D1062018

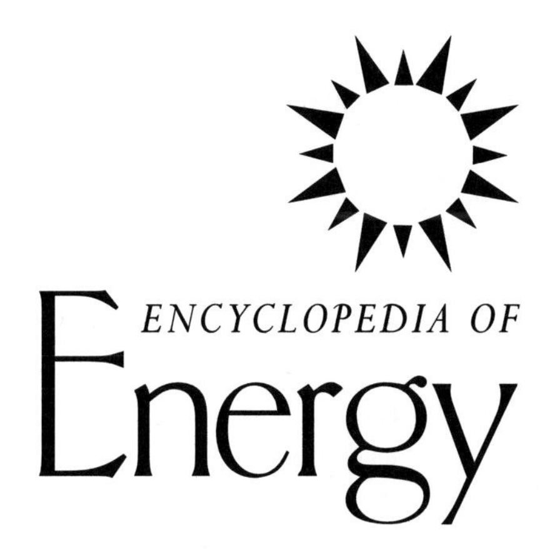

A–Ea

VOLUME 1

ENCYCLOPEDIA OF Energy

Editor-in-Chief

CUTLER J. CLEVELAND

Boston University, Boston, Massachusetts, United States

A – Ea

VOLUME 1

Amsterdam Boston Heidelberg London New York
Oxford Paris San Diego San Francisco Singapore Sydney Tokyo

This book is printed on acid-free paper.

Academic Press
An imprint of Elsevier Inc.
525 B Street, Suite 1900, San Diego, California 92101-4495, USA
http://www.academicpress.com

Academic Press
The Boulevard, Langford Lane, Kidlington, Oxford OX5 1GB, UK
http://www.academicpress.com

Library of Congress Catalog Card Number: 2003116610

International Standard Book Number: 0-12-176480-X (set)
International Standard Book Number: 0-12-176481-8 (Volume 1)
International Standard Book Number: 0-12-176482-6 (Volume 2)
International Standard Book Number: 0-12-176483-4 (Volume 3)
International Standard Book Number: 0-12-176484-2 (Volume 4)
International Standard Book Number: 0-12-176485-0 (Volume 5)
International Standard Book Number: 0-12-176486-9 (Volume 6)

Printed in the United States of America

04 05 06 07 08 09 MM 9 8 7 6 5 4 3 2 1

CONTENTS OF VOLUME 1

Contents of Volumes 1–6 ix
Contents by Subject Area xvii
Foreword xxvii
Preface xxxi
Guide to the Encyclopedia xxxiii

A

Acid Deposition and Energy Use 1
Jan Willem Erisman

Aggregation of Energy 17
Cutler J. Cleveland, Robert K. Kaufmann, and David I. Stern

Aircraft and Energy Use 29
Joosung J. Lee, Stephen P. Lukachko, and Ian A. Waitz

Air Pollution from Energy Production and Use 39
J. Slanina

Air Pollution, Health Effects of 55
Jonathan Levy

Alternative Transportation Fuels: Contemporary Case Studies 67
S. T. Coelho and José Goldemberg

Aluminum Production and Energy 81
H.-G. Schwarz

Aquaculture and Energy Use 97
M. Troell, P. Tyedmers, N. Kautsky, and P. Rönnbäck

Arid Environments, Impacts of Energy Development in 109
David Faiman

B

Batteries, Overview 117
Elton J. Cairns

Batteries, Transportation Applications 127
Michael M. Thackeray

Bicycling 141
Charles Komanoff

Biodiesel Fuels 151
Leon G. Schumacher, Jon van Gerpen, and Brian Adams

Biomass, Chemicals from 163
Douglas C. Elliott

Biomass Combustion 175
André P. C. Faaij

Biomass for Renewable Energy and Fuels 193
Donald L. Klass

Biomass Gasification 213
Ausilio Bauen

Biomass: Impact on Carbon Cycle and Greenhouse Gas Emissions 223
Carly Green and Kenneth A. Byrne

Biomass Resource Assessment 237
Marie E. Walsh

Bottom-Up Energy Modeling 251
Jayant Sathaye and Alan H. Sanstad

Business Cycles and Energy Prices 265
Stephen P. A. Brown, Mine K. Yücel, and John Thompson

C

Carbon Capture and Storage from Fossil Fuel Use 277
Howard Herzog and Dan Golomb

Carbon Sequestration, Terrestrial 289
R. Lal

Carbon Taxes and Climate Change 299
Marc Chupka

Cement and Energy 307
Ernst Worrell

City Planning and Energy Use 317
Hyunsoo Park and Clinton Andrews

Clean Air Markets 331
Alexander E. Farrell

Clean Coal Technology 343
Mildred B. Perry

Climate Change and Energy, Overview 359
Martin I. Hoffert and Ken Caldeira

Climate Change and Public Health: Emerging Infectious Diseases 381
Paul R. Epstein

Climate Change: Impact on the Demand for Energy 393
Timothy J. Considine

Climate Protection and Energy Policy 401
William R. Moomaw

Coal, Chemical and Physical Properties 411
Richard G. Lett and Thomas C. Ruppel

Coal Conversion 425
Michael A. Nowak, Anton Dilo Paul, Adrian Radziwon, and Rameshwar D. Srivastava

Coal, Fuel and Non-Fuel Uses 435
Anton Dilo Paul

Coal Industry, Energy Policy in 445
Richard L. Gordon

Coal Industry, History of 457
Jaak J. K. Daemen

Coal Mine Reclamation and Remediation 475
Robert L. P. Kleinmann

Coal Mining, Design and Methods of 485
Andrew P. Schissler

Coal Mining in Appalachia, History of 495
Geoffrey L. Buckley

Coal Preparation 507
Peter J. Bethell and Gerald H. Luttrell

Coal Resources, Formation of 529
Shreekant B. Malvadkar, Sarah Forbes, and Gilbert V. McGurl

Coal Storage and Transportation 551
James M. Ekmann and Patrick H. Le

Cogeneration 581
Doug Hinrichs

Combustion and Thermochemistry 595
Shen-Lin Chang and Chenn Qian Zhou

Commercial Sector and Energy Use 605
J. Michael MacDonald

Complex Systems and Energy 617
Mario Giampietro and Kozo Mayumi

Computer Modeling of Renewable Power Systems 633
Peter Lilienthal, Thomas Lambert, and Paul Gilman

Conservation Measures for Energy, History of 649
John H. Gibbons and Holly L. Gwin

Conservation of Energy Concept, History of 661
Elizabeth Garber

Conservation of Energy, Overview 673
Gordon J. Aubrecht II

Consumption, Energy, and the Environment 687
Simon Guy

Conversion of Energy: People and Animals 697
Vaclav Smil

Corporate Environmental Strategy 707
Bruce Piasecki

Cost–Benefit Analysis Applied to Energy 715
J. Peter Clinch

Crude Oil Releases to the Environment: Natural Fate and Remediation Options 727
Roger C. Prince and Richard R. Lessard

Crude Oil Spills, Environmental Impact of 737
Stanislav Patin

Cultural Evolution and Energy 749
Richard N. Adams

D

Decomposition Analysis Applied to Energy 761
B. W. Ang

Depletion and Valuation of Energy Resources 771
John Hartwick

Derivatives, Energy 781
Vincent Kaminski

Desalination and Energy Use 791
John B. Tonner and Jodie Tonner

Development and Energy, Overview 801
José Goldemberg

Diet, Energy, and Greenhouse Gas Emissions 809
Annika Carlsson-Kanyama

Discount Rates and Energy Efficiency Gap 817
Richard B. Howarth

Distributed Energy, Overview 823
Neil Strachan

District Heating and Cooling 841
Sven Werner

E

Early Industrial World, Energy Flow in 849
Richard D. Periman

Earth's Energy Balance 859
Kevin E. Trenberth

Easter Island: Resource Depletion and Collapse 871
James A. Brander

CONTENTS OF VOLUMES 1–6

Volume 1

Acid Deposition and Energy Use

Aggregation of Energy

Aircraft and Energy Use

Air Pollution from Energy Production and Use

Air Pollution, Health Effects of

Alternative Transportation Fuels: Contemporary Case Studies

Aluminum Production and Energy

Aquaculture and Energy Use

Arid Environments, Impacts of Energy Development in

Batteries, Overview

Batteries, Transportation Applications

Bicycling

Biodiesel Fuels

Biomass, Chemicals from

Biomass Combustion

Biomass for Renewable Energy and Fuels

Biomass Gasification

Biomass: Impact on Carbon Cycle and Greenhouse Gas Emissions

Biomass Resource Assessment

Bottom-Up Energy Modeling

Business Cycles and Energy Prices

Carbon Capture and Storage from Fossil Fuel Use

Carbon Sequestration, Terrestrial

Carbon Taxes and Climate Change

Cement and Energy

City Planning and Energy Use

Clean Air Markets

Clean Coal Technology

Climate Change and Energy, Overview

Climate Change and Public Health: Emerging Infectious Diseases

Climate Change: Impact on the Demand for Energy

Climate Protection and Energy Policy

Coal, Chemical and Physical Properties

Coal Conversion

Coal, Fuel and Non-Fuel Uses

Coal Industry, Energy Policy in

Coal Industry, History of

Coal Mine Reclamation and Remediation

Coal Mining, Design and Methods of

Coal Mining in Appalachia, History of

Coal Preparation

Coal Resources, Formation of

Coal Storage and Transportation

Cogeneration

Combustion and Thermochemistry
Commercial Sector and Energy Use
Complex Systems and Energy
Computer Modeling of Renewable Power Systems
Conservation Measures for Energy, History of
Conservation of Energy Concept, History of
Conservation of Energy, Overview
Consumption, Energy, and the Environment
Conversion of Energy: People and Animals
Corporate Environmental Strategy
Cost–Benefit Analysis Applied to Energy
Crude Oil Releases to the Environment: Natural Fate and Remediation Options
Crude Oil Spills, Environmental Impact of
Cultural Evolution and Energy
Decomposition Analysis Applied to Energy
Depletion and Valuation of Energy Resources
Derivatives, Energy
Desalination and Energy Use
Development and Energy, Overview
Diet, Energy, and Greenhouse Gas Emissions
Discount Rates and Energy Efficiency Gap
Distributed Energy, Overview
District Heating and Cooling
Early Industrial World, Energy Flow in
Earth's Energy Balance
Easter Island: Resource Depletion and Collapse

Volume 2

Ecological Footprints and Energy
Ecological Risk Assessment Applied to Energy Development
Economic Geography of Energy
Economic Growth and Energy
Economic Growth, Liberalization, and the Environment
Economics of Energy Demand
Economics of Energy Efficiency
Economics of Energy Supply
Economic Systems and Energy, Conceptual Overview
Economic Thought, History of Energy in
Ecosystem Health: Energy Indicators
Ecosystems and Energy: History and Overview
Electrical Energy and Power
Electricity, Environmental Impacts of
Electricity Use, History of
Electric Motors
Electric Power: Critical Infrastructure Protection
Electric Power Generation: Fossil Fuel
Electric Power Generation: Valuation of Environmental Costs
Electric Power Measurements and Variables
Electric Power Reform: Social and Environmental Issues
Electric Power Systems Engineering
Electric Power: Traditional Monopoly Franchise Regulation and Rate Making
Electric Power: Transmission and Generation Reliability and Adequacy
Electromagnetic Fields, Health Impacts of
Electromagnetism
Emergy Analysis and Environmental Accounting
Energy Development on Public Land in the United States
Energy Efficiency and Climate Change
Energy Efficiency, Taxonomic Overview
Energy Futures and Options
Energy in the History and Philosophy of Science
Energy Ladder in Developing Nations
Energy Services Industry

Entrainment and Impingement of Organisms in Power Plant Cooling
Entropy
Entropy and the Economic Process
Environmental Change and Energy
Environmental Gradients and Energy
Environmental Injustices of Energy Facilities
Environmental Kuznets Curve
Equity and Distribution in Energy Policy
Ethanol Fuel
European Union Energy Policy
Evolutionary Economics and Energy
Exergoeconomics
Exergy
Exergy Analysis of Energy Systems
Exergy Analysis of Waste Emissions
Exergy: Reference States and Balance Conditions
Experience Curves for Energy Technologies
External Costs of Energy
Fire: A Socioecological and Historical Survey
Fisheries and Energy Use
Flywheels
Food Capture, Energy Costs of
Food System, Energy Use in
Forest Products and Energy
Forms and Measurement of Energy
Fuel Cells
Fuel Cell Vehicles
Fuel Cycle Analysis of Conventional and Alternative Fuel Vehicles
Fuel Economy Initiatives: International Comparisons
Fuzzy Logic Modeling of Energy Systems
Gas Hydrates
Gasoline Additives and Public Health
Geographic Thought, History of Energy in
Geopolitics of Energy
Geothermal Direct Use
Geothermal Power Generation

Volume 3

Glass and Energy
Global Energy Use: Status and Trends
Global Material Cycles and Energy
Goods and Services: Energy Costs
Green Accounting and Energy
Greenhouse Gas Abatement: Controversies in Cost Assessment
Greenhouse Gas Emissions, Alternative Scenarios of
Greenhouse Gas Emissions from Energy Systems, Comparison and Overview
Ground-Source Heat Pumps
Gulf War, Environmental Impact of
Hazardous Waste from Fossil Fuels
Heat Islands and Energy
Heat Transfer
Heterotrophic Energy Flows
Human Energetics
Hunting and Gathering Societies, Energy Flows in
Hybrid Electric Vehicles
Hybrid Energy Systems
Hydrogen, End Uses and Economics
Hydrogen, History of
Hydrogen Production
Hydrogen Storage and Transportation
Hydropower Economics
Hydropower, Environmental Impact of
Hydropower, History and Technology of
Hydropower Resettlement Projects, Socioeconomic Impacts of

Hydropower Resources
Hydropower Technology
Indoor Air Quality in Developing Nations
Indoor Air Quality in Industrial Nations
Industrial Agriculture, Energy Flows in
Industrial Ecology
Industrial Energy Efficiency
Industrial Energy Use, Status and Trends
Industrial Symbiosis
Inflation and Energy Prices
Information Technology and Energy Use
Information Theory and Energy
Innovation and Energy Prices
Input–Output Analysis
Integration of Motor Vehicle and Distributed Energy Systems
Intelligent Transportation Systems
Internal Combustion Engine Vehicles
Internal Combustion (Gasoline and Diesel) Engines
International Comparisons of Energy End Use: Benefits and Risks
International Energy Law and Policy
Investment in Fossil Fuels Industries
Labels and Standards for Energy
Land Requirements of Energy Systems
Leisure, Energy Costs of
Life Cycle Analysis of Power Generation Systems
Life Cycle Assessment and Energy Systems
Lifestyles and Energy
Lithosphere, Energy Flows in
Livestock Production and Energy Use
Lunar–Solar Power System
Magnetic Levitation
Magnetohydrodynamics
Manufactured Gas, History of
Marine Transportation and Energy Use
Market-Based Instruments, Overview
Market Failures in Energy Markets
Markets for Biofuels
Markets for Coal
Markets for Natural Gas
Markets for Petroleum
Marx, Energy, and Social Metabolism
Material Efficiency and Energy Use
Materials for Solar Energy
Material Use in Automobiles

Volume 4

Mechanical Energy
Media Portrayals of Energy
Microtechnology, Energy Applications of
Migration, Energy Costs of
Modeling Energy Markets and Climate Change Policy
Modeling Energy Supply and Demand: A Comparison of Approaches
Motor Vehicle Use, Social Costs of
Multicriteria Analysis of Energy
National Energy Modeling Systems
National Energy Policy: Brazil
National Energy Policy: China
National Energy Policy: India
National Energy Policy: Japan
National Energy Policy: United States
Nationalism and Oil
National Security and Energy
Natural Gas, History of
Natural Gas Industry, Energy Policy in

Natural Gas Processing and Products
Natural Gas Resources, Global Distribution of
Natural Gas Resources, Unconventional
Natural Gas Transportation and Storage
Net Energy Analysis: Concepts and Methods
Neural Network Modeling of Energy Systems
Nongovernmental Organizations (NGOs) and Energy
Nuclear Engineering
Nuclear Fission Reactors: Boiling Water and Pressurized Water Reactors
Nuclear Fuel: Design and Fabrication
Nuclear Fuel Reprocessing
Nuclear Fusion Reactors
Nuclear Power Economics
Nuclear Power, History of
Nuclear Power Plants, Decommissioning of
Nuclear Power: Risk Analysis
Nuclear Proliferation and Diversion
Nuclear Waste
Obstacles to Energy Efficiency
Occupational Health Risks in Crude Oil and Natural Gas Extraction
Occupational Health Risks in Nuclear Power
Ocean, Energy Flows in
Ocean Thermal Energy
Oil and Natural Gas Drilling
Oil and Natural Gas: Economics of Exploration
Oil and Natural Gas Exploration
Oil and Natural Gas Leasing
Oil and Natural Gas Liquids: Global Magnitude and Distribution
Oil and Natural Gas: Offshore Operations
Oil and Natural Gas Resource Assessment: Classifications and Terminology
Oil and Natural Gas Resource Assessment: Geological Methods
Oil and Natural Gas Resource Assessment: Production Growth Cycle Models
Oil Crises, Historical Perspective
Oil Industry, History of
Oil-Led Development: Social, Political, and Economic Consequences
Oil Pipelines
Oil Price Volatility
Oil Recovery
Oil Refining and Products
Oil Sands and Heavy Oil
Oil Shale
OPEC, History of
OPEC Market Behavior, 1973–2003
Origin of Life and Energy
Passenger Demand for Travel and Energy Use
Peat Resources
Petroleum Property Valuation
Petroleum System: Nature's Distribution System for Oil and Gas

Volume 5

Philanthropy and Energy
Photosynthesis and Autotrophic Energy Flows
Photosynthesis, Artificial
Photovoltaic Conversion: Space Applications
Photovoltaic Energy: Stand-Alone and Grid-Connected Systems
Photovoltaic Materials, Physics of
Photovoltaics, Environmental Impact of
Physics and Economics of Energy, Conceptual Overview
Plastics Production and Energy
Polar Regions, Impacts of Energy Development
Population Growth and Energy

Potential for Energy Efficiency: Developing Nations
Prices of Energy, History of
Public Reaction to Electricity Transmission Lines
Public Reaction to Energy, Overview
Public Reaction to Nuclear Power Siting and Disposal
Public Reaction to Offshore Oil
Public Reaction to Renewable Energy Sources and Systems
Radiation, Risks and Health Impacts of
Rebound Effect of Energy Conservation
Recycling of Metals
Recycling of Paper
Refrigeration and Air-Conditioning
Remote Sensing for Energy Resources
Renewable Energy and the City
Renewable Energy in Europe
Renewable Energy in Southern Africa
Renewable Energy in the United States
Renewable Energy Policies and Barriers
Renewable Energy, Taxonomic Overview
Renewable Portfolio Standard
Reproduction, Energy Costs of
Research and Development Trends for Energy
Resource Curse and Investment in Energy Industries
Reuse and Energy
Risk Analysis Applied to Energy Systems
Rocket Engines
Rural Energy in China
Rural Energy in India
Service and Commerce Sector, Energy Use in
Sociopolitical Collapse, Energy and
Solar Cells
Solar Cookers
Solar Cooling, Dehumidification, and Air-Conditioning
Solar Detoxification and Disinfection
Solar Distillation and Drying
Solar Energy, History of
Solar Fuels and Materials
Solar Heat Pumps
Solar Ponds
Solar Thermal Energy, Industrial Heat Applications
Solar Thermal Power Generation
Solar Water Desalination
Steel Production and Energy
Stock Markets and Energy Prices
Storage of Energy, Overview
Strategic Petroleum Reserves
Subsidies to Energy Industries
Suburbanization and Energy
Sun, Energy from
Sustainable Development: Basic Concepts and Application to Energy
System Dynamics and the Energy Industry

Volume 6

Tanker Transportation
Taxation of Energy
Technology Innovation and Energy
Temperature and Its Measurement
Thermal Comfort
Thermal Energy Storage
Thermal Pollution
Thermodynamics and Economics, Overview
Thermodynamic Sciences, History of
Thermodynamics, Laws of
Thermoregulation

Tidal Energy
Trade in Energy and Energy Services
Transitions in Energy Use
Transportation and Energy, Overview
Transportation and Energy Policy
Transportation Fuel Alternatives for Highway Vehicles
Turbines, Gas
Turbines, Steam
Ultralight Rail and Energy Use
United Nations Energy Agreements
Uranium and Thorium Resource Assessment
Uranium Mining: Environmental Impact
Uranium Mining, Processing, and Enrichment
Urbanization and Energy
Value Theory and Energy
Vehicles and Their Powerplants: Energy Use and Efficiency
War and Energy
Waste-to-Energy Technology
Wave and Tidal Energy Conversion
Wetlands: Impacts of Energy Development in the Mississippi Delta
Wind Energy Economics
Wind Energy, History of
Wind Energy Technology, Environmental Impacts of
Wind Farms
Wind Resource Base
Women and Energy: Issues in Developing Nations
Wood Energy, History of
Wood in Household Energy Use
Work, Power, and Energy
World Environment Summits: The Role of Energy
World History and Energy

CONTENTS BY SUBJECT AREA

Basics of Energy

Batteries, Overview

Cogeneration

Combustion and Thermochemistry

Conservation of Energy, Overview

Conversion of Energy: People and Animals

Electrical Energy and Power

Electric Motors

Electromagnetism

Entropy

Flywheels

Forms and Measurement of Energy

Fuel Cells

Heat Transfer

Internal Combustion (Gasoline and Diesel) Engines

Magnetohydrodynamics

Mechanical Energy

Refrigeration and Air-Conditioning

Rocket Engines

Storage of Energy, Overview

Sun, Energy from

Temperature and Its Measurement

Thermal Energy Storage

Thermodynamics, Laws of

Turbines, Gas

Turbines, Steam

Work, Power, and Energy

Coal

Coal Industry, Energy Policy in

Coal Industry, History of

Coal Mine Reclamation and Remediation

Coal Mining in Appalachia, History of

Clean Coal Technology

Coal, Chemical and Physical Properties

Coal Conversion

Coal, Fuel and Non-Fuel Uses

Coal Mining, Design and Methods of

Coal Preparation

Coal Resources, Formation of

Coal Storage and Transportation

Markets for Coal

Peat Resources

Conservation and End Use

Aircraft and Energy Use

Alternative Transportation Fuels: Contemporary Case Studies

Aquaculture and Energy Use
Batteries, Transportation Applications
Bicycling
Commercial Sector and Energy Use
Conservation Measures for Energy, History of
Conservation of Energy, Overview
Diet, Energy, and Greenhouse Gas Emissions
Discount Rates and Energy Efficiency Gap
Distributed Energy, Overview
District Heating and Cooling
Economics of Energy Efficiency
Energy Efficiency, Taxonomic Overview
Fisheries and Energy Use
Food System, Energy Use in
Fuel Cell Vehicles
Fuel Cycle Analysis of Conventional and Alternative Fuel Vehicles
Hybrid Electric Vehicles
Hydrogen, End Uses and Economics
Industrial Ecology
Industrial Energy Efficiency
Industrial Energy Use, Status and Trends
Information Technology and Energy Use
Integration of Motor Vehicle and Distributed Energy Systems
Intelligent Transportation Systems
Internal Combustion Engine Vehicles
International Comparisons of Energy End Use: Benefits and Risks
Lifestyles and Energy
Livestock Production and Energy Use
Lunar–Solar Power System
Magnetic Levitation
Marine Transportation and Energy Use
Obstacles to Energy Efficiency
Passenger Demand for Travel and Energy Use
Potential for Energy Efficiency: Developing Nations
Rebound Effect of Energy Conservation
Service and Commerce Sector, Energy Use in
Thermal Comfort
Transportation and Energy, Overview
Transportation Fuel Alternatives for Highway Vehicles
Ultralight Rail and Energy Use
Vehicles and Their Powerplants: Energy Use and Efficiency

Economics of Energy

Aggregation of Energy
Business Cycles and Energy Prices
Corporate Environmental Strategy
Derivatives, Energy
Economic Geography of Energy
Economic Growth and Energy
Economic Growth, Liberalization, and the Environment
Economics of Energy Demand
Economics of Energy Efficiency
Economics of Energy Supply
Economic Thought, History of Energy in
Energy Futures and Options
Energy Services Industry
Entropy and the Economic Process
Evolutionary Economics and Energy
Exergoeconomics
External Costs of Energy
Hydrogen, End Uses and Economics
Hydropower Economics
Inflation and Energy Prices
Innovation and Energy Prices

Investment in Fossil Fuels Industries

Market Failures in Energy Markets

Markets for Biofuels

Markets for Coal

Markets for Natural Gas

Markets for Petroleum

Marx, Energy, and Social Metabolism

Nuclear Power Economics

Oil and Natural Gas: Economics of Exploration

Oil Price Volatility

OPEC Market Behavior, 1973–2003

Petroleum Property Valuation

Physics and Economics of Energy, Conceptual Overview

Prices of Energy, History of

Rebound Effect of Energy Conservation

Resource Curse and Investment in Energy Industries

Stock Markets and Energy Prices

Subsidies to Energy Industries

Taxation of Energy

Thermodynamics and Economics, Overview

Trade in Energy and Energy Services

Wind Energy Economics

Electricity

Electrical Energy and Power

Electricity, Environmental Impacts of

Electricity Use, History of

Electric Motors

Electric Power: Critical Infrastructure Protection

Electric Power Generation: Fossil Fuel

Electric Power Generation: Valuation of Environmental Costs

Electric Power Measurements and Variables

Electric Power Reform: Social and Environmental Issues

Electric Power Systems Engineering

Electric Power: Traditional Monopoly Franchise Regulation and Rate Making

Electric Power: Transmission and Generation Reliability and Adequacy

Electromagnetic Fields, Health Impacts of

Electromagnetism

Hybrid Electric Vehicles

Public Reaction to Electricity Transmission Lines

Energy Flows

Conversion of Energy: People and Animals

Earth's Energy Balance

Ecosystem Health: Energy Indicators

Ecosystems and Energy: History and Overview

Environmental Gradients and Energy

Food Capture, Energy Costs of

Heat Transfer

Heterotrophic Energy Flows

Human Energetics

Industrial Agriculture, Energy Flows in

Industrial Symbiosis

Lithosphere, Energy Flows in

Migration, Energy Costs of

Ocean, Energy Flows in

Origin of Life and Energy

Photosynthesis and Autotrophic Energy Flows

Reproduction, Energy Costs of

Sun, Energy from

Thermoregulation

Environmental Issues

Acid Deposition and Energy Use
Air Pollution from Energy Production and Use
Air Pollution, Health Effects of
Aquaculture and Energy Use
Arid Environments, Impacts of Energy Development in
Biomass: Impact on Carbon Cycle and Greenhouse Gas Emissions
Carbon Capture and Storage from Fossil Fuel Use
Carbon Sequestration, Terrestrial
Clean Air Markets
Clean Coal Technology
Climate Change and Energy, Overview
Climate Change and Public Health: Emerging Infectious Diseases
Climate Change: Impact on the Demand for Energy
Climate Protection and Energy Policy
Coal Mine Reclamation and Remediation
Consumption, Energy, and the Environment
Crude Oil Releases to the Environment: Natural Fate and Remediation Options
Crude Oil Spills, Environmental Impact of
Desalination and Energy Use
Economic Growth, Liberalization, and the Environment
Ecosystem Health: Energy Indicators
Ecosystems and Energy: History and Overview
Electricity, Environmental Impacts of
Electric Power Generation: Valuation of Environmental Costs
Electric Power Reform: Social and Environmental Issues
Energy Efficiency and Climate Change
Entrainment and Impingement of Organisms in Power Plant Cooling
Environmental Change and Energy
Environmental Gradients and Energy
Environmental Injustices of Energy Facilities
Fisheries and Energy Use
Global Material Cycles and Energy
Greenhouse Gas Emissions, Alternative Scenarios of
Greenhouse Gas Emissions from Energy Systems, Comparison and Overview
Gulf War, Environmental Impact of
Hazardous Waste from Fossil Fuels
Heat Islands and Energy
Hydropower, Environmental Impact of
Indoor Air Quality in Developing Nations
Indoor Air Quality in Industrial Nations
Land Requirements of Energy Systems
Nuclear Power Plants, Decommissioning of
Nuclear Waste
Photovoltaics, Environmental Impact of
Polar Regions, Impacts of Energy Development
Thermal Pollution
Uranium Mining: Environmental Impact
Wetlands: Impacts of Energy Development in the Mississippi Delta
Wind Energy Technology, Environmental Impacts of
World Environment Summits: The Role of Energy

Global Issues

Climate Change and Energy, Overview
Cultural Evolution and Energy
Development and Energy, Overview
Economic Geography of Energy
Economic Growth and Energy
Geopolitics of Energy

Global Energy Use: Status and Trends
International Comparisons of Energy End Use: Benefits and Risks
International Energy Law and Policy
Nationalism and Oil
Nongovernmental Organizations (NGOs) and Energy
Nuclear Proliferation and Diversion
Population Growth and Energy
Technology Innovation and Energy
United Nations Energy Agreements
Women and Energy: Issues in Developing Nations
World Environment Summits: The Role of Energy

History and Energy

Coal Industry, History of
Coal Mining in Appalachia, History of
Conservation Measures for Energy, History of
Conservation of Energy Concept, History of
Early Industrial World, Energy Flow in
Economic Thought, History of Energy in
Ecosystems and Energy: History and Overview
Electricity Use, History of
Energy in the History and Philosophy of Science
Environmental Change and Energy
Fire: A Socioecological and Historical Survey
Geographic Thought, History of Energy in
Gulf War, Environmental Impact of
Hydrogen, History of
Hydropower, History and Technology of
Manufactured Gas, History of
Nationalism and Oil
Natural Gas, History of
Nuclear Power, History of
Oil Crises, Historical Perspective
Oil Industry, History of
OPEC, History of
OPEC Market Behavior, 1973–2003
Prices of Energy, History of
Sociopolitical Collapse, Energy and
Solar Energy, History of
Thermodynamic Sciences, History of
Transitions in Energy Use
War and Energy
Wind Energy, History of
Wood Energy, History of
World History and Energy

Material Use and Reuse

Aluminum Production and Energy
Cement and Energy
Forest Products and Energy
Glass and Energy
Global Material Cycles and Energy
Industrial Energy Efficiency
Material Efficiency and Energy Use
Materials for Solar Energy
Material Use in Automobiles
Plastics Production and Energy
Recycling of Metals
Recycling of Paper
Reuse and Energy
Steel Production and Energy
Uranium and Thorium Resource Assessment

Measurement and Models

Aggregation of Energy
Bottom-Up Energy Modeling

Computer Modeling of Renewable Power Systems

Cost–Benefit Analysis Applied to Energy

Decomposition Analysis Applied to Energy

Depletion and Valuation of Energy Resources

Ecological Risk Assessment Applied to Energy Development

Electric Power Generation: Valuation of Environmental Costs

Electric Power Measurements and Variables

Emergy Analysis and Environmental Accounting

Experience Curves for Energy Technologies

Forms and Measurement of Energy

Fuzzy Logic Modeling of Energy Systems

Green Accounting and Energy

Input–Output Analysis

Life Cycle Analysis of Power Generation Systems

Life Cycle Assessment and Energy Systems

Modeling Energy Markets and Climate Change Policy

Modeling Energy Supply and Demand: A Comparison of Approaches

Multicriteria Analysis of Energy

National Energy Modeling Systems

Net Energy Analysis: Concepts and Methods

Neural Network Modeling of Energy Systems

System Dynamics and the Energy Industry

Temperature and Its Measurement

Nuclear Power

Nuclear Engineering

Nuclear Fission Reactors: Boiling Water and Pressurized Water Reactors

Nuclear Fuel: Design and Fabrication

Nuclear Fuel Reprocessing

Nuclear Fusion Reactors

Nuclear Power Economics

Nuclear Power, History of

Nuclear Power Plants, Decommissioning of

Nuclear Power: Risk Analysis

Nuclear Proliferation and Diversion

Nuclear Waste

Occupational Health Risks in Nuclear Power

Public Reaction to Nuclear Power Siting and Disposal

Radiation, Risks and Health Impacts of

Uranium and Thorium Resource Assessment

Uranium Mining, Processing, and Enrichment

Oil and Natural Gas

Crude Oil Spills, Environmental Impact of

Gas Hydrates

Markets for Natural Gas

Markets for Petroleum

Natural Gas, History of

Natural Gas Processing and Products

Natural Gas Resources, Global Distribution of

Natural Gas Resources, Unconventional

Natural Gas Transportation and Storage

Occupational Health Risks in Crude Oil and Natural Gas Extraction

Oil and Natural Gas Drilling

Oil and Natural Gas: Economics of Exploration

Oil and Natural Gas Exploration

Oil and Natural Gas Leasing

Oil and Natural Gas Liquids: Global Magnitude and Distribution

Oil and Natural Gas: Offshore Operations

Oil and Natural Gas Resource Assessment: Classifications and Terminology

Oil and Natural Gas Resource Assessment: Geological Methods

Oil and Natural Gas Resource Assessment: Production Growth Cycle Models

Oil Crises, Historical Perspective

Oil Industry, History of

Oil-Led Development: Social, Political, and Economic Consequences

Oil Pipelines

Oil Price Volatility

Oil Recovery

Oil Refining and Products

Oil Sands and Heavy Oil

Oil Shale

OPEC, History of

Petroleum Property Valuation

Petroleum System: Nature's Distribution System for Oil and Gas

Public Reaction to Offshore Oil

Remote Sensing for Energy Resources

Strategic Petroleum Reserves

Tanker Transportation

Policy Issues

Carbon Taxes and Climate Change

City Planning and Energy Use

Clean Air Markets

Climate Protection and Energy Policy

Coal Industry, Energy Policy in

Corporate Environmental Strategy

Energy Development on Public Land in the United States

Equity and Distribution in Energy Policy

European Union Energy Policy

Fuel Economy Initiatives: International Comparisons

Geopolitics of Energy

Greenhouse Gas Abatement: Controversies in Cost Assessment

International Energy Law and Policy

Land Requirements of Energy Systems

Market-Based Instruments, Overview

National Energy Policy: Brazil

National Energy Policy: China

National Energy Policy: India

National Energy Policy: Japan

National Energy Policy: United States

National Security and Energy

Natural Gas Industry, Energy Policy in

Nuclear Proliferation and Diversion

Polar Regions, Impacts of Energy Development

Renewable Energy Policies and Barriers

Renewable Portfolio Standard

Research and Development Trends for Energy

Strategic Petroleum Reserves

Subsidies to Energy Industries

Taxation of Energy

Transportation and Energy Policy

Public Issues

City Planning and Energy Use

Climate Change and Public Health: Emerging Infectious Diseases

Consumption, Energy, and the Environment

Environmental Injustices of Energy Facilities

Hydropower Resettlement Projects, Socioeconomic Impacts of

Labels and Standards for Energy

Lifestyles and Energy

Media Portrayals of Energy

Motor Vehicle Use, Social Costs of

Oil-Led Development: Social, Political, and Economic Consequences

Passenger Demand for Travel and Energy Use

Philanthropy and Energy

Population Growth and Energy

Public Reaction to Electricity Transmission Lines

Public Reaction to Energy, Overview

Public Reaction to Nuclear Power Siting and Disposal

Public Reaction to Offshore Oil

Public Reaction to Renewable Energy Sources and Systems

Suburbanization and Energy

United Nations Energy Agreements

Urbanization and Energy

Renewable and Alternative Sources

Alternative Transportation Fuels: Contemporary Case Studies

Biodiesel Fuels

Biomass, Chemicals from

Biomass Combustion

Biomass for Renewable Energy and Fuels

Biomass Gasification

Biomass Resource Assessment

Computer Modeling of Renewable Power Systems

Ethanol Fuel

Forest Products and Energy

Geothermal Direct Use

Geothermal Power Generation

Ground-Source Heat Pumps

Hybrid Energy Systems

Hydrogen, End Uses and Economics

Hydrogen Production

Hydrogen Storage and Transportation

Hydropower Economics

Hydropower Resources

Hydropower Technology

Lunar–Solar Power System

Materials for Solar Energy

Microtechnology, Energy Applications of

Ocean Thermal Energy

Photosynthesis, Artificial

Photovoltaic Conversion: Space Applications

Photovoltaic Energy: Stand-Alone and Grid-Connected Systems

Photovoltaic Materials, Physics of

Public Reaction to Renewable Energy Sources and Systems

Renewable Energy and the City

Renewable Energy in Europe

Renewable Energy Policies and Barriers

Renewable Energy in Southern Africa

Renewable Energy in the United States

Renewable Energy, Taxonomic Overview

Renewable Portfolio Standard

Solar Cells

Solar Cookers

Solar Cooling, Dehumidification, and Air-Conditioning

Solar Detoxification and Disinfection

Solar Distillation and Drying

Solar Energy, History of

Solar Fuels and Materials

Solar Heat Pumps

Solar Ponds

Solar Thermal Energy, Industrial Heat Applications

Solar Thermal Power Generation

Solar Water Desalination

Tidal Energy

Transportation Fuel Alternatives for Highway Vehicles

Waste-to-Energy Technology

Wave and Tidal Energy Conversion

Wind Energy Economics

Wind Energy, History of

Wind Farms

Wind Resource Base

Risks

Air Pollution from Energy Production and Use

Air Pollution, Health Effects of

Climate Change and Public Health: Emerging Infectious Diseases

Ecological Risk Assessment Applied to Energy Development

Electromagnetic Fields, Health Impacts of

Gasoline Additives and Public Health

Hazardous Waste from Fossil Fuels

Nuclear Power: Risk Analysis

Nuclear Proliferation and Diversion

Nuclear Waste

Occupational Health Risks in Crude Oil and Natural Gas Extraction

Occupational Health Risks in Nuclear Power

Radiation, Risks and Health Impacts of

Risk Analysis Applied to Energy Systems

Tanker Transportation

Society and Energy

Cultural Evolution and Energy

Early Industrial World, Energy Flow in

Easter Island: Resource Depletion and Collapse

Electric Power Reform: Social and Environmental Issues

Goods and Services: Energy Costs

Hunting and Gathering Societies, Energy Flows in

Hydropower Resettlement Projects, Socioeconomic Impacts of

Industrial Agriculture, Energy Flows in

Leisure, Energy Costs of

Lifestyles and Energy

Motor Vehicle Use, Social Costs of

Population Growth and Energy

Renewable Energy and the City

Sociopolitical Collapse, Energy and

Suburbanization and Energy

Urbanization and Energy

War and Energy

Sustainable Development

Development and Energy, Overview

Ecological Risk Assessment Applied to Energy Development

Economic Growth and Energy

Economic Growth, Liberalization, and the Environment

Energy Ladder in Developing Nations

Environmental Kuznets Curve

Indoor Air Quality in Developing Nations

Oil-Led Development: Social, Political, and Economic Consequences

Potential for Energy Efficiency: Developing Nations

Rural Energy in China

Rural Energy in India

Sustainable Development: Basic Concepts and Application to Energy

United Nations Energy Agreements

Women and Energy: Issues in Developing Nations

Wood in Household Energy Use

Systems of Energy

Aggregation of Energy

Complex Systems and Energy

Ecological Footprints and Energy

Economic Systems and Energy, Conceptual Overview

Emergy Analysis and Environmental Accounting

Entropy and the Economic Process

Exergoeconomics

Exergy

Exergy Analysis of Energy Systems

Exergy Analysis of Waste Emissions

Exergy: Reference States and Balance Conditions

Fuzzy Logic Modeling of Energy Systems

Information Theory and Energy

Life Cycle Assessment and Energy Systems

National Energy Modeling Systems

Neural Network Modeling of Energy Systems

Physics and Economics of Energy, Conceptual Overview

Risk Analysis Applied to Energy Systems

System Dynamics and the Energy Industry

Thermodynamics and Economics, Overview

FOREWORD

Energy generation and use are strongly linked to all elements of sustainable development: economic, social, and environmental. The history of human development rests on the availability and use of energy, the transformation from the early use of fire and animal power that improved lives, to the present world with use of electricity and clean fuels for a multitude of purposes. This progress built on basic scientific discoveries, such as electromagnetism and the inventions of technologies such as steam engines, light bulbs, and automobiles.

It is thus abundantly clear that access to affordable energy is fundamental to human activities, development, and economic growth. Without access to electricity and clean fuels, people's opportunities are significantly constrained. However, it is really energy services, not energy *per se* that matters. Yet, today some 2 billion people lack access to modern energy carriers.

In addition to the great benefits, the generation, transportation, and use of energy carriers unfortunately come with undesired effects. The environmental impacts are multifaceted and serious, although mostly less evident. Emissions of suspended fine particles and precursors of acid deposition contribute to local and regional air pollution and ecosystem degradation. Human health is threatened by high levels of air pollution resulting from particular types of energy use at the household, community, and regional levels.

Emissions of anthropogenic greenhouse gases (GHG), mostly from the production and use of energy, are altering the atmosphere in ways that are affecting the climate. There is new and stronger evidence that most of the global warming observed over the last 50 years is attributable to human activities. Stabilization of GHG in the atmosphere will require a major reduction in the projected carbon emissions to levels below the present.

Dependence on imported fuels leaves many countries vulnerable to disruption in supply, which might pose physical hardships and economic burdens; the weight of fossil fuel imports on the balance of payments is unbearable for many poorer countries. The present energy system of countries heavily dependent on fossil fuels geographically concentrated in a few regions of the world adds security of supply aspects.

From the issues indicated here it is clear that major changes are required in energy system development worldwide. At a first glance, there appears to be many conflicting objectives. For example, is it possible to sustain poverty alleviation and economic growth while reducing GHG emissions? Can urban areas and transport expand while improving air quality? What would be the preferable trade-offs? Finding ways to expand energy services while simultaneously addressing the environmental impacts associated with energy use represents a critical challenge to humanity.

What are the options? Looking at physical resources, one finds they are abundant. Fossil fuels will be able to provide the energy carriers that the world is used to for hundreds of years. Renewable energy flows on Earth are many thousands of times larger than flows through energy markets. Therefore, there are no apparent constraints from a resource point of view. However, the challenge is how to use these resources in an environmentally acceptable way. The broad categories of options for using energy in ways that support sustainable development are (1) more efficient use of energy in all sectors, especially at the point of end use, (2) increased use of renewable energy sources, and (3) accelerated development and deployment of new and advanced energy technologies, including next-generation fossil fuel technologies that produce near-zero harmful emissions. Technologies are available in these areas to meet the challenges of sustainable development. In addition, innovation provides increasing opportunities.

Analysis using energy scenarios indicates that it is indeed possible to simultaneously address the sustainable development objectives using the available natural resources and technical options presented. A prerequisite for achieving energy futures compatible with sustainable development objectives is finding ways to accelerate progress for new technologies along the energy innovation chain, including research and development, demonstration, deployment, and diffusion.

It is significant that there already exist combinations of technologies that meet all sustainable development challenges at the same time. This will make it easier to act locally to address pollution problems of a major city or country while at the same time mitigating climate change. Policies for energy for sustainable development can be largely motivated by national concerns and will not have to rely only on global pressures.

However, with present policies and conditions in the marketplaces that determine energy generation and use such desired energy futures will not happen. A prerequisite for sustainable development is change in policies affecting energy for sustainable development. This brings a need to focus on the policy situation and understand incentives and disincentives related to options for options for energy for sustainable development.

Policies and actions to promote energy for sustainable development would include the following:

- Developing capacity among all stakeholders in all countries, especially in the public sector, to address issues related to energy for sustainable development.
- Adopting policies and mechanisms to increase access to energy services through modern fuels and electricity for the 2 billion without.
- Advancing innovation, with balanced emphasis on all steps of the innovation chain: research and development, demonstrations, cost buy-down, and wide dissemination.
- Setting appropriate market framework conditions (including continued market reform, consistent regulatory measures, and targeted policies) to encourage competitiveness in energy markets, to reduce total cost of energy services to end-users, and to protect important public benefits, including the following:
    - Cost-based prices, including phasing out all forms of permanent subsidies for conventional energy (now on the order of $250 billion a year) and internalizing external environmental and health costs and benefits (now sometimes larger than the private costs).
    - Removing obstacles and providing incentives, as needed, to encourage greater energy efficiency and the development and/or diffusion of new technologies for energy for sustainable development to wider markets.
    - Recent power failures on the North American Eastern Seaboard, in California, London (United Kingdom), Sweden, and Italy illustrate the strong dependence on reliable power networks. Power sector reform that recognizes the unique character of electricity, and avoids power crises as seen in recent years, is needed.
- Reversing the trend of declining Official Development Assistance and Foreign Direct Investments, especially as related to energy for sustainable development.

This is a long list of opportunities and challenges. To move sufficiently in the direction of sustainability will require actions by the public and the private sector, as well as other stakeholders, at the national, regional, and global levels. The decisive issues are not technology or natural resource scarcity, but the institutions, rules, financing mechanisms, and regulations needed to make markets work in support of energy for sustainable development. A number of countries, including Spain, Germany, and Brazil, as well as some states in the United States have adopted successful laws and regulations designed to increase the use of renewable energy sources. Some regions, including Latin America and the European Union, have set targets for increased use of renewable energy. However, much remains to be done.

Energy was indeed one of the most intensely debated issues at the United Nations World Summit on Sustainable Development (WSSD), held in Johannesburg, South Africa, in August/September, 2002. In the end, agreement was reached on a text that significantly advances the attention given to energy in the context of sustainable development. This was in fact the first time agreements could be reached on energy at the world level! These developments followed years of efforts to focus on energy as an instrument for sustainable development that intensified after the United Nations Conference on Environment and Development in 1992.

The United Nations General Assembly adopted the Millennium Development Goals (MDG) in 2000. These goals are set in areas such as extreme poverty and hunger, universal primary education, gender equality and empowerment of women, child mortality, maternal health, HIV/AIDS, malaria and other diseases, and environmental sustainability. However, more than 2 billion people cannot access affordable energy services, based on the efficient use of gaseous and liquid fuels and electricity. This constrains their opportunities for economic development and improved living standards. Women and children suffer disproportionately because of their relative dependence on traditional fuels. Although no explicit goal on energy was adopted, access to energy services is a prerequisite to achieving all of the MDGs.

Some governments and corporations have already demonstrated that policies and measures to promote energy solutions conducive to sustainable development can work, and indeed work very well. The renewed focus and broad agreements on energy in the Johannesburg Plan of Implementation and at the 18th World Energy Congress are promising. The formation of many partnerships on energy between stakeholders at WSSD is another encouraging sign. A sustainable future in which energy plays a major positive role in supporting human well-being is possible!

Progress is being made on many fronts in bringing new technologies to the market, and to widening access to modern forms of energy. In relation to energy, a total of 39 partnerships were presented to the United Nations Secretariat for WSSD to promote programs on energy for sustainable development, 23 with energy as a central focus and 16 with a considerable impact on energy. These partnerships included most prominently the DESA-led Clean Fuels and Transport Initiative, the UNDP/World Bank-led Global Village Energy Partnership (GVEP), the Johannesburg Renewable Energy Coalition (JREC), the EU Partnership on Energy for Poverty Eradication and Sustainable Development, and the UNEP-led Global Network on Energy for Sustainable Development (GNESD).

With secure access to affordable and clean energy being so fundamental to sustainable development, the publication of the *Encyclopedia of Energy* is extremely timely and significant. Academics, professionals, scholars, politicians, students, and many more will benefit tremendously from the easy access to knowledge, experience, and insights that are provided here.

Thomas B. Johansson
Professor and Director
International Institute for
Industrial Environmental Economics
Lund University
Lund, Sweden

Former Director
Energy and Atmosphere Programme
United Nations Development Programme
New York, United States

PREFACE

The history of human culture can be viewed as the progressive development of new energy sources and their associated conversion technologies. Advances in our understanding of energy have produced unparalleled transformations of society, as exemplified by James Watt's steam engine and the discovery of oil. These transformations increased the ability of humans to exploit both additional energy and other resources, and hence to increase the comfort, longevity, and affluence of humans, as well as their numbers. Energy is related to human development in three important ways: as a motor of economic growth, as a principal source of environmental stress, and as a prerequisite for meeting basic human needs. Significant changes in each of these aspects of human existence are associated with changes in energy sources, beginning with the discovery of fire, the advent of agriculture and animal husbandry, and, ultimately, the development of hydrocarbon and nuclear fuels. The eventual economic depletion of fossil fuels will drive another major energy transition; geopolitical forces and environmental imperatives such as climate change may drive this transition faster than hydrocarbon depletion would have by itself. There is a diverse palette of alternatives to meet our energy needs, including a new generation of nuclear power, unconventional sources of hydrocarbons, myriad solar technologies, hydrogen, and more efficient energy end use. Each alternative has a different combination of economic, political, technological, social, and environmental attributes.

Energy is the common link between the living and non-living realms of the universe, and thus provides an organizing intellectual theme for diverse disciplines. Formalization of the concept of energy and identification of the laws governing its use by 19th century physical scientists such as Mayer and Carnot are cornerstones of modern science and engineering. The study of energy has played a pivotal role in understanding the creation of the universe, the origin of life, the evolution of human civilization and culture, economic growth and the rise of living standards, war and geopolitics, and significant environmental change at local, regional, and global scales.

The unique importance of energy among natural resources makes information about all aspects of its attributes, formation, distribution, extraction, and use an extremely valuable commodity. The *Encyclopedia of Energy* is designed to deliver this information in a clear and comprehensive fashion. It uses an integrated approach that emphasizes not only the importance of the concept in individual disciplines such as physics and sociology, but also how energy is used to bridge seemingly disparate fields, such as ecology and economics. As such, this *Encyclopedia* provides the first comprehensive, organized body of knowledge for what is certain to continue as a major area of scientific study in the 21st century. It is designed to appeal to a wide audience including undergraduate and graduate students, teachers, academics, and research scientists who study energy, as well as business corporations, professional firms, government agencies, foundations, and other groups whose activities relate to energy.

Comprehensive and interdisciplinary are two words I use to describe the *Encyclopedia*. It has the comprehensive coverage one would expect: forms of energy, thermodynamics, electricity generation, climate change, energy storage, energy sources, the demand for energy, and so on. What makes this work unique, however, is its breadth of coverage, including insights from history, society, anthropology, public policy, international relations, human and ecosystem health, economics, technology, physics, geology, ecology, business management, environmental science, and engineering. The coverage and integration of the social sciences is a unique feature.

The interdisciplinary approach is employed in the treatment of important subjects. In the case of oil, as one example, there are entries on the history of oil, the history of OPEC, the history of oil prices, oil price volatility, the formation of oil and gas, the distribution of oil resources, oil exploration and drilling, offshore oil, occupational hazards in the oil industry, oil refining, energy policy in the oil industry, the geopolitics of oil, oil spills, oil transportation, public lands and oil development, social impacts of oil and gas development, gasoline additives and public health, and the environmental impact of the Persian Gulf War. Other subjects are treated in a similar way.

This has been a massive and extremely satisfying effort. As with any work of this scale, many people have contributed at every step of the process, including the staff of Academic Press/Elsevier. The project began through the encouragement of Frank Cynar and David Packer, with Frank helping to successfully launch the initiative. Henri van Dorssen skillfully guided the project through its completion. He was especially helpful with integrating the project formulation, production, and marketing aspects of the project. Chris Morris was with the project throughout, and displayed what I can only describe as an uncanny combination of vision, enthusiasm, and energy for the project. I owe Chris a great deal for his insight and professionalism. I spent countless hours on the phone with Robert Matsumura, who was the glue that held the project together. Chris and Robert were ably assisted by outstanding Academic Press/Elsevier staff, especially Nick Panissidi, Joanna Dinsmore, and Mike Early. Clare Marl and her team put together a highly effective and creative marketing plan.

At the next stage, the Editorial Board was invaluable in shaping the coverage and identifying authors. The Board is an outstanding collection of scholars from the natural, social, and engineering sciences who are recognized leaders in their fields of research. They helped assemble an equally impressive group of authors from every discipline and who represent universities, government agencies, national laboratories, consulting firms, think tanks, corporations, and nongovernmental organizations. I am especially proud of the international scope of the authors: more than 400 authors from 40 nations are represented from every continent and every stage of development. To all of these, I extend my thanks and congratulations.

Cutler Cleveland
Boston University
Boston, Massachusetts, United States

GUIDE TO THE ENCYCLOPEDIA

The *Encyclopedia of Energy* is a comprehensive and authoritative study of the subject of energy in all its various aspects, as well as the ways in which energy use involves or affects other areas such as environmental science, economics, public policy, international relations, and human development. The encyclopedia includes 380 different articles on various topics within the overall theme of energy, written by prominent experts from around the world who represent 40 different countries in all. The print version of this work consists of six separate volumes and about 5,400 pages.

Each entry in the encyclopedia provides a complete overview of the given topic, intended to inform a broad spectrum of readers, ranging from energy research professionals, energy policy makers, and scholars in energy-related fields to students and the interested general public. The entries are self-contained and can be read in isolation, but there is a general system linking related topics by means of cross referencing and by their placement in specific subject areas (see Organization, below).

In order that you, the reader, will derive the greatest possible benefit from your use of the *Encyclopedia of Energy,* we have provided this guide. It will explain how the encyclopedia has been formulated and how the information within it can be located.

ENTRY SELECTION

This encyclopedia was conceived with the goal of providing a complete description of all the issues contained within, or impacting upon, the field of energy. This approach defines energy not just in terms of its physical, chemical, and engineering aspects but with respect to all of its effects on society and the environment.

To that end, a thorough and systematic method of entry selection was devised for the work. To begin the selection process, the project's chief editor, Cutler Cleveland, prepared a thematic outline of the topic of energy. This thematic outline progressed through the entire scope of the subject from basic principles to peripheral issues, in the manner of a course curriculum or a textbook.

The reference staff of Academic Press/Elsevier then compared this original outline to a bibliography of leading source materials in the field, including books, journal articles, conference proceedings, Web sites, and so on. Professor Cleveland refined the original outline based on this research; at this point the number of possible entries was about 500, much larger than the eventual total would be for the published encyclopedia.

The outline was then cast in the form of a preliminary table of contents and was presented to members of the editorial board at a two-day conference in San Diego, California, United States. A number of proposed revisions for the contents list emerged from this forum, including suggestions for various topics worthy of being added and the identification of some existing topics that could be merged with others or dropped as nonessential.

Professor Cleveland then prepared another version of the thematic topic list in which he made certain revisions based on the editors' recommendations, and also other adjustments based on his own expert judgment. The Academic Press/Elsevier staff rearranged this thematic list into the alphabetical format that the actual encyclopedia would require and revised the wording of article titles as needed to accomplish this. The result was a working entry list of about 400 topics, which, after some attrition and the further combining of related topics, resulted in the final table of contents of 380 articles.

ORGANIZATION

For the purpose of this encyclopedia, the chief editor and the associate editors, in collaboration with the Academic Press/Elsevier staff, have defined the field of energy as consisting of 20 distinct subject areas, as follows:

Basics of Energy
Coal
Conservation and End Use
Economics of Energy
Electricity
Energy Flows
Environmental Issues
Global Issues
History and Energy
Material Use and Reuse
Measurement and Models
Nuclear Power
Oil and Natural Gas
Policy Issues
Public Issues
Renewable and Alternative Sources
Risks
Society and Energy
Sustainable Development
Systems of Energy

Every article in the encyclopedia is designated as part of one (or more) of these 20 subject areas. For example, various articles on energy pricing appear in the economics of energy subject area, and articles on climate change appear in the environmental issues section. (Please see p. xvii of this introductory section for a complete listing of the articles in the encyclopedia according to their subject area.) This table of contents by subject area is a good starting place for a reader who wants information on different facets of a particular issue, such as the relationship between energy use and sustainable development.

FORMAT

All the articles in the *Encyclopedia of Energy* are arranged in a single alphabetical sequence according to the wording of the article title. The placement of the articles by volume is as follows: articles whose titles begin with the letters A to Ea are in Volume 1, Ec to Ge in Volume 2, Gl to Ma in Volume 3, Me to Pe in Volume 4, Ph to S in Volume 5, and T to Z in Volume 6, along with the combined glossary, the appendix, and the subject index.

So that they can be easily located, article titles generally begin with the key word or phrase indicating the topic, with any generic terms following. Thus, for example, "Hydrogen, End Uses and Economics" is the article title rather than "End Uses and Economics of Hydrogen" and "Coal Mining, Design and Methods of" is the title rather than "Design and Methods of Coal Mining." This approach also allows related topics to be grouped together alphabetically, for example, a series of articles beginning with the word "Solar."

OUTLINE

Entries in the encyclopedia begin with a topical outline that indicates the general content of the article. This outline serves two functions. First, it provides a preview of the article, so that the reader can get a sense of what is contained there without having to leaf through the pages. Second, it serves to highlight important subtopics to be discussed within the article. For example, the article "Earth's Energy Balance" (by Kevin E. Trenberth) has this outline:

1. **The Earth and Climate System**
2. **The Global Energy Balance**
3. **Regional Patterns**
4. **The Atmosphere**
5. **The Hydrological Cycle**
6. **The Oceans**
7. **The Land**
8. **Ice**
9. **The Role of Heat Storage**
10. **Atmosphere–Ocean Interaction: El Niño**
11. **Anthropogenic Climate Change**
12. **Observed and Projected Temperatures**

The outline is intended as an overview and thus it lists only the major headings of the article. In addition, extensive second-level and third-level headings will be found within the article.

GLOSSARY

The glossary section appears before the beginning of the article text and is set off typographically from the narrative to follow. It contains terms that are important to an understanding of the article and that may be unfamiliar to the reader. Each term is defined within the context of the particular article in which it is used. Thus the same term may appear in more than one article with slightly varying definitions. The encyclopedia includes approximately 2,400 glossary terms. For example, the article "Oil and Natural Gas Liquids: Global Magnitude and Distribution" (by Thomas S. Ahlbrandt) includes the following glossary entries (among others):

continuous accumulations Petroleum that occurs in an extensive reservoir or reservoirs and is not necessarily related to conventional structural or stratigraphic traps. These accumulations of oil and/or gas lack well-defined down-dip petroleum/water contacts and thus are not localized by the buoyancy of oil or natural gas in water.

conventional accumulations Petroleum that occurs in structural or stratigraphic traps, commonly bounded by a down-dip water contact, and therefore affected by the buoyancy of petroleum in water.

crude oil A mixture of hydrocarbons that exists in a liquid phase in natural underground reservoirs and remains liquid at atmospheric pressure after passing through surface separation facilities. Crude oil may also contain some nonhydrocarbon components; referred to as oil in this article.

cumulative production Volumes of oil and natural gas liquids that have been produced.

endowment The sum of cumulative production, remaining reserves, mean undiscovered recoverable volumes, and mean additions to reserves by field growth.

field A contiguous area consisting of a single reservoir or multiple reservoirs of petroleum, all grouped on, or related to, a single geologic structural and/or stratigraphic feature.

future petroleum The sum of the remaining reserves, mean reserve growth, and the mean of the undiscovered volume. Cumulative production does not contribute to the future petroleum. The terms future oil, future liquid volume, or future endowment are sometimes used as variations of future petroleum to reflect those resources that are yet to be produced.

reserve The estimated quantities of petroleum expected to be commercially recovered from known accumulations relative to a specified date, under prevailing economic conditions, operating practices, and government regulations. Reserves are part of the identified (discovered) resources and include only recoverable materials.

reserve (field) growth The increases of estimated petroleum volume that commonly occur as oil and gas fields are developed and produced.

resource A concentration of naturally occurring solid, liquid, or gaseous hydrocarbons in or on the earth's crust, some of which is currently or potentially economically extractable.

DEFINING STATEMENT

The text of each article begins with a single introductory paragraph that precedes the body of the article. This introduction defines the topic under discussion and summarizes the content of the article. For example, the entry "Sociopolitical Collapse, Energy and" (by Joseph A. Tainter) begins with the following defining paragraph:

> Collapse is the rapid simplification of a society. It is the sudden, pronounced loss of an established level of social, political, or economic complexity. Widely known examples include the collapses of Mesopotamia's Third Dynasty of Ur (ca. 2100–2000 BC), the Mycenaean society of Greece (ca. 1650–1050 BC), the Western Roman Empire (last emperor deposed 476 AD), and Maya civilization of the lowlands of Guatemala (ca. 250–800 AD). There are at least two dozen cases of collapse that are known from history, archaeology, or both. States and empires are not the only types of institutions that may rapidly simplify. The entire spectrum of societies, from simple foragers to extensive empires, yields examples of collapse. Since all but a few human societies existed before the development of writing, there may be dozens or even hundreds of cases that are not yet recognized archaeologically. Collapse is therefore a recurrent process, and perhaps no society is invulnerable to it.

CROSS-REFERENCES

All the articles in the encyclopedia have cross-references to other articles. These appear at the end of the article, following the end of the narrative text and preceding the further reading section. The encyclopedia contains about 3,300 cross-references in all. The cross-references indicate related articles that can be consulted for further information on the same topic, or for information on a related topic. For example, the article "European Union Energy Policy" (by Felix C. Matthes) has been provided with the following cross-references:

> Equity and Distribution in Energy Policy • Fuel Economy Initiatives: International Comparisons • Geopolitics of Energy • National Energy Policy:

Brazil • National Energy Policy: China • National Energy Policy: India • National Energy Policy: Japan • National Energy Policy: United States • Research and Development Trends for Energy

FURTHER READING

The further reading section appears as the last element in an article. It consists of a selection of materials chosen by the author to provide readers with further information on the article topic. This section lists recent secondary sources to aid the reader in locating more detailed or more technical information. Review articles and research papers that are important to an understanding of the topic are also listed. For example, the article "OPEC, History of" (by Fadhil J. Chalabi) has the following suggested readings:

BP Amoco. (1970–2001). "Annual Statistical Reviews." BP Amoco, London.

Center for Global Energy Studies. (1990–2002). "Global Oil Reports." CGES, London.

Chalabi, F. J. (1980). "OPEC and the International Oil Industry: A Changing Structure." Oxford University Press, Oxford, UK.

Chalabi, F. J. (1989). "OPEC at the Crossroads." Pergamon, Oxford, UK.

Penrose, E. (1968). "The Large International Firm in Developing Countries: The International Petroleum Industry." Allen & Unwin, London.

Sampson, A. (1975). "The Seven Sisters." Hodder & Stroughton, London.

The further reading references are for the benefit of the reader; thus they consist of a limited number of entries. They do not represent a complete listing of all the sources consulted by the author in preparing the paper.

INDEX

A subject index is located at the end of Volume 6. This index is the most convenient way to locate a desired topic within the encyclopedia and thus it should be the first point of reference for any reader seeking to find a particular topic. The entries in the index are listed alphabetically and indicate the volume and page number where information on this topic can be found.

Acid Deposition and Energy Use

JAN WILLEM ERISMAN
Energy Research Centre of the Netherlands
Petten, The Netherlands

1. **Introduction**
2. **Acid Deposition, Its Effects, and Critical Loads**
3. **Processes in the Causal Chain: Emission, Transport, and Deposition**
4. **Emissions from Energy Use**
5. **Abatement and Trends in Emission**
6. **Acid Deposition**
7. **Benefits and Recovery of Ecosystems**
8. **Future Abatement**
9. **Conclusion**

Glossary

acid deposition The removal of acidic or acidifying components from the atmosphere by precipitation (rain, cloud droplets, fog, snow, or hail); also known as acid rain or acid precipitation.

acidification The generation of more hydrogen ions (H^+) than hydroxide ions (OH^-) so that the pH becomes less than 7.

critical level The maximum pollutant concentration a part of the environment can be exposed to without significant harmful effects.

critical load The maximum amount of pollutant deposition a part of the environment can tolerate without significant harmful effects.

deposition Can be either wet or dry. In dry deposition, material is removed from the atmosphere by contact with a surface. In wet deposition, material is removed from the atmosphere by precipitation.

emissions The release of primary pollutants directly to the atmosphere by processes such as combustion and also by natural processes.

eutrophication An increase of the amount of nutrients in waters or soils.

nitrification The conversion of ammonium ions (NH_4^+) to nitrate (NO_3^-).

nonlinearity The observed nonlinear relationship between reductions in primary emissions and in pollutant deposition.

pollutant Any substance in the wrong place at the wrong time is a pollutant. Atmospheric pollution may be defined as the presence of substances in the atmosphere, resulting from man-made activities or from natural processes, causing effects to man and the environment.

Acid deposition originates largely from man-made emissions of three gases: sulfur dioxide (SO_2), nitrogen oxides (NO_x), and ammonia (NH_3). It damages acid-sensitive freshwater systems, forests, soils, and natural ecosystems in large areas of Europe, the United States, and Asia. Effects include defoliation and reduced vitality of trees; declining fish stocks and decreasing diversity of other aquatic animals in acid-sensitive lakes, rivers, and streams; and changes in soil chemistry. Cultural heritage is also damaged, such as limestone and marble buildings, monuments, and stained-glass windows. Deposition of nitrogen compounds also causes eutrophication effects in terrestrial and marine ecosystems. The combination of acidification and eutrophication increases the acidification effects. Energy contributes approximately 82, 59, and 0.1% to the global emissions of SO_2, NO_x, and NH_3, respectively. Measures to reduce acid deposition have led to controls of emissions in the United States and Europe. Sulfur emissions were reduced 18% between 1990 and 1998 in the United States and 41% in Europe during the same period. At the same time, emissions in Asia increased 43%. In areas of the world in which emissions have decreased, the effects have decreased; most notably, lake acidity has decreased due to a decrease in sulfur emissions, resulting in lower sulfate and acid concentrations. However, systems with a long response time have not seen improvement yet and very limited recovery has also been observed. Emissions should therefore be further reduced and sustainable levels should be maintained to decrease the effects and to see recovery. This will require drastic changes in our energy consumption and the switch to sustainable energy sources. Before renewable energy sources can fulfill a large part of our energy needs, the use of zero-emission fossil fuels must be implemented. This should be done in such a way that different environmental impacts are addressed at the same time. The most cost-effective way is to reduce CO_2 emissions because this will result in decreases in SO_2 and NO_x emissions. When starting with SO_2 and NO_x emissions, an energy penalty compensates for the emission reductions.

1. INTRODUCTION

The alarm regarding the increasing acidification of precipitation in Europe and eastern North America was first sounded in the 1960s. Since then, most attention has focused on acid rain's effects, established and suspected, on lakes and streams, with their populations of aquatic life, and on forests, although the list of concerns is far broader: It includes contamination of groundwater, corrosion of man-made structures, and, recently, deterioration of coastal waters.

The processes that convert the gases into acid and wash them from the atmosphere began operating long before humans started to burn large quantities of fossil fuels. Sulfur and nitrogen compounds are also released by natural processes such as volcanism and the activity of soil bacteria. However, human economic activity has made the reactions vastly more important. They are triggered by sunlight and depend on the atmosphere's abundant supply of oxygen and water.

Wood was the first source for energy production. Wood was used for the preparation of food and for heating. After the discovery of fossil fuels in the form of coal as a source of energy, it was used for heating purposes. During the industrial revolution, coal, followed by crude oil and later natural gas, provided a large source of energy that stimulated technological development, resulting in the replacement of animals and people in the production of goods, increased possibilities for mobility, and the enabling of industries to process raw materials into a large variety of products. When the possibilities of energy production and use became apparent, the use of fossil fuels grew exponentially. Figure 1 provides an overview of the worldwide use of fossil fuels since 1900. It shows that the total global energy consumption is closely linked to the growth of the world's population. The share of oil and gas as energy

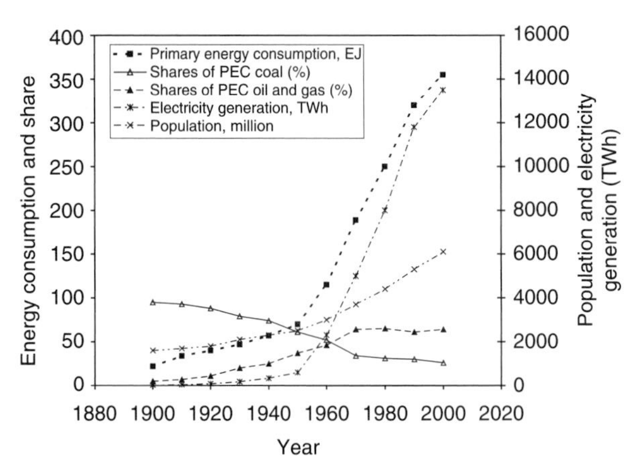

FIGURE 1 The worldwide use of fossil fuels in millions of tons per year. Adapted from Smil (2000).

sources has increased and has become higher than the share of coal.

The first environmental consequences of the use of fossil fuels were the problems of smoke in houses and urban air quality. Human health was affected by the use of coal with high sulfur content, producing sulfur dioxide and carbon monoxide. The history of London clearly shows the ever-increasing problems related to coal use and the struggle to abate this pollution, culminating in the famous London smog in 1952, from which thousands of people died. In the 1950s and 1960s, air pollution became an important issue but was mainly considered on the local scale. One of the abatement options was to increase the stack height in order to take advantage of atmospheric dispersion, leading to dilution of pollutant concentrations before humans could be exposed. This formed the basis of the transboundary nature of acid deposition as we know it today and increased the scale from local to regional and even continental. The use of fossil fuels increased, and the pollutants were transported to remote areas, which had very clean air until that time.

Acid rain was first recognized as a serious environmental problem by Robert Angus Smith, an alkali inspector based in Manchester. He started a monitoring network with approximately 12 sites at which rainfall was collected and analyzed. In 1852, he wrote about the correlation between coal burning and acid pollution in and around the industrial center of Manchester. He identified three types of air: "that with carbonate of ammonia in the fields at a distance ... that with sulphate of ammonia in the suburbs, ... and that with sulphuric acid, or acid sulphate, in the town." In 1872, he wrote the famous book, *Air and Rain: The Beginnings of a Chemical Climatology.* He referred to the acidity of rain and air and its effect on humans and buildings. This was the beginning of acidification research, although this was recognized only much later.

Acid deposition is one of the oldest transboundary environmental problems. Several components and thus emission sources contribute to acid deposition, making it a complex problem and difficult to abate. It is one of the most studied environmental problems and there is a large body of literature. This article gives an overview of the state of knowledge of energy production and use and acid deposition. First, an overview is given of acid deposition, the consequences of acid deposition, and the processes involved. The contribution of energy is discussed next, and an estimate of the progress made to decrease the emissions from energy sources is given. Finally, future options to decrease acid deposition and the options for decreasing energy emissions are discussed.

2. ACID DEPOSITION, ITS EFFECTS, AND CRITICAL LOADS

Acid deposition is the deposition of air pollutants to ecosystems leading to acidification and eutrophication, which cause negative effects in these ecosystems. The components contributing to acid deposition are sulfur compounds (SO_2 and SO_4), oxidized nitrogen compounds (NO, NO_2, HNO_2, HNO_3, and NO_3), and reduced nitrogen compounds (NH_3 and NH_4). Some rain is naturally acidic because of the carbon dioxide (CO_2) in air that dissolves with rain water and forms a weak acid. This kind of acid is actually beneficial because it helps dissolve minerals in the soil that both plants and animals need. Ammonia acts as a base in the atmosphere, neutralizing nitric and sulfuric acid, but in soil ammonia it can be converted by microorganisms to nitric acid, producing additional acid in the process.

The first sign of the effects of acid deposition was the loss of fish populations in Canadian, Scandinavian, and U.S. lakes in the early 1960s. It was found that the water in the lakes was acidified to a point at which fish eggs no longer produced young specimens. This was caused by acid deposition; precipitation had introduced so much acid in the lakes that pH levels had declined below the critical limit for egg hatching. A few years later, reports from Canada, the United States, and Germany indicated unusual damage in forests. Trees showed decreased foliage or needles, and this damage even progressed to the point that trees would die, a phenomenon that happened in the border area of former East Germany, Poland, and former Czechoslovakia. This tree dieback was also attributed to acidification of the soil. However, exposure to increased oxidant concentrations was determined to be another cause. In the mid-1980s, nitrogen deposition to terrestrial and aquatic ecosystems was found to cause negative effects through eutrophication. Acidification and eutrophication are both caused by atmospheric deposition of pollutants, and the combination increases the effects, causing too high nutrient concentrations in soil and groundwater. Nitrate and ammonium are beneficial, even essential, for vegetation growth, but in too high concentration they lead to the loss of diversity, especially in oligotrophic ecosystems (i.e., ecosystems adapted to low nutrient availability). This problem has mainly been encountered in central and northwestern Europe.

The effects of acid deposition include changes in terrestrial and aquatic ecosystems showing acidification of soils, shifts in plant community composition, loss of species diversity, forest damage, water acidification and loss of fish populations, reduction in growth yield of sensitive agricultural crops, and damage to cultural heritage and to building materials. The common factor for all these effects is that pollutants, or their precursors, are emitted, transported in the atmosphere, and deposited by way of precipitation (wet deposition) or as gases or particles (dry deposition). In some cases, acid deposition occurs 1000 km or more from where the responsible emissions were generated.

If deposition exceeds sustainable levels, effects can occur depending on the duration of the exposure and the level of exceedance. Sustainable levels can be represented by critical loads and critical levels. The critical load is formally defined by the United Nations Economic Commission for Europe (UNECE) as "a quantitative estimate of exposure to one or more pollutants below which significant harmful effects on sensitive elements of the environment do not occur according to present knowledge." This definition can be applied to a wide range of phenomena and not merely acid rain. Any process in which damage will occur if depositions exceed natural sink processes has a critical load set by those sink processes. The concept of critical loads or levels has been used to provide a scientific basis for the incorporation of environmental impacts in the development of national and international policies to control transboundary air pollutants, within the framework of the UNECE Convention on the Control of Transboundary Air Pollution. The concept is based on the precautionary principle: Critical loads and levels are set to prevent any long-term damage to the most sensitive known elements of any ecosystem. The critical loads approach is an example of an effects-based approach to pollution control, which considers that the impact of deposited material depends on where it lands. Similar industrial plants may have different pollution impacts depending on the different capacities of the receiving environments to absorb, buffer, or tolerate pollutant loads. The policy goal with effects-based strategies is ultimately to ensure that the abatement on emissions from all plants is adequate so that no absorptive capacities, nor air quality standards, are exceeded.

3. PROCESSES IN THE CAUSAL CHAIN: EMISSION, TRANSPORT, AND DEPOSITION

The causal chain represents the sequence of emissions into the atmosphere, transport and chemical reactions, deposition to the earth surface, and cycling within the ecosystems until eventually the components are fixed or transformed in a form having no effects. This causal chain is shown in Fig. 2. After the causal chain is understood and parameterized in models, it can be used to assess the possibilities for reducing the effects.

3.1 Emission

The use of fossil fuels for energy production is one of the main causes of acid deposition. Fossil fuels contain sulfur compounds, which are emitted into the atmosphere as SO_2 after combustion. Oxidized nitrogen has several sources but is formed mainly by combustion processes in which fuel N is oxidized or atmospheric N_2 is oxidized at high temperatures. These processes are relevant for industries, traffic, and energy production. Among the other sources of oxidized nitrogen, soils are most important. Ammonia is mainly emitted from agricultural sources. Fossil fuels seem less important for ammonia than for the other two pollutants. Through the use of fertilizer, which is produced from natural gas, there is a link with energy. Indirectly, the increased energy use from fossil fuels to produce fertilizer has increased nitrogen emissions from agriculture and contributed to acidification. Energy use first increased mobility. Food is therefore not necessarily produced in the area where it is consumed. Furthermore, relatively cheap nutrient resources (animal feed) are transported from one continent to another, where they are concentrated in intensive livestock breeding areas. Also, fossil fuels, first in the form of coal but now mainly from natural gas, provide the basis to produce fertilizers, which are used to produce the food necessary to feed the ever-growing population throughout the world and to fulfill the need for the growing demands for more luxurious food.

FIGURE 2 Emission, transport, and deposition of acidifying pollutants.

Base cations play an important role because they can neutralize acids in the atmosphere and the soil, limiting the effect of acid deposition. On the other hand, base cations contribute to the particle concentrations in the atmosphere, affecting human health when inhaled. Sources of base cations are energy production, agricultural practices, and road dust.

3.2 Transport and Deposition

The sulfur compounds are converted to sulfuric acid and NO_x to nitric acid, and ammonia reacts with these acids to form ammonium salts. Both precursors (SO_2, NO_x, and NH_3) and products (sulfuric acid, nitric acid, and ammonium salts) can be transported over long distances (up to 1500 km), depending on meteorological conditions, speed of conversion, and removal by deposition processes. The gases emitted, SO_2, NO_x, and NH_3, behave differently in the atmosphere. The residence time of NH_3 is relatively short because it is emitted at low level (near the ground), it converts quickly to NH_4^+, and the dry deposition rate is fairly high. NH_3 is transported 100–500 km. The horizontal and vertical concentration gradients are steep, so concentrations and depositions can vary significantly over small distances.

This means that a large proportion of the NH_3 emitted in The Netherlands is also deposited within this country. Once converted into NH_4^+, which has a much lower rate of deposition, the transport distances are much greater (up to 1500 km). SO_2 is mainly emitted into the atmosphere by high sources and can therefore be transported over long distances, despite its relatively high deposition rate. Some NO_x is also emitted by low sources (traffic). However, because of its low deposition rate and relatively slow conversion rate into rapidly deposited gases (HNO_3 and HNO_2), NO_x is transported over relatively long distances before it disappears from the atmosphere. SO_2 is quickly converted to sulfuric acid (H_2SO_4) after deposition, both in water and in soil. NO_x and NH_3 and their subsequent products contribute to the eutrophication and also the acidification of the environment as a result of conversion to nitric acid (HNO_3) in the air (NO_x) or in the soil (NH_3).

Since the three primary gases (SO_2, NO_x, and NH_3) can react and be in equilibrium with each other and with the different reaction products in the atmosphere, there is a strong and complex mutual relationship. If, for example, there were no NH_3 in the atmosphere, SO_2 would be converted less quickly to SO_4^{2-}. The environment, however, would also be "more acid" so that the deposition rate of acidifying compounds would be reduced (poor solubility of these compounds in acid waters). The net impact is difficult to determine, but it is certainly true that if the emission of one of the compounds increases or decreases relative to that of the others, this will also influence the transport distances and deposition rates of the other compounds. This is only partly taken into account in the scenario calculations because such links have not been fully incorporated into the models.

The base cations can neutralize the acidifying deposition and, after deposition to the soil, they act as a buffer both in terms of neutralization and in terms of uptake by plants and the prevention of nutrient deficiencies. This applies especially to the Mg^{2+}, Ca^{2+}, and K^+ compounds. The degree of deposition of base cations is therefore important in determining the critical loads that an ecosystem can bear or the exceedances thereof. Thus, accurate loads are also required for base cations on a local scale. These estimates are not available for The Netherlands nor for Europe.

The nature and size of the load of acidifying substances on the surface depend on the characteristics of the emission sources (height and point or diffuse source), the distance from the source, physical and chemical processes in the atmosphere, and the receptor type (land use, roughness, moisture status, degree of stomatal opening of vegetation, snow cover, etc.). When gases and/or particles are deposited or absorbed directly from the air, we speak of dry deposition. When they reach the surface dissolved in rain or another form of precipitation, we refer to wet deposition. If this occurs in mist or fog, it is cloud/fog deposition. The total deposition is the sum of the dry, wet, and cloud/fog deposition. Throughfall is washed from the vegetation by rain falling on the soil beneath the forest canopy. It is a measure of the load on the vegetation's surface, whereas the total deposition indicates the load on the whole system (soil + vegetation). Net throughfall is the difference between throughfall and wet deposition in the open field; it is a measure of the dry deposition plus cloud/fog deposition when no exchange takes place with the forest canopy (uptake or discharge of substances through leaves, microorganisms, etc.).

To understand the impact of emissions on the effects caused by acid deposition and eutrophication, the whole chain of emissions, transport, atmospheric conversion, dry and wet deposition, and the effects caused by the total deposition loads, including the role of groundwater and soils, must be understood. Wet and dry deposition are not only the processes by which pollutants are transported to the earth surface and humans and ecosystems are exposed to acid deposition and eutrophication but also play an essential role in cleaning of the atmosphere. If removal by wet and dry deposition stopped, the earth's atmosphere would be unsuitable to sustain life in a relatively short period—on the order of a few months.

Since acidifying deposition involves different substances, it is necessary to give these substances a single denominator in order to indicate the total load of acidifying substances. For this purpose, the total load is expressed as potential acid, calculated as follows:

$$2SO_x + NO_y + NH_x (\text{mol H}^+\text{ha/year}),$$

where SO_x are oxidized sulfur compounds, NO_y are oxidized nitrogen compounds, and NH_x are reduced nitrogen compounds. The concept of potential acid is used because NH_3 is considered to be a potentially acidifying substance. In the atmosphere, NH_3 acts as a base, which leads to the neutralization of acids such as HNO_3 and H_2SO_4. However, the NH_4^+ formed in the soil can be converted to NO_3^- so that acid is still produced via bacterial conversion (nitrification) according to

$$NH_4^+ + 2O_2 \rightarrow NO_3^- + H_2O + 2H^+.$$

Two moles of acid are finally formed via this process: one originating from the neutralized acid and one originating from NH_3. On balance, as for 1 mol NO_y, 1 mol NH_3 acts maximally to acidify 1 mol H^+ acid. One mole of the bivalent SO_4^{2-} can lead to the formation of 2 mol of H^+. The actual acidification depends on the degree to which NO_3^- and SO_4^{2-} leach out of the soil. Only when this occurs completely is the actual acidification equal to the potential acidification.

4. EMISSIONS FROM ENERGY USE

van Ardenne *et al.* provide a compilation of global sulfur emissions, mainly based on work by different authors. According to their estimates, anthropogenic contributions are on the order of 75% of the total sulfur emissions, 82% of which is related to energy production and use (Fig. 3). Anthropogenic emissions exceeded natural emissions as early as 1950. Important natural sources are volcanoes and DMS emissions from the oceans. Fossil fuel use in industry, fuel production in refineries, and electricity generation are the main activities responsible for anthropogenic emissions. According to EMEP (Co-operative Program for Monitoring and Evaluation of Air Pollutants in Europe) studies, the maximum sulfur emissions in western Europe occurred from 1975 to 1980, at a level of 40 million tons of SO_2. The emissions have decreased since then and are currently at a level of approximately 30 million tons of SO_2.

On a global scale, NO_x emissions are estimated to be equally divided between anthropogenic and biogenic sources. However, anthropogenic contributions are already much higher than the natural emissions in the Northern Hemisphere. Most natural emissions, through biomass burning, occur in agricultural areas of South America and Africa. In western Europe and the United States, traffic is the main source for NO_x, contributing 50% or more of the total NO_x emissions.

Globally, in 1995 the emissions from fossil fuels contributed 82, 59, and 0.1% to the total emissions of SO_2, NO_x, and NH_3, respectively. The contributions of fuel and other anthropogenic sources to the emissions of the three gases are shown in Fig. 4.

5. ABATEMENT AND TRENDS IN EMISSION

Figure 5 shows the change in global emissions of SO_2, NO_x, and NH_3 for the different source categories. The figure shows that there is a steady increase in emissions for the three gases between 1890 and 1995. The contribution of energy emissions is highest during the whole time period for SO_2 and NO_x; for NH_3, energy constitutes only a minor contribution. After World War II, the increase in emission is higher.

After 1990, there seems to be a smaller increase in global emissions. This is mainly due to the decrease

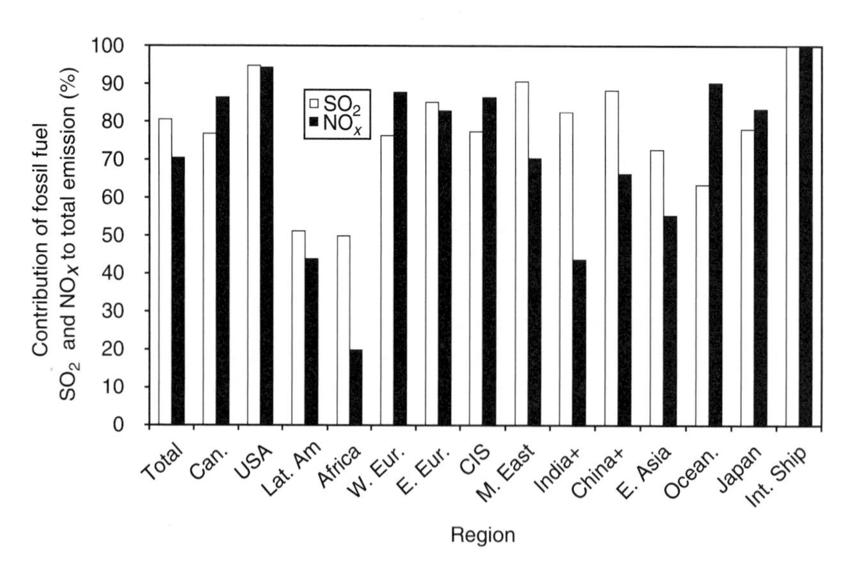

FIGURE 3 Share of fuel emissions of the total global sulfur and nitrogen emissions of 148.5 Tg SO_2 and 102.2 Tg NO_2. Data from van Ardenne *et al.* (2001).

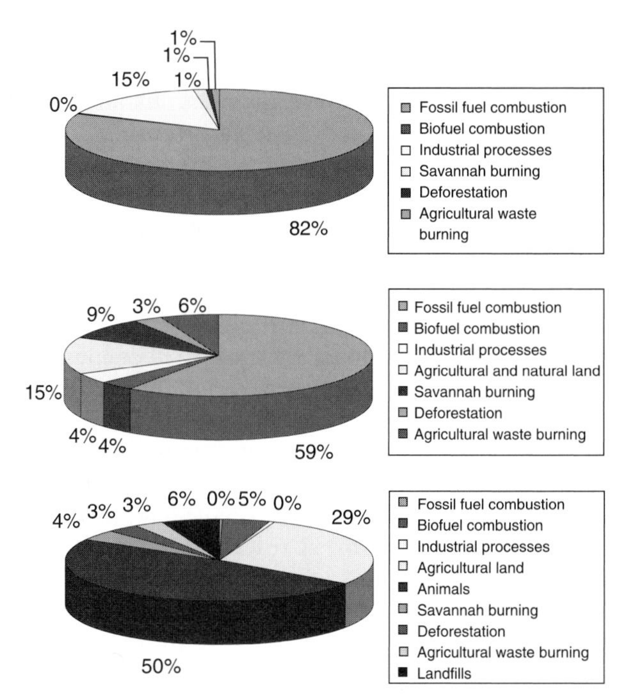

FIGURE 4 Contribution of different sources to the total anthropogenic emission of SO_2 (top of the figure), NO_x (middle), and NH_3 (bottom).

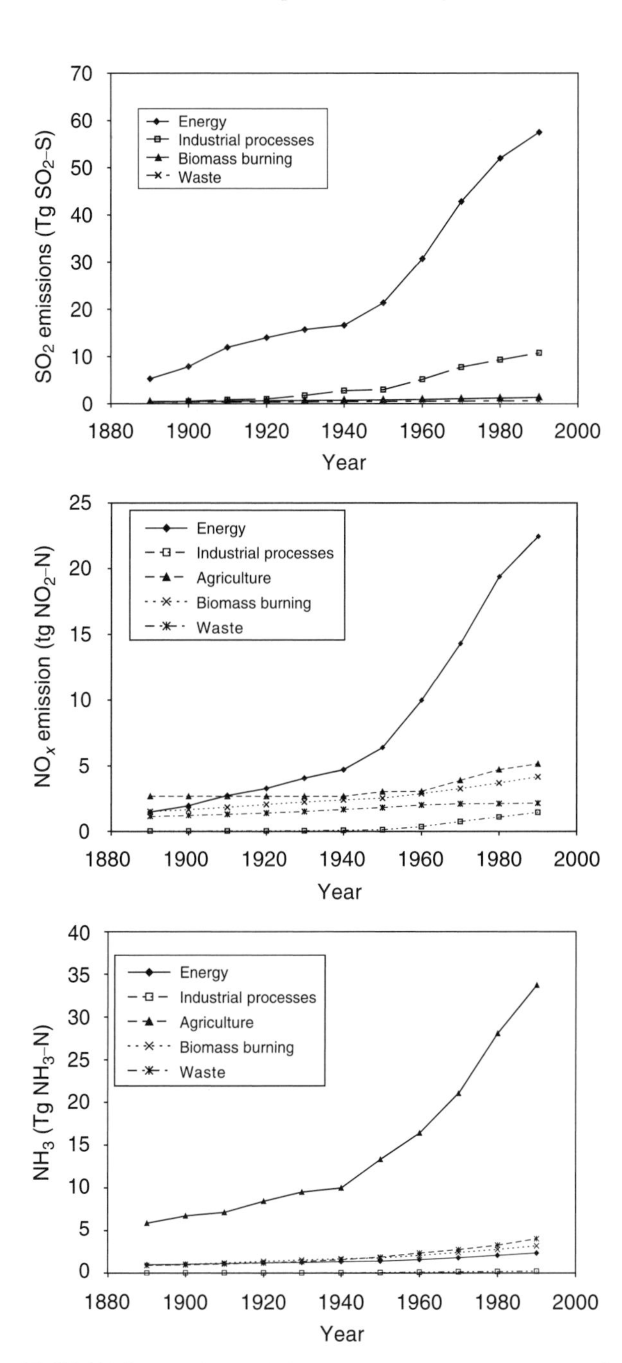

FIGURE 5 Trend in global SO_2, NO_x, and NH_3 emissions of different sources. Data from van Ardenne *et al.* (2001).

of emissions of SO_2 and, to a lesser extent, NO_x in the United States and Europe. This leveling is a net effect of decreased emissions in the industrialized areas of Europe and the United States and increased emissions in developing countries, especially Asia. Table I shows the estimated changes in emissions between 1990 and 1998 in Europe, the United States, and Asia. Europe shows the most positive trends in the world. Emission trends in the United States between 1990 and 1998 are somewhat upward for the nitrogen components, whereas SO_2 and volatile organic compounds (VOCs) show a downward trend of 18 and 14%, respectively. Emissions of SO_2, NO_x, NH_3, and VOCs show a downward trend of 41, 21, 14, and 24%, respectively. Clearly, policy has been much more successful in Europe than in the United States.

Table I shows that emissions of SO_2 and VOCs in the United States and Europe have decreased, whereas in Asia they have increased. The nitrogen components decreased only in Europe, whereas in the United States they remained about the same. In Asia, NO_x emissions more than doubled, whereas for NH_3 no data are available. There is a distinct difference between the policies in the United States and those in Europe to combat acidification. In the United States, the Clean Air Act is the driving force to reduce emissions, and in Europe the reductions are achieved through protocols under the Convention of Long-Range Transboundary Air Pollution, with the multipollutant–multieffect protocol signed in 1999 in Gothenburg being the most recent.

5.1 United States

For SO_2, the Acid Rain Program places a mandatory ceiling, or cap, on emissions nationwide from electric

TABLE I

Percentage Changes in Total Emissions of SO_2, NO_x, NH_3, and VOCs between 1990 and 1997 in Europe, the United States, and Asia

Component	Asia[a]	Europe	United States	Global
SO_2	+43	−41	−18	−3
NO_x	+29	−21	+1	+3
NH_3	n.a.[b]	−14	+4	n.a.
VOC	+12	−24	−14	+1

[a] Data from IPCC (2000) for the years 1990 to 1997. Streets *et al.* (2001) report increases of 16% for SO_2 and 53% for NO_x for northeast Asia.

[b] n.a., not available.

utilities and allocates emissions to these pollution sources in the form of allowances, or a permit to emit 1 ton of SO_2. Emissions from large electric power plants have been reduced through phase 1 of the Acid Rain Program. The emission trading has proven to be a cost-effective mechanism and has facilitated 100% compliance by affected sources and stimulated early emission reduction. Phase 2 was started in 2000 and will result in a reduction for a greater number of smaller plants, and emissions from larger plants will be further reduced.

Measures to limit NO_x emissions in the United States have largely focused on decreasing human health effects from tropospheric ozone, of which NO_x is a primary precursor. NO_x abatement has also been achieved as a means to address ecosystem acidification in conjunction with much larger decreases in SO_2 emissions. NO_x emissions from stationary and mobile sources are limited by efforts to comply with various current and future regulations, including National Ambient Air Quality Standards, New Source Performance Standards, Title IV of the 1990 Clean Air Act Amendments, the Ozone Transport Commission NO_x Budget Allowance Trading Program, state-implemented plans for NO_x emissions decreases, Section 126 of the 1990 Clean Air Act Amendments, and Mobile Source Emission Limits.

5.2 Europe

In Europe, the major efforts to decrease the effects of SO_2 and nitrogen emissions have been aimed at decreasing transfers to air, soil, and groundwater. Most of the measures in Europe are focused on decreasing human and plant exposure and decreasing ecosystem loads leading to acidification and eutrophication. Countries agreed to decrease air emissions by signing different protocols developed under the Convention on Long-Range Transboundary Air Pollution, including the first and second sulfur protocols, the first NO_x protocol, and the NMVOC (Non-Methane Volatile Organic Compounds) protocol. The last protocol, the Göteborg Protocol, is unique in the sense that it requires decreases in emissions of four pollutants with the objective of abating three specific effects—acidification, eutrophication, and the effects from tropospheric ozone on human health and vegetation. The protocol, which has been signed by 29 European countries together with the United States and Canada, is based on a gap-closure method aiming at decreasing the spatial exceedances of critical loads and levels in the most cost-efficient way. Critical loads for each European country are defined on the basis of information developed by each country. The agreed upon decreases in emissions for the European Union (EU) member states are expected to lead to overall decreases in the European (except Russia) emissions of 78% for SO_2 and approximately 44% for NO_x during the period 1990 to 2010. The corresponding figure for ammonia is 17%.

The protocols have had a major effect on the emission trends in Europe, especially for SO_2 (Table I). European emissions reductions are being made with the clear objective that environmental loads, exposures, and effects should be decreased—the so-called "effects-based approach," which was initiated under the Second Sulphur Protocol. For NO_x, the main measure to decrease emissions is exhaust gas regulations introduced in the EU countries in approximately 1990, resulting in the application of three-way catalysts in gasoline cars. Even regulations on heavy-duty vehicles have caused emission reductions of NO_x. Furthermore, selective catalytic reduction technologies (SCR) with ammonia or urea as a reductor have been implemented in many combustion plants. In eastern Europe, the main cause of decreases in NO_x emissions is the shutdown of a large number of industrial plants. Despite declining emissions rates, total NO_x emissions have remained steady or even increased during the same period in North America and Europe due to increases in vehicle kilometers/miles traveled, increases in electricity usage, and the sometimes differing regulatory frameworks applied to various sectors. Decreasing total NO_x emissions likely means that continuing technological advances need to be combined with regulatory approaches (e.g., emissions caps).

6. ACID DEPOSITION

Within the causal chain, the activities and resulting emissions form the basis of the effects and thus the abatement should be focused on the emissions. However, the effects are linked primarily with acid deposition. Therefore, the real progress in reducing the effects should be derived from trends in effects or in acid deposition. Erisman *et al.* evaluated the progress in acid and nitrogen policies in Europe. Potential acid deposition, the molar sum of SO_2, NO_x, and NH_3, has decreased 2.5% per year between 1980 and 1999. This is mainly due to the reduction in sulfur emissions, which have decreased 59% averaged throughout Europe. The decrease in emissions has resulted in improved environmental quality. Ambient concentrations of SO_2 have decreased well below critical levels. NO_2 exceedances of critical levels are now only observed near sources and in industrialized areas. Deposition measurements are limited mostly to wet deposition. Trends in measurements show a decrease in wet deposition. However, the decrease is smallest in remote areas, where the most vulnerable ecosystems are located. This is due to nonlinearity in emissions and depositions. Consequently, appreciable emission reductions are still needed to protect ecosystems in remote areas. Although acidity has been decreased to a large extent, nitrogen pollution is still excessive and needs to be reduced to reach critical loads.

In the northeast and mid-Atlantic regions of the United States, where ecosystems are most sensitive to acidic deposition, sulfate levels in precipitation (wet deposition) have declined by up to 25%, mirroring the reductions in SO_2 emissions achieved through the implementation of the Acid Rain Program. No distinct regional trends in nitrate have been detected, consistent with NO_x emissions, which have remained approximately the same. Trends in dry deposition of sulfur and nitrogen are comparable.

Because sulfur emissions decreased more than nitrogen emissions, the relative contribution of nitrogen deposition compared to sulfur deposition has increased. The focus of policies should therefore be on nitrogen emissions, both oxidized and reduced nitrogen.

The unbalanced reduction in emissions can have unexpected consequences. The large decrease in base cation emissions, which was much faster than the decrease in sulfur emissions, has caused precipitation pH to decrease rapidly in several areas. Another example is the nonlinearity in sulfur and nitrogen emission and the respective deposition in remote areas. Nonlinearity in the relationship between the emission and deposition patterns for sulfur and oxidized nitrogen has been detected, which leads to slower than expected reductions in critical loads exceedance in remote areas. There is evidence that NO_3 in rain and wet deposition is increasing at remote, mainly West Coast locations, whereas close to the major sources a small reduction has been detected. This is the result of the nonlinearity.

In relation to acidification, little emphasis is usually placed on the role of deposition of base cations such as Na^+, Mg^{2+}, Ca^{2+}, and K^+. Besides their ability to neutralize acid input, base cations are important nutrient elements for ecosystems. From a variety of measurements of Ca in precipitation in Europe and North America, Hedin *et al.* detected a long-term decline in 1994. The decline in Ca is due to the abatement of particle emissions from coal combustion and industrial sources. The reduction has resulted in a decrease in the neutralizing capacity of the atmosphere, leading to a decrease in pH. Furthermore, because some abatement of SO_2 at industrial sources coincided with a decrease in Ca, the net effect on atmospheric acidity was small; thus, recovery of ecosystems would not proceed as expected. At the time, no data on long-term Ca aerosol concentrations were available. In 1998, Lee *et al.* reanalyzed previously unreported long-term data sets of the atmospheric aerosol composition for the United Kingdom and showed a clear decline in Ca during the period 1970–1995. Paralleling this decline were trends in scandium and arsenic. Scandium is often used as a tracer for noncombustion sources, whereas arsenic is often used as a tracer for combustion. Declines of both metals clearly show a strong decline in fly-ash from industrial and combustion sources; hence, the decline in Ca can be understood.

7. BENEFITS AND RECOVERY OF ECOSYSTEMS

Policies to reduce emissions and deposition are ultimately aimed at decreasing negative effects and/or increasing recovery of affected ecosystems. The response of effects to decreases in emissions depends on the type of effect. Figure 6 shows relative response times to changes in emissions. The effect is fastest for air concentrations and the resulting improvement in visibility, the direct plant response, and acute human health. However, it takes longer for chronic health, aquatic systems, or plant responses to show the results of reduced emissions. It will take several years

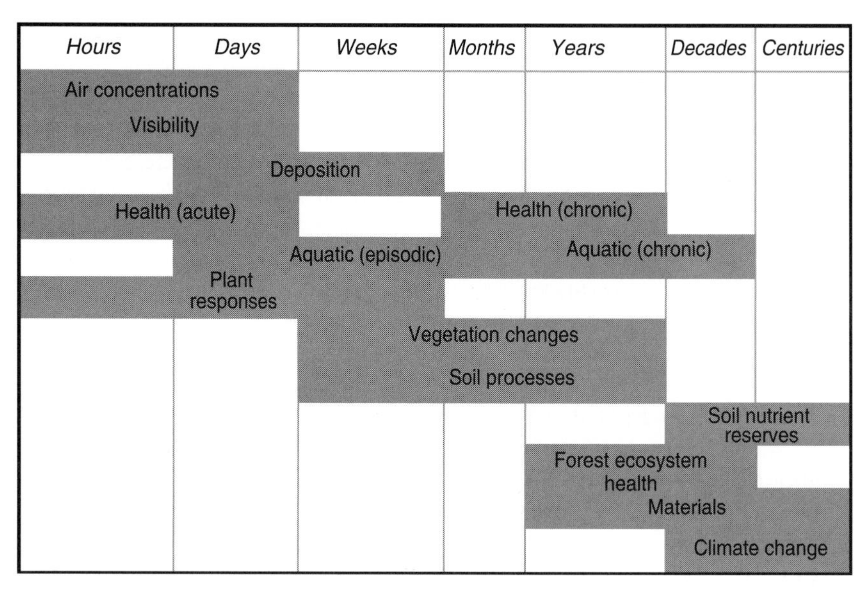

FIGURE 6 Relative response times to changes in emissions. Data from U.S. National Acid Precipitation Assessment Program (2000) and Galloway (2001).

to centuries before the forest ecosystem or soil nutrient status respond to the reductions. This difference in response times has to be taken into account when evaluating the changes in effect parameters, especially because the decrease in emissions started in the 1980s and 1990s.

7.1 Surface Water

Lake ecosystems have improved in northern Europe and North America, where emission reductions occurred. In general, SO_4 concentrations have decreased following the emission trends, but nitrogen concentrations have not shown changes. Data from 98 sites were tested for trends in concentrations over the 10-year period 1989–1998. The (grouped) sites clearly showed significant decreases in SO_4 concentrations. Nitrate, however, showed no regional patterns of change, except possibly for central Europe: Decreasing trends occurred in the Black Triangle. Concentrations of base cations declined in most regions. All regions showed tendencies toward increasing dissolved organic carbon. Recovery from acidification reflected by an increase in surface water acid neutralizing capacity (ANC) and pH was significant in the Nordic countries/United Kingdom region. In central Europe, there was a regional tendency toward increasing ANC, but large spatial differences were found with the low ANC sites showing the largest recovery. Nonforested sites showed clear and consistent signals of recovery in ANC and pH and appropriate (relative to SO_4 trends) rates of base cation declines. Hence, it was concluded that the observed recovery was associated with declining SO_4.

In the most acidic sites in central Europe, improvements in water quality have not yet resulted in improvements in biology. Biological improvements of these sites require considerable improvements in water quality with respect to acidification.

In many lakes in Scandinavia, there is evidence of a small but significant recovery and many species that died because of acidification are returning. The positive signs are mainly observed in lakes and streams with limited acidification. For the most acidified waters, the signs of recovery are still small and unclear.

In a study by Stoddard *et al.*, data for 205 sites in eight regions in North America and Europe between 1980 and 1995 were used to test trends. The data they used were primarily from the International Co-operative Program (ICP) Waters study. They found trends of decreasing SO_4 concentrations in all regions except the United Kingdom and no or very small changes in NO_3. SO_4 levels declined from 0 to -4 μeq/liter/year in the 1980s to -1 to -8 μeq/liter/year in the 1990s. Recovery of alkalinity was associated with the decrease in SO_4, especially in the 1990s.

7.2 Forests

Forested catchments in Europe are studied within the ICP Integrated Monitoring (IM). These sites are located in undisturbed areas, such as natural parks or comparable areas. The network currently covers

50 sites in 22 countries. Fluxes and trends of sulfur and nitrogen compounds were recently evaluated for 22 sites. The site-specific trends were calculated for deposition and runoff water fluxes and concentrations using monthly data and nonparametric methods. Statistically significant downward trends of SO_4 and NO_3 bulk deposition (fluxes and concentrations) were observed at 50% of the sites. Sites with higher nitrogen deposition and lower carbon/nitrogen ratios clearly showed an increased risk of elevated nitrogen leaching. Decreasing SO_4 and base cation trends in output fluxes and/or concentrations of surface/soil water were commonly observed at the ICP IM sites. At several sites in Nordic countries, decreasing NO_3 and H^+ trends (increasing pH) were also observed. These results partly confirm the effective implementation of emission reduction policy in Europe. However, clear responses were not observed at all sites, showing that recovery at many sensitive sites can be slow and that the response at individual sites may vary greatly.

Defoliation and discoloration of stands have been monitored in Europe on a regular 16×16-km grid since 1986—the so-called level 1 program [ICP-Forests (F)]. The network currently covers 374,238 sample trees distributed on 18,717 plots in 31 countries. At approximately 900 sites in Europe, key parameters, such as deposition, growth, soil chemistry, and leaf content, have been monitored since 1994 to study cause–effect relationships and the chemical and physical parameters determining forest ecosystem vitality (level 2). The duration of the level 2 program is too short for trend detection. These data are more relevant for studying cause–effect relationships. Figure 7 shows the annual variation in defoliation for different tree species in Europe as reported in the extended summary report by ICP-F. It is concluded by ICP-F that in all parts of Europe there is defoliation to a various extent. Of the 1999 total transnational tree sample, mean defoliation is 19.7%. Of the main tree species, *Quercus robur* has the highest defoliation (25.1%), followed by *Picea abies* (19.7%), *Fagus sylvatica* (19.6%), and *Pinus sylvestris* (18.9%). The trend in defoliation over 14 years for continuously observed plots shows the sharpest deterioration of *Pinus pinaster* and *Quercus iles* in southern Europe. *Fagus sylvatica* deteriorated in the sub-Atlantic, mountainous (south), and continental regions. *Picea abies* deteriorated in several parts of Europe but improved in the main damage areas of central Europe since the mid-1990s.

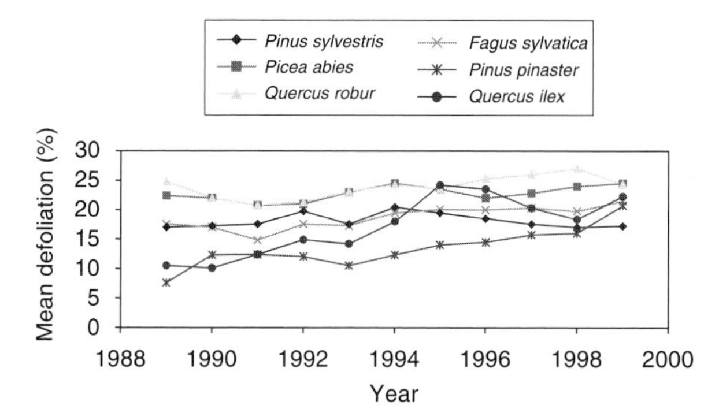

FIGURE 7 Development of mean defoliation of the six most abundant species in Europe. Data from ICP (2000).

Because there are no specific symptoms of individual types of damage, defoliation reflects the impact of many different natural and anthropogenic factors. Weather conditions and biotic stresses are most frequently cited. Several countries refer to air pollution as a predisposing, accompanying, or triggering factor, but the degree to which air pollution explains the spatial and temporal variation of defoliation on the large scale cannot be derived from crown condition assessment alone. Statistical analysis of 262 of the 860 level 2 plots indicates that defoliation is influenced mainly by stand age, soil type, precipitation, and nitrogen and sulfur deposition. These factors explain 30–50% of the variation. Nitrogen deposition correlated with defoliation of spruce and oak, and sulfur deposition was found to correlate with defoliation of pine, spruce, and oak. Calculated drought stress was found to impact nearly all tree species, whereas calculated ozone mainly impacted broadleaves, specifically the common beech trees.

8. FUTURE ABATEMENT

There are basically two ways of reducing acid deposition. Emission control technologies can be attached to smokestacks at power plants and other industries, removing the acid gases before they are emitted into the atmosphere. In coal-fired power plants, sulfur emissions are removed with a "scrubber," in which a limestone slurry is injected into the flue gas to react with the SO_2. The resulting gypsum slurry can eventually be used in other industrial processes. The main problem with scrubbers is that they are expensive, and they decrease the overall operating efficiency of a power plant. The decreased efficiency results in increased emissions of carbon dioxide, a greenhouse gas.

The other alternative to reduce acid deposition is to burn less high-sulfur fossil fuel. This can be accomplished by switching to alternative sources of energy or improving the efficiency of energy-consuming technologies. Coal-fired power plants can reduce SO_2 emissions by burning coal with a lower sulfur content. Another alternative is for these power plants to switch to fuels with lower acid gas emissions, such as natural gas. Ultimately, the most effective ways to reduce acid rain are the use of renewable energy and improving energy efficiency. Such measures will decrease the use of fossil fuels and therewith reduce the emissions. Renewable energy technologies, such as solar and wind energy, can produce electricity without any emissions of SO_2 or NO_x. There are also many ways to decrease consumption of energy by improving the efficiency of end-use technologies. Both renewable energy and energy efficiency have the added benefit that they also result in reduced emissions of carbon dioxide, the greenhouse gas most responsible for global warming.

It is technically feasible and likely economically possible to further decrease NO_x emissions from fossil fuel combustion to the point at which they become only a minor disturbance to the nitrogen cycle at all scales. Clean electric generation and transportation technologies are commercially available today, or will be commercially available within one or two decades, that have the potential to further decrease NO_x emissions in industrialized countries and to decrease NO_x emissions from projected business-as-usual levels in developing countries. In addition, technologies currently under development, such as renewable energy and hydrogen-based fuel cells, could operate with zero NO_x emissions.

During the 1970s, major attention was given to energy conservation measures designed to avoid depletion of coal, crude oil, and gas reserves. Since then, the known reserves of natural gas have increased from 40,000 to 146,400 billion m^3 in 1998. The current world coal, crude oil, and natural gas reserves are estimated to be 143,400, 500, and 131,800 Mtoe, respectively (Mtoe = million ton oil equivalents). Thus, there appears to be enough fossil fuel available for at least 50–100 years, and it is expected that this will increase as new technologies become available with which to discover and add new sources of fossil fuels. The current world concern about fossil fuel combustion is driven by the apparent necessity to decrease CO_2 emissions. The increase in CO_2 content of the atmosphere and the associated climate change will call for a drastic change in energy production and use. Focusing only on SO_2 and NO_x as pollutants from energy production and use will not lead to drastic changes in the energy systems. In order to abate these emissions from energy production, a multiple-pollutant approach should be used; that developed within the Göteborg Protocol appears to be very promising. Unless yet-to-be developed carbon sequestration technology emerges, decreasing CO_2 emissions will automatically decrease NO_x and SO_2, although this has to be proved. On the contrary, NO_x and SO_2 abatement almost always leads to increased energy use and thus increased CO_2 emissions. These technologies are focused on enhanced combustion, producing less NO_x (lean burn), or using end of pipe technologies, such as SCR or scrubbers. Both technologies decrease energy efficiency, and the majority of SCR technologies require ammonia or urea as a reductor, which is produced from natural gas.

8.1 Consequences of Carbon Management: Addressing Greenhouse Gases for Energy Nitrogen and Sulfur

If the energy system of the world remains based on fossil fuels throughout the 21st century and little is done to target the atmospheric emissions of CO_2, it is plausible that the atmospheric CO_2 concentration may triple its “preindustrial” concentration of approximately 280 parts per million by the year 2100. Strategies to slow the rate of buildup of atmospheric CO_2 are being developed worldwide, and all appear to have a positive effect of reducing the production of energy nitrogen and sulfur. First among these strategies is an increase in the efficiency of energy use throughout the global economy, resulting in less energy required to meet the variety of amenities that energy provides (e.g., mobility, comfort, lighting, and material goods). Greater energy efficiency directly decreases energy nitrogen and sulfur.

Other CO_2 emission-reduction strategies address the composition of energy supply. First, nuclear energy and renewable energy are nonfossil alternatives, resulting in CO_2 emissions only to the extent that fossil fuels are used to produce the nonfossil energy production facilities. Neither nuclear nor renewable energy sources produce energy nitrogen and sulfur, but nuclear waste is another source of concern.

Second, the mix of coal, oil, and natural gas in the energy supply affects CO_2 emissions because they have different carbon intensities (carbon content per unit of thermal energy). Specifically, coal has the highest and natural gas the lowest carbon intensity so that shifts from coal to oil or natural gas and shifts from oil to natural gas, other factors held constant, reduce the greenhouse effect of the fossil fuel system. The nitrogen and sulfur intensity of fossil fuels (nitrogen and sulfur content per unit of thermal energy) differ in the same way (i.e., on average, the intensity of coal is highest and the intensity of natural gas is lowest). Thus, fuel shifts within the fossil fuel system that reduce greenhouse effects will also reduce fossil nitrogen and sulfur.

Third, many countries throughout the world are investing in research, development, and demonstration projects that explore the various forms of capturing carbon from combustion processes before it reaches the atmosphere and sequestering it on site or off. Several available technologies can be used to separate and capture CO_2 from fossil-fueled power plant flue gases, from effluents of industrial processes such as iron, steel, and cement production, and from hydrogen production by reforming natural gas. CO_2 can be absorbed from gas streams by contact with amine-based solvents or cold methanol. It can be removed by absorption on activated carbon or other materials or by passing the gas through special membranes. However, these technologies have not been applied at the scale required to use them as part of a CO_2 emissions mitigation strategy. The goal is to sequester the carbon in a cost-effective way, for example, as CO_2 injected deep below ground in saline aquifers. This is a relatively new area of research and development, and little attention has been given to the consequences of fossil carbon sequestration for energy nitrogen and sulfur, but decreases are a likely result. For example, to capture and sequester the carbon in coal will require the gasification of coal and the subsequent production of hydrogen and a CO_2 gas stream. Coal nitrogen or sulfur should be amenable to independent management, with results that include extraction as a saleable product or cosequestration below ground with CO_2. The first of these results is one of many "polygeneration" strategies for coal, in which products may include electricity, hydrogen, process heat, hydrocarbons, dimethyl ether, and nitrogen. The long-term goal is to run the economy on noncarbon secondary energy sources, specifically electricity and hydrogen, while sequestering emissions of CO_2.

9. CONCLUSION

The atmosphere functions as a pool and chemical reaction vessel for a host of substances. Many of the most important ones (e.g., oxygen, carbon dioxide, nitrogen, and sulfur compounds) are released by the activity of organisms. Often with the help of the water cycle, they pass through the atmosphere and are eventually taken up again into soil, surface water, and organic matter. Through technology related to energy production and use, humans have added enormously to the atmospheric burden of some of these substances, with far-reaching consequences for life and the environment. The evidence is clearest in the case of acid deposition: gases, particles, and precipitation depositing to the surface causing acidification.

In North America and in some European nations, public concern regarding the effects of acid rain has been transformed into regulations restricting the amount of SO_2 and NO_x released by electric utilities and industries. The result has been a decrease in annual acidic deposition in some areas, especially due to the reduction in sulfates. There is also evidence that when acid deposition is reduced, ecosystems can recover. The chemistry of several lakes has improved during the past 20 years, but full biological recovery has not been observed. Nitrogen has become the main concern because emissions have been reduced much less than those of SO_2. In the future, it will be very important for the industrialized world to transfer its technology and experience to the developing world in order to ensure that the same acid rain problems do not occur as these countries consume more energy during the process of industrialization. Whereas in the past few years the acid deposition decreased in Europe and the United States, an increase has been observed in Asia. It is essential to establish or optimize monitoring programs aimed at following the trends in acid deposition and recovery.

The acid deposition levels in the industrialized areas of the world are well above the critical thresholds of different ecosystems. Because energy emissions are the largest source for acid deposition, abatement is necessary. Multipollutant–multieffect approaches are necessary to ensure that measures are cost-effective and do not create problems for other areas. It has to be determined which effects limit the emissions in different areas most: Is it because of the air quality and the effects on humans, or is it the ecosystem loading? In this way, regional emissions can be optimized and caps can be set to prevent any effects. Furthermore, it is probably most effective to follow CO_2 emission reduction options and to then consider SO_2 and NO_x, instead of the reverse. In most cases, when CO_2 emission is decreased, SO_2

and NO_x are also decreased, whereas if measures to reduce SO_2 and NO_x are taken first, CO_2 is increased because of the energy penalty or additional energy use of abatement options.

SEE ALSO THE FOLLOWING ARTICLES

Air Pollution from Energy Production and Use • Air Pollution, Health Effects of • Biomass: Impact on Carbon Cycle and Greenhouse Gas Emissions • Clean Air Markets • Ecosystem Health: Energy Indicators • Ecosystems and Energy: History and Overview • Greenhouse Gas Emissions from Energy Systems, Comparison and Overview • Hazardous Waste from Fossil Fuels

Further Reading

Breemen, N. van, Burrough, P. A., Velthorst, E. J., Dobben, H. F. van, Wit, T. de, Ridder, T. B., and Reinders, H. F. R. (1982). Soil acidification from atmospheric ammonium sulphate in forest canopy throughfall. *Nature* **299**, 548–550.

Brimblecombe, P. (1989). "The Big Smoke." Routledge, London.

Chadwick, M. J., and Hutton, M. (1990). "Acid Depositions in Europe. Environemental Effects, Control Strategies and Policy Options." Stockholm Environmental Institute, Stockholm, Sweden.

Cowling, E., Galloway, J., Furiness, C., Barber, M., Bresser, T., Cassman, K., Erisman, J. W., Haeuber, R., Howarth, R., Melillo, J., Moomaw, W., Mosier, A., Sanders, K., Seitzinger, S., Smeulders, S., Socolow, R., Walters, D., West, F., and Zhu, Z. (2001). Optimizing nitrogen management in food and energy production and environmental protection: Proceedings of the 2nd International Nitrogen Conference on Science and Policy. *Sci. World* 1(Suppl. 2), 1–9.

EMEP/MSC-W (1998). Transboundary acidifying air pollution in Europe. Estimated dispersion of acidifying and eutrophying compounds and comparison with observations, MSC-W (Meteorological Synthesizing Centre–West) status report. Norwegian Meteorological Institute, Oslo, Norway.

Erisman, J. W. (2000). Editorial: History as a basis for the future: The environmental science and policy in the next millennium. *Environ. Sci. Pollut.* **3**, 5–6.

Erisman, J. W., and Draaijers, G. P. J. (1995). Atmospheric deposition in relation to acidification and eutrophication. *Stud. Environ. Res.* **63**.

Erisman, J. W., Grennfelt, P., and Sutton, M. (2001). Nitrogen emission and deposition: The European Perspective. *Sci. World* **1**, 879–896.

European Environmental Agency (1999). "Environment in the European Union at the Turn of the Century." European Environmental Agency, Copenhagen.

European Environmental Agency (2000). "Environmental Signals 2000." European Environmental Agency, Copenhagen.

Galloway, J. N. (2001). Acidification of the world: Natural and anthropogenic. *Water Soil Air Pollut.* **130**, 17–24.

Gorham, E. (1994). Neutralising acid rain. *Nature* **367**, 321.

Hedin, L. O., Granat, L., Likens, G. E., Buishand, T. A., Galloway, J. N., Butler, T. J., and Rodhe, H. (1994). Steep declines in atmospheric base cations in regions of Europe and North America. *Nature* **367**, 351–354.

Heij, G. J., and Erisman, J. W. (1997). Acidification research in the Netherlands; Report of third and last phase. *Stud. Environ. Sci.* **69**.

ICP (2000). Forest condition in Europe, 2000 executive report. *In* "UN ECE Convention on Long-Range Transboundary Air Pollution International Co-operative Programme on Assessment and Monitoring of Air Pollution Effects on Forests." Federal Research Centre for Forestry and Forest Products (BFH), Hamburg, Germany.

ICP Waters (2000). The 12-year report: Acidification of surface water in Europe and North America; Trends, biological recovery and heavy metals. Norwegian Institute for Water Research, Oslo.

IPCC (2000). Third assessment report on climate change. UN/FCCC, Bonn, Germany.

Lee, D. S., Espenhahn, S. E., and Baker, S. (1998). Evidence for long-term changes in base cations in the atmospheric aerosol. *J. Geophys. Res.* **103**, 21955–21966.

Matyssek, R., Keller, T., and Günthardt-Goerg, M. S. (1990). Ozonwirkungen auf den verschiedenen Organisationsebenen in Holzpflanzen. *Schweizerische Z. Forstwesen* **141**, 631–651.

Mylona, S. (1993). Trends of sulphur dioxide emissions, air concentrations and depositions of sulphur in Europe since 1880, EMEP/MSC-W Report 2/93. Norwegian Meteorological Institute, Oslo.

Nihlgård, B. (1985). The ammonium hypothesis—An additional explanation for the forest dieback in Europe. *Ambio* **14**, 2–8.

Nilles, M. A., and Conley, B. E. (2001). Changes in the chemistry of precipitation in the United States, 1981–1998. *Water Air Soil Pollut.* **130**, 409–414.

Nilsson, J. (1986). Critical deposition limits for forest soils. *In* "Critical Loads for Nitrogen and Sulphur" (J. Nilsson, Ed.), Miljørapport 1986, Vol. 11, pp. 37–69. Nordic Council of Ministers, Copenhagen.

Nilsson, J., and Grennfelt, P. (1988). "Critical Loads for Sulphur and Nitrogen," Miljørapport 1988, Vol. 15. Nordic Council of Ministers, Copenhagen.

Smil, V. (2000). Energy in the twentieth century. *Annu. Rev. Energy Environ.* **25**, 21–51.

Smith, R. A. (1852). On the air and rain of Manchester. *Manchester Literary Philos. Soc.* **10**, 207–217.

Smith, R. A. (1872). "Air and Rain. The Beginnings of a Chemical Climatology." Longmans, Green, London.

Stoddard, J. L., Jeffries, D. S., Lukewille, A., Clair, T. A., Dillon, P. J., Driscoll, C. T., Forsius, M., Johannessen, M., Kahl, J. S., Kellogg, J. H., Kemp, A., Mannio, J., Monteith, D. T., Murdoch, P. S., Patrick, S., Rebsdorf, A., Skjelkvale, B. L., Stainton, M. P., Traaen, T., van Dam, H., Webster, K. E., Wieting, J., and Wilander, A. (1999). Regional trends in aquatic recovery from acidification in North America and Europe. *Nature* **401**, 575–578.

Streets, D. G., Tsai, N. Y., Akimoto, H., and Oka, K. (2001). Trends in emissions of acidifying species in Asia, 1985–1997. *Water Air Soil Pollut.* **130**, 187–192.

Ulrich, B. (1983). Interaction of forest canopies with atmospheric constituents: SO_2, alkali and earth alkali cations and chloride.

In "Effects of Accumulation of Air Pollutants in Forest Ecosystems" (B. Ulrich and J. Pankrath, Eds.), pp. 33–45. Reidel, Germany.

U.S. National Acid Precipitation Assessment Program (2000). "Acid Deposition: State of Science and Technology Progress Report 2000." U.S. National Acid Precipitation Assessment Program, Washington, DC.

Van Ardenne, J. A., Dentener, F. J., Olivier, J. G. J., Klein Goldewijk, C. G. M., and Lelieveld, J. (2001). A 1×1 resolution data set of historical anthropogenic trace gas emissions for the period 1890–1990. *Global Biogeochem. Cycl.* **15**, 909–928.

WGE (2000). 9th annual report 2000. *In* "UN ECE Convention on Long-Range Transboundary Air Pollution International Co-operative Programme on Integrated Monitoring of Air Pollution Effects on Ecosystems" (S. Kleemola and M. Forsius, Eds.). Working Group on Effects of the Convention on Long-Range Transboundary Air Pollution, Helsinki.

Aggregation of Energy

CUTLER J. CLEVELAND and ROBERT K. KAUFMANN
Boston University
Boston, Massachusetts, United States
DAVID I. STERN
Rensselaer Polytechnic Institute
Troy, New York, United States

1. Energy Aggregation and Energy Quality
2. Economic Approaches to Energy Quality
3. Alternative Approaches to Energy Aggregation
4. Case Study 1: Net Energy from Fossil Fuel Extraction in the United States
5. Case Study 2: Causality in the Energy/GDP Relationship
6. Case Study 3: The Determinants of the Energy/GDP Relationship
7. Conclusions and Implications

Glossary

Divisia index A method of aggregation used in economics that permits variable substitution among material types without imposing a priori restrictions on the degree of substitution.

emergy The quantity of solar energy used directly and indirectly to produce a natural resource, good, or service.

energy quality The relative economic usefulness per heat equivalent unit of different fuels and electricity.

energy/real gross domestic product (GDP) ratio (E/GDP ratio) The ratio of total energy use to total economic activity; a common measure of macroeconomic energy efficiency.

energy return on investment (EROI) The ratio of energy delivered to energy costs.

exergy The useful work obtainable from an energy source or material is based on the chemical energy embodied in the material or energy based on its physical organization relative to a reference state. Exergy measures the degree to which a material is organized relative to a random assemblage of material found at an average concentration in the crust, ocean, or atmosphere.

Granger causality A statistical procedure that tests whether (i) one variable in a relation can be meaningfully described as a dependent variable and the other variable as an independent variable, (ii) the relation is bidirectional, or (ii) no meaningful relation exists. This is usually done by testing whether lagged values of one of the variables add significant explanatory power to a model that already includes lagged values of the dependent variable and perhaps also lagged values of other variables.

marginal product of energy The value marginal product of a fuel in production is the marginal increase in the quantity of a good or service produced by the use of one additional heat unit of fuel multiplied by the price of that good or service.

net energy analysis Technique that compares the quantity of energy delivered to society by an energy system to the energy used directly and indirectly in the delivery process.

Investigating the role of energy in the economy involves aggregating different energy flows. A variety of methods have been proposed, but none is accepted universally. This article shows that the method of aggregation affects analytical results. We review the principal assumptions and methods for aggregating energy flows: the basic heat equivalents approach, economic approaches using prices or marginal product for aggregation, emergy analysis, and thermodynamic approaches such as exergy analysis. We argue that economic approaches such as the index or marginal product method are superior because they account for differences in quality among different fuels. We apply economic approaches to three case studies of the U.S. economy. In the first, we account for energy quality to assess changes in the energy surplus delivered by the extraction of fossil fuels from 1954 to 1992. The second and third case studies examine the effect of energy quality on statistical analyses of the relation between energy use and gross domestic product

(GDP). First, a quality-adjusted index of energy consumption is used in an econometric analysis of the causal relation between energy use and GDP from 1947 to 1996. Second, we account for energy quality in an econometric analysis of the factors that determined changes in the energy/GDP ratio from 1947 to 1996. Without adjusting for energy quality, the results imply that the energy surplus from petroleum extraction is increasing, that changes in GDP drive changes in energy use, and that GDP has been decoupled from aggregate energy. These conclusions are reversed when we account for changes in energy quality.

1. ENERGY AGGREGATION AND ENERGY QUALITY

Aggregation of primary-level economic data has received substantial attention from economists for a number of reasons. Aggregating the vast number of inputs and outputs in the economy makes it easier for analysts to discern patterns in the data. Some aggregate quantities are of theoretical interest in macroeconomics. Measurement of productivity, for example, requires a method to aggregate goods produced and factors of production that have diverse and distinct qualities. For example, the post-World War II shift toward a more educated workforce and from nonresidential structures to producers' durable equipment requires adjustments to methods used to measure labor hours and capital inputs. Econometric and other forms of quantitative analysis may restrict the number of variables that can be considered in a specific application, again requiring aggregation. Many indexes are possible, so economists have focused on the implicit assumptions made by the choice of an index in regard to returns to scale, substitutability, and other factors. These general considerations also apply to energy.

The simplest form of aggregation, assuming that each variable is in the same units, is to add up the individual variables according to their thermal equivalents (Btus, joules, etc.). Equation (1) illustrates this approach:

$$E_t = \sum_{i=1}^{N} E_{it}, \qquad (1)$$

where E is the thermal equivalent of fuel i (N types) at time t. The advantages of the thermal equivalent approach are that it uses a simple and well-defined accounting system based on the conservation of energy and the fact that thermal equivalents are easily measured. This approach underlies most methods of energy aggregation in economics and ecology, such as trophic dynamics national energy accounting, energy input–output modeling in economies and ecosystems, most analyses of the energy/gross domestic product (GDP) relationship and energy efficiency, and most net energy analyses.

Despite its widespread use, aggregating different energy types by their heat units embodies a serious flaw: It ignores qualitative differences among energy vectors. We define energy quality as the relative economic usefulness per heat equivalent unit of different fuels and electricity. Given that the composition of energy use changes significantly over time (Fig. 1), it is reasonable to assume that energy quality has been an important economic driving force. The quality of electricity has received considerable attention in terms of its effect on the productivity of labor and capital and on the quantity of energy required to produce a unit of GDP. Less attention has been paid to the quality of other fuels, and few studies use a quality-weighting scheme in empirical analysis of energy use.

The concept of energy quality needs to be distinguished from that of resource quality. Petroleum and coal deposits may be identified as high-quality energy sources because they provide a very high energy surplus relative to the amount of energy required to extract the fuel. On the other hand, some forms of solar electricity may be characterized as a

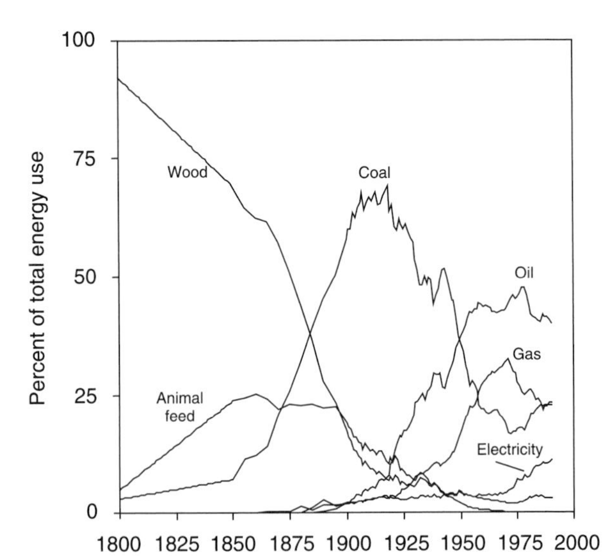

FIGURE 1 Composition of primary energy use in the United States. Electricity includes only primary sources (hydropower, nuclear, geothermal, and solar).

low-quality source because they have a lower energy return on investment (EROI). However, the latter energy vector may have higher energy quality because it can be used to generate more useful economic work than one heat unit of petroleum or coal.

Taking energy quality into account in energy aggregation requires more advanced forms of aggregation. Some of these forms are based on concepts developed in the energy analysis literature, such as exergy or emergy analysis. These methods take the following form:

$$E_t^* = \sum_{i=1}^{N} \lambda_{it} E_{it}, \tag{2}$$

where λ represents quality factors that may vary among fuels and over time for individual fuels. In the most general case that we consider, an aggregate index can be represented as

$$f(E_t) = \sum_{i=1}^{N} \lambda_{it} g(E_{it}), \tag{3}$$

where $f(\,)$ and $g(\,)$ are functions, λ_{it} are weights, the E_i are the N different energy vectors, and E_t is the aggregate energy index in period t. An example of this type of indexing is the discrete Divisia index or Tornquist–Theil index described later.

2. *ECONOMIC APPROACHES TO ENERGY QUALITY*

From an economic perspective, the value of a heat equivalent of fuel is determined by its price. Price-taking consumers and producers set marginal utilities and products of the different energy vectors equal to their market prices. These prices and their marginal productivities and utilities are set simultaneously in general equilibrium. The value marginal product of a fuel in production is the marginal increase in the quantity of a good or service produced by the use of one additional heat unit of fuel multiplied by the price of that good or service. We can also think of the value of the marginal product of a fuel in household production.

The marginal product of a fuel is determined in part by a complex set of attributes unique to each fuel, such as physical scarcity, the capacity to do useful work, energy density, cleanliness, amenability to storage, safety, flexibility of use, and cost of conversion. However, the marginal product is not uniquely fixed by these attributes. Rather, the energy vector's marginal product varies according to the activities in which it is used; how much and what form of capital, labor, and materials it is used in conjunction with; and how much energy is used in each application. As the price rises due to changes on the supply side, users can reduce their use of that form of energy in each activity, increase the amount and sophistication of capital or labor used in conjunction with the fuel, or stop using that form of energy for lower value activities. All these actions raise the marginal productivity of the fuel. When capital stocks have to be adjusted, this response may be somewhat sluggish and lead to lags between price changes and changes in the value marginal product.

The heat equivalent of a fuel is just one of the attributes of the fuel and ignores the context in which the fuel is used; thus, it cannot explain, for example, why a thermal equivalent of oil is more useful in many tasks than is a heat equivalent of coal. In addition to attributes of the fuel, marginal product also depends on the state of technology, the level of other inputs, and other factors. According to neoclassical theory, the price per heat equivalent of fuel should equal its value marginal product and, therefore, represent its economic usefulness. In theory, the market price of a fuel reflects the myriad factors that determine the economic usefulness of a fuel from the perspective of the end user.

Consistent with this perspective, the price per heat equivalent of fuel varies substantially among fuel types (Table I). The different prices demonstrate that end users are concerned with attributes other than heat content. Ernst Berndt, an economist at MIT, noted that because of the variation in attributes among energy types, the various fuels and electricity are less than perfectly substitutable, either in production or in consumption. For example, from the point of view of the end user, 1 Btu of coal is not perfectly substitutable with 1 Btu of electricity; since the electricity is cleaner, lighter, and of higher quality, most end users are willing to pay a premium price per Btu of electricity. However, coal and electricity are substitutable to a limited extent because if the premium price for electricity were too high, a substantial number of industrial users might switch to coal. Alternatively, if only heat content mattered and if all energy types were then perfectly substitutable, the market would tend to price all energy types at the same price per Btu.

Do market signals (i.e., prices) accurately reflect the marginal product of inputs? Empirical analysis of the relation between relative marginal product and price in U.S. energy markets suggests that this is

TABLE I

U.S. Market Price for Various Energy Types[a]

Energy type	Market price ($/10^6 btu)
Coal	
Bituminous	
Mine-mouth	
Consumer cost	
Anthracite	
Mine-mouth	
Oil	
Wellhead	2.97
Distillate oil	7.70
Jet fuel	4.53
LPG	7.42
Motor gasoline	9.73
Residual fuel oil	2.83
Biofuels	
Consumer cost	
Natural gas	
Wellhead	2.10
Consumer cost	20.34

[a] *Source*. Department of Energy (1997). Values are 1994 prices.

TABLE II

Marginal Product of Coal, Oil, Natural Gas, and Electricity Relative to One Another

	Minimum	Year	Maximum	Year
Oil : coal	1.83	1973	3.45	1990
Gas : coal	1.43	1973	2.76	1944
Electricity : coal	4.28	1986	16.42	1944
Oil : gas	0.97	1933	1.45	1992
Electricity : oil	1.75	1991	6.37	1930
Electricity : gas	2.32	1986	6.32	1930

indeed the case. In the case of the United States, there is a long-term relation between relative marginal product and relative price, and several years of adjustment are needed to bring this relation into equilibrium. The results are summarized in Table II and suggest that over time prices do reflect the marginal product, and hence the economic usefulness, of fuels.

Other analysts have calculated the average product of fuels, which is a close proxy for marginal products. Studies indicate that petroleum is 1.6–2.7 times more productive than coal in producing industrial output, and that electricity is 2.7–18.3 times more productive than coal.

2.1 Price-Based Aggregation

If marginal product is related to its price, energy quality can be measured by using the price of fuels to weight their heat equivalents. The simplest approach defines the weighting factor (λ) in Eq. (2) as

$$\lambda_{it} = \frac{P_{it}}{P_{1t}}, \tag{4}$$

where P_{it} is the price per Btu of fuel. In this case, the price of each fuel is measured relative to the price of fuel type 1.

The quality index in Eq. (4) embodies a restrictive assumption—that fuels are perfect substitutes—and the index is sensitive to the choice of numeraire. Because fuels are not perfect substitutes, an increase in the price of one fuel relative to the price of output will not be matched by equal changes in the prices of the other fuels relative to the price of output. For example, the increase in oil prices in 1979–1980 would cause an aggregate energy index that uses oil as the numeraire to decline dramatically. An index that uses coal as the numeraire would show a large decline in 1968–1974, one not indicated by the oil-based index.

To avoid dependence on a numeraire, a discrete approximation to the Divisia index can be used to aggregate energy. The formula for constructing the discrete Divisia index E^* is

$$\ln E_t^* - \ln E_{t-1}^* = \sum_{i=1}^{n}\left(\left(\frac{P_{it}E_{it}}{2\sum_{i=1}^{n} P_{it}E_{it}} + \frac{P_{it-1}E_{it-1}}{2\sum_{i=1}^{n} P_{it-1}E_{it-1}}\right) \times(\ln E_{it} - \ln E_{it-1})\right) \tag{5}$$

where P are the prices of the n fuels, and E are the quantities of Btu for each fuel in final energy use. Note that prices enter the Divisia index via cost or expenditure shares. The Divisia index permits variable substitution among material types without imposing a priori restrictions on the degree of substitution. This index is an exact index number representation of the linear homogeneous translog production function, where fuels are homothetically weakly separable as a group from the other factors of production. With reference to Eq. (3), $f() = g() = \Delta\ln()$, whereas λ_{it} is given by the average cost share over the two periods of the differencing operation.

2.2 Discussion

Aggregation using price has its shortcomings. Prices provide a reasonable method of aggregation if the aggregate cost function is homothetically separable in the raw material input prices. This means that the elasticity of substitution between different fuels is not a function of the quantities of nonfuel inputs used. This may be an unrealistic assumption in some cases. Also, the Divisia index assumes that the substitution possibilities among all fuel types and output are equal.

Another limit on the use of prices is that they generally do not exist for wastes. Thus, an economic index of waste flows is impossible to construct.

It is well-known that energy prices do not reflect their full social cost due to a number of market imperfections. This is particularly true for the environmental impact caused by their extraction and use. These problems lead some to doubt the usefulness of price as the basis for any indicator of sustainability. However, with or without externalities, prices should reflect productivities. Internalizing externalities will shift energy use, which in turn will change marginal products.

Moreover, prices produce a ranking of fuels (Table I) that is consistent with our intuition and with previous empirical research. One can conclude that government policy, regulations, cartels, and externalities explain some of the price differentials among fuels but certainly not the substantial ranges that exist. More fundamentally, price differentials are explained by differences in attributes such as physical scarcity, capacity to do useful work, energy density, cleanliness, amenability to storage, safety, flexibility of use, and cost of conversion. Eliminate the market imperfections and the price per Btu of different energies would vary due to the different combinations of attributes that determine their economic usefulness. The different prices per Btu indicate that users are interested in attributes other than heat content.

3. ALTERNATIVE APPROACHES TO ENERGY AGGREGATION

Although we argue that the more advanced economic indexing methods, such as Divisia aggregation, are the most appropriate way to aggregate energy use for investigating its role in the economy, the ecological economics literature proposes other methods of aggregation. We review two of these methods in this section and assess limits on their ability to aggregate energy use.

3.1 Exergy

Other scientists propose a system of aggregating energy and materials based on exergy. Exergy measures the useful work obtainable from an energy source or material, and it is based on the chemical energy embodied in the material or energy based on its physical organization relative to a reference state. Thus, exergy measures the degree to which a material is organized relative a random assemblage of material found at an average concentration in the crust, ocean, or atmosphere. The higher the degree of concentration, the higher the exergy content. The physical units for exergy are the same as for energy or heat, namely kilocalories, joules, Btus, etc. For fossil fuels, exergy is nearly equivalent to the standard heat of combustion; for other materials, specific calculations are needed that depend on the details of the assumed conversion process.

Proponents argue that exergy has a number of useful attributes for aggregating heterogeneous energy and materials. Exergy is a property of all energy and materials and in principle can be calculated from information in handbooks of chemistry and physics and secondary studies. Thus, exergy can be used to measure and aggregate natural resource inputs as well as wastes. For these reasons, Ayres argues that exergy forms the basis for a comprehensive resource accounting framework that could "provide policy makers with a valuable set of indicators." One such indicator is a general measure of "technical efficiency," the efficiency with which "raw" exergy from animals or an inanimate source is converted into final services. A low exergy efficiency implies potential for efficiency gains for converting energy and materials into goods and services. Similarly, the ratio of exergy embodied in material wastes to exergy embodied in resource inputs is the most general measure of pollution. Some also argue that the exergy of waste streams is a proxy for their potential ecotoxicity or harm to the environment, at least in general terms.

From an accounting perspective, exergy is appealing because it is based on the science and laws of thermodynamics and thus has a well-established system of concepts, rules, and information that are available widely. However, like enthalpy, exergy should not be used to aggregate energy and material inputs aggregation because it is one-dimensional. Like enthalpy, exergy does not vary with, and hence does not necessarily reflect, attributes of fuels that determine their economic usefulness, such as energy density, cleanliness, and cost of conversion. The same is true for materials. Exergy cannot explain, for example, impact resistance, heat resistance, corrosion resistance, stiffness, space maintenance, conductivity, strength, ductility, or other properties of metals that determine their usefulness. Like prices, exergy does not reflect all the environmental costs of fuel use. The exergy of coal, for example, does not reflect coal's contribution to global warming or its impact on human health relative to natural gas. The exergy of wastes is at best a rough first-order approximation of environmental impact because it does not vary with the specific attributes of a waste material and its receiving environment that cause harm to organisms or that disrupt biogeochemical cycles. In theory, exergy can be calculated for any energy or material, but in practice the task of assessing the hundreds (thousands?) of primary and intermediate energy and material flows in an economy is daunting.

3.2 Emergy

The ecologist Howard Odum analyzes energy and materials with a system that traces their flows within and between society and the environment. It is important to differentiate between two aspects of Odum's contribution. The first is his development of a biophysically based, systems-oriented model of the relationship between society and the environment. Here, Odum's early contributions helped lay the foundation for the biophysical analysis of energy and material flows, an area of research that forms part of the intellectual backbone of ecological economics. The insight from this part of Odum's work is illustrated by the fact that ideas he emphasized—energy and material flows, feedbacks, hierarchies, thresholds, and time lags—are key concepts of the analysis of sustainability in a variety of disciplines.

The second aspect of Odum's work, which we are concerned with here, is a specific empirical issue: the identification, measurement, and aggregation of energy and material inputs to the economy, and their use in the construction of indicators of sustainability. Odum measures, values, and aggregates energy of different types by their transformities. Transformities are calculated as the amount of one type of energy required to produce a heat equivalent of another type of energy. To account for the difference in quality of thermal equivalents among different energies, all energy costs are measured in solar emjoules, the quantity of solar energy used to produce another type of energy. Fuels and materials with higher transformities require larger amounts of sunlight to produce and therefore are considered more economically useful.

Several aspects of the emergy methodology reduce its usefulness as a method for aggregating energy and/or material flows. First, like enthalpy and exergy, emergy is one-dimensional because energy sources are evaluated based on the quantity of embodied solar energy and crustal heat. However, is the usefulness of a fuel as an input to production related to its transformity? Probably not. Users value coal based on it heat content, sulfur content, cost of transportation, and other factors that form the complex set of attributes that determine its usefulness relative to other fuels. It is difficult to imagine how this set of attributes is in general related to, much less determined by, the amount of solar energy required to produce coal. Second, the emergy methodology is inconsistent with its own basic tenant, namely that quality varies with embodied energy or emergy. Coal deposits that we currently extract were laid down over many geological periods that span half a billion years. Coals thus have vastly different embodied emergy, but only a single transformity for coal is normally used. Third, the emergy methodology depends on plausible but arbitrary choices of conversion technologies (e.g., boiler efficiencies) that assume users choose one fuel relative to another and other fuels based principally on their relative conversion efficiencies in a particular application. Finally, the emergy methodology relies on long series of calculations with data that vary in quality. However, little attention is paid to the sensitivity of the results to data quality and uncertainty, leaving the reader with little or no sense of the precision or reliability of the emergy calculations.

4. CASE STUDY 1: NET ENERGY FROM FOSSIL FUEL EXTRACTION IN THE UNITED STATES

One technique for evaluating the productivity of energy systems is net energy analysis, which compares the quantity of energy delivered to society by an energy system to the energy used directly and indirectly in the delivery process. EROI is the ratio of energy delivered to energy costs. There is a long debate about the relative strengths and weaknesses of net energy analysis. One restriction on net energy analysis' ability to deliver the insights it promises is its treatment of energy quality. In most net energy analyses, inputs and outputs of different types of energy are aggregated by their thermal equivalents. This case study illustrates how accounting for energy quality affected calculations for the EROI of the U.S. petroleum sector from 1954 to 1992.

4.1 Methods and Data

Following the definitions in Eq. (2), a quality-corrected EROI* is defined by

$$\mathrm{EROI}_t^* = \frac{\sum_{i=1}^{n} \lambda_{i,t} E_{i,t}^{\mathrm{o}}}{\sum_{i=1}^{n} \lambda_{i,t} E_{i,t}^{\mathrm{c}}}, \qquad (6)$$

where $\lambda_{i,t}$ is the quality factor for fuel type i at time t and E^{o} and E^{c} are the thermal equivalents of energy outputs and energy inputs, respectively. We construct Divisia indices for energy inputs and outputs to account for energy quality in the numerator and denominator. The prices for energy outputs (oil, natural gas, and natural gas liquids) and energy inputs (natural gas, gasoline, distillate fuels, coal, and electricity) are the prices paid by industrial end users for each energy type.

Energy inputs include only industrial energies: the fossil fuel and electricity used directly and indirectly to extract petroleum. The costs include only those energies used to locate and extract oil and natural gas and prepare them for shipment from the wellhead. Transportation and refining costs are excluded from this analysis. Output in the petroleum industry is the sum of the marketed production of crude oil, natural gas, and natural gas liquids.

The direct energy cost of petroleum is the fuel and electricity used in oil and gas fields. Indirect energy costs include the energy used to produce material inputs and to produce and maintain the capital used to extract petroleum. The indirect energy cost of materials and capital is calculated from data for the dollar cost of those inputs to petroleum extraction processes. Energy cost of capital and materials is defined as the dollar cost of capital depreciation and materials multiplied by the energy intensity of capital and materials (Btu/$). The energy intensity of capital and materials is measured by the quantity of energy used to produce a dollar's worth of output in the industrial sector of the U.S. economy. That quantity is the ratio of fossil fuel and electricity use to real GDP produced by industry.

4.2 Results and Conclusions

The thermal equivalent and Divisia EROI for petroleum extraction show significant differences

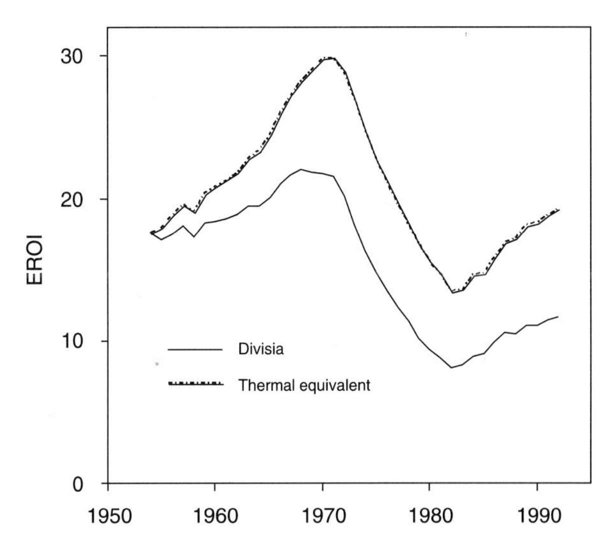

FIGURE 2 Energy return on investment (EROI) for petroleum extraction in the United States, with energy inputs and outputs measured in heat equivalents and a Divisia index.

(Fig. 2). The quality-corrected EROI declines faster than the thermal-equivalent EROI. The thermal-equivalent EROI increased by 60% relative to the Divisia EROI between 1954 and 1992. This difference was driven largely by changes in the mix of fuel qualities in energy inputs. Electricity, the highest quality fuel, is an energy input but not an energy output. Its share of total energy use increased from 2 to 12% during the period; its cost share increased from 20 to 30%. Thus, in absolute terms the denominator in the Divisia EROI is weighted more heavily than in the thermal-equivalent EROI. The Divisia-weighted quantity of refined oil products is larger than that for gas and coal. Thus, the two highest quality fuels, electricity and refined oil products, comprise a large and growing fraction of the denominator in the Divisia EROI compared to the thermal-equivalent EROI. Therefore, the Divisia denominator increases faster than the heat-equivalent denominator, causing EROI to decline faster in the former case.

5. CASE STUDY 2: CAUSALITY IN THE ENERGY/GDP RELATIONSHIP

One of the most important questions about the environment–economy relationship regards the strength of the linkage between economic growth and energy use. With a few exceptions, most analyses ignore the effect of energy quality in the assessment of this relationship. One statistical approach to address this question is Granger causality and/or cointegration analysis. Granger causality tests whether (i) one variable in a relation can be meaningfully described as a dependent variable and the other variable as an independent variable, (ii) the relation is bidirectional, or (iii) no meaningful relation exists. This is usually done by testing whether lagged values of one of the variables add significant explanatory power to a model that already includes lagged values of the dependent variable and perhaps also lagged values of other variables.

Although Granger causality can be applied to both stationary and integrated time series (time series that follow a random walk), cointegration applies only to linear models of integrated time series. The irregular trend in integrated series is known as a stochastic trend, as opposed to a simple linear deterministic time trend. Time series of GDP and energy use are usually integrated. Cointegration analysis aims to uncover causal relations among variables by determining if the stochastic trends in a group of variables are shared by the series so that the total number of unique trends is less than the number of variables. It can also be used to test if there are residual stochastic trends that are not shared by any other variables. This may be an indication that important variables have been omitted from the regression model or that the variable with the residual trend does not have long-term interactions with the other variables.

Either of these conclusions could be true should there be no cointegration. The presence of cointegration can also be interpreted as the presence of a long-term equilibrium relationship between the variables in question. The parameters of an estimated cointegrating relation are called the cointegrating vector. In multivariate models, there may be more than one such cointegrating vector.

5.1 Granger Causality and the Energy GDP Relation

A series of analyses use statistical tests to evaluate whether energy use or energy prices determine economic growth, or whether the level of output in the United States and other economics determine energy use or energy prices. Generally, the results are inconclusive. Where significant results are obtained, they indicate causality from output to energy use.

One analysis tests U.S. data (1947–1990) for Granger causality in a multivariate setting using a vector autoregression model of GDP, energy use, capital, and labor inputs. The study measures energy use by its thermal equivalents and the Divisia aggregation method discussed previously. The relation among GDP, the thermal equivalent of energy use, and the Divisia energy use indicates that there is less "decoupling" between GDP and energy use when the aggregate measure for energy use accounts for qualitative differences (Fig. 3). The multivariate methodology is important because changes in energy use are frequently countered by substitution with labor and/or capital and thereby mitigate the effect of changes in energy use on output. Weighting energy use for changes in the composition of the energy input is important because a large portion of the growth effects of energy is due to substitution of higher quality energy sources such as electricity for lower quality energy sources such as coal (Fig. 1).

Bivariate tests of Granger causality show no causal order in the relation between energy and GDP in either direction, regardless of the measure used to qualify energy use (Table III). In the multivariate model with energy measured in primary Btus, GDP was found to "Granger cause" energy use. However, when both innovations—a multivariate model and energy use adjusted for quality—are employed, energy Granger causes GDP. These results show that adjusting energy for quality is important, as is considering the context within which energy use is occurring. The conclusion that energy use plays an important role in determining the level of economic

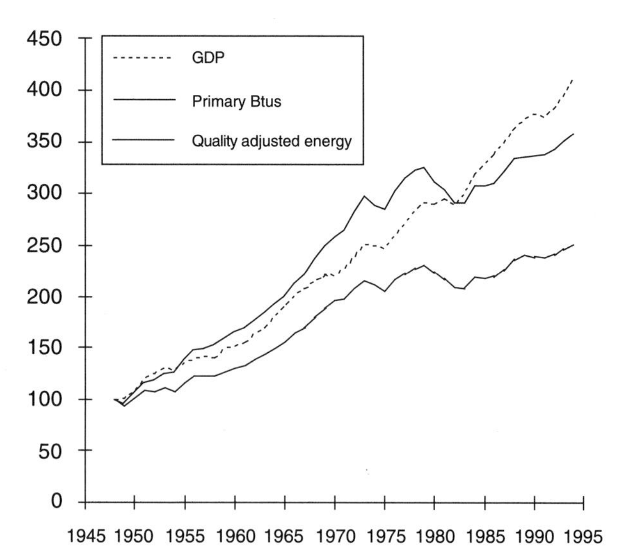

FIGURE 3 Energy use and GDP in the United States, with energy use measured in heat equivalents and a Divisia index. From Stern (1993).

TABLE III

Energy GDP Causality Tests for the United States, 1947–1990[a]

	Bivariate model		Multivariate model	
	Primary Btus	Quality-adjusted energy	Primary Btus	Quality-adjusted energy
Energy causes GDP	0.8328	0.9657	0.5850	3.1902
	0.4428	*0.4402*	*0.5628*	*0.3188E−01*
GDP causes Energy	0.3421	0.7154	9.0908	0.8458
	0.7125	*0.5878*	*0.7163E−03*	*0.5106*

[a]The test statistic is an F statistic. Significance levels in italics. A significant statistic indicates that there is Granger causality in the direction indicated.

activity is consistent with results of price-based studies of other energy economists.

6. CASE STUDY 3: THE DETERMINANTS OF THE ENERGY/GDP RELATIONSHIP

One of the most widely cited macroeconomic indicators of sustainability is the ratio of total energy use to total economic activity, or the energy/real GDP ratio (E/GDP ratio). This ratio has declined since 1950 in many industrial nations. There is controversy regarding the interpretation of this decline. Many economists and energy analysts argue that the decline indicates that the relation between energy use and economic activity is relatively weak. This interpretation is disputed by many biophysical economists, who argue that the decline in the E/GDP ratio overstates the ability to decouple energy use and economic activity because many analyses of the E/GDP ratio ignore the effect of changes in energy quality (Fig. 1).

The effect of changes in energy quality (and changes in energy prices and types of goods and services produced and consumed) on the E/GDP ratio can be estimated using Eq. (7):

$$\frac{E}{\mathrm{GDP}} = \alpha + \beta_1 \ln\left(\frac{\text{natural gas}}{\mathrm{E}}\right) + \beta_2 \ln\left(\frac{\text{oil}}{\mathrm{E}}\right) + \beta_3 \ln\left(\frac{\text{primary electricity}}{\mathrm{E}}\right) + \beta_4\left(\frac{\mathrm{PCE}}{\mathrm{GDP}}\right) + \beta_5(\text{product mix}) + \beta_6 \ln(\text{price}) + \varepsilon, \quad (7)$$

where E is the total primary energy consumption (measured in heat units); GDP is real GDP; primary electricity is electricity generated from hydro, nuclear, solar, or geothermal sources; PCE is real personal consumption expenditures spent directly on energy by households; product mix measures the fraction of GDP that originates in energy-intensive sectors (e.g., chemicals) or non-energy-intensive sectors (e.g., services); and price is a measure of real energy prices.

The effect of energy quality on the E/GDP ratio is measured by the fraction of total energy consumption from individual fuels. The sign on the regression coefficients β_1–β_3 is expected to be negative because natural gas, oil, and primary electricity can do more useful work (and therefore generate more economic output) per heat unit than coal. The rate at which an increase in the use of natural gas, oil, or primary electricity reduces the E/GDP ratio is not constant. Engineering studies indicate that the efficiency with which energies of different types are converted to useful work depends on their use. Petroleum can provide more motive power per heat unit of coal, but this advantage nearly disappears if petroleum is used as a source of heat. From an economic perspective, the law of diminishing returns implies that the first uses of high-quality energies are directed at tasks that are best able to make use of the physical, technical, and economic aspects of an energy type that combine to determine its high-quality status. As the use of a high-quality energy source expands, it is used for tasks that are less able to make use of the attributes that confer high quality. The combination of physical differences in the use of energy and the economic ordering in which they are applied to these tasks implies that the amount of economic activity generated per heat unit diminishes as the use of a high-quality energy expands. Diminishing returns on energy quality is imposed on the model by specifying the fraction of the energy budget from petroleum, primary electricity, natural gas, or oil in natural logarithms. This specification ensures that the first uses of high-quality energies decrease the energy/GDP ratio faster than the last uses.

The regression results indicate that Eq. (7) can be used to account for most of the variation in the E/GDP ratio for France, Germany, Japan, and the United Kingdom during the post-World War II period and in the United States since 1929. All the variables have the sign expected by economic theory and are statistically significant, and the error terms have the properties assumed by the estimation technique.

Analysis of regression results indicate that changes in energy mix can account for a significant portion of the downward trend in E/GDP ratios. The change from coal to petroleum and petroleum to primary electricity is associated with a general decline in the E/GDP ratio in France, Germany, the United Kingdom, and the United States during the post-World War II period (Fig. 4). The fraction of total energy consumption supplied by petroleum increased steadily for each nation through the early 1970s. After the first oil shock, the fraction of total energy use from petroleum remained steady or declined slightly in these four nations. However, energy mix continued to reduce the E/real GDP ratio after the first oil shock because the fraction of total energy use from primary electricity increased steadily. The effect of changes in energy mix on the E/GDP ratio shows no trend over time in Japan, where the fraction of total energy consumption supplied by primary electricity declined through the early 1970s and increased steadily thereafter. This U shape offsets the steady increase in the fraction of total energy use from petroleum that occurred prior to 1973.

These regression results indicate that the historical reduction in the E/GDP ratio is associated with shifts in the types of energies used and the types of goods and services consumed and produced. Diminishing returns to high-quality energies and the continued consumption of goods from energy-intensive sectors such as manufacturing imply that the ability of changes in the composition of inputs and outputs to reduce the E/real GDP ratio further is limited.

7. CONCLUSIONS AND IMPLICATIONS

Application of the Divisia index to energy use in the U.S. economy illustrates the importance of energy quality in aggregate analysis. The quality-corrected index for EROI indicates that the energy surplus delivered by petroleum extraction in the United States is smaller than indicated by unadjusted EROI. The trend over time in a quality-adjusted index of total primary energy use in the U.S. economy is significantly different, and declines faster, than the standard heat-equivalent index. Analysis of Granger causality and cointegration indicates a causal relationship running from quality-adjusted energy to GDP but not from the unadjusted energy index. The econometric analysis of the E/real GDP ratio indicates that the decline in industrial economies has been driven in part by the shift from coal to oil, gas, and primary electricity. Together, these results suggest that accounting for energy quality reveals a relatively strong relationship between energy use and economic output. This runs counter to much of the conventional wisdom that technical improvements and structural change have decoupled energy use from economic performance. To a large degree, technical change and substitution have increased the use of higher quality energy and reduced the use of lower quality energy. In economic terms, this means that technical change has been "embodied" in the fuels and their associated energy converters. These changes have increased energy efficiency in energy extraction processes, allowed an apparent decoupling between energy use and economic output, and increased energy efficiency in the production of output.

The manner in which these improvements have been largely achieved should give pause for thought. If decoupling is largely illusory, any increase in the cost of producing high-quality energy vectors could have important economic impacts. Such an increase might occur if use of low-cost coal to generate electricity is restricted on environmental grounds, particularly climate change. If the substitution process cannot continue, further reductions in the E/GDP ratio would slow. Three factors might limit future substitution to higher quality energy. First, there are limits to the substitution process. Eventually, all energy used would be of the highest quality variety—electricity—and no further substitution could occur. Future discovery of a higher quality energy source might mitigate this situation, but it would be unwise to rely on the discovery of new physical principles. Second, because different energy sources are not perfect substitutes, the substitution process could have economic limits that will prevent full substitution. For example, it is difficult to imagine an airliner running on electricity. Third, it is likely that supplies of petroleum, which is of higher quality than coal, will decline fairly early in the 21st century.

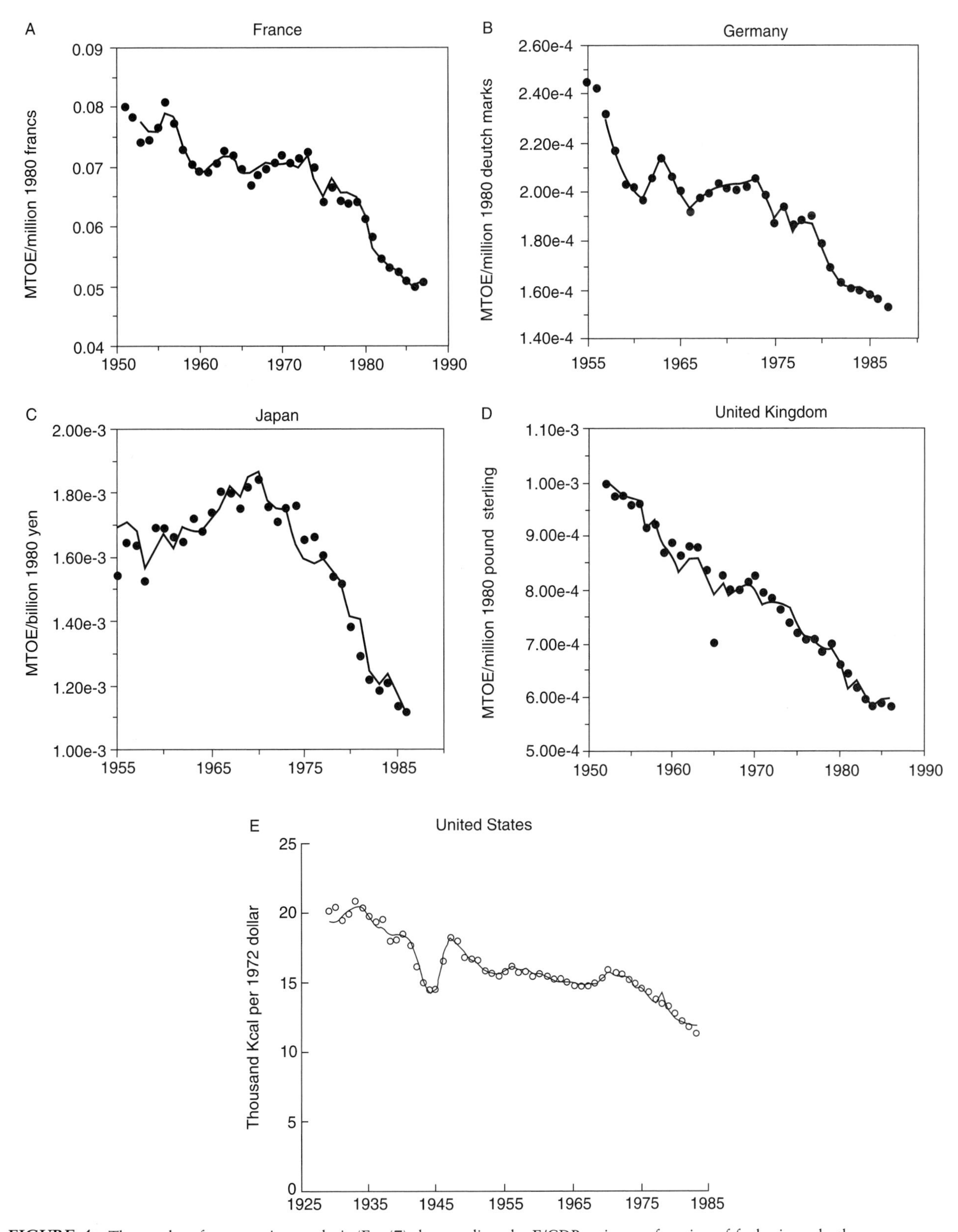

FIGURE 4 The results of a regression analysis (Eq. (7) that predicts the E/GDP ratio as a function of fuel mix and other variables. •, actual values; solid lines, values predicted from the regression equation; ○, actual values for the United States. From Kaufmann (1992) and Hall *et al.* (1986).

Finally, our conclusions do not imply that one-dimensional and/or physical indicators are universally inferior to the economic indexing approach we endorse. As one reviewer noted, ecologists might raise the problem of Leibig's law of the minimum, in which the growth or sustainability of a system are constrained by the single critical element in least supply. Exergy or mass are appropriate if the object of analysis is a single energy or material flux. Physical units are also necessary to valuate those flows. Integrated assessment of a material cycle within and between the environment and the economy is logical based on physical stocks and flows. However, when the question being asked requires the aggregation of energy flows in an economic system, an economic approach such as Divisa aggregation or a direct measure of marginal product embody a more tenable set of assumptions than does aggregation by one-dimensional approaches.

SEE ALSO THE FOLLOWING ARTICLES

Economic Thought, History of Energy in • Emergy Analysis and Environmental Accounting • Entropy and the Economic Process • Exergy • Exergy Analysis of Energy Systems • Exergy: Reference States and Balance Conditions • Net Energy Analysis: Concepts and Methods • Thermodynamics and Economics, Overview

Further Reading

Ayres, R., and Martiñas, K. (1995). Waste potential entropy: The ultimate ecotoxic? *Econ. Appliqueé* **48,** 95–120.
Berndt, E. (1990). Energy use, technical progress and productivity growth: A survey of economic issues. *J. Productivity Anal.* **2,** 67–83.
Cleveland, C. J. (1992). Energy quality and energy surplus in the extraction of fossil fuels in the U.S. *Ecol. Econ.* **6,** 139–162.
Cleveland, C. J., Costanza, R., Hall, C. A. S., and Kaufmann, R. (1984). Energy and the U.S. economy: A biophysical perspective. *Science* **255,** 890–897.
Cottrell, W. F. (1955). "Energy and Society." McGraw-Hill, New York.
Darwin, R. F. (1992). Natural resources and the Marshallian effects of input-reducing technological changes. *J. Environ. Econ. Environ. Management* **23,** 201–215.
Gever, J., Kaufmann, R., Skole, D., and Vorosmarty, C. (1986). "Beyond Oil: The Threat to Food and Fuel in the Coming Decades." Ballinger, Cambridge, UK.
Hall, C. A. S., Cleveland, C. J., and Kaufmann, R. K. (1986). "Energy and Resource Quality: The Ecology of the Economic Process." Wiley Interscience, New York.
Hamilton, J. D. (1983). Oil and the macroeconomy since World War II. *J. Political Econ.* **91,** 228–248.
Kaufmann, R. K. (1992). A biophysical analysis of the energy/real GDP ratio: Implications for substitution and technical change. *Ecol. Econ.* **6,** 35–56.
Odum, H. T. (1996). "Environmental Accounting." Wiley, New York.
Rosenberg, N. (1998). The role of electricity in industrial development. *Energy J.* **19,** 7–24.
Schurr, S., and Netschert, B. (1960). "Energy and the American Economy, 1850–1975." Johns Hopkins Univ. Press, Baltimore.
Stern, D. I. (1993). Energy use and economic growth in the USA: A multivariate approach. *Energy Econ.* **15,** 137–150.

Aircraft and Energy Use

JOOSUNG J. LEE, STEPHEN P. LUKACHKO, and IAN A. WAITZ
Massachusetts Institute of Technology
Cambridge, Massachusetts, United States

1. Introduction
2. Economic Growth, Demand, and Energy Use
3. Energy Use, Emissions, and Environmental Impact
4. Trends in Energy Use
5. Energy Consumption in an Aircraft System
6. Historical Trends in Technological and Operational Performance
7. Technological and Operational Outlook for Reduced Energy Use in Large Commercial Aircraft
8. Industry Characteristics, Economic Impacts, and Barriers to Technology Uptake

Glossary

bypass ratio The ratio of air passed through the fan system to that passed through the engine core.

contrail The condensation trail that forms when moist, high-temperature air in a jet exhaust, as it mixes with ambient cold air, condenses into particles in the atmosphere and saturation occurs.

drag The aerodynamic force on an aircraft body; acts against the direction of aircraft motion.

energy intensity (E_I) A measure of aircraft fuel economy on a passenger-kilometer basis; denoted by energy used per unit of mobility provided (e.g., fuel consumption per passenger-kilometer)

energy use (E_U) A measure of aircraft fuel economy on a seat-kilometer basis (e.g., fuel consumption per seat-kilometer).

great circle distance The minimum distance between two points on the surface of a sphere.

hub-and-spoke system Feeding smaller capacity flights into a central hub where passengers connect with flights on larger aircraft that then fly to the final destination.

lift-to-drag ratio (L/D) A measure of aerodynamic efficiency; the ratio of lift force generated to drag experienced by the aircraft.

load factor The fraction of passengers per available seats.

radiative forcing A measure of the change in Earth's radiative balance associated with atmospheric changes; positive forcing indicates a net warming tendency relative to preindustrial times.

structural efficiency (OEW/MTOW) The ratio of aircraft operating empty weight (OEW) to maximum takeoff weight (MTOW); a measure of the weight of the aircraft structure relative to the weight it can carry (combined weights of structure plus payload plus fuel).

thrust A force that is produced by engines and propels the aircraft.

thrust specific fuel consumption (SFC) A measure of engine efficiency as denoted by the rate of fuel consumption per unit thrust (e.g., kilograms/second/Newton).

turbofan engine The dominant mode of propulsion for commercial aircraft today; a turbofan engine derives its thrust primarily by passing air through a large fan system driven by the engine core.

An aircraft is composed of systems that convert fuel energy to mechanical energy in order to perform work—the movement of people and cargo. This article describes how aircraft technology and operations relate to energy use. Historical trends and future outlook for aircraft performance, energy use, and environmental impacts are discussed. Economic characteristics of aircraft systems as they relate to energy use are also presented.

1. INTRODUCTION

The first powered passenger aircraft were developed at the turn of the 20th century. Since then, there has been rapid growth in aviation as a form of mobility and consequently significant growth in energy use. In 2002, aviation accounted for 3 trillion revenue passenger-kilometers (RPKs), or approximately 10% of world RPKs traveled on all transportation modes and 40% of the value of world freight shipments. Among all modes of transport, demand for air travel has grown fastest. If, as expected, strong growth in air travel demand continues, aviation will become the dominant mode of transportation, perhaps surpassing the mobility provided by automobiles within a century. This evolution of transportation demand also suggests an increase in per-person energy use for transportation. Minimizing energy use has always been a fundamental design goal for commercial aircraft. However, the growth of air transportation renders ever-increasing pressures for improvements in technology and operational efficiency to limit environmental impacts.

In the analysis presented here, trends in aviation transportation demand, energy use, and associated environmental impacts are examined (Sections 2–4). In Sections 5 and 6, aircraft systems from an energy conversion perspective are introduced and key performance parameters of aircraft technology and operation are discussed. A technology and operational outlook for reduced aircraft energy use is presented in Section 7, followed by a summary of industry characteristics and economic impacts that affect energy use of individual aircraft and the fleet as a whole in Section 8.

2. *ECONOMIC GROWTH, DEMAND, AND ENERGY USE*

On a per capita basis, rising demand for mobility is well correlated with growth in gross domestic product (GDP)—a measure of national economic activity—across a wide variety of economic, social, and geographic settings. One reason for this may be found in the roughly constant shares of income and time people dedicate to transportation. A fixed budget for travel leads to an increase in total travel demand per capita (e.g., RPK per capita) in approximate proportion to income. In addition, a person spends an average of 1.0–1.5 hours/day traveling. One key implication of such invariance is that as demand for movement increases, travelers tend to shift toward faster modes of transportation. Consequently, continuing growth in world population and income levels can be expected to lead to further demand for air travel, both in terms of market share and RPKs. As a result, high-speed transportation, in which aviation is anticipated to be the primary provider, will play an increasingly important role and may account for slightly more than one-third of world passenger traffic volume within the next 50 years. In general, among industry and government predictions, growth in passenger air transportation has been typically projected to be between 3 and 6%/year as an average over future periods of 10–50 years.

Aviation fuel consumption today corresponds to 2–3% of the total fossil fuel use worldwide, more than 80% of which is used by civil aviation operation. Energy use in the production of aircraft is relatively minor in comparison to that consumed in their operation. Although the majority of air transportation demand is supplied by large commercial aircraft, defined as those aircraft with a seating capacity of 100 or more, smaller regional aircraft have emerged as an important component of both demand and energy use within air transportation. For example, in the United States, although regional aircraft currently perform under 4% of domestic RPKs, they account for almost 7% of jet fuel use and for 40–50% of total departures. Future growth in demand for regional aircraft RPKs could be up to double the rate for large commercial aircraft. Cargo operations account for some 10% of total revenue ton-kilometers and fuel use within the aviation sector. Economic activity, as measured by world GDP, is the primary driver for the air cargo industry growth. World air cargo traffic is expected to grow at an average annual rate of over 6% for the next decade.

3. *ENERGY USE, EMISSIONS, AND ENVIRONMENTAL IMPACT*

The growth in air transportation volume has important global environmental impacts associated with the potential for climate change. On local to regional scales, noise, decreased air quality related primarily to ozone production and particulate levels, and other issues, such as roadway congestion related to airport services and local water quality, are all recognized as important impacts. In this section, the focus is on emissions-related impacts; because of its relative importance, some additional detail on the aviation role in climate change is provided.

The total mass of emissions from an aircraft is directly related to the amount of fuel consumed. Of the exhaust emitted from the engine core, 7–8% is composed of carbon dioxide (CO_2) and water vapor (H_2O); another 0.5% composed of nitrogen oxides (NO_x), unburned hydrocarbons (HC), carbon monoxide (CO), and sulfur oxides (SO_x); there are other trace chemical species that include the hydroxy family (HO_x) and the extended family of nitrogen compounds (NO_y), and soot particulates. Elemental species such as O, H, and N are also formed to an extent governed by the combustion temperature. The balance (91.5–92.5%) is composed of O_2 and N_2.

Emissions of CO_2 and H_2O are products of hydrocarbon fuel combustion and are thus directly related to the aircraft fuel consumption, which in turn is a function of aircraft weight, aerodynamic design, engine design, and the manner in which the aircraft is operated. Emissions of NO_x, soot, CO, HC, and SO_x are further related to details of the combustor design and, to some extent, to postcombustion chemical reactions occurring within the engine. These emissions are thus primarily controlled by the engine design, but total emissions can be reduced through improvements in fuel efficiency. Such emissions are therefore typically quoted relative to the total amount of fuel burned as an emission index (e.g., grams of NO_x/kilogram of fuel). A host of minor constituents exist in very small, trace amounts.

The climate effects of aviation are perhaps the most important of the environmental impacts, both in terms of economic cost and the extent to which all aspects of the aviation system, operations, and technology determine the impact. Because a majority of aircraft emissions are injected into the upper troposphere and lower stratosphere (typically 9–13 km in altitude), resulting impacts on the global environment are unique among all industrial activities. The fraction of aircraft emissions that is relevant to atmospheric processes extends beyond the radiative forcing effects of CO_2. The mixture of exhaust species discharged from aircraft perturbs radiative forcing two to three times more than if the exhaust was CO_2 alone. In contrast, the overall radiative forcing from the sum of all anthropogenic activities is estimated to be a factor of 1.5 times CO_2 alone. Thus the impact of burning fossil fuels at altitude is approximately double that due to burning the same fuels at ground level. The enhanced forcing from aircraft compared with ground-based sources is due to different physical (e.g., contrails) and chemical (e.g., ozone formation/destruction) effects resulting from altered concentrations of participating chemical species and changed atmospheric conditions. However, many of the chemical and physical processes associated with climate impacts are the same as those that determine air quality in the lower troposphere.

Estimates of the radiative forcing by various aircraft emissions for 1992 offered by the Intergovernmental Panel on Climate Change (IPCC) and the 1999 projections from Penner *et al.* for the year 2050 are shown in Fig. 1. The estimates translate to 3.5% of the total anthropogenic forcing that occurred in 1992 and to an estimated 5% by 2050 for an all-subsonic fleet. Associated increases in ozone levels are expected to decrease the amount of ultraviolet radiation at the surface of the earth. Future fleet composition also impacts the radiative forcing estimate. A supersonic aircraft flying at 17–20 km would have a radiative forcing five times greater than a subsonic equivalent in the 9- to 13-km range. It is important to note that these estimates are of an uncertain nature. Although broadly consistent with

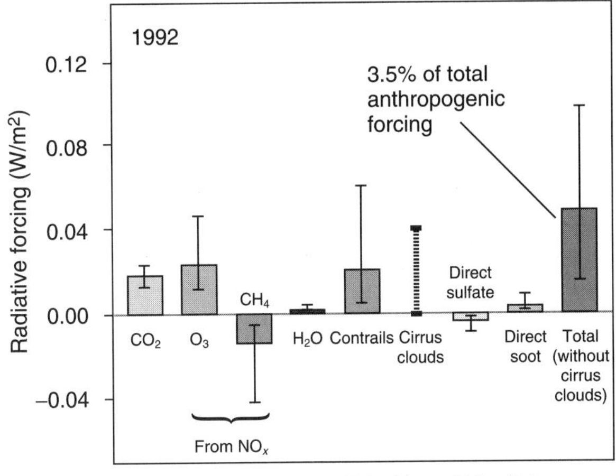

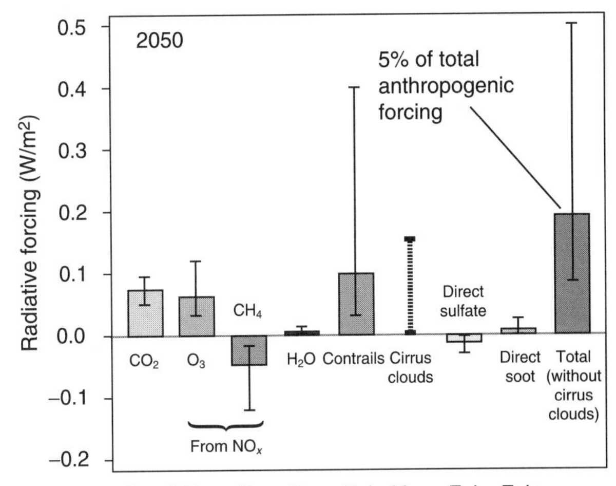

FIGURE 1 Radiative forcing estimated for 1992 (0.05 W/m^2 total) and projected to 2050 (0.19 W/m^2 total). Note differences in scale. Note also that the dashed bars for aviation-induced cirrus cloudiness describe the range of estimates, not the uncertainty. The level of scientific understanding of this potential impact is very poor and no estimate of uncertainty has been made. Cirrus clouds are not included in the total radiative forcing estimate. Reproduced from Penner *et al.* (1999), with permission.

these IPCC projections, subsequent research reviewed by the Royal Commission on Environmental Protection (RCEP) in the United Kingdom has suggested that the IPCC reference value for the climate impact of aviation is likely to be an underestimate. In particular, although the impact of contrails is probably overestimated in Fig. 1, aviation-induced cirrus clouds could be a significant contributor to positive radiative forcing; NO_x-related methane reduction is less than shown in Fig. 1, reducing the associated cooling effect, and growth of aviation in the period 1992–2000 has continued at a rate larger than that used in the IPCC reference scenario.

4. TRENDS IN ENERGY USE

Fuel efficiency gains due to technological and operational change can mitigate the influence of growth on total emissions. Increased demand has historically outpaced these gains, resulting in an overall increase in emissions over the history of commercial aviation. The figure of merit relative to total energy use and emissions in aviation is the energy intensity (E_I). When discussing energy intensity, the most convenient unit of technology is the system represented by a complete aircraft. In this section, trends in energy use and E_I are elaborated. In the following section, the discussion focuses on the relation of E_I to the technological and operational characteristics of an aircraft.

Reviews of trends in technology and aircraft operations undertaken by Lee *et al.* and Babikian *et al.* indicate that continuation of historical precedents would result in a future decline in E_I for the large commercial aircraft fleet of 1.2–2.2%/year when averaged over the next 25 years, and perhaps an increase in E_I for regional aircraft, because regional jets use larger engines and replace turboprops in the regional fleet. When compared with trends in traffic growth, expected improvements in aircraft technologies and operational measures alone are not likely to offset more than one-third of total emissions growth. Therefore, effects on the global atmosphere are expected to increase in the future in the absence of additional measures. Industry and government projections, which are based on more sophisticated technology and operations forecasting, are in general agreement with the historical trend. Compared with the early 1990s, global aviation fuel consumption and subsequent CO_2 emissions could increase three- to sevenfold by 2050, equivalent to a 1.8–3.2% annual rate of change. In addition to the different demand growth projections entailed in such forecasts, variability in projected emissions also originates from different assumptions about aircraft technology, fleet mix, and operational evolution in air traffic management and scheduling.

Figure 2 shows historical trends in E_I for the U.S. large commercial and regional fleets. Year-to-year variations in E_I for each aircraft type, due to different operating conditions, such as load factor, flight speed, altitude, and routing, controlled by different operators, can be $\pm 30\%$, as represented by the vertical extent of the data symbols (Fig. 2A). For large commercial aircraft, a combination of technological and operational improvements led to a reduction in E_I of the entire U.S. fleet of more than 60% between 1971 and 1998, averaging about 3.3%/year. In contrast, total RPK has grown by 330%, or 5.5%/year over the same period. Long-range aircraft are $\sim 5\%$ more fuel efficient than are short-range aircraft because they carry more passengers over a flight spent primarily at the cruise condition. Regional aircraft are 40–60% less fuel efficient than are their larger narrow- and wide-body counterparts, and regional jets are 10–60% less fuel efficient compared to turboprops. Importantly, fuel efficiency differences between large and regional aircraft can be explained mostly by differences in aircraft operations, not technology.

Reductions in E_I do not always directly imply lower environmental impact. For example, the prevalence of contrails is enhanced by greater engine efficiency. NO_x emissions also become increasingly difficult to limit as engine temperatures and pressures are increased—a common method for improving engine efficiency. These conflicting influences make it difficult to translate the expected changes in overall system performance into air quality impacts. Historical trends suggest that fleet-averaged NO_x emissions per unit thrust during landing and takeoff (LTO) cycles have seen little improvement, and total NO_x emissions have slightly increased. However, HC and CO emissions have been reduced drastically since the 1950s.

5. ENERGY CONSUMPTION IN AN AIRCRAFT SYSTEM

Energy intensity can be related to specific measures of technological and operational efficiency in the air transportation system. The rest of this article takes a more detailed look at trends in these efficiencies and

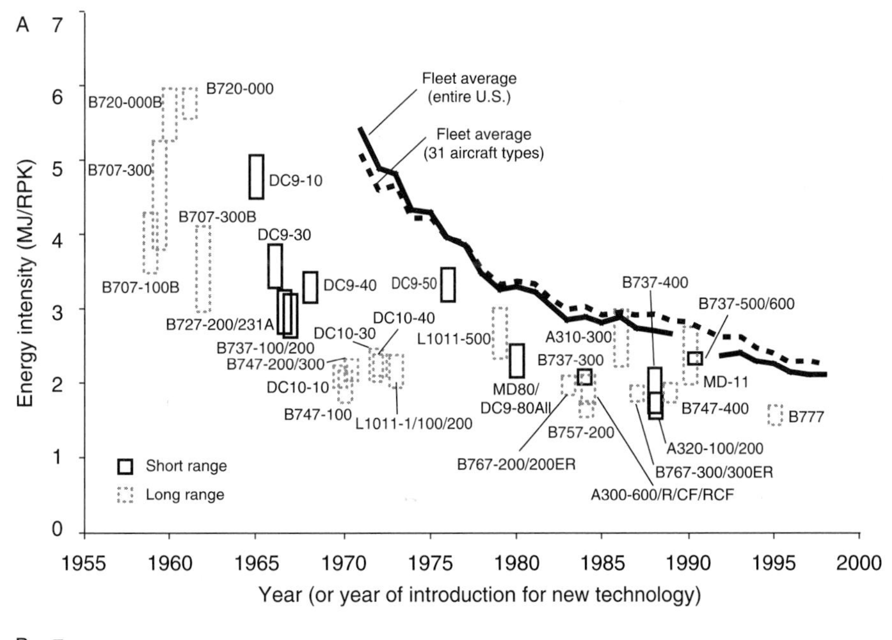

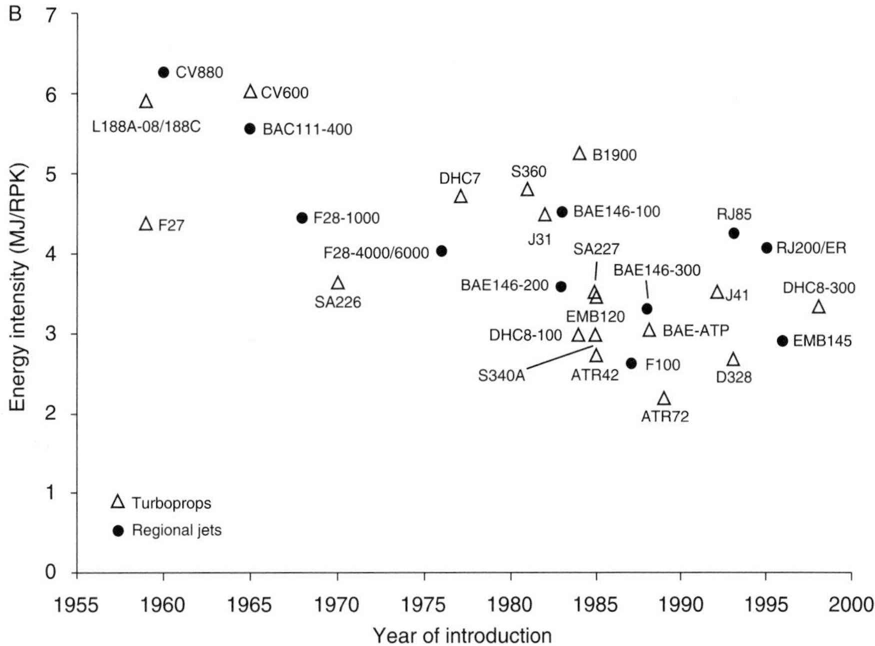

FIGURE 2 (A) Historical trends in energy intensity of the U.S. large commercial fleets. Individual aircraft E_I based on 1991–1998 operational data with the exception of the B707 and B727, which are based on available operational data prior to 1991. Fleet averages were calculated using a revenue passenger-kilometer (RPK) weighting. Data were not available for the entire U.S. fleet average during 1990 and 1991. Reproduced from Lee *et al.* (2001), with permission. (B) Historical trends in energy intensity of the U.S. regional fleets.

options for controlling energy use. The first step is a simplified description of the energy conversion within an aircraft engine. An aircraft engine converts the flow of chemical energy contained in aviation fuel and the air drawn into the engine into power (thrust multiplied by flight speed). Overall engine efficiency is defined by the ratio of power to total fuel energy flow rate. Only one-fourth to one-third of fuel energy is used to overcome drag and thus propel the aircraft. The remaining energy is expelled as waste heat in the engine exhaust. A parameter that is closely related to the overall engine efficiency is the specific fuel consumption (SFC). When judging the efficiency of an aircraft system, however, it is more relevant to consider work in terms of passengers or payload carried per unit distance. Energy intensity is an appropriate measure when comparing efficiency and environmental impact to other modes. E_I consists of two components—energy use, E_U, and load factor, α, as described by Eq. (1). Energy use is energy consumed by the aircraft per seat per unit distance traversed and is determined by aircraft technology parameters, including engine efficiency. E_U observed in actual aircraft operations reflects operational inefficiencies such as ground delays and airborne holding. The fleet average E_U is of interest because it is the fleet fuel efficiency that determines the total energy use. Load factor is a measure of how efficiently aircraft seats are filled and aircraft kilometers are utilized for revenue-generating purposes. Increasing load factor leads to improved fuel consumption on a passenger-kilometer basis.

$$E_I = \frac{\text{MJ}}{\text{RPK}} = \frac{\text{MJ}}{\text{ASK}} \Big/ \frac{\text{RPK}}{\text{ASK}} = \frac{E_U}{\alpha}, \qquad (1)$$

where MJ is megajoules of fuel energy, RPK is revenue passenger-kilometers, ASK is available seat-kilometers, and α is load factor. To show E_I as a function of the engine, aerodynamic, and structural efficiencies of an aircraft system as well as load factor, it is necessary to have a model of aircraft performance. Because a major portion of aircraft operation is spent at cruise, the Breguet range (R) equation, which describes aircraft motion in level, constant-speed flight, is a relevant model. In the Breguet range equation [Eq. (2)], engine thrust is balanced by drag, and lift balances aircraft weight. Propulsion, aerodynamic, and structural characteristics are represented by three parameters: SFC, lift-to-drag ratio (L/D), and structural weight ($W_{\text{structure}}$). Given these technological characteristics as well as other operability parameters, including the amount of payload (W_{payload}) and fuel on board (W_{fuel}), the Breguet range equation can be used to determine maximum range for a level, constant-speed flight.

$$R = \frac{V(\text{L/D})}{g \cdot \text{SFC}} \times \ln\left(1 + \frac{W_{\text{fuel}}}{W_{\text{payload}} + W_{\text{structure}} + W_{\text{reserve}}}\right), \qquad (2)$$

where g is the gravitational acceleration constant, W_{reserve} is the reserve fuel, and V is flight speed. By rearranging Eq. (2), a relationship between aircraft energy use and technology parameters can be derived as shown in Eq. (3). As implied by Eq. (3), aircraft system efficiency improves with lower SFC, greater L/D, and lighter structural weight.

$$E_U \equiv \alpha E_I \equiv \frac{1}{\eta_U}$$

$$E_U = \frac{Q W_{\text{fuel}}}{S} \cdot \frac{g \cdot \text{SFC}}{V(\text{L/D})} \times \frac{1}{\ln\left(\dfrac{W_{\text{fuel}}}{W_{\text{payload}} + W_{\text{structure}} + W_{\text{reserve}}}\right)}, \qquad (3)$$

where η_U is fuel efficiency (e.g., seat-kilometers/kilogram of fuel consumption), Q is the lower heating value of jet fuel, and S is the number of seats.

6. *HISTORICAL TRENDS IN TECHNOLOGICAL AND OPERATIONAL PERFORMANCE*

6.1 Technological Performance

As shown in Eq. (3), engine, aerodynamic, and structural efficiencies play an important role in determining the energy intensity of an aircraft. Engine efficiency in large commercial aircraft, as measured by the cruise SFC of newly introduced engines, improved by approximately 40% over the period 1959–1995, averaging an annual 1.5% improvement. Most of this improvement was realized prior to 1970, with the introduction of high-bypass turbofan engines. However, as bypass ratios have increased, engine diameters have also become larger, leading to an increase in engine weight and aerodynamic drag. Other routes to engine efficiency improvement include increasing the peak pressure and temperature within the engine, which is limited by materials and cooling technology, and improving engine component efficiencies. Aerodynamic efficiency in large commercial aircraft has increased by approximately 15% historically, averaging 0.4%/year for the same period. Better wing design and improved propulsion/airframe integration, enabled by improved computational and experimental design tools, have been the primary drivers. Historical improvements in structural efficiency are less evident. One reason is that over the 35-year period between the introduction of the B707 and the B777, large commercial aircraft have been constructed almost exclusively of aluminum and are currently about 90% metallic by weight. Composites are used for a limited number of components. Another reason is that improvements in aircraft structural efficiency have been largely traded for other technological improvements, such as larger, heavier engines and increased passenger comfort.

6.2 Operational Performance

Infrastructure characteristics also impact efficiency. In particular, delays on the ground and in the air can increase energy intensity. Extra fuel is burned on the ground during various non-flight operations, and hours spent in the air (airborne hours) do not account for more than 75–90% of the total operational hours of the aircraft (block hours). The ratio of airborne to block hours can be treated as ground-time efficiency, η_g. Similarly, non-cruise portions of the flight, poor routing, and delays in the air constitute inefficiencies related to spending fuel during the flight beyond what would be required for a great circle distance trip at constant cruise speed. This inefficiency can be measured by the ratio of minimum flight hours to airborne hours, η_a. Minimum flight hours are calculated with the assumption that all aircraft fly the entire route at Mach 0.80 and at an altitude of 10.7 km (no climbing, descending, or deviation from the minimum distance, the great circle route). Minimum flight hours represent the shortest time required to fly a certain stage length and reveal any extra flight time due to nonideal flight conditions. The product of η_g and η_a gives the flight time efficiency, η_{ft}. Both η_a and η_g increase with stage length. The lower η_{ft} associated with short-range aircraft is related to the more than 40% of block time spent in non-cruise flight segments. Long-range aircraft operate closer to the ideal as total flight time efficiency approaches 0.9. The impact of operational differences on E_U is evident in Fig. 3, which shows the variation of E_U with stage length for turboprop and jet-powered aircraft (both regional and large jets) introduced during and after 1980. Aircraft flying stage lengths below 1000 km have E_U values between 1.5 and 3 times higher compared to aircraft flying stage lengths above 1000 km. Regional aircraft, compared to large aircraft, fly shorter stage lengths, and therefore spend more time at airports taxiing, idling, and maneuvering into gates, and in general spend a greater fraction of their block time in non-optimum, non-cruise stages of flight. Turboprops show a pattern distinct from that of jets and are, on average, more efficient at similar stage lengths. The energy usage also increases gradually for stage lengths above 2000 km because the increasing amount of fuel required for increasingly long stage lengths leads to a heavier aircraft and a higher rate of fuel burn.

Aircraft E_I is also improved through better utilization (e.g., load factor) and greater per-aircraft capacity (e.g., number of seats). Historically, the load factor on domestic and international flights operated by U.S. carriers climbed 15% between 1959 and 1998, all of which occurred after 1970 at an average of 1.1%/year. Figure 4 shows historical load factor evolution for both U.S. large commercial and regional aircraft. Load factor gains have been attributed to deregulation in the United States and global air travel liberalization, both of which contributed to the advent of hub-and-spoke systems. As airlines have sought greater route capacity, the

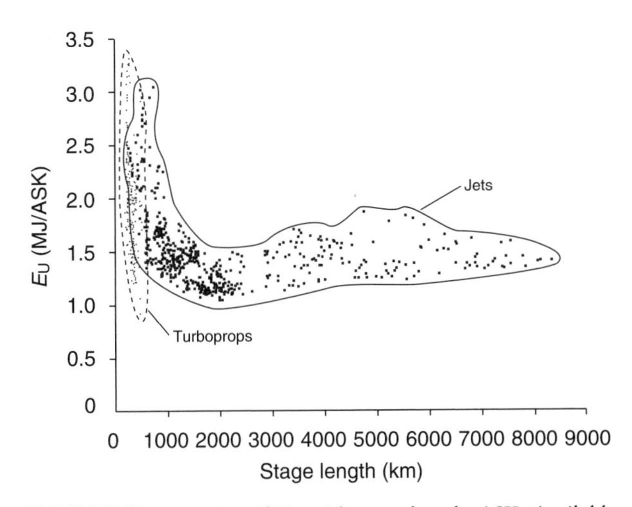

FIGURE 3 Variation of E_U with stage length. ASK, Available seat-kilometers.

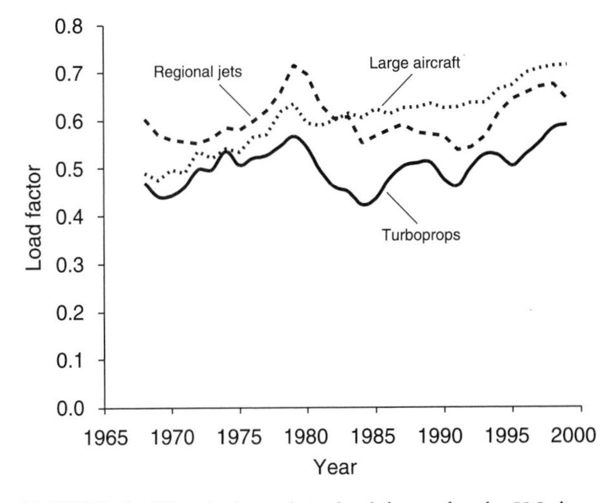

FIGURE 4 Historical trends in load factor for the U.S. large commercial and regional fleets.

average number of seats has also increased, by 35% between 1968 and 1998, or from 108 to 167 seats (an average of 1.4%/year), most of which occurred prior to 1980.

7. TECHNOLOGICAL AND OPERATIONAL OUTLOOK FOR REDUCED ENERGY USE IN LARGE COMMERCIAL AIRCRAFT

The outlook for future reductions in energy use is necessarily based on the potential for increased technological and operational efficiencies. In this section, the outlook for such improvements in large commercial aircraft over the next quarter century is examined.

Engine efficiencies may be improved by between 10 and 30% with further emphasis on moving more mass through engines that operate at higher temperatures and higher pressures. A continuation of the historical trend would lead to a 10% increase in L/D by 2025, and further improvements in the reduction of parasitic drag may extend these savings to perhaps 25%. However, the technologies associated with these improvements have weight and noise constraints that may make their use difficult. For example, the lack of historical improvement in structural efficiency suggests that weight reductions will be offset by added weight for other purposes (e.g., engines, entertainment). However, weight represents an area wherein major improvements may be found without the constraints that may hinder improvement in engine or aerodynamic efficiency. If lighter weight, high-strength materials can be substituted into the predominantly metallic aircraft structures of today, the potential exists for 30% savings through the use of composites to the extent they are currently employed for some military applications.

Although some studies suggest that non-optimum use of airspace and ground infrastructure will be reduced through congestion control, such improvements may only maintain current efficiencies because air traffic growth remains strong. Historical trends for η_g, η_a, and η_{ft} also show constant air traffic efficiencies since 1968. Improved scheduling and equipment commitment can improve load factor, but congestion and low load factor during early morning/late evening flights may limit improvements to the historical rate. This represents a continuation of recent historical trends. At this rate of improvement, about 0.2%/year, the worldwide average load factor is expected to reach around 0.77 by 2025. For an individual aircraft, a seating arrangement that utilizes a small amount of floor space per seat is beneficial to E_I. This trend also applies to the fleet as a whole. It has been estimated that the average number of seats per aircraft may grow by 1.0% annually over the next 20 years.

Based on this outlook, a 25–45% reduction in E_U would be possible by 2025. This is equivalent to a change in E_U of 1.0–2.0%/year, i.e., 40–75% of the average rate over the previous 35 years. In terms of E_I, the addition of load factor results in an estimated improvement of 1.2–2.2%/year, or a 30–50% reduction in E_I by 2025. As was shown in Fig. 2, over the period 1971–1985, airlines found profitable an average 4.6%/year reduction in fleet average E_I on an RPK basis, which translates into a slower 2.7% improvement on a seat-kilometer basis when the contribution of load factor is removed (E_U). Over the period 1985–1998, however, the rate of change was slower, at approximately 2.2% in E_I and 1.2%/year in E_U. Fleet average projections in the literature suggest a 1.3–2.5% annual change in fleet average E_I and a 0.7–1.3%/year change in E_U. These studies are consistent with recent historical trends.

Beyond the evolution of the current aircraft platform, hydrogen and ethanol have been proposed as alternative fuels for future low-emission aircraft. Hydrogen-fueled engines generate no CO_2 emissions at the point of use, may reduce NO_x emissions, and greatly diminish emissions of particulate matter. However, hydrogen-fueled engines would replace CO_2 emissions from aircraft with a threefold increase in emissions of water vapor. Considering uncertainties over contrails and cirrus cloud formation, and the radiative impact of water vapor at higher altitudes (Fig. 2), it is not clear whether use of hydrogen would actually reduce the contribution of aircraft to radiative forcing. In addition, several issues must be resolved before a new fuel base is substituted for the existing kerosene infrastructure. The actual usefulness of such alternative fuels requires a balanced consideration of many factors, such as safety, energy density, availability, cost, and indirect impacts through production. Renewable biomass fuels such as ethanol have much lower energy density than does kerosene or even hydrogen, requiring aircraft to carry more fuel. They would again increase water vapor emissions from aircraft in flight. Hence, kerosene is likely to remain the fuel for air travel for the foreseeable future.

8. INDUSTRY CHARACTERISTICS, ECONOMIC IMPACTS, AND BARRIERS TO TECHNOLOGY UPTAKE

Although reducing energy intensity tends to reduce overall emissions, factors inherent to air transportation can act to counter the potential benefits. Reductions in emissions are hindered by the relatively long life span and large capital and operating costs of individual aircraft, and the resulting inherent lag in the adoption of new technologies throughout the aviation fleet. In improving the performance of technologies that are adopted, trade-offs are inevitable. For example, increasing the efficiency of new engines may potentially increase NO_x emissions as a result of higher peak engine temperatures. Further, the impact of any efficiency improvements is diminished by fuel wasted in airborne or ground travel delays or in flying partially empty aircraft. Perhaps most importantly, we do not know the cost of change. In this section, some industry characteristics and the economic impact of introducing energy-saving aircraft technologies are examined.

The lag in technology introduction is apparent in Fig. 2. It has typically taken 15–20 years for the U.S. fleet to achieve the same fuel efficiency as that of newly introduced aircraft. Apart from in-use aircraft performance improvements, the rate of improvement in the average E_I is determined by the gradual adoption of new, more fuel-efficient aircraft into the existing fleet. This process of technology uptake depends on various cost factors and market forces. The models described below consider the limitations on this process imposed by cost factors. In assessing future aviation fuel consumption and emissions, it is important to consider the time delay between technology introduction and its full absorption by the world fleet.

Figure 5 shows the relationship between how much it costs to carry a passenger 1 km, in terms of aircraft operation (direct operating cost/revenue passenger-kilometer) and the fuel efficiency of the aircraft (here in RPK/kilogram) for 31 selected aircraft types during the period 1968–1998. In addition to fuel efficiency, stage length has a strong influence on operating costs of regional (very short stage length) flights. The direct operating cost (DOC)–fuel efficiency relationship is indicative of the use of technological advances for the purposes of lowering operating costs. However, reductions in the future cost stream are purchased through higher capital and investment costs. That is, airlines are willing to pay a higher acquisition cost if they can gain from savings in DOC, mainly through lower fuel and maintenance costs during the lifetime of aircraft. The plot of aircraft price-per-seat versus DOC/RPK in Fig. 6 shows that aircraft price is inversely proportional to DOC.

The DOC–fuel efficiency and price–DOC relationships imply a potential constraint for energy use and emissions reduction in the aviation sector. If the relative changes in DOC and price with respect to technological improvements occur at historically accepted levels, airlines will continue to adopt newer and more efficient technologies at a higher price, balanced by the promise of sufficient future revenue.

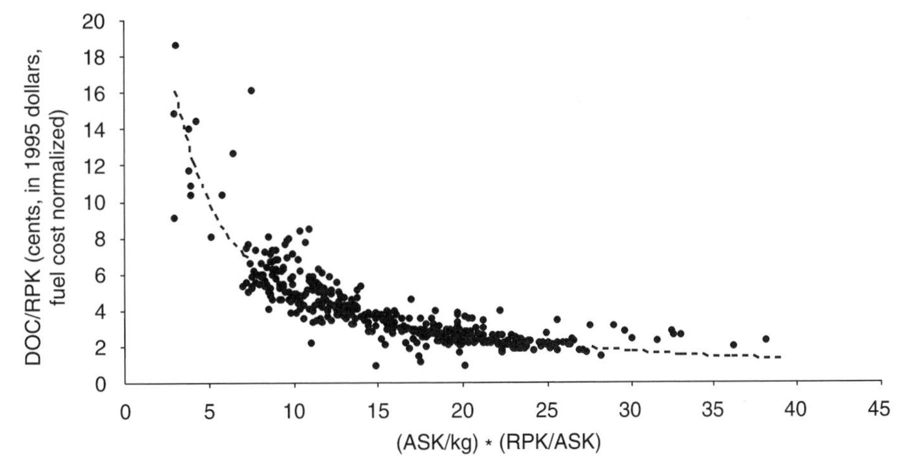

FIGURE 5 Direct operating cost (DOC) and fuel efficiency relationship. The DOC here is composed of crew, fuel, and maintenance costs. RPK, revenue passenger-kilometers; ASK, available seat-kilometers. Reproduced from Lee *et al.* (2001), with permission.

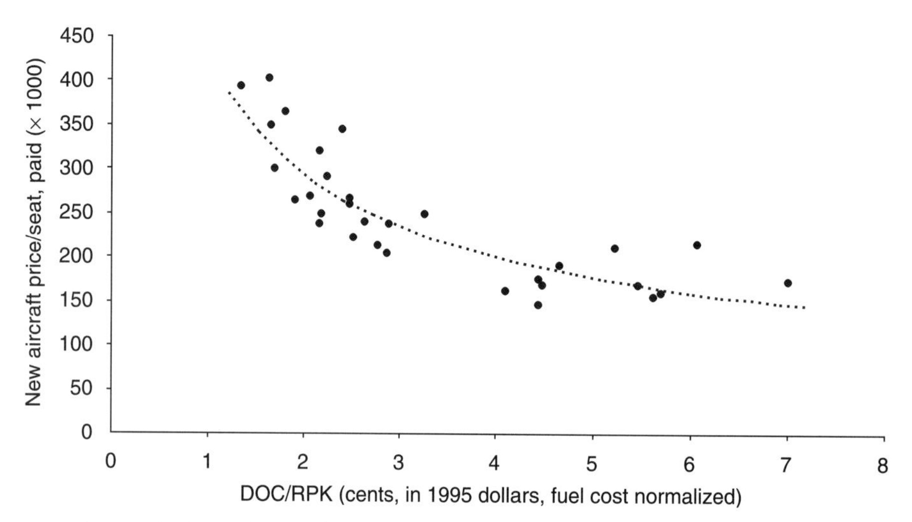

FIGURE 6 Direct operating cost (DOC) and price relationship. Reported prices are average market values paid in then-year (1995) dollars for new airplanes at the time of purchase. All prices were adjusted to 1995 dollars using gross domestic product deflators. RPK, revenue passenger-kilometers. Reproduced from Lee *et al.* (2001), with permission.

However, it is unclear whether future technologies can be delivered at an acceptable price-to-DOC ratio. If the price is too high, airlines may not choose to pay more for energy-saving technologies, in which case further improvements in energy use for the aviation sector may be limited.

SEE ALSO THE FOLLOWING ARTICLES

Internal Combustion Engine Vehicles • Marine Transportation and Energy Use • Transportation and Energy, Overview • Transportation and Energy Policy • Transportation Fuel Alternatives for Highway Vehicles

Further Reading

Anderson, J. D. (2000). "Introduction to Flight." 4th ed. McGraw-Hill, Boston.

Babikian, R., Lukachko, S. P., and Waitz, I. A. (2002). Historical fuel efficiency characteristics of regional aircraft from technological, operational, and cost perspectives. *J. Air Transport Manage.* 8(6), 389–400.

Balashov, B., and Smith, A. (1992). ICAO analyses trends in fuel consumption by world's airlines. *ICAO J.* August, 18–21.

Greene, D. L. (1992). Energy-efficiency improvement potential of commercial aircraft. *Annu. Rev. Energy Environ.* **17**, 537–573.

International Civil Aviation Organization (ICAO). (2002). 2001 annual civil aviation report. *ICAO J.* August, 12–20.

Kerrebrock, J. L. (1992). "Aircraft Engines and Gas Turbines." 2nd ed. MIT Press, Cambridge, Massachusetts.

Lee, J. J., Lukachko, S. P., Waitz, I. A., and Schafer, A. (2001). Historical and future trends in aircraft performance, cost and emissions. *Annu. Rev. Energy Environ.* **26**, 167–200.

National Research Council, Aeronautics and Space Engineering Board. (1992). "Aeronautical Technologies for the Twenty-First Century." National Academies, Washington, D.C.

Penner, J. E., Lister, D. H., Griggs, D. J., Dokken, D. J., and McFarland, M. (eds.). (1999). "Aviation and the Global Atmosphere: A Special Report of the Intergovernmental Panel on Climate Change." Cambridge Univ. Press, Cambridge, U.K.

Royal Commission on Environmental Pollution (RCEP). (2002). "Special Report. The Environmental Effects of Civil Aircraft in Flight." RCEP, London.

Schafer, A., and Victor, D. (2000). The future mobility of the world population. *Transp. Res. A* **34**(3), 171–205.

U.S. Federal Aviation Administration (FAA). (2000). "FAA Aerospace Forecasts Fiscal Years 2000–2011." FAA, Washington, D.C.

Air Pollution from Energy Production and Use

J. SLANINA
Netherlands Energy Research Foundation
Petten, The Netherlands

1. Introduction
2. Emissions
3. The Atmosphere, Meteorology, and Transport
4. Chemical Conversion in the Troposphere and Stratosphere
5. Deposition
6. Lifetimes of Pollutants
7. Impact of Air Pollutants
8. Abatement Measures

Glossary

acid deposition The total effects caused by the deposition of acids and acid-generating substances on ecosystems.

aerosol The system of small particles suspended in air. The term is also loosely used to describe the particles as such. Aerosol particles range from 0.003 to 100 μm in aerodynamic diameter. They are partly emitted as particles (e.g., by dust storms) and partly generated by conversion of gaseous emissions in the atmosphere.

anthropogenic emissions Caused, directly or indirectly, by human activities. The emission of sulfur dioxide due to use of fossil fuels is an example of a direct cause of emissions, and the emission of nitrogen oxides from farmland as a function of fertilizer application is an example of an indirect cause.

eutrophication The total effects on ecosystems, terrestrial as well as aqueous, due to high input of nutrients, mainly nitrogen and phosphorus compounds.

global air pollution Increased air pollutant concentrations of long-lived (>5 years) compounds, which affect air quality on a global scale regardless of the location of the emissions.

local air pollution Increased air pollutant concentrations and the effect caused by this increase near sources of emission, up to distances of a few tens of kilometers.

natural emissions Occur without any interaction of human activities. Generation of sea spray aerosols by waves or the emission of sulfur dioxide by volcanoes are examples.

oxidants The products of the reactions of volatile organic compounds with nitrogen oxides, which impact human health, agriculture, and ecosystems. The most important oxidant is ozone, but hydrogen peroxide and peroxy acetyl nitrate are other representatives.

regional air pollution Increased air pollutant concentrations and the effects thereof caused by emissions at distances up to 1000 km.

The increased use of fossil energy since the industrial revolution, and especially since 1950, has been the major cause of increased emissions of air pollutants and, correspondingly, many environmental problems. Emissions due to the use of energy are major sources of sulfur dioxide, nitrogen oxides, carbon dioxide, and soot and constitute a large contribution of methane, nonmethane volatile organic compounds, and heavy metals. Depending on conversion due to atmospheric chemical reactions, on meteorological transport, and on deposition processes, air pollution can be transported from hundreds of kilometers (ammonia) to several thousands kilometers (aerosols) on a truly global scale (CO_2 and CFCs). The adverse effects of emissions due to use of energy range from very local to regional and global. In cities, traffic can cause very high concentrations of nitrogen oxides and carbon monoxide, but also secondary products such as ozone and aerosols, especially under conditions of stagnant air. The adverse effects of ozone on human health at concentrations higher than $200\,\mu g\,m^{-3}$ are well documented. Investigations indicate that aerosol concentrations of $50–100\,\mu g\,m^{-3}$ are harmful and in The Netherlands (16 million inhabitants) cause the same number of deaths as traffic accidents per year, approximately 1500. On a regional scale (1000 km), acid deposition and photochemical smog (ozone) cause well-documented harm to vegetation and human health. Eutrophication (too high concentrations of nutrients such as phosphate, ammonium, and nitrate) is also a regional problem, quite manifest in Western Europe. Loss of visibility due to backscatter of light by aerosols is also largely a regional phenomenon. A truly global problem is the changes in the radiative balance of the earth due to increased concentrations of greenhouse gases and aerosols.

1. INTRODUCTION

Air pollution is difficult to define because many air pollutants (at low concentrations) are essential nutrients for sustainable development of ecosystems. Therefore, air pollution can be defined as a state of the atmosphere that leads to exposure of humans and/or ecosystems to such high levels or loads of specific compounds or mixtures thereof that damage is caused. With very few exceptions, all compounds that are considered air pollutants have both natural and man-made origins, mainly from energy production and use. Also, air pollution due to energy use is not a new phenomenon; it was forbidden in medieval times to burn coal in London while Parliament was in session. Air pollution problems have dramatically increased in intensity and scale due to the increased use of fossil fuel since the industrial revolution.

All reports on air pollution in the 19th and early 20th centuries indicated that the problems were local, in or near the industrial centers and the major cities. Even the infamous environmental catastrophes in the area of Liege in the 1930s and in London in the 1950s were essentially local phenomena. In the London smog episode, stagnant air accumulated such extremely high sulfur dioxide and sulfuric acid concentrations of approximately 1900 and 1600 $\mu g\,m^{-3}$, respectively—some 20 times the current health limit—that 4000 inhabitants died as a result. The main causes were the emissions from coal stoves used for heating and the fact that all emissions were trapped in a layer of air probably only a few hundred meters high, with no exchange of air within the city.

During the second half of the 20th century, the effects of air pollution due to energy use and production were detected on regional (>500 km), continental, and even global scales. In approximately 1960, the first observed effects from acid deposition were observed on regional and continental scales. Fish populations in lakes in Scandinavia and North America declined as the lakes were acidified by acid deposition to such a degree that fish eggs would not hatch and no young fish were produced. Approximately 10 years later, damage to forests, the loss of vitality of trees, also contributed to acid deposition. Smog episodes in cities such as Los Angeles were reported during the same period. Reactions of volatile organic compounds (VOCs) and nitrogen oxides, emitted by traffic, produced high concentrations of ozone and peroxides, which are harmful for humans and ecosystems. During the same period, high oxidant concentrations (the complex mixture of ozone, peroxides, and other products of the reactions of organics and nitrogen oxides) were becoming increasingly more frequent in Europe during stagnant meteorological conditions. Also, severe eutrophication (damage and changes to ecosystems due to the availability of large amounts of nutrients) occurred in Europe and the United States. Deposition of ammonium and nitrates (partly caused by fossil energy use) was shown to contribute substantially to high nutrient concentrations in soil and groundwater, leading to large-scale dying off of fish in the United States and Europe. It also caused extremely high nitrate concentrations in groundwater, with the result that a large part of the superficial groundwater in The Netherlands is now unfit for human consumption.

Since 1990, the increased concentrations of radiative active substances (compounds that alter the radiative balance of the earth—greenhouse gases, aerosols, and water in liquid form as clouds) and the resulting climatic change have received a lot of attention. Greenhouse gases absorb long-wave infrared radiation emitted from the earth, thereby retaining heat in the atmosphere and increasing the total radiative flux on the surface of the earth. Aerosols and clouds reflect incoming short-wave sunlight and influence the optical properties of clouds toward more reflection of sunlight; hence, increasing aerosol concentration leads to a decrease in the radiative flux on the surface.

The destruction of stratospheric ozone by chlorofluorocarbon compounds (CFCs) is one of the few environmental problems not related to energy use. Epidemiological research has demonstrated the effects of aerosols on the respiratory tract (inducing asthma and bronchitis), and a large part of the ambient aerosol is caused by emissions due to energy use and production.

This timing of air pollution problems could give the impression that there were sudden increases in air pollutant concentration, but this is probably not correct, as can be demonstrated in the case of ozone. By carefully characterizing old methodologies, it has been possible to reconstruct ozone concentrations in the free troposphere (the air not directly influenced by processes taking place on the earth surface) during the past 125 years (Fig. 1). The ozone concentrations in Europe slowly increased at a rate of 1% or 2% per year from 10 parts per billion (ppb) to more than 50 ppb as a result of the use of fossil energy. It is well documented that the effects of ozone start at levels of approximately 40 ppb (ppb is a mixing ratio of one molecule of ozone in 1 billion molecules of air). Therefore, it is not surprising that the effects of ozone were detected in the 1970s because the background concentration of continental ozone was 30 ppb and additional oxidant formation would increase the ozone concentrations locally or regionally. However, the increase in the continental background concentrations, mainly caused by fossil fuel, had been occurring for a long period.

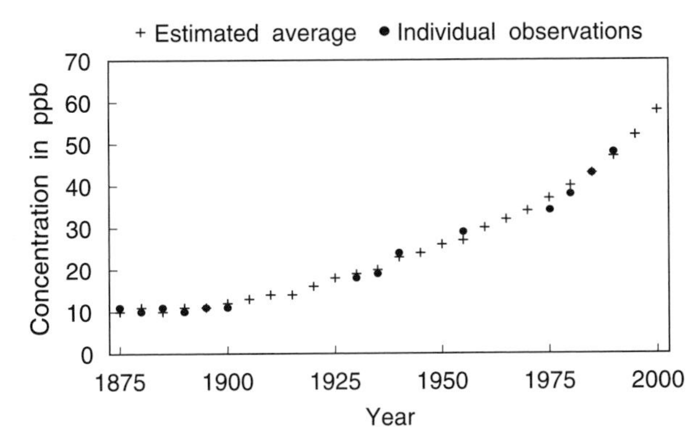

FIGURE 1 Ozone concentration in Europe since 1870. After Voltz-Thomas and Kley (1988).

In general, effects of pollution, and those of air pollution, are a function of the degree of transgression of the limits over which effects can be expected. Figure 2 shows that the use and production of energy is the major cause of most of the current environmental problems, with the exceptions of eutrophication and stratospheric ozone loss.

2. EMISSIONS

Most pollutants are emitted by both natural and anthropogenic sources. Natural sources are not influenced by man or man-induced activities. Volcanoes are a good example. Many emissions are biogenic (i.e., produced by living organisms), but these emissions are often influenced by man's activities. N_2O is a greenhouse gas that is largely emitted during nitrification and denitrification processes (the conversion of ammonium to nitrate and nitrate to N_2 and ammonium, respectively) that occur in the soil. However, the largest N_2O emissions are observed where nitrogen-containing fertilizer is applied in agriculture. The ratio between total anthropogenic, energy-related and natural emissions is very important in order to assess the impact of the use of fuel.

The SO_2 emissions from the use of fossil fuel (Table I) are by far the largest anthropogenic source and are much higher compared to natural emissions

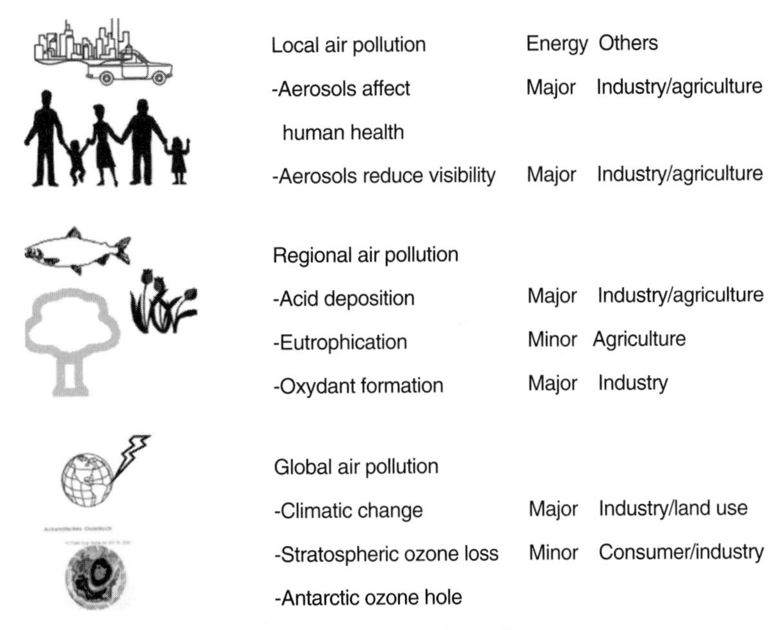

FIGURE 2 Impact of energy use on the environment.

TABLE I

Global Natural and Anthropogenic Emissions of SO_2 and NO_x

	Emissions (Tg/year)	
	SO_2[a]	No_x[b]
Industrial and utility activities	76	22
Biomass burning	2.2	6
Volcanoes	9.3	
Lightning		5
Biogenic emissions from land areas	1.0	15
Biogenic emissions from oceans	24	
Total anthropogenic emissions	78.2	27
Total natural emissions	34.3	21
Total emissions	112.5	48

[a]Data from Albritton and Meira Filho (2001).
[b]Data from Graedel and Crutzen (1993).

and exceeded these natural emissions as early as 1950. On a global scale, the use of fossil fuel may contribute approximately 60% of acid deposition.

The emissions of NO_x due to fossil fuels are on the same order of magnitude as the natural emissions. NO_x emissions, together with the emissions of VOCs, are responsible for the formation of oxidants such as ozone and peroxide acetyl nitrate (PAN). Some NO_x sources, especially the natural ones, are uncertain, but the contribution of natural sources to ambient VOCs is not very well-known either. An overview of the emissions of VOCs on the global scale is given in Table II.

Assuming that ozone formation is generally NO_x limited, energy use is responsible for at least 50% of oxidant formation on the global scale. The actual contribution of fossil fuel consumption to acid deposition and ozone is probably much higher because these environmental problems are most pronounced in areas where high emissions by fossil fuel use occur.

The estimates of VOC emissions, especially for the natural/biogenic emissions, are quite uncertain but make clear that the natural sources are predominant. The contribution of fossil fuels is 45% of the total anthropogenic emissions. Biomass burning does not include the use of biomass for energy but denotes the burning of forests and savannahs for agricultural purposes.

If it is assumed (not completely true) that the ambient concentrations vary linearly with the emissions and that only natural emissions are responsible for concentrations before the year 1800, the ratio of natural versus anthropogenic sources can be derived from old trapped air (e.g., in glaciers and polar ice masses). The ratio of natural and anthropogenic contributions varies considerably for the different greenhouse gases: It is highest for N_2O, approximately a factor 10; for carbon dioxide, approximately a factor of 3; for methane, approximately a factor of 1; and lowest for CFCs, a factor of 0, because there are no known natural sources. An overview of the emissions of greenhouse gases is given in Table III. Natural sources for CO_2 are not mentioned because large carbon cycles (e.g., decay of biomass in fall and uptake during leaf formation and growth in spring) occur, which are in equilibrium.

Emissions due to fossil fuel use contribute approximately 80% of the increased carbon dioxide (the most important greenhouse gas, responsible for

TABLE II

Global Emissions of VOCs

	Anthropogenic (Tg/year)	Natural (Tg/year)
Fossil fuel use	45	
Biomass burning, etc.	45	
Solvents	15	
Oceans		25
Isoprene by vegetation		350
Terpenes by vegetation		480
Total	105	855

TABLE III

Overview of the Emissions of Greenhouse Gases

	CO_2 (Pg C/year)	CH_4 (Tg/year)	N_2O (Tg N/year)
Fossil fuel	5.5	100	
Biomass burning	1.6	40	
Ruminants		80	
Waste treatment		80	
Rice fields		60	
Total anthropogenic	7.1	360	4.4 (sources not well quantified)
Natural Wetlands/continents		120	6
Oceans		10	2
Other		20	
Total natural	C cycles very large	150	8

50% of the greenhouse effect) and approximately 20% of the increase in methane, which is thought to contribute approximately 20% of the greenhouse effect. Obviously, fossil fuel use also contributes to N_2O emissions (e.g., emissions of cars equipped with three-way catalysts), but this contribution is not well-known.

From the overview provided in this section, it is clear that the use of fossil fuel is the basis of most environmental problems.

3. THE ATMOSPHERE, METEOROLOGY, AND TRANSPORT

Emitted pollutants are diluted by mixing with ambient air, and they are transported by the general displacement of the air as a function of meteorological conditions. Nearly all pollutants are converted to other species by means of atmospheric reactions. The effect of emissions of pollutants is strongly dependent on dilution, transport, and conversion.

Vertical and horizontal mixing is a function of the lifetime of pollutants. If pollutants have a short lifetime, they are not transported very far and vertical mixing in the atmosphere is very limited. If they have a long lifetime, mixing on a global scale occurs.

The basis of all transport of air masses in the atmosphere is movements of air from areas with high pressure to areas with lower pressure. Unequal heating of the earth's surface is the main reason for these pressure differences. The solar heat flux is much higher in the tropics compared to the polar regions. Therefore, heat is transferred by air movement as well as ocean currents (such as the warm Gulf Stream) from the equator to the poles. Also, the characteristics of the surface (sand is heated very quickly, whereas ocean surfaces are not) are very important and lead to phenomena that differ very much in scale, from a local see breeze to the distribution of high and low pressures over oceans and continents. Variation in the earth's rotation speed from the poles to the equator prevents the straightforward movement of air from low- to high-pressure areas and induces the so-called Coriolis force, resulting in circular movements of air around high-pressure and low-pressure areas. The friction induced by mountains and smaller objects, such as cities or forests, is another factor that influences wind direction and speed. All these factors together result in an average wind pattern as shown in Fig. 3.

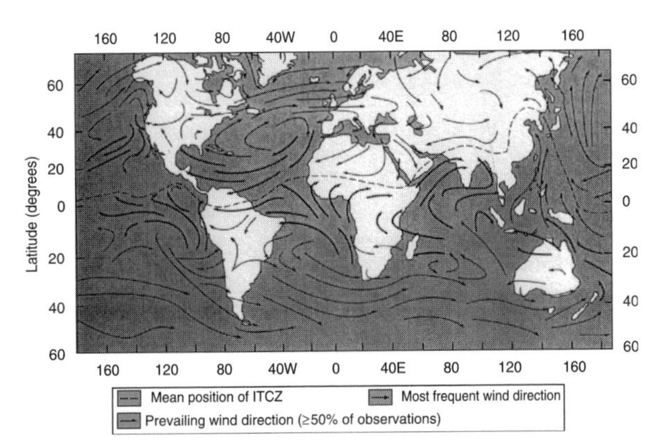

FIGURE 3 Average surface winds (geostrophic winds).

The average wind direction are obviously very important for transport of pollutants: The western winds at the coast in Europe will transport the emissions of the United Kingdom to The Netherlands and Germany, whereas a relatively small amount of German emissions will reach the United Kingdom. All winds converge near the equator, and this phenomenon delays transport from one hemisphere to the other. Consequently, whereas mixing in either hemisphere takes approximately 1 year, a 5-year period is required for global mixing.

In the troposphere (Fig. 4), the temperature decreases as a function of adiabatic expansion (temperature in a gas decreases if the pressure is lowered without heat exchange, approximately 1°C/100 m) and less heat transport from the surface. Rising heated air (called sensible heat) is not the only factor for vertical heat transport. At the surface, over water bodies, and due to evaporation from vegetation, large amounts of water vapor are added to the air. This evaporation takes up much energy, which is freed as water vapor is condensed to water droplets in clouds at higher altitude. However, latent heat transport becomes weaker higher in the troposphere because cooling results in less water vapor in the air.

The heated surface of land, together with turbulence created by objects such as trees and houses, creates a well-mixed layer—the mixing layer or planetary boundary layer. This layer varies from a few hundred meters during the night to 1.5 km during the day under conditions of solar heating. Emissions into this layer are dispersed quickly, especially during the day. If a warm layer is formed over cold air in the troposphere, it can act like a barrier for vertical dispersion: Air masses moving up under these conditions are colder and heavier than this layer and cannot pass this layer. This situation is

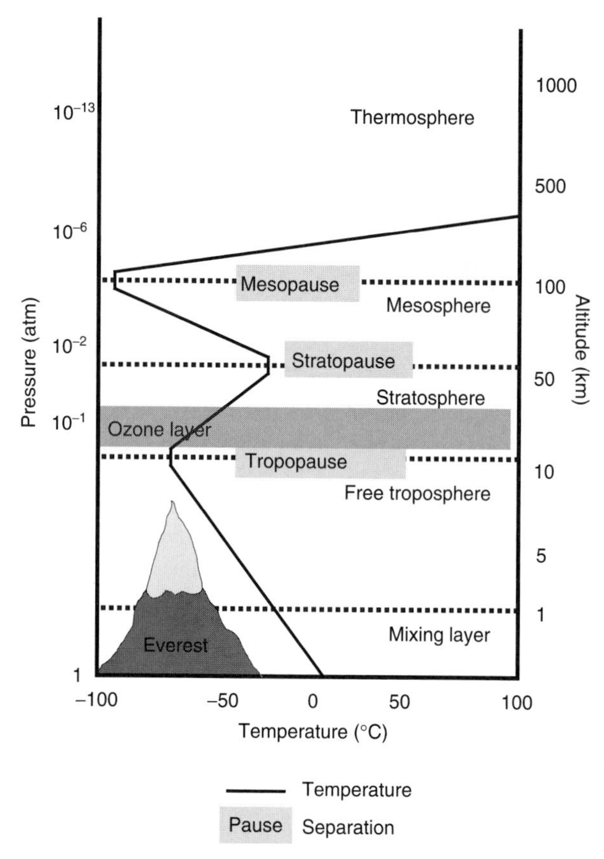

FIGURE 4 Vertical structure of the atmosphere.

called inversion (temperature increasing with height), and all emissions are trapped under this warm layer of air. A strong inversion was an important factor in the infamous London fog episode and is also important for photochemical smog in cities such as Los Angeles, Athens, and Mexico City.

Pollutants are transported by turbulence to the free troposphere, the part of the troposphere not in contact with the earth surface. Movement of air masses in this layer is the main factor for long-range transport of air pollutants, as discussed later. Temperatures decrease as a function of altitude in the troposphere until a minimum of −50 to −80°C is reached at the tropopause. The height of the tropopause varies with latitude. Due to much higher solar fluxes, the tropopause is at a height of 17 km in the tropics and approximately 10 km in the polar regions. At greater height, the atmosphere is heated by absorption of ultraviolet (UV) light by oxygen and ozone,

$$O_2 + h\nu \rightarrow 2\ O^{\bullet} \text{ (wavelength} < 240 \text{ nm)}$$
$$O^{\bullet} + O_2 \rightarrow O_3$$
$$\text{Net } 3\ O_2 + h\nu \rightarrow 2\ O_3,$$

and also destroyed again:

$$O_3 + h\nu \rightarrow O^{\bullet} + O_2 \text{ (wavelength} < 1140 \text{ nm)}$$
$$O^{\bullet} + O_3 \rightarrow 2\ O_2$$
$$\text{Net } 2\ O_3 + h\nu \rightarrow 3\ O_2,$$

where $O^{\bullet}$ is an oxygen atom with an unpaired electron, a radical.

The net result is that an ozone concentration on the order of 5000 ppb is encountered, whereas at the surface concentration levels of approximately 10–50 ppb are found. The warm layer above the tropopause is a barrier to exchange between the troposphere and stratosphere. Hence, complete vertical mixing over the troposphere and stratosphere takes many years.

Short-lived compounds (a few hours) are transported a few hundred kilometers, mostly in the mixing layer. Compounds with lifetimes of a few weeks are transported in the free troposphere and reach distances of 1000 km. It takes several months to 1 year for a compound emitted at the surface to reach the stratosphere. In some areas, less time is required; cold fronts in the northern Atlantic mix the atmosphere to a height of 20 km, which leads to relatively rapid exchange (weeks) locally between the troposphere and stratosphere. Compounds with lifetimes of 5 years or longer are globally mixed. An overview of all mixing processes, including mixing in the ocean, is given in Fig. 5.

Some air pollutants (e.g., ozone) have such a long lifetime that intercontinental transport occurs. Ozone originating in the United States, for instance, can reach Europe with the predominant western winds. Oxidants of European origin contribute to the oxidant concentrations in Asia. Currently, there is concern regarding the Asiatic contribution to ozone problems in the United States. Even though the contribution is relative small, less than 10% of the average concentrations according to recent estimates, it will become significant as precursor concentrations increase in one region while stringent abatement reduces the emissions of these precursors in others.

4. CHEMICAL CONVERSION IN THE TROPOSPHERE AND STRATOSPHERE

The competition between meteorological transport and losses due to conversion and removal by deposition determines how far pollutants are transported. The main atmospheric reactant is the OH

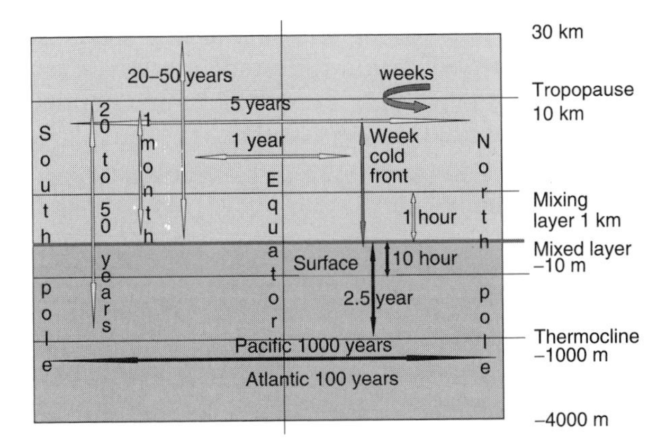

FIGURE 5 Transport times for different scales.

radical, generated by reaction of ozone, water, and short-wave solar radiation:

$$O_3 \overset{\text{Uv rad}}{\Rightarrow} O_2 + O$$

$$H_2O + O^{\bullet} \Rightarrow 2\ OH^{\bullet}.$$

The dot following OH is used to indicate that the species is a radical; it has only one electron instead of the electron pair, which normally constitutes a chemical bond.

The main emissions due to the use of fossil fuels are sulfur and nitrogen oxides and organic compounds. Sulfur and nitrogen oxides are oxidized in the atmosphere to sulfates and nitrates, and organic compounds are transformed into carbon dioxide and water. Some of the emitted organic compounds and the intermediate products of atmospheric conversion are toxic and affect human health(e.g., benzene, toluene, and formaldehyde).

SO_2 is oxidized to sulfuric acid (H_2SO_4) via gas phase reactions and in cloud droplets. The gas phase conversion starts with reaction with the OH radical, leading to the formation of sulfuric acid:

$$SO_2 + OH^{\bullet} \Rightarrow HOSO_3 + O_2 \Rightarrow HO^{\bullet} + SO_3$$

$$SO_3 + H_2O \Rightarrow H_2SO_4.$$

The speed of gas phase conversion can vary from a few tenths of a percent to a few percent per hour, depending on OH radical concentrations and other factors. The conversion in the water phase starts with uptake of SO_2 in droplets and the formation of sulfurous acid:

$$SO_2 \Leftrightarrow SO_{2(aq)} + H_2O \Leftrightarrow H_2SO_3 \Leftrightarrow H^+ + HSO_3^-.$$

Thus, SO_2 uptake is dependent on the next equilibria. If the H^+ formed by dissociation of sulfurous acid is removed (e.g., by reaction with ammonia or carbonates), more sulfur dioxide is dissolved. If, however, high H^+ concentrations are present (e.g., due to uptake of nitric or hydrochloric acid or the formation of sulfuric acid), the uptake of SO_2 is inhibited.

Next, H_2SO_3 is oxidized:

$$H_2SO_3 + O_3/H_2O_2/O_2 \Rightarrow H_2SO_4 \Rightarrow 2H^+ + SO_4^{2-}.$$

Oxidation by oxygen is only possible when suitable catalysts, such as Fe(III) or MN(II), are present. The reaction with ozone is fast at low H^+ concentrations but slow in acid conditions. Sulfur dioxide reacts fast with hydrogen peroxide independent of the pH of the droplets. The pH of the cloud droplets determines the uptake of SO_2 in the droplets; lower pH means that the equilibrium is displaced toward SO_2.

Nitrogen dioxide is oxidized only in the gas phase because both NO and NO_2 are very sparingly soluble in water. The basic reaction in daytime is

$$NO_2 + OH^{\bullet} \Rightarrow HNO_3.$$

The reaction rate of NO_2 with OH radicals in the summer is higher compared to that of SO_2.

A very important series of reactions involving OH radicals, nitrogen oxides, and reactive organic compounds leads to the formation of oxidants. Again, the initial reaction involves OH radicals:

$$RH + OH^{\bullet} \Rightarrow RO + H(A),$$

where RH is a organic aliphatic or olefenic compound, such as ethane, butane, or propene:

$$RO + OH^{\bullet} \Rightarrow RO_2 + H^{\bullet}.$$

Here, a peroxide is formed, such as PAN:

$$H^{\bullet} + O_2 \Rightarrow HO_2^{\bullet}$$

$$HO_2^{\bullet} + NO \Rightarrow OH^{\bullet} + NO_2.$$

The role of NO is to convert HO_2 to OH radicals:

$$NO + RO_2 \Rightarrow NO_2 + RO.$$

NO, NO_2, O_3, and the total light flux are in equilibrium:

$$O_3 + NO \Rightarrow NO_2 + O_2$$

$$NO_2 + h\nu \Rightarrow NO + O^{\bullet}(<380\ \text{nm})$$

$$O^{\bullet} + O_2 \Rightarrow O_3$$

Net:

$$RO_2 + h\nu + O_2 \rightarrow RO + O^{\bullet} + O_3.$$

This means that the hydrocarbon compound is oxidized, ultimately, to carbon dioxide and water; that ozone and OH radicals are formed; but that nitrogen oxide is formed again. Therefore, nitrogen oxide acts as a catalyst but is not really a reactant.

The next important observation is that radicals are actually formed during oxidant formation as long as sufficient NO_x is present. The positive aspect is that the oxidation by OH radicals of organic compounds cleans the atmosphere; in this way, many toxic organic compounds are eliminated.

Species that are not destroyed by reactions with the OH radical or the longer wave UV radiation encountered in the lower layers of the atmosphere ultimately pass the tropopause and enter the stratosphere.

5. DEPOSITION

Deposition is the process that brings air pollutants down to vegetation, soil, and water surfaces and exposes ecosystems to acid deposition. Deposition of acid substances and eutrophication (excessive deposition of nutrients, especially nitrogen compounds) damage ecosystems. Deposition, together with conversion, is a sink for pollutants and, hence, a factor that determines the range of transport.

Three main processes for deposition can be distinguished:

1. Wet deposition: Compounds are transported to the surface by means of precipitation, rain or snow.
2. Occult deposition: Compounds dissolved in fog or cloud droplets are deposited on the surface by means of deposition of these cloud or fog droplets.
3. Dry deposition: Deposition of compounds to the surface directly, not involving the liquid phase.

Wet deposition is the main deposition process in "background" areas—those regions where low concentrations of gases and aerosols in the atmosphere occur, such as Amazonia in Brazil, northern Norway, and the Tarim Basin in the Gobi desert.

Occult deposition is important at locations frequently exposed to fog and clouds (e.g., some hills in Scotland and the Po Valley in Italy) but a minor deposition mechanism at most places. Dry deposition is the dominate process in areas characterized by high air concentrations of gases and aerosols because, in general, dry deposition fluxes increase linearly with concentration.

Whereas wet deposition in northwestern Norway (a background area) is responsible for more than 80% of the deposition load (load is deposition flux integrated over a time period, typically expressed in units of moles per hectare per year) and occult deposition contributes more than 40% of the pollution load over Great Dunn Fell in Scotland (which is covered with low-altitude clouds 50% of the time), dry deposition is responsible for more than 75% of the total deposition load in The Netherlands (typical for European industrial countries that are moderately polluted).

5.1 Wet Deposition

Wet deposition is directly linked with cloud formation. Cloud droplets are formed on aerosol particles. If no particles are present, very high supersaturation of water vapor is needed to form droplets. Particles with a diameter of 0.05–2 μm, consisting of water-soluble material, are the favorite cloud condensation nuclei—the particles on which droplets are formed. This pathway is very important for the uptake of aerosols in clouds and precipitation. Water-soluble gases are also scavenged by cloud droplets.

NO_2 and NO are examples of compounds that are not readily incorporated in the water phase due to limited solubility. Thus, they are not scavenged effectively. SO_2 is scavenged as a function of the pH of the droplet, and the resulting sulfite is oxidized to sulfate.

Maps of wet deposition are generally directly based on measurements: Precipitation is sampled in collectors of wet deposition monitoring networks and analyzed in the laboratory. Figure 6 provides an overview of the wet deposition of sulfate (corrected for the sulfate produced by sea salt) per hectare per year in Europe.

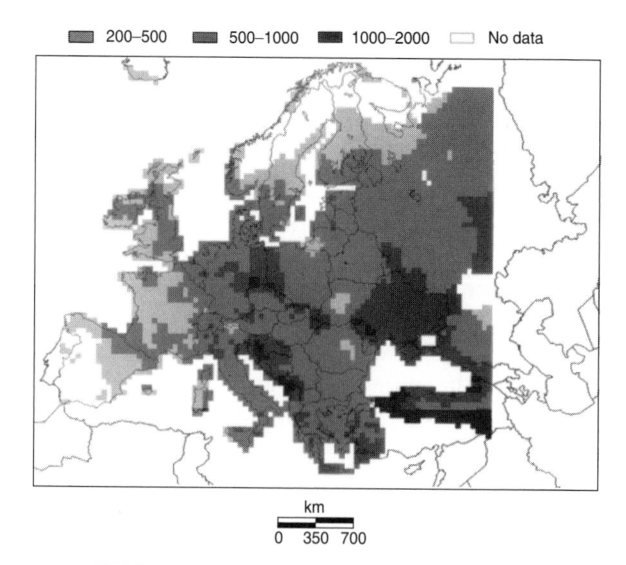

FIGURE 6 Wet deposition of sulfate in Europe.

5.2 Dry Deposition

Aerosol particles and gases are also deposited directly on vegetation and soil without prior uptake in the water phase. The deposition flux is calculated as the product of vertical deposition velocity (Vd) and ambient concentration:

$$F_x = \mathrm{Vd}_x \cdot C_x,$$

where F_x is the dry deposition flux of compound X, Vd_x is the deposition velocity of compound X, and C_x is the concentration of compound X.

The deposition velocity, the vertical speed of compounds directed downward, is dependent on a number of factors:

- Transport in troposphere occurs by means of turbulence in the form of eddies. If the troposphere is turbulent (e.g., due to high wind speed or strong mixing due to solar heating), the vertical transport of gaseous compounds and aerosol in the boundary layer is quite fast
- Each surface dampens turbulent transport, and very near the surface the turbulent transport is completely stopped.
- Transport through this stagnant layer is only possible by means of diffusion (in the case of gases) or by impaction or interception (for aerosols).
- The compound or aerosol particle reaches the surface next. It can adhere to the surface or bounce back to the atmosphere. Very polar gases, such as HNO_3, NH_3, or HCl, are adsorbed strongly by nearly all surfaces, so they are retained quantitatively. The same is true for particles, which are in fact very small droplets. Other compounds are not adsorbed and can leave the surface again.

The dry deposition load (deposited flux over a specified time period per unit area) is a linear function of the concentration. Thus, dry deposition loads are dominant in areas with higher air pollution concentration.

Vd_x levels vary depending on circumstances. The surface of vegetation in Western Europe is often covered with a water layer due to precipitation or dew. Sulfur dioxide dissolves easily in water, especially at a pH of 7 or higher, so the combination of high ambient ammonia concentrations and the frequent presence of water layers means that sulfur dioxide is easily taken up by the surface of vegetation. In southern Spain, on the other hand, water layers are rarely present on vegetation. Thus, the Vd in The Netherlands is at least twice the Vd for SO_2 in Spain, so the acid deposition fluxes in The Netherlands are two times higher at the same ambient concentration.

Because deposition fluxes and Vd are difficult to measure, dry deposition fluxes are generally modeled, and the models are parameterized and validated using dry deposition flux measurements. The flux is either derived from variations in the concentration as a function of vertical wind speed or from the gradient in concentration of the compound of interest as a function of height.

5.3 Modeling of Acid Deposition

Models are used to derive the concentrations as well as wet and dry deposition fluxes based on descriptions of emissions, atmospheric transport, and conversion and deposition processes. One of the major challenges in atmospheric chemistry is to formulate these models and to provide the necessary data to derive descriptions and to validate the model results. This complicated exercise is shown in Fig. 7.

Emissions and meteorological data are used to develop concentrations maps (Fig. 7, left). Based on high-resolution land use maps (Fig. 7, right) and meteorological data, the vertical deposition velocity, Vd, is derived for every compound. Next, the dry deposition flux is calculated by multiplying the concentration with the vertical deposition velocity. An example of the results is provided in Fig. 8, which shows the NO_y load in Europe in moles/ha/year (NO_y is the sum of HNO_3, HNO_2, NO_2, and NO).

The wet deposition flux is added to derive the total deposition flux (Fig. 6). The total deposition (Fig. 9) is used to derive the impact of the deposition and to calculate the loss of compounds by deposition, necessary to calculate lifetime and transport of emissions. The largest deposition loads are found near the regions where most fossil fuels are used, but some of the emissions are transported up to 1000 km.

6. *LIFETIMES OF POLLUTANTS*

Conversion in the troposphere and stratosphere and removal by deposition processes determine the fate and the lifetime of pollutants. A simplified approach is used here. Analogous to a resistance model, the overall lifetime can be expressed as

$$1/\tau_{(\mathrm{resultant})} = 1/\tau_{\mathrm{w}} + 1/\tau_{\mathrm{d}} + 1/\tau_{\mathrm{c}},$$

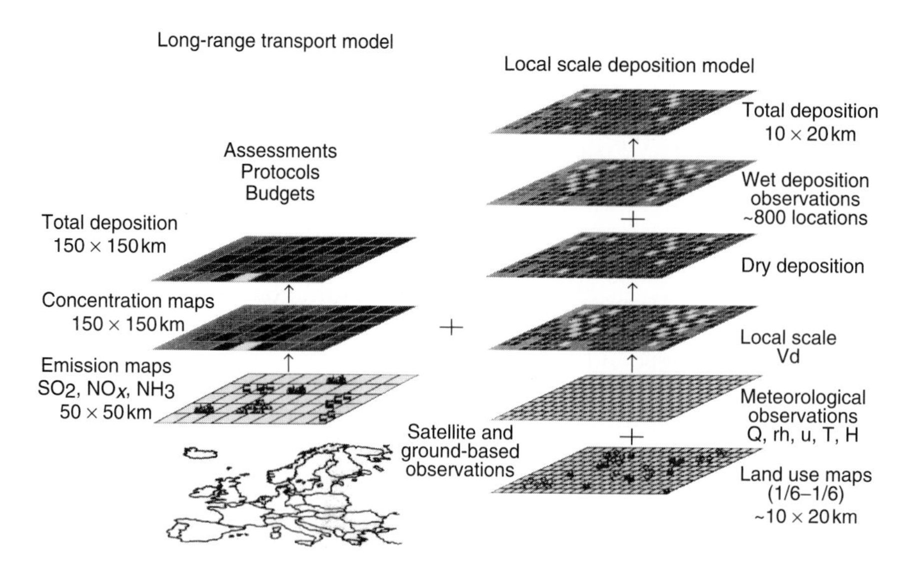

FIGURE 7 Scheme for calculation of deposition loads by modeling.

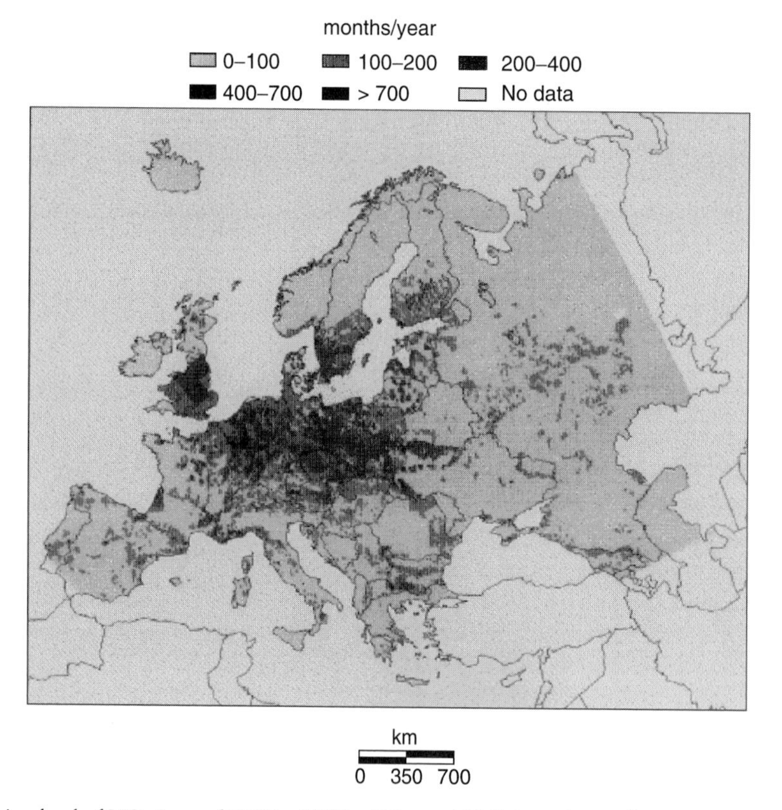

FIGURE 8 Dry deposition load of NO_y (sum of HNO_3, HNO_2, NO_2, and NO) in moles per hectare per year for Europe.

where $\tau_{(resultant)}$ denotes the total lifetime, τ_c is lifetime as a function of conversion, τ_d is lifetime as a function of removal by dry deposition, and τ_w lifetime as a function of removal by wet deposition. An overview of lifetimes of some important pollutants is given in Table IV. In addition to estimates of τ_c, τ_w, and τ_d, Table IV also provides estimates of the total lifetime [$\tau_{(resultant)}$] and of the transport distance. If the lifetime of a compound is so long that a global distribution is observed, this is indicated by "global." "Stratosphere" means that the compound is not destroyed by OH radicals in the troposphere but only by short-wave UV radiation in the stratosphere.

Very small aerosol particles, as generated by combustion engines, have very short lifetimes because they more or less behave like gas molecules, including transport by Brownian diffusion and rapid coagulation. The large ones are removed by gravity; they fall out of the atmosphere. But particles of approximately 1 μm, which contain most anthropogenic-produced sulfates, nitrates, and organics, live quite long and are generally removed by cloud formation only.

For sulfur dioxide, deposition processes and conversion are of equal weight, whereas nitrogen oxides are removed by conversion. The transport distance of ammonia is very limited due to short deposition and conversion lifetimes. The reactivity of different classes of organic compounds varies considerably; hence, there are major differences in lifetimes and transport distance.

Conversion of ozone is important near sources because it reacts rapidly with NO, such as emitted by traffic. Far from sources, dry deposition is the main loss term.

Carbon dioxide is taken up by vegetation (e.g., in spring) and again released as litter decomposes (e.g., in fall), under condition that the vegetation is in steady state. The total cycle of carbon, including vegetation and ocean uptake and release, is so large that the lifetime of emitted CO_2 is estimated to be 100–200 years.

Methane is first converted to CO, and next to CO_2 in the reaction with OH radicals. Uptake of methane occurs in soils, but this probably constitutes only a small fraction compared with conversion to CO and CO_2. N_2O and CFCs are not destroyed in the troposphere but are decomposed in the stratosphere.

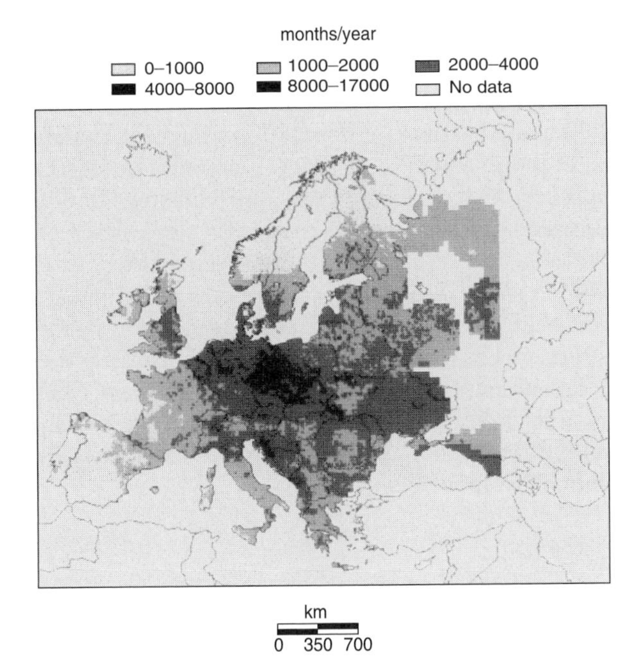

FIGURE 9 Total acid deposition in Europe in moles of acid rain per hectare per year.

7. IMPACT OF AIR POLLUTANTS

The effects of energy-related air pollutants on human health, ecosystems, and essential requirements such

TABLE IV

Lifetimes and Transport Distances of Air Pollutants

Compound	τ_c	τ_w	τ_d	τ (resultant)	Distance (km)
Aerosol					
<0.1 μm	Minutes–hours	—	Hours	Hour	10
>10 μm	—	Large	Hours	Hours	100
(≈1 μm)	—	Week	Week	Week	1000
Sulfur dioxide	Days	Days	Days	Days	1000
Nitrogen oxides	Days	—	Days	Days	1000
Ammonia	Hours	Hours	Hours	Hours	250
VOC reactive	Hours–days	Large	—	Hours–days	200–1000
Ozone	Hours–days	—	Days	Hours	200–1000
Carbon dioxide	—	—	200 Years	200 Years	Global
Methane	10 Years	—	Large	12 Years	Global
N_2O	Stratosphere	—	—	200 Years	Global
CFCs	Stratosphere	—	—	100–200 Years	Global

as drinking water vary widely in scale, ranging from very local to global. Here, an overview is given as a function of scale, starting with local effects and proceeding to regional impacts and global issues.

7.1 Local Pollution and Effects

Local pollution is generally caused by too high ambient concentrations as found near sources. Historic pollution problems, such as the London smog periods in the 1950, fall into this category. Most local air pollution problems in cities in developed nations are associated with traffic or, rather, the use of combustion engines in traffic. Cities in the United States and Europe, but also those in Mexico and east Asia, such as Beijing and Delhi, have problems with CO and NO_x concentrations that are too high, directly traffic related. In developing countries in which sulfur-rich coal is used and in which many industries are located in towns, local SO_2 concentrations also present environmental problems. Most developed countries have successfully eliminated this environmental hazard.

Countries and also multinational institutions, such as the European Union, have developed different air quality standards to ensure that the population is not affected by emissions in cities. These standards vary between different countries; an example is given in Table V for nitrogen dioxide. Some cities have limited exchange of ambient air due to the fact that they are surrounded by mountains. In this stagnant air, emissions from traffic can cause high concentrations of nitrogen oxides and reactive organic compounds, and oxidants can form, with high ozone concentrations as the result. Ozone affects human health and also vegetation. This is reflected in the air quality standards for ozone. Table VI includes different ozone air quality standards for protection of vegetation and human health. In some cities in developing countries, the limit at which the population should be warned (e.g., to stay indoors as much as possible) is frequently exceeded.

Epidemiological investigations have provided evidence that particulate matter, aerosols, induce increased mortality, ranging from 4% for U.S. cities to 2% for Western European cities and less than 1% for Eastern European cities for a level of 100 $\mu g\,m^{-3}$. In addition to increased mortality (which for a country such as The Netherlands amounts to 1500–4000 acute deaths in a population of approximately 16 million, equal to the total number of deaths caused by traffic incidents and approximately 10,000 per year due to chronic effects), an increased incidence of afflictions of the respiratory tract (asthma and bronchitis) has been found (e.g., amounting to 30,000 extra cases of bronchitis and asthma in The Netherlands). No mechanisms for this increased mortality have been established, but research indicates that very small particles (mainly emitted by traffic and other fossil fuel use) are probably responsible.

TABLE V

Air Quality Standards for NO_2

Value ($\mu g\,m^{-3}$)	Average time	Remarks
135	98% (1 hour)[a]	The Netherlands standard
175	99.5% (1 hour)	The Netherlands standard
25	50% (1 hour)	Target value, The Netherlands
80	98% (1 hour)	Target value, The Netherlands
200	98% (1 hour)	EU standard
50	50% (1 hour)	Target value, EU
135	98% (1 hour)	Target value, EU

[a]Ninety-eight percent indicates that the standard may be exceeded only 2% of the time. Target value is the air quality standard that is seen as desirable and that probably will be implemented in the future because the necessary technical means have been developed to reach this standard.

TABLE VI

Ozone Air Quality Standards[a]

Value ($\mu g\,m^{-3}$)	Average time	Remarks
240	1 hour	NL[b] standard (not to exceed 2 days/year)
160	8 hours	NL standard (not to exceed 5 days/year)
100	Growing season	March 1–September 1, 9 am to 5 pm
240	1 hour	NL target value
160	8 hours	NL target value
100	Growing season	NL target, March 1–September 1, 9 am to 5 pm
120	1 hour	NL long-term goal
50	Growing season	NL long-term goal
110	8 hours	EU standard (health protection)
200	1 hour	EU standard (protection of vegetation)
65	24 hours	EU standard (protection of vegetation)
180	1 hour	EU information of civilians
360	1 hour	EU warning of civilians

[a]See Table V footnote for description.
[b]NL, The Netherlands.

Aerosols scatter light, and this scattering is very effective if the diameter of the aerosol particle is on the same order as the wave length of the scattered light. This backscattering process is detected by the human eye as haze, limiting visibility. Aerosols with a diameter of 0.1 to approximately 2 μm scatter visible solar light effectively and hence impede visibility. In some polluted cities, aerosols limit visibility to 500 m, especially under conditions of high humidity. In Fig. 9, the left picture was taken when aerosol concentrations where approximately $20\,\mu g\,m^{-3}$. Visibility is quite good and probably more than 30 km. The picture on the right was taken during a dust storm. Dust storms contain large concentrations of not only large particles but also small particles, which reduce visibility to less than 1 km.

7.2 Regional Problems

Regional problems are caused by emissions within a distance of 1000 km; acid deposition, eutrophication, and regional ozone formation fall into this category. The first signs of the effects of acid deposition were encountered when losses of fish populations in Canadian, Scandinavian, and U.S. lakes were reported in the early 1960s. A few years later, there were reports of unusual damage to forests from Canada, the United States, and Germany. Trees were observed to have less foliage or needles than normal, and this damage even progressed to the point that trees died, a phenomenon that occurred on the border of former East Germany, Poland, and former Czechoslovakia. In the case of tree dieback, the cause–effect relation was much more complicated than that for the loss of fish in lakes. In fact, much later it was determined that not only acid deposition but also increased oxidant concentrations (e.g., ozone, PAN, and hydrogen peroxide) contribute to tree damage.

A third type of effect of acid deposition was reported in the mid-1980s. The same deposition processes that were responsible for the high loads (load is a time-integrated flux) of acid compounds were also the cause of high nutrient concentrations in soil, groundwater, and surface waters. Nitrate and ammonium are beneficial, even essential, for development in vegetation, but in excessively high concentrations they lead to the loss of biodiversity, especially in oligotrophic (adapted to low nutrient availability) ecosystems. This problem originated in The Netherlands, Belgium, and parts of Germany, Denmark, and southern Sweden but is now clearly affecting large areas of the United States, Canada, and China. In most countries, agricultural activities play a much larger role in eutrophication than fossil fuel use.

The common factor in all these cases is that pollutants, or their precursors, are emitted, transported in the atmosphere, and deposited by either dry or wet deposition. In some cases, acid deposition takes place 1000 km or more from the place where the responsible emissions originated. The most important acid-forming species are sulfur dioxide (which is partially converted to sulfuric acid in the atmosphere), nitrogen oxides (converted to nitric acid during atmospheric transport), and ammonia. Ammonia may act as a base in the atmosphere, neutralizing nitric and sulfuric acid, but in soil or groundwater it is largely converted by microorganisms to nitric acid, producing additional acid in the process.

Both precursors (SO_2 and NO_x) and products (sulfuric acid, nitric acid, and ammonium) can be transported large distances (up to 1500 km) depending on meteorological conditions, the speed of conversion, and removal by deposition processes. Ammonia, on the other hand, is transformed quickly in ammonium by reactions with sulfuric acid and nitric acid and deposited rapidly in the form of both wet and dry deposition; it is generally transported 250 km or less.

Oxidant problems are not only a local but also a regional phenomenon. An example is the buildup of high concentrations of oxidants, mainly ozone, in Europe when the weather is dominated by large high-pressure systems. The air within these high-pressure systems, which often have a diameter of 1000 km, is not exchanged quickly. So precursor concentrations of NO_x and reactive organic compounds increase due to ongoing emissions mainly related to fossil fuel, and very high ozone concentrations, often exceeding $150\,\mu g\,m^{-3}$, are the result. These ozone concentrations affect not only human health but also vegetation (Fig. 10).

The loss of production of wheat due to ozone starts at a threshold concentration of approximately $40\,\mu g\,m^{-3}$ and increases linearly with the ozone concentration. (Fig. 11). Not only do ozone problems occur in Europe and the United States, but there is also increasing concern in developing countries. For instance, high ozone concentrations occur in Hong Kong and the Chinese province of Guangdong, which includes cities such as Guangzhou and Shenzhen.

Aerosols generally derive from both local and regional sources. The backscatter by aerosols is receiving much attention because backscatter of solar radiation is a very important but uncertain factor in determining the balance of incoming and

FIGURE 10 Visibility on the campus of Peking University. (Left) Photograph taken when the aerosol concentration was approximately 20 μg m^{-3}. (Right) Photograph taken during a dust storm, with a high concentration of small particles.

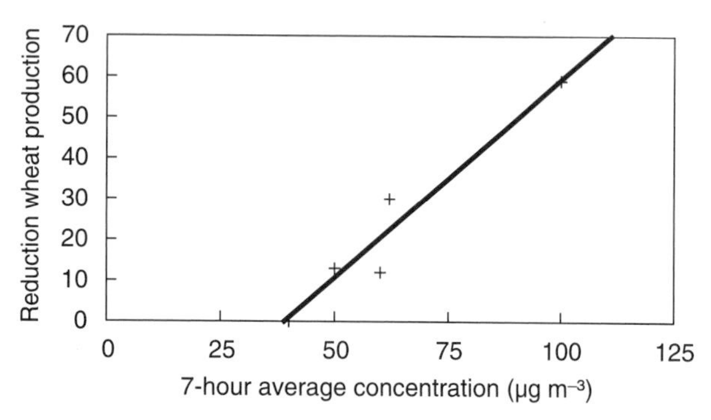

FIGURE 11 Reduction of wheat production as a function of daytime ozone concentrations.

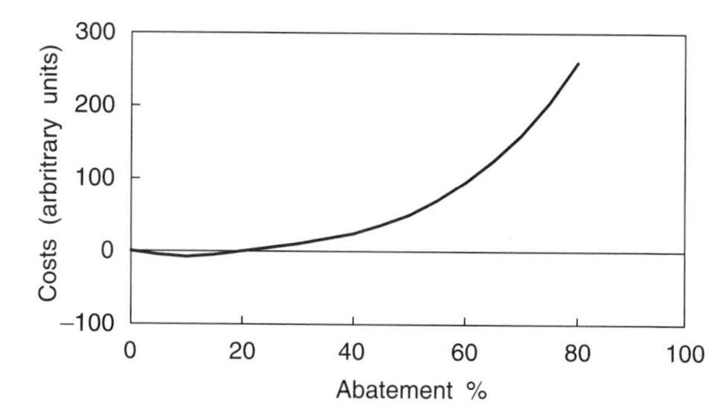

FIGURE 12 Costs of abatement in relative cost units as a function of reduction (abatement).

outgoing radiation of the earth and hence in assessments of climatic change.

7.3 Pollution on a Global Scale

Pollutants that are not converted in the troposphere or deposited rapidly and have lifetimes of 10 years or more can have an impact on a global scale. The greenhouse effect and the Antarctic ozone hole are caused by long-lived air pollutants. Because climatic change is discussed elsewhere in this encyclopedia and the destruction of the stratospheric ozone, leading to the Antarctic ozone hole, is not primarily caused by the use of fossil fuels, these subjects are not treated here.

8. *ABATEMENT MEASURES*

In view of the major environmental problems caused by the use of fossil fuel, on the one hand, and economic development, on the other hand, it is very important to evaluate the effects of possible abatement measures. The costs of abatement measures increase exponentially with the degree of reduction (Fig. 12).

The scientific basis for abatement measures is obviously emission–effect relations: How are undesirable environmental effects related to specific emissions, and how will changes in these emissions influence the effects of air pollution? It is very important to have as detailed as possible descriptions of emission–effect relations in order to have a scientific and solid fundament for environmental measures. Not only is this necessary to assess the cost-effectiveness of measures but also national consensus of economic groups and international consensus between countries are very dependent on the acceptance of emission–effect relations.

It should be kept in mind that smaller emission reductions, in general, are not very costly. Problems arise when very large emission reductions have to be made. The first hurdle is to reach a generally accepted view on emission–effect relations. Then, a national or international consensus must be reached and so a basis is constituted for reaching agreements. Examples are the protocols under the aegis of UN-EMEP

(Cooperative Programme for Monitoring and Evaluation for Long Range Transport in Europe) for abatement of acid deposition, eutrophication, and impact of oxidants in Europe or the Montreal Protocol, which stringently minimizes emissions of CFCs.

These emission–effect relations are generally so complex that such a description is seldom based on observations only, and mathematical models are needed to obtain these relations. The first stage of this process is to develop a description of the critical limits or critical loads (load = integrated deposition fluxes). Critical concentrations and loads are the maximum concentrations and deposition fluxes above which damage is likely to occur. These descriptions are based on information on the effects of acid deposition, eutrophication, or ozone on ecosystems, taking into account the buffer capacity of the soils and the sensitivity of ecosystems. The result is a map of critical loads in Europe, as shown in Fig. 13 for acid deposition. This description of loads of acid deposition (Fig. 9) is then compared to the description of critical loads, and the transgression (the term exceedence is generally used) can be derived (Fig. 14). The difference between the actual deposition load and the critical load is the central steering mechanism for European environmental policy. These descriptions are the internationally accepted basis of the European protocols for abatement of acid deposition, eutrophication, and the impact of oxidants. If a quantitative and validated scientific description is generally accepted, then a good scientific fundament is present for abatement measures. Therefore, EMEP must describe all processes from emissions, natural and anthropogenic, transport and chemical conversion, and deposition to effects and also provide an assessment of the costs for abatement in order to obtain scientifically defendable conclusions.

Next, individual countries must meet the targets using appropriate measures, optimized to their specific conditions. The following measures have proven to be very effective:

• Not only will the use of wet desulfurization plants reduce the emissions of sulfur dioxide of, for

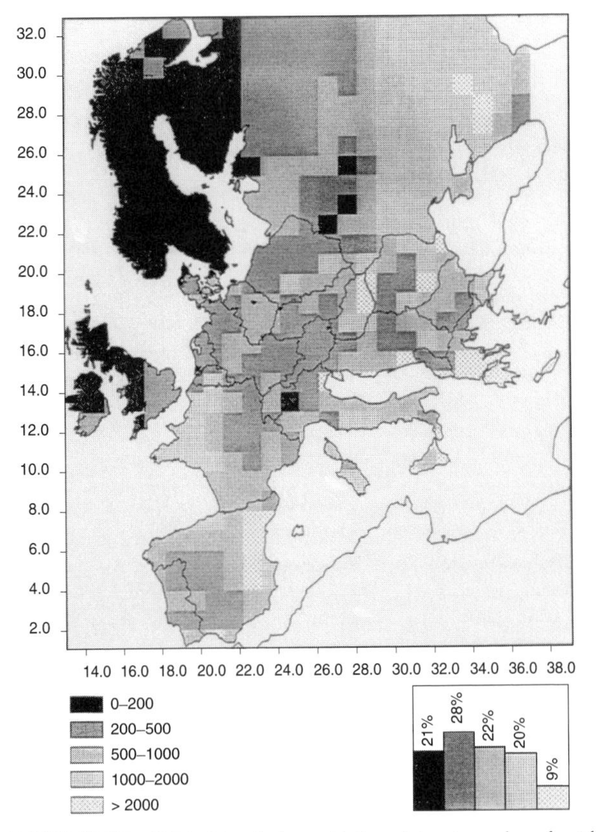

FIGURE 13 Critical loads for acid deposition in moles of acid per hectare per year. The very dark areas are extremely sensitive; damage occurs at loads as low as 200 mol/ha/year. The light areas can accept deposition loads of at least 2000 mol/ha/year without unacceptable damage.

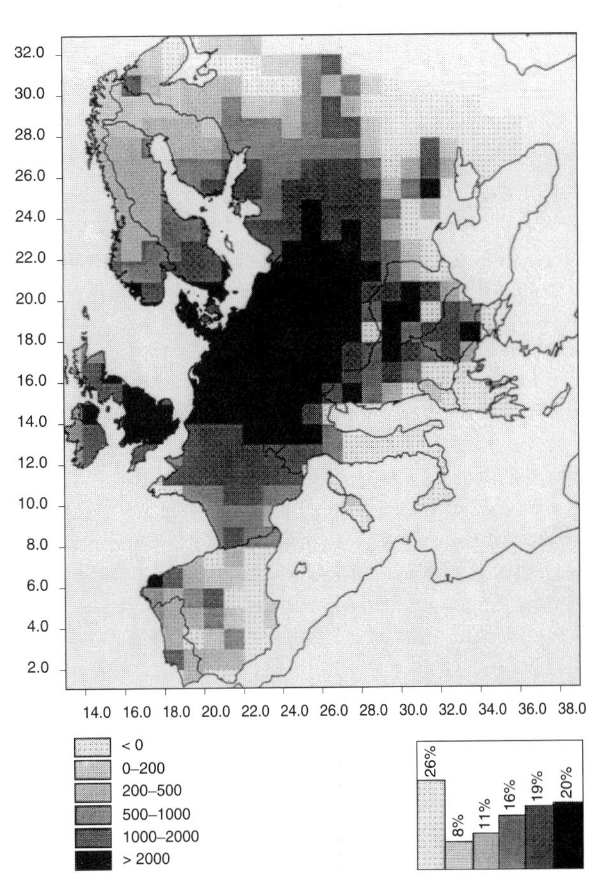

FIGURE 14 Exceedence of critical loads in Europe. The light areas denote no or little transgressions, and the dark areas indicate regions with serious acid deposition problems.

example, electricity-generation plants, by approximately 95% but these installations will also reduce emissions of compounds such as hydrochloric acid and hydrofluoric acid by the same amount.

• The three-way catalyst, now in general use, reduces the emissions of nitrogen oxides and reactive volatile organic compounds by 90–95%, depending on traffic conditions.

• Low NO_x burners, burning fossil fuel in two stages, can reduce NO_x emissions by power plants and industry by 70% or more. Low NO_x burners are even applied in household central heating systems using natural gas.

• Increasing the efficiency of the use of fossil fuels is a very effective way to decrease most emissions.

A very important issue for which such a general acceptance of the scientific description of emission–effect relations has not been reached is the reduction of emissions of greenhouse gases (Kyoto Protocol). On the one hand, the costs of such reductions play an important role, and costs of abatement measures increase exponentially at higher reduction rates. On the other hand, the perceived uncertainty in the assessment of the environmental effects is seen as very important:

• The measures necessary to reach the agreed abatement are judged to be very expensive by a number of countries.

• The effects of climatic change are potentially very damaging and far-reaching. However, the uncertainties regarding regional impacts as well as estimates of the time before these changes are detected give countries reasons to defer measures.

• Uncertainties exist with regard to sources and sinks of greenhouse gases, especially regarding the role of vegetation in the release and sequestration (uptake) of carbon in vegetation and soil.

• Because most of the current increases in ambient carbon dioxide concentrations must be allotted to the use of fossil fuel by developed countries, undeveloped countries voice quite strongly the opinion that they should be exempt from measures that could hamper their economic development. Some developed countries counter this argument by pointing out that developing countries will be responsible in the near future for a large percentage of greenhouse gas emissions and that measures will not be effective unless developing countries also take abatement measures.

• The main impact of climatic change will be experienced, in the opinion of many scientists, by developing countries, whereas most of the costs must be covered by developed countries. This obviously is a very complicating factor.

Although part of the necessary solutions to the debate must be provided by atmospheric chemistry research, particularly information on the fluxes of different sources and sinks, reduction in the use of fossil fuel seems to be unavoidable in order to mitigate the current environmental problems.

SEE ALSO THE FOLLOWING ARTICLES

Acid Deposition and Energy Use • Air Pollution, Health Effects of • Clean Air Markets • Ecosystems and Energy: History and Overview • Gasoline Additives and Public Health • Hazardous Waste from Fossil Fuels • Indoor Air Quality in Developing Nations • Indoor Air Quality in Industrial Nations • Thermal Pollution

Further Reading

Albritton, D. L., Meira Filho, L. G., and the IPCC. (2001). Technical summary. www.ipcc.ch.

EMEP Summary Report. (2000). Transboundary acidification and eutrophication in Europe, CCC and MSC-W Research Report No. 101. Norwegian Meteorological Institute, Oslo.

Finlayson-Pitts, B. J., and Pitts, J. N. (2000). "Chemistry of the Upper and Lower Atmosphere." Academic Press, New York.

Graedel, T. E., and Crutzen, P. J. (1993). "Atmospheric Change; An Earth System Perspective." Freeman, New York.

Jacob, D. J. (1999). "Introduction to Atmospheric Chemistry." Princeton Univ. Press, Princeton, NJ.

Slanina, J. (ed.). (1997). "Biosphere–Atmosphere Exchange of Trace Gases and Pollutants in the Troposphere." Springer-Verlag, Berlin.

Volz-Thomas, A. and Kley, D. (1988). *Biogeochemistry* **23**, 197–215.

Air Pollution, Health Effects of

JONATHAN LEVY
Harvard University School of Public Health
Boston, Massachusetts, United States

1. Historical Air Pollution Episodes
2. Types of Studies Used to Evaluate Health Impacts
3. Key Pollutants and Health Outcomes
4. Health Effects of Indoor Air Pollution
5. Conclusions

Glossary

cohort study An epidemiological study in which a group of individuals is enrolled and tracked over time to evaluate whether more highly exposed individuals are more likely to develop disease or to die prematurely.

confounder A variable independently associated with both the exposure and the health outcome in an epidemiological study, and could therefore be responsible for the observed relationship.

epidemiological study An observational study of human populations (either occupational or general); designed to determine the association between exposure and disease.

inhalation unit risk The excess lifetime cancer risk estimated to result from continuous lifetime exposure to a substance at a concentration of 1 $\mu g/m^3$ in the air.

particulate matter (PM) A solid or liquid suspended in the air; typically defined by the aerodynamic diameter of the particle and categorized as fine or respirable particles ($PM_{2.5}$, particles less than 2.5 μm in diameter), inhalable particles (PM_{10}, particles less than 10 μm in diameter), coarse particles ($PM_{2.5-10}$, particles between 2.5 and 10 μm in diameter), or total suspended particles (TSPs; suspended particles of all sizes).

reference dose An estimate (with uncertainty spanning perhaps an order of magnitude) of the daily exposure to the human population (including sensitive subgroups) that is likely to be without an appreciable risk of adverse effects during a lifetime; used for the evaluation of noncancer risks of toxic chemicals.

time-series study An epidemiological study in which changes in pollution levels over time are correlated with changes in health outcomes over time, capturing the short-term effects of pollution on health.

toxicological study A controlled laboratory experiment in which a small number of animals is given a gradient of doses of compounds with suspected health impacts, and the relationship between dose and disease rate is estimated.

Understanding the health impacts of air pollution is a crucial step in the determination of appropriate energy policy. Although it has been well established that large increases in pollution exposures can lead to substantial increases in death and disease, it is somewhat more complex to quantify the precise impacts of an incremental change in exposures to a specific pollutant at the levels commonly faced by individuals. This is in part because there is an array of different air pollutants in outdoor and indoor air; pollutants can come from different sources and have different types of health effects, thus different forms of evidence are used to draw conclusions regarding pollutant health impacts. Broadly, air pollution causes health impacts through a pathway in which emissions lead to concentrations of key pollutants in the air, human activity patterns influence the exposures people face from these concentrations, and characteristics of the exposed populations help determine the types and magnitudes of health impacts.

1. *HISTORICAL AIR POLLUTION EPISODES*

Although there have been a number of documented events whereby extreme levels of air pollution led to elevated death and disease rates, three episodes occurring near the middle of the 20th century are among the most well documented and were the first pollution events studied in great detail. In December 1930, an atmospheric inversion led to a thick pollution fog settling over the Meuse Valley in

Belgium. During a 4-day period, 60 deaths were attributed to the pollution (10 times more than normal), and many individuals suffered from respiratory distress, with the elderly or those with respiratory or cardiovascular disease at greatest risk. Levels of a number of pollutants were increased during this episode, with extremely high concentrations of sulfur-based compounds. In the second event, in October 1948, a similar set of circumstances (temperature inversion in an area with substantial industrial activity) led to an air pollution episode in Donora, Pennsylvania. Nearly half of the residents of the town became ill, with 20 deaths attributed to the multiday episode. Finally, perhaps the most damaging and well-known pollution episode occurred in December 1952 in London, where a temperature inversion and associated elevated pollution levels led to 4000 deaths during the episode and to another 8000 deaths during the subsequent months (demonstrating a lagged or cumulative effect of exposure). As previously, elderly individuals and those with respiratory and cardiovascular disease were at greatest risk of health impacts.

These and other similar episodes provided clear evidence of the effects of extremely high levels of air pollution, which generally occurred when significant emissions from low stacks were trapped near the ground due to atmospheric conditions. Because of increased emission controls and taller stacks in many settings at the start of the 21st century, the pollution levels seen in the early studies are currently rarely seen in either developed or developing countries. Evidence from additional sources of information is therefore needed to understand health impacts of lower levels of air pollution, as well as to determine precisely which pollutants are responsible for mortality and morbidity impacts.

2. *TYPES OF STUDIES USED TO EVALUATE HEALTH IMPACTS*

The evidence used to determine the magnitude of mortality and morbidity effects of air pollution comes from two general types of studies. Toxicological studies are controlled laboratory experiments in which relatively high doses of compounds are administered to animals and rates of diseases (often cancer) are estimated. As controlled experiments, toxicological studies avoid some of the issues in observational studies, wherein a number of potential causal factors can be changing at the same time. However, in order to determine health effects in humans, results from toxicological studies must be extrapolated in at least two important ways—results from animals must be applied to humans, and results from high doses necessary to see effects in a small number of animals must be used to draw conclusions about lower doses typically seen in humans. Because of these often substantial uncertainties, epidemiological studies, when they are available and interpretable, are generally preferred for quantifying the health impacts of air pollution, with toxicological studies providing supporting evidence for a causal effect.

Epidemiological studies are observational studies of human populations; they are designed to show the relationship between exposure to an agent or set of agents and the development or exacerbation of disease. Although the focus on human populations and observed exposures removes some of the extrapolation issues associated with toxicological studies, epidemiological studies run a greater risk of yielding results that do not reflect a true causal relationship between exposure and disease (because correlation does not necessarily imply causation). This would occur if there were confounders, which are variables independently associated with both the exposure and the disease in question, that could actually be responsible for the observed relationship. For an epidemiological study to adequately determine the health effects of air pollution, it must use statistical techniques or a restrictive study design to remove the possibility of confounding by nonpollution variables. Also, unlike toxicological studies, which generally focus on a single pollutant or a defined list of pollutants, epidemiological studies measure health effects that could be associated with a number of pollutants that are often correlated with one another in time or space. Epidemiological studies must also be able to provide accurate exposure estimates, which can be more difficult than in toxicological studies, in which the doses can be measured in a controlled setting, particularly for large populations, for long-term exposures, or for pollutants that vary greatly in time or space.

There are three primary types of epidemiological studies used to understand the health effects of air pollution, each of which has different strengths and different issues related to potential confounders. The first is the cross-sectional study, in which disease rates in different locations are correlated with pollution levels in those settings, looking at a single point in time. Although these studies can generally be conducted using only publicly available information, they can suffer both from confounding issues and from the ecological fallacy, which is the incorrect assumption that relationships evaluated from large groups are applicable to individuals. A classic example of the ecological fallacy involved a study that showed that whereas there was a correlation between race and illiteracy rates across states, this relationship was drastically reduced when considering individuals rather than states.

In terms of confounding issues, individual behaviors with important linkages to health (such as smoking status or occupation) cannot be incorporated into a cross-sectional study, making it difficult to interpret the causal association between air pollution and health. Therefore, although many of the original air pollution epidemiological studies were cross-sectional because of the availability of information, these studies are conducted less often at present.

Air pollution health effects are also determined through time-series studies, which are studies that compare changes in daily or multiday air pollution levels with changes in daily or multiday health outcomes (such as emergency room visits or deaths). These studies therefore yield information about the influence of air pollution on health over a relatively short number of days (known as acute health effects). Time-series studies have the strength of having a relatively limited set of potential confounding variables, because any confounder must be correlated with both air pollution levels and health impacts on a temporal basis. For example, cigarette smoking could only confound the relationship in a time-series study if more people smoked on high-pollution days than on low-pollution days, which is unlikely. However, cigarette smoking could act as an effect modifier, a variable that influences the relationship between exposure and disease but does not distort the true relationship between exposure and disease. Plausible confounders include weather variables such as temperature and relative humidity, which can affect both pollutant formation and population health, as well as air pollutants other than those studied or measured. Because time-series studies rely largely on publicly available information and are informative about the relationship between air pollution and health, the majority of air pollution epidemiological studies are time-series studies.

Despite the numerous advantages of time-series studies, they cannot provide information about the effects of longer term exposures to air pollution (known as chronic health effects). Time-series studies can expand the time window under consideration beyond periods of days, but seasonal trends in health risks and statistical limitations of time-series studies make it impossible to evaluate the effects of daily exposure beyond about a couple of months. Information about long-term health risks is primarily taken from retrospective or prospective cohort studies, in which a group of individuals is tracked for a significant period of time (often decades) to evaluate whether their air pollution exposures are linked to higher risks of mortality or chronic disease. Because individual-level data about nonpollution risk factors can be collected, confounders can be dealt with in a more substantial way than in cross-sectional studies. However, there remain more plausible confounders for a cohort study than for a time-series study (including lifestyle factors such as smoking, diet, or occupation, which can theoretically be spatially correlated with air pollution exposures). Statistical methods are used to evaluate the independent effects of air pollution given the levels of other risk factors, which addresses this issue to an extent. The long time horizon to conduct a cohort study implies that fewer of these studies can be done, but the long-term exposure focus means that the health impacts estimated in cohort studies would be expected to be greater than the health impacts in time-series studies (which do not capture the effects of long-term exposures).

In addition, it should be noted that other types of studies provide evidence about the health effects of air pollution. Human chamber studies, in which people are exposed to a pollutant for a short period of time in a controlled laboratory setting, can help corroborate relationships observed in epidemiological studies (particularly for short-term and reversible health effects). Other epidemiological study designs can also be informative, such as case-crossover studies (in which individuals act as their own controls, comparing exposures when the health outcome of concern occurred with exposure levels during a comparable point in time) or panel studies (in which subjects are monitored intensively over a short period of time to evaluate the effects of exposure on outcomes such as daily symptoms or lung function).

Regardless of the type of toxicological or epidemiological information, a crucial final step in determining the health effects of an air pollutant is to evaluate whether the evidence is sufficient to infer a causal relationship rather than a simple association. Although proper statistical control of confounders is an important first step, inferences about causality require evidence that is outside of the domain of individual studies. To make this evaluation more systematic, Sir Austin Bradford Hill proposed nine "causal criteria" that could be applied. These include strength of the association, consistency across studies or techniques, specificity of the health effect, temporality (exposure preceding disease), biological gradient (higher exposures leading to greater health risks), biological plausibility, coherence with known facts about the disease, supporting experimental evidence, or analogy with other exposures. Although these provide a useful basis for analysis, Hill believed that these should not be interpreted as rules, all of which must be followed, but rather as a framework in which evidence about causality can be interpreted.

3. *KEY POLLUTANTS AND HEALTH OUTCOMES*

A number of air pollutants have been connected with human health impacts in epidemiological and/or toxicological studies. These pollutants are conventionally divided into two categories:

1. Criteria pollutants: particulate matter (PM), ozone (O_3), sulfur dioxide (SO_2), nitrogen dioxide (NO_2), carbon monoxide (CO), and lead (Pb), as defined by the United States Environmental Protection Agency (U.S. EPA) in the Clean Air Act. These pollutants are generally ubiquitous and have a range of health impacts at typical ambient levels.
2. Toxic air pollutants: also known as "hazardous air pollutants," substances considered to cause cancer or to lead to other potential noncancer effects that might include reproductive, developmental, or neurological impacts. The U.S. EPA maintains a list of hundreds of toxic air pollutants that are anticipated to impact human health.

As might be expected, there is substantially more epidemiological evidence for criteria air pollutants than for toxic air pollutants, given the generally higher levels and shorter term effects (as well as the existence of monitoring networks necessary for most epidemiological investigations). In general, criteria air pollutants have been associated with respiratory or cardiovascular outcomes ranging from symptom exacerbation to premature death, with some limited evidence for lung cancer effects. Toxic air pollutants have largely been associated with cancer through a combination of toxicological and occupational epidemiology studies, along with miscellaneous neurological or reproductive end points. Another important difference between the two pollutant categories is that although it is assumed that cancer risks from air toxics have no threshold (meaning that any level of exposure would lead to some risk of cancer), criteria air pollutants or noncancer risks from air toxics are assumed to have thresholds (levels below which no population health impacts would occur) unless it can be demonstrated otherwise.

3.1 Particulate Matter

Of the criteria pollutants, particulate matter has received the greatest attention within the scientific community to date. PM can be simply defined as any solid or liquid substance suspended in the air. As such, PM can include many different substances, ranging from sea salt, to elemental carbon directly emitted by diesel vehicles, to sulfate particles secondarily formed from SO_2 emissions. Air quality regulations have focused historically on the size of particles rather than on the chemical composition, given the complexity of particulate matter, the difficulty of measuring numerous constituents, and the physiological importance of particle size.

Originally, regulations in the United States and elsewhere focused on total suspended particles (TSPs), which included airborne particles of all sizes. However, the largest particles seemed unlikely to cause significant health impacts, because they remain suspended in the air for brief periods of time and cannot travel into the lungs (see Fig. 1). In 1987, the U.S. EPA modified the particulate matter National Ambient Air Quality Standard (NAAQS) to focus on PM_{10}, defined as the fraction of particles less than $10\,\mu m$ in aerodynamic diameter. PM_{10} has also been referred to as inhalable particles, because these particles are sufficiently small to enter the respiratory system. The size fraction of interest was further constrained by the U.S. EPA in 1997, when they proposed to change the focus of the NAAQS to $PM_{2.5}$, the fraction of particles less than $2.5\,\mu m$ in aerodynamic diameter. These smaller particles, known as fine particles or respirable particles, are most able to reach the lower portions of the lungs, where gas exchange occurs.

Historically, a large amount of the evidence of particulate matter health effects has come from epidemiological studies, in part because of the regulatory emphasis on total particulate matter rather than chemical constituents. A number of time-series studies have evaluated the relationship

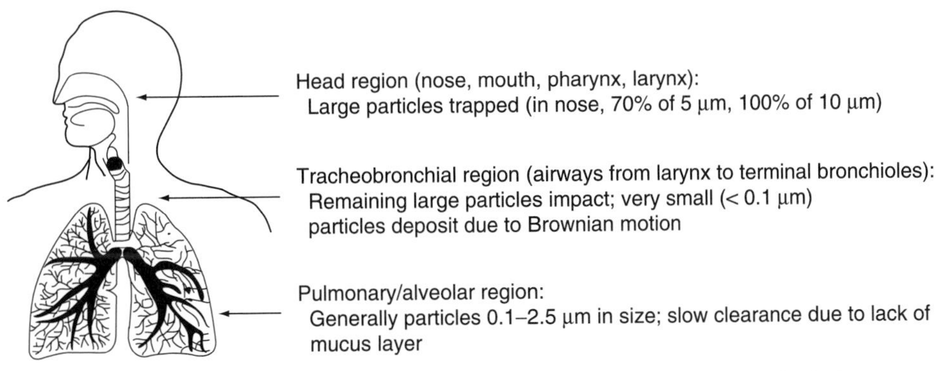

FIGURE 1 Relationship between particle size and site of particle deposition in the lung.

between daily changes in particulate matter concentrations and daily changes in number of deaths, generally within a few days of the pollution increment. These studies have been conducted in both developed and developing country settings and in cities with a range of pollution types or levels, but they have largely demonstrated a consistent relationship between PM and mortality even when the effects of weather or other air pollutants are taken into account. Across the numerous studies that evaluated the relationship between PM_{10} and daily mortality, estimates generally have been on the order of a 0.5–1% increase in deaths per $10\,\mu g/m^3$ increase in daily average PM_{10} concentrations (although with selected studies yielding either nonsignificant or greater effects). These estimates have been similar in cities with different levels of other air pollutants, different sources of particulate matter, and different correlations between other pollutants and particulate matter, lending support to the hypothesis that particulate matter rather than other pollutants or confounders is responsible for this observed relationship. As anticipated, in most studies, the effects were greater for respiratory or cardiovascular causes of death than for other causes of death.

A crucial issue for interpretation of time-series air pollution studies is whether they represent a significant loss of life expectancy or whether the deaths are simply a function of "harvesting," whereby individuals who would have otherwise died the next day are dying one day earlier due to acute pollution exposure. If this were the case, then the death rate might increase above the expected level on one day but drop below the expected level the next day, with no net increase in deaths over the week following the pollution increase. Studies that have investigated the harvesting question have found that the relationship between PM and mortality does not vanish as the time period after the exposure increases, indicating that harvesting is unlikely.

A final question from the time-series mortality literature is whether a threshold exists, and, if so, at what concentration. Studies have generally not found evidence of a population threshold within the range of ambient concentrations measured, although a threshold could exist at levels lower than those observed in the epidemiological studies.

In contrast, there have been relatively fewer cohort mortality studies in the published literature, given the resource intensity of these investigations. Although multiple cohort mortality studies exist, two studies published in the mid-1990s were among the earliest completed and provided an important basis for subsequent regulatory decisions and assessments of the health effects of air pollution. The Six Cities Study tracked 8111 white adults who lived in six cities in the eastern half of the United States for 15–17 years. In each city, at the start of the study, the research team set up a centrally located monitoring station, where concentrations of both particulate matter of various sizes and gaseous pollutants were measured. As a cohort study, this study was able to collect detailed information about individual behaviors that could influence health, such as smoking, obesity, and occupational exposures.

Controlling for these and other plausible confounders, the investigators found that mortality rates were significantly higher in the high-pollution cities compared with the low-pollution cities, with the strongest associations found for three different measures of particulate matter (PM_{10}, $PM_{2.5}$, and sulfate particles). The relationship for SO_2 was weaker than the PM relationships, and there was no association between mortality and ozone. Air pollution was significantly associated with death from cardiovascular and respiratory disease, was positively associated with lung cancer (although not statistically significantly so), and was not associated with other causes of death. The relative risk for mortality was reported to be 1.26 for an 18.6 $\mu g/m^3$ increase in $PM_{2.5}$ concentrations (95% confidence interval: 1.08, 1.47), with no apparent threshold at the pollution levels documented in the study (between 11.0 and 29.6 $\mu g/m^3$ of $PM_{2.5}$).

To determine if the conclusions from the Six Cities Study were robust, the American Cancer Society cohort study was conducted, in which individuals enrolled in the American Cancer Society Cancer Prevention Study II were matched with the nearest ambient air pollution monitoring data. This cohort contained over 500,000 United States subjects living in all 50 states, with detailed individual risk factor data collected. As in the Six Cities Study, the American Cancer Society study found a significant increase in mortality risks as levels of both $PM_{2.5}$ and sulfate particles increased across cities. Cardiopulmonary mortality was significantly elevated for both pollutants, whereas lung cancer mortality was significantly associated with sulfate particles but not $PM_{2.5}$. The relative risk for all-cause mortality was 1.17 for a 24.5 $\mu g/m^3$ increase in median annual $PM_{2.5}$ concentrations (95% confidence interval: 1.09, 1.26), slightly lower than the relative risk from the Six Cities Study. No threshold for mortality risks was apparent across the range of fine particle concentrations in the study (between 9.0 and 33.5 $\mu g/m^3$ of median annual $PM_{2.5}$ concentrations).

Additional analyses of these two cohorts emphasized the robustness of the findings across different statistical formulations and additional confounders. A detailed reanalysis of both studies by the Health Effects Institute concluded that the findings did not appear to be confounded by non-air-pollution variables and that there was an association between mortality and $PM_{2.5}$ and sulfate particles, as well as with gaseous SO_2. A follow-up to the American Cancer Society study used more monitoring data and more years of population follow-up on the original cohort, along with a refined statistical approach and an increased number of behavioral confounders. The estimated effect of air pollution on mortality was similar to the earlier findings, but the follow-up study documented a statistically significant association between PM exposure and lung cancer (potentially due to the increased sample size given more years of follow-up).

Thus, these two seminal cohort mortality studies documented an effect of air pollution on mortality that is somewhat larger than the effect from the time-series studies (even when both types of studies consider the same particle size) and was associated in large part with particulate matter. Furthermore, the cohort mortality studies likely represent a greater loss of life expectancy potentially related to induction rather than exacerbation of disease (such as lung cancer, which cannot plausibly be related to air pollution in a time-series study). Although inherent limitations of epidemiology make it difficult to disentangle the effects of individual pollutants, and cohort studies have numerous behavioral confounders that must be addressed statistically, the cohort mortality studies supported the time-series mortality studies regarding the role of particulate matter and demonstrated a potentially greater effect for fine particles than for larger particles.

Along with the mortality effects, the epidemiological literature for PM has documented morbidity outcomes with a wide range of severity. Hospital admissions and emergency room visits for cardiovascular or respiratory causes have been studied extensively in time-series investigations, in part due to the publicly available information and the relative ease of defining the health end points. Other acute health end points that have been associated with particulate matter exposure include upper and lower respiratory symptoms, asthma attacks, and days with restricted activities. Long-term particulate matter exposure has also been associated with the development of chronic bronchitis, defined as inflammation of the lining of the bronchial tubes on a chronic basis, with the presence of mucus-producing cough most days of the month for at least 3 months of the year for multiple years, without other underlying disease. These morbidity end points are consistent with the types of premature deaths found in the mortality studies and are therefore supportive of the relationship between particulate matter exposure and respiratory and cardiovascular health.

3.2 Ozone

In contrast with the health evidence for PM, the primary studies documenting ozone health effects have been a combination of toxicological, human chamber, and epidemiological studies. This is in part because, unlike PM, ozone is a single substance with well-understood chemical properties. Ozone is an oxidant gas that is poorly water soluble. Ozone is therefore able to travel throughout the respiratory tract, reacting with molecules on the surface of the lung and leading to pulmonary edema, inflammation, and the destruction of epithelial cells that line the respiratory tract. These and other physical changes to the lung, many of which are transient but some of which may be permanent, could plausibly lead to a number of health outcomes documented in the epidemiological literature, given sufficient ozone exposure. As in any epidemiological analysis, the crucial question is what level of ozone exposure is sufficient to induce health effects (or whether population health effects are found for any level of ozone exposure) and what the magnitude of the relationship is between ozone exposure and various health outcomes.

As anticipated, epidemiological and human chamber studies have shown that ozone has an array of acute effects on the respiratory system. Short-term exposures to ozone have been associated with significant but reversible decreases in lung function as well as increases in pulmonary inflammation. As would be expected, these physiological changes have resulted in increases in both mild (respiratory symptoms and restrictions of activity) and severe (emergency room visits and hospitalizations) morbidity outcomes, depending on the health of the individual and the magnitude of the exposure.

Ozone has also been associated with premature death in some time-series studies; however, because high-ozone days tend to correspond with hot and humid days (which are associated with higher mortality rates), it can be difficult to determine the relative contributions of air pollution and weather. Ozone may also be correlated with particulate matter and other air pollutants in some settings, impairing the ability to determine the independent role of each pollutant. Finally, ozone does not penetrate indoors as readily as PM, making indoor exposures lower and outdoor monitor-based health evidence harder to interpret. Regardless, considering a subset of published studies that controlled both for other air pollutants (primarily PM) and for the nonlinear effects of weather on health, there appears to be an association between daily average ozone concentrations and premature death, with an approximate 1% increase in deaths per 10 parts per billion (ppb) increase in daily average ozone. Because fewer studies have focused on ozone mortality in detail than have focused on particulate matter mortality, both the existence and the magnitude of this effect are somewhat uncertain.

Ozone may also be associated with chronic, nonreversible health outcomes. Children who exercise outdoors in high-ozone areas have been shown to have increased risk of asthma development, demonstrating potential long-term effects of ozone exposure as well as the importance of breathing rate and time spent outdoors (because ozone levels indoors are low, given the reactivity of ozone). Ozone exposure has also been linked with decreased lung function in young adults who lived for long periods of time in cities with high ozone levels. Although the Six Cities Study and the American Cancer Society study did not document an influence of long-term exposure to ozone on premature death, there is both epidemiological and physiological evidence supporting chronic pulmonary inflammation and other chronic health effects due to low-level ozone exposure.

3.3 Other Criteria Air Pollutants

Although PM and ozone are the two pollutants studied most extensively in many past epidemiological investigations (especially studies of mortality risk), there is also evidence of health effects of other criteria air pollutants. It has been well established that extremely high levels of carbon monoxide (CO) can lead to damage to the central nervous system, eventually leading to death at high enough concentrations. This is because hemoglobin preferentially reacts with CO over oxygen (forming carboxyhemoglobin), thereby reducing the oxygen-carrying capacity of the blood. Symptoms at lower levels of carboxyhemoglobin in the blood (below 30%) include headache, fatigue, and dizziness, with coma, respiratory failure, and subsequent death occurring at levels above 60%.

There has also been evidence that low levels of exposure to CO (corresponding to typical ambient levels in some settings) can lead to cardiovascular impairment. Time-series investigations have correlated cardiovascular hospital admissions with exposure to CO even when controlling for other pollutants, although the degree to which these effects are independent of the effects of other pollutants is difficult to determine. However, controlled human chamber studies have demonstrated that carboxyhemoglobin levels as low as 2% are associated with cardiovascular effects such as anginal pain or changes in electrocardiogram measures, providing support for the epidemiological findings.

Similarly, there is clear evidence that exposure to extremely high levels of nitrogen dioxide can lead to respiratory distress, with symptoms such as cough, shortness of breath, and chest tightness occurring given short-term exposures on the order of 1 part per million (ppm; a level rarely seen in outdoor air but occasionally found in poorly ventilated indoor settings containing combustion sources).

These effects occur for reasons similar to those involved in ozone health impacts, with the insoluble nature of nitrogen oxides coupled with their oxidative properties leading to damage to epithelial cells and other forms of cytotoxicity.

At lower levels of air pollution, epidemiological studies have documented linkages between NO_2 exposures and outcomes such as decreased lung function or increased respiratory infections. A number of studies have investigated the relationship between NO_2 exposure and respiratory illness in children, often using the presence of gas stoves as a proxy for elevated exposures. These studies generally found that acute lower respiratory symptoms in children increased with incremental increases in NO_2 concentrations, down to fairly low levels. Children and individuals with preexisting respiratory disease have been most strongly linked with respiratory outcomes associated with NO_2 exposure.

As mentioned previously, there was an association between sulfur dioxide and premature mortality in the American Cancer Society cohort study. However, it was difficult to determine the relative importance of gaseous SO_2 versus $PM_{2.5}$ or secondary sulfate particles, and this finding has generally not been interpreted as a definitive causal relationship, given the lack of supporting evidence for the lethality of low-level SO_2 exposure. Sulfur dioxide has been established as an irritant to the eyes and upper respiratory tract at high concentrations (multiple ppm). The site of the documented respiratory tract effect differs somewhat from the sites hypothesized for ozone or nitrogen oxides. This is because SO_2 is highly water soluble and can therefore react in the upper respiratory tract, making that the site of its irritant effects.

Considering acute SO_2 exposures at typical ambient concentrations, high-risk populations include children with mild to moderate asthma and others with preexisting respiratory disease. Short-term exposures to SO_2 concentrations on the order of 0.25–0.5 ppm have been associated with bronchoconstriction and subsequent decreases in lung function and increases in respiratory symptoms, for those with asthma or other respiratory diseases. In agreement with these findings, some epidemiological studies have associated respiratory hospital admissions or emergency room visits with SO_2, although the evidence supporting these more severe outcomes has not been substantial. However, it is clear that sulfur dioxide and its related secondary by-products have potential health impacts at higher pollution levels.

Finally, although the health effects of airborne lead have been drastically reduced in much of the developed world due to the use of unleaded fuels, outdoor airborne lead can have significant health impacts near industrial sites (such as lead smelters) or in countries that still use a significant portion of leaded fuel. Although extremely high levels of blood lead have been associated with severe central nervous system problems, the more prevalent health impacts, which have been documented at blood lead levels as low as 10 μg/dl or lower, include incremental increases in blood pressure and incremental decreases in the IQ of children. Although both of these health end points are only incrementally affected by lead at generally seen blood lead levels, the population health implications of small shifts in blood pressure or IQ can be significant.

3.4 Toxic Air Pollutants

Given the numerous toxic air pollutants (189 compounds are listed in Section 112 of the 1990 U.S. Clean Air Act Amendments) with both cancer and noncancer health effects, the anticipated health effects of each of these pollutants cannot be described in detail here. However, the severity of the carcinogenic effects of a toxic air pollutant can generally be described in one of three ways—its cancer classification, its inhalation unit risk, or its population cancer risk.

Because evidence about carcinogenicity can arise from a number of sources, the U.S. EPA, the International Agency for Research on Cancer (IARC), and other similar organizations have developed methods to categorize the weight of evidence for carcinogenicity. As indicated in Table I, both the U.S. EPA and IARC evaluate evidence from animal and human studies, as well as available information about the mechanism of action of the compound, to draw conclusions about the likelihood that the substance is carcinogenic. In general, few substances have been categorized as known human carcinogens (EPA Category A, IARC Category 1), because this requires statistically significant epidemiological evidence, which is difficult to obtain, given the need for studies that follow exposed populations for decades, accurately characterize long-term exposures, and adequately control for potential confounders for cancer development (such as smoking). Similarly, few substances have been determined to be noncarcinogens, because most substances tested (which is only a subset of those substances to which humans are exposed) have some positive evidence of carcinogenicity in some test or species.

Once the likelihood of carcinogenicity has been established, the inhalation unit risk is calculated, given evidence from animal and human studies. The

TABLE I

Cancer Classification Systems by the U.S. Environmental Protection Agency (EPA) and International Agency for Research on Cancer (IARC)

Category	Description	Animal evidence	Human evidence
U.S. EPA			
A	Known carcinogen	Any	Sufficient
B1	Probable carcinogen	Sufficient	Limited
B2	Probable carcinogen	Sufficient	Inadequate or none
C	Possible carcinogen	Limited	Inadequate or none
D	Not classifiable	Inadequate	Inadequate or none
E	Noncarcinogen	None	None
IARC			
1	Known carcinogen	Any Sufficient	Sufficient Strong relevant mechanism
2A	Probable carcinogen	Sufficient	Limited Inadequate + strong relevant mechanism
2B	Possible carcinogen	<Sufficient Sufficient	Limited Inadequate
3	Not classifiable	Inadequate or limited	Inadequate
4	Probable noncarcinogen	Negative	Negative Inadequate

inhalation unit risk refers to the excess lifetime cancer risk estimated to result from continuous lifetime exposure to the given substance at a concentration of 1 $\mu g/m^3$ in the air. For most compounds, the inhalation unit risk must be calculated from animal studies that use much higher doses than are commonly found for human populations. Statistical techniques based in part on hypothesized mechanisms of action for cancer development are used to determine appropriate inhalation unit risks at low doses, but this analytical process contains multiple steps that contribute uncertainty to the final quantitative estimate.

As is clear from the definition of inhalation unit risk, the population cancer risk from a toxic air pollutant will be a function of both the inhalation unit risk and the level of population exposure to the substance. Investigations by the U.S. EPA provide some indication of the outdoor air toxics that contributed most substantially to cancer risk in the United States in 1990, based on estimated inhalation unit risks and modeled outdoor concentrations of the pollutants. In total, five substances contributed more than 75% of the total estimated cancer risk from outdoor air toxics—polycyclic organic matter, 1,3-butadiene, formaldehyde, benzene, and chromium. Along with their relatively high inhalation unit risks and ambient concentrations, these substances are all considered probable or known human carcinogens.

Although this captures some of the most substantial health effects from air toxics, there are a few important omissions. The analytical approach just described focused on cancer risks; however, many air toxics also have noncancer risks that can be substantial. A prominent example is mercury, for which there is substantial evidence of neurological impairment (such as impaired motor skills or cognitive function) but limited evidence of carcinogenic effects. For these noncancer effects, the general analytical approach is to determine whether an adverse health effect is likely to occur, rather than quantifying a unit risk factor. This is done by determining a reference dose from animal or human studies, i.e., an estimate (with uncertainty spanning perhaps an order of magnitude) of the daily exposure to the human population (including sensitive subgroups) that is likely to be without an appreciable risk of adverse effects during a lifetime. As indicated by the careful language in the definition, this reference dose is not meant to be a definitive bright line, in part because of the number of uncertainties embedded in its calculation (i.e., extrapolation from short-term studies of genetically homogeneous rats to health effects from long-term exposures in heterogeneous human populations).

IV. HEALTH EFFECTS OF INDOOR AIR POLLUTION

The information discussed thus far was obtained from studies that largely focused on the effects of pollution generated outdoors (such as from power plants or motor vehicles) on population health, which includes exposure during time spent outdoors as well as exposure indoors from pollution penetrating from the outdoors. However, most of these studies did not consider indoor-generated pollution. Indoor-generated air pollution can be extremely important in both developing and developed country settings, both because of the amount of time people spend at home and because, in a confined space, pollution levels can quickly reach extremely high levels when a highly emitting fuel is burned indoors.

The uncontrolled combustion of biofuels, ranging from wood, to animal waste, to crop residues, to coal, is prevalent among poorer households in developing countries, with about 50% of the global population relying on these energy sources for heating and cooking. Numerous pollutants are emitted at high rates from biofuel combustion, including particulate matter, carbon monoxide, nitrogen oxides, and carcinogens such as formaldehyde or benzo[*a*]pyrene. The high emission rates into confined spaces result in extremely high indoor concentrations, with daily average levels as high as 50 ppm of CO and 3000 $\mu g/m^3$ of PM_{10} (versus U.S. EPA primary health standards of 9 ppm of CO and 150 $\mu g/m^3$ of PM_{10}). As a result, rates of multiple diseases are elevated in homes burning biofuels, including acute respiratory infections, chronic obstructive pulmonary disease, cancer, and low birth weight. It has been estimated that indoor air pollution is responsible for nearly 3 million deaths per year globally, approximately two-thirds of which occur in rural areas of developing countries.

Although the health effects of indoor air quality and its linkage with energy are predominantly a developing country story, the developed world is not immune to these problems. Homes that have been tightened to improve energy efficiency can have reduced ventilation rates and subsequent increases in indoor pollution concentrations. As already mentioned, the use of gas stoves can increase indoor concentrations of nitrogen dioxide, which has been associated with increased respiratory disease in children. In addition, for many air toxics, exposure to indoor sources can exceed exposure to outdoor sources, due to the use of cleaning products, off-gassing from building materials, and other emission sources.

V. CONCLUSIONS

Although the evidence supporting the health impacts of air pollution is varied and evolving over time, some general conclusions can be drawn based on existing information:

1. At current levels of air pollution in most developed and developing country settings, there is evidence of both mortality and morbidity effects. However, extreme episodes are rare and the relative risk of health effects of air pollution is low enough to necessitate incorporating evidence from a number of types of studies (including epidemiological and toxicological studies) to determine the magnitude of the effect and the likelihood of causality.
2. For criteria air pollutants, the most extensive evidence is related to particulate matter, which has been associated with premature death and cardiopulmonary morbidity due to both short-term and long-term exposures. However, uncertainties exist regarding the components of particulate matter causally linked with health outcomes and the relative contributions of other criteria pollutants to the public health burden of air pollution. The fact that long-term cohort studies have been conducted only in developed countries also contributes uncertainty in extrapolating the findings to locations with much higher levels of pollution.
3. In most settings, a subset of toxic air pollutants contributes a substantial portion of the population cancer risk from outdoor air toxic emissions, with polycyclic organic matter, 1,3-butadiene, formaldehyde, benzene, and chromium among the most substantial contributors in the United States at present.
4. Indoor air pollution, compared to outdoor air pollution, could contribute more significantly to health impacts, given the amount of time spent indoors and the higher concentrations of combustion-related pollutants in indoor settings with limited ventilation. This problem is most severe in rural settings in developing countries, where the direct burning of biofuels has been associated with respiratory disease and other health end points.

SEE ALSO THE FOLLOWING ARTICLES

Acid Deposition and Energy Use • Air Pollution from Energy Production and Use • Clean Air Markets • Climate Change and Public Health: Emerging Infectious Diseases • Gasoline Additives and Public Health • Indoor Air Quality in Industrial Nations • Radiation, Risks and Health Impacts of • Thermal Pollution

Further Reading

Bruce, N., Perez-Padilla, R., and Albalak, R. (2000). Indoor air pollution in developing countries: A major environmental and public health challenge. *Bull. World Health Org.* **78,** 1078–1092.

Dockery, D. W., Pope, C. A. 3rd, Xu, X., Spengler, J. D., Ware, J. H., Fay, M. E., Ferris, B. G. Jr., and Speizer, F. E. (1993). An association between air pollution and mortality in six U.S. cities. *N. Engl. J. Med.* **329,** 1753–1759.

Hill, A. B. (1965). The environment and disease: Association or causation? *Proc. R. Soc. Med.* **58,** 295–300.

Holgate, S. T., Samet, J. M., Koren, H. S., and Maynard, R. L. (Eds.). (1999). "Air Pollution and Health." Academic Press, San Diego.

Krewski, D., Burnett, R. T., Goldberg, M. S., Hoover, K., Siemiatycki, J., Jarrett, M., Abrahamowicz, M., and White, W. H. (2000). "Particle Epidemiology Reanalysis Project. Part II: Sensitivity Analyses." Health Effects Institute, Cambridge, MA.

Levy, J. I., Carrothers, T. J., Tuomisto, J., Hammitt, J. K., and Evans, J. S. (2001). Assessing the public health benefits of reduced ozone concentrations. *Environ. Health Perspect.* **109,** 1215–1226.

Pope, C. A. III, Burnett, R. T., Thun, M. J., Calle, E. E., Krewski, D., Ito, K., and Thurston, G. D. (2002). Lung cancer, cardiopulmonary mortality, and long-term exposure to fine particulate air pollution. *J. Am. Med. Assoc.* **287,** 1132–1141.

Pope, C. A. III, Thun, M. J., Namboodiri, M. M., Dockery, D. W., Evans, J. S., Speizer, F. E., and Heath, C. W. Jr. (1995). Particulate air pollution as a predictor of mortality in a prospective study of U.S. adults. *Am. J. Respir. Crit. Care Med.* **151,** 669–674.

Thurston, G. D., and Ito, K. (2001). Epidemiological studies of acute ozone exposures and mortality. *J. Expo. Anal. Environ. Epidemiol.* **11,** 286–294.

Woodruff, T. J., Caldwell, J., Cogliano, V. J., and Axelrad, D. A. (2000). Estimating cancer risk from outdoor concentrations of hazardous air pollutants in 1999. *Environ. Res. (Sect. A)* **82,** 194–206.

Alternative Transportation Fuels: Contemporary Case Studies

S. T. COELHO and J. GOLDEMBERG
University of São Paulo
São Paulo, Brazil

1. Introduction
2. General Overview
3. Case Studies: Ethanol Use Worldwide
4. Flex-Fuel Vehicles

Glossary

biodiesel An ester that can be made from substances such as vegetable oils and animal fats. Biodiesel can either be used in its pure state or blended with conventional diesel fuel derived from petroleum.

ethanol Ethyl alcohol (CH_3CH_2OH); one of a group of chemical compounds (alcohols) composed of molecules that contain a hydroxyl group (OH) bonded to a carbon atom. Ethanol is produced through the fermentation of agricultural products such as sugarcane, corn, and manioc. Most of the ethanol produced in the world is from sugarcane, mainly in Brazil. In the United States, ethanol is made from corn.

gasohol A mixture of 20–26% anhydrous ethanol (99.6° Gay-Lussac and 0.4% water) and gasoline; used in most vehicles in Brazil.

methyl *tert*-butyl ether (MTBE) A chemical compound that is manufactured by the chemical reaction of methanol and isobutylene. MTBE is produced in very large quantities (over 3.18×10^6 liters/day in the United States in 1999) and is almost exclusively used as a fuel additive in motor gasoline. It is one of a group of chemicals commonly known as "oxygenates" because they raise the oxygen content of gasoline. At room temperature, MTBE is a volatile, flammable, and colorless liquid that dissolves rather easily in water.

Organization for Economic Cooperation and Development (OECD) An international body composed of 30 member nations. Its objective is to coordinate economic and development policies of the member nations.

transesterification The reaction of a fat or oil with an alcohol in the presence of a catalyst to produce glycerin and esters or biodiesel. The alcohol, which carries a positive charge to assist in quick conversion, is recovered for reuse. The catalyst is usually sodium hydroxide or potassium hydroxide.

Alternative transportation fuels are those fuels not derived from petroleum. They include not only those derived from renewable sources but also natural gas and its derived fuels. A general overview of the existing status of development of alternative fuels for transportation throughout the world is presented in this article. Long-term experiences, such as the Brazilian Alcohol Program and several programs in the United States, as well as those in other countries, are also discussed.

1. INTRODUCTION

The perspective that envisions exhaustion of the world's oil reserves is not the main reason for researching and developing alternative fuels; rather, it is for environmental and strategic reasons that alternative fuel technology is of importance. For example, big cities such as São Paulo, Mexico City, and Delhi (among others) currently face serious environmental problems due to the pollutant emissions from vehicles. Some countries have proposed and enacted solutions for these problems. Ambient concentrations of lead in the São Paulo metropolitan region dropped more than 10 times from 1978 to 1991, far below the air quality standard maximum, due to the mandatory use of either an ethanol/gasoline blend or straight ethanol in all cars, legislated through the Brazilian Alcohol Program (further details are discussed in Section 3.1). Strategic aspects of alternative fuel development are also significant due to the fact that most oil reserves are

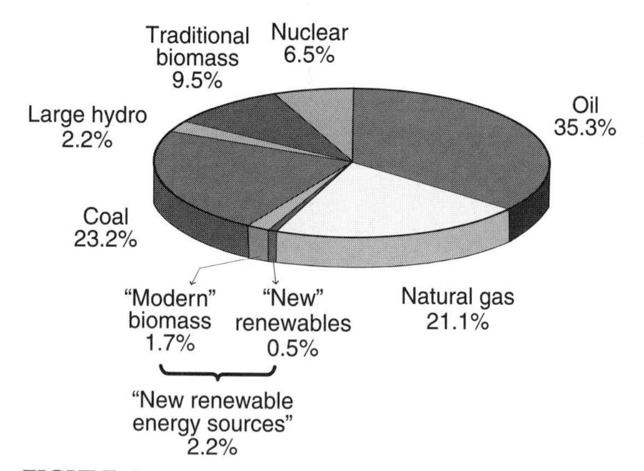

FIGURE 1 World consumption of primary energy and renewables, by energy type, 1998.

in the Middle East, a region facing complex political conflicts. The current trends show that the world will continue to depend on fossil fuels for decades; however, because the largest share of the world's oil resources is concentrated in regions with potential or active political or economic instabilities, alternative fuels ease the complexity of worldwide distribution of necessary energy resources. Nuclear energy plants, although an alternative to fossil fuels, are also concentrated only in a few countries and nuclear technology raises numerous concerns on physical security grounds.

Organization for Economic Cooperation and Development (OECD) countries, which account for 80% of the world economic activity, are quite dependent on oil imports, with a 63% share of global oil consumption (expected to rise to 76% in 2020). Asian Pacific countries are expected to increase the external dependence of their energy requirements to 72% by 2005. Compared to fossil and nuclear fuels, renewable energy resources are more evenly distributed, although only 2.2% of the world energy supply in 1998 was from new renewable sources. New renewable sources include modern biomass, small hydropower, geothermal energy, wind energy, solar (including photovoltaic) energy, and marine energy. Natural gas accounts for 21.1% of this supply (Fig. 1).

2. GENERAL OVERVIEW

The alternative fuels being used in the world today include biodiesel, electricity, ethanol, hydrogen, methanol, natural gas, propane, and solar energy. In 1999, the U.S. Department of Energy finalized an amendment to the Energy Policy Act of 1992 that added certain blends of methyltetrahydrofuran (MeTHF), ethanol, and light hydrocarbons to the list of approved "alternative fuels." These liquid products for spark-ignited engines have come to be known under the registered trademark, "P-series." The discussion here is limited to the fuels already being commercialized.

2.1 Biodiesel

Biodiesel is an ester that can be made from substances such as vegetable oils and animal fats. Biodiesel can either be used in its pure state or blended with conventional diesel fuel derived from petroleum. Vegetable oil was used as a diesel fuel as early as 1900, when Rudolf Diesel demonstrated that a diesel engine could run on peanut oil. For vegetable oils to be used as fuel for conventional diesel engines, the oils must be further processed, primarily because of their high viscosity. Transesterification (production of the ester) of vegetable oils or animal fats, using alcohol in the presence of a catalyst, is the most popular process. For every 100 units of biodiesel fuel produced using this method, there are 11 units of glycerin as a by-product. Glycerin is used in such products as hand creams, toothpaste, and lubricants. Another biodiesel production process in limited use involves cold-pressed rapeseed oil, but no glycerin by-product is produced. Alternatively, unprocessed vegetable oils can be used in modified diesel engines. Such engines have limited production and are therefore more expensive, although their numbers are increasing in Europe.

The main benefits of biodiesel can be categorized as strategic (increased energy self-sufficiency for oil-importing countries), economic (increased demand for domestic agricultural products), and environmental (biodegradability and improved air quality, particularly lower sulfur emissions and almost null carbon balance). Exhaust emission improvements include substantial reduction in carbon monoxide, hydrocarbons, and particulates, although the production of nitrogen gases can be similar to that of regular diesel fuel (depending of the diesel quality). The United States, New Zealand, Canada, and several European Union countries have conducted extensive tests of biodiesel in trucks, cars, locomotives, buses, tractors, and small boats. Testing has included the use of pure biodiesel and various blends with conventional diesel. Among the developing countries, the Biodiesel Program started in Brazil in 2002 was notable for its goal to replace part of the diesel consumption in the Brazilian transportation sector. Presently, the major disadvantage of biodiesel is its high production cost. As discussed in Section 3.1, in the case of Brazil, biodiesel forecasted costs are higher than diesel costs.

2.1.1 Biodiesel in the United States

Interest in biodiesel in the United States was stimulated by the Clean Air Act of 1990, combined with regulations requiring reduced sulfur content in diesel fuel and reduced diesel exhaust emission. The Energy Policy Act of 1992 established a goal of replacing 10% of motor fuels with petroleum alternatives by the year 2000 (a goal that was not reached) and increasing to 30% by the year 2010.

2.1.2 Biodiesel in Europe

Two factors have contributed to an aggressive expansion of the European biodiesel industry. Reform of the Common Agricultural Policy to reduce agricultural surpluses was of primary importance. This policy, which provides a substantial subsidy to non-food crop production, stimulated the use of land for non-food purposes. Secondarily, high fuel taxes in European countries normally constitute 50% or more of the retail price of diesel fuel. In 1995, Western Europe biodiesel production capacity was 1.1 million tons/year, mainly produced through the transesterification process. This added over 80,000 tons of glycerin by-products to the market annually, creating created a large surplus. Germany thus decided to limit the production of biodiesel using the transesterification process. When it is not possible to market the glycerin by-product, one method of disposal of the excess is incineration; however, this creates an environmental risk and results in additional costs. Germany is now focusing on biodiesel production using the cold-pressed rapeseed method to avoid the problem of excess glycerin.

2.1.3 Biodiesel in Japan

In early 1995, Japan decided to explore the feasibility of biodiesel by initiating a 3-year study. Biodiesel plants using recycled vegetable oils collected in the Tokyo were planned; 10% of federal vehicles were expected to use alternative fuels to set an example for the private automotive and fuel industries. The 3-year study indicated production costs for biodiesel in Japan are 2.5 times that of petroleum diesel. The program has only recently reach its objectives.

2.1.4 Biodiesel in Canada

In the early 1990s, Canadian canola production increased in response to higher market prices relative to cereal grains and increased grain handling and transportation costs. Canola production peaked in 1994 and 1995, limited by suitable land base and crop rotational requirements. However, higher yield due to new *Brassica juncea* varieties and improved chemical weed control may further increase production in the medium term. There is a potential for the use of lower quality canola oils derived from overheated or frost-damaged seed without any ill effects on biodiesel quality.

2.1.5 Biodiesel in Brazil

In 1998, several initiatives were implemented in Brazil, aiming to introduce biodiesel into the Brazilian energy matrix. The initiatives included (1) tests performed in South Brazil, using the so-called B20 blend (20% ester and 80% diesel oil), in specific fleets of urban buses, (2) the building of a small-scale pilot plant for biodiesel production from fat and palm oil (largely produced in North Brazil), and (3) laboratory-scale production and tests of biodiesel using soybean oil/sugarcane ethanol. The Brazilian federal government subsequently decided to establish a work group of specialists from all involved sectors, creating the National Biodiesel Program in 2002. This program will mainly analyze the use of surplus of soybean oil, which is produced on a large scale in Brazil and is presently facing, export barriers.

The economic competitiveness of biodiesel and diesel oil has been evaluated; studies in Brazil show that biodiesel production costs are higher than diesel costs (Table I).

TABLE I

Production Cost for Methyl Ester (1 Ton) from Soy Oil in Brazil[a]

	Amount			
Input	Kilograms	Tons	Price (R $/ton)	Cost
Soy oil	1015	1.015	169.64	172.19
Methanol[b]	140	0.14	93.57	13.10
Catalyzer[c]	12	0.012	125.00	1.50
Input cost				186.79
Production cost (115% of input cost)				401.59
Total cost (U.S. $/ton)				588.38

[a]Data from the Brazilian Reference Center on Biomass (CENBIO). Exchange rate, Brazilian real (R)/U.S. dollar, $2.8/1 (July 2002). Diesel price in São Paulo pump stations is around R $1.00/liter (about U.S. $350/ton).

[b]Methanol cost based on prices at pump stations in California (U.S. $0.88–1.10/gallon).

[c]Catalyzer cost based on Brazilian market price.

2.2 Electricity

Electricity is not, technically speaking, a fuel, but it is used in existing alternative technologies for powering transportation. Electricity can be used to power electric and fuel cell (discussed later) vehicles. When used to power electric vehicles (EVs), electricity is stored in an energy storage battery, which must be replenished by plugging the vehicle into a recharging unit. The electricity for recharging the batteries can come from the existing power grid (from large hydroelectric or thermoelectric power plants), or from distributed renewable sources, such as biomass, small hydro, solar, or wind energy. The main benefits to electric-powered vehicles are the lower pollutant emissions at the point of use, although the emissions generated in the electricity production process at the power plants or from the fuel reform reaction (when the fuel cells use hydrogen produced by the reform reaction, as discussed later) can be indirectly attributed to EVs. Economic aspects of using EVs include the high initial capital cost, which can be partially offset by the lower maintenance costs. When compared to the cost of gasoline, the cost of an equivalent amount of fuel for an EV is lower. Maintenance costs for EVs are lower because EVs have fewer moving parts to service and replace.

2.3 Ethanol

Ethanol (ethyl alcohol, CH_3CH_2OH) is one of a group of chemical compounds (alcohols) with molecules that contain a hydroxyl group (OH) bonded to a carbon atom. Ethanol is produced through the fermentation of agricultural products such as sugarcane, corn, and manioc, among others. Most ethanol produced worldwide is from sugarcane, mainly in Brazil. In the United States, ethanol is made from corn.

Ethanol is used as a high-octane fuel in vehicles. More than 4 million cars run on pure, hydrated ethanol in Brazil, and all gasoline in the country is blended with anhydrous ethanol (20–26% ethanol), as a result of a government program to make ethanol from sugarcane, in place since the 1970s. In the United States, there is a similar program being started and the number of vehicles using ethanol is increasing.

Ethanol makes an excellent motor fuel: it has a motor octane number that exceeds that of gasoline and a vapor pressure that is lower than that of gasoline, which results in lower evaporative emission. Ethanol's flammability in air is also much lower than that of gasoline, which reduces the number and severity of vehicle fires. Anhydrous ethanol has lower and higher heating values of 21.2 and 23.4 megajoules (MJ)/liter, respectively; for gasoline the values are 30.1 and 34.9 MJ/liter. Because ethanol in Brazil is produced from sugarcane, it has the lowest production cost in the world. This is due not only to high agricultural and industrial productivity levels, but also to the extremely favorable energy balance of the alcohol production. In the United States, ethanol is produced from corn and represents a large consumption of fossil fuels, with much lower energy balance, despite the existing controversy among specialists. Table II shows a comparison of the ethanol energy balance in Brazil (from sugarcane) and the United States (from corn). Ethanol can also be produced from cellulose feedstock (corn stalks, rice straws, sugarcane bagasse, etc.), through a process still under development.

In the United States, because of support from corn-growing states and the U.S. Departments of Energy and Agriculture, ethanol-fueled vehicle production is increasing. Auto manufacturers began in 1997 to produce cars and pickup trucks that could use either ethanol or gasoline. These flexible fuel (or flex-fuel) vehicles are discussed in Section 4.

2.4 Hydrogen

Hydrogen (H_2) is a gas that has considerable potential as an alternative fuel for transportation but, at this point, little market presence. The most important use of hydrogen is expected to be in electric fuel cell vehicles in the future. Fuel cell cars are in development by most major manufacturers, but hydrogen still lacks a wide distribution infrastructure. The current emphasis is on the use of hydrogen to supply the fuel cells that power electric vehicles. Fuel cells produce electricity. Similar to a battery, a fuel cell converts energy produced by a chemical reaction directly into usable electric power. However, unlike a battery, a fuel cell needs an external fuel source—typically hydrogen gas—and generates electricity as long as fuel is supplied, meaning that it never needs electrical recharging. Inside most fuel cells, oxygen and hydrogen from a fuel tank combine (in an electrochemical device) to produce electricity and warm water. As a simple electrochemical device, a fuel cell does not actually burn fuel, allowing it to operate pollution free. This also makes a fuel cell quiet, dependable, and very fuel efficient.

An impediment to the use of fuel cells is the hydrogen production and storage. The predominant

TABLE II

Energy Balance of Ethanol[a] from Sugarcane and from Corn

Product	Source	Fossil fuel consumption in ethanol production (MJ/liter of ethanol)[b]	Final energy balance[c]
Ethanol from sugarcane[d]	Macedo (2000)	1.89[h]	11.2:1
Ethanol from corn			
Existing plants[e]	Pimentel (1991)	42.6	Negative
Laboratory tests[f]	Pimentel (1991)	25.6	Negative
Existing plants[e]	Unnasch (2000)	11.8	1.8:1
With allocation to coproducts[g]	Unnasch (2000)	7.19	2.9:1

[a] Lower ethanol energy value: 21.2 MJ/liter.

[b] Fossil fuel consumption in ethanol production corresponds to diesel oil use in the agricultural phase and during transport, to coal and natural gas use in corn-based ethanol plants, and to natural gas use for fertilizer production.

[c] Final energy balance corresponds to the low energy content in 1 liter of ethanol divided by the total fossil fuel consumption to produce 1 liter of ethanol.

[d] Includes bagasse surplus production (Brazil); see footnote b.

[e] Large plants in the United States.

[f] Using membrane technology.

[g] Considering the coproduction of corn oil and animal feed products, the allocation of energy inputs corresponds to 1096 kcal/liter of ethanol.

[h] In sugarcane-origin ethanol plants, there is no fossil fuel consumption in the plant (all fuel consumed in the plant is sugarcane bagasse, the by-product of sugarcane crushing); fossil fuel consumption corresponds to the agricultural phase and fertilizer production.

method of making hydrogen today involves using natural gas as a feedstock. Petroleum-based fuels, including gasoline and diesel, can also be used, but this may compromise a major objective behind alternative fuels, i.e., to reduce oil consumption. Hydrogen production is through the "reform reaction" of an existing fuel (natural gas, methanol, ethanol, or naphtha), a chemical reaction that "extracts" the hydrogen from the fuel, producing a gas mixture of carbon monoxide (CO) and hydrogen. The hydrogen must be separated from the CO to be fed into the fuel cell. This reaction can be performed in a stationary system (and the vehicle will carry a high-pressure hydrogen storage tank) or in an on-board reformer/fuel cell system (and the vehicle will carry a conventional fuel tank to feed the system). Both possibilities are now under study in several countries. According to some specialists, ethanol should be the fuel of choice for fuel cells due to its lower emissions and because it is produced from a renewable source (biomass).

Though fuel cells have been widely publicized in recent years, they are not new: the first one was produced in 1839. Fuel cells powered the Gemini spacecraft in the 1960s, continue to power the Space Shuttle, and have been used by the National Aeronautics and Space Administration (NASA) on many other space missions. Although their operation is simple, they have been quite expensive to make. Extensive research and development have promised the widespread use of fuel cells in the near future. All major auto companies have fuel cell-powered vehicles in the works, and a nascent fuel cell industry is growing rapidly. By the kilowatt, fuel cells still cost more today than do conventional power sources (from $1000 upto $5000/kW, U.S. dollars), but an increasing number of companies are choosing fuel cells because of their many benefits, and large-scale production is expected to make fuel cell costs decline.

According to the U.S. National Renewable Energy Laboratory, hydrogen prices vary widely, depending on the transport distance and the type of hydrogen (from 17 to 55 cents/100 cubic feet), but cost figures for hydrogen use as transportation fuel are not yet available.

2.5 Methanol

Methanol (or methyl alcohol) is an alcohol (CH_3OH) that has been used as alternative fuel in flexible fuel vehicles that run on M85 (a blend of 85% methanol and 15% gasoline). However, methanol is not commonly used nowadays because car manufacturers are no longer building methanol-powered vehicles. Methanol can also be used to make methyl-*tert*-butyl ether (MTBE), an oxygenate that is blended with gasoline to enhance octane and reduce pollutant emissions. However, MTBE production and use have declined due to the fact that MTBE contaminates groundwater. In the future, methanol may be an important fuel, in addition to ethanol, to produce the hydrogen necessary to power fuel cell vehicles; such a process is now under development.

Methanol is predominantly produced by steam reforming of natural gas to create a synthesis gas, which is then fed into a reactor vessel in the presence of a catalyst to produce methanol and water vapor. Although a variety of feedstocks other than natural gas can and have been used, today's economics favor natural gas. Synthesis gas refers to combinations of carbon monoxide (CO) and hydrogen (H_2). Similar to ethanol vehicles, methanol-powered vehicles emit smaller amounts of air pollutants, such as hydrocarbons, particulate matter, and NO_x, than do gasoline-powered vehicles. However, the handling of methanol is much more dangerous compared to ethanol due to its high negative impacts on health.

World demand for methanol is around 32 million tons/year and is increasing modestly by about 2–3%/year, but with significant changes in the industry profile. Since the early 1980s, larger plants using new, efficient low-pressure technologies have replaced less efficient small facilities. Demand patterns have been changing in Europe; methanol was once blended into gasoline (when its cost was around one-half of that of gasoline), but now is not competitive with lower oil prices. Offsetting this was the phasing out of leaded gasoline in developed countries, mandating the use of reformulated gasoline. The United States promoted the use of MTBE derived from methanol. The United States produces almost one-quarter of the world's supply of methanol, but there is a significant surplus of methanol in the world. Figure 2 shows the world methanol supply, indicating this significant excess capacity.

The largest market for methanol in the United States is for the production of MTBE; there are nearly 50 U.S. plants. It is estimated that 3.3 billion gallons of MTBE was used in 1996 for blending into clean, reformulated gasoline serving 30% of the U.S. gasoline market. MTBE displaces 10 times more gasoline than all other alternative vehicle fuels combined.

Methanol prices in the United States have varied significantly since 1989; they doubled by 1994 and returned to the 1993 levels by 1996. In 2002, methanol prices reached 64 cents/gallon, slightly higher than European prices (61.7 cents/gallon) and Asian prices (55.9 cents/gallon).

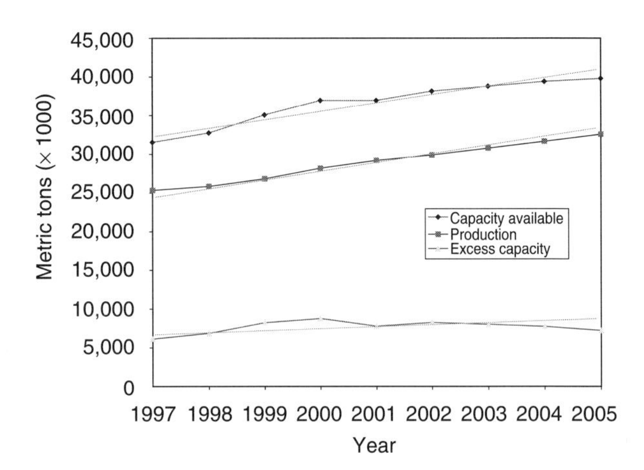

FIGURE 2 World methanol supply. Data from the Methanol Institute (www.methanol.org).

2.6 Natural Gas

Natural gas is a mixture of hydrocarbons, mainly methane (CH_4), but it also contains hydrocarbons such as ethane (C_2H_4) and propane (C_3H_8) and other gases such as nitrogen, helium, carbon dioxide, hydrogen sulfide, and water vapor. It is produced either from gas wells or in conjunction with crude oil production (associated natural gas). Natural gas is consumed in the residential, commercial, industrial, and utility markets. The interest in natural gas as an alternative fuel comes mainly from its clean burning properties, with extremely low sulfur and particulate contents. Also, carbon emissions from natural gas are lower compared to those from other fossil fuels, such as oil and coal. Because of the gaseous nature of natural gas, it must be stored on board a vehicle in either a compressed gaseous state (compressed natural gas) or in a liquefied state (liquefied natural gas). Natural gas is mainly delivered through pipeline systems.

Natural gas reserves are more evenly distributed worldwide than are those of oil. According to BP Global, reserves at the end of 2000 totaled 5304 trillion cubic feet, with 37.8% of the reserves located in the former Soviet Union, 35% in the Middle East, 6.8% in Asia Pacific, 6.8% in North America, 4.6% in South and Central America, and 3.5% in Europe. By contrast, crude oil reserves are heavily concentrated in the Middle East, which alone held 65.3% of oil reserves at the end of 2000. Global proved natural gas reserves available in 2000 would last more than 60 years at the 2000 rate of production. Although production and consumption are likely to increase over time, past trends suggest that reserves will increase as well, especially with improvements to the technology employed to find and produce natural gas.

Consumption of natural gas for fuel increased more than consumption of any other fuel in 2000, with global consumption rising by 4.8%, the highest rate since 1996. This was driven by growth in consumption of 5.1% in the United States and Canada, which together represent more than 30% of world demand. Chinese consumption increased by 16%, although China still represents only 1% of world consumption. In the former Soviet Union, growth of 2.9% was the highest for a decade.

Gas production rose by 4.3% worldwide, more than double the average of the last decade. This growth is mainly due developments in the United States and Canada, where production rose by 3.7%, and the fastest since 1994. Growth exceeded 50% in Turkmenistan, Nigeria, and Oman, and 10% in 11 other countries. Russia, the second largest producer, saw output decline by 1.1%. North America produced and consumed in 2000 more than 70 billion cubic feet (bcf)/day. The former Soviet union produced about 68 bcf/day and consumed about 53 bcf/day. Europe produced less than 30 bcf/day but consumed about 1.5 times more. The balance is practically even in South and Central America (with a little less than 10 bcf/day) and is almost even in Asia Pacific (about 30 bcf/day; consumption higher than production) and the Middle East (about 20 bcf/day; production a little higher than consumption). Africa produced more than 10 bcf/day but consumed less than half of it.

In the U.S. market, natural gas prices are about $1.5/million British thermal units (MMBtu) above oil spot prices, which ranged from $2.1 to $3.5/MMBtu by the end of 2002 (according to the U.S. Department of Energy; information is available on their Web site at www.eia.doe.gov).

2.7 Propane

Liquefied petroleum gas (LPG) consists mainly of propane (C_3H_8) with other hydrocarbons such as propylene, butane, and butylene, in various mixtures. However, in general, this mixture is mainly propane. The components of LPG are gases at normal temperatures and pressures. Propane-powered vehicles reportedly have less carbon build-up compared to gasoline- and diesel-powered vehicles.

LPG is a by-product from two sources: natural gas processing and crude oil refining. When natural gas is produced, it contains methane and other light hydrocarbons that are separated in a gas-processing plant. The natural gas liquid components recovered during processing include ethane, propane, and butane, as well as heavier hydrocarbons. Propane and butane, along with other gases, are also produced during crude refining as a by-product of the processes that rearrange and/or break down molecular structure to obtain more desirable petroleum compounds.

The propane market is a global market. Approximately 1.3 billion barrels of propane are produced worldwide. Although the United States is the largest consumer of propane, Asian consumption is growing fast. According to the World LP Gas Association, during 1999 China achieved a growth rate of over 20% in their consumption of propane, largely in the residential/commercial sector. Other notable increases were recorded in India, Iran, and South Korea, which are rebounding from the Asian economic crisis.

2.8 Solar Energy

Solar energy technologies use sunlight to produce heat and electricity. Electricity produced by solar energy through photovoltaic technologies can be used in conventional electric vehicles. Using solar energy directly to power vehicles has been investigated primarily for competition and demonstration vehicles. Solar vehicles are not available to the general public, and are not currently being considered for production. Pure solar energy is 100% renewable and a vehicle run on this fuel emits no pollutants.

3. *CASE STUDIES: ETHANOL USE WORLDWIDE*

3.1 The Brazilian Experience

The Brazilian Alcohol Program (PROALCOOL) to produce ethanol from sugarcane was established during the 1970s, due to the oil crises, aiming to reduce oil imports as well as to be a solution to the problem of the fluctuating sugar prices in the international market. The program has strong positive environmental, economic, and social aspects, and has become the most important biomass energy program in the world. In 1970, some 50 million tons of sugarcane were produced, yielding approximately 5 million tons of sugar. In 2002, sugarcane production reached 300 million tons, yielding 19 million tons of sugar and 12 billion liters of alcohol (ethanol). In 2002, the total land area covered by sugarcane plantations in Brazil was approximately 4.2 million hectares (60% in the state of São Paulo, where sugarcane has replaced, to a large extent, traditional coffee plantations). The average productivity of sugarcane crops in Brazil is 70 tons/hectare, but in São Paulo State there are mills with a productivity of 100 tons of cane per hectare.

Ethanol production costs were close to $100/barrel in the initial stages of the program in 1980. After that, they fell rapidly, to half that value in 1990, due to economies of scale and technological progress, followed by a slower decline in recent years. Considering the hard currency saved by avoiding oil importation through the displacement of gasoline by ethanol, it is possible to demonstrate that the Alcohol Program has been an efficient way of exchanging dollar debt for national currency subsidies, which were paid by the liquid fossil fuel users.

The decision to use sugarcane to produce ethanol in addition to sugar was a political and economic one that involved government investments. Such decision was taken in Brazil in 1975, when the federal government decided to encourage the production of alcohol to replace gasoline, with the idea of reducing petroleum imports, which were putting great constraints in the external trade balance. At that time, sugar price in the international market was declining very rapidly and it became advantageous to shift from sugar to alcohol production. Between 1975 and 1985, the production of sugarcane quadrupled and alcohol became a very important fuel used in the country. In 2002, there were 321 units producing sugar and/or alcohol (232 in central–south Brazil and 89 in northeast Brazil). An official evaluation of the total amount of investments in the agricultural and industrial sectors for production of ethanol for automotive use concluded that in the period 1975–1989, a total of $4.92 billion (in 2001 U.S. dollars) was invested in the program. Savings on oil imports reached $43.5 billion (2001 U.S. dollars) from 1975 to 2000.

In Brazil, ethanol is used in one of two ways: (1) as an octane enhancer in gasoline in the form of 20–26% anhydrous ethanol (99.6° Gay-Lussac and 0.4% water) and gasoline, in a mixture called gasohol, or (2) in neat-ethanol engines in the form of hydrated ethanol at 95.5° Gay-Lussac. Increased production and use of ethanol as a fuel in Brazil were made possible by three government actions during the launching of the ethanol program. First, it was decided that the state-owned oil company, Petrobrás, must purchase a guaranteed amount of ethanol; second, economic incentives were offered to agro-

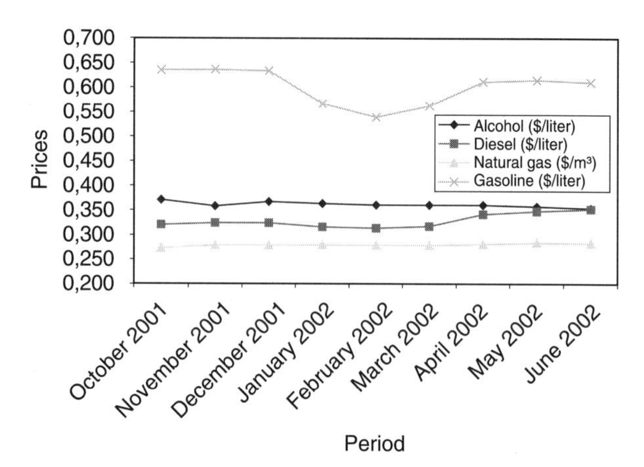

FIGURE 3 Transportation fuel prices in Brazil (in U.S. dollars).

industrial enterprises willing to produce ethanol, in the form of loans with low interest rates, from 1980 to 1985; third, steps were taken to make ethanol attractive to consumers, by selling it at the pump for 59% of the price of gasoline. This was possible because the government at that time set gasoline prices.

The subsidies for ethanol production have been discontinued and ethanol is sold for 60–70% of the price of gasoline at the pump station, due to significant reduction of production costs. These results show the economic competitiveness of ethanol when compared to gasoline. Considering the higher consumption rates for net-ethanol cars, ethanol prices at the station could be as much as 80% of gasoline prices. Fig. 3 shows a comparison of the different transportation fuels in Brazil.

3.1.1 Impact of Alcohol Engines on Air Pollution

All gasoline used in Brazil is blended with 25% anhydrous ethanol, a renewable fuel with lower toxicity, compared to fossil fuels. In addition to the alcohol–gasoline (gasohol) vehicles, there is a 3.5 million-vehicle fleet running on pure hydrated ethanol in Brazil, 2.2 million of which are in the São Paulo metropolitan region. Initially, lead additives were reduced as the amount of alcohol in the gasoline was increased, and lead was completely eliminated by 1991. Aromatic hydrocarbons (such as benzene), which are particularly toxic, were also eliminated and the sulfur content was reduced as well. In pure ethanol cars, sulfur emissions were eliminated, which has a double dividend. The simple substitution of alcohol for lead in commercial gasoline has dropped the total carbon monoxide, hydrocarbon, and sulfur emissions by significant numbers. Due to the ethanol blend, ambient lead concentrations in the São Paulo metropolitan region dropped from 1.4 $\mu g/m^3$ in 1978 to less than 0.10 μg/m^3 in 1991, according to CETESB (the Environmental Company of São Paulo State), far below the air quality standard maximum of 1.5 μg/m^3. Alcohol hydrocarbon exhaust emissions are less toxic, compared to those from gasoline. They present lower atmospheric reactivity and null greenhouse emissions balance (9.2 million tons of carbon dioxide avoided in 2000 due only to the gasoline replacement by ethanol).

One of the drawbacks of the use of pure ethanol is the increase in aldehyde emissions as compared to gasoline or gasohol use. It can be argued, however, that the acetaldehyde from alcohol use is less detrimental to human health and to the environment, compared to the formaldehyde produced when gasoline is used. Total aldehyde emissions from alcohol engines are higher than emissions from gasoline engines, but it must be noted that these are predominantly acetaldehydes. Acetaldehyde emissions produce fewer health effects compared to the formaldehydes emitted from gasoline and diesel engines. Aldehyde ambient concentrations in São Paulo present levels substantially below the reference levels found in the literature. In addition, carbon monoxide (CO) emissions have been drastically reduced: before 1980, when gasoline was the only fuel in use, CO emissions were higher than 50 g/km; they went down to less than 5.8 g/km in 1995.

3.1.2 Social Aspects of the Brazilian Alcohol Program

Social considerations are the real determinants of the impact of the Brazilian alcohol program. Presently, ethanol production generates some 700,000 jobs in Brazil, with a relatively low index of seasonal work. The cost of creating a single job in the ethanol agroindustry is around $15,000 (U.S. dollars), according to recent studies by the São Paulo Sugarcane Agroindustry Union (UNICA). In Brazil, job generation in most other industries requires higher investments, as shown in Fig. 4.

Although Brazilian workers have low incomes relative to workers in developed countries, there has been a significant increase in agricultural mechanization, particularly in the São Paulo region. There are two main reasons for this trend: social changes such as urban migration have reduced the number of workers available to the sugarcane industry, promoting mechanization, and mechanization has proved cheaper than hand labor. Thus, the delicate balance between mechanization and the number and quality of new jobs created by the ethanol industry is likely to remain a key issue for several years.

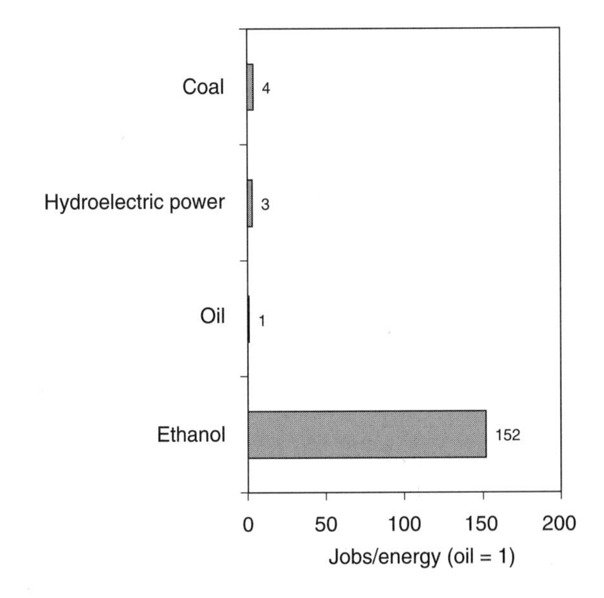

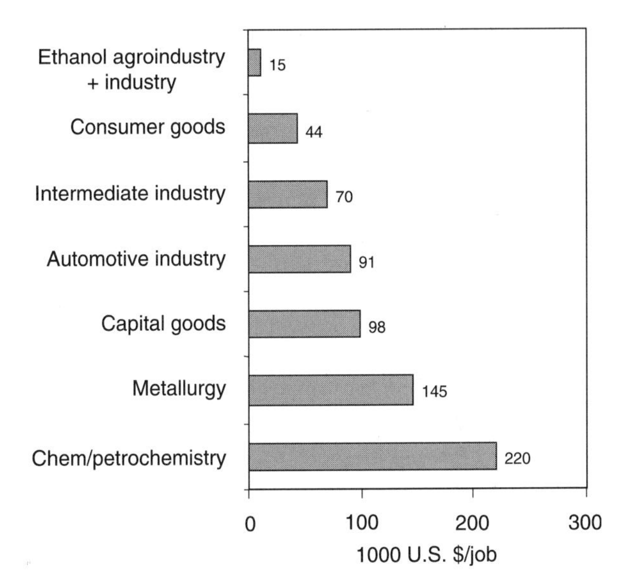

FIGURE 4 Employment numbers (2001) for the Brazilian Ethanol Program. Data from UNICA (2002).

3.1.3 Outlook for the Brazilian Alcohol Program

The use of alcohol as a fuel in Brazil was an extraordinary success until 1990. In 1988, sales of ethanol-powered cars represented 96% of the market; by the end of the decade, 4.5 million automobiles were sold. Since then, however, there has been a precipitous drop in the sale of new ethanol-powered cars, due to the following reasons:

- The price of pure alcohol was set in 1979 at 64.5% of the price of gasoline but increased gradually to 80%, reducing the economic competitiveness.

- The tax on industrialized products (IPI), initially set lower for alcohol-fueled cars, was increased in 1990, when the government launched a program of cheap "popular cars" (motors with a cylinder volume up to 1000 cm^3) for which the IPI tax was reduced to 0.1%. According to the local manufacturers, the "popular cars" could not be easily adapted to use pure alcohol because this would make them more expensive and would take time. Competition between manufacturers required an immediate answer to benefit from the tax abatement.
- Significant lack of confidence in a steady supply of alcohol and the need to import ethanol/methanol to compensate for a reduction in local production due to the increase on sugar exports (high sugar prices in international market).

As a result, the sales of pure ethanol cars dropped almost to zero in 1996. However, in 2001, this scenario started to change: 18,335 units of pure ethanol car were sold, showing an increase in consumer interest. Use of blended ethanol–gasoline has not been affected due the existence of a federal regulation that requires addition of ethanol in all gasoline sold in Brazil.

Other countries adopt special taxes (mainly on gasoline) that are quite substantial (around 80% in Italy and France, around 60% in the United Kingdom, Belgium, Germany, and Austria, and around 30% in the United States). In addition, ethanol as an additive is made viable in some countries by differentiated taxes and exemptions. In the United States, federal taxes on gasoline are 13.6 cents/gallon (3.59 cents/liter). The addition of 1 gallon of ethanol leads to a credit of 54 cents/gallon, or $26.68/barrel (U.S. dollars). Some states have also introduced tax exemptions that are the equivalent of up to $1.10/gallon, or $46.20/barrel. The rationale for these subsidies and exemptions is the same as was used in Brazil in the past, such as the social benefits of generating employment and positive environmental impacts.

An interesting option regarding the future for the ethanol program is its expansion. To guarantee ethanol market expansion, there are two distinct alternatives: increase national demand or create an international alternative liquid fuel market. However, international trade of ethanol still faces several difficulties regarding import policies of developed countries. Finally, it is also important to consider the effect the use of ethanol has in the balance of payments made by Brazil. Since the beginning of the alcohol program in 1975, and until 2000, at least 216 million m^3 of ethanol have been produced, displacing the consumption of 187 million m^3 of gasoline. This means that around $33 billion (U.S. 1996 dollars) in hard currency expenditures were avoided.

The international alternative liquid fuel market is currently an interesting option if developed countries decide effectively to limit CO_2 emissions to satisfy the goals established in the 1992 Rio de Janeiro conference, in the Climate Convention adopted in the United Nations World Summit for Sustainable Development. There are also reasons to develop an alternative to fossil fuels for economic and strategic reasons while promoting better opportunities for farmers. Food production efficiency has improved so much in the past 30 years that around 40–60 million hectares (Mha) of land is now kept out of production to maintain food product prices at a reasonable level able to remunerate farmers in the OECD countries. With the introduction of biomass-derived liquid fuels, a new opportunity exists for utilization of such lands and preservation of rural jobs and employment in industrialized countries. Not all industrial countries have free agricultural areas to produce biofuels, and even those that do will not be able to satisfy fully the demand for biofuel, at a reasonable price and in an environmentally sound way. The total automobile fleet in Western Europe exceeds 200 million units, demanding more than 8 million barrels/day (bl/day) of gasoline. To satisfy such a level of demand with a 10% ethanol blend, production of 0.8 million bl/day is necessary, of which probably two-thirds would be imported. This means an international market of 530,000 bl/day for ethanol from sugarcane, which could be shared mainly by a few developing tropical countries, such as Brazil, India, Cuba, and Thailand. With the creation of such a market, some developing countries could redirect their sugarcane crops to this market instead of to sugar. Sugar is a well-established commodity with a declining real average price in the international market. This is not the case for gasoline, for which the average real price is increasing.

Finally it is worthwhile to add that the international fuel market option has the potential to increase ethanol demand to a level above the level presented here, because (depending on the seriousness of environmental problems) ethanol blends of up to 25% may be marketed in the future, as is already happening in Brazil, as well as the possibility that such a practice will be followed by the diesel oil industry. Brazil's sugarcane industry is recovering from a significant decrease in production in recent

TABLE III

Comparative Prices of Transportation Fuels in Brazil and the United States[a]

	Prices[b]			
	Brazil		United States	
Fuel	Low	High	Low	High
Gasoline	20.92	24.69	11.49	20.38
Natural gas	7.63	7.94	6.21	6.48
Diesel	8.67	9.75	8.81	10.41
Ethanol[c]	16.39	17.76	17.92	21.90
Methanol	na[d]	na[d]	12.93	16.14

[a] Data from Brazilian Petroleum National, the Alternative Fuels Data Center, the U.S. Gas Price Watch, and the Ministry of Science and Technology. Calculations by the Brazilian Reference Center on Biomass.

[b] All prices in U.S. dollars ($/m^3); average prices taken between January and July, 2002. Exchange rate, Brazilian real (R)/U.S. dollar, 2.8/1.

[c] Brazilian ethanol from sugarcane and U.S. ethanol from corn.

[d] na, Not available.

years; in 2002, a sugarcane surplus allowed Brazil to increase ethanol exports. Table III shows comparative prices of gasoline, natural gas, diesel oil, methanol, and ethanol in both Brazil and the United States. The low prices of gasoline in the United States are remarkable compared to Brazilian prices. It also must be noted that ethanol prices in Brazil, in contrast to the United States, do not include any subsidies.

3.2 The U.S. Experience

In 2000, 19.7 million barrels of crude oil and petroleum products were consumed in the United States per day (25% of world production); more than half of this amount was imported. The United States has the lowest energy costs in the world. In 2001, ethanol-blended fuels represented more than 12% of motor gasoline sales, with 1.77 billion gallons produced from corn. Corn is used as the principal feedstock, in contrast to practices in Brazil, where all ethanol is produced from sugarcane, which considerably less expensive than corn. The U.S. Congress established the Federal Ethanol Program in 1979 to stimulate rural economies and reduce the U.S. dependence on imported oil. In 1992, the Energy Policy Act established a goal of replacing 10% of motor fuels with petroleum alternatives by the year 2000, increasing this to 30% by the year 2010.

In 2002, the U.S. Senate passed an energy bill that includes a provision that triples the amount of ethanol to be used in gasoline in the next 10 years. This policy, combined with regulations requiring reduced sulfur content in diesel fuel and reduced diesel exhaust emission, also was designed to foster an interest in biodiesel. Ethanol blended fuel is now marketed throughout the United States as a high-quality octane enhancer and oxygenate, capable of reducing air pollution and improving vehicle performance. The federal administration denied California an oxygenate waiver in 2002, thus ethanol fuel production is expected to grow strongly. According to the Renewable Fuels Association, by the end of 2003, U.S. annual ethanol production capacity will reach 3.5 billion gallons (13.25 billion liters).

California currently uses 3.8 billion gallons/year of MTBE, compared with total U.S. use of 4.5 billion gallons/year. MTBE, like ethanol, is an oxygenate that allows lower pollutant emissions from gasoline. Under federal law, in urban areas with the worst pollution (as in many California cities), gasoline must contain at least 2% oxygen by weight. This requirement applies to about 70% of the gasoline sold in California. The use of MTBE will be discontinued by 2003 in California, being replaced by ethanol, which does not present risks of contamination of water supplies.

In the United States, fuel ethanol production increased from about 150 million gallons in 1980 to more than 1700 million gallons in 2001, according to the Renewable Fuels Association. In the period 1992–2001, the U.S. demand for fuel ethanol increased from 50 to 100 thousand barrels/day, whereas MTBE demand increased from 100 to 270 thousand barrels/day. According to the Renewable Fuels Association, ethanol production in the United States affords several advantages:

- Ethanol production adds $4.5 billion to U.S. farm income annually.
- More than 900,000 farmers are members of ethanol production cooperatives; these cooperatives have been responsible for 50% of new production capacity since 1990.
- Ethanol production provides more than 200,000 direct and indirect jobs.
- Ethanol production reduces the U.S. negative trade balance by $2 billion/year.

In June 2002, ethanol and MTBE prices in the United States were practically equivalent ($1/gallon). Gasoline prices were $0.8/gallon.

3.3 Other Countries

3.3.1 Europe

In the European Union, currently only about 0.01 billion liters of ethanol is used as fuel. In 1994, the European Union decided to allow tax concessions for the development of fuel ethanol and the other biofuels, and as result a number of ethanol projects have been announced in The Netherlands, Sweden, and Spain. France has one of the most developed fuel alcohol programs in the European Union. A 1996 law requires the addition of oxygenate components to fuel, and ethanol was given an early exemption from the gasoline excise tax.

The European Commission (EC) is currently drafting a directive that could force member states to require 2% biofuels in all motor fuels. The directive, if eventually confirmed by the European Parliament and Council, could be enforced as early as 2005. France, Austria, and Germany have already experimented with biodiesel, and other countries, including Sweden, Spain, and France, use ethanol or ethyl *tert*-butyl ether (ETBE) in gasoline. Sweden is the first country in Europe that has implemented a major project introducing E85 fuel and vehicles. Approximately 300 vehicles have been in use in Sweden since the mid-1990s and 40 E85 fueling stations have been established from Malmo, in the south, to Lulea, in the north. In 1999, a consortium composed of the Swedish government, many municipal administrations, companies, and private individuals contacted over 80 different car factories inquiring about the production of an ethanol-fueled vehicle for use in Sweden. A major criterion was that it should not be much more expensive than a gasoline-powered vehicle. Car manufacturers are presently marketing these vehicles. The ethanol fuel will be produced and distributed by all major gasoline companies in Sweden. The production of the blend, 4–5% ethanol in gasoline from the first large-scale plant, will occur at the oil depots in Norrköping, Stockholm, and Södertälje. In the region near these depots, all unleaded 95-octane gasoline will contain ethanol. In 1998, Swedish gasoline consumption was 5.4 million m^3; a 5% ethanol blend means consumption of 270,000 m^3 ethanol annually, i.e., more than five times the capacity of the plant in Norrköping.

3.3.2 Australia

There is a plan under development in Australia that will enable Australia to produce 350 million liters of ethanol annually by 2010. The increase in production, equivalent to approximately 7% of the current petrol market, would total 1% of the liquid fuel market. The Agriculture Minister also intends to ensure that biofuels contribute to 2% of Australia's transportation fuel use by 2010. Under the new plan, a capital subsidy of A$0.16 for each liter of new or expanded biofuel production capacity constructed would be presented to investors. This is equivalent to a subsidy of about 16% on new plant costs. At least five new ethanol distilleries are expected to be set up under the new program, generating a possibility of 2300 construction and 1100 permanent jobs.

The Australian Environment Ministry recently announced that the federal government is trying to determine an appropriate level of ethanol in fuel and would not enforce mandatory levels of ethanol. The federal government has been asked by Australian sugarcane growers to create a plan that would develop a variable and sustainable ethanol industry throughout the country. A producers' working group would assist the government in the production of the plan. The sugarcane growers have also announced plans to launch an educational campaign that will focus on the benefits of ethanol. The cane growers hope to win the support of other farmers throughout the country. Manufacturers announced they would build a 60-million gallon/year ethanol plant in Dalby, Australia, making it the country's largest. The project is expected to be completed by the end of 2003. The plant will process sorghum, wheat, and maize into ethanol that will be blended with petrol sold through 40 or more of the company's service stations.

3.3.3 New Zealand

The Queensland Government has announced that its vehicle fleet will switch to petrol blended with ethanol. Although the switch was welcomed by the Queensland Conservation Council, concerns were expressed about the environmental risks and problems associated with sugarcane farming. The Council noted that they would work to ensure that sugarcane growing is environmentally sound. The Cane Growers' Association believes it is unlikely that the sugar industry will expand due to the government's decision. The Association will not encourage the planting of more cane, because it is likely that the ethanol will be made from the by-product molasses or that it will be sourced out of cereal crops.

3.3.4 South Africa

There are four major ethanol producers in South Africa. The largest ones produce up to 400 million liters/year of ethanol from coal. This capacity was developed during the 1950s to reduce South Africa's dependence on oil imports during the apartheid era. Synthetic ethanol production can top 400 million liters/year, but usually fluctuates with demand on the world market in general, and on the Brazilian market in particular. Since the mid-1990s, the efforts to introduce coal-derived ethanol into gasoline have repeatedly failed because of petroleum and automobile industry complaints about the low quality of the ethanol. Coal-derived alcohols are not pure ethanol. The original alcohol additive contained only 65% ethanol, which caused significant engine problems. In 1990, the quality issues seem to have been resolved through development of an 85% ethanol blend, and the coal-derived ethanol is used in South Africa as a 12% blend with gasoline. South Africa also produces ethanol from natural gas, about 140 million liters/year, and it is quite expensive. In addition, there are other plants that use molasses as a feedstock.

3.3.5 *Thailand*

In 2000, Thailand launched a program to mix 10% ethanol with gasoline. Ethanol would be produced from molasses, sugarcane, tapioca, and other agricultural products. The goal was the production of 730 million liters/year by 2002. One of the major sugar groups in Thailand announced that it would invest Bt800 in an ethanol plant in 2002. This group would apply for a license to produce ethanol as an alternative fuel for automobiles. On approval from the National Ethanol Development Committee, the building of the plant was scheduled to begin immediately and it was expected to take 12 to 18 months to complete. The group expects to use molasses, a by-product supplied by the group's sugar mills, to produce 160,000 liters of ethanol/day.

3.3.6 *Japan*

In order to reduce automobile emissions, Japan is considering introducing a policy that will support the use of blending ethanol with gasoline. The policy is a result of pressures to cut greenhouse gas emissions that lead to global warming. As the second largest consumer of gasoline in the world, Japan has no extra agricultural produce to use for fuel output. Thus, the use of biofuels could create a big export opportunity for ethanol-producing countries. The trading house of Mitsui & Co. is backing ethanol use in Japan. Mitsui estimates Japan's ethanol market at approximately 6 million kiloliters/year at a blending ratio of 10%. Due to an ample supply of low-cost gasoline produced by imported crude oil, the Japanese have yet to use ethanol as a fuel. Japan's interest in green energy increased with the 1997 Kyoto Protocol, which went into effect in 2002. The Kyoto Protocol aims to cut carbon dioxide emissions by 5.2% by 2012.

In order to implement mandatory use of ethanol in Japan, the Environment Ministry must win the support of the oil-refining industry as well as the energy and transportation arms of the government.

3.3.7 *Malawi*

Malawi has very favorable economic conditions for ethanol. Like Zimbabwe, Malawi was on the forefront of fuel alcohol development; Malawi has blended ethanol with its gasoline continuously since 1982, and has thereby eliminated lead. Because of high freight costs, the wholesale price of gasoline is about 56 cents/liter retail. Moreover, Malawi's molasses has a low value because the cost of shipping it to port for export typically exceeds the world market price. Malawi's Ethanol Company Ltd. produces about 10–12 million liters/year, providing a 15% blend for gasoline.

3.3.8 *China*

China is also interested in introducing an alcohol program. A pilot program was introduced in the Province of Jilin in 2001, and, in 2002, a Chinese delegation from the Province of Heilonjiand visited Brazil. The Chinese government is concerned about the increase in oil consumption in the country (around 7–7.5%/year between 2000 and 2005) and the alcohol–gasoline blend appears to be an interesting option due to job generation and to the potential for reducing the pollution in large Chinese cities. China finds itself short on fuel and producing only 70% of the nation's demand, and the country is also faced with a sagging economy for farmers. In order to fight both problems, China is considering a new program to launch its first ethanol plant. Despite the numerous projected benefits, the program is not expected to be implemented right away. Although ethanol is environmentally friendly, it is still expensive to produce in China and there are difficulties in transporting it. China hopes that the cost will come down in the next 5 years, making ethanol a viable option for the country.

Despite no laws mandating the use of greener fuels, China has continued to phase out petroleum products since 1998. China is encouraging the use of alternative fuels in their "Tenth Five-Year Plan," which includes trial fuel ethanol production (2001–2005).

3.3.9 India

India is one of the largest producers of sugarcane in the world (300 million tons of sugarcane, similar to Brazilian production) and has vast potential to produce ethanol fuel, which would significantly reduce air pollution and the import of petroleum and its products. The Indian distillery industry has an installed capacity of over 3 billion liters (Brazilian production is in average 13 billion liters of alcohol per year), but operates only at 50% capacity. Also, the Indian sugar industry currently faces an uncertain future, with high stock valuations, unattractive export balances, and narrow operating margins. There are also the issues related to the balance of payments due to oil imports. Therefore, there is a growing interest in ethanol as major business opportunity, mainly to produce anhydrous ethanol to be used in a 5% ethanol–gasoline blend.

The New Delhi government is currently reviewing the excise duty structure on ethanol. The government hopes to make ethanol more attractive to oil companies that blend ethanol with petrol. Government officials are currently concerned about India's sugar farmers, who have found it difficult to sell their crop. As the demand for ethanol-blended petrol increases, so will the demand for sugarcane increase. India produced 18.5 million tons of sugar in the past year. It is expected that, if a trial run proves successful, ethanol-blended petrol (5% ethanol) will be sold throughout the country. The government plans to free all market pricing of petrol.

4. FLEX-FUEL VEHICLES

Flex-fuel vehicles are vehicles that can operate with multiple fuels (or fuel blends). Such technology was created 1980 and there are around 2 million flex-fuel vehicles in the United States today. The main fuels used include gasoline and several alternative fuels, such as pure ethanol (already used in Brazil in automotive vehicles) and blends of ethanol and gasoline. Already in use in Brazil is an ethanol–gasoline; blend at a percentage of 20–26% ethanol. Another blend of ethanol–gasoline (E85), with 85% ethanol, is used in the United States; a blend of methanol–gasoline also used in the United States has 85% methanol. The methanol–gasoline blend has a limited potential for widespread use, because most automobile manufacturers do not build fuel systems compatible with methanol blends. Flex-fuel vehicles have a small processor placed inside the fuel system; the processor detects the fuel blend being used and automatically adjusts the ignition time and the mixture of air and fuel. The greatest advantage of the flex-fuel vehicles is that they can operate with regular gasoline when alternative fuels are not available or are not economically competitive.

Flex-fuel cars in the United States are built to utilize natural gas, pure gasoline, and gasoline blended with a small percentage of ethanol. In Brazil, flex-fuel motors have to be built to accept a much larger percentage of ethanol; the larger percentage of ethanol does negatively affect the life span of the fuel tank and other parts of the engine system, and the design requirements and shortened system life span add to the overall expense of the vehicles.

Acknowledgments

The authors acknowledge the contribution of specialists from the Brazilian Reference Center on Biomass (CENBIO), the University of São Paulo, and the Environmental Company of São Paulo State (CETESB).

SEE ALSO THE FOLLOWING ARTICLES

Biodiesel Fuels • Biomass for Renewable Energy and Fuels • Ethanol Fuel • Hybrid Electric Vehicles • Hydrogen, End Uses and Economics • Internal Combustion (Gasoline and Diesel) Engines • National Energy Policy: Brazil • Natural Gas Processing and Products • Solar Fuels and Materials • Transportation Fuel Alternatives for Highway Vehicles

Further Reading

Goldemberg, J. (2002). "The Brazilian Energy Initiative." World Summit on Sustainable Development, São Paulo, June 2002.

Macedo, I. (2000). O ciclo da cana-de-açúcar e reduções adicionais nas emissões de CO_2 através do uso como combustível da palha da cana. Inventário de Emissões de Gases Efeito Estufa, Report to the International Panel on Climate Change (IPCC), Ministério de Ciência e Tecnologia, March 2000, Brazil.

Moreira, J. R., and Goldemberg, J. (1997). The Alcohol Program, Ministério de Ciência e Tecnologia, June 1997, Brazil.

Pimentel, D. (1991). Ethanol fuels: Energy security, economics and the environment. *J. Agricult. Environ. Ethics* 4, 1–13.

Pioneira (ed.). (2002). *Energia e Mercados*, March 16, 2002.

São Paulo Sugarcane Agroindustry Union (UNICA). (2001). *Informação Unica* 5(43), October 2001.

Unnasch, S. (2000). Fuel cycle energy conversion efficiency. Report prepared for the California Energy Commission, June 2000. California Energy Commission, Sacramento.

Winrock International India (WII). (2001). "Ethanol, Sustainable Fuel for the Transport Sector," Vol. 3, October. WII, New Delhi.

Aluminum Production and Energy

H.-G. SCHWARZ
University of Erlangen
Erlangen, Germany

1. Aluminum Process Chain and Production Figures
2. Specific Final Energy Consumption
3. Specific Primary Energy Consumption
4. Energy as Cost Factor
5. Summary

Glossary

alumina (aluminum oxide) Made from bauxite via the Bayer process. Alumina is smelted using the Hall–Héroult process to produce primary aluminum.

aluminum Second most abundant metallic element in the earth's crust after silicon. It is a comparatively new industrial metal that has been produced in commercial quantities for slightly more than 100 years. It is made from alumina via the Hall–Héroult process (primary aluminum). In addition, aluminum is produced from secondary raw materials—that is, from old and/or new scrap (secondary aluminum). Aluminum is used in a wide range of applications, such as automobiles, airplanes, buildings tools, and packaging.

aluminum alloy Aluminum can be alloyed with other materials to make an array of metals with different properties. The main alloying ingredients are silicon, zinc, copper, and magnesium. Aluminum alloys are classified into two groups: cast aluminum and wrought aluminum. Cast aluminum cannot be reshaped plastically and must therefore be cast. The content of alloying ingredients is relatively high. In contrast, wrought aluminum can be reshaped plastically. The content of alloying ingredients is low for wrought aluminum.

bauxite Naturally occurring, heterogeneous material composed primarily of one or more aluminum hydroxide minerals plus various mixtures of silica, iron oxide, titania, and other impurities in trace amounts. Bauxite is converted to alumina via the Bayer process. Bauxite occurs mainly in tropical and subtropical areas, such as Africa, the West Indies, South America, and Australia.

Bayer process Wet chemical caustic leach process to convert bauxite to alumina.

Hall–Héroult process Process of electrolytic reduction in a molten bath of natural and synthetic cryolite to convert alumina to aluminum. The process is energy intensive. The specific electricity consumption is 13–17 MWh per ton of primary aluminum.

recycling share and quota Recycling share is the share of secondary aluminum production of the total aluminum production of 1 year (aluminum, secondary aluminum). In contrast, recycling quota is the share of the actual recycled secondary material of the theoretical available secondary material. It describes how much aluminum is recovered at the end of the use phase of products containing aluminum. The calculation of the recycling quota for one product is based on its aluminum content, lifetime, the consumption for one reference year, and the quantity of recycled aluminum. The total recycling quota of a metal is a weighted average of the quotas of the different products containing the metal. The lifetime is very different for the different aluminum-consuming sectors. The range is 6 months for containers and packaging to approximately 20–40 years for building and construction materials. The average lifetime is approximately 15 years. Only an average lifetime of less than 1 year leads to an identicalness of recycling share and total recycling quota.

secondary aluminum (recycling aluminum) Aluminum produced from secondary raw materials (i.e., from old and/or new scrap). Depending on the type of scrap and the desired product quality, different types of furnaces for melting aluminum scrap are used. Scrap for the production of casting alloys is commonly melted in rotary furnaces under a layer of liquid melting salt. Producers of wrought alloys prefer open-hearth furnaces in varying designs (aluminum alloys).

Aluminum is a comparatively new industrial metal that has been produced in commercial quantities for slightly more than 100 years. A necessary precondition was the (independent) discovery of molten-salt electrolysis by American Charles Martin Hall and Frenchman Paul Toussaint Héroult in the late 19th century. World production in 1900 totaled 7339 tons and took place in only four countries. The United States was the leading producer, with a production share of 44%, followed by Switzerland (34%), France (14%), and the United Kingdom (8%). One hundred years later, primary aluminum production has multiplied: In 2000, approximately 24.5 million tons of primary aluminum was produced in 43 countries. The United States remained the world's largest producer and accounted for 15% of the world's production, followed by the former Soviet Union (15%), China (12%), and Canada (10%). Since the inception of the aluminum industry, metal has been recovered from scrap. After World War II, so-called secondary production expanded, especially since the 1960s. In 2000, approximately 8.5 million tons of secondary aluminum was produced. Secondary production accounted for one-fourth of the total global aluminum consumption of 33 million tons.

Aluminum is commonly seen as coagulated energy. In fact, specific energy requirements for primary production of aluminum are high. In the mid- to late 1990s, the specific worldwide average electricity consumption for electrolysis was 15.5 MWh per ton of primary aluminum. The electricity supply for electrolysis represented more than three-fourths of total specific primary energy consumption of approximately 200 GJ per ton of primary aluminum. The specific primary energy consumption of primary aluminum production is at least eight times higher than the consumption caused by the primary production of iron and at least seven times higher than that for the primary production of copper.

The absolute energy consumption figures are also impressive. In 2000, the global production of 24.5 million tons of primary aluminum required 380 TWh of electric power for electrolysis, which represented 2.8% of the total global power generation. The primary energy consumption of 4.90 EJ for global primary aluminum production accounted for 1.2% of the total worldwide primary energy consumption.

This article focuses on the remarkable role of the production factor "energy" in aluminum production. First, the process chain of aluminum production is discussed and the production figures for the Year 2000 are given. Second, the determinants of specific energy consumption and primary energy consumption are presented. Finally, the importance of cost factor energy is analyzed.

1. ALUMINUM PROCESS CHAIN AND PRODUCTION FIGURES

1.1 Survey of Process Chain

The process of primary aluminum production takes place in three stages (Fig. 1). Initially, bauxite is extracted by mining. Then, it is converted to alumina via the Bayer process. According to the worldwide average, 2.3 tons of bauxite is necessary to produce 1 ton of alumina. Primary aluminum is produced via the Hall–Héroult process, a process of electrolytic reduction in a molten bath of natural and synthetic cryolite. Two tons of alumina is necessary to produce 1 ton of primary aluminum.

The primary aluminum produced is alloyed in the cast house; here, cast aluminum and wrought aluminum are produced. The content of alloying ingredients is relatively high for cast aluminum. Cast aluminum cannot be reshaped plastically and must therefore be cast. In contrast, wrought aluminum can be reshaped plastically and its content of alloying ingredients is low. The alloys are converted to castings and semis (sheet plates and profiles). These are used for the production of final products. Figure 2 provides an overview of the consumption of aluminum by domestic end users in the United States, the world's most important aluminum consumer.

Old scrap and new scrap are the initial products of secondary aluminum production. Old scrap consists of

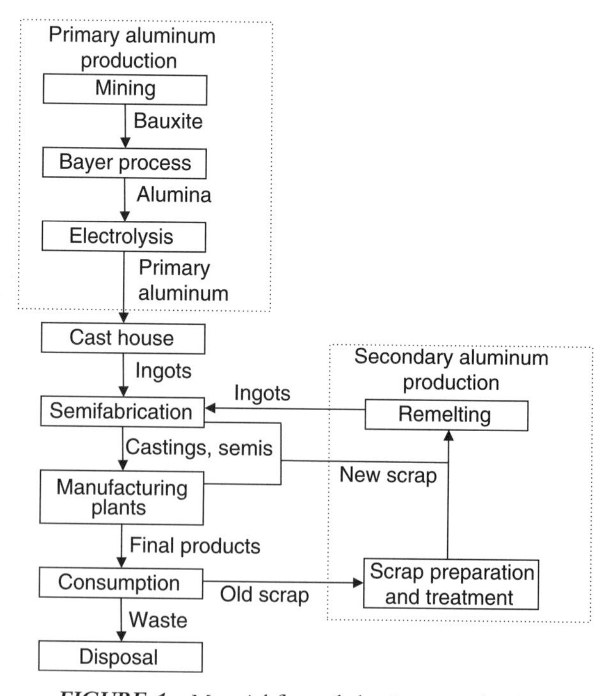

FIGURE 1 Material flow of aluminum production.

products made from aluminum that at the end of their service lives enter the recycling chain as secondary raw materials. New scrap (production scrap) arises during the processing of semifabricated products, mostly clean and sorted, in the form of processing residues (e.g., turnings and residues from punching or stamping). Secondary aluminum production results in predominantly cast aluminum but, increasingly, also wrought aluminum for further processing.

Currently, the alloy elements (e.g., silicon, zinc, copper, and magnesium) within the aluminum scrap cannot be cost-effectively removed. Therefore, in contrast to primary production, the composition of the initial product (scrap) determines to a great extent the possible range of applications. Nevertheless, secondary aluminum is a substitute for primary aluminum, at least as long as cast aluminum, which is predominantly produced, can be sold profitably on the market and therefore the input of secondary aluminum can replace the input of primary aluminum. Thus, it seems reasonable to regard an increase in the recycling share as an important instrument to reduce specific energy consumption in aluminum production.

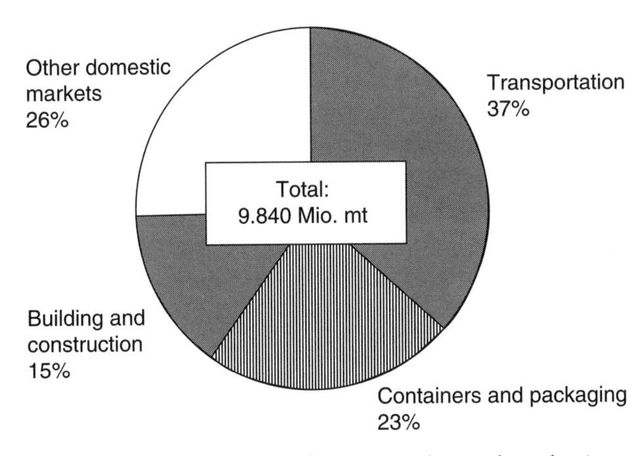

FIGURE 2 Consumption of primary and secondary aluminum by domestic end users in the United States in 2000. Mio mt., million metric tons. From U.S. Geological Survey (2000). "U.S. Geological Survey Minerals Yearbook." U.S. Geological Survey, Reston, VA.

1.2 Primary Aluminum Production

1.2.1 Mining

Bauxite is the most common aluminum ore. Approximately 98% of primary aluminum production is based on bauxite. In the former Soviet Union, aluinite and kaolinite are also used in small proportions for aluminum production. Bauxite is a naturally occurring, heterogeneous material composed primarily of one or more aluminum hydroxide minerals plus various mixtures of silica, iron oxide, titania, and other impurities in trace amounts. The principal aluminum hydroxide minerals found in varying proportions within bauxite are gibbsite (trihydrated) and boehmite and diaspore (both monohydrated). The content of aluminum oxide is 30–60%. Bauxite is generally extracted by open cast mining from strata, typically 4–6 m thick under a shallow covering of topsoil and vegetation (Table I).

Bauxite occurs mainly in tropical and subtropical areas. In 2000, Australia was the leading worldwide bauxite producer with a production share of 39%, followed by Guinea (13%), Brazil (10%), and Jamaica (8%) (Table II). Bauxite reserves are estimated to be 23,000 million tons. They are concentrated in areas that currently dominate production: Australia (share of 24%), Guinea (24%), Brazil (12%), and Jamaica

TABLE I

Characteristics of Important Bauxite Deposits[a]

Bauxite deposits	Reserves (million metric tons)	Content of Al_2O_3 (%)	Average ore thickness (m)	Average surface thickness (m)
Weipa (Australia)	230 reliable + 150 likely	55	2.4	0.8
Gove (Australia)	150 reliable	50	3.5	0.8
Boddington (Australia)	330 reliable + 140 likely	31	6	0.9
Boké (Guinea)	300 reliable	56	7	2
Trombetas (Brazil)	780 reliable	49	6	8
Los Pijiguaos (Venezuela)	177 reliable + 500 likely	49	5	0.5
Linden/Berbice (Guyana)	500 reliable + likely	55–61	3–10	20–60
Dry Harbour Mountain (Jamaica)	110 reliable	48	8	0.3

[a] Adapted with permission from Mori and Adelhardt (2000).

TABLE II

Geography of Aluminum Production and Consumption in 2000[a]

	Primary production (%)			Secondary aluminum production (%)	Aluminum total consumption (%)
	Bauxite	Alumina	Primary aluminium		
Europe	9	22	33	31	>21
France	0	1	2	3	3
Germany	0	2	3	7	8
Italy	0	2	3	7	8
Norway	0	0	4	3	2
United Kingdom	0	0	1	3	3
Former USSR	6	11	15	0	n.a.[b]
Asia	13	11	21	17	>15
China	6	8	12	0	n.a.
India	5	0	3	0	n.a.
Japan	0	2	0	14	12
Korea	0	0	0	1	3
Africa	13	1	5	1	1
Guinea	13	1	0	0	0
South Africa	0	0	3	1	1
America	26	33	34	50	37
Brazil	10	7	5	3	2
Canada	0	2	10	1	3
Jamaica	8	7	0	0	0
Suriname	3	4	0	0	0
United States	0	9	15	41	29
Australia and Oceania	39	31	9	2	2
Australia	39	31	7	1	1
World (1000 Metric tons)	137,547	50,916	24,485	8490	32,401

[a] Adapted with permission from the World Bureau of Metal Statistics (2001), "Metal Statistics." World Bureau of Metal Statistics, Ware, UK.

[b] n.a., not applicable.

(9%). With a global production of 135 million tons in 2000, the static range of coverage is approximately 170 years. Only 4% of mined bauxite is not converted to alumina, and this is used, among other things, in the chemical and cement industry.

1.2.2 Alumina Production

Due to high specific transport costs of bauxite, alumina production often takes place near bauxite mines. In 2000, Australia manufactured 31% of the 50.9 million tons of alumina produced worldwide, follow by the United States (9%), China (8%), and Jamaica (7%) (Table II). Approximately 10% of the alumina produced is not converted to aluminum and is used, among others things, in the ceramic industry.

Approximately 98% of the alumina produced worldwide is manufactured via the Bayer process, which consists of digestion and calcination. Initially, the bauxite is washed and ground. During digestion, the bauxite is dissolved in caustic soda (sodium hydroxide) at high pressure (40–220 bar) and at high temperature (120–300°C). The resulting liquor contains a solution of sodium aluminates and undissolved bauxite residues containing iron, silicon, and titanium. These residues sink gradually to the bottom of the tank and are removed. They are known as red mud. The clear sodium aluminate solution is pumped into a huge tank called a precipitator. Aluminum hydroxide is added to seed the precipitation of aluminum hydroxide as the liquor cools. The aluminum hydroxide sinks to the bottom of the tank and is removed. It is then passed through a rotary or fluidized calciner at up to 1250°C to drive off the chemically combined water. The result is alumina (aluminum oxide), a white powder.

1.2.3 Electrolysis

The basis for all primary aluminum smelting plants is the Hall–Héroult process (Table II). Alumina is dissolved in an electrolytic bath of molten cryolithe (sodium aluminum fluoride) at 960°C inside a large carbon- or graphite-lined steel container known as a pot. An electric current is passed through the electrolyte at low voltage (3.9–4.7 V) and at very high amperage (100–320 kA). The electric current flows between a carbon anode (positive), made of petroleum coke and pitch, and a cathode (negative), formed by the thick carbon or graphite lining of the pot. The anodes are used up during the process when they react with the oxygen from the alumina. Molten aluminum is deposited at the bottom of the pot and is siphoned off periodically (Fig. 3).

1.3 Secondary Aluminum Production

Secondary aluminum production is concentrated in Western industrial countries, the main aluminum consumers. In 2000, the United States was the world's leading producer with a production share of 41%, followed by Japan (14%), Italy (7%), and Germany (7%) (Table II). Secondary production consists of two process steps: scrap preparation, mixing, and charging, on the one hand, and the processes within the remelter/refiner (remelting, refining, and salt slag preparation), on the other hand.

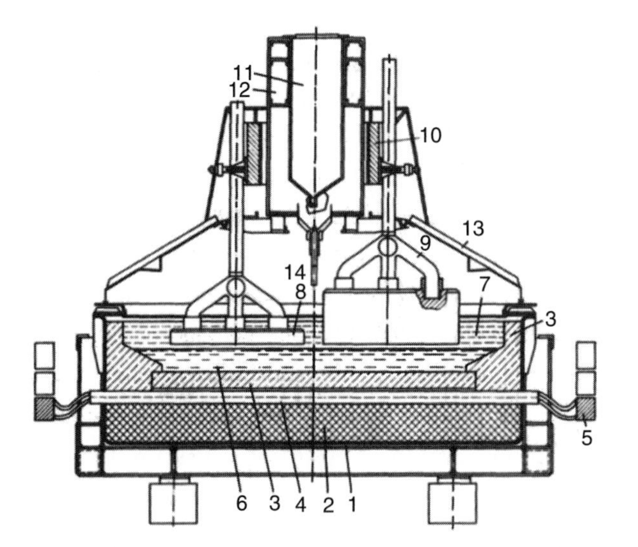

FIGURE 3 Profile of a modern electrolytic cell. 1, steel shell; 2, insulation; 3, cathode (carbon in base and sides; 4, iron cathode bar; 5, cathodic current entry; 6, liquid aluminum; 7, electrolyte; 8, prebaked anodes (carbon); 9, steel nipple; 10, anode beam; 11, aluminum oxide hopper; 12, gas suction unit; 13, detachable covering; 14, crust crusher. From Kammer, C. (1999). "Aluminium Handbook. Volume 1: Fundamentals and Materials." Aluminium-Verlag, Düsseldorf. Reproduced with permission.

1.3.1 Scrap Preparation, Mixing, and Charging

Scraps recovered are treated according to their quality and characteristics. Common treatment processes are sorting, cutting, baling, or shredding. Free iron is removed by magnetic separators. The different (prepared) scrap types are selected and mixed in such a way that their chemical composition is as close as possible to that of the required alloy.

1.3.2 Melting, Refining, and Salt Slag Preparation

Depending on the type of scrap and the desired product quality, different types of furnaces for melting aluminum scrap are used. Scrap for the production of casting alloys is commonly melted in rotary furnaces under a layer of liquid melting salt. A company producing casting alloys from old and new scrap is commonly called a refiner. Producers of wrought alloys prefer open hearth furnaces in varying designs. These furnaces are normally used without salt. Wrought aluminum from mainly clean and sorted wrought alloy scrap is produced in a remelter.

The alloy production in rotary furnaces is followed by a refining process. The molten alloy is fed into a holding furnace (converter) and purified through the addition of refining agents.

After the melting process, the liquid melting salt used in rotary furnaces is removed as salt slag. In the past, the salt slag was land filled. Today, the salt slag is prepared as a rule. The aluminum and the salt within the salt slag are recovered.

2. SPECIFIC FINAL ENERGY CONSUMPTION

2.1 Survey

Table III shows the specific final energy consumption of world aluminum production in the mid- to late 1990s. The electricity consumption for primary production is almost 35 times higher than the consumption for secondary production. The fuel consumption is more than 2.5 times higher. Electrolysis accounts for the main portion (96%) of electricity requirements,

TABLE III

Specific Final Energy Consumption of World Aluminum Production, Mid- to Late 1990s[a]

	Primary production		Total production (75% PA + 25% SA)[b]		Secondary production		
	Electricity (MWh/t_{pa})	Fuels (GJ/t_{pa})	Electricity (MWh/t_{ai})	Fuels (GJ/t_{ai})		Electricity (MWh/t_{sa})	Fuels (GJ/t_{sa})
Transport	<0.1	1.6				0.0	1.0
Processes							
Mining	<0.1	1.0			Scrap preparation and treatment	0.05	2.5
Refining	0.5	27.1			Remelting plant	0.42	8.4
Smelting	15.5	0.0			Remelting	0.32	4.5
					Salt slag treatment	0.09	2.1
					Refining	<0.01	1.8
Main physical inputs							
Caustic soda production	0.1	0.2					
Anode product	<0.1	0.9					
Total	16.2	30.8	12.3	26.1		0.47	11.9

[a] Adapted with permission from Schwarz (2000) and Krone (2000).
[b] PA, primary aluminum; SA, secondary aluminum.

and alumina production accounts for the main portion of fuel consumption (88%) of primary production. The processes within the refiner/remelter dominate in terms of electricity (89%) and fuel consumption (71%) for secondary production.

2.2 Recycling Share

In addition to the specific final energy consumption for the different process steps, the share of secondary production in total production (recycling share) is of great importance for the specific final energy consumption for aluminum production. This is due to the very different final energy requirements for primary and secondary production. Table IV shows the development of total, primary, and secondary production as well as the development of the recycling share since 1950. Aluminum production expanded at high rates in the 1950s and 1960s. Since the 1970s, there has been a noticeable slowdown in growth rates. The recycling share began to increase again in the 1960s and 1970s. In 2000, the recycling share was 26%. This percentage seems to be rather low, but one has to take into consideration the lifetime of aluminum consumer goods—an estimated 15 years. Compared to a total production of 19.9 million tons in 1985, secondary production of 8.5 million tons in 2000 represents a recycling quota of approximately 43%.

(Optimistic) projections estimate a further expansion of aluminum production. In 2030, a total production of 50 million tons is expected, with a recycling share of 44–48% (Table IV). There are two reasons for the projected increase in recycling share: the predicted growth in the recycling quota and the expected continued slowdown in the growth rates of total production. The presented scenarios assume a future growth rate of 1.4% per year compared with 2.6% in the 1980s and 1990s. The decline in the growth rate automatically results in an increase in the recycling share.

If a fixed level of final energy consumption is assumed, a duplication of the recycling share from approximately 25 to 50% would lead to a decrease in specific electricity consumption of 32% and a decrease in fuel consumption of 18%.

TABLE IV
Primary, Secondary, and Total Aluminum Production, 1950–2030(p)[a]

	1950	1960	1970	1980	1990	2000	2030p
Primary production (million tons)	1.5	4.5	10.3	16.1	19.4	24.5	26–28
Secondary production (million tons)	0.4	0.8	2.3	3.7	6.1	8.4	22–24
Total (million tons)	1.9	5.3	12.6	19.7	25.5	32.9	50
Recycling share (%)	22	14	18	19	24	26	44–48
Growth rates (% per year)		1951–1960	1961–1970	1971–1980	1981–1990	1991–2000	2001–2030p
Primary production		11.7	8.5	4.6	1.9	2.4	0.2–0.4
Secondary production		5.8	11.9	4.6	5.3	3.3	3.2–3.5
Total		10.6	9.0	4.6	2.6	2.6	1.4

[a]Adapted with permission from the World Bureau of Metal Statistics, "Metal Statistics" (various volumes), World Bureau of Metal Statistics, Ware, UK; and Altenpohl and Paschen (2001).

2.3 Specific Final Energy Consumption in Primary Production

2.3.1 *Transport, Mining, and Alumina Production*

The final energy consumption for transport is of little importance. Transport accounts for approximately 5% of the total fuel consumption for primary production. The importance of final energy consumption for bauxite extraction is also minimal. Three percent of the total fuel consumption for primary production is due to mining. Diesel is used as fuel for trucks, scrappers, and excavators. The belt conveyers are responsible for the very low electricity consumption.

Substantial amounts of thermal energy are required for alumina production—for digestion as well as for calcinations. Autoclaves, in almost equal shares of low and high temperature, were predominantly (98%) used for digestion of bauxite in the mid- to late 1990s. Tube reactor use was very low (2%). High-temperature autoclaves and the tube reactor are able to digest all types of bauxite, whereas low-temperature autoclaves can only digest trihydrated bauxite economically. The digestion of monohydrated bauxite would cause a great loss of alumina. On average, the tube reactor offers the best value for energy requirements (and consumption of caustic soda), followed by low-temperature autoclaves. Autoclaves and tube reactors can be combined with a rotary kiln or fluidized bed for calcinations. A fluidized bed is considered the superior technology with higher energy efficiencies. Therefore, rotary kilns are increasingly being replaced by fluidized beds. In the late 1990s, approximately three-fourths of worldwide capacity was based on the fluidized bed.

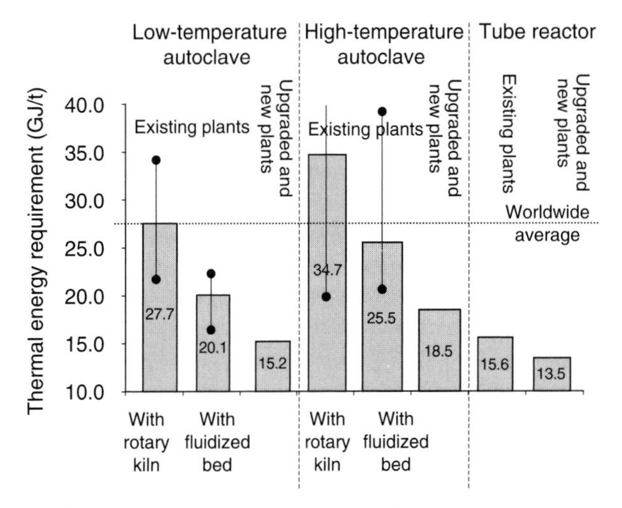

FIGURE 4 Average worldwide specific thermal energy consumption of existing alumina-producing plants (inclusive range for 16 world regions) and for upgrading and new plants (GJ/t primary aluminum). From Schwarz (2000). Reproduced with permission.

Figure 4 provides a review of thermal energy requirements for the different technologies for existing plants on a worldwide average and for upgraded and new plants. Upgraded and new plants require approximately 50% of the thermal energy necessary for existing plants. The upgrading of existing plants and the building of new plants mean that a noticeable decline in specific final energy requirement can be estimated for the next few years. Estimates indicate a decrease in specific thermal energy consumption of approximately 30% in the 15 years from 1995 to 2010.

2.3.2 *Electrolysis*

The Hall–Héroult process has been the basis for all primary smelting plants for more than 100 years. Indeed, cell design has been improved since the beginning. Modern electrolysis cells work at very high amperage. They use prebaked anodes and are point fed. In addition, process computers are used. Table V shows typical parameters of modern electrolysis cells for the years 1895, 1950, and 2000. At least in the next few years, it is not expected that greater process changes (e.g., the use of inert anodes) will prove to be economically beneficial.

Usually, existing electrolysis cells are characterized in terms of the kinds of anodes used (Soederberg and prebaked technology) and the method of feeding alumina to the smelter (side-worked, center-worked, and point-feeder technology). Soederberg technology uses self-baking anodes. Liquid electrode mass in a steel tank is put into the electrolytic bath. Prebaked technology uses prebaked anodes in which the anodes are baked in a separate furnace. The main advantage of prebaked anodes is their homogeneous structure resulting from better conditions during baking. Due to the homogeneous structure, prebaked anodes have lower electrical resistance and lower specific consumption of anodes in comparison to self-baking anodes.

Side-worked means that the alumina is replenished from the side of the furnace. The crust over the electrolyte is opened by hammer crushers approximately every 3 h until hundreds of kilograms of alumina have been fed into the furnace. Center-worked plants open the crust between the anode lines with beam crushers. The intervals of feeding new alumina can be less than 1 h. Point feeding is the most modern technology. The voltage of the cell is monitored continuously. The measured values allow assessment of how much additional alumina is necessary. The crust is opened by a ram and alumina is added automatically to the smelting pot. This operation is repeated every few minutes. Only a few kilograms of alumina is put into the pot each time to ensure optimal concentration of alumina.

In the mid- to late 1990s, slightly more than 30% of smelting capacity was based on (mostly side-worked) Soederberg technology. The modern point-feeder prebaked (PFPB) technology accounted for more than 40% of smelting capacity. The older center-worked prebaked and side-worked prebaked technologies represented the remaining 30% of smelting capacity. Figure 5 shows the electricity consumption for the different technologies for existing plants as a worldwide average and for upgraded and new plants. The electricity consumption for new plants is only 14% lower than the average consumption for existing plants. Due to the upgrading of old plants with PFPB and the building of new PFPB plants, a moderate decline in specific electricity consumption of electrolysis can be expected. A 10% decline is estimated to occur from 1995 to 2010.

TABLE V

Typical Parameters of Aluminum Reduction Cells—1895, 1950, and 2000[a]

Parameter	1895[b]	1950	2000
Current rating (kA)	9	50–60	300–325
Aluminum production (kg Al/day)	59	385	2475
Unit energy (DC kWh/ kg Al)	31	18.5–19.0	12.9–13.5
Potroom worker hours (per ton of Al)	n.a.[c]	5–8	1.7
Interval of Al_2O_3 additions (min)	n.a.	80–240	0.7–1.5

[a]Adapted with permission from Morel (1991) and Æye *et al.* (1999).

[b]Data for a plant of the Pittsburgh Reduction Company (Niagara Falls).

[c]n.a., not applicable.

2.4 Specific Final Energy Consumption in Secondary Production

2.4.1 *Transport and Scrap Preparation*

Transport accounts for approximately 8% of the total fuel consumption for secondary production due to decentralized scrap recovery and the resulting transport distances. Scrap recovery represents 21% of total

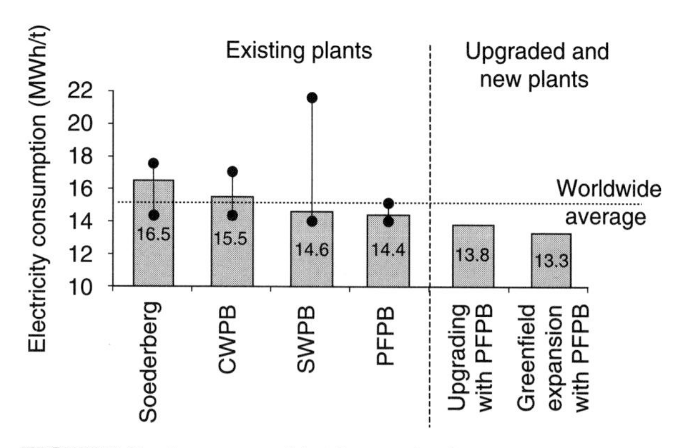

FIGURE 5 Average worldwide specific electricity consumption of existing smelters (inclusive range for 16 world regions) and for upgrading and new plants (MWh/t primary aluminum). From Schwarz (2000). Reproduced with permission.

TABLE VI

***Final Energy Consumption of Preparation and Treatment of Different Important Scrap Types in Germany, Mid- to Late 1990s*[a]**

Scrap	Processes	Thermal energy (GJ/t_{SA})	Electric energy (MWh/t_{SA})
Old cars	Disassembly, shredder, sink-float-separation	—	0.13
Packaging	Sorting, with part pyrolysis, mechanical treatment	0.097	0.07
Building and mixed scrap	Shredder, with part DesOx-granulating, sink-float-separation	—	0.12
Electrical	Shredder, with part sink-float-separation	—	0.18

[a] Adapted with permission from Wolf (2000), "Untersuchungen zur Bereitstellung von Rohstoffen für die Erzeugung von Sekundäraluminium in Deutschland. ["Studies on the Preparation of Raw Materials for Secondary Aluminum Production in Germany"]. Shaker, Aachen, Germany.

fuel consumption and 11% of total electricity consumption for secondary production. Different treatment processes are used depending on the initial products (scrap types). These treatment processes have very different energy requirements. Table VI presents the specific final energy consumption of scrap preparation for different scrap types in Germany.

2.4.2 Melting, Refining, and Salt Slag Preparation

Melting is responsible for 38% of the overall specific fuel consumption and 69% of overall specific electricity consumption for secondary consumption. Additionally, the requirements for refining (15 and 1%, respectively) and salt slag preparation (18 and 19%, respectively) have to be considered.

These data are only valid as averages. As previously mentioned, depending on the type of scrap and the desired product quality, different types of furnaces for melting aluminum scrap are used. The selection of the most appropriate melting process is determined by the metal content of the scarp (oxide content), type and content of impurity (annealing loss), geometry of the scrap, frequency of change in alloy composition, and operating conditions (Fig. 6).

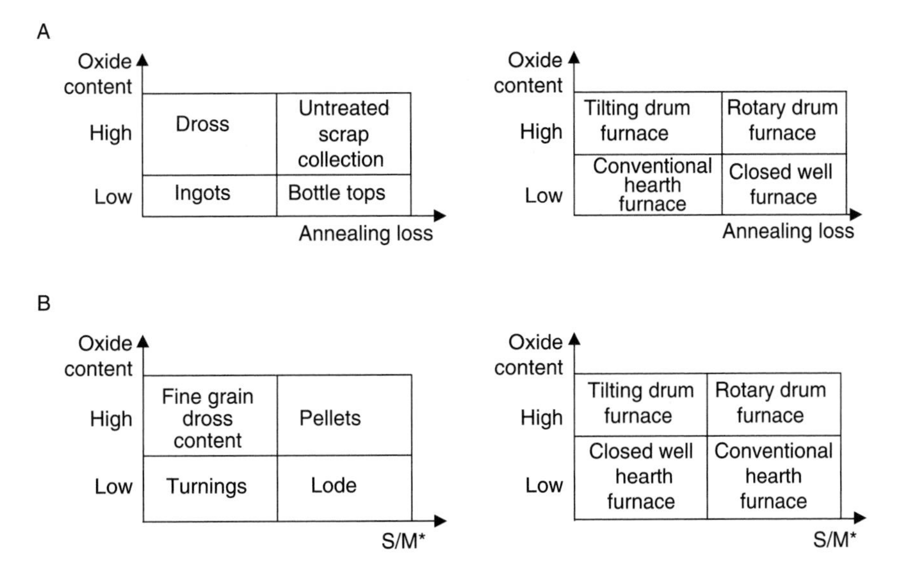

FIGURE 6 Classification of scrap types and furnace selection criteria by (A) oxide content and annealing loss and (B) oxide content and ratio of surface area to intermediate part size. *Surface area: mass of scrap pieces. From Boin, U., Linsmeyer, T., Neubacher, F., and Winter, B. (2000). "Stand der Technik in der Sekundäraluminiumerzeugung im Hinblick auf die IPPC-Richtlinie" ["Best Available Technology in Secondary Aluminum Production with Regard to the IPCC Guideline"]. Umweltbundesamt, Wien, Austria. Reproduced with permission.

TABLE VII

Typical Final Energy Consumption of Modern Secondary Aluminum Furnaces[a]

Furnace type	Scrap	Typical yield of metal (%)	Assumed yield of metal (%)	Electricity consumption (MWh/t_{SA})	Thermal energy consumption (GJ/t_{SA})
Induction	Pigs, ingots, turnings	95–99	98	0.70–0.93	—
Closed-well	Clean scrap, organically contaminated scrap	88–95	92	—	2.52–4.30
Rotary drum (static)	Turnings, pellets, shredder scrap, new scrap	75–92	74	—	4.41–4.70
Rotary drum (static + O_2 burner)	Turnings, pellets, shredder scrap, new scrap	75–92	74	—	1.87–2.05
Rotary drum (tilting)	Dross, pellets	50–80	78	—	2.67

[a] Adapted with permission from Boin *et al.* (2000). "Stand der Technik in der Sekundäraluminiumerzeugung im Hinblick auf die IPPC-Richtlinie" ["Best Available Technology in Secondary Aluminum Production with Regard to the IPCC-Guideline"]. Umweltbundesamt, Wien, Austria.

The appropriate melt aggregate for organically contaminated scrap with a low oxide content is the closed-well hearth furnace. The higher the fine and/or oxide content, the more appropriate it becomes to use a rotary drum furnace. In secondary melting plants, induction furnaces are only used in individual cases due to the necessity for clean, oxide-free scrap. Hearth and rotary drum furnaces are heated by fuel oil and natural gas. Only induction furnaces use electricity for melting. Table VII gives an indication of the very different final energy consumption for the most commonly used melting furnaces.

3. *SPECIFIC PRIMARY ENERGY CONSUMPTION*

3.1 Survey

Table VIII shows the specific primary energy consumption for global aluminum production in the mid- to late 1990s. If, as is usually the case, the regional energy carrier mix of electricity production is used for all processes, including electrolysis, to convert electricity consumption into primary energy consumption, the specific primary energy consumption of primary production is more than 11 times higher than the requirements for secondary production. If the contract mix is used for electrolysis, the consumption of primary production is nine times higher. Electrolysis accounts for the main portion (80 and 75%, respectively) of primary energy consumption for primary production, the processes within the refiner/remelter dominate (75%) in terms of the primary energy consumption for secondary production.

3.2 Determinants, Especially the Overall Efficiency of Electricity Production

The overall efficiency of both fuel supply and electricity production must be determined to convert final energy consumption into primary energy consumption. In our case (Table VIII), the same overall efficiencies of fuel supply (diesel, 88%; oil, 90%; natural gas, 90%; and coal, 98%) are assumed worldwide. The overall efficiency of electricity production depends on the efficiencies of power plants, the mix of energy carriers for electricity production, and the distribution losses. The allocation of the primary energy carriers to electricity production is usually determined by the regional mix of energy carriers. The regional mix is based on the average shares of electricity production of energy carriers within a region. In our case, the overall efficiencies of electricity production for 16 world regions were calculated. The resulting world overall efficiency of electricity production is different for the separate process stages (e.g., mining and alumina production) and depends on the associated production shares of the regions, which are used as weighted factors.

Additionally, the contract mix of energy carriers is used for the calculation of the primary energy consumption for electrolysis. The contract mix reflects the mix of energy carriers according to the

TABLE VIII

Specific Primary Energy Consumption of World Aluminum Production, Mid- to Late 1990s[a]

Primary production (GJ/t_{PA})		Total production (75% PA + 25% SA)[b](GJ/t_{ai})	Secondary production (GJ/t_{SA})	
Transport	2.0			1.1
Processes				
Mining	1.3		Scrap preparation and treatment	3.3
Refining	34.2		Remelting plant	13.3
Smelting	159.8[c] (120.1)[d]		Remelting	8.1
			Salt slag treatment	3.2
			Refining	2.0
Main physical inputs				
Caustic soda production	2.2			
Anode product	1.4			
Total	200.9[c] (161.1)[d]	155.1[c](125.3)[d]		17.7

[a] Adapted with permission from Schwarz (2000) and Krone (2000).
[b] PA, primary aluminum; SA, secondary aluminum.
[c] Regional energy carrier mix for electrolysis.
[d] Contract energy carrier mix for electrolysis.

contracts of the smelters with utilities. Whereas the regional mix provides a purely statistical allocation of smelters to power-generating units, the contract mix permits a more causal allocation. The higher the electricity demand of a smelter, the more electricity has to be supplied by the contractual generating units. Despite this advantage, the calculations based on the contract mix for electrolysis are seen merely as additional information (and therefore the data are given in brackets; see Table VIII). The reason is simple; for a consistent calculation of primary energy consumption for total aluminum production based on contract mix, it would be necessary to assess the contract mix of energy carriers for all production stages in all regions. This is simply not possible because of a lack of data.

Table IX shows the share of hydropower for the 16 world regions based on the regional mix and on the contract mix for electrolysis. In most regions, a larger share of electricity from hydropower is used for the contract mix. This is not surprising because low-price electricity from hydropower is one of the deciding factors for the siting of electrolysis. The larger share of hydropower leads mostly to a higher overall efficiency of electricity production because of the relatively high efficiency of hydroelectric power plants (80% is assumed). The efficiencies of power generation based on the other energy carriers are much lower and in the mid- to late 1990s were quite different for the individual regions (the following are assumed: natural gas, 22–33%; oil, 23–41%; coal, 24–36%; and nuclear power plants, 33%).

For most regions, the lower overall efficiency of electricity production of the regional mix leads to a higher primary energy consumption for electrolysis (Table IX). This is also true on a worldwide average. The specific primary energy consumption for electrolysis based on a regional mix is 160 GJ per ton of primary aluminum, with an overall efficiency of 35% and a share of hydropower of 29% of total electricity production. Based on contract mix, the data are 120 GJ per ton of primary aluminum, 46% efficiency, and 60% share of hydropower.

3.3 Future Primary Energy Consumption

The specific primary energy consumption of world aluminum production should decrease noticeable in the future. One main reason is the expected increase in the recycling share. An increase in the recycling share from 25 to 50% as projected for 2030 would lead to a decrease of 30% in specific primary consumption. In addition, there should be a decrease in final energy consumption for primary and secondary production and an improvement in overall efficiency of electricity production. Estimates for electrolysis, the deciding energy-consuming process, present an interesting picture: A 26% decrease in

TABLE IX

Regional Primary Energy Consumption for Electrolysis and Regional Overall Efficiency and Share of Hydropower for Electricity Production[a]

	Regional mix			Contract mix		
	Primary energy consumption electrolysis (GJ/t_{PA})	Overall efficiency (%)	Share of hydropower (%)	Primary energy consumption electrolysis (GJ/t_{PA})	Overall efficiency (%)	Share of hydropower (%)
Europe						
Germany	175.0	32	6	173.7	32	8
EU without Germany	142.0	36	17	125.6	41	29
Rest of Western Europe	81.3	67	86	68.1	80	10
Former USSR	233.3	28	21	118.8	55	80
Rest of Eastern Europe	212.4	26	2	139.8	39	53
Asia						
China	195.9	28	19	195.0	28	18
India	190.0	29	20	194.4	28	20
Near and Middle East	180.3	30	5	131.4	42	50
Rest of Asia	148.7	36	10	67.1	80	100
Africa	152.1	35	15	139.1	38	29
America						
Brazil	75.1	78	95	68.1	79	100
Canada	65.9	51	61	103.5	80	100
United States	163.4	34	12	127.0	44	44
Rest of Latin America	153.7	36	30	69.5	79	100
Australia and Oceania	150.8	35	11	139.1	38	23
World[b]	159.8	(35)	(29)	120.1	(46)	(60)

[a] Adapted with permission from Schwarz (2000).
[b] Regional primary aluminium production shares as weighted factor.

primary energy consumption is estimated for the years 1995–2010. Interestingly, the main reason for this decline is not the decrease in specific electricity consumption, which accounts for less than 40% of the decline. The main reason is the improvement in overall efficiency of electricity production globally, which accounts for the remaining 60% or more. This improvement in worldwide overall efficiency is the result of increased efficiency of fossil-fired power plants, especially in the former Soviet Union, China, and the United States. These are regions with high primary aluminum world production shares and with high shares of fossil fuels for electricity production. In addition, the relocation of primary aluminum production to hydropower-rich regions accounts for the increase in worldwide overall efficiency.

4. ENERGY AS COST FACTOR

4.1 Survey

Table X presents the average cost structure of world aluminum production. Less reliable are the data for secondary production (only the processes within the refiner/remelter). These are based on rough estimations. However, it is clear that the share of energy costs in total operating costs of secondary production is very low in comparison with the costs of all processes of primary production. The energy cost shares of alumina and primary aluminum production are approximately the same. Nevertheless, the energy costs of electrolysis are more important with respect to choosing location.

4.2 Primary Production

4.2.1 Mining and Alumina Production

Energy costs represent 11% of total operating costs for bauxite mining. Labor is the dominant cost factor, with a share of more than 40%. Deposits with bauxite, which is easy to extract (poor surface and good ore thickness) and of high quality (e.g., high aluminum oxide content), are preferred for mining. The conditions of extraction influence the necessary labor input and energy requirements. The higher the quality of the bauxite, the higher the attainable price. Furthermore, general political conditions play a major role. An exorbitant bauxite levy, for example, hindered the bauxite industry in Jamaica for many years.

Energy costs are relatively high for alumina production. On worldwide average, they account for 25% of total operating costs. Bauxite costs are the dominant cost factor. They represent more than one-third of the total operating costs. Bauxite costs are heavily dependent on the location of alumina production. They vary considerably between regions. For example, in Latin America the cost of bauxite per ton of alumina was approximately $40 in the mid- to late 1990s, whereas in Western Europe (with the exception of Greece) the cost was $80. Approximately half of this difference is due to high transport costs, which amounted to $15–20 per ton for transport to Western Europe. The longer the transport distances are (and the lower the alumina content of the bauxite), the higher the specific transport costs. This explains the trend for locating refineries close to mines (and also why only bauxite with a high alumina content is traded between regions). The bauxite-producing countries (with a production >2 million tons in 2000) increased their alumina production by approximately 36% from 1990 to 2000, whereas that in countries with no significant bauxite production declined by approximately 14%. The production share of countries with significant mining activities increased from 64% in 1990 to 74% in 2000.

The location of alumina production is determined by transport costs for bauxite. The energy costs play only a minor role in this respect. On the one hand, upgraded and new plants have almost the same technical parameters, including energy requirements for alumina production. On the other hand, the energy carriers coal, oil, and natural gas can increasingly be obtained in similar conditions worldwide, with the result that energy costs (at least for upgraded and new plants) do not differ much between regions.

TABLE X

World Average Total Operating Costs for Primary and Secondary Aluminium Production, Mid- to Late 1990s[a] (USD/t_{output})

	Primary production			
Total operating costs (%)	Mining	Refining	Smelting	Secondary production[b]
	15	150	1350	(1400–1500)
Main raw material	—	34	33	(>65)
Labour	43	13	10	(>8)
Electricity	—	3	24	(1)
Other energy	11	22	—	(1–2)
Caustic soda	—	13	—	—
Other operating costs	46	15	33	(<25)

[a] Adapted with permission from Schwarz (2000).
[b] Remelting, inclusive salt slag treatment and refining.

4.2.2 Electrolysis

For smelting, alumina costs were dominant, with a share of 33% of total operating costs. Nevertheless, electricity costs, which were responsible for one-fourth of the total costs, are more important with regard to determining where to locate smelters because the variation in price of electricity is much greater than that of alumina. This is not surprising because electricity is only tradable to a limited extent. The necessary net infrastructure and the distribution losses limit regional interchange. Typically, primary aluminum smelters that operate with low-electricity tariffs obtain power from nearby hydroelectric power stations. Smelters that operate in regions with a high potential for cheap hydroelectric power have a natural competitive advantage. The abundance of energy is practically "stored" in aluminum and therefore becomes tradable.

There is a moderate trend of shifting smelting capacity to regions with low electricity prices. The building of new plants and capacity expansions of existing plants are usually done in these regions. Nevertheless, regions with high electricity prices for smelting, such as in the European Union, still command a relatively large share of world aluminum production (approximately 10% in 2000). The status of old plants in disadvantageous regions is usually not endangered because the high initial investment costs of these plants can be characterized as sunk costs due to the high degree of factor immobility and the limited scope of alternative uses of plants and machinery in the smelting industry. Therefore, old capacity will only be superseded by new capacity when the specific total costs of new capacity are lower than the specific operating costs of old capacity.

4.3 Secondary Aluminum Production

Energy costs for secondary aluminum production are of little relevance. The share of energy costs in total operating costs is 3% maximum. The cost of the scrap containing aluminum is the dominant factor (>65%), followed by labor costs (>8%). The profitability of aluminum recycling is determined by the difference (margin) between the scrap price and the attainable price of the different alloys. The prices for secondary and primary aluminum are interdependent. Assuming that initially both markets are in equilibrium, if the price of secondary aluminum were to increase, secondary aluminum would be increasingly substituted by primary aluminum. The decreasing demand for secondary aluminum would lead to a declining demand for scrap containing aluminum and therefore to lower scrap prices until the initial market equilibrium is again attained.

The major aluminum-consuming countries are still the major secondary aluminum producers. Obviously, the close proximity to initial products (scrap) and purchasers is still the decisive factor with regard to location. In the future, many expect that producers in South Asia, South America, and the Gulf states will gain production shares because of lower labor, environmental, and other costs.

5. SUMMARY

The determinants of specific primary energy consumption for world aluminum production are the recycling share, final energy requirements of the relevant process steps, and the overall efficiencies of fuel supply and electricity production. If, as expected, recycling share increases, final energy consumption decreases, and overall efficiencies improve, primary energy consumption will decline noticeably in the next few years.

Market processes are responsible for the expected growth in recycling share and the decrease in final energy requirements. The importance of energy costs in primary production makes both secondary production (and therefore the increase in recycling share) and efforts to reduce final energy requirements (especially for alumina production and primary aluminum smelting) attractive.

One question remains: What is the future absolute primary energy consumption of world aluminum production? Let us assume an increase in world aluminum production from 33 million tons in 2000 to 50 million tons in 2030 (an increase of 52%). Unchanged absolute primary energy consumption would require a decrease in specific primary energy consumption of 34%. Assuming an increase in the recycling share from 26 to 46–48% (Table IV), this would lead to a 22–26% decrease in specific consumption. In this scenario, it is very likely that the increase in the recycling share, together with the decrease in final energy requirements of the relevant process steps and improvements in overall efficiencies of fuel supply and electricity production, will lead to a decline in specific consumption of >34% and therefore to a decreasing absolute primary energy requirement for world aluminum production. The question is whether the slowdown in growth rates of primary production will in fact be as strong as assumed here.

SEE ALSO THE FOLLOWING ARTICLES

Cement and Energy • Cost–Benefit Analysis Applied to Energy • External Costs of Energy • Plastics Production and Energy • Steel Production and Energy

Further Reading

Altenpohl, D., and Paschen, P. (2001). Secondary aluminium is the raw material highlight of the 21st century. *Aluminium* 77(1/2), 8–13.

Bielfeldt, K., and Grojtheim, K. (1988). "Bayer and Hall–Héroult Process—Selected Topics." Aluminium-Verlag, Düsseldorf.

Kammer, C. (1999). "Aluminium Handbook. Volume 1: Fundamentals and Materials." Aluminium-Verlag, Düsseldorf.

Krone, K. (2000). "Aluminiumrecycling. Vom Vorstoff zur fertigen Legierung" ["Aluminum Recycling. From Initial Product to Aluminum Alloy"]. Vereinigung deutscher Schmelzhütten, Düsseldorf.

Æye, H. A., Mason, N., Peterson, R. D., Rooy, E. L., Stevens McFadden, F. J., Zabreznik, R. D., Williams, F. S., and Wagstaff, R. B. (1999). Aluminum: Approaching the new millennium. *J. Metals* **51**(2), 29–42.

Morel, P. (1991). "Histoire technique de la production d'aluminium" ["History of Technologies in Aluminum Production"]. Presses Universitaires de Grenoble, Grenoble, France.

Mori, G., and Adelhardt, W. (2000). "Stoffmengenflüsse und Energiebedarf bei der Gewinnung ausgewählter mineralischer Rohstoffe, Teilstudie Aluminium" ["Material Mass Flows and Energy Consumption Caused by the Production of Selected Mineral Raw Materials, Section: Aluminum"]. Bundesanstalt für Geowissenschaften und Rohstoffe, Hannover, Germany

Pawlek, P. (1999). 75 Years development of aluminium electrolysis cells. *Aluminium* 75(9), 734–743.

Schwarz, H.-G. (2000). "Grundlegende Entwicklungstendenzen im weltweiten Stoffstrom des Primäraluminiums" ["Fundamental Trends in the Global Flow of Primary Aluminum"]. Forschungszentrum, Zentralbibliothek, Jülich.

Schwarz, H.-G., Kuckshinrichs, W., and Krüger, B. (2000). Studies on the global flow of primary aluminium. *Aluminium* **76**(1/2), 60–64.

Schwarz, H.-G., Briem, S., and Zapp, P. (2001). Future carbon dioxide emissions in the global flow of primary aluminium. *Energy* **26**(8), 775–795.

Thonstad, J., Fellner, P., Haarberg, G.M., Híves, J., Kvande, H., and Sterten, A. (2001). "Aluminium Electrolysis. Fundamentals of the Hall–Héroult Process." Aluminium-Verlag, Düsseldorf.

Aquaculture and Energy Use

M. TROELL
The Beijer Institute and Stockholm University
Stockholm, Sweden

P. TYEDMERS
Dalhousie University
Halifax, Nova Scotia, Canada

N. KAUTSKY
Stockholm University and The Beijer Institute
Stockholm, Sweden

P. RÖNNBÄCK
Stockholm University
Stockholm, Sweden

1. Introduction
2. Overview of Aquaculture
3. Energy Analysis in Aquaculture
4. Energy Performance of Aquaculture
5. Biophysical Efficiency and Sustainability

Glossary

aquaculture The farming of aquatic organisms, including fish, mollusks, crustaceans, and aquatic plants. Farming implies some form of intervention in the rearing process to enhance production, such as regular stocking, feeding, and protection from predators. Farming also implies individual or corporate ownership of the stock being cultivated.

embodied energy density The accumulated energy use per area of production. Can include both direct and indirect industrial energy inputs as well as fixation of solar energy in ecological systems.

emergy The sum of the available energy (exergy) of one kind previously required directly and indirectly through input pathways to make a product or service.

energy intensity (EI) The accumulated energy inputs required to provide a given quantity of a product or service of interest. In the current context, energy intensity is expressed as the total joules of energy required to produce a live-weight tonne of fish, shellfish, or seaweed.

fossil fuel equivalents An expression of auxiliary energy inputs from the economy, embodied in goods and services (e.g., 1 unit of electricity = 3–4 units of fossil fuels).

industrial energy analysis (IEA) The sum of the energy (fossil fuel equivalents) needed for manufacturing machinery and products used as inputs to the system, and the fuel energy needed for operation.

life-support systems The parts of the earth that provide the physiological necessities of life, namely, food and other energies, mineral nutrients, oxygen, carbon dioxide, and water; the functional term for the environment, organisms, processes, and resources interacting to provide these physical necessities.

solar energy equivalents A unit making it possible to combine ecological and economic energy requirements, based on the energy from solar fixation. It takes about 10 times of solar fixation per 1 unit of fossil fuel.

In this article describing the various techniques used to assess the energy performance of aquaculture, the discussion focuses on the results of different energy analyses that have been undertaken. The main focus is on industrial energy analysis, i.e., embodied fossil fuel energy. However, analyses within a wider ecosystem context are also discussed. An analysis of the broad range of aquaculture practices is not within the scope of this general overview. Instead, the focus is on some selected culture systems, representing intensities and methodologies dominating global production.

1. INTRODUCTION

The need for indicators to assess sustainable development of aquaculture has been high on the research agenda for the past decade. Useful indicators preferably focus on both biophysical and socioeconomic dimensions. An example of a biophysical indicator used in aquaculture is the measure of the dependence of aquaculture on different forms of energy. Even though energy consumption (i.e., of auxiliary energy) within the aquaculture sector is small compared with global economic activities, and energy analyses generally provide only a partial tool for assessing sustainability of an industry, it is of interest to increase our understanding of aquaculture's dependence on an external finite resource. In addition, energy dependence provides a basis on which we can evaluate the efficiency characteristics of different types of aquaculture production systems and processes; thus, this indicator can be used to measure the extent to which alternative options move toward efficiency goals implied by sustainability. Further, the energy costs of production not only involve sustainability issues of ecosystem resource efficiency and nonrenewable resource depletion, but also the potential cost to future societies through environmental change from pollution and global climate change.

2. OVERVIEW OF AQUACULTURE

2.1 Definition and Forms

Aquaculture is the farming of aquatic organisms, including fish, mollusks, crustaceans, and aquatic plants. By definition, aquaculture implies some sort of intervention in the rearing process to enhance production. From an industrial perspective, the objective is to maximize production of some demanded species to achieve maximum economic benefits. This is the objective for most recently developed operations, but not necessarily for the many traditional systems whose outputs are primarily directed toward rural and household subsistence consumption. Aquaculture is a highly diverse activity, and species choice is important from an energy perspective because different species have different characteristics that limit the circumstances under which they can be cultured.

Farming takes place in fresh, brackish, and marine waters and encompasses two main subsectors, commercial/industrial- and rural/subsistence-based operations. The former usually target species generating high income, and the latter target species largely used for subsistence consumption or local markets. Culturing may be practiced either as monoculture or polyculture. Traditional freshwater farming systems in China have been, and to a large extent still are, practiced as polycultures using low stocking densities. By contrast, modern-day operations are usually monoculture systems employing high stocking densities. Freshwater pond farming accounts for the majority of global cultured finfish production. Only a small proportion of freshwater production is conducted in land-based tanks and these are almost exclusively found in developed countries. The most common open-water culture technologies, conducted primarily in marine waters, include either cages or pens for finfish, and rafts, long lines, or poles for seaweed and mussel culture.

Farming practices can be classified into extensive, semiintensive, and intensive farming. The classification is often based on resource (e.g., feed) and energy use, but rearing densities, maintenance costs or inputs, chemical use, etc. may also be considered. Intensive culture systems include the following general characteristics:

- High stocking/rearing densities.
- High dependence on exogenous direct and indirect energy inputs, with the latter in the form of feed and chemicals.
- Low inputs of labor energy.
- Production of relatively high-valued species.

Intensive culture is typically conducted in tanks, ponds, and in open-water cages. In most of these systems, with the possible exception of unlined earthen ponds, capital investments are usually high. Carnivorous finfish species such as salmon, seabass, groupers, eel, halibut, and cod are typically farmed intensively and depend almost exclusively on commercially prepared feeds or aquafeeds. This is also true for some omnivorous penaeid shrimp species (e.g., tiger shrimps). In contrast, extensive and semiintensive systems in which most detrivorous, herbivorous, and omnivorous fish species (carps, milkfish, tilapia, etc.) are cultured, rely to varying degrees on food produced within the culture environment. However, many of these systems are fertilized, using either industrial fertilizers or organic by-products from agriculture, to stimulate primary production within the pond environment. In recent years, many of these traditional extensive systems have been moving along a path toward greater intensification and compound aquafeed usage in order to increase production.

2.2 Importance of Aquaculture

A wide range of plants and animals are farmed globally, with more than 220 different species of finfish and shellfish being produced. Although most aquaculture is undertaken to produce food, other applications include fodder, fertilizers, extracted products for industrial purposes (e.g., seaweed compounds used in manufacturing processes, food processing, and microbiology), goods for decorative purposes (shell, pearls, and ornamental fish), and recreation (sport fishing, ornamentals, etc.). The demand for seafood is steadily increasing worldwide, but most of the world's fishing areas have reached their maximal potential for capture fisheries production (Fig. 1). It is therefore logical that aquaculture, currently one of the fastest growing food production sectors in the world ($\sim 10\%\ yr^{-1}$), will be increasingly called on to compensate for expected shortages in marine fish and shellfish harvests. Currently, aquaculture accounts for approximately 30% of total food fish production globally, of which roughly 90% originates from Asia. Production histories of some major aquaculture species groups from 1970 to 2000 are presented in Fig. 1.

Geographically, China contributes more than two-thirds of total aquaculture production (mainly from traditional freshwater cultures). Europe, North America, and Japan collectively produce just over one-tenth of global output but consume the bulk of farmed seafood that is traded internationally.

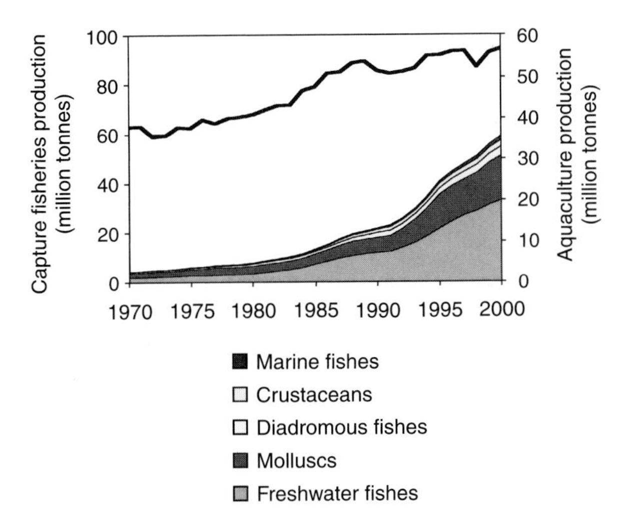

FIGURE 1 Production development of global capture fisheries (thick line, including both marine and freshwater fishes) and major aquaculture groups from 1970 to 2000. Figures for aquaculture (but not for capture fisheries) are cumulative.

On a tonnage basis, extensive and semiintensive freshwater aquaculture dominates global fish production (56% by weight). Currently, within these systems about 80% of carp and 65% of tilapia worldwide are farmed without the use of modern compound feeds. It should be noted, however, that during the 1990s, the more capital- and energy-intensive production of shrimp and marine (including diadromous) fish species increased significantly.

3. *ENERGY ANALYSIS IN AQUACULTURE*

3.1 Forms of Energy Used in Aquaculture

3.1.1 Ecosystem Support

From an ecological perspective, most aquaculture systems involve the redirection and concentration of energy and resource flows from natural ecosystems to the culture environment. Consequently, very few systems depend solely on the solar radiation incident on the production facility. This is easily appreciated when one considers the relatively small surface area of many high-density culture systems that are exposed to the sun. Exceptions, however, include extensive culture of autotrophic species such as seaweeds. Similarly, filter-feeding organisms such as mussels are dependent on phytoplankton produced in adjacent waters that are brought to the farm by currents. For most farming systems, though, productivity is enhanced by supplementing auxiliary energy and resource inputs produced in other ecosystems (Fig. 2). For example, a fish cage or shrimp pond is dependent on large ecosystem support areas outside the farm border (i.e., a culture system's "ecological footprint" is much larger than the area occupied by the farm), where solar energy is being transferred through food webs to produce feed. Currently, fish meal and fish oil remain the dominant ingredients in most compound feeds for carnivorous finfish and marine shrimps. As a result, many aquaculture systems that utilize compounded feeds containing fish meal and fish oil actually appropriate more fish biomass as inputs to the feed than is represented by the farmed product produced. And as will be shown, this relative dependence on capture fisheries products is also important when discussing energy consumption, because the energy associated with providing the feed is often the main input to many aquaculture systems.

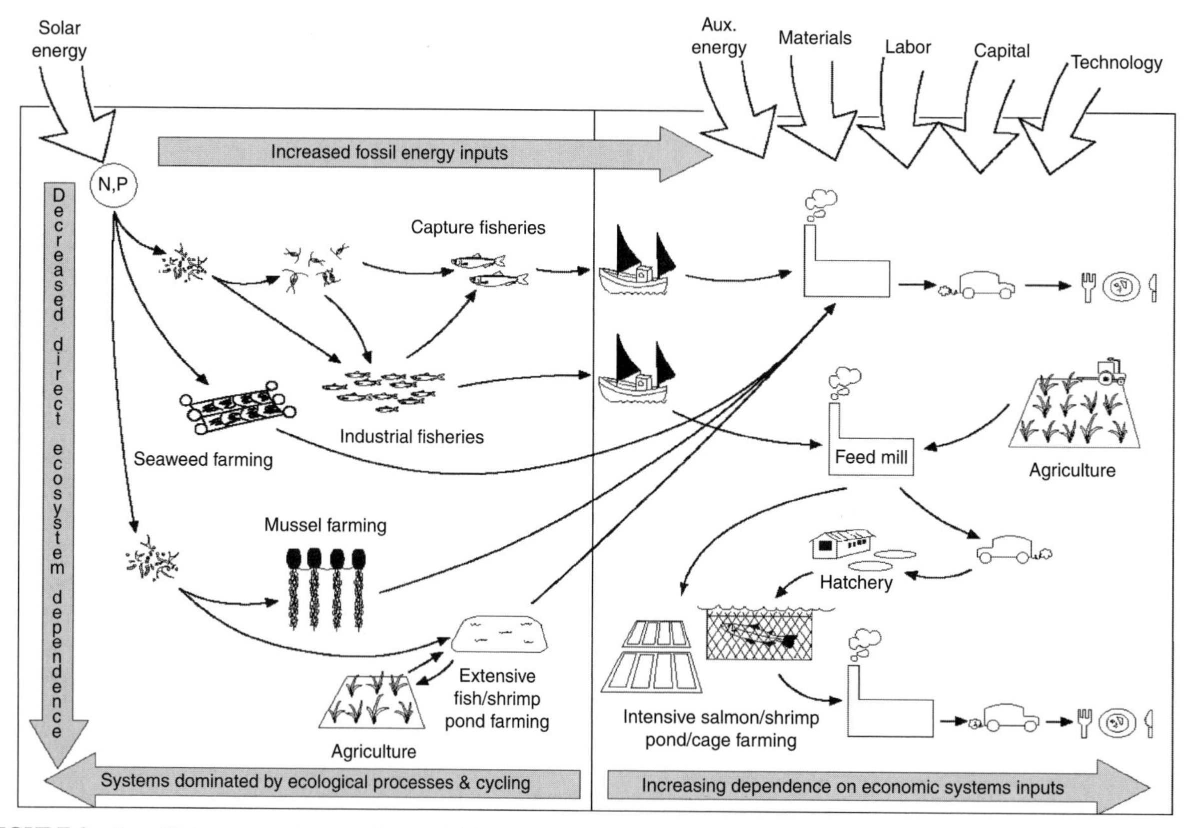

FIGURE 2 Simplified energy and matter flow model for various types of seafood production from solar fixation by plants in the natural system and until the product reaches the consumer. In natural ecosystems, as well as in extensive farming systems, solar energy and recirculation of nutrients constitute the limiting factor for biological production. Due to energy losses in each transfer between trophic levels, the harvest per area will decrease when a higher level of the food chain is exploited. Extensive seaweed and mussel farming may require additional inputs of materials for support structures (buoys, ropes, etc.), and extensive fish and shrimp production usually requires ponds and inputs of seed, fertilizers, and/or agricultural by-products to maintain production. In contrast, intensive cultures of carnivorous species such as salmon and shrimp require high industrial energy intensity (fossil fuel input per weight of product), mainly due requirements for formulated feed made from fish- and crop-derived products, sophisticated cages and ponds, energy for aeration and pumping, inputs of hatchery-reared smolts/fry, and medication. Moreover, each step in the feed production chain from fisheries and crop harvesting, processing, transportation, etc. requires fossil energy, materials, labor, technology, and capital to a varying degree. Consequently, it is clear that there are significantly more energy-demanding and pollution-generating steps involved in intensive shrimp/salmon farming than in extensive aquaculture or capture fisheries.

3.1.2 Direct Energy Inputs

Direct energy inputs to aquaculture operations encompass a range of activities, including the collection/production of juveniles, general system operations, and the harvesting, processing, and transport of the product. A first-order approximation of the industrial energy dependence of a culture system can be obtained by considering the intensity of the operation, the degree of mechanization, the type or means of production, and the quality/quantity of feed inputs. Provisions of feed and juveniles for stocking are the main material inputs, and fuels and electricity are needed for various grow-out operations.

Energy inputs associated with harvesting, processing, and transporting the various feed components are implicit in the reliance of intensive forms of aquaculture on remote ecosystems (Fig. 2). For example, fishing fleets and reduction plants consume large amounts of fossil and electrical energy in order to provide fish meal and fish oil. Similarly, a wide range of direct and indirect energy inputs are required by the agriculture industry to supply plant- and livestock-derived components of feed pellets. This includes the indirect energy cost for fuel and labor, together with the direct energy-demanding processes. Further energy inputs to aquaculture operations are associated with either the production of juveniles in hatcheries or their collection from the wild. Hatchery production is also dependent on energy-consuming activities for the provision of resources such as feed and broodstock (adult individuals, either wild captured or reared entirely within the culture environment).

By comparison, most extensive culture systems receive no formulated feeds. Instead, inputs to these systems typically comprise agricultural products or wastes (e.g., supplemented manure) and chemical fertilizers. This typically results in smaller industrial energy inputs, particularly if crop residuals and manure, essentially by-products of local agricultural production, are treated as biophysically free with only marginal energy inputs associated with further processing. Similarly, feed inputs to semiintensive systems may consist, either wholly or in part, of fish from local fisheries. In these cases, the unprocessed fish may be used directly or supplied as feed after minor transformation (i.e., as moist pellets). Thus, utilization of locally available fisheries resources may involve less fossil fuel energy, particularly if drying- and transportation-related energy inputs are limited. However, depending on the fishing technology used and the relative abundance of local harvestable stocks, fuel inputs may be surprisingly high when compared with more modern, large-scale reduction fisheries. Moreover, because raw fish products can spoil quickly if unused, energy inputs to a culture system that utilizes these products will increase as a result of any waste that occurs. Alternatively, fish processing wastes can, to some extent, replace fish protein inputs from whole fish. Such a substitution might again imply "no energy cost" for extraction, but only for transportation and any secondary processing that is required.

The methodological appropriateness of allocating a zero biophysical cost for by-products from fish processing and agriculture production is debatable. It can be argued that treating the energy costs associated with making a given by-product available as free, simply because they are designated as "wastes," is arbitrary and potentially misleading. From this perspective, the energy inputs associated with any production process, regardless of how its many outputs may be regarded, should be fully included in any energy analysis. Alternatively, for some products the energy costs could be based on the proportion to the relative weights of product and by-product. However, such an accounting system may not be applicable to all products (e.g., manure production). A third possible accounting convention might be based on the relative prices of coproducts. This approach only works, however, if a market for the by-product exists, or if alternatives to the by-product exist, from which an opportunity cost could be estimated. The degree to which these alternative approaches capture the true energy costs of different inputs to aquafeeds is open for debate. In practice, however, to date, most, if not all, of these techniques have been used by various analysts when evaluating the energy costs of aquaculture.

Accounting for labor inputs is also a complicated issue within the context of energy analysis. In most aquaculture analyses, labor has been excluded, with the justification that its contribution to the energy profile of the system is negligible when compared to fossil fuel inputs. Again, this may be accurate for some types of systems but not for others. Methods that have been used to account for the energy costs of labor either incorporate the nutritional energy content of the food required to sustain the requisite labor inputs, or, alternatively, incorporate the industrial energy needed to provide the food. It is, however, questionable if these methods, based as they are on standard energy conversion values, should be applied widely, because there are large discrepancies with respect to energy production costs of food and consumption levels between developed and developing countries. To address these context-specific concerns, some analysts incorporate the total industrial energy associated with wages paid for labor inputs, using the industrial energy to gross national product (GNP) ratio for the country within which labor is supplied.

3.1.3 Indirect (Embodied) Energy Inputs

Fixed capital inputs may include different forms of enclosures (tanks, nets, etc.), feeders, aerators, pumps, boats, vehicles, etc. The energy embodied in building and maintaining these capital inputs may be substantial, particularly in intensive systems. Methodologically, a common practice for estimating the energy inputs associated with fixed capital has been to use input–output analysis of the country's economy. The resulting energy equivalents provide, after accounting for depreciation, estimates of the related energy costs per year. However, not all studies of aquaculture systems have accounted for fixed costs in this way. Instead of applying an annualized energy cost over a defined time period, some studies treat fixed costs as a single lump energy input. Alternatively, some studies have simply excluded fixed costs completely from the analysis. To what extent these alternative approaches affect the overall energy analysis depends on the characteristics of the system being analyzed. Generally, the fixed energy cost in intensive systems is insignificant compared to the overall energy profile, due to the high embodied energy cost for feed (Table I). In culture systems that depend predominantly on naturally available productivity and with low inputs of supplemented feed, the energy embodied in capital infrastructure may play a more important role in the overall industrial energy analysis (IEA) (Table I). The variation in the methodological approaches used and the rigor with which they are applied between different studies can make the direct comparison of their results difficult. For example, in some cases, total energy inputs are reported as a single value without any specification of the various inputs. In other instances, only one form of energy may be accounted for (for example, only the electrical energy consumption in the grow-out phase is included, whereas the energy inputs associated with feed, hatchery rearing, fixed infrastructure, harvesting, etc. are left out). In other seemingly more comprehensive analyses, a specific source of energy input may be completely overlooked (for example, not accounting for the energy consumed within the fishery sector that is required to sustain raw feed materials). Similarly, energy inputs to hatchery operations may be left out.

3.2 Analytical Techniques

Relatively few energy analyses of aquaculture systems have been conducted to date. A variety of analytical approaches have been used, the most frequent being quantification of direct and indirect auxiliary (industrial) energy inputs (e.g., electric energy, oil, gas, or coal) dissipated in the myriad supporting economic processes. These apply regular process and input–output analysis normally using land, labor, and capital as independent primary factors of economic production. This method sets the "system boundary" at the point of material extraction. Thus, the method is limited in the sense that it does not consider the impact of extraction on the future availability of resources, and does not value the wider resource base on which the production depends.

In a second approach, the boundaries of the system are expanded to include processes and the energy transformed by renewable resources. In doing so, the focus of an analysis shifts to an understanding of the solar energy dependency of economic activities. In the context of aquaculture, solar energy enters mainly through agriculture and fisheries, together with the energy embodied in fuels and raw materials (Fig. 2). The most highly developed form of this approach has given rise to the concept of "emergy." This method includes all forms of energy such as material and human inputs and environmental services (including energy of hydrological and atmospheric processes) being converted into their solar energy equivalents. Thus, it evaluates the work done by the environment and by the human economy on a common basis.

Not surprisingly, different approaches result in different kinds of information. Ideally, it is necessary to consider the total energy use from pond to plate, expressed per unit of edible produced protein. Within the context of sustainability, it can be argued that an analysis that also captures the work by nature, i.e., the energy flows through dynamic ecosystems, more accurately reflects dependencies on the overall resource base. All economic activities depend on these life-supporting systems. Often, however, this support is taken for granted and is not included in any efficiency calculations.

4. *ENERGY PERFORMANCE OF AQUACULTURE*

4.1 Extensive Systems

Extensive culture systems generally have relatively low dependence on direct auxiliary and indirect embodied energy inputs due to farming practice and species choice. The culture of species groups such as mussels, carp, and tilapia (Table I) have generally low energy intensity (EI) input per unit of production and also per edible protein energy return. These organisms are able to utilize natural primary production (phytoplankton or macrophytes) or low-quality agriculture by-products and thereby require little if any artificial feed. The various means for enhancing natural production, i.e., applying fertilizers or organic inputs (e.g., manure), are in most cases not costly from an energy perspective. Human labor inputs are typically high in countries where extensive culturing methods dominate, both at the farm site and in activities generating different inputs to the farm (Table I).

The classification into extensive culture systems, based on energy consumption, may not always be as straightforward as it might seem. For example, the same type of culture system may depend on very different forms and quantities of energy inputs depending on where the system is situated (e.g., whether in industrialized or nonindustrialized countries). For some activities, fossil fuels may be substituted for human labor. Similarly, the energy embodied in various material inputs that could be

TABLE I

Direct and Indirect Energy Inputs and Resulting Energy Intensities of Semiintensive Shrimp Pond Farming, Intensive Salmon Cage Farming, Semiintensive Carp + Tilapia Pond Polyculture, Indian Carp Polyculture, Intensive Catfish Pond Farming, Semiintensive Tilapia Pond Farming, and Mussel Long-Line Culture[a]

	Species Farmed													
Parameter	**Shrimp**		**Salmon**		**Polyculture (carp recirc)[b]**		**Polyculture (Indian carp)**		**Catfish**		**Tilapia**		**Mussel**	
System Characteristics														
Site area (ha)	10		0.5		1		1.0		1.0		0.005		2	
Production/yr(t)	40		200		4		4.3		3.4		0.058		100	
Production/yr/ha(t)	4		400		4		4.3		3.4		12		50	
Energy Inputs[c]	EI	%	EI	%	EI	%	EI[d]	%	EI	%	EI	%	EI	%
Fixed Capital														
Structures/equipment	2500	2	5940	6	3556	7	2114	8	11,691	10	0	0	2700	58
Operting Inputs														
Fertilizer, chemicals	22,750	15	0	0	9316	18	633	2	369	0	1000	3	0	0
Seed	18,750	12	2970	3	15	0	0	0	11,076	10	0	0	0	0
Feed	58,250	37	78,210	79	15,451	31	287	1	86,389	75	23,280	97	0	0
Electricity, fuel	54,250	35	11,880	12	21,928	44	26,176	97	5415	5	0	0	1900	42
Total	156,750		99,000		50,265		27,096		114,940		24,000		4600	
Labor Inputs[e]														
Person-days/t	n.d		10.0		6.5		66.7		3.5		172.4		5.0	
Percentage	6.4[f]		14		n.d		n.d		n.d		35[g]		63	
Energy Ratio														
Edible product(MJ/kg)[h]	275	***	142		84	***	45	***	192	***	40		12	
Edible protein(MJ/kg)	784	**	688		272		135	**	575	**	199		116	

[a]Data from Larsson *et al.* (1994) (shrimp), Stewart (1994) (salmon), Bardach (1980) (recirc. Asian carp), Singh and Pannu (1998) (Indian carp), Waldrop and Dillard (1985) (catfish), and Stewart (1994) (tilapia and mussels).

[b]One-third of water recirculated.

[c]Energy intensities expressed in kilojoules/kilogram (wet weight).

[d]Calculating with 20 yrs of deprivation time for pond construction.

[e]Labor inputs expressed as person-days per tonne of production and as a percentage of the total capital cost; n.d., no data available.

[f]Due to lack of data this value is based on % labor energy consumption (calculated form GNP and wages) of total IE inputs.

[g]Labor for pond construction dominating input.

[h]Per kilogram of wet weight.**, Calculating with 20% protein of product wet weight; ***, calculating with 60% of product being edible.

used to construct a culture system may differ due to variations in the energy associated with the input extraction, processing, and transport. To illustrate the potential challenges associated with an energy-based classification scheme, consider mussel farming, widely regarded as an extensive culture system in that it involves purely filter-feeding organisms that grow well without supplemented feeds or fertilizers. Although indirect energy inputs may be relatively small, the forms and quantities of direct energy for harvesting, sorting, and cleaning can vary widely. For example, mussel farming in the developed world is typically more dependent on industrial energy inputs for these operations. In contrast, in developing countries, industrial energy inputs are often low because most operational work is carried out by hand, because labor is relatively abundant and inexpensive. Similarly, the farming of herbivorous fish species in which agricultural inputs are used to enhance production may, in more industrialized countries, result in higher energy consumption, because the underlying agricultural production system is more fossil fuel dependent. Even though the discussion here includes no case requiring change in the general application of the intensity classification, it is necessary to be aware that site-specific factors can significantly influence the energy analysis, even if the same types of systems and species are being used.

Increased pressure for expansion of global aquaculture production, together with increased scientific knowledge concerning the optimization of growth performance, has resulted in the incremental intensification of many extensive culture systems. This intensification process typically entails the increased use of artificial feeds containing fish meal. Not only does this transformation reduce the total global supply of fish potentially available for human consumption, but it also increases the industrial energy intensity of the culture system. For example, relatively small increases in the use of artificial feeds, in the context of extensive systems, can quickly dominate the total industrial energy inputs to the system Table I.

4.2 Intensive Systems

It becomes obvious from comparing extractive species, such as seaweeds and mussels, with carnivorous species, such as salmon, that the inherent or natural trophic level of the species being farmed plays a significant role in the energy intensity of the culture system. However, Table I also reveals that there are other production factors that can play a significant role in the overall industrial energy costs of a given system. These include intensity level, dependence on artificial feed, degree of mechanization, and environmental control needed in the husbandry system.

Direct and indirect energy inputs to operational activities can represent more than 90% of the total EI inputs to intensive culture systems. Of this, the energy required to provide feed is usually the single most important item (Fig. 3A). Other inputs of importance, however, include direct electricity and fuel inputs, along with the energy required to supply juveniles. In herbivorous/omnivorous fish culture systems, the energy required to supply fertilizer and green fodder can also be substantial (Fig. 3B). In contrast, the embodied and direct construction-related energy costs associated with the fixed capital infrastructure of many intensive farming systems are relatively insignificant (Fig. 3A, Table I). Similarly, regardless of type of culture system, labor in Western countries is generally negligible in comparison to fossil fuel inputs.

It is no surprise that intensive land-based and open-water aquaculture activities differ significantly with respect to their energy costs associated with

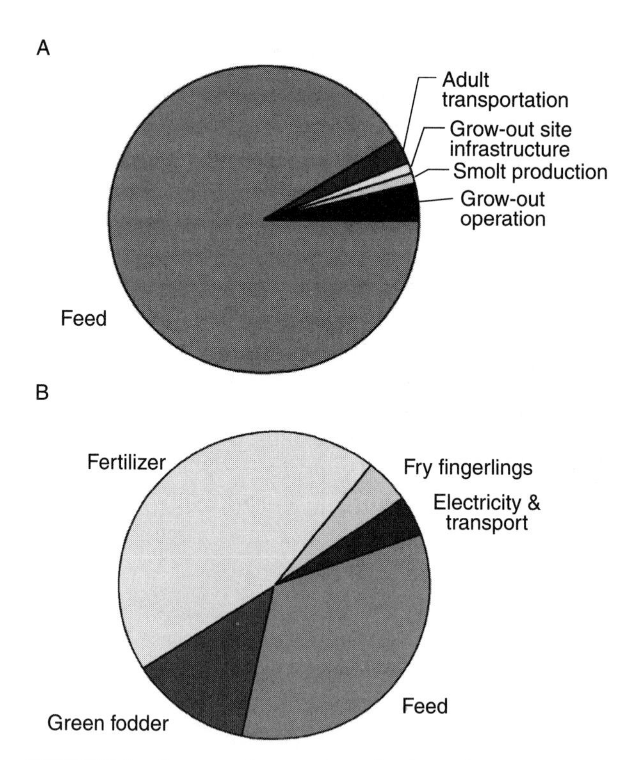

FIGURE 3 Source and percentage of industrial energy inputs to produce (A) Atlantic salmon (intensive cage farming) (from Tyedmers, 2000) and (B) silver and bighead carps (semi-intensive ponds) (from Jingsong and Honglu, 1989).

water quality maintenance (i.e., for circulation and water aeration). Thus, cages in open water benefit from natural water circulation that removes waste metabolites and reoxygenates the water. Conversely, tank and pond systems on land can have high energy requirements for replacing, circulating, and aerating the water. Consequently, a semiintensive shrimp pond farm can consume as much as five times more energy for pumping and aeration activities compared to a salmon cage farm energy consumption for building and maintaining the cage structures (Table I). Although feed-related energy inputs typically account for a large proportion of the overall energy profile of an intensive culture system, the absolute scale of these inputs is very sensitive to the specific components of the feed being used. In general, fish- and livestock-derived inputs to feeds typically have much higher associated energy costs than do crop-derived components. Moreover, even within the context of a specific type of feed input, say fish meal and oils, energy inputs can vary widely between different sources and through time as conditions change. In the case of these fish-derived inputs, major sources of energy input variations result from differences in the fishing methods used, and in the relative abundance of the fish stocks being targeted. Similarly, the energy costs of agriculture products can also vary depending on the specific crops produced and the agriculture system they are derived from. This is of specific importance when aquaculture systems in industrialized and nonindustrialized countries are being compared.

To illustrate the relative energy costs associated with providing the major forms of feed inputs, Fig. 4 shows the energy inputs associated with producing a typical salmon feed used in Canada. By comparing Figs. 4A and 4B, we see that although over 50% of the mass of the feed is derived from fish, these inputs only represent about 35% of the total energy inputs. What is more interesting to note, however, is that although livestock-derived inputs represent only 16% of the feed by mass, they account for well over 40% of the total energy costs of this feed. In contrast, crop-derived inputs comprise over 25% of the feed but only account for approximately 5% of the total energy cost.

5. BIOPHYSICAL EFFICIENCY AND SUSTAINABILITY

Aquaculture is becoming an increasingly important component of the food production sector. As such, it is of interest to compare its energy performance with other food production systems. Table II provides an overview of the energy performance of selected food production systems, including fisheries and terrestrial crop and animal production, using a ratio of total industrial energy invested in the system relative to the edible protein energy return. By this measure, we find that the energy performance of intensive shrimp and salmon aquaculture is of a magnitude similar to that of shrimp and flatfish fisheries, as well as feedlot beef production. Seaweed, mussels, and some extensive fish aquaculture systems are comparable to sheep and rangeland beef farming. Interestingly, intensive rainbow trout cage culture has a surprisingly low energy ratio, similar to that of tilapia and mussel farming. Lobster and oyster farming systems in tanks represent the most energy-demanding aquaculture system for which data are available. It is important to note, however, in the case of both of these systems, the energy data represent experimental production systems from the late 1970s and early 1980s. As far as any

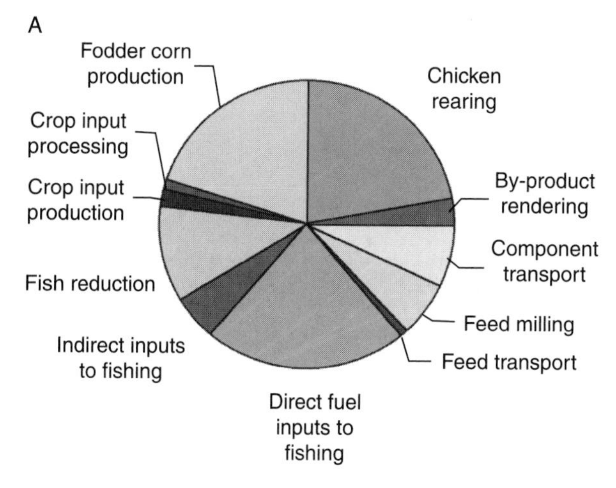

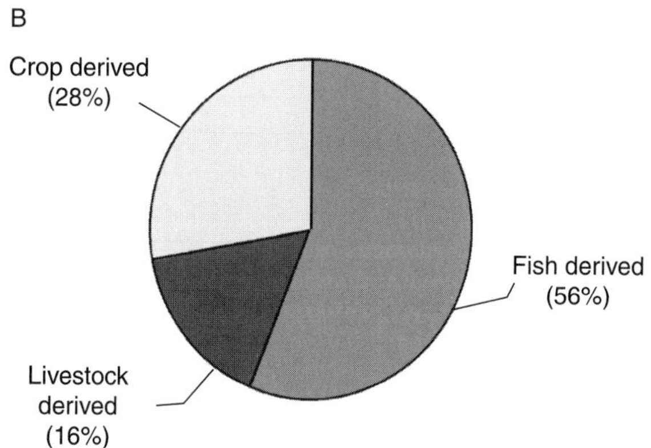

FIGURE 4 (A) Breakdown of industrial energy inputs to produce a generic salmon feed; (B) the major components of this feed. From Tyedmers (2000).

TABLE II

Ranking of Industrial Energy Inputs Excluding Labor per Protein Output in Aquaculture Compared with Capture Fisheries and Agriculture[a]

Food type	Industrial energy input/protein energy output (J/J)
Herring (purse, seine, NE Atlantic)	2–3
Vegetable crops	2–4
Carp farming (ponds, Indonesia, China, Israel)	1–9
Tilapia culture (extensive ponds, Indonesia)	8
Seaweed culture (West Indies)	7–5
Sheep farming	10
Rangeland beef farming	10
Cod fisheries (trawl and long line, North Atlantic)	10–12
Salmon fisheries (purse seine, gillnet, troll, NE Pacific)	7–14
Milk (USA)	14
Mussel culture (long line, Northern Europe)	10–20
Tilapia culture (semiintensive ponds, Africa)	18
Rainbow trout culture (cages, Finland, Ireland)	8–24
Catfish culture (ponds, USA)	25
Eggs (USA)	26
Broiler farming	34
Shrimp culture (Semiintensive Ponds, Ecuador)	40
Atlantic salmon culture (cages, Canada–Sweden)	40–50
Shrimp fisheries	17–53
Flatfish (trawl, NE Atlantic)	53
Beef feed-lot (USA)	53
Lobster fisheries (trawl, NE Atlantic, NW Pacific)	38–59
Seabass culture (Cages, Thailand)	67
Shrimp culture (intensive ponds, Thailand)	70
Oyster culture (tanks, USA)	136
Lobster culture (tanks, USA)	480

[a]Data for oyster farming and carp tank farming from the late 1970s; lobster farming data from the early 1980s. All other data mainly from the 1990s and subsequent years, with some few exceptions from late 1980s. Data from Tyedmers (2001), Watanabe and Okubo (1989), Tyedmers (2000), Larsson *et al.* (1994), Folke and Kautsky (1992), Pimentel *et al.* (1996), Seppälä *et al.* (2001), Stewart (1994), Berg *et al.* (1996), Smith (1990), Ackefors (1994), Rawitscher (1978), Bardach (1980), and Sing and Pannu (1996).

potential changes in performance from experimental systems in the 1970s and 1980s to now, it would be expected that the energy performance of these systems has improved. This is due to decreased fish meal/oil content in the feed, to increased feeding efficiency (methods, feed characteristics, etc.), and probably also to increased technological efficiency (in machinery production, increased machinery efficiency, etc.).

The data in this table should not be discussed in isolation from other environmental considerations, i.e., various negative externalities, but instead should be seen as adding important pieces of information to such discussions. By expressing energy performance on the basis of protein energy output, an important aspect of food production is captured. However, it is necessary to acknowledge that fish make particularly valuable contributions to the human diet because they provide complete protein and are low in saturated and high in polyunsaturated fats compared to other animal protein sources. A further complicating factor, and one not usually addressed when using maximum protein yield as a basis for comparing energy performance, is the degree to which potentially available protein is actually utilized. This issue is potentially most problematic when comparisons are being made between food production systems set in different cultural contexts. For example, a catfish produced and consumed in the United States may actually result in a smaller portion of the available protein of the fish being consumed when compared to catfish consumed in many Asian countries, where a much larger proportion of the animal may be regarded as edible.

It has been argued that without a clear recognition of the aquaculture industry's large-scale dependency and impact on natural ecosystems and traditional societies, the aquaculture industry is unlikely to either develop to its full potential or continue to supplement ocean fisheries. Thus, only by defining and quantifying linkages between the aquaculture operation and its supporting systems, together with socioeconomic considerations, will it be possible to ensure sustainable development. The extent to which measurement of energy inputs to aquaculture operations contributes to our understanding of its sustainability is an important issue for further consideration. Does less need for industrial energy inputs imply increased sustainability? If a reduction in dependence on fossil fuels is considered the most important or only indicator of sustainability, the answer is clearly yes. No doubt, reduced pressure on valuable finite resources is important. However, from a more holistic perspective on aquaculture, it may not be the most important factor for achieving sustainability. In other words, although energy analysis can certainly help select forms and types of culture systems that reduce demand for industrial energy, it cannot capture a wider range of negative environmental or socio-economic externalities and benefits that result from production. For example, a comparison of EI inputs between a system that depends on renewable energy resources and a system more dependent on industrial energy does not reflect whether the extraction of renewable resources is being performed in a sustainable way (i.e., sources, extraction methods, socio-economic conflicts). To do this requires additional tools and analysis. Industrial energy analysis should instead be viewed as a measure of one aspect of sustainability–not as a complete tool for evaluating sustainability. Its narrow focus on nonrenewable resource use, and in particular fossil fuel use, functionally sets the analytical boundaries at the point of material extraction. It does not value the wider resource base on which the activity depends.

Comparing results from different energy analyses of aquaculture systems is also not as straightforward as the literature may indicate. A common problem is the lack of detail regarding what has been included in the analysis. Frequently, information is missing regarding (a) how transportation has been addressed (e.g., including also transport from farm to processing), (b) detailed descriptions of the feed components used, (c) how the energy required to provide various feed components has been calculated, (d) if inputs to hatchery operations prior to grow-out are included, (e) how fixed infrastructure costs been accounted for, if at all, (f) what if any harvest-related energy costs are included (processing, storage, packing, and transportation), and (g) the size of the operation that EI usages are based on.

Further, there is a general problem when comparing EI inputs between different studies due to spatial and temporal variations in how raw materials are obtained and processed (different sources, extraction processes, distances and means of transport, and processing technology). This makes it difficult to generalize about a specific culture system. Also, comparisons of farming systems in industrialized countries with developing countries may result in a skewed picture because labor is much cheaper in developing countries, resulting in less dependence on industrial energy-consuming technologies. Finally, information regarding the size of the production unit being studied is important given that larger culture units may require less energy input per unit of production—essentially giving rise to "energy economies of scale".

Acknowledgments

The authors thank Alan Stewart, Craig Tucker, and Peter Edwards for help with data access, and Robert Kautsky for help with illustrations.

SEE ALSO THE FOLLOWING ARTICLES

Ecological Footprints and Energy • Ecological Risk Assessment Applied to Energy Development • Ecosystem Health: Energy Indicators • Emergy Analysis and Environmental Accounting • Fisheries and Energy Use • Industrial Agriculture, Energy Flows in • Livestock Production and Energy Use • Ocean, Energy Flows in • Sustainable Development: Basic Concepts and Application to Energy • Wetlands: Impacts of Energy Development in the Mississippi Delta

Further Reading

Ackefors, H., Huner, J. V., and Konikoff, M. (1994). Energy use in aquaculture production. *In* "Introduction to the General Principles of Aquaculture" (R. E. Gough, Ed.), pp. 51–54. Food Products Press, New York.

Andersen, O. (2002). Transport of fish from Norway: energy analysis using industrial ecology as the framework. *J. Cleaner Prodn.* **10**, 581–588.

Bardach, J. (1980). Aquaculture. *In* "Handbook of Energy Utilisation in Agriculture" (D. Pimentel, Ed.), pp. 431–437. CRC Press, Boca Raton.

Berg, H., Michélsen, P., Troell, M., and Kautsky, N. (1996). Managing aquaculture for sustainability in tropical lake Kariba, Zimbabwe. *Ecol. Econ.* **18**, 141–159.

Boyd, C. E., and Tucker, C. S. (1995). Sustainability of channel catfish production. *World Aquacult.* **26**(3), 45–53.

Costa-Pierce, B. A. (ed.). (2002). "Ecological Aquaculture." Blackwell Science, Oxford, UK.

Edwardson, W. (1976). The energy cost of fishing. *Fish. News Int.* **15**(2), 2.

Folke, C., and Kautsky, N. (1992). Aquaculture with its environment: prospects for sustainability. *Ocean Coast. Manage.* **17**, 5–24.

Larsson, J., Folke, C., and Kautsky, N. (1994). Ecological limitations and appropriation of ecosystem support by shrimp farming in Columbia. *Environ. Manage.* **18**(5), 663–676.

Mathews, S. B., Mock, J. B., Willson, K., and Senn, H. (1976). Energy efficiency of Pacific salmon aquaculture. *Progress. Fish-Cultur.* **38**(2), 102–106.

Naylor, L., Goldburg, R. J., Primavera, J., Kautsky, N., Beveridge, M., Clay, J., Folke, C., Lubchenco, J., Mooney, H., and Troell, M. (2000). Effect of aquaculture on world fish supplies. *Nature* **405**, 1017–1024.

Pimentel, D., Shanks, R. E., and Rylander, J. C. (1996). Bioethics of fish production: energy and the environment. *J. Agricult. Environ. Ethics* **9**(2), 144–164.

Rawitscher, M. A. (1978). Energy cost of nutrients in the American diet. Ph.D. thesis (unpubl.). University of Connecticut, Storrs, Connecticut.

Seppälä, J., Silvenius, F., Grönroos, J., Mäkinen, T., Silvo, K., and Storhammar, E. (2001). "Rainbow Trout Production and the Environment. The Finnish Environment" [in Finnish]. Suomen ympäristö 529. Technical report.

Sing, S., and Pannu, C. J. S. (1998). Energy requirements in fish production in the state of Punjab. *Energy Convers. Manage.* **39**, 911–914.

Stewart, J. A. (1995). "Assessing Sustainability of Aquaculture Development." Ph.D. thesis. University of Stirling, Stirling, Scotland.

Tyedmers, P. (2000). Salmon and sustainability: The biophysical cost of producing salmon through the commercial salmon fishery and the intensive salmon culture industry. Ph.D. dissertation. University of British Columbia, Vancouver, Canada.

Tyedmers, P. (2001). Energy consumed by north atlantic fisheries. Fisheries impacts on North Atlantic ecosystems: Catch, effort and national/regional datasets (D. Zeller, R. Watson, and D. Pauly, Eds.). *Fisheries Centre Res. Rep.* 9(3), 12–34.

Watanabe, H., and Okubo, M. (1989). Energy input in marine fisheries of Japan. *Bull. Jpn. Soc. Scientif. Fisheries/Nippon Suisan Gakkaishi* 53(9), 1525–1531.

Arid Environments, Impacts of Energy Development in

DAVID FAIMAN
Ben-Gurion University of the Negev
Beer-Sheva, Israel

1. Arid Lands as a Key to Energy Stability
2. The Desert Ecosystem
3. Danger from Unbridled Development of Deserts
4. Technological Milestones for Safe Development of Arid Zones
5. Potential of Photovoltaic Technology
6. Other Available Technologies
7. Prospects and Conclusions

Glossary

desertification The conversion to true desert, as a result of human activities, of dryland located on the margin between desert and fertile regions.

ecosystem The totality of interdependent living organisms in a given physical environment.

megawatt (MW) Generally, a unit of power ($= 10^6$ W). When used in the context of a photovoltaic system, megawatt units refer to the maximum power output of the photovoltaic modules under specific (and somewhat artificial) test conditions. For this reason, the power ratings of photovoltaic and nonphotovoltaic power plants are not directly comparable.

performance ratio The annual energy output of a photovoltaic system, at a particular site, divided by the power rating of its photovoltaic modules. This figure of merit is, to a large extent, independent of the specific kind of photovoltaic modules employed.

photovoltaic module A panel assembled from a number of individual photovoltaic cells connected in series and parallel so as to provide the required voltage and current specifications of the module. Photovoltaic cells use the energy of incoming light to generate an electric current via the so-called photovoltaic effect. A single silicon photovoltaic cell, with typical dimensions 10×10 cm, would typically give a voltage of 0.5 V, and a current of 2 A when exposed to strong sunlight.

solar thermal A method of generating power from solar energy; in contrast to the photovoltaic method, solar thermal produces heat as an intermediate stage (which may then be used to drive a turbogenerator).

terawatt-hour (TWh) A unit of electrical energy equal to 10^{12} watt-hours, or 3.6×10^{-3} exajoules (EJ).

There is a need to develop the world's arid areas; dual and related incentives to do so involve both raising living standards in the developing world and reducing fossil fuel consumption in the developed world. The focus here is on reasons why this may be possible and proposals on how it may come about. A number of possible impacts of development of arid regions are emphasized, although the list of such impacts cannot be exhaustive at the present time, given the enormous area of Earth's arid ecosystems and the huge range of variables that must be considered in proposing major changes. The purpose here is to outline the nature of the principal impacts and how it might be possible to control them.

1. ARID LANDS AS A KEY TO ENERGY STABILITY

The world is suffering increasingly from environmental problems, many of which, if not all, may be traced to the uncontrolled use of energy and related resources by an ever-increasing global population. On the one hand, the industrially developed part of the world consumes fossil fuel to generate the energy needs of its inhabitants and industries, and to provide transportation both on the ground and in the air. At the other end of the social spectrum, the developing world suffers from an acute dearth of the energy needed for development and, not unnaturally, clamors to right this imbalance. If this situation is not to lead to a globally catastrophic conclusion, it will be necessary to solve a number of urgent energy-related problems during the next few generations. First, the standard of living among all people on Earth must be raised to a high enough level to promote political stability. Second, the worldwide consumption of energy and the concomitant production of waste products must level off at a rate commensurate with environmental stability. Third, worldwide population growth must be reduced to zero.

Of course, each of these problems, though easy to state, encompasses a veritable Pandora's box of social and technical subproblems. As such, even to approach their solution will be a far from simple matter. However, it is clear that a start must be made, and for all three problems the energy development of the world's deserts may hold an important key. First, much of the world's poverty is located in arid regions. Second, solar energy is locally available for the development of those regions. Third, the deserts receive sufficient solar energy that, if it is harnessed, the deserts could provide nonpolluting power station replacements for the rest of the world. Fourth, there is sufficient available land area within the world's deserts to accommodate a large population increase during the time it will take to achieve the ultimately desired goal of global environmental stability.

Naturally, and this is often forgotten, there are many possible impacts (both positive and negative) of energy development in arid environments. It is therefore of vital importance to promote the positive impacts while minimizing the negative ones.

2. THE DESERT ECOSYSTEM

Certain regions on Earth are arid due to the natural circulation of the atmosphere. Warm, moist air, rising from the equatorial belt, cools and releases its precipitation in the tropics, and the resulting dry air returns to Earth, forming two desert regions at approximate latitudes of $\pm 30^{\circ}$. As a result of the dearth of rainfall in these regions, they have little natural vegetation and are consequently sparsely populated both by animals and humans. In spite of an apparent almost complete absence of life in deserts, these regions actually support a wealth of species, both floral and faunal. However, in absolute numbers, their abundances are low and they coexist as a very frail ecosystem. In fact, so frail is the desert ecosystem that even though the species have evolved to be able to withstand arid conditions, extended periods of drought periodically bring about extinctions. It is also possible, and even likely, that additional extinction of species in the arid regions comes about as a result of anthropogenic activities elsewhere on Earth. For example, unnatural damage to the ozone layer, caused by the worldwide use of chlorofluorocarbons (CFCs) and other chemicals, can result in the exposure of living species in the desert belts to elevated and deadly levels of solar ultraviolet-B (UV-B) radiation. Also, sparse as the rainfall in these regions is, the introduction of externally caused acid rain may also be expected to present a serious threat to the natural desert ecosystem.

Naturally, the encroachment of humans on the arid regions, albeit, in the past, in relatively small numbers, also causes serious perturbations to the natural desert ecosystem; desertification, the conversion of marginal desert areas to true deserts, is probably the most notorious example of the impact of human encroachment. It is apparent that today, although the deserts of the world are still relatively undeveloped, energy development elsewhere on Earth can have a negative impact on arid environment ecosystems. Of course, with the massive development of deserts that inevitably may be expected to occur, their natural ecosystems will vanish into history. However, whatever replaces them will clearly need to be taken into consideration because, as we are learning from the phenomenon of global warming, Nature has a tendency to fight back.

In the light of the relatively limited knowledge we possess today on desert ecosystems, it is obvious that there are some rather nontrivial questions that science would do well to try and answer, before unbridled technological development occurs in the arid regions. Specifically, it is important to ask what the respective effects are on a desert ecosystem of introducing various kinds of technology, what should be the criteria for assessing whether the results are good or bad, and what kinds of resulting ecosystems could be considered desirable.

3. DANGER FROM UNBRIDLED DEVELOPMENT OF DESERTS

In addition to being concerned about desert ecosystems, the entire global ecosystem must also be considered. There is already evidence that anthropogenic activities outside arid regions can have a negative impact on the desert ecosystem. However, the reverse can also be true. If the development of arid regions was traditionally limited in the past by the nature of their aridity, modern technology has proved that this need no longer be the case. In fact, the real limiting factor is not water but, rather, energy. After all, were enough energy to be available, then, following the Saudi Arabian example, seawater could be desalinated and transported large distances, allowing food to be grown in even the most arid of deserts. So, suppose, for example, that solar energy could suddenly become available for the practical development of arid regions. The next step is to examine what the overall consequences could be for the world at large.

The first effect would be a massive improvement in the local standard of living. For example, a recent study employing matrices of input/output economic factors for Morocco calculated some of the effects of constructing a single photovoltaic module manufacturing facility in the Sahara Desert. The facility would, by supposition, be capable of producing 5 MW of photovoltaic modules per year. The authors of the study concluded that such a manufacturing facility would create more than 2500 new jobs. In order to place this result in its correct perspective, it is important to point out that the study was part of a scenario in which five such factories would be constructed initially, to be followed by four much larger factories, each with an annual module production of 50 MW. The combined enterprise would provide (a) enough electricity for operating the successive factories and (b) enough photovoltaic modules to enable the construction of a 1.5-GW photovoltaic electric power generating plant, a so-called very large solar photovoltaic plant (VLS-PV). The entire project would take, by calculation, 43 years to complete. The resulting annual electricity output would be 3.5 TWh, at a cost perhaps as low as $0.05/kWh (U.S. dollars)—enough electricity to provide nearly 1 million people with living standards comparable to those enjoyed in today's industrially developed countries. Of course, the concomitant cost reduction that would accompany such a large-scale manufacture of photovoltaic panels would also lead to their more widespread use on an individual family basis in outlying desert areas. Such improved living conditions may be expected to result in a corresponding population growth fueled by the migration of poor people into this new paradise on Earth. Housing and feeding these people should present little problem thanks to the plentiful availability of energy. This much is clearly desirable, because the large-scale relief of poverty may be expected to lead to greater political stability. But what would be achieved at the global environmental level?

Although the local generation of electric power from solar energy would be accompanied by a negligible generation of carbon dioxide, this is not necessarily the case for the associated growth in infrastructure that would result. In the first place, surface and air transport, both within the newly developed arid region and between it and the other developed parts of the world, would result in the release of some 3 kg of CO_2 for each 1 kg of fuel that was burned. If we take octane (C_8H_{18}) as being a typical constituent of motor fuel, then the complete combustion of one molecule of this hydrocarbon produces eight molecules of CO_2, or, in mass units, 1 kg of octane produces 3.1 kg of carbon dioxide. Second, there would be a corresponding increase in the worldwide production of construction materials and consumer goods, transported from places of manufacture where solar energy might not be available at prices that could compete with fossil fuel. Third, the benefits of newfound prosperity in the arid regions would likely lead to lower infant mortality and fewer deaths from disease, resulting in larger, longer lived families, hence increased energy consumption. Clearly, all of these energy-related side effects could lead to a serious overall increase in CO_2 production even though the original wherewithal—solar-generated electricity—is environmentally benign. Of course, solar-generated electricity may lead to a net per capita decrease in CO_2 production, but unless world population growth declines at a sufficiently rapid rate, this will still lead to an absolute increase in CO_2 production.

4. TECHNOLOGICAL MILESTONES FOR SAFE DEVELOPMENT OF ARID ZONES

In order avoid the various spin-off problems of arid region development, a number of parallel technological developments will have to reach fruition:

1. Solar energy (particularly photovoltaic) technology will have to reach an appropriate level of mass-producibility to enable large-area systems to be constructed in all major desert locations.
2. High-temperature (actually, ambient-temperature) superconductive cables will need to be developed, so as to enable the lossless transmission of solar-generated power from desert regions to the other, more industrially developed, parts of the world. This will enable large numbers of fossil-fueled power plants to be made redundant. Of course, environmentally clean electricity could then become a valuable export commodity for desert dwellers.

3. Hydrogen fuel cells and hydrogen-burning combustion engines will have to reach an appropriate level of development and scale of mass-production, to enable the replacement of today's hydrocarbon-based transportation fuel by solar-generated hydrogen. In this way, the present CO_2 waste product (and many others) would be replaced by water vapor. (H_2O is, of course, the major greenhouse gas in Earth's atmosphere, being some three to four times more potent than CO_2. However, it is important to realize that water vapor released by hydrogen combustion engines is generated from water in the first place, so it merely constitutes a part of the natural hydrological cycle of nature. It is here assumed that hydrogen is generated via the electrolysis of water, using solar-generated electricity.)

4. Nonpolluting air transport will need to be developed. This could be based (in situations where time considerations are not of pressing urgency) on the return of a more streamlined and safety-conscious form of hydrogen-based airship technology. Alternatively, new, hydrogen-fueled jet engines will need to be developed.

With the possible exception of pollution-free aeroengines, the technologies itemized here are already under intensive study.

5. POTENTIAL OF PHOTOVOLTAIC TECHNOLOGY

At this stage, it is worth pausing to examine some numbers. Table I lists the total world consumption of primary energy for the year 2001, according to geographic region. Also shown is the potential for photovoltaic power generation of the principal deserts in those regions. A number of clarifying comments may facilitate a better understanding of the tabulated data. First, primary energy consumption includes only commercially traded fuels. It excludes many of the important local fuels such as wood, peat, and animal waste, for which consumption statistics are hard to document. Second, the consumption of primary energy is given in the original source in units of tonnes of oil equivalent (TOE, which has here been converted to Système International units at the conversion rate of 1 TOE = 42 GJ). This begs the important question of how the primary energy is actually used. For low-grade heating purposes, the available energy from each TOE is comparatively high, but it is much lower if used for transportation fuel or electricity production. Thus, each of the tabulated entries for primary consumption should really be broken down into usage categories before comparison with solar availability. On the other hand, the tabulated photovoltaic potential of various desert areas assumes the generation of alternating current (AC) electricity via the use of photovoltaic cells. This is a relatively low-efficiency method (typically 10%) of harnessing solar energy compared with using solar collectors to generate thermal energy (with typical efficiency of 50%). Thus, the tabulated photovoltaic potential is an underestimate of the total solar potential of these deserts. Finally, the tabulated photovoltaic potential has been calculated assuming the use of 14% efficient photovoltaic modules covering each desert surface with a 0.5 spacing factor. What this means is that, in effect, only 50% of the total ground area is covered by photovoltaic panels, the remaining area being needed for access, for ancillary system components, and to prevent mutual shading among the panels (Fig. 1). A system performance ratio of 0.7 has been adopted as being typical of the range of values indicated by the detailed computer studies.

TABLE I

Regional Comparison of Primary Energy Consumption in 2001[a]

Geographic region	Primary energy consumption for the year 2001 (EJ)	Estimated photovoltaic energy capability of local deserts (EJ)
North America	110.9	458.6
South and Central America	19.0	209.2
Europe	79.6	—
Former Soviet Union	39.9	190.4
Middle East	16.7	1115.3
Africa	11.8	4005.4
Asia Pacific	105.5	1013.4
World total	383.2	6992.3

[a] Data from British Petroleum Company (2002) and the estimated photovoltaic electricity production capability of deserts in those regions (Kurokawa, 2003).

Turning now to Table I, we may distinguish between two sets of energy needs: the need to provide energy for the developing world and the need to replace the present usage of fossil fuel in the developed world by renewable energy. These two needs may be vividly contrasted on a map of the world (Fig. 2), on which light areas represent those regions that require energy for development and dark areas represent regions in which today's heavy energy

FIGURE 1 An array of photovoltaic panels at kibbutz Samar in the Negev Desert. Note the spacing between panels, necessary to prevent mutual shading and to allow access for cleaning. Photo courtesy of Bryan Medwed.

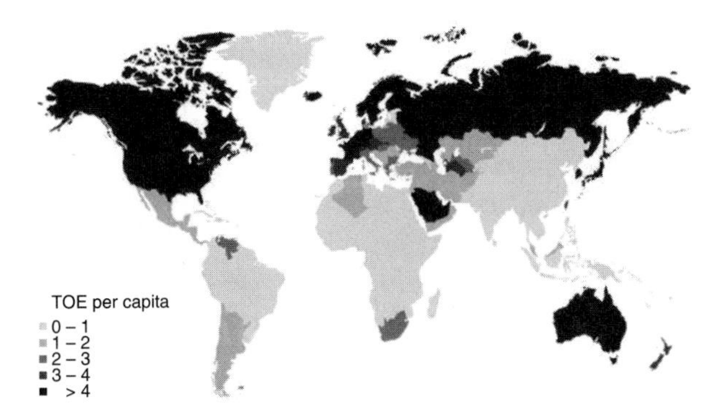

FIGURE 2 World map showing primary energy consumption per capita (2001). TOE, Tonnes of oil equivalent. From British Petroleum Company (2002).

users reside. Regarding the former, one may notice, from Table I, the startling difference between energy consumption and solar availability in both Africa and the Middle East. It is not impossible that much of the instability that has traditionally been associated with these regions could be alleviated by the systematic development of solar energy. Take just one of the Middle East's present-day complex issues, the Israeli–Palestinian conflict, and consider it purely within the context of solar energy development. Although Israel's Negev Desert (12,000 km^2) is minuscule by world, or even Middle Eastern, standards, it occupies approximately 43% of the total land area of the state but houses only a negligible proportion of the country's 6 million inhabitants. Most of that population enjoys a high standard of living due, in large measure, to 9 GW of fossil-fueled electrical generating capacity. On the other hand, the Palestinians have only a poorly developed electrical infrastructure and a correspondingly low standard of living. If, however, a mere 1% of the Negev Desert were to be covered with photovoltaic panels, they could generate the electrical equivalent of a 2- to 3-GW fossil-fueled power plant, enough electrical energy for 2–3 million people. This small example from the Negev provides a vivid indication of the enormous clean energy potential that resides in the vast expanses of desert that cover the entire Middle East.

As regards Africa, there are two promising features in Table I. First, we note that Europe does not, at first glance, appear to have any desert energy-generating potential. However, that continent is only a short distance, by superconducting cable, from the Sahara, where the solar electricity generation potential is clearly enormous. Therefore, a mere 2% of Africa's photovoltaic potential could replace the equivalent of 100% of Europe's present consumption of primary fuel. This would require an area of desert only 400×400 km. Superconducting cable links to the European electricity grid could be made from Morocco to Spain or from Libya to Italy. Thus, solar-generated electricity could constitute an important future export commodity for the African continent. Second, the Sahara Desert also contains sufficient wasteland and solar irradiance to provide energy for all the developmental needs of the African continent in addition to those of Europe. The Sahara Desert is thus a perfect place for the large-scale development of solar energy,

Continuing with the need to replace fossil fuel consumption in the developed world, Table I indicates that North America, the world's major energy consumer, "may" not have sufficient desert area to enable solar energy to provide for all of its future needs. The word "may" is placed in quotation marks because, modulo the previously mentioned caveats, the photovoltaic potential of the North American deserts is a factor of four times higher than their present energy usage. However, it is clearly unfeasible to use all or most of these arid areas for such a purpose. This touches on a far less quantifiable impact of energy development in arid regions than has been hitherto discussed, related, of course, to the problem of severely changing the natural ecosystem, and to all of those scientific questions that need to be studied. But it is also a question of aesthetics. Many people justifiably regard deserts as places of exquisite, unspoiled beauty, sanctuaries for periodic retreat from the hustle and bustle of everyday life. Thus, covering the entire desert, or even 25% of it, with solar collectors would probably be regarded as socially unacceptable. Therefore, to the various ecological questions suggested in Section II, there should be added a number of important sociological questions that will need to be answered before the energy development of arid regions can take place within the developed world.

6. OTHER AVAILABLE TECHNOLOGIES

The need to develop certain renewable technologies for widespread adoption both in arid regions and worldwide was mentioned in Section 4. These technologies were power transmission cables that would superconduct at ambient temperatures, hydrogen fuel cells and hydrogen combustion engines for surface and air travel, and low-cost photovoltaic solar electricity generators. In Section 5 the discussions elaborated the promise of very large-scale photovoltaic systems for two reasons. On the one hand, such discussions serve as a useful indicator as to the manner in which arid regions could serve as power stations of the future, and what some of the associated developmental impacts might be. But on the other hand, photovoltaics represents a technology that has been in existence for approximately half a century and is, consequently, relatively mature, albeit costly compared to fossil fuel alternatives. Nevertheless, it is important to remember that there are a number of alternative renewable technologies that have also been successfully demonstrated at the multimegawatt scale, and which may also be expected to play an important role in the development of arid regions.

Solar-thermal power stations have been built on the scale of tens of megawatts apiece, and some 350 MW of total electrical generating capacity have been operating in the California desert for the past 15–20 years. These power plants employ long rows of troughlike parabolic mirrors in order to concentrate sunlight onto a central tube. Oil flowing within the tube is heated to high temperatures and is used to convert water into steam that, in turn, drives a turbogenerator. Although the economic climate that prevailed in California when these systems were first constructed favored their use for electricity generation, it is important to realize that the thermal energy these systems generate as an intermediate stage represents a far more efficient use of solar energy. In these systems, the oil is heated to approximately 350°C, at an efficiency close to 50%. This heat is then used to power a steam turbine at an efficiency of about 30%. Thus, the combined solar-to-electrical efficiency is only around 15%. It was previously mentioned that for many situations in which thermal energy is required (e.g., industrial processes), the use of photovoltaic electricity would be needlessly inefficient. It is understandable, therefore (Fig. 3), how, in addition to providing electricity, such rows of parabolic trough reflectors could provide solar-powered thermal energy for the industries of desert towns in the future.

Wind power represents another technology that has been successfully demonstrated on a large scale (Fig. 4). This technology could usefully supplement the employment of solar energy in arid zones because its generating capacity is not limited to the daylight hours. In fact, at certain locations and with careful system design, it may be possible to use the same parcels of land for power generation using both solar and wind technologies.

Of course, as the variety of energy-producing technologies increases, so too, to a certain extent,

FIGURE 3 Solar-thermal electric power plant at Kramer Junction, California, United States. Photo courtesy of Solel Solar Systems Ltd.

FIGURE 4 Wind turbines at Jaisalmer, India. Photo courtesy of Safrir Faiman.

will be the variety of additional impacts that will have to be studied. The possibility of oil leakage and fire hazard will be among topics of relevance if solar-thermal systems are to be employed on a very large scale, as will the possible effects of wind turbines on bird life and human hearing.

7. PROSPECTS AND CONCLUSIONS

Historically, technological advances have occurred piecemeal, being driven, by and large, by motives of greed. It is only comparatively recently that the importance of the larger picture has come to be realized: namely, the effects of technological development on the environment. Hand in glove with such technological advances, important social issues are also more appreciated today than in the past.

The development of deserts might be allowed to come about driven purely by the profit motive. For example, photovoltaic manufacturers will naturally be interested in pursuing the potentially large markets that exist in the world's desert regions. With capital investment from the outside, such actions will certainly improve local living standards. However, as was indicated previously, if this occurs too rapidly, on a large enough scale, it could lead to an alarming increase of induced CO_2 production at the global level, as the industrial countries increase their output of consumer goods. Ideally, what is needed is an orchestrated approach by industry and government among the wealthier nations. The first priority should be to raise living standards in the developing world by the provision of environmentally clean electricity. Not only will this help stabilize many of those regions in which poverty is rife, but it will also help bring on that time when all the world's population can, together, discuss what is best for everybody. Of course, as the use of consumer goods increases among the newly wealthy parts of the world, it may be necessary for the previously developed nations to take ever more stringent measures to check their own usage of energy. This will not necessarily mean reducing living standards, but it will mean heavier investment into energy efficiency and research into new methods of alternative energy usage. Mention was made of the importance of hydrogen fuel cells and the need to develop hydrogen combustion engines for surface and air transport. But, no doubt, many other developments may be expected to occur as the research proceeds.

SEE ALSO THE FOLLOWING ARTICLES

Aircraft and Energy Use • Fuel Cell Vehicles • Global Energy Use: Status and Trends • Hydrogen, End Uses and Economics • Photovoltaic Energy: Stand-Alone and Grid-Connected Systems • Photovoltaics, Environmental Impact of • Population Growth and Energy • Solar Thermal Energy, Industrial Heat Applications • Solar Thermal Power Generation

Further Reading

British Petroleum Company (BP). (2002). "BP Statistical Review of World Energy," 51st Ed. BP Corporate Communication Services, London. Available at the corporate Web site at http://www.bpamoco.com.

Johansson, T.B., Kelly, H., Reddy, A.K.N., and Williams, R.H. (eds.). (1993). Renewable Energy: Sources for Fuels and Electricity. Island Press, Washington, D.C.

Kurokawa, K. (ed.). (2003). "Energy from the Desert: Feasibility of Very Large Scale Photovoltaic Power Generation (VLS-PV) Systems." James & James, London.

Rabl, A. (1985). "Active Solar Collectors and Their Applications." Oxford Univ. Press, New York.

Batteries, Overview

ELTON J. CAIRNS
Lawrence Berkeley National Laboratory and
University of California, Berkeley
Berkeley, California, United States

1. How Batteries Work
2. History
3. Types of Batteries and Their Characteristics
4. Applications
5. Environmental Issues
6. Future Outlook

Glossary

battery An assemblage of cells connected electrically in series and/or parallel to provide the desired voltage and current for a given application.

cell An electrochemical cell composed of a negative electrode, a positive electrode, an electrolyte, and a container or housing.

electrode An electronically conductive structure that provides for an electrochemical reaction through the change of oxidation state of a substance; may contain or support the reactant or act as the site for the electrochemical reaction.

electrolyte A material that provides electrical conduction by the motion of ions only; may be a liquid, solid, solution, polymer, mixture, or pure substance.

primary cell A cell that cannot be recharged and is discarded after it is discharged.

secondary cell A rechargeable cell.

Batteries are an important means of generating and storing electrical energy. They are sold at a rate of several billions of dollars per year worldwide. They can be found in nearly all motor vehicles (e.g., automobiles, ships, aircraft), all types of portable electronic equipment (e.g., cellular phones, computers, portable radios), buildings (as backup power supplies), cordless tools, flashlights, smoke alarms, heart pacemakers, biomedical instruments, wristwatches, hearing aids, and the like. Batteries are so useful and ubiquitous, it is difficult to imagine how life would be without them. Batteries, strictly speaking, are composed of more than one electrochemical cell. The electrochemical cell is the basic unit from which batteries are built. A cell contains a negative electrode, a positive electrode, an electrolyte held between the electrodes, and a container or housing. Cells may be electrically connected to one another to form the assembly called a battery. In contrast to the alternating current available in our homes from the electric utility company, batteries deliver a direct current that always flows in one direction. There are a few different types of batteries: Primary batteries can be discharged only once and then are discarded; they cannot be recharged. Secondary batteries are rechargeable. Forcing current through the cells in the reverse direction can reverse the electrochemical reactions that occur during discharge. Both primary and secondary batteries can be categorized based on the type of electrolyte they use: aqueous, organic solvent, polymer, ceramic, molten salt, and so on.

1. HOW BATTERIES WORK

The electrical energy produced by batteries is the result of spontaneous electrochemical reactions. The driving force for the reactions is the Gibbs free energy of the reaction, which can be calculated easily from data in tables of thermodynamic properties. The maximum voltage that a cell can produce is calculated from this simple relationship:

$$E = -\Delta G/nF, \qquad (1)$$

where E is the cell voltage, ΔG is the Gibbs free energy change for the cell reaction (joules/mole), n is the number of electrons involved in the reaction (equivalents/mole), and F is the Faraday constant (96487 coulombs/equivalent).

It is clear from Eq. (1) that a large negative value for the Gibbs free energy of the cell reaction is desirable if we wish to have a significant cell voltage.

In practical terms, cell voltages of 1 to 2 V are achieved when using aqueous electrolytes, and cell voltages up to approximately 4 V are achieved when using nonaqueous electrolytes. When larger voltages are required, it is necessary to connect a number of cells electrically in series. For example, 12-volt automotive batteries are composed of six 2-volt cells connected in series.

An important property of batteries is the amount of energy they can store per unit mass. This is the specific energy of the battery, usually expressed in units of watt-hours of energy per kilogram of battery mass (Wh/kg). The maximum value of the specific energy is that which can be obtained from a certain mass of reactant materials, assuming the case that any excess electrolyte and terminals have negligible mass. This is called the theoretical specific energy and is given by this expression:

$$\text{Theoretical Specific Energy} = -\Delta G/\Sigma Mw, \quad (2)$$

where the denominator is the summation of the molecular weights of the reactants.

It can be seen from Eq. (2) that the theoretical specific energy is maximized by having a large negative value for ΔG and a small value for ΣMw. The value of ΔG can be made large by selecting for the negative electrode those reactant materials that give up electrons very readily. The elements with such properties are located on the left-hand side of the periodic chart of the elements. Correspondingly, the positive electrode reactant materials should readily accept electrons. Elements of this type are located on the right-hand side of the periodic chart. Those elements with a low equivalent weight are located toward the top of the periodic chart. These are useful guidelines for selecting electrode materials, and they help us to understand the wide interest in lithium-based batteries now used in portable electronics. Equations (1) and (2) give us a useful framework for representing the theoretical specific energy and the cell voltage for a wide range of batteries, as shown in Fig. 1. The individual points on the plot were calculated from the thermodynamic data. These points represent theoretical maximum values. The practical values of specific energy that are available in commercial cells are in the range of one-fifth to one-third of the theoretical values. As expected, the cells using high equivalent weight materials (e.g., lead, lead dioxide) have low specific energies, whereas those using low equivalent weight materials (e.g. lithium, sulfur) have high specific energies. The lines in Fig. 1 represent the cell voltages and simply represent the relationships given by Eqs. (1) and (2).

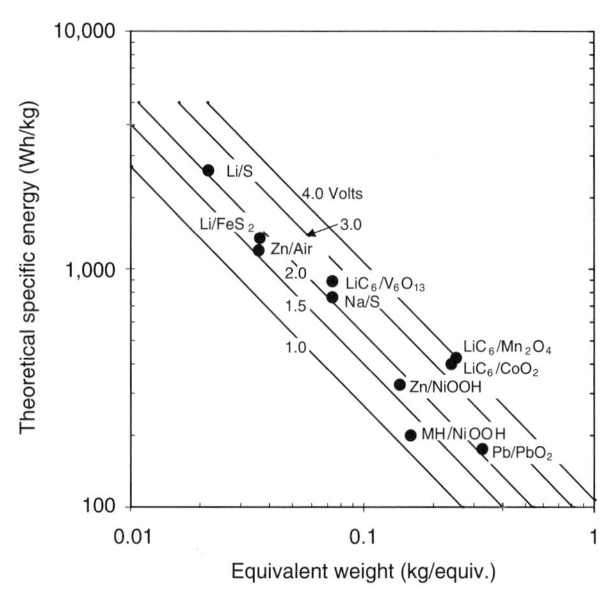

FIGURE 1 Theoretical specific energy for various cells as a function of the equivalent weights of the reactants and the cell voltage.

The details of the electrochemical reactions in cells vary, but the principles are common to all. During the discharge process, an electrochemical oxidation reaction takes place at the negative electrode. The negative electrode reactant (e.g., zinc) gives up electrons that flow into the electrical circuit where they do work. The negative electrode reactant is then in its oxidized form (e.g., ZnO). Simultaneously, the positive electrode reactant undergoes a reduction reaction, taking on the electrons that have passed through the electrical circuit from the negative electrode (e.g., MnO_2 is converted to MnOOH). If the electrical circuit is opened, the electrons cannot flow and the reactions stop.

The electrode reactions discussed in the preceding can be written as follows:

$$Zn + 2OH^- = ZnO + H_2O + 2e^-$$

and

$$2MnO_2 + 2H_2O + 2e^- = 2MnOOH + 2OH^-.$$

The sum of these electrode reactions is the overall cell reaction:

$$Zn + 2MnO_2 + H_2O = ZnO + 2MnOOH.$$

Notice that there is no net production or consumption of electrons. The electrode reactions balance exactly in terms of the electrons released by the negative electrode being taken on by the positive electrode.

2. HISTORY

Volta's invention of the Volta pile in 1800 represents the beginning of the field of battery science and engineering. The pile consisted of alternating layers of zinc, electrolyte soaked into cardboard or leather, and silver. Following Volta's report, many investigators constructed electrochemical cells for producing and storing electrical energy. For the first time, relatively large currents at high voltages were available for significant periods of time. Various versions of the pile were widely used. Unfortunately, there was a lack of understanding of how the cells functioned, but as we now know, the zinc was electrochemically oxidized and the native oxide layer on the silver was reduced. The cells were "recharged" by disassembling them and exposing the silver electrodes to air, which reoxidized them. Inevitably, many other electrochemical cells and batteries were developed.

John F. Daniell developed a two-fluid cell in 1836. The negative electrode was amalgamated zinc, and the positive electrode was copper. The arrangement of the cell is shown in Fig. 2. The copper electrodes were placed in (porous) porcelain jars, which were surrounded by cylindrical zinc electrodes and placed in a larger container. A copper sulfate solution was placed in the copper electrode's compartment, and sulfuric acid was put in the zinc electrode compartment.

The electrode reactions were as follows:

$$Zn = Zn^{+2} + 2e^-$$

and

$$Cu^{+2} + 2e^- = Cu.$$

FIGURE 2 A set of three Daniell cells connected in series.

Sir William R. Grove, a lawyer and inventor of the fuel cell, developed a two-electrolyte cell related to the Daniell cell in 1839. Grove used fuming nitric acid at a platinum electrode (the positive) and zinc in sulfuric acid (the negative). Variants on this formulation were popular for a number of years. Of course, all of these cells were primary cells in that they could be discharged only once and then had to be reconstructed with fresh materials.

Gaston Planté invented the first rechargeable battery in 1860. It was composed of lead sheet electrodes with a porous separator between them, spirally wound into a cylindrical configuration. The electrolyte was sulfuric acid. These cells displayed a voltage of 2.0 V, an attractive value. During the first charging cycles, the positive electrode became coated with a layer of PbO_2. The charging operation was carried out using primary batteries—a laborious process. Because of the low cost and the ruggedness of these batteries, they remain in widespread use today, with various evolutionary design refinements.

The electrode reactions during discharge of the lead–acid cell are as follows:

$$Pb + H_2SO_4 = PbSO_4 + 2H^+ + 2e^-$$

and

$$PbO_2 + H_2SO_4 + 2H^+ + 2e^- = PbSO_4 + 2H_2O.$$

Notice that sulfuric acid is consumed during discharge and that the electrolyte becomes more dilute (and less dense). This forms the basis for determining the state of charge of the battery by measuring the specific gravity of the electrolyte. The theoretical specific energy for this cell is 175 Wh/kg, a rather low value compared with those of other cells.

Waldemar Jungner spent much of his adult life experimenting with various electrode materials in alkaline electrolytes. He was particularly interested in alkaline electrolytes because there generally was no net consumption of the electrolyte in the cell reactions. This would allow for a minimum electrolyte content in the cell, minimizing its weight. Jungner experimented with many metals and metal oxides as electrode materials, including zinc, cadmium, iron, copper oxide, silver oxide, and manganese oxide.

In parallel with Jungner's efforts in Sweden, Thomas Edison in the United States worked on similar ideas using alkaline electrolytes and many of the same electrode materials. Patents to these two inventors were issued at nearly the same time in 1901.

During the period since the beginning of the 20th century, a wide variety of cells have been investigated, with many of them being developed into commercial products for a wide range of applications. Representative cells are discussed in the next section.

3. TYPES OF BATTERIES AND THEIR CHARACTERISTICS

Batteries are usually categorized by their ability to be recharged or not, by the type of electrolyte used, and by the electrode type. Examples of various types of electrolytes used in batteries are presented in Table I.

Primary (nonrechargeable) cells with aqueous electrolytes are usually the least expensive and have reasonably long storage lives. Historically, the most common primary cell is the zinc/manganese dioxide cell used for flashlights and portable electronics in sizes from a fraction of an amp-hour to a few amp-hours. Various modifications of this cell have been introduced, and now the most common version uses an alkaline (potassium hydroxide) electrolyte. The materials in this cell are relatively benign from an environmental point of view. The electrode reactions in an alkaline electrolyte may be represented as follows:

$$Zn + 2OH^- = ZnO + H_2O + 2e^-$$

and

$$2MnO_2 + 2H_2O + 2e^- = 2MnOOH + 2OH^-.$$

The cell potential is 1.5 V, and the theoretical specific energy is 312 Wh/kg. Practical cells commonly yield 40 to 50 Wh/kg.

Another common primary cell is the zinc/air cell. It uses an aqueous potassium hydroxide electrolyte, a zinc negative electrode, and a porous catalyzed carbon electrode that reduces oxygen from the air. The overall electrode reactions are as follows:

$$Zn + 2OH^- = ZnO + H_2O + 2e^-$$

and

$$\tfrac{1}{2}O_2 + H_2O + 2e^- = 2OH^-.$$

TABLE I

Electrolyte Types and Examples

Aqueous electrolyte	H_2SO_4, KOH
Nonaqueous electrolyte	
Organic solvent	Li salt in ethylene carbonate-diethyl carbonate
Solid	
Crystalline	$Na_2O \times 11Al_2O_3$
Polymeric	Li salt in polyethylene
Molten salt (high temperature)	LiCl–KCl

The sum of these reactions gives this overall cell reaction:

$$Zn + \tfrac{1}{2}O_2 = ZnO.$$

The cell voltage is 1.6 V, and the theoretical specific energy is 1200 Wh/kg. Practical specific energy values of 200 Wh/kg have been achieved for zinc/air cells. This cell has a very high energy per unit volume, 225 Wh/L, which is important in many applications. The materials are inexpensive, so this cell is very competitive economically. This cell is quite acceptable from an environmental point of view.

The zinc/mercuric oxide cell has the unique characteristic of a very stable and constant voltage of 1.35 V. In applications where this is important, this is the cell of choice. It uses an aqueous potassium hydroxide electrolyte and is usually used in small sizes. The electrode reactions are as follows:

$$Zn + 2OH^- = ZnO + H_2O + 2e^-$$

and

$$HgO + H_2O + 2e^- = Hg + 2OH^-.$$

The high equivalent weight of the mercury results in the low theoretical specific energy of 258 Wh/kg. Because mercury is toxic, there are significant environmental concerns with disposal or recycling.

The zinc/silver oxide cell has a high specific energy of about 100 Wh/kg (theoretical specific energy = 430 Wh/kg), and because of this, the cost of the silver is tolerated in applications that are not cost-sensitive. Potassium hydroxide is the electrolyte used here. The electrode reactions are as follows:

$$Zn + 2OH^- = ZnO + H_2O + 2e^-$$

and

$$AgO + H_2O + 2e^- = Ag + 2OH^-.$$

This gives the following overall cell reaction:

$$Zn + AgO = Ag + ZnO.$$

The cell voltage ranges from 1.8 to 1.6 V during discharge. Because of the high density of the silver oxide electrode, this cell is quite compact, giving about 600 Wh/L.

Primary (nonrechargeable) cells with nonaqueous electrolytes have been under development since the 1960s, following the reports of Tobias and Harris on the use of propylene carbonate as an organic solvent for electrolytes to be used with alkali metal electrodes. It has long been recognized that lithium has very low equivalent weight and electronegativity (Table II). These properties make it very attractive

TABLE II

Some Properties of Lithium and Zinc

	Lithium	Zinc
Equivalent weight (g/equiv.)	6.94	32.69
Reversible potential (V)	−3.045	−0.763
Electronegativity	0.98	1.65
Density (g/cm^3)	0.53	7.14

for use as the negative electrode in a cell. Because lithium will react rapidly with water, it is necessary to use a completely anhydrous electrolyte. There are many organic solvents that can be prepared free of water. Unfortunately, organic solvents generally have low dielectric constants, making them poor solvents for the salts necessary to provide electrolytic conduction. In addition, organic solvents are thermodynamically unstable in contact with lithium, so solvents that react very slowly and form thin, conductive, protective films on lithium are selected.

Perhaps the most common lithium primary cell is that of Li/MnO_2. It has the advantage of high voltage (compared with aqueous electrolyte cells), 3.05 V, and a high specific energy (~170 Wh/kg). The electrode reactions are as follows:

$$Li = Li^+ + e^-$$

and

$$MnO_2 + Li^+ + e^- = LiMnO_2.$$

The electrolyte is usually $LiClO_4$ dissolved in a mixture of propylene carbonate and 1,2-dimethoxyethane. In general, the specific power that can be delivered by these cells is less than that available from aqueous electrolyte cells because the ionic conductivity of the organic electrolyte is much lower than that of aqueous electrolytes.

An interesting cell is the one using fluorinated carbon as the positive electrode material. Because this material is poorly conducting, carbon and titanium current collection systems are used in the positive electrode. The electrode reaction is as follows:

$$(CF)_n + nLi^+ + ne^- = nLiF + nC.$$

The electrolyte commonly is $LiBF_4$ dissolved in gamma butyrolactone.

An unusual primary cell is that of lithium/thionyl chloride ($SOCl_2$). The thionyl chloride is a liquid and can act as the solvent for the electrolyte salt ($LiAlCl_4$) and as the reactant at the positive electrode. This cell can function only because of the relatively stable thin protective film that forms on the lithium electrode, protecting it from rapid spontaneous reaction with the thionyl chloride. The cell reaction mechanism that is consistent with the observed products of reaction is the following:

$$4Li + 2SOCl_2 = 4LiCl + 2SO'$$

and

$$2SO' = SO_2 + S,$$

where SO′ is a radical intermediate that produces SO_2 and sulfur. This cell provides a high cell voltage of 3.6 V and a very high specific energy but can be unsafe under certain conditions. Specific energies of up to 700 Wh/kg have been achieved compared with the theoretical value of 1460 Wh/kg.

Table III summarizes the characteristics of several representative primary cells.

Primary cells with molten salt electrolytes are commonly used in military applications that require a burst of power for a short time (a few seconds to a few minutes). These batteries are built with an integral heating mechanism relying on a chemical reaction to provide the necessary heat to melt the electrolyte (at 400–500°C). A typical cell uses lithium as the negative electrode and iron disulfide as the positive electrode. The following are example electrode reactions:

$$Li = Li^+ + e^-$$

and

$$FeS_2 + 4Li = 2Li_2S + Fe.$$

The molten salt electrolyte (a mixture of alkali metal chlorides) has a very high conductivity, allowing the cell to operate at very high specific power levels in excess of 1 kW/kg.

Rechargeable cells with aqueous electrolytes have been available for more than 140 years, beginning with the Planté cell using lead and sulfuric acid as discussed previously. Modern Pb/PbO_2 cells and batteries have received the benefit of many incremental improvements in the design and optimization of the system. Current versions are very reliable and inexpensive compared with competitors. Depending on the design of the cell, lifetimes vary from a few years to more than 30 years. Specific energy values of up to approximately 40 Wh/kg are available.

Alkaline electrolyte systems are available with a variety of electrodes. Perhaps the most common alkaline electrolyte cell is the Cd/NiOOH cell. It offers very long cycle life (up to thousands of charge–discharge cycles) and good specific power (hundreds

TABLE III

Properties of Some Lithium Primary Cells

	Open circuit voltage (V)	Practical specific energy (Wh/kg)	Theoretical specific energy (Wh/kg)
Li/MnO_2	3.0	280	750
$Li/(CF)_n$	2.8	290	2100
$Li/SOCl_2$	3.6	450–700	1460
Li/FeS_2	1.7	220	760

of W/kg), albeit with a modest specific energy (35–55 Wh/kg). The cell voltage, 1.2 V, is lower than most and can be somewhat of a disadvantage. The electrode reactions are as follows:

$$Cd + 2OH^- = Cd(OH)_2 + 2e^-$$

and

$$2NiOOH + 2H_2O + 2e^- = 2Ni(OH)_2 + 2OH^-.$$

Of course, the reverse of these reactions takes place on recharge.

A very robust rechargeable cell is the Edison cell, which uses iron as the negative electrode, nickel oxyhydroxide as the positive electrode, and an aqueous solution of 30 w/o potassium hydroxide as the electrolyte. During the early years of the 20th century, Edison batteries were used in electric vehicles. They proved themselves to be very rugged and durable, although they did not have high performance. The iron electrode reaction is as follows:

$$Fe + 2OH^- = Fe(OH)_2 + 2e^-.$$

The positive electrode reaction is the same as for the Cd/NiOOH cell described earlier.

The NiOOH electrode has proven itself to be generally the best positive electrode for use in alkaline electrolytes and has been paired with many different negative electrodes, including Cd, Fe, H_2, MH, and Zn. Recently, the metal hydride/nickel oxyhydroxide cell (MH/NiOOH) has been a significant commercial success and has captured a large fraction of the market for rechargeable cells. It offers a sealed, maintenance-free system with no hazardous materials. The performance of the MH/NiOOH cell is very good, providing up to several hundred watts/kilogram peak power, up to about 85 Wh/kg, and several hundred cycles. The negative electrode operates according to the following reaction:

$$MH + OH^- = M + H_2O + e^-.$$

The very high theoretical specific energy and low materials cost of the zinc/air cell make it attractive for consumer use. There have been many attempts to develop a rechargeable zinc/air cell, but with limited success due to the difficulties of producing a high-performance rechargeable air electrode.

The electrode reactions during discharge for this cell are as follows:

$$Zn + 2OH^- = ZnO + H_2O + 2e^-$$

and

$$O_2 + 2H_2O + 4e^- = 4OH^-.$$

During recent years, the cycle life of the air electrode has been improved to the point where more than 300 cycles are now feasible. Development efforts continue. In the meantime, various versions of a “mechanically rechargeable” zinc/air cell have been tested. These cells have provision for removing the discharged zinc and replacing it with fresh zinc. This approach avoids the difficulties of operating the air electrode in the recharge mode but creates the need for recycling the spent zinc to produce new zinc electrode material.

The status of some rechargeable cells with aqueous electrolytes is shown in Table IV.

Rechargeable cells with nonaqueous electrolytes have been under development for many years, although they have been available on the consumer market for only a decade or so. All of the types of nonaqueous electrolytes shown in Table I have been used in a variety of systems.

Organic solvent-based electrolytes are the most common of the nonaqueous electrolytes and are found in most of the rechargeable lithium cells available today. A typical electrolyte consists of a mixture of ethylene carbonate and ethyl-methyl carbonate, with a lithium salt such as $LiPF_6$ dissolved in it. Various other solvents, including propylene carbonate, dimethyl carbonate, and diethyl carbonate, have been used. Other lithium salts that have been used include lithium perchlorate, lithium hexafluoro arsenate, and lithium tetrafluoro borate. All of these combinations of solvents and salts yield electrolytes that have much lower conductivities than do typical aqueous electrolytes. As a result, the electrodes and electrode spacing in these lithium cells are made very thin to minimize the cell resistance and maximize the power capability.

Lithium is difficult to deposit as a smooth compact layer in these organic electrolytes, so a host material, typically carbon, is provided to take up the

TABLE IV

Rechargeable Aqueous Battery Status

System	Cell voltage (V)	Theoretical specific energy (Wh/kg)	Specific energy (Wh/kg)	Specific power (W/kg)	Cycle life	Cost (dollars/kWh)
Pb/PbO_2	2.10	175	30–45	50–100	>700	60–125
Cd/NiOOH	1.20	209	35–55	400	2000	>300
Fe/NiOOH	1.30	267	40–62	70–150	500–2000	>100
H_2/NiOOH	1.30	380	60	160	1000–2000	>400
MH/NiOOH	1.20	200	60–85	200+	500–600	>400
Zn/NiOOH	1.74	326	55–80	200–300	500+	150–250 (est.)
Zn/Air	1.60	1200	65–120	<100	300	100 (est.)

Note. est., estimate.

Li on recharge and deliver it as Li ions to the electrolyte during discharge. Many types of carbon, both graphitic and nongraphitic, have been used as the lithium host material. In addition, a variety of intermetallic compounds and metals have been used as Li host materials. All of the commercial Li rechargeable cells today use a carbon host material as the negative electrode.

The positive electrodes of Li cells are usually metal oxide materials that can intercalate Li with a minimal change in the structure of the material. The most common positive electrode material used in commercial Li cells is $LiCoO_2$. This material is very stable toward the removal and reinsertion of Li into its structure. It can be cycled more than 1000 times with little change in capacity.

The electrode reactions during discharge for this cell are as follows:

$$xLi(C) = xLi^+ + C + xe^-$$

and

$$Li_{1-x}CoO_2 + xLi^+ + xe^- = LiCoO_2.$$

Excellent engineering of this cell has resulted in a specific energy of more than 150 Wh/kg and a peak specific power of more than 200 W/kg, making it very desirable for applications requiring high specific energy. This cell has been manufactured extensively first in Japan and now in Korea and China, and it has been used worldwide.

Even though the $Li(C)/CoO_2$ cell has many desirable features, it has some limitations and disadvantages. The use of Co represents an environmental hazard because Co is toxic. It is also too expensive to use in large batteries suitable for low-cost applications such as electric and hybrid vehicle propulsion. In addition, there are safety problems associated with overcharge of this cell. Significant overcharge can cause the cell to vent and burn. Operation at elevated temperatures can also cause fire. Special microcircuits mounted on each cell provide control over the cell, preventing overcharge and overdischarge.

During the past several years, there has been a large effort to replace the $LiCoO_2$ with a less expensive, more environmentally benign material that can withstand overcharge and overdischarge while retaining the excellent performance of $LiCoO_2$. Many metal oxide materials have been synthesized and evaluated. Some of them are promising, but none has surpassed $LiCoO_2$ so far. Popular candidate materials include lithium manganese oxides such as $LiMn_2O_4$, $LiMnO_2$, and materials based on them, with some of the Mn replaced by other metals such as Cr, Al, Zn, Ni, Co, and Li. Other candidates include $LiNiO_2$, $LiFePO_4$, and variants on these, with some of the transition metal atoms replaced by other metal atoms in an attempt to make the materials more stable to the insertion and removal of the Li during discharge and charge. Progress is being made, and some Mn-based electrode materials are appearing in commercial cells.

Some of the instabilities that have been observed are traceable to the electrolyte and its reactivity with the electrode materials. This has resulted in some research on more stable solvents and Li salts. The Li salts must be stable to voltages approaching 5 V and must be very soluble and highly conductive. The solvents must have a wide temperature range of liquidity so that cells can be operated from very cold to very warm temperatures (-40 to $+60$°C for some applications).

The emphasis on lower cost electrode materials has continued, and some encouraging results have been obtained for materials based on $LiFePO_4$. This material is very stable toward repeated insertion and removal of Li and has shown hundreds of cycles with little capacity loss. One problem with $LiFePO_4$ is its low electronic conductivity, making full use and high rate operation difficult, even with enhanced current collection provided by added carbon.

Some unusual electrode materials are under investigation in the quest to reduce cost. These include elemental sulfur and organo-disulfide polymers. The Li/S cell has been investigated several times during the past 20 years or more. From a theoretical specific energy point of view, it is very attractive at 2600 Wh/kg. The cost of sulfur is extremely low, and it is environmentally benign. A significant difficulty with the sulfur electrode in most organic solvents is the formation of soluble polysulfides during discharge. The soluble polysulfides can migrate away from the positive electrode and become inaccessible, resulting in a capacity loss.

The overall reaction for the sulfur electrode is as follows:

$$S + 2e^{-} = S^{-2}.$$

And the overall cell reaction is as follows:

$$2Li + S = Li_2S.$$

Recent work on this system has involved isolating the soluble polysulfides with an ionically conductive membrane or using solvents that have a low solubility for the polysulfides that are formed during discharge. These approaches have resulted in a large increase of cycle life, up to several hundred cycles.

The organo–disulfides offer a lower specific energy than does elemental sulfur, but they do not exhibit the same relatively high solubility. Long cycle lives have been demonstrated for these electrodes. They are usually coupled with a lithium metal negative electrode, which has a short cycle life. Current efforts are under way to improve the cycle life of the lithium metal electrode.

The status of some rechargeable cells with nonaqueous electrolytes is presented in Table V. The performance of several rechargeable batteries is shown in Fig. 3.

There are a few rechargeable batteries with molten salt or ceramic electrolytes that operate at elevated temperatures, but these have found only specialty (demonstration) applications such as stationary energy storage and utility vehicle propulsion and are not yet in general use. Examples of these are

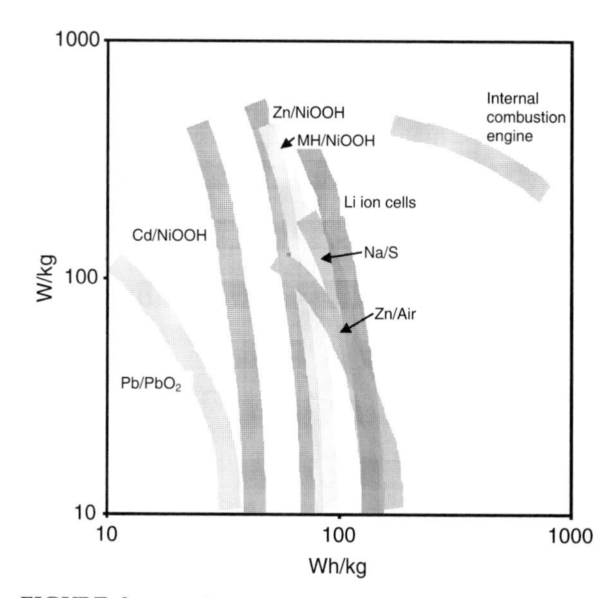

FIGURE 3 Specific power, watts/kilogram (W/kg), as a function of specific energy, watt-hours/kilogram (Wh/kg), for various cells. A curve for an internal combustion engine is included for reference.

TABLE V

Nonaqueous Rechargeable Battery Status

System	Cell voltage (V)	Equivalent weight (kg)	Theoretical specific energy (Wh/kg)	Specific energy (Wh/kg)	Specific power (W/kg)	Cost (dollars/kWh)
Li/CoO_2	3.60	0.1693	570	125	>200	>>100
Li/Mn_2O_4	4.00	0.1808	593	150	200	~200
$Li/FePO_4$	3.5	0.151	621	120	100(est.)	~200
Li/V_6O_{13}	2.40	0.0723	890	150	200	>200
Li/TiS_2	2.15	0.1201	480	125	65	>200
Li/SRS	3.00	0.080	1000	200(est.)	400(est.)	N.A.
Li/S	2.10	0.0216	2600	300	200	<100

Note. est., estimate; N.A., not available.

TABLE VI

High-Temperature Battery Status

System	Cell voltage (V)	Theoretical specific energy (Wh/kg)	Specific energy (Wh/kg)	Specific power (W/kg)	Cycle life	Cost (dollars/kWh)	Temperature (°C)
Na/S	2.08	758	180	160	500–2000	>100	350
$Na/NiCl_2$	2.58	787	90	90–155	600	>300	350
$LiAl/FeS_2$	1.80	514	180–200	>200	1000	N.A.	400

Note. N.A., not available.

the lithium/iron sulfide cell with a molten alkali halide electrolyte, the sodium/sulfur cell with a beta-alumina ceramic ($Na_2O \cdot Al_2O_3$) electrolyte, and the sodium/nickel chloride cell with both molten salt and beta-alumina electrolytes. These cells operate at 350 to 400°C. The characteristics of these cells are shown in Table VI.

4. *APPLICATIONS*

Current applications of batteries are very wide-ranging. It is difficult to identify many technologies that do not rely on either primary or secondary batteries. Batteries are part of our everyday lives and are essential to many fields with which most people do not come into contact.

Consumer applications of batteries include flashlights, portable electronics (e.g., cell phones), radios, televisions, MP3 players, personal digital assistants, notebook computers, portable tools, power garden tools, burglar alarms, smoke detectors, emergency lighting, remote control devices for television, stereos, electronic games, remote-controlled toys, and other entertainment equipment. Medical applications include heart pacemakers, medical diagnostic devices (e.g., glucose monitors), pumps for administering drugs, and nerve and muscle stimulators. Automobiles depend on batteries for starting, lighting, ignition, and related functions. Golf carts, utility vehicles, and forklift trucks represent a variety of propulsion-power applications.

There are many government and military applications with stringent requirements for batteries, including communications radios and phones, portable lasers for range finding, field computers, night vision equipment, sensors, starting for all types of vehicles, power sources for fuses, timers, various missiles, aircraft starting and electrical systems, electrically powered vehicles (e.g., submarines), and off-peak energy storage for utility power.

The space program depends critically on batteries for all manner of power supplies for communication, operation of the electrical systems in shuttle craft and space stations, power for the small vehicles and sensors deployed on the moon and planets, life support systems, power for space suits, and many others.

The batteries that are used in these applications come in many sizes, shapes, and types. The sizes of individual cells range from a couple of millimeters or less to a couple of meters. Depending on the application, many cells are connected in series-parallel arrangements to provide the required voltage and current for the necessary time.

Future applications for batteries are certain to include those listed here as well as a variety that are difficult to imagine at this time. We are now seeing the appearance of electric hybrid vehicles with a battery as part of the propulsion system and an electric motor. These vehicles are much more efficient than the standard vehicles and deliver approximately 50 miles per gallon. In this application, metal hydride/nickel oxide cells and lithium cells are used. We are certain to see many more of these hybrid vehicles with relatively large batteries providing much of the propulsion power. These batteries store a few kilowatt-hours of energy. We have also seen a small number of all-battery electric vehicles on the road. They use larger batteries of up to 20 kWh storage capability. As batteries of higher specific energy and lower cost become available, more electrically powered vehicles (both battery and hybrid) will be sold.

Systems for the remote generation and storage of electrical energy depend on batteries. These include solar- and wind-powered systems for remote buildings. We can expect to see ever-smaller cells being used as power sources on microcircuit chips in computers and other electronic devices. Biomedical applications will expand with a wide variety of battery-powered sensors, stimulators, transmitters, and the like. A battery-powered artificial heart is a distinct possibility.

Future household applications include a battery-powered robot capable of performing routine household tasks. As computer-controlled homes become more common, batteries will be used in many sensors, transmitters, and receivers, either because they are mobile or to be independent of the power grid. As fuel cells are introduced, it is likely that in most applications the fuel cell will be accompanied by a battery to store energy and meet peak power requirements, lowering the total cost of the system for a given peak power requirement.

5. ENVIRONMENTAL ISSUES

Some types of batteries contain hazardous materials that must be kept out of the environment. Examples of these are the lead in lead–acid batteries and the cadmium in cadmium/nickel oxide batteries. As batteries continue to experience growth in their use, developers and manufacturers must be cognizant of environmental issues relating to the materials used in the batteries. For example, more than 80% of the lead in a new lead–acid battery has been recycled. With the use of clean technology in the recycling process, this hazardous material can be used and reused safely.

6. FUTURE OUTLOOK

We can expect to see continuing growth in the variety of electrochemical couples available and the sizes and shapes of cells to satisfy a growing variety of applications, from microwatt-hours to megawatt-hours of energy. The world market for batteries is a multi-billion-dollar per year market and has been growing very rapidly, especially during the past decade or so. This growth will continue, especially as the developing countries install their communications systems and more extensive infrastructures for electrical power and transportation. Lower cost batteries will be in demand, as will batteries for various specialty applications. Batteries capable of fitting into oddly shaped spaces will be needed, and cells with polymer electrolytes can satisfy this unique requirement. Such batteries are being introduced now.

As batteries show an increase in specific energy in the range of 300 to 400 Wh/kg, battery-powered electric cars will become quite attractive. Batteries could power buses and other vehicles with ranges of 200 to 300 miles.

In the development of new cells, it will be important to continue to reduce the equivalent weight of the electrodes. At this time, the highest theoretical specific energy of any cell being developed is that of Li/S at 2600 Wh/kg. If a lithium/oxygen, lithium/phosphorus, or lithium/fluorine cell could be developed, this would lead to an even higher specific energy.

SEE ALSO THE FOLLOWING ARTICLES

Batteries, Transportation Applications • Electrical Energy and Power • Electric Motors • Storage of Energy, Overview

Further Reading

Linden, D., and Reddy, T. B. (eds.). (2002). "Handbook of Batteries," 3rd ed. McGraw–Hill, New York.

Socolow, R., and Thomas, V. (1997). The industrial ecology of lead and electric vehicles. *J. Industr. Ecol.* **1**, 13.

Vincent, C. A., Bonino, F., Lazzari, M., and Scrosati, B. (1984). "Modern Batteries: An Introduction to Electrochemical Power Sources." Edward Arnold, London.

Batteries, Transportation Applications

MICHAEL M. THACKERAY
Argonne National Laboratory
Argonne, Illinois, United States

1. Battery Systems for Electric Transportation
2. Performance Requirements
3. Lead–Acid Batteries
4. Nickel (Alkaline) Batteries
5. High-Temperature Batteries
6. Lithium Batteries
7. Regulations for Transporting Batteries

Glossary

battery A direct-current voltage source consisting of two or more electrochemical cells connected in series and/or parallel that converts chemical energy into electrical energy.

capacity The current output capability of an electrode, cell, or battery over a period of time, usually expressed in ampere-hours.

conductivity A measure of the ability of a material to conduct an electrical current, equal to the reciprocal of resistivity, $S = 1/R$.

current The rate of transfer of electrons, positive ions or negative ions, measured in amperes.

electric vehicle A vehicle that is powered solely by an electrochemical power source, such as a battery or fuel cell.

electrochemical cell A device that converts chemical energy into electrical energy by passing current between a negative electrode and a positive electrode through an ionically conducting electrolyte.

electrode A conductor by which an electrical current enters or leaves an electrochemical cell.

electrolyte Any liquid or solid substance that conducts an electric current by the movement of ions.

energy density The amount of energy per unit volume, usually expressed in watt-hours per liter.

fuel cell An electrochemical cell in which electrical energy is produced by the electrochemical oxidation of a fuel, usually hydrogen.

hybrid electric vehicle A vehicle that is powered by a combination of power sources such as a battery with an internal combustion engine or a fuel cell.

impedance The effective resistance to the flow of electric current.

power The rate at which electrical energy can be supplied, measured by the mathematical product of voltage and current, usually expressed in watts.

power density The amount of power per unit volume, usually expressed in watts per liter.

specific capacity The amount of capacity per unit mass, usually expressed in ampere-hours per kilogram.

specific conductivity The conductivity of a material per unit length, usually expressed in siemens per centimeter.

specific energy The amount of energy per unit mass, usually expressed in watt-hours per kilogram.

specific power The amount of power per unit mass, usually expressed in watts per kilogram.

voltage The potential difference between two electrodes, measured in volts.

Batteries provide a means for storing energy and as such they have become an indispensable entity in modern-day life. They are used for powering a myriad of devices, both large and small. The use of batteries as a source of energy for electrically powered vehicles was recognized early. Electric cars powered by nickel–iron batteries were manufactured during the early 20th century, but this mode of transportation could not withstand or compete against the rapid and very successful development of the internal combustion engine. Nevertheless, electric transportation continued to exist for specialized applications, such as mining and underground vehicles. Although the nickel–iron battery was extremely rugged, long-lived, and durable and played a significant role from its introduction in 1908 through the 1970s, it lost market share to the industrial lead–acid batteries. Lead–acid batteries have continued to dominate these specialized transportation sectors, which now include applications such as forklift trucks, delivery vehicles, and golf carts.

The oil crisis in the early 1970s, when the world experienced a temporary shortage of petroleum fuels, sparked a renewed interest in finding alternative sources of power for mass transport. At the same time, there was a growing awareness that electric vehicles (EVs) with "zero gas" emissions and hybrid electric vehicles (HEVs) with "low gas" emissions would significantly reduce the levels of pollution in urban areas. Since then, several new battery technologies with significantly superior performance properties have emerged—for example, nickel–metal hydride batteries, high-temperature molten-salt systems with sodium or lithium electrodes, and, most recently, nonaqueous lithium ion and lithium–polymer systems. These developments in battery technology have been driven by the need for improved high-energy, high-power systems and by environmental, safety, and cost issues.

1. *BATTERY SYSTEMS FOR ELECTRIC TRANSPORTATION*

1.1 Historical Background

Batteries have been known for two centuries. The first significant discovery has been credited to Volta who, between 1798 and 1800, demonstrated that a "voltaic pile" (an electrochemical cell) in which two unlike metal electrodes (silver and zinc), separated by a liquid electrolyte, provided an electrochemical potential (the cell voltage). At the same time, he demonstrated that when the area of the electrodes was increased, the cell could deliver a significantly higher current. In so doing, he unequivocally proved that there was a strong relationship between chemical and electrical phenomena. Scientists were quick to capitalize on this discovery, with the result that during the 19th century great progress was made in advancing electrochemical science. Many notable names appear in the chronology of battery development during this period. For example, Humphrey Davy (1807) showed that a battery could be used to electrolyze caustic soda (NaOH); Michael Faraday (1830) developed the basic laws of electrochemistry; the copper–zinc Daniell cell was discovered in 1836; William Grove (1839) demonstrated the principle of a fuel cell; Planté (1859) discovered the lead–acid battery; LeClanché (1868) introduced the zinc–carbon "dry" cell; and, by the turn of the century, Junger (1899) had patented the nickel–cadmium cell and Thomas Edison (1901) had patented other nickel-based electrochemical couples, notably nickel–iron. These pioneering discoveries of battery systems that used aqueous electrolytes laid the foundation for many of the battery systems in use today and opened the door to the discovery and development of alternative battery chemistries, particularly the nonaqueous sodium- and lithium-based systems that revolutionized battery technology toward the end of the 20th century.

1.2 Electrochemical Reactions and Terminology

Batteries consist of an array of electrochemical cells, connected either in series to increase the battery voltage or in parallel to increase the capacity and current capability of the battery. The individual cells consist of a negative electrode (the anode) and a positive electrode (the cathode) separated by an ionically conducting electrolyte. During discharge, an oxidation reaction occurs at the anode and ionic charge is transported through the electrolyte to the cathode, where a reduction reaction occurs; a concomitant transfer of electrons takes place from the anode to the cathode through an external circuit to ensure that charge neutrality is maintained. For secondary (rechargeable) batteries, these processes are reversed during charge.

The voltage of an electrochemical cell (E^0_{cell}) is determined by the difference in electrochemical potential of the negative and positive electrodes, i.e.,

$$E^0_{\text{cell}} = E^0_{\text{positive electrode}} - E^0_{\text{negative electrode}}.$$

From a thermodynamic standpoint, the cell voltage can also be calculated from the free energy of the electrochemical reaction (ΔG):

$$\Delta G = nFE^0_{\text{cell}},$$

where n is the number of electrons transferred during the reaction and F is Faraday's constant (96,487 coulombs). During the discharge of electrochemical cells, it is a general requirement that if maximum energy and power are to be derived from the cell, particularly for heavy-duty applications such as electric vehicles, then the voltage of the cell should remain as high and as constant as possible. This requirement is never fully realized in practice because the internal resistance (impedance) of cells and polarization effects at the electrodes lower the practical voltage and the rate at which the electrochemical reactions can take place. These limitations are controlled to a large extent not only by the cell design but also by the type of electrochemical reaction that occurs at the individual electrodes and by the ionic conductivity of the electrolyte. For example, the following electrochemical reactions take place at the individual electrodes of a sodium–nickel chloride cell:

$$\text{Anode}: \ 2Na \rightarrow 2Na^{+} + 2e^{-}$$

$$\text{Cathode}: \ NiCl_2 + 2e^{-} \rightarrow Ni + 2Cl^{-}$$

$$\text{Overall reaction}: 2Na + NiCl_2 \rightarrow 2NaCl + Ni.$$

At the anode, the sodium provides a constant potential throughout the reaction. The cathode also operates as a constant potential electrode because it contains three components (C: Na, Ni, and Cl) and three phases (P: $NiCl_2$, NaCl, and Ni). The Gibbs phase rule,

$$F = C - P + 2,$$

dictates that the number of degrees of freedom (F) for such an electrode reaction is two (temperature and pressure), thereby keeping the potential invariant. Therefore, sodium–nickel chloride cells provide a constant potential (under thermodynamic equilibrium) throughout charge and discharge.

In contrast, lithium ion cells operate by insertion reactions, whereby lithium ions are electrochemically introduced into, and removed from, host electrode structures during discharge and charge. For a Li_xC_6/$Li_{1-x}CoO_2$ cell with $x_{max} \approx 0.5$, the electrode reactions are

$$\text{Anode}: \ Li_xC_6 \rightarrow xLi^{+} + C_6 + xe^{-}$$

$$\text{Cathode}: \ xLi^{+} + Li_{1-x}CoO_2 + xe^{-} \rightarrow LiCoO_2$$

$$\text{Overall reaction}: Li_xC_6 + Li_{1-x}CoO_2 \rightarrow LiCoO_2 + C_6.$$

The lithiated graphite electrode operates by a series of staging reactions, each of which constitutes a two-component, two-phase electrode that provides a constant potential a few tens of millivolts above that of metallic lithium. On the other hand, the $Li_{1-x}CoO_2$ electrode is a two-component (Li and CoO_2) and single-phase ($Li_{1-x}CoO_2$) system. Therefore, according to the Gibbs phase rule, the number of degrees of freedom for the cathode reaction is three, and the potential of the $Li_{1-x}CoO_2$ electrode varies with composition (x) as it is discharged and charged. Lithium ion cells therefore do not operate at constant potential; they show a sloping voltage profile between the typical upper and lower limits of operation—4.2 and 3.0 V, respectively.

The calendar (shelf) life and cycle life of cells, which depend on the chemical stability of the charged and discharged electrodes in the electrolyte environment, are also important characteristics that have to be taken into account when considering battery systems for electric vehicles. For transportation applications, it is necessary for the battery to be able to store a large quantity of energy per unit mass and also to deliver this energy at rapid rates, particularly during the acceleration of the vehicles. The specific energy of the battery in watt-hours per kilogram (Wh/kg), which is proportional to the distance that a vehicle will be able to travel, is defined by the product of the battery voltage, V, and the electrochemical capacity of the electrodes, the unit of which is ampere-hour per kilogram (Ah/kg). The volumetric energy density is given in terms of watt-hours per liter (Wh/liter), which is an important parameter when the available space for the battery is a prime consideration. The specific power, which defines the rate at which the battery can be discharged, is given as watts per kilogram (W/kg) and the power density as watts per liter (W/liter). The power parameters are largely defined by the kinetics of the electrochemical reaction, the surface area and thickness of the electrodes, the internal resistance of the individual cells, and the size and design of the cells. Therefore, it is clear that in order to be able to travel large distances and for vehicles to be able to quickly accelerate, the batteries should be as powerful and light as possible.

Tables I and II provide a list of the battery systems that have received the most attention during the past 25 years and hold the best promise for transportation applications. Table I defines the chemistry of each system and provides theoretical voltages, capacities, and specific energies based on the masses of the electrochemically active components of the cell reactions. Table II provides typical performance characteristics that have been reported specifically for EV or HEV applications. The order in which the systems are given approximates the chronology of their development.

1.3 Electric Vehicles vs Hybrid Electric Vehicles

Pure EVs are limited by the range they can travel before the batteries become depleted and need to be recharged. Vehicle range is determined by the specific energy of the batteries. Table I shows that the practical specific energy of heavy-duty batteries varies widely from ~30 Wh/kg for lead–acid to ~150Wh/kg for advanced lithium batteries. Although lead–acid batteries are relatively inexpensive and have a rugged construction, the range they

TABLE I

Battery Systems and Theoretical Electrochemical Properties[a]

System	Negative electrode (anode)	Positive electrode (cathode)	Open circuit voltage of reaction (V)	Theoretical capacity (Ah/kg)	Theoretical specific energy (Wh/kg)
Lead–acid	Pb	PbO_2	2.1	83	171
Nickel–cadmium	Cd	NiOOH	1.35	162	219
Nickel–metal hydride	Metal hydride alloy	NiOOH	1.35	~178[b]	~240[b]
Sodium–sulfur	Na	S	2.1–1.78(2.0)[c]	377	754
Sodium–metal chloride	Na	$NiCl_2$	2.58	305	787
Lithium ion	Li_xC_6	$Li_{1-x}CoO_2$	4.2–3.0(4.0)[c]	79[d]	316[d]
Lithium–polymer	Li	VO_x	~3.0–2.0(2.6)[c]	~340	~884

[a]Theoretical values of capacity and specific energy are based on the masses of electrode and electrolyte components that participate in the electrochemical reaction. For example, for lead–acid cells the masses of Pb, PbO_2, and H_2SO_4 are included in the calculation.
[b]Values are dependent on the composition of the selected metal hydride alloy.
[c]Average voltages for the reaction in parentheses.
[d]Calculations are based on $x_{max} = 0.5$.

TABLE II

Typical Performance Characteristics of Various Battery Systems[a]

System	Specific energy (Wh/kg)	Energy density (Wh/liter)	Specific power (W/kg)	Power density (W/liter)	No. of cycles
Lead–acid					
EV	35–50	90–125	150–400	400–1000	500–1000
HEV	27	74	350	900	?
Nickel–cadmium					
EV	60	115	225	400	2500
Nickel–metal hydride					
EV	68	165	150 (80% DOD)	430 (80% DOD)	600–1200
HEV	50	125	750 (50% DOD)	>1600 (50% DOD)	?
Sodium–sulfur					
EV	117	171	240	351	800
Sodium–metal chloride					
EV	94	147	170	266	>1000
Lithium–ion[b]					
EV battery module	93	114	350	429	800 (40°C)
Lithium–polymer					
EV battery module	155	220	315	447	600

[a]Abbreviations used: EV, electric vehicle; HEV, hybrid electric vehicle; DOD, depth of discharge of the battery.
[b]C/$LiMn_2O_4$ system.

offer (on average, 50 miles between charges) limits their application to utility fleets, such as forklift trucks, delivery vehicles, and golf carts, for which short driving distances are acceptable. (Because lead–acid batteries are heavy, much of the energy is used to transport the mass of the battery.) For more energetic and advanced systems, such as nickel–metal hydride (68 Wh/kg), sodium–nickel chloride (94 Wh/kg), lithium ion (93 Wh/kg), and lithium–polymer batteries (155 Wh/kg), greater attention has to be paid to factors such as safety and cost. Although high-energy density battery systems that provide 100 Wh/kg can offer a typical range of 100 miles or more, this is still too short a distance for pure EVs to be considered attractive, particularly in the United States, where driving distances can be considerably longer. In the absence of built-in infrastructures to cater to the demands of recharging EVs and the time that it may take to charge them, it seems that this mode of consumer transportation will be initially restricted to urban travel and introduced particularly in cities that are severely overpolluted by gas emissions from internal combustion vehicles.

Currently, HEVs, in which a battery and an internal combustion engine (or a fuel cell) are coupled, are the most attractive option for reducing the demands on fossil fuels and for optimizing gasoline consumption. Such systems still make use of the current infrastructure of the gasoline industry and do not require independent battery charging stations. In HEVs, the battery is used predominantly to assist in acceleration; the gasoline engine takes over during cruising, when the battery can be recharged. Energy from regenerative braking can also be used to charge the batteries. Unlike pure EVs in which the batteries may become completely discharged, which can negatively impact cycle life, batteries for HEVs are subject to only shallow charge and discharge reactions during operation, thereby enabling a very high number of cycles and significantly longer calendar lifetimes.

2. PERFORMANCE REQUIREMENTS

The growing concern about pollution led the California Air Resources Board in the early 1990s to mandate that 2% of all vehicles sold in the state of California should be zero-emission vehicles by 1998 and 10% by 2003. Although these mandates have not been successfully implemented, they led to the creation of the U.S. Advanced Battery Consortium (USABC), which consists of U.S. automobile manufacturers, electric utility companies, and government authorities, to support and manage the development of batteries for pure EVs. The emphasis of the development efforts has been redirected toward developing high-power batteries for HEVs through the Program for a New Generation of Vehicles, now called the FreedomCAR Partnership. Selected guidelines and performance requirements established by the USABC in 1996 for mid- and long-term EV batteries and by Freedom CAR in 2001 for HEV batteries are provided in Tables III and IV, respectively. HEV batteries are classified into two types: those for "power-assist" operation, where the battery is always used in conjunction with the internal combustion engine, and those for "dual-mode" operation, where the battery can provide energy and power to the vehicle either on its own or together with the engine. The energy requirement for dual mode (15 Wh/kg) is considerably higher than that for power assist (7.5 Wh/kg) so that the battery can provide sufficient energy and power to the vehicle on occasions when it is used alone.

TABLE III

Selected USABC Mid-Term and Long-Term Battery Performance Goals for Electric Vehicles (1996)

Performance criteria	Mid-term (5 years)	Long term (10 years)
Specific energy (Wh/kg)	80	200
Energy density (Wh/liter)	135	300
Specific power (W/kg)	150	400
Power density (W/liter)	250	600
Life (years)	5	10
Cycle life (cycles, 80% depth of discharge)	600	1000
Operating temperature (°C) (environment)	−30 to 65	−40 to 85
Recharge time (h)	<6	3–6

3. LEAD–ACID BATTERIES

Lead–acid batteries have existed as commercial products for more than 100 years and have found application in a large number of diverse systems. The most notable application has been in the automotive industry for starting internal combustion engines and for providing the energy to power electrical equipment in vehicles, such as lights, windows, and displays. These batteries are commonly called SLI batteries (starting, lighting, and ignition) and are capable of providing the high currents required for initially cranking the engine. Other applications include photovoltaic energy storage, standby energy storage (particularly for providing backup emergency power and uninterrupted power supply for strategic equipment, such as lighting, telephone, and communication systems), and transportation devices such as underground mining conveyances, submarines (particularly for silent maneuverability when submerged),

TABLE IV

Selected FreedomCAR Battery Performance Goals for Hybrid Electric Vehicles (2002)[a]

Performance criteria	Power assist	Dual mode
Total available energy over DOD range in which power goals are met (Wh/kg)	7.5 (at C/1 rate)[b]	15 (at 6-kW constant power)
Pulse discharge power (W/kg)	625 (18 s)	450 (12 s)
Calendar life (years)	15	15
Cycle life (cycles, for specified SOC increments)	300,000	3750
Operating temperature (°C)	−30 to 50	−30 to 50

[a] Abbreviations used: DOD, depth of discharge of the battery; SOC, state of charge.
[b] C/1 rate, 1-h rate.

golf carts, milk and postal delivery trucks, and small electrically powered cars and trucks. Lead–acid batteries constitute approximately 40% of the world's total battery sales, which can be attributed to their well-developed and robust technology and significant cost advantage.

3.1 General Chemistry, Construction, and Operation

Lead–acid batteries consist of a metallic lead (Pb) negative electrode, a lead dioxide (PbO_2) positive electrode, and a sulfuric acid electrolyte. The overall cell reaction is

$$Pb + PbO_2 + 2H_2SO_4 \rightarrow 2PbSO_4 + 2H_2O.$$

The voltage of lead–acid cells on open circuit is approximately 2 V; a standard 12-V (SLI) battery therefore consists of six individual cells connected in series. During discharge, lead sulfate ($PbSO_4$) is produced as both Pb and PbO_2 electrodes, with water (H_2O) as a by-product of the reaction. Because the specific gravity (SG) or density of H_2SO_4 differs markedly from that of H_2O, the SG of the electrolyte can be conveniently used to monitor the state of charge of the cells and battery and to identify "dead" cells. The SG of lead–acid cells depends on the type of battery construction; fully charged cells typically have an SG of approximately 1.3, whereas the SG of fully discharged cells is typically 1.2 or lower.

Lead–acid batteries are assembled in the discharged state from electrodes that are manufactured by reacting PbO, Pb, and sulfuric acid to form "tribasic" ($3PbO{\cdot}PbSO_4$) and "tetrabasic" ($4PbO{\cdot}PbSO_4$) salts, which are either pasted onto flat lead grids or compacted into porous, tubular electrode containers with a central current-collecting rod, and then they are cured under controlled conditions of humidity and temperature. A formation (initial charge) process is required to convert the salts to Pb and PbO_2 at the negative and positive electrodes, respectively. Red lead (Pb_3O_4), which is more conductive than PbO, is sometimes added to the positive electrode to promote the formation of PbO_2.

Because Pb is a soft metal (melting point, 327°C), small amounts of additives, such as antimony, calcium, and selenium, have been used to increase the stress resistance of the lead grids. Antimony, in particular, significantly increases the mechanical strength of the grids but also increases their electrical resistance. An additional disadvantage of antimony is that toxic stibene (SbH_3) can form during charge, which precludes the use of lead–acid batteries with lead–antimony electrodes in poorly ventilated areas, such as in underground mining operations and submarines.

Charging of lead–acid batteries can also be hazardous. The voltage of lead–acid batteries (2 V) is higher than that required for the electrolysis of water, with the result that hydrogen and oxygen are released while batteries are being charged. Improvements have been made with the introduction of valve-regulated lead–acid ("maintenance-free") batteries in which evolved oxygen is recombined with lead at the negative electrode and in which evolved hydrogen, which can be minimized by the addition of tin to the lead grids, is vented. Nevertheless, hydrogen evolution can still pose a safety hazard in some lead–acid battery designs. Another unavoidable and inherent safety hazard of lead–acid batteries is the potential risk of spillage of the sulfuric acid electrolyte, which is corrosive and can cause severe chemical burns.

3.2 Transportation Applications

Despite their relatively low energy density, lead–acid batteries have found widespread application in numerous transportation devices, as previously noted, where the range of the vehicle is not a prime consideration. Disadvantages that will ultimately count against lead–acid batteries for transportation applications in the face of alternative emerging technologies are their relatively low specific energy and decay during long-term storage if kept in a discharged state. Nonetheless, great strides have been made by General Motors during the past few years to develop a compact, limited-production electric vehicle, known as EV1, which is now in its second generation. The estimated driving range is 55–95 miles, depending on driving conditions, which is made achievable through an aerodynamic, lightweight design and regenerative braking of the battery. The greatest advantage that lead–acid batteries have over competitive technologies is cost; they are as much as 10 times less expensive than advanced nickel–metal hydride, lithium ion, and lithium–polymer batteries that are being developed for high-performance vehicles.

4. *NICKEL (ALKALINE) BATTERIES*

Several nickel-based batteries with a NiOOH positive electrode and an alkaline (KOH) electrolyte were developed during the 20th century, including the nickel–iron (Edison), nickel–zinc, nickel–cadmium, and nickel–metal hydride systems. Although both nickel–iron and nickel–zinc systems commanded significant attention as possible candidates for electric vehicles, they fell from favor because of insurmountable problems. For example, despite their ability to provide a high number of deep discharge cycles, nickel–iron batteries were limited by low power density, poor low-temperature performance, and poor charge retention and gas evolution on stand. A major limitation that has slowed the development of nickel–zinc batteries is the solubility of the zinc electrode in the alkaline electrolyte. The severe shape change of the zinc electrode that occurs during cycling, and the formation of dendrites, which can cause internal short circuits, are problems that have not been entirely overcome. Moreover, corrosion of the zinc electrode generates hydrogen gas. Because of these limitations and the lack of major interest in these technologies today, the nickel–iron and nickel–zinc systems are not discussed further.

4.1 Nickel–Metal Hydride and Nickel–Cadmium Batteries

Nickel–metal hydride (NiMH) cells were developed during the last quarter of the 20th century as an environmentally acceptable alternative to nickel–cadmium (NiCad) cells; the toxic cadmium (negative) electrode was replaced with a metal hydride electrode. During this period, the rapid growth of NiMH technology was fueled by the need for advanced batteries for early consumer electronics devices, such as cellular phones and laptop computers.

Both NiMH and NiCad systems use an alkaline (KOH) electrolyte and a NiOOH positive electrode, and they both have a cell voltage of 1.35 V. NiMH cells can therefore be substituted directly for NiCad cells in many applications. The electrochemical reaction that occurs in a NiCad cell is

$$Cd + 2NiOOH + 2H_2O \rightarrow Cd(OH)_2 + 2Ni(OH)_2.$$

Water is consumed during the reaction to form $Cd(OH)_2$ and $Ni(OH)_2$ at the negative and positive electrodes, respectively; however, in practice, the concentration of the alkaline electrolyte is such that it does not change much during the reaction. The insolubility of the electrodes in the electrolyte at all states of charge and discharge contributes to the long cycle and standby life of NiCad cells, even though the cells lose capacity on storage and require charging to be restored to their full capacity. Particular advantages of NiCad cells compared to NiMH cells are lower cost, higher rate capability, and better low-temperature operation. However, the toxicity of cadmium and the lower specific energy and energy density delivered by practical NiCad batteries compared to NiMH batteries (Table II) are major factors preventing further growth of the NiCad battery market.

For NiMH cells, the overall electrochemical reaction is

$$MH + NiOOH \rightarrow M + Ni(OH)_2.$$

During discharge, hydrogen is released from the MH electrode into the electrolyte, which is inserted into the NiOOH electrode in the form of protons (H^+ ions) with the concomitant reduction of the nickel ions from a trivalent to divalent state to form $Ni(OH)_2$; water is therefore not consumed during the overall reaction. Because the electrolyte is water based, an inherent problem of MiNH cells is that electrolysis of water can occur during charge to form hydrogen and oxygen at the negative and positive electrodes, respectively. In sealed NiMH cells, however, the inherent dangers of overcharge are reduced because the oxygen evolved at the positive electrode can diffuse through the electrolyte and permeable separator to the MH electrode, where it can react to form water and thus reduce a pressure increase in the cell:

$$2MH + 1/2O_2 \rightarrow 2M + H_2O.$$

An additional advantage of this reaction, which depletes the negative electrode of hydrogen, is that the negative electrode never becomes fully charged, thereby suppressing the evolution of hydrogen gas. Nickel–metal hydride cells are therefore constructed with a "starved-electrolyte" design, which promotes and exploits the advantages of the oxygen-recombination reaction.

The MH negative electrodes of NiMH batteries have complex formulae; they are based on the overall composition and structure types of AB_5 or AB_2 alloys. AB_5 alloys contain several rare earth metals (Misch metals) and tend to be preferred over the AB_2-type alloys even though they have slightly lower specific and volumetric capacities. The NiOOH positive electrodes commonly contain minor quantities of Co and Zn to improve the electronic conductivity of the NiOOH electrode (Co) and to control oxygen evolution and microstructural features of the electrode (Co and Zn).

NiMH batteries have made a significant contribution to improving the range of pure EVs and to the fuel consumption of HEVs. Replacement of lead–acid batteries with NiMH batteries in the General Motors electric vehicle EV1 increased the driving range from 55–95 to 75–130 miles. NiMH batteries were introduced into the first generation of modern-day HEVs by Toyota (the Prius) and Honda (the Insight) in the late 1990s.

Despite the very significant progress that has been made, the future of NiMH batteries is uncertain, particularly in the consumer electronics industry, in which they have lost significant market share to lithium–ion batteries, which offer a significantly higher cell voltage and superior specific energy (Tables I and II).

5. HIGH-TEMPERATURE BATTERIES

The possibility of using high-temperature batteries with molten electrodes or molten salts was first seriously explored in the early 1960s. Early attempts to construct an electrochemical lithium–sulfur cell with a LiCl–KCl molten salt electrolyte (melting point, 352°C) were unsuccessful because both electrodes were also molten at the operating temperature of the cell (400°C). In order to overcome the problems of electrode containment, researchers replaced the two liquid electrodes with a solid lithium alloy and a metal sulfide. This approach resulted in the development of a highly successful primary $LiAl/FeS_x$ battery ($1 < x < 2$), which has been used predominantly in military devices such as missile systems. Attempts to develop a viable rechargeable system for EVs were thwarted by several problems, such as cycle life, corrosion, and cost, with the result that the USABC abandoned its support for the project in 1995 and opted for the emerging lithium battery technologies. Nevertheless, research on lithium–iron sulfide batteries continues, but for operation at lower temperature (90–135°C) using a metallic lithium negative electrode with a polymer electrolyte. Time will tell whether this new approach will be successful.

Two other high-temperature batteries were developed in the latter part of the 20th century, namely, the sodium–sulfur battery and the sodium–nickel chloride battery, both of which use molten sodium as the positive electrode. These advanced sodium batteries were made possible by the discovery in 1967 of the sodium ion-conducting solid electrolyte, β-alumina. This solid electrolyte, which has a composition between $Na_2O \cdot 5.33Al_2O_3$ and $Na_2O \cdot 11Al_2O_3$, has an anomalously high sodium ion specific conductivity of 0.2 S/cm at 300°C, comparable to that of liquid electrolytes (e.g., the specific conductivity of 3.75 M H_2SO_4 at room temperature is 0.8 S/cm). The structure of β-alumina consists primarily of Al_2O_3 spinel blocks, which are structurally related to the mineral "spinel" $MgAl_2O_4$. These spinel blocks are separated by oxygen-deficient layers, in which sodium ions can diffuse rapidly when an electrochemical potential difference exists across the ceramic electrolyte. The conventional design of the β-alumina solid electrolyte for sodium batteries is tubular. The β-alumina tubes are brittle and must be thin to permit the electrochemical reaction to occur at an acceptable rate. The manufacture of high-quality tubes, which requires a sintering temperature of 1600°C to ensure a high-density ceramic, is a complex process and great care must be taken to minimize flaws that might lead to fracture of the tubes during battery operation.

5.1 Sodium–Sulfur Batteries

In most conventional sodium–sulfur cells, which have the configuration Na/β-alumina/S, the molten sodium electrode is contained within the β-alumina tube, typically called the central sodium design. The molten sulfur electrode is contained on the outside between the tube and the inner walls of the metal container, which are usually coated with chromium to resist attack from the highly corrosive sulfur electrode. Cells can also be constructed with a central sulfur design, but this design provides lower power and is less tolerant to freeze–thaw cycling (i.e., cooling and heating of the battery during which the electrodes are allowed to solidify and melt repeatedly).

During discharge of Na/β-alumina/S cells, the sodium reacts electrochemically with sulfur to form a series of polysulfide compounds according to the general reaction

$$2Na + xS \rightarrow Na_2S_x$$

over the range $3 \leq x \leq 5$. The reaction can be considered to take place in two steps. The initial reaction produces Na_2S_5, which is immiscible with the molten sulfur. This reaction is therefore a two-phase process; it yields a constant cell voltage of 2.08 V. Thereafter, the sulfur electrode discharges in a single-phase process to form Na_2S_x compositions between Na_2S_5 and Na_2S_3, during which the cell voltage decreases linearly. The open circuit voltage at the end of discharge is 1.78 V. The composition Na_2S_3 represents the fully discharged composition because further discharge yields Na_2S_2, which is a solid, insoluble product that leads to an increase in the internal resistance of the cell. Moreover, it is difficult to recharge a Na_2S_2 electrode. In practice, therefore, it is common for the end voltage of sodium–sulfur cells to be restricted to between 1.9 and 1.78 V not only because it prevents over-discharge in local areas of the cells but also because it reduces the severity of the corrosion reactions that occur for Na_2S_x products with lower values of x.

Because sodium and sulfur are relatively light elements, the sodium–sulfur battery offers a high theoretical specific energy (754 mAh/g). However, the robust battery construction, heating systems, and thermal insulation that are required for efficient and safe operation place a penalty on the specific energy that can be delivered in practice (117 Wh/kg). The molten state of the electrodes and high-temperature operation ensure an acceptable rate capability; a specific power of 240 W/kg for electric vehicle batteries has been reported (Table I).

One of the major disadvantages of sodium–sulfur cells is that they are inherently unsafe; any fracture of the β-alumina tube that allows the intimate contact of molten sodium and molten sulfur can result in a violent reaction that is difficult to control. Furthermore, damaged cells that fail in an open-circuit mode need to be electrically bypassed to ensure continued operation of the battery. These limitations, the need for hermetic seals in a very corrosive environment, and the difficulty of meeting cost goals were among the major reasons why support for the development of sodium–sulfur batteries for electric vehicle applications was terminated worldwide in the mid-1990s. Nevertheless, considerable development of this technology is under way, particularly in Japan, for stationary energy storage applications.

5.2 Sodium–Nickel Chloride Batteries

The sodium–nickel chloride battery, commonly referred to as the Zebra battery, had its origins in South Africa in the late 1970s. The early research on these batteries capitalized on the chemistries and designs of two high-temperature battery systems, sodium–sulfur and lithium–iron sulfide, that had been made public several years before. The first generation of Zebra cells used an iron dichloride electrode with the configuration,

$$Na/\beta\text{-alumina}/FeCl_2,$$

in which the sodium electrode and β-alumina electrolyte components mimic the sodium–sulfur cell and the iron dichloride electrode acts as a substitute for the FeS_x electrode in the LiAl/KCl, LiCl/FeS_x cell. Because $FeCl_2$ is a solid, a molten salt electrolyte must be added to the positive electrode compartment of Na/β-alumina/$FeCl_2$ cells. $NaAlCl_4$ is an ideal candidate for this purpose because it has a low melting point (152°C) and because it dissociates into two stable ionic species, Na^+ and $AlCl_4^-$, thereby providing good Na^+ ion conductivity at elevated temperatures. Moreover, it is probable that the chloride ions of the molten electrolyte play a significant role in the dechlorination and rechlorination reactions that occur at the iron electrode during discharge and charge.

The electrochemical reaction of a Na/β-alumina/$FeCl_2$ cell occurs in two distinct steps:

$$6Na + 4FeCl_2 \rightarrow Na_6FeCl_8 + 3Fe$$
$$2Na + Na_6FeCl_8 \rightarrow NaCl + 4Fe.$$

The overall reaction is

$$8Na + 4FeCl_2 \rightarrow NaCl + 4Fe.$$

The initial reaction provides an open-circuit voltage of 2.35 V; the second reaction occurs only a few tens of millivolts lower. The intermediate phase that is formed, Na_6FeCl_8, has a defect rock salt structure in which one iron divalent atom replaces two Na atoms in the NaCl structure. Moreover, despite a volume increase, the closely packed arrangement of the chloride atoms of the $FeCl_2$ structure remains essentially unaltered during the transitions to Na_6FeCl_8 and NaCl. Crystallographically, the reaction at the positive electrode is therefore a relatively simple one. Moreover, all the compounds formed during charge and discharge are essentially insoluble in the $NaAlCl_4$ electrolyte, provided that the electrolyte is kept basic [i.e., slightly NaCl rich (or $AlCl_3$ deficient)]. This prevents ion-exchange reactions from occurring between the Fe^{2+} in the electrode and Na^+ in the β-alumina structure, which would severely degrade the conductivity of the solid electrolyte membrane. The simplicity of the electrochemical reaction and the insolubility of the metal chloride electrode in the molten salt electrolyte are key factors that contribute to the excellent cycle life of Zebra cells.

The second generation of Zebra cells used a nickel chloride electrode because it provided a slightly higher cell voltage (2.58 V), resulting in cells with a higher power capability. Unlike $FeCl_2$ electrodes, $NiCl_2$ electrodes do not form any intermediate compositions but react directly with sodium to form NaCl:

$$2Na + NiCl_2 \rightarrow 2NaCl + Ni.$$

Zebra cells are usually designed with the positive electrode housed inside the β-alumina tube. Therefore, during charge, sodium is electrochemically extracted from the NaCl/Ni composite electrode to form $NiCl_2$; the metallic sodium is deposited on the outside of the tube and fills the void between the tube and inner wall of the cylindrical metal container. Although the power capability of the early Zebra cells was limited to approximately 100 W/kg, significantly improved performance was achieved by changing the shape of the cylindrical β-alumina tube to a fluted-tube design, thereby increasing the surface area of the solid electrolyte considerably. Fully developed Zebra batteries now have a specific energy of approximately 100 Wh/kg and a specific power of 170 W/kg.

Zebra cells have several major advantages over sodium–sulfur cells:

1. Zebra cells have a higher operating voltage (2.58 vs 2.08 V).
2. Unlike sodium–sulfur cells, which are assembled in the charged state with highly reactive sodium that requires special handling conditions, Zebra cells can be assembled in the discharged state by blending common salt with nickel powder. (Note that state-of-the-art Zebra electrodes contain a small proportion of iron, which significantly improves the power capability of the cells toward the end of discharge.)
3. From a safety standpoint, if the β-alumina tubes fracture, the molten $NaAlCl_4$ electrolyte is reduced according to the reaction

   $$NaAlCl_4 \rightarrow 4NaCl + Al,$$

   resulting in solid products; the deposited aluminum metal also provides an internal short circuit that allows for failed cells to be electrically bypassed.
4. Zebra cells can be safely overdischarged by electrochemical reduction of the electrolyte at 1.58 V (the reaction is identical to that given for point 3), provided that the reaction does not go to completion when the electrode solidifies completely.
5. There is less corrosion in Zebra cells.
6. Zebra cells offer a wider operating temperature range (220–450°C) compared with sodium–sulfur cells (290–390°C).
7. Zebra cells are more tolerant to freeze–thaw cycles.

Zebra batteries have been tested extensively in EVs by Daimler-Benz in Europe, for example, in a Mercedes A Class sedan. These batteries have met all the USABC mid-term performance goals listed in Table III, except for operating temperature. Although development work on Zebra batteries is ongoing, the market for pure EVs for which Zebra batteries are best suited is lacking, particularly because of the greater attention that is currently being paid to HEVs. Also, high-temperature batteries have been criticized because they require sophisticated thermal management systems for both heating and cooling. Nonetheless, because they are temperature controlled, these systems have been shown to work exceptionally well in both extremely hot and extremely cold climates.

6. *LITHIUM BATTERIES*

6.1 Lithium Ion Batteries

Although primary, nonaqueous, lithium batteries that operate at room temperature have been in existence since the 1970s for powering small devices such as watches and cameras, it was the introduction of rechargeable lithium ion batteries by Sony Corporation in 1990 that brought about a revolution in the battery industry. These batteries were introduced by Japan in anticipation of the boom in electronic devices such as cellular phones and laptop computers, which took the world by storm during the next decade and was responsible for the technology "dot-com" industry.

Lithium ion batteries derive their name from the electrochemical processes that occur during charge and discharge of the individual cells. During charge of these cells, lithium ions are transported from a positive host electrode to a negative host electrode with concomitant oxidation and reduction processes occurring at the two electrodes, respectively; the reverse process occurs during discharge.

State-of-the-art lithium ion cells consist of a graphite negative electrode and a $LiCoO_2$ positive electrode. The electrolyte consists of a lithium salt, such as $LiPF_6$, dissolved in an organic solvent, typically a blend of organic carbonates, such as ethylene carbonate and diethyl carbonate. The cell reaction is

$$Li_xC_6 + Li_{1-x}CoO_2 \rightarrow C_6 + LiCoO_2$$

Lithium ion cells, like Zebra cells, are conveniently assembled in the discharged state, which avoids the necessity of having to handle metallic lithium. The electrochemically active lithium is therefore initially contained in the $LiCoO_2$ positive electrode. Lithium extraction from $LiCoO_2$ occurs at a potential of approximately 4 V vs a metallic lithium (or lithiated graphite) electrode. This voltage is three times the voltage of a nickel–metal hydride cell. Therefore, one lithium ion cell provides a voltage equivalent of three nickel–metal hydride cells in series. Four-volt lithium ion cells are high-energy power sources. From a safety standpoint, it is fortunate that a protective passivating layer forms at the lithiated graphite–electrolyte interface, which prevents the Li_xC_6 electrode from reacting continuously and vigorously with the electrolyte. The high voltage and the light electrodes give lithium ion cells a high theoretical specific energy ($\sim$400 Wh/kg). However, despite this advantage, these high-voltage cells contain highly reactive electrodes, particularly when they are fully charged. If cells are overcharged, the possibility of exothermic reactions of a highly delithiated and oxidizing $Li_{1-x}CoO_2$ electrode and an overlithiated and reducing Li_xC_6 electrode (which may contain metallic lithium at the surface of the lithiated graphite particles) with a flammable organic electrolyte poses a severe safety hazard. In the event of thermal runaway and the buildup of gas pressure, cells can vent, rupture, and catch fire. Individual lithium ion cells must therefore be protected electronically from becoming overcharged, which contributes significantly to the cost of lithium ion cells. With built-in microprocessors, millions of small lithium ion cells, now considered safe, are produced (mainly by Japanese manufacturers) and sold every month. For larger scale applications, such as EVs and HEVs, in which considerably more energy is contained in the batteries, the safety issue is of paramount importance and continues to receive much attention.

Compared with all other systems, lithium ion batteries are the most versatile in terms of their chemistry. They can offer a diverse range of electrochemical couples because there are a large number of structures that can act as host electrodes for lithium. For example, various metals, such as Al, Si, Sn, and Sb, and intermetallic compounds, such as SnSb, Cu_6Sn_5, and Cu_2Sb, have received considerable attention because they react with lithium, either by lithium insertion or by a combination of lithium insertion into and metal displacement from a host structure at a potential above that of a Li_xC_6 electrode. These metal/intermetallic electrodes should therefore be potentially safer than Li_xC_6 electrodes. The electrochemical performance of these metal and intermetallic electrodes is limited by large crystallographic volume increases during the uptake of lithium and by a large irreversible capacity loss that commonly occurs during the initial cycle of the lithium cells.

Although $LiCoO_2$ has been the positive electrode of choice for more than a decade, lithium extraction is limited to approximately 0.5 Li per formula unit, which corresponds to a practical specific capacity of 137 mAh/g. Moreover, cobalt is a relatively expensive metal. Therefore, there has been a great motivation in the lithium battery community to find alternative positive electrodes to increase the capacity and specific energy of lithium ion cells and to reduce their cost. In particular, much effort has been focused on developing alternative lithium–metal oxide electrodes that are isostructural with $LiCoO_2$, such as $LiNi_{0.8}Co_{0.15}Al_{0.05}O_2$, which has a higher capacity (170–180 mAh/g) without compromising the cell voltage significantly. A $LiMn_{0.33}Ni_{0.33}Co_{0.33}O_2$ system has been developed that offers an electrode capacity in excess of 200 mAh/g. These advances bode well for the next generation of lithium ion batteries.

Other positive electrode materials have also commanded attention. One example is $LiMn_2O_4$, which has a spinel-type structure. This electrode is considerably less expensive than $LiCoO_2$ and is more stable to the extraction of lithium, a reaction that also occurs at approximately 4 V vs a metallic lithium electrode. Moreover, the Mn_2O_4 spinel framework provides a stable three-dimensional interstitial space for lithium diffusion and is more tolerant to high-rate discharge and charge compared with the layered electrode structures. Although $LiMn_2O_4$ has an inferior capacity compared to that of $LiCoO_2$, the spinel remains a very attractive low-cost and high-rate electrode for HEV batteries, which are subjected to only shallow levels of charge and discharge during operation. Spinel electrodes suffer from solubility problems if the temperature of the battery reaches 40–50°C as a result of trace quantities of hydrofluoric acid that can be generated in the cell by hydrolysis of commonly used fluorinated lithium salts such as $LiPF_6$. It has been reported that this problem can be overcome by using nonfluorinated lithium salts that contain boron.

Another possible alternative positive electrode to $LiCoO_2$ is the family of lithium transition metal phosphates, which have an olivine-type structure, that is, $LiFePO_4$. This low-cost material, although it provides a relatively low theoretical electrode capacity (170 mAh/g), is significantly more stable to lithium extraction than layered and spinel transition metal oxide electrode structures. Unlike the layered $LiMO_2$ (M = Co, Ni, and Mn) and spinel LiM_2O_4 (M = Mn and Ni) electrodes that, on lithium extraction, operate off either a $M^{4+/3+}$ or a $M^{4+/2+}$ couple at 4 V or higher, the $LiFePO_4$ electrode provides a significantly more stable $Fe^{3+/2+}$ couple at an attractive potential of 3.5 V. $Li_xC_6/Li_{1-x}FePO_4$ cells therefore show very little capacity decay and provide exceptionally good cycling stability. An intrinsic limitation of phosphate electrodes is that they are inherently poor electronic and ionic conductors, two essential features for a high-performance insertion electrode. Significant progress has been made in improving the electronic conductivity of these electrodes, for example, by applying a coating of electronically conducting carbon to the surface of the phosphate particles.

Spinel and olivine electrode structures can be exploited by selecting various transition metals as the electrochemically active component to tailor the potential of the electrode and hence the voltage of the cell. For example, in the family of spinel compounds, $Li_4Ti_5O_{12}$ (alternatively, in spinel notation, $Li[Li_{0.33}Ti_{1.67}]O_4$) provides a constant 1.5 V vs Li^0 when reacted with lithium to the rock salt composition $Li_7Ti_5O_{12}$. Therefore, $Li_4Ti_5O_{12}$ can be used as a negative electrode against a high-potential positive electrode such as $LiMn_{1.5}Ni_{0.5}O_4$ (4.5 V vs Li^0) to yield a 3.0-V cell. Lithium ion cells with metal oxide negative and positive electrodes should be inherently safer compared with state-of-the-art $Li_xC_6/LiCoO_2$ cells, but they are limited by relatively low voltages and specific energy. With respect to the family of olivine electrodes, replacement of Fe in $LiFePO_4$ by Mn and Co yields lithium ion cells with significantly higher voltages of 4.1 and 4.8 V, respectively.

With respect to electrolytes, considerable progress has been made in developing "polymer-gel" systems in which a liquid electrolyte component is immobilized in a polymer matrix. Although the Li^+ ion conductivity of these gel electrolytes is slightly inferior to that of pure liquid electrolyte systems, the ability to manufacture cells with laminated electrodes and electrolytes has provided much flexibility in the design of lithium ion cells. Moreover, cells with polymer-gel electrolytes are construed to be safer than those containing liquid electrolytes; they can also be easily constructed with lightweight packaging materials such as aluminum foil, thereby significantly improving the energy density of the cells.

Lithium ion battery technology has not reached maturity, and there is little doubt that significant advances will continue to be made in the years to come with regard to improving energy, power, and safety. Although the emphasis on the immediate development of a pure EV has been reduced, the future of lithium ion batteries for HEVs and in the shorter term for other selected modes of transportation, such as electric bikes and scooters, remains bright. In this respect, the first generation of lithium ion batteries containing a manganese spinel electrode has been developed for both EV and HEV applications by Shin-Kobe in Japan.

6.2 Lithium–Solid Polymer Electrolyte Batteries

Batteries that contain a metallic lithium negative electrode coupled via solid polymer electrolyte to a transition metal oxide positive electrode (usually a vanadium oxide, for example, V_2O_5, V_6O_{13}, and LiV_3O_8, commonly denoted VO_x) have been under development since the mid-1970s. A major advantage of using lithium instead of a lithiated graphite electrode is that a metallic lithium electrode offers a substantially higher capacity (3863 mAh/g) than LiC_6 (372 mAh/g). However, the performance of these batteries is compromised by the relatively low room-temperature ionic conductivity of polymer electrolytes, which necessitates an elevated operating temperature, typically 80–120°C. Another limitation of these batteries is that dendritic growth of lithium may occur during the charging of the battery, leading to electrolyte penetration and internal short circuits. This problem has not been entirely overcome, thereby restricting the cycle life of this type of battery. Despite this limitation, excellent progress has been made during the past few years, and development efforts are ongoing. Battery modules are being evaluated in demonstration EVs and HEVs, such as the Ford Motor Company's Th!nk *city* vehicle and the Honda Insight.

7. REGULATIONS FOR TRANSPORTING BATTERIES

International regulations exist for transporting hazardous materials by land, sea, and air. Batteries that contain highly reactive elements, such as lithium and sodium, therefore fall under the scope of these regulations. In particular, since the early 1990s, the regulations for lithium ion, lithium–polymer, sodium–sulfur, and sodium–metal chloride batteries have been continually revised to allow their transport, particularly for large and heavy-duty applications, such as EVs and HEVs. The "United Nations Recommendations on the Transport of Dangerous Goods" provides regulatory guidance for the transport of dangerous goods both internationally and domestically. Regulations stipulate that lithium cells and batteries of any size can be shipped if they pass a set of safety tests described under the section titled "Manual of Tests and Criteria." Because both sodium–sulfur and sodium–metal chloride batteries operate at elevated temperatures, the regulations also define specific conditions for the transportation of cold and hot sodium batteries.

Acknowledgments

I thank Ralph Brodd, Dennis Dees, and Gary Henriksen for useful discussions. Financial support from the U.S. Department of Energy, Offices of FreedomCAR & Vehicle Technologies, is gratefully acknowledged. This article was created under Contract No. W-31-109-ENG-38.

SEE ALSO THE FOLLOWING ARTICLES

Batteries, Overview • Electric Motors • Fuel Cells • Fuel Cell Vehicles • Fuel Cycle Analysis of Conventional and Alternative Fuel Vehicles • Hybrid Electric Vehicles • Internal Combustion Engine Vehicles

Further Reading

Besenhard, J. O. (ed.). (1999). "Handbook of Battery Materials," Wiley-VCH, Weinheim.

Linden, D., and Reddy, T. B. (eds.). (2002). "Handbook of Batteries," 3rd ed. McGraw-Hill, New York.

McNicol, B. D., and Rand, D. A. J. (1984). "Power Sources for Electric Vehicles." Elsevier, Amsterdam.

Morris, C. (ed.). (1992). "Dictionary of Science and Technology." Academic Press, London.

Nakai, K., Aiba, T., Hironaka, K., Matsumura, T., and Horiba, T. (2000). Development of manganese type lithium ion battery for pure electric vehicles (PEV). *In* "Proceedings of the 41st Japan Battery Symposium, The Electrochemical Society of Japan, November 20–22, Nagoya."

Ovshinsky, S. R., Fetcenko, M. A., Young, K., Fierro, C., Reichman, B., and Koch, J. (2002). High performance NiMH battery technology. *In* "Proceedings of the 19th International Seminar and Exhibit on Primary and Secondary Batteries, Florida Educational Seminars, Inc., March 11–14, Fort Lauderdale."

Pistoia, G. (ed.). (1994). "Lithium Batteries, New Materials, Developments and Perspectives." Elsevier, Amsterdam.

Rand, D. A. J., Woods, R., and Dell, R. M. (1998). "Batteries for Electric Vehicles." Research Studies Press, Taunton, UK.

St.-Pierre, C., Rouillard, R., Belanger, A., Kapfer, B., Simoneau, M., Choquette, Y., Gastonguay, L., Heiti, R., and Behun, C. (1999). Lithium–metal–polymer battery for electric vehicle and hybrid electric vehicle applications. *In* "Proceedings of EVS16 Symposium, China Electrotechnical Society, Beijing."

Sudworth, J., and Tilley, R. (1985). "The Sodium/Sulfur Battery." Chapman & Hall, London.

United Nations (1997). "UN Recommendations on the Transport of Dangerous Goods." 10th rev. ed. United Nations Sales Section, New York.

Van Schalkwijk, W. A., and Scrosati, B. (eds.). (2002). "Advances in Lithium-Ion Batteries." Kluwer Academic/Plenum, New York.

Vincent, C. A., and Scrosati, B. (1997). "Modern Batteries." 2nd ed. Arnold, London.

Yamanaka, K., Hata, K., Noda, T., Yamaguchi, K., and Tsubota, M. (2001). Development of 36 V valve regulated lead–acid batteries for hybrid vehicle application. *In* "Proceedings of the 42nd Japan Battery Symposium, The Electrochemical Society of Japan, November 21–23, Yokohama."

Bicycling

CHARLES KOMANOFF
Komanoff Energy Associates
New York, New York, United States

1. Introduction
2. Bicycle Development
3. The Bicycle During the Auto Age
4. The Bicycle and Human Power
5. Bicycle Variety
6. Bicycles Around the World
7. Bicyclist Safety and Danger
8. Bicycle Policies
9. Bicycle Prospects

Glossary

automobile Any self-guided, motorized passenger vehicle used for land transport, usually with four wheels and an internal combustion engine.

bicycle Any pedal-driven, two-wheeled conveyance propelled by human power.

bike lane A linear portion of a roadway demarcated for the predominant or exclusive use of bicycles.

bike path A linear path or pathway physically separated from conventional motorized roadways for the predominant or exclusive use of bicycles.

efficiency For transport vehicles, ratio of distance traversed to energy consumed.

gear For bicycles, relationship between pedaling and movement, often expressed as distance traveled in one revolution of the pedals (metric gear ratio).

sprawl Land-use patterns characterized by low population density, high automobile density, extensive road network, and little or no opportunity to travel via transit, cycling, or walking.

traffic calming Any roadway design feature or environmental intervention intended to reduce speeds and volumes of vehicular traffic.

A bicycle is any pedal-driven, two-wheeled conveyance propelled by human power. Bicycles revolutionized transport during the 1800s, becoming the first mass-produced personal transportation device, only to be literally pushed aside by automobiles during the 20th century. Nevertheless, the bicycle remains the world's most numerous transport vehicle, outnumbering cars by a two-to-one margin. Bicycles are the predominant transport vehicle in China and are a staple of urban transit throughout Northern Europe. They are the most energy-efficient machine of any kind in widespread use and are three orders of magnitude less energy-consuming than automobiles per distance traversed. Moreover, bicycles conserve not just on a per-trip basis but also at the social level by encouraging the substitution of proximity for distance and adding to the efficiency advantage of dense urban settlements over sprawling, suburbanized land-use patterns. In addition, cycling provides the opportunity to obtain healthful physical activity in the course of daily life while enhancing personal autonomy vital to mental health. Accordingly, preserving and indeed expanding the bicycle's role in urban transport is increasingly viewed as a global priority for social cohesion, urban viability, oil conservation, and protection of the climate against greenhouse gases.

1. *INTRODUCTION*

The bicycle—any pedal-driven, two-wheeled conveyance propelled by human power—is the world's most numerous transport vehicle and the most energy-efficient machine in widespread use. The bicycle was also the first mass-produced personal transportation device and created the preconditions for the development of the automobile a century ago. It is ironic, then, that the automobile, wherever it has been widely adopted, has largely driven the bicycle from the roads, causing global energy use to skyrocket along with a multitude of other social ills.

Consider that to travel 1 km on flat terrain, a cyclist operating a sturdy "roadster" bicycle at 10 mph expends a mere 60,000 joules (14.5 kilocalories). To cover the same distance with a typical U.S. passenger car rated at 20 mpg requires 115 million joules or nearly 2000 times more energy.

To be sure, the car/bicycle energy ratio varies with the number of people carried (cars' "load factors" average more than 1, but so do bicycles' load factors in developing countries), with vehicle weight and efficiency (small cars use 33–50% less fuel than so-called light trucks, and similarly lightweight "10-speed" bicycles require less energy than do roadster bikes), and with energy losses in processing gasoline and obtaining food. Still, under most combinations of assumptions, bicycles can cover a given distance using one-thousandth of the fuel that automobiles use.

Moreover, as social theorist Ivan Illich observed, the distance one travels is a product of the dominant mode of transport. Bicycles go hand in hand with short distances and urban density, whereas cars' voracious need for space both demands and feeds suburban sprawl. In short, bicycles serve proximity, whereas cars create distance.

Partly as a result of this vicious circle, cars have come to account for nearly 30% of world petroleum consumption and to produce nearly 15% of emissions of carbon dioxide, the most prominent greenhouse gas. Thus, expanding the role of bicycles vis-à-vis automobiles seems to be an obvious prescription for a world riven by conflict over petroleum and facing ecological upheaval from climate change.

Yet despite concerted efforts by thousands of cycle advocates in scores of countries, cycling appears to be losing ground, or at best running in place, in most nations' transport mix. Bikes outnumber cars worldwide, but they are used for fewer "trips" overall and probably cover less than one-tenth as many person-miles as do autos. How can this quintessentially human scale and efficient machine flourish again during the 21st century?

2. BICYCLE DEVELOPMENT

The essential elements of the bicycle are two wheels in line connected by a chain drive mechanism, with a rider simultaneously balancing and pedaling. Some bicycle historians attribute the first sketch of such a device to Leonardo da Vinci or, as argued by author David Perry, to an assistant in Leonardo's studio. A drawing with two large eight-spoked wheels, two pedals, a chain drive, saddle supports, a frame, and a tiller bar appears in the *Codex Atlanticus,* a volume of Leonardo's drawings and notations dating from around 1493 and assembled during the 16th century. However, the drawing was not discovered until 1966, during a restoration, leaving its authenticity open to question.

European inventors produced a number of self-propelled, hand- or foot-powered conveyances over the subsequent several centuries, all employing four wheels for stability. A practical two-wheeled device was first produced by the German Karl von Drais in 1816 and was patented 2 years later. His 40-kg Laufmaschine ("running machine") was propelled not by a mechanical drive but rather by pushing the feet against the ground, like a hobby horse, yet was capable of 13- to 14-kph speeds on dry firm roads. Similar machines with foot-operated drive mechanisms, using treadle cranks, appeared in Scotland during the late 1830s. True bicycles—two-wheel vehicles with pedal cranks located on the hub of the drive wheel—finally emerged in 1863 in France and spread quickly throughout Europe and to the United States.

Because these "pedal velocipedes" lacked gearing, each turn of the pedals advanced them only a distance equal to the circumference of the front wheel. To achieve high speeds, designers resorted to enormous front-drive wheels, reaching diameters of 4 feet for ordinary use and up to 6 feet for racing models. Although ungainly and difficult to operate, these "high-wheelers" proliferated and led to important innovations such as tangent-spoked wheels to resist torque, tubular diamond-shaped frames to absorb stress, hand-operated "spoon" brakes that slowed the wheels by pushing against them, and hubs and axles with adjustable ball bearings.

The final two steps in the evolution of the modern bicycle came during the 1880s: gearing to allow the use of smaller, more manageable wheels and chain-and-sprocket drives that transferred the drive mechanism to the rear wheel. These advances led to the so-called safety bicycle, the now-familiar modern design in which the cyclist sits upright and pedals between two same-sized wheels—the front for steering and the rear for traction.

As noted by Perry, this modern machine revolutionized cycling and is widely considered the optimal design. Innovations making bicycles safer, easier to use, and more comfortable followed in quick succession; these included pneumatic tires, "free-wheels" allowing coasting, multispeed gearing, and lever-actuated caliper brakes operating on rims rather than on tires. The safety bike transformed bicycling from a sport for athletic young men to a transport vehicle for men, women, and children alike. By 1893, safety bicycles had replaced velocipedes, and by 1896, Americans owned more than 4 million bicycles—1 per 17 inhabitants.

The bicycle boom of the late 1800s swept through the industrialized world, and bicycle manufacture became a major industry in Europe and America. One census found 1200 makers of bicycles and parts, along with 83 bicycle shops, within a 1-mile radius in lower Manhattan. The pace of invention was so frenetic that during the mid-1890s the United States had two patent offices: one for bicycles and another for everything else. The lone urban traffic count in the United States to include bicycles, taken in Minneapolis, Minnesota, in 1906 after bicycling levels had peaked, found that bicycles accounted for more than one-fifth of downtown traffic—four times as much as did cars.

This "golden age of bicycling" proved to be short-lived. Following the classic pattern of corporate capitalism, a wave of mergers and buy-outs consolidated small shops run by enthusiasts and financed on the cheap into factories whose assembly-line efficiencies entailed high fixed costs. Overproduction followed, and then came market saturation, price wars, stock manipulations, and bankruptcies. Never universally popular, particularly in dense urban areas where swift and stealthy bicycles frightened pedestrians, the bicycle industry found its public image tarnished.

Of course, reversals of fortune were standard fare in late-19th century capitalism, and many industries, particularly those employing advanced technology, bounced back. Unfortunately for the bicycle business, on the heels of the shakeout in bike manufacture came the automobile.

3. THE BICYCLE DURING THE AUTO AGE

The bicycle catalyzed development of the car. A number of technical advances essential to the fledgling automobile industry, from pneumatic tires and ball bearings to factory-scale production engineering and a functional network of paved urban roads, owe their emergence to bicycles. No less important, the bicycle's ethos of independent, self-guided travel helped to split open the railroad-based paradigm of travel as mass transport along a fixed linear track.

But once the car took hold, it imposed its own ideology, one antithetical to bicycles. For one thing, cars used the same roads as did bicycles (the very roads that were paved as a result of bicyclists' campaigning), and through incessant noise and fumes, superior speed, and sheer physical force, cars literally pushed bicyclists aside. What is more, as recounted by social historian Wolfgang Sachs, the engine-driven car proved to be a more alluring cultural commodity than did the self-propelled bicycle. Although the bicycle leveraged bodily energy and broadened the individual's arena of direct activity many times over, the substitution of mechanical power for muscular exertion conveyed a sense of joining the leisure class and became equated with progress.

Thus, the bicycle's "defect of physicality," as Sachs termed it, put it at a disadvantage compared with the new technologies of internal combustion, electric motor drive, and flying machines. Rather than defend their right to cycle, the masses aspired to abandon the bicycle in favor of the auto. And abandon it they did, as fast as rising affluence and the advent in 1908 of the mass-produced, affordable car, Henry Ford's Model T, permitted. Although reliable data are lacking, by the end of the 1920s, bicycles probably accounted for only 1 to 2% of U.S. urban travel, an order-of-magnitude decline in just a few decades.

A similar devolution occurred in Europe during the long boom after World War II, albeit less steeply and with important exceptions. However, even now, bicycles outnumber cars by a two-to-one margin around the world, primarily due to economics. Cars cost roughly 100 times as much to buy as do bicycles and require fuel as well as maintenance, putting them out of reach of a majority of the world's people.

4. THE BICYCLE AND HUMAN POWER

According to data compiled by Vance A. Tucker of Duke University, a walking human consumes approximately 0.75 calorie of energy per gram of body weight for each kilometer traveled. This is less than the rate for birds, insects, and most mammals but is more than that for horses and salmon. However, atop a bicycle, a human's energy consumption falls fivefold, to a rate of roughly 0.15 calorie per gram per kilometer. As S. S. Wilson noted, "Apart from increasing his unaided speed by a factor of three or four, the cyclist improves his efficiency rating to No. 1 among moving creatures and machines."

Wilson attributed the bicycle's high efficiency mainly to its effective use of human muscles.

TABLE I

Cyclist Performance Factors

Factor	Average	Sport	Pro
Resting heart rate (breaths/min)	70	50	30
Anaerobic threshold (breaths/min)	160	175	185
Maximum heart rate (breaths/min)	180	190	195
Blood volume (ml)	10	25	50
Lung capacity (L)	5	6.5	8
VO_2 maximum (O_2 uptake, ml/kg-minute)	40	60	85
Thrust (Foot-pounds)	15	30	55
Cadence (rpm)	70	90	100
Watts	50	200	500
Calories (kcal)	135	750	2,150
Speed (kph)	15	30	50

Whereas a walker expends energy raising and lowering the entire body as well as accelerating and decelerating the lower limbs, the cyclist's sitting posture relieves the leg muscles of their supporting function. Because the cyclist's feet rotate smoothly at a constant speed and the rest of the body is still, the only reciprocating parts of the cyclist's body are the knees and thighs. Even the acceleration and deceleration of the legs is optimized given that one leg is raised by the downward thrust of the other. Wind resistance, the main constraint on the racing cyclist (because it varies with the square of the cyclist's velocity relative to the wind), is less significant at ordinary utilitarian speeds.

According to Perry, during a 1-h ride, an average person on a touring bike covering 15 km burns approximately 135 calories for an average power output of 50 W; over the same hour, a professional racing cyclist covers 50 km, burning 2150 calories and producing approximately 500 W (0.67 horsepower) (Table I).

"It is because every part of the design must be related to the human frame," wrote Wilson, "that the entire bicycle must always be on a human scale." He concluded, "Since the bicycle makes little demand on material or energy resources, contributes little to pollution, makes a positive contribution to health, and causes little death or injury, it can be regarded as the most benevolent of machines."

5. BICYCLE VARIETY

There is an enormous variety of bicycle types, reflecting human ingenuity, technical evolution, and a broad range of design criteria such as comfort, roadworthiness, speed, durability, and economy. As noted by Perry, a 22-pound road bicycle contains some 1275 parts in two dozen functional systems (e.g., wheels, chains, derailleurs, crank sets). Bicycle components require an unusual mix of lightness and durability, rigidity, and flexibility to provide a range of functions such as steering, braking, balancing, and climbing over a variety of terrains in various weather conditions. Moreover, in much of the world, bicycles must withstand the stress of carrying several passengers or cargoes weighing several hundred kilograms.

Most bikes fit into one of five broad categories:

- *Safety bicycles.* These bikes are the standard design established during the late 19th century and still employed throughout Asia and Africa. They have wide upright handlebars, medium-width tires, heavy construction for stability and durability, and up to three internal hub gears.
- *Racing or touring (10-speed or road) bicycles.* These bikes have the lightweight aerodynamic design initially developed for racing during the early 20th century and widely used today for sport. They have narrow "dropped" handlebars and skinny tires and achieve 10 gears through a double front chain ring and a 5-speed rear derailleur.
- *Mountain bikes.* These bikes are a recent (circa 1980) design, adding lightweight racing and touring components to the classic upright posture safety bicycle, with wide knobby tires for off-road use and a triple chain ring for steep inclines. A "hybrid" variant with slimmer tires is widely used in industrial countries for urban commuting.
- *Human-powered vehicles (or HPVs).* HPVs are an entire class of machines combining aspects of bicycles, tricycles, and even cars developed by engineers and enthusiasts to "push the envelope" of nonmotorized transportation. They include recumbent bicycles in which riders recline against a backrest and the pedals are placed far forward. Using streamlined "fairing" to minimize wind resistance, HPVs have achieved 60-min speeds of 82 kph versus a maximum of 56 kph for standard racing bikes.
- *Utility cycles.* These bikes, with dedicated compartments and/or trailers for carrying large and heavy loads, are common in Asia and also are used in industrial nations in settings ranging from factory floors to urban food delivery. Pedicabs conveying passengers in separate compartments are widely used in China, Bangladesh, and parts of Africa, although authorities in Indonesia and elsewhere

have confiscated millions in forced motorization campaigns.

6. *BICYCLES AROUND THE WORLD*

Along with 1.2 billion bicycles, the world's 6.1 billion people possess 600 million motorized passenger vehicles (cars and light trucks), making for roughly 1 bike per 5 persons and 1 automobile per 10 persons. However, only a handful of countries actually show this precise two-to-one ratio because most of the world's motor vehicles are in the industrial nations, whereas most bicycles are in the developing world.

Although precise data are not available, it is likely that the average car is driven approximately 10,000 miles per year, whereas the average bicycle probably logs fewer than 500 miles per year. Based on these rough figures, the world's bicycles collectively travel less than one-tenth as many miles as do cars, although their share of trips is somewhat larger.

Three countries or regions are of particular interest: China, the world's most populous nation and still a bicycling stronghold, in spite of policies designed to encourage car use; the United States, the *ne plus ultra* of automobile use; and Northern Europe, where public policy restraining automobile use has brought about a bicycle renaissance amid affluence (Table II).

4.1 China

Large-scale bicycle manufacture and use have been a centerpiece of Chinese industrialization and urbanization since shortly after the 1949 revolution. By the 1980s, production for both domestic use and export had reached 40 million bikes per year, outnumbering total world car output. Today, China's 1.3 billion people own a half-billion bicycles, 40% of the world's total, and the bicycle is the mainstay of urban transportation throughout the country. Not just individuals but also whole families and much cargo are conveyed on bicycles, bike manufacture and servicing are staples of China's economy, and mass urban cycling is an indelible part of China's image in the world. Traffic controllers in the largest cities have counted up to 50,000 cyclists per hour passing through busy intersections; in comparison, a four-lane roadway has a maximum throughput of 9000 motor vehicles per hour.

This is now changing, perhaps irreversibly, as China invests heavily in both automobiles and mechanized public transport. Although domestic auto sales in 2002 numbered just 800,000 versus bicycle sales of 15 to 20 million, the auto sector is growing rapidly at 10 to 15% per year. "Bicycle boulevards" in Beijing and other major cities have been given over to cars, with bikes excluded from 54 major roads in Shanghai alone. The safety and dignity enjoyed by generations of Chinese are beginning to crumble under the onslaught of motorization.

Although the number of bicycles in China is still growing, sales have dropped by one-third since the early 1990s. With cyclists increasingly forced onto buses and subways, the bicycle's share of trips in Beijing and Shanghai has fallen precipitously to 40% and 20%, respectively, from more than 50% a decade or so ago. Indeed, the incipient conversion of the world's premier bicycle nation into a car-cum-transit society is eerily reminiscent of America a century ago.

TABLE II

Bicycles and Automobiles in Selected Countries and Regions (circa 2000)

Country or region	Bicycles	Autos	Bicycle/auto ratio	Bicycles per 1000	Autos per 1000
China	500,000,000	18,000,000	28	392	14
India	60,000,000	10,000,000	6	59	10
Japan	60,000,000	40,000,000	1.5	472	315
Germany	60,000,000	40,000,000	1.5	732	488
The Netherlands	12,000,000	6,000,000	2	750	375
United States	120,000,000	180,000,000	0.7	421	632
Argentina	5,000,000	5,000,000	1	135	135
Africa	40,000,000	20,000,000	2	50	25
World totals	1,200,000,000	600,000,000	2	198	99

The difference is that China's population is an order of magnitude larger than that of the United States in 1900, making the fate of cycling in China a matter of global moment. Profligate burning of fossil fuels is recognized to threaten humanity through global climate change, and a simple calculation demonstrates that if China were to match the U.S. per capita rate of auto use, world carbon dioxide emissions would rise by one-quarter, a catastrophic defeat in the effort to limit greenhouse gases.

6.2 United States

Like no other society in history, the United States is dominated spatially, economically, and psychologically by automobiles. Registered autos outnumber bikes by a two-to-one ratio, but more important, more than 90% of individuals' transportation trips are by car. A mere one-hundredth as many, or 0.9%, are by bicycle, and a majority of these are for recreational riding rather than for "utilitarian" transport. Some reasons follow:

- *Car culture.* Under America's cultural triumvirate of mobility, physical inactivity, and speed, the car has been enshrined as the norm and cycling is consigned to the margins. In turn, the perception of cycling as eccentric or even deviant contributes to an unfavorable climate for cycling that tends to be reinforcing.
- *Sprawling land use.* A majority of Americans live in suburbs, increasingly in the outermost metropolitan fringe whose streets and roads are suitable only for motorized vehicles. Cities are more conducive to cycling, with smaller distances to cover and congestion that limits motor vehicle speeds, but they too are engineered around autos, and few have marshaled the fiscal and political resources to ensure safety, much less amenity, for cyclists.
- *Subsidized driving.* Low gasoline taxes, few road tolls, and zoning codes requiring abundant free parking are a standing invitation to make all journeys by car, even short trips that could be walked or cycled.
- *Poor cycling infrastructure.* Spurred by federal legislation letting states and localities apply transportation funds to "alternative" modes, the United States has invested $2 billion in bicycle facilities since the early 1990s. Nevertheless, provision of on-street bike lanes, separated bike paths, cycle parking, and transit links has been haphazard at best and is often actively resisted.
- *Cycling danger.* A bike ride in the United States is three times more likely to result in death than is a trip in a car, with approximately 800 cyclists killed and 500,000 injured annually. Not surprisingly, the prospect of accident and injury is a powerful impediment to bicycling in the United States. Moreover, cycling's actual risks are compounded by cultural attitudes that attribute cycle accidents to the supposedly intrinsic perils of bicycles, unlike motorist casualties, which are rarely considered to imply that driving as such is dangerous.

These inhibiting factors are mutually reinforcing. Particularly with America's pressure group politics, the lack of broad participation in cycling severely limits support for policies to expand it. The lack of a consistent visible presence of cyclists on the road exacerbates the inclination of drivers to see cyclists as interlopers and to treat them with active hostility. "Feedback loops" such as these keep cycling levels low.

6.3 Northern Europe

There is one region in which bicycling coexists with affluence and automobiles. Despite high rates of car ownership, more than a half-dozen nations of Northern Europe make at least 10% of urban trips by bike, surpassing the U.S. mode share at least 10-fold.

The highest bike share, and the steadiest, is in The Netherlands, with 26% of urban trips in 1978 and 27% in 1995. Cycling's modal share rose sharply in Germany during the same period, from 7 to 12%—still less than half the Dutch level but impressive given Germany's rapid growth in auto ownership and use. Also notable is cycling's high share of trips made by seniors: 9% in Germany and 24% in The Netherlands.

Robust bicycling levels in Northern Europe are not accidental but rather the outcome of deliberate policies undertaken since the 1970s to reduce oil dependence and to help cities avoid the damages of pervasive automobile use. Not only are road tolls, taxes, and fees many times higher than those in the United States, but generously funded public transit systems reduce the need for cars, increasing the tendency to make short utilitarian trips by bicycle. Particularly in Germany, Denmark, and The Netherlands, comprehensive cycle route systems link "traffic-calmed" neighborhoods in which cycling alongside cars is safe and pleasant.

Indeed, the same kind of feedback loops that suppress cycling in the United States strongly nurture it in Northern Europe. Both density and bicycling are encouraged by policies ranging from provision of transit and cycle infrastructures to "social pricing" of driving; and Northern European states refrain from subsidizing sprawl development. Not only do a majority of Europeans live in cities as a result, but population densities in urban areas are triple those in the United States; correspondingly, average trip distances are only half as great, a further inducement to cycle.

7. BICYCLIST SAFETY AND DANGER

The bicycle's marvelous economy and efficiency have negative corollaries. First, unlike three- or four-wheeled conveyances, bikes are not self-balancing. Continuous motion is required to keep them upright, and they can tip and crash due to road defects, mechanical failure, or operator error. Second, crash mitigation measures such as seat belts, air bags, and crumple zones that have become standard in automobiles are not feasible for bicycles; only helmets offer a modicum of protection, and perhaps less than is commonly believed.

The exposed position of cyclists on the road makes them vulnerable to motor vehicles. Surpassing bicycles several-fold in velocity and at least 100-fold in mass, automobiles have approximately 1000 times more kinetic energy to transfer to a bicycle in a collision than is the case vice versa. Not surprisingly, although most of the total injury-accidents to bicyclists occur in falls or other bike-only crashes, severe injuries and fatalities are due mostly to being hit by cars. Approximately 90% of bicycle fatalities (95% for child cyclists) in the United States have motorist involvement, and the percentages elsewhere are probably as high.

Ironically, bicycles pose little danger for other road users and far less than do automobiles. In the United States, fewer than 5 pedestrians die each year from collisions with bikes, whereas 5000 are killed by motor vehicles. Motor vehicle users exact an enormous toll on themselves as well as on each other. Worldwide, total road deaths are estimated at 1 million people each year, with millions more becoming disabled in accidents. Based on "disability-adjusted life years," a statistic incorporating permanent injuries and the relative youth of victims, road deaths were ranked by the World Health Organization as the world's ninth-leading health scourge in 1990 and were predicted to rank third by 2020.

7.1 Worldwide Differences

Bicycle safety policies differ widely around the world, as illustrated in the regions profiled earlier. China and other developing countries are too poor to invest in bicycle safety programs, but until recently they were also too poor for the cars that make such programs necessary. Historically, cyclists in China and other Asian nations have been able to rely on their sheer numbers to maintain their rights-of-way. In the United States, bicycle safety measures focus on changing cyclist behavior, primarily increasing helmet use (especially by children) and training cyclists to emulate motor vehicles through "effective cycling" programs. Little effort is made to address the nature and volume of motor vehicle traffic or the behavior of drivers toward cyclists.

In contrast, bicycle safety in Europe is promoted holistically as part of policies to encourage widespread cycling and universal road safety. Germany and The Netherlands promote both through provision of elaborate and well-maintained cycling infrastructures, urban design oriented to cycling and walking rather than to motor traffic, disincentives for and restrictions on car use, and enforcement of traffic regulations that protect pedestrians and cyclists.

The European safety model appears to be validated by low fatality rates. Despite stable or rising cycling levels from 1975 to 1998, cycle fatalities fell 60% in Germany and The Netherlands. The U.S. decline was less than half as great (25%) and may have been largely an artifact of reduced bicycling by children. Currently, bicycle fatalities per kilometer cycled are two to three times lower in Germany and The Netherlands than in the United States, and pedestrian fatalities per kilometer walked are three to six times less. Perhaps most tellingly, although Germany and The Netherlands have higher per-kilometer fatality rates for auto users than does the United States, their overall per capita rates of road deaths are lower—by one-third in Germany and by one-half in The Netherlands—in large part because cars are used less in both countries.

7.2 Cycle Helmets

Helmets have become an intensely polarized subject in bicycling during recent years. Many observers trace the origins of the debate to a 1989 epidemiological study in Seattle, Washington, associating helmet use with an 85% reduction in brain and head injuries. The authors subsequently employed better statistical methods and scaled back their results considerably, to a mere 10% reduction in severe injuries when body and not just head trauma is taken into account. But the initially reported connection between helmet use and injury reduction sparked campaigns in the United States and Australia to compel helmet use by child cyclists (later extended to skateboards, roller skates, and scooters) and to promote helmet wearing by adult cyclists.

Whether these campaigns have been useful or not is difficult to say. Child cycling fatalities have decreased in the United States, but that may be because fewer children now cycle. Adult fatalities have risen, possibly due to growth in the more dangerous kinds of motor traffic such as sport utility vehicles and drivers' use of mobile phones. Nevertheless, although links between helmet promotion and cycling injury prevention are inconclusive, the "helmet paradigm" is firmly established in U.S. policy.

In Europe, where injury prevention is subordinated to the larger goal of health promotion and where social responsibility is emphasized alongside individual accountability, helmets are considered irrelevant or even counterproductive to health. The influential British social scientist Mayer Hillman contends that cardiovascular and other physiological and psychological gains from cycling far outweigh the rider's crash risk, even in unsatisfactory present-day road environments. Therefore, Hillman's paradigm of cycle encouragement holds that cycling is so beneficial to individuals and society that no interferences should be tolerated, not even the inconvenience and unattractiveness of a helmet or the subliminal message that helmets may send about the dangers of cycling.

7.3 Safety in Numbers

Anecdotal evidence has long suggested that the per-cyclist rate of bicycle–motor vehicle crashes declines as the amount of cycling on a road or in a region increases. This "safety in numbers" effect is thought to occur because as cyclists grow more numerous and come to be an expected part of the road environment, motorists become more mindful of their presence and more respectful of their rights. The implication is that adding more cyclists to the road makes it less likely that a motorist will strike an individual cyclist and cause serious injury. Conversely, removing cyclists from the traffic stream raises the risk to those who continue to cycle.

This safety in numbers effect offers a plausible explanation for the fact that per-kilometer cycling fatality rates in Germany and The Netherlands are four times less than that in the United States, even though cycling percentages are more than 10 to 20 times higher in these European countries. Now, time-series estimates of this effect, although preliminary and site specific, are pointing intriguingly toward a "power law" relationship of approximately 0.6 between cyclist numbers and cyclist safety. According to this relationship, the probability that a motorist will strike an individual cyclist on a particular road declines with the 0.6 power of the number of cyclists on that road. Say the number of cyclists doubles. Because 2 raised to the 0.6 power is 1.5, each cyclist would be able to ride an additional 50% without increasing his or her probability of being struck. (The same phenomenon can be expressed as a one-third reduction in per-cyclist crash risk per doubling in cycling volume given that the reciprocal of 1.5 is 0.67.)

The implications for cycling are profound. Countries that have based safety promotion on cyclist behavior modification (e.g., the United States) might reconstruct safety in a social context. One consequence would be to deemphasize helmet use in favor of jump-starting broader participation in cycling so as to stimulate a "virtuous circle" in which more cycling begets greater safety, which in turn encourages more cycling. In addition, countries such as China might reconsider policies that threaten to erode large-scale cycling, lest safety in numbers in reverse leads to a downward spiral as in the current U.S. situation, where bike riding is limited to small numbers of enthusiasts and to others who have no alternatives.

8. *BICYCLE POLICIES*

Policies to support and "grow" bicycling fall into three categories: cycling infrastructure, cyclists' rights, and disincentives to driving.

Cycling infrastructure policies aim to attract cycle trips by providing "facilities" such as on-street bicycle lanes, off-street bicycle paths (e.g., "greenways") with separate rights-of-way, bicycle parking, and integration with the metropolitan or regional transit system. Constructing and maintaining such facilities has proven to be politically popular in some states and localities in the United States and absorbed most of the $2 billion spent on bicycle programs from 1992 to 2002.

Cyclists' rights initiatives seek to improve the legal standing of cycling and, thus, to make cycling safer and more socially validated. It is believed that exerting closer authority over driver conduct through the legal system, police enforcement, and cultural shifts would directly reduce the threat to bicyclists and so encourage cycling.

Disincentives to driving are policies to make driving less attractive economically and logistically and, therefore, to reduce the level of motor traffic. Measures falling under the rubric of "social pricing" of automobiles include gasoline taxes, "carbon" taxes on fossil fuels, road pricing (fees on each kilometer driven), and restructuring auto insurance and local road taxes to pay-per-use.

That these three kinds of initiatives are complementary is seen by examining Germany and The Netherlands, which have used all three to maintain and restore bicycling since the early 1970s. No single approach is sufficient, and each supports the others by increasing opportunities and rewards for cycling and establishing a social context in which cycling is "valorized" as appropriate and praiseworthy rather than viewed as deviant behavior.

9. BICYCLE PROSPECTS

Strong societal currents are pushing bicycling forward, yet equally mighty forces are suppressing it. Much hangs in the balance, both for billions of the earth's peoples who may wish to master their own mobility through cycling and for our planet's ecological and political well-being.

Each trip cycled instead of driven conserves gasoline and stops the addition of climate-altering carbon dioxide to the atmosphere. The potential effects are large given that passenger vehicles account for nearly one-third of global petroleum consumption and generate more than one-eighth of carbon dioxide emissions. "Green cars" are no more than a palliative; a world in which everyone drove at the U.S. per capita rate would emit more carbon dioxide than is currently the case, even if autos could be made to be five times more efficient.

Thus, sustaining the earth's climate and political equilibrium requires robust alternatives to the American model of one car per journey. Moreover, bicycles conserve several times over, and not just at the per-trip level, by encouraging the substitution of proximity for distance and adding to the efficiency advantage of dense urban settlements over sprawling suburbanized land-use patterns. Therefore, ecological imperatives are a potent reason to maintain bicycling in China and other developing countries as well as to foster it in automobile-dependent societies such as the United States.

Health promotion is a major rationale as well. As noted, road traffic accidents are or soon will be among the world's half-dozen leading causes of death and disability. Sedentary lifestyles, including substitution of motorized transport for walking and cycling, are also recognized as a cause of fast-growing obesity and of ill health generally. In contrast, cycling provides the opportunity to obtain physical activity in the course of daily life while enhancing personal autonomy and aiding mental health.

Yet motorization itself is a powerful suppressant to cycling, and not just in the often-lethal competition between cars and bikes. Just as pernicious is the automobile's grip on transportation's "mind-share"—the automatic equating of mobility with motor vehicles and of motor vehicles with success—that leaves bicyclists, both individually and institutionally, on the outside looking in.

Large-scale cycling seems reasonably assured in the countries of Northern Europe that view it as integral to national objectives of reducing greenhouse gases, sustaining urban centers, and promoting health and self-guided mobility. Preserving mass cycling in China and developing it in the United States will probably require dethroning the automobile as the symbol and engine of prosperity and sharply reducing its enormous financial and political power—a tall order.

The bicycle—a pinnacle of human efficiency and an icon of vernacular culture for more than a century—will survive. Whether it will again flourish may make a difference in how, and whether, humanity itself survives.

SEE ALSO THE FOLLOWING ARTICLES

Alternative Transportation Fuels: Contemporary Case Studies • Development and Energy, Overview • Fuel Cycle Analysis of Conventional and Alternative Fuel Vehicles • Global Energy Use: Status and Trends • Internal Combustion Engine Vehicles • Lifestyles and Energy • Motor Vehicle Use, Social Costs of • Passenger Demand for Travel and Energy Use • Vehicles and Their Powerplants: Energy Use and Efficiency

Further Reading

Carlsson, C. (ed.). (2002). "Critical Mass: Bicycling's Defiant Celebration." AK Press, Edinburgh, UK.

Herman, M., Komanoff, C., Orcutt, J., and Perry, D. (1993). "Bicycle Blueprint: A Plan to Bring Bicycling into the Mainstream in New York City." Transportation Alternatives, New York.

Hillman, M. (1992). "Cycling: Towards Health and Safety." British Medical Association, London.

Illich, I. (1974). "Energy and Equity." Harper & Row, New York.

Jacobsen, P. (2003). Safety in numbers: More walkers and bicyclists, safer walking and bicycling. *Injury Prevention* **9**, 205–209.

Komanoff, C., and Pucher, J. (2003). Bicycle transport in the U.S.: Recent trends and policies. *In* "Sustainable Transport: Planning for Walking and Cycling in Urban Environments" (R. Tolley, Ed.). Woodhead Publishing, Cambridge, UK.

Lowe, M. D. (1989). "The Bicycle: Vehicle for a Small Planet." Worldwatch Institute, Washington, DC.

McShane, C. (1994). "Down the Asphalt Path: The Automobile and the American City." Columbia University Press, New York.

Perry, D. B. (1995). "Bike Cult." Four Walls Eight Windows, New York.

Pucher, J. (1997). Bicycling boom in Germany: A revival engineered by public policy. *Transport. Q.* **51**(4), 31–46.

Pucher, J., and Dijkstra, L. (2000). Making walking and cycling safer: Lessons from Europe. *Transport. Q.* **54**(3), 25–50.

Rivara, F. P., Thompson, D. C., and Thompson, R. S. (1997). Epidemiology of bicycle injuries and risk factors for serious injury. *Injury Prevention* 3(2), 110–114.

Sachs, W. (1992). "For Love of the Automobile." University of California Press, Berkeley.

Tolley, R. (ed.). (2003). "Sustainable Transport: Planning for Walking and Cycling in Urban Environments." Woodhead Publishing, Cambridge, UK.

Whitt, F. R., and Wilson, D. G. (1989). "Bicycling Science." MIT Press, Cambridge, MA.

Wilson, S. S. (1973). Bicycle technology. *Sci. Am.* **228**, 81–91.

Biodiesel Fuels

LEON G. SCHUMACHER
University of Missouri, Columbia
Columbia, Missouri, United States
JON VAN GERPEN
Iowa State University
Ames, Iowa, United States
BRIAN ADAMS
University of Missouri, Columbia
Columbia, Missouri, United States

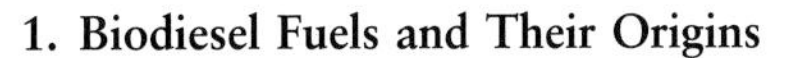

1. Biodiesel Fuels and Their Origins
2. Chemical Properties
3. Physical Properties
4. Biodiesel Production
5. Advantages of Biodiesel
6. Disadvantages
7. Biodiesel Storage and Use
8. Biodiesel Economic Considerations

Glossary

aromatics Any unsaturated hydrocarbon containing resonance stabilized carbon-to-carbon bonds characterized by benzene-type ring structure.

biodiesel A renewable energy fuel from vegetable or animal derived oil (triglyceride) that has been chemically modified to reduce its viscosity. It can be used in any concentration with petroleum-based diesel fuel in existing diesel engines with little or no modification. Biodiesel is not the same thing as raw vegetable oil. It is produced by a chemical process, which removes the glycerol from the oil.

distillation curve A measure of volatility of a fluid. More specifically, it is the temperature of the fluid (fuel) when a specific volume of the fluid has evaporated.

feedstock Raw or processed material that is chemically reacted to produce biodiesel. Note that the raw material must first be processed before the oil can be used to produce biodiesel.

gas chromatograph A chemical is heated to a gaseous state, and then the gas is passed through a cylinder and the different chemicals that make up the gas "stick" to the wall of the cylinder at different intervals. This information is then used to determine the chemical makeup of the gas.

olefin Any unsaturated hydrocarbon containing one or more pairs of carbon atoms linked by a double bond.

triglyceride A naturally occurring ester formed from glycerol and one to three fatty acids. Triglycerides are the main constituent of vegetable and animal derived fats and oils.

The United States depends heavily on imported oil to fuel its transportation infrastructure. The use of alternative fuel derived from plant oils was examined by researchers in the mid-1970s to determine if internal combustion engines could be fueled from sources other than petroleum. The initial research on pure vegetable oils as a replacement for petroleum diesel fuel was met with mostly negative results. Researchers determined that transesterification of these plant- and animal-derived oils reduced the viscosity of the oil without any other significant changes to the oil. Since the new fuel was bio-derived and was used to fuel a diesel engine, the name "biodiesel" was selected to refer to the new fuel. This article focuses more specifically on how biodiesel fuel was developed, its chemical and physical properties, advantages, disadvantages, and how biodiesel is used and stored. The article concludes by reviewing the economic issues associated with its use.

1. BIODIESEL FUELS AND THEIR ORIGINS

1.1 What Is Biodiesel?

Biodiesel is made from a number of feedstocks including vegetable oil, tallow, lard, and waste cooking oils (yellow grease). "Biodiesel is defined as mono-alkyl esters of long chain fatty acids derived from vegetable oils or animal fats which conform to American Society of Testing Materials (ASTM D5453) International specifications for use in diesel engines." Biodiesel contains no petroleum, but it can be blended at any level with petroleum diesel to create a biodiesel blend. Feedstocks can be transesterified to make biodiesel using an alcohol that has been mixed with a catalyst such as potassium hydroxide or sodium hydroxide. The most commonly used alcohol for transesterification is methanol. Methanol reacts easily and is less expensive to use than most other alcohols.

Soybean derived biodiesel is the most commonly used biodiesel in the United States. The most commonly used feedstock for biodiesel production in Europe is rapeseed. Biodiesel is biodegradable, nontoxic, and essentially free of sulfur and aromatics. Biodiesel is considered a renewable resource due to the fact that it is derived from products that can be grown and produced domestically.

1.2 What Biodiesel Is Not

In 1898 when Rudolph Diesel's compression ignition engine was demonstrated at the World's Exhibition in Paris, it ran on virgin peanut oil. Some document this event as the first use of biodiesel. However, biodiesel is not vegetable oil or animal fats. Biodiesel refers to the alkyl esters produced from a transesterification reaction between the oil or fat and an alcohol. Others refer to mixtures of biodiesel and petroleum diesel fuel as "biodiesel." This mixture is referred to as a biodiesel blend and is commonly designated in much the same way as a blend of gasoline and alcohol (E85). For example, B100 is 100% biodiesel; B20 is a blend of 20% biodiesel and 80% diesel fuel.

1.3 Beginnings

Vegetable oils were transesterified prior to the mid-1800s. Transesterification is the process of reacting a triglyceride molecule with an excess of alcohol in the presence of a catalyst (KOH, NaOH, $NaOCH_3$, etc.) to produce glycerol and fatty esters as shown in the chemical reaction in Fig. 1.

Companies such as Proctor & Gamble have used this process to make soap for years. According to Knothe, the earliest known use of alkyl esters for fuel appears in a Belgian patent granted in 1937 to G. Chavanne.

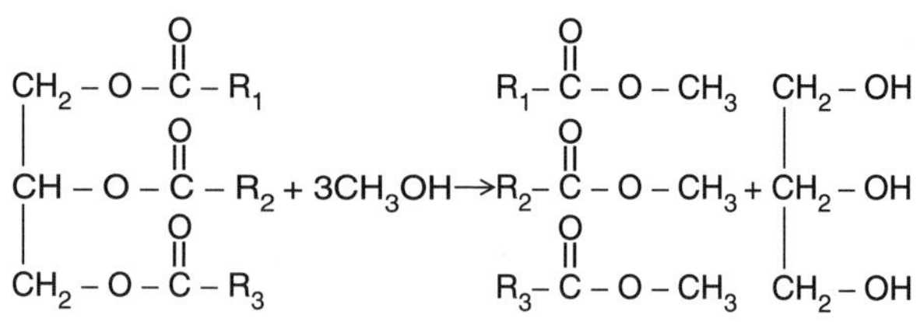

$$
\begin{array}{l}
CH_2-O-\overset{O}{\overset{\|}{C}}-R_1 \\
| \\
CH-O-\overset{O}{\overset{\|}{C}}-R_2 + 3CH_3OH \rightarrow \\
| \\
CH_2-O-\overset{O}{\overset{\|}{C}}-R_3
\end{array}
\begin{array}{l}
R_1-\overset{O}{\overset{\|}{C}}-O-CH_3 \quad CH_2-OH \\
\qquad\qquad\qquad\quad\;\; | \\
R_2-\overset{O}{\overset{\|}{C}}-O-CH_3 + CH_2-OH \\
\qquad\qquad\qquad\quad\;\; | \\
R_3-\overset{O}{\overset{\|}{C}}-O-CH_3 \quad CH_2-OH
\end{array}
$$

FIGURE 1 Chemical structure of biodiesel.

Methyl and ethyl esters are essentially by-products of this process. Ethyl esters are made using ethanol and methyl esters are made using methanol. Ethanol is made from grain such as corn. The methanol is either wood based or derived from natural gas (methane).

Peanut oil, hemp oil, corn oil, and tallow are typically transesterified in the soap-making process. Soybeans, industrial or edible rapeseed (or its cousin, canola oil), corn, recycled fryer oil, lard, and tallow are common resources for the complex fatty acids and their by-products.

It is important to find other sources of oil to enhance our ability to produce biodiesel. Research has been conducted concerning oil production from algae according to sources at the National Renewable Energy Laboratory (NREL). This oil source could have yields greater than any feedstock now known. Most biodiesel researchers worldwide agree that it would be difficult to replace more than 10% of the diesel fuel that is used for transportation purposes with biodiesel.

2. CHEMICAL PROPERTIES

Researchers have determined that the following properties characterize the chemical properties of biodiesel: fatty acid content, aromatics, olefins, paraffins, carbon, hydrogen, oxygen, and sulfur content, acid neutralization number, iodine number, and Conradson carbon residue number.

Note the chemical structure of a triglyceride (Fig. 2). R_1, R_2, and R_3 represent the hydrocarbon chain of the fatty acid elements of the triglyceride. There is a three-carbon chain called the glycerol backbone that runs along the left side of the molecule. Extending away from this backbone are the three long fatty acid chains. The properties of the triglyceride and the biodiesel fuel will be determined by the amounts of each fatty acid that are present in the molecules.

Fatty acids are designated by two numbers: the first number denotes the total number of carbon

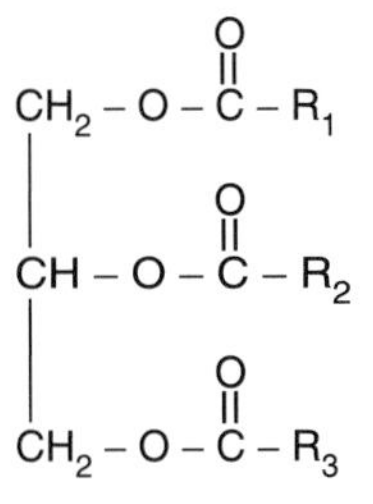

$$
\begin{array}{l}
CH_2-O-\overset{\overset{\large O}{\|}}{C}-R_1 \\
| \\
CH-O-\overset{\overset{\large O}{\|}}{C}-R_2 \\
| \\
CH_2-O-\overset{\overset{\large O}{\|}}{C}-R_3
\end{array}
$$

Triglyceride

FIGURE 2 Chemical structure of a triglyceride.

Chain	Percentage by weight
C14:0	0.278
C16:0	10.779
C18:0	4.225
C18:1	20.253
C18:2	54.096
C18:3	9.436
C20:0	0.395

FIGURE 3 Fatty acid profile for soybean derived biodiesel.

atoms in the fatty acid, and the second is the number of double bonds. For example, 18:1 designates oleic acid, which has 18 carbon atoms and one double bond. A typical sample of soybean oil based biodiesel would have the fatty acid profile shown in Fig. 3.

Biodiesel is essentially free of sulfur and aromatics. This is an advantage for biodiesel because sulfur poisons catalytic converter technology that is used to reduce engine exhaust emissions. The sulfur levels in biodiesel by ASTM D5453 are found to be as low as 0.00011% by mass (1 ppm), where petroleum diesel is often no lower than 0.02% (200 ppm). The lack of aromatic hydrocarbons is also an advantage for biodiesel, as many of these compounds are believed to be carcinogenic. The test procedure normally used to measure aromatics in petroleum fuel (ASTM D1319) should not be used to determine the aromatics of biodiesel. This analytical procedure mistakenly identifies the double bonds commonly found in biodiesel as the resonance stabilized bonds normally associated with aromatics.

Paraffins are hydrocarbon compounds that are normally associated with petroleum diesel fuel. These compounds help increase the cetane value of the diesel fuel. However, they also typically increase cold flow problems of petroleum diesel fuel. Olefins are hydrocarbons that contain carbon-carbon double bonds. Molecules having these types of bonds are called unsaturated. Biodiesel from common feedstocks is usually 60 to 85% unsaturated. Some olefins are present in petroleum-based diesel fuel. However, the amount of olefins is usually small as they contribute to fuel oxidation.

The carbon content of biodiesel is nearly 15% lower than petroleum diesel fuel on a weight basis. Conversely, biodiesel has approximately 11% oxygen, on a weight basis, while petroleum diesel has almost no oxygen. Very little differences exist between biodiesel and petroleum diesel fuel concerning the weight percentage of hydrogen.

The neutralization number is used to reflect the acidity or alkalinity of an oil. This number is the weight in milligrams of the amount of acid (hydrochloric acid [HCL]) or base (potassium hydroxide [KOH]) required to neutralize one gram of the oil, in accordance with ASTM test methods. If the neutralization number indicates increased acidity (i.e., high acid number) of an oil, this may indicate that the oil or biodiesel has oxidized or become rancid. Biodiesel is allowed to have a neutralization number up to 0.8mg KOH/g.

The iodine value is a measure of the unsaturated fatty acid content of biodiesel, and reflects the ease with which biodiesel will oxidize when exposed to air. The iodine value for petroleum diesel fuel is very low, but the iodine value of biodiesel will vary from 80 to 135.

A weighed quantity of fuel is placed in a crucible and heated to a high temperature for a fixed period to determine the Conradson carbon residue of a fuel. The crucible and the carbonaceous residue is cooled in a desiccator and weighed. The residue that remains is weighed and compared to the weight of the original sample. This percentage is reported as the Conradson carbon residue value. This procedure provides an indication of relative coke forming properties of petroleum oils. No real differences are to be expected when comparing biodiesel with petroleum diesel fuel (0.02 versus 0.01).

3. PHYSICAL PROPERTIES

The following properties reflect the physical properties of biodiesel: distillation curve, density/specific gravity, API gravity, cloud point, pour point, cold filter plug point, flash point, corrosion, viscosity, heat of combustion, and cetane number.

Each fuel has its own unique distillation curve. Some compare this to a fingerprint. This curve tells a chemist which components are in the fuel, their molecular weights by identifying their boiling points and their relative amounts. The temperature at which a specific volume boils off is used to establish this curve (0, 10, 50, 90, and 100%). The range of boiling points for biodiesel is much narrower than petroleum diesel fuel. For example, the initial boiling point of petroleum diesel fuel is 159°C. and the end boiling point is 336°C. The initial boiling point of biodiesel is 293°C. and the end boiling point is 356°C. Thus, the distillation range for diesel fuel is 177° versus 63°C for biodiesel fuel.

The specific gravity of a product is the weight of the product compared to an equal volume of water. The specific gravity of biodiesel is slightly higher than petroleum diesel fuel. For example, the specific gravity of No. 2 petroleum diesel fuel is approximately 0.84 (7.01 pounds per gallon). The specific gravity of biodiesel is approximately 0.88 (7.3 pounds per gallon).

ASTM test procedures are used to determine cloud and pour point of biodiesel. Cloud point is the temperature at which the first wax crystals appear. Pour point is the temperature when the fuel can no longer be poured. According to ASTM specifications for diesel fuel, the cloud points of petroleum diesel fuel are determined by the season (temperature) that the fuel is used. For example, ASTM standards essentially require that the fuel distributor treat number 2 petroleum diesel fuel with cold flow improvers (CFI) for winter operation. Cold weather operation with 100% biodiesel also forces the distributor to use a CFI. However, the CFI of choice for petroleum diesel fuel is usually not as effective when used with biodiesel. The cloud point for biodiesel is higher than petroleum diesel fuel (1.6°C for biodiesel compared with −9.4° to −17.7°C for diesel fuel). Chemical companies (Lubrizol, Octell Starreon) are experimenting with CFI chemicals that work effectively with biodiesel.

An alternative method used to determine how the fuel will perform during cold weather operation is the cold filter plugging point. The cold filter plugging point is the lowest temperature at which the fuel, when cooled under specific conditions, will flow through a filter during a given period of time. The filter is a defined wire mesh. This procedure indicates the low temperature operability of a fuel with/without cold flow improver additives when cooled below the cloud point temperature. The cold filter plugging point for fuels has become an important issue due to the reduction in the size of the openings in the fuel filters (2–5 microns versus 45 microns).

The flash point of a fuel is defined as the temperature at which the air/fuel vapor mixture above the product will ignite when exposed to a spark or flame. A sample is heated and a flame is passed over the surface of the liquid. If the temperature is at or above the flash point, the vapor will ignite and a detectable flash (unsustained or sustained flame) will be observed. The flash point of biodiesel is substantially higher (159°C versus 58°C) than for petroleum diesel fuel. Biodiesel is categorized as a combustible fuel, not a flammable fuel, as is the standard classification for petroleum diesel fuel. Biodiesel is a safer fuel to transport due to its higher flash point.

As fuel deteriorates it becomes acidic. Copper is particularly susceptible to corrosion by the acids in the fuel. As a result, test procedures have been developed to detect the fuel's corrosiveness to copper. A polished copper strip is immersed in a heated sample of fuel. After a prescribed period of time, the strip is removed and examined for evidence of corrosion. A standardized system is used to assign a value between 1 and 4. This number is assigned based on a comparison with the ASTM Copper Strip Corrosion Standards. No differences have been detected using this methodology between petroleum diesel fuel and biodiesel.

The viscosity of a fluid is a measure of its resistance to flow. The greater the viscosity, the less readily a liquid will flow. Test procedures are used to measure the amount of time necessary for a specific volume of fuel to flow through a glass capillary tube. The kinematic viscosity is equal to the calibration constant for the tube multiplied by the time needed for the fuel to move through the tube. The viscosity of biodiesel is approximately 1.5 times greater than petroleum diesel fuel (4.01 versus 2.6 cSt@40°C).

The heat of combustion is the amount of energy released when a substance is burned in the presence of oxygen. The heat of combustion, also known as the heating value, is reported in two forms. The higher, or gross, heating value assumes that all of the water produced by combustion is in the liquid phase. The lower, or net, heating value assumes the water is vapor. The lower heating value for biodiesel is less than for number 2 diesel fuel (37,215 kJ/kg versus 42,565 kJ/kg).

The cetane number reflects the ability of fuel to self-ignite at the conditions in the engine cylinder. In general, the higher the value, the better the performance. The cetane number of biodiesel varies depending on the feedstock. It will be 48 to 52 for soybean oil based biodiesel and more than 60 for recycled greases. This is much higher than the typical

TABLE I

ASTM Biodiesel Specification—D6751 versus ASTM LS #2 Diesel Fuel—D975

Property	ASTM Method	Limits D6751	Units D6751	Limits D975	Units D975
Flash point	D3	130.0 min.	°C	52.0 min.	°C
Water and Sediment	D2709	0.050 max.	% vol.	0.050 max.	% vol.
Carbon residue, 100% sample	D4530	0.050 max.	% mass	N/A	N/A
Ramsbottom carbon residue	D524	N/A	N/A	0.35	% mass
Sulfated ash	D874	0.020 max.	% mass	N/A	N/A
Ash	D482	N/A	N/A	0.01	% mass
Kinematic viscosity, 400°C	D445	1.9–6.0	mm2/s	1.9–4.1	mm2/s
Sulfur	D5453	0.05 max.	% mass	N/A	N/A
Sulfur	D2622	N/A	N/A	0.05 max.	% mass
Cetane	D613	47 min.		40 min.	
Cloud point	D2500	By customer	°C	By customer	°C
Copper strip corrosion	D130	No. 3 max.		No. 3 max.	
Acid number	D664	0.80 max.	mg KOH/gm	N/A	N/A
Free glycerol	D6584	0.020 max.	% mass	N/A	N/A
Total glycerol	D6584	0.240 max.	% mass	N/A	N/A
Phosphorus content	D4951	0.001	% mass	N/A	N/A
Distillation temperature, 90% recovered	D1160	360 max.	°C	N/A	N/A
Distillation temperature, 90% recovered	D86	N/A	N/A	338 max.	°C

43 to 47 observed for petroleum diesel fuel. The cetane number is typically estimated for petroleum diesel fuel using the cetane index. However, due to the fact that this index was developed for diesel fuel and not biodiesel, this ASTM test procedure provides faulty information for biodiesel. A cetane engine must be used to determine the cetane number for biodiesel. See Table I.

4. BIODIESEL PRODUCTION

4.1 Introduction

Biodiesel consists of the monoalkyl esters of fatty acids derived from vegetable oils or animal fats. It is most commonly produced through a process known as transesterification, which is a chemical reaction where an alkoxy group of an ester is exchanged with that of another alcohol to form the new ester product. The chemical reaction with methanol is shown schematically in Fig. 1.

This reaction is reversible so to force the equilibrium in the direction of the products, from 60% to 200% excess methanol is added. The reaction requires a catalyst and strong bases such as potassium hydroxide and sodium hydroxide. The actual reaction using 100% excess methanol is shown in Fig. 4.

Soybean oil (885 g) + 2X Methanol ($6 \times 32.04 = 192.24$ g)

$\xrightarrow{NaOH}$ Methyl soyate ($3 \times 296.5 = 889$ g) + Glycerol (96.12 g) + XS Methanol (92.10 g)

FIGURE 4 Basic transesterification reaction for soybean oil.

The alkali catalysts usually result in the reaction proceeding to completion in 4 to 8 hours at ambient conditions and in 1 hour at 60°C. In general, the catalyst is dissolved in the methanol before addition to the oil to prevent direct contact between the concentrated catalyst and the oil. Since the methanol is only slightly soluble in the soybean oil, agitation is required during the early part of the reaction. The reaction proceeds through a sequence of steps involving the removal of fatty acid chains from the triglyceride to produce diglycerides, monoglycerides, and, ultimately, free glycerol. When the reaction has proceeded to the point where substantial amounts of di- and monoglycerides have been produced, agitation is less important. To further drive the equilibrium to products, the reaction is often conducted in steps. During the first step, only a portion, say 80%, of the methanol and catalyst are added. The reaction proceeds substantially to equilibrium and then the reactants are allowed to settle so the resulting glycerol can be removed. Then, the remaining methanol and catalyst are added and the reaction continued. Removal of the glycerol forces the reaction to the product side and since the alkali catalyst is selectively attracted to the glycerol, the presence of the glycerol can limit the speed of the reaction.

The most important characteristics of the triglyceride feedstocks are low water content (preferably less than 0.2%) and low free fatty acids (less than 0.5%). High free fatty acid (FFA) feedstocks can be processed but pretreatment to remove the FFAs or to convert them to biodiesel is required. This pretreatment will be described later. Water should be excluded from the reaction because it contributes to soap formation as the fatty acid chains are stripped from the triglycerides. The soap sequesters the alkali catalyst and inhibits the separation of the glycerol from the biodiesel. Excessive soap also contributes to the formation of emulsions when water washing is used at a later stage of the production process.

Another important characteristic of the feedstock is the presence of saturated and polyunsaturated triglycerides. Excessive levels of saturated fatty acid chains can produce biodiesel with a high pour point making it difficult to use at low temperatures. High levels of polyunsaturates can provide poor oxidative stability requiring the resulting fuel to be treated with an antioxidant.

While most biodiesel is made using methanol, because of its low price (and quick conversion), other alcohols, such as ethanol and isopropanol, can also be used. Higher alcohols provide superior cold flow properties but are generally more difficult to produce, requiring higher temperatures, lower levels of water contamination, and more complex alcohol recycling due to the formation of azeotropes.

As mentioned earlier, strong alkalis such as sodium hydroxide and potassium hydroxide are common catalysts. These bases form the corresponding methoxides when dissolved in methanol. Water is also formed in this reaction and is probably responsible for some soap formation although not enough to inhibit the transesterification reaction. A more desirable option is to use sodium methoxide formed using a water-free process. This catalyst is available as a concentrated solution in methanol (25% and 30%), which is easier to use because it is a liquid. Research is underway to develop heterogeneous catalysts for biodiesel production that would minimize soap formation and provide cleaner glycerol. Since most of the catalyst ends up as a contaminant in the glycerol, either as soap or free alkali, a heterogeneous catalyst would simplify glycerol refining. Research is also being conducted to find reaction conditions that do not require a catalyst. However, these conditions are at very high temperature (>250°C) and appear to produce undesirable contaminants in the biodiesel.

Figure 5 shows a schematic of a biodiesel production process. Oil enters the reactor where it is mixed with methanol and catalyst. Usually, the catalyst has been mixed with the methanol before contacting the oil to prevent direct contact of the concentrated catalyst and the oil to minimize soap formation. The reactor can be either a batch process or, as is more common with larger plants, a continuously stirred tank reactor (CSTR) or plug flow reactor. When a CSTR is used, it is common to use more than one stage to ensure complete reaction. After the reaction is complete, the glycerol is separated from the biodiesel. This separation can be accomplished with a gravity decanter or using a centrifuge. The unreacted methanol will split between the biodiesel and glycerol giving about 1 to 3% methanol in the biodiesel and 30 to 50% in the glycerol. The methanol in the biodiesel should be recovered for reuse. It may be as much as half of the excess methanol. This is usually accomplished by a vacuum flash process, but other devices such as falling film evaporator have also been used. The methanol-free biodiesel is then washed with water to remove residual methanol, catalyst, soap, and free glycerol. If the biodiesel contains excessive soap, this washing process can be problematic, as the soap will cause an emulsion to form between the water and the biodiesel. To minimize the formation of emulsions, a strong acid is sometimes added to the biodiesel to split the soap into free fatty acids (FFA) and salt. Without the soap, water consumption is greatly reduced and the salt, methanol, catalyst, and glycerol are removed with as little as 3 to 10% water. The water should be heated to 60°C to assist in the removal of free glycerol and should be softened to minimize the transfer of calcium and magnesium salts to the biodiesel. The final step in the process is to heat the biodiesel to remove water that may be dissolved in the biodiesel or entrained as small droplets. This is accomplished with a flash process.

Also shown in the diagram is the preliminary processing of the co-product glycerol. This glycerol contains virtually all of the catalyst and a considerable amount of soap and unreacted methanol. Usually the glycerol will be acidulated to split the soaps into FFA and salt. The FFAs are not soluble in the glycerol and rise to the top where they can be

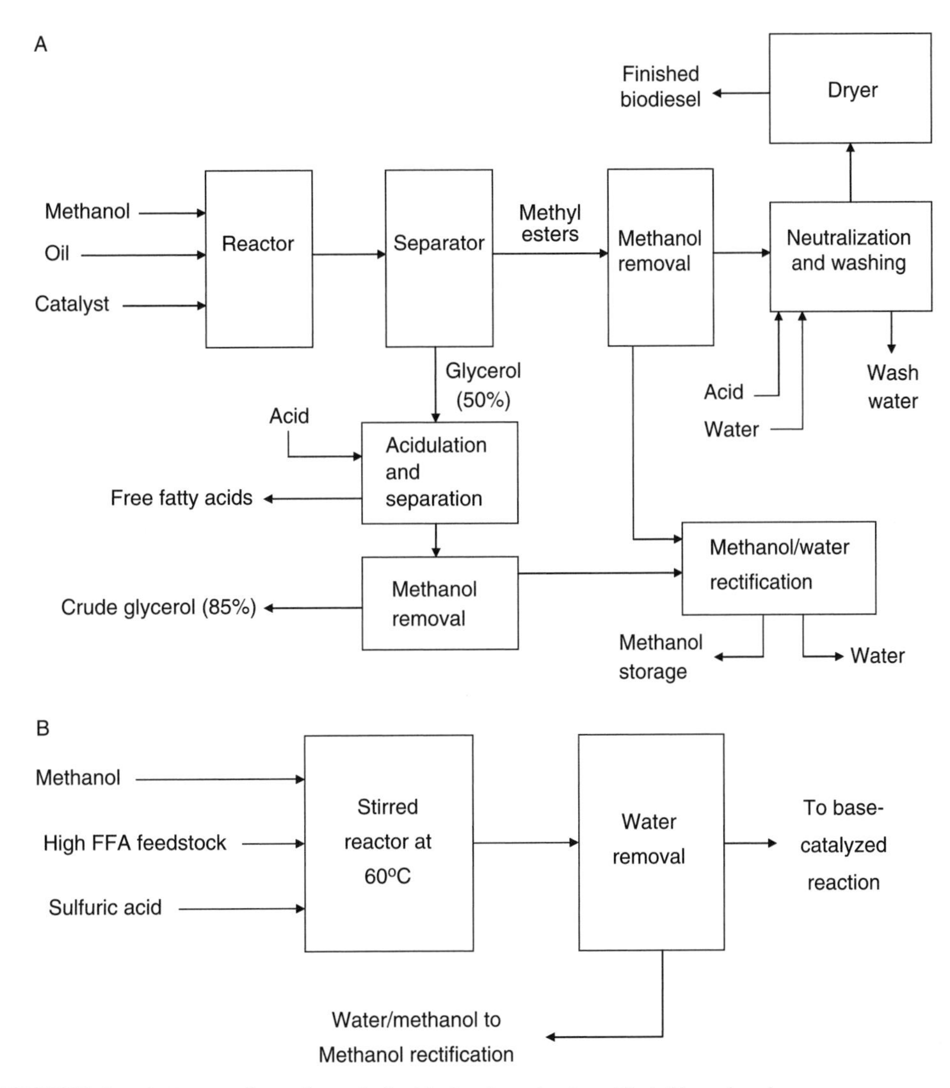

FIGURE 5 (A) Process flow schematic for biodiesel production. (B) Acid-catalyzed pretreatment process.

decanted and returned to the biodiesel process after pretreatment. After FFA removal, the methanol is removed by a flash process or by a thin-film evaporator leaving a crude glycerol product that is 80 to 90% pure. The balance will be salts, residual FFAs, water, and phosphotides, color-bodies, and other contaminants from the oil.

Although water is not deliberately added to the process until after the methanol has been removed, small amounts of water will enter the system as a contaminant in the oil, alcohol, and catalyst. This water will tend to accumulate in the methanol, so before it can be returned to the process, the methanol should undergo fractional distillation.

Biodiesel plants can use either batch or continuous flow processing. Batch processing is most common in small plants of less than 4 million liters/year. Batch processing provides the ability to modify the process for variations in feedstock quality. Continuous flow requires greater uniformity in the feedstock quality, generally requires 24 hour operation, 7 days per week, increases labor costs, and is most suitable for larger operations of greater than 40 million liters/year.

4.2 Pretreatment

Small amounts of FFAs can be tolerated by adding enough alkali catalyst to neutralize the FFAs while still leaving enough to catalyze the reaction. When an oil or fat has more than 3 to 5% FFA, the amount of soap formed with the use of an alkali catalyst will be large enough that separation of the glycerol from the oil may not be possible. In addition, when excess alkali is added, the FFAs are lost and not available for conversion to biodiesel.

An alternative is to convert the FFAs to methyl esters using an acid catalyst such as sulfuric acid with a large excess of methanol ($>20:1$ molar ratio based on the FFA content) (Fig. 5B). The acid-catalyzed reaction is relatively fast (1 hour at 60°C), converting the FFAs to methyl esters with water produced as a by-product. The water eventually stops the reaction before all of the FFAs have been converted and must be removed from the system, either by decanting with the methanol or by vaporization. After the FFAs have been converted to methyl esters, an alkali catalyst can be used to convert the triglycerides. Acid catalysis can be used to complete the transesterification reaction, but the time required is prohibitive.

4.3 Product Quality

Modern diesel engines require high-quality fuels. The fuel injection system, which is often the most expensive element of the engine, can be damaged by fuel contaminants. Water and solid particles are the largest problem.

The contaminants most frequently found in biodiesel are the products of incomplete reaction and residual alcohol, catalyst, and free glycerol. Incompletely reacted biodiesel will contain monoglycerides, diglycerides, and triglycerides. These compounds are usually detected using a gas chromatograph and then the glycerol portion is summed to yield a total glycerol quantity for the fuel. ASTM standards require that the total glycerol be less than 0.24%. This means that more than 98% of the original glycerol portion of the triglycerides feedstock must be removed. Excessive amounts of monoglycerides, especially for saturated compounds, may precipitate from the fuel and plug fuel filters.

If the biodiesel is not washed with water, it may contain some unreacted alcohol. The amount will usually be small enough that it does not adversely affect the operation of the engine, but it can lower the flash point of the fuel to where it must be considered flammable and accorded the same safety requirements as gasoline. The residual catalyst can cause excessive ash formation in the engine. Free glycerol can separate from the fuel and collect in the bottom of storage tanks. This glycerol layer can extract mono- and diglycerides from the biodiesel and produce a sludge layer that may plug filters and small passages in the fuel system.

5. *ADVANTAGES OF BIODIESEL*

Table II shows the changes that were observed in the regulated exhaust emissions of three diesel engines that were tested to produce emissions characterization data for the U.S. Environmental Protection Agency's Fuels and Fuel Additives registration program. The reductions in unburned hydrocarbons (HC) and carbon monoxide (CO) are dramatic although these specific pollutants are not generally a concern with diesel engines. The particulate matter (PM) reductions are also quite striking. Oxides of nitrogen (NO_x) were found to increase with the use of biodiesel. The increase varied depending on the engine tested, but it is clear that biodiesel may produce a NO_x increase of 5 to 13%. The reasons for this increase are still under investigation but appear to be a combination of several effects, including biodiesel's higher speed of sound and isentropic bulk modulus and the tendency of many engine fuel injection systems to advance the injection timing when greater volumes of fuel are injected. Due to biodiesel's lower energy content, a typical test protocol may demand a higher fuel flow rate when biodiesel is used, causing an inadvertent timing advance and resulting NO_x increase.

A comprehensive Life-Cycle Inventory of biodiesel conducted by the National Renewable Energy Laboratory showed that biodiesel provided 3.2 units of fuel energy for every unit of fossil energy consumed in its life cycle. Further, although some fossil-based CO_2 is released during biodiesel production, mainly from the methanol consumed, the net production of CO_2 is reduced by 78%.

6. *DISADVANTAGES*

6.1 Economics

One of the largest factors preventing the adoption of biodiesel is cost. The feedstock costs for biodiesel

TABLE II

Changes in Regulated Emissions with Biodiesel

Engine	HC	CO	PM	NO_x
Cummins N-14	−95.6	−45.3	−28.3	+13.1
DDC S-50	−83.3	−38.3	−49.0	+11.3
Cummins B5.9	−74.2	−38.0	−36.7	+4.3

Derived from Sharp *et al.*

tend to be high in comparison to the cost of petroleum diesel fuel. The end result is that the cost of biodiesel is higher than that of petroleum diesel fuel. In addition, transportation costs are significantly greater for biodiesel due to the fact that the transportation infrastructure for biodiesel is in its infancy. The costs associated with the production and transportation of biodiesel fuels are discussed in more detail in the economics section of this article.

6.2 NO_x and Other Exhaust Emissions

Biodiesel produces more NO_x emissions than diesel fuel. If B100 is used, NO_x production may be increased by 13%. If a B20 blend is used, NO_x production is only increased by 2%, and the engine will typically satisfy the EPA engine exhaust emissions requirements under the Clean Air Act. To meet the EPA emissions requirements in 2006, engine manufacturers will likely use exhaust after-treatment technology that will reduce NO_x emissions. The low sulfur levels in biodiesel fuels make them a good candidate for use with the exhaust after-treatment technologies that are available. Even though biodiesel fuels produce more NO_x emissions, they have been shown to reduce carbon monoxide, particulate matter, unburned hydrocarbons, and other pollutants.

6.3 Fuel Quality

Many problems associated with biodiesel stem from poor fuel quality from the supplier. Most often this is related to the completeness of the production reaction. The ASTM has developed a quality standard for biodiesel. At this point in time, fuel manufacturer compliance with the standard is voluntary. Generally, it is a good idea to ensure that the biodiesel manufacturer sells biodiesel that meets or exceeds the ASTM specifications.

6.4 Energy Content

Biodiesel fuels contain about 12.5% less energy per unit of weight than petroleum diesel fuel (37,215 kJ/kg vs. 42,565 kJ/kg). However, since biodiesel has a higher density, the energy content per unit of volume is only 8% less. As a result, the fuel economy of the diesel engine that is powered with biodiesel tends to be slightly less than when powered with petroleum diesel fuel.

6.5 Cold Weather

The cloud point and cold filter plugging point are much higher for biodiesel than diesel fuel. This means that the fuel will not work in the engine as well as diesel fuel at lower temperatures. Additives can be used to reduce the cloud point and CFPP of biodiesel fuel. The cold-weather properties of biodiesel can also be improved by using a lower level blend of biodiesel fuel (i.e., B5 instead of B20). Additionally, the fuel may be blended with number 1 diesel instead of number 2 diesel to improve the cold-weather properties.

6.6 Material Compatibility

Biodiesel fuel will react with some plastics and some metals in a negative manner. The plastics that seem to be compatible with biodiesel include nylon, teflon, and viton. When in contact with nonferrous metals, such as copper and zinc, biodiesel fuel can cause precipitates to form. Some of these materials can be found in fuel tank liners, fuel lines, transfer pump diaphragms, injector seals, and injection pump seals (among others).

6.7 Solvency

Biodiesel can also act as a solvent. This creates some problems when used in existing systems. The biodiesel can dissolve existing residues in fuel tanks and lines and carry them to the fuel system. Generally, after a few tanks of fuel have been used, these problems tend to be reduced.

6.8 Stability

The oxidative stability of biodiesel fuel is a major factor in determining the allowable storage time for biodiesel fuel. The iodine number can be used to estimate the oxidative stability of the fuel before any stabilizers are added. Typically biodiesel fuels can be stored for up to 6 months without problems. If biodiesel fuel needs to be stored longer, antioxidants can be added to the fuel to improve the stability. If the fuel is not stabilized, biodiesel can form gums and sediments that clog filters or form deposits on fuel system components, including fuel pumps and injectors.

Additionally, as with diesel fuel, some climatic conditions promote biological growth (such as algae), in the fuel. If this occurs, the problem can be treated with a biocide. Reduction in water contamination also reduces the amount of biological growth in the fuel since the algae grows on the water.

6.9 Warranties

Most engine manufacturers do not warranty their engines for use with a specific fuel. Consumers with engine problems that can be traced to the fuel are directed to their fuel supplier. Many engine manufacturers have developed policy statements for biodiesel that allow the use of up to 5% biodiesel but indicate that more experience is needed before fueling with higher level blends. Most require that biodiesel meet the ASTM standard. The best practice is to check with the engine and vehicle manufacturer before using biodiesel.

7. BIODIESEL STORAGE AND USE

7.1 Blending

Blending of biodiesel is not recommended if the temperature of either fuel is below 4.4°C. Low temperatures impact how easily the biodiesel mixes with petroleum diesel fuel. In most situations, splash blending works effectively (i.e., splashing or pouring the biodiesel into the diesel fuel), as biodiesel mixes readily with petroleum diesel fuel. Once mixed, the biodiesel tends to remain blended.

If splash blending biodiesel in a large tank, the biodiesel should be introduced after the diesel fuel has been placed in the tank or the blend should be prepared before placing the blended fuel into storage. This is due to the fact that the biodiesel is heavier than diesel fuel and will essentially rest at the bottom of the tank until some type of agitation is provided. Bulk plants or terminals may also use pumps but will most likely rely on electronic injection or in-line blending to prepare the required blend.

7.2 Transportation

Biodiesel, due to its high flash point, is *not* considered flammable. The fuel is considered combustible, just as is vegetable oil (feedstock). As such, the transportation of "neat" biodiesel may be handled in the same manner as vegetable oil (Code of Federal Regulations 49 CFR 171-173). This is not the case for a low-level blend or a B20 blend. These blends exhibit flash point tendencies that essentially mirror diesel fuel. Blends of biodiesel should be handled in the same manner as petroleum diesel fuel.

7.3 Storage Tanks

Storage tanks for biodiesel can be constructed from mild steel, stainless steel, fluorinated polyethylene, fluorinated polypropylene and teflon. Biodiesel, like petroleum diesel fuel, should be stored in a clean, dry, dark environment. In the event that the container selected is made from polyethylene or polypropylene, the container should be protected from sunlight.

Some authors suggest that aluminum is suitable for use as a storage tank. However, nonferrous metals, such as aluminum, tend to react unfavorably with biodiesel by shortening the shelf life of the fuel. Much is the same for tin and zinc. Concrete lined tanks, varnish lined tanks, or tanks lined with PVC cannot be used to store biodiesel. Biodiesel reacts with each of these products, breaking down the chemical structure of each.

As with any fuel, steps must be taken to prevent water from entering the tank. Algae can grow in biodiesel just as it does with petroleum diesel fuel.

Measures should be taken to ensure that the biodiesel will flow in cold weather. This is often accomplished by mixing the fuel with either number 1 or number 2 diesel fuel. Cold flow improvers (CFI) can also be added to enhance the cold flow characteristics of biodiesel.

7.4 Material Compatibility

Essentially the same materials that are used to construct a biodiesel storage tank can be used with biodiesel (stainless steel, mild steel, viton, some forms of teflon, and fluorinated polyethylene/polypropylene). Rubber elastomers cannot be used, as pure biodiesel will dissolve the rubber. The effect is lessened with lower percentage blends, but little research has been conducted to determine the long-term material compatibility of biodiesel blends.

7.5 Safety

Biodiesel is nontoxic, biodegradable, and much less irritating to the skin than petroleum diesel. However, the same safety rules that pertain to petroleum diesel fuel also apply to the use of biodiesel. The following list summarizes several of these issues:

- Store in closed, vented containers between 10°C and 50°C.
- Keep away from oxidizing agents, excessive heat, and ignition sources.
- Store, fill, and use in well-ventilated areas.

- Do not store or use near heat, sparks, or flames; store out of the sun.
- Do not puncture, drag, or slide the storage tank.
- A drum is not a pressure vessel; never use pressure to empty.
- Wear appropriate eye protection when filling the storage tank.
- Provide crash protection (i.e., large concrete-filled pipe near the storage tank).

8. BIODIESEL ECONOMIC CONSIDERATIONS

The cost of biodiesel consists of five major components: the cost of the feedstock, the cost of the biodiesel, the price of glycerol by-product, and availability of biodiesel. We will examine each of these components.

8.1 Cost of the Feedstock

The cost of the feedstock (oil/tallow/lard, etc.) will vary from one season to the next. The cost will also vary from one crop to the next (soybeans versus rapeseed) as well as from one year to the next, depending on supply and demand. For example, soybean oil has fluctuated from 14.15 cents/pound to 25.8 cents/pound during 1995–2002. Peanut oil has varied from 27 to 49 cents/pound during the same time period. At the same time, lard has sold for 12.5 to 23.02 cents/pound. During 2001–2002, the price of cottonseed oil varied from 14.4 to 22.3 cents per pound and corn oil varied from 11.38 to 23.02 cents/pound. During 2001–2002, the high for corn was in December and the low was in May. The high for cottonseed oil occurred in September and the low occurred in October. Some biodiesel producers have been able to purchase used restaurant grease for as little as 3 to 4 cents/pound (McDonalds, etc.), while some have been paid to haul the waste oil away (transportation was the only feedstock cost).

8.2 Cost of Biodiesel

Economies of scale are a factor when buying biodiesel. Since biodiesel is not yet available on the pipeline network, most deliveries are made by truck or rail. Transportation costs can be a significant portion of the product cost.

The most significant contributor to the cost of the fuel is the cost of the oil itself. As noted in the feedstock section, the price of the oil has varied from as little as 3 to 4 cents/pound to a high of 25.8 cents/pound. If soybeans were used as the feedstock, the cost could range from \$1.73 to \$3.10/gallon. If a less expensive feedstock were used, the price range would even be greater anywhere from \$0.50 to \$63.10/gallon. (See Fig. 6.)

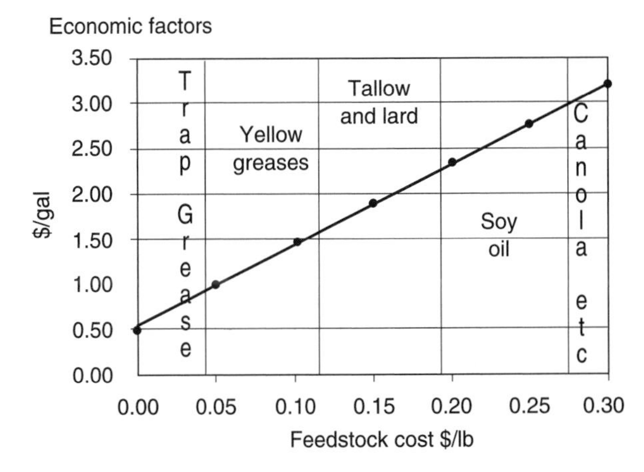

FIGURE 6 Production cost per gallon. Reprinted the from National Renewable Energy Laboratory, Kansas Cooperative Development Center/KSU.

The next highest cost when producing biodiesel is the cost to convert the feedstock from a pure oil to a transesterified fuel. Some researchers report that this amounts to approximately 30% of the total cost (as nearly 70% of the cost is tied up in the raw materials [soybean oil feedstock, methanol, catalyst]).

The conversion costs for biodiesel can range from \$0.30 to \$0.60/gallon. One study reported that it cost \$0.58/gallon for transesterification and \$0.33/gallon for overhead. A credit of \$0.39/gallon was applied for the resale of the glycerol, bringing the total for transesterification to \$0.52/gallon.

8.3 Price of Glycerol

As noted in the previous paragraph, the glycerol is a valuable by-product that must be disposed of after transesterification (credit = \$0.39/gallon). Glycerol is used in pharmaceuticals, cosmetics, toothpaste, paints, and other products. Most researchers believe that a flood of glycerol on the market (which would result from increased biodiesel production) would reduce the net credit. Alternative uses for the glycerol would develop and these uses should help maintain a solid price for the glycerol. Presently, refined glycerol prices stand in the United States at approximately \$0.65/pound.

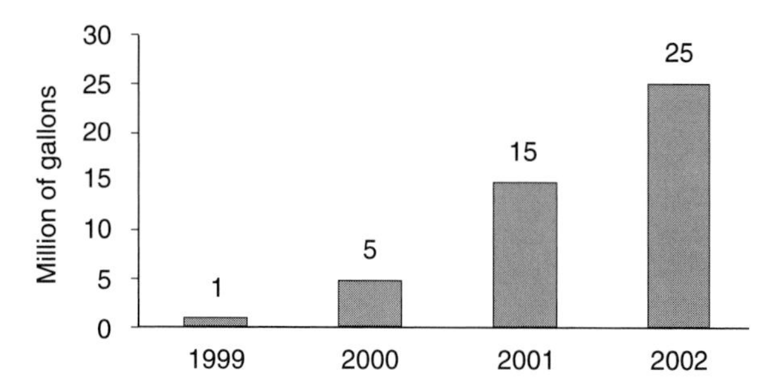

FIGURE 7 U.S. biodiesel production. Reprinted from the National Biodiesel Board, Kansas Cooperative Development Center/KSU.

8.4 Availability

Most economists are of the opinion that biodiesel has the capability to replace about 2 to 5% of the diesel fuel used for transportation in the United States. Although this may seem small in comparison to the volume of diesel fuel used, this would indicate that we would need to produce roughly 6 billion gallons of biodiesel each year. However, Brazil has become a major exporter in the soybean market, and large carryover stocks of soybeans have resulted. The estimated soybean oil carryover for the year 2002 was 1.2 million metric tons. The United States produces 25 million gallons of biodiesel a year. (See Fig. 7.)

SEE ALSO THE FOLLOWING ARTICLES

Alternative Transportation Fuels: Contemporary Case Studies • Biomass, Chemicals from • Biomass for Renewable Energy and Fuels • Ethanol Fuel • Fuel Cycle Analysis of Conventional and Alternative Fuel Vehicles • Internal Combustion (Gasoline and Diesel) Engines • Life Cycle Analysis of Power Generation Systems • Renewable Energy, Taxonomic Overview

Further Reading

Ash, M., and Dohlman, E. (2003). Bumper South American soybean crop buoys world consumption growth. *In* "Oil Crops Outlook," a USDA report in February of 2003. Found at www.ers.usda.gov.

"Biodiesel: Fueling Illinois." (2003). Illinois Soybean Association. Bloomington, IL. Found at www.ilsoy.org.

Coltrain, J. (2002). Biodiesel: Is it worth considering? Kansas State University. Found at www.agecon.ksu.edu.

National Renewable Energy Laboratory (2003). Kansas Cooperative Development Center. Department of Agricultural Economics. Kansas State University. Manhattan, KS 66506.

Sharp, C. A., S.A. Howell, and J. Jobe (2000).The effect of biodiesel fuels on transient emissions from modern diesel engines, part I: Regulated emissions and performance. Society of Automotive Engineers Paper 2000-01-1967.

Sheehan, J., Carnobreco, V., Duffield, J., Graboski, M., and Shapouri, H. (1998). Urban bus operation. *In* "Life Cycle Inventory of Biodiesel and Petroleum Diesel for Use in an Urban Bus," pp. 171–188. Found at www.nrel.gov.

Sigmon, M. (1997). Demonstrations and economic analysis of biodiesel in fleet vehicle and marine applications. For the City of Chicago and Department of Energy. Found at www.biodiesel.org.

"Standard Specification for Diesel Fuel Oils." (2002). ASTM International. West Conshohocken, PA. Found at www.astm.org.

Biomass, Chemicals from

DOUGLAS C. ELLIOTT
Pacific Northwest National Laboratory
Richland, Washington, United States

1. Historical Developments
2. The "Biorefinery" Concept
3. Biomass-Derived Chemical Products
4. The Future Biorefinery

Glossary

biomass Material generated by living organisms; typically the cellular structure of a plant or animal or chemical products generated by the organism.

biorefinery A factory incorporating a number of processing steps, including pretreatments, separations, and catalytic and biochemical transformations, for the production of chemical and fuel products from biomass.

carbohydrate Chemicals composed of carbon, hydrogen, and oxygen and typically having a hydrogen-to-oxygen ratio of 2:1. Representatives include sugars, such as glucose or xylose, starches, and cellulose or hemicelluloses.

catalyst An additive to a chemical reaction that increases the rate of reaction without being consumed in the reaction.

chemurgy The philosophy of using agriculturally derived feedstocks for production of chemical products.

fermentation The production of simple chemical products by microbial action on a complex feedstock, often a carbohydrate.

hydrogenation The chemical addition of hydrogen into a molecule, usually performed in the presence of a metal catalyst. Special types of hydrogenation include hydrodeoxygenation, in which oxygen is removed from a molecule by the reaction of hydrogen, and hydrogenolysis, in which a molecule is broken into smaller molecules by the reaction of hydrogen.

hydrolysis A chemical reaction in which water is chemically combined into a reactant, and in the process the reactant is broken down into smaller molecules.

Refined processing methods and advanced technologies for biomass conversion enable low-value byproducts or waste materials to be transformed into value-added products. Such products include the same chemicals produced from petroleum that are used, for example, as plastics for car components, food additives, clothing fibers, polymers, paints, and other industrial and consumer products. Other replacement products that can be produced from biomass include different chemicals with similar or better properties. Currently, 17% of the volume of the products derived from petroleum in the United States is chemicals. If that petroleum, used for chemical synthesis, were displaced by biomass, it would thus be available for the energy market. In 1998, the U.S. Department of Energy stated as its goal the production of at least 10% of the basic chemical building blocks from plant-derived renewables by 2020 (a fivefold increase) and 50% by 2050 (a subsequent fivefold increase). The chemicals and products discussed here are capable of reducing the 17% petroleum utilization number and making a significant contribution toward meeting those goals.

The discussion on chemicals from biomass is limited to large-scale commodity products for which a noticeable impact on the energy market would be made by displacing petroleum. Specialty chemical products whose limited market would have minimal impact on energy markets are not addressed, nor is the use of biomass for material products such as cellulosic fiber for paper or lignocellulosic material for construction materials. Using biomass to produce a vast array of small-market specialty chemicals based on unique biochemical structures is another part of the overall concept of a sustainable economy but is outside the scope of this article. Instead, it focuses on the chemical structure of various commodity products and their fabrication from biomass. A history of the use of biomass as the basis for chemical products is included for background.

1. HISTORICAL DEVELOPMENTS

Biomass has been used to meet the needs of civilization since prehistory. Several biomass chemical products can be traced to a time when neither the chemical mechanistic transformation process nor the chemical identity of the product were understood. These early products include chemicals still widely used, such as ethanol and acetic acid. A more scientific understanding of chemistry in the 19th century led to the use of biomass as a feedstock for chemical processes. However, in the 20th century, the more easily processed and readily available petroleum eventually changed the emphasis of chemical process development from carbohydrate feedstock to hydrocarbon feedstock. Through major investments by industry and government, utilization of petroleum as a chemical production feedstock has been developed to a high level and provides a wide array of materials and products. Now, the expansion of synthetic chemical products from petroleum provides the framework in which to reconsider the use of renewable biomass resources for meeting society's needs in the 21st century. However, only since the 1980s has interest returned to the use of biomass and its chemical structures for the production of chemical products; therefore, much research and development of chemical reactions and engineered processing systems remains to be done (see Table I)

1.1 Early Developments

The earliest biomass uses often involved food products. One of the first was ethanol, which was limited to beverage use for centuries. Ethanol's origin is lost in antiquity, but it is generally thought to be an accidental development. Only with an improved understanding of chemistry in the mid-1800s was ethanol identified and its use expanded for solvent applications and as a fuel. Acetic acid, the important component in vinegar, is another accidental product based on fermentation of ethanol by oxidative organisms. A third fermentation product from antiquity, lactic acid, also an accidental product of oxidative fermentation, is commonly used in sauerkraut, yogurt, buttermilk, and sourdough.

Another old-world fermentation method is the production of methane from biomass, a multistep process accomplished by a consortium of bacteria, which is an important part of the cyclical nature of carbon on Earth. These processes, now generally referred to as anaerobic digestion, are well-known in ruminant animals as well as in stagnant wet biomass. The methane from these processes has only been captured at the industrial scale for use as a fuel since the 1970s.

Additional early uses of biomass involved its structural components. The fibrous matrix (wood) was first used in constructing shelter. Later, the fibrous component was separated for paper products. As a result, two major industries, the wood products industry and the pulp and paper industry, were developed at sites where biomass was available. Biomass also comprised materials used for clothing. Cotton fibers, consisting mainly of cellulose, and flax fiber are examples.

At the beginning of the 20th century, cellulose recovered from biomass was processed into chemical products. Cellulose nitrate was an early synthetic polymer used for "celluloid" molded plastic items and photographic film. The nitrate film was later displaced by cellulose acetate film, a less flammable and more flexible polymer. Cellulose nitrate lacquers

TABLE I

Petroleum-Derived Products in the United States 2001 [a]

	Net production (thousands of barrels per day)	Percentage of net production
Finished motor gasoline(total)	8560	44.6
Finished aviation gasline(total)	22	0.1
Jet fuel	1655	8.6
Kerosene	72	0.4
Distillate fuel oil	3847	20.0
Residual fuel oil	811	4.2
Asphalt and road oil	519	2.7
Petroleum coke	437	2.3
Chemical products	3283	17.1
Liquefied petroleum gases (net[b])	1778	
Ethane/ethylene	695	
Propane/Propylene	1142	
Butane/Butylene	−5	
Isobutane/Isobutylene	−28	
Pentanes plus	−32	
Naphtha for petrochemicals	258	
Other oils for petrochemicals	312	
Special naphthas	41	
Lubricants	153	
Waxes	18	
Still gases	670	
Miscellaneous products	59	
Total net production	19,206	

[a] Source: Energy Information Administration, Petroleum Supply Annual 2001, Vol. 1, Table 3.

[b] Net, products minus natural gas liquids input.

retained their market, although acetate lacquers were also developed. Cellulose acetate fiber, called Celanese, grew to one-third of the synthetic market in the United States by 1940. Injection-molded cellulose acetate articles became a significant component of the young automobile industry, with 20 million pounds produced in 1939. Today, cellulose-based films and fibers have largely been supplanted by petroleum-based polymers, which are more easily manipulated to produce properties of interest, such as strength, stretch recovery, or permeability.

Wood processing to chemicals by thermal methods (pyrolysis), another practice from antiquity, was originally the method used for methanol and acetic acid recovery but has been replaced by petrochemical processes. One of the earliest synthetic resins, the phenol-formaldehyde polymer Bakelite, was originally produced from phenolic wood extractives. However, petrochemical sources for the starting materials displaced the renewable sources after World War I. More chemically complex products, such as terpenes and rosin chemicals, are still recovered from wood processing but only in a few cases as a by-product from pulp and paper production.

1.2 Chemurgy

In the early 20th century, the chemurgy movement blossomed in response to the rise in the petrochemical industry and in support of agricultural markets. The thrust of the movement was that agricultural products and food by-products could be used to produce the chemicals important in modern society. A renewable resource, such as biomass, could be harvested and transformed into products without depleting the resource base. Important industrialists, including Henry Ford, were participants in this movement, which espoused that anything made from hydrocarbons can be made from carbohydrates.

Supply inconsistencies and processing costs eventually turned the issue in favor of the petrochemical industry. The major investments for chemical process development based on hydrocarbons were more systematically coordinated within the more tightly held petrochemical industry, whereas the agricultural interests were too diffuse (both geographically and in terms of ownership) to marshal much of an effort. Furthermore, the chemical properties of petroleum, essentially volatile hydrocarbon liquids, allowed it to be more readily processed to chemical products. The large scale of petroleum refining operations, developed to support fuel markets, made for more efficient processing systems for chemical production as well.

2. *THE "BIOREFINERY" CONCEPT*

The vision of a new chemurgy-based economy began evolving in the 1980s. The biorefinery vision foresees a processing factory that separates the biomass into component streams and then transforms the component streams into an array of products using appropriate catalytic or biochemical processes, all performed at a scale sufficient to take advantage of processing efficiencies and resulting improved process economics. In essence, this concept is the modern petroleum refinery modified to accept and process a different feedstock and make different products using some of the same unit operations along with some new unit operations. The goal is to manufacture a product slate, based on the best use of the feedstock components, that represents higher value products with no waste. Overall process economics are improved by production of higher value products that can compete with petrochemicals. Some biomass processing plants already incorporate this concept to various degrees.

2.1 Wood Pulp Mill

The modern wood pulp mill represents the optimized, fully integrated processing of wood chips to products. The main process of pulping the wood allows the separation of the cellulosic fiber from the balance of the wood structure. The cellulose has numerous uses, ranging from various paper and cardboard types to chemical pulp that is further processed to cellulose-based chemicals. The lignin and hemicellulose portions of the wood are mainly converted to electrical power and steam in the recovery boiler. In some cases, the lignin is recovered as precipitated sulfonate salts, which are sold based on surfactant properties. In other cases, some lignin product is processed to vanillin favoring. The terpene chemicals can be recovered, a practice limited primarily to softwood processing plants in which the terpene fraction is larger.

2.2 Corn Wet Mill

In the corn wet mill, several treatment and separation processes are used to produce a range of food and chemical products. Initial treatment of the dry corn kernel in warm water with sulfur dioxide results in a softened and protein-stripped kernel and a protein-rich liquor called corn steep liquor. The corn is then milled and the germ separated. The germ is pressed and washed with solvent to recover the corn oil. Further milling and filtering separate the cornstarch from the corn fiber (essentially the seed coat). The starch is the major product and can be refined to a number of grades of starch product. It can also be hydrolyzed to glucose. The glucose, in turn, has various uses, including isomerization to fructose for high-fructose corn syrup; hydrogenation to sorbitol; or fermentation (using some of the steep liquor for nutrients) to a number of products, such as ethanol, lactic acid, or citric acid. The balance of the corn steep liquor is mixed with the corn fiber and sold as animal feed.

2.3 Cheese Manufacturing Plant

Cheese manufacturing has evolved into a more efficient and self-contained processing operation in which recovering the whey by-product (as dried solids) is the industry standard instead of disposing of it, which was the practice for many centuries. The whey is a mostly water solution of proteins, minerals, and lactose from the milk following coagulation and separation of the cheese. To make the whey more valuable, ultrafiltration technology is used in some plants to separate the protein from the whey for recovery as a nutritional supplement. Further processing of the ultrafiltration permeate is also used in a few plants to recover the lactose for use as a food ingredient. Developing methods for recovering the mineral components is the next stage in the development process.

3. *BIOMASS-DERIVED CHEMICAL PRODUCTS*

For production of chemicals from biomass, there are several processing method categories, which are used based on the product desired: fermentation of sugars to alcohol or acid products; chemical processing of carbohydrates by hydrolysis, hydrogenation, or oxidation; pyrolysis of biomass to structural fragment chemicals; or gasification by partial oxidation or steam reforming to a synthesis gas for final product formation. Various combinations of these processing steps can be integrated in the biorefinery concept. Table II illustrates the wide range of biomass-derived chemicals. These products include the older, well-known fermentation products as well as new products whose production methods are being developed. Development of new processing technology is another key factor to making these products competitive in the marketplace.

3.1 Modern Fermentation Products

Fermentation products include ethanol, lactic acid, acetone/butanol/ethanol, and citric acid. Ethanol and lactic acid are produced on a scale sufficient to have an impact on energy markets. Ethanol is primarily used as a fuel, although it has numerous potential derivative products as described later. Lactic acid is new as an industrial chemical product; the first lactate derivative product, poly lactic acid, came on stream in 2002.

3.1.1 Ethanol

In 1999, ethanol was produced at 55 plants throughout the United States at a rate of 1.8 billion gallons per year. The feedstock for the fermentation ranges from corn starch-derived glucose, wheat gluten production by-product starches, and cheese whey lactose to food processing wastes and, until recently, wood pulping by-products. Although some of this ethanol is potable and goes into the food market, the vast majority is destined for the fuel market for blending with gasoline. However, other possible chemical products derived from ethanol include acetaldehyde, acetic acid, butadiene, ethylene, and various esters, some of which were used in the past and may again become economically viable.

3.1.2 Lactic Acid

Lactic acid, as currently utilized, is an intermediate-volume specialty chemical used in the food industry for producing emulsifying agents and as a nonvolatile, odor-free acidulant, with a world production of approximately 100 million pounds per year in 1995. Its production by carbohydrate fermentation involves any of several *Lactobacillus* strains. The fermentation can produce a desired stereoisomer, unlike the chemical process practiced in some areas of the world. Typical fermentation produces a 10% concentration lactic acid broth, which is kept neutral by addition of calcium carbonate throughout the 4- to 6-day fermentation. Using the fermentation-derived lactic acid for chemical products requires complicated separation and purification, including the reacidification of the calcium lactate salt and resulting disposal of a calcium sulfate sludge by-product. An electrodialysis membrane separation method, in the development stage, may lead to an economical process. The lactic acid can be recovered as lactide by removing the water, which results in the dimerization of the lactic acid to lactide that can then be recovered by distillation.

TABLE II
Chemicals Derived from Biomass

Chemical	Derivative chemicals	Uses
Fermentation products		
Ethanol		Fuel
Lactic acid	Poly lactic acid	Plastics
Acetone/butanol/ethanol		Solvents
Citric acid		Food ingredient
Carbohydrate chemical derivatives		
Hydrolysate sugars	Fermentation feedstocks	Fermentation products
Furfural, hydroxymethylfurfural	Furans, adiponitrile	Solvents, binders
Levulinic acid	Methyltetrahydrofuran, δ-amino lactic acid, succinic acid	Solvents, plastics, herbicide/pesticide
Polyols	Glycols, glycerol	Plastics, formulations
Gluconic/glucaric acids		Plastics
Pyrolysis products		
Chemicals fractions	Phenolics, cyclic ketones	Resins, solvents
Levoglucosan, levoglucosenone		Polymers
Aromatic hydrocarbons	Benzene, toluene, xylenes	Fuel, solvents
Gasification products		
Synthesis gas	Methanol, ammonia	Liquid fuels, fertilizer
Tar chemicals		Fuels
Development fermentations		
2,3-Butanediol/ethanol		Solvents
Propionic acid		Food preservative
Glycerol, 1,3-propanediol		C3 plastics, formulations
3-Dehydroshikimic acid	Vanillin, catechol, adipic acid	Flavors, plastics
Catalytic/bioprocessing		
Succinic acid	Butanediol, tetrahydrofuran	Resins, solvents
Itaconic acid	Methyl-1,4-butanediol and -tetrahydrofuran (or methyltetrahydrofuran)	Resins, solvents
Glutamate, lysine	Pentanediol, pentadiamine	C5 plastics
Plant-derived		
Oleochemicals	Methyl esters, epoxides	Fuel, solvents, binders
Polyhydroxyalkanoates		Medical devices

Poly lactic acid (PLA) is the homopolymer of biomass-derived lactic acid, usually produced by a ring-opening polymerization reaction from its lactide form. It is the first generic fiber designated by the Federal Trade Commission in the 21st century and the first renewable chemical product commercially produced on an industrial scale. Cargill–Dow, which produces PLA, opened its plant rated at 300 million pounds per year capacity in April 2002.

Other chemical products that can be derived from lactic acid include lactate esters, propylene glycol, and acrylates. The esters are described as nonvolatile, nontoxic, and biodegradable, with important solvent properties. Propylene glycol can be produced by a simple catalytic hydrogenation of the lactic acid molecule. Propylene glycol is a commodity chemical with a 1 billion pound per year market and can be used in polymer synthesis and as a low-toxicity antifreeze and deicer. Dehydration of lactate to acrylate appears to be a straightforward chemical process, but a high-yield process has yet to be commercialized.

3.1.3 Acetone/Butanol/Ethanol

Acetone/butanol/ethanol fermentation was widely practiced in the early 20th century before being displaced by petrochemical operations. The two-step bacterial fermentation is carried out by various species of the *Clostridium* genus. The solvent product ratio is typically 3:6:1, but due to its toxicity, the butanol product is the limiting component. The yield is only 37%, and final concentrations are approximately 2%. The high energy cost for recovery by distillation of the chemical products from such a dilute broth is a major economic drawback. Alternative recovery methods, such as selective membranes, have not been demonstrated.

3.1.4 Citric Acid

More than 1 billion pounds per year of citric acid is produced worldwide. The process is a fungal fermentation with *Aspergillus niger* species using a submerged fermentation. Approximately one-third of the production is in the United States, and most of it goes into beverage products. Chemical uses will depend on the development of appropriate processes to produce derivative products.

3.2 Chemical Processing of Carbohydrates

3.2.1 Hydrolysate Sugars

Monosaccharide sugars can be derived from biomass and used as final products or further processed by catalytic or biological processes to final value-added products. Disaccharides, such as sucrose, can be recovered from plants and used as common table sugar or "inverted" (hydrolyzed) to a mixture of glucose and fructose. Starch components in biomass can readily be hydrolyzed to glucose by chemical processes (typically acid-catalyzed) or more conventionally by enzymatic processes. Inulin is a similar polysaccharide recovered from sugar beets that could serve as a source of fructose. More complex carbohydrates, such as cellulose or even hemicellulose, can also be hydrolyzed to monosaccharides. Acid-catalyzed hydrolysis has been studied extensively and is used in some countries to produce glucose from wood. The processing parameters of acidity, temperature, and residence time can be controlled to fractionate biomass polysaccharides into component streams of primarily five- and six-carbon sugars based on the relative stabilities of the polysaccharides. However, fractionation to levels of selectivity sufficient for commercialization has not been accomplished.

The development of economical enzymatic processes for glucose production from cellulose could provide a tremendous boost for chemicals production from biomass. The even more complex hydrolysis of hemicellulose for monosaccharide production provides a potentially larger opportunity because there are fewer alternative uses for hemicellulose than for cellulose.

3.2.2 Furfural, Hydroxymethylfurfural, and Derived Products

More severe hydrolysis of carbohydrates can lead to dehydrated sugar products, such as furfural from five-carbon sugar and hydroxymethylfurfural from six-carbon sugar. In fact, the careful control of the hydrolysis to selectively produce sugars without further conversion to furfurals is an important consideration in the hydrolysis of biomass when sugars are the desired products. However, furfural products are economically valuable in their own right. The practice of processing the hemicellulose five-carbon sugars in oat hulls to furfural has continued despite economic pressure from petrochemical growth, and it is still an important process, with an annual domestic production of approximately 100 million pounds. Furfural can be further transformed into a number of products, including furfural alcohol and furan. Even adiponitrile was produced from furfural for nylon from 1946 to 1961.

3.2.3 Levulinic Acid and Derived Products

Levulinic acid (4-oxo-pentanoic acid) results from subsequent hydrolysis of hydroxymethylfurfural. It is relatively stable toward further chemical reaction under hydrolysis conditions. Processes have been developed to produce it from wood, cellulose, starch, or glucose.

Levulinic acid is potentially a useful chemical compound based on its multifunctionality and its many potential derivatives. In the latter half of the 20th century, its potential for derivative chemical products was reviewed numerous times in the chemical literature. Some examples of the useful products described include levulinate esters, with useful solvent properties; δ-aminolevulinic acid, identified as having useful properties as a herbicide and for potentially controlling cancer; hydrogenation of levulinic acid, which can produce γ-valerolactone, a potentially useful polyester monomer (as hydroxyvaleric acid); 1,4-pentanediol, also of value in polyester production; methyltetrahydrofuran, a valuable solvent or a gasoline blending component; and diphenolic acid, which has potential for use in polycarbonate production.

3.2.4 Hydrogenation to Polyols

Sugars can readily be hydrogenated to sugar alcohols at relatively mild conditions using a metal catalyst. Some examples of this type of processing include sorbitol produced from glucose, xylitol produced from xylose, lactitol produced from lactose, and maltitol produced from maltose. All these sugar alcohols have current markets as food chemicals. New chemical markets are developing; for example, sorbitol has been demonstrated in environmentally friendly deicing solutions.

In addition, the sugar alcohols can be further processed to other useful chemical products. Isosorbide is a doubly dehydrated product from sorbitol produced over an acid catalyst. Isosorbide has uses in polymer blending, for example, to improve the rigidity of polyethylene terephthalate in food containers such as soda pop bottles.

Hydrogenation of the sugar alcohols under more severe conditions can be performed to produce lower molecular-weight polyols. By this catalytic hydrogenolysis, five- or six-carbon sugar alcohols have been used to produce primarily a three-product slate consisting of propylene glycol, ethylene glycol, and glycerol. Although other polyols and alcohols are also produced in lower quantities, these three are the main products from hydrogenolysis of any of the epimers of sorbitol or xylitol. Product separations and purifications result in major energy requirements and costs. Consequently, controlling the selectivity within this product slate is a key issue in commercializing the technology. Selectivity can be affected by processing conditions as well as catalyst formulations.

3.2.5 Oxidation to Gluconic/Glucaric Acid

Glucose oxidation can be used to produce six-carbon hydroxyacids—either the monoacid, gluconic, or the diacid, glucaric. The structures of these chemicals suggest opportunities for polyester and polyamide formation, particularly polyhydroxypolyamides (i.e., hydroxylated nylon). Oxidation is reportedly performed economically with nitric acid as the catalyst. Specificity to the desired product remains to be overcome before these chemicals can be produced commercially.

3.3 Potential Pyrolysis Products

Pyrolysis of wood for chemicals and fuels production has been practiced for centuries; in fact, it was likely the first chemical production process known to humans. Early products of charcoal and tar were used not only as fuels but also for embalming, filling wood joints, and other uses. Today, pyrolysis is still important in some societies, but it has largely been displaced by petrochemical production in developed countries. At the beginning of the 20th century, important chemical products were methanol, acetic acid, and acetone. Millions of pounds of these products were produced by wood pyrolysis until approximately 1970, when economic pressure caused by competition with petroleum-derived products became too great. Due to new developments in "flash" pyrolysis beginning in the 1980s, there are new movements into the market with wood pyrolysis chemicals. However, these products are specialty chemicals and not yet produced on a commodity scale.

3.3.1 Chemical Fractions

Most new development work in pyrolysis involves separating the bulk fractions of chemicals in the pyrolysis oil. Heavy phenolic tar can be separated simply by adding water. A conventional organic chemical analytical separation uses base extraction for acid functional types, including phenolics. Combining the water-addition step and the base extraction with a further solvent separation can result in a stream of phenolics and neutrals in excess of 30% of the pyrolysis oil. This stream has been tested as a substitute for phenol in phenol–formaldehyde resins commonly used in restructured wood-based construction materials, such as plywood and particle board. This application allows for the use of the diverse mixture of phenolic compounds produced in the pyrolysis process, which typically includes a range of phenols, alkyl-substituted with one- to three-carbon side chains, and a mix of methoxyphenols (monomethoxy from softwoods and a mix of mono- and dimethoxy phenols from hardwood) with similar alkyl substitution. In addition, there are more complex phenolics with various oxygenated side chains, suggesting the presumed source—lignin in the wood.

The lower molecular-weight nonphenolics comprise a significant but smaller fraction of pyrolysis oil as produced in flash processes. Methanol and acetic acid recovery has potential. Hydroxyacetaldehyde, formic acid, hydroxyacetone, and numerous small ketones and cyclic ketones, which may have value in fragrances and flavorings, also have potential.

3.3.2 Levoglucosan and Levoglucosenone

Cellulose pyrolysis can result in relatively high yields of levoglucosan or its subsequent derivative levoglucosenone. Both are dehydration products from glucose. Levoglucosan is the 1,6-anhydro product, and levoglucosenone results from removal of two of the three remaining hydroxyl groups of levoglucosan with the formation of a conjugated olefin double bond and a carbonyl group. Levoglucosenone's chiral nature and its unsaturated character suggest uses for this compound in pharmaceuticals and in polymer formation.

3.3.3 Aromatic Hydrocarbon Compounds

The large fraction of phenolic compounds in pyrolysis oil suggests the formation of aromatics for chemical or fuel use. Catalytic hydrogenation (in this case, hydrodeoxygenation) can be used for removing the hydroxyl group from the phenolic compounds to produce the respective phenyl compound. For example, hydrogenation of phenol or methoxyphenol (guaiacol) would give benzene, and methyl guaiacol would give toluene. This chemistry was studied in the 1980s, when aromatics were being added to gasoline to improve octane number. Due to increased restriction on aromatics in gasoline to meet emission guidelines, this process concept was shelved. Production of such hydrocarbons (and other saturated cyclic hydrocarbons) remains an option for chemical production from wood pyrolysis; however, the economics are a drawback due to the extensive amount of hydrogenation required to convert the oxygenate to a hydrocarbon.

3.4 Potential Gasification Products

Whereas the pyrolysis process results in a product that retains much of the chemical functional character of the biomass feedstock, gasification operates under more severe process conditions and results in products of a relatively simple chemical structure but with wide applications. Gasification usually entails high-temperature partial oxidation or steam reforming of the biomass feedstock with typical products of hydrogen, carbon monoxide and dioxide, and light hydrocarbons. Other gasification processes using catalysis or bioconversion methods can also result in a primarily methane and carbon dioxide product gas.

3.4.1 Synthesis Gas to Produce Chemicals

Synthesis gas produced from biomass can be used for chemical production in much the same way as synthesis gas from steam reforming of natural gas or naphtha. Certain process configurations can be used with proper optimization or catalysis to control the important ratio of hydrogen to carbon monoxide in the synthesis gas. The synthesis gas can then be used in catalyzed reactions to produce methanol or hydrocarbons or even ammonia. Methanol can subsequently be converted by further catalytic steps to formaldehyde, acetic acid, or even gasoline.

3.4.2 Tar Chemicals

By-products formed in incomplete gasification include an array of organics ranging from the pyrolysis chemicals to further reacted products, including aromatics and particularly polycyclic aromatics, if a sufficient amount of time at high temperature is applied. As a function of temperature and time at temperature, the composition of the tar can vary from a highly oxygenated pyrolysis oil produced at lower temperature and short residence time to a nearly deoxygenated polycyclic aromatic hydrocarbon produced at high temperature. Because of these high-temperature tar components, including four- and five-aromatic ring structures, mutagenic activity can be significant.

3.5 Developmental Fermentations

New fermentation processes for additional biomass-derived chemical products are in various developmental stages. These processes range from improvements to known fermentations to new chemical products produced in genetically engineered organisms.

3.5.1 Acetic Acid

Acetic acid is a major organic chemical commodity product, with an annual domestic demand in excess of 6 billion pounds. The acetic acid currently available for chemical uses is petrochemically derived. Acetic acid fermentation, known from antiquity to form vinegar, is an *Acetobacter* partial oxidation of ethanol. More recently developed is the homofermentative conversion of glucose to acetic acid by "acetogenic" *Clostridium* bacteria. In some cases, 3 mol of acetate can be produced from each mole of six-carbon sugar feedstock, and concentrations of acetate of up to 1.5% can be accumulated. This type of fermentation effectively increases the theoretical yield of acetic acid from glucose by 50% and provides a more reasonable pathway for renewable resources.

3.5.2 Propionic Acid

All chemical production of propionic acid, best known as a food preservative, is currently derived from petroleum as a coproduct or by-product of several processes. However, propionic acid can readily be fermented with propionibacterium. Domestic production is approximately 220 million pounds per year and increasing only 1 or 2% annually. The production of esters for use as environmentally benign solvents represents a small but growing market.

3.5.3 Extremophilic Lactic Acid

As an economical production process, lactic acid fermentation from glucose has led the way into the marketplace for chemicals production. Although the lactic acid process used for PLA production is based on conventional fermentation, the recovery of the lactide product for polymerization eliminates the neutralization and reacidification steps typically required. Efforts to improve the process are focused on the development of new organisms that can function under acidic conditions and produce higher concentrations of lactic acid. These organisms, called extremophiles, are expected to operate at a pH less than 2 in a manner similar to citric acid fermentations. Other work to develop higher temperature-tolerant organisms may also result in reduced production costs.

3.5.4 2,3-Butanediol/Ethanol

For this fermentation, a variety of bacterial strains can be used. An equimolar product mix is produced under aerobic conditions, whereas limited aeration will increase the production of the 2,3-butanediol to the level of exclusivity. Either optically active 2,3-butanediol or a racemic mixture can be produced, depending on the bacterial strain. Similarly, the final product concentration, ranging from 2 or 3 to 6–8%, depends on the bacterial strain. Recovering the high-boiling diol by distillation is problematic.

3.5.5 Succinic Acid

Succinate production from glucose and other carbohydrates has been well characterized for numerous bacterial organisms, including several succinogenes and the *Anaerobiospirillum succiniproducens*. The biological pathways for these organisms are similar and usually result in significant acetate production. The pathways also include a carbon dioxide insertion step such that potential theoretical yield is greater than 100% based on glucose feedstock. Mutated *Escherichia coli* has also been developed for succinic acid production, with yields as high as 90% shown on the pilot scale and a final succinic acid concentration of 4 or 5%. As with lactic acid production, maintaining the pH of the fermentor at approximately neutral is vital. The resulting disposal of mineral by-products from neutralization and reacidification is problematic. Recent advances in membrane separation methods or acid-tolerant fermentation organisms are important for the commercial development of succinic acid production.

3.5.6 Itaconic Acid

Itaconic acid is succinic acid with a methylene group substituted onto the carbon chain. It is typically produced in a fungal fermentation at relatively small scale and high cost ($1.2–2/lb). It is feasible that the economics of this fermentation could be improved in a manner similar to that for citric acid fermentation. As a result, itaconic acid could become available for use in polymer applications as well as for further processing to value-added chemical products.

3.5.7 Glycerol and 1,3-PDO

Bacterial fermentation of glycerol to 1,3-propanediol (PDO) has been studied as a potential renewable chemical production process. The glycerol feedstock could be generated as by-product from vegetable oil conversion to diesel fuel (biodiesel). The anaerobic conversion of glycerol to PDO has been identified in several bacterial strains, with *Clostridium butyricum* most often cited. The theoretical molar yield is 72%, and final product concentration is kinetically limited to approximately 6 or 7%. Genetic engineering of the bacteria to include a glucose to glycerol fermentation is under development and would result in a single-step bioprocess for PDO directly from glucose. The PDO product is a relatively new polyester source, only recently economically available from petroleum sources. Its improved properties as a fiber include both stain resistance and easier dye application.

3.5.8 3-Dehydroshikimic Acid

This interesting building block has been identified as a potential renewable resource. It is a hydroaromatic intermediate in the aromatic amino acid biosynthetic pathway. Its formation in mutated *E. coli* and its recovery in significant yields have been reported. It can be produced from six-carbon sugars with a theoretical yield of 43%, whereas a 71% yield is theoretically possible from five-carbon sugars based on a different metabolic pathway. Further processing of 3-dehydroshikimic acid can lead to protocatechuic acid, vanillin, catechol, gallic acid, and adipic acid.

3.6 Catalytic/Bioprocessing Combinations

Combining bioprocessing systems with catalytic processing systems provides some important opportunities for biomass conversion to chemicals. Although fermentation of biomass feedstocks to useful chemicals can be a direct processing step, in many cases the fermentation product can, in turn, be transformed into a number of value-added chemical products. Thus, the fermentation product acts as a platform chemical from which numerous final products can be derived. Some of these combined bioprocessing platform chemicals and catalytically derived families of value-added chemicals are described here.

3.6.1 Succinic Acid Derivatives

Two branches of the succinic acid family tree have been described. The first, involving hydrogenation, is similar to hydrogenation of maleic anhydride performed in the petrochemical industry based on butane oxidation. The hydrogenation products from succinic acid include γ-butyrolactone, 1,4-butanediol (BDO), and tetrahydrofuran (THF). Currently, 750 million pounds of both BDO and THF are produced per year. Lactone and THF have solvent markets, whereas BDO is an important polyester resin monomer (polybutylene terephthalate). THF is also used for Spandex fiber production.

The second pathway for succinic acid chemical production is through the succinamide to form pyrrolidones. *N*-methylpyrrolidone is an important low-toxicity, environmentally benign solvent with a growing annual market of more than 100 million pounds, displacing chlorinated hydrocarbon solvents. 2-Pyrrolidone can form the basis for several polymers, including polyvinyl pyrrolidone.

3.6.2 Itaconic Acid Derivatives

The structural similarity of itaconic acid to that of succinic acid suggests that it could be used to produce similar families of chemical products. Typically, the products would be methylated versions of the succinic-derived products, such as 2-methyl-1,4-butanediol, 3-methyltetrahydrofuran, or 3-methyl-*N*-methylpyrrolidone. Therefore, the final products would include methylated polyesters or methylated Spandex.

3.6.3 Glutamic Acid and Lysine

Glutamic acid and lysine are major chemical products, with glutamate sold as the sodium salt, monosodium glutamate, and lysine used as an important animal feed supplement. Both compounds are produced industrially by fermentation.

Glutamate is produced as L-glutamic acid at 680 million pounds per year, with 85% of the production in Asia. In the first half of the 20th century, glutamate was produced by hydrolysis of protein, primarily wheat gluten. The current fermentation process, using molasses or another cheap glucose source, had replaced protein hydrolysis processing by 1965. Because the current cost of glutamate is more than $1.4/lb, it is too expensive for use as a platform chemical, and its use for chemical production is dependent on improved fermentation, recovery method modification, and integration of catalytic processing with the existing fermentation process.

Worldwide lysine production is only approximately one-third that of glutamate. Thus, the cost of lysine is approximately twice that of glutamate. Consequently, chemical production based on glutamate rather than lysine appears to be more likely.

Glutamate and lysine are interesting as renewable chemical feedstocks because each provides both multifunctionality and a five-carbon backbone. The five-carbon-based polymers are not widely used and are less well-known or understood since there is no easy way to produce five-carbon petroleum products. Processing steps can be envisioned for direct conversion of glutamate either to a five-carbon diacid by deamination (or thereafter to 1,5-pentanediol) for polyester production or to a five-carbon terminal amine/acid for nylon production. Lysine could be decarboxylated to produce a five-carbon diamine or could be deaminated to the amine/acid (the same as that for glutamate) for nylon production.

3.7 Plant-Derived Chemicals

One strategy for the use of biomass for chemicals production is the recovery of chemical products produced directly by plants. This strategy can involve either recovery from existing plants, such as the oils from certain seed crops, or recovery from genetically modified plants in which the chemical is produced specifically for harvesting.

3.7.1 Oleochemicals

The oil recovered from certain seed crops has been an important food product for many centuries. The growth of scale of the processing industry in the case of corn and soybeans has resulted in a reduction of the cost of the oil to the point that it can be considered for use in commodity chemical products. The use of vegetable oil for fuel as the transesterified methyl esters of the fatty acids derived from the triglycerides, known as biodiesel, has made some market inroads but primarily to meet alternative fuel use requirements and it is not based on demand. An important market breakthrough for renewable resources was the use of soybean oils in printing inks, instituted in the 1990s, which has been growing steadily. They are also used in toners for photocopy machines and in adhesive formulations. Other uses, such as plasticizers or binders based on epoxide formation from the unsaturated fatty acids, have also been investigated.

3.7.2 Polyhydroxyalkanoates

Polyhydroxyalkanoate (PHA) production has been demonstrated by both fermentative and plant growth methods. PHA is a biodegradable polymer with applications in the biomedical field. Originally, PHA was limited to polyhydroxybutyrate (PHB), but it was found to be too brittle. A copolymer (PHBV) of 75% hydroxybutyrate and 25% hydroxyvalerate is now a commercial product with applications in medical devices. However, the fermentative process has some potentially significant drawbacks that, when considered on a life cycle basis, could result in poorer performance of PHA production from a renewable perspective than that of petroleum-derived polystyrene. To address these drawbacks, which include high energy requirements for cell wall rupture and product recovery, plant-based production of PHA has been investigated using genetic modification to engineer the PHB production pathway into *Arabidopsis* as the host plant. Alternatively, PHBV might be produced by other organisms that can use multiple sugar forms from a less-refined biomass feedstock than corn-derived glucose.

4. THE FUTURE BIOREFINERY

The future biorefinery (Fig. 1) is conceptualized as an optimized collection of the various process options described previously. In addition to the chemical processing steps for producing value-added chemicals, it will likely incorporate components from existing biorefinery-type operations, such as wood pulp mill, wet corn mill, and cheese manufacturing processes.

Carbohydrate separation and recovery will likely involve both the five- and six-carbon compounds. These compounds will respond to separations by selective hydrolysis because they have different activities in either chemical or enzymatic hydrolyses. Any oil or protein components liberated through this processing will be considered as potential high-value chemicals. The lignin component, the least developed of the biomass fractions, is often viewed as a fuel to drive the processing systems, but it is also a potential source of aromatic chemical products. Mineral recovery is most likely to be important for maintaining agricultural productivity, with the return of the minerals to the croplands as fertilizer.

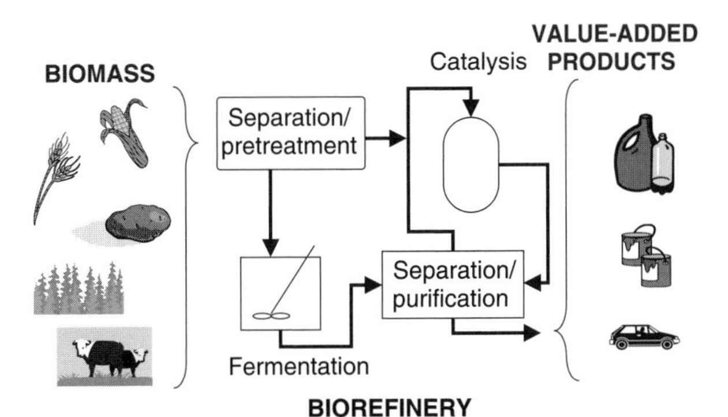

FIGURE 1 Biorefinery concept for production of value-added chemicals from biomass.

In order to achieve a true biorefinery status, each biomass component will be used in the appropriate process to yield the highest value product. Therefore, a combination of processing steps, tailored to fit the particular operation, will be used to produce a slate of chemical products, which will be developed in response to the market drivers to optimize feedstock utilization and overall plant income. In this manner, the return on investment for converting low-cost feedstock to high-value chemical products can be maximized.

The future biorefinery will likely have a major impact on society as fossil-derived resources become scarce and more expensive. Indeed, use of biomass is the only option for maintaining the supply of new carbon-based chemical products to meet society's demands. The biorefinery can have a positive environmental impact by deriving its carbon source (via plants) through reducing atmospheric levels of carbon dioxide, an important greenhouse gas. With appropriate attention to the biomass-derived nitrogen and mineral components in the biorefinery processes, a life cycle balance may be achieved for these nutrient materials if their recovery efficiencies can be maintained at high levels.

SEE ALSO THE FOLLOWING ARTICLES

Biomass Combustion • Biomass for Renewable Energy and Fuels • Biomass Gasification • Biomass: Impact on Carbon Cycle and Greenhouse Gas Emissions • Biomass Resource Assessment

Further Reading

Arntzen, C. J., Dale, B. E., *et al.* (2000). "Biobased Industrial Products." National Academy Press, Washington, DC.

Bozell, J. J. (ed.). (2001). "Chemicals and Materials from Renewable Resources," ACS Symposium Series No.784. American Chemical Society, Washington, DC.

Bozell, J. J., and Landucci, R. (eds.). (1993). "Alternative Feedstocks Program, Technical and Economic Assessment: Thermal/Chemical and Bioprocessing Components." National Renewable Energy Laboratory, Golden, CO.

Donaldson, T. L., and Culberson, O. L. (1983). "Chemicals from Biomass: An Assessment of the Potential for Production of Chemical Feedstocks from Renewable Resources, ORNL/TM-8432." Oak Ridge National Laboratory, Oak Ridge, TN.

Executive Steering Committee, Renewables Vision 2020 (1999). "The Technology Roadmap for Plant/Crop-Based Renewable Resources 2020," DOE/GO-10099-706. U.S. Department of Energy, Washington, DC.

Fuller, G., McKeon, T. A., and Bills, D. D. (eds.). (1996). "Agricultural Materials as Renewable Resources—Nonfood and Industrial Applications, ACS Symposium Series No.647." American Chemical Society, Washington, DC.

Goldstein, I. S. (ed.). (1981). "Organic Chemicals from Biomass." CRC Press, Boca Raton, FL.

Morris, D., and Ahmed, I. (1992). "The Carbohydrate Economy: Making Chemicals and Industrial Materials from Plant Matter." Institute for Local Self-Reliance, Washington, DC.

Pierre, St., L.E., and Brown, G. R. (eds.). (1980). "Future Sources of Organic Raw Materials: CHEMRAWN I." Pergamon, Oxford, UK.

Rowell, R.M., Schultz, T.P., Narayan, R. (eds.). (1992). "Emerging Technologies for Materials and Chemicals from Biomass," ACS Symposium Series No.476. American Chemical Society, Washington, DC.

Saha, B. C., and Woodward, J. (eds.). (1997). "Fuels and Chemicals from Biomass, ACS Symposium Series No.666." American Chemical Society, Washington, DC.

Scholz, C., and Gross, R. A. (eds.). (2000). "Polymers from Renewable Resources—Biopolyesters and Biocatalysis, ACS Symposium Series No.764." American Chemical Society, Washington, DC.

Biomass Combustion

ANDRÉ P. C. FAAIJ
Utrecht University
Utrecht, The Netherlands

1. Introduction
2. Basics of Biomass Combustion: Principles and Fuels
3. Key Technological Concepts: Performance and Status
4. The Role of Biomass Combustion to Date: Different Contexts and Markets
5. Future Outlook and Closing Remarks

Glossary

bubbling fluidized bed/circulating fluidized bed (BFB/CFB) Boiler designs in which the fuel is fed to a furnace where an inert bed material (like sand) is kept in motion by blowing (combustion) air from underneath.

co-combustion and co-firing Co-combustion or co-firing implies the combined combustion of different fuels (e.g., biomass and coal) in one furnace.

combined heat and power generation (CHP) Combustion facilities that produce both electricity (e.g., by driving a steam turbine and generator) and heat (e.g., process steam or district heating).

excess air Amount or ratio of air needed in surplus of oxygen theoretically needed for stochiometric complete combustion; excess air ratios can differ substantially between differ boiler concepts for effective combustion.

heating value (lower heating value [LHV]/higher heating value [HHV]) Energy content of a fuel released at complete combustion. Higher heating value gives the energy content for dry fuel. For the lower heating value, the energy needed to evaporate water present in the fuel and formed during combustion is substracted from the HHV.

municipal solid waste (MSW) Heterogeneous mixture of organic fractions, plastics, paper, and so on, as collected in urban areas from households and the service sector.

Biomass combustion means burning fuels from organic origin (e.g., wood, organic wastes, forest, and agriculture residues) to produce heat, steam, and power. Biomass combustion is responsible for over 95% of the production of secondary energy carriers from biomass (which contributes some 10 to 15% of the world's energy needs). Combustion for domestic use (heating, cooking), waste incineration, use of process residues in industries, and state-of-art furnace and boiler designs for efficient power generation all play their role in specific contexts and markets. This chapter gives an overview of the role, performance, technological concepts, and markets of biomass combustion.

1. INTRODUCTION

Biomass combustion is as old as civilization. Wood and other biomass fuels have been used for cooking and industrial processes, like iron production, for thousands of years.

Energy supplies in the world are dominated by fossil fuels (some 80% of the total use of more than 400 EJ per year). About 10 to 15% (or 45 ± 10 EJ) of this demand is covered by biomass resources (see Table I). On average, in the industrialized countries biomass contributes some 9 to 13% to the total energy supplies, but in developing countries this is as high as one-fifth to one-third. In quite a number of countries, biomass covers even over 50 to 90% of the total energy demand. It should be noted, though, that a large part of this biomass use is noncommercial and used for cooking and space heating, generally by the poorer part of the population. This also explains why the contribution of biomass to the energy supply is not exactly known; noncommercial use is poorly mapped. In addition, some traditional use is not sustainable because it may deprive local soils of needed nutrients, cause indoor and outdoor pollution, and result in poor health. It may also contribute to greenhouse gas (GHG) emissions and affect ecosystems. Part of this use is commercial though—that is, the household fuel wood in industrialized countries and charcoal and firewood in urban and

TABLE I

World Primary Energy Use (1998)

	ExaJoule	ExaJoule	%
Fossil fuels		320	80
Oil	142		
Natural Gas	85		
Coal	93		
Renewables		56	14
Large hydropower	9		
Traditional biomass	38		
'Modern' renewables	9		
Nuclear		26	6
Total		402	100

industrial areas in developing countries—but there are almost no data on the size of those markets. An estimated 9 ± 6 EJ is covered by this category.

Poor statistics on bioenergy use even occur for industrial countries because of differences in definitions and fuels in or excluded (such as MSW and peat). Modern, commercial energy production from biomass (such as in industry, power generation, or transport fuels) makes a lower, but still significant contribution (some 7 EJ/year in 2000), and this share is growing. By the end of the 1990s, an estimated 40 GWe biomass-based electricity production capacity was installed worldwide (good for 160 TWh/year) and 200 GW heat production capacity ($>$700 TWh/year).

Nevertheless, by far the largest contribution of renewables to the world's energy supply comes from biomass, and combustion plays the key role as conversion route for producing useful energy from it. Besides domestic and industrial use of biomass via combustion, waste incineration is another major market where biomass combustion technology is widely applied, and a diversity of concepts has emerged over time.

In the past, a considerable part of the combustion capacity for electricity production (e.g., at the sugar and paper and pulp industry for utilization of process residues) was installed for solving disposal problems of biomass residues. The use of low-cost fuels has resulted in plants with low efficiencies. In the past decade or so, attention has begun to shift toward more efficient biomass combustion power plants, which is partly explained by increased utilization of more expensive biomass resources (such as agricultural and forest residues, which are more costly to collect and transport) and environmental and energy policies in various countries that stimulate the use of low- or zero-carbon energy sources. A range of advanced combustion concepts and technologies has been and is being developed over time, such as fluid bed technology, fully automated plants, flexible fuel concepts capable of dealing with both fossil and a diversity of biomass fuels, and a variety of co-combustion options. Another generic trend is that such plants are built with increasing capacities, thus resulting in economies of scale.

Section 2 goes into the basics of biomass combustion in relation to the properties of biomass fuels. Section 3 gives an overview of technological concepts deployed for various applications. Section 4 discusses the role biomass combustion plays in a number of example countries and sector. This article concludes with a look at the future of biomass combustion in Section 5.

2. BASICS OF BIOMASS COMBUSTION: PRINCIPLES AND FUELS

2.1 Basics: The Process of Biomass Combustion

Combustion can ideally be defined as a complete oxidation of the fuel. The actual full oxidation of the biomass fuel consist of several basic steps: drying, followed by pyrolysis, gasification, and finally full combustion. Biomass is never 100% dry, and any water in the fuel will be evaporated. The higher the moisture content, the more energy is needed to evaporate water. Given that many biomass resources have high moisture contents, this is an important parameter in the overall system performance. At moisture contents over 60%, it is generally impossible to maintain the combustion process. The heating value of a (biomass) fuel can be defined by the higher heating value, which is basically the energy content on dry basis, and the lower heating value, which subtracts the energy needed for the evaporation of water during the combustion process. Lower heating values are calculated by the following formula:

$$\mathrm{LHV_{wet}} = \mathrm{HHV^{*}_{dry}}(1 - \mathrm{W}) - \mathrm{E^{*}_{w}}(\mathrm{W} + \mathrm{H^{*}m_{H_2O}}),$$

in which HHV is the energy content of the dry biomass released during complete combustion, E_w is the energy required for evaporation of water (2.26 MJ/kg), W is the moisture content, H is the hydrogen content (weight percent of wet fuel), and

m_{H2O} is the weight of water created per unit of hydrogen (8.94 kg/kg).

This formula explains that the LHV can drop below zero at given moisture contents. Drying prior to the combustion process (e.g., with waste heat) is a possible way to raise the heating value to acceptable levels and this is sometimes deployed in practice. The heating value and moisture content of the fuel therefore also heavily affect the net efficiency of the combustion process in question. Some boiler concepts deploy condensation of the water in present the flue gases, which enables recovery of the condensation energy from the combustion process. This is an expensive option though.

Drying is followed by pyrolysis, thermal degradation in absence of oxygen and devolatization, at which tars, charcoal, and combustible gases are formed. The subsequent gasification results in formation of combustible gases (e.g., through char oxidation). Finally, at complete combustion, all gases are converted to CO_2 and H_2O and a range of contaminants, which partly depend on fuel composition and partly on the conditions of the combustion process. Incomplete combustion can, for example, lead to emissions of CO, soot, methane, polycyclic aromatic hydrocarbons (PAH), and volatile organic components (VOC), which is typical for lower temperature combustion and poor combustion air control. Dust emissions are caused by the mineral fraction (e.g., sand) of the fuel. Fuel bound nitrogen as well as thermal conversion of molecular nitrogen can be converted to NO_x, in particular at higher temperatures. N_2O (a strong greenhouse gas) can be formed in small amounts during combustion as well, depending on process conditions. Sulfur in the fuel is partly converted to SO_x. Both NO_x and SO_x contribute to acidification. SO_2 can also lead to corrosion in downstream equipment. The same is true for chlorine, which results in HCl emissions. Presence of chlorine may, under specific conditions, also lead to formation small amounts of the highly toxic dioxins and furans. Alkali metals (in particular K and Na) also contribute to corrosion problems. Those metals, as well as the silicon content and Ca, influence ash melting points as well. In case heavy metals are present, these can partly evaporate and end up in the flue gases as well. In particular, cadmium and mercury are volatile and evaporate already at lower combustion temperatures.

Combustion temperatures can vary between 800 and 2000°C and are influenced by moisture and excess air ratio. The latter can typically be controlled by the design of the combustion equipment—that is, with various air feeding steps (primary and secondary air) and boiler or furnace geometry (e.g., to create specific primary and secondary combustion zones, dominated by one of the basic processes described above). The higher the air ratio, the lower the maximum (adiabatic) temperature, but also the more complete the combustion.

The control and performance of the entire combustion process (including costs, efficiency, and emission profile) is always a trade-off. On the one hand, there are various technical variables such as design of the equipment, materials used, air and fuel feeding methods, and control strategies; on the other hand, a number of process variables (e.g., heat transfer, residence times, excess air, insulation) and fuel properties (moisture, mineral fraction, composition) must all be balanced to obtain the desired performance. Clearly, with bigger facilities, more advanced design, materials, and equipment such as emission control devices become more feasible; in general, better quality fuels (clean wood, pellets, etc.) are used for firing smaller combustion installations (i.e., below some 1 MWth) and poorer quality fuels (such as straw, waste fuels) for bigger installations.

2.2 Characteristics of Biomass Resources

Biomass resources that can be used for energy are diverse. A distinction can be made between primary, secondary, and tertiary residues (and wastes), which are available already as a by-product of other activities and biomass that is specifically cultivated for energy purposes:

- Primary residues are produced during production of food crops and forest products. Examples are residues from commercial forestry management and straw from various food crops (cereals and maize). Such biomass streams are typically available in the field and need to be collected to be available for further use.
- Secondary residues are produced during processing of biomass for production of food products or biomass materials. Such residues typically are available in larger quantities at processing facilities, such as the food- and beverage industry, saw- and paper mills, and the like.
- Tertiary residues become available after a biomass derived commodity has been used, meaning a diversity of waste streams is part of this category, varying from the organic fraction of municipal solid waste (MSW), waste and demolition wood, sludges, and so on.

Properties like moisture content, mineral fraction, density, and degree of contamination (e.g., heavy and alkali metals, nitrogen, sulfur, chlorine) differ widely depending on the source of the biomass.

Biomass is a highly reactive fuel compared to coal and has a much higher oxygen content, higher hydrogen/carbon ratio, and higher volatile content. The bulk composition of biomass in terms of carbon, hydrogen, and oxygen (CHO) does not differ much among different biomass sources though; typical (dry) weight percentages for C, H, and O are 45 to 50%, 5 to 6%, and 38 to 45%, respectively. The volatile content (lighter components that are particularly released during the pyrolysis stage) can vary between some 70 and 85%. The latter is typically higher for nonwoody and 'younger' greener biomass sources. CHO shares can be different for atypical fuels as sludges or MSW (the higher C content is due to biological breakdown and plastic fraction). Table II summarizes some characteristics for a range of biomass resources in terms of indicative ranges. Clearly, the diversity of characteristics (also in relation to price) of biomass resources is a challenge for combustion processes, and in practice a wide diversity of technical concept is deployed to convert different feedstocks to energy carriers for different markets and capacities.

Characteristics of the biomass are influenced by the origin of the biomass, but also by the entire supply system preceding any conversion step. Depending on the combustion process technology chosen, various pretreatment steps such as sizing (shredding, crushing, chipping) and drying are needed to meet process requirements. Furthermore, certain fuel properties exclude certain biomass fuels for specific combustion options, partially for technical and partially for environmental reasons.

Storage, processing (e.g., densification) before transport, and pretreatment influence moisture content, ash content (e.g., by means of sieving), particle size, and sometimes even degree of contamination (an example being straw washing, which reduces the salt content (chlorine and alkali content).

2.3 Power Generation

The resulting hot gases from the combustion can be used for a variety of applications such as the following:

- Smaller scale heat production for direct space heating, cooking, and water heating.
- Larger scale heat production for district and process heating (such as furnaces and drying).

TABLE II

Some Example Values on the Composition and Characteristics of a Selection of Biomass Fuels

Biomass feedback	Moisture content (percentage wet basis)	Ash content (percentage dry basis)	Heating value (LHVGJ/ wet ton)	Nitrogen (mg/kg, dry basis)	Sulfur (mg/kg dry basis)	Chlorine (mg/kg, dry basis)	Heavy metals	Costs (indications) (Euro/GJ)
Wood (chips)	30–50	1–2%	8–13	900–2000	70–300	50–60	Very low	From 1 to 5, when cultivated
Bark	40–60	3–5%	7–10	3000–4.500	300–550	150–200	Very low	From some 0.5 to 1
Straw	15	10%	14–15	3000–5000	500–1100	2500–4000	Low	From zero to high market prices
Grass	15–60	10%	6–15	4000–6000	200–1400	500–2000	Low	As residue low, cultivated Miscanthus: 1.5 to 4
Demolition wood	10–15	1%	15–17	300–3000	40–100	10–200	Low to very high	From negative to market prices
Organic domestic waste	50–60	20%	4–5	1000–4000	400–600	100–1000	Low	Negative prices due to tipping fees to zero
Sewage sludge	Over 70 in wet form	40%	1–10 (when dried)	2000–7000	600–2000	50–200	Very high	Strongly negative prices due to processing costs, up to zero.

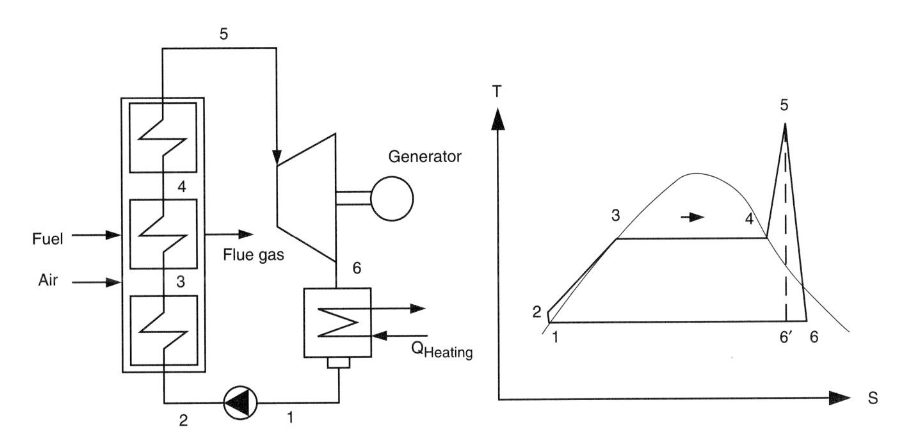

FIGURE 1 Basic principle and T/S diagram for a steam cycle including cogeneration. Steps from 1 to 6 cover pressure increase (1–2), heating up to evaporation temperature (2–3), evaporation in the boiler (3–4), superheating of vapor (4–5), expansion in the turbine (5–6) and finally condensation (6–1).

- Raising (high temperature) steam for industrial processes or subsequent electricity production using a steam turbine or steam engines.

Various devices can be used to convert heat from combustion to power. The most important and widely deployed are steam turbines (in the range of 0.5 to 500 MWe). Furthermore, steam engines for capacities below 1 MWe or so are used. Alternatives are the use of closed gas turbines (hot air turbine), Stirling engines, and an Organic Rankine Cycle, but those are either under development or to date only used in a few specific projects.

Steam turbines are mature, proven technology, which, when applied at larger scales, can reach high efficiencies when high pressure and steam is available. Steam pressures deployed in steam turbines range from 20 to 200 bars for superheated steam. Raising such high pressures with biomass fuels can be problematic when corrosive components (such as chlorine, sulfur, and alkali metals) are present in high concentrations in the flue gas. This is a reason why, for example, waste incinerators generally maintain low steam pressures (with resulting lower efficiencies).

Combined heat and power generation (CHP) configurations and delivery of process steam are made easy by the use of back-pressure turbines, a technique widely deployed in the cane-based sugar and paper-and-pulp industry for processing residues and supplying both power and process steam to the factory. Steam turbines (and the required steam generator) are expensive on a smaller scale, however, and high-quality steam is needed. Superheating steam (for high efficiency) is generally only attractive at larger scales. The same is true for efficiency-increasing measures like feed water preheating and steam reheating. For a condensing plant (implying exhausted steam that exits the turbine is cooled to ambient temperatures), a simple process scheme and T/S, temperature/entropy diagram, is depicted in (Fig. 1). Typical net overall electrical efficiencies of power plants range from less than 20% electrical efficiency at smaller scales to over 40% for very large systems (e.g., obtained with co-firing in a coal plant).

Steam engines play a much smaller role in current markets. Advantages are their flexibility (part load operation) and suitability for smaller scales. Electrical efficiencies are limited to 6% to up to 20% for the most advanced designs.

2.4 Ash

Ash remains after complete combustion, containing the bulk of the mineral fraction of the original biomass, but so does unburned carbon (which strongly depends on the technology deployed). High alkali and Si content typically give low ash melting temperatures (while Mg and Ca increase ash melting temperature). Ash sintering, softening, and melting temperatures can differ significantly among biofuels, and this characteristic is essential in determining temperature control to avoid sintering (e.g., on grates or in fluid beds) or slagging (resulting in deposits on, e.g., boiler tubes). Typically, grasses and straw have mineral fractions giving low melting temperatures.

Potential utilization of ash is also influenced by contaminants (such as heavy metals) and the extent to which the ash is sintered or melted. In case clean biomass is used and the ash contains much of the minerals (such as phosphor) and important trace elements, it can be recycled to forest floors, for example. This is demonstrated in Sweden and Austria. Heavy metals typically concentrate in fly ash, which itself presents the possibility of removing a smaller ash fraction with most of the pollutants. This is the practice in Austria, particularly for removing cadmium.

Co-firing can give specific issues, since biomass additions to, for example, coal firing can change ash characteristics (which are of importance for reuse in cement production, for example). More important is that biomass ash can result to increased fouling and reducing the fusion temperatures of coal ash. The extent to which those mechanisms become serious issues depends on the fuel properties, both of the coal and biomass, as well as the specific boiler design.

3. *KEY TECHNOLOGICAL CONCEPTS: PERFORMANCE AND STATUS*

As argued, a wide range of combustion concepts is developed and deployed to serve different markets and applications. To structure this range, a distinction can first be made between small-scale domestic applications (generally for heating and cooking) and larger scale industrial applications. First, domestic applications will be discussed, distinguishing between the use of stoves in (in particular in developing countries) and the use of more advanced furnaces, boilers, and stoves utilized in industrialized countries. For industrial applications, a basic distinction is made between standalone systems and co-firing.

3.1 Domestic Biomass Use in Developing Countries

The aforementioned (and largely noncommercial) use of biomass for cooking and heating in particular in developing countries is applied with simple techniques, varying from open fires to simple stoves made from clay or bricks. Typically, the efficiency of such systems is low, to less than 10% when expressed as energy input in wood fuel versus actual energy needed for cooking food. Improvements in the energy efficiency of cooking in particular can therefore lead to significant savings of fuel demand. Improved stoves, the use of high-pressure cookers, and in some cases the shift to liquid or gaseous fuels (like bio-ethanol, fischer-tropsch liquids, or biogas) can have a major positive impact in both overall energy efficiency and emission control.

The use of agricultural residues is common in regions were forest resources are lacking. The use of straw or dung is typically found in countries like China, India, and parts of Africa. Wood is definitely preferred over agricultural residues because the latter have poor combustion properties, low density complicating transport, and cause a lot of smoke when used.

3.2 Modern Biomass Combustion for Domestic Applications

Typical capacities for combustion systems in domestic applications for production of heat and hot water range from only a few kWth to 25 kWth. This is still a major market for biomass for domestic heating in countries like Austria, France, Germany, Sweden, and the United States. Use of wood in open fireplaces and small furnaces in houses is generally poorly registered, but estimated quantities are considerable in countries mentioned. Use of wood in classic open fireplaces generally has a low efficiency (sometimes as low as 10%) and generally goes with considerable emissions of, for example, dust and soot. Technology development has led to the application of strongly improved heating systems, which are, for example automated, have catalytic gas cleaning, and make use of standardized fuel (such as pellets). The efficiency benefit compared to, for example, open fireplaces is considerable: open fireplaces may even have a negative efficiency over the year (due to heat losses through the chimney), while advanced domestic heaters can obtain efficiencies of 70 to 90% with strongly reduced emissions. The application of such systems is especially observed in, for example, Scandinavia, Austria, and Germany. In Sweden in particular, a significant market has developed for biomass pellets, which are fired in automated firing systems.

Concepts for production of heat vary from open fireplaces to fireplace inserts, heat storing stoves (where a large mass of stone or tiles is used to store heat and again gradually dissipate that heat for a period of up to several days) to catalytic combustion systems, and use of heat exchangers that avoid losses of the produced heat through the chimney. The transfer of heat takes place by convection and direct radiation.

Boilers are used for hot water production (for either space heating or tap water). Basic concepts cover over-fire, under-fire, and downdraught boilers, which primarily differ with the direction of the combustion air flow through the fuel. Such systems are usually linked to water heating systems with a capacity of some 1 to 5 m^3, including the possibility for temporal heat storage. Equipment is commercially available and deployed with advanced emission control systems. All systems mentioned generally rely on good quality wood, like logs, which are not too wet and uncontaminated.

Pellet burners are also available and the way combustion is performed shows some similarities with oil firing, since the pellets are crushed prior to combustion, resulting in a constant flow of fine particle to the burner, which is easily controlled. Fully automated systems are on the market that rely on a large storage of pellets in, for example, a basement which is filled once every half year or so. Control is steered by the heat demand of the hot water system. Infrastructure with pellet distribution is present in countries like Sweden.

Wood chip stoves can have similar characteristics in terms of automation, but require more voluminous storage and an oven or firebox for combustion of the chips. Those are called stoker burners. Temperatures can rise up to 1000°C when good quality, dry fuel is used.

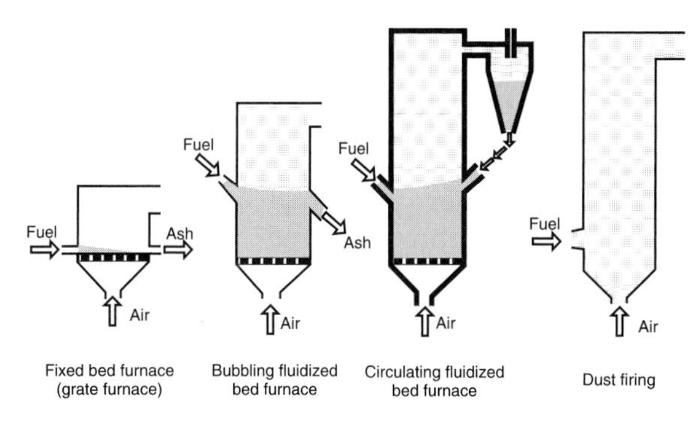

FIGURE 2 Basic combustion concepts for industrial applications.

3.3 Biomass Combustion for Industrial Applications: Stand Alone

Biomass combustion concepts exceeding a capacity of some 1 MWth are generally equipped with automated feeding and process control. Basic combustion concepts include fixed bed or grate furnaces, as well as bubbling fluidized bed (BFB) and circulating fluidized bed (CFB) furnaces. Furthermore, dust firing is deployed, but this is less common (see Fig. 2). Table III summarizes some of the main characteristics of the combustion concepts discussed.

- *Fixed bed.* Fixed bed furnaces (e.g., stoker furnaces) are typically suited for smaller capacities (e.g., up to 6 MWth). Grate-fired systems are deployed in a wide scale range. They can be divided

TABLE III

Some Main Characteristics of Key Combustion Concepts Currently Deployed

Combustion concept	Key advantages	Key disadvantages	Key applications
Underfeed stokers	Low cost at smaller capacities, easy control	Needs high-quality fuel with respect to composition and size	< 6 MWth heating; buildings, district heating
Grate furnaces	Relatively cheap and less sensitive to slagging	Different fuels cannot be mixed for certain grate-type; nonhomogeneous combustion conditions (resulting in higher emissions)	Wide range; widely applied, especially for waste fuels and residues; typical capacity some 20–30 MWe; waste incinerators up to 1–2 Mtonne capacity/year
Dust combustion	High efficiency, good load control and low Nox	Small particle size required, wearing of lining	Less common (especially suited for sawdust)
BFB furnaces	Highly flexible for moisture contents, low Nox, high efficiency and decreased flue gas flow	High capital and operational costs, sensitive to particle size (requires pretreatment), high dust load in flue gas, some sensitivity for ash slagging; erosion of equipment	> 20 MWth; deployed for flexible fueled plants and more expensive fuels (e.g., clean wood)
CFB furnaces	High fuel flexibility with respect to moist and other properties, low Nox, easy adding additives (e.g., for emission control), high efficiency and good heat transfer	High capital and operational costs, sensitivity to particle size (requires pre-treatment), high dust loads in flue gas, inflexible for part load operation; very sensitive to ash slagging, erosion of equipment	> 30 MWth; deployed for flexible fueled plants and more expensive fuels; of CFB and BFB combined, some 300 installations build worldwide

into co-, cross-, and counter-current fired systems, which principally differ in the direction of the combustion air-flow in relation to the biomass flow. While the fuel is moving over the grate, the combustion process develops starting with drying, devolatization, formation of charcoal, and finally ash on the grate. Typically, primary and secondary combustion chambers are found, leading to staged combustion. Grate furnaces are very flexible with respect to fuel properties (e.g., moisture and size). Depending on the fuel properties, different grate types are deployed. Traveling grates are used meaning the grate is working like a transport belt, while the biomass is gradually converted along the way. Ash typically falls through the grate and is removed with collection systems. Ash removal can be both wet and dry. Horizontal, vibrating, and inclined grates are also applied. Typically, MSW incineration (which is a very inhomogeneous fuel) makes use of inclined grates, which result in longer residence times of the fuel particles. Another distinction can be made between air and water cooled grates. The latter is typically used for dryer fuels that give a higher combustion temperature; the first is deployed for wetter fuels.

- *Bubbling fluidized bed.* In a BFB, the fuel is fed to a furnace where an inert bed material (like sand) is kept in motion by blowing (combustion) air from underneath. Temperatures are typically 800 to 900°C, and bed temperatures are homogeneous. Low excess air is needed, minimizing flue gas flows. Homogenous conditions allow for good emission control and adding of additives to the bed (like limestone) can be deployed to reduce emissions further. Fuel bed (FB) technology is flexible for different fuel properties. In contrast, this technology is inflexible to particle size, resulting in the need for more extensive pretreatment.
- *Circulating fluidized bed.* Key characteristics are rather similar to BFB technology, but fluidizing velocities of the bed are higher, leading to higher turbulence. This results in better heat transfer (and higher efficiency), but requires a particle collector (or cyclone) to feed bed material that is blown out of the furnace back to the bed. Dust loads in the flue gas are high therefore. Due to high capital and operation costs, both BFB and CFB become attractive at larger scales (i.e., over 30 MWth). At such capacities, the advantages for smaller downstream equipment (boiler and flue gas) cleaning start to pay off.
- *Other concepts.* Some concepts that do not really fit the division given here are the dust combustion (e.g., deployed for dry fine fuels as sawdust), the rotating cone furnace (involving a rotating vessel in which the fuel is fed; in particular suited for waste fuels), and the whole tree energy concept. This design is capable of firing compete logs, which are dried over a longer period time in a covered storage using waste heat from the combustion process, and the resulting net energy efficiency is high. The system has not yet been commercially applied though.

3.4 Co-combustion

Co-combustion involves the use of biomass in existing fossil fuel fired power plants next to coal, wastes, peat, or other fuels. Co-combustion can be categorized in direct co-firing (directly fed to the boiler), indirect co-firing (fuel gas produced via gasification is fed to the boiler), and parallel combustion (use of biomass boiler and use of the resulting steam in an existing power plant).

Direct co-firing can be done by mixing with coal, separate pretreatment and handling of the biomass fuel and subsequent joint firing with coal, separate biomass fuel burners, and the use of biofuel as a reburn fuel (for NO_x reduction), but the latter is not commercially applied to date.

The advantages of co-firing are apparent: the overall electrical efficiency is high due to the economies of scale of the existing plant (usually around 40% in the case of pulverized coal boilers of some 500 to 1000 MWe), and investments costs (in particular for pretreatment) are low to negligible when high-quality fuels as pellets are used. Also, directly avoided emissions are high due to direct replacement of coal. Combined with the fact that many coal-fired power plants in operation are fully depreciated, this makes co-firing usually an attractive GHG mitigation option. In addition, biomass firing generally leads to lowering sulfur and other emissions, although ash utilization and increased fouling can become serious problems in some cases.

3.5 Environmental Performance and Emission Control

Both the biomass properties and composition and the combustion process influence the environmental performance of combustion processes. Section 2 listed the main contaminants. Table IV summarizes some generic data on major emissions of key combustion options for West European conditions.

TABLE IV

Generic Emission Data for Key Biomass Combustion Options

Combustion option	NO_x (mg/MJ)	Particulates (mg/MJ)	Tar (mg/MJ)	CO (mg/MJ)	UHC as CH_4 (mg/MJ)	PAH (μg/MJ)
FB boiler	170	2	n.m.	0	1	4
Grate-fired boiler	110	120	n.m.	1850	70	4040
Stoker burner	100	60	n.m.	460	4	9
Modern wood stove	60	100	70	1730	200	30
Traditional wood stove	30	1920	1800	6960	1750	3450
Fireplace	n.m.	6050	4200	6720	n.m.	110

Values concern wood firing only and West European conditions (and emissions standards). N.m., not measured.

Those data clearly reflect the better emission profiles of FB technology (except fro NOx) and the poor performance of small-scale, traditional combustion.

Depending on applicable emission standards, various technical measures may need to be taken to meet those standards. Measures can be divided into primary measures (covering design and control of the combustion process and technology) and secondary measures that remove contaminants after the actual combustion. These typically include particle control (e.g., by means of cyclones, filters, and scrubbers), SO_2 removal (additives as limestone and scrubbing), and No_x reduction measures, which include primary measures as staged combustion but also selective (Non-) catalytic reduction. Generally, secondary measures add significantly to the total capital and operational costs, which becomes even more apparent at smaller scales.

3.6 Efficiency and Economics

Table V shows a generic overview of energy efficiency and costs of the main conversion options discussed. For comparison, gasification for electricity production is also included.

Overall economy of a plant or system strongly depends on fuel prices (and quality), emissions standards, load factors, and economic factors as interest rates and feed-in tariffs for power and heat (when applicable). Clearly, as will also be discussed in the next section, various countries installed very different policy measures, financial instruments, and standards that sometimes resulted in very specific markets for certain technologies. The next section discusses a number of countries and markets and the specific deployment of biomass combustion technology to date.

4. THE ROLE OF BIOMASS COMBUSTION TO DATE: DIFFERENT CONTEXTS AND MARKETS

This section discusses some main developments in the role of biomass combustion in various countries, contexts, and sectors, covering domestic heating, district heating and CHP, large-scale combustion and industrial use, waste incineration, and co-firing. Table VI gives an overview of some countries in the bioenergy field, as well as their specific policies, resources, and the role of combustion for production of power and heat.

It is important to understand the interlinkages between biomass resource availability and the use of specific technologies. The biomass resource base is a determining factor for which technologies are attractive and applied in relation to its characteristics (as discussed in Section 2) but also in relation to the growing biomass market. To take Europe as an example, organic wastes (in particular MSW) and forest residues represent the largest available potential and deliver the largest contribution to energy production from organic (renewable) material. Market developments, natural fluctuations in supply (e.g., due to weather variations), as well as the economics and various other factors (such as accessibility of areas) of bioenergy all influence the technical and economic potential of such resources, and it is impossible to give exact estimates of the potential. However, *available* resources are increasingly utilized. When the use of bioenergy expands further in the future (as projected by many scenario studies and policy targets), such larger contributions of biomass to the energy supply can only be covered by dedicated production systems (e.g., energy crops). Over time, several stages can be observed in biomass

TABLE V

Global Overview of Current and Projected Performance Data for the Main Conversion Routes of Biomass to Power and Heat, and Summary of Technology Status and Deployment in the European Context

Conversion option		Typical capacity range	Net efficiency (LHV basis)	Investment cost ranges (Euro/kW)	Status and deployment in Europe
Combustion	*Heat*	Domestic 1–5 MWth	From very low (classic fireplaces) up to 70–90% for modern furnaces	~100/kWth 300–700/kWth for larger furnaces	Classic firewood use still widely deployed in Europe, but decreasing. Replacement by modern heating systems (i.e., automated, flue gas cleaning, pellet firing) in, for example, Austria, Sweden, Germany, ongoing for years
	CHP	0.1–1 MWe	60–90% (overall) ~10% (electrical) 80–100% (overall) 15–20% (electrical)		Widely deployed in Scandinavia countries, Austria, Germany, and to a lesser extent France; in general increasing scale and increasing electrical efficiency over time
	Stand-alone	20–100's MWe	20–40% (electrical)	2.500–1600	Well-established technology, especially deployed in Scandinavia; various advanced concepts using fluid bed technology giving high efficiency, low costs, and high flexibility commercially deployed; mass burning or waste incineration goes with much higher capital costs and lower efficiency; widely applied in countries like the Netherlands and Germany
	Co-combustion	Typically 5–20 MWe at existing coal fired stations; higher for new multifuel power plants.	30–40% (electrical)	~250 + costs of existing power station	Widely deployed in many EU countries; interest for larger biomass cofiring shares and utilization of more advanced options (e.g., by feeding fuel gas from gasifiers) is growing in more recent years
Gasification	*Heat*	Usually smaller capacity range around 100's kWth	80–90% (overall)	Several 100's/kWth, depending on capacity	Commercially available and deployed; but total contribution to energy production in the EU is very limited
	CHP gas engine	0.1–1 MWe	15–30%	3.000–1.000 (depends on configuration)	Various systems on the market; deployment limited due to relatively high costs, critical operational demands, and fuel quality
	BIG/CC	30–100 MWe	40–50% (or higher; electrical efficiency)	5.000–3.500 (demos) 2.000–1.000 (longer term, larger scale)	Demonstration phase at 5–10 MWe range obtained, rapid development in the 1990s has stalled in recent years; first generation concepts prove capital intensive

Based on a variety of literature sources. Due to the variability of data in the various references and conditions assumed, all cost figures should be considered as indicative. Generally they reflect European conditions.

utilization and market developments in biomass supplies. Different countries seem to follow these stages over time, but clearly differ in the stage of development (see also Table V):

1. Waste treatment. Waste treatment, such as MSW and the use of process residues (paper industry, food industry), onsite at production facilities is generally the starting phase of a developing bioenergy market and system. Resources are available and often have a negative value, making utilization profitable. Utilization is usually required to solve waste management problems.
2. Local utilization. Local utilization of resources from forest management and agriculture is generally next. Such resources are more expensive to collect and transport, but usually still economically attractive. Infrastructure development is needed.
3. Biomass market development on regional scale. Larger scale conversion units with increasing fuel flexibility are deployed, increasing average transport distances and further improved economies of scale. Increasing costs of biomass supplies make more energy-efficient conversion facilities necessary as well as feasible. Policy support measures such as feed-in tariffs are usually needed to develop into this stage.
4. Development of national markets with an increasing number of suppliers and buyers, creation of a marketplace, increasingly complex logistics. There is often increased availability, due to improved supply systems and access to markets. Price levels may therefore even decrease.
5. Increasing scale of markets and transport distances, including cross border transport of biofuels; international trade of biomass resources (and energy carriers derived from biomass). In some cases, conflicts arise due to profound differences in national support schemes (subsidies, taxes, and environmental legislation; an example is the export of waste wood from the Netherlands and Germany to Sweden). Biomass is increasingly becoming a globally traded energy commodity.
6. Growing role for dedicated fuel supply systems (biomass production largely or only for energy purposes). There are complex interlinkages with existing markets for pulp- and construction wood, use of agricultural residues, and waste streams. So far, dedicated crops are mainly grown because of agricultural interests and support (subsidies for farmers, use of set-aside subsidies), which concentrates on oil seeds (like rapeseed) and surplus food crops (cereals and sugar beet). Sweden has largest experience with SRC-Willow. There is interest in agricultural crops such as sweet sorghum in Southern Europe. Perennial grasses, poplar, and plantation forest receive attention in various countries.

4.1 Domestic Heating

Combustion on a domestic scale for heat production (space heating, cooking) is widely applied, as argued. In various developing countries in Asia and Africa, biomass use accounts for 50% to even 90% of the national energy supply, which is almost exclusively deployed through often inefficient combustion. Improved technologies, often meaning relatively simple stove designs, are available and could make a huge difference in fuel wood consumption and emissions. It is clear though that biomass is not a preferred fuel once people can afford more modern energy carriers, such as kerosene, gas, or electricity. Typically, developing regions follow the so-called fuel ladder, which starts with biomass fuels (first agricultural residues, then wood, charcoal, kerosene, propane, gas, and finally electricity). Given the inefficiency of direct biomass use for cooking in particular, the future use of available biomass resources in developing countries may well be the production of modern energy carriers as power and fuels.

In industrialized countries, where more modern combustion technology for domestic heating is available, production of heat with biomass is not increasing as fast when compared to biomass-based power generation or, more recently, production of fuels for the transport sector. This is illustrated by development in the European Union where in 1990 the production of heat from biomass amounted about 1500 PJ rising to over 1800 PJ in 1999 (an increase of 2% per year), while electricity production from biomass amounted 15 TWhe in 1990 and rose up to 46 TWhe in 1999 (an increase of 9% per year). This growth is almost exclusively realized by application of modern biomass combustion technology. Furthermore, despite the modest role of biofuels in energy terms, the production and use of biofuels rapidly increased over the past 10 years. Biodiesel production increased from 80 ktonne in 1993 up to 780 ktonnes in 2001. Germany and France produce the bulk, with minor contributions from Italy and Austria. Ethanol production in the EU increased from 48 up to 216 ktonne in the same period.

TABLE VI

Global Overview of Bioenergy Use, Policy, and Developments Over Time in a Selection of Countries

Country	Status of bioenergy, main achievement and goals	Main policy instruments deployed	Biomass supplies; use and potential	Heat and power
Austria	Biomass accounts for 11% of the national energy supply. Forest residues are used for (district) heating, largely in systems of a relatively small scale. Some of the policy measures deployed are province oriented.	Financial support for individual heating systems and gasifiers. Subsidies for farmers producing biomass (as rape). Tax reduction biofuels and tax on mineral fuels.	Mainly forest residues and production of rape for biodiesel.	Heat production declined end of nineties due to declining number of classic wood boilers. Wood boilers for domestic heating still important. About one-fifth consists of modern concepts. Bioelectricity steady grower.
Denmark	Running program for utilisation of 1.2 million tonnes of straw as well as the utilization of forest residues. Various concepts for cofiring biomass in larger scale CHP plants, district heating, and digestion of biomass residues.	Long-term energy program; energy crops to contribute from 2005. Subsidies for bioenergy projects up to 50% are possible. Feed in tariffs guaranteed for 10 years	Use of straw and wood residue, waste, and biogas production (digestion) and a significant role for energy crops in 2010 (total ~150 PJ).	1990–2000, heat production more or less constant, electricity production increased 10-fold.
Finland	Obtains 20% of its primary energy demand from biomass (one third being peat). Especially the pulp and paper industry makes a large contribution by efficient residue and black liquor utilization for energy production. Strong government support for biomass; a doubling of the contribution is possible with available resources.	Exemption from energy tax. Financial support for forest owners for producing forest residues. Targets for waste to energy production.	Forest residues, black liquor (paper industry; two-thirds of the total) and to a lesser extent peat play a key role. Further use of forest residues likely to result in more expensive supplies in the future.	In 2010 over 1000 MWe biomass electricity generation capacity should be installed. Heat production increased about 60% between 1990 and 2000, electricity some 70%.
France	There is no specific bioenergy policy in France, but the role of biofuels (ethanol and biodiesel) and biomass for heating is significant: some 400 PJ heat, 500 PJ of fuel and 10 PJe are produced. Growth rates have been very low, however, in recent years. Biomaterials explicitly included in the strategy. Electricity has a low priority. Biomass contributes about two-thirds of the total energy production for renewables (in turn, 6% of the total).	100% tax exemption for biodiesel; 80% for bioethanol. Investment subsidies, R&D. Feed-in tariffs waste incineration and landfill gas; probably to be extended to other options.	Mainly focus on available resources, residues, and wastes for heat production. Surplus cereal production and production rape. Specific attention for increasing the use of biomaterials.	Major role for domestic heating, stable market, no real technological developments. Smaller contribution for collective and industrial heating systems. Modern boiler concepts for heat production key are pursued. Furthermore, considerable biogas utilization.
Germany	Renewables contribute some 3% to the total energy supply. The contribution of biomass is to increase from about 300 PJ (fuel input) to over 800 PJ in 2010. Waste policies relevant due to stringent targets in reuse and prohibiting landfilling. Targets for biofuels are ambitious; about 10-fold more compared to current levels.	Tax exemption on biofuels. Subsidies up to 100% and support for higher fuel costs. Various R&D programmes and support measures. Important role for the ministry of agriculture.	Waste and residue utilisation of key importance. Rapeseed production to biodiesel.	About a doubling of bioelectricity production between 1990 and now (total some 19 Pje). Similar rate for heat (now over 200 PJ). Significant technical developments in smaller scale heat and electricity production concepts (e.g., gasification based).

Netherlands	Biomass and waste contribute over 50 PJ in 2000. Target for 2007 in 85 PJ, going up to about 150 PJ in 2020. In long-term strategy formulation biomass envisaged to play a key role in the energy supply (600–1000 PJ in 2040).	Feed-in tariffs for various technologies (differentiating between scales and fuels). Tax measures and subsidies for investments.	Waste and residues main resources. Import of biomass current practice for green electricity production. Import a key role to play for meeting long-term targets.	Waste incineration and cofiring in coal-fired power stations play a prominent role. Widespread application of digestion. Electricity production to increase.
New Zealand	Renewables cover some 30% of the national energy supply; bioenergy and waste over 6%. Natural resource base (forest in particular) only partly used.	National strategy on renewable energy, but so far not covering quota or energy taxes.	Woody biomass (forest residues) and black liquor account for the bulk of currently used resources.	25 PJ in wood processing industries, 21 PJ in pulp and paper industry. Growth foreseen in domestic heat markets.
Spain	Contribution of biomass in total to increase from some 160 PJ (about 50% of total renewable energy) at present to 430 PJ in 2010 (about two-thirds of renewable energy). RES contribute some 6% in 2000 with targets for 2010 set at 30%.	Tax deduction for investments (10%) and direct subsidies (up to 30%) and discount on interest (up to 5 points) on bioenergy projects. Reduced rate excise duty.	Use of waste and residues but also major attention for energy crops, both classic (rape, cereals) as new options such as sweet sorghum.	Ambitious targets for increased electricity production and biogas production and use. Heat production constant.
Sweden	Biomass accounts for 17% of the national energy demand. Use of residues in the pulp and paper industry and district heating (CHP) and the use of wood for space heating and dominant. Biomass projected to contribute 40% to the national energy supply in 2020.	Taxation and administrative measures, most notably a CO_2 tax and energy tax. Subsidies for CHP facilities. Tax exemption (energy, environmental, and fees) for biofuels; no direct subsidies.	Wood fuels (forest residues) dominate the supplies (about 70% of the total). Wood market expanded over time with logistics covering the whole country and international trade; 14.000 ha SRC-Willow.	Bioelectricity not a very fast grower. Heat production steady increase from 160 PJ in 1990 to some 270 PJ in 2000.
United Kingdom	Renewable energy in total contributes about 1% to the total energy supply of the United Kingdom. Biomass accounts for about two-thirds of that. Rapid growth in electricity production from biomass and waste is observed. Wastes but also larger scale use of energy crops to play main role.	Bioenergy part of general GHG mitigation policy. Waste policy of major relevance. Innovative bids for biofuel projects supported. Duty reduction on biofuels. No direct subsidies. Support for farmers growing SRC.	MSW large part of total current supplies. Energy crops (SRC-Willow in particular) seen as important for longer term.	MSW incineration accounts for about 50% of total energy from biomass. Electricity production increased eight-fold between 1990 and 2000 (up to 4500 GWhe). Heat production doubled to almost 40 PJ.
United States	Approximately 9000 MWe biomass fired capacity installed; largely forest residues. Production of 4 billion litres of ethanol In total some 2600 PJ supplied from biomass (over 3% of primary energy use)	Past decade emphasis on efficient power generation. For 2020 bioenergy is identified to become major cost effective supplier of heat and power. The biorefinery concept (combined production of products, fuels, and power) deserves considerable attention.	Forestry residues, waste, and agricultural residues are all important. Considerable use of (surplus) corn for ethanol production. Various crop production systems tested and considered.	Electricity production and CHP cover some 1600 PJ of biomass use. Mainly through combustion of residue and agricultural waste processing. An additional 600 PJ for domestic heating. Increase in installed capacity has stalled.

4.2 District Heating and CHP

The application of biomass fired district heating is widely applied in Scandinavian countries and Austria. In Scandinavia, biomass fired CHP really took off in the 1980s as a result from national climate and energy policies. In the first stages, retrofits of existing coal-fired boilers were popular. Over time, the scale of CHP systems shows an increasing trend, with apparent advantages as higher electrical efficiencies and lower costs. This was also combined with a developing biomass market, allowing for more competitive and longer distance supplies of biomass resources (especially forest residues). During the 1990s, Denmark deployed a major program for utilizing straw. Various technical concepts were developed and deployed, such as the so-called cigar burners combined with efficient straw baling equipment, transport, and storage chains. Other innovations were needed to deal with the difficult combustion characteristics of straw, such the high alkali and chlorine content. This led to complex boiler concepts, such as those involving two-stage combustion, but also new pretreatment techniques such as straw washing. Austria, another leading country in deploying biomass-fired CHP, focused on smaller scale systems on village level, generally combined with local fuel supply systems. All countries mentioned have colder climates, making CHP economically attractive. Furthermore, involvement of local communities has proven important. Municipalities and forest owners are often the owners of the CHP-plants. Energy costs of those systems are usually somewhat higher. Local societal support is generally strong though, especially due to the employment and expenditures that benefit the local community. High labor costs also led to high degrees of automation though, with unmanned operation typical for many of the newer facilities. In Eastern European countries, district heating is widely deployed (but with little use of biomass). Opportunities in Central and Eastern Europe come from the energy system requiring further modernization and replacement of outdated power (and heat) generating capacity. Retrofitting and repowering schemes for coal-fired power and CHP generation capacity allowing for co-combustion of biomass streams are already successful in various Central and Eastern European countries. In the past 5 years or so, in particular Scandinavian companies have become active in this field in the context of joint implementation. Clearly, due to the combination of available district heating networks and abundant forest resources, the potential for biomass fired CHP in countries like Poland, the Baltic region, and Russia is high.

4.3 Larger Scale Combustion and Industrial Use for Power Generation and Process Heat

Larger scale combustion of biomass for the production electricity (plus heat and process steam) is applied commercially worldwide. Many plant configurations have been developed and deployed over time. Basic combustion concepts include pile burning, various types of grate firing (stationary, moving, vibrating), suspension firing, and fluidized bed concepts. Typical capacities for standalone biomass combustion plants (most using woody materials, such as forest residues, as fuel) range between 20 and 50 MWe, with related electrical efficiencies in the 25 to 30% range. Such plants are only economically viable when fuels are available at low costs or when a carbon tax or feed-inn tariff for renewable electricity is in place. Advanced combustion concepts have penetrated the market. The application of fluid bed technology and advanced gas cleaning allows for efficient and clean production of electricity (and heat) from biomass. On a scale of about 50 to 80 MWe, electrical efficiencies of 30 to 40% are possible. Finland is on cutting edge of the field with development and deployment of BFB and CFB boilers with high fuel flexibility, low costs, high efficiency, and deployed at large scale. One of the latest plants realized in Finland has a capacity of 500 MWe and is co-fired with a portfolio of biomass fuels, partly supplied by water transport.

Thailand and Malaysia are two countries in Asia that are active in deploying state-of-the-art combustion technology (typically plants of 20 to 40 MWe) for utilizing forest, plantation, and agricultural residues. The potential in the southeast Asian region for such projects is large and fits in the rapid growth of the electricity demand in the region giving ample room for investments.

Two large industrial sectors offer excellent opportunities to use available biomass resources efficiently and competitively worldwide. Those are the paper and pulp industry and the sugar-industry (particularly the one using sugar cane as feed). Traditionally, those sectors have used biomass residues (wood waste and bagasse) for their internal energy needs, which usually implies inefficient conversion to low pressure steam and some power. Plants are usually laid out to meet internal power needs only. Older plants often use process steam to drive steam turbines for direct mechanical drive of equipment (such as the sugar mills). Although in general the power production potential is much larger, most plants are not equipped with power generation units to export a surplus power. This situation is particularly caused by the absence of regulations that ensure reasonable electricity tariffs for independent power producers, thus making it unattractive for industries to invest in more efficient power generation capacity. However, liberalization of the energy markets in many countries removed this barrier (or is close to implementation). Combined with a need to reduce production costs and the need for modernization of, often old, production capacity, this provides a window of opportunity for these sectors worldwide.

In the sugar industry, the improved use of bagasse for energy production in sugar mills has gained some momentum. In Africa, Asia (e.g., through a World Bank supported project in Indonesia), and Latin America, retrofits have been carried out or are considered at many mills. Retrofits generally include the installation of efficient boilers and in some cases energy efficiency measures in the mills themselves. In combination, such retrofits could easily double the electricity output of a sugar mill, although projects are always site and plant specific.

Power output and exports can be considerably increased at low investment costs. As examples, in Nicaragua, electricity production from available bagasse by using improved boilers could meet the national electricity demand. Countries like Cuba, Brazil, and others offer similar opportunities. The other key sector applying biomass combustion for power generation is the paper-and-pulp industry. Combustion of black liquor, a main residue from the pulping process containing various chemicals, and, to a lesser extent, wood residues are the key fuels. Conventional boilers for combined production of power and process steams and recovery of pulping chemicals (so-called Tomlinson boilers) constitute common technology for the paper-and-pulp sector. More advanced boilers or gasification technology (BIG/CC) could offer even further efficiency gains and lower costs, such as when applied for converting black liquor, since recovery of pulping chemicals is easier. Generally, the power generation is competitive to grid prices, since the biomass is available anyway.

4.4 Waste Incineration

Waste incinerators, combined with very stringent emission standards, were widely deployed starting in the 1980s in countries like Germany and the Netherlands. Typical technologies deployed are large-scale (i.e., around 1 Mtonne capacity per plant per year) moving grate boilers (which allow mass burning of very diverse waste properties), low steam pressures and temperatures (to avoid corrosion), and extensive flue gas cleaning. Typical electrical efficiencies are between 15 to over 20%. Mass burning became the key waste-to-energy technology deployed in Europe, but it is also relatively expensive with treatment costs in the range of 50 to 150 Euro/tonne (off-set by tipping fees). Emission standards have a strong impact on the total waste treatments costs. In the Netherlands, flue gas cleaning equipment represents about half of the capital costs of a typical waste to energy plant. The more advanced BFB and CFB technology may prove to be a more competitive alternative in the short to medium term, because emission profiles are inherently better. On the other hand, more complex pretreatment (e.g., production of refuse derived fuel from municipal solid waste, basically an upgrading step) is required.

4.5 Co-combustion

Co-combustion of biomass, in particular in pulverized coal-fired power plants, is the single largest growing conversion route for biomass in many OECD countries (e.g., in the United States, Scandinavia, Spain, Germany, and the Netherlands, to name a few). Generally, relatively low co-firing shares are deployed with limited consequences for boiler performance and maintenance. For some countries, biomass co-firing also serves as a route to avoid the need for investing in additional flue gas cleaning equipment that would be required by national emission standards for 100% coal firing.

Because many plants in mentioned countries are now equipped with some co-firing capacity, interest for higher co-firing shares (e.g., up to 40%) is rising. Technical consequences for feeding lines and boiler performance, for example, as well as ash compositions (ash is generally used and changing characteristics can cause problems in established markets) are more severe though and development efforts focus on those issues. This is a situation where gasification becomes attractive as co-firing option, since feeding fuel gas avoids many of aforementioned consequences. Two examples of such configurations are Lahti in Finland (60 MWth gasification capacity for a 200 MWe and 250 MWth CHP plant, using natural gas, biomass, and other solid fuels) and AmerGas in the Netherlands (over 80 MWth gasification capacity for a 600 MWe coal-fired power station). Power plants capable of firing, for example, natural gas or coal with various biomass streams, have been built in Denmark (e.g., the Avedore plant) and Finland with the benefit of gaining economies of scale as well as reduced fuel supply risks. In Denmark straw is a common fuel. The chlorine and alkaline-rich straw caused problems in conventional combustion systems through increased corrosion and slagging. In multi-fuel systems, however, straw can be used for raising low-temperature steam, after which the steam is superheated by fossil fuels. This approach nearly eliminates the problems mentioned.

5. FUTURE OUTLOOK AND CLOSING REMARKS

To date, the use of biomass for energy production in the world (which is considerable with some 45 EJ) is dominated by combustion. In terms of net energy produced, gasification, pyrolysis, fermentation (for ethanol production), extraction (e.g., biodiesel), and other options play a marginal role. Although various technological options mentioned may have better economic and efficiency perspectives in the long term, in general, it is likely that biomass combustion will have a major role to play for a long time to come. Currently, biomass combustion is the work-horse for biomass as a primary and renewable energy source. Looking ahead, many technological improvements can be deployed and it is likely the principle of technological learning will lead to further improvements in energy efficiency, environmental performance, reliability, and flexibility and economy.

Better stoves in developing countries can have a tremendous positive impact on both efficiency improvements of wood fuel use as well as improving indoor health quality and resulting improvement of health, in particular for women who are generally responsible for preparing meals and maintaining the household. Simply reducing the need for fuel wood collection can save large amounts of time in many situations and adds to increased well-being and welfare as well.

Possibilities for further improvements of economy, energy efficiency, emissions, and availability for biomass combustion at larger scales include optimized and novel boiler designs lowering emissions, better materials, such as for steam generators and new power-generation concepts (closed turbine, etc.), and Stirling engines. Economies of scale have been discussed as a key development, as well as advanced co-firing concepts. With respect to co-firing, the distinction between combustion and gasification is becoming less relevant, since biomass gasifiers are now successfully deployed to deliver fuel gas to larger boilers, which are fired with coal or other fuels. Furthermore, the organization and optimization of the entire fuel supply chain (including combining different fuels and biomass pretreatment in direct relation to the conversion technology) has improvement potential in most current markets.

Clearly, new, advanced technologies will replace the older generation combustion technology, both in domestic and industrial markets. Furthermore, specific markets (such as those that affect fuels, economic conditions, and the energy system) require specific solutions, and dedicated designs are required to find optimal solutions. Another key trend, as described, is the increasing scale of bioenergy projects and the bioenergy market. The increased availability of biomass, which is also more expensive, requires conversion units to be more efficient and reach lower cost. Deploying larger scale conversion units is therefore a likely continuing trend, which is already observed in various markets. In addition, advanced co-firing systems have a key role to play, also a trend from the 1990s. Increased fuel flexibility, reducing supply risks, and allowing for cost minimization while at the same time allow for gaining economies of scale are important pathways for bioenergy as a whole.

In turn, biomass markets become increasingly intertwined, and specific national approaches (which have led to many innovations over time) may be replaced by increased international collaboration between countries and companies, resulting in products and systems for the world market. Nevertheless, new and improved technologies in the field of gasification (such as biomass integrated gasification combined cycle [BIG/CC] technology) and production of biofuels (such as methanol, fischer-tropsch liquids, and hydrogen, as well as bio-ethanol production from ligno-cellulosic biomass) have the potential over the long term to make the use of (more expensive and cultivated) biomass competitive or at least more attractive in efficiency and economic terms than direct combustion for power and heat. This also implies increased competition among different options. At this stage, it is likely various technologies will have their own specific markets and development trajectories for some time to come.

Furthermore, developments in energy systems as a whole, like increased deployment of renewables (such as wind energy), efficient use of low carbon fuels (natural gas), or CO_2 removal and storage can affect the attractiveness of biomass use for power generation once power production becomes less carbon intensive. Similar reasoning holds for increased use of biomaterials, which may have the combined benefit of saving fossil fuels in production of reference materials (e.g., plastics from mineral oil versus bioplastics) and which can be converted to energy when turned to waste. Over time, this may shift the attractiveness of applying biomass for heat and power to production of transportation fuels or bio-materials.

SEE ALSO THE FOLLOWING ARTICLES

Biodiesel Fuels • Biomass, Chemicals from • Biomass for Renewable Energy and Fuels • Biomass Gasification • Biomass: Impact on Carbon Cycle and Greenhouse Gas Emissions • Biomass Resource Assessment • Combustion and Thermochemistry • Peat Resources • Renewable Energy, Taxonomic Overview • Waste-to-Energy Technology • Wood Energy, History of

Further Reading

The Bio-energy Agreement of the International Energy Agency Web site, found at www.ieabioenergy.com, and more specifically the Task on Biomass Combustion and Co-firing, found at www.ieabioenergy-task32.com.

Broek, R. van den, Faaij, A., and van Wijk, A. (1996). Biomass combustion for power generation. *Biomass & Bioenergy* **11**(4), 271–281.

Dornburg, V., and Faaij, A. (2001). Efficiency and economy of wood-fired biomass energy systems in relation to scale regarding heat and power generation using combustion and gasification technologies. *Biomass & Bioenergy* Vol. 21(2), 91–108.

Faaij, A. (2003). Bio-energy in Europe: Changing technology choices. Prepared for a special of the *J. Energy Policy on Renewable Energy in Europe,* August.

Faaij, A., van Wijk, A., van Doorn, J., Curvers, A., Waldheim, L., Olsson, E., and Daey-Ouwens, C. (1997). Characteristics and availability of biomass waste and residues in the netherlands for gasification. *Biomass and Bioenergy* **12**(4), 225–240.

Harmelinck, M., Voogt, M., Joosen, S., de Jager, D., Palmers, G., Shaw, S., and Cremer, C. (2002). "PRETIR, Implementation of Renewable Energy in the European Union until 2010." Report executed within the framework of the ALTENER program of the European Commission, DG-TREN. ECOFYS BV, 3E, Fraunhofer-ISI, Utrecht, The Netherlands, plus various country reports.

Hillring, B. (2002). Rural development and bioenergy—experiences from 20 years of development in sweden. *Biomass and Bioenergy* **23**(6), 443–451.

Hoogwijk, M., Faaij, A., van den Broek, R., Berndes, G., Gielen, D., and Turkenburg, W. Exploration of the ranges of the global potential of biomass for energy. *Biomass and Bioenergy* **25**(2), 119–133.

Loo van, S., and Koppejan, J. (eds.). (2002). "Handbook Biomass Combustion and Co-firing." Twente University Press, Enschede, The Netherlands.

Nikolaisen, L., *et al.* (eds.). (1998). Straw for energy production. Centre for Biomass Technology, Denmark, available at www.videncenter.dk.

Serup, H., *et al.* (eds.) (1999). Wood for energy production. Centre for Biomass Technology, Denmark, available at www.videncenter.dk.

Solantausta, Y., Bridgewater, T., and Beckman, D. (1996). "Electricity Production by Advanced Biomass Power Systems." VTT Technical Research Centre of Finland, Espoo, Finland (report no. 1729).

Turkenburg, W. C., and Faaij, A., *et al.* (2001). Renewable energy technologies. "World Energy Assessment of the United Nations," UNDP, UNDESA/WEC, Chapter 7, September. UNDP, New York.

U.S. Department of Energy, Office of Utility Technologies (1998). "Renewable Energy Technology Characterizations." Washington, DC, January.

Biomass for Renewable Energy and Fuels

DONALD L. KLASS
Entech International, Inc.
Barrington, Illinois, United States

1. Fundamentals
2. Biomass Conversion Technologies
3. Commercial Biomass Energy Markets and Economics
4. Environmental Impacts

Glossary

barrels of oil equivalent (boe) The total energy content of a non-petroleum-based product or fuel in GJ divided by 5.904–6.115 GJ/boe.

biodiesel The methyl or ethyl esters of transesterified triglycerides (lipids, fats, cooking greases) from biomass.

biofuel A solid, gaseous, or liquid fuel produced from biomass.

biogas A medium-energy-content gaseous fuel, generally containing 40 to 80 volume percent methane, produced from biomass by methane fermentation (anaerobic digestion).

biomass All non-fossil-based living or dead organisms and organic materials that have an intrinsic chemical energy content.

biorefinery A processing plant for converting waste and virgin biomass feedstocks to energy, fuels, and other products.

gasohol A blend of 10 volume percent ethanol and 90 volume percent gasoline.

independent power producer (IPP) A nonutility generator of electricity, usually produced in a small capacity plant or industrial facility.

integrated biomass production conversion system (IBPCS) A system in which all operations concerned with the production of virgin biomass feedstocks and their conversion to energy, fuels, or chemicals are integrated.

landfill gas (LFG) A medium-energy-content fuel gas high in methane and carbon dioxide produced by landfills that contain municipal solid wastes and other waste biomass.

methyl *t*-butyl ether (MTBE) An organic compound used as an oxygenate and octane-enhancing additive in motor gasolines.

oxygenated gasoline Gasolines that contain soluble oxygen-containing organic compounds such as fuel ethanol and MTBE.

quad One quad is 10^{15} (1 quadrillion) Btu.

refuse-derived fuel (RDF) The combustible portion of municipal solid wastes.

tonnes of oil equivalent (toe) The total energy content of a non-petroleum-based product or fuel in GJ divided by 43.395–44.945 GJ/toe.

The world's energy markets rely heavily on the fossil fuels coal, petroleum crude oil, and natural gas as sources of thermal energy; gaseous, liquid, and solid fuels; and chemicals. Since millions of years are required to form fossil fuels in the earth, their reserves are finite and subject to depletion as they are consumed. The only natural, renewable carbon resource known that is large enough to be used as a substitute for fossil fuels is biomass. Included are all water- and land-based organisms, vegetation, and trees, or virgin biomass, and all dead and waste biomass such as municipal solid waste (MSW), biosolids (sewage) and animal wastes (manures) and residues, forestry and agricultural residues, and certain types of industrial wastes. Unlike fossil fuel deposits, biomass is renewable in the sense that only a short period of time is needed to replace what is used as an energy resource.

1. *FUNDAMENTALS*

1.1 The Concept

The capture of solar energy as fixed carbon in biomass via photosynthesis, during which carbon dioxide (CO_2) is converted to organic compounds, is the key initial step in the growth of virgin biomass and is depicted by the following equation:

$$CO_2 + H_2O + \text{light} + \text{chlorophyll} \rightarrow (CH_2O) + O_2. \quad (1)$$

Carbohydrate, represented by the building block (CH_2O), is the primary organic product. For each gram mole of carbon fixed, about 470 kJ (112 kcal) is absorbed.

The upper limit of the capture efficiency of the incident solar radiation in biomass has been estimated to range from about 8% to as high as 15%, but under most conditions in the field, it is generally less than 2% as shown in Table I. This table also lists the average annual yields on a dry basis and the average insolation that produced these yields for a few representative biomass species.

The global energy potential of virgin biomass is very large. It is estimated that the world's standing terrestrial biomass carbon (i.e., the renewable, above-ground biomass that could be harvested and used as an energy resource) is approximately 100 times the world's total annual energy consumption. The largest source of standing terrestrial biomass carbon is forest biomass, which contains about 80 to 90% of the total biomass carbon (Table II). Interestingly, marine biomass carbon is projected to be next after the forest biomass carbon in terms of net annual production, but is last in terms of availability because of its high turnover rates in an oceanic environment.

The main features of how biomass is used as a source of energy and fuels are schematically illustrated in Fig. 1. Conventionally, biomass is harvested for feed, food, fiber, and materials of construction or is left in the growth areas where natural decomposition occurs. The decomposing biomass or the waste products from the harvesting and processing of biomass, if disposed on or in land, can in theory be partially recovered after a long period of time as fossil fuels. This is indicated by the dashed lines in the figure. The energy content of biomass could be diverted instead to direct heating applications by collection and combustion. Alternatively, biomass and any wastes that result from its processing or consumption could be converted directly into synthetic organic fuels if suitable conversion processes were available. Another route to energy products is to grow certain species of biomass such as the rubber tree *(Hevea braziliensis)*, in which high-energy hydrocarbons are formed within the species by natural biochemical mechanisms, or the Chinese tallow tree *(Sapium sebiferum)*, which affords high-energy triglycerides in a similar manner. In these cases, biomass serves the dual role of a carbon-fixing apparatus and a continuous source of high-energy organic products without being consumed in the process. Other biomass species, such as the herbaceous guayule bush *(Parthenium argentatum)* and the gopher plant *(Euphorbia lathyris)*, produce hydrocarbons too, but must be harvested to recover them. Conceptually, Fig. 1 shows that there are several pathways by which energy products and synthetic fuels can be manufactured.

Another approach to the development of fixed carbon supplies from renewable carbon resources is to convert CO_2 outside the biomass species to synthetic fuels and organic intermediates. The ambient air, which contains about 360 ppm by volume of CO_2, the dissolved CO_2 and carbonates in the oceans, and the earth's large carbonate deposits, could serve as renewable carbon resources. But since CO_2 is the final oxidation state of fixed carbon, it contains no chemical energy. Energy must be supplied in a chemical reduction step. A convenient method of supplying the required energy and of simultaneously reducing the oxidation state is to reduce CO_2 with hydrogen. The end product, for example, can be methane (CH_4), the dominant component in natural gas and the simplest hydrocarbon known, or other organic compounds. With all components in the ideal gas state, the standard

$$CO_2 + 4H_2 \rightarrow CH_4 + H_2O \quad (2)$$

enthalpy of the process is exothermic by −165 EJ (−39.4 kcal) per gram mole of methane formed. Biomass can also serve as the original source of hydrogen via partial oxidation or steam reforming to yield an intermediate hydrogen-containing product gas. Hydrogen would then effectively act as an energy carrier from the biomass to CO_2 to yield a substitute or synthetic natural gas (SNG). The production of other synthetic organic fuels can be carried out in a similar manner. For example, synthesis gas (syngas) is a mixture of hydrogen and carbon oxides. It can be produced by biomass gasification processes for subsequent conversion to a wide range of chemicals and fuels as illustrated in Fig. 2. Other renewable sources of hydrogen can also be utilized. These include continuous water splitting by electrochemical, biochemical, thermochemical, microbial, photolytic, and biophotolytic processes.

The basic concept then of using biomass as a renewable energy resource consists of the capture of solar energy and carbon from ambient CO_2 in growing biomass, which is converted to other fuels (biofuels, synfuels, hydrogen) or is used directly as a

TABLE I

Examples of Biomass Productivity and Estimated Solar Energy Capture Efficiency

Location	Biomass community	Annual yield dry matter (t/ha-year)	Average insolation (W/m^2)	Solar energy capture efficiency (%)
Alabama	Johnsongrass	5.9	186	0.19
Sweden	Enthrophic lake angiosperm	7.2	106	0.38
Denmark	Phytoplankton	8.6	133	0.36
Minnesota	Willow and hybrid poplar	8–11	159	0.30–0.41
Mississippi	Water hyacinth	11.0–33.0	194	0.31–0.94
California	*Euphorbia lathyris*	16.3–19.3	212	0.45–0.54
Texas	Switchgrass	8–20	212	0.22–0.56
Alabama	Switchgrass	8.2	186	0.26
Texas	Sweet sorghum	22.2–40.0	239	0.55–0.99
Minnesota	Maize	24.0	169	0.79
New Zealand	Temperate grassland	29.1	159	1.02
West Indies	Tropical marine angiosperm	30.3	212	0.79
Nova Scotia	Sublittoral seaweed	32.1	133	1.34
Georgia	Subtropical saltmarsh	32.1	194	0.92
England	Coniferous forest, 0-21 years	34.1	106	1.79
Israel	Maize	34.1	239	0.79
New South Wales	Rice	35.0	186	1.04
Congo	Tree plantation	36.1	212	0.95
Holland	Maize, rye, two harvests	37.0	106	1.94
Marshall Islands	Green algae	39.0	212	1.02
Germany	Temperate reedswamp	46.0	133	1.92
Puerto Rico	*Panicum maximum*	48.9	212	1.28
California	Algae, sewage pond	49.3–74.2	218	1.26–1.89
Colombia	Pangola grass	50.2	186	1.50
West Indies	Tropical forest, mixed ages	59.0	212	1.55
Hawaii	Sugarcane	74.9	186	2.24
Puerto Rico	*Pennisetum purpurcum*	84.5	212	2.21
Java	Sugarcane	86.8	186	2.59
Puerto Rico	Napier grass	106	212	2.78
Thailand	Green algae	164	186	4.90

Note. Insolation capture efficiency calculated by author from dry matter yield data of Berguson, W., *et al.* (1990). "Energy from biomass and Wastes XIII" (Donald L. Klass, Ed.). Institute of Gas Technology, Chicago; Bransby, D. I., and Sladden, S. E. (1991). "Energy from Biomass and Wastes XV" (Donald L. Klass, Ed.). Institute of Gas Technology, Chicago; Burlew, J. S. (1953). "Algae Culture from Laboratory to Pilot Plant," Publication 600. Carnegie Institute of Washington, Washington, DC; Cooper, J. P. (1970). "Herb." *Abstr. m,* **40**, 1; Lipinsky, E. S. (1978). "Second Annual Fuels from Biomass Symposium" (W. W. Shuster, Ed.), p. 109. Rensselaer Polytechnic Institute, Troy, New York; Loomis, R. S., and Williams, W. A. (1963). *Crop. Sci.* **3**, 63; Loomis, R. S., Williams, W. A., and Hall, A. E. (1971). *Ann. Rev. Plant Physiol.* **22**, 431; Rodin, I. E., and Brazilevich, N. I. (1967). "Production and Mineral Cycling in Terrestrial Vegetation." Oliver & Boyd, Edinburgh, Scotland; Sachs, R. M., *et al.* (1981). *Calif. Agric.* **29**, July/August; Sanderson, M. A., *et al.* (1995). "Second Biomass Conference of the Americas," pp. 253–260. National Renewable Energy Laboratory, Golden, CO; Schneider, T. R. (1973). *Energy Convers.* 13, 77; and Westlake, D. F. (1963). *Biol. Rev.* 38, 385.

source of thermal energy or is converted to chemicals or chemical intermediates.

The idea of using renewable biomass as a substitute for fossil fuels is not new. In the mid-1800s, biomass, principally woody biomass, supplied over 90% of U.S. energy and fuel needs, after which biomass energy usage began to decrease as fossil fuels became the preferred energy resources. Since the First Oil Shock of 1973–1974, the commercial utilization of biomass energy and fuels has increased slowly but steadily. The contribution of biomass energy to U.S. energy consumption in the late 1970s was more than

TABLE II

Estimated Distribution of World's Biomass Carbon

	Forests	Savanna and grasslands	Swamp and marsh	Remaining terrestrial	Marine
Area (10^6 km^2)	48.5	24.0	2.0	74.5	361
Percentage	9.5	4.7	0.4	14.6	70.8
Net C production (Gt/year)	33.26	8.51	2.70	8.40	24.62
Percentage	42.9	11.0	3.5	10.8	31.8
Standing C (Gt)	744	33.5	14.0	37.5	4.5
Percentage	89.3	4.0	1.7	4.5	0.5

Note. Adapted from Table 2.2 in Klass, D. L. (1998). "Biomass for Renewable Energy, Fuels, and Chemicals." Academic Press, San Diego, CA.

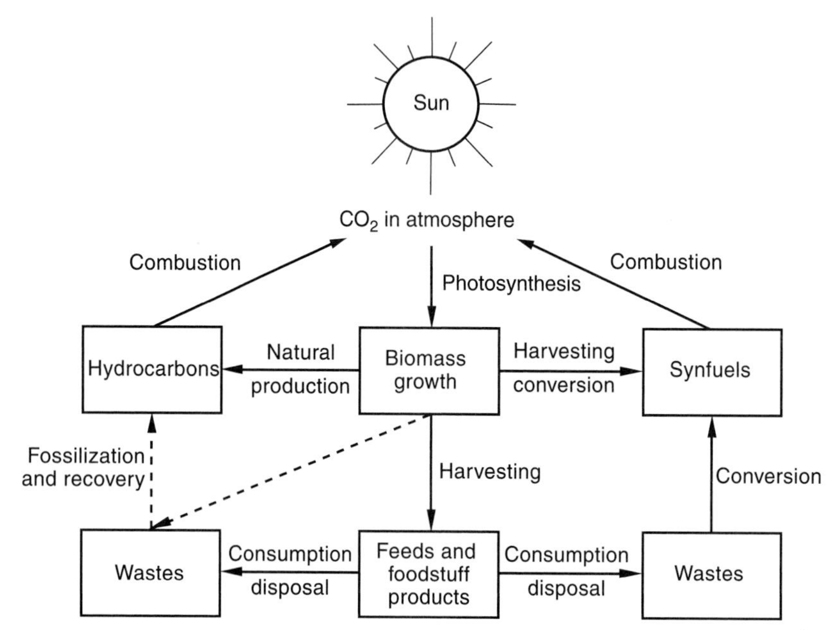

FIGURE 1 Main features of biomass energy technology. From Klass (1998).

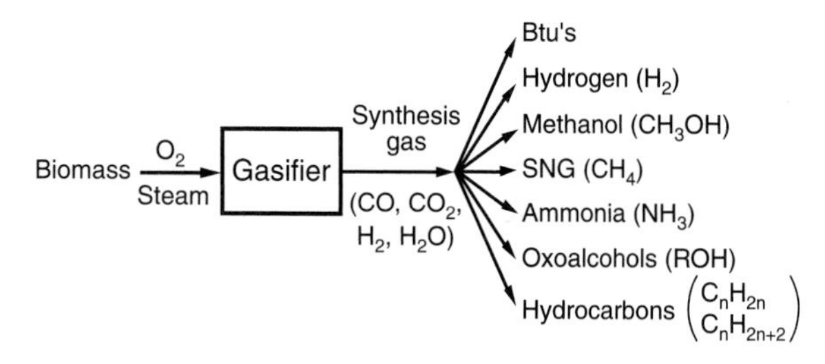

FIGURE 2 Chemicals from syngas by established processes. From Klass (1998).

850,000 barrels of oil equivalent per day (boe/day), or more than 2% of total primary energy consumption at that time. In 1990, when total U.S. primary energy consumption was about 88.9 EJ (84.3 quad), virgin and waste biomass resources contributed about 3.3% to U.S. primary energy demand at a rate of about 1.4 Mboe/day, as shown in Table III. By 2000, when total primary energy consumption had increased to 104.1 EJ (98.8 quad), virgin and waste biomass resources contributed about 23% more to primary energy demand, 1.60 Mboe/day, although the overall percentage contribution was about the same as in 1990 (Table IV).

According to the United Nations, biomass energy consumption was about 6.7% of the world's total energy consumption in 1990. For 2000, the data compiled by the International Energy Agency (IEA) from a survey of 133 countries indicate that biomass' share of total energy consumption, 430 EJ (408 quad), for these countries is about 10.5% (Table V). Although the IEA cautions that the quality and

TABLE III

Consumption of Biomass Energy in United States in 1990

Biomass resource	EJ/year	boe/day
Wood and wood wastes		
Industrial sector	1.646	763,900
Residential sector	0.828	384,300
Commercial sector	0.023	10,700
Utilities	0.013	6,000
Total	2.510	1,164,900
Municipal solid wastes	0.304	141,100
Agricultural and industrial wastes	0.040	18,600
Methane		
Landfill gas	0.033	15,300
Biological gasification	0.003	1,400
Thermal gasification	0.001	500
Total	0.037	17,200
Transportation fuels		
Ethanol	0.063	29,200
Other biofuels	0	0
Total	0.063	29,200
Grand total	2.954	1,371,000
Percentage primary energy consumption	3.3	

Note. From Klass, D. L. (1990). *Chemtech* **20**(12), 720–731; and U.S. Department of Energy (1991). "Estimates of U.S. Biofuels Consumption DOE/EIA-0548," October. Energy Information Administration, Washington, DC.

TABLE IV

Consumption of Biomass Energy in United States in 2000

Biomass resource	EJ/year	boe/day
Wood	2.737	1,270,100
Waste	0.570	264,800
Alcohol fuels	0.147	68,000
Total	3.454	1,602,900

Note. Adapted from Energy Information Administration (2002). *Monthly Energy Review,* August, Table 10.1. Washington, DC. Wood consists of wood, wood waste, black liquor, red liquor, spent sulfite liquor, wood sludge, peat, railroad ties, and utility poles. Waste consists of MSW, LFG, digester gas, liquid acetonitrile waste, tall oil, waste alcohol, medical waste, paper pellets, sludge waste, solid by products, tires, agricultural by-products, closed-loop biomass, fish oil, and straw. Alcohol fuels consist of ethanol blended into motor gasoline.

reliability of the data they compiled on biomass may be limited, which makes comparison between countries difficult, and that the proper breakdown between renewables and nonrenewables is often not available, it is clear that a significant portion of global energy consumption is based on biomass resources. It is also evident that the largest biomass energy consumption occurs among both industrialized and developing countries. Some countries meet a large percentage of their energy demands with biomass resources such as Sweden, 17.5%; Finland, 20.4%; Brazil, 23.4%; while many other countries in South America, Africa, and the Far East use biomass energy resources that supply much higher percentages of total energy demand. As expected, most countries in the Middle East where large proved crude oil and natural gas reserves are located and where dedicated energy crops might be difficult to grow meet their energy and fuel needs without large contributions from biomass resources.

The IEA reports that the share of energy consumption for renewables in 2000 was 13.8% of total energy consumption, of which 79.8% is combustible renewables and waste, most of which is biomass, and that the balance of 20.2% consists of hydroelectric power, 16.5%, and 3.7% other renewables.

Despite some of the inconsistencies that can occur because of data reliability, specific comparisons for total and biomass energy consumption in Table VI, and for total electricity and biomass-based electricity generation in Table VII, are shown for the eleven largest energy-consuming countries including the United States. The data for seven countries in these tables are for 2000; data for four countries are for 1999. The United States is the largest energy consumer, but China and India are the largest biomass energy consumers. Primary biomass solids as described in the footnote for Table VI are the largest biomass resources for these countries as well as the other countries listed in this table. In the case of electricity generation from biomass, the United States, Japan, and Germany are the largest producers of the 11 countries listed in Table VII, and the biomass resource most utilized is primary biomass solids for the United States and Japan, while Germany uses much less of that resource. It is surprising that China and India are each reported to use "0" biogas for power generation, since it is well known that each these countries operate millions of small-scale and farm-scale methane fermentation units, while many major urban cities utilize the high-methane fuel gas produced during wastewater treatment by anaerobic digestion. The lack of data is probably the cause of this apparent inconsistency.

It is noteworthy that some energy analysts have predicted that the end of seemingly unlimited petroleum crude oil and natural gas resources is in

TABLE V

Total Energy Consumption and Biomass' Share of Total Consumption for 133 Countries in 2000

Region	Country	Total consumption		Biomass' share total consumption		
		(Mtoe)	(EJ/year)	(%)	(EJ/year)	(Mboe/day)
North America	Canada	251	10.9	4.5	0.491	
	Cuba	13.2	0.573	21.1	0.121	
	Dominican Republic	7.8	0.34	17.4	0.059	
	Haiti	2	0.09	75.4	0.068	
	Jamaica	3.9	1.5	12.1	0.182	
	Mexico	153.5	6.664	5.2	0.347	
	Panama	2.5	0.11	18.1	0.020	
	Trinidad and Tobago	8.7	0.38	0.4	0.002	
	United States	2300	99.85	3.4	3.395	
	Subtotal:	2742.6	120.41	3.9	4.685	2.100
South America	Argentina	61.5	2.67	4.4	0.117	
	Bolivia	4.9	0.21	14.7	0.031	
	Brazil	183.2	7.953	23.4	1.861	
	Chile	24.4	1.06	17.4	0.184	
	Colombia	28.8	1.25	18.3	0.229	
	Ecuador	8.2	0.36	8.5	0.031	
	El Salvador	4.1	0.18	34.0	0.061	
	Guatemala	7.1	0.31	54.5	0.169	
	Honduras	3	0.1	44.0	0.044	
	Netherlands Antilles	1.1	0.048	0.0	0.000	
	Nicaragua	2.7	0.12	51.6	0.062	
	Paraguay	3.9	0.17	58.2	0.099	
	Peru	12.7	0.551	17.6	0.097	
	Uruguay	3.1	0.13	13.7	0.018	
	Venezuela	59.3	2.57	0.9	0.023	
	Subtotal:	408.0	17.7	17.7	3.026	1.356
Europe	Albania	1.6	0.069	3.6	0.002	
	Austria	28.6	1.24	10.9	0.135	
	Belgium	59.2	2.57	1.2	0.031	
	Bosnia and Herzegovina	4.4	0.19	4.2	0.008	
	Bulgaria	18.8	0.816	3.1	0.025	
	Croatia	7.8	0.34	4.8	0.016	
	Cyprus	2.4	0.10	0.4	0.000	
	Czech Republic	40.4	1.75	1.5	0.026	
	Denmark	19.5	0.847	8.8	0.075	
	Finland	33.1	1.44	20.4	0.294	
	France	257.1	11.16	4.5	0.502	
	Germany	339.6	14.74	2.5	0.369	
	Gibraltar	0.2	0.009	0.0	0.000	
	Greece	27.8	1.21	3.7	0.045	
	Hungary	24.8	1.08	1.5	0.016	
	Iceland	3.4	0.15	0.0	0.000	
	Ireland	14.6	0.634	1.2	0.008	
	Italy	171.6	7.449	4.9	0.365	
	Luxembourg	3.7	0.16	0.8	0.001	

continues

Table V continued

Region	Country	Total consumption (Mtoe)	Total consumption (EJ/year)	Biomass' share total consumption (%)	Biomass' share total consumption (EJ/year)	Biomass' share total consumption (Mboe/day)
	Macedonia	2.8	0.12	7.7	0.009	
	Malta	0.8	0.03	0.0	0.000	
	Netherlands	75.8	3.29	2.3	0.076	
	Norway	25.6	1.11	5.3	0.059	
	Poland	90	3.9	4.5	0.176	
	Portugal	24.6	1.07	8.3	0.089	
	Romania	36.3	1.58	7.9	0.125	
	Russia	614	26.7	1.1	0.294	
	Slovak Republic	17.5	0.760	0.5	0.004	
	Slovenia	6.5	0.28	6.5	0.018	
	Spain	124.9	5.422	3.6	0.195	
	Sweden	47.5	2.06	17.5	0.361	
	Switzerland	26.6	1.15	6.0	0.069	
	United Kingdom	232.6	10.10	0.9	0.091	
	Fed. Rep. of Yugoslavia	13.7	0.595	1.8	0.011	
	Former Yugoslavia	35.1	1.52	1.8	0.027	
	Subtotal:	2432.9	105.6	3.3	3.522	1.579
Former USSR	Armenia	2.1	0.091	0.0	0.000	
	Azerbaijan	11.7	0.508	0.1	0.001	
	Belarus	24.3	1.05	4.1	0.043	
	Estonia	4.5	0.20	11.1	0.022	
	Georgia	2.9	0.13	2.5	0.003	
	Kazakhstan	39.1	1.70	0.1	0.002	
	Kyrgystan	2.4	0.10	0.2	0.000	
	Latvia	3.7	0.16	22.4	0.036	
	Lithuania	7.1	0.31	8.7	0.027	
	Republic of Moldova	2.9	0.13	2.0	0.003	
	Tajikistan	2.9	0.13	0.0	0.000	
	Turkmenistan	13.9	0.169	0.0	0.000	
	Ukraine	139.6	6.060	0.2	0.012	
	Uzbekistan	50.2	2.18	0.0	0.000	
	Subtotal:	307.3	13.34	1.1	0.149	0.067
Africa	Algeria	29.1	1.26	0.3	0.004	
	Angola	7.7	0.33	74.5	0.246	
	Benin	2.4	0.10	75.5	0.076	
	Cameroon	6.4	0.28	78.4	0.220	
	Congo	0.9	0.04	65.6	0.026	
	Cote d'lvoire	6.9	0.30	60.9	0.183	
	Egypt	46.4	2.01	2.8	0.056	
	Eritrea	0.7	0.03	70.9	0.021	
	Ethiopia	18.7	0.812	93.1	0.756	
	Gabon	1.6	0.072	59.2	0.043	
	Ghana	7.7	0.33	68.8	0.227	
	Kenya	15.5	0.239	76.1	0.182	
	Libya	16.4	0.712	0.8	0.006	
	Morocco	10.3	0.447	4.3	0.019	
	Mozambique	7.1	0.31	92.7	0.287	

continues

Table V continued

Region	Country	Total consumption		Biomass' share total consumption		
		(Mtoe)	(EJ/year)	(%)	(EJ/year)	(Mboe/day)
	Namibiae	1	0.04	16.8	0.007	
	Nigeria	90.2	3.92	80.2	3.144	
	Senegal	3.1	0.13	55.8	0.073	
	South Africa	107.6	4.671	11.6	0.542	
	Sudan	16.2	0.703	86.9	0.611	
	United Rep. of Tanzania	15.4	0.669	93.6	0.626	
	Togo	1.5	0.065	67.7	0.044	
	Tunisia	7.9	0.34	15.7	0.053	
	Zambia	6.2	0.27	82.2	0.222	
	Zimbabwe	10.2	0.443	54.8	0.243	
	Subtotal:	437.1	19.00	41.7	7.917	3.548
Middle East	Bahrain	6.4	0.28	0.0	0.000	
	Iran	112.7	4.893	0.7	0.034	
	Iraq	27.7	1.20	0.1	0.001	
	Israel	20.2	0.877	0.0	0.000	
	Jordan	5.2	0.23	0.1	0.000	
	Kuwait	20.9	0.907	0.0	0.000	
	Lebanon	5.1	0.22	2.5	0.006	
	Oman	9.8	0.43	0.0	0.000	
	Qatar	15.7	0.682	0.0	0.000	
	Saudi Arabia	105.3	4.571	0.0	0.000	
	Syria	18.4	0.799	0.1	0.001	
	Turkey	77.1	3.35	8.4	0.281	
	United Arab Emirates	29.6	1.28	0.1	0.001	
	Yemen	3.5	0.15	2.2	0.003	
	Subtotal:	457.6	19.87	1.6	0.327	0.147
Far East	Bangladesh	18.7	0.812	40.8	0.331	
	Brunei	2	0.09	0.9	0.001	
	China	1142	49.58	18.7	9.271	
	Taiwan	83	3.6	0.0	0.000	
	Hong Kong (China)	15.5	0.673	0.3	0.002	
	India	501.9	21.79	40.2	8.760	
	Indonesia	145.6	6.321	32.6	2.061	
	Japan	524.7	22.78	1.1	0.251	
	North Korea	46.1	2.00	1.1	0.022	
	South Korea	193.6	8.405	5.8	0.487	
	Malaysia	49.5	2.15	5.1	0.110	
	Myanmar	12.5	0.543	73.3	0.398	
	Nepal	7.9	0.34	85.2	0.290	
	Pakistan	64	2.8	37.6	1.053	
	Philippines	42.4	1.84	22.5	0.414	
	Singapore	24.6	1.07	0.0	0.000	
	Sri Lanka	8.1	0.35	52.8	0.185	
	Thailand	73.6	3.20	19.4	0.621	
	Vietnam	37	1.6	61.2	0.979	
	Subtotal:	2992.7	129.92	19.4	25.236	11.311

continues

Table V continued

Region	Country	Total consumption (Mtoe)	Total consumption (EJ/year)	Biomass' share total consumption (%)	Biomass' share total consumption (EJ/year)	Biomass' share total consumption (Mboe/day)
Oceania	Australia	110.2	4.784	4.9	0.234	
	New Zealand	18.6	0.807	6.5	0.052	
	Subtotal:	128.8	5.591	5.1	0.286	0.128
	Total:	9907	430.1	10.5	45.148	20.236

Note. Total energy consumption in Mtoe for each country listed was compiled by the International Energy Agency (2002). IEA's data for total energy consumption were converted to EJ/year (for 2000) in this table using a multiplier of 0.043412. The multiplier for converting EJ/year to Mboe/day is 0.4482×10^6. For each country, the IEA reported the total share of renewables as a percentage of the total consumption and as a percentage of the total consumption excluding combustible renewables and waste (CRW). Since CRW is defined to contain 97% commercial and noncommercial biomass, the percentage share of biomass for each country listed here is calculated as the difference between the percentage of total consumption and the percentage of CRW.

TABLE VI

Total Energy Consumption, Total Biomass Energy Consumption, and Biomass Energy Consumption by Biomass Resource in EJ/Year for United States and Top 10 Energy-Consuming Countries

Country	Total	Total biomass	Renewable MSW	Industrial wastes	Primary biomass solids	Biogas	Liquid biomass
United States	99.85	3.373	0.308	0.166	2.616	0.143	0.140
China*	47.25	9.244	0	0	9.191	0.054	0
Russia*	26.18	0.326	0	0.111	0.216	0	0
Japan	22.78	0.242	0.044	0	0.198	0	0
India*	20.84	8.596	0	0	8.596	0	0
Germany	14.74	0.366	0.076	0.045	0.213	0.024	0.007
France	11.16	0.496	0.079	0	0.399	0.008	0.011
Canada	10.9	0.487	0	0	0.487	0	0
United Kingdom	10.10	0.093	0.012	0.002	0.036	0.035	0
South Korea	8.405	0.092	0.065	0.015	0.007	0.002	0
Brazil*	7.80	1.862	0	0	1.547	0	0.322

Note. The energy consumption data for each country listed here were adapted from the International Energy Agency (2002). The data presented in Mtoe were converted to EJ/year using a multiplier of 0.043412. The data for those countries marked with an asterisk are for 1999; the remaining data are for 2000. Data reported as "0" by the IEA are shown in the table (see text). The sum of the energy consumption figures for the biomass resources may not correspond to total biomass energy consumption because of rounding and other factors (see text). The nomenclature used here is IEA's, as follows: biomass consists of solid biomass and animal products, gas/liquids from biomass, industrial waste, and municipal waste any plant matter that is used directly as fuel or converted into fuel, such as charcoal, or to electricity or heat. Renewable MSW consists of the renewable portion of municipal solid waste, including hospital waste, that is directly converted to heat or power. Industrial waste consists of solid and liquid products such as tires, that are not reported in the category of solid biomass and animal products. Primary biomass solids consists of any plant matter used directly as fuel or converted into other forms before combustion, such as feedstock for charcoal production. This latter category includes wood, vegetal waste including wood wastes and crops used for energy production. Biogas consists of product fuel gas from the anaerobic digestion of biomass and soild wastes—including landfill gas, sewage gas, and gas from animal wastes—that is combusted to produce heat or power. Liquid biomass includes products such as ethanol.

sight. Irreversible shortages of these fossil fuels are expected to occur before the middle of the 21st century because their proved reserves have been projected to be insufficient to meet demands at that time. Supply disruptions are expected to start first with natural gas. This is illustrated by using a reserves availability model to plot global proved natural gas reserves and five times the proved reserves versus year as shown in Fig. 3. Presuming this model provides results that are more valid over the long term than reserves-to-consumption ratios, the trend in the curves indicates that shortages of natural gas would be expected to occur in the early years of the 21st century and then begin to cause

TABLE VII

Total Electricity Generation, Total Biomass-Based Electricity Generation, and Biomass-Based Electricity Generation by Biomass Resource in TWh/Year for United States and Top 10 Energy-Consuming Countries

Country	Total	Total biomass	Renewable MSW	Industrial wastes	Primary biomass solids	Biogas
United States	4003.5	68.805	15.653	6.552	41.616	4.984
China*	1239.3	1.963	0	0	1.963	0
Russia*	845.3	2.075	0	2.045	0.030	0
Japan	1081.9	16.518	5.209	0	11.309	0
India*	527.3	0	0	0	0	0
Germany	567.1	10.121	3.688	3.946	0.804	1.683
France	535.8	3.290	1.995	0	0.949	0.346
Canada	605.1	7.379	0	0	7.379	0
United Kingdom	372.2	4.360	0.695	0	0.700	2.556
South Korea	292.4	0.396	0.361	0	0	0
Brazil*	332.3	8.519	0	0	8.519	0

Note: The electricity generation data for each country listed here were adapted from the International Energy Agency (2002). The data for those countries marked with an asterisk are for 1999; the remaining data are for 2000. Data reported as "0" by the IEA are shown in the table (see text). The data compiled by the IEA is defined as "electricity output." The sum of the electricity generation figures for the biomass resources may not correspond to total biomass electricity generation because of rounding and other factors (see text). The nomenclature used here is IEA's, as follows: biomass consists of solid biomass and animal products, gas/liquids from biomass, industrial waste and municipal waste, and any plant matter that is used directly as fuel or converted into fuel, such as charcoal, or to electricity or heat. Renewable MSW consists of the renewable portion of municipal solid waste, including hospital waste, that is directly converted to heat or power. Industrial waste consists of solid and liquid products, such as tires, that are not reported in the category of solid biomass and animal products. Primary biomass solids consists of any plant matter used directly as fuel or converted into other forms before combustion, such as feedstock for charcoal production. This latter category includes wood, vegetal waste including wood wastes and crops used for energy production. Biogas consists of product fuel gas from the anaerobic digestion of biomass and soild wastes—including landfill gas, sewage gas, and gas from animal wastes—that is combusted to produce heat or power. Liquid biomass includes products such as ethanol.

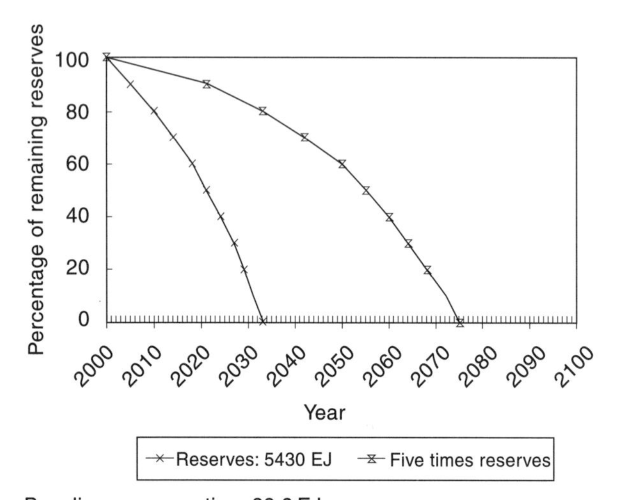

FIGURE 3 Global natural gas reserves remaining at annual growth rate in consumption of 3.2%. From Klass (2003). *Energy Pol.* **31**, 353.

serious supply problems in the next 20 to 30 years. Large-scale price increases for fossil fuels are probable because of what has been called the first derivative of the law of supply and demand, the law of energy availability and cost. This eventuality coupled with the adverse impacts of fossil fuel consumption on the environment are expected to be the driving forces that stimulate the transformation of virgin and waste biomass and other renewable energy resources into major resources for the production of energy, fuels, and commodity chemicals.

1.2 Biomass Composition and Energy Content

Typical organic components in representative, mature biomass species are shown in Table VIII, along with the corresponding ash contents. With few exceptions, the order of abundance of the major organic components in whole-plant samples of terrestrial biomass is celluloses, hemicelluloses, lignins, and proteins. Aquatic biomass does not appear to follow this trend. The cellulosic components are often much lower in concentration than the hemicelluloses as illustrated by the data for water hyacinth *(Eichhornia crassipes)*. Other carbohydrates

TABLE VIII

Organic Components and Ash in Representative Biomass

Biomass type	Marine	Fresh water	Herbaceous	Woody	Woody	Woody	Waste
Name	Giant brown kelp	Water hyacinth	Bermuda grass	Poplar	Sycamore	Pine	RDF
Component (dry wt %)							
Celluloses	4.8	16.2	31.7	41.3	44.7	40.4	65.6
Hemicelluloses		55.5	40.2	32.9	29.4	24.9	11.2
Lignins		6.1	4.1	25.6	25.5	34.5	3.1
Mannitol	18.7						
Algin	14.2						
Laminarin	0.7						
Fucoidin	0.2						
Crude protein	15.9	12.3	12.3	2.1	1.7	0.7	3.5
Ash	45.8	22.4	5.0	1.0	0.8	0.5	16.7
Total	100.3	112.5	93.3	102.9	102.1	101.0	100.1

Note. All analyses were performed by the Institute of Gas Technology (Gas Technology Institute). The crude protein content is estimated by multiplying the nitrogen value by 6.25. RDF is refuse-derived fuel (i.e., the combustible fraction of municipal solid waste).

and derivatives are dominant in marine species such as giant brown kelp *(Macrocystis pyrifera)* to almost complete exclusion of the celluloses. The hemicelluloses and lignins have not been found in *M. pyrifera.*

Alpha-cellulose, or cellulose as it is more generally known, is the chief structural element and major constituent of many biomass species. In trees, it is generally about 40 to 50% of the dry weight. As a general rule, the major organic components in woody biomass on a moisture and ash-free basis in weight percent are about 50 cellulosics, 25 hemicellulosics, and 25 lignins. The lipid and protein fractions of plant biomass are normally much less on a percentage basis than the carbohydrate components. The lipids are usually present at the lowest concentration, while the protein fraction is somewhat higher, but still lower than the carbohydrate fraction. Crude protein values can be approximated by multiplying the organic nitrogen analyses by 6.25. The sulfur contents of virgin and waste biomass range from very low to about 1 weight percent for primary biosolids. The sulfur content of most woody species of biomass is nil.

The chemical energy content or heating value is of course an important parameter when considering energy and fuel applications for different biomass species and types. The solid biomass formed on photosynthesis generally has a higher heating value on a dry basis in the range of 15.6 to 20.0 MJ/kg (6,700 to 8,600 Btu/lb), depending on the species. Typical carbon contents and higher heating values of the most common classes of biomass components are shown on a dry basis in Table IX. The higher the carbon content, the greater the energy value. It is apparent that the lower the degree of oxygenation, the more hydrocarbon-like and the higher the heating value. When the heating values of most waste and virgin biomass samples are converted to energy content per mass unit of carbon, they usually

TABLE IX

Typical Carbon Content and Heating Value of Selected Biomass Components

Component	Carbon (wt %)	Higher heating value (MJ/kg)
Monosaccharides	40	15.6
Disaccharides	42	16.7
Polysaccharides	44	17.5
Crude proteins	53	24.0
Lignins	63	25.1
Lipids	76–77	39.8
Terpenes	88	45.2
Crude carbohydrates	41–44	16.7–17.7
Crude fibers	47–50	18.8–19.8
Crude triglycerides	74–78	36.5–40.0

Note. Adapted from Klass, D. L. (1994). "Kirk-Othmer Encyclopedia of Chemical Technology," 4th ed., vol. 12, pp. 16–110. John Wiley & Sons. New York. Carbon contents and higher heating values are approximate values for dry mixtures; crude fibers contain 15 to 30% lignins.

fall within a narrow range. The energy value of a sample can be estimated from the carbon and moisture analyses without actual measurement of the heating values in a calorimeter. Manipulation of the data leads to a simple equation for calculating the higher heating value of biomass samples and also of coal and peat samples. One equation that has been found to be reasonably accurate is

$$\text{Higher heating value in MJ/dry kg} = 0.4571\ (\%\text{C on dry basis}) - 2.70. \quad (3)$$

2. *BIOMASS CONVERSION TECHNOLOGIES*

2.1 Processes

The technologies include a large variety of thermal and thermochemical processes for converting biomass by combustion, gasification, and liquefaction, and the microbial conversion of biomass to obtain gaseous and liquid fuels by fermentative methods. Examples of the former are wood-fueled power plants in which wood and wood wastes are combusted for the production of steam, which is passed through a steam turbine to generate electricity; the gasification of rice hulls by partial oxidation to yield a low-energy-value fuel gas, which drives a gas turbine to generate electric power, and finely divided silica coproduct for sale; the rapid pyrolysis or thermal decomposition of wood and wood wastes to yield liquid fuel oils and chemicals; and the hydrofining of tall oils from wood pulping, vegetable oils, and waste cooking fats to obtain high-cetane diesel fuels and diesel fuel additives. Examples of microbial conversion are the anaerobic digestion of biosolids to yield a relatively high-methane-content biogas of medium energy value and the alcoholic fermentation of corn to obtain fuel ethanol for use as an oxygenate and an octane-enhancing additive in motor gasolines.

Another route to liquid fuels and products is to grow certain species of biomass that serve the dual role of a carbon-fixing apparatus and a natural producer of high-energy products such as triglycerides or hydrocarbons. Examples are soybean, from which triglyceride oil coproducts are extracted and converted to biodiesel fuels, which are the transesterified methyl or ethyl esters of the fatty acid moieties of the triglycerides having cetane numbers of about 50, or the triglycerides are directly converted to high-cetane value paraffinic hydrocarbon diesel fuels having cetane numbers of about 80 to 95 by catalytic hydrogenation; the tapping of certain species of tropical trees to obtain liquid hydrocarbons suitable for use as diesel fuel without having to harvest the tree; and the extraction of terpene hydrocarbons from coniferous trees for conversion to chemicals. A multitude of processes thus exists that can be employed to obtain energy, fuels, and chemicals from biomass. Many of the processes are suitable for either direct conversion of biomass or conversion of intermediates. The processes are sufficiently variable so that liquid and gaseous fuels can be produced that are identical to those obtained from fossil feedstocks, or are not identical but are suitable as fossil fuel substitutes. It is important to emphasize that virtually all of the fuels and commodity chemicals manufactured from fossil fuels can be manufactured from biomass feedstocks. Indeed, several of the processes used in a petroleum refinery for the manufacture of refined products and petrochemicals can be utilized in a biorefinery with biomass feedstocks. Note also that selected biomass feedstocks are utilized for conversion to many specialty chemicals, pharmaceuticals, natural polymers, and other higher value products.

2.2 Integrated Biomass Production-Conversion Systems

The energy potential of waste biomass, although of significant importance for combined waste disposal, energy-recovery applications, is relatively small compared to the role that virgin biomass has as an energy resource. The key to the large-scale production of energy, fuels, and commodity chemicals from biomass is to grow suitable virgin biomass species in an integrated biomass-production conversion system (IBPCS) at costs that enable the overall system to be operated at a profit. Multiple feedstocks, including combined biomass–fossil feedstocks and waste biomass, may be employed. Feedstock supply, or supplies in the case of a system that converts two or more feedstocks, is coordinated with the availability factor (operating time) of the conversion plants. Since growing seasons vary with geographic location and biomass species, provision is made for feedstock storage to maintain sufficient supplies to sustain plant operating schedules.

The proper design of an IBPCS requires the coordination of numerous operations such as biomass planting, growth management, harvesting, storage, retrieval, transport to conversion plants, drying, conversion to products, emissions control, product separation, recycling, wastewater and waste solids treatment and disposal, maintenance, and transmission or transport of salable products to market. The design details of the IBPCS depend on the feedstocks involved and the type, size, number, and location of biomass growth and processing areas needed. It is evident that a multitude or parameters are involved. In the idealized case, the conversion plants are located in or near the biomass growth areas to minimize the cost of transporting biomass to the plants, all the nonfuel effluents of which are recycled to the growth areas (Fig. 4). If this kind of plantation can be implemented in the field, it would be equivalent to an isolated system with inputs of solar radiation, air, CO_2, and minimal water; the outputs consist of the product slate. The nutrients are kept within the ideal system so that addition of external fertilizers and chemicals is not necessary. Also, the environmental controls and waste disposal problems are minimized.

It is important to understand the general characteristics of IBPCSs and what is required to sustain their operation. Consider an IBPCS that produces salable energy products at a rate of 10,000 boe/day from virgin biomass. This is a small output relative to most petroleum refineries, but it is not small for an IBPCS. Assume that the plant operates at an availability of 330 day/year at an overall thermal efficiency of converting feedstock to salable energy products of 60%, a reasonable value for established thermochemical conversion technologies. Equivalent biomass feedstock of average energy content of 18.60 GJ/dry tonne would have to be provided at the plant gate to sustain conversion operations at a rate of 5291 dry tonne/day, or a total of 1,746,000 dry tonne/year. This amount of feedstock, at an average biomass yield of 25 dry tonne/ha-year, requires a biomass growth area of 69,840 ha (270 square miles), or a square area 26.4 km (16.4 miles) on each edge. For purposes of estimation, assume the product is methanol and that no coproducts are

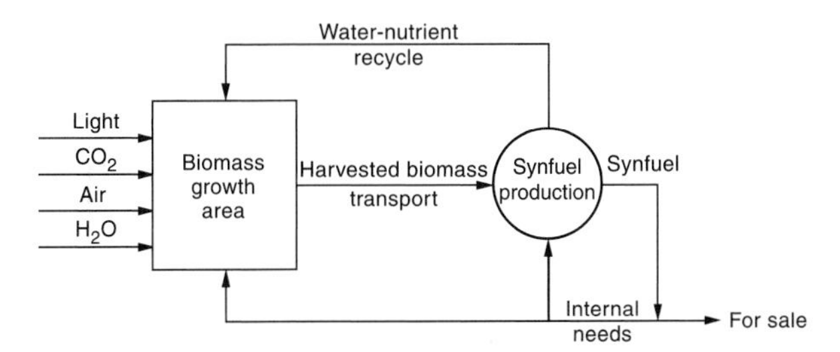

FIGURE 4 Idealized biomass growth and manufacturing system. From Klass (1998).

formed. The total annual methanol production is then approximately 1.237 billion liters/year (327 million gallons/year). Fifty-four IBPCSs of this size are required to yield 1.0 quad of salable methanol energy per year, and the total growth area required is 3,771,400 ha (14,561 square miles), or a square area 194.2 km (120.7 miles) on each edge. Again exclusive of infrastructure and assuming the conversion facilities are all centrally located, the growth area is circumscribed by a radial distance of 101.4 km (68.1 miles) from the plants.

This simplistic analysis shows that the growth areas required to supply quad blocks of energy and fuels would be very large when compared with conventional agricultural practice, but that 10,000-boe/day systems are not quite so large when compared with traditional, sustainable wood harvesting operations in the forest products industry. The analysis suggests that smaller, localized IBPCSs in or near market areas will be preferred because of logistics and product freight costs, and multiple feedstocks and products will have advantages for certain multiproduct slates. For example, commercial methanol synthesis is performed mainly with natural gas feedstocks via synthesis gas. Synthesis gas from biomass gasification used as cofeedstock in an existing natural gas-to-methanol plant can utilize the excess hydrogen produced on steam reforming natural gas. Examination of hypothetical hybrid methanol plants shows that they have significant benefits such as higher methanol yields and reduced natural gas consumption for the same production capacity.

Sustainable virgin biomass production at optimum economic yields is a primary factor in the successful operation of IBPCSs. Methodologies such as no-till agriculture and short-rotation woody crop (SWRC) growth have been evaluated and are being developed for biomass energy and combined biomass energy-coproduct applications. Most of the IBPCSs that have been proposed are site specific—that is, they are designed for one or more biomass species, in the case of a multicropping system, for specific regions. Field trials of small IBPCSs or modules of IBPCSs are in progress in the United States, but no full-scale systems have yet been built. Some of the large, commercial forestry operations for tree growth, harvesting, and transport to the mills can be considered to be analogous in many respects to the biomass production phase of managed IBPCSs. The growth, harvesting, and transport of corn to fermentation plants for fuel ethanol manufacture in the U.S. Corn Belt is perhaps the closest commercial analog to an IBPCS in the United States. Most of the other large IBPCSs that have been announced for operation outside the United States are either conceptual in nature or have not been fully implemented.

The historical development of IBPCSs shows that large-scale biomass energy plantations must be planned extremely carefully and installed in a logical scale-up sequence. Otherwise, design errors and operating problems can result in immense losses and can be difficult and costly to correct after construction of the system is completed and operations have begun. It is also evident that even if the system is properly designed, its integrated operation can have a relatively long lag phase, particularly for tree plantations, before returns on investment are realized. The financial arrangements are obviously critical and must take these factors into consideration.

3. COMMERCIAL BIOMASS ENERGY MARKETS AND ECONOMICS

3.1 Some Examples

The United States only has about 5% of the world's population, but is responsible for about one-quarter of total global primary energy demand. The markets for biomass energy in the United States are therefore already established. They are large, widespread, and readily available as long as the end-use economics are competitive.

To cite one example, petroleum crude oils have been the single largest source of transportation fuels since the early 1900s in the United States. Because of undesirable emissions from conventional hydrocarbon fuels, the U.S. Clean Air Act Amendments of 1990 included a requirement for dissolved oxygen levels in unleaded gasoline of at least 2.7 weight percent during the four winter months for 39 so-called nonattainment areas. The Act also required a minimum of 2.0 weight percent dissolved oxygen in reformulated gasoline in the nine worst ozone nonattainment areas year-round. The largest commercial oxygenates for gasolines in the United States have been fermentation ethanol, mainly from corn, and MTBE, which is manufactured from petroleum and natural gas feedstocks. Oxygenated gasolines are cleaner burning than nonoxygenated gasolines, and the oxygenates also serve as a replacement for lead additives by enhancing octane value in gasoline blends.

U.S. motor gasoline production was about 473 billion liters (125 billion gallons) in 2000, during which total U.S. gasohol production was about 61.7 billion liters (16.3 billion gallons) if it is assumed that all domestically produced fuel ethanol from biomass feedstocks in 2000, 6.17 billion liters (1.63 billion gallons), was blended with motor gasolines as gasohol. MTBE has been a major competitor of fuel ethanol as an oxygenate and octane-improving additive for unleaded gasolines since the phase-out of leaded gasolines began in the 1970s and 1980s. Without reviewing the detailed reasons for it other than to state that leakage from underground storage tanks containing MTBE-gasoline blends has polluted underground potable water supplies, federal legislation is pending that would eliminate all MTBE usage in gasolines and establish a renewables energy mandate in place of an oxygenate mandate. The provisions of this mandate are expected to include the tripling of fuel ethanol production from biomass by 2012. MTBE would be prohibited from use in motor gasoline blends in all states within a few years after enactment of the legislation. If the mandate does not become federal law, the replacement of MTBE by fuel ethanol is still expected to occur because many states have already announced plans to prohibit MTBE usage. Other states are exploring the benefits and logistics of removing it from commercial U.S. gasoline markets.

The cost of fermentation ethanol as a gasoline additive has been reduced at the pump by several federal and state tax incentives. The largest is the partial exemption of $0.053/gallon for gasohol-type blends ($0.53/gallon of fuel ethanol), out of a federal excise gasoline tax of $0.184/gallon, and a small ethanol producers tax credit of $0.10/gallon. The purpose of these incentives is to make fermentation ethanol-gasoline blends more cost competitive. Without them, it is probable that the market for fuel ethanol would not have grown as it has since it was re-introduced in 1979 in the United States as a motor fuel component. With corn at market prices of $2.00 to $2.70/bushel, its approximate price range from 1999 to 2002, the feedstock alone contributed 20.3 to 27.4 cents/liter ($0.769 to $1.038/gallon) to the cost of ethanol without coproduct credits. In 1995–1996, when the market price of corn was as high as $5.00/bushel, many small ethanol producers had to close their plants because they could not operate at a profit.

An intensive research effort has been in progress in the United States since the early 1970s to improve the economics of manufacturing fermentation ethanol using low-grade, and sometimes negative-cost feedstocks such as wood wastes and RDF, instead of corn and other higher value biomass feedstocks. Significant process improvements, such as large reductions in process energy consumption, have been made during this research, but the target production cost of \$0.16/liter (\$0.60/gallon) has not yet been attained. It is believed by some that fuel ethanol production will exhibit even larger increases than those mandated by legislation, possibly without the necessity for tax incentives, when this target is attained.

Some cost estimates indicate that fuel ethanol as well as the lower molecular weight C_3 to C_6 alcohols and mixtures can be manufactured by thermochemical non-fermentative processing of a wide range of waste biomass feedstocks at production costs as low as 25 to 50% the cost of fermentation ethanol from corn. The C_3+ alcohols and mixtures with ethanol also have other advantages compared to ethanol in gasoline blends. Their energy contents are closer to those of gasoline; their octane blending values are higher; the compatibility and miscibility problems with gasolines are small to nil; excessive Reid vapor pressures and volatility problems are less or do not occur; and they have higher water tolerance in gasoline blends, which facilitates their transport in petroleum pipelines without splash blending. Splash blending near the point of distribution is necessary for fuel ethanol-gasoline blends.

Another example of a commercial biomass-based motor fuel is biodiesel fuel. It is utilized both as a cetane-improving additive and a fuel component in diesel fuel blends, and as a diesel fuel alone. Biodiesel is manufactured from the triglycerides obtained from oil seeds and vegetable oils by transesterifying them with methanol or ethanol. Each ester is expected to qualify for the same renewable excise tax exemption incentive on an equal volume basis as fuel ethanol from biomass in the United States. Unfortunately, the availability of biodiesel is limited. The main reason for the slow commercial development of biodiesel is the high production cost of \$75 to \$150/barrel caused mainly by the relatively low triglyceride yields per unit growth area, compared to the cost of conventional diesel fuel from petroleum crude oils.

In Europe, where the costs of motor fuels including diesel fuel are still significantly higher than those in the United States, the commercial scale-up of biodiesel, primarily the esters from the transesterification of rape seed triglycerides, has fared much better. Production is targeted at 2.3 million tonnes (5.23 billion liters, 1.38 billion gallons) in 2003, and 8.3 million tonnes (18.87 billion liters, 4.98 billion gallons) in 2010. Several European countries have established zero duty rates on biodiesel to increase usage and reduce greenhouse gas emissions from diesel-fueled vehicles.

Still another example of the commercial application of biomass energy is its use for the generation of electricity. U.S. tax incentives have been provided to stimulate and encourage the construction and operation of biomass-fueled power generation systems. Most of them are operated by independent power producers or industrial facilities, not utilities. The installed, nonutility electric generation capacity fueled with renewables in the United States and the utility purchases of electricity from nonutilities generated from renewable resources including biomass in 1995 by source, capacity, and purchases are shown in Table X. Note that the sums of the biomass-based capacities—wood and wood wastes, MSW and landfills, and other biomass—and purchases are about 43% and 57% of the totals from all renewable energy resources.

Unfortunately, several of the federal tax incentives enacted into law to stimulate commercial power generation from biomass energy have expired or the qualifying conditions are difficult to satisfy. In 2002, there were no virgin biomass species that were routinely grown as dedicated energy crops in the United States for generating electricity. There are many small to moderate size power plants, however, that are fueled with waste biomass or waste biomass-fossil fuel blends throughout the United States. These plants are often able to take credits such as the tipping fees for accepting MSW and RDF for disposal via power plants that use combustion or gasification as the means of energy recovery and disposal of these wastes, the federal Section 29 tax credit for the

TABLE X

Installed U.S. Nonutility Electric Generation Capacity from Renewable Energy Resources and Purchases of Electricity by Utilities from Nonutilities by Resource in 1995

Renewable resource	Capacity (GW)	Purchases (TWh)
Wood and wood wastes	7.053	9.6
Conventional hydro	3.419	7.5
MSW and landfills	3.063	15.3
Wind	1.670	2.9
Geothermal	1.346	8.4
Solar	0.354	0.8
Other biomass	0.267	1.5
Total	17.172	46.0

Note. Adapted from Energy Information Administration (1999). Renewable Energy 1998: Issues and Trends, DOE/EIA-0628(98), March, Washington, DC.

conversion to electricity of the LFG collected from landfills, the equivalent cost of purchased power generated from biogas in a wastewater treatment plant for on-site use, or the sale of surplus power to utilities by an IPP at the so-called avoided cost, which is the cost the utility would incur by generating the power itself.

3.2 Advanced Technologies

The research programs funded by the public and private sectors in the United States to develop renewable energy technologies since the First Oil Shock have led to numerous scientific and engineering advances for basically all renewable energy resources. Some of the advanced biomass-related technologies are listed here. Many of them have already been or will be commercialized.

- The development of hybrid trees and special herbaceous biomass species suitable for use as dedicated energy crops in different climates.
- Advanced plantation designs for the managed multicropping of virgin biomass species in integrated biomass production-conversion systems.
- Advanced biorefinery system designs for the sustained production of multiple product slates.
- Practical hardware and lower cost installation methods for recovering LFG from sanitary landfills for power generation and mitigation of methane emissions.
- Safety-engineered, unmanned LFG-to-electricity systems that operate continuously.
- High rate anaerobic treatment processes for greater destruction of pathogens and biosolids in wastewaters at higher biogas yields and production rates.
- Zero-emissions waste biomass combustion systems for combined disposal-energy recovery and recycling.
- Genetically engineered microorganisms capable of simultaneously converting all pentose and hexose sugars from cellulosic biomass to fermentation ethanol.
- Catalysts for thermochemical gasification of biomass feedstocks to product gases for conversion to preselected chemicals in high yields.
- Processes for the thermochemical conversion of waste and virgin biomass feedstocks to ethanol and lower molecular weight alcohols and ethers.
- Close-coupled biomass gasification-combustion systems for the production of hot water and steam for commercial buildings and schools.
- Advanced biomass gasification processes for the high-efficiency production of medium-energy-content fuel gas and power.
- Short-residence-time pyrolysis processes for the production of chemicals and liquid fuels from biomass.
- Catalytic processes for the direct conversion of triglycerides and tall oils to "super cetane" diesel fuels and diesel fuel additives having cetane numbers near 100.

3.3 Economic Impacts and Barriers

When full-scale, well-designed IBPCSs are in place in industrialized countries and are supplying energy, organic fuels, and commodity chemicals to consumers, conventional fossil fuel production, refining, and marketing will have undergone major changes. Numerous economic impacts are expected to occur. Biomass energy production and distribution will be a growth industry, while the petroleum and gas industries will be in decline. Because of the nature of IBPCSs, employment in agriculture and forestry and the supporting industries will exhibit significant increases over many different areas of the country. Unlike petroleum refineries, which are geographically concentrated in relatively few areas, and are therefore dependent on various long-distance modes of transporting refined products to market, the biomass energy industry will be widely dispersed in rural areas. Most IBPCSs will incorporate their own biorefineries. The transport distances of refined products to market will be relatively short, and the logistics of supplying energy demands will change. It is apparent that there will be many national and international impacts of the Renewable Energy Era.

The regional economic impact of biomass energy alone is illustrated by an assessment for the U.S. Southeast from which it was concluded that industrial wood energy generated 71,000 jobs and 1 billion dollars of income annually. It was estimated in another study that 80 cents of every dollar spent on biomass energy in a given region stays in the region, while almost all expenditures on petroleum products leave the region. Still another assessment conducted for the state of Wisconsin in the Midwest Corn Belt indicates the economic impacts of shifting a portion of Wisconsin's future energy investment from fossil fuels to biomass energy. This study assumed a 75% increase in the state's renewable energy use by 2010: 775 MW of new electric generating capacity to supply electricity to 500,000 Wisconsin homes and 379 million liters

(100 million gallons) per year of new ethanol production to supply gasohol to 45% of Wisconsin's automobiles. This scenario generated about three times more jobs, earnings, and sales in Wisconsin than the same level of imported fossil fuel usage and investment and was equivalent to 62,234 more job-years of net employment, $1.2 billion in higher wages, and $4.6 billion in additional output. Over the operating life of the technologies analyzed, about $2 billion in avoided payments for imported fuels would remain in Wisconsin to pay for the state-supplied renewable resources, labor, and technologies. Wood, corn, and waste biomass contributed 47% of the increase in net employment.

Nationwide, the projected economic impacts of biomass energy development are substantial. In 2001, with petroleum crude oil imports at about 9.33 million barrels/day, consumption of biomass energy and fuels corresponds to the displacement of 1.64 Mboe/day, or 17.6% of the total daily imports. This effectively reduces expenditures for imported oil, and beneficially impacts the U.S. trade deficit. Since agricultural crops and woody biomass as well as industrial and municipal wastes are continuously available throughout the United States, biomass also provides a strategic and distributed network of renewable energy supplies throughout the country that improve national energy security.

Conservatively, the energy and fuels available in the United States for commercial markets on a sustainable basis from virgin and waste biomass has been variously estimated to range up to about 15 quad per year, while the energy potentially available each year has been estimated to be as high as 40 quad. This is made up of 25 quad from wood and wood wastes and 15 quad from herbaceous biomass and agricultural residues. Utilization of excess capacity croplands of up to 64.8 million hectares (160 million acres) estimated to be available now for the growth of agricultural energy crops could open the way to new food, feed, and fuel flexibility by providing more stability to market prices, by creating new markets for the agricultural sector, and by reducing federal farm subsidy payments. Based on the parameters previously described for one-quad IBPCSs, this acreage is capable of producing about 17 quad of salable energy products from herbaceous feedstocks each year. Other opportunities to develop large IBPCSs exist in the United States using federally owned forest lands. Such IBPCSs would be designed for sustainable operations with feedstocks of both virgin and waste wood resources such as thinnings, the removal of which would also reduce large-scale forest fires that have become commonplace in the dryer climates, particularly in the western states.

Because of the multitude of organic residues and biomass species available, and the many different processing combinations that yield solid, liquid, and gaseous fuels, and heat, steam, and electric power, the selection of the best feedstocks and conversion technologies for specific applications is extremely important. Many factors must be examined in depth to choose and develop systems that are technically feasible, economically and energetically practical, and environmentally superior. These factors are especially significant for large-scale biomass energy plantations where continuity of operation and energy and fuel production are paramount. But major barriers must be overcome to permit biomass energy to have a large role in displacing fossil fuels.

Among these barriers are the development of large-scale energy plantations that can supply sustainable amounts of low-cost feedstocks; the risks involved in designing, building, and operating large IBPCSs capable of producing quad blocks of energy and fuels at competitive prices; unacceptable returns on investment and the difficulties encountered in obtaining financing for first-of-a-kind IBPCSs; and the development of nationwide biomass energy distribution systems that simplify consumer access and ease of use. These and other barriers must ultimately be addressed if any government decides to institute policies to establish large-scale biomass energy markets.

Without IBPCSs, biomass energy will be limited to niche markets for many years until oil or natural gas depletion starts to occur. The initiation of depletion of these nonrenewable resources may in fact turn out to cause the Third Oil Shock in the 21st century.

4. *ENVIRONMENTAL IMPACTS*

Several environmental impacts are directly related to biomass energy production and consumption. The first is obviously the environmental benefit of displacing fossil fuel usage and a reduction in any adverse environmental impacts that are caused by fossil fuel consumption. In addition, the use of a fossil fuel and biomass together in certain applications, such as electric power generation with coal and wood or coal and RDF in dual-fuel combustion or cocombustion plants, can result in reduction of undesirable emissions. The substitution of fossil fuels and their derivatives by biomass and biofuels also helps to conserve depletable fossil fuels.

Another beneficial environmental impact results from the combined application of waste biomass disposal and energy recovery technologies. Examples are biogas recovery from the treatment of biosolids in municipal wastewater treatment plants by anaerobic digestion, LFG recovery from MSW landfills, which is equivalent to combining anaerobic digestion of waste biomass and LFG "mining," and the conversion of MSW, refuse-derived fuel (RDF), and farm, forestry, and certain industrial wastes, such as black liquor generated by the paper industry, to produce heat, steam, or electric power. Resource conservation and environmental benefits certainly accrue from such applications.

Another environmental impact is more complex. It concerns the growth and harvesting of virgin biomass for use as dedicated energy crops. By definition, sustainable, biomass energy plantations are designed so that the biomass harvested for conversion to energy or fuels is replaced by new biomass growth. If more biomass is harvested than is grown, the system is obviously not capable of continued operation as an energy plantation. Furthermore, the environmental impact of such systems can be negative because the amount of CO_2 removed from the atmosphere by photosynthesis of biomass is then less that that needed to balance the amount of biomass carbon removed from the plantation. In this case, virgin biomass is not renewable; its use as a fuel results in a net gain in atmospheric CO_2. Energy plantations must be designed and operated to avoid net CO_2 emissions

TABLE XI

Estimated Annual Global Carbon Dioxide and Carbon Exchanges with the Atmosphere

	Carbon dioxide		Carbon equivalent	
Source or sink	To atmosphere (Gt/year)	From atmosphere (Gt/year)	To atmosphere (Gt/year)	From atmosphere (Gt/year)
Terrestrial				
Cement production	0.51		0.14	
Other industrial processes	0.47		0.13	
Human respiration	1.67		0.46	
Animal respiration	3.34		0.91	
Methane emissions equivalents	1.69		0.46	
Natural gas consumption	3.98		1.09	
Oil consumption	10.21		2.79	
Coal consumption	8.15		2.22	
Biomass burning	14.3		3.90	
Gross biomass photosynthesis		388		106
Biomass respiration	194		53	
Soil respiration and decay	194		53	
Total terrestrial	432	388	118	106
Oceans				
Gross biomass photosynthesis		180		49
Biomass respiration	90		25	
Physical exchange	275	202	75	55
Total oceans	365	382	100	104
Total terrestrial and oceans	797	770	218	210

Note. The fossil fuel, human, and animal emissions were estimated by Klass (1998). Most of the other exchanges are derived from exchanges in the literature or they are based on assumptions that have generally been used by climatologists. It was assumed that 50% of the terrestrial biomass carbon fixed by photosynthesis is respired and that an equal amount is emitted by the soil. The total uptake and emission of carbon dioxide by the oceans were assumed to be 104 and 100 Gt C/year (Houghton, R. A., and Woodwell, G. M. (1989). *Sci. Am.* **260**(4), 36) and biomass respiration was assumed to emit 50% of the carbon fixed by photosynthesis. The carbon dioxide emissions from cement production and other industrial processes are process emissions that exclude energy-related emissions; they are included in the fossil fuel consumption figures.

to the atmosphere. A few biomass plantations are now operated strictly to offset the CO_2 emissions from fossil-fired power plants, particularly those operated on coal. Sometimes, the fossil-fired power plant and the biomass plantation are geographically far apart. It is important to emphasize that established IBPCSs that utilize dedicated energy crops will normally involve the harvesting of incrementally new virgin biomass production.

Finally, there is the related issue of the causes of increasing concentrations of atmospheric CO_2, which is believed to be the greenhouse gas responsible for much of the climatic changes and temperature increases that have been observed. Most climatologists who have studied the problem portray atmospheric CO_2 buildup to be caused largely by excessive fossil fuel usage. Some assessments indicate that biomass contributes much more to the phenomenon than formerly believed, possibly even more than fossil fuel consumption. Because terrestrial biomass is the largest sink known for the removal of atmospheric CO_2 via photosynthesis, *the accumulated loss in global biomass growth areas with time and the annual reduction in global CO_2 fixation capacity* are believed by some to have had a profound adverse impact on atmospheric CO_2 buildup. The population increase and land use changes due to urbanization, the conversion of forest to agricultural and pasture lands, the construction of roads and highways, the destruction of areas of the rainforests, large-scale biomass burning, and other anthropological activities appear to contribute to atmospheric CO_2 buildup at a rate that is much larger than fossil fuel consumption. This is illustrated by the estimated annual global CO_2 exchanges with the atmosphere shown in Table XI. Despite the possibilities for errors in this tabulation, especially regarding absolute values, several important trends and observations are apparent and should be valid for many years. The first observation is that fossil fuel combustion and industrial operations such as cement manufacture emit much smaller amounts of CO_2 to the atmosphere than biomass respiration and decay and the physical exchanges between the oceans and the atmosphere. The total amount of CO_2 emissions from coal, oil, and natural gas combustion is also less than 3% of that emitted by all sources. Note that human and animal respiration are projected to emit more than five times the CO_2 emissions of all industry exclusive of energy-related emissions. Note also that biomass burning appears to emit almost as much CO_2 as oil and natural gas consumption together. Overall, the importance of the two primary sinks for atmospheric CO_2—terrestrial biota and the oceans—is obvious. No other large sinks have been identified.

Somewhat paradoxically then, it is logical to ask the question: How can large-scale biomass energy usage be considered to be a practical application of virgin biomass? The answer, of course, has already been alluded to. At a minimum, all virgin biomass harvested for energy and fuel applications must be replaced with new growth at a rate that is at least equal to the rate of removal. Even more desirable is the creation of additional new biomass growth areas, most likely forests, because they are the largest, long-lived, global reserve of standing, terrestrial biomass carbon. New biomass growth in fact seems to be one of the more practical routes to remediation of atmospheric CO_2 buildup.

SEE ALSO THE FOLLOWING ARTICLES

Alternative Transportation Fuels: Contemporary Case Studies • Biodiesel Fuels • Biomass, Chemicals from • Biomass Combustion • Biomass Gasification • Biomass: Impact on Carbon Cycle and Greenhouse Gas Emissions • Biomass Resource Assessment • Ethanol Fuel • Forest Products and Energy • Renewable Energy, Taxonomic Overview • Waste-to-Energy Technology

Further Reading

"Bioenergy '96, Proceedings of the Seventh National Bioenergy Conference," Vols. I–II. (1996). The Southeastern Regional Biomass Energy Program (and previous and subsequent biennial proceedings published by the U.S. Regional Biomass Energy Program). Tennessee Valley Authority, Muscle Shoals, AL.

Bisio, A., Boots, S., and Siegel, P. (eds.). (1997). "The Wiley Encyclopedia of Energy and the Environment, Vols. I–II." John Wiley & Sons, New York.

Bridgwater, A. V., and Grassi, G. (eds.). (1991). "Biomass Pyrolysis Liquids Upgrading and Utilization." Elsevier Science, Essex, United Kingdom.

Chartier, P., Beenackers, A. A. C. M., and Grassi, G. (eds.). (1995). "Biomass for Energy, Environment, Agriculture, and Industry, Vols. I-III (and previous and subsequent biennial books)." Elsevier Science, Oxford, United Kingdom.

Cross, B. (ed.). (1995). "The World Directory of Renewable Energy Suppliers and Services 1995." James & James Science, London, United Kingdom.

Cundiff, J. S., *et al.* (eds.) (1996). "Liquid Fuels and Industrial Products from Renewable Resources." American Society of Agricultural Engineers, St. Joseph, MI.

Directory of U.S. "Renewable Energy Technology Vendors: Biomass, Photovoltaics, Solar Thermal, Wind." (1990). Biomass Energy Research Association, Washington, DC.

"First Biomass Conference of the Americas: Energy, Environment, Agriculture, and Industry," Vols. I–III (1993, 1942). NREL/CP-200-5768, DE93010050 (and subsequent biennial books). National Renewable Energy Laboratory, Golden, CO.

Hogan, E., Robert, J., Grassi, G., and Bridgwater, A. V. (eds.). (1992). "Biomass Thermal Processing. Proceedings of the First Canada/European Community Contractors Meeting." CPL Press, Berkshire, United Kingdom.

Klass, D. L. (ed.). (1981). "Biomass as a Nonfossil Fuel Source, American Chemical Society Symposium Series 144." American Chemical Society, Washington, DC.

Klass, D. L. (ed.). (1993). "Energy from Biomass and Wastes XVI (and previous annual books)." Institute of Gas Technology, Chicago, IL.

Klass, D. L. (1998). "Biomass for Renewable Energy, Fuels, and Chemicals." Academic Press, San Diego, CA.

Klass, D. L., and Emert, G. H. (eds.). (1981). "Fuels from Biomass and Wastes. Ann Arbor Science." The Butterworth Group, Ann Arbor, MI.

Myers, R. A. (ed.). (1983). "Handbook of Energy Technology & Economics." John Wiley & Sons, New York.

Nikitin, N. I. (1966, translated). "The Chemistry of Cellulose and Wood." Academy of Sciences of the USSR, translated from Russian by J. Schmorak, Israel Program for Scientific Translations.

Wilbur, L. C. (ed.). (1985). "Handbook of Energy Systems Engineering, Production and Utilization." John Wiley & Sons, New York.

Biomass Gasification

AUSILIO BAUEN
Imperial College London
London, United Kingdom

1. Introduction
2. Fundamentals of Gasification
3. Gasification and Gasifier Types
4. Activities Characteristic of Biomass Gasification Systems for Heat and Electricity Generation
5. Economics of Biomass Gasification for Heat and Electricity Generation
6. Constraints and Issues Affecting the Development of Biomass Gasification Systems

Glossary

biomass Organic matter of vegetable and animal origin.

gasification Thermochemical process that converts a solid or liquid hydrocarbon feedstock of fossil or renewable origin to a gaseous fuel.

Gasification is a thermochemical process that has been exploited for more than a century for converting solid feedstocks to gaseous energy carriers. The first gasifier patent was issued in England at the end of the 18th century and producer gas from coal was mainly used as lighting fuel throughout the 19th century. At the turn of the 20th century, the main use of producer gas, obtained essentially from coal, switched to electricity generation and automotive applications via internal combustion engines. The use of producer gas was gradually supplanted by the use of higher energy density liquid fuels and as a result confined to areas with expensive or unreliable supplies of petroleum fuels. Efforts in gasification technology research have persisted, however, driven mainly by the need for cleaner and more efficient electricity generation technologies based on coal. In the past decade, biomass and municipal solid waste gasification have attracted increasing interest.

1. INTRODUCTION

Biomass gasification allows the conversion of different biomass feedstocks to a more convenient gaseous fuel that can then be used in conventional equipment (e.g., boilers, engines, and turbines) or advanced equipment (e.g., fuel cells) for the generation of heat and electricity. The conversion to a gaseous fuel provides a wider choice of technologies for heat and electricity generation for small- to large-scale applications. Furthermore, electricity generation from gaseous fuels is likely to be more efficient compared to the direct combustion of solid fuels. Efficiency is a particularly important issue for biomass systems because of the possible energy and cost implications of the production and transport of biomass fuels, which are generally characterized by a low energy density. The upgrading of biomass feedstocks to gaseous fuels is also likely to lead to a cleaner conversion. In addition to the production of heat and electricity, the product gas could be used to produce transport fuels, such as synthetic diesel or hydrogen.

The coupling of biomass gasification with gas and steam turbines can provide a modern, efficient, and clean biomass system for the generation of heat and electricity. A number of small- to large-scale biomass gasification systems integrated with power generation equipment are being developed and commercialized. Their market penetration will depend on a number of factors:

- Successful demonstration of the technology
- Economic competitiveness with other energy conversion technologies and fuel cycles
- Environmental performance of the biomass fuel cycle as well as the influence of environmental factors on decision making
- Other socioeconomic factors (e.g., energy security and independence, job creation, and export potential of the technology)

This article provides an overview of the state-of-the-art of biomass gasification with a focus on heat and electricity generation. The product gas, however, could also be used for the production of a range of transport fuels.

2. FUNDAMENTALS OF GASIFICATION

Thermochemical processing of biomass yields gaseous, liquid, and solid products and offers a means of producing useful gaseous and/or liquid fuels. Gasification is a total degradation process consisting of a sequence of thermal and thermochemical processes that converts practically all the carbon in the biomass to gaseous form, leaving an inert residue. The gas produced consists of carbon monoxide (CO), hydrogen (H_2), carbon dioxide (CO_2), methane (CH_4), and nitrogen (N_2) (if air is used as the oxidizing agent) and contains impurities, such as small char particles, ash, tars, and oils. The solid residue will consist of ash (composed principally of the oxides of Ca, K, Na, Mg, and Si) and possibly carbon or char. Biomass ash has a melting point of approximately 1000°C; thus, it is important to keep the operating temperature below this temperature to avoid ash sintering and slagging. At temperatures higher than 1300°C, the ash is likely to melt and may be removed as a liquid.

The following sequence is typical of all gasifiers: drying, heating, thermal decomposition (combustion and pyrolysis), and gasification. Table I illustrates the reactions occurring in gasifiers. In directly heated gasifiers, the energy (heat) necessary for the endothermic reactions is provided by combustion and partial combustion reactions within the gasifier. In indirectly heated gasifiers, the heat is generated outside the gasifier and then exchanged with the gasifier. Gasifiers can operate at low (near-atmospheric) or high (several atmospheres) pressure.

The Badouard, steam–carbon, and the methanization reactions are favored with increasing temperature and decreasing pressure; the hydrogasification reaction is favored with decreasing temperature and increasing pressure; the water–gas shift reaction is favored at low temperatures but is independent of pressure. Temperature and pressure operating conditions as well as residence time are key factors in determining the nature of the product gas. Data from existing gasifiers indicate that the heating value of the product gas varies little between pressurized and atmospheric gasification for similar operating temperatures. Gas quality is usually measured in terms of CO and H_2 quantity and ratio.

TABLE I

Reactions Occurring in Gasifiers

Reactions	Enthalpy of reaction (kJ/mol)
Heterogeneous reactions	
Combustion	
$C + \frac{1}{2}O_2 = CO$	−123.1
$C + O_2 = CO_2$	−405.9
Pyrolysis	
$4C_nH_m = mCH_4 + (4n - m)C$	Exothermic
Gasification	
$C + CO_2 = 2CO$ (Badouard)	159.7
$C + H_2O = CO + H_2$ (steam–carbon)	118.7
$C + 2H_2 = CH_4$ (hydrogasification)	−87.4
Homogeneous reactions	
Gas phase reactions	
$CO + H_2O = CO_2 + H_2$ (water–gas shift)	−40.9
$CO + 3H_2 = CH_4 + H_2O$ (methanization)	−206.3

2.1 Effect of Feedstock Properties

Water vapor is an essential component of gasification reactions. However, a high moisture content of the feedstock has an adverse effect on the thermal balance. A low ash content improves the thermal balance, reduces the occlusion and loss of carbon in the residue, and reduces operating problems due to sintering and slagging. Sintering and slagging depend on the gasifier temperature and, in the case of biomass, are likely to be related to the presence of SiO_2, which possesses the lowest melting point among the ash components. The size of the fuel particles will affect the rate of heat and mass transfer in the gasifier. Elements such as sulfur and chlorine lead to the formation of corrosive gas components such as H_2S and HCl. Alkali metals are also a major concern with regard to corrosion, especially when combined with sulfur. Nitrogen in the feedstock leads to the formation of ammonia (NH_3), which can act as a major source of NO_x emissions when combusted in engines or gas turbines.

2.2 Effect of Operating Parameters

The operating temperature will determine the equilibrium composition of the gas. High operating temperatures increase to different extents the intrinsic rate of all chemical and physical phenomena and result in leaner gases. Gasifier temperatures should be sufficiently high to produce noncondensable tars in order to avoid problems in downstream conversion equipment. Condensable tars must be avoided if the product gas is to be used in engine or gas turbine applications and they will have to be cracked or removed prior to the operation of engines or gas turbines. Exceedingly high temperatures ($>1000^{\circ}C$) may lead to ash sintering and slagging. High operating pressures increase the absolute rate of reaction and the heat and mass transfer, and they shift the gas equilibrium composition in favor of CH_4 and CO_2. The air factor (air flow rate into the gasifier) is a key regulating parameter of a fluidized bed gasifier using air as the gasifying agent. Excessive air factors lead to low heating values (nitrogen dilution and excessive combustion). Insufficient air factors lead to low reactor temperatures and low rates of gasification.

3. *GASIFICATION AND GASIFIER TYPES*

Three main types of gasification can be distinguished based on the gasifying agent and the way in which the heat for the gasification reactions is provided: directly heated air gasification, directly heated oxygen gasification, and indirectly heated gasification.

In the first two cases, the injected gasifying agent burns part of the feedstock to provide the heat necessary to gasify the remainder of the feed in an air-poor environment. Air gasification leads to a product gas rich in nitrogen (50–65%) and consequently low in calorific value (4–8 MJ/Nm^3). Small-scale gasifiers are usually of the air gasification type, but air gasification may also be the choice for larger scale gasification systems. Oxygen gasification requires an oxygen-producing plant, which increases costs and energy consumption but leads to a producer gas low in nitrogen content and of medium calorific value (10–18 MJ/Nm^3). Steam can be added to both air and oxygen gasification processes to act as a thermal moderator and as a reagent in the gasification process, and it enhances the calorific value of the gas. Oxygen and steam gasification are not required for biomass gasification but are used for the gasification of less reactive fuels such as coal.

Indirectly heated gasifiers do not require air or oxygen input because the heat necessary for gasification is generated outside the gasifier. For example, the bed material can be heated in a separate reactor by burning the char from the gasifier, as in the Battelle process, or the heat can be generated by pulse burners and transferred by in-bed tubes, as in the MTCI process. Indirectly heated gasifiers produce a medium calorific value gas, and steam can be input to the gasifier in order to favor the gasification reaction.

The oxygen or indirectly heated gasification route may be preferred in cases in which the dilution of the product gas with nitrogen may have adverse effects on its downstream uses, for example, in relation to the production of other fuels, possibly for transport applications.

Gasifier design influences the gaseous product with respect to gas composition, condensed liquids, suspended solids, and water content. A number of reactor types are available: fixed bed, fluidized bed, circulating fluidized bed, entrained flow, and molten bath. There are numerous examples of small-scale fixed bed biomass gasification applications throughout the world, mainly coupled with engines for electricity generation. Larger gasifiers, of the fluidized bed type, are in use in niche industrial applications, and there is little operating experience with gasifiers coupled to gas turbines.

4. *ACTIVITIES CHARACTERISTIC OF BIOMASS GASIFICATION SYSTEMS FOR HEAT AND ELECTRICITY GENERATION*

In general, the conversion of biomass to heat and electricity via gasification involves the following steps: biomass storage and transport, size reduction, drying, feeding, gasification, fuel gas cleaning and cooling, power generation, flue gas cleaning, and ash disposal or recycling. Most of these activities, apart from power generation and flue gas cleaning, are also typical of gasification systems in which the product gas is destined to other end uses (e.g., transport fuel production). This section focuses on equipment and activities typical of medium- to large-scale systems.

4.1 Storage, Transfer, and Pretreatment of the Feedstock

4.1.1 Storage

Biomass storage is required to ensure the continuous operation of the facility. To limit the space required for storage at the plant site, biomass must be stored in relatively high piles. Two main problems associated with fuel storage are decomposition and self-heating. Self-heating increases the rate of decomposition and fire risk, and it encourages the growth of thermophilic fungi whose spores can cause a respiratory condition in humans similar to farmers lung. Some small biomass losses may occur at the storage stage, but they are likely to be negligible. For intermediary storage of the fuel between the pretreatment (e.g., drying and sizing) and gasification stage, storage silos may be used.

4.1.2 On-Site Biomass Transfer

On-site biomass transfer occurs between storage facilities, pretreatment equipment, and gasification equipment. Such transport technology is readily available, and cranes and enclosed conveyor belt systems may be used. The material flow characteristics of the biomass fuel need to be considered in the design of the on-site biomass transfer and intermediary storage system in order to avoid flow problems.

4.1.3 Size Control

Biomass particle size affects gasification reaction rates and gas composition. Since size control operations are expensive and energy intensive, there is a trade-off, in terms of cost and energy, between particle size reduction and reactor design and the yield and characteristics of the product gas. In practical terms, the size of the feedstock particles is dependent on the biomass requirements imposed by the adopted gasification system.

In the case of wood chips, the acceptable size of the chips fed to the gasifier is usually 2–5 cm, and size reduction may be achieved by using crushers. Size reduction generally makes drying, transfer, and intermediate storage of the biomass easier.

4.1.4 Drying

The moisture content of the feedstock affects the gas composition and the energy balance of the process since gasification is an endothermic process. Water vapor, however, is an essential component of gasification reactions. Therefore, there is a trade-off between the extent of fuel drying and the quality of product gas. Drying of the feedstock to a moisture content of approximately 15% is commonly adopted.

Fuel drying is likely to be the most energy-intensive activity in the gasification process. Important contributions can be made to the energy balance by using flue gases or steam to dry the biomass. The heat used for drying does not have to be high temperature, and a low temperature level is actually desired because it will prevent the evaporation of undesirable organic components. Direct heating systems, in which the heating medium (e.g., flue gas) is in direct contact with the fuel to be dried, based on a rotary drum or fluidized bed design may be used. The systems are also likely to be open, meaning that the heating medium is discharged to the atmosphere.

Drying activities will result in dust emissions. In most cases, a simple baghouse filter is employed to satisfactorily reduce dust emissions. However, in the case of a large drying facility, considerable quantities of water vapor (likely containing significant quantities of organic compounds) could result in addition to dust. A wet gas scrubber followed by a flue gas condensation system may be used to clean the flue gas (mainly of dust and organic compounds such as terpenes) and to recover the heat from the water vapor present in the flue gas. The condensed water requires a biological treatment before it can be discharged to the sewage system. Condensed organic compounds from the fuel-drying activity possess a fuel value; hence, the energy can be recovered from their combustion. Particular care is required in the design of drying installations to avoid fire and explosion risks.

4.2 Feeding the Biomass

Physical properties of the biomass fuel, such as size and density, affect the performance of feeding systems and have often been a source of technical problems. The choice of a feeder depends on the gasifier design, mainly the pressure against which it has to operate. Although some simple small-scale gasifiers may have an open top through which the biomass is fed, screw feeders are commonly used in low-pressure closed-vessel gasifiers.

A screw feeder, in which the screw forms a compact, pressure-retaining plug from the feedstock in the feed channel, is suitable for atmospheric gasifiers. For pressurized gasifiers, a lock-hopper feeder or a lock-hopper/screw-piston feeder is required. The lock-hopper feeder uses a screw feeder but is more complex than a simple screw feeder because of the need to pressurize the feedstock prior to its input into the gasifier. Pressurization of the feedstock is generally achieved using lock-hopper devices, which may require large quantities of inert gases (e.g., liquid nitrogen). Care is required in the design of feeding systems to limit blockages as well as fire and explosion risks.

4.3 Circulating Fluidized Bed Gasification

Air-blown circulating fluidized bed gasifiers are of interest because they produce a good quality, low calorific value (LCV) gas (4–6 MJ/Nm^3) and possess a very high carbon conversion efficiency while allowing high capacity, good tolerance to variations in fuel quality, and reliable operation. The high and homogeneously distributed temperatures and the use of particular bed materials, such as dolomite, favor tar cracking. Successful tar cracking can also be achieved using secondary circulating fluidized bed reactors. Also, successful tests on catalytic tar cracking have been performed, for example, by introducing nickel compounds into the gasifier. Sulfur control is made easier because of the significant reduction that can be achieved by adding limestone or dolomite to the gasifier bed. However, biomass feedstocks are not likely to require sulfur control because of their very low sulfur content, which is likely to meet turbine requirements indicated in Table II. The high fluid velocity entrains large amounts of solids with the product gas that are recycled back to the gasifier via the cyclones to improve conversion efficiency. Carbon conversion efficiencies for circulating fluidized bed gasifiers can reach approximately 98%.

A fluidized bed gasifier consists of a plenum chamber with a gas distributor, a refractory lined steel shell enclosing the granular bed and the freeboard zone, and a feeding device. The bed material consists of a clean and graded fraction of heat-resistant material, generally sand, alumina, limestone, dolomite, or fly ash. It is fluidized by the upward stream of the gasifying agent rising in the form of bubbles, and the continuous motion causes a mixing of the solids and results in a fairly uniform temperature distribution. High superficial velocities of the gas cause elutriation of solid particles leading to a circulating bed design.

The gas composition and heating values do not differ significantly between pressurized and atmospheric operation for similar operating temperatures. The LCV gas (4–6 MJ/Nm^3) consists mainly of inert nitrogen and carbon dioxide gases and the combustible gases carbon monoxide, hydrogen, and methane. Contaminant concentrations (e.g., sulfur, chlorine, alkali and heavy metals) depend on the quality and composition of the biomass and the bed material used.

Pressurized gasifiers require a more complex and costly feeding system (i.e., lock-hopper feeder and inert gas) compared to near-atmospheric gasifiers. Also, pressurized systems require a higher degree of process control. However, if the product gas is to be used for power generation in a gas turbine, pressurized gasifiers considerably reduce the product gas compression requirements. Pressurization is obtained by compressing the gasifying air using the turbine compressor. This results in the compression of a much lower volume of air compared to the product gas volume that would otherwise have to be compressed prior to combustion in the gas turbine. Also, in pressurized gasification systems using hot gas cleanup equipment, there is less chance that tars will condense and cause damage to equipment; thus, they need not be removed from the gas, which allows use of the energy content of tars present in the product gas. Generally, the reduced energy requirement of hot gas cleanup and reduced compression needs of pressurized systems may lead to efficiency gains on the order of 3% compared to systems based on atmospheric gasification and wet gas scrubbing. Hot gas cleanup may also be applied to atmospheric gasification systems.

TABLE II

Notional Gas Turbine Fuel Specifications

Gas heating value (LHV, MJ/Nm^3)	4–6
Gas inlet temperature (°C)	<425
Gas hydrogen content (vol. %)	10–20
Alkali metals (Na + K) (ppmw)	<0.1
H_2S (ppmv)	<100
HCl (ppmw)	<0.5
Naphthalene and tars (ppmv)	<100
Particulates (99% below 10 μm) (ppmw)	<1
Vanadium (ppmw)	<0.1
Combinations: Alkali metals + sulfur (H_2S) (ppmw)	<0.1

4.4 Product Gas Cleaning

The product gas contains a series of impurities (e.g., organic compounds, alkali metals, ammonia, char, and ash). These need to be removed to varying degrees depending on the downstream conversion process and on potential environmental impacts.

Gasification systems in which the product gas is used in engines generally comprise a series of cyclones and physical filters (e.g., baghouse filter). Gasification systems in which the product gas demands more stringent cleanup (e.g., for use in gas turbines) generally comprise catalytic or thermal tar cracking, cyclones, and high-efficiency hot or cold gas cleanup equipment. Physical filtration, in the form of hot gas filtration systems using ceramic or sintered metal filters, is likely to offer a simpler and less costly option than wet gas scrubbing. It also considerably reduces the water consumption and liquid effluents of the plant. However, unlike gas scrubbing, hot gas filtration systems are not a fully demonstrated and commercial technology. The removal of some polluting and corrosive compounds (e.g., ammonia and chlorine) and of excessive water vapor from the product gas may be problematic when using hot gas filters. However, preliminary testing using hot gas cleanup systems indicates that environmental and gas turbine fuel requirements can be met with hot gas filtration systems. These results are very much fuel dependent.

4.5 Heat and Electricity Generation

The product gas can be burned in boilers to generate heat and raise steam, in internal combustion engines to generate electricity and heat at small to medium scale (from a few kilowatts to a few megawatts), and in gas turbines to generate electricity (Brayton cycle) and heat at small to large scale. In large-scale systems using gas turbines, the exhaust gas from the gas turbine can be used to raise steam in a heat recovery steam generator to generate additional electricity using a steam turbine (Rankine cycle), resulting in combined cycle operation. In a combined heat and power plant, designed for district heating, the flue gas from the combustion of the product gas goes through a heat exchange system to raise the temperature of a heat transport fluid, generally water, circulating in a district heating system. Residual heat in the flue gas can be used to dry the biomass prior to its discharge to the atmosphere.

Factors such as scale, technical performance, capital costs, efficiencies, and emissions will determine the preferred generating technology. Also, the relative demand for heat and electricity will influence the technology choice.

4.5.1 Combustion in Engines

Engines require input of a clean gas to minimize engine wear, avoid tar deposition, and reduce coking in the engine, particularly in the valve area. The gas temperature should be as low as possible to inject a maximum amount of energy into the cylinders.

Engines have the advantage that they can be run on a variety of fuels or fuel combinations with relatively minor adjustments. Gas engines for power generation are commercially available for small- to large-scale applications (from tens of kilowatts to tens of megawatts). Their efficiencies are estimated to range between 25 and 40% depending on the capacity.

There is considerable experience with coupling boilers and engines to gasifiers at small to medium scale ($<10\,MW_e$). The provision of a product gas of suitable quality is the major problem leading to frequent maintenance and shortened engine lifetimes. Engines are commercially available, there is much experience with running engines on a wide variety of gases, and they are generally more tolerable to contaminants than turbines. However, turbines hold promise of higher efficiencies (especially at scales allowing for combined cycle operation), lower costs, and cleaner operation. Also, turbines are likely to be more suited to combined heat and power application because of higher grade heat generation compared to engines.

4.5.2 Combustion in Gas Turbines

Gas turbines cover a wide range of electrical capacities ranging from a few hundred kilowatts to tens of megawatts, with recent developments in microturbines of a few tens to a few hundred kilowatts capacity. However, there is little experience with operating gas turbines in product gas from biomass gasification. Gas turbines have more stringent fuel requirements but may have higher efficiency and lower emissions compared to reciprocating engines. Due to the stringent gas quality requirements of gas turbines, care must be taken in considering the quality and composition of the feedstock, the gasification process, and the gas cleaning system used. The main areas of concern are contamination of the fuel gas by alkali metals, tars, sulfur, chlorine, particulates, and other trace elements; the fuel gas heating value and therefore its composition and volume; the flame properties of the gas within the combustion chamber; and the presence in the fuel gas of fuel-bound nitrogen leading to NO_x formation during combustion. The minimum allowable gas heating value depends on turbine design (heating value affects air-to-fuel ratio and therefore affects the inlet requirements based on total mass flow). Care must be taken to ensure that the fuel gas is supplied to the turbine at a temperature greater than the gas dew point temperature in order to avoid droplet formation. Also, the fuel gas must be unsaturated with water to avoid the formation of acids that could result mainly from the presence of H_2S and CO_2.

Gas turbine fuel requirements are shown in Table II. Table III provides an indicative comparison of fuel gas requirements for different applications.

TABLE III

Product Gas Requirements

Application	Fuel gas temperature	Maximum particulates	Maximum tars	Contaminants
Boilers	High to use sensible heat	Low–moderate	Moderate	Low–moderate
Engine	As low as possible	Very low	Very low	Low
Turbine	As high as possible	None	None	None–low

The strict gas specifications result in low emission levels for most pollutants. Thermal NO_x formation for LCV gas combustion may achieve very low levels (less than 10 ppmv), particularly compared to emissions from natural gas-fired turbines, in which low NO_x burners achieve emissions of approximately 25 ppmv. Low emissions result from the lower combustion temperature and a leaner combustion.

The following factors contribute to the achievement of gas turbine fuel requirements:

1. Biomass feedstock properties:
   - Low (10–15%) moisture content so as not to hamper efficient gasification process and adversely affect the gas heating value.
   - Low ash feedstock to reduce filtering demand and potential slagging.
   - Low alkalinity feedstock to reduce fouling.
2. Gasification process:
   - Choice of gasifying agent (air, oxygen, or steam) influences heating value and can also reduce or eliminate formation of problem tars.
   - Choice of gasifier type influences carryover of particulates.
   - Higher operating temperatures vaporize alkali metals, which must be condensed and filtered out [e.g., vaporization occurs in circulating fluidized beds (900–1100°C) but not in bubbling fluidized beds (600–800°C)], and tars are also vaporized and cracked to a certain extent (e.g., circulating fluidized beds crack more tar compounds than do bubbling fluidized beds).
3. Gas cooling and cleaning system:
   - Special bed materials and catalysts in the gasifier or in a separate reactor can be used for sulfur and chlorine removal, to avoid the formation of ammonia, and for tar cracking.
   - Gas cooling is necessary to protect gas turbine components from heat damage, and the degree of cooling depends on the gas cleaning technique adopted (e.g., hot gas filter or wet gas scrubbing).
   - Hot gas filter systems remove particulates and alkali metals.
   - Tars may not need to be removed from the gas if the temperature remains higher than the gas condensation temperature.
   - Wet gas scrubbing removes most fuel gas contaminants (ammonia is not removed by hot gas filters but can be washed out of the gas by an acidic solution in a wet gas scrubber).

Pressurized gasifiers are well suited for hot gas cleanup (below 600°C) prior to direct combustion in the turbine. At temperatures below 600°C, alkali metals precipitate onto the particulates present in the gas and are removed by the gas cleaning system. Tar cracking is not required for pressurized gasification systems because of the elevated gas temperature (the tars are likely to be in noncondensable form). In the case of hot gas cleaning, fuel-bound nitrogen, in the form of ammonia, may cause concern regarding NO_x emissions since approximately 60% of the ammonia in the gas may be converted to NO_x during combustion. Atmospheric gasifiers generally require tar cracking or removal, cold (<200°C) gas cleaning, and compression of the gas prior to turbine combustion.

4.5.2.1 Combined Cycle Operation Single cycle efficiencies of approximately 36% and combined cycle efficiencies of 47% or higher (up to approximately 52%) are typical of natural gas-fueled plants. Integrated biomass gasification and single or combined cycle efficiencies, relative to the energy content of the fuel input, will not match the efficiencies of gas turbines fueled with natural gas because combustion temperatures are lower and part of the energy content in the biomass fuel is dissipated in the production of the fuel gas. Table IV shows notional gas and steam turbine generating efficiencies and Table V provides an indication of biomass integrated gasification/combined cycle (BIG/CC) efficiencies.

4.5.3 Comparison with Biomass Direct Combustion

Gasification has a number of advantages compared to direct combustion, particularly at the relatively

TABLE IV

Notional Generating Equipment Efficiencies[a]

Equipment	Efficiency (%)
GT running on NG	36
GT running on LCV gas	31
ST generating system (>50 MW_e)	35
ST generating system (<50 MW_e)	15–35

[a]Abbreviations used: GT, gas turbine; NG, natural gas; LCV, low calorific value; ST, steam turbine.

TABLE V

Notional BIG/CC Electrical Efficiencies

	Capacity (%)	
	<100 MW_e	>100 MW_e
Atmospheric system	43	50
Pressurized system	46	53

small scales typical of biomass-to-energy systems. Engines and single or combined cycle gas turbines are likely to have higher electrical conversion efficiencies compared to steam cycles. At small scale, steam turbines have lower efficiencies compared to engines and gas turbines, and at a larger scale gasification offers the possibility of combined cycle operation. Also, direct combustion and steam turbine systems are characterized by important economies of scale, possibly making gasification systems coupled with engines and gas turbines a more likely economically viable option at small to medium scale (from a few hundred kilowatts to a few tens of megawatts).

4.5.4 Comparison with Coal Gasification

Coal has a number of advantages and disadvantages with respect to biomass. On the one hand, it is easier and less costly to handle and process; it is easier to feed, particularly in pressurized systems; and it may require less complex supply logistics. On the other hand, coal has a much lower level of volatiles than biomass (typically 30% compared to more than 70% for biomass); its char is significantly less reactive than biomass char, implying greater reactor sizes, residence times, and the use of oxygen as a gasifying agent; it possesses a higher ash content than biomass (the pro and it often contains significant quantities of sulfur, requiring more complex and costly gas cleaning systems. Coal also has a major environmental disadvantage with respect to biomass in that it produces large CO_2 emissions, contributing to global warming.

4.6 Waste Disposal

Wastewater and solid waste result from power plant operation and require treatment prior to disposal or recycling. Liquid effluents resulting, for example, from wet gas scrubbing and flue gas condensation must be treated. The wastewater treatment is conventional, although oxygenated organics such as phenols (derived from tars) and ammonia may create problems. Systems using hot gas filtration reduce liquid effluents and may potentially reduce costs and environmental impacts.

Solid residues consist of inert ash. The ash is likely to be disposed of conventionally through landfilling or recycled, for example, as a soil nutrient. Care must be taken, however, because in some cases dust from fly ash may be toxic due to the absorption of chemicals such as benzo[*a*]pyrene. Also, heavy metal concentrations in the ash may in some cases be too high for direct recycling of the ash. Research programs in Sweden have obtained promising results with regard to ash recycling from wood gasification and combustion systems. In most cases, ash can be recycled without any processing other than wetting and crushing it.

5. ECONOMICS OF BIOMASS GASIFICATION FOR HEAT AND ELECTRICITY GENERATION

The economics of biomass gasification for electricity and heat generation is the principal factor influencing its future market penetration. The Costs of semi-commercial gasification systems integrated with engines at a scale of approximately 300 kW_e are estimated to be approximately \$1800/kW_e. Costs could be reduced by approximately 20% with continued commercialization. Larger scale BIG/CC systems are in the demonstration stage and characterized by high costs in the range of \$5000–6,000/kW_e. However, commercialization of the technology is expected to reduce systems costs to \$1500/kW_e for plants with capacities between 30 and 60 MW_e.

Capital costs alone do not provide sufficient basis for comparison between different technologies and energy sources. Variable costs, particularly fuel costs, are of key importance. These are also affected by the efficiency of the plants. Biomass fuel costs can vary considerably depending on type and location. For biomass energy to be competitive, an indicative cost of biomass of $2/GJ or less is often cited. However, biomass costs may range from negative (i.e., in the case of waste products) to significantly in excess of the suggested cost.

6. CONSTRAINTS AND ISSUES AFFECTING THE DEVELOPMENT OF BIOMASS GASIFICATION SYSTEMS

Gasification systems for power generation are in the semicommercial or demonstration stage, and their market development will depend on a number of factors ranging from technical developments to energy market structure and government policies with regard, for example, to energy security and the environment.

A number of technical issues must be addressed with regard to gasification systems. Demonstration projects are valuable in this respect and technical problems do not appear to be a major obstacle to the development of the systems. More operating experience is required, particularly with different types of biomass fuels.

Gasification systems are characterized by relatively high investment costs and, in some cases, high biomass fuel costs. However, gasification offers advantages in terms of scale, efficiency, and emissions that make it an attractive renewable energy technology. Technology development, replication, learning by doing, and competition will lead to cost reductions. The identification of opportunities for early market penetration is key for the successful commercialization of the technology.

Efficient biomass supply logistics and low-cost biomass resources are very important in the development of biomass energy systems. The issue of security of fuel supply (i.e., ensuring a fuel supply for the lifetime of the plant) needs to be addressed and is of particular importance in the case of energy crops.

The regulatory environment is also a key factor in the successful development of biomass gasification facilities. Regulations that facilitate decentralized power generation are necessary. Also, regulations or incentives aimed at reducing emissions may favor integrated gasification systems, which are likely to be characterized by low emission levels of regulated pollutants and no CO_2 emissions from the conversion of biomass.

Biomass gasification may also provide a source of clean transport fuels ranging from synthetic diesel to hydrogen.

SEE ALSO THE FOLLOWING ARTICLES

Biodiesel Fuels • Biomass, Chemicals from • Biomass Combustion • Biomass for Renewable Energy and Fuels • Biomass: Impact on Carbon Cycle and Greenhouse Gas Emissions • Biomass Resource Assessment • Combustion and Thermochemistry • Renewable Energy, Taxonomic Overview

Further Reading

Bridgwater, A. V. (1995). The technical and economic feasibility of biomass gasification for power generation. *Fuel* **74**(5), 631–653.

Bridgwater, A. V., and Evans, G. D. (1993). "An Assessment of Thermochemical Conversion Systems for Processing Biomass and Refuse." Energy Technology Support Unit, UK.

Brown, A. E., and van den Heuvel, E. J. M. T. (1996). "Producer Gas Quality Requirements for IGCC Gas Turbine Use—A State-of-the-Art Review." Biomass Technology Group, Enschede, The Netherlands.

Consonni, S., and Larson, E. D. (1996). Biomass gasifier/aeroderivative gas turbine combined cycles: Part B—Performance calculations and economic assessment. *Trans. ASME* **118**.

International Energy Agency Bioenergy Program. www.ieabioenergy.com.

Prabhu, E., and Tiangco, V. (1999). The Flex-Microturbine for biomass gases—A progress report. In "Biomass: A Growth Opportunity in Green Energy and Value-Added Products. Proceedings of the 4th Biomass Conference of the Americas, August 29–September 2, Oakland CA" (R. Overend, *et al.*, Eds.). Elsevier.

Stassen, H. E. (1995). Small-scale biomass gasifiers for heat and power: A global review, World Bank Technical Paper No. 296. World Bank, Washington, DC.

Sydkraft. (1998). Värnamo demonstration plant: A demonstration plant for biofuel-fired combined heat and power generation based on pressurised gasification—Construction and commissioning 1991–1996. Sydkraft/Elforsk/Nutek, Sweden.

Wilen, C., and Kurkela, E. (1997). Gasification of biomass for energy production—State of the technology in Finland and global market perspectives, Research Note No. 1842. Technical Research Centre of Finland (VTT), Espoo, Finland.

Biomass: Impact on Carbon Cycle and Greenhouse Gas Emissions

CARLY GREEN and KENNETH A. BYRNE
University College Dublin
Dublin, Ireland

1. Biomass and Carbon Cycling
2. Climate Change and International Policy
3. Bioenergy and CO_2 Emissions
4. Bioenergy Sources
5. Managing Biomass as a Carbon Sink
6. Potential Role of Wood Products
7. Increasing Global Importance
8. Concluding Remarks

Glossary

afforestation Direct, human-induced conversion of land that has not been forested for a period of at least 50 years to forested land through planting, seeding, or the human induced promotion of natural seed sources.

autotrophic respiration Carbon loss as a result of internal plant metabolism, which typically amounts to half of that fixed through photosynthesis.

bioenergy Renewable energy produced from biomass. Biomass is organic nonfossil material of biological origin, for example, forest residues, wood and wood wastes, agricultural crops and wastes, municipal and industrial wastes. Biomass may be used directly as a fuel or processed into solid, liquid, or gaseous biomass fuels.

biological carbon sequestration The uptake and storage of carbon in the biosphere. Trees and other plants absorb carbon dioxide through photosynthesis, release the oxygen, and store the carbon in their biomass.

carbon dioxide (CO_2) A naturally occurring gas, produced through respiration, decomposition, and combustion. The rise in atmospheric CO_2 concentration is attributed to burning of fossil fuels and deforestation. It is the principal anthropogenic greenhouse gas that affects the earth's temperature and it is the reference gas against which other greenhouse gases are indexed.

carbon sink Any process, activity or mechanism that removes CO_2 from the atmosphere.

carbon source Any process or activity that releases CO_2 to the atmosphere.

CO_2 equivalent A measure used to compare the emissions of different greenhouse gases based upon their global warming potential (GWP). The CO_2 equivalent for a non-CO_2 greenhouse gas is calculated by multiplying the quantity of the gas by the associated GWP. For example, the GWP for carbon dioxide is 1 and for methane is 21. Therefore 1 ton of methane is equivalent to 21 tons of carbon dioxide.

CO_2 flux The rate of transfer of CO_2 between carbon reservoirs. Positive values are fluxes to the atmosphere; negative values represent removal from the atmosphere.

coppice A method of regenerating a stand in which trees in the previous stand are cut and the majority of new growth is associated with vigorous sprouting from stumps, roots, or root suckers.

gigatonne C (GtC) The unit of measure for stock and fluxes of carbon in the global carbon cycle. 1 GtC = 1000 million tonnes of carbon.

greenhouse gas (GHG) Atmospheric gases, both natural and anthropogenic, that absorb and reemit infrared radiation. Greenhouse gases include water vapor, carbon dioxide (CO_2), methane (CH_4), nitrous oxide (N_2O), halogenated fluorocarbons (HCFCs), ozone (O_3), perfluorinated carbons (PFCs), and hydrofluorocarbons (HFCs).

net primary production (NPP) The difference between photosynthesis and autotrophic respiration and is referred to as net primary production (NPP). It constitutes the total annual growth increment (both above and below ground) plus the amounts grown and shed in senescence (the growth phase from full maturity to death), reproduction or death of short-lived individuals, plus the amounts consumed by herbivores.

reference energy system Used as a base case in a life-cycle assessment, a reference energy system should consider how electricity, heat, or transportation fuels would have been produced in the absence of the bioenergy system.

reservoir A component of the climate system where a greenhouse gas is stored.

thinning Removal of trees to reduce stand density primarily to concentrate growth on fewer stems, thereby increasing production of high-value timber.

Large amounts of carbon are stored in the earth's biosphere as biomass. Humans have long made use of the various forms of biomass carbon for energy; however, measures to mitigate the enhanced greenhouse effect have stimulated renewed interest in its utilization. The role of biomass in modern energy production and its carbon sink potential can significantly influence the contemporary global carbon cycle resulting in potential reductions in greenhouse gas (GHG) emissions to the atmosphere. The various application of biomass in this context and its potential contribution to stabilizing GHG emissions into the future is explored.

1. *BIOMASS AND CARBON CYCLING*

Biomass accumulates in the earth's biosphere in vegetation following the process of photosynthesis. Powered by energy from sunlight, plants use carbon dioxide (CO_2) from the atmosphere and water from the soil to manufacture carbohydrates, a storable energy source. Oxygen (O_2) and water vapor are released to the atmosphere as by-products. CO_2 is cycled back to the atmosphere as the plant material decays, is respired or is burned, and therefore becomes available again for photosynthesis.

The average annual amount of solar energy potentially available at the earth's surface is estimated at $180\,Wm^{-2}$. The proportion available for photosynthesis varies both temporally and spatially, being dependent on latitude and season. The photosynthetic conversion efficiency of land plants is estimated at 3.3 to 6.3%, varying between plant species. Taking these constraints into consideration, the energy stored in biomass globally is substantial with photosynthetic carbon fixation comprising a major component of the global carbon cycle.

1.1 The Global Carbon Cycle

The global carbon cycle is made up of carbon reservoirs (stocks) and the dynamic transfer of carbon between them (fluxes). The complex system is summarized in a simplified carbon cycle (Fig. 1).

The total carbon stored in the earth's surface is estimated at more than 650,000,000 GtC. The majority is stored in glacial deposits and minerals, fossil fuels, and the deep oceans. Atmospheric carbon accounts for a further 750 GtC. Due to their heterogeneity, biomass carbon stocks are difficult to measure empirically. The carbon stored in soil (which includes detritus, soil organic matter, and inert carbon) are 1400 GtC with plants accounting for a further 550 GtC.

Various natural and anthropogenic processes stimulate CO_2 fluxes between the reservoirs. The overall net flux of the global carbon cycle indicates that carbon is accumulating in the atmosphere at an average rate of $3.2\,GtCyr^{-1}$. Natural carbon fluxes are driven primarily by plant photosynthesis, respiration, and decay as well as oceanic absorption and release of CO_2. Anthropogenic processes such as burning fossil fuels for energy, as well as deforestation contribute a significant net positive flux to the atmosphere.

The largest anthropogenic flux within the global carbon cycle is caused by the anthropogenic burning of fossil fuels. During the 1990s, this source was reportedly $6.3\,GtCyr^{-1}$ and is considered the main cause of large increases in atmospheric CO_2 concentrations over the past 100 to 150 years. The largest natural flux is experienced between the atmosphere and the ocean where the ocean acts as a net sink of $1.7\,GtCyr^{-1}$. The terrestrial biosphere is a slightly smaller net sink. Incorporating both natural processes and human-induced activities, the flux from atmosphere to biosphere is estimated at $1.4\,GtCyr^{-1}$, representing the annual net accumulation of carbon in the biosphere.

1.2 Net Primary Production

Net primary production (NPP) is a measure of the annual productivity of the plants in the biosphere. The 550 GtC of carbon in the global reservoir of plant biomass has a corresponding NPP of $60\,GtC\,yr^{-1}$, which indicates that globally the average carbon residence time on land is approximately 9 years with an average biomass production (NPP) of $4\,tCha^{-1}\,yr^{-1}$. However, average NPP, as reported by the Intergovernmental Panel on Climate Change (IPCC) (Table I) and calculated residence time varies greatly at ecosystem level and is dependent on plant species and management, ranging from 1 year for croplands to approximately 15 years for forests.

Production forests are comparatively stable ecosystems, experiencing a longer growth cycle than food and energy crops. Food and energy crops are usually harvested at the end of a relatively short rapid growth phase, leading to a higher average NPP per unit area. In contrast to stable forest ecosystems, the majority of the NPP associated with energy and

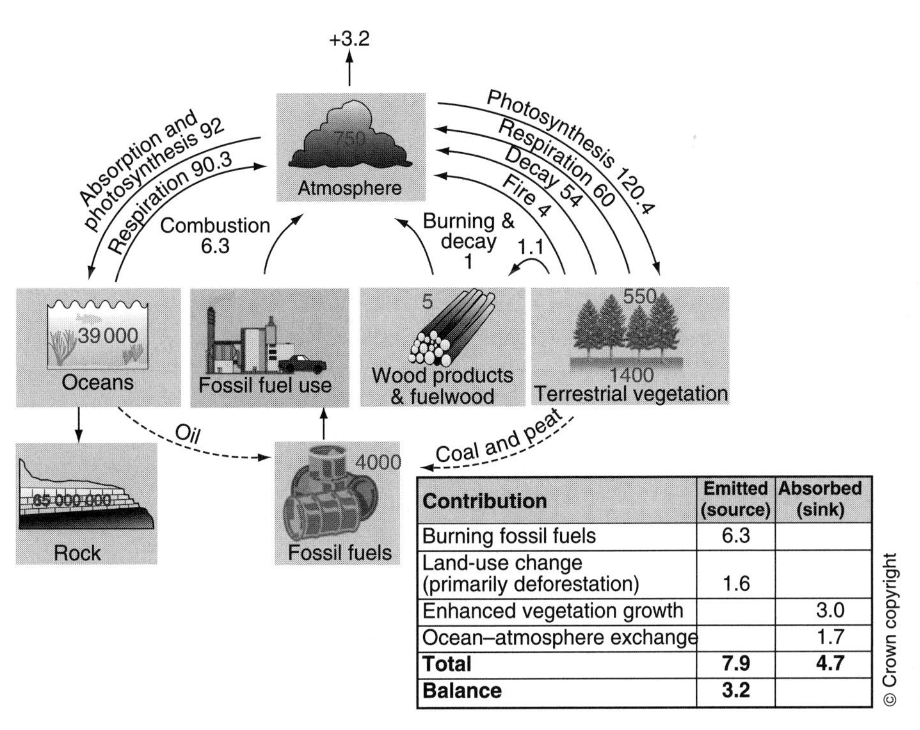

Contribution	Emitted (source)	Absorbed (sink)
Burning fossil fuels	6.3	
Land-use change (primarily deforestation)	1.6	
Enhanced vegetation growth		3.0
Ocean–atmosphere exchange		1.7
Total	**7.9**	**4.7**
Balance	**3.2**	

FIGURE 1 The simplified global carbon cycle; carbon reservoirs (GtC) represented within the boxes and associated fluxes ($GtCyr^{-1}$) indicated by solid arrows between the reservoirs. The broken lines represent long-term (millennia) movement of carbon within the system. A simplified summary of the global carbon budget is also given. Values are for the 1990s. The most important natural fluxes in terms of the atmospheric CO_2 balance are the land-atmosphere exchange ($+1.4\,GtCyr^{-1}$), which includes land-use change and in enhanced vegetation growth, and the physical ocean-atmosphere exchange ($+1.7\,GtCyr^{-1}$). Fossil fuel burning ($-6.3\,GtCyr^{-1}$) represents the largest anthropogenic release of CO_2 to the atmosphere. The net result of these processes is an increase in atmospheric CO_2 concentration of $+3.2\,GtCyr^{-1}$. From Broadmeadow and Matthews (2003).

TABLE I

Terrestrial Carbon Stock Estimates, NPP and Carbon Residence Time Globally Aggregated by Biome

Biome	Area (10^9 ha)	Global carbon stocks (GtC)	Carbon density (tC/ha)	NPP (GtC/yr)	Residence time[a]
Tropical forests	1.75	340	194	21.9	15.5
Temperate forests	1.04	139[b]	134	8.1	17.2
Boreal forests	1.37	57	42	2.6	22.1
Tropical savannas and grasslands	2.76	79	29	14.9	5.4
Temperate grasslands and Mediterranean shrublands	1.78	23	13	7.0	3.3
Deserts and semi-deserts	2.77	10	4	3.5	3.2
Tundra	0.56	2	4	0.5	4.5
Croplands	1.35	4	3	4.1	1.0
Total	14.93[c]				

Source. Modified from Mooney *et al*., 2001.
[a] Average carbon residence time = (carbon density × area)/NPP.
[b] Based on mature stand density.
[c] 1.55×10^9 ha of ice are included.

food crops ends up in products exported from the site. Soil carbon density is generally lower and may become depleted as a result.

1.3 The Available Biomass Resource

In determining the extent of the biomass resource, an understanding of the global allocation to land use is required. As the values shown in Table I are globally aggregated by biome, they do not identify competing land use functions such as areas for human settlement, recreation, and conservation of biodiversity, and ecosystem function. Estimates of the possible contribution of biomass in the future global energy supply vary from 100 to 400 $EJ\,yr^{-1}$ by 2050. The variation in estimates can be largely attributed to land availability and energy crop yield levels. In terms of GHG emission benefits, such an increase in bioenergy production could lead to an overall reduction of between 2 and 6.2 GtC per year as a result of the direct substitution of fossil fuels. Projections suggest that annual GHG emissions associated with fossil fuel combustion will exceed 11.4 GtC by the year 2050 if consumption continues in accordance with trends. Therefore biomass can potentially provide GHG emissions avoidance in the order of 17.5 to 54% by 2050.

In contrast, the potential carbon sink through vegetation management has been estimated at between 60 and 87 GtC by 2050. This is equivalent to the sequestration of 14 to 20% of the average fossil fuel emissions for the period 2000 to 2050.

2. *CLIMATE CHANGE AND INTERNATIONAL POLICY*

Climate change caused by the enhanced greenhouse effect is one of the most significant global environmental issues. Increased emissions of GHG to the atmosphere, most notably CO_2, are considered the main cause of global climate change. Increasing energy consumption, a reliance on fossil fuels to meet these needs, and deforestation related to land use change are the main sources of increasing atmospheric CO_2. The ability of biomass to make a significant contribution to the stabilization of atmospheric CO_2 concentrations through its utilization as an energy source and the preservation of terrestrial biomass stocks has lead to its inclusion in global policy developments.

2.1 Global Environmental Policy

Global concern over the increase in atmospheric GHG concentrations and the acknowledgment that human activity is leading to climate change led to nations recognizing the need for international effort. The United Nations Framework Convention on Climate Change (UNFCCC) was agreed at the Rio Earth Summit in 1992 and entered into force in 1994. Parties agreed to report on GHG emissions and to take action to mitigate climate change by limiting those emissions and to protect and enhance GHG sinks and reservoirs. The Kyoto Protocol to the UNFCCC, developed in 1997, set the relatively short-term target for developed countries and countries with economies in transition to stabilize their combined GHG emissions, in CO_2 equivalents, to 5.2% below 1990 levels during the first commitment period 2008–2012. Actions that enhance carbon stock in biomass and soil, such as reforestation and altered management in forestry and agriculture, can be used as credits to offset GHG emissions to meet commitments under the Kyoto Protocol.

Biomass can address the issue of global climate change in three ways: (1) bioenergy produced from biomass can replace fossil fuels, directly reducing emissions; (2) biomass can sequester carbon as it is growing, increasing the terrestrial carbon stock; and (3) biomass products can substitute for more energy intensive building products, reducing the emissions associated with production of these materials and additionally increasing the carbon stock in wood products.

Offsetting fossil fuel emissions with sequestration in biomass and reducing fossil fuel emissions through substitution of renewable bioenergy are likely to be key factors in the ability of countries to achieve their reduction commitments under the Kyoto Protocol.

2.2 Carbon Neutrality

Bioenergy produced from biomass is sometimes called a carbon-neutral energy source, because the same quantity of carbon released when the biomass is burned is sequestered again when the crop or forest is regrown (Fig. 2). Referring to bioenergy as carbon neutral or having zero net emissions may be misleading; there are emissions associated with producing the biomass, such as from fossil fuel used in cultivation, harvest, processing and transport, and in manufacture and construction of fuel conversion technology. Then there are less obvious emissions, such as the fossil fuel emissions in producing fertilizer and

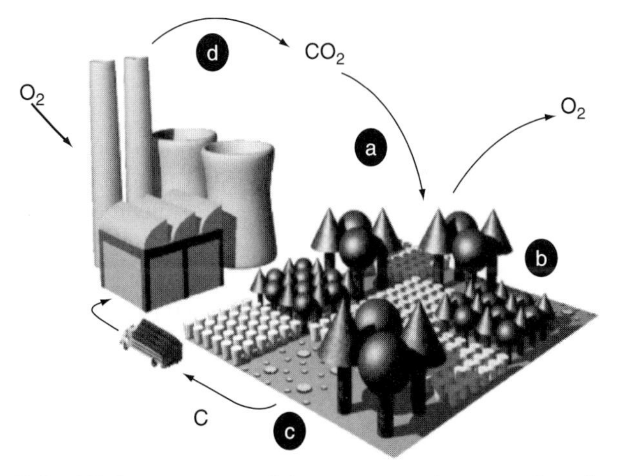

FIGURE 2 Illustration of the apparent carbon neutrality of a bioenergy system. (a) CO_2 is captured by the growing crops and forests; (b) oxygen (O_2) is released and carbon (C) is stored in the biomass of the plants; (c) carbon in harvested biomass is transported to the power station; (d) the power station burns the biomass, releasing the CO_2 captured by the plants back to the atmosphere. Considering the process cycle as a whole, there are no net CO_2 emissions from burning the biomass. From Matthews and Robertson (2002).

herbicide. Although it is clearly not a zero-emissions process, the greenhouse impact of bioenergy is much lower than that of fossil fuel energy sources.

The production, processing, transport, and conversion of fossil fuels also requires additional fossil fuel inputs; the significant difference between fossil fuels and biomass in terms of GHG emissions is the ability of biomass to "recycle" the emissions released in fuel conversion acting as a closed loop process.

GHG emissions and removals are reported to the United Nations Framework Convention on Climate Change (UNFCCC) by sector. Fuel use is reported separately from agricultural activities and forestry removals. The emissions associated with particular forestry and agricultural operations are not explicit in national GHG inventories; nevertheless, they are reported in consumption figures. Recommendations for future reporting suggest that full life cycle accounting be undertaken for projects, so that credits will be earned only for net removals. Therefore policy is being formulated that will make such auxiliary emissions completely transparent.

3. BIOENERGY AND CO_2 EMISSIONS

Substituting fossil fuels with renewable energy sources offers a significant opportunity for reducing GHG emissions. Renewable bioenergy has a number of benefits over other renewable technologies such as wind and solar as it is capable of supplying the full range of energy markets (i.e., heat, electricity and transport fuel).

3.1 Role of Bioenergy

Biomass has the longest history of any energy source and still provides approximately 44 EJyr^{-1} to meet 11% of the worlds primary energy needs. Most of this is utilized in developing countries where traditional low efficiency applications of fuel wood for cooking and heating account for 38 EJ. The remaining 6 EJ is utilized in the developed countries for heat and power generation with modern technologies.

In the past, the most common way to capture energy from biomass was through combustion. However, advancements in technology mean that more efficient and cleaner methods for energy conversion are readily available. Solid, liquid, and gaseous biofuels can replace fossil fuels in almost every application.

The extent and way that biofuels are utilized varies greatly. Only 3% of the total primary energy needs in industrialized counties are supplied by biomass compared, with over 40% in the developing world. A comparison of the contribution that biomass makes to various countries' energy requirements, Table II shows a wide range, from 1% in Japan, Ireland, and The Netherlands to 94% in Tanzania and Ethiopia. Of the industrialized countries, Finland, Sweden, and Austria source relatively high levels of their energy requirements from bioenergy, which can be largely attributed to the use of wood chips, industrial wood waste, and straw as fuel for district heating.

3.2 Bioenergy Life Cycles

The quantification of the actual reduction in GHG emissions resulting from the substitution of fossil fuels with biomass requires a complete lifecycle assessment (LCA). A systematic framework for estimating the net GHG emissions from bioenergy systems and comparing them against the fossil fuel reference system that it would replace has been developed (Fig. 3). The major considerations of the life cycle assessment approach to quantifying the greenhouse impacts of bioenergy are as follows:

- *Carbon stock dynamics*. Changes in the carbon stock of plants, plant debris, and soils result from

TABLE II

Bioenergy Consumption and Resulting Greenhouse Gas Benefits

Country	Total energy consumption (TJ)	Bioenergy consumption (TJ)	Energy supply share (%)	Emission avoidance ($GtCO_2$)[a]
Austria	1162	104	9	7
Japan	21558	312	1	20
Sweden	2174	348	16	22
Finland	1385	246	18	18
Ireland	523	7	1	0
Netherlands	3136	41	1	3
Australia	4255	219	5	14
United States	90527	2841	3	183
Brazil	7203	1693	24	109
China	46010	8720	19	562
Egypt	1657	52	3	3
India	19302	8072	42	520
Mozambique	321	290	90	19
Tanzania	597	561	94	36
Ethiopia	717	675	94	43

[a]Emissions factor = 64.4 $kgCO_2/GJ$. Values are for 1997. (From World Resources Institute, 2003.)

biomass growth and harvest. This necessitates time-dependent analysis, as such changes might extend over long periods of time until a new equilibrium is reached.

• *Trade-offs and synergies*. The trade-off between competing land uses, namely biomass production, food production, and carbon sequestration, should be considered in determining the benefits of bioenergy. A timber plantation, thinned to maximize value of wood products, where thinning residues are used for bioenergy, is an example of synergy.

• *Leakage*. The use of biomass fuels does not always avoid the use of fossil fuels to the full extent suggested by the amount of bioenergy actually used and therefore the reduction in fossil fuel use that can be attributed to the project will be reduced. This is commonly referred to as leakage.

• *Permanence*. Irreversible GHG mitigation offered by the system is required to permanently offset the CO_2 emissions. The greenhouse mitigation benefit from substituting bioenergy for fossil fuel use is irreversible. In contrast, the mitigation benefit of reforestation will be lost if the forest biomass is reduced by harvest or natural disturbances.

• *Emissions factors*. The quantity of GHG produced per unit of fossil fuel energy consumed. This factor influences the net benefit of a bioenergy project: if bioenergy displaces natural gas, the benefit is lower than if it displaces coal, as coal has a higher emissions factor than natural gas.

• *By-products*. The emissions and offsets associated with both products and by-products of a biomass production system must be considered.

• *Efficiency*. In determining the quantity of fossil fuel energy displaced by a quantity of biomass feedstock, it is important to know the energy output per unit of that feedstock.

• *Upstream and downstream GHG emissions*. Auxiliary (energy inputs) emissions resulting from the bioenergy and reference systems should be considered, such as from fossil fuels used in production/extraction, processing, and transport of the feedstock.

• *Other greenhouse gases*. The emissions of other GHG associated with biomass and fossil fuel chains, such as CH_4 and N_2O, expressed in CO_2 equivalents, should be included in the calculation of net greenhouse mitigation benefits.

LCA assessments have been applied to a variety of fuel systems to generate an understanding of their contribution to atmospheric CO_2 concentrations. For example, full fuel cycle CO_2 emissions per GWh for coal have been calculated as 1142 t compared to 66 to 107 t for biomass.

4. BIOENERGY SOURCES

Biomass used directly or indirectly for the production of bioenergy can be sourced from any organic matter,

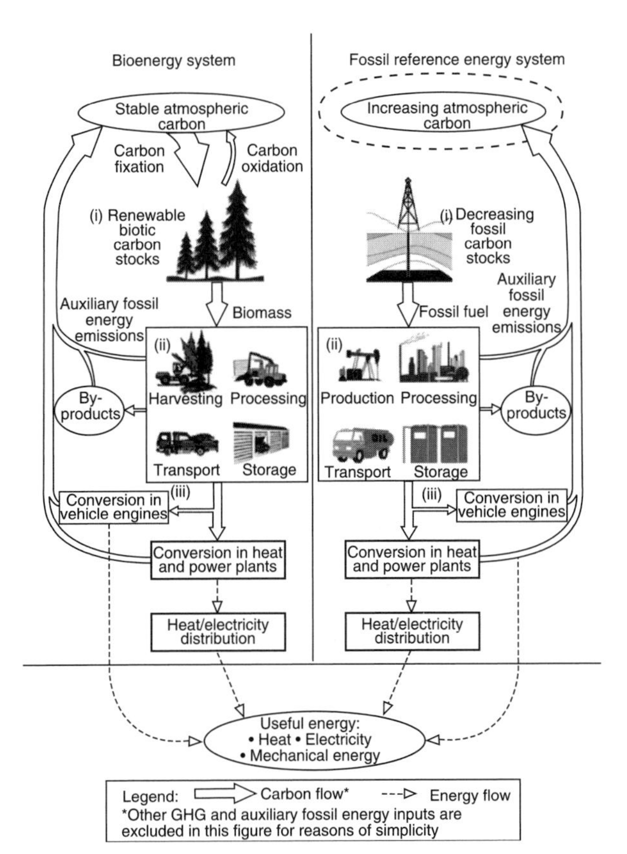

FIGURE 3 Life cycle assessment (LCA) approach for calculation of GHG balances. The components of the comparison between the (a) Bioenergy system and (b) fossil reference system to reach the (c) useful energy stage are shown. The system boundary incorporates the (i) carbon flows related to the fuel source *in situ*, (ii) carbon flows involved in supplying the fuel for energy conversion including any by-products or auxiliary energy requirements, and (iii) carbon and energy flows associated with conversion technologies to useful energy. Reproduced from IEA Bioenergy (2002).

of plant or animal origin, and can be categorized into two types: dependent or dedicated biomass resources. Dependent biomass resources are by-products of other activities (e.g., wood residues and recovered wood waste, straw, forestry residues, livestock manure, sugarcane, and recoverable waste cooking oil). Dedicated biomass resources are energy crops grown specifically for fuel (e.g., short rotation tree crops, coppice, herbaceous grasses, and whole crop cereals).

4.1 Bioenergy from Dependent Resources

Agricultural and forestry residues provide the largest proportion of biomass used for the production of bioenergy. Some estimates suggest that globally available biomass in the form of recoverable residues represents about 40 $EJyr^{-1}$, enough to meet 10% of the total present energy use of 406 $EJyr^{-1}$. However, realizing this potential is limited by factors such as ease and cost of recovery and environmental concerns relating to sustainable land use practices.

In developed countries, sawmill and pulp mill residues comprise the largest proportion of residues in the existing bioenergy mix. Industries make use of recovered operational waste such as sawdust and off cuts in timber mills and black liquor in pulp and paper production to supply process energy needs. Additionally, techniques such as densification can produce a usable fuel from sawmill sawdust and bark as well as shavings from furniture manufacturing.

Substantial opportunities exist in the forestry sector. Forest management residues including branches, tree tops, and small diameter stems from thinning and harvest of commercial plantations are a potentially large source of biomass, which is underutilized. Full utilization of thinnings, which may represent approximately 25% of the biomass produced, permits energy recovery from a resource that may otherwise decay on the forest floor. In both cases, the carbon in the biomass is returned to atmosphere, but where bioenergy is produced, greenhouse mitigation benefit is obtained from displacement of fossil fuel emissions. Technological advancements are being made in this area, making use of bundling technology for the densification of harvest residues. This process, largely being developed in Finland, involves recovery of the residues and bundling at source to produce an easily transportable and storable fuel.

Used vegetable oils and animal fats can be reprocessed into biodiesel. Biodiesel has similar physical properties to conventional diesel to which it can be blended or alternatively used in its pure form. Biodiesel is commercially available at service stations in countries such as the United States, Brazil, France, Germany, and Austria. In 1996, the International Energy Agency (IEA) reported that 21 countries produced some 591,000 ton of biodiesel. Direct substitution of diesel oil would result in the mitigation of over 500,000 ton of CO_2 emissions. As approximately 20% of global CO_2 emissions can be attributed to the transportation sector, further development of biofuels, including biodiesel, can contribute significantly to reduction in GHG emissions from this sector.

Wet waste streams such as domestic, agricultural, and industrial sludge and animal manure emit methane (CH_4), as a product of anaerobic decomposition of organic matter. CH_4 has a greenhouse warming potential 21 times higher than that of CO_2 and therefore, avoidance of CH_4 emissions produces significant greenhouse benefits. Anaerobic digestion is a waste management and energy recovery process highly suited to wet waste streams. The bacterial fermentation of organic material produces biogas (65% CH_4, 35% CO_2), which is captured and utilized to produce heat and power. Although anaerobic digestion of sewage, industrial sludge and wastewater are fully commercialized, systems involving the processing of animal manure and organic wastes are still being developed. Technology can generate approximately 200 kilowatt-hours (kWh) of electricity from 1 ton of organic waste. If this was used to replace electricity from coal production, direct GHG emissions could be reduced by approximately 220 $kgCkWh^{-1}$.

The energy contribution of animal sludge and municipal solid waste globally was estimated at 5300 to 6300 MW in 1995, 95% of which can be attributed to Asia. Globally, anaerobic digestion of organic wastes is predicted to increase to 8915 to 20130 MW in 2010. This projection is largely attributed to the developed world increasing its contribution.

Energy production from solid municipal waste offers a potentially sustainable approach to a growing environmental problem. Such sources include urban wood waste (i.e., furniture, crates, pallets) and the biomass portion of garbage (i.e., paper, food, green waste.). The biogas emitted when waste decomposes in landfill can be extracted for use in the generation of electricity. Maximizing the potential of such waste streams is hampered by logistical issues relating to recovery, negative public perception of "incineration," and opposition from environmental groups to the establishment of a market for waste.

4.2 Bioenergy from Dedicated Resources

The future development of energy crops, to the level at which they would replace residues as the major bioenergy fuel source, will be largely dependent on regional factors such as climate and local energy requirements and emission factors, which will determine their environmental and financial viability.

Energy production per hectare is determined by crop yields. This is in turn dependent on climate, soil, and management practices in addition to the number of harvests per year, the fraction of the biomass harvested, and the usable fraction of the harvest. The conversion efficiency to a useable energy form is a further important consideration.

Generally perennial crops produce higher net energy yields than annual crops, due to the lower management inputs, and therefore lower production costs, both economic and ecological. Management techniques, fertilizer application, and genetic improvements may increase the productivity by a factor of 10 compared with no inputs; however, this can result in increased environmental problems. Eutrophication of water can be associated with fertilizer runoff; soil erosion with intensive cultivation, and concerns regarding consequences for biodiversity with genetically improved crops.

There is sometimes a misconception that it takes more energy to produce biomass for bioenergy than is generated from the biomass itself. However, it has been repeatedly shown that such systems are capable of producing significant energy ratios (Table III).

For example, the energy production per hectare of woody biomass from a commercial plantation is 20 to 30 times greater than the energy input for agricultural operations such as fertilizer and harvesting. However, liquid fuel production from energy crops may experience conversion rates that are substantially less. In Europe and the United States for example, achieving a high-energy output per unit input requires careful management at the production stage. Ethanol produced in the U.S. yields about 50% more energy than it takes to produce, whereas in Brazil where ethanol is produced from sugarcane, higher yields are possible largely as a result of greater rates of biomass growth and lower auxiliary energy inputs.

4.3 Modernizing Bioenergy

Estimation of the future technical potential of biomass as an energy source is dependent on assumptions with respect to land availability and productivity as well as conversion technologies. With the emergence of energy crops as the major source of biomass fuel, land use conflicts, especially in relation to food production, may arise. However, with efficient agricultural practices, plantations and crops could supply a large proportion of energy needs, with residues playing a smaller role without compromising food production or further intensifying agricultural practices.

Considering future projections of increasing energy and food demands associated with increasing standard of living and population, pressure on the availability of land is likely to increase. In the European Union (EU) it has been proposed that

TABLE III

Current and Feasible Biomass Productivity, Energy Ratios, and Energy Yields for Various Crops and Conditions

Crop and conditions	Yield (dry $t\,ha^{-1}yr^{-1}$)	Energy ratio[a]	Net energy yield[b] ($GJ\,ha^{-1}yr^{-1}$)
Short rotation woody crops (willow, hybrid poplar; United States, Europe)			
Current	10–12	10:1	180–200
Feasible	12–15	20:1	220–260
Miscanthus/ switchgrass			
Current	10–12	12:1	180–200
Feasible	12–15	20:1	220–260
Sugarcane (Brazil, Zambia)	15–20	18:1	400–500
Wood (commercial forestry)	1–4	20/30:1	30–80
Sugar beet (northwest Europe)			
Current	10–16	10:1	30–100
Feasible	16–21	20:1	140–200
Rapeseed (including straw yields; northwest Europe)			
Current	4–7	4:1	50–90
Feasible	7–10	10:1	100–170

[a]Energy ratio = Unit of bioenergy produced per unit of fossil fuel input. Fuel for processing and transporting the biomass is excluded, except for the ration of sugarcane in Brazil and Zambia.

[b]Net energy yield = Energy output − Energy input (i.e., agricultural operations, fertilizer, and harvest operations).

From World Energy Assessment (2002).

biomass could contribute an additional 3.8 EJ annually by 2010, compared with the contribution of 1.9 EJ. Energy crops are expected to provide this extra capacity on 4% of the total land area of the EU. If all of this 1.9 EJ replaced energy generated from coal in Western Europe, reductions of net CO_2 in the order of 50 $MtCyr^{-1}$ or more than 7% of the current anthropogenic CO_2 emissions from the region could be expected.

Heat and power generation through combustion is still the most common method of generating bioenergy, using the same technology used in power plants burning solid fossil fuels. Co-firing biomass with coal in existing electricity generation plants is becoming more widespread and may be one of the most promising near-term options for increasing market share of bioenergy. With low capital input, fuels such as waste wood, crop debris, forestry residues, and agricultural wastes can be mixed with coal to reduce net GHG emissions. The technique is being developed in the European Union and the United States where biomass has been successfully co-fired with coal at around 10% without compromising efficiency. If 10% of the 50,000 PJ of electricity generated each year from coal were to be replaced with biomass in only 10% of installations, approximately 350 $MtCO_2yr^{-1}$ emitted to the atmosphere could be offset. Co-firing offers additional environmental benefits associated with reduced NO_x and SO_x emissions from conventional generating plants.

Technological advancements in the production of liquid biofuels such as pyrolysis are at the pilot stage, and new processes for the production of ethanol from woody biomass have been developed, however, they are not yet commercial. In developing countries, efficiency of conversion technologies is also being increased. For example, in India, traditional cooking fires are being replaced with improved higher efficiency cooking stoves, and biogas produced from anaerobic digestion of animal and other wastes is now used for cooking, lighting, and power generation. Modernizing biomass production and bioenergy conversion technologies will transform this traditional energy source to one that is feasible and widely acceptable and ultimately offer significant GHG reductions.

Although the majority of energy sourced from biomass is converted through combustion, and some through liquification to alcohol, the adoption of new technologies such as gasification can improve the power production efficiency of biomass. Increasing efficiencies in conversion and the global economy will determine to what extent biofuels can displace CO_2 emissions into the future.

5. MANAGING BIOMASS AS A CARBON SINK

Another possible role for biomass as a means for mitigating atmospheric GHG is through carbon sequestration in the growing biomass and soil. The rapid rates of assimilation of CO_2 by vegetation provide opportunities for sequestering carbon in the relatively short term. Forests can capture and retain a more significant stock of carbon in biomass (40 to

120 $tCha^{-1}$) compared with grassland (7 to 29 $tCha^{-1}$) or cropland (3 $tCha^{-1}$), underlying the importance of the role of forests preserved as a carbon reservoir.

Afforestation and reforestation activities have a role in reducing atmospheric CO_2 concentrations. The Kyoto Protocol requires countries to report such activities and allows carbon stock increases due to post-1990 afforestation and reforestation to be offset against emissions. If the forest is maintained, it does not continue to sequester CO_2 indefinitely; however, it provides a perpetual sink for the quantity of carbon represented by the difference in average carbon density between the previous and the forested land use. Utilizing forests to sequester carbon provides some relatively short-term solutions to rising atmospheric CO_2 concentrations.

5.1 Carbon Dynamics in Forest Ecosystems

The accumulation of carbon within a forest growth cycle can be considered in four stages. The initial establishment stage involves low carbon accumulation and may even experience net carbon loss (particularly from soil) as a result of site preparation and low biomass inputs. A rapid uptake of carbon is then experienced during the second phase, known as the full vigor stage, which subsequently levels off as the stand reaches the mature stage. Finally, the forest reaches old growth and the carbon is in steady state with accumulation associated with new growth balanced by mortality and disturbances.

Modeling management for forest conservation in the United Kingdom indicates that the maximum accumulation of carbon, and the subsequent removal of CO_2 from the atmosphere, over the life cycle of the stand is approximately 200 $tCha^{-1}$ (Fig. 4a). Forest stands managed for commercial production experience periodic harvesting and generally have lower carbon stocks than stands that are not harvested. Under such conditions, commercial stands rarely experience growth phases past the full vigor phase. Over a typical 50-year rotation, commercially managed Sitka spruce stands yield average carbon accumulation rates of 70 $tCha^{-1}$ (Fig. 4b). Although these figures indicate that the carbon stock increases even under periodic harvesting, the longer term C stock depends on a balance between the impacts of harvesting and the rate of forest regeneration. Consideration of the long-term average C stocks in these two examples indicates that even though young, fast-growing trees experience high sequestration rates, over the longer term, average carbon stocks are higher in old growth forests.

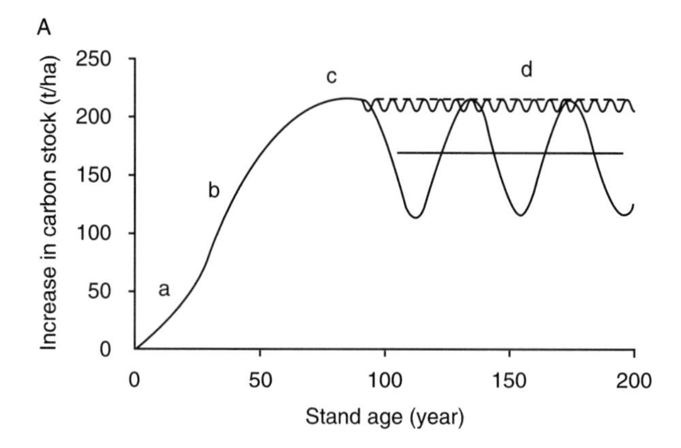

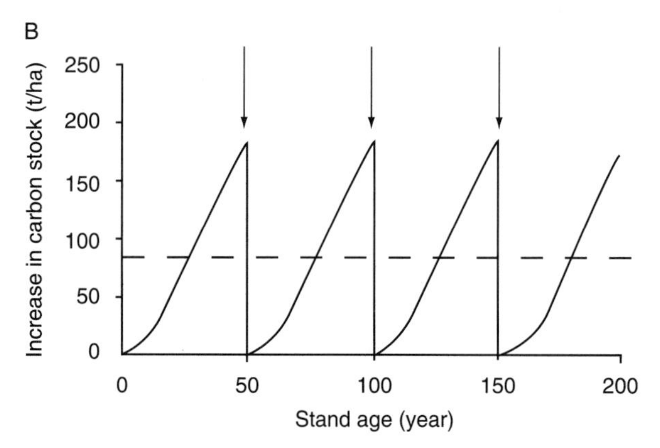

FIGURE 4 (A) Carbon accumulation in a newly created stand of trees managed as a carbon sink, based on an average stand of Sitka Spruce in Britain. Four stages of growth are shown: (a) establishment phase, (b) full-vigor phase, (c) mature phase, (d) long-term equilibrium phase. Following an initial increase in carbon stock due to the establishment of the stand, carbon stocks neither increase nor decrease because accumulation of carbon in growing trees is balanced by losses due to natural disturbances and oxidation of dead wood onsite. Two examples of carbon dynamics with low (dotted line) and high (dashed line) long-term equilibrium stocks are illustrated. Carbon dynamics in soil, litter, and coarse woody debris are ignored. (B) Carbon accumulation in a newly created commercial forest stand. Periodically the stand is felled (indicated by vertical arrows) and subsequently replanted. Over several rotations, carbon stocks neither increase nor decrease because accumulation of carbon in growing trees is balanced by removals due to harvesting. In practice, a forest usually consists of many stands, all established and harvested at different times. Averaged over a whole forest, therefore, the accumulation of carbon stocks is more likely to resemble the time-averaged projection shown as a dashed line. Carbon dynamics in soil, litter, coarse woody debris, and wood products are ignored. From Matthews and Robertson (2002).

Based on these two different management examples, between 5000 and 14,000 ha of newly established forests would be required to sequester 30 years of CO_2

emissions from a 30 MW power station burning fossil fuels. In contrast, approximately 11,250 ha of dedicated energy crops could supply such a power plant and indefinitely replace fossil fuels, providing permanent GHG emission reductions.

There may be potential to promote carbon sinks globally over the next 100 years or more to the extent of sequestering 90 GtC, but increasing carbon stocks in vegetation will ultimately reach ecological or practical limits and other measures will need to be adopted. Forests have a finite capacity to remove CO_2 from the atmosphere. The provisions of the Kyoto Protocol with respect to sinks can be seen as a valuable incentive to protect and enhance carbon stock now, while possibly providing the biomass resources needed for continued substitution of fossil fuels into the future.

5.2 Permanence

If a forest is harvested and not replanted (deforestation) or is permanently lost as a result of fire or disease, then the carbon reservoir is lost. Whatever safeguards may be in place to protect against future loss of stored carbon, it is impossible to guarantee indefinite protection, and therefore any emissions reductions are potentially reversible and therefore not guaranteed to be permanent.

For a sink to offset the emissions from the factory, the forest would have to be maintained in perpetuity. The issue of permanence has become increasingly prominent in discussions on measures to reduce national and international GHG emissions. These discussions focus around the use of accounting systems to trace land areas ensuring that any loss in carbon storage is identified and reported. This raises the issue of intergenerational equity, where land use choice may be foreseen as a liability to future generations.

In contrast, the GHG benefits resulting from the utilization of biomass as an energy source are irreversible. Even if the bioenergy system operates for a fixed period, the offset effects are immediate and are not lost even if the energy system reverts back to fossil fuels.

6. *POTENTIAL ROLE OF WOOD PRODUCTS*

To date, most discussion and research relating to the various roles of biomass in mitigating CO_2 emissions has been focused around its use as a fuel or as a sink. However, full utilization of the potential of biomass products, particularly from woody biomass, may provide significant opportunities.

Wood products themselves are a carbon pool and can be a sink if increasing demand for wood products results in increased carbon storage. The potential carbon sink in wood products is estimated to be small compared to the carbon sink in living vegetation and biomass, or compared to the potential of wood products to displace fossil fuel consumption; however, utilizing the full range of options may enable such products to play a significant role.

Indirectly, biomass could displace fossil fuels by providing substitute products for energy intensive materials such as steel, aluminum, and plastics (Fig. 5). Although, product substitution will be dependent on practical and technically feasible applications at a local to regional level, there is a large potential to replace a range of materials in domestic and industrial applications.

The construction industry is one example where maximizing the use of biomass materials may provide GHG emissions reductions of between 30 and 85% in the construction of housing for example. However, over its life cycle, house-heating requirements contribute 90% of the total associated GHG emissions, therefore the use of bioenergy for domestic heating may provide a more significant role.

Disposal options of wood products at the end of their usable life, such as landfill or combustion, can influence CO_2 emissions. Calculations suggest that a

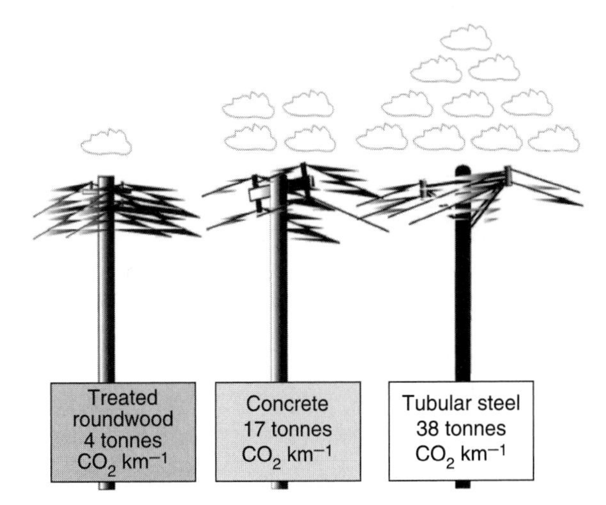

FIGURE 5 Biomass can be used as product substitutes such as in the construction of transmission lines. In CO_2-equivalents, 1 km using treated round wood would emit 4 tonnes $CO_2\,km^{-1}$, compared to concrete 17 tonnes $CO_2\,km^{-1}$ and 38 tonnes $CO_2\,km^{-1}$ for tubular steel. This analysis was based on a 60-year life cycle and includes the impact of disposal. From Matthews and Robertson (2002).

significant sink may exist should wood products be disposed to landfill. Such products decompose to a very limited extent in landfill with over 90% of the carbon apparently stored permanently. Alternatively, wood products could be collected and used as a combustible "carbon neutral" fuel.

7. INCREASING GLOBAL IMPORTANCE

The IPCC considers that increasing the use of biomass across all scenarios is important in the long-term mitigation of greenhouse gas emissions. Increasing the share of biomass in the total energy market to between 25 to 46% by 2100 could be possible. In the biomass intensive scenario (46%), cumulative emissions between 1990 and 2100 could be as low as 448 GtC compared with 1300 GtC in the business-as-usual case.

As discussed in Section 5, vegetation, particularly forests, can be managed to sequester carbon. Through global afforestation and management, the forestry sector could have the potential to sequester up to 90 GtC over the 21st century with policy developments encouraging such activity. However, each use need not compete and preventative and reactive measures to mitigating atmospheric CO_2 emissions could be jointly applied.

7.1 Possible Synergistic Opportunities

Synergistic opportunities exist, often on a regional scale, between bioenergy, wood production, and management as a carbon sink. For example, establishing a forest or crop on former degraded land can provide a source of biomass, which can be harvested for bioenergy or wood products. As a direct result of this establishment, the aboveground, belowground, and soil carbon stock may be increased to create a sink. Alternatively, management approaches aimed at increasing productivity, such as thinning a commercial forest stand, may provide specified amounts of fuel wood for bioenergy. The wood harvested from the stand could be used and reused in a number of products, being available as a fuel at the end of its useful life. Depending on the objectives, the choice of biomass crop and application is likely to be regionally specific based on environmental, economic, and social factors (Fig. 6).

In the production of biomass, high yields are important; therefore fast-growing crops and short rotation forestry are best suited, which in some cases also have a small carbon sink capacity by increasing soil carbon. Such crops are more suited to providing pulp, wood chip, and some timber production. Residues provide some options for bioenergy and a small sink capacity. However, if the intention were long-term storage of the carbon stock, maximizing the carbon ultimately attained and retaining it indefinitely, rather than the rapid capture of carbon in the initial stages, then enduring tree species and the promotion of regeneration and succession would be preferable.

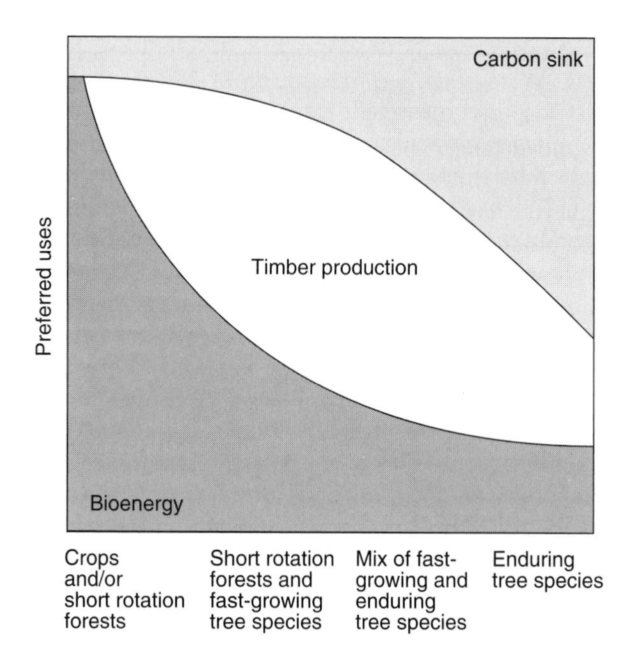

FIGURE 6 The choice of crops, type of tree species, and management regime can be selected to achieve a mix of bioenergy production, wood production, and carbon sink. From Matthews and Robertson (2002).

Although sequestration of carbon in biomass sinks may be promoted under climate change policy, a preventative approach, reducing total GHG emissions by substituting fossil fuels either directly with bioenergy or indirectly as an alternative to energy intensive materials, could yield more stable and less problematic reductions.

7.2 Balance of Benefits

The future potential of biomass to provide a substantial contribution to reducing GHG emissions will depend on a number of environmental, technical, social, and economic factors.

Land availability and competition may be the largest limiting factor to maximizing the potential of biomass. Offsetting CO_2 emissions using biomass, either through carbon storage or directly by fuel substitution, requires large areas of land. Competition, especially for food production, is likely to occur and vary widely from region to region. Large areas of degraded land may be available in developing countries, and through sustainable planning, such areas could be restored as energy crops while still encouraging biodiversity and enhancing the local environment.

Biomass is often seen as a fuel of the past and one that is associated with poverty and poor health as traditional uses have not been sustainable. However, when managed in a sustainable manner ensuring that all issues related to its practical exploitation are carefully considered, bioenergy can contribute significantly to the development of renewable energy markets in both the developed and less developed countries.

Bioenergy is competing in a heavily subsidized market. Unless cheap or negative cost wastes and residues are used, bioenergy is rarely competitive. For large-scale energy crops to make a substantial contribution to the energy market and subsequently to reducing GHG emissions, they need to be competitive. Carbon and energy taxes may encourage this transition. More efficient conversion technologies may also contribute through the provision of higher conversions at lower cost. Further development of combined heat and power as well as gasification technologies will be crucial. The development of dedicated fuel supply systems, including higher yields, greater pest resistance, better management techniques combined with reduced inputs, and improvements in harvesting, storage, and logistics will be required to lower costs and raise productivity without contributing to nutrient and organic matter depletion.

8. CONCLUDING REMARKS

As energy demands grow and pressure from international conventions and protocols lead to the requirement for lower GHG emissions, biomass may provide an important part of an integrated renewable energy future, through direct supply or as a substitute for energy intensive materials. Biomass will also provide options for increasing the biospheric sink capacity, helping to reduce atmospheric concentrations of CO_2 in the coming decades.

Forests, forest resources and energy crops can affect the global cycling of carbon in several ways. Carbon dynamics in a number of areas such as carbon storage in terrestrial reservoirs, fossil fuel displacement, auxiliary energy requirements associated with biomass management and conversion processes and the various applications of biomass products can influence the concentration of atmospheric CO_2. To fully appreciate the effect that biomass can have in terms of the carbon cycle and GHG emissions, a detailed assessment of global, regional, and local conditions taking a life-cycle approach should determine whether it is better utilized as a sink, as fuel, or as long-lived timber products.

The success of biomass, in the future of renewable energy and sustainable production, will be strongly influenced by policy and economic incentives to stimulate reductions in GHG emissions. Evaluating the options of carbon management through the use of biomass will require a number of considerations including prospective changes in land use, sustainable management practices, forest and crop productivity rates, as well as the manner and efficiency with which biomass is used. Regardless of the focus, biomass has a major role in the global energy system of the future.

Acknowledgments

Financial support for this study was provided by the International Energy Agency (IEA) Implementing Agreement Task 38—Greenhouse Gas Balances of Biomass and Bio-energy Systems. We are grateful to Annette Cowie and to Gregg Marland for reviewing the manuscript, Robert Matthews for assisting in the preparation of the figures, and for the many useful comments provided by the National Task Leaders of IEA Bioenergy Task 38.

SEE ALSO THE FOLLOWING ARTICLES

Biomass, Chemicals from • Biomass Combustion • Biomass for Renewable Energy and Fuels • Biomass Gasification • Biomass Resource Assessment • Carbon Sequestration, Terrestrial • Climate Change and Energy, Overview • Diet, Energy, and Greenhouse Gas Emissions • Greenhouse Gas Emissions, Alternative Scenarios of • Greenhouse Gas Emissions from Energy Systems, Comparison and Overview

Further Reading

Berndes, G., Hoogwijk, M., and van de Broek, R. (2003). The contribution of biomass in the future global energy supply: A review of 17 studies. *Biomass and Bioenergy* **25**(1), 1–28.

Broadmeadow, M., and Matthews, R. (2003). The Carbon Dynamics of Woodland in the UK. Information Note. UK Forestry Commission, Edinburgh.

Gustavsson, L., Karjalainen, T., Marland, G., Savolainen, I., Schlamadinger, B., and Apps, M. (2000). Project based greenhouse-gas accounting: Guiding principles with a focus on baselines and additionality. *Energy Policy* **28**, 935–946.

Hall, D. O., and Scrase, J. I. (1998). Will biomass be the environmentally friendly fuel of the future? *Biomass and Bioenergy* **15**(4 and 5), 357–367.

Hoogwijk, M., Faaij, A., van den Broek, R., Berndes, G., Gielen, D., and Turkenburg, W. (2003). Exploration of the ranges of the global potential of biomass for energy. *Biomass and Bioenergy* **25**(2), 119–123.

IEA Bioenergy Task 38 (2002). "Greenhouse Gas Balances of Biomass and Bioenergy Systems." http://www.joanneum.ac.at/iea-bioenergy-task38/publications/

IPCC (2001). Climate change 2001. IPCC Third Assessment Report. Intergovernmental Panel on Climate Change.

IPCC (2002). "Climate Change 2001: The Scientific Basis. Contribution of Working Group I to the Third Assessment Report of the Intergovernmental Panel on Climate Change" (J. T. Houghton, Y. Ding, D. J. Griggs, M. Noguer, P. J. van der Linden, X. Dai, K. Maskell, and C. A. Johnson, Eds.). Cambridge, UK, and New York, NY.

Kheshgi, H. S., Prince, R. C., and Marland, G. (2000). The Potential of Biomass Fuels in the Context of Global Climate Change: Focus on Transportation Fuels. *Annual Rev. Energy and the Environment* **25**, 199–244.

Kirschbaum, M. (2003). To sink or burn? A discussion of the potential contributions of forests to greenhouse gas balances through storing carbon or providing biofuels. *Biomass and Bioenergy* **24**(4 and 5), 297–310.

Matthews, R., and Robertson, K. (2002) Answers to ten frequently asked questions about bioenergy, carbon sinks and their role in global climate change. IEA Bioenergy Task 38. http://www.joanneum.ac.at/iea-bioenergy-task38/publications/faq/

Mooney, H., Roy, J., and Saugier, B. (eds.) (2001). "Terrestrial Global Productivity: Past, Present, and Future." Academic Press, San Diego, CA.

Schlamadinger, B., and Marland, G. (1996). The role of forest and bioenergy strategies in the global carbon cycle. *Biomass and Bioenergy* **10**, 275–300.

Schlamadinger, B., and Marland, G. (2001). The role of bioenergy and related land use in global net CO_2 emissions. *In* "Woody Biomass as an Energy Source: Challenges in Europe" (P. Pelkonen, P. Hakkila, T. Karjalainen, and B. Schlamadinger, Eds.) EFI Proceedings No. 39, pp. 21–27. European Forest Institute, Finland.

United Nations Development Programme (UNDP) (2000). World Energy Assessment: Energy and the Challenge of Sustainability. United Nations Publications, New York.

World Resources Institute (2003). Earth Trends. The Environmental Information Portal. Found at http://earthtrends.wri.org

Biomass Resource Assessment

MARIE E. WALSH
Oak Ridge National Laboratory
Oak Ridge, Tennessee, United States

1. Introduction
2. Approaches to Biomass Resource Assessment
3. Example of an Integrated Supply Curve Approach for Biomass Resource Assessment
4. Discussion of Limitations and Conceptual Differences in Biomass Resource Assessments
5. Summary and Conclusions

Glossary

agricultural residues Include the stems, leaves, cobs, husks, some roots, and chaff of agricultural crops such as corn, wheat, and other small grains as well as rice, sugarcane, and oilseed crops.

animal wastes Mostly manure (dung), but may include carcasses and industrial processing wastes (whey from milk separation).

biomass resource assessment An estimate of the quantities of biomass resources available by location and price levels given existing and potential physical, technical, environmental, economic, policy, and social conditions.

biomass supply curve An estimate of the quantity of biomass resources that could be available as a function of the price that can be paid for them; can be static (model variables are independent and do not vary in response to changes in each other) or dynamic (model variables change in response to changes in each other) as well as deterministic (there is no variability in model parameters) or stochastic (model parameters display a distribution of values).

energy crops Include a number of grass resources (herbaceous crops) such as switchgrass (*Panicum virgatum*), *Miscanthus,* reed canary grass, *Arundo donax,* and bamboo, as well as fast-growing short-rotation trees such as hybrid poplar (*Populus spp*), *Eucalyptus,* and willows (*Salix spp*).

forest industry residues Include those residues collected from forest operations such as logging residues, forest thinning, and removal of cull, dead, and dying material; also may include those residues produced at primary mills (e.g., saw mills, pulp mills, veneer mills) and secondary mills (e.g., furniture and cabinet makers).

fossil fuels Include coal, oil, and natural gas and comprise approximately 86% of the primary energy used in the world.

other miscellaneous biomass resources Include captured gases from decomposing plant and animal materials (biogas) that occur in landfills, sewage treatment plants, and feedlots; food processing wastes (e.g., olive pits, nut shells), other fibers (e.g., cotton gin, clothing remnants), and the wood component of the municipal solid waste and construction and demolition waste streams; also may include oils from oilseed crops (e.g., soybeans, rapeseed) and industry (e.g., restaurant cooking fats) and starch derived from agricultural food and feed crops (e.g., corn, wheat, other small grains).

Biomass resource assessments estimate the quantities of resources available by location and price levels given existing and potential physical, technical, environmental, economic, policy, and social conditions. Factors that affect the quantities and prices of biomass resources include physical limitations (e.g., soil and water limitations for energy crops, access to forests in remote areas), technical limitations (e.g., efficiency of harvest machinery to collect the resource, energy crop yields), environmental limitations (e.g., quantities of agricultural residues needed to control soil erosion and soil carbon for long-term productivity, wildlife diversity needs), economic considerations (e.g., competition for land between energy crops and traditional crops, alternative uses for biomass resources), policy considerations (e.g., crop subsidies, renewable portfolio standards that require a certain percentage of energy generation to come from biomass, emission regulations), and social factors (e.g., ethical values, aesthetic values, NIMBY [not in my back yard]).

1. INTRODUCTION

Biomass resources are widely used throughout the world, mostly for home cooking and heat but also for industrial uses such as heat, steam, and electricity. Estimates indicate that for the year 2000, 11% (~ 1095 million tons oil equivalent) of the world's total primary energy supply (TPES) came from biomass. Regional percentages of TPES ranged from a low of 0.3% in the Middle East to a high of 48% in Africa, with biomass accounting for 18% of TPES in China, 31% in the rest of Asia, and 17% in Latin America. Biomass as a percentage of TPES in non-Organization for Economic Cooperation and Development (OECD) Europe, the former Soviet Union, and OECD countries accounted for an estimated 5, 1, and 3%, respectively.

Biomass resources include a wide range of plant and animal materials available on a renewable or recurring basis. Most assessments focus on a few key resources such as forest industry residues, agricultural residues, and animal wastes—the most widely available and used resources in the world. Many assessments also evaluate potential new energy crops such as grass crops (herbaceous resources, e.g., switchgrass [*Panicum virgatum*], *Miscanthus*, reed canary grass, *Arundo donax*, bamboo) and fast-growing short-rotation trees (e.g., hybrid poplar [*Populus spp*], *Eucalyptus*, willows [*Salix spp*]). Forest industry residues include those collected from forest operations such as logging residues, forest thinning, and removal of cull, dead, and dying material. Additional forest industry residues include those produced at primary mills (e.g., saw mills, pulp mills, veneer mills) and secondary mills (e.g., furniture and cabinet makers). Agricultural residues include the stems, leaves, cobs, husks, some roots, and chaff of agricultural crops such as corn (stover), wheat, and other small grains (straw) as well as rice, sugarcane, and oilseed crops. Animal wastes are mostly manure (dung).

In addition to these biomass resources, food processing wastes (e.g., olive pits, nut shells); other fibers (e.g., cotton gin, clothing remnants); the wood component of municipal solid waste, construction and demolition waste, and yard trimmings; oils from oilseed crops (e.g., soybeans, rapeseed) and industry (e.g., restaurant cooking fats); and starch from agricultural food and feed crops (e.g., corn, wheat, other small grains) are available. Captured gases from decomposing plant and animal materials (i.e., biogas) in landfills, sewage treatment plants, and feedlots are also considered to be biomass resources.

The focus on energy crops (particularly trees), wood from forest and mill operations, agricultural residues, and animal manure in most studies is due not only to their worldwide availability but also to their assumed uses as steam, heat, and/or electricity. These resources are readily combustible and are well suited to these uses. However, biomass resources can be used to produce numerous other products, such as liquid and gaseous transportation fuels (e.g., ethanol, methanol, compressed natural gas, bio-diesel, hydrogen for fuel cells), and bioproducts currently produced from fossil fuels (e.g., coal, oil, natural gas), such as organic chemicals (e.g., plastic precursors, acetic acid, lactic acid, ethylene), new composite fiber materials (e.g., wood-plastic, wood-fiberglass), lubricants, pharmaceuticals, cosmetic ingredients, coatings (e.g., lacquers, varnishes, paints), and printing inks, among others. The key biomass resources can be used to produce many of these products as well, but some of the described products use biomass resources other than energy crops, forest industry residues, agricultural residues, or animal dung.

Interest in bioenergy and bioproducts is increasing because biomass is an abundant renewable resource that, when used in high-efficiency systems, can reduce greenhouse gas emissions (e.g., SO_x, N_2O, CO_2) relative to fossil fuel systems. Its use can also enhance economic development, particularly in rural areas, and can provide increased farm income, improve soil quality, be a viable option (e.g., energy crops) for production on degraded and abandoned lands, and offer the greatest flexibility (among renewable resources) in terms of the products that can be produced. This flexibility provides opportunities to establish highly diverse bio-based industries that can displace imported oil from politically unstable regions, enhancing energy security for some countries.

Three basic approaches are used for biomass resource assessments: an inventory approach that estimates quantities only, a quantity and associated cost approach that estimates either an average or range of quantities with an associated average or range of production and/or collection costs, and a supply curve approach that estimates biomass quantities as a function of the price that can be paid. The supply curve approach is generally the best suited to conduct integrated assessments that evaluate changes in physical, technical, environmental, economic, policy, and social factors. The potential availability of biomass resources for bioenergy and bioproducts generally decreases as one goes from the inventory approach to the supply curve approach.

The complexity of the models (integrated supply curves) and data limitations become more severe as the analysis becomes more geographically disaggregated, that is, as it goes from the global level to the local level. However, biomass resource quantities and prices are influenced significantly by local conditions (e.g., soil quality and weather for agricultural residues, forest residues, energy crops) and the low energy density of biomass resources (i.e., energy available vs volume and/or weight of the resource) that substantially increase transportation costs as distance increases. Conducting biomass resource assessments at the smallest geographic unit feasible is preferable.

2. *APPROACHES TO BIOMASS RESOURCE ASSESSMENT*

The inventory, quantity and associated cost, and supply curve approaches can be applied to all biomass resources. However, throughout this article, one resource category—energy crops—is used to demonstrate each approach. Hypothetical examples are presented to facilitate simplicity and transparency; however, the costs and yields used in the examples are within the yield and production cost ranges that can be realistically achieved.

2.1 Inventory Approach

The inventory approach is the most common method used to evaluate biomass resources. This approach generally estimates physical quantities only and does not include costs or prices that can be paid. It is used for local, regional, national, and global analysis. The degree of sophistication ranges from simple local surveys and/or observation of local use, to GIS satellite imaging, to solar radiation combined with soil type, water and nutrient availability, and weather conditions to estimate biomass growth potential. Lack of available data (especially in developing countries) is a major limitation and constrains the types and geographic scale on which assessments can be made. Inventory approaches are highly useful in establishing theoretical maximum quantities of biomass resources.

Although most inventory studies estimate the maximum physical quantities of biomass, some adjust these quantities to estimate available resources by including physical, technical, economic, and/or environmental factors. Physical limitations include areas where temperature, soil, and water conditions limit crop production opportunities. Technical considerations include limits to collection machinery efficiency and physical inaccessibility such as forest harvest at high altitudes. Environmental constraints include leaving a portion of agricultural and forest residues to maintain soil quality and reduce erosion, although most studies use very general rather than detailed local approximations to make this adjustment. Economic constraints generally include subtracting quantities of biomass currently being used.

A study by Hall and colleagues illustrates this approach. They estimated reasonable current yields of short-rotation woody crops that can be produced in the midwestern United States. Theoretical maximum quantities based on photosynthetic efficiency (a function of the amount of visible light that can be captured, converted to sugars, and stored) are first estimated. Losses from plant respiration are subtracted. Biomass production rates are estimated next by using the energy contained in the visible light that reaches an area (e.g., square meters) and multiplying by the photosynthetic efficiency. For short-rotation trees, this results in approximately 104 dry metric tons per hectare per year (dMT/ha/year) (where dry is 0% moisture). The theoretical maximum is then adjusted for temperature, and this reduces biomass yields to 60 dMT/ha/year. Adjustments for the parts of the trees that will be used for energy (e.g., trunk, branches) further reduce biomass yields to 35 dMT/ha/year. Given additional adjustments for the light intercepted during the growing season, and assuming sufficient water, nutrients, and pest and disease control, expected yields of 10 to 15 dMT/ha/year are reasonable. Genetic improvements could potentially double these yields. Total biomass is estimated as the expected yields multiplied by the land available to support these yield levels. Results from selected inventory studies are shown in Table I. Units are those reported by the authors of the study, with no attempt to adjust to a common measure.

2.2 Quantity and Associated Cost Approach

The quantity and associated cost approach includes an inventory approach component but extends the assessment to include biomass cost estimates in addition to physical quantities. This approach has been used in local to global assessments and varies substantially in complexity and sophistication. Studies range from examining one biomass resource (e.g., an energy crop) to evaluating

TABLE I

Selected Biomass Quantities: Inventory Approach

Reference number	Biomass resources included	Geographic region	Biomass quantities
1	Forest industry residues, agricultural residues, wood energy crops, animal wastes	Global	266.9EJ/Year
2	Forest industry residues, agricultural residues, animal wastes, biogases	Non-OECD countries	879.6 billion tons oil-equivalent
3	Forest industry residues, agricultural residues	Latin America, Caribbean	664,225 Kboe
4	Forest industry residues, agricultural residues, energy crops	Asia	2.51 billion tons
5	Forest industry residues(fuel wood)	World	472.3 million tons oil-equivalent

Sources. (1) Hall, D., Rosillo-Calle, F., Williams, R., and Woods, J. (1993). Biomass for energy: supply prospects. *In* "Renewable Energy: Sources for Fuels and Electricity" (T. Johansson, H. Kelly, A. Reddy, and R. Williams, Eds.), pp. 593–651. Island Press, Washington, DC. (2) Denman, J. (1998). Biomass energy data: System methodology and initial results. *In* "Proceedings of Biomass Energy: Data, Analysis, and Trends" Paris, France. (3) Hernandez, G. (1998). Biomass energy in Latin America and the Caribbean: Estimates, analysis, and prospects. *In* "Proceedings of Biomass Energy: Data, Analysis, and Trends." Paris, France. (4) Koopmans, A., and Koppejan, J. (1998). "Agriculture and Forest Residues: Generation, Utilization, and Availability," Food and Agricultural Organization Regional Wood, Energy Development Programme in Asia. (5) World Energy Council, (2001). "Survey of Energy Resources." London (www.worldenergy.org)

multiple resources (e.g., energy crops, forest residues, agricultural residues, animal wastes) in the geographic area examined.

As a first step, these studies create an inventory of available biomass resources using techniques similar to those in the inventory approach and often include some physical, technical, economic, and environmental adjustments. Biomass resource quantities are estimated generally as an average amount or a range. This approach can be applied to all of the biomass resources, but energy crops are used to illustrate it.

The average approach estimates the amount of resource available as the number of hectares that can be used to produce energy crops times the average yield per hectare. That is, if 10,000 ha are available in a region and the average yield is 10 dMT/ha, the total quantity of biomass available is 100,000 dMT. The range approach generally estimates available quantities as a minimum and a maximum (with a minimum expected yield of 5 dMT/ha and a maximum of 15 dMT/ha, for 10,000 ha, energy crop quantities range from 50,000 to 150,000 dMT).

Biomass resource costs are estimated to correspond to the average, minimum, and maximum quantities. For the average approach, the costs of management and production practices associated with achieving the expected yield level are estimated. If expected yields are 10 dMT/ha, estimated cost is U.S. $40/dMT, and available land is 10,000 ha, the cost of supplying 100,000 dMFT of the energy crops is U.S. $40/dMT. For the range approach, if expected yields are 5 to 15 dMT/ha and corresponding production costs are U.S. $50 and $30/dMT, respectively, the cost of supplying 50,000 to 150,000 dMT is U.S. $30 to $50/dMT.

Production and/or collection costs generally include all variable cash costs, but the extent to which fixed costs and owned resource costs are included varies substantially by study. For energy crops, variable cash costs include items such as seeds (e.g., grass crops); cuttings (e.g., tree crops); fuel, oil, and lubrication; machinery repair; fertilizers and pesticides; hired labor; custom harvest; hired technical services; and interest associated with purchasing inputs used during production but not repaid until the crop is harvested. Fixed costs include general overhead (e.g., bookkeeping, maintaining fences and roads, equipment storage), property taxes, insurance, management, and either rental payments for leasing land or interest for purchasing land. Owned resource costs include a producer's own labor, the value of owned land, machinery depreciation costs, and either the interest rate for loans to purchase machinery or the forgone earnings that could be obtained if capital to purchase machinery was invested in other opportunities. Future changes in technologies (e.g., machinery efficiency, crop productivity) require additional assumptions as to how costs will change over time. No consensus international production costs methodology exists, but some countries and regions have standard recommended methodologies (e.g., the American Agricultural Economics Association for methodology, the American Society of Agricultural Engineers for methodology and machinery data). Table II presents results from selected studies. Units are those reported by the study authors, with no attempt to adjust to a common measure.

2.3 Supply Curve Approach

In the supply curve approach, biomass resource quantities are estimated as a function of the price that can be paid for the resources. Estimates of several such price/quantity combinations can be used to construct the supply curve. Supply curve estimates can use simple approaches based largely on the quantity and associated cost approach for several quantity/cost combinations. More sophisticated approaches generally use an optimization approach (either profits are maximized or costs to produce a given quantity of resource are minimized). Integrated supply curve analyses also use optimization approaches but generally conduct the analyses within a dynamic framework (i.e., allow variables to change in response to changes in other variables), allocate land in the case of energy crops or biomass resources to different end-use products based on relative economics, and often try to incorporate as many physical, technical, environmental, economic, policy, and social factors as is reasonably feasible. In general, the simpler approaches are used to estimate global supply curves, whereas the more integrated approaches are used for smaller geographic scales (e.g., a national level) due to the complexity of the models and data limitations. Most supply curve analyses estimate supply curves independently for each biomass resource, although attempts to integrate multiple biomass resources that interact with each other simultaneously are being pursued. An increasing number of integrated supply curve analyses are either under way or planned.

For the energy crops example in the previous section, expected yields are 5, 10 (average), and 15 dMT/ha, with respective production costs of U.S. $50, $40, and $30/dMT. A simple supply curve can be estimated by increasing the number of yield/production cost points (e.g., yields of 5.0, 7.5, 10.0, 12.5, and 15.0 dMT/ha and corresponding production costs of U.S. $50, $45, $40, $35, and $30/dMT). At this point, a typical return to the producer (profit) and/or a typical transportation cost (for defined distances) may be added to the production cost estimates but are excluded from this example. The land available for energy crop production is further delineated to include that which can support each production level (of the 10,000 ha available, yields of 5.0 dMT/ha can be achieved on 1000 ha, 7.5 dMT/ha on 2000 ha, 10.0 dMT/ha on 4000 ha, 12.5 dMT/ha on 2000 ha, and 15.0 dMT/ha on 1000 ha). Thus, at U.S. $50, $45, $40, $35, and $30/dMT, quantities of energy crops that can be produced are 5000, 15,000, 40,000, 25,000, and 15,000 dMT, respectively. The

TABLE II

Selected Biomass Quantities and Costs: Quantity and Associated Cost Approach

Reference number	Geographic region	Biomass resources	Quantity	Cost
1	United States	Secondary mill residues	1.22 million dry tons/year	$U.S. 20/dry ton
2	Finland	Forest	3.6 cubic m/year	FIM 12.5/GJ
			8.8 cubic m/year	FIM 15.3/GJ
3	Belarus	SRWC	12 tons/ha	40 Euros/ton[a]
			6 tons/ha	100 Euros/ton[a]
4	Nicaragua	Eucalyptus	12.7 dMT/ha/year	$1.7/GJ[b]

Sources. (1) Rooney, T. (1998). "Lignocellulose Feedstock Resource Assessment," NEOS Corporation, Lakewood, CO.
(2) Malinen, J., Pesonen, M., Maatta, T., and Kajanus, M. (2001). Potential harvest for wood fuels (energy wood) from logging residues and first thinnings in southern Finland. *Biomass and Bioenergy* **20**, 189–196.
(3) Vandenhove, H., Goor, F., O' Brien, S., Grebenkov, A., and Timofeyev, H. (2002). Economic viability of short rotation coppice for energy production for reuse of Caesium-contaminated land in Belarus. *Biomass and Bioenergy* **22**, 421–431.
(4) van den Broek, R., van den Burg, T., van Wijk, A., and Turkenburg, W. (2000). Electricity generation from eucalyptus and bagasse by sugar mill in Nicaragua: A comparison with fuel oil electricity generation on the basis of costs, macroeconomic impacts, and environmental emissions. *Biomass and Bioenergy* **19**, 311–335.

[a] Price paid to farmer (not production cost).

[b] Case study for one plant.

supply curve is constructed by ordering the costs from lowest to highest value (U.S. $30 to $50/dMT) and using cumulative quantities at each price level. Thus, for U.S. $35/dMT, the total quantity is that at U.S. $30/dMT (15,000 dMT) plus that between U.S. $30 and $35/dMT (25,000 dMT) for a total of 40,000 dMT. Figure 1 illustrates the supply curve for this example.

3. EXAMPLE OF AN INTEGRATED SUPPLY CURVE APPROACH FOR BIOMASS RESOURCE ASSESSMENT

A joint U.S. Department of Agriculture and U.S. Department of Energy project conducted by Oak Ridge National Laboratory (ORNL) and the University of Tennessee (UT) is used to illustrate the integrated supply curve approach. Work completed is presented, and possible extensions to the analysis are discussed.

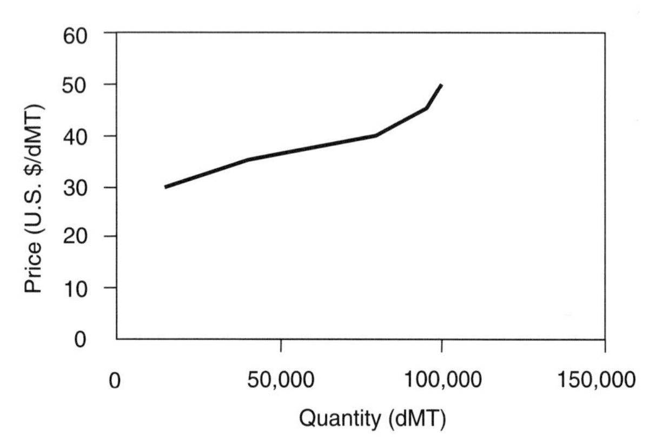

FIGURE 1 Example supply curve for energy crops: simple supply curve approach.

3.1 Model Description

A dynamic agricultural sector model (POLYSYS) is used to estimate the potential for energy crop production and its impacts on U.S. agriculture. POLYSYS is structured as a system of interdependent modules simulating crop supply and demand, livestock supply and demand, and farm income. Figure 2 provides a schematic of the POLYSYS model.

POLYSYS can simulate impacts on the U.S. agricultural sector resulting from changes in market, policy, technical, economic, or resource conditions. It is anchored to published baseline projections and estimates deviations from the baseline for planted and harvested acres, yields, production quantities, exports, variable costs, demand quantities for each use, market price, cash receipts, government payments, and net income.

The crop supply module is composed of 305 independent regions with relatively homogenous production characteristics (agricultural statistical districts [ASDs]). Cropland in each ASD that can be brought into crop production, shifted to production of a different crop, or moved out of production is first identified The extent to which lands can shift among crops is a function of whether the profit for

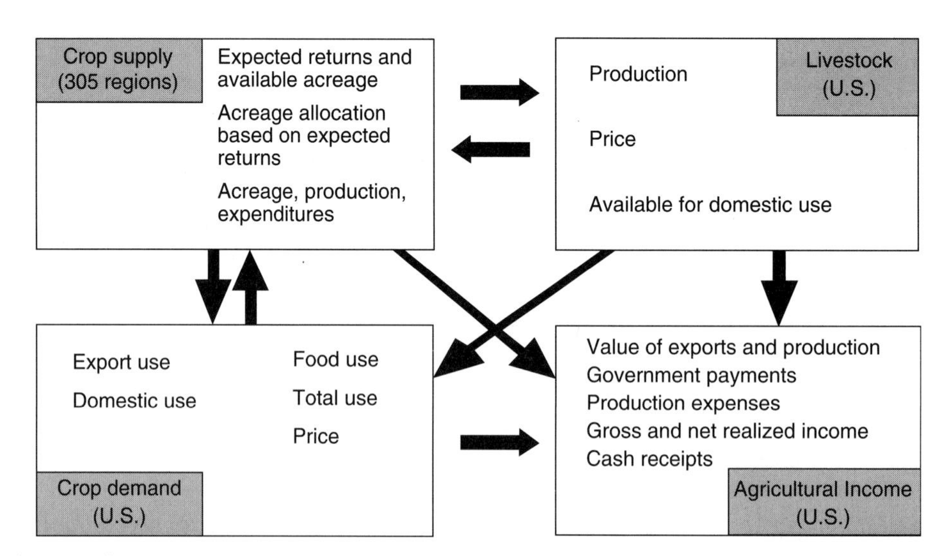

FIGURE 2 Schematic of the POLYSYS model showing the linkages among the submodules. From Ray, D., De La Torre Ugarte, D., Dicks, M., and Tiller, K. (1998). "The POLYSYS Modeling Framework: A Documentation" (unpublished report). Agricultural Policy Analysis Center, University of Tennessee, Knoxville. (http://apacweb.ag.utk.edu/polysys.html.)

each crop is positive, negative, or a mixture for the preceding 3 years, with larger land shifts permitted for crops with negative profits. The amount of idle and pasture lands that can be shifted is limited to 20% of the land in an ASD in any given year. Once the land that can be shifted is determined, the supply module allocates available cropland among competing crops based on maximizing returns above costs (profits).

The crop demand module estimates demand quantities and market prices for each crop and use (e.g., food, feed, industrial). Crop demand is a function of each crop's own price, prices of other crops that compete for the same end-use markets, and nonprice variables (e.g., quality, policy). The livestock module uses statistically derived equations that interact with the demand and supply modules to estimate production quantities and market prices. The primary link between the livestock and crop sectors is through livestock feed demand. Livestock quantities are a function of current and previous years' production and feed prices. The income module uses information from the supply, demand, and livestock modules to estimate cash receipts, production costs, government outlays, net returns to farmers, and net income.

Each module is self-contained but works interdependently to perform a multiperiod simulation. The demand module generates market prices used to form price expectations in the supply module that are used to allocate land to crops by ASD and to form national supply quantities. National supply quantities are then returned to the demand module to estimate prices, demand quantities, and surplus quantities that are carried over to the next year (i.e., carryover stocks). The livestock sector is linked to the crop sector through feed grains where livestock quantities affect feed grain demand and prices and feed grain prices and supplies affect livestock production decisions. Livestock and crop production quantities and prices are used by the income module.

POLYSYS includes the major U.S. crops, that is, corn, soybeans, wheat, grain sorghum, oats, barley, cotton, rice, alfalfa, and other hay. The livestock sector includes beef, pork, lamb and mutton, broiler chickens, turkeys, eggs, and milk. Food, feed, industrial, and export demands are considered as well as carryover stocks. The model includes all U.S. lands classified as cropland (174.7 million ha, 431.4 million acres) currently in crop production, idled, in pasture, or in the Conservation Reserve Program (CRP).

Crop production costs are developed using detailed field operation schedules and include variable input costs (e.g., seed, fertilizer, pesticides, custom operations, fuel and lube, repairs, paid labor, drying, irrigation, technical services, interest), fixed costs (e.g., insurance and taxes, storage, general overhead), capital returns on machinery, and the value of the producer's own labor and management. The analysis uses recommended methodology.

3.2 Modification of Model to Include Energy Crops

POLYSYS was modified to include three energy crops: switchgrass (*Panicum virgatum*), hybrid poplar (*Populus spp.*), and willow (*Salix spp.*). Switchgrass is a native perennial grass and a major species of the U.S. tall grass prairies. Hybrid poplar is a cross between two or more *Populus spp.* that include aspens and cottonwoods. Willows used for energy are improved varieties of native shrubs (not landscaping ornamentals). Other trees and grasses can be used and may be preferred under some conditions; however, many of these alternatives have similar expected yields and production practices, so switchgrass, poplar, and willow can serve as reasonable representations of other energy crops. U.S. cropland identified as suitable for energy crop production (given current development status) and used in the analysis is 149 million ha (368 million acres).

The production location, yields, and management practices of each energy crop differ across each ASD in POLYSYS. Switchgrass is planted for 10 years and is harvested annually as large round bales. Varieties appropriate to each region are assumed, and fertilizer (quantity and type) varies by region. Hybrid poplar is planted for 6- to 10-year rotations (depending on region) and is harvested once at the rotation end as whole tree chips. Fertilizer applications vary by region. Willows are planted in a 22-year rotation, harvested every third year with regrowth from the stump following harvest (coppicing), and delivered as whole tree chips. Fertilizer is applied during the year following harvest. Herbicide applications for all crops are limited to the first 2 years of production. All assumed management practices are based on research results, demonstration of commercial field experience where available, and expert opinion.

Achievable energy crop yields on land currently producing traditional agricultural crops are used as the base yields in the analysis. Energy crop yields on idled and pasture lands are assumed to be 85% of the base yields. Production practices used on idle and pasture lands are similar to those on cropped lands except for additional herbicide applications to kill existing cover and increased tillage to prepare the sites for planting.

Energy and traditional crop production costs are estimated using the POLYSYS budget generator model except for hybrid poplar and willow harvesting costs. These costs are estimated using the ORNL BIOCOST model. Because of the multiyear production rotations of energy crops, annual profits for energy and traditional crops are discounted using a 6.5% real discount rate (a net present value approach). A common planning horizon of 40 years is used to account for the different rotation lengths of the energy crops.

Cropland idled or in pasture is assumed to be so due to economic reasons; at given prices, production costs, and yields, converting the land to pasture or idling it is the most economic use. To return these lands to production, the present value profits of crops must be higher than the most profitable crop under the baseline assumptions. To account for possible inertia to keep the land in its current use and/or the value of pasture land in livestock operations, a premium of 10% above the baseline present value returns for idled lands and 15% for pasture lands is required.

Energy crops compete for land with both traditional crops and each other; at an equivalent energy price (same cost/gigajoule [U.S. $/GJ]) for each energy crop, land is allocated to the most profitable energy crop given expected yields and production costs. Switchgrass is the most profitable energy crop, followed by hybrid poplar and willow in most ASDs; thus, nearly all (99%) of the currently planted, idle, and pasture land shifted to energy crop production goes to switchgrass production.

The analysis estimates that at a price of U.S. $2.44/GJ (U.S. $44/dMT [U.S. $40/dry ton]), 11.7 million ha (29 million acres) producing 120.4 million dMT (132.5 million dry tons) of energy crops can be more profitable than existing agricultural uses for cropped, idled, and pasture lands. The biomass produced is equivalent to 2.016 exajoules (2.05 quads) of primary energy and, given existing conversion rates, can produce 44.7 billion L (11.8 billion gallons) of ethanol or 216 billion kilowatt-hours of electricity equivalent to approximately 5% of current U.S. production levels. Traditional crop prices increase by 9 to 14% depending on crop. Net farm income increases by U.S. $6.0 billion/year (U.S. $2.3 billion from energy crops and the rest from higher traditional crop prices). Figure 3 summarizes estimated quantities of energy crops that are more profitable than alternative land uses for several price levels. Figure 4 shows the regional distribution of energy crop production at U.S. $2.44/GJ.

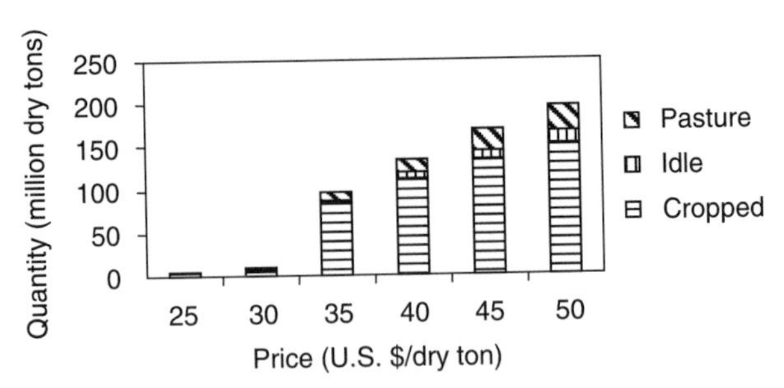

FIGURE 3 Energy crop production on U.S. cropped, idle, and pasture acres at selected crop prices.

3.3 Analysis of the Conservation Reserve Program

Modifications to the CRP have been suggested as a means to introduce energy crops to agriculture and reduce their prices to end users. The CRP, established in the 1985 Farm Bill, sets asides environmentally sensitive lands under 10- to 15-year contracts. CRP lands are planted to conservation crops, such as perennial grasses and trees, and farmers receive an annual rental payment. Harvest is prohibited except under emergency conditions. Policy changes in this agricultural program were analyzed with the modified POLYSYS model.

Enrolled lands considered in the analysis (12.1 million ha [29.7 million acres]) were adjusted for geographic suitability and environmental sensitivity. Lands that are enrolled as buffer strips, classified as wetlands, or critical to watershed management, or that provide critical habitat in wildlife conservation priority areas were excluded. CRP land available for energy crop production in the analysis amounts to 6.8 million ha (16.9 million acres).

Two energy crop management practices were analyzed: one to achieve high biomass production (production scenario) and one to achieve high wildlife diversity (wildlife scenario). The wildlife scenario uses fewer fertilizer and chemical inputs than does the production scenario, and the former restricts switchgrass harvest to alternating years. Cover crops to minimize erosion are required during the establishment years for all energy crops. Yields of energy crops produced on CRP lands are adjusted by an index of traditional crop yields obtained on CRP lands prior to being enrolled in the program.

When existing CRP contracts expire, farmers can renew their current contracts or produce energy crops under modified contracts (the analysis is simplified by not enrolling new lands in the program or by switching existing CRP lands back to traditional crop production). In exchange for being

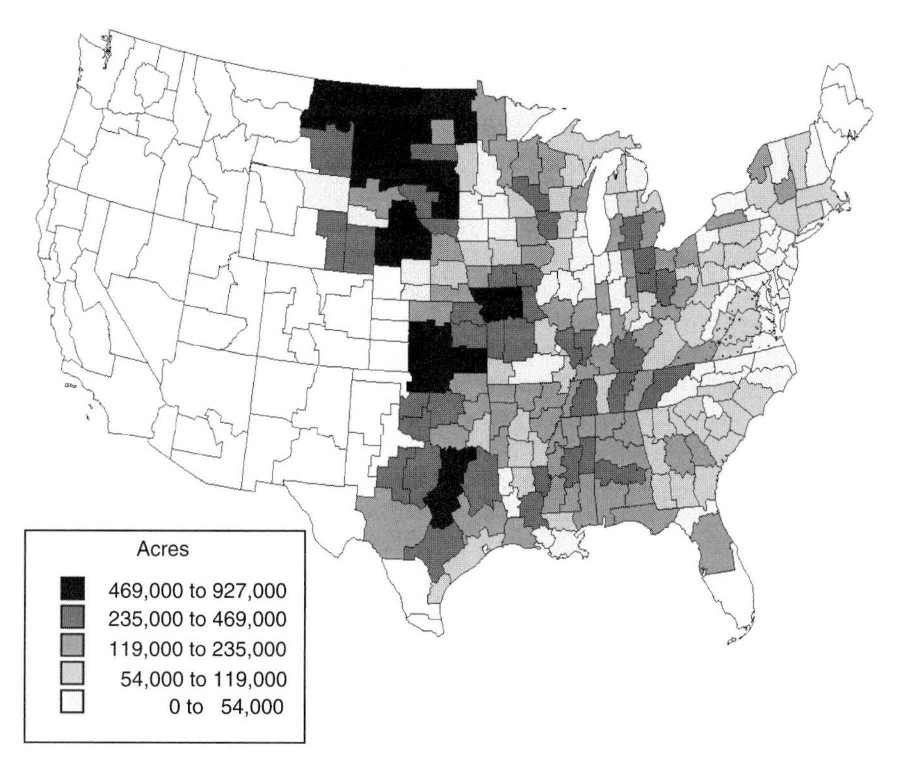

FIGURE 4 Distribution of energy crops on cropland currently in traditional crop production, idled, or in pasture at a price of U.S. $2.44/GJ (U.S. $40/dry ton). From De La Torre Ugarte, D., Walsh, M., Shapouri, H., and Slinsky, S. (2003). "The Economic Impacts of Bioenergy Crop Production on U.S. Agriculture," Agricultural Economic Report No. 816. U.S. Department of Agriculture, Washington, DC.

allowed to produce and harvest energy crops on CRP lands, 25% of the current rental rate is forfeited. Thus, the decision to allocate CRP lands to energy crop production is reduced to a comparison of their present value profits and the present value of the forgone CRP rental payments.

Under the wildlife scenario at U.S. $2.44/GJ, CRP lands are split between switchgrass (annual production of 1.0 million dMT) and hybrid poplar (35.1 million dMT when harvested). Under the production scenario, switchgrass dominates, with 5.2 million ha (12.91 million acres) producing 50.3 million dMT annually. The U.S. Congress used the analysis to establish a pilot program (based on the wildlife scenario) to allow production and harvest of energy crops on CRP lands (Public Law 106-78, section 769). Figure 5 shows the distribution of CRP lands under the production scenario at U.S. $2.44/GJ.

3.4 Potential Extensions of the Integrated Supply Curve Analysis

The POLYSYS model provides a solid decision framework to anchor additional analyses. For the analysis described previously, a deterministic enterprise version of POLYSYS was used. A stochastic crop rotation version also exists, and future plans are to incorporate energy crops into this version. Because POLYSYS is an agricultural sector model, additional agricultural-based biomass resources (e.g., corn stover, wheat straw) can be added. The removal of agricultural residues raises many soil quality issues such as erosion, carbon levels, moisture, and long-run productivity. POLYSYS was originally designed to be able to evaluate environmental constraints and impacts and includes several major soil types, tillage, and management options by crop rotation and ASD. Work is currently under way to estimate the amounts of agricultural residues that must be left on the field to maintain soil erosion and carbon at determined levels. The work evaluates needed residue levels considering grain yields, crop rotation, tillage and other management practices, soil type, topology, and climate. Response curves developed from this analysis will be incorporated into POLYSYS, allowing biomass supply curves for agricultural residues, energy crops, and traditional crops to be estimated simultaneously.

The supply curves from POLYSYS, as well as the underlying data used to produce them, can be used

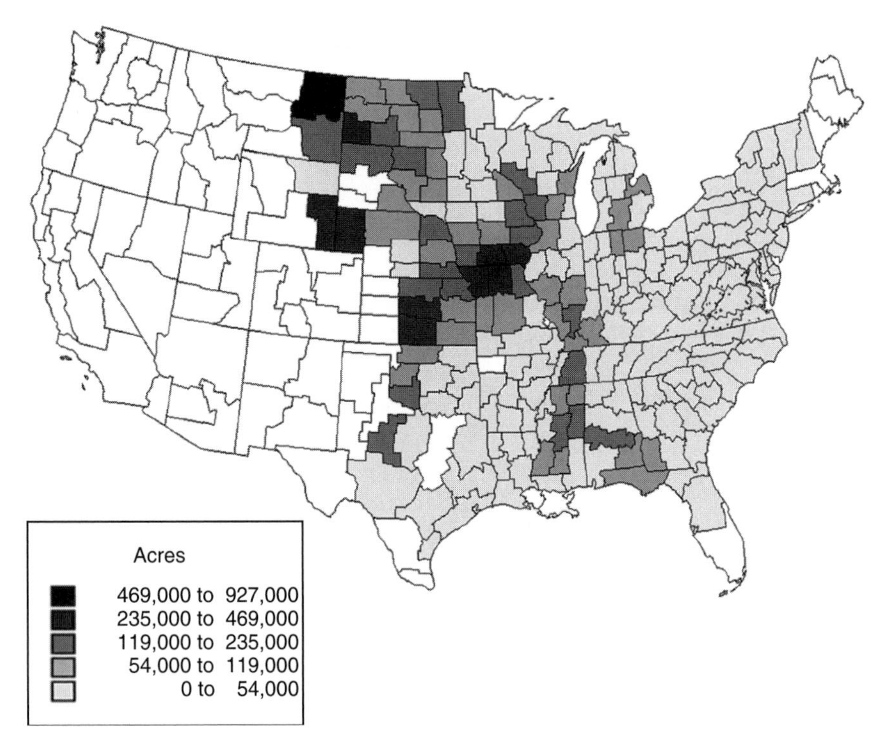

FIGURE 5 Distribution of energy crops on Conservation Reserve Program (CRP) acres at a price of U.S. \$2.44/GJ (U.S. \$40/dry ton) and assuming the production CRP scenario. From De La Torre Ugarte, D., Walsh, M., Shapouri, H., and Slinsky, S. (2003). "The Economic Impacts of Bioenergy Crop Production on U.S. Agriculture," Agricultural Economic Report No. 816. U.S. Department of Agriculture, Washington, DC.

with other models to conduct additional analyses, including developing industry-level bioenergy and bioproduct expansion curves, life cycle analyses, and economic impact assessments. Product expansion curves can be estimated using a transportation and siting model to identify location and delivered biomass cost given user facility demand levels. Addition of conversion costs and market prices for the product(s)—multiple products can be examined simultaneously—permits estimation of the size, location, and cost of the industry as a whole (industry expansion curves). Life cycle analyses can be used to estimate the potential to displace fossil fuels, energy balances, impact on greenhouse gas emissions, and impact on other air pollutants relative to existing products (e.g., ethanol vs gasoline as a transportation fuel). Economic impacts (e.g., jobs, value added, total product output, farm income) of a national bioenergy or bioproduct industry can be estimated using appropriate economic models. Once the entire system is established, the impacts of changes in technology, economic and environmental policy, and regulatory conditions can be evaluated. Figure 6 illustrates how the component pieces relate to each other.

4. DISCUSSION OF LIMITATIONS AND CONCEPTUAL DIFFERENCES IN BIOMASS RESOURCE ASSESSMENTS

International consensus regarding the definition of biomass resource assessment, the terminology and methodologies used to conduct assessments, and the units used to present results does not currently exist. Disciplinary training, historical conditions, and cultural differences are among the factors that complicate reaching a consensus.

This article defines biomass resource assessment as estimating the quantity of biomass that can be realistically supplied by price and location under existing and potential physical, technical, environmental, economic, policy, and social conditions. This definition includes the inventory, average quantity and cost, and simple to integrated supply curve approaches. However, not all biomass analysts concur with this view, and a number would limit their definition to inventory analysis only.

Terminology to define biomass resource assessment approaches is not standard. For example, the

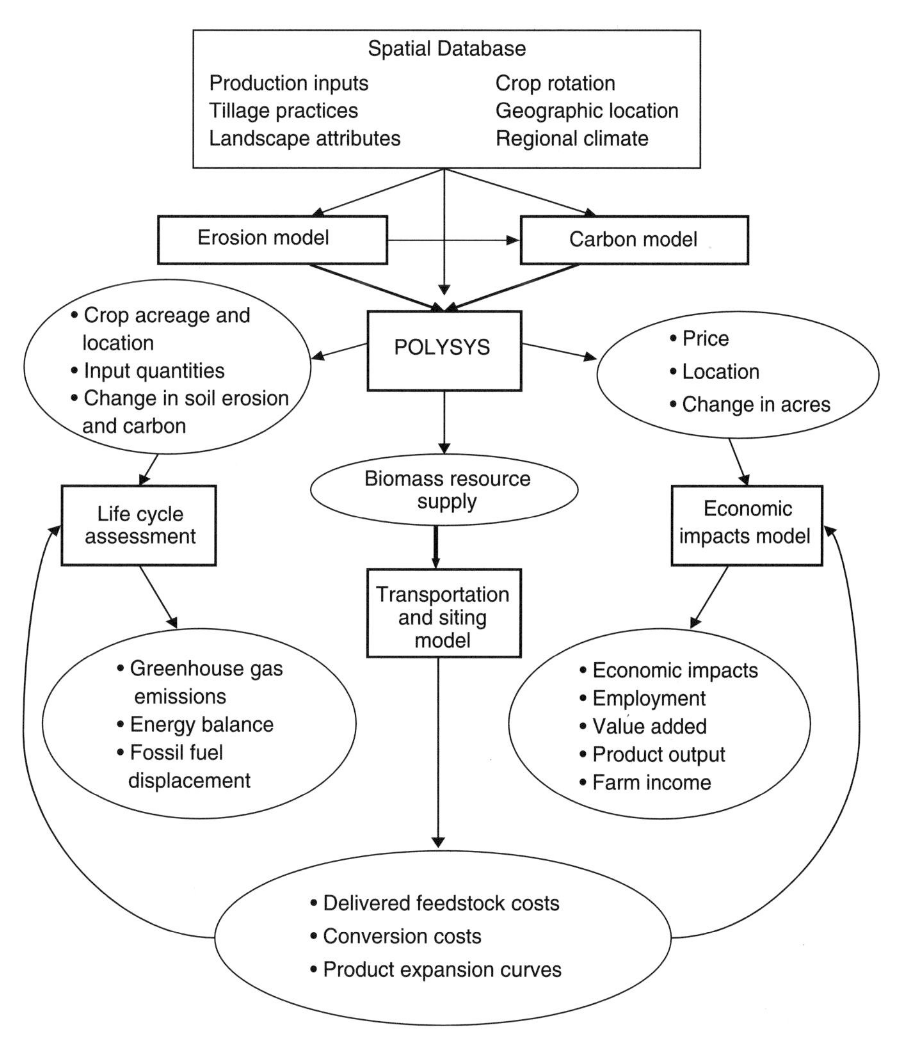

FIGURE 6 Schematic of an integrated biomass system. The figure illustrates the relationship among potential extensions of the integrated supply curve analysis (POLYSYS modeling).

International Institute for Applied Systems Analysis defines assessments in terms of theoretical, technical, and economic approaches. This article combines the theoretical and technical approaches into the inventory approach and divides the economic approach into the quantity and associated cost and supply curve approaches. Recommended methodologies are available for selected components of biomass resource assessments in some regions, but an international consensus of approaches flexible enough to be applied to different situations for all components does not exist.

As illustrated in Tables I and II, units to express results vary substantially by study. The various units used are appropriate to the purpose of each study but make it difficult to compare studies. With respect to biomass resource assessments, dry metric tons are applicable to most solid biomass resources. Inclusion of this measure would facilitate comparisons across studies. It is important to note that trying to develop one standard for all situations might not be an appropriate goal. Rather, a set of standards applicable to different situations may be better. Alternative methods might be needed to accommodate data limitations.

Inclusion/exclusion of biomass resources currently used for low-efficiency energy or nonenergy purposes is a major conceptual difference among biomass resource assessments. Most studies exclude biomass resources currently used that can reduce available quantities to as little as 10% of those produced in some studies. The rationale for exclusion is that these alternative uses are more economically attractive relative to energy use. However, the relative attractiveness is based on current technologies, assumed energy product, and existing economic, market, policy, and regulatory conditions. Under different conditions, the relative attractiveness can differ radically. Some studies assume that currently used resources are still available for energy use, but at higher prices. This approach allows greater flexibility to evaluate biomass potential under alternative situations. Studies using this approach show that under plausible technology, product, market, economic, policy, and regulatory conditions, prices paid for biomass can be sufficient to divert these resources from existing uses to bioenergy and bioproduct uses.

A second major conceptual difference is the value of idled land used to produce energy crops. Most studies assume zero to low value for idled lands. This assumption is reasonable for severely degraded lands that are physically unable to support most forest or food crop production such as previously forested areas that were clear cut, planted to food crops until the soils were depleted, and then abandoned. Alternative uses for these lands are severely limited. However, in many industrialized nations, lands are idle due to low profits or policy and can be returned to production under different conditions. Under these circumstances, the assumption of zero to low value is questionable; valuing idle lands based on their best alternative uses is more appropriate.

A third conceptual difference involves whether to include environmental factors as constraints to biomass quantities that can be collected or as impacts that occur after collection. This difference is particularly apparent in the way in which available agricultural residues are estimated, particularly with respect to soil erosion and carbon. Most studies either ignore these issues altogether or view them as impacts resulting from collection. Some studies acknowledge the erosion concerns and try to account for them, albeit usually in simple ways (e.g., 50% of residues must remain). Soil carbon implications are rarely considered. In the United States, studies are under way to estimate the quantities of agricultural residues that can be removed and still maintain soil erosion and carbon at defined levels under different crop rotation, soil type, tillage and management practices, crop yields, and geo-climatic conditions.

In addition to conceptual differences and no standard definitions and methodologies, biomass resource assessments are significantly affected by data availability. In many cases, data are unavailable to support even simple inventory approaches, much less the extremely data-intensive integrated supply curve approaches. Also, the complexity and difficulty of creating integrated dynamic and stochastic models can be severely limiting.

For those interested in finding out more about biomass resource assessments, bioenergy, and bioproducts, good starting points include biomass-related journals, Web sites at institutions conducting biomass research, and biomass conference proceedings.

5. SUMMARY AND CONCLUSIONS

Biomass resource assessments estimate the quantities of resources available by location and price levels given existing and potential physical, technical, environmental, economic, policy, and social conditions. Three basic approaches are used to estimate biomass resources: the inventory approach, the quantity and associated cost approach, and the supply curve approach. The inventory approach estimates physical quantities of biomass resources only, with no consideration of price. The quantity and associated cost approach typically estimates a minimum–maximum range or average quantities along with corresponding production and/or collection costs. The supply curve approach estimates biomass quantities as a function of the price that can be paid. Each is inclusive of the previous approach and results in fewer available resources and higher prices.

International consensus regarding the definition of biomass resource assessment, biomass terminology, methodology, and units to express results is lacking. Major differences in conceptual issues lead to substantially different ways in which to conduct biomass resource assessments and to substantially different results of the analyses.

Data availability is a major constraint limiting the types of biomass resource assessments that can be done. Integrated supply curve approaches incorporating as many physical, technical, environmental, economic, policy, and social factors as are reasonably feasible are extremely data intensive and involve creating complex models. However, they provide maximum flexibility to examine changes in any given variable, and if they are conducted in a dynamic framework, they allow variables within a model to change in response to each other. An example of an integrated supply curve analysis in the United States, with proposed extensions, was presented.

SEE ALSO THE FOLLOWING ARTICLES

Biomass, Chemicals from • Biomass Combustion • Biomass for Renewable Energy and Fuels • Biomass

Gasification • Biomass: Impact on Carbon Cycle and Greenhouse Gas Emissions • Forest Products and Energy • Renewable Energy, Taxonomic Overview • Waste-to-Energy Technology

Further Reading

Biomass and Bioenergy. (n.d.). Elsevier Science, New York.

Hall, D. O., Rosillo-Calle, F., Williams, R. H., and Woods, J. (1993). Biomass for energy: Supply prospects. *In* "Energy Sources for Fuels and Electricity" (T. Johansson, H. Kelly, A. Reddy, and R. Williams, Eds.), pp. 593–651. Island Press, Washington, DC.

Overend, R., and Chornet, E. (eds.). (1999). "Proceedings of the Fourth Biomass Conference of the Americas: Biomass—A Growth Opportunity in Green Energy and Value-Added Products." Elsevier Science, Oakland, CA.

"Proceedings of Bioenergy 2000: Moving Technology into the Marketplace." (2000). Northeast Regional Biomass Program, Buffalo, NY.

"Proceedings of Biomass Energy: Data, Analysis, and Trends Conference." (1998). International Energy Agency, Paris.

"Proceedings of the First World Conference on Biomass for Energy and Industry." (2001). James & James, Sevilla, Spain.

Walsh, M., De La Torre Ugarte, D., Shapouri, H., and Slinsky, S. (2003). Bioenergy crop production in the United States: potential quantities, land use changes, and economic impacts on the agricultural sector. *Environ. Resource Econ.* **24,** 313–333.

Bottom-Up Energy Modeling

JAYANT SATHAYE and ALAN H. SANSTAD
Lawrence Berkeley National Laboratory
Berkeley, California, United States

1. Goals, Context, and Use: Bottom-Up Approaches
2. Structure of an Energy Sector Bottom-Up Assessment
3. Typical Models for a Bottom-Up Energy Sector Assessment
4. Key Challenges in the Bottom-Up Modeling Approach
5. Data Sources for Bottom-Up Analysis
6. Developing Scenarios for Use in a Bottom-Up Assessment
7. Results from a Bottom-Up Approach

Glossary

base year The year for which the inventory is to be taken. In some cases (such as estimating CH_4 from rich production), the base year is simply the last year of a number of years over which an average must be taken.

bottom-up modeling A modeling approach that arrives at economic conclusions from an analysis of the effect of changes in specific parameters on narrow parts of the total system.

carbon tax A tax on fossil fuels based on the individual carbon content of each fuel. Under a carbon tax, coal would be taxed the highest per MBtu, followed by petroleum and then natural gas.

demand-side management The planning, implementation, and monitoring of utility activities designed to encourage customers to modify their pattern of electricity usage.

discount rate The rate at which money grows in value (relative to inflation) if it is invested.

dynamic In the field of modeling, a dynamic model includes intertemporal relations between variables. A model that does not include such relations is called static.

energy forms and levels Primary energy is energy that has not been subjected to any conversion or transformation process. Secondary energy (derived energy) has been produced by the conversion or transformation of primary energy or of another secondary form of energy. Final energy (energy supplied) is the energy made available to the consumer before its final conversion (i.e., before utilization). Useful energy is the energy made usefully available to the consumer after its final conversion (i.e., in its final utilization).

energy intensity The amount of energy required per unit of a particular product or activity.

energy services The service or end use ultimately provided by energy. For example, in a home with an electric heat pump, the service provided by electricity is not to drive the heat pump's electric motor but rather to provide comfortable conditions inside the house.

engineering approach A particular form of bottom-up modeling in which engineering-type process descriptions (e.g., fuel efficiency of end-use devices) are used to calculate a more aggregated energy demand. This term is particularly used in contrast to econometric models.

exogenous variables Variables determined outside the system under consideration. In the case of energy planning models, these may be political, social, or environmental, for example.

feedback When one variable in a system (e.g., increasing temperature) triggers changes in a second variable (e.g., cloud cover), which in turn ultimately affects the original variable (i.e., augmenting or diminishing the warming). A positive feedback intensifies the effect. A negative feedback reduces the effect.

fossil fuel Coal, petroleum, or natural gas or any fuel derived from them.

general equilibrium analysis An approach that considers simultaneously all the markets in an economy, allowing for feedback effects between individual markets. It is particularly concerned with the conditions that permit simultaneously equilibrium in all markets and with the determinants and properties of such an economy-wide set of equilibrium.

income elasticity The expected percentage change in the quantity demand for a good given a 1% change in income. An income elasticity of demand for electricity of 1.0 implies that a 1% increase in income will result in a 1% increase in demand for electricity.

input-output analysis A method of investigating the interrelationship between the branches of a national economy in a specific time period. The representation, in the form of a matrix table, is called an input-output table. An input-output analysis allows the changes in total demand in related industrial branches to be estimated.

least-cost planning In energy planning, the practice of basing investment decisions on the least costly option for providing energy services. It is distinguished from the more traditional approach taken by utilities, which focuses on the least costly way to provide specific types of energy, with little or no consideration of less costly alternatives that provide the same energy service at lower costs.

linear programming A practical technique for finding the arrangement of activities that maximizes or minimizes a defined criterion subject to the operative constraints. For example, it can be used to find the most profitable set of outputs that can be produced from a given type of crude oil input to a given refinery with given output prices. The technique can deal only with situations where activities can be expressed in the form of linear equalities or inequalities and where the criterion is also linear.

macroeconomics The study of economic aggregates and the relationships between them. The targets of macroeconomic policy are the level and rate of change of national income (i.e., economic growth), the level of unemployment, and the rate of inflation. In macroeconomics, the questions about energy are how its price and availability affect economic growth, unemployment, and inflation, and how economic growth affects the demand for energy.

marginal costs In linear programming, this term has the very specific meaning of change of the objective function value as a result of a change in the right-hand-side value of a constraint. If, for example, the objective is to minimize costs, and if the capacity of a particular energy conversion facility, such as a power plant, is fully utilized, the marginal cost in the linear planning sense expresses the (hypothetical) reduction of the objective function value (i.e., the benefit) of an additional unit of capacity.

market clearing The economic condition of supply equaling demand.

optimization model A model describing a system or problem in such a way that the application of rigorous analytical procedures to the representation results in the best solution for a given variable(s) within the constraints of all relevant limitations.

price elasticities The expected percentage change in quantity demand for a good given a 1% change in price. A price elasticity of demand for electricity of −0.5 implies that a 1% increase in price will result in a half percent decrease in demand for electricity.

renewable energy Energy obtained from sources that are essentially inexhaustible (unlike, for example, the fossil fuels, of which there is a finite supply). Renewable sources of energy include wood, waste, wind, geothermal, and solar thermal energy.

retrofit To update an existing structure or technology by modifying it, as opposed to creating something entirely new from scratch. For example, an old house can be retrofitted with advanced windows to slow the flow of energy into or from the house.

scenario Coherent and plausible combination of hypotheses, systematically combined, concerning the exogenous variables of a forecast.

sensitivity analysis A method of analysis that introduces variations into a model's explanatory variables to examine their effects on the explained.

simulation model Descriptive model based on a logical representation of a system and aimed at reproducing a simplified operation of this system. A simulation model is referred to as static if it represents the operation of the system in a single time period; it is referred to as dynamic if the output of the current period is affected by evolution or expansion compared with previous periods. The importance of these models derives from the impossibility of excessive cost of conducing experiments on the system itself.

top-down modeling A modeling approach that proceeds from broad, highly aggregated generalizations to regionally or functionally disaggregated details.

Two general approaches have been used for the integrated assessment of energy demand and supply: the so-called bottom-up and top-down approaches. The bottom-up approach focuses on individual technologies for delivering energy services, such as household durable goods and industrial process technologies. For such technologies, the approach attempts to estimate the costs and benefits associated with investments in increased energy efficiency, often in the context of reductions in greenhouse gas (GHG) emission or other environmental impacts. The top-down method assumes a general equilibrium or macroeconomic perspective, wherein costs are defined in terms of losses in economic output, income, or gross domestic product (GDP), typically from the imposition of energy or emissions taxes.

1. GOALS, CONTEXT, AND USE: BOTTOM-UP APPROACHES

The fundamental difference between the two approaches is in the perspective taken by each on consumer and firm behavior and the performance of markets for energy efficiency. The bottom-up approach assumes that various market "barriers" prevent consumers from taking actions that would be in their private self-interest—that is, would result in the provision of energy services at lower cost. These market barriers include lack of information about energy efficiency opportunities, lack of access to capital to finance energy efficiency investment, and misplaced incentives that separate responsibilities for making capital investments and paying operating costs. In contrast, the top-down approach generally assumes that consumers and firms correctly perceive, and act in, their private self-interest (are utility and profit maximizers) and that unregulated markets serve to deliver optimal investments in energy efficiency as a function of prevailing prices. In this view, any market inefficiencies pertaining to energy efficiency result solely from the presence of environmental externalities that are not reflected in market prices.

In general, an assessment carried out using the bottom-up approach will very likely show significantly lower costs for meeting a given objective (e.g., a limit on carbon emissions) than will one using a top-down approach. To some extent, the differences may lie in a failure of bottom-up studies to accurately account for all costs associated with implementing specific actions. Top-down methods, on the other hand, can fail to account realistically for consumer and producer behavior by relying too heavily on aggregate data, as noted by Krause *et al.* In addition, some top-down methods sacrifice sectoral and technology detail in return for being able to solve for general equilibrium resource allocations. Finally, Boero *et al.* noted that top-down methods often ignore the fact that economies depart significantly from the stylized equilibria represented by the methods. Each approach, however, captures costs or details on technologies, consumer behavior, or impacts that the other does not. Consequently, a comprehensive assessment should combine elements of each approach to ensure that all relevant costs and impacts are accounted for.

The two approaches have been used in the development of national energy plans or policies that require identification and analysis of different actions that governments could take to encourage adoption of energy technologies and practices. Based on such analyses, policymakers can decide which options not only satisfy specific policy objectives but are also within institutional, political, and budget constraints. Typically, the analytic process will follow a series of steps, each of which produces information for decision makers. The manner in which these steps are performed will reflect each country's resources, objectives, and decision-making process.

Both approaches have been used extensively in the assessment of costs of climate change mitigation. The earlier literature dating to 1970s focused primarily on evaluation of the energy and particularly the petroleum sector, and categorized approaches into sectoral models, industry-market models, energy system models, and energy economy models. The first three approaches would be referred to as bottom-up approaches and the last one as the top-down approach today. From the late 1980s to date, much of the attention in the application of these two energy-sector approaches has focused on climate change mitigation, that is, the long-term stabilization of climate change and the short-term reduction of GHG emissions. The following sections focus on the evolution and use of energy-sector bottom-up approaches as they have been applied for mitigation assessment, particularly the short-term reduction of GHG emissions.

2. *STRUCTURE OF AN ENERGY SECTOR BOTTOM-UP ASSESSMENT*

The energy sector comprises the major energy demand sectors (industry, residential and commercial, transport, and agriculture) and the energy supply sector (resource extraction, conversion, and delivery of energy products). GHG emissions occur at various points in the sector, from resource extraction to end use, and, accordingly, options for mitigation exist at various points.

The bottom-up approach involves the development of scenarios based on energy end uses and evaluation of specific technologies that can satisfy demands for energy services. One can compare technologies based on their relative cost to achieve a unit of GHG reduction and other features of interest. This approach gives equal weight to both energy supply and energy demand options. A variety of screening criteria, including indicators of cost-effectiveness as well as noneconomic concerns, can be used to identify and assess promising options, which can then be combined to create one or more mitigation scenarios. Mitigation scenarios are evaluated against the backdrop of a baseline scenario, which simulates the events assumed to take place in the absence of mitigation efforts. Mitigation scenarios can be designed to meet specific emission reduction targets or to simulate the effect of specific policy interventions. The results of a bottom-up assessment can then be linked to a top-down analysis of the impacts of energy sector scenarios on the macroeconomy.

Energy-sector bottom-up assessments require physical and economic data about the energy system, socioeconomic variables, and specific technology options, and GHG emissions if these are targeted. Using these data, a model or accounting system of the energy sector is designed to suit local circumstances.

The manner in which an assessment is performed reflects each country's resources, objectives, and decision-making process as well as the type of modeling approach employed. Figure 1 depicts the basic steps of a typical mitigation assessment and how they relate to one another. Some of the steps are interlinked, so they are not necessarily sequential, and require iterations.

An initial step is to assemble data for the base year on energy demand by sector, energy supply by type of energy source, and energy imports and exports. The disaggregated energy data is normalized to match the national energy supply totals for the base year. One then calibrates base year emissions with the existing GHG inventory, as needed. The analyst also assembles data for the base year on the technologies used in end-use sectors and in energy supply.

The data for the base year is used as a starting point for making projections of future parameters and developing integrated scenarios of energy demand and supply. On both the demand and supply side, one identifies and screens potential technology options to select those that will be included in the analysis. The screening is guided by information from an assessment of energy resources, as well as the potential for energy imports and exports.

Once the list of technologies has been made manageable by the screening process, the analyst characterizes potential technology options in end-use sectors and in energy supply with respect to costs, performance, and other features. On the demand side, this characterization will assist in projecting end-use energy demands in the various sectors. Projecting energy demand also requires one to project activity levels in each subsector for indicators such as tons of various industrial products, demand for travel and freight transport, and number of urban and rural households. These projections are based on the assumptions for growth in key parameters such as GDP and population. Assumptions about sectoral policies with respect to energy pricing and other factors are also important in developing projections of energy demand.

The data from the energy demand and supply analyses are then entered into an energy sector model or accounting framework that allows for integrated analysis of the various options that can meet energy requirements. This analysis calculates costs and impacts over the time horizon considered, the results are reviewed for reasonableness, and uncertainty is taken into consideration. This step involves combining technology options to meet the objectives of each scenario. The selection of technologies may be made directly by the analyst or performed by the model (as with an optimization model).

The baseline scenario projects energy use and emissions over the time horizon selected, reflecting the development of the national economy and energy system under the assumption that no policies are introduced to reduce GHG emissions. The baseline scenario must include sufficient detail on future energy use patterns, energy production systems, and technology choices to enable the evaluation of specific policy options. An alternative baseline

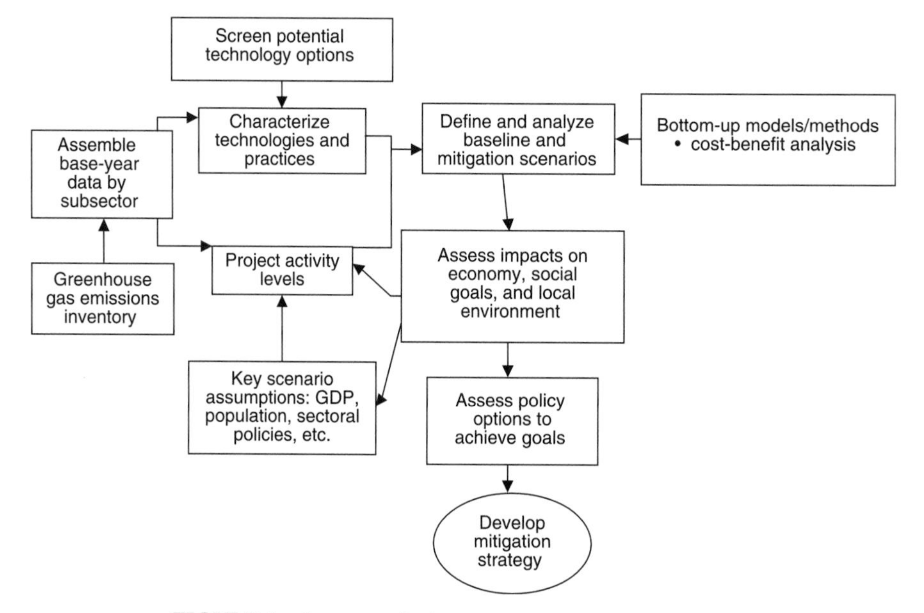

FIGURE 1 Structure of a bottom-up assessment.

scenario can be developed if desired (for example, to reflect different assumptions about GDP growth).

Policy scenarios can be defined to meet particular emission reduction targets, to assess the potential impact of particular policies or technologies, or to meet other objectives. The comparisons of policy and baseline scenarios reveal the net costs and impacts of the policy options. The results are assessed with respect to reasonableness and achievability, given barriers to implementation and the policy instruments that might be used, such as taxes, standards, or incentive programs.

For both baseline and policy scenarios, the analyst assesses the impacts on the macroeconomy, social goals (such as employment), and the national environment. One approach is to integrate bottom-up assessment with a macroeconomic model. Decision analysis methods that allow for consideration of multiple criteria may also be appropriate.

After scenarios have been analyzed and options have been ranked in terms of their attractiveness (on both quantitative and qualitative terms), it is desirable to conduct a more detailed evaluation of policies that can encourage adoption of selected options. Such an evaluation can play an important role in the development of a national strategy. The latter step requires close communication between analysts, policymakers, and other interested parties.

3. TYPICAL MODELS FOR A BOTTOM-UP ENERGY SECTOR ASSESSMENT

Bottom-up assessments of the energy sector typically use an accounting or modeling framework to capture the interactions among technologies and to ensure consistency in the assessment of energy, emission, and cost impacts. Accounting and modeling methods can vary greatly in terms of their sophistication, data intensiveness, and complexity. This section provides an overview of key concepts and capabilities of some models that have been used for this purpose in energy/environmental studies in developing and industrialized countries.

As discussed earlier, it is common to divide energy models into two types, so-called bottom-up and top-down, depending on their representation of technology, markets, and decision making. In practice, there is a continuum of models, each combining technological and economic elements in different ways. At one extreme are pure bottom-up energy models, which focus on fuels and energy conversion or end-use technologies and treat the rest of the economy in an aggregated fashion. At the other extreme are pure top-down models, which treat energy markets and technologies in an aggregated manner and focus instead on economy-wide supply-demand relations and optimizing behavior. Between these two cases are a number of models that combine elements of both extremes with various degrees of emphasis and detail.

The description of the future varies among the models. Some models can only analyze a snapshot year and compare this to another year, without any representation of the transition between them. Dynamic models, on the other hand, allow for time-dependent descriptions of the different elements of the energy system. While the snapshot models enable great detail in the representation of the system, dynamic models allow for representation of technology capacity transfer between time periods and thus time-dependent capacity expansion, time-dependent depletion of resources, and abatement costs as they vary over time.

In dynamic modeling, information about future costs and prices of energy is available through two diametrically different foresight assumptions. With myopic foresight, the decisions in the model are made on a year-by-year basis, reflecting the assumption that actors expect current prices to prevail indefinitely. Assuming perfect foresight, the decisions at any year are based on the data for the entire time horizon. The model thus reflects the activities of market participants as if they use the model itself to predict prices.

Table I summarizes the key design features of some of the models. Most of the models listed in Table I can be used to integrate data on energy demand and supply. The models can use this information for determining an optimal or equilibrium mix of energy supply and demand options. The various models use cost information to different degrees and provide for different levels of integration between the energy sector and the overall economy.

3.1 Energy Accounting Models

Energy accounting models reflect an engineering or input-output conception of the relations among energy, technology, and the services they combine to produce. This view is based on the concept of energy services that are demanded by end users. Schematically, this can be represented as follows:

Energy inputs > Technology > Energy services.

TABLE I

Key Design Features of Types of Bottom-Up Energy Sector Models

Model characteristics	Energy accounting	Engineering optimization	Iterative equilibrium	Hybrid
Energy supply representation	Process analysis	Process analysis	Supply curve	Process analysis
Energy demand representation	Exogenous	Exogenous	Exogenous	Utility maximization
Multiperiod	Yes	Yes	Yes	Yes
Consumer/producer foresight	Not applicable	Perfect/myopic	Myopic	Perfect/myopic
Solution algorithm	Accounting	Linear or nonlinear optimization	Iteration	Nonlinear optimization

For policy purposes, the significance of this approach is that a given type and level of energy service can be obtained through various combinations of energy inputs and technologies. In particular, holding the service constant while increasing the energy efficiency of the technology allows decrease in the required level of energy input. In a range of cases, when other factors are held equal, this lowers the overall cost of the energy service. With accounting models, the evaluation and comparison of policies is performed by the analyst external to the model itself.

These models are essentially elaborations on the following accounting identity describing the energy required for satisfying a given level of a specific energy service:

$$E = AI,$$

where E indicates energy, A indicates activity, and I indicates intensity. With multiple end uses, aggregate energy demand is simply the sum as follows:

$$E = \text{Sum of } (A_i I_i).$$

Accounting models are essentially spreadsheet programs in which energy flows and related information such as carbon emissions are tracked through such identities. The interpretation of the results, and the ranking of different policies quantified in this manner, is external to the model and relies primarily on the judgment of the analyst.

Note that these calculations assume that a number of factors are held constant, including energy service level and equipment saturations. Also, these expressions represent a quasi-static view; in actual practice, such calculations would be performed over time paths of costs, activities, intensities, and prices developed in scenario construction. Finally, it is easy to see that factors for carbon savings from the shift to efficient technologies can be easily included in such calculations.

In energy accounting models, macroeconomic factors enter only as inputs in deriving demand-side projections—that is, there is no explicit representation of feedback from the energy sector to the overall economy. While different models contain different levels of detail in representing the supply sector, supply-demand balancing in this type of model is accomplished by back calculation of supply from demand projections.

In contrast to optimization models, accounting models cannot easily generate a least-cost mitigation solution. They can be used to represent cost-minimizing behavior estimated by the analyst, however. They tend to require less data and expertise and are simpler and easier to use than optimization models.

3.2 Engineering Optimization Models

In engineering optimization models, the model itself provides a numerical assessment and comparison of different policies. These models are linear programs in which the most basic criterion is total cost of providing economy-wide energy services under different scenarios; when this criterion is used, the structure of this type of model as used in mitigation analysis can be represented schematically as follows:

Minimize total cost of providing energy
and satisfying end-use demand subject to

1. Energy supplied, energy demanded
2. End-use demands satisfied
3. Available resource limits not exceeded

In addition to the overall optimization structure of these models, perhaps the key distinction between these and the accounting models is that, within the model structure itself, trade-offs are made among different means of satisfying given end-use demands for energy services.

The intertemporal structure of these linear programming models varies. Some are constructed to perform a target year analysis: the model is first parameterized and run for the base year, then the procedure is repeated for a single designated future year (typically 2005, 2010, or 2020). Others perform a more elaborate dynamic optimization, in which time paths of the parameters are incorporated and the model generates time paths of solutions.

In engineering optimization models, macroeconomic factors enter in two ways. First, they are used to construct forecasts of useful energy demands. Second, they can be introduced as constraints. For example, the overall cost minimization can be constrained by limits on foreign exchange or capital resources. In both cases, the models do not provide for the representation of feedbacks from the energy sector to the overall economy.

Supply and demand are balanced in engineering optimization models by the presence of constraints, as indicated earlier. The engineering detail and level of disaggregation used in both the supply and demand side are at the discretion of the user, and in practice these vary widely among models.

This type of model allows several means of analyzing GHG emissions and the effects thereupon of various policy options. For example, as an alternative to minimizing energy costs, criteria such as minimizing carbon output subject to the constraints can be employed. In addition, an overall cap on carbon emissions can be entered as a constraint in the model and the cost minimization performed with this restriction. Each such approach allows the comparison of different policy intervention.

3.3 Iterative Equilibrium Models

These models incorporate the dynamics of market processes related to energy via an explicit representation of market equilibrium—that is, the balancing of supply and demand. These are used to model a country's total energy system and do not explicitly include an economy model integrated with the energy system model. Thus, macroeconomic factors enter the model exogenously, as in the previous model types discussed. (That is, demands for energy services are derived from macroeconomic drivers rather than being obtained endogenously.) These models thus occupy an intermediate position between engineering, energy-focused models, and pure market equilibrium models.

The methodology employed to solve the model is a process network wherein individual energy processes are represented with standard model forms, with application specific data, and linked together as appropriate. Prices and quantities are then adjusted iteratively until equilibrium is achieved. This iterative approach makes it much easier to include noncompetitive-market factors in the system than in the optimization approach.

3.4 Hybrid Models

In hybrid models, the basic policy measure is the maximization of the present value of the utility of a representative consumer through the model planning horizon. Constraints are of two types: macroeconomic relations among capital, labor, and forms of energy, and energy system constraints. The model generates different time paths of energy use and costs and macroeconomic investment and output. The energy submodel contains representations of end-use technologies, with different models containing different levels of detail. Schematically, this type of model can be represented as follows:

Maximize (discounted) utility of consumption
subject to

1. Macroeconomic relations among output, investment, capital, labor, and energy
2. Energy system and resource constraints (as in engineering optimization models)

The constraints in this case are also dynamic: they represent time paths for the model variables.

In this type of model, energy demands are endogenous to the model rather than imposed exogenously by the analyst. In addition, this optimization structure indicates the difference we noted earlier in the way the different models incorporate macroeconomic data. Specifically, in accounting and engineering optimization models, these data—on GDP, population growth, capital resources, and so on—enter essentially in the underlying constructions of the baseline and policy scenarios. In the hybrid model, however, such data enter in the macroeconomic relations (technically, the aggregate production function) as elasticities and other parameters. Within this model framework, changes in energy demand and supply can feed back to affect macroeconomic factors. It should be noted that, despite their inclusion of engineering optimization subcomponents, these models typically do not contain as much detail on specific end-use technologies as many purely engineering models.

4. KEY CHALLENGES IN THE BOTTOM-UP MODELING APPROACH

A number of key challenges arise in the bottom-up modeling approach. These include (1) incorporating efficiency versus equity; (2) aggregation over time, regions, sectors, and consumers; (3) representing decision rules used by consumers; (4) incorporating technical change; (5) capturing facility retirement dynamics; (6) avoiding extreme solutions; and (7) accounting for carbon flows.

4.1 Incorporating Efficiency versus Equity

None of the models discussed earlier provide explicitly for making trade-offs between efficiency and equity. Different models, however, have different implications for the analyst's consideration of this important issue. Nonoptimization models do not themselves choose among different policies in an explicit way but can allow for the ranking of policies according to criteria specified by the analyst, including considerations of equity. Engineering optimization models, since they focus on least-cost solutions to the provision of energy services, leave to the analyst the judgment of how to trade-off the importance of energy with that of other economic and social priorities. The models that optimize the utility of a representative consumer, in a sense, constrain consideration of the issue the most. Embedded in this modeling structure is a view of the economic system that equates social optima with competitive economic equilibria; the appropriateness of this perspective in the application at hand must be weighed carefully.

4.2 Aggregation over Time, Regions, Sectors, and Consumers

Perhaps the most fundamental formulation issue is the level of aggregation at which costs and benefits are calculated. Economic efficiency is generally insured if discounted net benefits are maximized in the aggregate; any desired income redistribution is handled subsequently. Decision makers, however, are fundamentally interested in how the costs and benefits fall on various income, industry, and regional groups. Coupled with the relative emphasis of the analysis on equity versus efficiency is the desired level of disaggregation of the model by region, time periods, industry, and income group. Obviously, the level of disaggregation must be sufficient to allow reporting of results at the desired level, but in some cases the projection of an aggregate variable can be improved by some level of disaggregation to capture the heterogeneity in decision making objects on the part of the different groups. These decision rules themselves are critical elements of the models and range from minimizing discounted future costs (or maximizing benefits) over a 40- or 50-year time horizon to picking investments that come close to minimizing costs based on conditions for a single year only.

4.3 Representing Decision Rules Used by Consumers

Accounting models contain no explicit representation of consumer decision making. In practice, however, their use often reflects the view that certain market barriers constrain consumers from making optimal decisions with respect to energy efficiency. At the other extreme, the use of the representative consumer in the optimization models rests on strong assumptions regarding consumer behavior, serving primarily to ensure mathematical and empirical tractability. Key among these are perfect foresight and essential homogeneity among consumers or households.

4.4 Incorporating Technical Change

Another set of key assumptions about inputs are those made about the costs and efficiencies of current and future technologies, both for energy supply and energy use. Most analysts use a combination of statistical analysis of historical data on the demand for individual fuels and a process analysis of individual technologies in use or under development to represent trends in energy technologies. At some point these two approaches tend to look quite similar though, as the end-use process analysis usually runs out of new technology concepts after some years or decades, and it is then assumed that the efficiency of the most efficient technologies for which there is an actual proposed design will continue to improve as time goes on. Particularly important, but difficult, here is projecting technological progress. Almost all top-down models have relied on the assumption of "exogenous" or "autonomous" technical change, that is, productivity trends (relating to energy as well as other factors of production) that are a function of time only, and in particular not affected by either changes in relative factor prices or by policy interventions. Bottom-up approaches, by contrast, assume (usually implicitly) that technological progress can be accelerated by policy intervention. This difference constitutes another significant contrast between the two approaches. There has been an acceleration of research, among top-down modelers, on incorporating technological change that is "endogenous"—that is, responsive to price changes, policies, or both. This research may eventually contribute to a partial reconciliation of the two approaches in the treatment of technological progress.

4.5 Capturing Facility Retirement Dynamics

Most modeling approaches focus on investments in new energy producing and consuming equipment, which is typically assumed to have a fixed useful lifetime. In scenarios where conditions change significantly (either through external factors or explicit policy initiatives), it may be economic to retire facilities earlier or later than dictated purely by physical depreciation rates. This endogenous calculation of facility retirement dates can be handled analytically in most models, but it represents a major increase in data and computational requirements.

4.6 Avoiding Extreme Solutions

Another typical problem, particularly with models that assume optimizing behavior on the part of individual economic agents, is the danger of knife edge results, where a small difference in the cost of two competing technologies can lead to picking the cheaper one only. This is generally handled by disaggregating consumers into different groups who see somewhat different prices for the same technology (e.g., coal is cheaper in the coal fields than a thousand miles away), modeling the decision process as somewhat less than perfect, or building appropriate time lags into the modeling structure.

4.7 Accounting for Carbon Flows

Finally, estimating carbon flows for a given energy system configuration can be complicated. It is more accurate to measure emissions as close to the point of combustion as possible so types of coal and oil product can be distinguished and feedstocks (which do not necessarily produce carbon emissions) can be netted out. However, a point-of-use model requires far more data and computation than the models described here, which aggregate several fuel types and use average carbon emissions factors for each fossil fuel.

5. DATA SOURCES FOR BOTTOM-UP ANALYSIS

Regardless of the approach taken and analysis tool used, the collection of reliable data is a major and relatively time-consuming aspect of bottom-up analysis. To keep data constraints from becoming a serious obstacle to the analysis, two points are essential. First, the bottom-up model should be sufficiently flexible to adapt to local data constraints. Second, the data collection process should be as efficient as possible. Efficiency can be maximized by focusing the detailed analysis on sectors and end uses, where the potential for GHG mitigation is most significant, and avoiding detailed data collection and analysis in other sectors.

Data collection generally begins with the aggregate annual energy use and production figures typically found on a national energy balance sheet. The remaining data requirements depend largely on (1) the disaggregated structure of the analysis, (2) the specific technology and policy options considered, and (3) local conditions and priorities.

Table II shows the typical types of data needed for a bottom-up approach to mitigation analysis. They tend to fall within five general categories: macroeconomic and socioeconomic data, energy demand data, energy supply data, technology data, and emission factor data. The full listing of potential data requirements may appear rather daunting. In practice, however, much of the data needed may already be available in the form of national statistics, existing analytical tools, and data developed for previous energy sector studies. The development and agreement on baseline projections of key variables, the characterization of mitigation options relevant to local conditions, and, if not already available, the compilation of disaggregated energy demand data are typically the most challenging data collection tasks facing the analyst.

In general, emphasis should be placed on locally derived data. The primary data sources for most assessments will be existing energy balances, industry-specific studies, household energy surveys, electric utility company data on customer load profiles, oil company data on fuel supply, historical fuel price series maintained by government departments, vehicle statistics kept by the transportation department,

TABLE II

Data Sources for a Bottom-Up Mitigation Analysis

Data categories	Types of data	Common data sources
Macroeconomic variables		
Aggregate driving variables	GDP/value added, population, household size	National statistics and plans; macroeconomic studies
More detailed driving variables	Physical production for energy intensive materials; transportation requirements; agricultural production and irrigated area; changes in income distribution, etc.	Macroeconomic studies; transport sector studies; household surveys, etc.
Energy demand		
Sector and subsector totals	Fuel use by sector/subsector	National energy statistics, national energy balance, energy sector yearbooks (oil, electricity, coal, etc.)
End-use and technology characteristics by sector/subsector	Energy consumption breakdown by enduse and device: e.g., energy use characteristics of new versus existing building stock; vehicle stock; breakdown by type, vintage, and efficiencies; or simpler breakdowns	Local energy studies; surveys and audits; studies in similar countries; general rules of thumb from end-use literature
Response to price and income changes (optional)	Price and income elasticities	Local econometric analyses; energy economics literature
Energy supply		
Characteristics of energy supply, transport, and conversion facilities	Capital and O&M costs, performance (efficiencies, unit intensities, capacity factors, etc.)	Local data, project engineering estimates, Technical Assessment Guide; IPCC Technology Characterization Inventory
Energy prices		Local utility or government projections; for globally traded energy products
Energy supply plans	New capacity online dates, costs, characteristics	National energy plans; electric utility plans or projections; other energy sector industries (refineries, coal companies, etc.)
Energy resources	Estimated, proven recoverable reserves of fossil fuels; estimated costs and potential for renewable resources	Local energy studies
Technology options		
Technology costs and performance	Capital and O&M costs, performance (efficiencies, unit intensities, capacity factors, etc.)	Local energy studies and project engineering estimates; technology suppliers; other mitigation studies
Penetration rates	Percent of new or existing stock replaced per year; overall limits to achievable potential	
Emission factors	Kg GHG emitted per unit of energy consumed, produced, or transported	National inventory assessments; IPCC Inventory Guidelines IPCC Technology Characterization Inventory

and so on. The main thrust of the data collection effort is not so much on collecting new primary data but on collating secondary data and establishing a consistent data set suitable for analysis using the model of choice.

Where unavailable, particularly in developing country analyses, local data can be supplemented with judiciously selected data from other countries. For example, current and projected cost and performance data for some mitigation technologies (e.g., high-efficiency motors or combined cycle gas units) may be unavailable locally, particularly if the technologies are not presently in wide use. For this purpose, technology data from other countries can provide indicative figures and a reasonable starting point. For data on energy use patterns, such as the fraction of electricity used for motor drive in the textile industry, the use of extemal data can be somewhat more problematic. In general, it may be possible to use estimates and general rules of thumb suggested by other country studies, particularly data from other countries with similar characteristics.

6. *DEVELOPING SCENARIOS FOR USE IN A BOTTOM-UP ASSESSMENT*

A bottom-up assessment typically compares the energy and environmental consequences of one future scenario against another. The scenarios may be composed of widely different views of the world (e.g., a rapidly growing economic world versus a slowly growing one). These types of scenarios allow companies to prepare for operations in either type of future. Often, governments though may wish to analyze the consequences of their policy or programmatic actions and in such a case it is desirable to compare a policy scenario against a reference one. The latter is often referred to as a business-as-usual or baseline scenario. Developing these scenarios is a complex process. It requires the combining of both analyses with some judgment of the future evolution of variables that are likely to affect the energy and environment system.

6.1 Developing a Baseline Scenario

Developing a baseline scenario that portrays social, demographic, and technological development over a 20- to 40-year or longer time horizon can be one of the most challenging aspects of a bottom-up analysis. The levels of projected future baseline emissions shape the amount of reductions required if a specific policy scenario target is specified and the relative impacts and desirability of specific mitigation options. For instance, if many low-cost energy efficiency improvements are adopted in the baseline scenario, this would yield lower baseline emissions and leave less room for these improvements to have an impact in a policy scenario.

Development of a baseline scenario begins with the definition of scenario characteristics (e.g., business as usual). Changes in exogenous driving variables are then specified and entered into the model, which is run to simulate overall energy use and emissions over the time horizon selected. The baseline scenario is evaluated for reasonableness and consistency and revised accordingly. Uncertainty in the evolution of the baseline scenario can be reflected through a sensitivity analysis of key parameters such as GDP growth.

The procedure will vary somewhat depending on the modeling approach used and the nature of a baseline scenario. In an optimization model, the use of different technologies is to a certain degree decided within the model, dependent on how much one wants to constrain the evolution of the baseline scenario. For example, the analyst might choose to construct a baseline scenario in which the energy supply system closely reflects or extrapolates from published plans. Alternately, the analyst might choose to give the model more flexibility to select a future energy supply system based on specific criteria. If one is using an optimization model, it is necessary to introduce certain constraints if one wishes to force the model toward a solution that approximates a business-as-usual future.

The rate of economic growth and changes in domestic energy markets are among the most important assumptions affecting projected baseline emissions. Official government GDP projections may differ from other macroeconomic projections. In terms of domestic energy markets, the removal of energy price subsidies could greatly affect fuel choice and energy efficiency, and thus baseline emissions and the impacts of mitigation options.

A preparatory step in developing a baseline scenario is to assemble available forecasts, projections, or plans. These might include national economic development plans, demographic and economic projections, sector-specific plans (e.g., expansion plans for the iron and steel industry), plans for transport and other infrastructure, studies of trends in energy use (economy wide, by sector, or by end use), plans for investments in energy facilities (electricity expansion plans, new gas pipelines, etc.), studies of resource availability, and projections of future resource prices. In short, all studies that attempt to look into a country's future—or even the future of a region—may provide useful information for the specification of a baseline scenario. However, it is unlikely that every parameter needed to complete the baseline scenario will be found in national documents or even that the documents will provide a consistent picture of a country's future. As with much of the modeling process, the judgment of the analyst in making reasonable assumptions and choices is indispensable.

6.1.1 Developing Projections of Energy Demand

In bottom-up approaches, projections of future energy demand are based on two parameters for each subsector or end use considered: a measure of the activity that drives energy demand and a measure of the energy intensity of each activity, expressed in energy units per unit of activity.

Measures of activity include data on household numbers, production of key industrial products, and demand for transport and services. Activity may be measured in aggregate terms at the sectoral level (e.g., total industrial value added, total passenger-km or ton-kin) and by using indicators at the subsector level. These two measures need not be identical. For example, total industrial value added is a common indicator for aggregate activity for the industrial sector, but for specific subsectors such as steel or cement one often uses tons of production as a measure of activity. In general, physical measures of activity are preferable, but they are not appropriate in all cases (such as in light industry, where there is no aggregate measure of physical production).

In bottom-up approaches, future values for driving activities are exogenous—that is, based on external estimates or projections rather than being estimated by the model itself. Future values can be drawn from a variety of sources or estimated using various forecasting methods. Estimates of the future levels of activity or equipment ownership depend on local conditions and the behavioral and functional relationships within each sector.

Projections of the future development of energy intensities in each subsector or end use can be expressed in terms of final energy or useful energy. When the choice of end-use options is conducted within the model, however, the energy demand should be given in terms of useful energy to allow the model to select among technologies for meeting the energy requirements.

Projections should start from the base year values, such that the sum of the product of energy intensity and activity level in each subsector add up to total final or useful energy use in the base year. In energy statistics, the data are normally presented in terms of final energy. If the useful energy refers to the amount of energy required to meet particular demands for energy services. It is typically estimated by multiplying final energy consumption by the average conversion efficiency of end-use equipment (e.g., of oil-fired boilers). Applying the concept of useful energy is more difficult in the transport sector, although one can use the estimated conversion efficiency of generic engine types in various vehicle classes. If the projections are to be given in useful energy units, the statistical data should be converted using estimated base year efficiencies for each end use.

Ideally, the projections of energy intensities are based in part on historical developments. To the extent statistical data on a disaggregated level are available, they give limited information for economies that have undergone significant structural changes or changes in taxation/subsidies on energy. Even if reliable historical data are available, assumptions on the development of the energy intensities have to be made using careful judgment about the status of existing end-use technologies and future efficiency improvements that should be included in the projections.

It is important to distinguish between improvements included in the exogenous demand projections and improvements that are explicitly included as technology options in the model. For example, future improvements of building insulation standards can either be directly included in the demand projections as an improvement of intensity for space heating or modeled as technology options that can be chosen individually in the assessment, depending on their attractiveness in the different scenarios. The distinction is especially important in optimization models. One can assume that the insulation standards will be implemented (e.g., through use of regulations) or allow the model to choose the implementation. In the latter case, the insulation option will be implemented if its cost is less than the cost of supplying the heat using available options for space heating.

Once the initial baseline scenario is prepared, it is reviewed to assess whether it presents a comprehensive and plausible future for the country in light of real-world constraints. Some specific questions might include the following:

- Can the indicated growth rates in energy demand be sustained over the study period? Is it a reasonable rate, given recent experience in the country and region?
- Is the level of capital investment needed to sustain the indicated levels of industrial growth likely to be forthcoming?
- Will the country be able to afford the bill for fuel imports that is implied by the baseline scenario?
- Will the capital investment needed for energy supply system expansion be available, given competition for limited financial resources?
- Is the indicated increase in transportation use plausible, given current and planned transportation infrastructure?
- Are the emission factors in use appropriate for future technologies?

Answers to these types of questions might indicate the need for adjustments to the baseline scenario or sensitivity analyses of key parameters.

6.2 Developing Policy Scenarios

The process of developing a policy scenario or scenarios involves establishing a scenario objective and combining specific options in an integrated scenario. Integrated scenario analysis is essential for developing accurate and internally consistent estimates of overall cost and emissions impacts since the actual emissions reduction from employing a specific option can depend on the other options included in a scenario. For instance, the level of reduction in GHG emission associated with an option that saves electricity is dependent on the electricity generation resources whose use would be avoided (e.g., coal, oil, hydro, or a mix). In reality, the type of electricity generation that the efficiency option would avoid will change over time, and, if lower GHG emitting electricity resources are also introduced in the scenario, the GHG savings of the efficiency option may be reduced. Integrated scenario analyses are intended to capture these and other interactive effects.

Where using an optimization model, the difference in input data for the baseline scenario and the policy scenario(s) is typically less than in an accounting model, where the choice of technologies is exogenous to the model. An optimization model chooses from the whole set of available technologies to satisfy the given constraints.

6.2.1 Objectives for a Policy Scenario

Several objectives are possible for designing a policy scenario. The objective depends on political and practical considerations. Types of objectives include the following:

- *Emission reduction targets.* For example: 12.5% reduction in CO_2 emissions by 2010, and 25% reduction by 2030, from baseline levels. An alternative is to specify reductions from base year levels, which avoids making the amount of reduction dependent on the specification of the baseline scenario.
- *Identification of options up to a certain cost per ton of energy or emissions reduction.* The energy or emissions reduction given by the resulting technology mix would reflect the level of reduction that could be achieved at a certain marginal cost.
- *No-regrets scenario.* This scenario is a common variant of the previous type of objective, where the screening threshold is essentially zero cost per tonne of energy or GHG reduced.
- *Specific technology options or packages of options.* Examples of this type of scenario might be a natural gas scenario, a renewable energy scenario, or a nuclear scenario.

7. RESULTS FROM A BOTTOM-UP APPROACH

Following are some key questions that an analyst would seek to answer from a bottom-up energy sector assessment:

1. What is the economic cost of providing energy services in a baseline or policy scenario? What is the incremental cost between scenarios?

1a. What are the capital and foreign exchange implications of pursuing alternative scenarios?

2. What is the economic cost of pursuing particular mitigation (policy and technology) options, such as high efficiency lighting or a renewable technology, and what are its local and global environmental implications?
3. What are the costs of reducing emissions to a predetermined target level? (Target may be annual or cumulative).
4. What is the shape of the marginal cost curve for reducing carbon emissions?

4a. How do alternative technologies rank in terms of their carbon abatement potential?

Any one of the three types—accounting, optimization, and iterative equilibrium—of models discussed here can address questions 1 and 2. Question 1a is important for developing country planners, since these two cost components are often insufficient in these countries.

Question 3 is easiest to address using an optimization model. Question 4 requires that energy supply and demand be evaluated in an integrated manner. A demand-side mitigation measure may change the energy supply configuration measure, which will affect the GHG emissions of the energy system. The optimization and iterative models are capable of capturing the integrated effect and deriving the changes in environmental emissions. The accounting models may not capture the changes in the energy supply mix and the consequent GHG emissions and thus may show higher emissions reduction for a mitigation measure. Only the optimization models can calculate marginal costs directly. In the other models, an approximation of the marginal cost curve can be constructed by performing a large number of carefully selected runs.

SEE ALSO THE FOLLOWING ARTICLES

Economics of Energy Demand • Economics of Energy Supply • Modeling Energy Markets and Climate Change Policy • Modeling Energy Supply and Demand: A Comparison of Approaches • Multi-criteria Analysis of Energy • National Energy Modeling Systems

Further Reading

IPCC (2001). Third Assessment Report, Chapter 8.

Krause, F., Baer P., and DeCanio, S. (2001). Cutting Carbon emissions at a Profit: Opportunities for the US. IPSEP, El Cerrito CA.

LBNL and ORNL (2000). Clean Energy Futures for the U.S.

Sathaye, J., and Ravindranath, N. (1998). Climate change mitigation in the energy and forestry sectors of developing countries. *Annual Rev. Energy and Environment* **23**, 387–437.

Business Cycles and Energy Prices

STEPHEN P. A. BROWN, MINE K. YÜCEL, and JOHN THOMPSON
Federal Reserve Bank of Dallas
Dallas, Texas, United States

1. Definition of Business Cycles and the Role of Energy Prices
2. Understanding the Basic Response
3. Amplification of the Basic Response
4. Asymmetry of the Response
5. A Weakening Response
6. Conclusions and Implications for Policy

Glossary

aggregate channels Changes in economic activity that work through changes in aggregate supply or demand.

aggregate economic activity The sum total of economic activity in a country, which is commonly measured by gross domestic product (GDP).

allocative channels Changes in economic activity that work through changes in the composition of output.

asymmetry A process in which aggregate economic activity is reduced more by an oil price increase than it is increased by an oil price decrease.

business cycles Fluctuations in aggregate economic activity.

consumption smoothing The process of adjusting consumption spending to lifetime income by holding consumption relatively constant when there are short-term fluctuations in income.

core inflation A measure of inflation that is thought to provide a better signal of underlying inflationary pressure because it excludes food and energy prices, which are quite volatile.

federal funds rate The interest rate at which commercial banks borrow from each other's reserves that are on deposit with the Federal Reserve System.

net oil price An oil price series that captures no price decreases and only those increases that take the price of oil higher than it has been during the past year.

neutral monetary policy The conduct of monetary policy in such a way that it has no effect on economic activity. According to Robert Gordon of Northwestern University, a neutral monetary policy leaves nominal GDP unchanged. According to Milton Friedman, who is retired from the University of Chicago, a neutral monetary policy requires that the money supply be unchanged. Many other economists regard a constant federal funds rate as neutral monetary policy. In the absence of aggregate supply shocks, all three definitions of neutrality are consistent with each other. Aggregate supply shocks can cause a divergence between these three measures of monetary neutrality.

oil price shock A sharp change in the price of oil.

putty–clay technology Technology in which the energy-to-output, capital-to-output, and labor-to-output ratios can be varied over the long term as capital is purchased and installed but cannot be changed in the short term because they are embedded in the capital stock.

real interest rate The market interest rate minus expected inflation.

sticky wages and prices The inability of market forces to change wages and prices. This stickiness usually occurs when market forces would reduce nominal wages and prices.

Oil price shocks have figured prominently in U.S. business cycles since the end of World War II, although the relationship seemed to weaken during the 1990s. In addition, the economy appears to respond asymmetrically to oil price shocks; increasing oil prices hurt economic activity more than decreasing oil prices help it. This article focuses on an extensive economics literature that relates oil price shocks to aggregate economic activity. It examines how oil price shocks create business cycles, why they seem to have a disproportionate effect on economic activity, why the economy responds asymmetrically to oil prices, and why the relationship between oil prices and economic activity may have weakened. It also addresses the issue of developing energy policy to mitigate the economic effects of oil price shocks.

1. DEFINITION OF BUSINESS CYCLES AND THE ROLE OF ENERGY PRICES

Despite occasional periods of notable recession, the U.S. economy has generally traveled an expansionary path throughout its history. In fact, U.S. gross domestic product (GDP) has averaged a 3.6% annual rate of increase since 1929. Population increases account for some of the growth, but productivity gains have also been essential to long-term national economic growth.

Although the U.S. economy has generally expanded over time, its upward trend has not been free of interruption. Shocks have occasionally hit the U.S. economy, disrupting the expansionary forces and creating business cycles. Since World War II, oil price shocks have played a significant role in U.S. business cycles. In fact, increasing oil prices preceded 9 of the 10 recessions that occurred from the end of World War II through 2002 (Fig. 1). Research conducted by James Hamilton strongly suggests that the recessions that followed sharply rising oil prices were not the result of other business cycle variables, such as aggregate demand shocks or contractionary monetary policy.

By examining the influence of oil price shocks on U.S. economic activity in the six decades since the end of World War II, economists have found that oil price shocks seem to have a disproportionate effect on economic activity. In addition, the economy appears to respond asymmetrically to oil price movements—that is, the gains in economic activity that follow oil price declines are not commensurate with the losses in economic activity that follow oil price increases. In addition, oil price shocks seem to have had less effect on economic activity during the 1990s than in previous decades.

These observations suggest a number of questions. What basic factors account for the negative influence that sharply rising oil prices have on U.S. economic activity? Why do sharply rising oil prices have a disproportionate effect in weakening economic activity? Why doesn't the economy respond as favorably to declining oil prices as it responds unfavorably to increasing oil prices? Why has the economy become less sensitive to oil price shocks? How does our understanding of the effect of oil price shocks on economic activity influence desired energy policies?

2. UNDERSTANDING THE BASIC RESPONSE

The oil price shock of 1973 and the subsequent recession, which (at the time) was the longest of the post-World War II recessions, gave rise to many studies on the effects of oil price increases on the economy. The early research documented and sought to explain the inverse relationship between oil price increases and aggregate economic activity. Subsequent

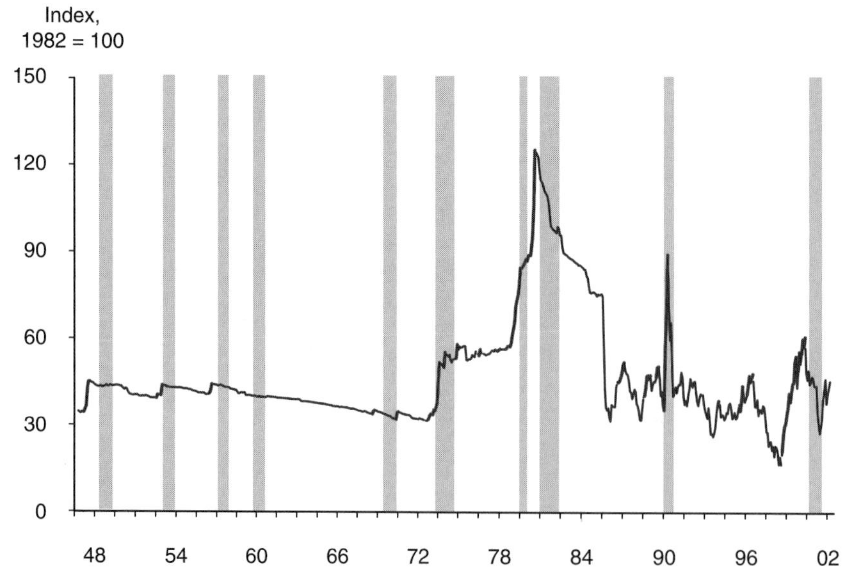

FIGURE 1 Real oil prices and U.S. business cycles. Episodes of sharply rising oil prices have preceded 9 of the 10 post-World War II recessions in the United States (bars).

empirical studies confirmed an inverse relationship between oil prices and aggregate economic activity for the United States and other countries. The latter research included James Hamilton's finding that other economic forces could not account for the negative effect that rising oil prices had on U.S. economic activity.

Economists have offered a number of explanations for why rising oil prices hurt aggregate U.S. economic activity. The most basic explanation is the classic supply shock, in which rising oil prices are indicative of the reduced availability of an important input to production. Another explanation is that rising oil prices result in income transfers from oil-importing nations to oil-exporting nations, which reduces U.S. aggregate demand and slows economic activity.

The remaining explanations sought to attribute the effects of oil price shocks to developments in the financial markets. One is simply that the monetary authorities responded to rising oil prices with a contractionary monetary policy that boosted interest rates. Another is that rising oil prices led to increased money demand as people sought to rebalance their portfolios toward liquidity. A failure of the monetary authority to meet growing money demand with an increased money supply boosted interest rates. In either case, rising interest rates retarded economic activity.

Sorting through the explanations that economists have offered for the inverse relationship between oil prices and economic activity requires a comparison between how the economy has responded to oil price shocks historically and how the various explanations suggest the economy should respond. In the United States, past episodes of rising oil prices have generally resulted in declining GDP, a higher price level, and higher interest rates (Table I). As discussed later, of the various explanations that have been offered for why rising oil prices hurt economic activity, only the classic supply-side effect can account for declining GDP, a rising price level, and higher interest rates.

TABLE I

Expected Responses to Increasing Oil Price

	Real GDP	Price level	Interest rate
Historical record	Down	Up	Up
Classic supply shock	Down	Up	Up
Aggregate demand shock	Down	Down	Down
Monetary shock	Down	Down	Up
Real balance effect	Down	Down	Up

2.1 A Classic Supply Shock

Many economists consider an oil price shock to be illustrative of a classic supply shock that reduces output. Elevated energy prices are a manifestation of increased scarcity of energy, which is a basic input to production. With reduced inputs with which to work, output and labor productivity are reduced. (In more mild cases, the growth of output and productivity are slowed.) In turn, the decline in productivity growth reduces real wage growth and increases the unemployment rate.

If consumers expect such a rise in oil prices to be temporary, or if they expect the short-term effects of output to be greater than the long-term effects, they will attempt to smooth their consumption by saving less or borrowing more. These actions boost the real interest rate. With reduced output and a higher real interest rate, the demand for real cash balances declines and the price level increases (for a given rate of growth in the monetary aggregate). Therefore, higher oil prices reduce real GDP, boost real interest rates, and increase the price level (Table I). The expected consequences of a classic supply shock are consistent with the historical record.

2.2 Income Transfers and Reduced Aggregate Demand

When oil prices increase, there is a shift in purchasing power from the oil-importing nations to the oil-exporting nations. Such a shift is another avenue through which oil price shocks could affect U.S. economic activity. Rising oil prices can be thought of as being similar to a tax that is collected from oil-importing nations by oil-exporting nations. The increase in oil prices reduces purchasing power and consumer demand in the oil-importing nations. At the same time, rising oil prices increase purchasing power and consumer demand in the oil-exporting nations. Historically, however, the increase in consumer demand occurring in oil-exporting nations has been less than the reduction in consumer demand in the oil-importing nations. On net, world consumer demand for goods produced in the oil-importing nations is reduced.

Reduced consumer demand for goods produced in the oil-importing nations increases the world supply of savings, which puts downward pressure on interest rates. Lower interest rates could stimulate investment, partially offsetting the lost consumption spending and partially restoring aggregate demand. The net result, however, is a reduction in aggregate demand.

The reduction in aggregate demand puts downward pressure on the price level. Economic theory suggests that real prices will continue to decline until aggregate demand and GDP are restored to preshock levels. However, if nominal prices are sticky downward, as most economists believe, the process of adjustment will not take place, and aggregate demand and GDP will not be restored—unless unexpected inflation increases as much as GDP growth declines.

The reduction in aggregate demand necessitates a lower real price level to yield a new equilibrium. If the real price level cannot decline, consumption spending will decline more than investment spending increases. Consequently, aggregate demand and output are reduced. With nominal prices downward sticky, the only mechanism by which the necessary reduction in real prices can occur is through unexpected inflation that is at least as great as the reduction in GDP growth. Domestically, the necessary change in inflation can be accomplished through a monetary policy that holds the growth of nominal GDP constant. A monetary policy that allows nominal GDP to decline will not generate enough unexpected inflation to restore aggregate demand and output.

To the extent that income transfers and reduced aggregate demand account for the aggregate effects of oil price shocks, monetary policy would have to allow nominal GDP to decline. The aggregate effects would be lower interest rates, reduced real GDP, and inflation that decreased or increased by less than the reduction in real GDP growth (Table I). These effects are inconsistent with the historical record for the United States, which shows that interest rates rise, real GDP declines, and the price level increases by as much as real GDP declines. Such a record seems to indicate that the effects of income transfers are negligible at best.

2.3 Monetary Policy

Monetary policy was prominent among the early explanations of how oil price shocks affected aggregate economic activity, but it was gradually supplanted by real business cycle theory, which attributed the effects to classic supply shocks rather than monetary policy. Nevertheless, an apparent breakdown in the relationship between oil and the economy during the 1980s and 1990s led researchers to question the pure supply shock theory of real business cycle models and to revisit other channels through which oil could affect the economy, such as changes in monetary policy.

However, it is difficult to explain the basic effects of oil price shocks on aggregate economic activity using monetary policy. Restrictive monetary policy will result in rising interest rates, reduced GDP growth, reduced inflation, and reduced nominal GDP growth (Table I). This is inconsistent with the historical record.

2.4 The Real Balance Effect

The real balance effect was one of the first explanations that economists offered for the aggregate economic effects of an oil price. In this theory, an increase in oil prices led to increased money demand as people sought to rebalance their portfolios toward liquidity. The failure of the monetary authority to meet growing money demand with an increased money supply would boost interest rates and retard economic growth, just like a reduction in the money supply. Adjustment would put downward pressure on the price level. The effects would be higher interest rates, lower GDP, and a reduced price level, inconsistent with the historical record (Table I).

2.5 Sorting through the Basic Theories

Of the explanations offered for the inverse relationship between oil price shocks and GDP growth, the classic supply-side shock argument best explains the historical record (Table I). The classic supply shock explains the inverse relationship between oil prices and real GDP and the positive relationship between oil price shocks and measured increases in inflation and interest rates. Income transfers that reduce aggregate demand can explain reduced GDP but cannot explain rising interest rates. Neither monetary policy nor the real balance effect can explain both slowing GDP growth and increased inflationary pressure.

3. *AMPLIFICATION OF THE BASIC RESPONSE*

A number of economists have recognized that basic supply-shock effects can account for only a portion of the intense effect that oil price shocks have on aggregate economic activity. Consequently, additional explanations for the intensity of the response are important. Among the possible explanations are restrictive monetary policy, adjustment costs, coordination externalities, and financial stress.

3.1 Monetary Policy

Monetary policy can influence how an oil price shock is experienced. When the monetary authorities hold the growth of nominal GDP constant, the inflation rate will accelerate at the same rate at which real GDP growth slows. To the extent that the market is slow to adjust to monetary surprises, a more accommodative monetary policy, which is accomplished by reducing interest rates, would temporarily offset or partially offset the losses in real GDP while it increased inflationary pressure. A counterinflationary (restrictive) monetary policy, which is accomplished by increasing interest rates, would temporarily intensify the losses in real GDP while it reduced inflationary pressure.

In the case of counterinflationary monetary policy, adjustment could be lengthy. If nominal wages and prices are sticky downward, real wages and prices would fail to decrease as is necessary to clear the markets. Consequently, unemployment would increase, aggregate consumption would decline, and GDP growth would be slowed beyond that which would arise directly from the supply shock.

If wages are nominally sticky downward, the reduction in GDP growth will lead to increased unemployment and a further reduction in GDP growth, unless unexpected inflation increases as much as GDP growth decreases. The initial reduction in GDP growth is accompanied by a reduction in labor productivity. Unless real wages decrease by as much as the reduction in labor productivity, firms will lay off workers, which will generate increased unemployment and exacerbate GDP losses. Therefore, if nominal wages are sticky downward, the only mechanism by which the necessary wage reduction can occur is through unexpected inflation that is at least as great as the reduction in GDP growth.

Several lines of research assert that restrictive monetary policy accounts for much of the decline in aggregate economic activity following an oil price increase. Douglas Bohi reasoned that energy-intensive industries should be most affected if a classic supply shock explains the principal effects of oil price shocks. Using industry data for four countries, Bohi found no relationship between the industries affected and their energy intensity. He also found inconsistent effects of oil price shocks across countries and time. Asserting that monetary policy was tightened in each of the countries he examined, Bohi concluded that much of the negative impact of higher oil prices on output must be due to restrictive monetary policy.

Similarly, Ben Bernanke, Mark Gertler, and Mark Watson showed that the U.S. economy responds differently to an oil price shock when the federal funds rate is constrained to be constant than when it is unconstrained. In their unconstrained case, a sharp increase in oil prices leads to a higher federal funds rate and a reduction in real GDP. With the federal funds rate held constant, they found that an oil price increase leads to an increase in real GDP. Defining neutral monetary policy as holding the federal funds rate constant, they found that monetary policy has tightened in response to increased oil prices, and they conclude that this monetary tightening accounts for the fluctuations in aggregate economic activity.

Despite some findings to the contrary, other research casts doubt on the idea that monetary policy accounts for much of the response of aggregate economic activity to oil price shocks. James Hamilton and Ana Maria Herrera show that the findings of Bernanke *et al.* are sensitive to specification. Using a longer lag time, Hamilton and Herrera found that oil price shocks have a substantially larger direct effect on the real economy. Furthermore, with the longer lag time, Hamilton and Herrera found that even when the federal funds rate is kept constant, an oil price shock still yields a sizeable reduction in output, which implies that monetary policy has little effect on easing the real consequences of an oil price shock.

Hamilton and Herrera's findings are consistent with research conducted by others that showed that counterinflationary monetary policy was only partly responsible for the real effects of oil price shocks from 1970 to 1990. Some researchers agree that monetary policy has become more restrictive following an oil price shock but conclude that oil price shocks have a stronger and more significant impact on real activity than monetary policy.

Brown and Yücel argue that U.S. monetary policy likely has had no role in aggravating the effects of past oil price shocks. Using a specification similar to that of Bernanke *et al.*, Brown and Yücel found that oil price shocks lead to an increased federal funds rate, reduced real GDP, and an increased price level. The increase in the price level is approximately the same as the reduction in real GDP, which means that nominal GDP is unchanged. Such a finding conforms to Robert Gordon's definition of monetary neutrality, which is achieved when monetary policy is adjusted to hold nominal GDP constant.

Brown and Yücel also criticize the assertion that an unchanged federal funds rate necessarily represents a neutral monetary policy. They found that holding the federal funds rate constant (in a counterfactual experiment) after an oil price shock boosts real GDP, the price level, and nominal GDP, which is consistent with Gordon's definition of accommodative monetary policy. In short, monetary policy can cushion the real effects of an oil price shock but at the expense of accelerating inflation.

The use of the federal funds rate to gauge the effects of monetary policy can be faulty when the economy is adjusting to a supply shock. As explained previously, the attempt by consumers to smooth consumption in response to a basic supply shock accounts for the higher interest rates. In a market with rising interest rates, a policy of holding the federal funds rate constant will accelerate money growth, contributing to short-term gains in real GDP and increased inflationary pressure.

In summation, oil price shocks create the potential for a monetary policy response that exacerbates the basic effects of an oil supply shock. The research assessing whether monetary policy has amplified the basic effects of a supply shock is contradictory, but the most compelling evidence suggests monetary policy has had a relatively small or no role in amplifying the basic effects in the United States. Furthermore, interest rates are not a good way to assess the effect of monetary policy when there is a supply shock, and measured in other ways, monetary policy has remained neutral in response to past oil price shocks.

3.2 Adjustment Costs

Adjustment costs are another way in which the basic response to rising oil prices might be amplified. Adjustment costs can arise from either capital stock that embodies energy technology or sectoral imbalances. In either case, adjustment costs can further retard economic activity.

To a great extent, the technology that a firm chooses for its production is embedded in the capital equipment it purchases. (Economists refer to this characteristic of technology and capital as "putty–clay" because the firm can vary its energy-to-output, capital-to-output, and labor-to-output ratios over the long term as it purchases capital but not in the short term.) With production technology embedded in the capital stock, a firm must change its capital stock to respond to rising energy prices. The consequences are slow adjustment and a disruption to economic organization when energy prices increase, with stronger effects in the short term compared to the long term.

In a similar way, changes in oil prices can also create sectoral imbalances by changing the equilibrium between the sectors. For example, rising oil prices require a contraction of energy-intensive sectors and an expansion of energy-efficient sectors. These realignments in production require adjustments that cannot be achieved quickly. The result is increased unemployment and the underutilization of resources.

3.3 Coordination Problems

Coordination problems are a potential outgrowth of sectoral imbalances. Coordination problems arise when individual firms understand how changing oil prices affect their output and pricing decisions but lack enough information about how other firms will respond to changes in oil prices. As a consequence, firms experience difficulty adjusting to each other's actions and economic activity is further disrupted when oil prices increase.

3.4 Uncertainty and Financial Stress

Uncertainty about future oil prices increases when oil prices are volatile, and such uncertainty reduces investment. When firms are uncertain about future oil prices, they find it increasingly desirable to postpone irreversible investment decisions. When technology is embedded in the capital, the firm must irreversibly choose the energy intensity of its production process when purchasing its capital. As uncertainty about future oil prices increases, the value of postponing investment decision increases, and the net incentive to invest decreases. In addition, uncertainty about how firms might fare in an environment of higher energy prices is likely to reduce investor confidence and increase the interest rates that firms must pay for capital. These two effects work to reduce investment spending and weaken economic activity.

3.5 Sorting through the Explanations

A consensus about why the effects of rising oil prices are so strong has not been reached. Recent research casts doubt on the idea that monetary policy is the primary factor amplifying the aggregate economic effects of oil price shocks. No empirical research has attempted to sort through the other effects—adjustment costs, coordination problems, and financial stress. In fact, these effects are consistent with observed facts and are mutually reinforcing rather than mutually exclusive. All three effects may be at work.

4. *ASYMMETRY OF THE RESPONSE*

Prior to the 1980s, the large shocks in oil prices were increases. The 1980s brought the first major decrease in oil prices, and it gradually became evident that U.S. economic activity responded asymmetrically to oil price shocks. That is, rising oil prices hurt U.S. economic activity more than declining oil prices helped it. Although all but one of the post-World War II recessions followed sharp rises in oil prices, accelerated economic activity did not follow the sharp price declines of the 1980s and 1990s.

4.1 The Discovery of Asymmetry

Initially, the weak response of economic activity to oil price decreases was seen as a breakdown in the relationship between oil price movements and the economy, and researchers began to examine different oil price specifications in their empirical work to reestablish the relationship between oil price shocks and economic activity. Knut Mork found that when he grouped oil price changes into negative and positive oil price changes, oil price increases had more effect on economic activity than oil price decreases. Later research conducted by others found that oil price increases had a significant effect on economic activity, whereas oil price decreases did not. Similar asymmetry was also found at a more detailed industry level, and further research established that economic activity in seven industrialized countries responded asymmetrically to oil price movements.

Hamilton further refined the analysis by creating what he called a "net oil price." This measure of oil prices attempts to capture episodes of oil price increases that are outside normal market fluctuations. This measure reflects only those increases in oil prices that take the price higher than it has been in the past 12 months. (Technically, the net oil price series is the price of oil for all periods in which the price is higher than it has been during the past 12 months and zero in all other months.) Hamilton's research and subsequent research by others found a statistically significant and stable negative relationship between the net oil price series and output, whereas various series constructed to reflect episodes of sharp oil price declines do not seem to have any explanatory power. In a related vein, Steven Davis and John Haltiwanger constructed an oil price series that combined asymmetry with persistence and also found an asymmetric relationship between oil price shocks and economic activity.

4.2 Understanding Asymmetry

Classic supply-side effects cannot explain asymmetry. Operating through supply-side effects, reductions in oil prices should help output and productivity as increases in oil prices hurt economic output and productivity. Accordingly, economists have begun to explore the channels through which oil prices affect economic activity. Monetary policy, adjustment costs, coordination problems, uncertainty and financial stress, and asymmetry in the petroleum product markets have been offered as explanations. Of these, adjustment costs, coordination problems, and financial stress seem most consistent with the historical record.

4.3 Monetary Policy and Asymmetry

Monetary policy may contribute to an asymmetric response in aggregate economic activity to oil price shocks in two ways. Monetary policy may respond to oil price shocks asymmetrically. Another possibility is that nominal wages are sticky downward but not upward, and monetary policy is conducted in such a way that nominal GDP declines when oil prices are rising and nominal GDP increases when oil prices are declining.

When oil prices increase, real wages must decline to clear markets and restore equilibrium. If real wages do not decline, the economic displacement will be greater. When oil prices decline, real wages must increase to clear markets and restore equilibrium. If real wages do not increase, the gains in economic activity will be greater.

If nominal wages are sticky downward, an increase in unexpected inflation that is at least as great as the decline in real GDP is necessary to yield the necessary reduction in real wages. If the monetary authority maintains a neutral monetary policy (nominal GDP is unchanged), prices will increase as much as real GDP declines, and real wages will adjust sufficiently. If the monetary authority conducts policy such that nominal GDP declines, however, prices will increase by less than the amount that real GDP declines, and because nominal wages are sticky downward, real wages will not adjust sufficiently to restore equilibrium.

Because nominal wages can adjust upward freely, however, unexpected disinflation is not required for adjustments in real wages to occur, and the conduct of monetary policy is not as important. If the monetary authority conducts policy such that nominal GDP increases, prices will increase by more than the real GDP declines, but because nominal wages adjust upward freely, real wages will increase enough to restore equilibrium. Consequently, a monetary policy that boosts nominal GDP when oil prices fall would have less effect on real activity than a policy that decreases nominal GDP when oil prices rise.

John Tatom provided early evidence that monetary policy responded asymmetrically to oil price shocks by showing that the economy responded symmetrically to oil price shocks if the monetary policy is taken into account. In a later contribution, Peter Ferderer showed that monetary policy cannot account for the asymmetry in the response of real activity to oil price shocks. Nathan Balke, Stephen Brown, and Mine Yücel found that the Federal Reserve's response to oil price shocks does not cause asymmetry in real economic activity. In their model, the asymmetry does not disappear, and is in fact enhanced, when monetary policy (as measured by either the federal funds rate or the federal funds rate plus expectations of the federal funds rate) is held constant.

4.4 Adjustment Costs and Coordination Problems

James Hamilton contributed the idea that adjustment costs may lead to an asymmetric response to changing oil prices. Rising oil prices retard economic activity directly, and decreasing oil prices stimulate economic activity directly. The costs of adjusting to changing oil prices retard economic activity whether oil prices are increasing or decreasing. Consequently, rising oil prices result in two negative effects on economic activity that reinforce each other. On the other hand, declining oil prices result in both positive and negative effects, which tend to be offsetting.

As described previously, these adjustment costs can be the result of the energy-to-output ratio being embedded in the capital or the result of sectoral imbalances. Coordination problems may reinforce adjustment costs whether oil prices are increasing or decreasing.

4.5 Uncertainty and Financial Stress

Uncertainty and financial stress may also contribute to an asymmetric response in aggregate economic activity to oil price shocks. As explained by Peter Ferderer, uncertainty about future oil prices is reflected in increased interest rates and reduced investment demand, which adversely affect aggregate economic activity. In addition, if the energy-to-output ratio is embedded in the capital stock, firms will find it increasingly desirable to postpone irreversible investment decisions until they are more certain about future oil prices.

Volatile oil prices contribute to oil price uncertainty and weaker economic activity, whether oil prices are increasing or decreasing. As is the case for adjustment costs, uncertainty and financial stress augment the basic supply-side effect when oil prices are increasing and offset the basic supply-side effect when oil prices are decreasing. The result is that aggregate economic activity responds to oil price shocks asymmetrically.

4.6 Asymmetry in Petroleum Product Prices

Petroleum product prices are another avenue through which the economy may respond asymmetrically to fluctuations in crude oil prices. A considerable body of research shows that petroleum product prices respond asymmetrically to fluctuations in oil prices, and most of the price volatility originates from crude oil prices rather than product prices. It is a relatively short leap to suggest that asymmetry in the response of product prices may account for the asymmetry between crude oil prices and aggregate economic activity.

Hillard Huntington found that the economy responds symmetrically to changes in petroleum product prices, but that petroleum product prices respond asymmetrically to crude oil prices. The consequence is an asymmetric relationship between crude oil prices and aggregate economic activity. These findings need further examination, but substantial research shows that the asymmetric response of aggregate economic activity to oil price shocks arises through channels that cannot be explained by the asymmetric response of product prices alone.

4.7 Sorting through the Explanations of Asymmetry

Although asymmetry is now fairly well accepted, relatively few studies have attempted to distinguish empirically through what channels oil price shocks might yield an asymmetric response in aggregate economic activity. The available research seems to rule out the likelihood that asymmetry is the result of monetary policy, but the findings are consistent with explanations of asymmetry that rely on adjustment costs, coordination problems, or uncertainty and financial risk.

Prakash Loungani found that oil price shocks lead to a reallocation of labor across sectors, which increases the unemployment rate, whether oil prices are increasing or decreasing. These sectoral shifts are consistent with adjustment costs and coordination problems, but they do not preclude uncertainty and financial risk.

Steven Davis and John Haltiwanger assessed the channels through which oil price shocks affect economic activity. By examining job elimination and creation data, they found that allocative channels (e.g., changes in the desired distributions of labor and capital) contribute to the asymmetric response of aggregate economic activity to oil price shocks. Again, sectoral shifts are consistent with adjustment costs and coordination problems, but they do not preclude uncertainty and financial risk. In addition, Davis and Haltiwanger showed that the allocative effects are relatively strong and reinforce the basic supply-side effect when oil prices are increasing and cancel the basic supply-side effect when prices are declining.

Balke, Brown, and Yücel found that monetary policy responds asymmetrically because the economy has responded asymmetrically to oil price shocks and that monetary policy cannot account for the asymmetry. They also found that output and interest rates respond asymmetrically to oil price shocks, with the shocks transmitted through asymmetric movements in market interest rates. Such interest rate movements are consistent with several explanations for asymmetry. Asymmetric movements in the interest rates may be indicative of the increased uncertainty and financial risk that result from oil price shocks. Alternatively, movements in market interest rates may be a reflection of expectations that oil price shocks will necessitate costly adjustments in the future.

5. *A WEAKENING RESPONSE*

By the mid-1990s, the quantitative relationship between oil prices and economic activity seemed fairly robust and reasonably well understood, even if the exact channels through which the effects operated were not known with certainty. During the latter half of the 1990s, however, the relationship seemed to weaken. In the late 1990s and early 2000s, rising oil prices had less effect on economic activity than previous research suggested might have been expected.

Rising oil prices led to increases in the unemployment rate from the early 1970s through the early 1990s, and unemployment declined with oil prices from 1982 through 1990 and in the late 1990s (Fig. 2). Nonetheless, the relationship began to weaken in the late 1990s. Although oil prices were relatively strong, the unemployment rate continued to decline. One explanation is that high world oil prices were a result of a strong world economy, and strong demand rather than a supply shock accounted for rising oil prices.

The data also show a weaker relationship between rising oil prices and core inflation in the late 1990s (Fig. 3). Mark Hooker reevaluated the oil price–inflation relationship and found that since approximately 1980, oil price changes have not seemed to affect core measures of inflation. Prior to 1980, however, oil price shocks contributed substantially to core inflation in the United States. One possibility is that U.S. monetary policy in the Volcker and Greenspan eras was significantly less accommodative to oil price shocks than it had been under their predecessors, and thus monetary policy no longer contributed to high inflation.

Nonetheless, the relationship between oil prices and interest rates was relatively unchanged through the 1990s. Rising oil prices led to higher interest rates, which is the expected consequence of supply shocks that have greater short-term effects than

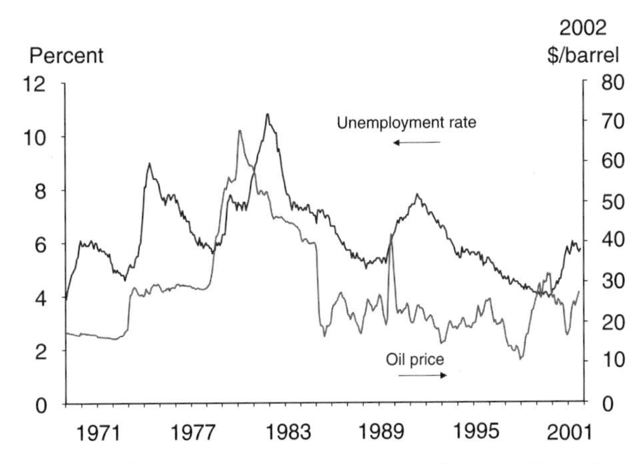

FIGURE 2 Real oil prices and U.S. unemployment. Unemployment declined with oil prices from 1982 to 1990 and in the late 1990s, but the relationship began to weaken in the late 1990s.

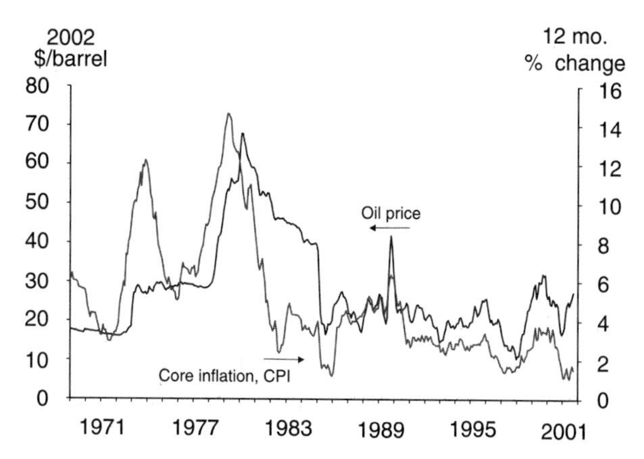

FIGURE 3 Real oil prices and core inflation. Prior to 1980, real oil prices and core inflation were more closely related.

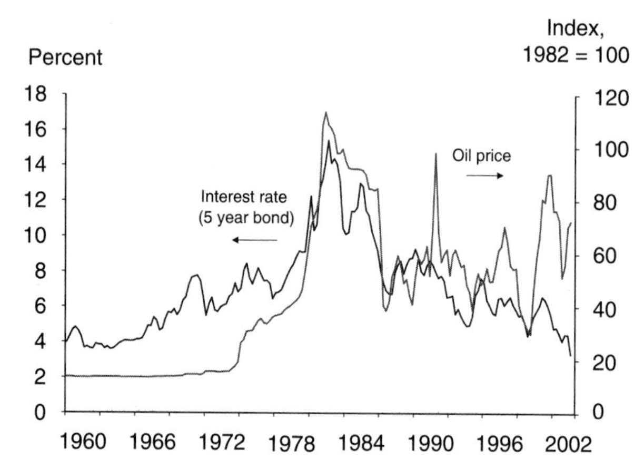

FIGURE 4 Real oil prices and U.S. interest rates. Oil prices and market interest rates moved closely together until mid-2000.

long-term effects (Fig. 4). Brown and Yücel have shown that some of the increases in the U.S. federal funds rate that occurred in 1999 and 2000 may have been part of a general increase in market interest rates that resulted from higher oil prices. Since approximately mid-2000, however, interest rates have not increased with oil prices. This changing relationship may be further evidence of an economy that is becoming less sensitive to oil price shocks.

5.1 Factors Contributing to a Weakening Relationship

Several factors have likely contributed to a weakening relationship between oil prices and economic activity during the past three decades, including a reduced energy-to-GDP ratio, the fact that oil price increases were the result of increased demand rather

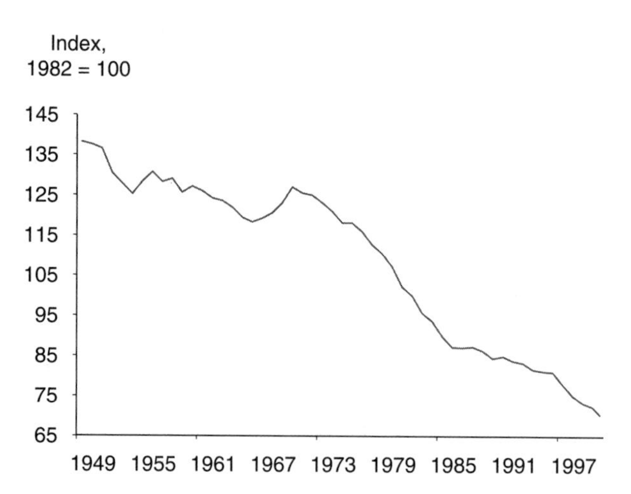

FIGURE 5 Decrease in U.S. energy-to-GDP ratio. The energy-to-GDP ratio has decreased since the end of World War II. Higher energy prices contributed to strong reductions in the energy-to-GDP ratio from the mid-1970s to the mid-1980s.

than oil supply shocks, and prior experience with oil price shocks. In addition, the 1990s boom was marked by strong productivity gains that may have simply obscured the relationship between oil prices and aggregate economic activity.

5.2 A Reduced Energy-to-GDP Ratio

Although it represents a continuing trend, the energy consumption-to-GDP ratio declined from the 1970s to the early 2000s (Fig. 5). On the basis of this decline, Brown and Yücel estimate that the U.S. economy may have been approximately one-third less sensitive to oil price fluctuations in 2000 than it was in the early 1980s and approximately half as sensitive as it was in the early 1970s.

5.3 Increased Demand Rather Than a Supply Shock

Another factor that could have contributed to a weakening relationship between oil prices and economic activity was the fact that rising oil prices in the late 1990s were largely the result of economic expansion. Price increases that are the result of increased demand rather than supply shocks may be less disruptive to economic activity.

Due to increased energy efficiency, U.S. oil consumption increased only moderately during the 1990s, but oil consumption in the industrialized nations increased steadily (Fig. 6). World oil consumption was further boosted by the dramatic gains in oil

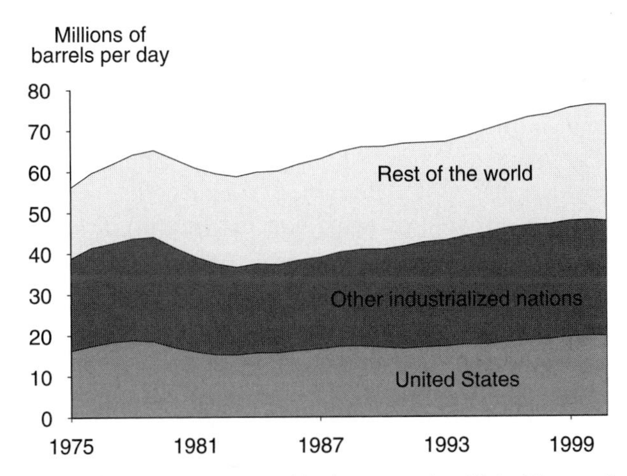

FIGURE 6 Increase in world oil consumption. United States oil consumption increased moderately in the 1990s. Oil consumption in the industrialized countries combined increased steadily during the 1990s. The largest increase in oil consumption occurred in the newly industrializing economies, such as China and Korea.

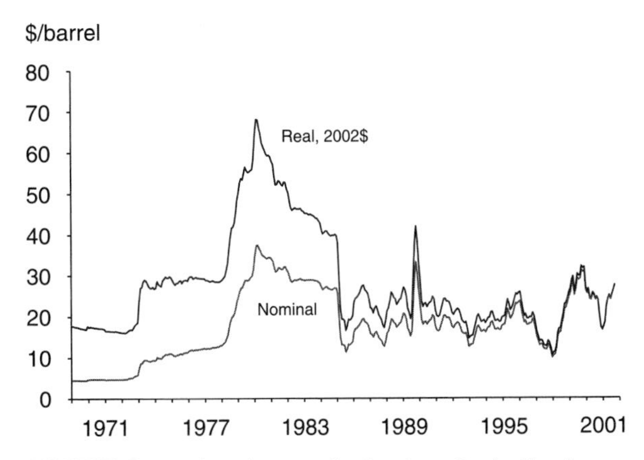

FIGURE 8 Real and nominal oil prices. Real oil prices are adjusted for inflation, whereas nominal oil prices are not. The real oil price indicates that the oil price gains in 2000 were moderate by historical standards.

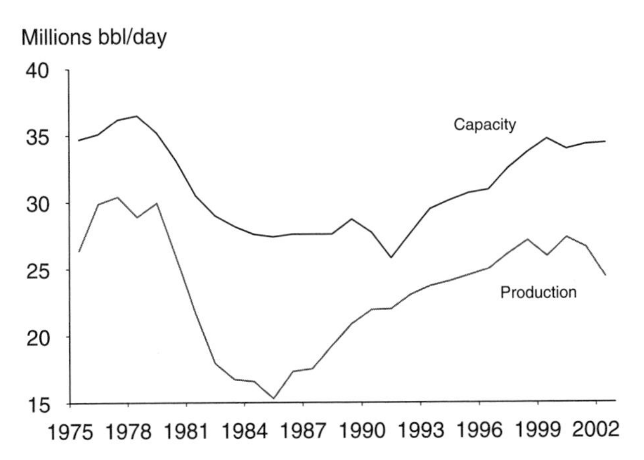

FIGURE 7 Production and capacity of the Organization of Petroleum Exporting Countries (OPEC). OPEC operated fairly close to capacity in the 1990s.

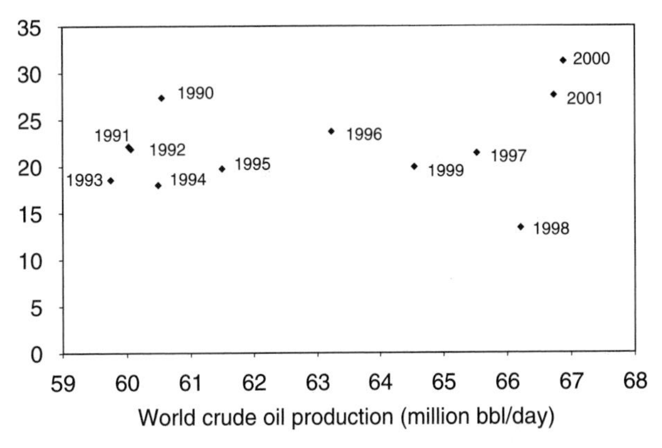

FIGURE 9 Oil demand increased sharply in 2000. The plotted points show the market prices and quantities of oil for the years 1991–2001. Increases in quantity and price are indicative of increased demand.

consumption outside countries belonging to the Organization for Economic Cooperation and Development, with the strongest gains in consumption occurring in Asian countries, such as China and Korea.

In addition, world capacity to produce oil did not keep pace with growing consumption throughout much of the 1990s. Consequently, the gap between capacity and production of the Organization of Petroleum Exporting Countries was relatively small throughout most of the 1990s (Fig. 7). With demand growing faster than supply, oil prices increased in the latter part of the decade, although the real increase was relatively small by 1980s standards (Fig. 8). In 2000, world oil demand increased sharply (Fig. 9), which led to greater consumption and an increase in prices. Because the increase in oil prices was the result of strong demand due to economic expansion, the usual decline in economic activity did not follow.

5.4 Prior Experience with Energy Price Shocks

Prior experience with oil price shocks may have also contributed to the muted response of U.S. economic activity to oil price shocks. Such experience may have resulted in reduced adjustment costs, coordination problems, and uncertainty and financial stress. In addition, monetary authorities have increased experience with oil price shocks, which may have resulted in reduced inflationary pressure caused by oil price shocks.

6. CONCLUSIONS AND IMPLICATIONS FOR POLICY

Economic research provides considerable evidence that rising oil prices contribute to real GDP losses. Economic theory suggests that a number of channels may account for this phenomenon. The most basic is a classic supply-side effect in which rising oil prices are indicative of the reduced availability of a basic input to production. Other channels include reduced aggregate demand and contractionary monetary policy. Of these channels, the empirical evidence is most consistent with a classic supply-side effect. This effect best explains the reduced aggregate output, increased price level, and higher interest rates.

Given the current understanding of how oil price shocks affect economic activity, analysts are relatively well positioned to prescribe the proper fiscal and monetary policy responses to oil price shocks. Also, given that rising oil prices reduce potential GDP and the reductions in aggregate demand are likely to be negligible, a fiscal policy response seems unnecessary. Regarding monetary policy, a neutral response to rising oil prices, which consists of holding nominal GDP constant, will neither aggravate the GDP losses nor offset them. Such a policy will also lead to an increase in the price level that is the same as the loss in real GDP.

In addition to taking a neutral stance, monetary policy can shape the aggregate effects of an oil price shock. If the monetary authority is willing to accept higher inflation, it can temporarily boost real GDP through an expansionary policy. If the monetary authority wants to lessen the inflationary consequences of rising oil prices, it can tighten policy, which will temporarily aggravate the losses in real GDP.

Economists have variously suggested that monetary policy, adjustment costs, coordination problems, and increased uncertainty and financial stress account for an amplification of the basic supply-side effect when oil prices are increasing and a lessening of the aggregate response when oil prices are decreasing. Although research indicates that the conduct of U.S. monetary policy has not resulted in such effects, monetary policy could have such effects, and the issue is not completely resolved. Researchers have provided evidence that adjustment costs, coordination problems, or increased uncertainty and financial stress amplify basic supply-side effects when oil prices are increasing. They have not been able to reliably distinguish between these effects.

Because economic research has not been able to distinguish between the contributions of adjustment costs, coordination problems, and uncertainty and financial stress, analysts are less able to prescribe the best course of action for energy policy. In particular, research has not determined whether the private sector is capable of providing the optimal level of insurance against price shocks. Given the possibility of coordination problems across industries and the asymmetric response of aggregate economic activity to oil price shocks, an energy policy that counters movements in international oil prices seems justified. Nonetheless, considerably more research must be conducted before economists can provide sound guidance regarding the extent to which policymakers should go to reduce an economy's vulnerability to oil price shocks.

SEE ALSO THE FOLLOWING ARTICLES

Economics of Energy Demand • Economics of Energy Supply • Energy Futures and Options • Inflation and Energy Prices • Innovation and Energy Prices • Oil Price Volatility • Prices of Energy, History of • Stock Markets and Energy Prices

Further Reading

Balke, N. S., Brown, S. P. A., and Yücel, M. K. (2002). Oil price shocks and the U.S. economy: Where does the asymmetry originate? *Energy J.* **23**(3), 27–52.

Bernanke, B. S., Gertler, M., and Watson, M. (1997). Systematic monetary policy and the effects of oil price shocks. *Brookings Papers Econ. Activity* **1997**(1), 91–157.

Brown, S. P. A., and Yücel, M. K. (1999). Oil prices and U.S. aggregate economic activity: A question of neutrality. *Econ. Financial Rev.* [Federal Reserve Bank of Dallas], Second Quarter, 16–23.

Brown, S. P. A., and Yücel, M. K. (2002, Second quarter). Energy prices and aggregate economic activity: An interpretative survey. *Q. Rev. Econ. Finance* 42(2).

Davis, S. J., and Haltiwanger, J. (2001). Sectoral job creation and destruction responses to oil price changes and other shocks. *J. Monetary Econ.* **48**, 463–512.

Ferderer, J. P. (1996). Oil price volatility and the macroeconomy: A solution to the asymmetry puzzle. *J. Macroecon.* **18**, 1–16.

Hamilton, J. D. (1983). Oil and the macroeconomy since World War II. *J. Political Econ.* **91**, 228–248.

Hamilton, J. D., and Herrera, A. M. (2004). Oil shocks and aggregate macroeconomic behavior: The role of monetary policy. *J. Money Credit Banking* 35(6).

Huntington, H. G. (2003). Oil security as a macroeconomic externality. *Energy Econ.* **25**, 119–136.

Mork, K. A. (1994). Business cycles and the oil market. *Energy J.* **15**, 15–38.

Carbon Capture and Storage from Fossil Fuel Use

HOWARD HERZOG and DAN GOLOMB
Massachusetts Institute of Technology
Cambridge, Massachusetts, United States

1. Introduction
2. Carbon Sources
3. Capture Processes
4. CO_2 Storage
5. Economics
6. Alternate Approaches

Glossary

carbon capture The separation and entrapment of CO_2 from large stationary sources.

carbon sequestration The capture and secure storage of carbon that would otherwise be emitted to or remain in the atmosphere.

carbon sources Large stationary sources of CO_2, e.g., fossil-fueled power plants, cement manufacturing and ammonia production plants, iron and nonferrous metal smelters, industrial boilers, refineries, and natural gas wells.

CO_2 storage The injection of CO_2 into geologic or oceanic reservoirs for timescales of centuries or longer.

One of the approaches for mitigating potential global climate change due to anthropogenic emissions of CO_2 and other greenhouse gases is to capture CO_2 from fossil fuel-using sources, and to store it in geologic or oceanic reservoirs. The capture technologies are described, and their efficiencies, cost, and energy penalties are estimated. Storage capacities and effectiveness are estimated, as well as transportation costs and possible environmental impacts.

1. INTRODUCTION

Carbon sequestration can be defined as the capture and secure storage of carbon that would otherwise be emitted to, or remain, in the atmosphere. The focus of this article is the removal of CO_2 directly from industrial or utility plants and subsequently storing it in secure reservoirs. This is called carbon capture and storage (CCS). The rationale for carbon capture and storage is to enable the use of fossil fuels while reducing the emissions of CO_2 into the atmosphere, and thereby mitigating global climate change. The storage period should exceed the estimated peak periods of fossil fuel exploitation, so that if CO_2 reemerges into the atmosphere, it should occur after the predicted peak in atmospheric CO_2 concentrations. Removing CO_2 from the atmosphere by increasing its uptake in soils and vegetation (e.g., afforestation) or in the ocean (e.g., iron fertilization), a form of carbon sequestration sometimes referred to as enhancing natural sinks, is only addressed briefly.

At present, fossil fuels are the dominant source satisfying the global primary energy demand, and will likely remain so for the rest of the century. Fossil fuels supply over 85% of all primary energy; nuclear power, hydroelectricity, and renewable energy (commercial biomass, geothermal, wind, and solar energy) supply the remaining 15%. Currently, nonhydroelectric renewable energy satisfies less than 1% of the global energy demand. Although great efforts and investments are made by many nations to increase the share of renewable energy to satisfy the primary energy demand and to foster conservation and efficiency improvements of fossil fuel usage, addressing climate change concerns during the coming decades will likely require significantly increasing the contributions from carbon capture and storage.

2. CARBON SOURCES

Pathways for carbon capture derive from three potential sources (see Fig. 1). Several industrial

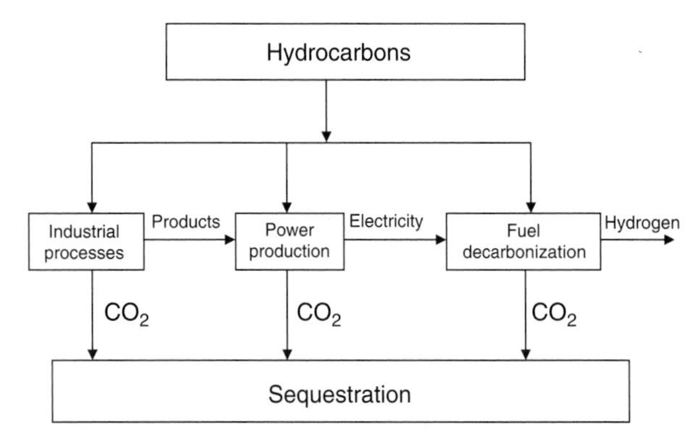

FIGURE 1 Sources of CO_2 for sequestration, as industrial by-product, captured from power plants, or as by-product of future fuel decarbonization plants.

processes produce highly concentrated streams of CO_2 as a by-product. Although limited in quantity, these by-products make a good capture target, because the captured CO_2 is integral to the total production process, resulting in relatively low incremental capture costs. For example, natural gas ensuing from wells often contains a significant fraction of CO_2 that can be captured and stored. Other industrial processes that lend themselves to carbon capture are ammonia manufacturing, fermentation processes, and hydrogen production (e.g., in oil refining).

By far the largest potential sources of CO_2 today are fossil-fueled power production plants. Power plants emit more than one-third of the CO_2 emissions worldwide. Power plants are usually built in large centralized units, typically delivering 500–1000 MW of electrical power. A 1000-MW pulverized coal-fired power plant emits between 6 and 8 Mt/yr of CO_2, an oil-fired single-cycle power plant emits about two-thirds of that, and a natural gas combined-cycle power plant emits about one-half of that.

Future opportunities for CO_2 capture may also arise from decarbonization, i.e., producing hydrogen fuels from carbon-rich feedstocks, such as natural gas, coal, and biomass. The CO_2 by-product will be relatively pure and the incremental costs of carbon capture will be relatively low. The hydrogen can be used in fuel cells and other hydrogen fuel-based technologies, but there are major costs involved in developing a mass market and infrastructure for these new fuels.

3. CAPTURE PROCESSES

CO_2 capture processes from power production fall into three general categories: (1) flue gas separation, (2) oxyfuel combustion in power plants, and (3) precombustion separation. Each of these technologies carries both an energy and an economic penalty. The efficiencies and economics of several technologies are discussed in Section 5.

3.1 Flue Gas Separation

Currently, flue gas separation and CO_2 capture are practiced at about a dozen facilities worldwide. The capture process is based on chemical absorption. The captured CO_2 is used for various industrial and commercial processes, e.g., urea production, foam blowing, carbonation of beverages, and dry ice production. Because the captured CO_2 is used as a commercial commodity, the absorption process, although expensive, is profitable because of the price realized for the commercial CO_2.

Chemical absorption refers to a process in which a gas (in this case, CO_2) is absorbed in a liquid solvent by formation of a chemically bonded compound. When used in a power plant to capture CO_2, the flue gas is contacted with the solvent in a packed absorber column, where the solvent preferentially removes the CO_2 from the flue gas. Afterward, the solvent passes through a regenerator unit, where the absorbed CO_2 is stripped from the solvent by counterflowing steam at 100–120°C. Water vapor is condensed, leaving a highly concentrated (over 99%) CO_2 stream, which may be compressed for commercial utilization or storage. The lean solvent is cooled to 40–65°C and is recycled into the absorption column. The most commonly used absorbent for CO_2 absorption is monoethanolamine (MEA). The fundamental reaction for this process is

$$C_2H_4OHNH_2 + H_2O + CO_2 \leftrightarrow C_2H_4OHNH_3^+ + HCO_3^-$$

During the absorption process, the reaction proceeds from left to right; during regeneration, the reaction proceeds from right to left. Cooling and heating of the solvent, pumping, and compression require power input from the power plant thermal cycle, derating the thermal efficiency (heat rate) of the power plant. A schematic of a chemical absorption process for power plant flue gas is depicted in Fig. 2.

In order to reduce the capital and energy costs and the size of the absorption and regenerator (stripper) columns, new processes are being developed. One example is the membrane-absorption process, whereby a microporous membrane made of poly(tetrafluoroethylene) separates the flue gas from the

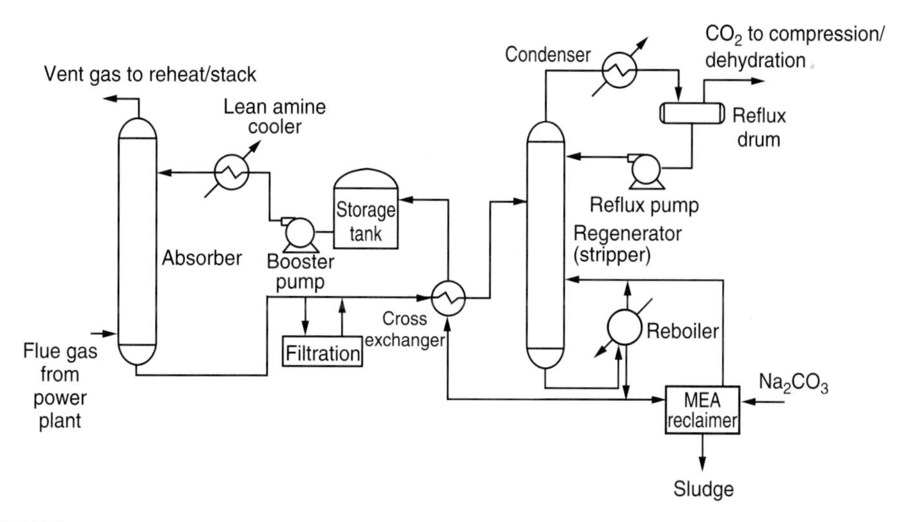

FIGURE 2 Process flow diagram for the amine separation process. MEA, monoethanolamine.

solvent. The membrane allows for greater contacting area within a given volume, but by itself the membrane does not perform the separation of CO_2 from the rest of the flue gases. It is the solvent that selectively absorbs CO_2. The use of a gas membrane has several advantages: (a) high packing density, (b) high flexibility with respect to flow rates and solvent selection, (c) no foaming, channeling, entrainment, and flooding, all common problems in packed absorption towers, (d) ease of transport of the unit, e.g., offshore, and (e) significant savings in weight.

It is possible to design a once-through scrubbing process (i.e., no regeneration step). For example, the CO_2 could be scrubbed from flue gas with seawater, returning the whole mixture to the ocean for storage. However, to date these approaches are not as practical compared to those using a regenerable solvent. In the seawater scrubbing example, the large volumes of water that are required result in large pressure drops in the pipes and absorber.

Other processes have been considered to capture CO_2 from power plant and industrial boiler flue gases, e.g., membrane separation, cryogenic fractionation, and adsorption using molecular sieves. Generally, these processes are less energy efficient and more expensive compared to the absorption methods.

3.2 Oxyfuel Combustion

When a fossil fuel (coal, oil, and natural gas) is combusted in air, the fraction of CO_2 in the flue gas ranges from 3 to 15%, depending on the carbon content of the fuel and the amount of excess air necessary for the combustion process. The separation of CO_2 from the rest of the flue gases (mostly N_2) by chemical or physical means is capital and energy intensive. An alternative is to burn the fossil fuel in pure or enriched oxygen. In such a fashion, the flue gas will contain mostly CO_2 and H_2O. A part of the flue gas needs to be recycled into the combustion chamber in order to control the flame temperature. From the nonrecycled flue gas, water vapor can be readily condensed, and the CO_2 can be compressed and piped directly to the storage site.

Of course, the separation process has now shifted from the flue gas to the intake air: oxygen must be separated from nitrogen in the air. The air separation unit (ASU) alone may consume about 15% of a power plant's electric output, requiring a commensurate increase of fossil fuel to be consumed for achieving the rated electric output of the plant. In the ASU, air is separated into liquid oxygen, gaseous nitrogen, argon, and other minor ingredients of air. The latter are marketable by-products of the oxyfuel plant. Pilot scale studies indicate that the oxyfuel method of capturing CO_2 can be retrofitted to existing pulverized coal (PC) plants.

3.3 Precombustion Capture

Capturing CO_2 before combustion offers some advantages. First, CO_2 is not yet diluted by the combustion air. Second, the CO_2-containing stream is usually at elevated pressure. Therefore, more efficient separation methods can be applied, e.g., using pressure-swing absorption in physical solvents, such as methanol or poly (ethylene glycol) (commercial brands are called Rectisol and Selexol). Precombustion capture is usually applied in integrated coal gasification combined cycle power plants. This process includes gasifying the coal to produce a synthesis gas composed of CO and H_2, reacting the CO with water (water–gas shift reaction) to produce CO_2 and H_2, capturing the CO_2, and sending the H_2 to a turbine to produce electricity. Because the primary fuel sent to the gas turbine is now hydrogen, some can be bled off as a fuel for separate use, such as in hydrogen fuel cells to be used in transportation vehicles. One of the biggest barriers to this pathway is that currently electricity generation is cheaper in pulverized coal power plants than in integrated coal gasification combined cycle plants. The precombustion process could be utilized when natural gas is the primary fuel. Here, a synthesis gas is formed by reacting natural gas with steam to produce CO_2 and H_2. However, it is unproved whether precombustion capture is preferable to the standard postcombustion capture for the case of using natural gas.

Worldwide, gasification facilities currently exist; most do not generate electricity, however, but produce synthesis gas and various other by-products of coal gasification. In these facilities, after the gasification stage, CO_2 is separated from the other gases, such as methane, hydrogen, or a mix of carbon monoxide and hydrogen. The synthesis gas or hydrogen is used as a fuel or for chemical raw material, e.g., for liquid-fuel manufacturing or ammonia synthesis. The CO_2 can also be used as a chemical raw material for dry ice manufacturing, carbonated beverages, and enhanced oil recovery (EOR). For example, the Great Plains Synfuel Plant, near Beulah, North Dakota, gasifies 16,326 metric tons per day of lignite coal into 3.5 million standard cubic meters per day of combustible syngas, and close to 7 million standard cubic meters of CO_2. A part of the CO_2 is captured by a physical solvent based on methanol. The captured CO_2 is compressed and 2.7 million standard cubic meters per day is piped over a 325-km distance to the Weyburn, Saskatchewan, oil field, where the CO_2 is used for enhanced oil recovery.

4. CO_2 STORAGE

Following the capture process, CO_2 needs to be stored, so that it will not be emitted into the atmosphere. Several key criteria must be applied to the storage method: (a) the storage period should be prolonged, preferably hundreds to thousands of years, (b) the cost of storage, including the cost of transportation from the source to the storage site, should be minimized, (c) the risk of accidents should be eliminated, (d) the environmental impact should be minimal, and (e) the storage method should not violate any national or international laws and regulations.

Storage media include geologic sinks and the deep ocean. Geologic storage options include deep saline formations (subterranean and subseabed), depleted oil and gas reservoirs, formations for enhanced oil recovery operations, and unminable coal seams. Deep ocean storage approaches include direct injection of liquid carbon dioxide into the water column at intermediate depths (1000–3000 m), or at depths greater than 3000 m, where liquid CO_2 becomes heavier than seawater, so CO_2 would drop to the ocean bottom and form a "CO_2 lake." In addition, other storage approaches are proposed, such as enhanced uptake of CO_2 by terrestrial and oceanic biota, and mineral weathering. The latter approaches refer to the uptake of CO_2 from the atmosphere, not CO_2 that has been captured from emission sources. Finally, captured CO_2 can be used as a raw material by the chemical industry. However, the prospective amounts of CO_2 that can be utilized are but a very small fraction of CO_2 emissions from anthropogenic sources. Table I lists the estimated worldwide capacities for CO_2 storage in the various media. As a comparison to the storage capacities, note that current global anthropogenic emissions amount to close to 7 Gt C per year (1 Gt C = 1 billion metric tons of carbon equivalent = 3.7 Gt CO_2).

4.1 Geologic Storage

Geological sinks for CO_2 include depleted oil and gas reservoirs, enhanced oil recovery operation formations, unminable coal seams, and deep saline

TABLE I

The Worldwide Capacity of Potential CO_2 Storage Reservoirs[a]

Sequestration option	Worldwide capacity (Gt C)[b]
Ocean	1000–10,000 +
Deep saline formations	100–10,000
Depleted oil and gas reservoirs	100–1000
Coal seams	10–1000
Terrestrial	10–100
Utilization	Currently < 0.1 Gt C/yr

[a] Ocean and land-based sites together contain an enormous capacity for storage of CO_2. The world's oceans have by far the largest capacity for carbon storage. Worldwide total anthropogenic carbon emissions are ~7 Gt C per year (1 Gt C = 1 billion metric tons of carbon equivalent).

[b] Orders of magnitude estimates.

formations. Together, these can hold hundreds to thousands of gigatons of carbon (Gt C), and the technology to inject CO_2 into the ground is well established. CO_2 is stored in geologic formations by a number of different trapping mechanisms, with the exact mechanism depending on the formation type.

4.1.1 Depleted Oil and Gas Reservoirs

Though a relatively new idea in the context of climate change mitigation, injecting CO_2 into depleted oil and gas fields has been practiced for many years. The major purpose of these injections was to dispose of "acid gas," a mixture of CO_2, H_2S, and other by-products of oil and gas exploitation and refining. In 2001, nearly 200 million cubic meters of acid gas was injected into formations across Alberta and British Columbia at more than 30 different locations. Acid gas injection has become a popular alternative to sulfur recovery and acid gas flaring, particularly in Western Canada. Essentially, acid gas injection schemes remove CO_2 and H_2S from the produced oil or gas stream, compress and transport the gases via pipeline to an injection well, and reinject the gases into a different formation for disposal. Proponents of acid gas injection claim that these schemes result in less environmental impact than alternatives for processing and disposing of unwanted gases. In most of these schemes, CO_2 represents the largest component of the acid gas, typically up to 90% of the total volume injected for disposal. Successful acid gas injection requires a nearby reservoir with sufficient porosity, amply isolated from producing reservoirs and water zones. Historically, depleted and producing reservoirs have proved to be extremely reliable containers of both hydrocarbons and acid gases over time.

4.1.2 Enhanced Oil Recovery Operations

Carbon dioxide injection into geological formations for enhanced oil recovery (EOR) is a mature technology. In 2000, 84 commercial or research-level CO_2 EOR projects were operational worldwide. The United States, the technology leader, accounts for 72 of the 84 projects, most of which are located in the Permian Basin. Combined, these projects produced 200,772 barrels (bbl) of oil per day, a small but significant fraction (0.3%) of the 67.2 million bbl per day total of worldwide oil production that year. Outside the United States and Canada, CO_2 floods have been implemented in Hungary, Turkey, and Trinidad.

In most CO_2 EOR projects, much of the CO_2 injected into the oil reservoir is only temporarily stored. This is because the decommissioning of an EOR project usually involves the "blowing down" of the reservoir pressure to maximize oil recovery. This blowing down results in CO_2 being released, with a small but significant amount of the injected CO_2 remaining dissolved in the immobile oil. The Weyburn Field in southeastern Saskatchewan, Canada, is the only CO_2 EOR project to date that has been monitored specifically to understand CO_2 storage. In the case of the Weyburn Field, no blow-down phase is planned, thereby allowing for permanent CO_2 storage. Over the anticipated 25-year life of the project, it is expected that the injection of some 18 million tons of CO_2 from the Dakota Gasification Facility in North Dakota will produce around 130 million BBL of enhanced oil. This has been calculated to be equivalent to approximately 14 million tons of CO_2 being prevented from reaching the atmosphere, including the CO_2 emissions from electricity generation that is required for the whole EOR operation.

4.1.3 Unminable Coal Seams

Abandoned or uneconomic coal seams are another potential storage site. CO_2 diffuses through the pore structure of coal and is physically adsorbed to it. This process is similar to the way in which activated carbon removes impurities from air or water. The exposed coal surface has a preferred affinity for adsorption of CO_2 rather than for methane, with a ratio of 2:1. Thus, CO_2 can be used to enhance the recovery of coal bed methane (CBM). In some cases, this can be very cost-effective or even cost free, because the additional methane removal can offset the cost of the CO_2 storage operations. CBM production has become an increasingly important component of natural gas supply in the United States during the past decade. In 2000, approximately 40 billion standard cubic meters (scm) of CBM was produced, accounting for about 7% of the nation's total natural gas production. The most significant CBM production, some 85% of the total, occurs in the San Juan basin of southern Colorado and northern New Mexico. Another 10% is produced in the Black Warrior basin of Alabama, and the remaining 5% comes from rapidly developing Rocky Mountain coal basins, namely, the Uinta basin in Utah, the Raton basin in Colorado and New Mexico, and the Powder River basin in Wyoming. Significant potential for CBM exists worldwide. A number of coal basins in Australia, Russia, China, India, Indonesia, and other countries have also been identified as having a large CBM potential. The total worldwide potential for CBM is estimated at around 2 trillion scm, with about 7.1 billion tons of associated CO_2 storage potential.

4.1.4 Deep Saline Formations

Deep saline formations, both subterranean and subseabed, may have the greatest CO_2 storage potential. These reservoirs, which are the most widespread and have the largest volumes, are very distinct from the more familiar reservoirs used for freshwater supplies. Research is currently underway in trying to understand what percentage of these deep saline formations could be suitable CO_2 storage sites.

The density of CO_2 depends on the depth of injection, which determines the ambient temperature and pressure. The CO_2 must be injected below 800 m, so that it is in a dense phase (either liquid or supercritical). When injected at these depths, the specific gravity of CO_2 ranges from 0.5 to 0.9, which is lower than that of the ambient aquifer brine. Therefore, CO_2 will naturally rise to the top of the reservoir, and a trap is needed to ensure that it does not reach the surface. Geologic traps overlying the aquifer immobilize the CO_2. In the case of aquifers with no distinct geologic traps, an impermeable cap rock above the underground reservoir is needed. This forces the CO_2 to be entrained in the groundwater flow and is known as hydrodynamic trapping. Two other very important trapping mechanisms are solubility and mineral trapping. Solubility and mineral trapping involve the dissolution of CO_2 into fluids, and the reaction of CO_2 with minerals present in the host formation to form stable, solid compounds such as carbonates. If the flow path is long enough, the CO_2 might all dissolve or become fixed by mineral reactions before it reaches the basin margin, essentially becoming permanently trapped in the reservoir.

The first, and to date only, commercial-scale project dedicated to geologic CO_2 storage is in operation at the Sleipner West gas field, operated by Statoil, located in the North Sea about 250 km off the coast of Norway. The natural gas produced at the field has a CO_2 content of about 9%. In order to meet commercial specifications, the CO_2 content must be reduced to 2.5%. At Sleipner, the CO_2 is compressed and injected via a single well into the Utsira Formation, a 250-m-thick aquifer located at a depth of 800 m below the seabed. About 1 million metric tons of CO_2 have been stored annually at Sleipner since October 1996, equivalent to about 3% of Norway's total annual CO_2 emissions. A total of 20 Mt of CO_2 is expected to be stored over the lifetime of the project. One motivation for doing this was the Norwegian offshore carbon tax, which was in 1996 about $50 (USD) per metric ton of CO_2 (the tax was lowered to $38 per ton on January 1, 2000). The incremental investment cost for storage was about $80 million. Solely on the basis of carbon tax savings, the investment was paid back in about 1.5 years. This contrasts to most gas fields worldwide, where the separated CO_2 is simply vented into the atmosphere.

Statoil is planning a second storage project involving about 0.7 Mt per year of CO_2 produced at the Snohvit gas field in the Barents Sea off northern Norway; injection will be into a deep subsea formation.

4.1.5 Environmental and Safety Concerns

Fundamentally, a geologic storage system can be broken down into two general subsystems, namely, operational and *in situ*. The operational subsystem is composed of the more familiar components of CO_2 capture, transportation, and injection, which have been successfully deployed in the previously discussed applications. Once CO_2 is injected in the reservoir, it enters an *in situ* subsystem in which the control of CO_2 is transferred to the forces of nature. Years of technological innovation and experience have given us the tools and expertise to handle and control CO_2 in the operational subsystem with adequate certainty and safety; however, that same level of expertise and understanding is largely absent once the CO_2 enters the storage reservoir. Direct environmental and human health risks are of utmost concern. As such, researchers are now conducting studies to evaluate the likelihood and potential impacts associated with leaks, slow migration and accumulation, and induced seismicity.

4.2 Ocean Storage

By far, the ocean represents the largest potential sink for anthropogenic CO_2. It already contains an estimated 40,000 Gt C (billion metric tons of carbon) compared with only 750 Gt C in the atmosphere and 2200 Gt C in the terrestrial biosphere. Apart from the surface layer, deep ocean water is unsaturated with respect to CO_2. It is estimated that if all the anthropogenic CO_2 that would double the atmospheric concentration were injected into the deep ocean, it would change the ocean carbon concentration by less than 2%, and lower its pH by less than 0.15 units. Furthermore, the deep waters of the ocean are not hermetically separated from the atmosphere.

Eventually, on a timescale of 1000 years, over 80% of today's anthropogenic emissions of CO_2 will be transferred to the ocean. Discharging CO_2 directly to the ocean would accelerate this ongoing but slow natural process and would reduce both peak atmospheric CO_2 concentrations and their rate of increase.

In order to understand ocean storage of CO_2, some properties of CO_2 and seawater need to be elucidated. For efficiency and economics of transport, CO_2 would be discharged in its liquid phase. If discharged above about 500 m depth (that is, at a hydrostatic pressure less than 50 atm), liquid CO_2 would immediately flash into a vapor and bubble up back into the atmosphere. Between 500 and about 3000 m, liquid CO_2 is less dense than seawater, therefore it would ascend by buoyancy. It has been shown by hydrodynamic modeling that if liquid CO_2 were released in these depths through a diffuser such that the bulk liquid breaks up into droplets less than about 1 cm in diameter, the ascending droplets would completely dissolve before rising 100 m. Because of the higher compressibility of CO_2 compared to seawater, below about 3000 m liquid CO_2 becomes denser than seawater, and if released there, would descend to greater depths. When liquid CO_2 is in contact with water at temperatures less than 10°C and pressures greater than 44.4 atm, a solid hydrate is formed in which a CO_2 molecule occupies the center of a cage surrounded by water molecules. For droplets injected into seawater, only a thin film of hydrate forms around the droplets.

There are two primary methods under serious consideration for injecting CO_2 into the ocean. One involves dissolution of CO_2 at middepths (1500–3000 m) by injecting it from a bottom-mounted pipe from shore or from a pipe towed by a moving CO_2 tanker. The other is to inject CO_2 below 3000 m, where it will form a "deep lake." Benefits of the dissolution method are that it relies on commercially available technology and the resulting plumes can be made to have high dilution to minimize any local environmental impacts due to increased CO_2 concentration or reduced pH. The concept of a CO_2 lake is based on a desire to minimize leakage to the atmosphere. Research is also looking at an alternate option of injecting the CO_2 in the form of bicarbonate ions in solution. For example, seawater could be brought into contact with flue gases in a reactor vessel at a power plant, and that CO_2-rich water could be brought into contact with crushed carbonate minerals, which would then dissolve and form bicarbonate ions. Advantages of this scheme are that only shallow injection is required ($>$200 m) and no pH changes will result. Drawbacks are the need for large amounts of water and carbonate minerals.

Discharging CO_2 into the deep ocean appears to elicit significant opposition, especially by some environmental groups. Often, discharging CO_2 is equated with dumping toxic materials into the ocean, ignoring that CO_2 is not toxic, that dissolved carbon dioxide and carbonates are natural ingredients of seawater, and, as stated before, atmospheric CO_2 will eventually penetrate into deep water anyway. This is not to say that seawater would not be acidified by injecting CO_2. The magnitude of the impact on marine organisms depends on the extent of pH change and the duration of exposure. This impact can be mitigated by the method of CO_2 injection, e.g., dispersing the injected CO_2 by an array of diffusers, or adding pulverized limestone to the injected CO_2 in order to buffer the carbonic acid.

5. ECONOMICS

Carbon capture and storage costs can be considered in terms of four components: separation, compression, transport, and injection. These costs depend on many factors, including the source of the CO_2, transportation distance, and the type and characteristics of the storage reservoir. In this section, costs associated with capture from fossil fuel-fired power plants and with subsequent transport and storage are considered. In this case, the cost of capture includes both separation and compression costs because both of these processes almost always occur at the power plant.

5.1 Cost of Capture

Technologies to separate and compress CO_2 from power plant flue gases exist and are commercially available. However, they have not been optimized for capture of CO_2 from a power plant for the purpose of storage. The primary difference in capturing CO_2 for commercial markets versus capturing CO_2 for storage is the role of energy. In the former case, energy is a commodity, and all we care about is its price. In the latter case, using energy generates more CO_2 emissions, which is precisely what we want to avoid. An energy penalty can be calculated as $(x-y)/x$, where x is the output in kilowatts of a reference power plant without capture and y is the output in kilowatts of the same plant with capture. The calculation requires that the same fuel input be used in both cases. For example, if the power plant output is reduced by 20%

because of the capture process ($y = 0.8x$), the process is said to have an energy penalty of 20%. We can account for the energy penalty by calculating costs on a CO_2 avoided basis. As shown in Fig. 3, due to the extra energy required to capture CO_2, the amount of CO_2 emissions avoided is always less than the amount of CO_2 captured. Therefore, capturing CO_2 for purposes of storage requires more emphasis on reducing energy inputs than is the case in traditional commercial processes.

Based on the results of major economic studies available in the literature adjusted to a common economic basis, Fig. 4 summarizes the present cost of electricity (COE) from three types of CO_2 capture power plants: integrated gasification combined cycle(IGCC), pulverized coal-fired single cycle (PC), and natural gas combined cycle (NGCC). The mean and range (± 1 standard deviation) are shown for each capture plant, along with a typical COE for a no-capture plant. This results in an increase in the cost of electricity of 1–2¢/kWh for an NGCC plant, 1–3¢/kWh for an IGCC plant, and 2–4¢/kWh for a PC plant.

The energy penalties for each of these processes have also been estimated. The energy penalty for an NGCC plant is about 16%, whereas for a PC plant it is 28%. Each of these plants uses the amine solvent process (see Section 3). The energy penalty for a PC plant is greater than for an NGCC plant because coal has a larger carbon content compared to gas. The major energy losses are associated with energy required to blow the flue gas through the amine absorption column, the heat required to strip off the CO_2 and regenerate the amine, and the energy required to compress the CO_2. The energy penalty for an IGCC plant is 14%, actually less than for a PC plant despite its use of coal. This is because the high CO_2 partial pressure in the IGCC plant allows the use of an energy-efficient physical absorption process instead of the chemical absorption process. However, some of these gains are offset by the energy loss associated with converting the coal into CO_2 plus H_2.

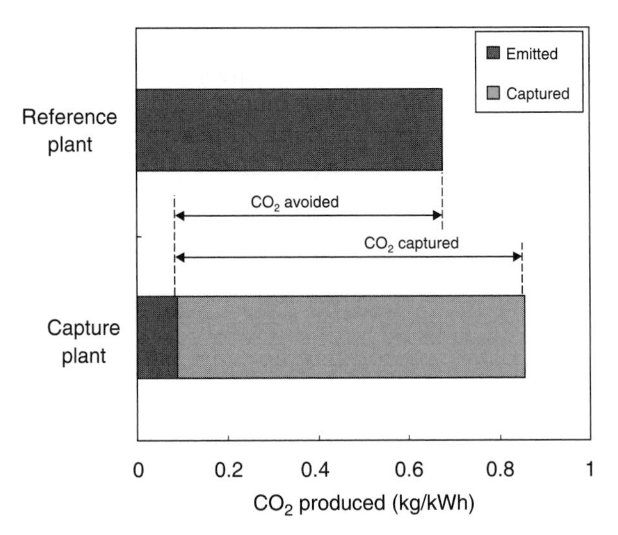

FIGURE 3 Graphical representation of avoided CO_2. The avoided emissions are simply the difference of the actual emissions per kilowatt-hour of the two plants. Note that due to the energy penalty, the emissions avoided are always less than the captured CO_2.

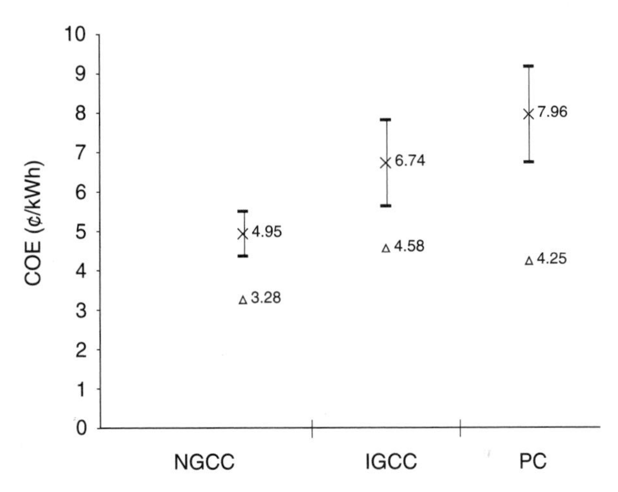

FIGURE 4 The cost of electricity (COE) with capture for various types of power plants. NGCC, Natural gas combined cycle; IGCC, integrated coal gasification combined cycle; PC, pulverized coal. The triangles represent a reference plant with no capture. The cost of electricity for CO_2 capture plants is based on a survey of the literature and is shown as a mean and a range of 1 standard deviation.

5.2 Cost of Transport

Figure 5 shows the cost of transporting CO_2 in large quantities by pipeline. Costs can vary greatly because pipeline costs depend on terrain, population density, etc. Economies of scale are realized when dealing with over 10 million metric tons per year (equivalent to about 1500 MW of coal-fired power). This cost is about \$0.50/t/100 km, compared to truck transport of \$6/t/100 km.

5.3 Cost of Injection and Storage

Figure 6 summarizes the cost of the various carbon storage technologies on a greenhouse gas-avoided basis. The points on the graphs are for a typical base case, and the bars represent the range between representative high- and low-cost cases. The ranges reflect the range of conditions found in the various reservoirs (depth, permeability, etc.), the distance

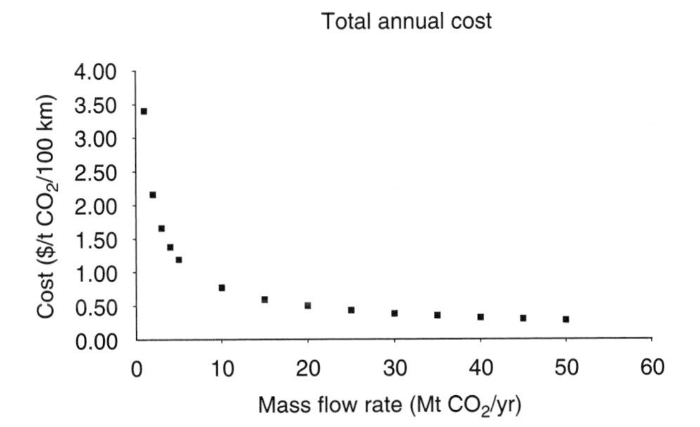

FIGURE 5 Total annual cost (capital and operation and maintenance) for CO_2 transport via pipeline as a function of CO_2 mass flow rate.

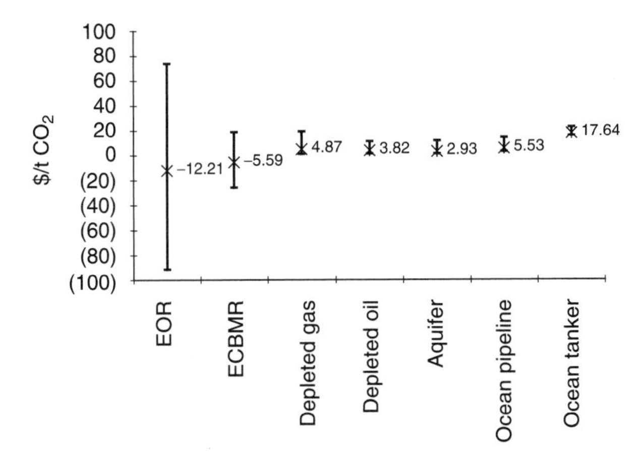

FIGURE 6 Range of costs for various carbon storage technologies on a greenhouse gas-avoided basis. EOR, Enhanced oil recovery; ECBMR, enhanced coal bed methane recovery.

between source and sink (a range of 0–300 km here), and the by-product prices (i.e., oil and gas). Excluding the more expensive ocean tanker option, the typical base case costs for CO_2 storage (transport + injection) without oil or gas by-product credit is in the range of $3–5.50/t CO_2 ($11–20/t C). The overall cost range can be characterized as $2–15/t CO_2 ($7–55/t C). With a by-product credit for the gas or oil, the credit will offset the storage costs in many instances. For example, in the base EOR case, one can afford to pay $12.21/t CO_2 and still break even (i.e., the costs equal the by-product credit).

5.4 Overall Costs

Economic models of the general economy (i.e., a general equilibrium model) can be used to estimate the market carbon price required for adaptation of CCS technologies in the electric power industry. Carbon prices must be established through government policy, such as a tax or a cap-and-trade system. Assuming the costs and technology level outlined in the preceding section, carbon prices must reach $100/t C in order for CCS technologies to start being adopted by the power industry on a significant scale ($>5\%$ market penetration). As the carbon price increases, CCS technologies will be adapted more quickly and achieve larger market penetration.

CCS technologies can be adopted at carbon prices much less than $100/t C. These targets of opportunity will either have very inexpensive capture costs (from nonpower sources such as natural gas processing and ammonia production) or will be able to claim a by-product credit (e.g., EOR). All the commercial-scale CO_2 storage projects either in operation (Sleipner, Weyburn) or planned (Snovit by Statoil in the North Sea and In Salah by BP in Algeria) can be classified as targets of opportunity. Finally, new technologies can reduce the costs associated with CCS.

6. *ALTERNATE APPROACHES*

In the previous sections, we addressed the technologies for separating CO_2 from fossil fuel streams before or after combustion and storing the captured CO_2 in geologic or oceanic sinks. In this section, we briefly identify some alternative approaches that have been proposed for CO_2 capture and/or storage. The topics that we have chosen to include in this section are ones that have received significant publicity and/or funding. Their inclusion is in no way an endorsement, just as the exclusion of any approach is not a rejection. The enhanced uptake of CO_2 by the terrestrial biosphere (e.g., afforestation) is currently a subject of intensive debate, but this approach falls outside the scope of this article.

6.1 Capture by Microalgae

The concept is to grow algae in artificial ponds, add the necessary nutrients, and fertilize the ponds with CO_2 from flue gas. Under these conditions, it is possible to enhance the growth of microalgae, harvest the algal biomass, and convert it to food, feed, or fuel. At present, about 5000 tons of food- and feed-grade microalgae biomass is produced annually in large open pond systems. As such, this approach cannot be considered as a sequestration method because the CO_2 will be returned to the atmosphere on digestion and respiration of the food or feed. What is even worse, when used as a feed for ruminating animals, some of the ingested carbon may be converted to methane, which, compared to carbon dioxide, is a stronger greenhouse gas. But if the biomass is converted to biofuel and subsequently combusted, then it replaces fossil fuel, and thus the commensurate emission of fossil fuel-generated CO_2 is avoided. However, for this approach to be viable as a greenhouse gas control method, it is necessary to significantly lower the cost from today's level. Despite some intensive efforts, primarily from Japan, little progress has been made toward this goal.

6.2 Ocean Fertilization

It has been hypothesized that by fertilizing the ocean with limiting nutrients such as iron, the growth of marine phytoplankton will be stimulated, thus increasing the uptake of atmospheric CO_2 by the ocean. The presumption is that a portion of the phytoplankton will eventually sink to the deep ocean. Researchers have targeted high-nutrient/low-chlorophyll (HNLC) ocean regions, specifically the eastern Equatorial Pacific, the northeastern Subarctic Pacific, and the Southern Ocean.

Four major open-ocean experiments have been conducted to test the "iron hypothesis," two in the Equatorial Pacific (IRONEX I in 1993 and IRONEX-II in 1995) and two in the Southern Ocean (SOIREE in 1999 and EISENEX in 2000). These experiments, funded through basic science programs (not sequestration programs), show conclusively that phytoplankton biomass can be dramatically increased by the addition of iron. However, although a necessary condition, it is not sufficient to claim iron fertilization will be effective as a CO_2 sequestration option.

The proponents of iron fertilization claim very cost-effective mitigation, on the order of $1–10/t C, but critical scientific questions remain unanswered. Although iron increases uptake of CO_2 from the atmosphere to the surface ocean, it needs to be exported to the deep ocean to be effective for sequestration. No experiments have yet attempted to measure export efficiency, which is an extremely difficult value to measure (some people claim that it cannot be measured experimentally). In addition, there are concerns about the effect on ecosystems, such as inducing anoxia (oxygen depletion) and changing the composition of phytoplankton communities.

6.3 Mineral Storage

Several minerals found on the surface of the earth uptake CO_2 from the atmosphere, with the formation of carbonates, thus permanently storing the CO_2. Such minerals are calcium and magnesium silicates. For example, the following reaction occurs with serpentine, a magnesium silicate:

$$Mg_3Si_2O_5(OH)_4 + 3CO_2(g) = 3MgCO_3 + 2SiO_2 + 2H_2O$$

Although the reaction is thermodynamically favored, it is extremely slow in nature (characteristic time on the order of a 100,000 years). The challenge is to speed up the reaction in order to be able to design an economically viable process. Many reaction pathways have been explored to varying degrees. Although some have shown progress, none has yet resolved all the issues necessary to make a commercial process.

6.4 Nonbiological Capture from Air

The terrestrial biosphere routinely removes CO_2 from air, primarily through photosynthesis. It has been suggested that CO_2 can also be removed from air via nonbiological means. Although some concept papers have been published, no viable methods to accomplish this goal have been proposed. The problem is that the partial pressure of CO_2 in air is less than 0.0004 atm, compared to about 0.1 atm in flue gas and up to 20 atm in synthesis gas. The difficulty in capture increases as the partial pressure of CO_2 decreases. Therefore, it is questionable whether CO_2 can be captured from air with acceptable energy penalties and costs. If so, it almost surely will take development of a capture process very different from those that exist today.

6.5 Utilization

CO_2 from fossil fuel could be utilized as a raw material in the chemical industry for producing commercial products that are inert and long-lived, such as vulcanized rubber, polyurethane foam, and polycarbonates. Only a limited amount of CO_2 can be stored in such a fashion. Estimates of the world's commercial sales for CO_2 are less than 0.1 Gt C equivalent, compared to annual emissions of close to 7 Gt C equivalent. It has been suggested that CO_2 could be recycled into a fuel. This would create a market on the same scale as the CO_2 emissions. However, to recycle CO_2 to a fuel would require a carbon-free energy source. If such a source existed, experience suggests that it would be more efficient and cost-effective to use that source directly to displace fossil fuels rather than to recycle CO_2.

SEE ALSO THE FOLLOWING ARTICLES

Biomass: Impact on Carbon Cycle and Greenhouse Gas Emissions • Carbon Sequestration: Terrestrial • Carbon Taxes and Climate Change • Climate Change and Energy, Overview • Climate Change: Impact on the Demand for Energy • Greenhouse Gas Abatement: Controversies in Cost Assessment • Greenhouse Gas Emissions, Alternative Scenarios of • Greenhouse Gas Emissions from Energy Systems, Comparison and Overview

Further Reading

Herzog, H. J. (2001). What future for carbon capture and sequestration? *Environ. Sci. Technol.* **35**, 148A–153A.

Herzog, H., Drake, E., and Adams. E. (1997). CO_2 capture, reuse and storage technologies for mitigating global climate change: A white paper. Energy Lab. Rep., DOE Order DE-AF22-96PC01257. Massachusetts Institute of Technology, Cambridge, MA.

U.S. Department of Energy. (1993). The capture, utilization and disposal of carbon dioxide from fossil fuel-fired power plants. DOE/ER-30194. Department of Energy, Washington, D.C.

U.S. Department of Energy. (1999). Carbon sequestration research and development. DOE/SC/FE-1. National Technical Information Service, Springfield, VA.

Williams, D. J., Durie, R. A., McMullan, P., Paulson, C. A. J., and Smith, A. Y. (eds.). (2001). Greenhouse gas control technologies. Proceedings of the Fifth International Conference on Greenhouse Gas Control Technologies, CSIRO Publ., Collingwood, Australia.

Carbon Sequestration, Terrestrial

R. LAL
The Ohio State University
Columbus, Ohio, United States

1. Introduction
2. Carbon Sequestration
3. Processes of Terrestrial Carbon Sequestration
4. Enhancing Terrestrial Carbon Sequestration
5. Ancillary Benefits
6. Restoration of Degraded Ecosystems and Terrestrial Carbon Sequestration
7. Potential and Challenges of Soil Carbon Sequestration
8. Climate Change and Terrestrial Carbon Sequestration
9. Nutrient Requirements for Terrestrial Carbon Sequestration
10. The Permanence of Carbon Sequestered in Terrestrial Ecosystems
11. Terrestrial Carbon Sequestration and Global Food Security
12. Conclusions

Glossary

ancillary benefits Additional benefits of a process.
anthropogenic Human-induced changes.
biomass/biosolids Material of organic origin.
carbon sequestration Transfer of carbon from atmosphere into the long-lived carbon pools such as vegetation and soil.
humus The decomposed organic matter in which remains of plants and animals are not identifiable.
mean residence time The average time an atom of carbon spends within a pool; calculated by dividing the pool by flux.
soil organic matter Sum of all organic substances in the soil.
soil quality Ability of a soil to perform functions of interest to humans such as biomass production, water filtration, biodegradation of pollutants, and soil carbon sequestration.
terrestrial carbon Carbon contained in the vegetation and soil.

Atmospheric concentration of carbon dioxide (CO_2) and other greenhouse gases (GHGs) has increased drastically since the industrial revolution and is still increasing with attendant risks of increase in global temperature. Thus, the need to identify options for mitigating atmospheric concentration of CO_2 is widely recognized. The Kyoto Protocol, ratified by several countries but not by the United States, mandates that industrialized nations reduce their net emissions with reference to the 1990 level by an agreed amount and recognizes a wide range of options to minimize the risks of global warming. The choice of specific options depends on the economic conditions, industrial infrastructure, land and water resources, agricultural conditions, and the like.

1. INTRODUCTION

Specific options of mitigating climate change can be grouped broadly into two categories: adaptive responses and mitigative responses (Fig. 1). Adaptive responses involve identification of options for sustainable management of natural resources (e.g., terrestrial ecosystems, water resources and wetlands, agricultural ecosystems, deserts) under changed climate with projected alterations in precipitation, frequency of extreme events, growing season duration, and incidence of pests and pathogens affecting health of human and other biota. Mitigative responses involve controlling atmospheric concentration of carbon dioxide (CO_2) either by removing it from the atmosphere or by avoiding its emission into the atmosphere. Important among mitigative strategies are reducing and sequestering emissions. Emissions reduction may involve improving energy use/conversion efficiency, developing alternatives to fossil fuel or developing carbon (C)-neutral fuel sources, using space reflectors to reduce the amount of radiation reaching the earth, absorbing CO_2 from the atmosphere using special extractants, and sequestering C in long-lived pools. Both adaptive and mitigative strategies are important to reducing

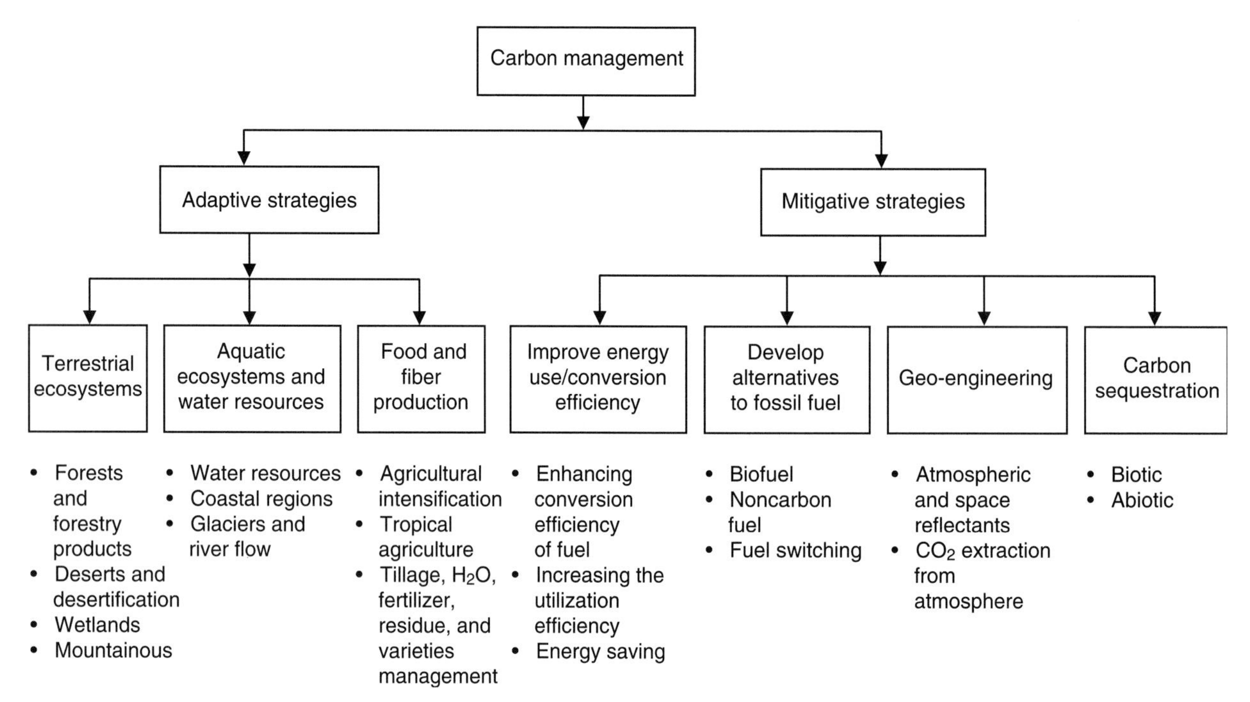

FIGURE 1 Carbon management strategies to mitigate the greenhouse effect.

the rate of enrichment of atmospheric concentration of CO_2. This article describes the principles, opportunities, and challenges of C sequestration in terrestrial pools composed of soils and vegetation.

2. CARBON SEQUESTRATION

Carbon sequestration implies capture and secure storage of C that would otherwise be emitted to or remain in the atmosphere. The objective is to transfer C from the atmosphere into long-lived pools (e.g., biota, soil, geologic strata, ocean). It is the net removal of CO_2 from the atmosphere or prevention of CO_2 emissions from terrestrial ecosystems. Strategies of C sequestration can be grouped under two broad categories: biotic and abiotic (Fig. 2). Biotic strategies are based on natural process of photosynthesis and transfer of CO_2 from atmosphere into vegetative, pedologic, and aquatic pools through mediation via green plants. Abiotic strategies involve separation, capture, compression, transport, and injection of CO_2 from power plant flue gases and effluent of industrial processes deep into ocean and geologic strata. The strategy is to use engineering techniques for keeping industrial emissions from reaching the atmosphere.

Terrestrial C sequestration is a natural process. It involves transfer of atmospheric CO_2 through photosynthesis into vegetative and pedologic/soil C pools. The terrestrial pool is the third largest among the five global C pools (Fig. 3). It includes 2300 petagrams (Pg, where 1 Pg $= 10^{15}$ g $=$ 1 billion metric tons) of C in soil to 1-m depth and 600 Pg in the vegetative pool, including the detritus material. Thus, the terrestrial C pool is approximately four times the size of the atmospheric pool and 60% the size of the fossil C pool. It is also a very active pool and is continuously interacting with atmospheric and oceanic pools.

The natural processes of terrestrial C sequestration are already absorbing a large fraction of the anthropogenic emissions. For example, oceanic uptake is about 2 Pg C/year, and unknown terrestrial sinks absorbed a total of 2.8 Pg C/year during the 1990s. Thus, 4.8 Pg/year out of a total anthropogenic emission of 8 Pg/year (or 59%) was absorbed by natural sinks during the 1990s. Therefore, it is prudent to identify and enhance the capacity of natural processes.

Carbon sequestration, both biotic and abiotic, is an important strategy for mitigating risks of global warming. It can influence the global C cycle over a short period and reduce the equilibrium level of atmospheric CO_2 until the alternatives to fossil fuel take effect. A hypothetical scenario in Fig. 4 shows that atmospheric concentration of CO_2 by 2100 may be 700 parts per million in terms of volume (ppmv)

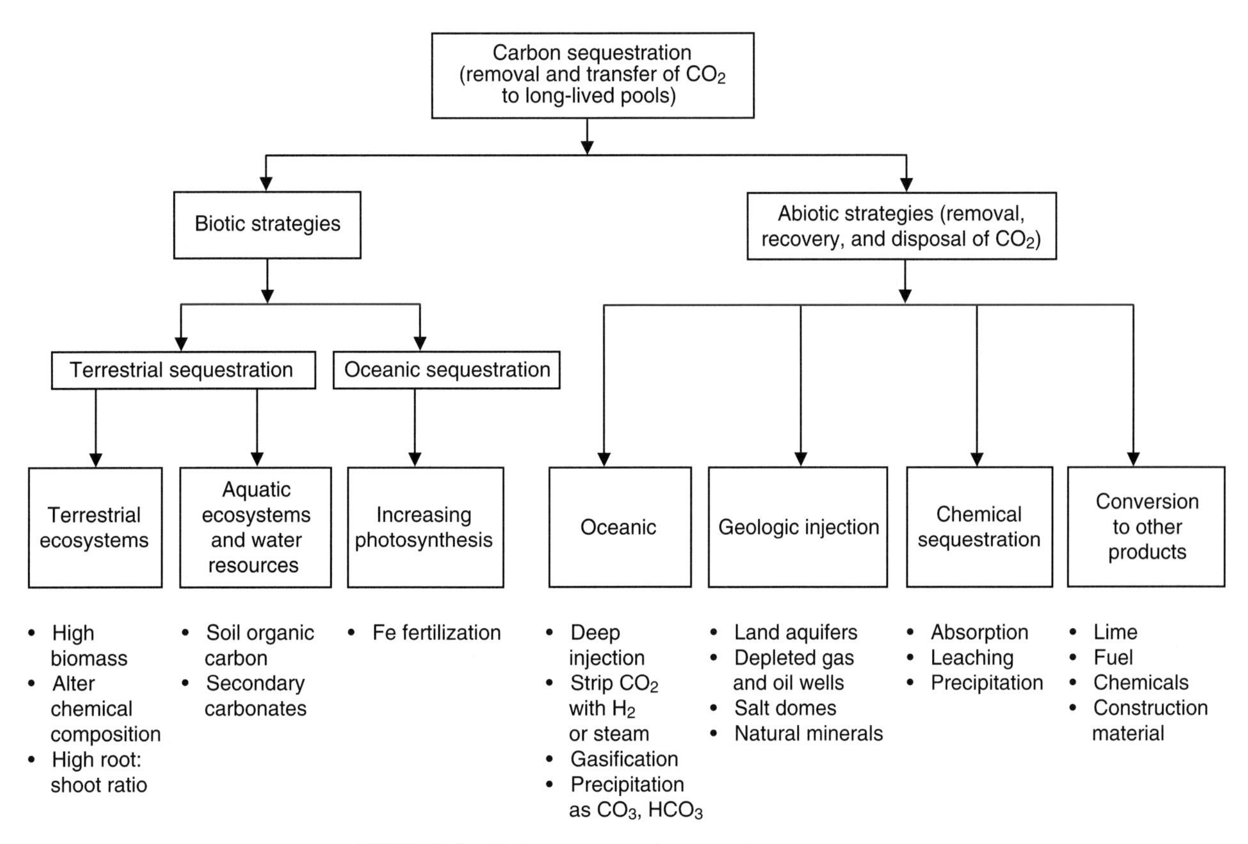

FIGURE 2 Carbon sequestration strategies.

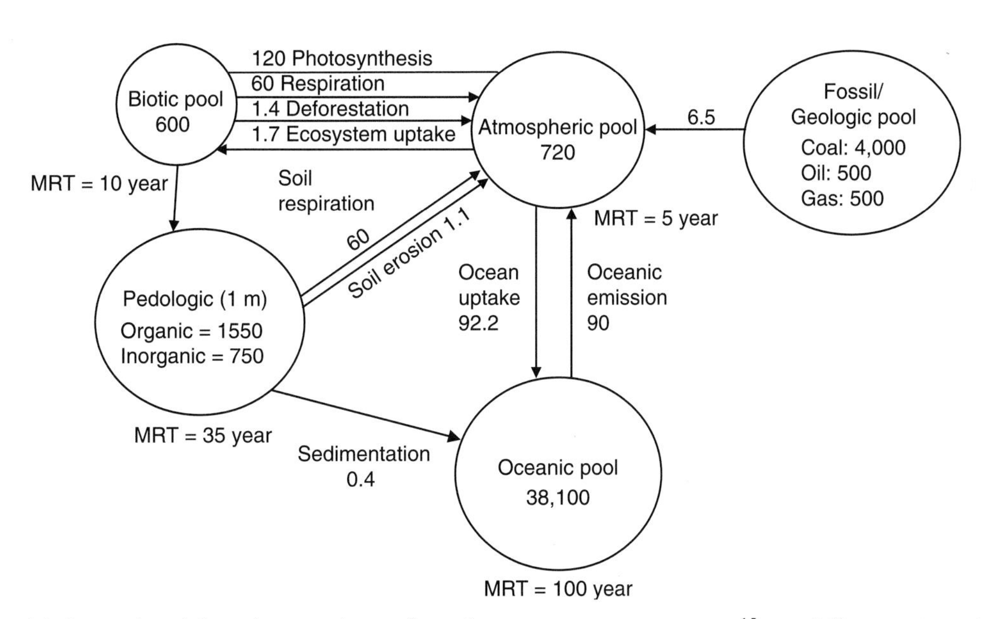

FIGURE 3 Global C pools and fluxes between them. All numbers are in petagrams (Pg = 10^{15} g = 1 billion metric tons). MRT, mean residence time.

with business as usual and only 550 ppmv by adoption of several C sequestration strategies. Feasibility of a C sequestration strategy depends on numerous factors: (1) sink capacity of the pool (e.g., soil, vegetation, geologic strata), (2) fate of the C sequestered and its residence time, (3) environmental impact on terrestrial or aquatic ecosystems, and (4) cost-effectiveness of the strategy.

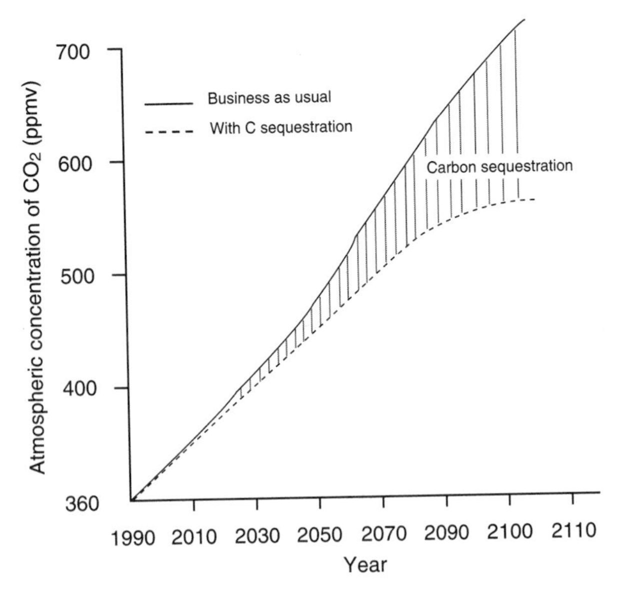

FIGURE 4 A hypothetical scenario about the impact of C sequestration by biotic and abiotic strategies on atmospheric concentration of CO_2.

The distribution of C pool in different biomes shows the high amount of terrestrial C in forest and savanna ecoregions and in northern peat lands (Table I). The terrestrial C pool in managed ecosystems (e.g., cropland, permanent crops) differs from others in terms of the biomass C pool that is nearly negligible because it is harvested for human and animal consumption. The data in Table II show the potential of terrestrial C sequestration at 5.7 to 10.1 Pg C/year, of which the potential of agricultural lands at 0.8 to 0.9 Pg C/year includes only soil C. Similar estimates for croplands and world soils have been made by others, yet there are numerous uncertainties in these estimates.

TABLE I

Distribution of Terrestrial C Pool in Various Ecosystems

Ecosystem	Area (Mha)	NPP (Pg C/year)	Biomass C (Pg)	Soil C to 1 m (Pg)
Forest				
Tropical	1480	13.7	244.2	123
Temperate	750	5.0	92.0	90
Boreal	900	3.2	22.0	135
Savanna				
Temperate woodlands	200	1.4	16.0	24
Chaparral	250	0.9	8.0	30
Tropical	2250	17.8	65.9	263
Temperate grassland	1250	4.4	9.0	295
Tundra, arctic & alpine	950	1.0	6.0	121
Deserts	3000	1.5	7.2	191
Lake and streams	200	0.4	0.0	0
Wetlands	280	3.3	12.0	202
Peat lands	340	0.0	0.0	455
Cultivated and permanent crops	1480	6.3	3.0	117
Perpetual ice	1550	0.0	0.0	0
Urban	200	0.2	1.0	10
Total	15,080	59.1	486.4	2056

Source. Adopted from Amthor and Huston (1998), Intergovernment Panel on Climate Change (2000), and U.S. Department of Energy (1999).

3. PROCESSES OF TERRESTRIAL CARBON SEQUESTRATION

The terrestrial sequestration involves biological transformation of atmospheric CO_2 into biomass and soil humus. Currently, the process leads to sequestration of approximately 2 Pg C/year. With improved management and natural CO_2 fertilization effect under the enhanced CO_2 levels, the potential of terrestrial C sequestration may be much larger (Table II). Principal among management options include the following:

- Conversion of marginal croplands to biomass energy crop production
- Restoration of degraded soils and ecosystems
- Agricultural intensification on cropland and pastures
- Afforestation and forest soil management
- Restoration of wetlands

If the rate of C sequestration by adopting these options can be increased by 0.2% per year over 25 years, it can transfer 100 Pg C from the atmosphere into the terrestrial ecosystem through natural processes of biological transformation.

4. ENHANCING TERRESTRIAL CARBON SEQUESTRATION

The natural processes governing terrestrial C sequestration are outlined in Fig. 5. The vegetative C pool can be enhanced by improving biomass production,

TABLE II
Global Potential of Terrestrial Carbon Sequestration

Biome	Potential of C sequestration (Pg C/year)
Agricultural lands	0.85–0.90
Biomass croplands	0.50–0.80
Grass lands	0.50
Range lands	1.2
Forest lands	1.0–3.0
Urban lands	—
Deserts and degraded lands	0.8–1.3
Terrestrial sediments	0.7–1.7
Boreal peat lands and other wetlands	0.1–0.7
Total	5.65–10.1

Source. U.S. Department of Energy (1999).

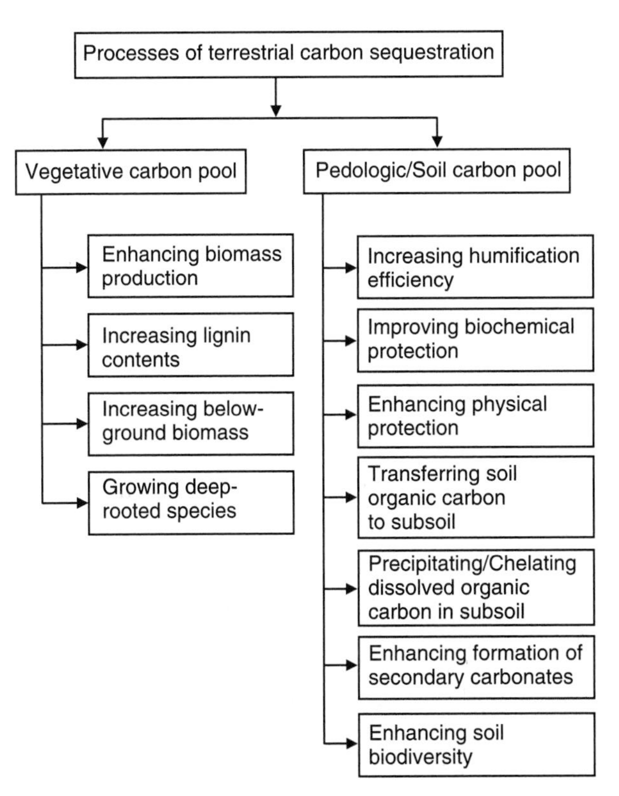

FIGURE 5 Processes of enhancing terrestrial carbon sequestration.

increasing the lignin content that reduces the rate of decomposition, and diverting biomass into a deep and prolific root system. Afforestation, conversion of agriculturally marginal soils into restorative land uses, choice of fast-growing species with deep root systems, and adoption of recommended management practices (RMPs) for stand, soil fertility, and pest management. In addition to the return of biomass to soil, there are several processes that moderate the soil C pool. The humification efficiency, the proportion of biomass converted to the stable humic fraction, depends on soil water and temperature regimes and the availability of essential elements (e.g., N, P, S, Zn, Cu, Mo, Fe). The amount and nature of clay, essential to formation and stabilization of aggregates, strongly influence the sequestration and residence time of C in soil. Effectiveness of chemical, biochemical, and physical processes of soil C sequestration depends on formation of stable micro-aggregates or organo–mineral complexes. Soil biodiversity, activity and species diversity of soil fauna such as earthworms and termites, plays an important role in bioturbation, C cycle, and humification of biosolids. Transfer of humus into the subsoil, precipitation of dissolved organic C (DOC), leaching and precipitation of carbonates, and formation of secondary carbonates are important processes that enhance soil C pool and increase its residence time.

Conversion of natural ecosystems to agricultural ones (e.g., cropland, grazing land) depletes soil organic C (SOC) pool by as much as 50 to 80%. The magnitude and rate of depletion are generally greater in warm climates than in cool climates, in coarse-textured soils than in heavy-textured soils, and in soils with high antecedent SOC pool than in low antecedent SOC pool. Therefore, most agricultural soils now contain lower SOC pool than their potential capacity for the specific ecoregion. Conversion to a restorative land use and adoption of RMPs can enhance SOC pool. Feasible RMPs of agricultural intensification include using no-till farming with frequent incorporation of cover crops (meadows) in the rotation cycle, using manure and other biosolids as a component of integrated nutrient management strategy, and improving pasture growth by planting appropriate species and controlling stocking rate and the like. Restoring eroded/degraded soils and ecosystems is extremely important to enhancing the terrestrial C pool.

5. ANCILLARY BENEFITS

There are numerous ancillary benefits of terrestrial C sequestration. Enhancing vegetative/ground cover and soil C pool reduces the risks of accelerated runoff, erosion, and sedimentation. The attendant decline in dissolved and suspended loads reduces risks of eutrophication of surface waters. Similarly, an increase in soil C pool decreases the transport of agricultural chemicals and other pollutants into the ground water. There are notable benefits of increased activity and species diversity of soil fauna and flora with an increase in SOC pool. In addition to enhancing soil structure, in terms of aggregation and its beneficial impacts on water retention and transmission, an increase in soil biodiversity also strengthens elemental/nutrient cycling. Soil is a living membrane between bedrock and the atmosphere, and its effectiveness is enhanced by an increase in soil C pool and its quality.

An increase in terrestrial C sequestration in general, and in soil C sequestration in particular, improves soil physical, chemical, and biological quality (Fig. 6). Improvements in soil physical quality lead to decreased risks of accelerated runoff and erosion, reduce crusting and compaction, and improve plant-available water capacity. Improvements in soil chemical quality lead to increased cation/anion exchange capacity, elemental balance, and favorable soil reaction. Improvements in soil biological quality lead to increased soil biodiversity and bioturbation, elemental/nutrient cycling, and biodegradation and bioremediation of pollutants. An integrative effect of improvement in soil quality is to increase biomass/agronomic productivity. On a global scale, soil C sequestration can have an important impact on food security.

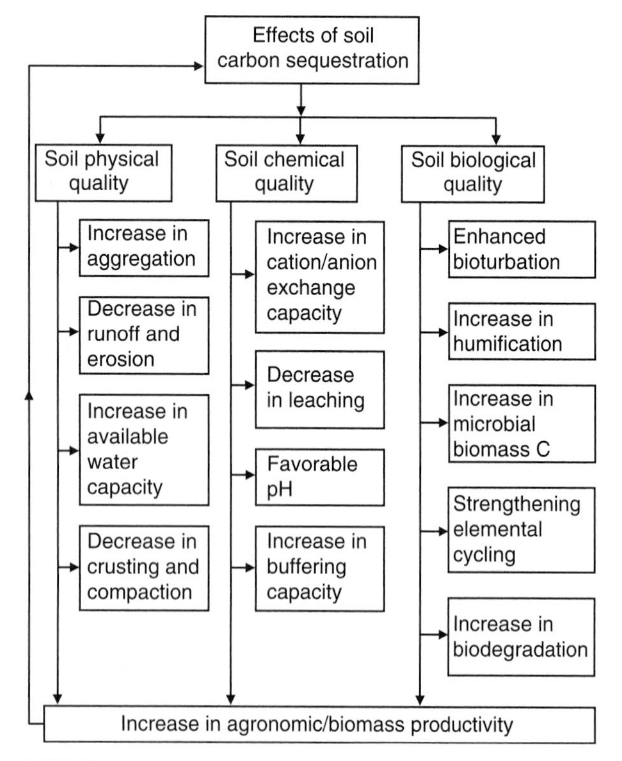

FIGURE 6 Favorable effects of soil carbon sequestration on soil quality and agronomic/biomass productivity.

6. RESTORATION OF DEGRADED ECOSYSTEMS AND TERRESTRIAL CARBON SEQUESTRATION

Accelerated soil erosion, soil degradation by other degradative processes (e.g., salinization, nutrient depletion, elemental imbalance, acidification), and desertification are severe global issues. Accelerated erosion by water and wind is the most widespread type of soil degradation. The land area affected by water erosion is 1094 million hectares (Mha), of which 751 Mha is severely affected. The land area affected by wind erosion is 549 Mha, of which 296 Mha is severely affected. In addition, 239 Mha is affected by chemical degradation and 83 Mha is affected by physical degradation. Thus, the total land area affected by soil degradation is 1965 Mha or 15% of the earth's land area. With a sediment delivery ratio of 13 to 20%, the suspended sediment load is estimated at 20×10^9 metric tons (Mg)/year. Soil erosion affects SOC dynamics by slaking and breakdown of aggregates, preferential removal of C in surface runoff or wind, redistribution of C over the landscape, and mineralization of displaced/redistributed C. The redistributed SOC is generally light or labile fraction composed of particulate organic carbon (POC) and is easily mineralized. As much as 4 to 6 Pg C/year is transported globally by water erosion. Of this, 2.8 to 4.2 Pg C/year is redistributed over the landscape and transferred to depressional sites, 0.4 to 0.6 Pg C/year is transported into the ocean, and 0.8 to 1.2 Pg C/year is emitted into the atmosphere. The historic loss of SOC pool from soil degradation and other anthropogenic processes is 66 to 90 Pg C (78 ± 12 Pg C), compared with total terrestrial C loss of 136 ± 55 Pg C. Of the total SOC loss, that due to erosion by water and wind is estimated at 19 to 32 Pg C (26 ± 9 Pg C) or 33% of the total loss.

There is also a strong link between desertification and emission of C from soil and vegetation of the dryland ecosystems. Desertification is defined as the diminution or destruction of the biological potential of land that ultimately can lead to desert-like conditions. Estimates of the extent of desertification are wide-ranging and are often unreliable. The land area affected by desertification is estimated at 3.5 to 4.0 billion ha. The available data on the rate of desertification is also highly speculative and is estimated by some to be 5.8 Mha/year.

Similar to other degradative processes, desertification leads to depletion of the terrestrial C pool. The historic loss of C due to desertification is estimated at 8 to 12 Pg from the soil C pool and 10 to 16 Pg from the vegetation C pool. Thus, the total historic C loss due to desertification may be 18 to 28 Pg C.

Therefore, desertification control and restoration of degraded soils and ecosystems can reverse the degradative trends and sequester a large fraction of the historic C loss. The potential of desertification control for C sequestration is estimated at 0.6 to 1.4 Pg C/year. Management of drylands through desertification control has an overall C sequestration potential of 1.0 Pg C/year. These estimates of high C sequestration potential through restoration of degraded/desertified soils and ecosystems are in contrast to the low overall potential of world soils reported by some.

7. POTENTIAL AND CHALLENGES OF SOIL CARBON SEQUESTRATION

Of the two components of terrestrial C sequestration—soil and vegetation—the importance of soil C sequestration is not widely recognized. The Kyoto Protocol has not yet accepted the soil C sink as offset for the fossil fuel emission. Yet soil C sequestration has vast potential and is a truly win–win situation.

Assuming that 66 Pg C has been depleted from the SOC pool, perhaps 66% of it can be resequestered. The global potential of soil C sequestration in the world's croplands is estimated at 0.6 to 1.2 Pg C/year. This potential can be realized over 50 years or until the sink capacity is filled.

Some have questioned the practicability or feasibility of realizing the potential of soil C sequestration. The caution against optimism is based on the hidden C costs of fertilizers, irrigation, and other input that may produce no net gains by soil C sequestration. However, it is important to realize that SOC sequestration is a by-product of adopting RMPs and agricultural intensification, which are needed to enhance agronomic production for achieving global food security. The global population of 6.2 billion is increasing at a rate of 73 million persons (or 1.3%) per year and is projected to reach 7.5 billion by 2020 and 9.4 billion by 2050. Consequently, the future food demand will increase drastically, and this will necessitate adoption of restorative measures to enhance soil quality and use of off-farm input (e.g., fertilizers, pesticides, irrigation, improved varieties) to increase productivity. Thus, fertilizers and other inputs are being used not for soil C sequestration but rather for increasing agricultural production to meet the needs of the rapidly increasing population. Soil C sequestration is an ancillary benefit of this inevitability and global necessity.

There is also a question of the finite nature of the terrestrial C sink in general and that of the soil C sink in particular. The soil C sink capacity is approximately 50 to 60 Pg C out of the total capacity of the terrestrial C sink of 136 Pg. In contrast, the capacity of geologic and terrestrial sinks is much larger. However, soil C sequestration is the most cost-effective option in the near future, and there are no adverse ecological impacts. Engineering techniques for abiotic (e.g., deep injection of CO_2 into geologic strata and ocean) strategies remain a work in progress.

The question of permanence also needs to be addressed. How long can C sequestered in soil and biota remain in these pools? Does an occasional plowing of a no-till field negate all of the gains made so far in soil C sequestration? These are relevant questions, and the answers may vary for different soils and ecoregions.

Although ratified by neither the United States nor Russia, the Kyoto Protocol demands that C stored in terrestrial ecosystems be assessed by transparent, credible, precise, and cost-effective methods. Questions have often been raised about the assessment of soil C pool and fluxes on soilscape, farm, regional, and national scales. It is important to realize that standard methods are available to quantitatively assess SOC pool and fluxes on scales ranging from the pedon or soilscape level and then extrapolated to the ecosystem, regional, and national levels. Soil scientists and agronomists have studied changes in soil organic matter in relation to soil fertility and agronomic productivity since 1900 or earlier. They have adapted and improved the analytical procedures for assessing soil C sequestration for mitigating climate change since the 1980s. New procedures are being developed to assess soil C rapidly, precisely, and economically. Promising among new analytical procedures are accelerator mass spectrometry (AMS), pyrolysis molecular beam mass spectrometry (Py-MBMS), and laser-induced breakdown spectroscopy (LIBS). The LIBS technique presumably has good detection limits, precision, and accuracy.

8. CLIMATE CHANGE AND TERRESTRIAL CARBON SEQUESTRATION

The increase in global temperature during the 20th century has been estimated at 0.6°C and is projected to be 1.4 to 5.8°C by 2100 relative to the global temperature in 1990. The projected increase in global temperature is attributed to enrichment of the atmospheric concentration of CO_2 (which has increased from 280 to 365 ppm and is increasing at the rate of 0.4% or 1.5 ppm/year), CH_4 (which has increased from 0.80 to 1.74 ppm and is increasing at the rate of 0.75% or 0.015 ppm/year), and N_2O (which has increased from 288 to 311 ppb and is increasing at the rate of 0.25% or 0.8 ppb/year).

The implications of rapidly changing atmospheric chemistry are complex and not very well understood. The projected climate change may have a strong impact on terrestrial biomes, biomass productivity, and soil quality. It is estimated that for every 1°C increase in global temperature, the biome or vegetational zones (e.g., boreal forests, temperate forests, wooded/grass savannas) may shift poleward by 200 to 300 km. There may be a change in species composition of the climax vegetation within each biome. In northern latitudes where the temperature change is likely to be most drastic, every 1°C increase in temperature may prolong the growing season duration by 10 days. With predicted climate change, the rate of mineralization of soil organic matter may increase with adverse impacts on soil quality.

There is a general consensus that CO_2 enrichment will have a fertilization effect and will enhance C storage in the vegetation (both above and below ground) for 50 to 100 years. Numerous free-air CO_2 enrichment (FACE) experiments have predicted increases in biomass production through increased net primary productivity (NPP), particularly in high latitudes. It was predicted that with an increase in temperature, tundra and boreal biomass will emit increasingly more C to the atmosphere while the humid tropical forests will continue to be a net sink. Soils of the tundra biome contain 393 Pg C as SOC and 17 Pg C as soil inorganic C (SIC) (410 Pg total), which is 16.4% of the world soil C pool. Soils of the boreal biome contain 382 Pg of SOC and 258 Pg of SIC (640 Pg total), which is 25.6% of the world soil C pool. Together, soils of the tundra and boreal (taiga) biomes contain 772 Pg or 42% of the world soil C pool. Some soils of these regions, especially histosols and cryosols, have been major C sinks in the historic past. The projected climate change and some anthropogenic activities may disrupt the terrestrial C cycle of these biomes and render them a major source of CO_2, CH_4, and N_2O. There exists a real danger that an increase in global temperature may result in a long-term loss of the soil C pool.

There are numerous knowledge gaps and uncertainties regarding the impact of climate change on terrestrial C pool and fluxes. Principal knowledge gaps and uncertainties exist regarding fine root biology and longevity, nutrient (especially N and P) availability, interaction among cycles of various elements (e.g., C, N, P, H_2O), and below-ground response.

9. NUTRIENT REQUIREMENTS FOR TERRESTRIAL CARBON SEQUESTRATION

Carbon is only one of the building blocks of the terrestrial C pool. Other important components of the biomass and soil are essential macroelements (e.g., N, P, K, Ca, Mg) and micronutrients (e.g., Cu, Zn, B, Mo, Fe, Mn). A total of 1 Mg (or 1000 kg) of biomass produced by photosynthesis contains approximately 350 to 450 kg of C, 5 to 12 kg of N, 1 to 2 kg of P, 12 to 16 kg of K, 2 to 15 kg of Ca, and 2 to 5 kg of Mg. Therefore, biomass production of 10 Mg/ha/year requires the availability of 50 to 120 kg of N, 10 to 20 kg of P, 120 to 160 kg of K, 20 to 150 kg of Ca, and 20 to 50 kg of Mg. Additional nutrients are required for grain production. Nutrient requirements are lesser for wood and forages production than for grain and tuberous crops.

Biomass (e.g., crop residues, leaf litter, detritus material, roots) is eventually converted into SOC pool through humification mediated by microbial processes. The humification efficiency is between 5 and 15%, depending on the composition of the biomass, soil properties, and climate. Soil C sequestration implies an increase in the SOC pool through conversion of biomass into humus. Similar to biomass production through photosynthesis, humification of biomass into SOC involves availability of nutrients. Additional nutrients required are N, P, S, and other minor elements. Assuming an elemental ratio of 12:1 for C:N, 50:1 for C:P, and 70:1 for C:S, sequestering 10,000 kg of C into humus will require 25,000 kg of residues (40% C), 833 kg of N, 200 kg of P, and 143 kg of S. Additional nutrients may be available in crop residue or soil or may be supplied through fertilizers and amendments. Nitrogen may also be made available through biological nitrogen fixation (BNF) and atmospheric deposition. It is argued that sequestration of soil C can be achieved only by the availability of these nutrients.

Nutrient recycling, from those contained in the biomass and in the subsoil layers, is crucial to terrestrial C sequestration in general and to soil C sequestration in particular. For soil C sequestration of 500 kg C/ha/year, as is the normal rate for conversion from plow till to no till in Ohio, sequestration of 500 kg of C into humus (58% C) will require about 14 Mg of biomass (40% C and humification efficiency of 15%). This amount of residue from cereals will contain 65 to 200 kg of N, 8 to 40 kg of P, 70 to 340 kg of K, 22 to 260 kg of Ca, and 10 to 130 kg of Mg. In comparison, the nutrients required for sequestration of 500 kg of C into humus are 42 kg of N, 10 kg of P, and 7 kg of S. Indeed, most of the nutrients required are contained in the biomass. Therefore, soil C sequestration is limited mostly by the availability of biomass. However, the production of biomass is limited by the availability of nutrients, especially N and P.

10. THE PERMANENCE OF CARBON SEQUESTERED IN TERRESTRIAL ECOSYSTEMS

The mean residence time (MRT) of C sequestered in soil and biota depends on numerous factors, including land use, management system, soil type, vegetation, and climate. The MRT is computed as a ratio of the pool (Mg/ha) to the flux (Mg/ha/year). On this basis, the global MRT of C is 5 years in the atmosphere, 10 years in the vegetation, 35 years in the soils, and 100 years in the ocean. However, there is a wide range of MRT depending on the site-specific conditions and characteristics of the organic matter. For example, the MRT of soil organic matter is less than 1 year for the labile fraction, 1 to 100 years for the intermediate fraction, and several millennia for the passive fraction. The MRT also depends on management. Conversion from plow till to no till sequesters C in soil, and the MRT depends on continuous use of no till. Even occasional plowing can lead to emission of CO_2.

Biofuel production is another important strategy. The C sequestered in biomass is released when the biosolids are burned. However, C released from biofuel is recycled again and is in fact a fossil fuel offset. There is a major difference between the two. The fossil fuel combustion releases new CO_2 into the atmosphere, and burning of biofuels merely recycles the CO_2.

11. TERRESTRIAL CARBON SEQUESTRATION AND GLOBAL FOOD SECURITY

An important ancillary benefit of terrestrial C sequestration in general, and of soil C sequestration in particular, is the potential for increasing world food security. The projected global warming, with a likely increase in mean global temperature of 4 ± 2°C, may adversely affect soil temperature and water regimes and the NPP. The increase in soil temperature may reduce soil organic matter pool with an attendant decrease in soil quality, increase in erosion and soil degradation, and decrease in NPP and agronomic yields. In contrast, the degradative trends and downward spiral may be reversed by an increase in soil C sequestration leading to improvement in soil quality and an increase in agronomic/biomass productivity. There already exists a large intercountry/interregional variation in average yields of wheat (*Triticum aestivum*) and corn (*Zea mays*). Yields of wheat range from a high of 6.0 to 7.8 Mg/ha in European Union countries (e.g., United Kingdom, Denmark, Germany, France), to a middle range of 2.4 to 2.7 Mg/ha (e.g., United States, India, Romania, Ukraine, Argentina, Canada), to a very low range of 1.0 to 2.2 Mg/ha (e.g., Pakistan, Turkey, Australia, Iran, Russia, Kazakhstan). Similarly, average yields of corn range from a high of 8 to 10 Mg/ha (e.g., Italy, France, Spain, United States), to a middle range of 4 to 7 Mg/ha (e.g., Canada, Egypt, Hungary, Argentina), to a low range of 1 to 2 Mg/ha (e.g., Nigeria, Philippines, India). A concern regarding projected global warming is whether the agroecological yield may be adversely affected by the change in soil temperature and moisture regimes and the threat of increase in food deficit in Sub-Saharan Africa and elsewhere in developing countries.

The adverse impact of decline in SOC pool on global food security is also evident from the required increase in crop yield in developing countries to meet the demands of increased population. With no possibility of bringing new land area under cultivation in most developing countries, the yields of cereals in developing countries will have to increase from 2.64 Mg/ha in 1997–1998 to 4.4 Mg/ha by 2025 and to 6.0 Mg/ha by 2050. Can this drastic increase in yields of cereals in developing countries be attained by agricultural intensification in the face of projected global warming? An important factor that can help to attain this challenging goal is improvement in soil quality through terrestrial and soil C sequestration. Improvements in soil C pool can be achieved by increasing below-ground C directly through input of root biomass. The importance of irrigation, fertilizer and nutrient acquisition, erosion control, and use of mulch farming and conservation tillage cannot be overemphasized.

12. CONCLUSIONS

Terrestrial C sequestration is a truly win–win scenario. In addition to improving the much needed agronomic/biomass production to meet the needs of a rapidly increasing world population, terrestrial C sequestration improves soil quality, reduces sediment transport in rivers and reservoirs, enhances biodiversity, increases biodegradation of pollutants, and mitigates the risks of climate change. Besides, it is a natural process with numerous ancillary benefits. Although it has a finite capacity to absorb atmospheric CO_2, it is the most desirable option available at the onset of the 21st century. The strategy of terrestrial C sequestration buys us time until C-neutral fuel options take effect.

SEE ALSO THE FOLLOWING ARTICLES

Biomass for Renewable Energy and Fuels • Biomass: Impact on Carbon Cycle and Greenhouse Gas Emissions • Carbon Capture and Storage from Fossil Fuel Use • Carbon Taxes and Climate Change • Clean Coal Technology • Greenhouse Gas Abatement: Controversies in Cost Assessment • Greenhouse Gas Emissions, Alternative Scenarios of • Greenhouse Gas Emissions from Energy Systems, Comparison and Overview

Further Reading

Amthor, J. S., and Huston, M. A. (1998). "Terrestrial Ecosystem Responses to Global Change: A Research Strategy (ORNL/TM-1998/27)." Oak Ridge National Laboratory, Nashville, TN.

Eswaran, H., Vanden Berg, E., Reich, P., and Kimble, J. M. (1995). Global soil carbon resources. *In* "Soils and Global Change" (R. Lal, J. M. Kimble, E. Levine, and B. A. Stewart, Eds.), pp. 27–43. CRC Press, Boca Raton, FL.

Halmann, M. M., and Steinberg, M. (1999). "Greenhouse Gas Carbon Dioxide Mitigation: Science and Technology." Lewis Publishers, Boca Raton, FL.

Hao, Y., Lal, R., Owens, L., Izaurralde, R. C., Post, M., and Hothem, D. (2002). Effect of cropland management and slope position on soil organic carbon pools at the North Appalachian Experimental Watersheds. *Soil Tillage Res.* **68**, 133–142.

Himes, F. L. (1998). Nitrogen, sulfur, and phosphorus and the sequestering of carbon. *In* "Soil Processes and the Carbon Cycle" (R. Lal, J. M. Kimble, R. F. Follett, and B. A. Stewart, Eds.), pp. 315–319. CRC Press, Boca Raton, FL.

Intergovernment Panel on Climate Change. (1996). "Climate Change 1995: Impacts, Adaptations, and Mitigation of Climatic Change—Scientific–Technical Analyses." Cambridge University Press, Cambridge, UK.

Intergovernment Panel on Climate Change. (2000). "Land Use, Land Use Change, and Forestry: Special Report." Cambridge University Press, Cambridge, UK.

Intergovernment Panel on Climate Change. (2001). "Climate Change: The Scientific Basis." Cambridge University Press, Cambridge, UK.

Lal, R. (1995). Global soil erosion by water and carbon dynamics. *In* "Soils and Global Change" (R. Lal, J. M. Kimble, E. Levine, and B. A. Stewart, Eds.), pp. 131–141. CRC/Lewis Publishers, Boca Raton, FL.

Lal, R. (1999). Soil management and restoration for C sequestration to mitigate the greenhouse effect. *Prog. Environ. Sci.* **1**, 307–326.

Lal, R. (2001). Potential of desertification control to sequester carbon and mitigate the greenhouse effect. *Climate Change* **15**, 35–72.

Lal, R. (2001). World cropland soils as a source or sink for atmospheric carbon. *Adv. Agronomy* **71**, 145–191.

Lal, R. (2003a). Global potential of soil C sequestration to mitigate the greenhouse effect. *Critical Rev. Plant Sci.* **22**, 151–184.

Lal, R. (2003b). Soil erosion and the global carbon budget. *Environ. Intl.* **29**, 437–450.

Mainguet, M. (1991). "Desertification: Natural Background and Human Mismanagement." Springer-Verlag, Berlin.

Oberthür, S., and Ott, H. E. (2001). "The Kyoto Protocol: International Climate Policy for the 21st Century." Springer, Berlin.

Oldeman, L. R. (1994). The global extent of soil degradation. *In* "Soil Resilience and Sustainable Land Use" (D. J. Greenland and I. Szabolcs, Eds.), pp. 99–118. CAB International, Wallingford, UK.

Oldeman, L. R., and Van Lynden, G. W. J. (1998). Revisiting the GLASOD methodology. *In* "Methods for Assessment of Soil Degradation" (R. Lal, W. H. Blum, C. Valentine, and B. A. Stewart, Eds.), pp. 423–440. CRC Press, Boca Raton, FL.

Schlesinger, W. H. (1999). Carbon sequestration in soils. *Science* **286**, 2095.

U.S. Department of Energy. (1999). "Carbon Sequestration: Research and Development." U.S. DOE, National Technical Information Service, Springfield, VA.

Carbon Taxes and Climate Change

MARC CHUPKA
The Brattle Group
Washington, DC, United States

1. The Policy Issue
2. Environmental Taxes and Conventional Taxes
3. Carbon Taxes as Environmental Policy
4. Carbon Taxes as Fiscal Instruments
5. Carbon Tax Implementation
6. Conclusion

Glossary

distortionary tax A conventional tax levied on a good or service that has an economic burden in excess of the direct revenue generated.

environmental tax A tax levied on pollution or polluting behavior designed to reduce the amount of environmental damage.

equity The "fairness" of the distribution of tax burden or incidence.

incidence The measurement of the tax burden that accounts for shifts away from the nominal taxpayer to other parties through changes in prices and quantities transacted.

revenue neutrality Changes in tax structures or rates that increase revenues from one source and decrease revenues from another source by an equal amount.

revenue recycling Utilizing environmental tax revenues as a source of funds for reducing other taxes.

Carbon taxes, which are levies on fossil fuel in proportion to the carbon content, have been analyzed as a policy instrument to reduce carbon dioxide (CO_2) emissions. Economists have long recognized the economic advantages of emission taxes compared to more widely used instruments of environmental control such as standards. Environmental tax approaches also generate revenue that can be utilized to reduce distortionary taxes on labor, income, or consumption. In order to appreciably reduce CO_2 emissions, carbon taxes would be set at levels high enough to generate substantial revenues and significantly alter macroeconomic outcomes. This article explores some of the conceptual issues raised by imposing carbon taxes as instruments of environmental control and as tools of fiscal policy.

1. *THE POLICY ISSUE*

The bulk of greenhouse gas emissions from industrialized economies occurs as carbon dioxide (CO_2) released from burning fossil fuels, such as coal, petroleum products, and natural gas. Concern regarding the potential for global climate change from the accumulation of greenhouse gases in the atmosphere has spawned an assessment of policy tools to reduce greenhouse gas emissions. Among the vast array of potential policies to reduce CO_2 emissions, one frequently proposed and analyzed option would levy a tax on CO_2 emissions or, as a practical equivalent, tax fossil fuels in proportion to their carbon content. This is often called a carbon tax.

Analyses of the potential impacts of carbon taxes encompass a diverse set of economic disciplines, including environmental economics, energy economics, public finance, macroeconomic growth, and monetary and fiscal policy. The span of analytic approaches is necessary because a carbon tax is both an instrument of environmental control, with advantages and disadvantages relative to other potential measures; and a tax, which needs to be evaluated relative to alternative methods of raising revenues. The diversity of analyses also reflects the sheer magnitude of a carbon tax policy: A tax high enough to appreciably reduce aggregate CO_2 emissions could significantly alter energy prices, change patterns of economic activity, and generate a substantial tax revenue stream. These impacts are large enough to affect macroeconomic outcomes such as consumption, investment, fiscal balances, inflation, interest rates, and trade.

2. *ENVIRONMENTAL TAXES AND CONVENTIONAL TAXES*

The concept of an environmental tax is grounded in the basic intuition that an economic activity that imposes external cost (i.e., the social cost is greater than the private cost) will be pursued at a level higher than is socially optimal. Therefore, economic welfare can be improved by curtailing the activity, such as the production of a good that imposes pollution costs on individuals or other firms. A tax on the offending activity will discourage it, and if set optimally the tax will reduce the level of the activity to the point at which marginal social cost is equivalent to marginal social benefit. Casting the climate change issue in terms of external costs of fossil fuel consumption motivates the imposition of a carbon tax as a potential policy response. Assuming that CO_2 emissions cause economic harm, then a tax on CO_2 emissions represents a classic externality tax.

The assessment of carbon taxes raises issues similar to those that arise in the traditional analysis of conventional taxes (e.g., commodity excise taxes or income taxes), including efficiency, incidence, equity, and administrative burden. However, environmental taxes are fundamentally different from conventional taxes in one key respect: They are intended to discourage harmful behavior. Conventional taxes, in contrast, are levied on activities or transactions that normally create value or enhance economic welfare, such as providing labor or capital. To the extent that conventional taxes lower the returns from labor or investment, and thereby discourage employment, saving, or capital formation, they reduce overall economic welfare.

The distinction between conventional and environmental taxes is colloquially referred to as the difference between taxing economic "goods" and taxing economic "bads." The degree to which traditional economic analysis frameworks are modified to account for this distinction can significantly affect how one interprets the results of carbon tax analyses. The fundamental insight that environmental taxes can improve economic welfare while conventional taxes can reduce economic welfare also has motivated recent analyses of a "tax shift" in which the revenues from environmental taxes are used as a substitute for revenues from conventional distortionary taxes.

3. *CARBON TAXES AS ENVIRONMENTAL POLICY*

Evaluations of carbon taxes as an environmental policy instrument compare carbon taxes to other ways of reducing CO_2 emissions. These include such targeted measures as efficiency standards, including establishing appliance efficiency standards or fuel economy standards for vehicles, providing tax credits for specific investments, removing subsidies, or providing consumer information. An alternative economywide intervention into fossil fuel markets—tradable CO_2 permits—is related to a carbon tax and will be discussed in greater detail. The primary criteria used to judge the relative desirability of carbon taxes are the cost-effectiveness of CO_2 reductions, the predictability of CO_2 reductions, overall economic efficiency (including such dynamic impacts such as induced technology improvement), incidence (who ultimately pays the tax), and equity (fairness of the ultimate tax burden).

3.1 Carbon Taxes vs Direct Controls or Standards

Economists have long favored the use of economic incentives such as effluent taxes rather than various command-and-control instruments to reduce pollution because taxes are much more likely to induce efficient abatement responses from affected entities. If levied directly on the offending behavior, effluent taxes enable a broader set of economic adjustments that serve to reduce the costs of abatement.

Given the complexity of energy use in the economy, economists generally find that carbon taxes would produce more cost-effective CO_2 emission reductions than a comparable slate of more targeted interventions. This is primarily due to the inability of policymakers to accurately assess the marginal costs of carbon abatement in various sectors and across the myriad economic decisions that affect aggregate CO_2 emission levels. Firms and consumers responding to tax-enhanced price signals will reduce CO_2 emissions in the most efficient manner possible, assuming the absence of other significant market barriers or distortions. Subtle but widespread changes in behavior involving various substitutions—fuel choice in electricity generation, more investment in energy efficient technology, and reduced demand for energy services—all contribute to cost-effective carbon abatement.

Not all analyses of energy-related policies indicate the advantages of carbon taxes over other instruments. For example, "bottom-up" engineering analysis generally assumes that market barriers inhibit efficient outcomes, and therefore energy prices may not induce cost-effective abatement decisions compared to policies that might remove these barriers or simply mandate specific actions such as efficiency standards. The degree to which these market barriers exist and can be overcome by various policy measures is a matter of dispute among policy analysts.

In addition, emission reductions from efficiency standards can partially be eroded by offsetting behavioral adjustments, such as "snap back" or "rebound" effects and "new source bias." The former occurs when a mandated performance standard reduces energy-related variable costs and thus stimulates higher energy service demand, for example, when efficient air conditioners encourage homeowners to lower thermostats and fuel-efficient cars encourage more driving. The latter can occur if a performance standard raises the price of new technology, encouraging the continued use or repair of older, less efficient versions in order to avoid high replacement costs. This can retard capital replacement rates. If properly implemented, carbon taxes generally do not create offsetting incentives. Therefore, unless significant market barriers or distortions exist, carbon taxes are more likely to induce least-cost abatement.

3.2 Setting the Level of Carbon Taxes

Economic theory suggests that carbon taxes should be set at the level of the external cost of the activity in order to align marginal social cost to marginal social benefit. However, estimating the external cost of 1 ton of CO_2 or carbon emitted into the atmosphere is extremely difficult and controversial, particularly since the long-term nature of the problem raises significant discounting and intergenerational equity issues.

Instead, most analyses and recent climate policy proposals, including the Kyoto Protocol, would target a particular level of CO_2 emissions. This represents a challenge for and highlights a key disadvantage to the carbon tax approach: Namely, the inherent uncertainty regarding the emission reduction expected from a given level of tax. This uncertainty arises from the same lack of information regarding abatement costs and opportunities that contributes to the general cost-effectiveness of the carbon tax approach. Different models and analytic approaches used to estimate the emission impacts of carbon taxes demonstrate an exceptionally wide range of emission reduction results. Although carbon taxes would induce efficient abatement, for example, as measured on a $/ton of CO_2 reduced basis, uncertainty regarding market responses implies significant difficulty in predicting the resulting emission levels. To the extent that specific emission levels are desired, alternative instruments (e.g., capping CO_2 emissions and distributing tradable permits or allowances) may be preferred.

The primary trade-off between using price instruments such as carbon taxes and quantity targets such as emission caps emerges when supply and demand (or costs and benefits of reductions) are uncertain. Uncertainty in supply or demand of an activity subject to a tax or quantity restriction will manifest itself as relatively variable quantity outcomes under a tax policy and relatively variable price outcomes under a quantity restriction. In the present context, a carbon tax will result in uncertain CO_2 emission levels even though the overall costs of abatement are minimized. A CO_2 emission cap will yield an unpredictable price for CO_2 permits (and uncertainty in overall abatement costs) even while the cost per unit of emission reduced is still minimized through the tradable permit scheme. The inherently large uncertainty regarding the marginal benefit from CO_2 reductions would generally favor a tax approach over an emission cap policy.

3.3 Carbon Taxes, CO_2 Permits, and Revenues

Under emission "cap-and-trade" policies, the overall level of CO_2 is determined in advance, and the equivalent amount of permits to emit are distributed or auctioned to market participants. Under certain conditions, the resultant price of the CO_2 permit can be shown to be exactly the carbon tax that would achieve the same level of reductions, a result that holds whether or not permits are allocated freely or auctioned to market participants. The primary difference between a free allocation and an auction is the distribution of economic burden (incidence) and, in the case of the auction, a revenue stream comparable to the carbon tax revenues.

The revenues expected under a carbon tax and an auctioned permit system (even if set equivalently in terms of expected value of CO_2 emissions) would also be uncertain in ways that could differ between the two policies. In simple terms, revenues are tax rate (or permit price) multiplied by carbon quantity, and the variability in revenues under each policy depends on the relative range of uncertainty of the two terms of the equation. If the likely range of emissions (in percentage terms) were larger under the carbon tax than the likely range of permit prices (in percentage terms) under the permit auction, then the permit auction revenues would be more predictable. Conversely, carbon tax revenues would be more predictable if the range of emissions were smaller under the tax than the range of likely prices under the permit auction. Because the revenues from carbon taxes (or permit auctions) represent an opportunity to offset some of the economic burden of the policy through the reduction of other taxes, the relative predictability of revenues may be important in the selection of instrument choice.

3.4 Long-Term Efficiency and Technical Change

The economic literature on technical change generally concludes that the private market economy underinvests in R&D because the returns for successful R&D are not fully appropriated by those funding the research. This is the classic rationale for publicly funded R&D, especially at the basic research level. The presence of external costs creates an even larger divergence between social benefit of R&D and the private returns. For example, the private returns to an energy-saving innovation (to the extent that they accrue to the entity that undertook the R&D effort that led to the improvement) would not include the additional social benefit of reducing CO_2 emissions.

Because climate change is an inherently long-term problem, a key determinant of the cost of climate change policy is the potential ability of technology improvement to reduce the costs of shifting toward less carbon-intensive economic activities. A comparative evaluation of climate change policies, therefore, needs to consider the incentives created for research, development, and demonstration of new technologies. Alternative policy approaches may differ in their potential to encourage development of new, lower emitting technologies.

Carbon taxes may have advantages in the long term over other policy instruments to reduce CO_2 emissions because the taxes provide a continual incentive to reduce the costs of carbon abatement. This incentive would increase the private return on R&D into carbon-reducing technologies. Other policy instruments such as targeted standards would create more limited long-term dynamic incentives for improved technology.

4. CARBON TAXES AS FISCAL INSTRUMENTS

Because a carbon tax would collect revenue, it is also important to analyze and compare it with other taxes designed for the same purpose. Traditional environmental tax analysis did not concern itself with the collected revenues since the problems addressed generally were small in relationship to the economy. Because carbon taxes would collect a significant amount of revenue, the ultimate disposition of that revenue becomes a key determinant of the overall economic impact of the carbon tax policy. Therefore, economic analysis of carbon taxes must include explicit and realistic assumptions regarding the disposition of the revenue. There are several choices: Explicitly target revenues for particular public investments, reduce government fiscal deficits, or use carbon tax revenues to offset other taxes. The latter option is sometimes referred to revenue recycling and opens up the prospect for shifting taxes to create net economic gains.

Studies of revenue recycling typically examine the use of proceeds from carbon taxes to offset lower revenues from reducing or eliminating other taxes (i.e., a concept of revenue neutrality). Revenue neutrality is defined as a change in the tax code or spending policy that does not impact the net fiscal balances of government. In the context of carbon taxes, a revenue neutral shift would reduce other taxes by the amount of revenues expected from imposing a carbon tax.

Revenue neutrality does not imply welfare neutrality: Since conventional taxes on labor or investment impose economic penalties beyond their direct cost, there exists an opportunity to use carbon tax revenues to enhance economic efficiency. This is sometimes referred to as the double dividend hypothesis. In its strongest and most simple form, the double dividend hypothesis observes that a carbon tax that enhances economic welfare (e.g., when set at the level of external cost of CO_2 emissions) generates revenue that can be used to lower distortionary taxes on labor or investment. In this way, an environmental tax that cost-effectively reduces the economic damage from pollution (the first dividend) can also serve to increase economic welfare in another way when the revenues are used to reduce taxes that otherwise would impair economic welfare (the second dividend).

Although carbon tax revenues combined with some decrease in taxes that inhibit investment could, in theory, enhance overall economic outcomes, analyses of double dividend are quite complex. They must simultaneously account for the impact of carbon taxes on emissions and the interactions with all relevant existing features of the tax system to estimate the impact of reducing distortionary taxes. Some studies indicate the potential for this double dividend of decreased environmental harm from imposing carbon taxes and enhanced economic efficiency from using the collected revenues to reduce distortions from conventional taxes. Such results, however, depend on the specific taxes reduced by the carbon tax revenues, which preexisting tax distortions are measured, and the methods used to estimate overall economic welfare.

5. CARBON TAX IMPLEMENTATION

The ubiquitous use of fossil fuels in modern economies makes imposition of a tax directly on measured CO_2 emissions completely infeasible. The direct sources of CO_2, including homes, vehicles, buildings, and factories, are simply too numerous. The nature of fossil fuel combustion, however, suggests an indirect practical alternative that is equivalent to an actual emission tax. The carbon content of commercially available fossil fuels, such as coal, oil and petroleum products, and natural gas, is relatively easy to determine, does not vary significantly within defined classifications, and is essentially proportional to the CO_2 emissions released when combusted. Therefore, a direct tax on fossil fuels in proportion to carbon content will mimic the imposition of a tax on CO_2 emissions.

On a $/Btu basis, a tax on coal would be approximately twice that of natural gas, reflecting its nearly double carbon content, and that for oil would be approximately midway between the two. A carbon tax is a differential excise tax that is independent of the market price of these fuels (i.e., not an *ad valorem* tax based on a percentage of price). Because the tax would not change with the underlying market price, this construction assumes that no other taxes or subsidies affect the value of the carbon tax. Such an assumption does not necessarily reflect the reality of existing differential taxes and subsidies on fossil fuels in most industrialized economies.

5.1 Carbon Tax Base

Between initial extraction and end use, fossil fuels are widely transported, refined or converted into higher valued products such as vehicle fuels and electricity, and subjected to multiple commercial transactions. The issue of where to impose a carbon tax becomes one of balancing administrative efficiency and ensuring complete coverage in order to avoid introducing distortions that would arise from taxing some fossil fuels and not others. Similar issues arise with regard to the distribution of tradable carbon permits under a cap-and-trade policy. Generally, this leads to proposals to tax fossil fuels "upstream" near the point of extraction, importation, or initial commercial transaction. Thus, taxing coal at the mine or processing plant, oil at the wellhead or refinery, and natural gas at the collection field or processing level represents points in the domestic value chain at which the fewest commercial entities would be subject to tax liabilities while ensuring universal coverage. Carbon taxes on imported fuels could be levied at the point of importation, such as harbors and pipelines. Finally, nonfuel uses, such as oil and natural gas used as chemical production feedstocks, can be exempted or subject to tax rebates. Likewise, fossil fuel exports would generally remain untaxed because their disposition does not entail CO_2 emissions from the exporting country.

5.2 Carbon Tax Incidence

Traditional analysis suggests that carbon tax incidence will shift both backward (reducing net prices received by extraction industries) and forward into energy and product prices (increasing prices for fossil fuel users). The degree of forward incidence depends on the elasticity of demand, the cost share of taxed fossil fuels, and pricing rules and regulations in downstream markets. To the extent that intermediate users (e.g., electric generators) and/or end users (e.g., consumers) reduce their fossil fuel use in response to the carbon tax, incidence is avoided and emissions are reduced.

Electricity markets provide a good illustration of the complex responses and incidence impacts. Assuming a supply sector composed of coal- and natural gas-fired generators and nuclear and renewables, a carbon tax on utility fuels would have a variety of impacts. Under least-cost dispatch, coal-fired generation would decline, whereas natural gas generation may increase (i.e., substitute for coal) or decrease (i.e., nonfossil generation would substitute for coal and overall generation would decline as end-use consumers curtail electricity demand because of higher prices). The degree to which fuel price increases translate into end-use electric rate increases depends on wholesale trading rules and retail regulation. Finally, end users may respond to increased prices by reducing the demand for electricity services. These short-term responses will reduce emissions and shift the tax incidence.

In the longer term, the composition of investment in the generation sector would be influenced by the carbon tax, and electricity consumers will invest in more efficient equipment or substitute other energy forms for electricity. Manufacturers that use power and fuels will experience cost increases that they will attempt to pass on in higher product prices. However, the degree to which manufacturers can shift the tax burden forward depends on the elasticity of demand for their products. All these responses simultaneously will shift tax incidence throughout the economy and determine the amount of CO_2 reduction.

One interesting dimension of carbon tax policy concerns goods in international trade. Unless border adjustments are made (e.g., exempting exports from the estimated carbon tax embedded in the product price and levying taxes on imported products in proportion to their estimated carbon content), a domestic carbon tax can significantly impact trade balances and shift incidence across borders.

5.3 Macroeconomic Impacts

Studies that have estimated the economywide impact of carbon taxes are too numerous to cite or even usefully summarize here. To put the discussion into perspective, a carbon tax or tradable permit scheme to reduce U.S. CO_2 emissions by 10–20% from projected levels approximately a decade in the future is typically estimated to cost a few percent of gross domestic product under generally unfavorable assumptions regarding revenue recycling. Revenue recycling that encourages capital formation tends to lower these figures and in some cases shows a net economic benefit over time. These assessments assume a baseline of no action (i.e., there is no presumed economic savings compared with pursuing CO_2 reductions with less efficient policy instruments).

Some general observations are in order. First, as implied previously, the overall outcomes depend significantly on the assumed mechanism for revenue recycling. Second, the degree of substitutability (between fossil and non-fossil fuels, among different products, between capital and energy, etc.) assumed by the model will determine the amount of CO_2 reduction expected by a given carbon tax and the measured economic impacts. Third, the model structure, temporal horizon, and the manner in which expectations of future economic conditions affect current behavior will influence the estimated level of a carbon tax needed to achieve a given CO_2 reduction target level and the types of economic impacts estimated.

Short-term macroeconomic models often do not represent detailed substitutions in energy markets while tending to emphasize transitional imbalances, such as unemployment or myopic monetary policy reactions, that might flow from a carbon tax. Thus, these models typically estimate high carbon taxes needed to reduce CO_2 emissions and commensurately large economic costs. Despite these limitations, such models shed important light on some of the issues involved in a rapid shift toward a less carbon-intensive economy. Longer term general equilibrium models often provide richer substitution possibilities and assume that the economy will rapidly adjust to the new energy prices. Thus, smaller carbon taxes suffice to reduce CO_2 emissions and the tax policy inflicts less economic harm. Although these models generally gloss over the potential short-term dislocations from imposing an aggressive carbon tax policy, they are helpful in showing how the economy might eventually adjust to a less carbon-intensive economic structure with modest long-term economic penalties. Given the extraordinary complexity involved with estimating how billions of economic decisions might be influenced by carbon taxes and revenue recycling policies, it is probably wise to examine the results of different models to gain understanding of how actual policies might affect real economic outcomes.

5.4 Equity

Environmental policies of this magnitude, especially ones that involve tax structures, naturally raise equity issues. Although economics as a discipline can only make limited moral or ethical judgments, some concepts of equity have economic dimensions. A few observations are offered here.

First, there is a tension between the measurement of contemporaneous cost burden of CO_2 reductions and the potential for significant climate change damages that might be borne by future generations. These do not easily reduce to arguments over discount rates; perhaps ethicists or philosophers are better equipped to analyze these issues.

Second, carbon taxes are typically considered to be regressive since in most industrialized economies energy costs represent a higher portion of expenditures or income for lower income families. However, a complete analysis must consider both the impact of existing tax structure and expenditure patterns, and possibly the impact on various income classes of alternative policies to reduce CO_2 emissions.

Finally, equity issues are especially sensitive to revenue recycling assumptions. In many economic analysis frameworks, reducing existing distortionary taxes on capital provides a greater boost to economic growth than equivalent reductions on labor taxes or devising a more neutral transfer based on expanding the personal exemption (e.g., shielding the initial portion of income from income tax). In any case, the revenues from a carbon tax might be used to offset perceived inequities in the current tax system. The equity implications of any overall carbon tax and revenue recycling scheme must be evaluated carefully and completely before drawing conclusions.

6. CONCLUSION

Although any policy prescription to combat climate change has advantages and disadvantages, on balance, carbon taxes appear to have several distinct advantages over other approaches to deal with climate change. Economic incentive approaches, such as taxes or tradable permits, will induce cost-effective abatement responses. The incidence of such policies is complicated but generally operates to reduce CO_2 emissions in cost-effective ways in both the short and long term. The revenues generated from carbon tax (or permit systems) can be used to reduce distortions or inequities or both in current tax schemes; such revenue recycling will lower and possibly eliminate the economic cost associated with reducing CO_2 emissions.

SEE ALSO THE FOLLOWING ARTICLES

Biomass for Renewable Energy and Fuels • Biomass: Impact on Carbon Cycle and Greenhouse Gas Emissions • Carbon Capture and Storage from Fossil Fuel Use • Carbon Sequestration, Terrestrial • Climate Change and Energy, Overview • Climate Change: Impact on the Demand for Energy • Energy Efficiency and Climate Change • Greenhouse Gas Abatement: Controversies in Cost Assessment • Greenhouse Gas Emissions from Energy Systems, Comparison and Overview • Taxation of Energy

Further Reading

Bohm, P., and Russell, C. (1985). Comparative analysis of alternative policy instruments. *In* "Handbook of Natural Resource and Energy Economics" (A. V. Kneese and J. L. Sweeney, Eds.), Vol. 1, pp. 395–460. North-Holland, Amsterdam.

Bovenberg, A. L., and de Mooij, R. A. (1994). Environmental levies and distortionary taxation. *Am. Econ. Rev.* **84,** 1085–1089.

Bovenberg, A. L., and Goulder, L. H. (1996). Optimal environmental taxation in the presence of other taxes: general equilibrium analyses. *Am. Econ. Rev.* **86,** 985–1000.

Boyd, R., Krutilla, K., and Viscusi, W. K. (1995). Energy taxation as a policy instrument to reduce CO_2 emissions: A net benefit analysis. *J. Environ. Econ. Management* **29,** 1–24.

Dewees, D. N. (1983). Instrument choice in environmental policy. *Econ. Inquiry* **21,** 53–71.

Gaskins, D., and Weyant, J. (eds.). (1995). "Reducing Global Carbon Emissions: Costs and Policy Options." Energy Modeling Forum, Stanford, CA.

Goulder, L. H. (1995). Environmental taxation and the "double dividend": A reader's guide. *Int. Tax Public Finance* **2,** 157–183.

Goulder, L. H., and Schneider, S. H. (1999). Induced technological change and the attractiveness of CO_2 abatement policies. *Resour. Energy Econ.* **21**(3/4), 211–253.

Goulder, L. H., Parry, I. W. H., and Burtraw, D. (1997). Revenue-raising vs. other approaches to environmental taxation: The critical significance of pre-existing tax distortions. *Rand J. Econ.* **28,** 708–731.

Goulder, L. H., Parry, I. W. H., Williams, R. C., and Burtraw, D. (1999). The cost effectiveness of alternative instruments for environmental protection in a second-best setting. *J. Public Econ.* **72,** 329–360.

Grübler, A., Nakicenovic, N., and Nordhaus, W. D. (eds.). (2002). "Technological Change and the Environment." Resources for the Future, Washington, DC.

Parry, I. W. H. (1995). Pollution taxes and revenue recycling. *J. Environ. Econ. Management* **29,** S64–S77.

Parry, I. W. H. (1997). Environmental taxes and quotas in the presence of distorting taxes in factor markets. *Resour. Energy Econ.* **19,** 203–220.

Parry, I. W. H., and Bento, A. M. (2000). Tax deductions, environmental policy, and the "double dividend" hypothesis. *J. Environ. Econ. Management* **39,** 67–96.

Parry, I. W. H., Williams, R. C., and Goulder, L. H. (1999). When can carbon abatement policies increase welfare? The fundamental role of distorted factor markets. *J. Environ. Econ. Management* **37,** 52–84.

Pizer, W. A. (1999). The optimal choice of climate change policy in the presence of uncertainty. *Resour. Energy Econ.* **21**(3/4), 255–287.

Repetto, R., and Austin, D. (1997). "The Costs of Climate Protection: A Guide for the Perplexed." World Resources Institute, Washington, DC.

Repetto, R., Dower, R. C., Jenkins, R., and Geoghegan, J. (1992). "Green Fees: How a Tax Shift Can Work for the Environment and the Economy." World Resources Institute, Washington, DC.

Sterner, T. (2002). "Policy Instruments for Environmental and Natural Resource Management." Resources for the Future, Washington, DC.

Terkla, D. (1984). The efficiency value of effluent tax revenues. *J. Environ. Econ. Management* **11**(2), 107–123.

Cement and Energy

ERNST WORRELL
Lawrence Berkeley National Laboratory
Berkeley, California, United States

1. Introduction
2. Cement
3. Cement Production
4. Energy Use in Cement Making
5. Energy Conservation Opportunities in Cement Making

Glossary

blended cement Cement containing less clinker than typical portland cement and in which the clinker is replaced with other cementious materials, such as blast furnace slag or fly ash.

calcination The reaction of limestone at high temperatures that produces calcium oxide and carbon dioxide.

cement A powder that consists of clinker and other materials and that can be mixed with sand and water to make concrete.

clinker A hard substance produced in a high-temperature process that consists of a mixture of calcined limestone and other minerals.

rotary kiln A rotating tube in which limestone and other ingredients are heated to high temperatures to calcine the limestone and produce clinker.

Cement is one of the most important building materials in the world. It is mainly used for the production of concrete. Cement is produced in almost every country, although a few countries dominate the global production (e.g., China, India, the United States, and Japan). Cement production is a highly energy-intensive process. The cement industry consumes more than 2% of world energy use and contributes approximately 5% of global anthropogenic CO_2 emissions. Cement is made in two basic steps: the making of clinker from limestone and the mixing of clinker with other materials. Energy is mainly consumed in the first production step because of the energy-intensive calcination process. Energy conservation opportunities include energy efficiency improvement, new processes, application of waste fuels, and increased use of additives in cement making.

1. INTRODUCTION

Cement is one of the most important building materials in the world. It is mainly used for the production of concrete. Concrete is a mixture of inert mineral aggregates (e.g., sand, gravel, crushed stones, and cement). Cement consumption and production are closely related to construction activity and therefore to general economic activity. Global cement production is estimated at 1650 million tonnes, and it is one of the most produced materials in the world.

Due to the importance of cement as a construction material and the geographic abundance of the main raw material (i.e., limestone), cement is produced in virtually all countries. The widespread production is also due to the relative low price and high density of cement, which limits ground transportation due to the relatively high cost. Ground deliveries generally do not exceed distances of more than 150–200 km. Bulk sea transport is possible but very limited. Generally, the international trade (excluding plants located on borders) is limited compared to the global production. The major producers are large countries (e.g., China and India) or regions and countries with a developed economy, such as the European Union, the United States, and Japan.

Cement production is a highly energy-intensive process. Energy consumption by the cement industry is estimated to be approximately 2% of the global primary energy consumption or approximately 5% of the total global industrial energy consumption. Due to the dominant use of carbon-intensive fuels such as coal in clinker making, the cement industry is a major source of CO_2 emissions. In addition to energy consumption, during clinker making CO_2 is emitted from the calcining process. Due to both emission sources and the emissions from electricity production, the cement industry is a major source of carbon emissions, and it should be considered in the assessment of carbon emission-reduction options.

The high temperatures in the combustion process also contribute to relatively high NO_x emissions, whereas sulfur in the coal may be emitted as SO_x. Furthermore, the cement industry is a source of particulate matter emissions from both quarrying and material processing. However, in modern cement plants NO_x emissions are reduced by the use of precalciners, and filters reduce particulate and SO_x emissions.

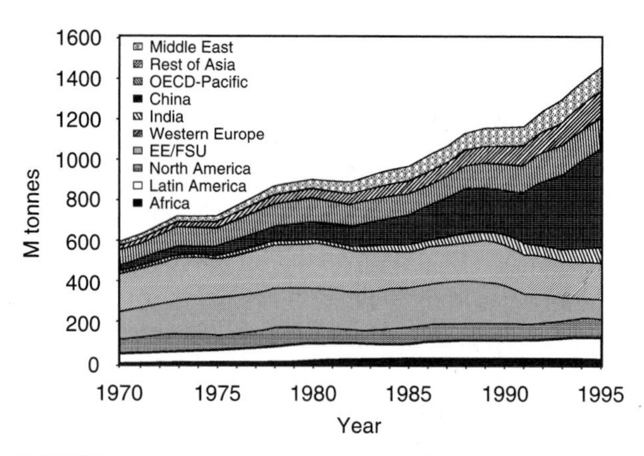

FIGURE 1 Historical production trends for cement in 10 regions of the world.

2. CEMENT

Cement is an inorganic, nonmetallic substance with hydraulic binding properties. Mixed with water, it forms a paste that hardens due to the formation of hydrates. After hardening, the cement retains its strength. In the common cement types (e.g., portland), the hardening is mainly due to the formation of calcium silicate hydrates. Cement is made from a mixture of raw materials, mainly calcium oxide, silicon dioxide, aluminum oxide, and iron (III) oxide. These compounds react during a heating process, forming the intermediate "clinker." This clinker is ground to a desired fineness together with additives to form cement. There are numerous cement types due to the use of different sources of calcium oxide and different additives to regulate properties. The most important sources of calcium oxide are limestone, blast furnace slag, and fly ash.

Global cement production increased from 594 Mt in 1970 to 1650 Mt in 2001. Consumption and production of cement are cyclical, concurrent with business cycles. Historical production trends for 10 world regions are shown in Fig. 1. The regions with the largest production levels are China, Europe, the Organization for Economic Cooperation and Development (OECD)-Pacific region, the rest of Asia, and the Middle East.

As a region, China clearly dominates current world cement production, manufacturing 595 Mt in 2001, more than twice as much as the next largest region. Cement production in China has increased dramatically, with an average annual growth rate of 12%. In many respects, China's cement industry is unique due to the large number of plants, the broad range of ownership types, and the variety of production technologies. By late 1994, there were more than 7500 cement plants throughout the country. Chinese plants tend to be small, approximately one-10th the size of the average plant in the United States. In India, cement production has increased at an average annual rate of 6.6%. Currently, India is the second largest cement producer in the world, producing approximately 100 Mt annually. In 1995, the OECD-Pacific region produced 154 Mt of cement, predominantly in Japan and South Korea. South Korean cement production increased at a high rate to 52 Mt in 2001. The rest of Asia experienced a high average annual growth of 8% between 1970 and 1995, increasing to 130 Mt in 1995. The largest producing countries in this region are Thailand, Indonesia, and Taiwan.

Cement production in North America increased slightly to 91 Mt in 2001. Recent economic growth has led to increased cement demand. Brazil (40 Mt) and Mexico (35 Mt) dominate cement production in Latin America; together they are responsible for more than 50% of the production in this region.

Cement production in Western Europe was relatively stable. The largest cement-producing countries in this region are Germany (40 Mt), Italy (36 Mt), and Spain (30 Mt). In the Eastern Europe/former Soviet Union region, cement production increased at an average rate of 2.3% per year between 1970 and 1988. After the breakup of the Soviet Union and the major restructuring that began in that region in 1988, production levels decreased by 13% per year on average between 1990 and 1995. Countries of the former Soviet Union with the highest production levels are the Russian Federation, Ukraine, and Uzbekistan.

Production of cement in the Middle East also increased rapidly, averaging 7% per year. The largest cement-producing countries in this region are

Turkey, Egypt, Iran, and Saudi Arabia. Africa showed relatively high growth, with an increase in production to approximately 50 Mt at an average annual rate of 4 to 5%. The largest cement-producing African countries are South Africa, Algeria, and Morocco, although none are among the top 20 cement-producing countries worldwide.

3. CEMENT PRODUCTION

3.1 Mining and Quarrying

The most common raw materials used for cement production are limestone, chalk, and clay. Most commonly, the main raw material, limestone or chalk, is extracted from a quarry adjacent or very close to the plant. Limestone and chalk provide the required calcium oxide and some of the other oxides, whereas clay, shale, and other materials provide most of the silicon, aluminum, and iron oxides required for portland cement. Limestone and chalk are most often extracted from open-pit quarries, but underground mining can be employed. The collected raw materials are selected, crushed, ground, and proportioned so that the resulting mixture has the desired fineness and chemical composition for delivery to the pyroprocessing systems (Fig. 2). It is often necessary to increase the content of silicon oxides and iron oxides by adding quartz sand and iron ore. The excavated material is transported to a crusher. Normally, first a jaw or gyratory crusher, followed by a roller or hammer mill, is used to crush the limestone. The crushed material is screened and stones are returned. At least 1.5 tons of raw material is required to produce 1 ton of portland cement.

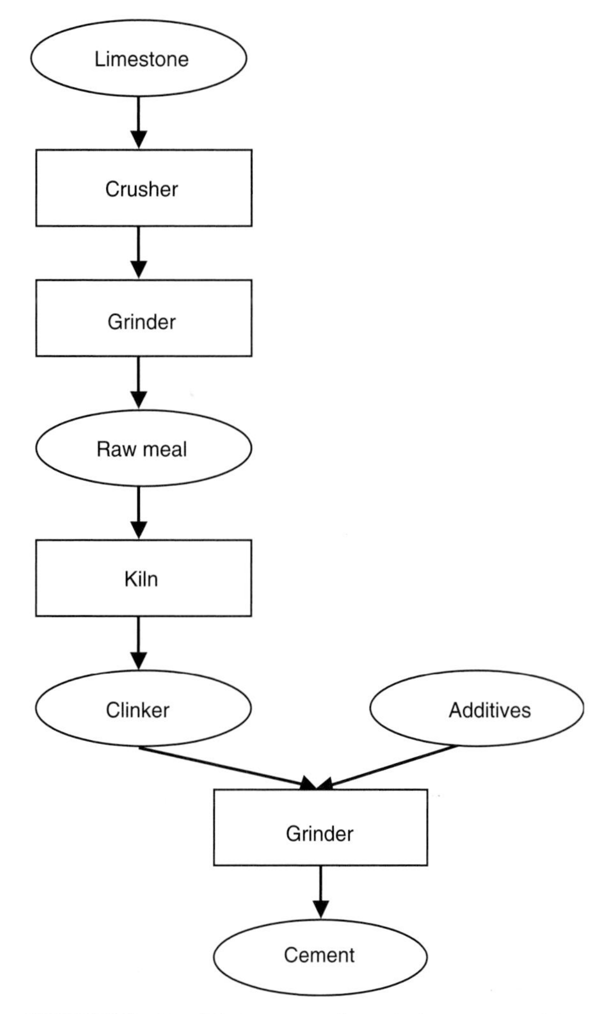

FIGURE 2 Simplified process schematic for cement making.

3.2 Kiln Feed Preparation

Raw material preparation is an electricity-intensive production step generally requiring approximately 25–35 kWh/tonne raw material, although it can require as little as 11 kWh/tonne. The raw materials are further processed and ground. The grinding differs with the pyroprocessing process used. The raw materials are prepared for clinker production into a "raw meal" by either dry or wet processing. In dry processing, the materials are ground into a flowable powder in ball mills or in roller mills. In a ball (or tube) mill, steel-alloy balls are used to decrease the size of the raw material pieces in a rotating tube. Rollers on a round table perform this task of comminution in a roller mill. The raw materials may be further dried from waste heat from the kiln exhaust, clinker cooler vent, or auxiliary heat from an air heater before pyroprocessing. The moisture content in the (dried) feed of the dry kiln is typically 0.5% (0–0.7%).

When raw materials contain more than 20% water, wet processing may be preferable. In the wet process, raw materials are ground with the addition of water in a ball mill to produce a slurry typically containing 24–48% water. Various degrees of wet processing [e.g., semiwet (moisture content of 17–22%] are used to reduce fuel consumption in the kiln.

3.3 Clinker Production (Pyroprocessing)

Clinker is the main constituent of portland cement. Clinker consists of calcium oxide and other mineral oxides (iron, aluminum, and silicon) and has cementious activity when reacted with water. Clinker is produced by pyroprocessing in large kilns. These kiln systems evaporate the free water in the meal, calcine the carbonate constituents (calcination), and form cement minerals (clinkerization). In the chemical transformation, calcium carbonate is transformed into calcium oxide, leading to a reduction in the original weight of raw materials used. Furthermore, cement kiln dust may be emitted to control the chemical composition of the clinker. Clinker production is the most energy-intensive stage in cement production, accounting for more than 90% of total industry energy use and virtually all the fuel use.

The main kiln type used throughout the world is the large-capacity rotary kiln. In these kilns, a tube with a diameter up to 8 m is installed at a 3–4° angle and rotates one to three times per minute. The ground raw material, fed into the top of the kiln, moves down the tube toward the flame. In the sintering (or clinkering) zone, the combustion gas reaches a temperature of 1800–2000°C. Although many different fuels can be used in the kiln, coal is the primary fuel in most countries.

In a wet rotary kiln, the raw meal typically contains approximately 36% moisture. These kilns were developed as an upgrade of the original long dry kiln to improve the chemical uniformity in the raw meal. The water is first evaporated in the kiln in the low-temperature zone. The evaporation step requires a long kiln. The length-to-diameter ratio may be up to 38, with lengths up to 230 m. The capacity of large units may be up to 3600 tonnes of clinker per day. Fuel use in a wet kiln can vary between 5.3 and 7.1 GJ/tonne clinker. The variation is due to the energy requirement for the evaporation and, hence, the moisture content of the raw meal. Originally, the wet process was preferred because it was easier to grind and control the size distribution of the particles in a slurry form. The need for the wet process was reduced by the development of improved grinding processes.

In a dry kiln, feed material with much lower moisture content (0.5%) is used, thereby reducing the need for evaporation and reducing kiln length. The dry process was first developed in the United States and consisted of a long dry kiln without preheating or with one-stage suspension preheating. Later, multi-stage suspension preheaters (i.e., a cyclone) or shaft preheaters were developed. Additionally, precalciner technology has recently been developed in which a second combustion chamber has been added to a conventional preheater that allows for further reduction of kiln energy requirements. The typical fuel consumption of a dry kiln with four- or five-stage preheating varies between 3.2 and 3.5 GJ/tonne clinker. A six-stage preheater kiln can theoretically use as low as 2.9–3.0 GJ/tonne clinker. The most efficient preheater, precalciner kilns, use approximately 2.9 GJ/tonne clinker. Kiln-dust bypass systems may be required in kilns in order to remove alkalis, sulfates, or chlorides. Such systems lead to additional energy losses since sensible heat is removed with the dust. Figure 3 depicts the typical specific energy consumption for different kiln types in use throughout the world.

Once the clinker is formed, it is cooled rapidly to ensure the maximum yield of alite (tricalcium silicate), an important component for the hardening properties of cement. The main cooling technologies are either the grate cooler or the tube or planetary cooler. In the grate cooler, the clinker is transported over a reciprocating grate passing through a flow of air. In the tube or planetary cooler, the clinker is cooled in a countercurrent air stream. The cooling air is used as combustion air for the kiln.

In developing countries, shaft kilns are still in use. These are used in countries with a lack of infrastructure to transport raw materials or cement or for the production of specialty cements. Today, most vertical shaft kilns are in China and India. In these countries, the lack of infrastructure, lack of capital, and power shortages favor the use of small-scale local cement plants. In China, this is also the consequence of the industrial development pattern, in which local township and village enterprises have been engines of rural industrialization, leading to a

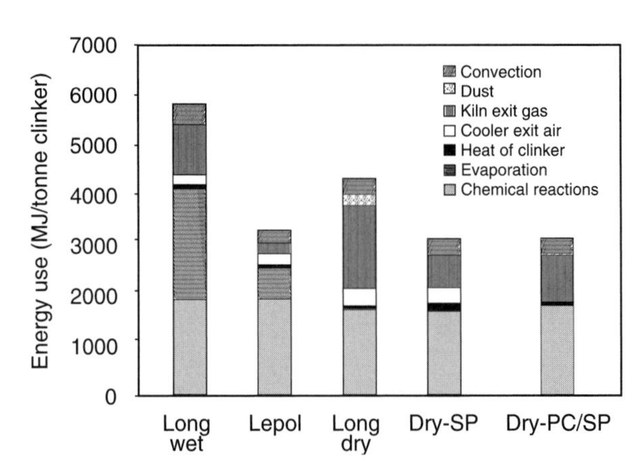

FIGURE 3 Energy consumption and losses for the major kiln types: wet process, Lepol or semiwet, long dry, dry process with four-stage suspension preheating (SP), and dry process with four-stage suspension preheating and precalcining (PC/SP).

substantial share of shaft kilns for the total cement production. Regional industrialization policies in India also favor the use of shaft kilns, in addition to the large rotary kilns in major cement-producing areas. In India, shaft kilns represent a growing portion of total cement production, approximately 10% of the 1996 production capacity. In China, the share is even higher, with an estimated 87% of output in 1995. Typical capacities of shaft kilns vary between 30 and 180 tonnes of clinker per day.

The principle of all shaft kilns is similar, although design characteristics may vary. The shaft kiln is most often cone shaped. The pelletized material travels from top to bottom, through the same zones as those of rotary kilns. The kiln height is determined by the time needed for the raw material to travel through the zones and by operational procedures, pellet composition, and the amount of air blown. Shaft kilns can achieve reasonably high efficiency through interaction with the feed and low energy losses through exhaust gases and radiation. Fuel consumption in China varies between 3.7 and 6.6 GJ/tonne clinker for mechanized kilns, with the average estimated at 4.8 GJ/tonne clinker. In India, presumably, shaft kilns use between 4.2 and 4.6 GJ/tonne clinker. The largest energy losses in shaft kilns are due to incomplete combustion. Electricity use in shaft kilns is very low.

3.4 Finish Grinding

After cooling, the clinker is stored in the clinker dome or silo. The material-handling equipment used to transport clinker from the clinker coolers to storage and then to the finish mill is similar to that used to transport raw materials (e.g., belt conveyors, deep bucket conveyors, and bucket elevators). To produce powdered cement, the nodules of cement clinker are ground. Grinding of cement clinker, together with additives (gypsum and/or anhydrite; 3–5% to control the setting properties of the cement) can be done in ball mills, roller mills, or roller presses. Combinations of these milling techniques are often applied. Coarse material is separated in a classifier and returned for additional grinding.

Power consumption for grinding depends on the required fineness of the final product and the additives used (in intergrinding). Electricity use for raw meal and finish grinding heavily depends on the hardness of the material (limestone, clinker, and pozzolan extenders) and the desired fineness of the cement as well as the amount of additives. Blast furnace slags are more difficult to grind and hence use more grinding power, between 50 and 70 kWh/tonne for a 3500 Blaine (expressed in cm^2/g). (Blaine is a measure of the total surface of the particles in a given quantity of cement, or an indicator of the fineness of cement. The higher the Blaine, the more energy required to grind the clinker and additives to the desired fineness.) Traditionally, ball or tube mills are used in finish grinding, but many plants use vertical roller mills. In ball and tube mills, the clinker and gypsum are fed into one end of a horizontal cylinder and partially ground cement exits the other end. Modern ball mills may use 32–37 kWh/tonne for cements with a Blaine of 3500.

Modern state-of-the-art concepts are the high-pressure roller mill and the horizontal roller mill (e.g., Horomill), which use 20–50% less energy than a ball mill. The roller press is a relatively new technology and is more common in Western Europe than in North America. Various new grinding mill concepts are under development or have been demonstrated.

Finished cement is stored in silos, tested and filled into bags, or shipped in bulk on bulk cement trucks, railcars, barges, or ships. Additional power is consumed for conveyor belts and packing of cement. The total consumption for these purposes is generally low and not more than 5% of total power use. Total power use for auxiliaries is estimated to be approximately 10 kWh/tonne clinker. The power consumption for packing depends on the share of cement packed in bags.

4. ENERGY USE IN CEMENT MAKING

The theoretical energy consumption for producing cement can be calculated based on the enthalpy of formation of 1 kg of portland cement clinker, which is approximately 1.76 MJ. This calculation refers to reactants and products at 25°C and 0.101 MPa. In addition to the theoretical minimum heat requirements, energy is required to evaporate water and to compensate for the heat losses. Heat is lost from the plant by radiation or convection and, with clinker, emitted kiln dust and exit gases leaving the process. Hence, in practice, energy consumption is higher. The kiln is the major energy user in the cement-making process. Energy use in the kiln basically depends on the moisture content of the raw meal. Figure 3 provides an overview of the heat requirements of different types of kilns. Most electricity is consumed in the grinding of the raw materials and finished cement. Power consumption for a rotary kiln is comparatively small, generally approximately 17–23 kWh/tonne clinker (including the cooler and preheater fans). Additional power is consumed for conveyor belts and packing of cement. Total power use for auxiliaries is estimated at approximately 10 kWh/tonne clinker. Table I summarizes the typical energy consumption for the different processing steps and processes used.

There is a relatively large difference in the energy consumption per tonne of cement produced in different countries. International comparisons of energy intensity have shown that part of the difference can be explained by the types of cement produced. However, a large part of the difference is due to the use of less energy-efficient equipment. For example, in Eastern Europe and the former Soviet Union countries, as well as the United States, a relatively large part of the clinker is produced in inefficient wet process kilns. The differences in energy efficiency, fuels used, and cement types produced also affect CO_2 emissions. Figure 4 shows the CO_2 emissions for cement production in various regions of the world.

TABLE I

Energy Consumption in Cement-Making Processes and Process Types[a]

Process step	Fuel use (GJ/tonne product)	Electricity use (kWh/tonne product)	Primary energy (GJ/tonne cement)[b]
Crushing			
Jaw crusher		0.3–1.4	0.02
Gyratory crusher		0.3–0.7	0.02
Roller crusher		0.4–0.5	0.02
Hammer crusher		1.5–1.6	0.03
Impact crusher		0.4–1.0	0.02
Raw meal grinding			
Ball mill		22	0.39
Vertical mill		16	0.28
Hybrid systems		18–20	0.32–0.35
Roller press, Integral		12	0.21
Roller press, pregrinding		18	0.32
Clinker kiln			
Wet	5.9–7.0	25	6.2–7.3
Lepol	3.6	30	3.9
Long dry	4.2	25	4.5
Short dry suspension preheating	3.3–3.4	22	3.6–3.7
Short dry, preheater and precalciner	2.9–3.2	26	3.2–3.5
Shaft	3.7–6.6	n.a.[d]	3.7–6.6
Finish grinding[c]			
Ball mill		55	0.60
Ball mill/separator		47	0.51
Roller press/ball mill/separator		41	0.45
Roller press/ separator/ball mill		39	0.43
Roller press/ separator		28	0.31

[a]Specific energy use is given per unit of throughput in each process. Primary energy is calculated per tonne of cement, assuming portland cement (containing 95% clinker), including auxiliary power consumption.

[b]Primary energy is calculated assuming a net power-generation efficiency of 33% (LHV).

[c]Assuming grinding of Portland cement (95% clinker and 5% gypsum) at a fineness of 4000 Blaine.

[d]n.a., not applicable.

5. ENERGY CONSERVATION OPPORTUNITIES IN CEMENT MAKING

Improvement in energy efficiency reduces the costs of producing cement. Improvement may be attained by applying more energy-efficient process equipment and by replacing old installations with new ones or shifting to completely new types of cement-production processes. By far the largest proportion of energy consumed in cement manufacture consists of fuel that is used to heat the kiln. Therefore, the greatest gain in reducing energy input may come from improved fuel efficiency. The dry process is generally more energy efficient than the wet process. The processes are exchangeable to a large extent but may be limited by the moisture content of the available raw material. The main opportunities for improving efficiency in the kiln are the conversion to more energy-efficient process variants (e.g., from a wet process to a dry process with preheaters and precalciner), optimization of the clinker cooler, improvement of preheating efficiency, improvement of burners, as well as improvement of process control and management systems. Electricity use can be reduced by improved grinding systems, high-efficiency classifiers, high-efficiency motor systems, and process control systems. Table II provides an overview of the different energy-efficient technologies available to the cement industry.

Several studies have demonstrated the existence of cost-effective potentials for energy efficiency

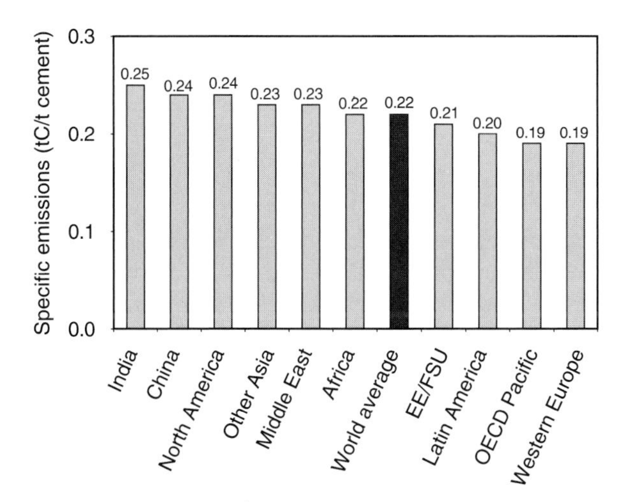

FIGURE 4 Carbon intensity of cement production in different regions.

improvement in the cement industry. In China, various programs have developed technologies to improve the efficiency of shaft kilns by increasing mechanization, insulation, bed distribution, as well as control systems. They demonstrated an energy efficiency improvement potential of 10–30% for all shaft kilns. A study of the Indian cement industry found a technical potential for energy efficiency improvement of approximately 33% with currently commercially available technology. Incorporating future technologies, the energy savings potential is estimated to be approximately 48%. However, the economic potential for energy efficiency improvement is estimated to be 24% of total primary energy use (using a discount rate of 30%). Focusing on commercially available technology, 29 energy-efficient technologies were identified that could be adopted to some extent by the U.S. cement industry. Together, these have a technical potential for energy efficiency improvement of 40%. However, the economic potential (using a discount rate of 30%) is estimated to be only 11% due to the high capital costs and low energy costs in the United States.

The production of clinker is the most energy-intensive step in the cement-manufacturing process and causes large process emissions of CO_2. In blended cement, a portion of the clinker is replaced with industrial by-products, such as coal fly ash (a residue from coal burning) or blast furnace slag (a residue from ironmaking), or other pozzolanic materials (e.g., volcanic material). These products are blended with the ground clinker to produce a homogeneous product—blended cement. Blended cement has different properties compared to portland cement (e.g., setting takes longer but ultimate strength is higher).

The current application of additives in cement making varies widely by country and region. Whereas in Europe the use of blended cements is quite common, it is less common in North America. The relative importance of additive use is indicated by the clinker:cement (C:C) ratio of cement production in a specific country. Portland cement has a C:C ratio of 0.95, whereas blast furnace slag cement has a C:C ratio as low as 0.35. Countries such as the United States, Canada, and the United Kingdom have high C:C ratios, indicating the dominance of portland cement, whereas countries such as Belgium, France, and the former Soviet Union have lower C:C ratios, indicating the relatively larger use of blended cements. Because no international sources collect clinker production data, it is not possible to accurately estimate the current practices in all cement-producing countries. The major barriers to further application of blended cements do not seem to be supply or environmental issues but rather existing product standards and specifications and building codes.

The potential for application of blended cements depends on the current application level, the availability of blending materials, and standards and legislative requirements. The global potential for CO_2 emission reduction by producing blended cement is estimated to be at least 5% of total CO_2 emissions from cement making (56 Mt CO_2) but may be as high as 20%. The potential savings will vary by country and region. The potential for energy savings for 24 countries in the OECD, Eastern Europe, and Latin America was estimated to be up to 29%, depending on the availability of blending materials and the structure of the cement market. The average emission reduction for all countries (producing 35% of world cement in 1990) was estimated to be 22%.

The costs of blending materials depend on the transportation costs and may be \$15–30 per ton for fly ash and approximately \$24 per ton for blast furnace slag. Shipping costs may significantly increase the price, depending on distance and shipping mode. The prices are still considerably lower than the production costs of cement, estimated to be approximately \$36 per ton (1990) in the United States.

Additives such as fly ash contain high concentrations of heavy metals, which may leach to the environment in unfavorable conditions. No negative environmental effects of slag and fly ash addition in cement have been found. The use of nonferrous slags seems to be limited to slag contents of 15% by mass. However, fly ash and blast furnace slag may be

TABLE II

Energy-Efficient Practices and Technologies in Cement Production

Raw materials preparation	
Efficient transport systems (dry process)	
Slurry blending and homogenization (wet process)	
Raw meal blending systems (dry process)	
Conversion to closed-circuit wash mill (wet process)	
High-efficiency roller mills (dry cement)	
High-efficiency classifiers (dry cement)	
Fuel preparation: roller mills	
Clinker production (wet)	Clinker production (dry)
Energy management and process control	Energy management and process control
Seal replacement	Seal replacement
Kiln combustion system improvements	Kiln combustion system improvements
Kiln shell heat loss reduction	Kiln shell heat loss reduction
Use of waste fuels	Use of waste fuels
Conversion to modern grate cooler	Conversion to modern grate cooler
Refractories	Refractories
Optimize grate coolers	Heat recovery for power generation
Conversion to preheater, precalciner kilns	Low-pressure drop cyclones for suspension preheaters
Conversion to semidry kiln (slurry drier)	Optimize grate coolers
Conversion to semiwet kiln	Addition of precalciner to preheater kiln
Efficient kiln drives	Long dry kiln conversion to multistage preheater kiln
Oxygen enrichment	Long dry kiln conversion to multistage preheater, precalciner kiln
	Efficient kiln drives
	Oxygen enrichment
Finish grinding	
Energy management and process control	
Improved grinding media (ball mills)	
High-pressure roller press	
High-efficiency classifiers	
General measures	
Preventative maintenance (insulation, compressed air system)	
High-efficiency motors	
Efficient fans with variable speed drives	
Optimization of compressed air systems	
Efficient lighting	
Product and feedstock changes	
Blended cements	
Limestone cement	
Low-alkali cement	
Use of steel slag in kiln (CemStar)	
Reducing fineness of cement for selected uses	

considered hazardous wastes under environmental legislation in some countries, limiting the use of fly ash to specified companies. In the United States, fly ash falls under the Resource Conservation and Recovery Act, which gives states the jurisdiction to define fly ash as a hazardous waste. In practice, the state regulation varies widely throughout the United States, limiting the reuse of fly ash.

SEE ALSO THE FOLLOWING ARTICLES

Aluminum Production and Energy • External Costs of Energy • Greenhouse Gas Emissions from Energy Systems, Comparison and Overview • Plastics Production and Energy • Steel Production and Energy

Further Reading

Alsop, P. A., and Post, J. W. (1995). "The Cement Plant Operations Handbook." Tradeship, Dorking, UK.

Cembureau (1996). "World Cement Directory." Cembureau, Brussels.

Cembureau (1997). "Best Available Techniques for the Cement Industry." Cembureau, Brussels.

Duda, W. H. (1985). "Cement Data Book, International Process Engineering in the Cement Industry." 3rd ed. Bauverlag, Wiesbaden, Germany.

Feng, L., Ross, M., and Wang, S. (1995). Energy efficiency of China's cement industry. *Energy* **20**, 669–681.

Hargreaves, D. (ed.). (1996). "The Global Cement Report," 2nd ed. Tradeship, Surrey, UK.

Hendriks, C. A., Worrell, E., Price, L., Martin, N., and Ozawa Meida, L. (1999). The reduction of greenhouse gas emissions from the cement industry, Report No. PH3/7. IEA Greenhouse Gas R&D Program, Gloucestershire, UK.

U.S. Geological Survey (various years). "Minerals Yearbook." U.S. Geological Survey, Washington, DC (http://minerals.er.usgs.gov/minerals).

Von Seebach, H. M., Neumann, E., and Lohnherr, L. (1996). State-of-the-art of energy-efficient grinding systems. *ZKG Int.* **49**, 61–67.

Worrell, E., Martin, N., and Price, L. (2000). Potentials for energy efficiency improvement in the U.S. cement industry. *Energy* **25**, 1189–1214.

Worrell, E., Price, L., Martin, N., Hendriks, C. A., and Ozawa Meida, L. (2001). Carbon dioxide emissions from the global cement industry. *Annu. Rev. Energy Environ.* **26**, 203–229.

City Planning and Energy Use

HYUNSOO PARK and CLINTON ANDREWS
Rutgers University
New Brunswick, New Jersey, United States

1. Scope of City Planning
2. Siting Energy Facilities
3. Financing Energy Infrastructure
4. Influence of Land Use Patterns on Energy
5. Influence of Urban/Architectural Design on Energy Policies and Practices
6. Centralized vs Distributed Energy Supplies
7. Integrated vs Segmented Energy Planning
8. Energy for the World's Burgeoning Megacities
9. Conclusion

Glossary

alternative dispute resolution (ADR) Any techniques, such as arbitration, mediation, early neutral evaluation, and conciliation, for the purpose of consensus building before litigation.

combined heat and power (CHP) Systems that generate electricity and thermal energy simultaneously in an integrated manner, so as to capture wasted energy.

distributed energy resources (DER) A variety of small, modular electric generation units located near the end user; used to improve the operation of the electricity network by complementing central power.

smart growth A development strategy emphasizing mixed land uses with convenient transportation choices in existing communities, thereby alleviating sprawl and preserving open spaces.

City planning is a future-oriented activity that shapes land use and the built environment by means of designs, regulations, and persuasion. The scope of city planning includes a range of interdependent decisions at the nexus of the private and public spheres: transportation and utility infrastructure investments, allowable land uses, and the locations and characteristics of housing, retail stores, offices, schools, hospitals, factories, and even energy-producing facilities. Although planners rarely focus explicitly on energy issues, planning decisions influence energy use and production in profound, long-lasting ways. This article focuses mostly on the U.S. case.

1. SCOPE OF CITY PLANNING

Although urban design has deep historical roots, modern city planning began with the Industrial Revolution and the associated rapid urbanization that led to a degradation of the local quality of life. City planning was a response to the growing ugliness and unhealthfulness of cities, and it became a movement to reform an old paradigm of cities, which in many countries had been based on laissez-faire principles. In the Anglo-American world, which paced the rest of northern Europe, early signs of the city planning impetus in the late 19th and early 20th centuries were the English Garden City movement led by Ebenezer Howard, the U.S. City Beautiful movement that started in Chicago, and the zoning movement that attempted to separate residential and industrial land uses. Decentralization advocate Frank Lloyd Wright, and "tower in a park" advocate Le Corbusier, utopians of very different stripes, provided intellectual fodder to planning debates.

After World War II, the agenda of United States city planners changed periodically in accordance with evolving economic, political, and ideological forces. Scientific planning, emphasizing optimization and rationalistic analysis, dominated during the 1950s and has never disappeared. Advocacy planning, emphasizing empowerment of disenfranchised slum dwellers, gained modest influence in the 1960s and 1970s. Equity planning, with a focus on improved fairness in land use decisions, became a central tenet of planning from the late 1960s onward. In the 1970s, environmental planning became an important component of city planning. Starting in the late 1980s, city planning increasingly emphasized participatory and communicative planning processes.

Encyclopedia of Energy, Volume 1.

Collaboration, community visioning, smarter growth, and sustainable development are the watchwords of the planning profession at the beginning of the 21st century.

Planning practice begins with how to regulate land use in urban areas. Zoning and subdivision regulations are basic tools in determining land use patterns. Land use planning is inextricably related to transportation planning, which analyzes the travel behavior of individuals and aggregates on current and prospective networks. Urban designers apply these tools to give cities unity and coherence, with clear and legible structures and hierarchies of centers. In the United States, however, low-density development in suburban areas during the post-World War II period created "sprawl." Planners devised growth management strategies to control sprawl, and starting in the 1990s, a "smart growth" movement advocated an integrated form of regional growth management to restore the community vitality of center cities and older suburbs. City planners increasingly rely on public decision-making processes to organize the built environment, and recent efforts at communicative planning encourage citizens' direct participation in policy-making processes, implementation, and monitoring of policies, so as to minimize conflicts and negative externalities.

The city planning profession, however, has paid little attention to energy use, even though most cities experienced significant energy crises in the 1970s. Most planners assume that cities are, above all, places to consume energy rather than produce it. They are increasingly being shown to be wrong.

2. *SITING ENERGY FACILITIES*

2.1 Regulations and Institutions

In the United States context, most land use and facility siting decisions take place at the local level. The federal government has reserved siting authority for only a few types of energy facilities: pipelines and nuclear waste dumps are the chief examples. Most states delegate facility siting tasks to municipalities, although statewide bodies have been created in a few states. Statewide decisions about siting facilities are made by a siting board, an energy facility siting council, or an energy commission. The members of the siting institutions vary from state to state; for instance, the members of the Energy Commission in California consist of four professionals and one citizen, and the members of the Energy Facility Siting Council in Oregon consist of seven geographically representative citizens appointed by the governor. These institutions issue a certification for the construction and operation of energy facilities.

The siting certification processes also vary from state to state. Generally, in the beginning of the process, these siting institutions or subcommittees oversee the submission of plans by the utilities. Then a potential applicant submits a notice of intent, which describes the proposed facilities and which allows the office to collect public inputs through public hearings and to identify laws, regulations, and ordinances. At the next stage, the applicants submit the application to the office and then the siting institution decides whether to issue a site certification.

General criteria for siting facilities are based on standards determined by each state. The environmental standards include considerations of noise, wetlands, water pollution, and water rights; soil erosion and environmentally sensitive areas are also considered. Structural standards consider public health and facility safety. The siting institutions consider the applicant's organizational, managerial, and technical expertise; the applicants are required to have financial assurance. Siting institutions must also evaluate socioeconomic effects such as expected population increases, housing, traffic safety, and so on. However, deciding where to site facilities is not easy due to the divergent interests of multiple stakeholders, and the process of siting facilities therefore often brings about locational conflicts.

2.2 Locational Conflicts

Siting facilities can be understood as an exercise in siting locally undesirable land uses (LULUs), many of which cause locational conflicts and thereby place government officials and policy makers in opposition to grassroots "not in my back yard" (NIMBY) groups. Locational conflicts can escalate from local disputes into broad protests against structural inequities. Major issues driving locational conflicts include health and safety concerns as well as environmental impacts. In addition, facilities may be sited in neighborhoods whose residents are poor or people of color, becoming an issue of environmental justice. Opponents of constructing facilities sometimes adopt grassroots techniques, such as protests, civil disobedience, and initiatives. Many conflicts lead to judicial review, promoting high transaction costs.

New approaches to the resolution of locational conflicts have emerged since the 1970s to reduce the high transaction costs. Alternative dispute resolution (ADR) is a method based on consensus building; this is also called principled negotiation or consensus-based negotiation. In their 1981 book, *Getting to Yes*, Roger Fisher and William Ury emphasize the importance of principled negotiation that applies scientific merit and objectivity to the problem in question. Planners engaged in siting facilities now routinely consider both efficiency and equity/fairness criteria.

2.3 Balancing Efficiency and Equity/Fairness Considerations

By and large, the criteria to evaluate the benefits and costs of siting energy facilities will be determined by three factors: risk from the facilities, fairness in siting facilities, and productivity or efficiency of the facilities after the decision on siting the facility. It is desirable to reduce risks related to health and safety issues. There is a fairness issue behind the risk issue. The NIMBY syndrome usually results from an unequal distribution of costs and benefits. Sites of energy facilities are correlated to a variety of economic factors, such as natural resources, land prices, and urban infrastructures. In his 1992 piece in the *Virginia Environmental Law Journal*, Robert Collin argues that generally racial and ethnic minorities and low-income groups are more likely to be exposed to hazardous waste and pollution because they are more likely to live near those treatment facilities. Thus, the balancing of efficiency and fairness becomes a major argument in siting energy facilities. Participation by the affected parties and negotiation are standard components of most current United States energy facility siting processes. During the negotiation process, each party reviews the plan to site facilities and pays close attention to the type and amount of compensation awarded to the community and the criteria used for siting facilities. If the compensation is not sufficient, renegotiation may be necessary.

3. *FINANCING ENERGY INFRASTRUCTURE*

3.1 Public or Private Ownership

Energy infrastructures include many components: generation, transmission, and distribution of electricity; physical networks of oil and natural gas pipelines; oil refineries; and other transportation elements such as marine and rail transportation. Historically, industrialized countries have financed the energy sector privately. Wood, coal, oil, natural gas, and electricity were produced and transported mostly by private firms well into the first decades of the 20th century. Thereafter, as network energy utilities became ubiquitous, public involvement increased. The new Soviet Union made electrification a national project; financially unstable private utilities of the United States begged for and received government regulation following World War I, and New Deal regional and rural policies made access to electricity a public goal justifying public investment and ownership. Before the Great Depression of the early 1930s, several holding companies controlled more than 75% of all U.S. generation. After the Great Depression, holding companies proved to be financially unstable and caused the price of electricity to increase, so support grew for government ownership of utilities, especially hydroelectric power facilities. Since the energy crisis in the early and late 1970s, the structure of electricity industry has changed; non-utilities have been added in the electricity market and the vertical structure has been unbundled. Because of the uncertainty over the future structure of the industry and recovery of investment costs in electricity infrastructures, financing electricity infrastructure has been in difficulty. In the United States as of 1999, federally owned utilities controlled 10.6% of the total generation, cooperatives controlled 5.2%, publicly owned utilities controlled 13.5%, investor-owned utilities controlled 70.7%, and nonutilities controlled 18.9%. The combination of private and public ownership differs from region to region. There are about 5000 power plants in the United States.

As electricity was transformed from a novelty to a necessity during the mid-20th century, many countries worldwide nationalized their electricity sectors. Only since the 1980s have privatization and liberalization reversed this trend. In the developing world, lack of public funds has forced many countries to turn to private, often transnational, financiers for energy sector investments.

In the oil and natural gas sectors, historically, major energy companies have possessed a vertically integrated oil and natural gas infrastructure, involving oil and natural gas exploration, development, and production operations and petroleum refining and motor gasoline marketing. Also, independent oil and natural gas companies take part in segments of vertically integrated structures as producers, petroleum refiners, and providers of transmission pipelines. Oil and natural gas pipelines may run across state boundaries or federal lands in order to link supply locations to market areas in cities. In the natural gas industry, there has been a structural change since the issue of Order 636 by the Federal Energy Regulatory Commission (FERC), which no longer permitted gas pipeline companies to engage in the sale of natural gas. Thus, several pipeline companies have been consolidated under single corporate umbrellas, such as El Paso Merchant Energy Company, the Williams Company, Duke Energy Corporation, and the now defunct Enron Corporation. Several of these companies have acquired interstate pipeline companies.

3.2 Vertical Disintegration

In the United States electricity sector, the industry structure throughout the 20th century consisted of three vertical components: generation, transmission, and distribution. With increasing demand, robust transmission networks, availability of new generating technologies, and an ideological shift, restructuring in electricity and natural gas industries has become significant. In the electricity sector, regulatory reform has changed the structure of the electricity industry; according to Order 888 issued by FERC, all utilities can have access to the U.S. transmission systems; in addition, as already mentioned, nonutilities can provide electricity through the transmission system. To reduce the costs resulting from constructing new power plants, the federal government is trying to remove the transmission constraints between regions, some of which are traceable to the 1935 Public Utility Holding Company Act (PUHCA) limiting interstate utility operations. Horizontal market power is replacing vertical integration as the dominant utility business strategy.

In the natural gas sector, business operations are going in a direction parallel to that of the electricity industry. In his 2001 report, *Natural Gas Transportation—Infrastructure Issues and Operational Trends*, James Tobin discusses that, to take a better position to handle the large growth in natural gas demand, natural gas companies have consolidated operations through major mergers.

3.3 Municipalization

In the early 20th century, local governments in the United States played key roles by awarding competing bidders electric service franchises. In fact, however, some local officials received bribes in return for granting franchises to utilities, and franchise holders imputed this burden to citizens. Citizens rebuked utility companies for causing the rates to be high, and called for local governments to take over some local electricity systems. Today, a significant minority of municipalities control electricity distribution to their residents, and a smaller number also generate and transmit power. In parallel, in the 20th century, strong state regulations were introduced to prevent corruption and guarantee the fair pricing among investor-owned utilities. From historical experience, municipalization has both strengths and weaknesses.

Nevertheless, the recent restructuring of the electricity sector might open new opportunities for public financial involvement. Municipal governments can serve as aggregators and use their market power to buy reasonably priced electricity on the open market. They can reduce rates for their residential customers or meet expectations of clean air and water by offering clean energy options. They can also use various renewable energy sources to meet their citizens' demand (this will be further discussed in Section 6).

4. *INFLUENCE OF LAND USE PATTERNS ON ENERGY*

4.1 Effects of Density, Grain, and Connectivity

Land use patterns can be parsimoniously characterized in four dimensions: degree of centralization or decentralization (urban form), ratio of population or jobs to area (density), diversity of functional land uses such as residential and industrial (grain), and extent of interrelation and availability of multiple modes of circulation for people and goods among local destinations (connectivity). The use of resources per capita diminishes as urban form becomes more centralized, density goes up, grain becomes finer, and connectivity shrinks. Metropolitan land use patterns in the United States after World War II show increased energy use due to increasing regional populations, decentralization, decreasing density, rougher grain, and increased connectivity.

Throughout the 19th century, most people in the United States lived in small towns and villages. With the advent of the 20th century, many people moved to industrial cities for jobs, a trend that peaked in the 1920s and was compounded by overseas immigration. Following the interruptions of the Great Depression and World War II, the pent-up demand for housing was met by a conscious process of suburbanization. Achieving the dream of home ownership became feasible for many Americans as new federal mortgage guarantee policies removed

TABLE I

Intensity of Land Use in Global Cities, 1990[a]

City	Metropolitan density		Central city density[b]		Inner-area density		Outer-area density	
	Pop.[c]	Jobs	Pop.	Jobs	Pop.	Jobs	Pop.	Jobs
American average[d]	14.2	8.1	50.0	429.9	35.6	27.2	11.8	6.2
Australian average[e]	12.2	5.3	14.0	363.6	21.7	26.2	11.6	3.6
Canadian average[f]	28.5	14.4	37.9	354.6	43.6	44.6	25.9	9.6
European average[g]	49.9	31.5	77.5	345.1	86.9	84.5	39.3	16.6
Asian average[h]	161.9	72.6	216.8	480.1	291.2	203.5	133.3	43.5

[a] From "Sustainability and Cities" by Peter Newman and Jeffrey Kenworthy. Copyright © 1999 by Peter Newman and Jeffrey Kenworthy. Adapted by permission of Island Press, Washington, D.C. Density expressed in persons per hectare and jobs per hectare.

[b] Central business district.

[c] Population.

[d] Average of Sacramento, Houston, San Diego, Phoenix, San Francisco, Portland, Denver, Los Angeles, Detroit, Boston, Washington, Chicago, and New York.

[e] Average of Canberra, Perth, Brisbane, Melbourne, Adelaide, and Sydney.

[f] Average of Winnipeg, Edmonton, Vancouver, Toronto, Montreal, and Ottawa.

[g] Average of Frankfurt, Brussels, Hamburg, Zurich, Stockholm, Vienna, Copenhagen, Paris, Munich, Amsterdam, and London.

[h] Average of Kuala Lumpur, Singapore, Tokyo, Bangkok, Seoul, Jakarta, Manila, Surabaya, and Hong Kong.

financial barriers to home ownership, as developers such as William Levitt perfected mass production of affordable housing units on greenfield sites, and as federal dollars poured into road building. Discrimination in lending and housing markets prevented most black Americans from participating in this exodus. Yet, by 1960, the suburban lifestyle was the conventional land use practice in the United States.

Suburban lifestyle has brought about sprawling, low-density suburban communities; according to the 1990 census, from 1970 to 1990, the density of urban population in the United States decreased by 23%. From 1970 to 1990, more than 30,000 square miles (19 million acres) of once-rural lands in the United States became urban areas, an area equal to one-third of Oregon's total land area. From 1969 to 1989, the population of the United States increased by 22.5%, and the number of miles traveled by that population ("vehicle miles traveled") increased by 98.4%.

Anthony Downs defines the term "sprawl" as (1) unlimited outward extension, (2) low-density residential and commercial settlements, (3) leapfrog development, (4) fragmentation of powers over land use among many small localities, (5) dominance of transportation by private automotive vehicles, (6) no centralized planning or control of land use, (7) widespread strip commercial development, (8) great fiscal disparities among localities, (9) segregation of types of land use in different zones, and (10) reliance mainly on the trickle-down, or filtering, process to promote housing to low-income households. The impacts of urban sprawl have caused increasing traffic congestion and commute times, air pollution, inefficient energy consumption and greater reliance on foreign oil, inability to provide adequate urban infrastructures, loss of open space and habitat, inequitable distribution of economic resources, and the loss of a sense of community.

When land use patterns in the United States are compared to other countries, they show significant differences. In comparisons of metropolitan density, U.S. cities are of low density in residential and business areas whereas European cities are three to four times denser. Asian cities are 12 times denser compared to American cities (Table I).

The average energy use for urban transportation in American cities is 64.3 gigajoules (GJ) of fuel per capita compared to 39.5 GJ in Australia, 39.2 GJ in Canada, 25.7 GJ in Europe, and 12.9 GJ in Asia. Energy use in American cities is five times more than in Asian cities. In addition, European cities consume two times more energy compared to Asian cities. This shows that energy use in transportation is closely related to land use patterns and income levels (Table II).

4.2 Environmental and Public Health Implications

Low-density development and urban sprawl are correlated with high-level air pollutant emissions from transportation. Emission rates of the greenhouse gas carbon dioxide in U.S. cities are higher than those of

TABLE II

Transportation Energy Use per Capita in Global Regions, 1990[a]

City	Private transportation			Public transportation			Total transportation energy (MJ)	Total Transportation energy/$ of GRP[b] (MJ/$)
	Gasoline (MJ)	Diesel (MJ)	Private % (of total)	Diesel (MJ)	Electricity (MJ)	Public % (of total)		
American average[c]	55,807	7764	99%	650	129	1%	64,351	2.38
Australian average[d]	33,562	4970	98%	764	159	2%	39,456	1.96
Canadian average[e]	30,893	6538	97%	1057	163	3%	39,173	?
European average[f]	17,218	7216	95%	604	653	5%	25,692	0.83
Asian average[g]	6311	5202	89%	1202	148	11%	12,862	3.81

[a] From "Sustainability and Cities" by Peter Newman and Jeffrey Kenworthy. Copyright © 1999 by Peter Newman and Jeffrey Kenworthy. Adapted by permission of Island Press, Washington, D.C. Use expressed in megajoules.

[b] GRP, gross regional product is the measure of all goods and services produced in the regional urban area.

[c] Average of Sacramento, Houston, San Diego, Phoenix, San Francisco, Portland, Denver, Los Angeles, Detroit, Boston, Washington, Chicago, and New York.

[d] Average of Canberra, Perth, Brisbane, Melbourne, Adelaide, and Sydney.

[e] Average of Winnipeg, Edmonton, Vancouver, Toronto, Montreal, and Ottawa.

[f] Average of Frankfurt, Brussels, Hamburg, Zurich, Stockholm, Vienna, Copenhagen, Paris, Munich, Amsterdam, and London.

[g] Average of Kuala Lumpur, Singapore, Tokyo, Bangkok, Seoul, Jakarta, Manila, Surabaya, and Hong Kong.

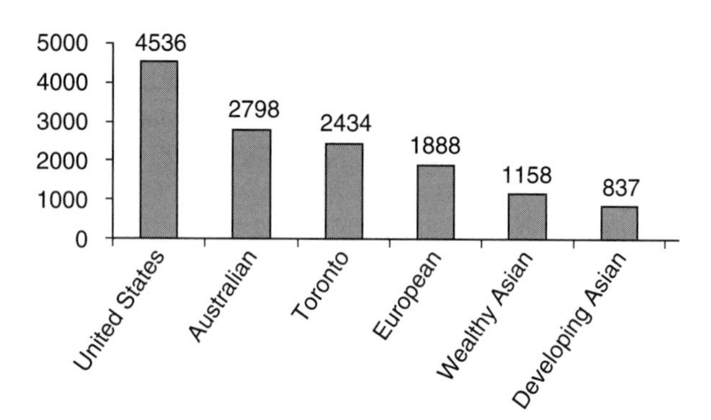

FIGURE 1 According to 1990 statistics, rates of annual CO_2 emissions (shaded bars, in kilograms/person) in the United States are 2.4 and 5.4 times higher than those of European cities and developing Asian cities, respectively. From "Sustainability and Cities" by Peter Newman and Jeffrey Kenworthy. Copyright © 1999 by Peter Newman and Jeffrey Kenworthy. Adapted by permission of Island Press, Washington, D.C.

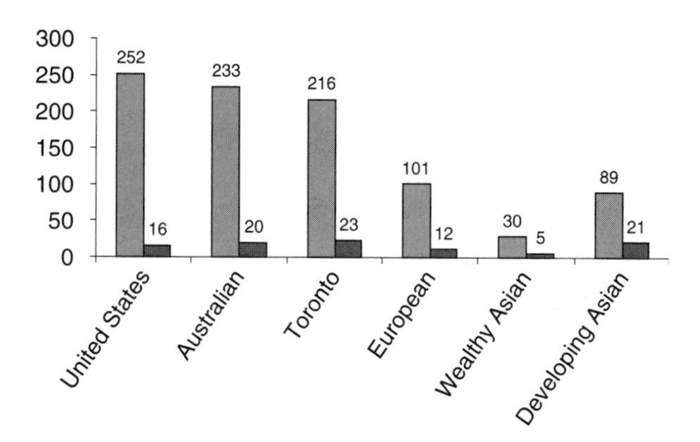

FIGURE 2 Rates of annual NO_x, SO_2, CO, and volatile hydrocarbon smog-related emissions (light shaded bars, in kilograms/person) in the United States are 2.5 and 2.8 times higher than those of European and Asian cities, respectively, in 1990. Volatile particulates are also shown (dark shaded bars, in grams/passenger-kilometers). From "Sustainability and Cities" by Peter Newman and Jeffrey Kenworthy. Copyright © 1999 by Peter Newman and Jeffrey Kenworthy. Adapted by permission of Island Press, Washington, D.C.

other cities (Fig. 1). Health problems are caused by smog-related emissions, involving nitrogen oxides (NO_x), sulfur dioxide (SO_2), carbon monoxide (CO), volatile hydrocarbons (VHCs), and volatile particulates (VPs). Per capita emission rates in the United States and Australia are higher than those in European and Asian cities (Fig. 2).

Los Angeles (LA) is famous for its smog caused by automobile emissions. The analysis in the 1996 report of the LA Air Quality Management District is that average ozone levels in LA, the South Coast Region, are twice the federal health standard. Ozone concentrations have exceeded standards on as many as 98 days per year. Because of the frequency of occurrence, the smog in this region is especially harmful to aged people and children. The symptoms caused by smog are usually aching lungs, wheezing, coughing, and headache.

Ozone also hurts the respiratory system's ability to fight infection. Other cities have similar air pollution problems; LA is no longer exceptional, especially among the growing cities of the American sunbelt.

4.3 Can Technological Innovations in Transportation Offset Impacts of Sprawling Land Use Patterns?

Optimists hope that innovations in transportation technology will solve the problems associated with sprawling land use patterns. Traffic congestion, air pollution, and burgeoning energy consumption are each the target of specific research initiatives. The U.S. government supports research to develop vehicles, such as hybrid electric and fuel cell cars, that will use energy sources (i.e., compressed natural gas, biodiesel fuel, ethanol, and hydrodiesel electricity) less harmful, compared to fossil fuels, to the environment. The government also promotes fuel efficiency through the Corporate Average Fuel Economy (CAFE) standards. In addition, with federal funds, some cities are introducing Intelligent Transportation Systems (ITSs), which increase the effectiveness of existing roadways by giving drivers real-time information about the best travel routes. However, many analysts argue that transportation energy efficiency and ITSs will not reduce the number of vehicles in use and air pollution emissions. Thus, technological innovations should be accompanied by nontechnological solutions. The Department of Energy suggests new land use planning strategies that will improve energy efficiency and will protect natural corridors and open space. These include transit-oriented design, mixed-use strategies, urban growth boundaries, infill development, greenways, brownfields redevelopment, transfer of development rights, open-space protection, urban forestry, land trusts, agricultural land protection, and solar access protection.

5. INFLUENCE OF URBAN/ ARCHITECTURAL DESIGN ON ENERGY POLICIES AND PRACTICES

Urban form at the metropolitan scale strongly affects transportation-related energy consumption, but smaller scale urban and architectural design decisions also influence energy use. Site layouts and building material choices affect microclimate and can create heat island effects.

Heat island effects arise from multiple sources. First, urbanized areas have residential, commercial, and industrial zones that displace trees and shrubs to varying extents. Trees and shrubs have the capacity to control their own temperature by releasing moisture, resulting in a natural cooling effect known as evapotranspiration. The displacement of trees and shrubs in the most built-up areas removes beneficial natural cooling. In addition, impervious surfaces, dumping excess heat from air-conditioning systems, and air pollution cause the ambient temperatures to rise, increasing energy use due to the large demand for air conditioning.

Site layouts and physical features of buildings and other impervious materials in a city have an influence on microclimate changes such as sunlight access, wind speed, temperature, and noise. Sunlight striking dark, impervious surfaces becomes sensible heat, causing the temperatures to rise. Building locations influence wind speed, sometimes impeding and other times redirecting the flow of wind on a site. Solar access, prevailing winds, tree location, and topographic modifications change microclimate. Wind facilitates good air circulation, thereby reducing build-up of heat. Effectively narrowing a street with street trees can reduce summer temperatures by 10°F. A modified site layout (street orientation, building placement, location on slope, and landscaping) can reduce the energy consumption of an ordinary residence by 20%.

Architects and, to a lesser extent, planners can influence building design and materials choices at the micro level. Energy efficiency in building design can be improved by maximizing solar access, by minimizing infiltration but taking full advantage of natural ventilation, by creating non-window spaces as buffers on north walls, and by utilizing natural convection and passive solar designs, for example. High-performance lighting and maximization of natural light by solar access can also improve energy efficiency, reducing energy costs. Institutionally, in the United States, the Energy Policy and Conservation Act (EPC Act) of 1975 and the Energy Policy Act (EP Act) of 1992 have improved the energy efficiency of household and commercial building appliances. Advanced building design techniques and associated technologies and appliances available today can cut in half the energy consumption of buildings.

6. CENTRALIZED VS. DISTRIBUTED ENERGY SUPPLIES

6.1 Implications for Planners

Energy supply, particularly electricity supply, needs to be analyzed in its regional context. It has long been assumed that increasing returns to scale result from increasing efficiency and technological specialization by means of large-scale supply. Thus, the single-minded pursuit of centralized energy supply led to a North American electric power system that was interconnected on a multistate basis and included gigawatt-scale generating plants. Centralized energy supply also brought with it certain inefficiencies associated with monopoly power, because utilities had few incentives to be operationally efficient and innovative. The recent restructuring efforts have been inspired by expectations of increased dynamic efficiency.

Another weakness of a centralized energy supply (characterized as "brittle" by Lovins) is the vulnerability of large power plants and transmission lines to disruptions, both natural and man-made, accidental and intentional. From the transmission system-related 1965 Northeast blackout in the United States, to the generation system-related 1986 Chernobyl accident in the former Soviet Union, to the utility distribution system disruptions caused by the 1993 and 2001 World Trade Center attacks in New York, to the 2003 cascading outages in the Northeast, there is accumulating evidence that the redundancies built into the large-scale grid provide inadequate reliability and security.

The introduction of smaller scale distributed energy (DE) power plants is one potential solution to these problems. Community-based energy supply falls within the planners' geographic sphere of influence. If this paradigm catches hold, energy planning could become a basic task for planners.

6.2 History of Centralization

In the United States, the electric power supply grew from the neighborhood scale in the 1880s to the municipal scale by the turn of the century. During the early 1900s, integrated large holding companies emerged, and interconnection of municipal systems proceeded apace, driven by a desire to increase load diversity and reliability of supply. Interstate transmission networks were entering use by the late 1920s. By that time, as mentioned earlier, 16 holding companies supplied most of the U.S. generation. Because of their nontransparent, unstable financial structure, the PUHCA was passed in 1935. This act mandated multistate holding companies to be subject to the state authority, and forbade a parent holding company from taking out loans from an operating utility. The state-based reorganization through PUHCA did not cause any serious problems until the 1960s. Following the 1965 Northeast blackout, the North American Electric Reliability Council (NERC) was formed to coordinate utility responses to regional disruptions and to avoid cascading failures.

Two oil crises in the 1970s and the restructuring of other regulated industries (such as telecommunication) brought about a rethinking of the electricity industry. During this period, technological innovations in gas turbines reduced generation costs and optimal generating-plant sizes, and thus made the large-scale power plants designed to exploit economies of scale no longer necessary. In addition, PUHCA exacerbated problems by placing the siting of transmission lines under state rather than federal authority, thereby constraining power flows from states with power surpluses to other states experiencing shortfalls. As a result, the vertically integrated structure is being unbundled, competition has been introduced in the electricity market, and a greater variety of new power generation techniques and resources are entering use. In particular, the DE technologies are gaining a toehold.

Most DE stations are located close to their ultimate customers. Supply-side distributed energy resources (DERs) include wind turbines, natural gas reciprocating engines, microturbines, photovoltaics, and fuel cells (Fig. 3). Demand-side DERs, for example, include schemes to reduce peak electricity demand and designs for high-efficiency buildings

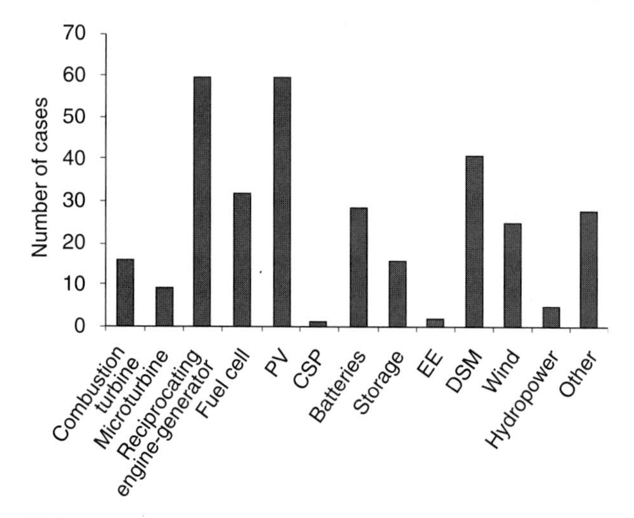

FIGURE 3 In a study of 275 distributed energy resources installations, reciprocating engines, solar electric power (PV, photovoltaic), and demand-side management (DSM) systems are the three most commonly used distributed energy resources technologies. In some cases, more than one technology was used per installation. CSP, Concentrating solar power; EE, energy efficiency. Reprinted from *DER Technologies*, the Office of Energy Efficiency and Renewable Energy, DOE.

and advanced motors and drives for industrial applications.

6.3 Special Case of District Energy Systems

Although centralized energy systems locate generators remote from consumers, some distributed energy systems locate generators on the consumer's premises, occupying valuable floor space and imposing significant installation and maintenance costs. District heating and/or cooling systems lie in-between these extremes. These systems produce steam, hot/chilled water, and electricity at a central location and distribute them to nearby buildings (Fig. 4).

A district heating or cooling system can reduce capital costs and save valuable building space, and can use not only conventional resources such as coal, oil, and natural gas, but also renewable fuels such as biomass and geothermal, thereby increasing fuel diversity. In particular, combined heat and power (CHP; also known as cogeneration), which reuses waste heat after producing electricity, can increase the overall process energy efficiency to more than 70%. Steam or hot water can be transformed to chilled water for refrigeration by means of absorption chiller technology.

City planning decisions strongly influence the economic feasibility of district heating and cooling systems. Such systems are most viable in compact downtown areas with high load density and a diversity of uses that support 24-hour operations. They are not viable in dispersed suburban settings.

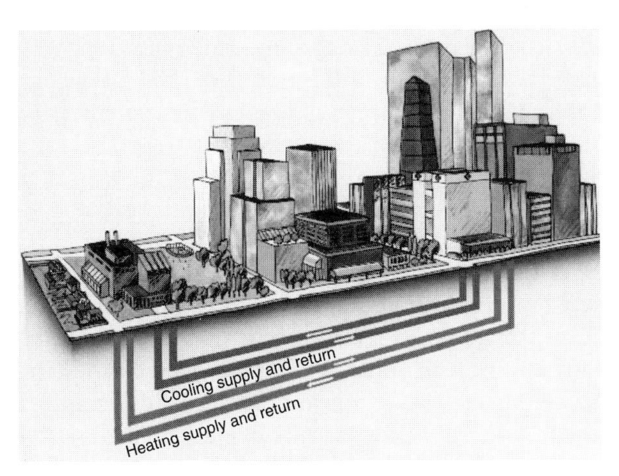

FIGURE 4 District energy systems distribute steam, hot/chilled water, and electricity from a central power plant to nearby buildings. With the technology of combined heat and power, they use 40% of input energy for electricity production and 40% for heating and cooling. Reprinted from *What Is District Energy?*, with permission of the International District Energy Association (2001).

6.4 Codes and Covenants Affecting Distributed Fossil Energy Systems and Distributed Renewable Energy Systems

The United States Public Utility Regulatory Policies Act of 1978 encouraged cogeneration and renewable (solar photovoltaic, wind) energy production. However, siting the facilities and interconnecting them with the grid involve significant problems, still unsolved today. There is no uniform standard code for interconnection between grids, and utilities have little incentive to modernize technical requirements for grid interconnection. Federal and state regulatory agencies share responsibility for setting the rules governing the development of distributed generation. The California Energy Commission argues that the main issues include paying off the utilities' stranded investments, siting and permitting hurdles, interconnection to the grid, environmental impacts, and transmission system scheduling and balancing.

At the state level, supportive policy-makers could revise the rules on siting facilities, financial incentive programs, interconnection, net-metering programs, and air quality standards to encourage the development of the DE market. At the local level, existing building codes, zoning ordinances, and subdivision covenants often forbid distributed energy system installations. For example, many communities have regulations that restrict homeowners' opportunities to install solar energy systems on their roof. Thus, a partnership among community groups, local governments, and developers is needed to address these restrictions before DERs can move significantly forward.

6.5 Prospects for Local Self-Sufficiency

The United States Department of Energy (DOE)sets forward one long-term vision of DER: "the United States will have the cleanest and most efficient and reliable energy system in the world by maximizing the use of affordable distributed energy resources." In the short term, DOE sets forward the plan to develop the technologies for the next-generation DE systems and to remove regulatory barriers. In the medium term, the goals are concentrated on reducing costs and emissions and on improving energy efficiency. In the long term, the target is to provide clean, efficient, reliable, and affordable energy generation and delivery systems. California is one of the few states to set a target: it has a plan to increase DE generation by 20% of state electricity generation. Yet municipal energy self-sufficiency is an unlikely prospect anywhere in the world during the next 20 years, if it is even desirable. More likely is an infiltration of DE technologies into a national grid that also contains many large central power plants. DE supply will depend on private investment in the DE industry. To encourage private investment, government will continue to have a role in setting market rules and helping investors overcome barriers associated with siting facilities, grid interconnection standards, and financing.

7. INTEGRATED VS. SEGMENTED ENERGY PLANNING

7.1 History of Debate

In his 2001 book, *Urban Development*, Lewis Hopkins argues that planning, when done well, is very much a big-picture exercise designed to improve coordination among distinct, yet interdependent, decision arenas. Good urban plans coordinate various private land use and public infrastructure system decisions, for example. In the energy arena, planning can have a similar integrating function.

7.1.1 Economies of Scope

A primary driver of integrated energy planning is the potential for economies of scope. This economic argument states that in certain circumstances, the average total cost decreases as the number of different goods produced increases. For instance, a company can produce both refrigerators and air conditioners at a lower average cost than what it would cost two separate firms to produce the same goods, because it can share technologies, facilities, and management skills in production, thereby reducing costs. In the energy industry, for over a century, many firms have realized economies of scope by integrating natural gas and electricity distribution and sales. Governmental energy planners have pursued similar economies since at least the 1970s by planning for the energy sector as a whole rather than keeping electricity, gas, oil, and other sectoral plans apart.

7.1.2 Integrated Resource Planning Movement of the 1980s/Early 1990s

The energy supply and financing crises of the 1970s inspired some U.S. utility regulators to demand a new kind of planning from regulated utilities. Rather than use only supply-side options and treat demand as exogenous, utility planners were directed to consider both supply- and demand-side options using a methodology known as least-cost planning, and later called integrated resource planning (IRP). In the United States, more than 30 state Public Utility Commissions (PUCs) adopted IRP procedures at the end of 1980s. The EP Act, passed in 1992, encouraged all electric utilities to exploit integrated resource planning.

Under IRP, utilities attempted to shape future energy supply and demand. In the process of planning, they considered factors such as energy efficiency and load-management programs, environmental and social aspects, costs and benefits, public participation, and uncertainties. In addition, demand-side resources were given the same weight as supply-side resources. Demand-side options included consumer energy efficiency, utility energy conservation, renewables installed on the customer side of the meter, and pricing signals. Supply-side options consisted of conventional power plants, non-utility-owned generation, power purchases from other suppliers, and remote renewables (Table III).

However, many states began to think about deregulating or restructuring their gas and electric power industries shortly after the EP Act was passed in 1992. This meant that state governments could no longer regulate planning practices and prices so closely, because utility companies were situated under competition in the energy supply market. Under restructuring, IRP became a strategic tool for the utility rather than a public planning process. Additionally, Kevin Kelly, in his 1995 chapter in the book *Regulating Regional Power Systems*, points out that there was a discrepancy between regional least-cost planning at the federal level and state least-cost planning at the state level. The need for public energy planning remained in the United States, but by the mid-1990s it no longer carried the force of regulation.

7.1.3 Sustainable Communities

Since the 1992 Earth Summit in Rio de Janeiro, and accelerating through the turn of the millennium, the term "sustainable development" has been a visible agenda item in every social sphere. Sustainable development does not simply regard development as growth; as Mark Roseland puts it in his 1998 book, *Toward Sustainable Communities*, it seeks to improve the well being of current and future generations while minimizing the environmental impact. There are different expressions of this idea: sustainable cities, sustainable communities, eco-cities, green communities, ecovillages, green villages,

TABLE III

Demand-Side and Supply-Side Options

Options	Example
Demand Side	
Consumer energy efficiency	Home weatherization, energy-effcient appliances for lighting, heating, air conditioning, water heating, duct repair, motor, refrigeration, energy-efficient construction programs, appliance timers and control, thermal storage, and geothermal heat pumps
Utility energy conservation	Load management, high-efficiency motors, and reduced transmission and distribution losses
Rates	Time-of-use, interruptible, and revenue decoupling
Renewables	Solar heating and cooling, photovoltaics, passive solar design, and daylighting
Supply side	
Conventional power plants	Fossil fuel, nuclear, life extensions of existing plants, hydro/pumped storage, repowering, and utility battery storage
Non-utility-owned generation	Cogeneration, independent power producers, and distributed generation
Purchase	Requirement transactions, coordination transmissions, and competitive bidding
Renewables	Biomass, geothermal, solar thermal, photovoltaics, and wind

Source. U.S. Department of Energy (2001).

econeighborhoods, and so on. One of the fundamental concepts is to minimize our consumption of essential natural capital, including energy resources.

In this sense, sustainable communities might perform IRP and could even give more weight to demand-side management than to supply options. However, in doing so, some argue that they push public policies counter to market forces, which currently make supply investments more profitable. Yet IRP could be performed from the consuming community's point of view, as a guide to aggregate expenditures on energy services. Some municipal electric utilities (in Sacramento, California, for example) have adopted this approach and they have maintained stable energy bills while shifting toward renewable energy supplies and more efficient energy usage.

Restructuring of energy industries is replacing regulated, firm-level IRP processes. Energy supply service is becoming more diversified under deregulation. Now communities are well placed to take on the challenge of integrative planning, and a very few are already becoming broad community businesses, extending the concept of the community energy utility to include recycling and reusing waste products within the community (Table IV).

7.2 Industrial Complexes and Ecoindustrial Parks

Integrated planning can clearly extend beyond energy into other realms. The Kalundborg industrial complex in Denmark is frequently cited as an

TABLE IV

Comparison of Conventional and Community Energy Utilities[a]

Parameter	Conventional energy utility	Community energy utility
Generating plant	Large scale and remote from customers	Small scale and locally based near customers
Customer base	Very large, with tens of thousands of customers	Relatively small, usually a few thousand customers and could be much smaller
Legal structure	Public limited company, subsidiary to multinational parent company	Variety of legal structures, including joint venture companies, cooperatives, and charities
Control	Main control lies with the multinational parent, but shareholders also influence decisions through their desire to maximize dividends	Day-to-day control could be in the hands of a commercial management; energy end users have a substantial stake in the utility; local authority may also be a stakeholder as an agent for the community
Technology	Conventional steam electricity, nuclear power, conventional hydroelectricity, etc.	Combustion turbine, microturbine, reciprocating engine-generator, fuel cell, photovoltaic, wind, etc.

[a] Adapted from Houghton (2000).

exemplar of industrial symbiosis that uses energy cascades and closed-loop materials cycling among several firms to yield economic benefits with less pollution, more efficient use of resources, and less need for environmental regulatory supervision, compared to traditional arrangements.

Firms in such ecoindustrial parks can share their environmental management infrastructure and increase ecoefficiency by rationalizing and optimizing aggregate materials and energy flows. Power plants, especially CHP systems, make natural anchor tenants. City planners specializing in industrial parks are just beginning to focus on such possibilities, although the examples and ideas have been around for decades.

7.3 Optimizing across Multiple Energy Sources

Energy planning must consider several factors: energy resources are unevenly distributed across Earth, technological innovations can disrupt existing equilibria, political and economic boundaries and regulations affect the use of energy resources, culturally and socioeconomically mediated human behavior can influence energy consumption patterns, and different places have different climates. Optimizing the use of energy resources will require different strategies in different places to reflect specific local conditions. For instance, a higher population density will make CHP more practical, but it may militate against other devices, such as passive solar design, due to the increased overshadowing.

7.4 What the Future Holds

Integrated energy planning remains an attractive proposition for firms and for governments, although deregulation of electricity and gas markets has disrupted previous regulatory drivers of this practice. Its scope may intersect in a significant way with that of city planning if current interest in distributed generation, industrial ecology, and sustainable communities persists.

8. ENERGY FOR THE WORLD'S BURGEONING MEGACITIES

8.1 Challenges of Scale and City–Hinterland Linkages

One of the trends at the beginning of the 21st century is the emergence of big cities. Currently, there are 30–35 cities in the world whose population size is more than 5 million. With the growing concern for the environmental impact of cities, people are beginning to pay attention to city systems from the perspective of sustainability. In his 2000 book, *Green Urbanism*, Timothy Beatley considers the ecological footprints of cities, which consume energy, materials, water, land, and food to support them, and then emit wastes. Urban growth and development have triggered the expansion of urban infrastructures, increasing energy consumption. For instance, in their 1997 *Nature* article, "The Value of the World's Ecosystem Service and Natural Capital," Robert Costanza and colleagues estimate that the energy consumed in urban areas is produced in coastal areas, forests, grasslands, and wetlands, and its total value is at least $721 billion per year. There is no primary energy production in urban areas. Rather, urban systems consume energy produced from the environment. The United Kingdom's International Institution of Environment and Development estimates that the entire productive land of the UK is necessary to maintain the population of London, whose population size is 7 million. The goal of managing urban areas is shifting from ensuring the unconstrained consumption of energy resources to encouraging their efficient use.

8.2 Cities as Systems

Cities are extremely complex combinations of physical subsystems, including those of urban infrastructure, various housing and other structures, transportation, communication, water transport, geology, ecosystem, solid waste, food and water distribution, economic zones, and demographics. In addition, social, political, economic, and cultural networks are interconnected, and create these built environments. To maintain and manage complex city systems, many agencies are involved. In the United States, these include the Department of Energy, Department of Defense, Department of Transportation, Environmental Protection Agency, Department of Housing and Urban Development, Federal Emergency Management Agency, United Nations, state and local planning agencies, and others. Yet none of these government agencies controls or plans cities to any significant extent: cities are self-organizing systems. Their top-down steering capacity is minimal. With continued urbanization worldwide, and with the development of high technology and the growing concerns for the limit of natural resources, there are severe challenges for cities. Adequate energy supplies rank high among these challenges. In the developing world, financial and environmental constraints will dictate more diversity in energy supplies, and more common use of demand management tools, than has been the case in the industrialized world to date.

8.3 Spaces of Flows

Cities are connected to one another in global transportation and communication networks that allow rapid flows of information, capital, people, and goods from one place to another. Under global capitalism, a hierarchy of cities exists in which a few cities (e.g., New York, London, and Tokyo) serve as corporate command centers while others specialize in manufacturing, back-office operations, market niches, and tourism, or serve primarily local needs. Cities likewise withdraw resources from and deliver products and residues to their hinterlands. In his 2001 book, *The Rise of the Network Society*, Manuel Castells refers to a shift from spaces of places to spaces of flows, arguing that cities are nodes through which flows occur. This conception implies that cities are influenced by a larger systemic logic, and that the nature of the flows and transformations the city imposes on those flows is relevant to city planners. The flows slow to a trickle in the absence of adequate energy supplies, and environmentally insensitive transformations of energy flows in urban areas are choking places such as Bangkok and Mexico City. Efforts by planners and local officials worldwide to ensure reliable, secure urban energy infrastructures have increased since the terrorist acts of September 11, 2001.

9. CONCLUSION

City planning evolved as a reaction to the ugliness and unhealthfulness of urban areas following the industrialization and urbanization in the late 19th century. In the 20th century, city planning broadened its scope and developed its tools, including land use planning, zoning, transportation planning, and growth management. Energy planning has not been the major agenda in city planning.

Since the oil crises in the 1970s and with increasing public awareness of the environmental impacts of fossil fuel consumption, there has been a fundamental change in energy policy. The security of the energy supply, the price of energy, and its environmental impacts have formed a volatile tripartite political agenda, and technically innovative, alternative energy resources have begun emerging.

Policy levers exist at several levels. For instance, although transit-oriented development is located in macro-level (or metropolitan) planning, walkable street development belongs to meso-level (or community) planning, and parking requirements play out at the micro (site) level of planning. At the meso (community) level are the placement of technological innovations such as distributed energy resources. At the micro (site) level, design decisions can alter the energy efficiency of building layouts and materials, lighting, and appliances. However, some innovations are restricted by current regulations and energy policies at each government level. There are especially severe energy-related challenges for city planning in the world's growing megacities.

Planning spans the gap between innovative technologies of energy production and their implementation. Hopkins argues that planning is a complicated and integrated process interconnecting regulation, collective choice, organizational design, market correction, citizen participation, and public sector action. Planners struggle with the vagueness of what their goals are, how their processes perform, and who can be involved in the planning process.

In sum, energy planning could be one of the major agendas in city planning, and it could conceivably be incorporated into city planning, although it has not been to date. Energy planning and city planning intersect at both the community and metropolitan levels, and thus planners will be involved in all attempts to provide more reliable, affordable, and environmentally sound energy for the world's cities.

SEE ALSO THE FOLLOWING ARTICLES

Ecological Footprints and Energy • Economic Growth and Energy • Heat Islands and Energy • Land Requirements of Energy Systems • Population Growth and Energy • Suburbanization and Energy • Urbanization and Energy

Further Reading

Andrews, C. J. (2002). Industrial ecology and spatial planning. *In* "A Handbook of Industrial Ecology" (R. U. Ayres and L. W. Ayres, Eds.), pp. 476–487. Edward Elgor, Cheltenham, UK.

Barton, H. (ed.). (2000). "Sustainable Communities the Potential for Eco-neighbourhoods." Earthscan Publ. Ltd., London.

Beatley, T. (2000). "Green Urbanism: Learning From European Cities." Island Press, Washington, D.C.

Calthorpe, P., and Fulton, W. (2001). "The Regional City." Island Press, Washington, D.C.
Campbell, S., and Fainstein, S. (eds.). (2000). "Readings in Planning Theory." Blackwell Publ., Cambridge, Massachusetts.
Castells, M. (2001). "The Rise of the Network Society." Blackwell Publ., Cambridge, Massachusetts.
Hopkins, L. D. (2001). "Urban Development: the logic of making plans." Island Press, Washington, D.C.
Newman, P., and Kenworthy, J. (1999). "Sustainability and Cities." Island Press, Washington, D.C.

Clean Air Markets

ALEXANDER E. FARRELL
University of California, Berkeley
Berkeley, California, United States

1. Introduction
2. Concepts
3. Examples
4. Looking Ahead

Glossary

banking The ability to save excess allowances in a Cap-and-Trade system from one period in order to use or trade them in a subsequent period.

cap-and-trade An emission trading program in which the government creates a fixed number of allowances, auctions or allocates these to regulated sources, and requires regulated sources to surrender allowances to cover actual emissions.

command-and-control regulation A form of regulation that is highly specific and inflexible, such as an emission standard (e.g., tons/year) or a performance standard (e.g., kg/unit of output).

discrete emission reduction A credit for a one-time reduction in pollution in the past below a regulatory standard. Often measured in mass (e.g., tons).

emission allowance The currency used in Cap-and-Trade programs. Allows the emission of one unit (e.g., ton) of pollutant.

emission reduction credit A credit for permanently reducing pollution below a regulatory standard or baseline. Often measured in mass emission rate (e.g., tons/year)

emission trading A generic term for environmental regulation that uses an allowances (or credit) for emissions (or emission reductions) that can be bought and sold.

market-based regulation Any regulatory method that use taxes, subsidies, tradable instruments (e.g., emissions allowances and emission reduction credits), or similar methods in stead of, or as a complement to command-and-control regulations.

verified emission reduction A credit for greenhouse gas emission reduction that has been verified by a third party but is not necessarily acknowledged by a government.

Government regulation can apply economic forces to achieve policy goals, as well as include mandatory standards and prohibitions. Such approaches are increasingly being applied to energy systems, especially for the control of air pollution and greenhouse gases, and the promotion of renewable energy. Most common are emission trading programs, which create clean air markets. This article discusses key concepts associated with clean air markets and the major implementation challenges associated with them. Several examples are used to illustrate these ideas.

1. INTRODUCTION

The use of market-based instruments (MBIs) for environmental policy is growing rapidly in regulating energy production and will likely be central in the still-evolving climate change policy regime. (See Farber in this volume for an overview.) Several features make MBIs attractive relative to more traditional command and control (CAC) regulation, but MBIs need to be designed and implemented carefully. This entry discusses some basic concepts and challenges of MBI and examines important examples of MBIs that feature air pollution allowances or credits, since these are of most importance in energy.

The most obvious benefit of MBIs is that they can help reduce the cost of environmental protection, which they do by providing significant flexibility to regulated organizations (typically firms). In the United States, programs that use tradable MBIs such as credits and allowances (e.g., emission trading) have enabled environmental protection while saving billions of dollars in pollution control costs. One important reason for the increasing interest in MBIs is that as environmental controls have become tighter and tighter, they have also become more expensive, making cost-effectiveness ever more important. The extreme case is climate change, which is likely to require the deployment of energy supply systems that are largely free of fossil fuels over the next several decades. Attempting such a massive change too fast or with CAC regulation alone would raise the cost and make it very difficult to achieve in democracies.

MBIs give firms incentives for environmental performance, rather than specifying emission rates or technical standards, as CAC regulations do. The basic logic behind MBIs is that firms differ in how much it costs them to control emissions, so if a high-cost company can pay a low-cost company to control emissions on its behalf, the net result is a savings. This approach assumes emissions from all sources are equivalent (i.e., uniform mixing); when this is not the case, additional rules can be added. Importantly, the use of MBIs can make environmental protection look more like an ordinary business issue to managers and can allow them to apply risk management tools (often in the form of financial derivatives). In addition, MBIs can be (but are not always) easier to implement since government often does not have use intrusive inspections or detailed permitting processes that require considerable industry knowledge, which may be hard for government to obtain. However, careful (and often expensive) monitoring may be needed, both at the plant level and in the marketplace, to ensure MBIs perform as expected. Finally, MBIs typically provide significant incentives for innovation, both managerially and technologically. However, the size and shape of these benefits vary greatly with the details of MBI design and implementation.

2. *CONCEPTS*

There are several types of MBIs, including taxes, subsidies, and different varieties of tradable instruments. Others include liability rules and information.

2.1 Taxes and Subsidies

The simplest MBI is a tax on pollution, which will give firms incentives to reduce emissions and to innovate to find new, cheaper ways to do so. Subsidies can work similarly, but, of course, they give positive incentives rather than negative. Economists point out that by making polluters pay for the damage they produce, and thus internalizing what they call externalities, emission taxes can increase economic efficiency. However, emission taxes place a cost on all firms, no matter how clean, and so are unpopular with firms. It is also difficult to achieve a specific environmental outcome with taxes, since the government will never know ahead of time what tax would be needed to achieve a desired level of emissions and requires governments to have detailed knowledge of regulated industries as well as the ability to forecast well. Two exceptions are carbon (or energy) taxes, which sometimes replace a previous energy tax, and auxiliary taxes in emission trading programs.

2.2 Emission Reduction Credits (ERC)

In places with poor air quality and a CAC regulatory framework, new, large sources (like a new power plant) or sources undergoing expansions are required to offset their new emissions. The instruments used for this purpose are emission reduction credits (ERCs), which are project based, rely on permanent reductions in emissions from specific sources compared to a baseline. For instance, ERCs can be created when a polluting facility goes beyond regulatory requirements in pollution control (overcontrol), permanently slows output (and thus emissions), or shuts down. Emission credits can be mass based (e.g., tons) or rate based (e.g., pounds per day), depending on the specifics of the underlying regulatory program to which the ERC has been added. However, the company that is trying to create emission credits must get government approval, which can be a slow process and raise transaction costs.

Typically, ERC systems are add-ons to preexisting CAC regulations, designed to provide flexibility and reduce costs. CAC regulations typically specify specific emission rates or control technologies and require regulated sources to submit detailed compliance plans explaining how they will meet these requirements. Regulators approve these plans and inspectors periodically verify compliance. Compliance planning is a significant burden for some firms since it is costly and limits their flexibility should market or technical conditions change. It also tends to reduce innovation since it is risky and time consuming to get approval for new technologies.

In the United States, ERC systems have had limited success, but they are not used much elsewhere. They have played a role in programs for federal air pollution control and are used by a few states. One problem is convincing environmental regulators of the size of emission reductions, because it is often impossible to measure emissions directly and because of disputes over appropriate baselines. Firms must prove that the ERCs are "surplus" or "additional"—that the actions that create the ERCs would not have taken place anyway. A further disincentive in ERC programs is the detailed oversight of ERC creation and trading.

An important feature of ERC systems is that they are entirely voluntary (although the underlying regulations are not), which, in practice, has meant that the incentives to create allowances and put them on the market have been weak. Firms generally do not go out of their way to create ERCs since they do not see themselves as being in the ERC business, do not want to invite greater scrutiny by regulators, and prefer to invest capital in their core businesses. When firms do create excess allowances, they often want to keep them to support possible expansions of their own facilities in the future rather than selling them into a market where potential competitors might buy them. Unfortunately, this problem can limit the number of available emission credits, making it hard for new entrants to buy credits to support new businesses. Additionally, the limited number of programs, the terms and conditions associated with ERCs, and the relatively small markets for them have hindered the development of risk management tools for them (e.g., futures and options), limiting their utility. One of the reactions to this problem has been for local governments to obtain ERCs as part of the process approving their creation, which can be used to foster growth (and job creation) by giving them to new companies entering the area.

2.3 Discrete Emission Reductions (DERs)

DERs are created by reducing emissions relative to a baseline and are project based. However, DERs are temporary and can only be used once. DERs can be generated by installation of pollution-control equipment, installation of control equipment with a higher-than-required efficiency, or prior to a compliance date for additional control or a process change. Importantly, DERs are wholly past reductions and are quantified after they have been created. In the United States, several states have rules that allow DERs, but they have different mechanisms for granting credits. For example, some states certify DERs, while others allow for self-certification with third-party verification. A similar term, verified emission reductions (VERs), is sometimes use to describe one-time reductions in greenhouse gas (GHG) emissions, but a key difference is that VERs are created *before* the rules for emission trading are in place. This makes their future validity uncertain.

2.4 Cap-and-Trade Systems (C/T)

The best known MBI is probably the emission allowance used in cap-and-trade (C/T) systems, which, as the name implies, create a permanent limit on emissions, a key virtue for environmental advocates. In a C/T system, the government defines the regulated sources and the total amount of pollution that they can emit, the "cap," over a set period of time, usually 1 year. Typically, the cap is set in mass units (e.g., tons), is lower than historical emissions, and declines over time. The government creates allowances equal to the size of the cap and then distributes them to the regulated sources, a process called allocation. All C/T existing systems allocate allowances based on historical emissions, which is problematic for new entrants, as discussed later.

The government then requires regulated facilities to surrender emission allowances equal to the emissions of the facilities on a periodic basis (sometimes called true-up or covering emissions). The government also sets standards for emissions monitoring, establishes rules for how allowances may be used, and defines enforcement. These are crucial choices, not just details. One particularly important decision is if and how allowances can be saved (or banked) from one period to another, which is discussed in detail later.

Because the allocation to each firm is smaller than its previous emissions, regulated firms have four basic options: (1) control emissions to exactly match their allocation, (2) undercontrol and buy allowances to cover their emissions, (3) overcontrol and then sell their excess, or (4) overcontrol and bank allowances for use in future years (when even fewer allowances will be allocated). The reason companies might buy or sell allowances is that facilities will have different emission control costs or they might change operations so that they needed more (or fewer) allowances. Companies with higher costs will be able to save money by undercontrolling and buying allowances from those with lower costs, which make money by overcontrolling and selling.

The government regulates the trading of emissions allowances differently in various C/T systems. The government usually acts as the accountant for C/T systems by establishing a registry for participants. Usually, participants are required to report the size of transactions and the names of buyer and seller. This can be facilitated by creating a serial number for each allowance. However, there is often no requirement that market participants disclose the price at which a sale was made, nor any requirement that they inform government of the trade in a timely manner. This lack of information can limit the transparency of the market as participants may delay reporting trades in order to conceal strategic information. Brokerage and consulting firms complete the picture by providing services to market participants, including markets in derivative commodities, and by increasing transparency by providing information (including price information) about the markets. Simplicity in market design and competition among brokers has tended to keep transaction costs low (up to a few percent of allowance prices) in emission allowance markets.

Several key features of C/T systems are worth noting. First, a cap on total emissions means that as an economy grows, new emissions-control technologies or emission-free production processes will be needed. Some observers worry that a fixed emission cap is a limit to economic growth; but so far the evidence shows no such effect. Second, the problem of getting allowances to new entrants can be minimized if an active, liquid market develops or if a small number of allowances are set aside by the government for this purpose. Third, the standardization of C/T allowances authorizes larger emissions markets and permit brokers to offer derivative securities based on them. This has proved important since the ability to use derivative securities like options and futures greatly enhances flexibility and reduces risk.

2.5 Open Market Trading

There have been attempts to allow DERs to be used in C/T systems, a concept called open market trading. Advocates of this approach typically look to create ERCs in the mobile source sector (by buying and scrapping old vehicles or paying for upgrades to cleaner vehicles) to sell to stationary sources. Although the open market trading approach has been attempted several times, it has usually failed, often due to disagreements over credit certification requirements or because this approach could nullify the cap. For example, a prominent open market trading program in New Jersey collapsed in late 2002 after years of work developing it. It is likely that the costs of adequate monitoring and verification for the use of DERs in C/T systems may be high enough to eliminate the value of doing so.

2.6 Renewable Energy Credits (RECs)

An addition to clean air markets is the renewable energy credit (REC), which is created by the production of electricity from renewable sources. RECs can be used to implement a renewable portfolio standard (RPS) cost-effectively. In such a program, RECs are created when renewable electricity is generated while companies that sell electricity are required to surrender RECs equal to a percentage of their sales as defined by the RPS. The logic here is similar to other MBIs—firms will vary a lot in their cost of generating renewable energy; an urban distribution company may find it more difficult to generate renewable energy, for instance, than a rural electric cooperative that has abundant land on which to site wind turbines.

2.7 Implementation Challenges

A number of significant challenges exist in the implementation of clean air market programs. Many of these are discussed in the cases that follow the descriptions presented next.

2.7.1 Environmental Goals

Ensuring that the environmental goals of a policy that uses an MBI (such as that the cap in a C/T program is not exceed) is important, since these are the essential reason for implementing such policies. A key threat is leakage, the migration of regulated activities to locations or sectors that are not controlled by the cap. For instance, if emissions from only from large electricity generators were regulated, the replacement of large power plants with smaller ones could constitute leakage. Migration of polluting industries from one country to another is a major concern of environmentalists and labor alike.

Another important issue is how MBIs might affect ambient air pollution and health risks since most pollutants are not uniformly mixed. A chief concern is over uneven spatial distribution of emissions and health risks (sometimes called the hotspots problem), which poorly designed emission trading systems may ignore or even exacerbate. A similar term is directionality, which is used when there are identifiable upwind and downwind locations. Another concern is that environmental equity goals may ignored by emissions trading systems. While such concerns have stopped some MBI proposals, they have been addressed in the design of others. With one minor exception (a part of the RECLAIM program, discussed later), there is no evidence that emission trading programs have caused environmental equity problems.

2.7.2 Regulatory and Price Uncertainty

Uncertainty is possibly the most important concern of firms in programs that use tradable instruments, and it comes from two basic sources. The first is that regulators will not be able to develop all the rules needed to operate a C/T or ERC program, or that they will change the rules. The second is whether allowances will be available and, if so, at what price. This uncertainty comes from the fact that most mangers (especially in small and medium firms) have little experience with allowance markets or resources to devote to the issue, while having enough allowances (or credits) may be crucial to continued operation. The implication of these uncertainties is that some firms regulated by MBIs may be unwilling to rely on entering the market as either buyer or seller, which makes concerns about a lack of allowances on the market self-fulfilling. However, it is not clear if emissions markets have more or fundamentally different uncertainties than do other markets (such as petroleum or computer chips) that have exhibited both dramatic price and regulatory changes in the past. If not, then the risk management tools developed for these markets can be adapted to manage uncertainty.

2.7.3 Banking

The ability to use allowances or credits left over from previous periods in the future is called banking and can be highly contentious. Note that DERs always involve banking, since they are created by retrospective analyses of one-time emission reductions. The ability to bank allowances is one of the major sources of flexibility in an emission trading program since it allows a firm to time its capital expenditures for process changes or emission controls. Banking also allows for the cheapest emissions reductions to occur first and for time to apply research to lower the costs of the more expensive cases.

2.7.4 Allowance Allocation

The creation of a C/T system produces a new type of asset, the allowance, which can have considerable value. For instance the SO_2 allowances given to the U.S. electric power industry each year has a value of more than $1.5 billion. However, a cost-free allocation based on historic emissions (called grandfathering) fails to internalize the external costs of pollution and give a weaker incentive for innovation. It also gives existing firms an inherent advantage over new entrants, who must buy allowances, often from potential competitors. Another important problem with cost-free allocation is that it adds to uncertainty about the market price of allowances.

Although all C/T systems in use so far in the United States have used grandfathered allocations, in principle allowances could be auctioned or distributed in other ways (e.g., distributed based on current production). Firms typically resist auctions, which will always cost more money, in effect arguing that their previous emissions entitles them to the new assets. This approach will also reduce the role of emissions brokers, since there would be less need for interfirm sales after such an auction.

One way to deal with the new entrant and price uncertainty problems is for the government to set a small portion of allowances aside for sale to new entrants and for early auctions to aid in what brokers call price discovery by providing evidence of what prices firms are willing to pay for allowances. Auctions could also be used if prices rise to levels that create significant hardships for a few market participants. These goals can also be accomplished with relatively small set-asides. A slightly different technique that has been proposed under the term "relief valve" is to have government create and sell as many allowances as firms wanted to buy for use that period at a predetermined value. The activation of such a safety valve would preclude banking for that year. This would break the cap of a C/T system for that year, obviously, but it would limit the costs of control and would have the desirable properties of an emissions tax.

2.7.5 Monitoring

Most emission trading programs are applied to stationary sources such as electric power plants, refineries, and manufacturers, partly because it is usually feasible to monitor the emissions of these sources accurately. Estimates of emissions are allowed in some cases, for instance, when calculating DER and ERC quantities since this always involves estimating a counterfactual baseline. Larger sources also have economies of scale in managing an emissions allowance account. Good monitoring and accounting practices reduce concerns about fraudulent (or simply flawed) sales of emission allowances and help ensure their environmental integrity. Fossil carbon in fuels is already accounted for and traded routinely in fuels markets, which may make it easier to monitor the withdrawal of fossil carbon from the ground than its release (as carbon dioxide, CO_2) to the atmosphere. However, such an upstream approach is probably politically impossible, especially internationally

2.7.6 Double Counting and Additionality
It is important for ERCs that the activities that create them are not already required by some law or regulation (additionality) and that they are not counted twice in two different ERC programs (double counting).

2.7.7 Market Transparency
It is not always easy for regulated firms to observe the operation of emission markets in detail, reducing their confidence in them. Unlike trading in securities (e.g., stocks, bonds, and futures contracts), emission trading markets are not regulated uniformly and often not very closely. In new or very small programs, emissions markets may be characterized by bilateral trades, generally put together by a consulting company and approved by government. Large emissions markets typically use computer systems that a handful of brokerage firms use, on which bids from regulated companies (which can number in the dozens to hundreds) to buy and sell allowances are posted. Regulated companies often do not have direct access to these markets, but work through brokers (who usually also trade in electricity, fuel, and sometimes other commodities). A few specialized newsletters survey brokers on market data and publish the results. In many cases, the environmental regulator will post some transaction information on the Internet, although just what information they put up varies.

Problems with market transparency include the lack of regulatory oversight of brokers (from which most of the price information comes), the potential for bilateral transactions to go unreported for long periods, and sometimes a difficulty in obtaining market information from government. These problems add to price uncertainty.

2.7.8 Incentives for Credit Creation
The creation of an ERC or C/T system does not automatically lead firms to reduce emissions for the purpose of selling credits or allowances. Managers may not see themselves as having any particular advantage in creating allowances, relative to their competitiveness in their product markets, or may simply not have the resources to devote to studying the problem. Those they do create may be considered vital to future expansion, so the firm may be unwilling to sell them.

2.7.9 Buyer/Seller Liability
There is some concern about what happens if an allowance or credit trade occurs and the traded instruments are later found to be invalid: Who is liable, the buyer or seller? What if the seller has gone bankrupt or disappeared? Questions associated with this issue are most important in international arenas, in ERC and DER programs, and in C/T systems without serialized allowances.

2.7.10 Enforcement
MBI programs typically include enforcement provisions. Enforcement for ERCs and DERs is based on the underlying CAC regulations, requiring action and judgment by the regulator, and often involves the courts. In contrast, C/T systems tend to have more automatic enforcement provisions, including such features as fines based on the cost of allowances, a requirement to surrender the next period with a multiple of the gap between actual emissions and allowances held at true-up. Such provisions reduce uncertainty and have led to extremely high compliance in the large C/T programs in the United States, where noncompliance has been limited to a handful of minor oversights, with one important exception (RECLAIM) discussed later.

3. *EXAMPLES*

Four emission trading systems for controlling air pollution in the United States will be examined, plus two climate change programs in Europe. This is by no means the entire set of emission trading programs in use. Others include pilot programs in Canada, China, and Europe, and regional efforts like the Houston/Galveston Area Mass Emission Cap & Trade program.

3.1 State ERC Programs in the United States

The original introduction of MBIs was the result of a U.S. Environmental Protection Agency (EPA) requirement that new facilities wishing to locate in areas with unacceptable air quality "offset" their emissions. The only way to do this was to obtain offsets from existing firms. This suddenly turned the legacy of emitting pollution into an asset that could be sold to firms entering or expanding in an area. This was late expanded to include a number of similar provisions called bubbling, netting, and banking, that improve the efficiency of CAC regulation. The banking provision enables many ERC markets.

State air pollution agencies typically manage these systems with guidance and oversight from the EPA.

Environmental groups have had an important role in ensuring the environmental integrity of these programs. Pollutants in these programs include volatile organic compounds (VOCs), nitrogen oxides (NO_X), sulfur dioxide (SO_2), and others. Price and availability for ERCs vary greatly by pollutant, location (ERCs cannot be transferred between different urban areas), and sometimes by other specific conditions attached to ERCs. These programs have created significant intrafirm trades (e.g., bubbling), but not many market transactions since they are somewhat cumbersome and have poor incentives for the generation and sale of credits.

3.2 U.S. Acid Rain Program (Title IV)

The best-known C/T system is the EPA's Acid Rain Program for SO_2 emissions from coal-fired power plants. Key features include the strict monitoring provisions for both SO_2 and NO_X; the national scope of the program; the relatively deep cuts in emissions (50%); completely unrestricted trading and banking; a small auction program in the early years of the program; and, most important, its success. Facilities regulated by the Acid Rain Program must still comply with health-based CAC SO_2 regulations that prevent hotspots from developing, although these restrictions have not affected the market.

The Acid Rain Program has been a success in several ways. First, substantial emission reductions have occurred. From 1990 to 2002, SO_2 emissions from regulated sources declined by about one-third. During the less-stringent Phase I (1995–1999), regulated sources overcontrolled and banked more than a year's worth of allowances, which they began to use in Phase II. This bank will be empty in about 2010.

Second, the program has greatly reduced the cost of SO_2 control compared to command-and-control polices. In the first 5 years, emissions trading reduced compliance costs by about one-third to one-half, and estimates of the savings range from $350 million to $1400 million. Allowance prices have thus been lower than forecast, ranging from $66/ton to about $200/ton. This is not to say that the cost of SO_2 control has been cheap. In 1995, annual costs were about $726 million, and capital costs for scrubbers in Phase I alone are estimated at $3.5 billion. However, most of these savings are not due to trading of allowances per se, but from the flexibility in compliance that allowed firms to find their own least-cost approach. An important development was that Midwestern power plants designed to burn high-sulfur local coal were adapted to burn low-sulfur Western fuel (from the Powder River Basin) just as it was becoming cheaper due to railroad deregulation.

Third, the SO_2 market has been a success. Prices in this market are relatively reliable because there are up to several dozen trades each day, resulting in from 20,000 to 100,000 allowances trading hands each week and considerable market information. All vintages of allowances are priced the same because there are no restrictions on banking, which helps smooth the operation of the market. An important feature of this market was auctions of set-aside allowances starting in 1992 and 1993, several years before the first compliance year. Although the design of these early auctions market has been criticized, they were extremely valuable because they helped with price discovery in an untested and highly uncertain market. The revenue from these auctions was returned to the regulated firms in proportion to their allowance allocation. These auctions also allowed for new entrants and the public to participate in the market. (Several school and public interest groups bought allowances in order to retire them.) This system also had an attractive program for early emission reductions, while some state programs encouraged early investment in emission control equipment (scrubbers), which made some allowances excess and created natural sellers. All of this helped the industry build up a bank of allowance before the first compliance year, reducing regulatory and price uncertainty.

Several important findings about the operation of allowance markets and industries regulated by a C/T system have emerged from the Acid Rain Program. First, market participation and compliance strategies have evolved, from an autarkic approach toward a greater and greater reliance on the market. Second, allowance prices have shown considerable volatility, and the lower bound of emissions prices have been shown to be equal to the marginal cost of operating emission control devices (scrubbers). Third, the relatively few units that did install scrubbers increased their utilization and lowered their emissions beyond original design specifications as a result of the incentives in the allowance market.

3.3 Ozone Transport Commission NO_x Budget

The first multilateral C/T system to go into effect is the Ozone Transport Commission (OTC) NO_X budget program, covering eleven states in the Eastern

Seaboard and the District of Columbia. The NO_x budget applies to power plants and large industrial facilities and covers emissions from May through September. There are more than 470 individual sources in the program, with about 100 owners. Beginning in 1999, the NO_x budget will eventually reduce emissions by 65 to 75% through a three-phase approach. Importantly, the OTC NO_X budget is not a centrally organized system (like the Acid Rain Program), but rather the result of coordinated laws and rules in each state, although the federal EPA provides some assistance (e.g., by tracking the allowances).

The states in the OTC were able to coordinate their individual laws and regulations to produce a successful program for a number of reasons, the most important of which were that they had similar interests in NO_X control, they had a history of cooperation in this area, and they engaged in a multiyear technical/political negotiation process that they all felt was fair. The result of this process was a model rule that could be adopted by each state with modifications to account for local differences without endangering the political or environmental integrity of the agreement.

This process illustrates how complex regulatory uncertainty can be. The development of the NO_X budget model rule began in late 1994, at which point the level and timing of emission reductions was set. The technical/political negotiation process that followed was designed to determine if emission trading would be used or not. In early 1996, the model rule was published and the OTC states began to develop their own rules, which were all final by the end of 1998 to allow the NO_X budget to start in May 1999 (except for Maryland, which entered a year late). Regulated sources felt this was actually rather short-fused, because the rules were not in place in several states until less than a year was left before the start of the first ozone season, while contracting for engineering and constructing NO_X control technologies can take several years. Regulators disagree with this assessment, noting that if the emission trading program had not come together, similar command-and-control regulation would have been the default. Further, much of the delay in developing state rules was due to the need to address the concerns of regulated industries, which then complained about the uncertainty.

The OTC NO_x budget has some distinctive features. First, it had no early auctions or other methods for price discovery before it went into effect. Second, the banking of allowances in the NO_x budget can be restricted in a way that can reduce the face value (in tons) of banked allowances, but firms do not know if or how much of a discount will be applied until after they have banked the allowances. This creates uncertainty since facilities cannot be sure exactly what their banked allowances will be worth in the future. Thus, banked allowances sell at a discount.

These factors led to a significant spike in allowance prices in the OTC NO_x budget, up to more than \$7000/ton, far above the cost of control for any regulated sources. Prices stayed high for several months, but by July they had fallen back to the predicted range and by the end of the year fell to around \$1000/ton. This volatility had nothing to do with the cost of emission controls, but was the result of the supply and demand changes in the allowance market. Importantly, neither state governments nor the regulated firms abandoned the allowance market by changing the rules or seeking regulatory relief in the courts. Since that first, difficult year, the OTC NO_x budget has matured; emission reductions have been substantial, compliance failures have been trivial, a large bank of NO_X allowances has been built up, and price discovery for allowances in the next, more stringent phase of the is now underway.

Subsequently, a failed attempt was made to develop a multilateral C/T program larger set of eastern states. This failure suggests several important factors are necessary. Probably the most important factor is that potential participants recognize that they have similar interests in controlling pollution cost-effectively. Secondly, they must be able to develop a process for creating and enforcing the C/T system that cannot be used against their interests.

3.4 California RECLAIM

The last U.S. example is California's Regional Clean Air Incentives Market (RECLAIM), a C/T program for SO_2 and NO_X emissions from industrial sources such as power plants, refineries, and metal fabricators. An important feature of RECLAIM is that it does not permit banking, since government regulators felt this would compromise its environmental integrity. When RECLAIM was first implemented in 1994, the cap was generous, allowing for an increase in emissions over historical levels for many sources, but it declined steadily each year, aiming at an overall reduction of about 75% by 2003. An example of an environmental equity problem in a C/T program is the mobile source opt-in provisions of the RECLAIM program. Residents living near several refineries were able to have these provisions overturned because they would have allowed the refineries to avoid reduce smog-forming emissions, which were also toxic pollutants.

For the first several years, the RECLAIM market functioned quite well, with readily available allowances at low prices. However, emissions in 1993 to 1998 did not decline nearly as fast as the cap due to a failure of many (but not all) participants to install emission control equipment. Although the state regulatory agency amply warned participants of a looming problem, many firms were unwilling to take appropriate actions because of a failure to consider future emission allowance markets and a belief that the government would bail them out in case of serious problems.

The result was a breakdown of the market and in response a temporary abandonment of the MBI approach by the state government. By early 2000 it had become clear to even the most shortsighted that emissions would exceed allocations, which was a problem because RECLAIM had no banking provision, and prices for NO_X allowances skyrocketed to over $40,000/ton. Electricity companies, which were making record profits at the time, could afford these prices, but other companies in the RECLAIM market could not. Thus, the RECLAIM cap was broken, and several firms were significantly out of compliance and paid record fines. This is the only failure of a C/T system to date. Facing significant political pressure, the state regulatory agency decided to essentially go back to a CAC approach for electric power plants by requiring them to submit compliance plans. This is particularly important in that most cost savings in C/T systems come from the ability to innovate in compliance strategy, not from buying or selling allowances. In addition, state regulators separated power companies from the rest of the RECLAIM market and subjected them to a high tax for emissions not covered by allowances. For other participants, RECLAIM proceeds as before and allowance prices have moderated.

Several key lessons emerge from the RECLAIM experience. First, because they force firms to gather more information and make more decisions, MBIs may be *more* difficult for firms to understand and manage than CAC programs, even if they have lower costs. This is especially true for smaller companies, some of which may even have increases in monitoring costs required by RECLAIM that were greater than the savings in control costs. Second, in some cases the optimal strategy may be noncompliance, placing more emphasis on the design of penalties. Third, emission markets are no different from others; they are volatile (especially when it is not possible to store the commodity, like electricity).

3.5 Denmark CO_2 Program

The first emission trading program to address climate change was a C/T system for CO_2 emissions adopted by Denmark in 1999 to achieve a national GHG emission reduction target of 5% in 2000 (relative to 1990) and 20% by 2005. The Danish program covers the domestic electricity sector (electricity imports and exports are treated separately), which is made up of eight firms, although two account for more than 90% of all emissions. Allocations were made on a modified historical basis and are not serialized. Banking is limited and there is some uncertainty about the validity of allowances beyond 2003. This program includes a tax of about $5.5/ton for emissions that are not covered by an allowance, which means the integrity of the cap is not guaranteed. It is possible to use VERs as well as credits created through provisions of the Kyoto Protocol (discussed later). Monitoring is accomplished through an analysis of fuel consumption—in-stack monitors are not required.

The C/T program was adopted to regulate the last major unregulated set of CO_2 emitters so that Denmark could reach its domestic CO_2 goals cost-effectively while also gaining experience with emission trading. In the first 2 years of the program, about two dozen trades were made in the $2 to $4/tonne range. This has resulted in more than half a million allowances changing hands. Some of these trades have been swaps of Danish allowances for VER credits, and one trade of Danish allowances for U.K. allowances (discussed later), the first instance of a trade of two government-backed MBIs. However, due to the small size of a market, it is unlikely that an "exchange" for Danish CO_2 allowances will emerge; instead most trades will probably be bilateral and the Danish system will probably be superseded by a European Union (EU) system.

3.6 United Kingdom Climate Change Levy and Emission Trading System

The first economy-wide GHG control policy was announced in November 2000 by the United Kingdom, and it contained a combination of MBIs, including taxes, subsidies, and ERCs. This program is designed both to achieve significant emission reductions and to provide U.K. organizations experience in emission trading, not least so that the City of London might become a leader in this activity.

The first part of this policy is the climate change levy (CCL) on fossil fuel energy supply for industrial users that came into effect in April 2001 and typically adds 15 to 20% to the cost of energy. The CCL has moderately increased the demand for energy efficiency, and the revenue it generates has been recycled into a fund that subsidizes consulting and capital purchases to reduce energy use. Moreover, it provides an important basis for voluntary participation and credit generation in the ERC program, possibly overcoming one of the key problems associated with past ERC programs.

The second component is the Emission Trading System (ETS), an ERC program open to reductions in all GHGs (measured in CO_2-equivalent, or CO_2e). To join the program, firms must enter into a climate change levy agreement (CCLA) in which they voluntarily accept an emissions cap in return for an 80% reduction in their climate change levy until 2013. Companies that adopt such a target may sell credits generated by exceeding their target.

The last component, direct entry, is a £215m ($310 million) subsidy that the U.K. government made available through an auction for voluntary actions by eligible firms to reduce GHG emissions in 2002–2006 from a 1998–2000 baseline and generate ETS credits. This auction is designed to obtain the maximum emission reduction and provide some price discovery by allowing eligible companies to bid a fixed amount of emission reduction for a given price and selecting (through repeated rounds) the price that yielded the maximum emission reduction given the government's budget. Most electricity and heat production were not eligible for this program, which made bids for reductions of non-CO_2 GHGs more likely and created a natural set of buyers (energy companies) and sellers (manufacturers and end users) for ETS credits. This auction was held via the Internet over 2 days in March 2002 and resulted in 34 bidders (of 38) winning subsidies at the level of about \$22/ton-CO_2e. Over half of the emissions will be non-CO_2e GHGs.

Organizations can also join the ETS through more traditional means by creating ETS credits through a specific project that meets all the necessary monitoring and verification requirements. Over time overseas-sourced ERCs (similar to those in anticipated in the Kyoto Protocol, as discussed later) may be allowed, but the U.K. government is going slowly to reduce uncertainty in the early stages of the market. By the end of 2002, more than 400 companies had opened accounts on the U.K. registry and about 1 million credits had been exchanged in several hundred transactions. Prices on this market are in the range of \$5 to \$10/ton-CO_2e.

4. *LOOKING AHEAD*

4.1 Multipollutant and Toxics

By 1999, European nations adopted an international treaty that deals with multiple pollutants and multiple impacts holistically, and many multipollutant emissions control proposals have been made in the United States. The reason for including multiple pollutants in one law is that this would greatly reduce the regulatory uncertainty for companies (especially electric power producers) that have been subject over the past 30 years to a long sequence of ever tighter control requirements on different pollutants. Such an approach is very expensive, especially for capital designed to last several decades. Capital investment in emission control retrofits and new facilities will be much less risky and probably less expensive for firms if they can count on flexible MBIs due to the flexibility in compliance strategy and risk management opportunities. Thus, industry may seek to replace uncoordinated and rigid CAC regulation with MBIs that will remain fixed for an extended period (a concept called the safe harbor). Typically, public and environmental groups press for more stringent regulations.

Major issues in considering multipollutant emission include whether or not to include toxic pollutants, such as mercury (Hg), and GHGs, such as CO_2, as well as what level of controls and predictability to include. In the United States, the term "3-P" applies to legislation that includes SO_2, NO_X, and Hg, and "4-P" applies to legislation that adds CO_2. Importantly, different environmental problem will almost always have different levels of consensus, both scientific and political, at any given time so the length of time a safe harbor may be different for various pollutants. Another is that there is no consensus exists about whether it is appropriate to allow emission trading to apply to toxics, since the issue of hotspots may be significant. Finally, achieving the emission reductions envisioned in 3-P and 4-P control will be serious technological challenges, especially for coal-fired electricity generation plants. The interaction of technologies to control all four emissions is highly complex. For instance, controlling SO_2 and NO_X tends to reduce power plant efficiency and thus increase Hg and CO_2 emissions, but some SO_2 and NO_X emission control technologies may also control Hg. The flexibility of MBIs will be even more important in multipollutant approaches since different facilities will likely have extremely varied least-cost compliance options.

4.2 Carbon Dioxide and the Kyoto Protocol

Probably the largest application of MBIs may be to control GHGs under the United Nations Framework Convention on Climate Change and the subsequent Kyoto Protocol (KP). The Framework Convention is an agreement of virtually all nations of the world to avoid "dangerous interference to the climate system," without committing signatories to any specific actions. Subsequently, negotiations produced the KP, which provides specific GHG emission caps for more than three dozen industrialized countries. These caps, combined with relatively lenient provisions for biological sequestration (e.g., credits for regrowing forests) and the decisions by the United States and Australia, create the background for international emissions trading in the near term. The original goal of the KP was to reduce emissions from industrialized countries to 5% below 1990 levels during 2008–2012. With the more recent provisions and participation, emission reductions of 1 to 2% are expected. Nonetheless, this is a substantial change from business as usual, which would have resulted in an increase of several percent.

The KP contains both C/T and ERC features. The C/T component addresses the trading of "assigned amounts" among industrialized countries, while the ERC components are mechanisms designed to allow trading between industrialized countries covered by an emissions cap and other countries that are not. These are called joint implementation (JI) and the clean development mechanism (CDM), and, like all ERC instruments, they are project specific. By the early 2000s, the rules for these provisions had still not been agreed upon, creating great regulatory uncertainty.

Despite uncertainties about future baselines, monitoring requirements, time frames, and other factors, many companies already began to engage in CO_2 emissions trading in 1996–2002. Most of these are VERs that have sold for \$2 to \$20/ton of CO_2. More than 1 billion such units have been traded. Although it is not clear that regulators will accept these credits in the future, there seem to be two values to these activities. First, companies that participate are learning how to measure their own CO_2 emissions and how to structure deals in this market. Second, although they are unofficial, these early CO_2 emission trades may be counted at least partially against future compliance obligations.

Although there is not yet agreement on emission trading rules for the KP as a whole, in October 2001 the European Union (EU) produced a directive to establish a framework for a mandatory EU-wide C/T system beginning in 2005 in order to implement the KP. Initially, only approximately 46% of estimated EU CO_2 emissions in 2010 were to be included. Between 4000 and 5000 installations from the following activities will be covered: electricity and heat generation (greater than 20 megawatts), petroleum refining, and the manufacture of iron, steel, cement clinker, ceramics, glass, and pulp and paper. Excluded sectors include oil and gas production, solid waste incineration, and chemical manufacturing, as well as transportation and residential energy consumption. The proposed system would allow for trading between companies, to be tracked by national governments that would establish registries for CO_2 emission allowances. A trade across a national boundary would require corresponding entries (one addition and one subtraction) in the two national registries involved. Banking from one year to the next is allowed, including from one commitment period to another. The EU CO_2 C/T system is also important because it sets a key precedent for international CO_2 emission trading. The EU would create the largest CO_2 emission market in the world, and other countries that wanted access to that market would have to follow the procedures set down by the EU. Expectations are that EU CO_2 credits would cost \$2.5 to \$10/ton before 2005 and \$5 to \$20/ton by 2010.

SEE ALSO THE FOLLOWING ARTICLES

Acid Deposition and Energy Use • Air Pollution from Energy Production and Use • Air Pollution, Health Effects of • Carbon Taxes and Climate Change • Fuel Economy Initiatives: International Comparisons • Greenhouse Gas Abatement: Controversies in Cost Assessment • Hazardous Waste from Fossil Fuels • Market-Based Instruments, Overview • Modeling Energy Markets and Climate Change Policy • Taxation of Energy

Further Reading

Ellerman, A. D., Joskow, P. L., and Harrison, D. (2003). "Emissions Trading in the U.S: Experience, Lessons, and Considerations for Greenhouse Gases." Pew Center on Global Climate Change: Arlington, VA.

Ellerman, A. D., *et al.* (2000). "Markets for Clean Air: the U.S. acid rain program." Cambridge University Press, Cambridge, UK.

Farrell, A. E. (2001). Multi-lateral emission trading: Lessons from inter-state NO_x control in the United States. *Energy Policy* **29**(13), 1061–1072.

Foster, V., and Hahn, R. W. (1995). Designing more efficient markets: Lessons from Los Angeles smog control. *J. Law and Economics* **38**(1), 19–48.

Grubb, M., Vrolijk, C., and Brack, D. (1999). "The Kyoto Protocol: A Guide and Assessment." Earthscan, London.

Israels, K., *et al.* (2002). "An Evaluation of the South Coast Air Quality Management District's Regional Clean Air Incentives Market—Lessons in Environmental Markets and Innovation." U.S. Environmental Protection Agency Region 9, San Francisco.

Raufer, R. (1996). Market-based pollution control regulation: implementing economic theory in the real world. *Environmental Policy and Law* **26**(4), 177–184.

Solomon, B., and Lee, R. (2000). Emissions trading systems and environmental justice: why are some population segments more exposed to pollutants than others? *Environment* **42**(8), 32–45.

Stavins, R. (2003). Experience with market-based environmental policy instruments. *In* "The Handbook of Environmental Economics" (K. Goren-Maler and J. Vincent, Eds.). North-Holland/Elsevier Science: Amsterdam.

Clean Coal Technology

MILDRED B. PERRY
U.S. Department of Energy, National Energy Technology Center
Pittsburgh, Pennsylvania, United States

1. Impact of Clean Coal Technology
2. Background of Development
3. International Clean Coal Program
4. United States Program
5. Benefits and Future of Clean Coal Technology

Glossary

acid rain When coal is burned, the sulfur and nitrogen in the coal are converted to sulfur dioxide (SO_2) and nitrogen oxides (NO_x). In the atmosphere, these substances are converted to sulfuric acid and nitric acid, which dissolve in raindrops to form acid rain.

carbon dioxide sequestration The storage of carbon dioxide (CO_2) in underground reservoirs or in deep-sea locations to prevent its release into the atmosphere. CO_2 can also be sequestered by enhancing the growth of forests and other vegetation.

clean coal technology Processes developed to reduce the emissions of pollutants from coal combustion; include technologies to remove pollutants and technology to improve efficiency, which also decreases emissions.

clean coal technology by-products Useful substances created during clean coal production; include sulfur and sulfuric acid, which have many applications; gypsum, used in the manufacture of wallboard; fly ash, used in the manufacture of cement; and aggregate, used in the construction industry.

flue gas desulfurization A clean coal technology consisting of a device, called a scrubber, fitted between a power plant's boiler and its stack, designed to remove the sulfur dioxide in the flue gas.

fluidized bed combustion (FBC) A process in which pulverized or granulated fuel and air are introduced into a fluidized bed of sand or some other material, where combustion takes place. Fluidized beds are classified as bubbling or circulating, depending on whether the bed material remains in place or is transported out of the vessel by the fluidizing gas, recovered, and returned to the bed. Plants may operate either at atmospheric pressure (AFBC) or at elevated pressure (PFBC).

fly ash Fine particles of ash that are entrained with the flue gas when coal is burned in a furnace.

greenhouse gases (GHGs) Gaseous compounds that act to trap heat in the atmosphere, thus tending to raise global temperature. The main GHG is carbon dioxide (CO_2). Other important GHGs are methane (CH_4) and nitrous oxide (N_2O).

gypsum A mineral having the chemical composition $CaSO_4 \cdot 2H_2O$, used in the manufacture of wallboard.

integrated gasification combined cycle (IGCC) A conversion process in which coal is first gasified to form a synthesis gas (a mixture of hydrogen and carbon monoxide), which is used as fuel for a combined-cycle power plant (gas turbine/steam turbine).

low-NO_x burner A burner that is designed to burn fuel in such a way so as to reduce the amount of NO_x produced.

NO_x The symbol for nitrogen oxides, consisting of a mixture of mainly NO and NO_2. NO_x is formed when coal is burned, partly from oxidation of the nitrogen in the coal and partly from reaction of nitrogen and oxygen in the combustion air.

overfire air The air introduced into a furnace above the combustion zone with the objective of completing combustion; used in conjunction with low-NO_x burners and reburning.

reburning A technique for reducing NO_x emissions consisting of introducing a hydrocarbon fuel into a reducing zone above the combustion zone to react with the NO_x and convert it to N_2.

repowering Rebuilding or replacing major components of a power plant as an alternative to building a new plant, often used to increase plant capacity.

scrubber A device designed to remove (scrub) sulfur oxides from flue gas, usually through contact with a limestone slurry to form calcium sulfate (gypsum).

selective catalytic reduction (SCR) A process installed downstream of a boiler, consisting of a catalytic reactor, through which the flue gas passes. A reducing agent, typically ammonia, is introduced into the flue gas upstream of the reactor and reacts with the NO_x in the flue gas to form nitrogen and water.

sulfur oxides When a fuel containing sulfur is burned, most of the sulfur in the fuel is converted to SO_2, with a small fraction being converted to sulfur trioxide (SO_3). In the atmosphere, sulfur oxides are converted to

Encyclopedia of Energy, Volume 1. Published by Elsevier Inc.

sulfuric acid, which is washed from the air and falls to the ground as acid rain, or converted to sulfate, which is a fine particulate.

synthesis gas (syngas) A mixture of hydrogen and carbon monoxide, used as an intermediate in the production of a number of chemicals. Syngas can also be used as a fuel.

trace element An element that appears in very small quantity (typically less than 10 parts per million) in coal. Because coal is a sedimentary rock, during its formation, it accumulated mineral matter, which is the source of most trace elements.

water–gas shift reaction The catalytic reaction of H_2O with CO to form H_2 and CO_2, used to increase the hydrogen content of synthesis gas.

Clean coal technologies can have a major impact on the world's economy by allowing the continued use of coal in an environmentally acceptable manner, thus reducing dependence on more expensive petroleum and natural gas, while avoiding economic problems caused by natural gas price fluctuations. Although natural gas is expected to fuel many new power plants and industrial facilities to meet growing demand over the next two decades, it is critical to maintain fuel flexibility, thus enabling the use of low-cost indigenous fuels, such as coal.

1. IMPACT OF CLEAN COAL TECHNOLOGY

Coal is one of the world's most abundant fossil fuels. It is also widely distributed, and many countries with little or no petroleum reserves have coal resources. Unfortunately, in addition to carbon and hydrogen, most coals contain sulfur, nitrogen, minerals, and chlorine, in various concentrations. Coals also contain trace elements, such as mercury, lead, arsenic, cadmium, selenium, thorium, and uranium, in very small concentrations, generally only a few parts per million. When coal is burned, the trace elements may be released to the environment in various, sometimes harmful, forms. In the past, coal combustion has contributed to acid rain, smog, forest die off, eutrophication of lakes, and other environmental problems. Because such impacts are no longer acceptable, new methods to reduce harmful emissions from plants utilizing coal are needed. Technology to accomplish this is generally referred to as clean coal technology (CCT) and includes technology designed either to remove or to reduce emissions, as well as technology designed to improve power plant efficiency. Efficiency improvements reduce pollution by decreasing the amount of coal that must be consumed to produce a given amount of power. Less coal burned means less pollution.

2. BACKGROUND OF DEVELOPMENT

Concern over environmental problems in the 20th century prompted a number of countries to address the problem of acid rain, resulting mainly from emissions of SO_2. In the United States in 1980, an independent organization, the National Acid Precipitation Assessment Program (NAPAP), was created, to coordinate acid rain research activities and periodically report to Congress on all aspects of the acid rain issue. In 1982, Canada proposed to reduce SO_2 emissions by 50%. In 1984, the United States, Poland, and the United Kingdom joined the "30% Club," a group of 18 European nations whose goal was to cut emissions of SO_2 by 30% or more by 1997. A similar agreement, the Helsinki Protocol, was later signed by 21 nations. In 1986, Canada and the United States assessed the international issues associated with transboundary air pollution and recommended actions for solving them, particularly with respect to acid rain. This entailed substantial research and financial commitments. Because of these efforts, SO_2 emissions were markedly reduced. For example, by the 1990s, emissions had dropped in most of Western Europe by 40% and in West Germany by 70%. Unfortunately, by that time, many forests and lakes worldwide had been damaged. Because of coal's importance to the economy, the United States government has been involved in the development and promotion of CCTs for over 20 years. In 1986, in response to regulatory requirements, the U.S. Department of Energy (DOE) initiated its original Clean Coal Technology Program, which consisted of five rounds of project funding. This program sponsored projects in the following areas: environmental control devices, advanced electric power generation, coal processing for clean fuels, and industrial applications.

3. INTERNATIONAL CLEAN COAL PROGRAM

Coal continues to be a vital fuel resource in numerous countries around the world. In order to maintain the viability of coal production, combustion, and conversion, a variety of international CCT research programs have been established to further enhance coal's utilization efficiency and environmental acceptability. Advanced technologies increase the amount of energy gained from each ton of coal consumed while reducing emissions and waste. This ongoing international research ensures that advances that focus on the needs of producers and end-users are continually being made.

Because of concern for coal's impact on the environment, many governments have been instrumental in funding, at least partially, the development of CCTs. Some of the more significant international programs and their goals are discussed in the following sections.

3.1 International Energy Agency Coal Research Clean Coal Center

The International Energy Agency (IEA) was established in 1974 within the framework of the Organization for Economic Cooperation and Development (OECD). A basic aim of the IEA is to foster cooperation among the IEA participating countries in order to increase energy security through diversification of energy supply and to promote energy conservation and cleaner and more efficient use of energy. This is achieved, in part, through a program of collaborative research and development of which IEA Coal Research, established in 1975, is by far the largest and the longest established project. IEA Coal Research is governed by representatives of its member countries. The IEA Clean Coal Center supports projects related to efficient coal supply and use, including technologies for reducing emissions from coal combustion.

3.2 United States

For more than 20 years, the U.S. DOE has had a very active program in the development and promotion of CCTs. This program is summarized in Section 4.

3.3 European Commission

The European Commission Clean Coal Technology Program encompasses projects performed in a series of research programs that promote the use of clean and efficient technologies in industrial plants using solid fuels, with an overall goal of limiting emissions, including CO_2, and encouraging the deployment of advanced clean solid fuel technologies, in order to achieve improved operations at affordable cost. A wide range of technologies is supported, including coal-cleaning plants for upgrading coal; handling, storage, and transport facilities; combustion and conversion plants; and waste disposal.

3.4 The United Kingdom

The U.K. Cleaner Coal Technology Program directly implements U.K. government policy for research and development (R&D) and technology transfer efforts related to environmentally friendly coal utilization technologies. This program was initiated through the Department of Trade and Industry (DTI) as a part of the government's Foresight Initiative. The DTI has established a 6-year collaborative program of activities that link R&D with technology transfer and export promotion activities. The overall aim of the program is to provide an incentive for U.K. industries to develop cleaner coal technologies and obtain an appropriate share of the growing world market for these technologies. The program objectives are to assist industry in meeting the technology targets for advanced power generation, to encourage fundamental coal science research in universities, to encourage the development of an internationally competitive clean coal industry and promote U.K. expertise and know-how in the main export markets, and to examine the potential for developing the U.K. coal bed methane resource and underground coal gasification technology.

3.5 The Netherlands

The Clean Fossil Fuels unit of the Energy Research Center of The Netherlands, the largest research center in The Netherlands concerned with energy, develops concepts and technology for cleaner and more efficient use of fossil fuels. The unit contributes to the transition toward sustainable energy based on three basic programs: climate-neutral energy carriers, emission reduction technologies (including CO_2 sequestration), and energy and environmental research and assessment. The climate-neutral energy supply program investigates decarbonization of fossil fuels, CO_2 capture technologies for power generation and hydrogen production systems, and novel hydrogen production technologies. The emissions reduction program is developing technologies, such as catalytic reduction of NO_x and N_2O, aimed at reducing the emissions of hazardous compounds to the environment. The objective of the energy and environmental research and assessments program is quantification of human influences on the environment as a result of fossil fuel use.

3.6 Australia

Several Cooperative Research Centers (CRCs) in Australia deal directly with the enhanced utilization of coal for power generation to decrease environmental impact and increase conversion efficiency. The CRC for Clean Power from Lignite has as its principal objective the development of technologies to reduce greenhouse gas emissions from lignite-fired power stations while enhancing Australia's international competitiveness by lowering energy costs. The technologies under development relate to both current power stations (pulverized coal-fired boilers) and to high-efficiency advanced cycles. Programs are in four areas: coal–water beneficiation, coal combustion and gasification, fluid bed process development, and ash and deposit formation.

The CRC for Coal in Sustainable Development, an unincorporated joint venture consisting of 18 participating organizations from the black coal-producing and -using industries and research organizations in Queensland, New South Wales, and Western Australia, has as one of its key goals to conduct research and development to improve the environmental performance of current technologies and reduce risks inherent in adopting emerging CCTs, such as pressurized fluidized bed combustion (PFBC) and the integrated gasification combined cycle (IGCC) process.

3.7 Japan

Since its establishment in June 1989, the Center for Coal Utilization, Japan (CCUJ) has been actively promoting the development of CCTs to utilize coal efficiently and cleanly, as well as the development of technical skills in the field of coal utilization. The activities of CCUJ have been extended to encompass sustainable economic growth and improvement of the global environment for the benefit of all industries that use coal, including power generation and the steel and cement industries. Efforts have been focused on development of more economical and efficient coal utilization systems with reduced emissions of CO_2, SO_2, NO_x, and particulates. CCUJ is promoting technical innovation in coal utilization and international cooperation in order to disseminate CCT information overseas, especially in Asian countries, in order to achieve a stable energy supply, sustainable economic growth, and an improved global environment. Some of the projects being sponsored are advanced PFBC, advanced coke making, coal gasification, and integrated coal gasification/fuel cell combined cycle (IGFC).

The New Energy and Industrial Technology Development Organization (NEDO) established the NEDO Clean Coal Technology Center in 1992 to carry out research and development of next-generation coal utilization technologies to reduce the amount of CO_2 and other emissions released to the atmosphere. CCT development areas include basic technology, next-generation coal utilization technology, low-emissions technologies, coal gasification to produce hydrogen for fuel cell applications, and coal liquefaction.

3.8 China

In April 1997, China's State Council approved the implementation of CCT in China. In August 1998, the State Electric Corporation established a long-term plan for controlling acid rain and SO_2 emissions during the next three Five-Year Plans, covering the period from 2001 to 2015. The Chinese CCT program covers key areas spanning coal processing, coal combustion, coal conversion, pollution control, and waste treatment. Power generation technologies being investigated are circulating fluidized bed combustion (CFBC), pressurized fluidized bed combustion, and integrated coal gasification combined cycle.

4. UNITED STATES PROGRAM

Coal accounts for over 94% of the proved fossil energy reserves in the United States and supplies the bulk of the low-cost, reliable electricity vital to the nation's economy and global competitiveness. Almost 90% of the coal mined in the United States is consumed in power plants to produce electricity. In 2000, over half of the U.S. electricity was produced with coal, and projections by the U.S. Energy Information Agency (EIA) predict that coal will continue to dominate electric power production well into the first quarter of the 21st century. However, if coal is to continue to be a major energy source for the United States, it must be used in an environmentally acceptable manner. The twin marketplace drivers of the need to continue burning coal and the simultaneous need to protect the environment led to the initiation in 1986 of the CCT Program in the United States. The CCT Program was established to demonstrate the commercial feasibility of CCTs to respond to a growing demand for a new generation of advanced coal-based units, characterized by enhanced operational, economic, and environmental performance. The CCT Program represents an investment of over \$5.2 billion in advanced coal-based technology. Industry and state governments provided an unprecedented 66% of the funding. Of 38 CCT projects selected in five separate solicitations, 29 projects, valued at $3.5 billion, address electric power generation applications (18 environmental control projects and 11 advanced power projects). The other nine projects involve coal processing for clean fuels and industrial applications.

Ten guiding principles have been instrumental in the success of the U.S. CCT Program:

- Strong and stable financial commitment for the life of a project.
- Multiple solicitations spread over a number of years.
- Demonstrations conducted at commercial scale in actual user environments.
- A technical agenda established by industry, not the government.
- Clearly defined roles of government and industry.
- A requirement for at least 50% cost-sharing.
- An allowance for cost growth, but with a ceiling and cost-sharing of the increase.
- Industry retention of real and intellectual property rights.
- A requirement for industry to commit to commercialization of the technology.
- A requirement to repay the government's cost-share from commercialization profits.

Typical clean coal technology projects sponsored in the initial U.S. CCT Program are described in the following sections. These examples, all completed between 1992 and 2003, illustrate the wide range of technologies developed to permit coal to be used in a more environmentally acceptable fashion. Information on the entire CCT Program is covered extensively in the Clean Coal Compendium on the Internet.

4.1 Environmental Control Devices—SO_2 Control

Pure Air's advanced flue gas desulfurization (AFGD) process (Fig. 1) was installed on a 528-MWe (MWe = megawatt electrical, as distinct from thermal) boiler at Northern Indiana Public Service Company's Bailly Generating Station. This process removes SO_2 from flue gas by contact with a limestone slurry in a scrubber vessel. Although installed on a large power plant, the scrubber occupied a relatively small plot area, because it incorporated three functions in one vessel: prequenching of the hot flue gas, absorbing SO_2 in a limestone slurry, and oxidizing the resultant sludge to gypsum. Because of the unit's high

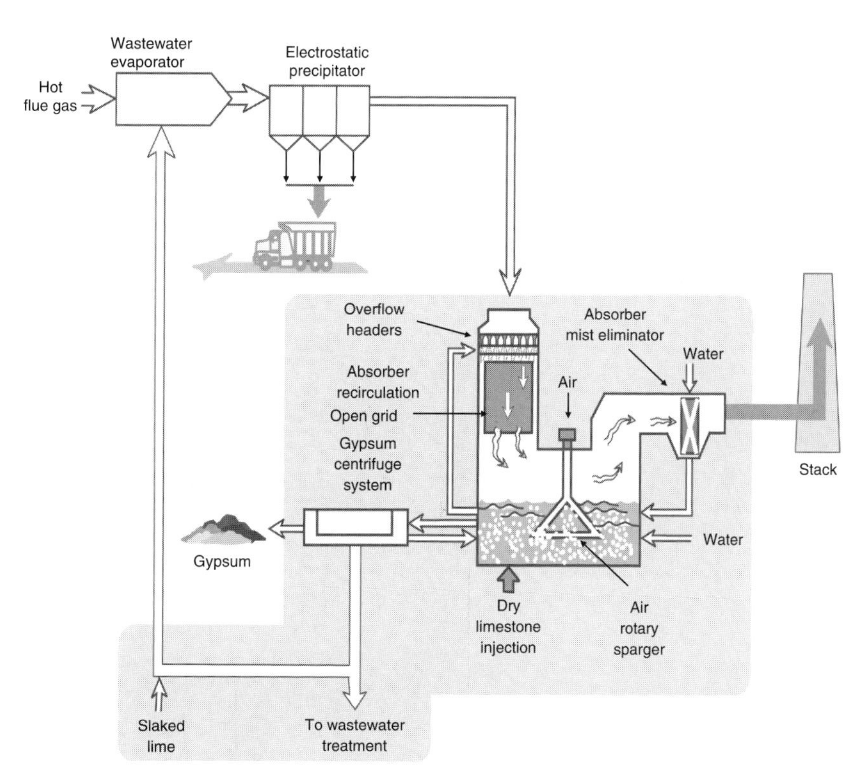

FIGURE 1 Pure Air's advanced flue gas desulfurization process.

availability of over 99%, no spare scrubber vessel was required. The AFGD unit successfully removed over 95% of the SO_2 in the flue gas and produced a wallboard-grade gypsum that was sold to a commercial producer, who used the gypsum for the manufacture of wallboard. Thus, this technology not only prevented sulfur from entering the atmosphere, but also generated a valuable by-product at the same time.

In another project, Chiyoda Corporation's Chiyoda Thoroughbred-121 (CT-121) AFGD process (Fig. 2), which uses a unique jet bubbling reactor (JBR), was installed at Georgia Power Company's Plant Yates. In this scrubber, the flue gas is injected below the surface of the limestone slurry in the JBR. Air is also injected into the slurry to oxidize the calcium sulfite initially formed into calcium sulfate. Over 90% SO_2 removal was achieved, and limestone utilization was over 97%. To prevent buildup of gypsum in the reactor, a slipstream is removed from the scrubber and sent to a collection pond. The scrubber also proved to be an effective particulate removal device.

4.2 Environmental Control Devices—NO_x Control

4.2.1 Selective Catalytic Reduction

A project at Gulf Power Company's Plant Christ consisted of testing a series of selective catalytic reduction (SCR) catalysts (Fig. 3). The test facility consisted of nine small reactors, each of which was loaded with a different SCR catalyst and fed a slipstream of flue gas from one of the plant's boilers. SCR catalyst is typically supported on the surface of ceramic honeycombs or plates, which are loaded into the reactor. Ammonia was injected into the flue gas before it passed over the catalyst. Ammonia injection needs to be carefully controlled to prevent the escape of unreacted ammonia, known as ammonia slip. The ammonia reacted with NO_x to form nitrogen gas and water. Although SCR had been used in Germany and Japan, it had not been proved to be applicable when burning higher sulfur U.S. coals. This study showed that SCR is capable of removing up to 90% of the NO_x in flue gas and that the process can be effective with U.S. coals.

4.2.2 Low-NO_x Burners

Several projects have demonstrated the effectiveness of low-NO_x burners for reducing NO_x emissions. These include demonstrations on wall-fired, tangentially fired, and cell burner-fired furnaces. Low-NO_x burners reduce NO_x by staged combustion. Figure 4 is an example of low-NO_x burners installed on a wall-fired boiler. Such burners are designed to mix the fuel initially with only part of the combustion air. This reduces NO_x formation both because the combustion temperature is reduced and because the concentration of oxygen is lower. The rest of the air, referred to as overfire air, is introduced at a higher

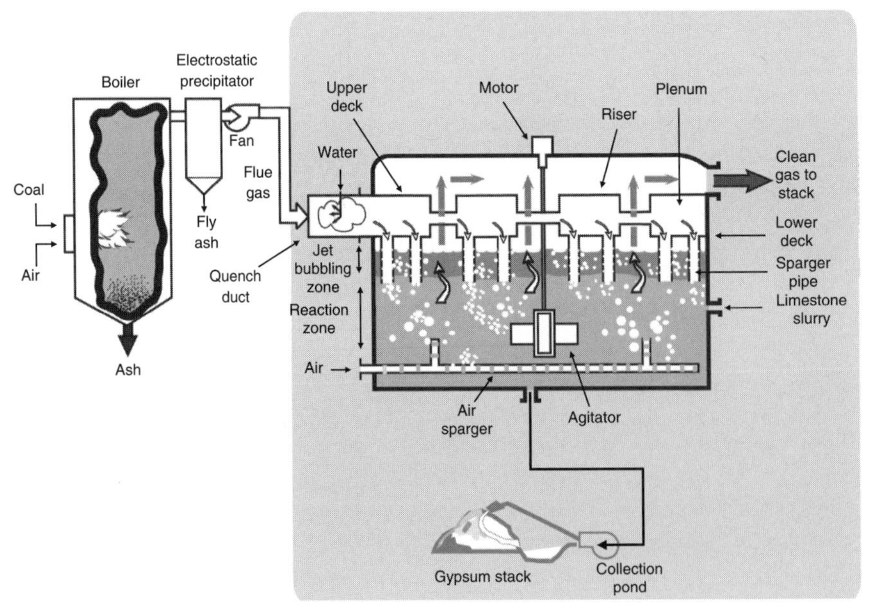

FIGURE 2 Chiyoda Corporation's Chiyoda Thoroughbred-121 advanced flue gas desulfurization process using the jet bubbling reactor.

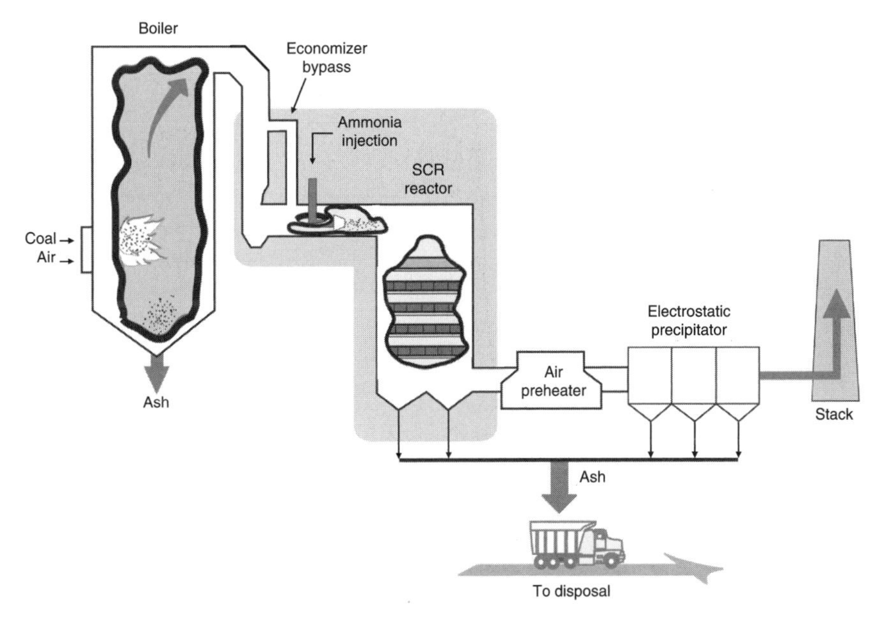

FIGURE 3 Selective catalytic reduction (SCR) installation at a power plant.

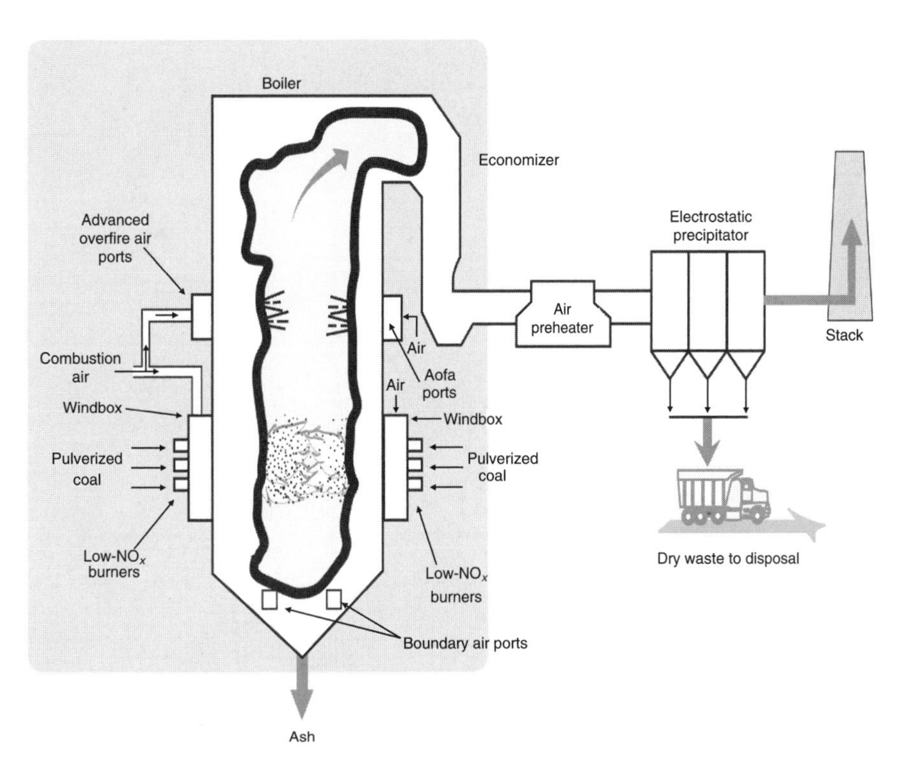

FIGURE 4 Low-NO_x burners with advanced overfire air (AOFA) installation on a wall-fired boiler.

elevation in the furnace to complete combustion. Low-NO_x burners typically reduce NO_x by 50–60%, which is generally not sufficient to meet current NO_x emissions regulations. Therefore, low-NO_x burners may need to be supplemented with SCR or another auxiliary process. Low-NO_x burners can result in increased unburned carbon in the ash, which may make the ash unsalable for cement production.

4.2.3 Reburning

Another NO_x reduction technology demonstrated in CCT projects is reburning (Fig. 5). In reburning, a hydrocarbon fuel, such as natural gas or finely pulverized coal, is injected into the furnace above the burners to create a reduction zone, where hydrocarbon fragments react with NO_x to form nitrogen gas. Overfire air is introduced above the reburning zone to complete combustion. Reburning, which typically achieves 50–60% NO_x reduction, is not as effective as SCR.

4.3 Environmental Control Devices—Combined SO_2 and NO_x Removal

In the SNOX (a trademarked name) process (Fig. 6), demonstrated at Ohio Edison's Niles Station in Niles, Ohio, flue gas first passes though a high-efficiency fabric filter baghouse to minimize fly ash fouling of downstream equipment. Ammonia is then added to

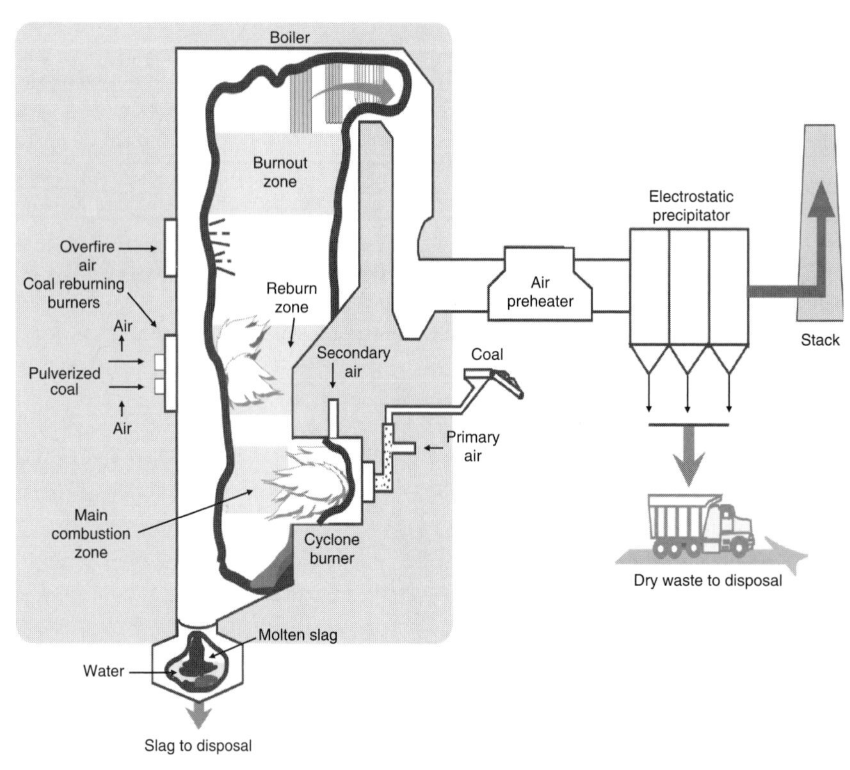

FIGURE 5 Coal reburning installation on a boiler with a cyclone burner.

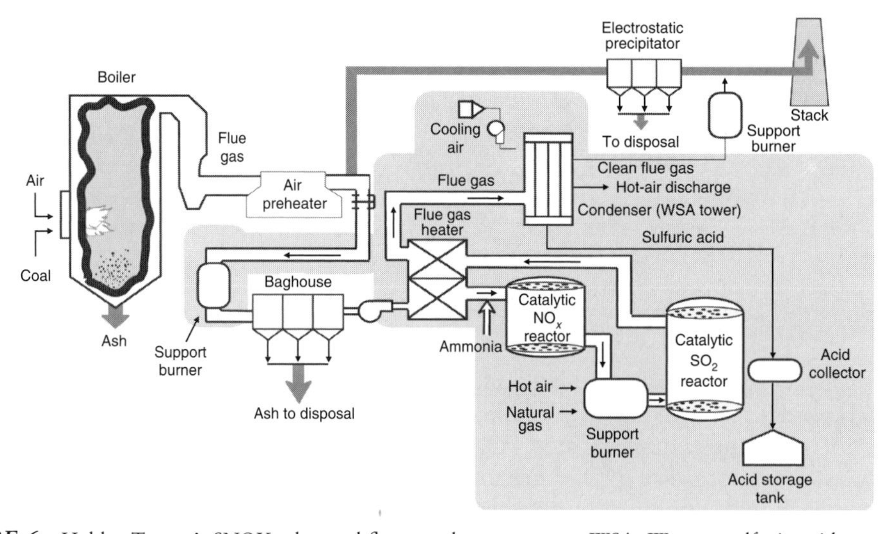

FIGURE 6 Haldor Topsoe's SNOX advanced flue gas cleanup system. WSA, Wet gas sulfuric acid.

the flue gas, which passes to a catalytic reactor where the ammonia reacts with NO_x to form nitrogen gas and water. The next step is a catalytic reactor, which oxidizes SO_2 to SO_3. The presence of this reactor oxidizes any unreacted ammonia to N_2 and water, thus preventing the formation of ammonium sulfate, which can form deposits that foul downstream equipment. The oxidation catalyst also virtually eliminates CO and hydrocarbon emissions. The gas then flows to a condenser for the recovery of sulfuric acid and then to the stack. Sulfur removal is about 95% and NO_x removal averages 94%. The sulfuric acid exceeds federal specifications for Class I acid. The fabric filter removes over 99% of the particulates.

Another demonstration combining SO_2 and NO_x removal was the SOx–NOx-Rox Box (SNRB) project at Ohio Edison's R.E. Burger Plant. The SNRB process combines the removal of SO_2, NO_x, and particulates in one unit, a high-temperature fabric filter baghouse (Fig. 7). SO_2 removal is effected by injecting a calcium- or sodium-based sorbent into the flue gas upstream of the baghouse. Ammonia is also injected into the flue gas to react with NO_x and to form N_2 and water over the SCR catalyst, which is installed inside the bags in the baghouse. Particulate removal occurs on the high-temperature fiber filter bags. Sulfur removal with calcium-based sorbents was in the 80–90% range at a Ca/S ratio of 2.0. NO_x removal of 90% was achieved.

4.4 Advanced Electric Power Generation

4.4.1 Integrated Gasification Combined Cycle

A project at PSI Energy's Wabash River Generating Station demonstrated the use of IGCC for the production of clean power from coal (Fig. 8). In IGCC, the coal is first partially combusted with air or oxygen to form a low to medium heating-value fuel gas, whose main combustible constituents are carbon monoxide and hydrogen, with some methane and heavier hydrocarbons. After being cleaned, the fuel gas is burned in a combustion turbine. The hot turbine flue gas is sent to a heat recovery steam generator (HRSG), which produces steam for a conventional steam turbine. A major advantage of IGCC is that, in the gasification process, the sulfur, nitrogen, and chlorine in the coal are converted to hydrogen sulfide, ammonia, and hydrogen chloride, respectively. It is much easier to remove these materials from the fuel gas than it is to remove SO_2 and NO_x from the flue gas of a conventional coal-fired boiler.

In the Wabash project, the gasifier used was Global Energy's E-Gas Technology. This is a two-stage gasifier, with a horizontal stage followed by a vertical stage. The ash in the coal is removed from the bottom of the gasifier as molten slag, which is quenched before being sent to disposal or sales. This has proved to be a very clean plant, with sulfur removal above 99% and NO_x levels at 0.15 lb/million Btu. Particulate levels were below detectable

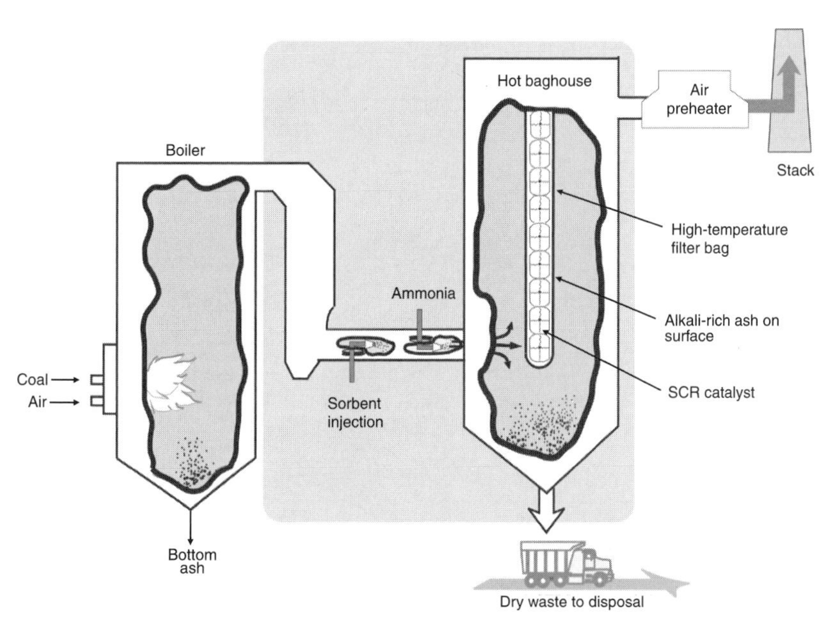

FIGURE 7 The Babcock & Wilcox Company's SOx–NOx-Rox Box process. SCR, Selective catalytic reduction.

limits. The recovered hydrogen sulfide was converted to liquid sulfur for sale.

Another IGCC demonstration was Tampa Electric Company's Polk Power Station Unit No. 1. This project used a Texaco gasifier with a radiant fuel gas cooler installed in the reactor vessel below the gasifier (Fig. 9). For this project, the recovered hydrogen sulfide was converted to sulfuric acid for sale. This plant also proved to be very clean, with SO_2, NO_x, and particulate emissions well below regulatory limits for the plant site.

4.4.2 Fluidized Bed Combustion

Foster Wheeler's atmospheric circulating fluidized bed (ACFB) combustion system (Fig. 10) was installed at the Nucla Station in Nucla, Colorado. In the combustion chamber, a stream of air fluidizes and entrains a bed of coal, coal ash, and sorbent (limestone). Relatively low combustion temperatures limit NO_x formation. Calcium in the sorbent combines with SO_2 to form calcium sulfite and sulfate. Solids exiting the combustion chamber with the flue gas are removed in a hot cyclone and recycled to the

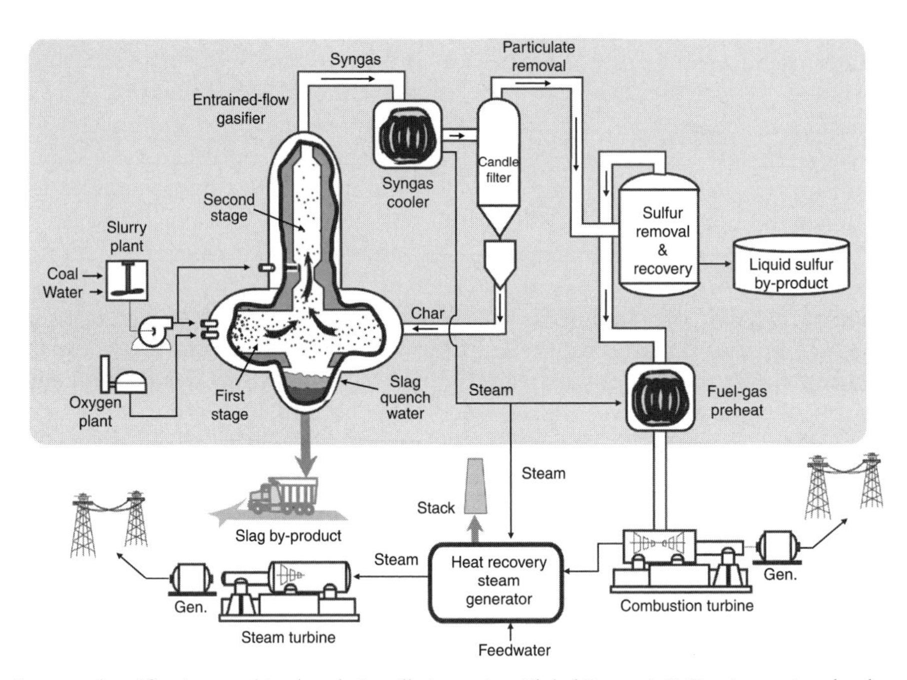

FIGURE 8 Integrated gasification combined cycle installation using Global Energy's E-Gas (syngas) technology.

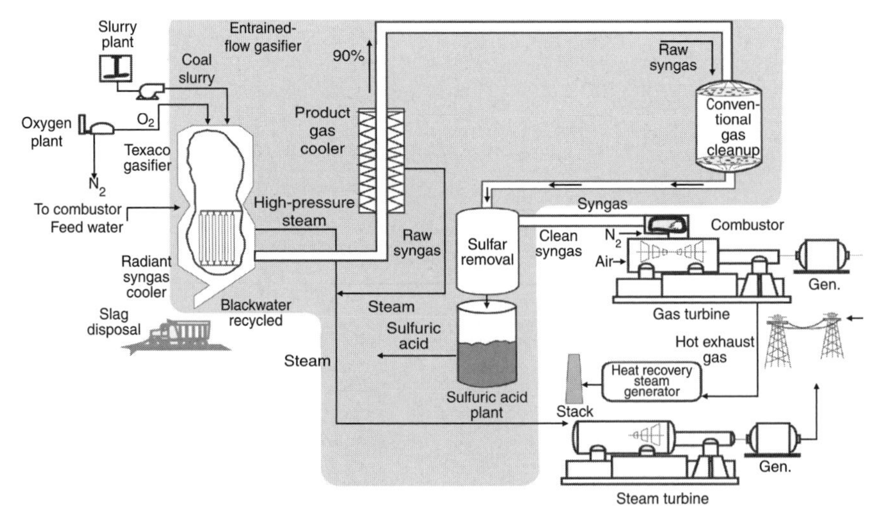

FIGURE 9 Integrated gasification combined cycle installation based on Texaco's pressurized, oxygen-blown, entrained-flow gasifier technology.

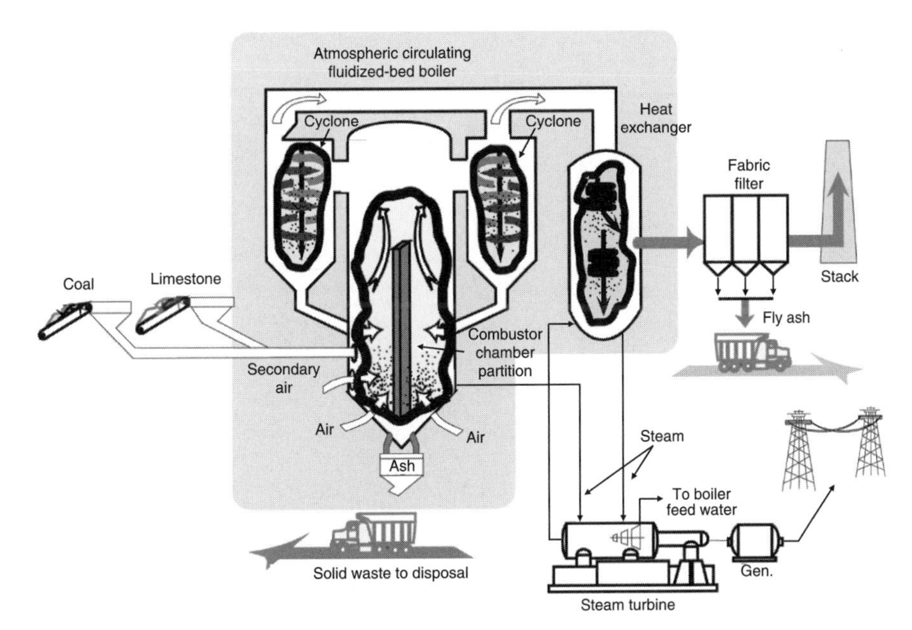

FIGURE 10 Foster Wheeler's atmospheric circulating fluidized bed combustion system.

combustion chamber. Continuous circulation of coal and sorbent improves mixing and extends the contact time of solids and gases, thus promoting high carbon utilization and high sulfur capture efficiency. Heat in the flue gas exiting the hot cyclone is recovered in the economizer. Flue gas passes through a baghouse, where particulate matter is removed. This project was the first repowering of a U.S. power plant with ACFB technology and demonstrated this technology's ability to burn a wide range of coals cleanly and efficiently.

4.5 Coal Processing for Clean Fuels

4.5.1 Methanol Production

A project at Eastman Chemical Company's Kingsport, Tennessee, facility demonstrated Air Products and Chemicals' liquid-phase methanol (LPMEOH) process (Fig. 11). This process converts synthesis gas (a mixture of CO and H_2) into methanol according to the reaction $CO + 2H_2 \rightarrow CH_3OH$. The process differs from conventional gas-phase processes in that instead of pelleted catalyst in beds or tubes, the LPMEOH process uses a finely divided catalyst suspended in a slurry oil. Use of a slurry reactor results in a uniform temperature throughout the reactor and improves heat removal by the internal heat exchanger. Thus, much higher per-pass conversions can be achieved. Furthermore, because the water–gas shift capabilities of the catalyst promote the conversion of CO to H_2, low H_2/CO ratio synthesis gas, typical of that produced by a coal gasifier, can be fed to the process. After some initial problems with catalyst poisons in the coal-derived synthesis gas were overcome, the LPMEOH process worked very well. All the methanol produced was used by Eastman Chemicals as an intermediate in the production of other chemicals.

One possibility for use of the LPMEOH process would be in conjunction with an IGCC system. In this concept, part of the synthesis gas from the gasifier in the IGCC unit would be diverted to the LPMEOH process for conversion to methanol. The amount diverted would depend on the load requirement of the power plant. This would permit the gasifier, the most expensive section of the IGCC unit, to operate at full load regardless of the power demand. The methanol produced could be sold or used as fuel in a combustion turbine during high electric power demand periods. During the demonstration project, tests showed that the LPMEOH process can rapidly ramp feed rate up and down, thus indicating its compatibility with the IGCC system concept.

4.6 Industrial Applications

4.6.1 Granulated Coal Injection into a Blast Furnace

One of the most expensive and pollution-prone aspects of blast furnace operation is the production of coke. To reduce coke usage, blast furnaces may inject natural gas or pulverized coal directly into the blast furnace. In a demonstration project at Bethlehem Steel's Burns Harbor Plant in Indiana, it was

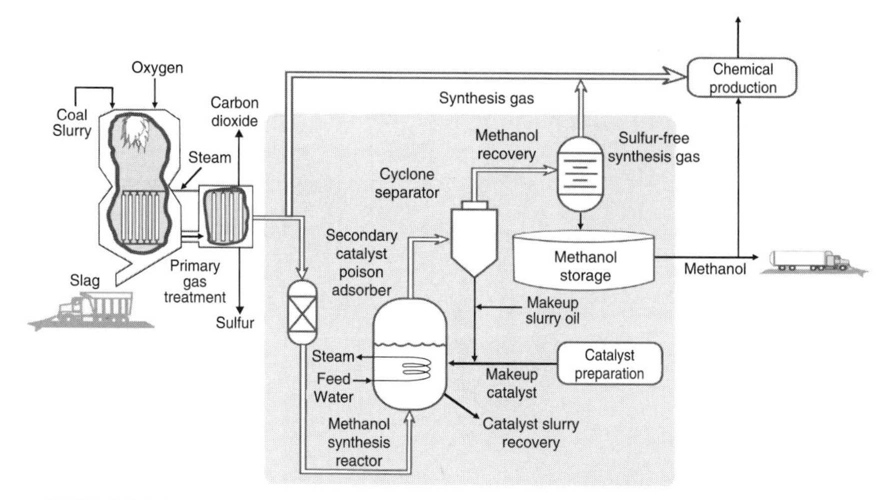

FIGURE 11 Air Products and Chemicals' liquid-phase methanol process.

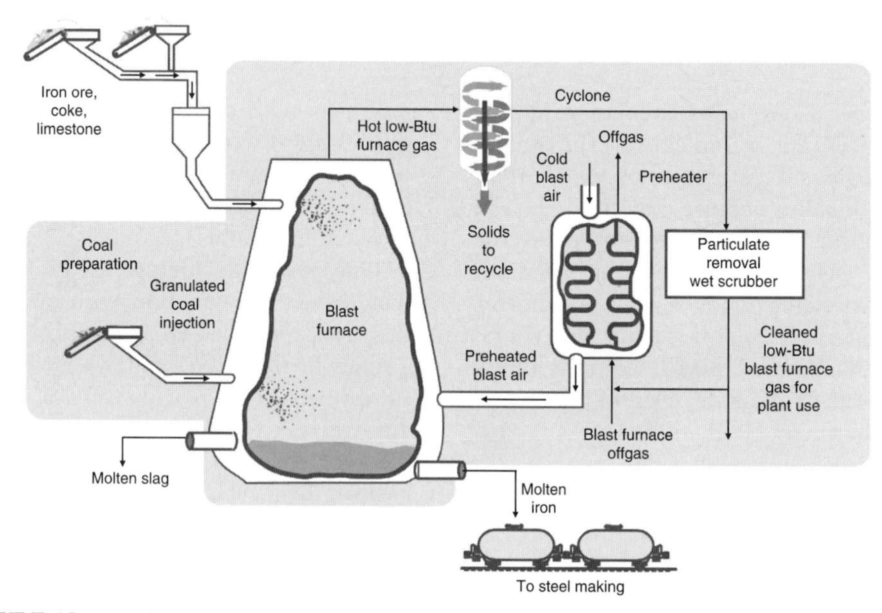

FIGURE 12 British Steel and Clyde Pneumatic blast furnace granular coal injection process.

shown that granular coal injection (Fig. 12) worked as well as pulverized coal injection and required about 60% less grinding energy. The coal replaced coke on almost a one-for-one basis. All the coke cannot be replaced because of the need to preserve porosity in the blast furnace, but enough coke can be displaced to improve economics significantly. Sulfur and ash in the injected coal end up in the slag removed from the furnace.

4.6.2 Cement Kiln Flue Gas Recovery Scrubber

A cement kiln flue gas recovery scrubber project was located at Dragon Products Company's coal-fired cement kiln in Thomaston, Maine. One of the problems at this plant was that about 10% of the kiln product was cement kiln dust (CKD), an alkali-rich waste that, because of its potassium content, cannot be blended with the cement product. Another problem was SO_2 in the kiln exhaust gas. The innovative cement kiln flue gas recovery scrubber (Fig. 13) demonstrated in this project solved both problems simultaneously by slurrying the CKD with water and using the slurry to scrub the kiln exhaust gas. The potassium in the CKD reacts with SO_2 in the kiln gas in the reaction tank to form potassium sulfate, which is water soluble. After liquid/solid separation, the low-potassium solid can be blended with the cement product, and potassium sulfate can

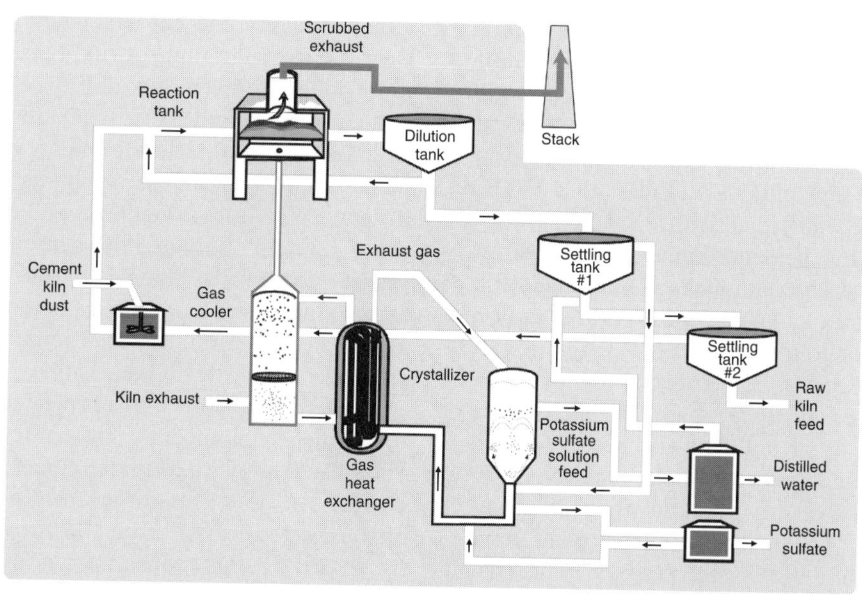

FIGURE 13 Passamaquoddy Technology recovery scrubber for treating cement kiln flue gas.

be recovered from the solution for sale as fertilizer. This project reduces SO_2 emissions by about 90%, while simultaneously reducing raw material requirements by about 10%, significantly improving economics.

4.7 Follow-On CCT Initiatives

The Clean Air Act of 1970 and subsequent amendments, especially the Acid Rain Program provisions of the 1990 amendments, and implementation of the original CCT Program have brought about major reductions in emissions of acid gases (SO_2 and NO_x) and particulates from coal-fired plants.

However, coal-fired power plants are increasingly being required to further cut emissions. Renewed concerns about fine particulates and their precursors (nitrogen and sulfur oxides), trace elements (especially mercury), and ozone (and its NO_x precursor) have created new requirements for cleaner plants.

Building on the success of the original CCT Program, the U.S. government has promoted two further solicitations, the Power Plant Improvement Initiative (PPII) in 2000 and the Clean Coal Power Initiative (CCPI) in 2002. PPII was established to promote the commercial-scale demonstration of technologies that would assure the reliability of the energy supply from existing and new electric-generating facilities. PPII is geared toward demonstrations of advanced near-term technologies to increase efficiency, lower emissions, and improve the economics and overall performance of coal-fired power plants. Most PPII projects focus on technologies enabling coal-fired power plants to meet increasingly stringent environmental regulations at the lowest possible cost. Proposed technologies must be mature enough to be commercialized within a few years, and the cost-shared demonstrations must be large enough to show that the technology is commercially viable.

The CCPI, a 10-year program, was initiated to further develop, improve, and expand the use of CCTs to increase national energy security and the reliability of the electric power supply and reduce costs while improving the environment. Candidate technologies are demonstrated at full scale to ensure proof of operation prior to widespread commercialization.

5. *BENEFITS AND FUTURE OF CLEAN COAL TECHNOLOGY*

The clean coal technology development effort has provided, and will continue to provide, significant economic, environmental, and health benefits. Economic benefits arise in a number of areas. The CCT Program has been instrumental in the commercialization of technologies such as AFBC and IGCC. The program has also demonstrated a variety of new options for the control of sulfur oxides, nitrogen oxides, and particulate emissions from electric power plants operating on coal. Considerable economic activity is generated through the sale, design, construction, and operation of these new technologies. Furthermore, these new power generation and pollution control options will permit coal to continue to be used as a fuel while minimizing environmental impacts. Because coal is the cheapest fossil fuel, its continued use will save billions of dollars that can be invested in other economic activities.

CCT benefits to the environment are obvious. Decreasing SO_2 and NO_x emissions reduces acid rain, which in turn reduces acidification and eutrophication of lakes and damage to forests and other vegetation. It also reduces damage to structures made of steel, limestone, concrete, and other materials. CCTs have reduced the amount of pollutants emitted by fossil fuel-fired power plants. For example, between 1970 and 2000, emissions of sulfur and nitrogen pollutants from the average U.S. coal-fired power plant have declined by 70 and 45%, respectively. This has enabled coal use to more than double while allowing the United States to meet its clean air objectives.

Another major benefit is in the area of human health. Decreases in smog precursors are very beneficial to the health of human beings, particularly people with respiratory problems, such as asthma. Decreases in emissions of mercury and other air toxics are expected to result in fewer cancers and other diseases. These benefits should decrease medical costs by tens, if not hundreds, of billions of dollars over the next couple of decades.

Another benefit of CCT efforts has been international cooperation. Because pollution does not respect national borders, all nations must be concerned, and this has fostered international efforts. Technologies developed in the United States and elsewhere are available internationally, and global cooperation on international energy issues is ongoing. However, new environmental constraints will require the development of new clean coal technologies. At the time the United States initiated its first CCT program, there were no regulations on mercury emissions or fine particulates (those smaller than 2.5 μm in diameter). NO_x and SO_2 are precursors to the atmospheric formation of fine particulates and, therefore, will likely undergo more stringent regulation. Further regulation of these pollutants will lead to the development of new clean coal technologies.

Although not currently regulated in the United States , there is growing concern about the buildup of carbon dioxide in the atmosphere. Decreasing CO_2 emissions from coal-burning plants will require the development of improved CO_2 capture technologies, as well as the development of CO_2 sequestration techniques. Of perhaps more importance in the short term is to improve the efficiency of power plants. Decreasing the amount of fuel that has to be burned to produce a given amount of electricity directly reduces the amount of CO_2 produced. An area that will see significant development will be advanced plant control systems, such as Electric Power Research Institute's Generic NO_x Control Intelligence System (GNOCIS), demonstrated at Gulf Power Company's Plant Hammond in the United States. Such systems allow a power plant to operate at or near optimum conditions, thus improving efficiency and lowering emissions.

In the United States, it is projected that reliance on fossil fuels will increase from the present level of 85% to 90% by 2020 under current price and usage trends. The use of fossil fuels to produce electricity is also expected to rise from the current 67% to 78% by 2020. Energy consumption in the developing world (Asia, Africa, the Middle East, and Central and South America) is expected to more than double by 2020, with the highest growth rates expected in Asia and Central and South America. The next generation of coal-fired power plants is emerging to meet this growth. These systems offer the potential to be competitive with other power systems from a cost and performance standpoint, while having improved environmental performance.

The Clean Coal Power Initiative, the newest CCT program in the United States, is aimed at technologies that will focus on development of multipollutant (SO_2, NO_x, and Hg) control systems. In the past, environmental regulations have been issued incrementally—generally, one pollutant was regulated at a time. This approach has proved to be costly to the electric power industry. The new multipollutant control approach should achieve equivalent reductions in pollutants at much lower cost. The CCPI will also focus on the development of high-efficiency electric power generation, such as gasification, advanced combustion, fuel cells and turbines, retrofitting/repowering of existing plants, and development of new plant technologies.

Because biomass combustion provides a means for recycling carbon, composite fuels consisting of coal and waste biomass can offset the release of greenhouse gases during combustion. Throughout U.S. coal-producing regions are large amounts of coal residing in waste ponds. Separation processes to recover this coal economically could make it available for power generation and, concurrently, eliminate existing and potential environmental problems associated with these waste sites. This type of waste coal is particularly applicable to fluidized bed combustion.

The examples discussed here illustrate the wide range of technologies included under the clean coal technology umbrella. As regulatory and economic drivers change, an even wider range of activities will need to be included in the future.

SEE ALSO THE FOLLOWING ARTICLES

Acid Deposition and Energy Use • Air Pollution, Health Effects of • Air Pollution from Energy Production and Use • Carbon Sequestration, Terrestrial • Clean Air Markets • Coal, Chemical and Physical Properties • Greenhouse Gas Emissions from Energy Systems, Comparison and Overview • Hazardous Waste from Fossil Fuels

Further Reading

Center for Coal Utilization, Japan (CCUJ). Program and project information is accessible on the Internet (http://www.ccuj.or.jp/index-e.htm). (Accessed May 20, 2003.)

Clean Coal Engineering & Research Center of Coal Industry (CCERC), China (http://www.cct.org.cn/eng/index.htm).

Cooperative Research Centre for Coal in Sustainable Development, Clean Coal Technology in Australia: Industry and Policy Initiatives (D .J. Harris and D. G. Roberts) (http://www.nedo.go.jp/enekan/ccd2.pdf). (Accessed September 9, 2003.)

Energy Information Administration. (2003). Annual Energy Outlook 2003 with Projections to 2025, January 2003. Energy Information Administration, Washington, D.C.

Energy, Security, and Environment in Northeast Asia (ESENA) Project. Program and project information is accessible on the Internet (http://www.nautilus.org/esena/). (Accessed January 7, 2000.)

European Commission Clean Coal Technology Programme. Program and project information is accessible on the Internet (http://europa.eu.int/comm/energy/en/pfs_carnot_en.html and http://www.euro-cleancoal.net/). (Accessed July, 2000.)

IEA Coal Research Clean Coal Centre. Program and project information is accessible on the Internet (http://www.iea-coal.org.uk/). (Accessed November 26, 2003.)

New Energy and Industrial Technology Development Organization (NEDO). Program and project information related to clean coal technology is accessible on the Internet (http://www.nedo.go.jp/enekan/eng/2-1.html). (Accessed September 9, 2003.)

UK Cleaner Coal Technology Program. Program and project information is accessible on the Internet (www.dti.gov.uk/cct). (Accessed March 14, 2003.)

U.S. Department of Energy. (2002). Clean Coal Technology Demonstration Program, Program Update 2001, July 2002. U.S. Department of Energy, Washington, D.C.

U.S. Department of Energy Clean Coal Power Initiative. Program and project information is accessible on the Internet (http://www.fossil.energy.gov/programs/powersystems/cleancoal/). (Accessed December 1, 2003.)

U.S. Dept. of Energy, Clean Coal Technology Compendium. Program and project information is available on the Internet (http://www.lanl.gov/projects/cctc). (Accessed October 31, 2003.)

U.S. Department of Energy Clean Coal Technology Program. Program and project information is accessible on the Internet (http://www.fossil.energy.gov/programs/powersystems/cleancoal/86-93program.html). (Accessed November 6, 2003.)

U.S. Department of Energy Power Plant Improvement Initiative. Program and project information is accessible on the Internet (http://www.lanl.gov/projects/ppii/). (Accessed October 20, 2003.)

Climate Change and Energy, Overview

MARTIN I. HOFFERT
New York University
New York, New York, United States

KEN CALDEIRA
U.S. Department of Energy, Lawrence Livermore
National Laboratory
Livermore, California, United States

1. Introduction: Global Warming as an Energy Problem
2. Efficiency
3. Fossil Fuels and CO_2 Emission Reduction
4. Renewables
5. Energy Carriers and Off-Planet Energy Production
6. Fission and Fusion
7. Geoengineering
8. Conclusion

Glossary

carbon intensity (i) Carbon intensity of economic productivity. Amount of carbon dioxide (CO_2) produced or emitted per unit economic output, as might be expressed, for example, in moles C per dollar. (ii) Carbon intensity of energy. Amount of CO_2 produced or emitted per unit energy produced or consumed, as might be expressed in moles C per joule. Care must be taken to distinguish between carbon intensity of economic productivity and carbon intensity of energy. In the case of carbon intensity of energy, care must be taken to clarify whether the reference is to primary or secondary energy. If mass units are used to measure CO_2, care must be taken to clarify whether the reference is to the mass of CO_2 or the mass of carbon in CO_2.

carbon sequestration The storage of carbon in a location or form that is isolated from the atmosphere. CO_2 can be separated from the combustion products of fossil fuels or biomass and stored in geologic formations or the ocean. CO_2 can be sequestered by removal from the atmosphere through photosynthesis or engineered processes. Carbon can be stored as CO_2, elemental carbon, carbonate minerals, or dissolved inorganic carbon. Some authors reserve the term sequestration to indicate removal from the atmosphere and use the word storage to refer to removal from industrial point sources, such as fossil-fuel-fired power plants.

climate sensitivity The change in global mean temperature that results when the climate system, or climate model, attains a new steady state in response to a change in radiative forcing, typically expressed in units of degree Celsius warming per watts per meter squared change in radiative forcing or degree Celsius warming per doubling of atmospheric CO_2 concentration. The new steady state is typically assumed to include equilibration of rapidly responding components of the climate system, such as clouds and snow cover, but not more slowly responding components, such as ice sheets and ecosystems.

economics A branch of behavioral biology dealing with the allocation of scarce resources by *Homo sapiens*, one of the many organisms found on Earth.

energy conservation A change in demand for services leading to a reduction in energy demand. The change in demand services may increase, decrease, or have no effect on the value of services provided. Typically, however, energy conservation attempts to contribute to a decline in energy intensity.

energy efficiency Useful energy output divided by energy input. The meaning of useful energy can vary with context. Energy input may mean primary or secondary energy sources, depending on context.

energy intensity Consumption of primary energy relative to total economic output (gross domestic, national, or world product), as may be expressed, for example, in joules per dollar. Changes in energy efficiency and structural and behavioral factors all contribute to changing energy intensity.

geoengineering (i) Approaches to climate stabilization involving alteration of the planetary energy balance, for example, through the scattering or reflection to space of sunlight incident on the earth. (ii) Any attempt to promote climate stabilization involving planetary-scale interference in natural systems (e.g., large-scale storage of CO_2 in the ocean interior).

primary energy Energy embodied in natural resources (e.g., coal, crude oil, sunlight, and uranium) that has not undergone any anthropogenic conversion or transformation.

radiative forcing The perturbation to the energy balance of the earth–atmosphere system following, for example, a change in the concentration of CO_2 or a change in the output of the sun. The climate system responds to the radiative forcing so as to reestablish the energy balance. A positive radiative forcing tends to warm the surface and a negative radiative forcing tends to cool the surface. It is typically measured in watts per meter squared.

secondary energy Energy converted from natural resources (e.g., coal, crude oil, sunlight, and uranium) into a form that is more easily usable for economic purposes (e.g., electricity and hydrogen). Typically, there are energy losses in the conversion from primary to secondary energy forms.

Stabilizing climate is an energy problem. Setting ourselves on a course toward climate stabilization will require a buildup within the following decades of new primary energy sources that do not emit carbon dioxide (CO_2) to the atmosphere in addition to efforts to reduce end-use energy demand. Mid-century CO_2 emissions-free primary power requirements could be several times what we now derive from fossil fuels ($\sim 10^{13}$ W), even with improvements in energy efficiency. In this article, possible future energy sources are evaluated for both their capability to supply the massive amounts of carbon emissions-free energy required and their potential for large-scale commercialization. Possible candidates for primary energy sources include terrestrial solar, wind, solar power satellites, biomass, nuclear fission, nuclear fusion, fission–fusion hybrids, and fossil fuels from which carbon has been sequestered. Non-primary-power technologies that could contribute to climate stabilization include conservation, efficiency improvements, hydrogen production, storage and transport, superconducting global electric grids, and geoengineering. For all, we find severe deficiencies that currently preclude cost-effective scale-up adequate to stabilize climate. A broad range of intensive research and development is urgently needed to develop technological options that can allow both climate stabilization and economic development.

1. INTRODUCTION: GLOBAL WARMING AS AN ENERGY PROBLEM

1.1 The Carbon Dioxide Challenge

More than 100 years ago, Svante Arrhenius suggested that CO_2 emissions from fossil fuel burning could raise the infrared opacity of the atmosphere enough to warm Earth's surface. During the 20th century, the human population quadrupled and humankind's energy consumption rate increased 16-fold, mainly from increased fossil fuel combustion. In recent years, observational confirmations of global warming from the fossil fuel greenhouse have accumulated, and we have come to better understand the interactions of fossil fuel burning, the carbon cycle, and climate change.

Atmospheric CO_2 concentration has been increasing during the past few centuries mainly from fossil fuel emissions (Fig. 1A). The buildup of atmospheric CO_2 so far has tracked energy use closely. Carbon cycle models recovering the observed concentration history (1900–2000) have most of the emitted CO_2 dissolved in the oceans (primarily as bicarbonate ions, the largest sink). As long as carbon emissions keep rising, approximately half the fossil fuel CO_2 will likely remain in the atmosphere, with the rest dissolving in the seas. Forests produce a two-way (photosynthesis–respiration) flux with residuals that temporarily add or subtract CO_2 but cancel in the long term unless the "standing crop" of biosphere carbon changes.

Stabilizing CO_2 concentration in the atmosphere at any specified level will require progressively more stringent emission reductions. However, even a total cessation of emissions would not restore preindustrial levels for more than millennia. The technology challenge of the century may be to reduce fossil fuel CO_2 emissions enough to prevent unacceptable climate change while population, energy use, and the world economy continue to grow.

1.2 Climate

A doubling of atmospheric CO_2 is estimated to produce 1.5–4.5°C surface warming based on paleoclimate data and models with uncertainties mainly from cloud–climate feedback. Unabated fossil fuel burning could lead to between two and three times this warming.

Climate change from the fossil fuel greenhouse is driven by radiative forcing—the change in solar

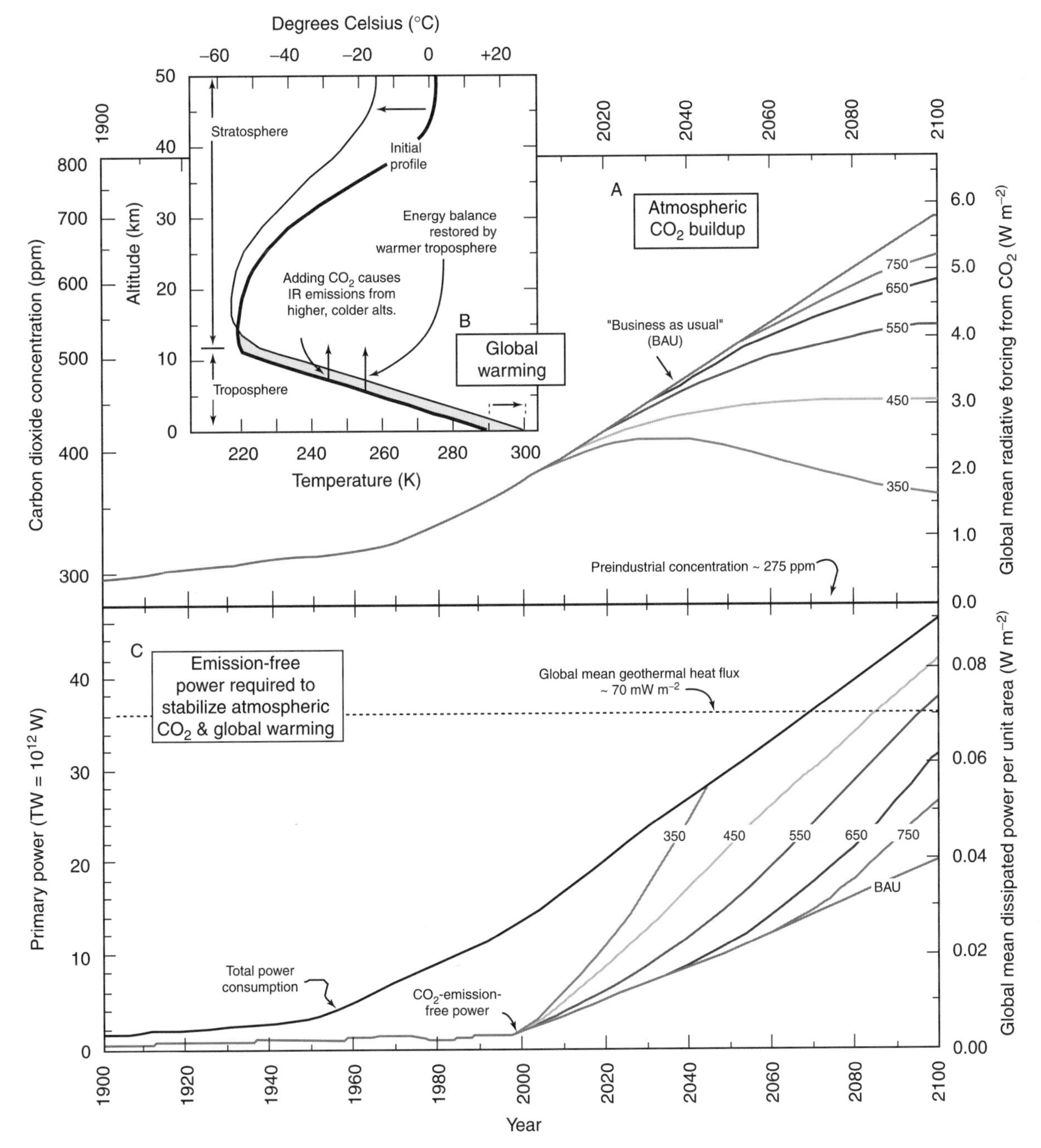

FIGURE 1 The trapping of outgoing long-wave radiation by CO_2 may mean that we need to produce vast amounts of power without releasing CO_2 to the atmosphere. (A) Atmospheric CO_2 concentration (left-hand scale) and radiative forcing (right-hand scale). The increase in CO_2 thus far is mainly from fossil fuel emissions. Future concentrations (2000–2100) are for a business as usual (BAU) emission scenario and the Wigley–Richels–Edmonds scenarios that stabilize atmospheric CO_2 (eventually) at concentrations in the range of 350–750 parts per million (ppm). (B) The lower atmosphere is rendered more opaque by IR-trapping greenhouse gases. Maintenance of the planetary energy balance requires that the earth's CO_2 radiate to space at earth's effective temperature from higher in the atmosphere. This results in surface warming. (C) Total and CO_2 emission-free power. Curves prior to 2000 are historical; curves from 2000 onward are BAU and CO_2-stabilization projections from a carbon cycle/energy model.

heating less infrared cooling to space per unit surface area relative to a preindustrial reference (Fig. 1B). Global warming by the "greenhouse effect" requires that (i) the lower atmosphere is rendered more opaque by infrared-trapping greenhouse gases, and (ii) the troposphere maintains (by convective adjustments) a nearly constant vertical temperature lapse rate ≈ 6 K/km.

The energy balance for Earth implies a subfreezing temperature for blackbody cooling to space, $T_{eff} \approx 255\,K\,(-18^{\circ}C)$. These infrared photons do not emanate, on average, from the surface but from a midtroposphere altitude of approximately 5.5 km. Natural greenhouse gases (H_2O, CO_2, and O_3) maintained a global surface temperature hospitable to life over Earth history that was 33 K warmer than the T_{eff} today: $T_{surface} \approx 255\,K + (5.5\,km \times 6\,K\,km^{-1}) \approx 288\,K\,(15^{\circ}C)$. More CO_2 in the atmosphere from fossil fuel burning makes the troposphere radiate from even higher, colder altitudes, reducing infrared cooling. Solar heating, now partly unbalanced by infrared cooling, generates greenhouse radiative forcing [$\approx 4\,W\,m^{-2}$ for doubling CO_2 to 550 parts per million (ppm)], driving the troposphere toward warmer temperatures, which depend on the climate sensitivity[= (temperature change)/(radiative forcing)].

Most of the infrared flux comes from the troposphere, but the center of the 15-μm CO_2 vibration band is so strongly absorbing that the atmosphere remains opaque near these wavelengths into the stratosphere. Here, temperature increases with altitude, and more CO_2 produces cooling.

Surface warming accompanied by stratospheric cooling from CO_2 increases was predicted in the 1960s and subsequently observed (ozone depletion combined with tropospheric soot might produce similar effects). Also predicted and observed is heat flow into the sea. Instrumental and paleotemperature observations are generally consistent with an emergent CO_2 greenhouse modulated by solar variability and by volcanic and anthropogenic aerosols. A marked global warming increase is predicted to occur in the following decades, the magnitude of which will depend on climate sensitivity and on our ability to limit greenhouse gas emissions as the global population and economy grow.

It is not known what concentrations will trigger the "dangerous anthropogenic interference with the climate system" that the United Nations' 1992 Rio de Janeiro Framework Convention on Climate Change aimed to prevent. A common reference case is doubling preindustrial CO_2, projected to produce global warming of 1.5–4.5 K, which is comparable in magnitude, but opposite in sign, to the global cooling of the last ice age 20,000 years ago. Evidently, dramatic increases in primary power from non-CO_2-emitting sources, along with improvements in the efficiency of primary energy conversion to end uses, could be needed to stabilize CO_2 at "only" twice preindustrial levels (Fig. 1C). This article assesses the ability of a broad range of advanced technologies to achieve CO_2 and climate stabilization targets.

1.3 Energy

Carbon dioxide is not a pollutant in the sense that it emerges from trace elements peripheral to the main business of energy. It is a product of combustion vital to how civilization is powered. The way to reduce CO_2 emissions significantly without seriously disrupting civilization is to identify and implement revolutionary changes in the way energy is produced, distributed, and lost to inefficiencies.

Factors influencing CO_2 emissions are evident in the Kaya identity, which illustrates the quandary of emission reductions with continued economic growth. The Kaya identity expresses the carbon emission rate, $\dot{C}$, as the product of population, N, per capita gross domestic product, GDP/N, primary energy intensity, E/GDP, and the carbon emission factor, C/E:

$$\dot{C} = N \times (\mathrm{GDP}/N) \times (E/\mathrm{GDP}) \times (C/E).$$

Few governments intervene in their nation's population growth, whereas continued economic growth is policy in virtually every country on Earth; thus, the first two terms of the equation are normally off limits to climate change policy. Their product is the growth rate of GDP. No nation advocates reducing its GDP to reduce carbon emissions. Stabilizing CO_2 with 2 or 3% per year global GDP growth requires comparable or greater reductions in the last two terms of the equation: the energy intensity and the carbon emission factor.

Figure 1C shows primary power required to run the global economy projected out 100 years for continued economic growth assuming a "business as usual" economic scenario. Figure 1C also shows the power needed from CO_2 emission-free sources to stabilize atmospheric CO_2 at a range of concentrations. Curves prior to 2000 are historical; curves from 2000 onward are business as usual and CO_2-stabilization projections run through a carbon cycle/energy model. Note the massive transition of the global energy system implied by all scenarios. Even business as usual (incorporating no policy incentives to limit emissions) ramps to 10 TW emission free by 2050. Emission-free primary power required by midcentury increased to 15 TW to stabilize at 550 ppm and to $\geq$30 TW to recover recent concentrations of 350 ppm (including energy to remove

CO_2 previously emitted). All scenarios assume aggressive continuing energy efficiency improvements and lifestyle changes such that global energy intensity (*E*/GDP) declines 1.0% per year during the entire 21st century. The main concern is CO_2, not heat dissipated by energy use. Projected to pass mean geothermal heating this century, heat dissipated by humankind's primary power consumption is approximately 1/100 the radiative forcing from the fossil fuel greenhouse.

Technologies capable of producing 100–300% of current total primary power levels, but CO_2 emission free, could be needed by midcentury (Fig. 1C). These do not exist operationally or at the pilot plant stage today. One need only consider that Enrico Fermi's "atomic pile" is more distant in time than the Year 2050, whereas nuclear power today represents less than 5% of primary power (17% of electricity) globally.

Can we produce enough emission-free power in time? We assess a broad range of advanced technologies for this job, along with pathways to their commercialization. Our goal is to explore possibilities that can revolutionize the global energy system. As Arthur C. Clarke stated, "The only way to discover the limits of the possible is to venture a little past them into the impossible." However, we cannot pick winners and losers, and we make no claim that our proposals are exhaustive. What is clear is that the future of our civilization depends on a great deal of research investment in energy options such as those considered here.

2. *EFFICIENCY*

2.1 Energy Efficiency and Energy Intensity

Reducing CO_2 emissions is sometimes considered synonymous with improvements in "efficiency." Energy efficiency is the ratio of useful energy output to energy input. Progress has been made with regard to improving energy efficiencies in building, manufacturing, and transportation end-use sectors of the United States and other nations. Technologies include compact fluorescents, low-emissivity window coatings, combined heat and power systems (cogeneration), and hybrid vehicles (internal combustion plus electric power). Improvements in energy efficiency may be adopted on purely economic grounds—the so-called "no regrets" strategy. However, cost-effectiveness is more important to market share than efficiency as such; for example, amorphous photovoltaic solar cells are less efficient, but cheaper per unit of electrical energy output, than crystalline cells.

Energy intensity (*E*/GDP, energy used per unit GDP generated) subsumes efficiency, lifestyle, and structural economic factors. Typically, the energy intensity of developing nations starts low, peaks during industrialization (when energy goes to infrastructure building rather than GDP), and then declines as service sectors dominate postindustrial economies. This final stage, modeled as a sustained percent decrease, can predict large emission reductions over century timescales. Current U.S. policy aims to continue the domestic trend of declining carbon intensity [$C/GDP = (E/GDP) \times (C/E)$]. However, even wealthy nations such as the United States that are on track to service economies consume goods manufactured elsewhere. A decline in global energy intensity adequate to stabilize the climate may be unrealistic given the massive industrialization, driven by globalization of capital and markets, in countries such as China, India, and Indonesia.

The stock of primary energy includes fossil fuels (coal, oil, and gas—a fraction of the organic carbon produced by plants from sunlight hundreds of millions of year ago), fission fuels (uranium and thorium—elements with metastable nuclei blown out of distant supernovae before the solar system formed in limited supply as high-grade ore deposits), and fusion fuels (cosmically abundant deuterium [D] and helium-three [^{3}He] relics of the big bang—D in seawater and ^{3}He in the lunar regolith and outer planet atmospheres). This energy is stored in chemical and nuclear bonds. "Renewables" are primary energy in natural fluxes (solar photons; wind, water, and heat flows). Converting this energy to human end uses always involves losses to heat dissipation—losses that engineers have in many cases already spent enormous effort reducing (Fig. 2). The efficiency of internal combustion engines, for example, is 18–23% after more than 100 years of development (Fig. 3).

Energy intensity reductions are often treated as a surrogate for energy efficiency improvements. With the "evolving economy" factor held constant, a decline of a few percent per year in *E*/GDP implies very large energy efficiency increases over long times: a factor of $(1.01)^{100} = 2.7$ at 1% per year and $(1.02)^{100} = 7.2$ at 2% per year after 100 years. Such increases will be physically impossible if they exceed thermodynamic limits.

Efficiency improvements and alternate energy sources need to be pursued in combination. It is a false dichotomy to pit efficiency against alternate

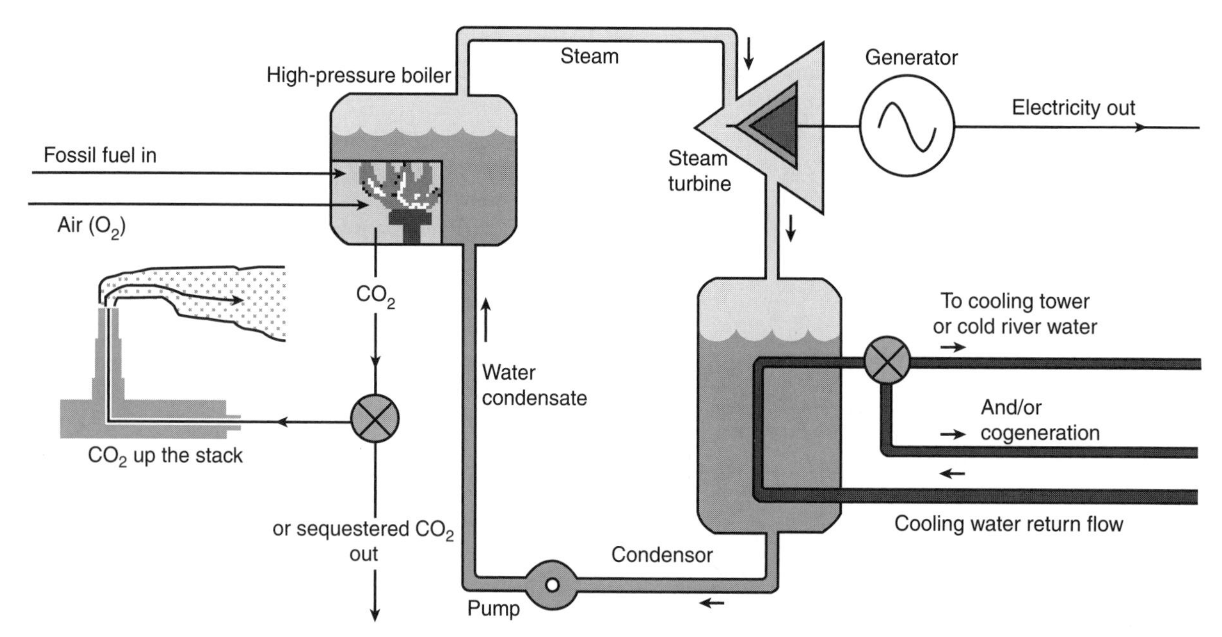

FIGURE 2 The world's electricity is generated mainly by heat engines employing hot water as the working fluid. Actual thermal efficiencies for power plant steam cycles are 39–50%, with primary energy-to-electricity efficiencies of 31–40%, roughly consistent with the conversion factor 3 kW (primary) $\approx$ 1 kW$_e$ (electrical). Because small temperature increases cause exponentially large increases in pressure (and hoop stress) in boilers, it is unlikely that turbine inlet temperature and turbine efficiency will increase significantly (and CO_2 emissions decline) without breakthroughs in materials strong enough to prevent explosions.

energy. The trade-off in which more of one requires less of the other to achieve a given CO_2 stabilization target was analyzed by Hoffert and colleagues in 1998. We have assumed a century-long 1% per year decrease in global *E*/GDP to estimate human primary power needed in Fig. 1C. Some believe even greater rates of decline are possible, perhaps as much as 2% per year. Our analysis indicates that even 1% per year will be difficult. For example, an effective strategy in the transportation sector would be to increase energy efficiency by a path leading to CO_2 emission-free hydrogen from renewable sources as the energy carrier.

2.2 The Transportation Sector

A frequently cited opportunity for emission reductions involves improving the efficiency of the transportation sector, which is responsible for 25% of CO_2 emissions in the United States, of which approximately half is by private cars. Figure 3 compares the efficiency of fossil-fueled cars of similar mass, aerodynamics, and driving cycles for four different propulsion systems: internal combustion (IC), battery–electric, IC–electric hybrid, and fuel cell–electric. A twofold improvement seems reasonable for a transition from the IC engine to fuel cell–electric. Analyses of transportation sector efficiency should consider the entire fuel cycle, "wells to wheels." Smaller cars requiring less fuel per passenger-kilometer, mass transit, and improvements in manufacturing efficiency can also reduce CO_2 emissions. According to some estimates, the manufacturing of a car can create as much emissions as driving it for 10 years.

In the United States, sport utility vehicles are driving emissions higher, whereas economic growth of China and India has led 2 billion people to the cusp of transition from bicycles to personal cars. (Asia already accounts for >80% of the growth in petroleum consumption.) CO_2 emissions from transportation depend on fuel carbon-to-energy ratios (*C*/*E*), fuel efficiency, vehicle mass, aerodynamic drag, and patterns of use. Storability and high energy density of liquid hydrocarbons have fostered crude oil derivatives as dominant motor and aviation fuels and massive capital investment worldwide in oil exploration, oil wells, supertankers, pipelines, refineries, and filling stations. Gasoline-powered IC technology today embodies a sophisticated and difficult to displace infrastructure. CO_2 emission-neutral fuels derived from biomass (e.g., ethanol) are limited in availability by low-yield photosynthesis and competition with agriculture and biodiversity for land. Natural gas

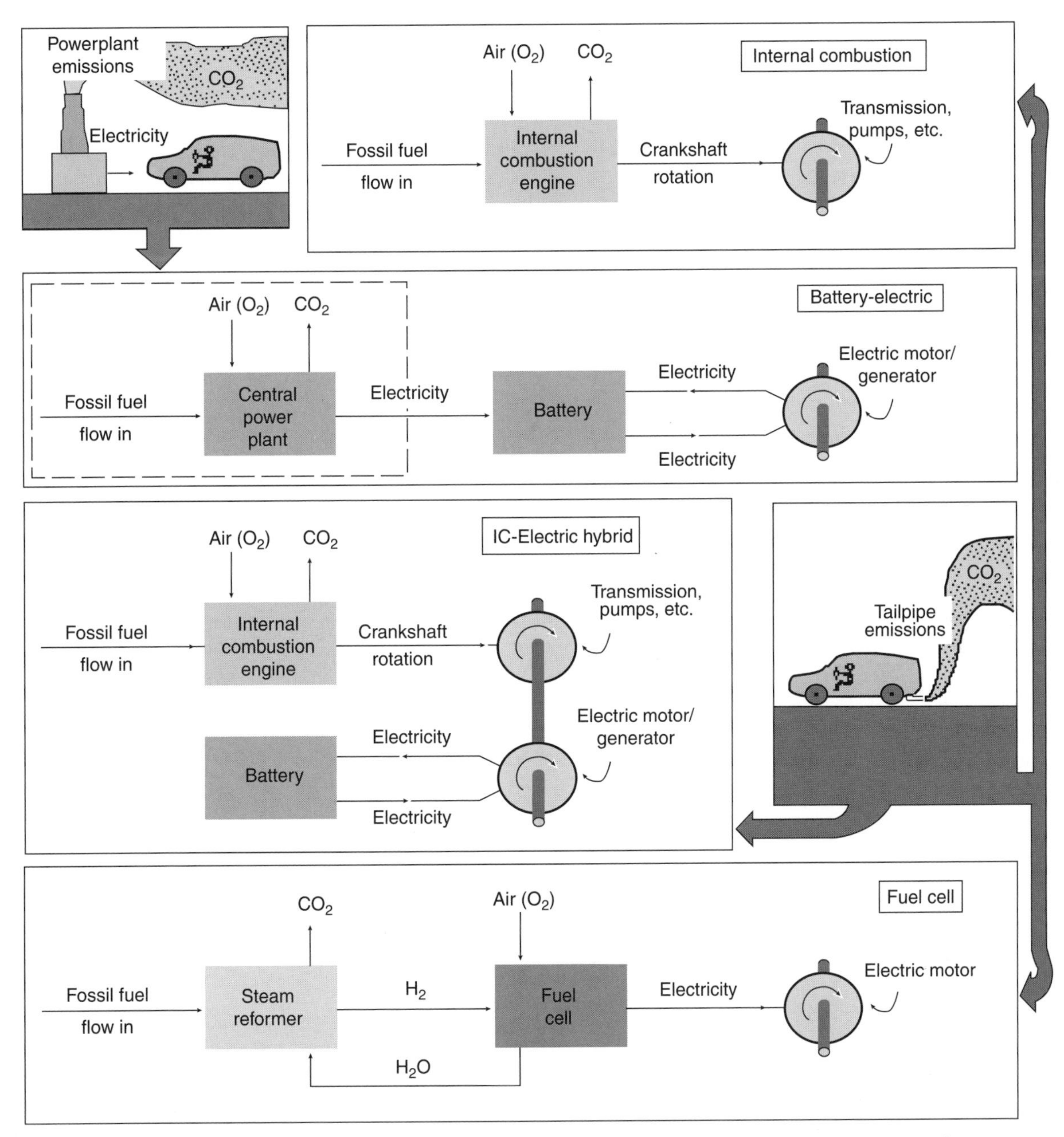

FIGURE 3 The transportation sector presents unique challenges for CO_2 emission reductions. CO_2 emissions from transportation depend on fuel carbon-to-energy ratios (C/E), fuel efficiency, vehicle mass, aerodynamic drag, and patterns of use. Fuel efficiencies [η = (mean torque × angular velocity)/(fossil fuel power in)] of the systems shown are as follows: internal combustion (IC), η = 18–23%; battery–electric, η = 21–27% (35–40% central power plant, 80–85% charge–discharge cycles, and 80–85% motor); IC–electric hybrid, η = 30–35% (higher efficiency from electric power recovery of otherwise lost mechanical energy); and fuel cell–electric, η = 30–37% (75–80% reformer, 50–55% fuel cell, and 80–85% motor). Cost-effective efficiencies of 40% may be possible with additional emission reductions per passenger-kilometer from lighter cars.

(methane, CH_4) is a possible transitional fuel, lower in CO_2 emissions than gasoline, but it requires high-pressure onboard storage (accompanied by cooling if stored as liquid natural gas).

Current batteries are prohibitively heavy for normal driving patterns unless combined with IC engines as hybrids. Fuel cells have high power densities; utilize various fuels, electrolytes, and electrode materials; and operate at various temperatures. The clear favorite for automobiles is the proton exchange membrane (PEM) cell—a compact unit converting H_2 to electricity at a relatively cool

350 K. Hydrogen (plus oxygen from the atmosphere with water as the reaction product) emits zero CO_2 from IC engines and fuel cells. A seamless transition to fuel cells is fostered by extraction of hydrogen from gasoline in a reformer, which is a miniature onboard chemical processing plant that does emit CO_2.

Fuel efficiencies [η = (mean torque × angular velocity)/(fossil fuel power in)] of the systems shown are as follows: IC, η = 18–23%; battery–electric, η = 21–27% (35–40% central power plant, 80–85% charge–discharge cycles, and 80–85% motor); IC–electric hybrid, η = 30–35% (higher efficiency from electric power recovery of otherwise lost mechanical energy); and fuel cell–electric, η = 30–37% (75–80% reformer, 50–55% fuel cell, and 80–85% motor). Cost-effective efficiencies of 40% may be possible with additional emission reductions per passenger-kilometer from lighter cars.

A scenario aimed at sustainable CO_2 emission-free vehicles would first replace IC engines with gasoline-powered fuel cells. In the second phase, a network of filling stations dispensing H_2 derived from fossil fuel with CO_2 sequestration is deployed as onboard hydrogen storage (metal hydride tanks, carbon nanotubes, etc.) becomes safe and cost-effective. In the third and presumably sustainable phase, H_2 is generated by water electrolysis from renewable or nuclear primary sources (Figs. 4–6). This scenario, and others emphasizing mass transit, requires technology breakthroughs from basic and targeted research. Also critical for CO_2 and climate stabilization are sequencing and diffusion of transportation sector transitions worldwide.

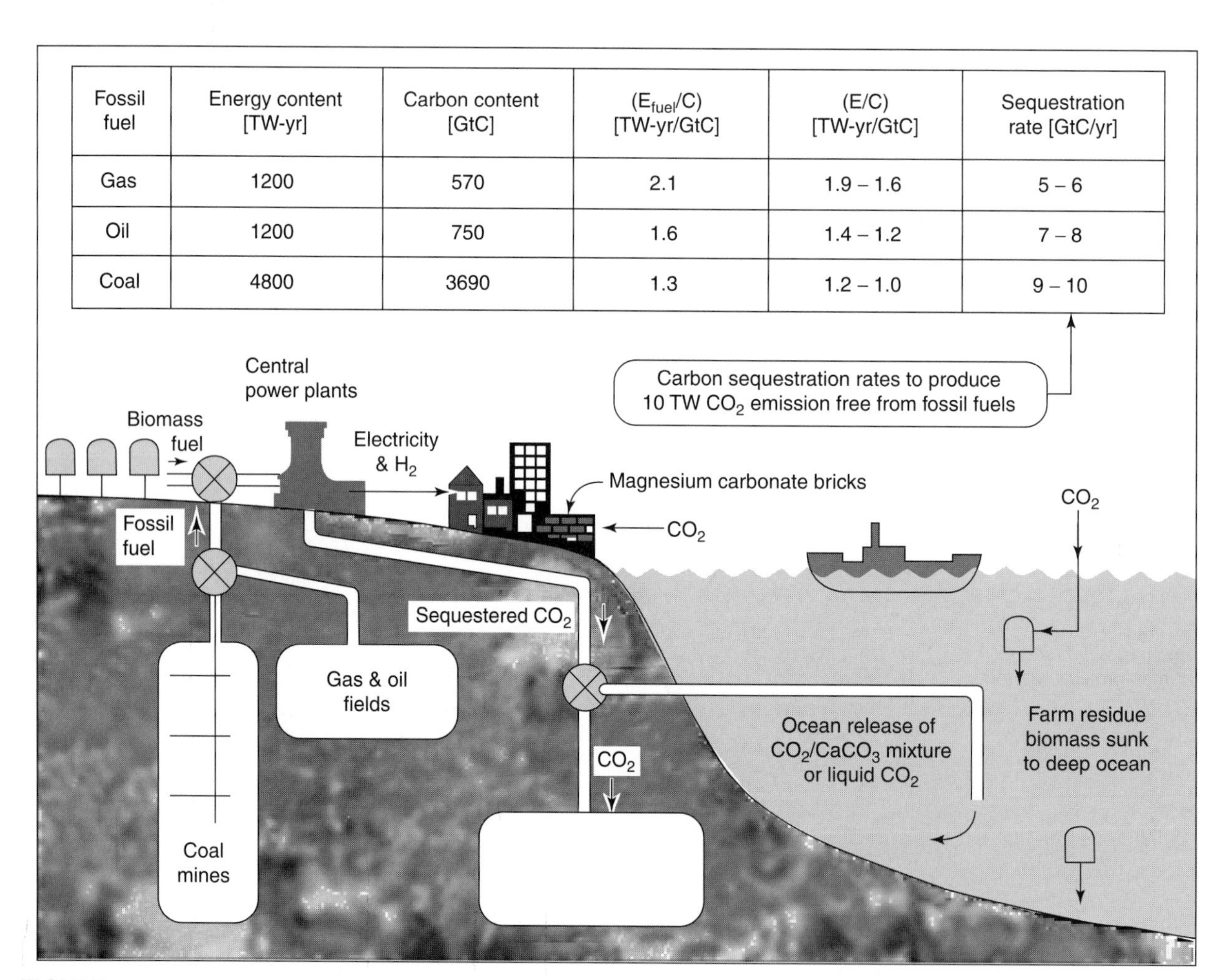

Fossil fuel	Energy content [TW-yr]	Carbon content [GtC]	(E_{fuel}/C) [TW-yr/GtC]	(E/C) [TW-yr/GtC]	Sequestration rate [GtC/yr]
Gas	1200	570	2.1	1.9 – 1.6	5 – 6
Oil	1200	750	1.6	1.4 – 1.2	7 – 8
Coal	4800	3690	1.3	1.2 – 1.0	9 – 10

FIGURE 4 Separating and sequestering the CO_2 or solid carbon by-product of fossil fuel or biomass oxidation in subsurface reservoirs while producing electricity in central power plants and/or hydrogen may have the potential to reduce anthropogenic CO_2 emissions and/or atmospheric concentrations. More CO_2 must be sequestered (and more fossil fuel consumed) per unit energy available for human power consumption than would be emitted as CO_2 to the atmosphere without sequestration, some of which may return later to the atmosphere from leaky reservoirs.

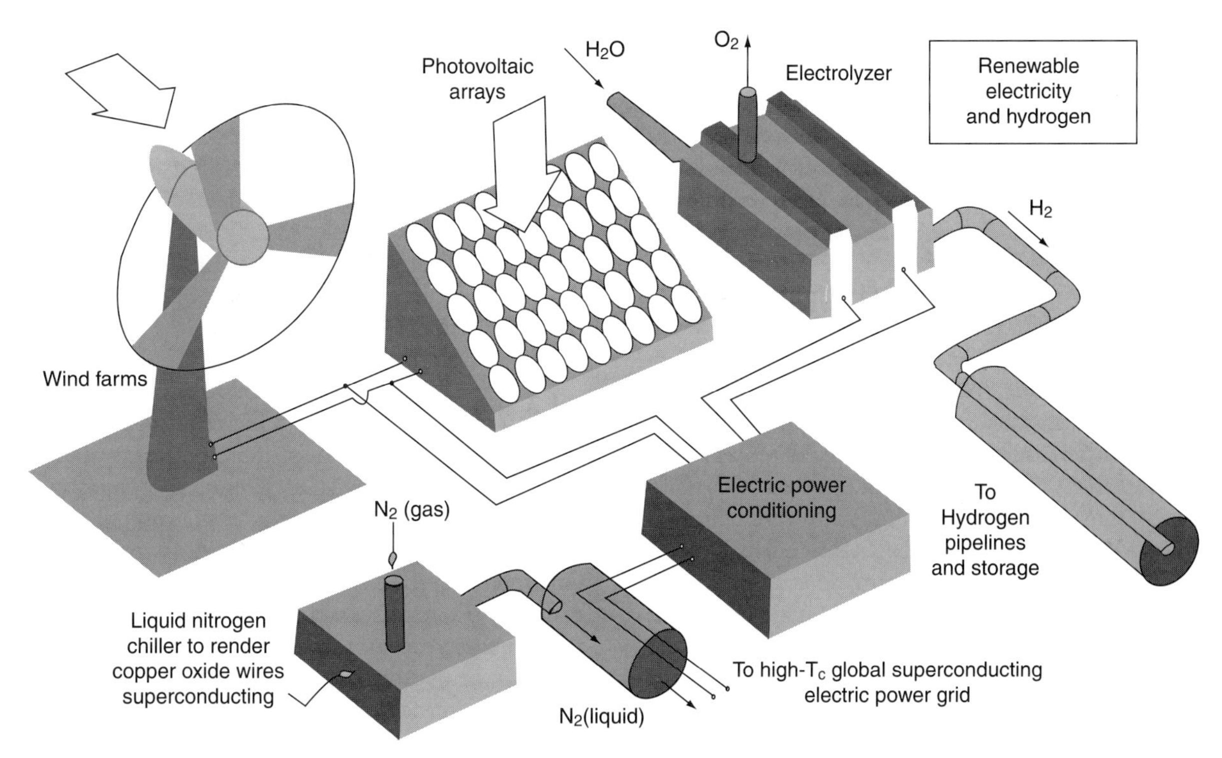

FIGURE 5 Renewable energy at Earth's surface. Greatest potential is from solar and wind with electric currents or H_2 as energy carriers.

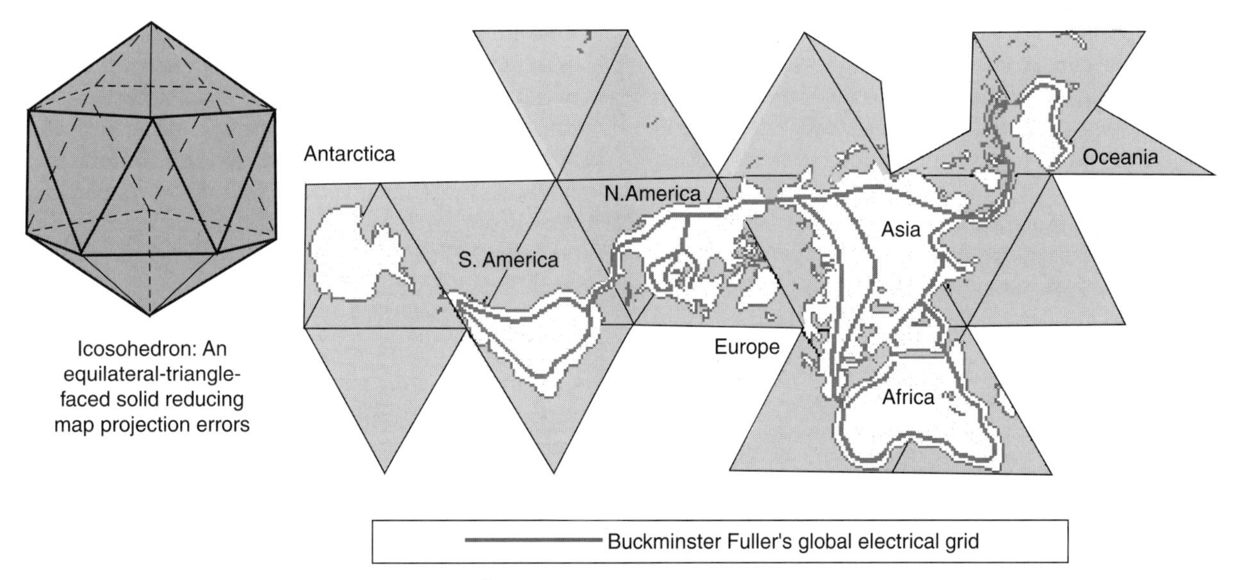

FIGURE 6 An alternative energy carrier is the low I^2R loss global electric grid envisioned by R. Buckminster Fuller. Modern computer technology may permit grids smart enough to match supply from intermittent renewable sources to baseload demands worldwide.

3. FOSSIL FUELS AND CO_2 EMISSION REDUCTION

Reductions in carbon intensity, *C*/*E*, the carbon emitted per unit of energy generated, reflect the degree to which societies decarbonize their energy sources. The long-term trend has been a shift from coal to oil to natural gas—hydrocarbons with decreasing C/H ratios emitting progressively less CO_2 per joule. However, the increasing use of clean low-carbon fuels is not sustainable without somehow disposing of excess carbon because it opposes the trend in the abundance of fossil fuels, with coal resources being the most abundant followed by oil and gas (Fig. 4).

Energy in recoverable fossil fuels ("reserves" plus "resources" from conventional and unconventional sources) is ~4800 TW-year for coal and ~1200 TW-year each for oil and gas (1 Gtoe = 44.8 EJ = 1.42 TW-year; natural gas includes liquids but excludes methane hydrates). Total energy in recoverable fossil fuels is in the range 6400–8000 TW-year, of which two-thirds is coal. Coal resources could roughly double if energy in shales could be extracted cost-effectively.

Separating and sequestering the CO_2 or solid carbon by-product of fossil fuel or biomass oxidation in subsurface reservoirs while producing electricity in central power plants and/or hydrogen may have large potential for CO_2 emission reductions. Industrial hydrogen is made primarily by partial (only the carbon is oxidized) oxidation of methane (net: $CH_4 + O_2 \rightarrow CO_2 + 2H_2$). Some energy loss is unavoidable for H_2 produced from fossil fuels even without CO_2 sequestration. Chemical energy transferred to hydrogen is typically 72% from natural gas, 76% from crude oil, and 55–60% from coal.

Primary energy available after sequestration ($E = E_{fuel} - E_{seq}$) per unit carbon content depends on the energy-to-carbon ratio of the fuel (E_{fuel}/C) and the ratio of sequestration energy to fuel energy (E_{seq}/E_{fuel}): $(E/C) = (E_{fuel}/C)[1-(E_{seq}/E_{fuel})]$. Carbon sequestration rates tabulated in Fig. 4 needed to produce 10 emission-free terawatts by midcentury from gas, oil, or coal [(10 TW)/(E/C)] are in the 5–10 GtC/year range, depending on the fossil fuel mix and assuming $(E_{seq}/E_{fuel}) = 10$–25%—an extrapolation of current technology judged necessary for cost-effectiveness. This would be a massive carbon transfer akin to the global extraction of fossil fuels in reverse. More CO_2 must be sequestered (and more fossil fuel consumed) per unit energy available for human power consumption than would be emitted as CO_2 to the atmosphere without sequestration, some of which may return later to the atmosphere from leaky reservoirs.

Potential carbon sequestration reservoirs include oceans, trees, soils, depleted natural gas and oil fields, deep saline reservoirs, coal seams, and solid mineral carbonates (Fig. 4). The advantage of sequestration is that it is compatible with existing fossil fuel infrastructure, including CO_2 injections for enhanced recovery from existing oil and gas fields and capture of CO_2 from power plant flue gas. Challenges include cost and energy penalties of capture and separation (more fuel is needed per joule than freely venting combustion products), understanding the long-term fate of CO_2 sequestered in various reservoirs, and investigating environmental impacts.

The simplest sequestration technology is growing trees. CO_2 uptake only works during net biomass growth. At best, it could mitigate 10–15% of emissions over a limited time. Approximately 2500 Gt C is organically fixed in land plants and soil. Carbon cycle models indicate that absorption by trees and soil exceeded emissions by 2.3 ± 1.3 Gt C per year in the 1990s. Model simulations matching observations are consistent with CO_2 uptake at midlatitudes; however, some models predict these lands will reverse from sinks to sources by midcentury as climate warms. The advantage of tree and soil sequestration is that they do not require separation of combustion products or more fuel. Although not a long-term solution, tree and soil uptake can buy time at low cost with current technology—time for learning, capital turnover, research, and development of advanced power technologies.

A biological sequestration scheme that might store fossil fuel CO_2 longer than trees and soils consists of sinking agricultural residues in bales to deep and/or anoxic zones of the oceans. This is conceptually similar to storing trees in underground vaults and to fertilizing high-latitude ocean plankton with iron. An early proposal by Cesare Marchetti being explored today is CO_2 disposal in the deep sea. Although most fossil fuel CO_2 emitted to the atmosphere dissolves in the oceans on decadal to century timescales, some persists in the atmosphere much longer because of a shift in the carbonate equilibrium of the air/sea system. For a given emission scenario, ocean injections can significantly decrease peak atmospheric CO_2 levels, although all cases eventually diffuse some CO_2 back to the atmosphere.

Carbon dioxide dissolves in water and slowly corrodes carbonate sediment, resulting in the storage of fossil fuel carbon primarily as bicarbonate ions dissolved in the ocean. Left to nature, the net reaction $CO_2 + H_2O + CaCO_3 \rightarrow Ca^{2+} + HCO_3^-$ would occur on the timescale of ~6000 years. Back diffusion to the atmosphere and pH impacts of ocean CO_2 disposal could be greatly diminished by accelerating carbonate mineral weathering reactions that would otherwise slowly neutralize the oceanic acidity produced by fossil fuel CO_2.

A potentially far-reaching removal scheme involves reacting CO_2 with the mineral serpentine to sequester carbon as a solid in magnesium carbonate "bricks." The core idea is to vastly accelerate silicate rock weathering reactions, which remove atmospheric CO_2 over geologic timescales. Atmospheric CO_2 is sequestered on geologic timescales by weathering calcium and magnesium silicate rocks. The net reaction with calcium silicate is schematically represented as $CO_2 + CaSiO_3 \rightarrow SiO_2 + CaCO_3$. Nature (rainfall, runoff, biology, and tectonics) precipitates the calcium carbonate product as seashell sediment and coral reefs. (Atmospheric CO_2 would be depleted in ~400,000 years were it not replenished by CO_2 emissions from volcanoes and hydrothermal vents.)

Carbon dioxide sequestration is a valuable bridge to renewable and/or nuclear energy. If emission-free primary power at 10–30 TW levels is unavailable by midcentury, then enormous sequestration rates could be needed for atmospheric CO_2 stabilization (Fig. 4). In light of these requirements, research and demonstration investments are required now for sequestration technology to be available at the required scale when needed.

4. *RENEWABLES*

4.1 Biomass

Biomass energy farms with tree regrowth can produce energy with zero net CO_2 emissions. Unfortunately, photosynthesis is too inefficient (~1%) for bioenergy to be the answer to CO_2 stabilization. Producing 10 TW from biomass would require at least 10% of Earth's land area, comparable to all human agriculture. Competition for land with agriculture and biological diversity is a likely showstopper for bioenergy as a dominant factor in climate stabilization, although it could be a notable contributor in some locales.

4.2 Wind and Solar

Solar and wind with electric currents or H_2 as energy carriers have great potential because their planetary-mean areal power densities are greater than those for biomass, hydroelectricity, ocean, and geothermal power. Considerable effort has been expended in recent years on the suite of renewable energy systems—solar thermal and photovoltaic (PV), wind, hydropower, biomass, ocean thermal, geothermal, and tidal. With the exception of hydroelectricity, these collectively represent <1% of global energy production. Limited by low areal power densities and intermittency of supply, renewables nonetheless have the potential to power human civilization. However, they require substantial investment in critical enabling technologies before they can be real CO_2 emission-free energy options at the global scale (Figs. 5 and 6).

Solar flux at Earth's orbital distance perpendicular to the sun's rays is $S_o \sim 1370\,W\,m^{-2}$. Sun power at the surface averaged over day–night cycles is reduced by the fraction of Earth's disk to surface areas ($f_1 = 0.25$), cloud fraction ($f_2 \sim 0.5$), and fraction of sunlight reaching the surface under clear skies ($f_3 \sim 0.8$): $S_{surf} = f_1 f_2 f_3 S_o \sim 0.1\,S_o \sim 140\,W\,m^{-2}$. [The factor of 10 solar flux reduction is a major argument for putting solar collectors in space.] Commercial photovoltaic arrays are 10–15% efficient, producing a long-term mean power ~10–20 $W_e\,m^{-2}$ (the theoretical peak for single bandgap crystalline cells is ~24%; it is higher for multi-bandgap cells and lower for the most cost-effective amorphous thin-film cells).

Wind power per surface area depends on the fraction of wind turbine disk to surface area minimizing interference from adjacent turbines (~0.1), air density ($\rho \sim 1.2\,kg\,m^{-3}$), and wind speed ($U \sim 5\,m\,s^{-1}$): $P_{wind} = 0.1 \times (1/2)\rho U^3 \sim 15\,W\,m^{-2}$. There are typically large local variations. Wind turbine efficiencies of 30–40% yield ~5 $W_e\,m^{-2}$ (theoretical peak is the Betz limit of ~59.3%; actual efficiency depends on wind spectrum and "cut-in" speeds).

The capital costs of PV arrays and wind turbines are rapidly decreasing. Locally, cost-effective wind power is available at many sites that tend to be remote and offshore high-wind locations. Distribution to consumers is limited by long-distance transmission costs. It is argued that the annualized total consumption of electric energy within the United States could be supplied by PV arrays covering a fraction of Nevada's relatively clear-sky deserts. It is similarly argued that the annualized consumption of global electric energy could be supplied by connected arrays of photovoltaics in the deserts of Asia, North Africa, the Americas, and Australia. However, arrays of terrestrial photovoltaics intended to be the sole supplier of large-scale commercial power face daunting problems, including the continental and global distribution of power and uncertainties in the supply of sunlight at the surface of Earth and the total energy storage requirements.

A challenge for demand-driven global energy systems is intermittency from variable clouds and winds. Solar and wind are intermittent, spatially dispersed sources unsuited to baseload electricity without breakthroughs in transmission and storage. Meeting urban electricity demand over day–night cycles and during overcast periods with locally sited PV arrays requires prohibitively massive pumped-storage or lead–acid batteries. A solar–hydrogen economy with electrolyzers, storage tanks, and pipelines could be self-sufficient in principle. However, solar–hydrogen system costs are so high that many advocates of H_2 as an energy carrier promote fossil fuel or nuclear as the primary energy source.

5. *ENERGY CARRIERS AND OFF-PLANET ENERGY PRODUCTION*

5.1 Hydrogen

A continued historical trend of decreasing carbon content of fuels points to the combustion fuel with zero carbon emissions: hydrogen. However, hydrogen does not exist in underground gas fields or Earth's atmosphere. It must be made from chemical processes requiring energy. If hydrogen is made from fossil fuels (as is commercial H_2 by steam-reforming natural gas), even more carbon or CO_2 must be disposed of than would be produced by burning these fuels directly.

A vision of coal power liberated from environmental constraints incorporates CO_2 capture and sequestration. In one such scheme, coal and/or other fossil fuel feedstocks along with biomass and some waste materials are gasified in an oxygen-blown gasifier, and the product is cleaned of sulfur and reacted with steam to form hydrogen and CO_2. After heat extraction, the CO_2 is sequestered from the atmosphere. The hydrogen can be used as a transportation fuel, or it could be oxidized in a high-temperature fuel cell, with reacted hot gases driving gas turbines or steam generators to make additional electricity. Electrochemical fuel cells are more efficient than heat engines subject to thermodynamic limits (steam turbines, gas turbines, and IC engines). This is somewhat offset by their need for hydrocarbon reforming to H_2 plus CO_2. Like conventional fossil fuel power plants, fuel cells are converters of chemical energy to electricity, not sources of primary power.

Alternately, hydrogen can be produced by electrolysis of water powered by renewable or nuclear sources. Decarbonization and fuel switching to natural gas and eventually H_2 will not mitigate global warming. The underlying problem is the need for non-CO_2-emitting primary power sources during the next 50 years capable of generating 10–30 TW.

Figure 5 depicts electricity routed directly to consumers via grid or employed to make hydrogen from water. (Photoelectrochemical cells combining PV and electrolysis cells in one unit are also being studied.) Water electrolysis employs H_2O as both feedstock and electrolyte (hydroxyl ions, OH^-, present in low concentrations in water are the charge carriers). Commercial electrolyzers employ various electrolytes (and ions) to raise ionic conductivity in alkaline water (K^+ and OH^-), solid polymer PEM (H^+), and ceramic solid oxide (O_2^-) cells. When a potential of approximately 1.5 V is applied, reactions at the electrodes evolve hydrogen and oxygen ($2H_2O + 2e^- \rightarrow H_2 + 2OH^-$ and $2OH^- \rightarrow (1/2)O_2 + H_2O + 2e^-$). Electrolyzers can operate reversibly as fuel cells like rechargeable batteries but with higher power densities because reactants and products flow to and from electrodes as opposed to diffusing though a solid matrix. Power densities per electrode area range from $2\,kW_e\,m^{-2}$ (commercial alkaline water electrolysis) to $20\,kW_e\,m^{-2}$ (PEM cells). The latter require platinum catalysts whose loading has been driven down by intensive research to $5 \times 10^{-3}\,kg\,Pt\,m^{-2}$ without significant loss of performance. However, the high power densities of PEM cells require $5 \times 10^{-4}\,kg\,Pt/kW$ assuming 50% overall efficiency. In other words, PEM cells (electrolyzers and/or fuel cells) need 5000 metric tonne Pt to produce and/or consume H_2 at our canonical 10 TW—30 times today's annual global platinum production rate. Another hurdle for renewable hydrogen is the need for 1 liter per day of chemically pure water per kilowatt, which may be an issue in deserts. Big bang hydrogen is the most abundant element in the universe, but it is too chemically reactive to exist freely on Earth and too light, even in its stable molecular form (H_2), to have been retained by the atmosphere. Hydrogen is a secondary fuel, and water splitting to get H_2 must be powered by primary fuels or energy fluxes in nature.

Renewable electric power systems could in principle evolve from natural gas-powered fuel cells intended to provide grid-independent electricity and cogenenerated heating/cooling to residential and commercial complexes. This path would replace reformers in a later phase with H_2 generated by fuel cells run backward as electrolyzers powered by local PV arrays and/or wind turbines (Fig. 5). With H_2

stored locally, dependence on grids and pipelines could be avoided. Such independence has perceived advantages, particularly during utility outages such as those experienced in California. A natural gas fuel cell path to renewables has the advantage of increasing fossil fuel energy efficiency on the way to renewable power, although it may require internalizing the costs of air capture of its emitted CO_2. Electricity costs from decentralized fuel cells are still many times more than those from utility grids, owing largely to precious metal catalysts employed by commercial reformers and fuel cell stacks.

5.2 Advanced Electric Grids

Advanced technology global electrical grids may be more promising for renewables. With current transmission lines, even if costs per kilowatt-hour of PV arrays and turbines declined drastically and production rates were increased to millions per year like mass-produced automobiles, the grids of the United States and the world could not handle the load-management demands. Power must flow where needed, when needed. Existing grids are hub-and-spoke networks designed for central power plants tens to hundreds of kilometers from users. These need to be reengineered into "smart grids"—a global Internet and superhighway system for electricity.

Expanding such grids to the global scale of tens of thousands of kilometers and undersea links between continents evokes the global electrical grid proposed by R. Buckminster Fuller (Fig. 6). Fuller envisioned electricity produced worldwide "wheeled" between day and night hemispheres and pole-to-pole to balance supply and demand with minimum power loss. Fuller's global grid had relatively short links between continents. A major technology issue may be maintaining the undersea grid segments.

Long-distance transmissions might be accomplished initially by high-voltage direct current lines using conventional aluminum/copper conductors. A more revolutionary technology exploits the discovery of "high-temperature" superconducting ceramic copper oxide wires. The electrical resistance of high-temperature superconductors vanishes below the boiling point of liquid nitrogen (77 K), which is much easier to produce from chilled air than from the 4 K liquid helium previously needed. Nitrogen-cooled cables have been fabricated as transmission lines. Long-term implications of high-temperature superconductors transmission lines for renewables and nuclear are profound.

Routing electricity over tens of thousands of kilometers loss-free by "smart grids" could permit solar energy to flow from the Sahara to sub-Saharan Africa, China, and India and nuclear energy to politically unstable regimes. Modern computer technology may permit grids smart enough to match supply from intermittent renewable sources to base-load demands worldwide. The ultimate electric utility deregulation would have buyers and sellers worldwide creating the market price of a kilowatt-hour on the grid.

5.3 Wireless Power Transmission and Solar Power Satellites

Approximately 40% of Earth's population is "off the grid," mainly in developing countries. Wireless power transmission envisioned by Nikola Tesla a century ago is feasible today. Microwave beams can propagate power efficiently along lines-of-sight over long distances. Orbiting microwave reflectors could form the basis of a global electric grid (Fig. 7).

An advanced technology path to electrification is the solar power satellite (SPS) proposed by Peter Glaser (Fig. 7). Solar flux is $\sim$10 times higher in space outside Earth's shadow cone than the long-term average at the surface of spinning, cloudy Earth, and power from space can be beamed by microwave efficiently through cloudy skies to the surface where it is needed.

Rockets routinely put PV-powered satellites in orbit, and the Moon, visited decades ago by astronauts, can provide a stable platform and materials. NASA and the Department of Energy studied a reference solar power satellite design in the 1970s with a solar array the size of Manhattan in geostationary orbit (35,000 km above the equator) beaming power to a 10×13-km rectenna on the surface at 35° latitude with 5 GW_e output to a grid. Today, the capital investment to build even one appears too mind-boggling to proceed, even if launch costs decline enough to make electricity cost-effective at 3 or 4¢ $(kW_e\text{-hr})^{-1}$. New ideas include satellite constellations providing integrated electric power and communication/Internet services with lower capital costs to "first power."

A solar power satellite in a lower orbit can be smaller and cheaper than one in geostationary orbit because it does not spread its beam as much, but it does not appear fixed in the sky and has a shorter duty cycle (the fraction of time power is received at a given surface site). When a line-of-sight exists the

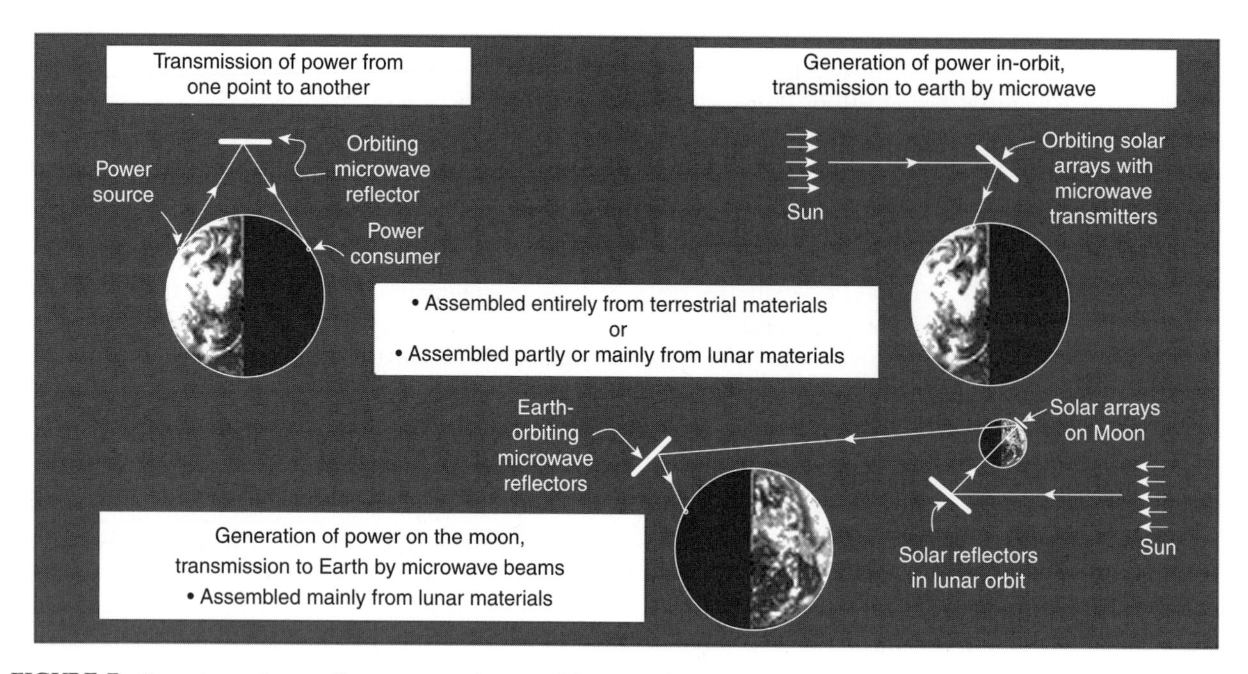

FIGURE 7 Capturing and controlling sun power in space. Microwave beams can propagate power efficiently along lines-of-sight over long distances. The power relay satellite, solar power satellite, and lunar power system all exploit unique attributes of space (high solar flux, lines-of-sight, lunar materials, and shallow gravitational potential well of the moon). In principle, a ring of these in geostationary orbit could power human civilization.

beam from a phased-array transmitter can be slued to track a surface rectenna. The flat transmitter array of aperture D_t is steered electronically with phase data from a computer homing on a pilot beam from the surface. Power is collected on the surface by a rectenna of aperture D_r of quarter-wave antenna and Shottky diode elements. Most of the rectenna is empty space, permitting vegetation and grazing below (Fig. 8). Biota and land use impacts could be minimal. A rectenna can produce the same power with 1/10 the areal footprint as PV panels for microwave intensities comparable to those routinely experienced by cell phone users (>100 million worldwide). Theoretically, transmission efficiency, $\eta_t = $(DC power out)/(EM power transmitted), is >90% when $D_tD_r/(\lambda d) \geq 2$, where d is distance and with $\eta_t = 60\%$ a reasonable target. Sunsats bypassing fossil-fueled central power stations offers a CO_2 emission-free path to developing nation electrification analogous to bypassing telephone lines with cell phones.

Important progress has been made in SPS enabling technologies: solar power generation (already at 30% efficiency and increasing), wireless power transmission, intelligent systems, and others. A major issue is the scale needed for net power generation. Beam diffraction drives component sizes up for distant orbits. (The "main lobe" where power is concentrated spreads by a small angle even in coherent diffraction-limited beams.) Unlike the case for fusion, no basic scientific breakthroughs are needed. Dramatic reductions in launch costs that NASA is pursuing independently could accelerate the technology.

A proposed short-term demonstration project would beam solar energy from low orbit. Relatively small equatorial satellites could supply electricity to developing nations within a few degrees of the equator. This appears feasible, but it would require a large number of satellites for orbits below the earth's radiation belts for continuous power. Papua New Guinea, Indonesia, Ecuador, and Columbia on the Pacific Rim, as well as Malaysia, Brazil, Tanzania, and the Maldives, have agreed to participate in a demonstration project.

One cost-reduction strategy is the building of solar power satellites from materials mined and launched from the moon. This requires less energy than launches from Earth but implies a long-term commitment to revisit the moon and exploit lunar resources. For power generation and transmission, geostationary orbit 36,000 km above the equator has the advantage of presenting a fixed point in the sky but has disadvantages as well. Alternate locations are 200- to 10,000-km altitude satellite constellations, the moon, and the earth–sun L_2 Lagrange point.

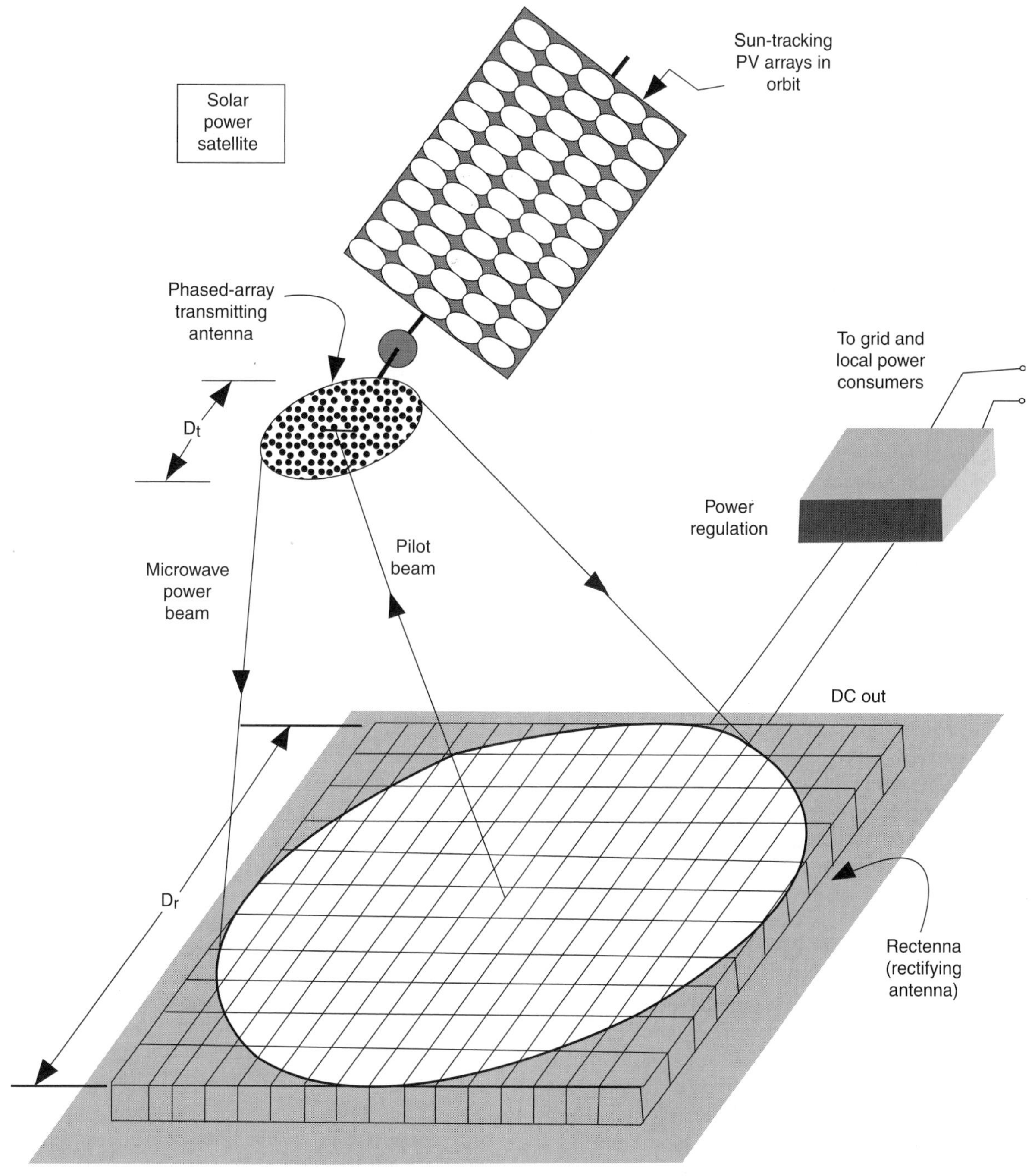

FIGURE 8 When a line-of-sight exists, the beam from a phased-array transmitter can be slued to track a surface rectenna. Most of the rectenna is empty space, permitting vegetation and grazing below. Biota and land use impacts could be minimal. A rectenna can produce the same power with 1/10 the areal footprint as photovoltaic (PV) panels for microwave intensities comparable to those routinely experienced by cell phone users.

With adequate research investments, space solar power could be matured and demonstrated in the next 15–20 years, with a full-scale pilot plant (producing $\sim$1–3 GW_e for a terrestrial market) in the 2025–2040 timeframe. Solar power beamed from space has the potential to deliver baseload power to global markets remote from electrical transmission lines and/or hydrogen pipelines. It could thus be an effective complement to terrestrial renewable electricity for developing nations. In the long term, space solar power could help stabilize global climate, provide a transition from military projects for aerospace industries, create new "orbital light and power" industries, and foster international cooperation.

6. *FISSION AND FUSION*

6.1 Fission

Nuclear electricity is fueled by fissioning the U-235 isotope—0.72% of natural uranium, usually slightly enriched in fuel rods. Neutrons liberated from fissioning nuclei (two or three per fission event) are moderator slowed to foster chain reactions in further uranium nuclei. Approximately 200 MeV of energy is released in each fission and appears as kinetic energy of the fission fragments. Reactors operate near steady states (criticality) in which heat is transferred by coolants to steam turbogenerators. Current nuclear reactor technology provides CO_2 emission-free electricity while posing unresolved problems of waste disposal and nuclear weapons proliferation.

Otto Hahn and Fritz Strassmann (working with Lisa Meitner) discovered that bombarding natural uranium with neutrons moving at a few eV split the nucleus, releasing a few hundred million eV: $^{235}U + n \rightarrow$ fission products $+ 2.43n + 202$ MeV. The ^{235}U isotope, 0.72% of natural uranium, is often enriched to 2 to 5% in reactor fuel rods. The ~500 nuclear power plants worldwide today are technology variants of ^{235}U thermal reactors: the light water reactor (LWR, in both pressurized and boiling versions); heavy water (CANDU); graphite-moderated, steam-cooled (RBMK), like the plant at Chernobyl; and gas-cooled graphite. A total of 85% are LWRs, based on Hyman Rickover's choice for the first nuclear submarine and commercial power reactor. The first electricity-generating commercial reactor (Shippingsport, PA) went online on December 2, 1957—18 years after the Hahn–Strassmann paper.

The LWR, primarily a "burner" powered by uranium slightly enriched in ^{235}U in zirconium alloy-clad fuel rods, employs H_2O as both coolant and working fluid (Fig. 9). Power is controlled by moderator rods that increase fission rates when lowered into the core. LWRs and their variants, and a smaller number of reactors employing D_2O (heavy water), produce CO_2 emission-free power (18% of global electricity; <5% primary power). However, they are susceptible to loss-of-coolant meltdowns and their efficiencies are limited by 600 K steam temperatures exiting the boiler. (Boiler pressures are set lower in nuclear plants [≤160 bar] than in fossil-fueled plants because of radioactivity danger from explosions.)

The "passively safe" helium-cooled, pebble-bed reactor (Fig. 9) is theoretically immune to meltdowns. Helium coolant exits the core at 1200 K, driving a 40% efficient gas turbine/generator. Its graphite-coated fuel pebbles are designed not to melt even if coolant flow stops.

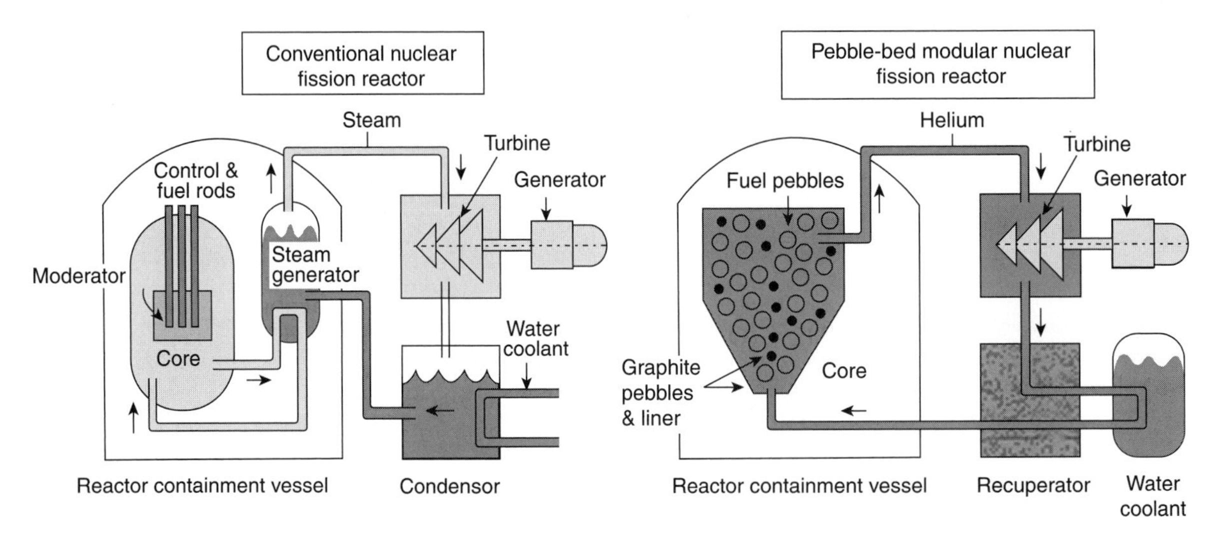

FIGURE 9 Fission. The light water reactor (LWR), primarily a burner powered by uranium slightly enriched in ^{235}U in zirconium alloy-clad fuel rods, employs H_2O as both coolant and working fluid. Power is controlled by moderator rods that increase fission rates when lowered into the core. They are susceptible to loss-of-coolant meltdowns and have efficiencies limited by 600 K steam temperatures exiting the boiler. The passively safe helium-cooled, pebble-bed reactor shown is theoretically immune to meltdowns. Helium coolant exits the core at 1200 K, driving a 40%-efficient gas turbine/generator.

6.2 Fuel Supply and Breeders

The overriding hurdle for fission as an emission-free power source is the mind-boggling scale of power buildup required: "Once through" reactors burning ^{235}U at 10 TW will exhaust identified ore deposits in 5–30 years, and mining the oceans for uranium requires seawater processing at flow rates more than five times global river runoff. Breeding fertile ^{232}Th and ^{238}U can increases fissile fuels ~ 100 times, but a massive commitment to neutron production is needed to make fuel fast enough. Moreover, diversion of nuclear fuel by terrorists and rogue regimes and disposal of high-level reactor wastes are unresolved issues.

Current best estimates of U-235 primary energy in crustal ores are 60–300 TW/year. Current estimates of uranium in proven and ultimately recoverable resources are 3.4 and 17 million tonnes, respectively. Deposits containing 500–2000 ppmw are considered recoverable. Vast amounts of uranium in shales at concentrations of 30–300 ppmw are unlikely to be recoverable because of high processing and environmental costs. The fission energy per tonne ^{235}U is

$$\frac{202\ \text{MeV}}{235\ \text{amu} \times 1.661 \times 10^{-27}\ \text{kg/amu}} \times \frac{1.602 \times 10^{-13}\ \text{J}}{1\ \text{MeV}} \times \frac{10^3\ \text{kg}}{1\ \text{tonne}} = 8.3 \times 10^{16} \frac{\text{J}}{\text{tU-235}}.$$

The energy content of the U-235 isotope in natural uranium is therefore $0.0072 \times 8.3 \times 10^{16}\ \text{J}\,\text{t}^{-1} \sim 5.7 \times 10^{14}\ \text{J}\,\text{t}^{-1}$, and the energy in proven uranium reserves and ultimately recoverable resources is ~ 61 and 300 TW/year, respectively (1 TW/year = 31.5 EJ). At 10 TW, reserves and ultimate resources of fissionable uranium last only 6 and 30 years, respectively. However, the recoverable uranium ore may be underestimated due to lack of exploration incentives.

What about the seas? Japanese researchers propose harvesting dissolved uranium with organic particle beds immersed in flowing seawater. The oceans contain 3.2×10^{-6} kg of dissolved uranium per cubic meter—a U-235 energy density of 1.8 million J/m^3. Multiplying by the oceans' huge volume ($1.37 \times 10^{18}\ \text{m}^3$) gives large numbers: 4.4 billion tonnes of uranium and 80,000 TW/year in U-235. Even with 100% U-235 extraction, the volumetric flow rate to make reactor fuel at the 10-TW rate is five times the outflow of all Earth's rivers, which is $1.2 \times 10^6\ \text{m}^3\,\text{s}^{-1}$. By comparison, the flow rate through a hypothetical seawater extraction plant to yield ^{235}U fuel at 10 TW is

$$10\ \text{TW} \times \frac{1.37 \times 10^{18}\text{m}^3}{8 \times 10^4\ \text{TW year}} \times \frac{1\ \text{year}}{3.15 \times 10^7\text{s}} = 5.4 \times 10^6\text{m}^3\text{s}^{-1},$$

approximately five times global river runoff.

Fissionable fuels could be made by breeding Pu-239 and/or U-233, which do not exist in nature. They must be bred from their feed stocks, U-238 ($\sim$99% of uranium) and Th-232 ($\sim$100% of thorium), using suitable neutron sources ($^{238}U + n \rightarrow {}^{239}Pu$; $^{232}Th + n \rightarrow {}^{233}U$). Fission energy per unit mass is comparable to U-235 ($^{239}Pu + n \rightarrow$ fission products $+ 2.9n + 210$ MeV; $^{233}U + n \rightarrow$ fission products $+ 2.5n + 198$ MeV). We estimate that breeding could increase fission energy resources by factors of 60 and 180 for Pu-239 and U-233, respectively. Breeding rates depend on feedstock mining rates and availability of neutrons.

Breeders might be more attractive if fuel cycles safely transmuting high-level wastes could be developed. The conclusion is inescapable that breeders have to be built very soon for fission to significantly displace CO_2 emissions. If one goes the breeder route, then thorium is a preferable feedstock to uranium because U-233 is more difficult than plutonium to separate from nuclear wastes and divert to weapons.

6.3 Fusion

The best hope for sustainable carbon emission-free nuclear power may not be fission but, rather, fusion. The focus so far has been on the D–tritium (T) reaction ($D + T \rightarrow {}^4He + n + 17.7$ MeV). Controlled fusion is difficult. Temperatures in the 30- to 300-million K range are needed, hence the skepticism about "cold fusion." To ignite the plasma, energy is normally input by neutral particle beams or lasers. In a reactor, energy would emerge as neutrons penetrating reactor walls to a liquid lithium blanket ultimately driving steam turbines.

The most successful path to fusion has been confining a D + T plasma with complex magnetic fields and heating it to ignition by neutral particle beams in a toroidal near-vacuum chamber (tokamak; Fig. 10). Classically, the 100- to 300-million K plasma should be contained by the magnetic fields, but in the real world "anomalous" diffusion (turbulence) has taken decades to suppress. "Breakeven" is when input power equals the power of fusion

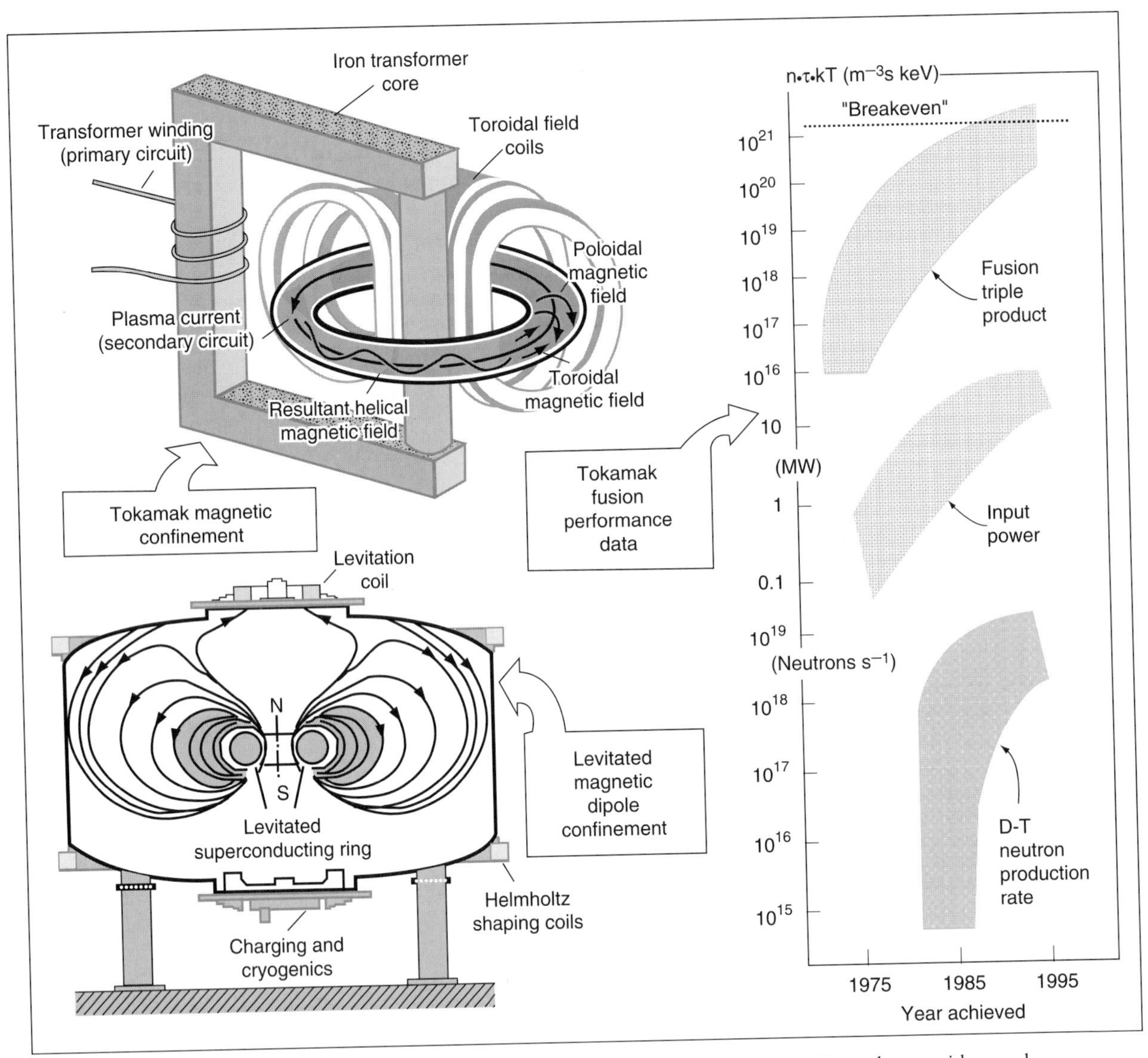

FIGURE 10 Fusion. The most successful path to fusion has been confining a deuterium + tritium plasma with complex magnetic fields and heating it to ignition by neutral particle beams (not shown) in a toroidal near-vacuum chamber (tokamak). Breakeven is when input power equals the power of fusion neutrons hitting the innermost reactor wall. This requires the fusion triple product (number density × confinement time × temperature) to equal a critical value. Despite progress, engineering breakthroughs are needed for fusion electricity to be viable. Experimental work on advanced fusion fuel cycles is being pursued at the National Ignition Facility and on simpler magnetic confinement schemes such as the levitated dipole experiment shown.

neutrons hitting the innermost reactor wall. This requires the fusion triple product (number density × confinement time × temperature) to equal a critical value. Recent tokamak performance improvements have been dramatic, capped by near breakeven. Commercial plants would surround the torus wall with neutron-absorbing molten lithium blankets where tritium is bred and heat exchanged with turbogenerators—the latter feeding electricity back to heaters and magnets, with excess power to consumers. Despite progress, engineering breakthroughs are needed for fusion electricity to be viable. Short-term experiments are advocated on accelerated breeding of fissile fuels from the high neutron flux in D–T fusion blankets. Fission–fusion hybrids could impact CO_2 stabilization by midcentury, even if pure fusion is delayed. (Current nuclear weapons, for example, use both fission and fusion nuclear reactions to create powerful explosions.) Experimental work includes simpler magnetic confinement schemes such as the levitated dipole experiment shown in Fig. 10.

Currently, there is no fusion-generated electricity. However, researchers have very nearly achieved breakeven [$Q =$ (neutron or charged particle energy out)/(energy input to heat plasma) $= 1$] with tokamak magnetic confinement proposed in the 1950s by Igor Tamm and Andrei Sakharov. Reactor power balances indicate that $Q = 1$ for the D–T reaction requires the product of number density n, confinement time τ, and temperature T to satisfy the Lawson criteria: $n \times \tau \times kT \sim 1 \times 10^{21}\,\mathrm{m^{-3}\,s\,keV}$, where $k = 0.0862$ keV/MK. Experimental results from the TFTR and JET tokamaks are within a factor of two of this value. The Lawson criteria for breakeven with D–D and D-^{3}He reactions requires a factor of 10 or higher triple product; net power-generating reactors require even higher Qs and triple products.

Tritium has to be bred from lithium ($n + {}^6\mathrm{Li} \rightarrow {}^4\mathrm{He} + \mathrm{T} + 4.8\,\mathrm{MeV}$). Reactors burning D–T mixtures are thus lithium burners. Lithium in crustal ores bred to tritium could generate 16,000 TW/year, approximately twice the thermal energy in fossil fuels. This is no infinite power source, although more lithium might be mined from the sea. Deuterium in the oceans is virtually unlimited, whether utilized in the D–T reaction or the more difficult to ignite D–D reactions ($\rightarrow {}^3\mathrm{He} + n + 3.2\,\mathrm{MeV}$ and $\rightarrow \mathrm{T} + p + 4.0\,\mathrm{MeV}$). D–T tokamaks give the best plasma confinement and copious neutrons.

New devices with either a larger size or a larger magnetic field strength are required for net power generation. Demonstrating net electric power production from a self-sustaining fusion reactor could be a breakthrough of overwhelming importance. A more near-term idea is fission–fusion hybrids. These could breed fissionable fuel from thorium mixed with lithium in tokamak blankets. Each 14-MeV fusion neutron, after neutron multiplication in the blanket, can breed one ^{233}U atom (burned separately in a nuclear power plant) and one T atom. Since $^{233}\mathrm{U} + n$ fission releases $\sim$200 MeV, the power output of a fission–fusion plant could be $\sim$15 times that of a fusion plant alone. A fusion breeder could support many more satellite reactors than a fission breeder. Fusion is energy poor but neutron rich, whereas fission is energy rich but neutron poor: Their marriage might breed fissionable fuel fast enough to help stabilize the fossil fuel greenhouse. Our analysis indicates that a fission–fusion hybrid plant based on existing tokamak technology but requiring major experimental confirmation might begin breeding fuel for 300-MW reactors in 10–15 years. However, the fission–fusion breeder path will also require (actually and perceived) safe ^{233}U fission reactor and fuel cycles by midcentury.

More difficult to ignite than D–T, there is renewed interest in the D–^{3}He reaction ($\mathrm{D} + {}^3\mathrm{He} \rightarrow {}^4\mathrm{He} + p + 18.3\,\mathrm{MeV}$), which yields charged particles directly convertible to electricity with little radioactivity (some neutrons may be produced by side reactions). Studies of D–^{3}He and D–D burning in inertial confinement fusion targets are being conducted at Lawrence Livermore National Laboratory. Computer modeling suggests that fast-ignited D–T spark plugs can facilitate ignition. Moreover, targets can be designed to release the vast majority of their fusion energy as high-energy charged particles leading to applications such as high-specific-impulse propulsion for deep-space missions and direct energy conversion for power generation. Experiments are also under way testing dipole confinement by a superconducting magnet levitated in a vacuum chamber. This method of plasma confinement was inspired by satellite observations of energetic plasma within the magnetospheres of the outer planets, and it may be a prototype for D–^{3}He fusion since theories predict low heat conduction at high plasma pressure. Rare on Earth, ^{3}He may be cost-effective to mine from the moon and is even more abundant in (and potentially recoverable from) atmospheres of gas giant planets. Deuterium in seawater and outer planet ^{3}He could provide global-scale power for a longer period than from any source besides the sun.

Fission and fusion offer several options to combat global warming. However, they can only become operational by midcentury if strategic technologies are pursued aggressively now through research and development and, in the case of nuclear fission, if outstanding issues of high-level waste disposal and weapons proliferation are satisfactorily resolved.

7. GEOENGINEERING

No discussion of global warming mitigation would be complete without mentioning geoengineering, which is also called climate engineering or planetary engineering on Earth and terraforming on other planets. CO_2 capture and sequestration having entered the mainstream, geoengineering today refers mainly to altering the planetary radiation balance to affect climate. Roger Revelle and Hans Suess recognized early on that mining and burning fossil fuels to power our industrial civilization was an

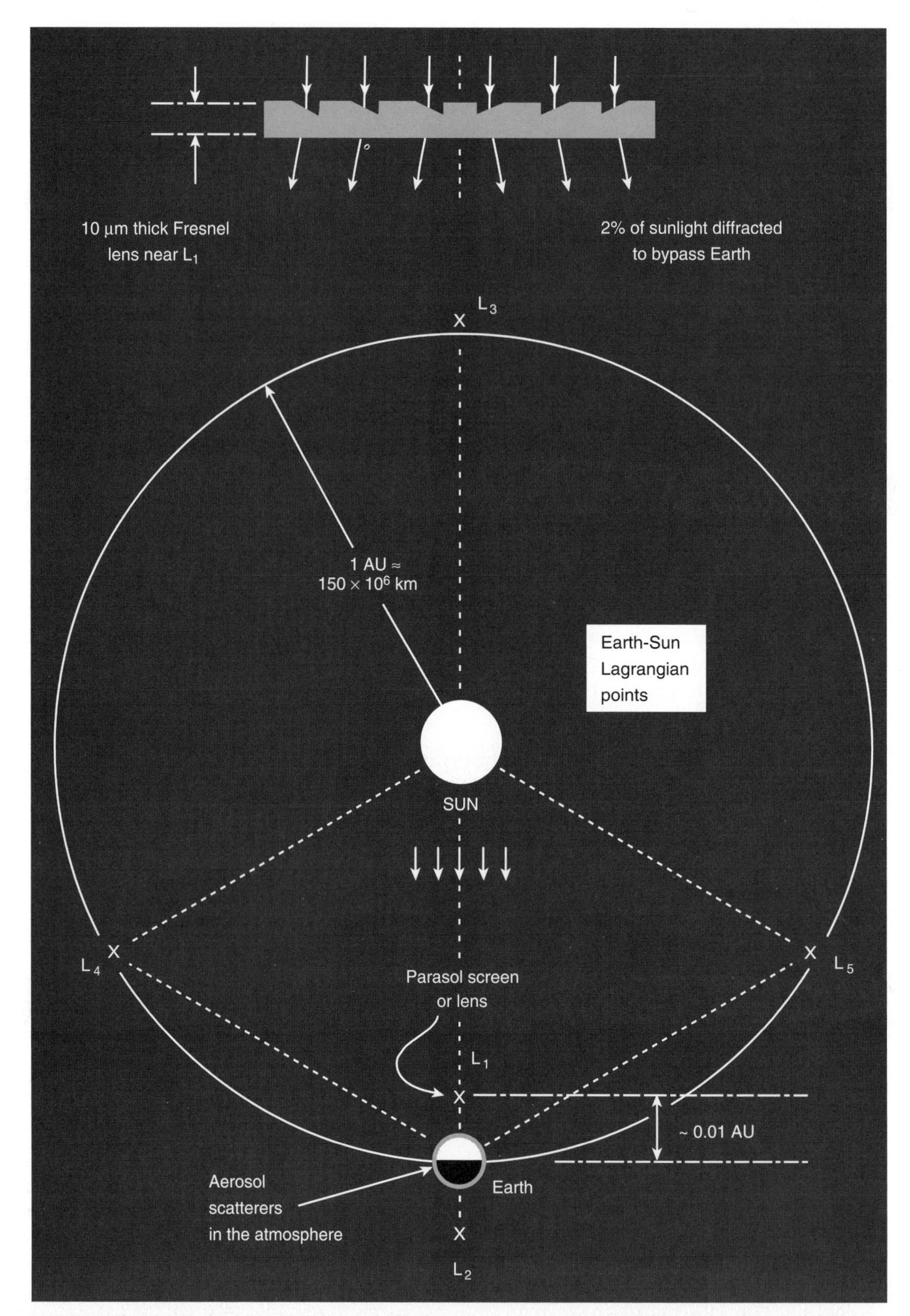

FIGURE 11 Space-based geoengineering. A 2000-km-diameter parasol near L1 could deflect 2% of incident sunlight, roughly compensating for the radiative forcing from CO_2 doubling. Deployment of stratospheric aerosols with optical properties engineered to compensate for adverse climatic change is also proposed, although continuous injections of upper-atmosphere particles to maintain 2% sunblocking could have unintended adverse consequences.

unprecedented technological intervention in the carbon cycle:

> *Human beings are now carrying out a large-scale geophysical experiment of the kind that could not have happened in the past nor be reproduced in the future. Within a few centuries we are returning to the atmosphere and oceans the concentrated organic carbon stored in sedimentary rocks over hundreds of millions of years.*

Geoengineering employs technologies to compensate for the inadvertent global warming produced by fossil fuel CO_2 and other greenhouse gases. An early idea by Soviet climatologist Mikhail Budyko would put layers of reflective sulfate aerosol in the upper atmosphere to counteract greenhouse warming. Variations on the sunblocking theme include injecting submicron dust to the stratosphere in shells fired by naval guns and increasing cloud cover by seeding and shadowing Earth by objects in space. Perhaps most ambitious is James Early's proposed 2000-km-diameter mirror of 10-μm glass fabricated from lunar materials at the L_1 Lagrange point of the sun–earth system illustrated in Fig. 11. The mirror's surface would look like a permanent sunspot, deflect 2% of solar flux, and roughly compensate for the radiative forcing of a CO_2 doubling. Climate model runs indicate that the spatial pattern of climate would resemble that without fossil fuel CO_2. Among other things, this scheme would require renegotiating existing treaties relating to human activities in space.

The Lagrange interior point L_1, one of the five libration points corotating with the earth–sun system, provides an opportunity for negative radiative forcing opposing global warming. However, solar parasols are currently more interesting as thought experiments than as real technology options. Deployment of stratospheric aerosols with optical properties engineered to compensate for adverse climatic change has also been proposed, although continuous injections of upper-atmosphere particles to maintain 2% sunblocking could have unintended adverse consequences. A better investment may be space power, which can produce $10\,TW_e$ (equivalent to $30\,TW$ primary) with $<3\%$ of the disk area needed for a $2 \times CO_2$ compensating parasol or aerosol layer and provide long-term renewable energy for Earth.

It seems prudent to pursue geoengineering research as an insurance policy should global warming prove worse than anticipated and conventional measures to stabilize CO_2 and climate fail or prove too costly. However, large-scale geophysical interventions are inherently risky and need to be approached with caution.

8. CONCLUSION

Even as reliable evidence for global warming accumulates, the fundamental dependence of our civilization on the oxidation of coal, oil, and gas for energy makes it difficult to respond appropriately. This condition will likely become more acute as the world economy grows and as ever larger reductions in CO_2-emitting energy relative to growing total energy demand are needed to stabilize atmospheric CO_2. Energy is critical to our aspirations for global prosperity and equity. If Earth continues to warm, as predicted, the people of Earth may turn to advanced, non-CO_2-emitting energy technologies for solutions. Combating global warming by radical restructuring of the global energy system may be a major research and development challenge of the century. It is likely to be a difficult but rewarding enterprise—one that could lead to sustainable global-scale power systems compatible with planetary ecosystems.

It is unlikely that a single "silver bullet" will solve this problem. We have identified a portfolio of promising advanced energy technologies here, many of which are radical departures from our current fossil fuel system. There are doubtless other good ideas. Inevitably, many concepts will fail, and staying the course will require leadership. Consider how John F. Kennedy voiced the goals of the Apollo program: "We choose to go to the moon in this decade, and to do the other things, not because they are easy, but because they are hard. Because that goal will serve to organize and measure the best of our energies and skills."

Stabilizing global climate change is not likely to be easy. At the very least, it will require political will, targeted research and development, and international cooperation. Most of all, it will require the recognition that, although regulation can play a role, the fossil fuel greenhouse is an energy problem that cannot be simply regulated away. Solutions need to be invented, tested, and implemented—an effort that could "organize and measure the best of our energies and skills."

SEE ALSO THE FOLLOWING ARTICLES

Biomass for Renewable Energy and Fuels • *Biomass: Impact on Carbon Cycle and Greenhouse Gas Emissions* • *Carbon Capture and Storage from Fossil Fuel Use* • *Carbon Sequestration, Terrestrial* • *Carbon*

Taxes and Climate Change • Climate Change and Public Health: Emerging Infectious Diseases • Climate Change: Impact on the Demand for Energy • Climate Protection and Energy Policy • Energy Efficiency and Climate Change • Greenhouse Gas Emissions from Energy Systems, Comparison and Overview

Further Reading

Department of Energy "The Vision 21 Energy Plant of the Future." Available at http://www.fe.doe.gov.

Fowler, T. K. (1997). "The Fusion Quest." Johns Hopkins Univ. Press, Baltimore.

Glaser, P. E., Davidson, F. P., and Csigi, K. I. (eds.). (1997). "Solar Power Satellites." Wiley–Praxis, New York.

Hoffert, M. I., Caldeira, K., Jain, A. K., Haites, E. F., Harvey, L. D. D., Potter, S. D., Schlesinger, M. E., Schneider, S. H., Watts, R. G., Wigley, T. M. L., and Wuebbles, D. J. (1998). Energy implications of future stabilization of atmospheric CO_2 content. *Nature* 395, 881–884.

Hoffert, M. I., Caldeira, K., Benford, G., Criswell, D. R., Green, C., Herzog, H., Katzenberger, J. W., Kheshgi, H. S., Lackner, K. S., Lewis, J. S., Manheimer, W., Mankins, J. C., Marland, G., Mauel, M. E., Perkins, L. J., Schlesinger, M. E., Volk, T., and Wigley, T. M. L. (2002). Advanced technology paths to global climate stability: Energy for a greenhouse planet. *Science* **295**, 981–987.

Johansson, T. B., Kelly, H., Reddy, A. K. N., and Williams, R. H. (eds.). (1993). "Renewable Energy." Island Press, Washington, DC.

Metz, B., Davidson, O., Swart, R., and Pan, J. (eds.). (2001). "Climate Change 2001: Mitigation: Contribution of Working Group III to the Third Assessment Report of the Intergovernmental Panel on Climate Change." Cambridge Univ. Press, New York.

Nakicenovic, N., and Swart, R. (eds.). (2001). "Emissions Scenarios: A Special Report of Working Group III of the Intergovernmental Panel on Climate Change." Cambridge Univ. Press, New York.

Nakicenovic, N., Grübler, A., and McDonald, A. (1998). "Global Energy Perspectives." Cambridge Univ. Press, New York.

National Laboratory Directors (1997). "Technology Opportunities to Reduce U.S. Greenhouse Gas Emissions." U.S. Department of Energy, Washington, D.C. Available at http://www.ornl.gov.

National Research Council (1996). "Nuclear Wastes: Technologies for Separations and Transmutations." National Academy Press, Washington, DC.

Reichle, D., *et al.* (1999). "Carbon Sequestration Research and Development." Office of Fossil Energy, Department of Energy, Washington, DC. Available at http://www.ornl.gov.

Smil, V. (1999). "Energies: An Illustrated Guide to the Biosphere and Civilization." MIT Press, Cambridge, MA.

Watts, R. G. (2002). "Innovative Energy Systems for CO_2 Stabilization." Cambridge Univ. Press, New York.

Climate Change and Public Health: Emerging Infectious Diseases

PAUL R. EPSTEIN
Harvard Medical School Center for Health and the Global Environment
Boston, Massachusetts, United States

1. Background on Climate Change
2. Climate and Infectious Disease
3. Pest Control: One of Nature's Services
4. Climate Change and Biological Responses
5. Sequential Extremes and Surprises
6. Case Study: West Nile Virus
7. Discontinuities
8. Conclusion
9. Next Steps

Glossary

climate change The heating of the inner atmosphere, oceans, and land surfaces of the earth. The warming is associated with more intense extreme weather events and the altered timing, intensity, and distribution of precipitation.

emerging infectious diseases Those diseases new to medicine since 1976, plus old diseases undergoing resurgence and redistribution.

Climate restricts the range of infectious diseases, whereas weather affects the timing and intensity of outbreaks. Climate change scenarios project a change in the distribution of infectious diseases with warming and changes in outbreaks associated with weather extremes, such as flooding and droughts. The hypothesis of this article is that the ranges of several key diseases or their vectors are already changing in altitude due to warming, along with shifts in plant communities and the retreat of alpine glaciers. In addition, more intense and costly weather events create conditions conducive to outbreaks of infectious diseases, as heavy rains leave insect breeding sites, drive rodents from burrows, and contaminate clean water systems. Conversely, drought can spread fungal spores and spark fires (and associated respiratory disease). In addition, sequences of extremes can destabilize predator/prey interactions, leading to population explosions of opportunistic, disease-carrying organisms. Advances in climate forecasting and health early warning systems may prove helpful in catalyzing timely, environmentally friendly public health interventions. If climate change continues to be associated with more frequent and volatile and severe weather events, we have begun to see the profound consequences that climate change can have for public health and the international economy.

Epidemics are like signposts from which the statesman of stature can read that a disturbance has occurred in the development of his nation—that not even careless politics can overlook.

—*Rudolf Virchow (1848)*

1. BACKGROUND ON CLIMATE CHANGE

The climate system can remain stable over millennia due to interactions and feedbacks among its basic components: the atmosphere, oceans, ice cover, biosphere, and energy from the sun. Harmonics among the six orbital (Milankovitch) cycles (e.g., tilt and eccentricity) of the earth about the sun, as revealed by analyses of ice cores and other "paleothermometers" (e.g., tree rings and coral cores), have governed the oscillations of Earth's climate between ice ages and warm periods until the 20th century. To explain the global warming during the 20th century of approximately 1°C, according to all studies reviewed by the Intergovernmental Panel on Climate Change, one must invoke the role of heat-trapping greenhouse gases (GHGs), primarily carbon dioxide (CO_2), oxides of nitrogen, methane, and chlorinated hydrocarbons. These gases have been steadily accumulating in the lower atmosphere (or troposphere), out to about 10 km, and have altered the heat budget of the atmosphere, the world ocean, land surfaces, and the cryosphere (ice cover).

For the past 420,000 years, as measured by the Vostok ice core in Antarctica, CO_2 has stayed within an envelope of between 180 and 280 parts per million (ppm) in the troposphere. Today, the level of CO_2 is 370 ppm and the rate of change during the past century surpassed that observed in ice core records. Ocean and terrestrial sinks for CO_2 have presumably played feedback roles throughout millennia. Today, the combustion of fossil fuels (oil, coal, and natural gas) is generating CO_2 and other GHGs, and the decline in sinks, primarily forest cover, accounts for 15–20% of the buildup.

1.1 Climate Stability

As important as the warming of the globe is to biological systems and human health, the effects of the increased extreme and anomalous weather that accompanies the excess energy in the system may be even more profound. As the rate of warming accelerated after the mid-1970s, anomalies and wide swings away from norms increased, suggesting that feedback, corrective mechanisms in the climate system are being overwhelmed. Indeed, increased variability may presage transitions: Ice core records from the end of the last ice age ($\sim$10,000 years ago) indicate that increased variability was associated with rapid change in state.

Further evidence for instability comes from the world ocean. Although the ocean has warmed overall in the past century, a region of the North Atlantic has cooled in the past several decades. Several aspects of global warming are apparently contributing.

Recent warming in the Northern Hemisphere has melted much North Polar ice. Since the 1970s, the floating North Polar ice cap has thinned by almost half. A second source of cold fresh water comes from Greenland, where continental ice is melting at higher elevations each year. Some meltwater is trickling down through crevasses, lubricating the base, accelerating ice "rivers," and increasing the potential for sudden slippage. A third source of cold fresh water is rain at high latitudes. Overall ocean warming speeds up the water cycle, increasing evaporation. The warmed atmosphere can also hold and transport more water vapor from low to high latitudes. Water falling over land is enhancing discharge from five major Siberian rivers into the Arctic, and water falling directly over the ocean adds more fresh water to the surface.

The cold, freshened waters of the North Atlantic accelerate transatlantic winds, and this may be one factor driving frigid fronts down the eastern U.S. seaboard and across to Europe and Asia in the winters of 2002–2004. The North Atlantic is also where deep-water formation drives thermohaline circulation, the "ocean conveyor belt," considered key to climate stabilization. In the past few years, the northern North Atlantic has freshened, and since the 1950s the deep overflow between Iceland and Scotland has slowed by 20%.

The ice, pollen, and marine fossils reveal that cold reversals have interrupted warming trends in the past. The North Atlantic Ocean can freshen to a point at which the North Atlantic deep-water pump—driven by sinking cold, salty water that is in turn replaced by warm Gulf Stream waters—can suddenly slow. Approximately 13,000 years ago, when the world was emerging from the Last Glacial Maximum and continental ice sheets were thawing, the Gulf Stream abruptly changed course and shot straight across to France. The Northern Hemisphere refroze—for the next 1300 years—before temperatures increased again, in just several years, warming the world to its present state.

Calculations (of orbital cycles) indicate that our hospitable climate regime was not likely to end due to natural causes any time soon. However, the recent buildup of heat-trapping greenhouse gases is forcing the climate system in new ways and into uncharted seas.

1.2 Hydrological Cycle

Warming is also accelerating the hydrological (water) cycle. As heat builds up in the deep ocean, down to 3 km, more water evaporates and sea ice melts. During the past century, droughts have lasted longer and heavy rainfall events (defined as >5 cm/day) have become more frequent. Enhanced evapotranspiration dries out soils in some regions, whereas the warmer atmosphere holds more water vapor, fueling more intense, tropical-like downpours elsewhere. Prolonged droughts and intense precipitation have been especially punishing for developing nations.

Global warming is not occurring uniformly. It is occurring twice as fast as overall warming during the winter and nighttime, a crucial factor in the biological responses, and the winter warming is occurring faster at high latitudes than near the topics. These changes may be due to greater evaporation and the increased humidity in the troposphere because water vapor is a natural greenhouse gas and can account for up to two-thirds of all the heat trapped in the troposphere. Warming nights and winters along with the intensification of extreme weather events have begun to alter weather patterns that impact the ecological systems essential for regulating the vectors, hosts, and reservoirs of infectious diseases.

Other climate-related health concerns include temperature and mortality, especially the role of increased variability in heat and cold mortality; synergies between climate change and air pollution, including CO_2 fertilization of ragweed and excess pollen production, asthma, and allergies; travel hazards associated with unstable and erratic winter weather; and genetic shifts in arthropods and rodents induced by warming. This article focuses on climate change and emerging infectious diseases.

2. CLIMATE AND INFECTIOUS DISEASE

Climate is a key determinant of health. Climate constrains the range of infectious diseases, whereas weather affects the timing and intensity of outbreaks. A long-term warming trend is encouraging the geographic expansion of several important infections, whereas extreme weather events are spawning "clusters" of disease outbreaks and a series of surprises. Ecological changes and economic inequities strongly influence disease patterns. However, a warming and unstable climate is playing an ever-increasing role in driving the global emergence, resurgence, and redistribution of infectious diseases.

The World Health Organization reports that since 1976, more than 30 diseases have appeared that are new to medicine, including HIV/AIDS, Ebola, Lyme disease, Legionnaires' disease, toxic *Escherichia coli*, a new hantavirus, and a rash of rapidly evolving antibiotic-resistant organisms. Of equal concern is the resurgence of old diseases, such as malaria and cholera. Declines in social conditions and public health programs underlie the rebound of diseases transmitted person to person (e.g., tuberculosis and diphtheria). The resurgence and redistribution of infections involving two or more species—mosquitoes, ticks, deer, birds, rodents, and humans—reflect changing ecological and climatic conditions as well as social changes (e.g., suburban sprawl).

Waves of infectious diseases occur in cycles. Many upsurges crest when populations overwhelm infrastructures or exhaust environmental resources, and sometimes pandemics can cascade across continents. Pandemics, in turn, can affect the subsequent course of history. The Justinian plague emerged out of the ruins of the Roman Empire in the 6th-century AD and arrested urban life for centuries. When the plague, carried by rodents and fleas, reappeared in the repopulated and overflowing urban centers of the 14th century, it provoked protests and helped end feudal labor patterns. In "Hard Times," Charles Dickens describes another transitional period in the overcrowded 19th-century England:

> *Of tall chimneys, out of which interminable serpents of smoke trailed themselves forever and ever, and never got uncoiled ... where the piston of the steam-engine worked monotonously up and down like the head of an elephant in a state of melancholy madness.*

These conditions bred smallpox, cholera, and tuberculosis. However, society responded with sanitary and environmental reform, and the epidemics abated. How will our society respond to the current threat to our health and biological safety?

2.1 An Integrated Framework for Climate and Disease

All infections involve an agent (or pathogen), host(s), and the environment. Some pathogens are carried by vectors or require intermediate hosts to complete their life cycle. Climate can influence pathogens, vectors, host defenses, and habitat.

Diseases carried by mosquito vectors are particularly sensitive to meteorological conditions. These relationships were described in the 1920s and quantified in the 1950s. Excessive heat kills mosquitoes. However, within their survivable range, warmer temperatures increase their reproduction and biting activity and the rate at which pathogens mature within them. At 20°C, falciparum malarial protozoa take 26 days to incubate, but at 25°C they develop in 13 days. *Anopheline* mosquitoes, which are carriers of malaria, live only several weeks. Thus, warmer temperatures permit parasites to mature in time for the mosquito to transfer the infection.

Temperature thresholds limit the geographic range of mosquitoes. Transmission of *Anopheline*-borne falciparum malaria occurs where temperatures exceed 16°C. Yellow fever (with a high rate of mortality) and dengue fever (characterized by severe headaches and bone pain, with mortality associated with dengue hemorrhagic fever and dengue shock syndrome) are both carried by *Aedes aegypti*, which is restricted by the 10°C winter isotherm. Freezing kills *Aedes* eggs, larvae, and adults. Thus, given other conditions, such as small water containers, expanding tropical conditions can increase the ranges and extend the season with conditions allowing transmission.

Warm nights and warm winters favor insect survival. Fossils from the end of the last ice age demonstrate that rapid, poleward shifts of insects accompanied warming, especially of winter temperatures. Insects, notably Edith's checkerspot butterflies today, are superb paleothermometers, outpacing the march of grasses, shrubs, trees, and mammals in response to advancing frost lines.

In addition to the direct impacts of warming on insects, volatile weather and warming can disrupt co-evolved relationships among species that help to prevent the spread of "nuisance" species.

3. PEST CONTROL: ONE OF NATURE'S SERVICES

Systems at all scales have self-correcting feedback mechanisms. In animal cells, errors in structural genes (mismatched base pairs) resulting from radiation or chemicals are "spell-checked" by proteins propagated by regulatory genes. Malignant cells that escape primary controls must confront an ensemble of instruments that comprise the immune surveillance system. A suite of messengers and cells also awaits invading pathogens—some that stun them, and others, such as phagocytes, that consume them.

Natural systems have also evolved a set of pheromones and functional groups (e.g., predators, competitors, and recyclers) that regulate the populations of opportunistic organisms. The diversity of processes provides resistance, resilience, and insurance, whereas the mosaics of habitat—stands of trees near farms that harbor birds and nectar-bearing flowers that nourish parasitic wasps—provide generalized defenses against the spread of opportunists. Against the steady background beat of habitat fragmentation, excessive use of toxins, and the loss of stratospheric ozone—all components of global environmental change—climate change is fast becoming a dominant theme, disrupting relationships among predators and prey that prevent the proliferation of pests and pathogens.

4. CLIMATE CHANGE AND BIOLOGICAL RESPONSES

Northern latitude ecosystems are subjected to regularly occurring seasonal changes. However, prolonged extremes and wide fluctuations in weather may overwhelm ecological resilience, just as they may undermine human defenses. Summer droughts depress forest defenses, increasing vulnerability to pest infestations. Also, sequential extremes and shifting seasonal rhythms can alter synchronies among predators, competitors, and prey, releasing opportunists from natural biological controls.

Several aspects of climate change are particularly important to the responses of biological systems. First, global warming is not uniform. Warming is occurring disproportionately at high latitudes, above Earth's surface and during winter and nighttime. Areas of Antarctica, for example, have already warmed over 1°C this century, and the temperatures within the Arctic Circle have warmed 5.5°C in the past 30 years. Since 1950, the Northern Hemispheric spring has occurred earlier and fall later.

Although inadequately studied in the United States, warm winters have been demonstrated to facilitate overwintering and thus northern migration of the ticks that carry tick-borne encephalitis and Lyme disease. Agricultural zones are shifting northward but not as swiftly as are key pests, pathogens, and weeds that, in today's climate, consume 52% of the growing and stored crops worldwide.

An accelerated hydrological cycle—ocean warming, sea ice melting, and rising atmospheric water vapor—is demanding significant adjustments from biological systems. Communities of marine species have shifted. A warmer atmosphere also holds more water vapor (for each 1°C warming 6%), insulates escaping heat, and enhances greenhouse warming. Warming and parching of Earth's surface intensifies the pressure gradients that draw in winds (e.g., winter tornadoes) and large weather systems.

Elevated humidity and lack of nighttime relief during heat waves directly challenge human (and livestock) health. These conditions also favor mosquitoes.

4.1 Range Expansion of Mosquito-Borne Diseases

Today, half of the world's population is exposed to malaria on a daily basis. Deforestation, drug resistance, and inadequate public health measures have all contributed to the recent resurgence. Warming and extreme weather add new stresses. Dynamic models project that the warming accompanying the doubling of atmospheric CO_2 will increase the transmission capacity of mosquitoes in temperate zones, and that the area capable of sustaining transmission will increase from that containing 45% of to world's population to that containing 60%, although recent statistical modeling projects less of a change. All these analyses rely on average temperatures, rather than the more rapid changes in minimum temperatures being observed, and thus may underestimate the biological responses.

In addition, historical approaches to understanding the role of temperature and infectious disease have argued that the relationships do not hold for periods such as the medieval warm period and the little ice age. It is important to note, however, that the changes in CO_2 and temperature, and their rates of change, during the 20th century are outside the bounds of those observed during the entire Holocene (the past 10,000 years).

Some of these projected changes may be under way. Since 1976, several vector-borne diseases (VBDs) have reappeared in temperate regions. *Anopheline* mosquitoes have long been present in North America and malaria circulated in the United States in the early 20th century. However, by the 1980s, transmission was limited to California after mosquito control programs were instituted. Since 1990, small outbreaks of locally transmitted malaria have occurred during hot spells in Texas, Georgia, Florida, Michigan, New Jersey, New York, and Toronto. Malaria has returned to South Korea, areas of southern Europe, and the former Soviet Union. Malaria has recolonized the Indian Ocean coastal province of South Africa, and dengue fever has spread southward into northern Australia and Argentina.

These changes are consistent with climate projections, although land clearing, population movements, and drug and pesticide resistance for malaria control have all played parts. However, a set of changes occurring in tropical highland regions are internally consistent and indicative of long-term warming.

4.2 Climate Change in Montane Regions

In the 19th century, European colonists sought refuge from lowland "mal arias" by settling in the highlands of Africa. These regions are now getting warmer. Since 1970, the height at which freezing occurs (the freezing isotherm) has increased approximately 160 m within the tropical belts, equivalent to almost 1°C warming. These measurements are drawn from released weather balloons and satellites.

Plants are migrating to higher elevations in the European Alps, Alaska, the U.S. Sierra Nevada, and New Zealand. This is a sensitive gauge because a plant shifting upward 500 m would have to move 300 km northward to adjust to the same degree of global warming.

Insects and insect-borne diseases are being reported at high elevations in eastern and central Africa, Latin America, and Asia. Malaria is circulating in highland urban centers, such as Nairobi, and rural highland areas, such as those of Papua New Guinea. *Aedes aegypti*, once limited by temperature to approximately 1000 m in elevation, has recently been found at 1700 m in Mexico and 2200 m in the Colombian Andes.

These insect and botanical trends, indicative of gradual, systematic warming, have been accompanied by the hardest of data: the accelerating retreat of summit glaciers in Argentina, Peru, Alaska, Iceland, Norway, the Swiss Alps, Kenya, the Himalayas, Indonesia, Irian Jaya, and New Zealand. Glaciers in the Peruvian Andes, which retreated 4 m annually in the 1960s and 1970s, were melting 30 m per year by the mid-1990s and 155 m per year by 2000. Many small ice fields may soon disappear, jeopardizing regional water supplies critical for human consumption, agriculture, and hydropower.

Highlands, where the biological, glacial, and isotherm changes are especially apparent, are sensitive sentinel sites for monitoring the long-term impacts of climate change.

4.3 Extreme Weather Events and Epidemics

Although warming encourages the spread of infectious diseases, extreme weather events are having the most profound impacts on public health and society. The study of variability also provides insights into the stability of the climate regime.

A shift in temperature norms alters the variance about the means, and high-resolution ice core records suggest that greater variance from climate norms indicates sensitivity to rapid change. Today, the enhanced hydrological cycle is changing the intensity, distribution, and timing of extreme weather events. Large-scale weather patterns have shifted. Warming of the Eurasian land surface, for example, has apparently intensified the monsoons that are strongly associated with mosquito- and water-borne diseases in India and Bangladesh. The U.S. southwest monsoons may also have shifted, with implications for disease patterns in that region.

Extremes can be hazardous for health. Prolonged droughts fuel fires, releasing respiratory pollutants. Floods foster fungi, such as the house mold *Stachybotrys atra*, which may be associated with an emerging hemorrhagic lung disease among children. Floods leave mosquito-breeding sites and also flush pathogens, nutrients, and pollutants into waterways, precipitating water-borne diseases (e.g., *Cryptosporidium*).

Runoff of nutrients from flooding can also trigger harmful algal blooms along coastlines that can be toxic to birds, mammals, fish, and humans; generate hypoxic "dead zones"; and harbor pathogens such as cholera.

4.4 The El Niño/Southern Oscillation

The El Niño/Southern Oscillation (ENSO) is one of Earth's coupled ocean–atmospheric systems that apparently helps to stabilize the climate system by undulating between states every 4 or 5 years. ENSO events are accompanied by weather anomalies that are strongly associated with disease outbreaks over time and with spatial clusters of mosquito-, water-, and rodent-borne illnesses. The ENSO cycle also affects the production of plant pollens, which are directly boosted by CO_2 fertilization, a finding that warrants further investigation as a possible contributor to the dramatic increase in asthma since the 1980s.

Other climate modes contribute to regional weather patterns. The North Atlantic Oscillation is a seesaw in sea surface temperatures (SSTs) and sea level pressures that governs windstorm activity across Europe. Warm SSTs in the Indian Ocean (that have caused bleaching of more than 80% of regional coral reefs) also contribute to precipitation in eastern Africa. A warm Indian Ocean added moisture to the rains drenching the Horn of Africa in 1997–1998 that spawned costly epidemics of cholera, mosquito-borne Rift Valley fever, and malaria, and warm SSTs catalyzed the southern African deluge in February 2000.

Weather extremes, especially intense precipitation, have been especially severe for developing nations, and the aftershocks ripple through economies. Hurricane Mitch, nourished by a warmed Caribbean, stalled over Central America in November 1998 for 3 days, dumping precipitation that killed more than 11,000 people and caused more than $5 billion in damages. In the aftermath, Honduras reported 30,000 cases of cholera, 30,000 cases of malaria, and 1000 cases of dengue fever. The following year, Venezuela suffered a similar fate, followed by malaria and dengue fever. In February 2000, torrential rains and a cyclone inundated large areas of southern Africa. Floods in Mozambique killed hundreds, displaced hundreds of thousands, and spread malaria, typhoid, and cholera.

Developed nations have also begun to experience more severe and unpredictable weather patterns. In September 1999, Hurricane Floyd in North Carolina afforded an abrupt and devastating end to an extended summer drought. Prolonged droughts and heat waves are also afflicting areas of Europe. In the summer of 2003, five European nations experienced about 35,000 heat-associated deaths, plus wildfires and extensive crop failures. Extreme weather events are having long-lasting ecological and economic impacts on a growing cohort of nations, affecting infrastructure, trade, travel, and tourism.

The 1990s was a decade of extremes, each year marked by El Niño or La Niña (cold) conditions. After 1976, the pace, intensity, and duration of ENSO events quickened, and extremes became more extreme. Accumulating heat in the oceans intensifies weather anomalies; it may be modifying the natural ENSO mode. Understanding how the various climate modes are influenced by human activities, and how the modes interact, is a central scientific challenge, and the results will inform multiple sectors of society. Disasters such as the $10 billion European heat waves and Hurricane Floyd suggest that the costs of climate change will be borne by all.

5. SEQUENTIAL EXTREMES AND SURPRISES

5.1 Hantavirus Pulmonary Syndrome

Extremes followed by subsequent extremes are particularly destabilizing for biological and physical systems. Light rains followed by prolonged droughts can lead to wildfires, and warm winter rains followed by cold snaps beget ice storms. Warm winters also create snowpack instability, setting the stage for avalanches, which are triggered by heavy snowfall, freezing rain, or strong winds.

Polar researchers suspect that melting at the base of the Greenland ice sheet may be sculpting fault lines that could diminish its stability. Shrinking of Earth's ice cover (cryosphere) has implications for water (agriculture, hydropower) and for albedo [reflectivity] that influences climate stability.

The U.S. Institute of Medicine report of 1992 on emerging infectious diseases warned that conditions in the United States were ripe for the emergence of a new disease. What it did not foresee was that climate was to play a significant role in the emergence and spread of two diseases in North America: the hantavirus pulmonary syndrome in the Southwest and the West Nile virus in New York City.

5.1.1 Hantavirus Pulmonary Syndrome, U.S. Southwest, 1993

Prolonged drought in California and the U.S. southwest from 1987 to 1992 reportedly reduced predators of rodents: raptors (owls, eagles, prairie falcons, red-tailed hawks, and kestrels), coyotes, and snakes. When drought yielded to intense rains in 1993 (the year of the Mississippi floods), grasshoppers and piñon nuts on which rodents feed flourished. The effect was synergistic, boosting mice populations more than 10-fold, leading to the emergence of a "new," lethal, rodent-borne viral disease: the hantavirus pulmonary syndrome (HPS). The virus may have already been present but dormant. Alterations in food supplies, predation pressure, and habitat provoked by sequential extremes multiplied the rodent reservoir hosts and amplified viral transmission.

Controlled experiments with rabbits demonstrate such synergies in population dynamics. Exclusion of predators with protective cages doubles their populations. With extra food, hare density triples. With both interventions, populations increase 11-fold.

By summer's end, predators apparently returned (indicating retained ecosystem resilience) and the rodent populations and disease outbreak abated. Subsequent episodes of HPS in the United States have been limited, perhaps aided by early warnings. However, HPS has appeared in Latin America, and there is evidence of person-to-person transmission.

6. CASE STUDY: WEST NILE VIRUS

West Nile virus (WNV) was first reported in Uganda in 1937. WNV is a zoonosis, with "spillover" to humans, which also poses significant risks for wildlife, zoo, and domestic animal populations. Although it is not known how WNV entered the Western Hemisphere in 1999, anomalous weather conditions may have helped amplify this *Flavivirus* that circulates among urban mosquitoes, birds, and mammals. Analysis of weather patterns coincident with a series of U.S. urban outbreaks of St. Louis encephalitis (SLE) (a disease with a similar life cycle) and four recent large outbreaks of WNV revealed that drought was a common feature. *Culex pipiens*, the primary mosquito vector (carrier) for WNV, thrives in city storm drains and catch basins, especially in the organically rich water that forms during drought and the accompanying warm temperatures. Because the potential risks from pesticides for disease control must be weighed against the health risks of the disease, an early warning system of conditions conducive to amplification of the enzootic cycle could help initiate timely preventive measures and potentially limit chemical interventions.

6.1 Background on WNV

WNV entered the Western Hemisphere in 1999, possibly via migratory or imported birds from Europe. Although the precise means of introduction is not known, experience with a similar virus, SLE, as well as the European outbreaks of WNV during the 1990s suggests that certain climatic conditions are conducive to outbreaks of this disease. Evidence suggests that mild winters, coupled with prolonged droughts and heat waves, amplify WNV and SLE, which cycle among urban mosquitoes (*Culex pipiens*), birds, and humans.

SLE and WNV are transmitted by mosquitoes to birds and other animals, with occasional spillover to humans. *Culex pipiens* typically breeds in organically rich standing water in city drains and catch basins as well as unused pools and tires. During a drought, these pools become even richer in the organic material that *Culex* needs to thrive. Excessive rainfall flushes the drains and dilutes the pools. Drought conditions may also lead to a decline in the number of mosquito predators, such as amphibians and dragonflies, and encourage birds to congregate around shrinking water sites, where the virus can circulate more easily. In addition, high temperatures accelerate the extrinsic incubation period (period of maturation) of viruses (and parasites) within mosquito carriers. Thus, warm temperatures enhance the potential for transmission and dissemination. Together, these factors increase the possibility that infectious virus levels will build up in birds and mosquitoes living in close proximity to human beings.

6.2 Outbreaks of St. Louis Encephalitis in the United States

SLE first emerged in the city of St. Louis in 1933, during the dust bowl era. Since 1933, there have been 24 urban outbreaks of SLE in the United States. SLE as an appropriate surrogate for study because of its similarity to WNV and because of the significant number of SLE outbreaks in the United States, along with accurate weather data (i.e., the Palmer Severity Drought Index [PSDI], a measure of dryness that is a function of precipitation and soil moisture compared with 30 years of data in the same location). The PSDI ranges from −4 (dry) to +4 (wet). From 1933 to the mid-1970s, 10 of the 12 urban SLE outbreaks—regionally clustered in Kentucky, Colorado, Texas, Indiana, Tennessee, and Illinois—were associated with 2 months of drought. (One of the remaining outbreaks was associated with 1 month of drought.) After the mid-1970s, the relationship shifted and outbreaks were associated with anomalous conditions that included droughts and heavy rains.

Note that outbreaks of SLE during the 1974–1976 period and after show a variable pattern in relation to weather. Once established in a region, summer rains may boost populations of *Aedes japonicus* and other *Aedes* spp. that function as "bridge vectors," efficiently carrying virus from birds to humans. The roles of "maintenance" (primarily bird-biting mosquitoes) and bridge vectors in WNV transmission are under study.

6.3 Outbreaks of WNV

6.3.1 Romania, 1996

A significant European outbreak of WNV occurred in 1996 in Romania, in the Danube Valley and in Bucharest. This episode, with hundreds experiencing neurological disease and 17 fatalities, occurred between July and October and coincided with a prolonged drought (May–October) and excessive heat (May–July). Human cases in Bucharest were concentrated in blockhouses situated over an aging sewage system in which C. *pipiens* were breeding in abundance.

6.3.2 Russia, 1999

A large outbreak of WNV occurred in Russia in the summer of 1999 following a drought. Hospitals in the Volgograd region admitted 826 patients; 84 had meningoencephalitis, of which 40 died.

6.3.3 United States, 1999

In the spring and summer of 1999, a severe drought (following a mild winter) affected the northeastern and mid-Atlantic states. The prolonged drought culminated in a 3-week July heat wave that enveloped the Northeast. Then the pendulum swung in the opposite direction, bringing torrential rains at the end of August (and, later, Hurricane Floyd to the mid-Atlantic states). *Culex* spp. thrived in the drought months; *Aedes* spp. bred in the late summer floodwaters. In the New York outbreak, 7 people died, and, of the 62 people who suffered neurological symptoms and survived, the majority reported chronic disabilities, such as extreme muscle weakness and fatigue.

6.3.4 Israel, 2000

WNV was first reported in Israel in 1951, and sporadic outbreaks followed. Israel, a major stopover for migrating birds, usually receives little precipitation from May to October. In 2000, the region was especially dry as drought conditions prevailed across southern Europe and the Middle East, from Spain to Afghanistan. Between August 1 and October 31, 2000, 417 cases of serologically confirmed WNV were diagnosed in Israel, and there were 35 deaths. *Culex pipiens* was identified as a vector.

6.3.5 United States, 2002

In 2002, much of the western and midwestern United States experienced a severe spring and summer drought. Lack of snowpack in the Rockies (warming winters leading to more winter precipitation falling as rain) contributed. Forest fires burned more than 7.3 million acres, and haze and respiratory disease affected several Colorado cities. There was also an explosion of WNV cases, with human or animal WNV being documented in 44 states and the District of Columbia, reaching to California. Drought conditions were present in June in Louisiana, the first epicenter of WNV in 2002. Widespread drought conditions and heat waves may have amplified WNV and contributed to its rapid spread throughout the continental United States.

(Health officials have also become convinced that WNV can be transmitted via organ transplant and blood transfusion.)

Of greatest concern, however, WNV spread to 230 species of animals, including 138 species of birds and 37 species of mosquitoes. Not all animals fall ill from WNV, but the list of hosts and reservoirs includes dogs, cats, squirrels, bats, chipmunks, skunks, rabbits, and reptiles. Raptors (e.g., owls and kestrels) have been particularly affected; WNV likely caused thousands of birds of prey to die in Ohio and other states in July 2002. Some zoo animals have died.

Note that the population impacts on wildlife and biodiversity have not been adequately evaluated. The impacts of the decline in birds of prey could ripple through ecological systems and food chains and could contribute to the emergence of disease.

Declines in raptors could have dramatic consequences for human health. These birds of prey are our guardians because they prey on rodents and keep their numbers in check. When rodent populations explode—when floods follow droughts, forests are clear-cut, or diseases affect predators—their legions can become prolific transporters of pathogens, including Lyme disease, leptospirosis, plague, hantaviruses and arenaviruses such as Lassa fever and Guaranito, Junin, Machupo, and Sabia, associated with severe hemorrhagic fevers in humans.

As of March 12, 2003, the Centers for Disease Control and Prevention reported the following for 2002:

Laboratory confirmed human cases nationally: 4156
WNV-related deaths: 284
Most deaths: Illinois (64), Michigan (51), Ohio (31), and Louisiana (25)

There were also 14,045 equine cases in 38 states reported to the U.S. Department of Agriculture APHIS by state health officials as of November 26, 2002. WNV has been associated with illness and death in several other mammal species, including squirrel, wolf, and dog in Illinois and mountain goat and sheep. The largest number of equine cases was reported in Nebraska. Because of the bird and mammal reservoirs for WNV, there is the potential for outbreaks in all eastern and Gulf States and into Canada in the future.

6.3.6 United States, 2003

In the summer of 2003, cases of WNV concentrated in Colorado, the Dakotas, Nebraska, Texas, and Wyoming—areas that experienced prolonged spring drought (and extensive summer wildfires) in association with anomalous conditions in the Pacific Ocean. The eastern part of the United States (where a cold, snowy winter occurred in association with the North Atlantic freshening and North Atlantic High along with continued warming of tropical waters) experienced a relatively calm summer/fall in relation to WNV. (Both the Pacific and the Atlantic Oceans were in anomalous states beginning in the late 1990s. The state of the Pacific created perfect conditions for drought in many areas of the world.)

6.4 Public Health Implications

Multimonth drought, especially in spring and early summer, was found to be associated with urban SLE outbreaks from its initial appearance in 1933 through 1973 and with recent severe urban outbreaks of WNV in Europe and the United States. Each new outbreak requires introduction or reintroduction of the virus, primarily via birds or wildlife, so there have been seasons without SLE outbreaks despite multimonth drought. Spread of WNV and sporadic cases may occur, even in the absence of conditions amplifying the enzootic cycling. In Bayesian parlance, drought increases the "prior probability" of a significant outbreak once the virus becomes established in a region. Other factors, such as late summer rains that increase populations of bridge vectors, may affect transmission dynamics.

Further investigation and modeling are needed to determine the role of meteorological factors and identify reservoirs, overwintering patterns, and the susceptibility of different species associated with WNV. The migration path of many eastern U.S. birds extends from Canada across the Gulf of Mexico to South America, and WNV has spread to Mexico, Central America, and the Caribbean.

Factors other than weather and climate contribute to outbreaks of these two diseases. Antiquated urban drainage systems leave more fetid pools in which mosquitoes can breed, abandoned tires and swimming pools are ideal breeding sites, and stagnant rivers and streams do not adequately support healthy fish populations to consume mosquito larvae. Such environmental vulnerabilities present opportunities for environmentally based public health interventions following early warnings of conducive meteorological conditions.

State plans to prevent the spread of and contain WNV have three components:

1. Mosquito surveillance and monitoring of dead birds

2. Source (breeding site) reduction though larviciding (*Bacillus sphaericus* and Altocid or methoprene) and neighborhood cleanups

3. Pesticide (synthetic pyrethrins) spraying, when deemed necessary.

The information covering predisposing climatic conditions and predictions of them may be most applicable for areas that have not yet experienced WNV but lie in the flyway from Canada to the Gulf of Mexico. Projections of droughts (e.g., for northeast Brazil during an El Niño event) could help focus attention on these areas, enhancing surveillance efforts (including active bird surveillance), public communication, and environmentally friendly, public health interventions. They may also help set the stage for earlier chemical interventions once circulating virus is detected.

Finally, in terms of the public perception and concerns regarding the risks of chemical interventions, understanding the links of WNV to climatic factors and mobilizing public agencies, such as water and sewage departments, to address a public health threat may prove helpful in garnering public support for the combined set of activities needed to protect public health.

The WNV may have changed because it took an unusually high toll on birds in the United States. Alternatively, North American birds were sensitive because they were immunologically naive. However, the unexpected outbreak of a mosquito-borne disease in New York City and rapid spread throughout the nation in 2002 also serve as a reminder of the potential for exponential spread of pests and pathogens, and that pathogens evolving anywhere on the globe—and the social and environmental conditions that contribute to these changes—can affect us all.

7. DISCONTINUITIES

Climate change may not prove to be a linear process. Polar ice is thinning and Greenland ice is retreating, and since 1976 several small stepwise adjustments appear to have reset the climate system. In 1976, Pacific Ocean temperatures warmed significantly; they warmed further in 1990 and cooled in 2000. The intensity of ENSO has surpassed the intensity it had 130,000 years ago during the previous warm interglacial period. Cold upwelling in the Pacific Ocean in 2000 could portend a multidecadal correction that stores accumulating heat at intermediate ocean layers. Meanwhile, two decades of warming in the North Atlantic have melted Arctic ice, plausibly contributing to a cold tongue from Labrador to Europe and enhancing the Labrador Current that hugs the U.S. east coast. Such paradoxical cooling from warming and ice melting could alter projections for climate, weather, and disease for northern Europe and the northeastern United States. It is the instability of weather patterns that is of most concern for public health and society.

Winter is a blessing for public health in temperate zones, and deep cold snaps could freeze *C. pipiens* in New York City sewers, for example, reducing the risk of WNV during those cold winters. Thus, the greatest threat of climate change lies not with year-to-year fluctuations but with the potential for a more significant abrupt change that would alter the life-support systems underlying our overall health and well-being.

8. CONCLUSION

The resurgence of infectious diseases among humans, wildlife, livestock, crops, forests, and marine life in the final quarter of the 20th century may be viewed as a primary symptom of global environmental and social change. Moreover, contemporaneous changes in greenhouse gas concentrations, ozone levels, the cryosphere, ocean temperatures, land use, and land cover challenge the stability of our epoch, the Holocene—a remarkably stable 10,000-year period that followed the retreat of ice sheets from temperate zones. The impacts of deforestation and climatic volatility are a particularly potent combination creating conditions conducive to disease emergence and spread. Given the rate of changes in local and global conditions, we must expect synergies and new surprises.

Warming may herald some positive health outcomes. High temperatures in some regions may reduce snails, the intermediate hosts for schistosomiasis. Winter mortality in the Northern Hemisphere from respiratory disease may decline. However, the consequences of warming and wide swings in weather are projected to overshadow the potential health benefits.

The aggregate of air pollution from burning fossil fuels and felling forests provides a relentless destabilizing force on the earth's heat budget. Examining the full life cycle of fossil fuels also exposes layers of damages. Environmental damage from their mining, refining, and transport must be added to direct health effects of air pollution and acid precipitation.

Returning CO_2 to the atmosphere through their combustion reverses the biological process by which plants drew down atmospheric carbon and generated oxygen and stratospheric ozone, helping to cool and shield the planet sufficiently to support animal life.

9. NEXT STEPS

Solutions may be divided into three levels. First-order solutions to the resurgence of infectious disease include improved surveillance and response capability, drug and vaccine development, and greater provision of clinical care and public health services.

Second is improved prediction. Integrating health surveillance into long-term terrestrial and marine monitoring programs—ecological epidemiology—can benefit from advances in satellite imaging and climate forecasts that complement fieldwork. Health early warning systems based on the integrated mapping of conditions, consequences, and costs can facilitate timely, environmentally friendly public health interventions and inform policies.

Anticipating the health risks posed by the extreme conditions facing the U.S. East Coast in the summer of 1999 could have (a) enhanced mosquito surveillance, (b) heightened sensitivity to bird mortalities (that began in early August), and (c) allowed treatment of mosquito breeding sites, obviating large-scale spraying of pesticides.

The third level is prevention, which rests on environmental and energy policies. Restoration of forests and wetlands, "nature's sponges and kidneys," is necessary to reduce vulnerabilities to climate and weather. Population stabilization is also necessary, but World Bank data demonstrate that this is a function of income distribution.

Development underlies most aspects of health and it is essential to develop clean energy sources. Providing basic public health infrastructure—sanitation, housing, food, refrigeration, and cooking—requires energy. Clean energy is needed to pump and purify water and desalinate water for irrigation from the rising seas. Meeting energy needs with nonpolluting sources can be the first step toward the rational use of Earth's finite resources and reduction in the generation of wastes, the central components of sustainable development.

Addressing all these levels will require resources. Just as funds for technology development were necessary to settle the Montreal Protocol on ozone-depleting chemicals, substantial financial incentives are needed to propel clean energy technologies into the global marketplace. International funds are also needed to support common resources, such as fisheries, and for vaccines and medications for diseases lacking lucrative markets.

Human and ecological systems can heal after time-limited assaults, and the climate system can also be restabilized, but only if the tempo of destabilizing factors is reduced. The Intergovernmental Panel on Climate Change calculates that stabilizing atmospheric concentrations of greenhouse gases requires a 60% reduction in emissions.

Worldviews can also shift abruptly. Just as we may be underestimating the true costs of "business as usual," we may be vastly underestimating the economic opportunities afforded by the energy transition. A distributed system of nonpolluting energy sources can help reverse the mounting environmental assaults on public health and can provide the scaffolding on which to build clean, equitable, and healthy development in the 21st century.

SEE ALSO THE FOLLOWING ARTICLES

Air Pollution, Health Effects of • Climate Change and Energy, Overview • Climate Change: Impact on the Demand for Energy • Climate Protection and Energy Policy • Development and Energy, Overview • Energy Efficiency and Climate Change • Environmental Injustices of Energy Facilities • Gasoline Additives and Public Health • Greenhouse Gas Emissions from Energy Systems, Comparison and Overview

Further Reading

Albritton, D. L., Allen, M. R., Baede, A. P. M., *et al.* (2001). "IPCC Working Group I Summary for Policy Makers, Third Assessment Report: Climate Change 2001: The Scientific Basis." Cambridge Univ. Press, Cambridge, U.K.

Daszak, P., Cunningham, A. A., and Hyatt, A. D. (2000). Emerging infectious diseases of wildlife—Threats to biodiversity and human health. *Science* **287**, 443–449.

Epstein, P. R. (1999). Climate and health. *Science* **285**, 347–348.

Epstein, P. R., Diaz, H. F., Elias, S., Grabherr, G., Graham, N. E., Martens, W. J. M., Mosley-Thompson, E., and Susskind, E. J. (1998). Biological and physical signs of climate change: Focus on mosquito-borne disease. *Bull. Am. Meteorol. Soc.* **78**, 409–417.

Harvell, C. D., Kim, K., Burkholder, J. M., Colwell, R. R., Epstein, P. R., Grimes, J., Hofmann, E. E., Lipp, E., Osterhaus, A. D. M. E., Overstreet, R., Porter, J. W., Smith, G. W., and Vasta, G. (1999). Diseases in the ocean: Emerging pathogens, climate links, and anthropogenic factors. *Science* **285**, 1505–1510.

Karl, T. R., and Trenbath, K. (2003). Modern climate change. *Science* **302,** 1719–1723.

Levitus, S., Antonov, J. I., Boyer, T. P., and Stephens, C. (2000). Warming of the world ocean. *Science* **287,** 2225–2229.

McCarthy, J. J., Canziani, O. F., Leary, N. A., Dokken, D. J., and White, K. S. (eds.). (2001). "Climate Change 2001: Impacts, Adaptation & Vulnerability Contribution of Working Group II to the Third Assessment Report of the Intergovernmental Panel on Climate Change (IPCC)." Cambridge Univ. Press, Cambridge, UK.

McMichael, A. J., Haines, A., Slooff, R., and Kovats, S. (eds.). (1996). "Climate Change and Human Health." World Health Organization, World Meteorological Organization, and United Nations Environmental Program, Geneva, Switzerland.

National Research Council, National Academy of Sciences (2001). "Abrupt Climate Change: Inevitable Surprises." National Academy Press, Washington, DC.

Petit, J. R., Jouze, J., Raynaud, D., Barkov, N. I., Barnola, J. M., Basile, I., Bender, M., Chapellaz, J., Davis, M., Delaygue, G., Delmotte, M., Kotlyakov, V. M., Legrand, M., Lipenkov, V. Y., Lorius, C., Peplin, L., Ritz, C., Saltzman, E., and Stievenard, M. (1999). Climate and atmospheric history of the past 420,000 years from the Vostok Ice Core, Antartica. *Nature* **399,** 429–436.

Rosenzweig, C., Iglesias, A., Yang, X. B., Epstein, P. R., and Chivian, E. (2001). Climate change and extreme weather events: Implications for food production, plant diseases, and pests. *Global Change Hum. Health* **2,** 90–104.

Climate Change: Impact on the Demand for Energy

TIMOTHY J. CONSIDINE
The Pennsylvania State University
University Park, Pennsylvania, United States

1. Impact of Climate Change on Energy Demand
2. Recent Climate Trends
3. Climate Impacts on Energy Demand
4. Sensitivity of Energy Demand to Temperature Changes
5. Numerical Simulation of Climate Impacts
6. Conclusion

Glossary

cooling degree days This measure is the maximum of zero or the difference between the average daily temperature at a given location and 65°F. For example, if average daily temperature is 85°, there are 20 heating degree days.

elasticity The proportionate change in one variable in response to a proportionate change in a causal factor. For example, an own-price elasticity of demand equal to −0.10 indicates that a 1% change in price reduces consumption by 0.1%.

heating degree days This measure is the maximum of zero or the difference between 65°F and the average daily temperature at a given location. For example, if average daily temperature is 35°, there are 30 heating degree days.

Although the focus of many policy studies of climate change is on establishing the causal links between anthropogenic systems, emissions of greenhouse gases, and climate, the line of causation also runs the other way. Short-term fluctuations in climate conditions, particularly in the temperate zones on the planet, affect energy consumption. If the popular expectation that the climate will become warmer becomes a reality, we can expect winters and summers that are warmer than those of the past. Warmer summers induce people to use more air-conditioning, which requires more electricity and greater consumption of fuels used to generate that electric power. On the other hand, warmer winters reduce heating requirements and the associated consumption of natural gas, distillate fuel oil, propane, and other heating fuels. The net effect of these two opposing forces on total annual energy demand depends on a number of factors, including local climate conditions, technology, and the behavioral response of consumers to weather conditions.

1. IMPACT OF CLIMATE CHANGE ON ENERGY DEMAND

Establishing the impact of climate change on energy demand requires a measure of heating and cooling requirements. In the United States, this measure is a degree day, which is defined in terms of an absolute difference between average daily temperature and 65°F, which is an arbitrary benchmark for household comfort. Heating degree days are incurred when outside temperatures are below 65°F, generally during the winter heating season from October through March. Cooling degree days occur when temperatures exceed 65°F, often during the summer cooling season from April through September. Degree days are intended to represent the latent demand for heating and air-conditioning services based on temperature departures from the comfort level of 65°F.

Heating and cooling degree days are reported by the National Oceanic and Atmospheric Administration at a weekly frequency for 50 cities throughout the United States. State, regional, and national population weighted averages of heating and cooling degree days are also reported and commonly used by the utility industry and energy commodity trading community.

Clearly, a higher number of degree days—heating, cooling, or both—should be associated with greater fuel consumption. Estimating how energy consumption rises for every degree day requires some form of model that links weather conditions to energy consumption. There are two basic modeling approaches. One approach uses engineering models that estimate energy consumption by multiplying a utilization rate for an energy-consuming durable by a fixed energy efficiency rate. For example, to compute the reduction in space heating energy from a warmer winter, an engineering approach would involve estimating the impact on the utilization rate and then multiplying the fixed unit energy efficiency by this new utilization rate. The main advantage of this approach is its high level of detail, although the accuracy is often in doubt given the limited data available to compute key parameters, such as utilization rates and unit energy efficiencies, particularly for older equipment.

Another approach develops statistical relationships often specified by economists on the basis of behavioral models that assume energy consumers, such as households and businesses, maximize their satisfaction conditional on heating and cooling degree days, relative fuel prices, income or sales levels, and the technical characteristics of energy-consuming durable equipment. Although this approach is more abstract than engineering models, it captures the behavioral dimensions of energy demand. Econometric models of energy demand are able to estimate the separate effects of price, income, weather variations, and other factors affecting energy demand.

The sensitivities of energy demand to price, weather, and other factors are expressed in terms of elasticities, defined as the percentage change in energy consumption associated with a given percentage change in one of these causal factors. The term associated is used to convey that the relationship is estimated based on a statistical model, which by definition allows a random error that reflects a number of uncertainties stemming from the inherent vagaries of human behavior. Such elasticities are a critical component in models of economic growth that are used to estimate the net social benefits of policies to mitigate or control greenhouse gas emissions. If a warmer climate is likely and if this results in a net reduction in energy demand, then the net social cost of greenhouse gas pollution would be lower by the energy expenditure savings. Determining the relevance and size of this potential effect has broad implications for climate change policies. Moreover, the feedback of climate change on energy demand also indirectly affects carbon emissions. Over the long term, this feedback effect may be an important element in understanding the global carbon cycle. In addition, future climate change agreements may need to allow some variance in carbon emissions due to normal climate fluctuations.

Many utilities and government agencies recognize the link between weather conditions and energy demand. Detecting the effects of climate trends on weather conditions is not without controversy, and the following section explains some of the nuances of detecting trends in heating and cooling degree days. Next, the formulation and construction of econometric models of energy demand that incorporate weather conditions are described in more detail. Estimates of price, income, and weather elasticities from these models are described along with numerical simulations of the impacts of climate change on U.S. energy demand. Finally, a discussion of the policy implications of the findings is presented.

2. *RECENT CLIMATE TRENDS*

Climatologists generally agree that there is accumulating evidence that a warming trend has been occurring since the mid-1960s. There are rather distinct seasonal and regional variations in the lower 48 states of the United States. Livezey and Smith determined that the average national warming trend has been 0.015°F per year. Since 1964, this implies that average annual temperature has increased by approximately one-half of one degree.

Nevertheless, there are some dissenting views, such as that of Balling, who argues that temperature has not changed in any meaningful way since 1932. His analysis tests for linear trends in temperature and finds no statistically significant trends in the data. Nonlinear trends, however, are possible and the evidence presented in Figs. 1 and 2 suggests that this indeed is the case. The smooth lines are predicted temperatures from a simple nonlinear statistical trend model in which degree days are specified as a function of a constant, the time period, and the time period squared. The hypothesis that these latter two terms are individually or jointly zero could not be rejected. The probability levels, or the chances that this hypothesis is wrong, are zero for cooling degree days and slightly less than 2% for the heating degree day trend terms. Notice that the trend for cooling degree days is downward from 1932 to approximately 1970 and then upward since then. Heating degree days display the inverse pattern, increasing during the early part of the sample but then decreasing during the 1980s and 1990s.

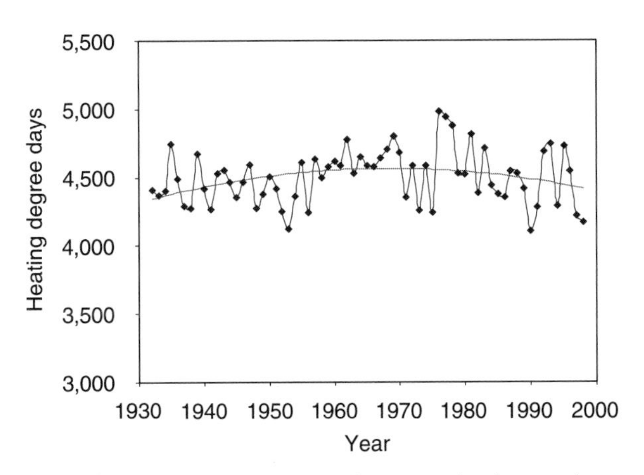

FIGURE 1 Actual and fitted nonlinear trend in heating degree days.

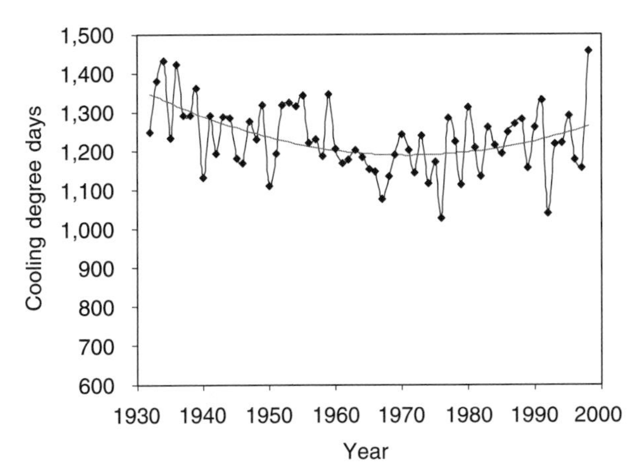

FIGURE 2 Actual and fitted nonlinear trend in cooling degree days.

The highly volatile nature of temperature as illustrated in Figs. 1 and 2 is cause for some concern because statistical models with a relatively small number of observations are prone to be heavily influenced or skewed by extreme observations, such as the severe cold during the winter of 1976–1977. Far more observations are required to place a high degree of confidence in detecting linear or nonlinear trends in temperature. With these caveats in mind, temperature trends during the past 30 years are consistent with the global warming hypothesis.

Morris found that climate trends vary by region and by season. Most regions of the eastern United States had average or slightly below average temperatures. In contrast, the western and Great Plains states experienced warmer temperatures. The western and upper Great Plains regions experienced the most warming during the winter months. Average temperatures during the fall are actually lower for most of the eastern and Midwest regions of the United States. This finding suggests that studies linking energy demand to temperature or degree days should capture some of these regional and seasonal patterns. The analysis presented next explicitly captures the seasonal dimensions. Although a regional energy demand model is not developed here, the findings of Enterline suggest that regional aggregation is not a serious problem.

3. *CLIMATE IMPACTS ON ENERGY DEMAND*

Energy is consumed by various segments of the economy, including households, commercial establishments, manufacturing enterprises, and electric power generators. Only a portion of total energy demand is sensitive to temperature. Consumption of gasoline and jet fuel has strong seasonal components but is not very sensitive to heating and cooling degree days. On the other hand, consumption of electricity, natural gas, propane, and heating oil is quite sensitive to weather conditions, particularly temperature. In addition, the derived demand for primary fuels used by electric utilities, such as coal, natural gas, and residual fuel oil, is also sensitive to weather because the demand for electricity changes with variations in temperature and other atmospheric conditions.

The challenge for understanding the importance of climate on energy demand is to identify and estimate the separate effects that the determinants of energy demand have on energy consumption. The demand for energy depends on a number of factors, including the price of energy relative to prices for other goods and services, household income and demographic features, output and technology adoption by businesses, and weather conditions. Using history as a guide, the challenge is to determine to what extent trends in energy consumption are influenced by trends in these basic drivers of energy demand.

Considine develops energy demand models that estimate the separate effects that these factors have on energy consumption trends. These models are derived from economic principles and then tested using econometric methods. For example, businesses are assumed to minimize energy expenditures subject to exogenously determined constraints, such as output levels, fuel prices, and weather conditions.

This optimization model provides a basis to formulate a set of energy demand equations that relate energy consumption and these exogenous factors. Moreover, this approach allows a consistent representation of substitution possibilities among fuels. For instance, if the demand for natural gas increases with an increase in fuel oil prices, suggesting that natural gas and fuel oil are substitutes, then the demand for fuel oil would increase with higher prices for natural gas. This consistency between demand equations is known as symmetry and follows from the mathematical conditions associated with economic optimization.

Another important property of demand systems is that if all fuel prices increase the same percentage, then total energy expenditures would increase the same amount and relative fuel use would not change. This property, known as zero-degree homogeneity in demand, essentially means that only relative prices matter in determining energy demand.

Given these considerations regarding the role of prices in determining energy demand, identifying the effect of climate on energy demand requires measurement of two components of the climate signal: the fixed seasonal effect and a random weather shock. Fixed seasonal effects include the length of day and other fixed monthly effects and can be represented by zero-one or dummy variables. For instance, household energy consumption in the United States may have a January effect, increasing a fixed amount each year as day length or consumer habits change in fixed seasonal ways. Energy demand during other months of the year may have similar fixed monthly or seasonal effects. The second effect, associated with random weather shocks, represents that portion of energy consumption directly associated with departures from normal temperatures. One good measure for this effect is heating and cooling degree day deviations from their 30-year means. The monthly deviation of degree days from its corresponding monthly mean measures the extent to which temperature is above or below normal. Therefore, for example, heating degree day deviations 30° below normal would imply a proportionately larger increase in energy demand than heating degree day deviations 10° below normal. This approach is similar to the study of natural gas demand by Berndt and Watkins.

The demand models developed by Considine assume that monthly energy consumption depends on economic forces, technological factors, fixed seasonal effects, and random weather shocks. Energy demand systems are estimated for four sectors of the U.S. economy: residential, commercial, industrial, and electric utilities. A two-step modeling approach is adopted. The first step determines the level of energy consumption in each sector, and the second determines the mix of fuels to satisfy that total demand.

In the first step, Considine specifies that aggregate energy demand is a simple log-linear function of real energy price, income or output, fixed seasonal effects, heating and cooling degree days, and a time trend as a proxy for exogenous technological change. This relationship is also dynamic so that shocks to demand during one period affect predictions of demand in future periods.

The second step, determining the fuel mix, assumes that the fuels in each sector's aggregate energy price index are weakly separable from other factors of production. This means that the marginal rate of substitution between two fuels is independent of the rate at which aggregate energy substitutes with other goods or factors of production. This specification is reasonable because substitution possibilities between energy and other factors of production are likely to be very limited over a month time span.

The combinations of fuels used in each sector are modeled using energy expenditure systems specified from a logistic function, which ensures adding-up and positive cost shares. Considine and Mount show how to impose constraints consistent with economic theory. Given the nature of the model, however, Considine imposes these conditions locally—either at a point with linear parameter restrictions or at each point in the sample using an iterative estimation procedure. The results presented next impose these conditions at the mean cost shares, which simplifies model simulation for policy analysis and forecasting.

4. SENSITIVITY OF ENERGY DEMAND TO TEMPERATURE CHANGES

Here, the sensitivity of energy demand to climate is measured two ways. The first method uses elasticities that provide simple summary measures of how departures from normal temperatures affect energy consumption. The second approach, reported in the following section, uses econometric simulation to estimate how climate changes affect energy demand.

The heating and cooling degree day elasticities of energy demand by sector are shown in Table I. The elasticities are interpreted as the percentage change in fuel consumption for a percentage change in heating or cooling degree days. For instance, total U.S. residential natural gas consumption increases 0.33%

TABLE I

Heating and Cooling Degree Day Elasticities of Energy Demand (with Probability Values)

	Residential	Commercial	Industrial	Electric utility
		Heating degree days		
Natural gas	0.333	0.296	0.114	0.05
Distillate oil	0.262	0.277	0.906	
Residual fuel oil			0.388	0.606
Coal			0.287	0.106
Electricity	0.148	0.08	0.055	
		Cooling degree days		
Natural gas	−0.022	0.038	−0.055	0.154
Distillate oil	−0.065	0.046	0.439	
Residual fuel oil			0.233	0.355
Coal			0.026	0.071
Electricity	0.141	0.059	−0.032	

for every 1% increase in heating degree days. With the exception of natural gas in the electric utility sector, all heating degree day elasticities are larger than the cooling degree day elasticities. In other words, heating degree days have a proportionately greater impact on energy consumption than cooling degree days. Overall, this finding suggests that global warming would reduce energy consumption because higher fuel use associated with hotter summers is offset by lower fuel use due to warmer winters.

This trade-off is clearly illustrated by the heating and cooling degree day elasticities for the residential sector reported in Table I. In addition to the previously natural gas elasticity, the heating degree day elasticities for distillate fuel oil and electricity are 0.262 and 0.148, respectively. In contrast, the cooling degree day elasticities for natural gas, distillate oil, and electricity are −0.022, −0.065, and 0.141, respectively. Higher cooling degree days reduce residential natural gas and heating oil consumption due to the reduced heating requirements during the spring and fall. Note that the heating and cooling degree day elasticities for electricity in the residential sector are approximately the same, suggesting that winter peaks from demands for electrical resistance heating are as sensitive to summer peak air-conditioning requirements. The commercial sector provides a similar set of elasticities, except that the cooling degree day elasticities are positive but extremely small for natural gas and distillate fuel oil.

Industrial demand for distillate and residual fuel oil is quite sensitive to temperature. For instance, for every percent change in heating degree days, industrial distillate use increases 0.9%. Residual fuel oil and coal demand increases with temperature, increasing 0.39 and 0.29%, respectively, for each percent change in heating degree days. Natural gas consumption is the next most temperature-sensitive fuel in the industrial sector, with a heating degree day eleasticity of 0.11. Electricity consumption is the least sensitive among the industrial fuels.

The demand for primary fuels in electric power generation is also quite sensitive to temperature. Like the other sectors, heating degree day elasticities are greater than the cooling degree day elasticities. Residual fuel oil consumption increases 0.66 and 0.36% for every 1% change in heating and cooling degree days, respectively. The natural gas cooling degree day elasticity is larger than the heating oil counterpart, which reflects the relatively greater use of natural gas to meet summer cooling demands. Note that the degree day elasticities assume that the demand for electricity is fixed. To capture the induced impact of degree days on primary fuel demand by electric utilities, a full model simulation is required that links the demand for electricity in the residential, commercial, and industrial sectors with electric power generation and fuel use.

5. NUMERICAL SIMULATION OF CLIMATE IMPACTS

To determine the effects of past climate trends on energy consumption, the econometric equations providing the degree day elasticities reported previously are combined into an econometric simulation model. The endogenous variables determined by the model include energy demand in the residential, commercial, and industrial sectors of the U.S. economy and the derived demand for primary fuels used in electric power generation. The demand and supply of electricity are determined in the model. The exogenous or predetermined variables include real personal disposable income, retail sales, industrial production, energy prices, and heating and cooling degree days. Like the degree day elasticities reported previously, the equations provide estimates of price and income elasticities of demand, which were reported by Considine in 2000. Hence, the model provides a tool for estimating the contributions of each determinant of energy demand, including climate, prices, and income.

To identify the effects of past climate trends on historical energy consumption, two simulations are required. The first simulation provides a baseline using the 30-year means for heating and cooling degree days, which gives an estimate of what energy demand would have been under average climate conditions. The second simulation uses actual degree days, which yields an estimate of predicted energy demand associated with actual weather. All other exogenous variables, such as energy prices and income, are the same in the two simulations. Consequently, the changes from the base simulation reported in Table II represent those changes in energy demands associated with deviations of weather conditions from their 30-year means.

As discussed previously, the climate in North America has been getting warmer since the early 1980s. Indeed, cumulative cooling degree days are 2.2% higher than normal and heating degree days are 1.65% lower. The impacts of these changes in

TABLE II

Historical Impacts of Weather on Energy Demand in Percentage Changes from Predictions Using Average Weather

		Fuel consumption					
Year	Cooling degree days	Electricity	Natural gas	Distillate	Residual	Coal	Propane
		Cooling season, April–September					
1983	6.57	0.79	1.15	2.03	6.01	1.35	1.16
1984	0.36	0.18	0.48	0.92	1.33	0.38	0.71
1985	−2.22	−0.47	−0.97	−1.26	−3.03	−0.94	−1.00
1986	3.58	0.10	−0.48	−0.28	0.59	0.14	−0.61
1987	8.23	0.31	−0.48	0.11	2.36	0.69	−0.93
1988	8.96	0.80	0.33	0.66	4.84	1.44	−0.14
1989	−3.91	−0.30	0.00	−0.17	−1.51	−0.56	0.16
1990	3.43	0.23	0.03	0.37	1.68	0.50	−0.05
1991	11.11	0.50	−0.63	0.21	3.56	1.10	−1.09
1992	−13.69	−1.03	−0.20	−0.97	−6.12	−2.18	0.31
1993	3.15	0.52	0.52	0.39	2.84	0.90	0.46
1994	1.56	−0.02	−0.36	−0.30	0.48	0.00	−0.41
1995	8.20	1.07	0.97	1.56	6.21	2.14	0.71
1996	0.16	0.06	0.23	0.36	0.49	0.20	0.31
1997	−7.82	−0.16	0.97	0.54	−1.22	−0.36	1.52

		Fuel consumption					
Year	Heating degree days	Electricity	Natural gas	Distillate	Residual	Coal	Propane
		Heating season, October–March					
1983–1984	3.16	0.17	0.88	2.18	2.43	0.60	1.64
1984–1985	−1.08	−0.04	0.00	0.33	−0.16	−0.04	−0.24
1985–1986	−3.24	−0.18	−1.15	−1.41	−1.28	−0.40	−1.20
1986–1987	−3.72	−0.41	−1.44	−2.04	−2.85	−0.79	−1.70
1987–1988	−0.18	−0.13	−0.22	−0.10	−0.88	−0.23	0.02
1988–1989	−1.06	0.02	−0.80	−0.99	−1.05	−0.05	−0.71
1989–1990	−4.55	−0.47	−1.86	−1.48	−3.01	−0.70	−1.50
1990–1991	−8.92	−1.01	−2.93	−3.65	−6.40	−2.00	−3.90
1991–1992	−7.88	−1.10	−3.12	−3.80	−7.01	−2.08	−3.66
1992–1993	2.05	0.27	0.40	0.64	1.32	0.48	0.53
1993–1994	4.63	0.64	1.64	2.54	3.84	1.21	2.21
1994–1995	−8.44	−1.23	−3.08	−3.53	−6.87	−2.48	−3.88
1995–1996	3.09	0.38	1.05	1.22	2.53	0.90	1.29
1996–1997	−3.62	−0.33	−1.63	−1.36	−2.66	−0.80	−1.63

temperature are illustrated in Table II, which presents the percentage changes in energy demand from the predictions using average and actual degree days during the cooling and heating seasons.

There is an unambiguous, positive relationship between heating degree days and energy demand (Table II). With warmer winters, the demand for energy declines. Notice the string of warm winters during the late 1980s and early 1990s and the associated reductions in energy consumption. For example, the winters of 1990–1991 and 1991–1992 were approximately 8% warmer than average. As a result, the consumption of natural gas was approximately 3% lower than it would have been under normal weather conditions. Distillate and residual fuel oil use was almost 4% and more than 6% lower than normal, respectively.

The effects of cooling degree days are more ambiguous because some months of the cooling season also contain heating degree days, which may offset the effects associated with cooling degree days. Nevertheless, there are several years with warmer than normal summers and slightly higher energy use. During the entire period, energy demand under the base simulation with actual weather is 0.2% lower than that under the alternative simulation using the 30-year mean. This suggests that a warmer climate would slightly lower energy consumption because lower fuel use associated with reduced heating requirements offsets higher fuel consumption to meet increased cooling demands.

Simulating the effects of a 3-month shock to heating and cooling degree days provides a more controlled experiment. First, consider the effects of a colder than normal winter with heating degree days 10% more than the 30-year mean for 3 months. All fuel demands increase (Fig. 3). Higher residential and commercial consumption is the principal reason for the approximately 7–10% increase in natural gas consumption. In addition, electric utility fuel use, particularly oil, increases due to the simulated 2–4% increase in electricity consumption. The large estimated elasticity of electric utility oil use with respect to electricity generation accounts for most of the increase in residual fuel consumption. This result may reflect the practice of electric utilities to use oil-fired generators to service peak power demands.

The last simulation estimates the effects of a 3-month, 10% increase in cooling degree days. The impacts on fuel demands and carbon emissions are shown in Fig. 4. As expected, electricity demand increases, leading to an increase in the demand for coal and oil. Natural gas consumption increases slightly due primarily to higher consumption in the electric utility sectors. Carbon emissions increase 1.5–3.7%.

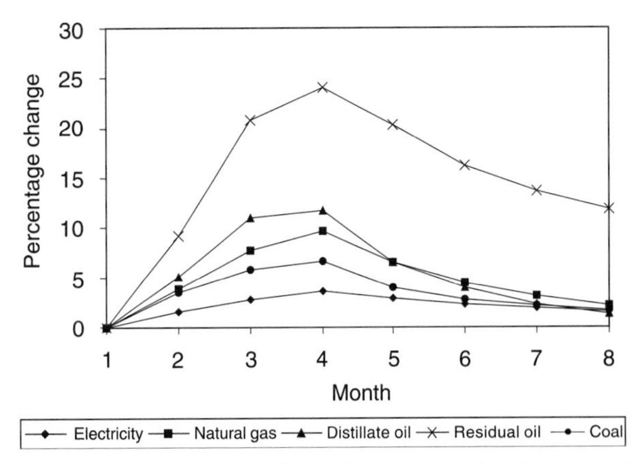

FIGURE 3 Simulated effects of a 10% colder than normal winter on fuel consumption.

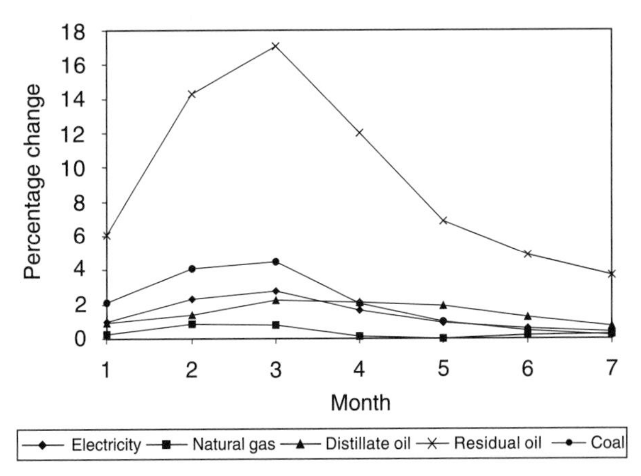

FIGURE 4 Simulated effects of a 10% warmer than normal winter on fuel consumption.

6. CONCLUSION

Overall, warmer climate conditions on balance slightly reduce U.S. energy demand. The estimated impact, however, is very small and pales in comparison to the 7% reduction from 1990 emission levels for the United States under the Kyoto Protocol. Short-term variations in temperature can have sizeable impacts on energy demand. Petroleum consumption is particularly sensitive to temperature, primarily through the induced effect of weather on electricity demand and the fuels in electric power generation. These short-term weather-related variations in energy demand suggest that any future international agreements on carbon emissions trading should have provisions that adjust compliance rules based on the variance in energy consumption due to weather fluctuations.

SEE ALSO THE FOLLOWING ARTICLES

Carbon Taxes and Climate Change • *Climate Change and Energy, Overview* • *Climate Protection and Energy Policy* • *Energy Efficiency and Climate Change* • *Modeling Energy Markets and Climate Change Policy*

Further Reading

Balling, R. C. (1999). Analysis of trends in United States degree days, 1950–1995. http://www.greeningearthsociety.org/Articles/1999/trends.htm.

Berndt, E. R., and Watkins, G. C. (1977). Demand for natural gas: Residential and commercial markets in Ontario and British Columbia. *Can. J. Econ.* **10,** 97–111.

Considine, T. J. (1990). Symmetry constraints and variable returns to scale in logit models. *J. Business Econ. Statistics* **8**(3), 347–353.

Considine, T. J. (2000). The impacts of weather variations on energy demand and carbon emissions. *Resource Energy Econ.* **22,** 295–312.

Considine, T. J., and Mount, T. D. (1984). The use of linear logit models for dynamic input demand systems. *Rev. Econ. Statistics* **66,** 434–443.

Enterline, S. (1999). The effect of weather and seasonal cycles on the demand for electricity and natural gas, M.S. thesis, Department of Energy, Environmental, and Mineral Economics, Pennsylvania State University, University Park.

Hansen, J., Sato, M., Glascoe, J., and Ruedy, R. (1998). A common sense climate index: Is climate changing noticeably? *Proc. Natl. Acad. Sci. USA* **95,** 4118–4120.

Livezey, R.E., and Smith, T. (1999). Covariability of aspects of North American climate with global sea surface temperatures on interannual and interdecadal timescales. *J. Climate*, January, 289–302.

Morris, M. (1999). The impact of temperature trends on short-term energy demand. http://www.eia.doe.gov/emeu/steo/pub/special/weather/temptrnd.html.

Climate Protection and Energy Policy

WILLIAM R. MOOMAW
The Fletcher School of Law and Diplomacy, Tufts University
Medford, Massachusetts, United States

1. Impact of Climate Change
2. Strategies for Energy Sources
3. Energy Strategies for Addressing Climate Change
4. Energy for Buildings
5. Energy for Industry
6. Energy for Transportation
7. Energy for Agriculture
8. Energy for Electric Power
9. Alternative and Integrated Strategies
10. Implications for the Future

Glossary

climate The 30- to 40-year average of weather measurements, such as average seasonal and annual temperature; day and night temperatures; daily highs and lows; precipitation averages and variability; drought frequency and intensity; wind velocity and direction; humidity; solar intensity on the ground and its variability due to cloudiness or pollution; and storm type, frequency, and intensity.

fossil fuels Energy-rich substances that contain carbon, hydrogen, and smaller amounts of other materials that have formed from long-buried plants. Fossil fuels, which include petroleum, coal, and natural gas, provide 85% of the energy that powers modern industrial society.

greenhouse effect The trapping of heat by greenhouse gases that allow incoming solar radiation to pass through the earth's atmosphere but prevent most of the outgoing infrared radiation from the surface and lower atmosphere from escaping into outer space. This process occurs naturally and has kept the earth's temperature approximately 33°C warmer than it would be otherwise.

greenhouse gases Any gas that absorbs infrared radiation in the atmosphere. Greenhouse gases include water vapor, carbon dioxide, methane, nitrous oxide, ozone, hydrochlorofluorocarbons, perfluorocarbons, and hydrofluorocarbons. Whereas the first three gases occur naturally and are increased by human activity, the latter gases are primarily synthesized by humans.

radiation Energy that is transmitted through space in the form of electromagnetic waves and their associated photons. Examples include visible light, ultraviolet light, and infrared light; also called radiant heat.

radiative forcing A measure of heat trapping by gasses and particles in the atmosphere (watts per square meter) that is directly proportional to the temperature increase at the earth's surface.

The earth's climate is determined by the interaction between the radiation received from the sun and the distribution and transformation of that energy by the atmosphere, oceans, land-based (terrestrial) ecosystems, and ice and snow and also by the differential absorption and reflection of solar energy by clouds, land, water, and snow. The one constant of climate over millions of years is its variability. The geological record reveals that at various times the earth has been either warmer or colder, and sometimes wetter or dryer, than it is now. The reason for the intense interest in climate and climate change is that there is evidence that the release of heat-trapping greenhouse gases from the energy sector, industry, and land use changes is directly and indirectly altering the earth's climate system. This article describes the technologies, policies, and measures that are proposed and that are being implemented to mitigate climate change for the energy sector.

1. IMPACT OF CLIMATE CHANGE

Since 1990, every 5 years a group of approximately 2000 natural and physical scientists, economists, social scientists, and technologists assemble under the auspices of the United Nations-sponsored Intergovernmental Panel on Climate Change (IPCC). These scientists spend 3 years reviewing all of the information on climate change and produce a voluminous report following a public review by others in the scientific community and by governments. “Climate Change 2001” is 2665 pages long and contains more than 10,000 references. Information for this article is based on this report, the references in it, as well as recent studies to ensure accurate, up-to-date information.

Climate is defined as the 30- to 40-year average of weather measurements, such as average seasonal and annual temperature; day and night temperatures; daily highs and lows; precipitation averages and variability; drought frequency and intensity; wind velocity and direction; humidity; solar intensity on the ground and its variability due to cloudiness or pollution; and storm type, frequency, and intensity.

Understanding the complex planetary processes and their interaction requires the effort of a wide range of scientists from many disciplines. Solar astronomers carefully monitor the intensity of the sun’s radiation and its fluxuations. The current average rate at which solar energy strikes the earth is 342 watts per square meter (W/m^2). It is found that 168 W/m^2 reaches the earth’s surface mostly as visible light: 67 W/m^2 is absorbed directly by the atmosphere, 77 W/m^2 is reflected by clouds, and 30 W/m^2 is reflected from the earth’s surface.

A number of trace gases in the atmosphere are transparent to the sun’s visible light but trap the radiant heat that the earth’s surface attempts to emit back into space. The net effect is that instead of being a frozen ball averaging $-19°C$, Earth is a relatively comfortable 14°C. This difference of 33°C arises from the natural greenhouse effect. Human additions of greenhouse gases appear to have increased the temperature an additional $0.6 \pm 0.2°C$ during the 20th century. The transmission of visible light from the sun and the trapping of radiant heat from the earth by gases in the atmosphere occur in much the same way as the windows of a greenhouse or an automobile raise the temperature by letting visible light in but trap outgoing radiant heat. The analogy is somewhat imperfect since glass also keeps the warm inside air from mixing with the cooler outside air.

The temperature rise at the earth’s surface is directly proportional to a measure of heat trapping called radiative forcing. The units are watts per square meter, as for the sun’s intensity. The radiative forcing from human additions of carbon dioxide since the beginning of the industrial revolution is 1.46 W/m^2. Methane and nitrous oxide additions have provided relative radiative forcings of 0.48 and 0.15 W/m^2, respectively. Other gases have individual radiative forcings of less than 0.1 W/m^2. The total radiative forcing of all greenhouse gases added by human activity to the atmosphere is estimated to be 2.43 W/m^2. This should be compared to the 342 W/m^2 that reaches the earth from the sun. Hence, the greenhouse gases added by human activity are equivalent to turning up the sun’s power by an extra 0.7%, which is enough to cause the global temperature to increase by the observed 0.6°C (1.1°F).

Carbon dioxide is removed from the atmosphere by dissolving in the ocean, by green plants through photosynthesis, and through biogeological processes, such as coral reef and marine shell formation. Approximately half of the carbon dioxide released today will remain in the atmosphere for a century, and one-fourth may be present 200 years from now. Some of the synthetic, industrial greenhouse gases have half-lives of hundreds, thousands, and, in at least one case (carbon tetrafluoride), tens of thousands of years. These long residence times are effectively irreversible changes in the composition of the atmosphere and in climate parameters on the timescale of human generations. This is why some argue that the continued release of greenhouse gases is not sustainable since it can adversely affect future generations. Even without complete information about the consequences of increasing greenhouse gases for climate change, many invoke the precautionary principle and urge the world to move toward reducing and eliminating them.

The post-ice age period in which human civilization has evolved is known as the Holocene. It has long been assumed that this represented an extended, stable, and generally warm climate period. Temperatures were slightly higher than those of today at the beginning of the Holocene and gradually declined until recently. Measurements from Greenland have shown that the general changes were subject to sudden shifts in temperature on timescales of a decade or a century. It is believed that these rapid temperature changes occurred as the result of sudden alterations in oceanic and atmospheric circulation.

Direct land and ocean instrumental temperature measurements began in 1861. These records show two periods of global temperature rise, from 1910 to 1945 and from 1976 to the present. In between, from 1946 to 1975, there was little change in global average temperature, although some regions cooled while others warmed. The total rise for land-based measurements from 1861 to 2000 is estimated by the IPCC to be 0.6°C±0.2°C. The warming rate across the entire earth's surface during the two periods has been 0.15°C per decade. Sea-surface temperatures are increasing at approximately one-third this rate, and it has been found that the heat content of the top 300 m of the ocean increased significantly during the second half of the 20th century. The lower atmosphere up to 8 km has also warmed, but at approximately one-third the land rate. Daily minimum temperatures are rising at 0.2°C per decade, approximately twice the rate of daily maximum temperatures. This has extended the frost-free season in northern areas, and indeed satellite data confirm that total plant growth is increasing at latitudes above 50°N.

There has been much discussion of whether the rise in the instrumental record during the 20th century is simply a return from a little ice age that occurred between the 15th and 19th centuries to conditions of a medieval warming period that predated it. An analysis of surrogate measurements of temperature utilizing data from tree rings, coral reefs, and ice cores that extends back 1000 years clearly shows that the temperature rise of the 20th century for the Northern Hemisphere lies outside the variability of the period. In fact, the current rate of increase, 0.6°C per century, reverses a downward trend of −0.02°C per century during the previous 900 years. The statistical analysis clearly shows that there is less than a 5% probability that the recent temperature increase is due to random variability in the climate system.

The IPCC concludes that the 20th century was the warmest in the past 1000 years, and the 1990s were the warmest decade in that period. The year 1998 may have been the warmest single year, 0.7°C above the 40-year average from 1961 to 1990 and nearly 1.1°C warmer than the projected average temperature extrapolated from the previous 900-year trend.

By the end of the 19th century, scientists had identified the major greenhouse gases in the atmosphere. In 1897, Svente Arhenius, a distinguished Swedish chemist, carried out the first calculation of the effect of greenhouse gases on the temperature of the earth. His estimate, made without a computer or even a good calculator, was surprisingly close to what we know today. Arhenius also performed a thought experiment in which he recognized that the industrial revolution was moving carbon from under the ground in the form of coal and oil and putting it into the atmosphere as carbon dioxide. What would happen, he asked, if human activity were to double the concentration of carbon dioxide in the atmosphere? His answer was approximately twice the upper limit of the range estimated by today's best climate models of 1.5–4.5°C. This is called the climate sensitivity. The actual increase in temperature depends on the amount of greenhouse gas released into the atmosphere. It is estimated that this will amount to approximately a doubling of carbon dioxide equivalent from preindustrial levels some time before 2100.

Systematic measurements of atmospheric composition begun in 1959 by Charles David Keeling in Hawaii have confirmed that carbon dioxide is indeed increasing along with the increase in fossil fuel consumption. Air trapped in glacial ice confirms that carbon dioxide in the atmosphere has risen one-third from 280 parts per million (ppm) in preindustrial times to 373 ppm in 2002.

Human activities are influencing climate in several ways. The most direct is through the release of greenhouse gases.

The greatest share of greenhouse gases comes from the combustion of fossil fuels, such as coal, oil, and natural gas, which release carbon dioxide when burned. Coal releases nearly twice as much carbon dioxide as does natural gas for each unit of heat energy released. Oil releases an amount in between.

Methane released from coal mining constitutes the largest source of anthropogenic methane. Methane releases during natural gas leaks during drilling and transportation are also significant. Long-burning coal mine and peat fires also contribute significant amounts of carbon dioxide, methane, and many air pollutants to the atmosphere.

Land clearance, deforestation, and forest fires release carbon dioxide (23% of total human emissions) and methane.

Land-use change alters the reflectivity or albedo of the land. Deforestation in the northern forests usually increases reflectivity by replacing dark evergreen trees with more white-reflecting snow. In the tropics, less solar energy is absorbed in most cases after forests are removed. Urbanization usually increases the absorption of solar energy. The net effect of land-use change is thought to offset warming by approximately $-0.2 \pm 0.2\ W/m^2$.

Combustion of fossil fuels and clearing of land also contribute dust, particulates, soot, and aerosol droplets. These either reflect sunlight away from the earth or absorb radiation. Some aerosols are also believed to increase the amount of cloud cover.

Agriculture releases carbon dioxide through fossil fuel energy use and from the oxidation of soils. This sector is the major producer of nitrous oxide from the bacterial breakdown of nitrogen fertilizer. Rice culture and livestock are also major producers of methane.

Industrial processes release a variety of greenhouse gases into the atmosphere either during manufacturing or through the use of products.

Air pollution that results in ozone and other gases near the earth's surface traps heat, whereas the depletion of the stratospheric ozone in the upper atmosphere allows more heat to escape to space. The pollution near the earth's surface is the larger effect.

The waste sector releases large quantities of methane from sewage and industrial wastewater, animal feedlots, and landfills.

2. STRATEGIES FOR ENERGY SOURCES

The two primary heat-trapping gases that are associated with energy are carbon dioxide from fossil fuel combustion and methane leakages from natural gas production, transport, and use. Ground-level ozone in smog is also an important secondary pollutant from the energy sector that is a significant heat-trapping gas. As temperatures rise on sunny days, chemical reactions in the atmosphere accelerate to produce more ozone, which in turn traps additional heat.

In the following sections, the sources of energy-related carbon dioxide are summarized by sector. It is customary to report only the carbon content of carbon dioxide that represents 12/44 or 27.3% of the mass of carbon dioxide. World total energy use and carbon dioxide increased rapidly between 1971 and 1995. Energy use increased by 67% to 319 exajoules (10^{18} J), whereas carbon dioxide emissions from energy use increased 54% during this period to 5.5 gigatons of carbon (GtC) (10^9 metric tons). The annual average growth rate between 1990 and 1995 was 0.7% for energy and 1.0% for carbon dioxide. In 1995, industrial countries released 49% of carbon dioxide from energy, with an average annual growth rate of 0.6% per year. Countries with economies in transition (i.e., the former Soviet Union and those of Eastern Europe) were well below their peak of 1990 with just 16% of emissions, an increase of 18% since 1971. Emissions were declining at a rate of 5.1% annually. Developing countries accounted for 35% of emissions, with growth rates ranging from 2.0% for Africa to 6.1% for Asia. There is considerable discussion among governments on responsibility for past contributions of greenhouse gases and the future potential for countries in which fossil fuel use is growing rapidly.

3. ENERGY STRATEGIES FOR ADDRESSING CLIMATE CHANGE

An important insight from Amory Lovins is to recognize that it is not energy we need but, rather, the services that energy can supply in differing forms of heat and work. Hence, in designing technologies, policies, and measures to reduce energy-related greenhouse gases, one must consider not only the present source of emissions and how to reduce them but also alternative ways to deliver the desired energy services.

A simple identity has been developed by Ehrlich and Holdren and extended by Kaya that simplifies this task:

$$\begin{aligned} CO_2\ (\text{tons}) = {} & (\text{population}) \\ & \times (\text{economic value/capita}) \\ & \times (\text{energy use/economic value}) \\ & \times (CO_2/\text{energy used}). \end{aligned}$$

Since greenhouse gas reduction policies do not address either population or wealth (economic value per capita), the major focus is on the last two factors. If one can either find a way to create large economic value with little energy use (factor 3) or use an energy source that produces little CO_2, then the total amount of CO_2 will be small. Note that using no energy or using a zero CO_2 energy technology drives the emissions to zero. Hence, strategies to control CO_2 focus on technological choice, energy efficiency, and the policies and practices that will promote them.

The strategy for reducing emissions from the energy sector consists of five basic options:

1. Increase the efficiency of providing the energy service by finding a less energy-demanding technology or practice.
2. Provide the energy in a more efficient manner.
3. Shift the technology to one that produces fewer or zero greenhouse gases.
4. Shift the fuel to one that produces lower or zero greenhouse gases.

5. Capture and store carbon dioxide and capture and burn methane for fuel.

In the following sections, we examine the opportunities for reducing climate-altering greenhouse gases by economic sector utilizing these strategies. In addition to being able to provide the appropriate energy services, alternative technologies and practices must be economically affordable and socially and politically acceptable.

4. ENERGY FOR BUILDINGS

Buildings use energy directly for heating and cooling and indirectly through electricity for lighting, appliances, office equipment, food storage and preparation, and motors in elevators, pumps, and ventilating systems. Globally, buildings account for 34.4% of energy and 31.5% of energy-related carbon dioxide emissions. The IPCC found many examples of opportunities for substantially reducing the energy use and carbon dioxide emissions from buildings, often at net negative cost. For example, in the United States, new refrigerators use only one-fourth as much electricity as they did in 1971. Improving windows, insulation, and heat exchangers and reducing air and moisture leaks in the building envelope combined with appropriate siting in an integrated building design have made it possible to lower the heating, cooling, and lighting requirements by an average of 40%, with demonstrated cases of 60–71% for residential and commercial buildings, respectively. The cost of achieving a 40% reduction is estimated to be \$3/GJ (10^9) saved in the United States compared to an average cost for building energy of \$14/GJ. Because of the rapid growth of appliance purchases in developing countries, there is enormous potential for major savings in electricity use at much lower cost than building additional power plants. For example, China is the largest producer, consumer, and exporter of refrigerators and air conditioners.

In the case of lighting, the shift from incandescent lamps, which still dominate because of their low initial cost and simplicity, to direct replacement compact fluorescent lamps lowers electricity use by a factor of four. Although the initial cost of compact fluorescent lamps is substantially higher than that of incandescant lamps, they last 10 times longer, resulting in a life cycle cost saving of a factor of two or more. However, compact fluorescents have not achieved universal acceptance because of their different color balance, less concentrated light for use in spotlighting, and higher initial cost. The impressive energy savings of compact fluorescent lamps may be eclipsed by solid-state lighting, such as light-emitting diodes. Because of their ability to be directed, it is possible to provide a solid-state reading light that uses only 1 W to produce the same illumination on a page as a 75-W incandescent. Light-emitting diodes are replacing traffic lights, producing brighter, easier to see signals with an 85% reduction in electricity and a 10 times longer life than incandescents.

It is estimated by the IPCC that there could be cost-effective savings in the building sector using available technology that would lower carbon dioxide emissions by 27% by 2010, 31% by 2020, and 52% by 2050. It is concluded that policies such as improved building codes will need to be put into place to achieve these cost-effective goals.

5. ENERGY FOR INDUSTRY

The industrial sector is the largest user of energy (41%) and releases the largest amount of energy-related carbon dioxide (42.5%) as well as other greenhouse gases. What is unusual in this sector is that for developed countries energy use has been growing slowly at only 0.9% per year, whereas carbon dioxide releases from all of the Organisation for Economic Cooperation and Development industrial countries combined was 8.6% lower in 1995 than in 1971. Developing country emissions amount to only 36.4% of total industrial carbon dioxide releases. Industrial emissions from countries with economies in transition are down from their peak in 1990 by 28%. The major growth has come from developing countries in Asia, where industrial carbon dioxide emissions now exceed those of the developed industrial countries. The IPCC estimates that there is the potential to reduce carbon dioxide emissions through energy efficiency gains by 10–20% by 2010 and by twice that amount by 2020. An additional 10% reduction could come from more efficient use of materials, including recycling, by 2010 and three times that amount by 2020.

Major industrial facilities and refineries are beginning to build fossil-fueled combined heat and electric power facilities that substantially reduce carbon dioxide emissions by 40–50%. The largest industrial combined heat and power facility in the world is the 1000-MW gas-fired facility run by Dow Chemical Company in the United States. The paper industry of Scandnavia and North America produces more than half of its steam and electricity from biomass waste, a renewable resource.

Energy required to produce 1 ton of iron or steel is actually less in some developing countries, such as Brazil or Korea, than it is the United Kingdom or the United States, but China and India remain substantially less efficient. Clearly, choosing improved technology can make a major difference as nations industrialize and as the economies in transition reindustrialize. With 75% of its energy being consumed in the industrial sector, it is important that China is improving its energy/gross domestic product intensity by an estimated 4% per year. There were even reports during the late 1990s that Chinese coal consumption and carbon dioxide emissions had actually declined. Japan continues to have the most energy-efficient industrial sector in the world, and industries in countries such as the United States, Canada, and several European nations have tried to match Japanese levels in specific industries.

6. *ENERGY FOR TRANSPORTATION*

The transport sector consumes just 21.6% of global primary energy and produces 22.0% of fossil fuel carbon dioxide. However, it is the fastest growing sector, with annual growth in carbon dioxide emissions of 2.4% per year. The world automobile fleet has been growing at a rate of 4.5% per year and now exceeds 600 million light vehicles. Unfortunately, the gains in efficiency that characterized the 1980s and early 1990s have begun to reverse.

On the technology side, two Japanese manufacturers have had hybrid gasoline electric cars on the market since 1997, approximately 5 years earlier than was anticipated by the IPCC report of 1996. These vehicles raise fuel economy by 50–100% relative to similar, gasoline-powered cars and provide a substantial clean air bonus as well without sacrificing performance. The second-generation hybrids introduced in the fall of 2003 boast a 22% improvement in fuel economy and a 30% decrease in emissions from already low levels. Europe diesel autos comprise 40% of the market and provide fuel economy gains over gasoline engines in the range of 30% or more. Fuel cell technology has also advanced rapidly: Several manufacturers have announced the introduction of fuel cell-powered vehicles by 2005, approximately 10–20 years earlier than was anticipated only recently. These innovations are expensive, and it remains to be seen if the price targets can be met.

In addition to new engines, there is the potential for major weight reduction with alternative materials. A prototype composite vehicle using a fuel cell has achieved fuel economy of 2.5 liters/100 km (100 miles per gallon), but the cost of a production model is uncertain. It is estimated by IPCC that the potential for carbon dioxide emission reductions from the transport sector is in the range of 10–30% by 2010, with reductions of twice that amount possible by 2020.

The major problem is that the amount of vehicle kilometers traveled continues to increase and outpace any technological gains in efficiency. European governments have kept fuel prices high through taxes on fuel and have negotiated a voluntary, improved fuel economy standard with manufacturers. In contrast, in the United States, the trend is toward larger, less fuel-efficient vehicles and the exploitation of loopholes that undermine the legal fuel economy standards.

The use of vehicle fuels derived from biomass could also reduce net carbon dioxide emissions substantially. Brazil has led the way with its extensive use of ethanol from the sugarcane industry to provide 40% of vehicle fuel mostly through a large fraction blend with gasoline. The Brazilian industry is the most efficient in the world, producing 10–12 times the energy value of alcohol fuel as it comsumes in fossil fuels. In contrast, the U.S. corn-based ehanol industry produces only approximately 1.3 times the energy content of biomass ethanol as it comsumes in fossil fuels. Approximately 140 countries produce sugarcane. Josè Morera of CENBIO proposed that this could become the basis for a major transportation fuel export industry that would boost rural development in poor tropical countries, provide transport fuel domestically, and lower balance of payments problems. Biodiesel from seed crops also has the potential to lower net carbon dioxide emissions and improve air quality by providing a zero-sulfur fuel for trucks and buses. The cost of Brazilian ethanol in recent years has been slightly higher than the price of petroleum-based gasoline, but when oil prices rise it becomes cheaper. The price of biodiesel is approximately twice as high as that of diesel fuel, but it is essentially a small-volume, handcrafted product in the United States and the price could decrease significantly in the future.

In addition to improving technological performance of vehicles, there is potential for alternative transport systems that build on more effective land-use planning, and there are alternatives to transporting people and goods. Low-energy, modern communications technology can substitute for travel to meetings and may be able to reduce travel for entertainment and other activities.

7. ENERGY FOR AGRICULTURE

Agriculture is responsible for only approximately 4% of energy-related carbon dioxide. Economies in transition are the heaviest users of fossil fuels in agriculture, followed by the developing countries of east Asia and the industrial nations. Other developing countries use very little fossil fuels. Rapid growth rates of 10% or higher in China and Latin America are offset by declines of 5% in economies in transition for an annual global growth rate in carbon dioxide of only 0.6%. Agriculture, however, is responsible for approximately 20% of global warming because of heavy emissions of methane from rice farming, livestock, and animal waste. Nitrous oxide is released from agricultural lands by bacterial denitrification of nitrogen fertilizer and the decomposition of animal waste, accounting for 65–80% of total human emissions of this gas. It is technically feasible to lower emissions of methane and nitrous oxide from the agriculture sector by 10–12% by 2010 and by approximately twice that amount by 2020.

Biogas (methane) from anaerobic manure and crop residue decomposition has long been used in India and China for on-farm fuel production. Recently, this technique has been introduced in Europe and North America as a substitute for natural gas and to assist in the management of animal waste and crop residues. Significant additional quantities of both methane and carbon dioxide are associated with rapid rates of deforestation and land-use change to produce new agricultural lands in developing countries. An unknown amount of soil carbon is also oxidized to carbon dioxide, especially from peat soils when they are exposed to air. Nontilling methods and other land management techniques can simultaneously reduce these sources of greenhouse gases and lower emissions from energy use.

Agriculture can also produce energy crops such as wood or sugar that can be burned directly or gasified or liquefied. Brazil has been at the forefront in converting sugar into ethanol for transport fuel, with the production of 10 units of energy for each unit of fossil fuel input. For U.S. corn production of ethanol for fuel, the conversion efficiency is only approximately 1.5. Brazilian industry representatives have realized that sugar constitutes only approximately one-third of the energy value of sugarcane and efforts are being made to gasify the rest of the plant to run gas turbines for electric power generation. Energy plantations can often utilize land that is unproductive for agricultural crops. The burning of biomass releases about as much carbon dioxide per unit of energy as does coal, but if energy crops are grown sustainably, the growing fuel crop absorbs as much carbon dioxide while growing as it will release when burned.

8. ENERGY FOR ELECTRIC POWER

The electric power sector is heavily dependent on coal and is responsible globally for 2.1 GtC per year, which is 38.2% of global energy and 37.5% of global carbon dioxide energy-related emissions. The emissions from this sector are incorporated into the buildings, industrial, transport, and agricultural sectors in calculating their contributions.

In many ways, electric power offers the largest range of possible technological options for reducing greenhouse gases. However, because of the rapid growth in electricity use, emission-reduction potential is estimated to be just 3–9% by 2010 and possibly two to four times that by 2020.

Efficient gas turbines have become the fossil fuel technology of choice since they achieve more than 50% conversion efficiencies from heat to electricity, have very low initial capital costs, and produce relatively few air pollutants. Natural gas also produces only approximately half the carbon dioxide emissions per unit of heat produced as does coal. Work is proceeding on coal gasification that could make use of this technology in an integrated fashion. Emissions from all fossil fuel power plants could be reduced by recovering a portion of the half to three-fourths of the heat that is simply dumped into the environment by utilizing combined heat and power systems. Fuel cells and microturbines promise a revolution of highly efficient distributed power systems that could change electric power generation by providing both electricity and heat on site with significant reductions in transmission and distribution losses.

Renewable energy also holds significant promise. On a percentage basis, wind turbines represent the fastest growing energy source, growing at a rate of more than 25% per year during the late 1990s; by the end of 2002, wind accounted for approximately 31 GW of installed power or 0.1% of installed electric capacity worldwide. The distribution is very uneven. Denmark currently meets 15% of its electric needs from wind. Germany is the global leader, with one-third of the world's total wind power, followed by Spain and the United States. India leads the developing countries in part because of Danish foreign assistance and investment. The potential for wind is particularly great in North America, Africa, and China and also in Russia and other economies in transition, and it substantially exceeds the total current use of electricity worldwide. The direct conversion of solar energy to electricity represents a small fraction of annually added energy, but conversion efficiency continues to increase and costs continue to decrease. Promising innovations such as solar roof tiles could enhance the potential for distributed roof-top power systems.

Burning or, even better, gasifying biomass would also substantially reduce net carbon dioxide to the atmosphere since the carbon is returned to growing plants on a sustained basis. Wood and wood waste in the forest products industry and bagass from sugarcane are among the leading biomass fuels used for electricity and steam production. Geothermal systems also make significant contributions in Central America, the Philippines, the United States, and Iceland. Hydropower continues to grow in many developing countries, including China, India, and Brazil, but few major projects are under development elsewhere.

A major effort has begun to explore the use of the hydrogen from fossil fuels and biomass for fuel cells and gas turbines. The carbon dioxide produced when hydrogen is removed could be sequestered in depleted oil and gas fields, injected into underground aquifers, or possibly injected into the deep ocean. This carbon dioxide storage option would greatly extend the use of fossil fuels without adding to the atmospheric loading of carbon dioxide. The economic and technical feasibility is being studied along with possible environmental consequences of carbon storage.

9. *ALTERNATIVE AND INTEGRATED STRATEGIES*

To stabilize carbon dioxide concentrations at current levels will require a reduction in annual emissions of 75–85%. To return atmospheric concentrations to preindustrial levels will require a shift to an energy system that produces zero emissions. If, in addition, the goal is to fully stabilize atmospheric concentrations of the other 40% of other greenhouse gases that contribute to global warming, it will be necessary to reduce the release of methane, nitrous oxide, and other industrial gases by 75–100%.

There are two alternative approaches for addressing climate change. The first is a gradualist approach that adjusts around the edges of existing technologies to improve efficiency or involves relatively modest shifts in technology and fuels. This approach can slow the rise of carbon dioxide and global warming but cannot by itself halt it. Nevertheless, it can effect significant reductions while awaiting major transformations of the energy sector. In the 2001 IPCC analysis, it was estimated that sufficient technologies currently exist to reduce energy-related carbon dioxide by 15% during each of the first two decades of the 21st century at zero or net negative cost. In other words, investing in these technologies would pay back in energy savings at reasonable social discount rates. An additional reduction of 15% could be achieved at a cost of between zero and $100 per metric tonne of carbon—the equivalent of 25¢ per gallon of gasoline or 6¢ per liter. Other practices, such as minimizing the use of lighting and appliances (turning off the lights, TV, computer, or copy machine, driving vehicles for maximum fuel economy, etc.), can yield substantial additional reductions of 10–50%.

The second strategy is to combine energy efficiency of existing technologies with a concerted effort to replace each existing technology with an alternative that emits little or no carbon dioxide. Such a strategy requires policies that shift the replacement of existing technologies to very low- or zero-emission technologies during the turnover of capital stock or as new additional facilities are built. For example, the light vehicle fleet is largely replaced every 8–12 years in industrial countries. After a new efficient auto technology, such as gasoline– or diesel–electric hybrids or fuel cell vehicles, is introduced, it might take a decade for it to penetrate the market depending on the cost and policies encouraging or mandating the shift. Hence, it might be possible to essentially replace the entire light vehicle fleet in approximately 35 years. For lighting, the change to compact fluorescent lighting and eventually to solid-state lighting could occur within a shorter time of approximately 10 years. Other end-use appliances such as office equipment are replaced on a 3- to 5-year cycle, whereas home appliances such as refrigerators and washers can be replaced every 15 years with devices that are twice as efficient.

Electric power plants have an operational lifetime of 40–50 years, so it would probably take a century to replace them all with some combination of zero-emitting renewable technologies (wind, solar, biomass, and geothermal), fuel cells, combined heat and power, a new generation of nuclear power, or a comprehensive system of carbon dioxide capture and storage technology. Transforming the electrical system to one of multiple, clean, distributed generators utilizing a combination of smaller combined heat and power turbines and fuel cells plus renewables could reduce current electric power generation emissions in half in less than 50 years. Such a distributed power system could also be much more robust and less likely to be disrupted by accidents, storms, or sabotage by terrorists.

Buildings are less permanent than they appear. Low-cost commercial structures often have a lifetime of 25 years or less, as does some housing. Homes and office buildings typically exceed 50–75 years. However, these are upgraded several times during their lifetime, and the addition of new replacement windows, improved insulation and moisture barriers, and advanced heating and cooling systems and controls can substantially lower energy use by 25–50%. Building rooftops are also ideal locations for solar hot water and photovoltaic electricity systems that can be either part of new construction or retrofitted to existing structures.

Hydrogen represents an energy carrier that can either be burned to provide heat or to fuel a gas turbine or it can react chemically in a fuel cell to produce electricity directly. The advantage of hydrogen is that it produces only water as it releases energy. It can be produced from water, fossil fuels, or biomass or solid waste. Producing hydrogen from any of these sources is very energy intensive. To achieve the benefits of zero carbon dioxide emissions from hydrogen on a life cycle basis will require the capture and storage of any carbon released when fossil fuels are the source. An advantage of obtaining hydrogen from biomass or organic solid waste is that if carbon capture and storage is used, there is a net removal of carbon dioxide from the atmosphere.

Finally, there are high-tech, long-term future scenarios that grip the imagination of some. Carbon dioxide capture and storage will require not only modifying all existing coal- and gas-burning power plants, and many industrial facilities, but also an enormous infrastructure of pipes and storage locations. It will take at least half a century to implement on a sufficient scale to make a major difference, but it does allow the current fossil fuel infrastructure to remain largely intact. Hopes for commercially available fusion power have been dashed numerous times, and it, like solar power satellites, remains in the realm of technological futurism. Costs are unknown but are likely to be quite high. Although such technologies may someday be a reality, waiting for these radical technological transformations when more modest technological changes to reduce greenhouse gases are available in the short to midterm does not appear to be a prudent strategy.

10. IMPLICATIONS FOR THE FUTURE

Reducing energy sector greenhouse gas emissions by the large amounts needed to restrain severe climate disruption requires nothing short of a dramatic shift in the delivery of energy services. Although there are advocates for different technologies, it appears that there is no single option that can provide those services in an acceptable manner at a cost that people are willing to pay. Since it is extremely difficult to pick winning future technology, it appears that a portfolio approach of multiple low- and zero-emission technologies is the most likely to succeed. Improving the efficiency of end-use devices, including vehicles, appliances, motors, and lighting, can not only reduce emissions using conventional fossil fuel energy supplies but also substantially lower the cost and many environmental impacts. Doubling the efficiency of lighting and appliances reduces the cost of zero-emission electricity supply whether photovoltaics, wind, fuel cells, or nuclear power plants provide the electricity. If fossil fuels continue to be utilized and carbon capture and storage is used, improved efficiency will lower the amount of carbon dioxide that must be removed, transported, and stored.

Because of the large current investment in existing capital stock of vehicles, buildings, appliances, and industrial facilities, it is essential to ensure that each generation of replacement technology provides substantially lower emissions. Also, since fossil fuels currently provide 85% of the world's energy, there is a strong "lock-in" effect that is technological, economic, and political. To shift course, major policy initiatives are needed that will require a combination of regulations and economic incentives. Until the cost of climate change and other environmental damage is included in the price of fossil fuels, it will be difficult to effect the revolution that is needed. In developing countries in which capital stock is just beginning to be accumulated, the opportunity exists to introduce low and zero carbon dioxide-emitting technology from the start of industrialization. Unfortunately, in most cases, governments and industries in these countries continue to opt for older, less efficient energy technology that has a lower initial cost. Companies and governments in industrial countries can best address this through a combination of technological sharing and investments. There are many low-cost opportunities to provide energy services to the people of developing countries with zero or low carbon dioxide emissions. Since climate change is truly global, it is in everyone's interest to reduce emissions throughout the world.

Since the shift to fossil fuels began in earnest more than 150 years ago, it may take a century or more to complete the next industrial revolution. A 2% annual reduction in carbon dioxide and methane emissions will lead to a net reduction in greenhouse gases necessary to stabilize gas concentrations during this century. It will require multiple policies and measures designed to deliver energy services in new ways if the world is to meet this reduction and avoid a dangerous increase in carbon monoxide levels to 550 ppm.

SEE ALSO THE FOLLOWING ARTICLES

Biomass: Impact on Carbon Cycle and Greenhouse Gas Emissions • Carbon Capture and Storage from Fossil Fuel Use • Carbon Sequestration, Terrestrial • Carbon Taxes and Climate Change • Climate Change and Energy, Overview • Climate Change: Impact on the Demand for Energy • Climate Change and Public Health: Emerging Infectious Diseases • Energy Efficiency and Climate Change • Greenhouse Gas Emissions from Energy Systems, Comparison and Overview

Further Reading

American Council for an Energy Efficient Economy Web site. at www.aceee.org.Fletcher School of Law and Diplomacy, Tufts University, Multilaterals treaty Web site at http://www.fletcher.tufts.edu.

Hawkins, P., Lovins, A., and Lovins, H. (1999). "Natural Capitalism: Creating the Next Industrial Revolution." Little, Brown, Boston.

Intergovernmental Panel on Climate Change (2001). "Climate Change 2001." Cambridge Univ. Press, Cambridge, UK.

Intergovernmental Panel on Climate Change Web site at http://www.ipcc.ch.

Romm, J. J. (1999). "Cool Companies." Island Press, Washington, DC.

Sawin, J. (2003). "Charting a New Energy Future State of the World 2003." Norton, New York.

Tata Energy Research Institute, India, Web site as http://www.teriin.org.

Tufts Climate Initiative, Tufts University, Web site at www.tufts.edu.

United Nations Development Program, United Nations Department of Economic and Social Affairs, World Energy Council (2000). "World Energy Assessment: Energy and the Challenge of Sustainability." United Nations Development Program, New York.

U.S. Environmental Protection Agency Energy Star Program Web site at http://www.energystar.gov.

Coal, Chemical and Physical Properties

RICHARD G. LETT
U.S. Department of Energy, National Energy Technology Laboratory
Pittsburgh, Pennsylvania, United States

THOMAS C. RUPPEL
National Energy Technology Laboratory Site Support Contractor, Parsons
South Park, Pennsylvania, United States

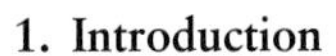

1. Introduction
2. Coal Sampling and Analysis
3. Coal Classification
4. Composition
5. Physical Properties
6. Chemical Properties
7. Coal Databases

Glossary

anthracite High-rank coal, usually possessing a bright luster and uniform texture, that is high in carbon content.

aromatic carbon Carbon atoms in unsaturated six carbon ring structures, exemplified by benzene, or in unsaturated poly-condensed ring structures.

ash The solid (oxidized) residue remaining after coal is subjected to specific combustion conditions.

bituminous coal Hard black coal of intermediate rank.

coalification The complex process of chemical and physical changes undergone by plant sediments to convert them to coal.

fixed carbon The solid residue, other than ash, obtained by heating coal under standardized conditions.

lignite/brown coal Nonfriable, low-rank coal, dark brown to nearly black in appearance.

mineral matter The inorganic compounds or minerals associated with a coal as opposed to the organic material.

rank The stage of geologic maturity or degree of coalification of a coal.

volatile matter The material, less moisture, evolved by heating coal in the absence of air under rigidly controlled conditions.

Coal is a sedimentary rock composed primarily of organic material, mineral matter, and water. It is formed from the partially decomposed remains of plant material that was originally deposited in a swampy environment and later covered by other sediments. The original plant material is transformed under primarily anaerobic conditions by a combination of microbial action, pressure, and heat over geological time periods in a process termed coalification. Peat is usually considered to be the precursor to coal.

1. INTRODUCTION

Coal is found widely distributed and occurs in seams layered between other sedimentary rocks. Owing to the diversity of the original organic debris and deposition environments and subsequent metamorphic processes, coal compositions and properties vary widely. Properties may vary considerably not only from seam to seam but also with location within a given seam. Since coal is heterogeneous, distinctions must be made between properties of some small size fraction of a coal of scientific interest and properties of representative bulk coal samples that are used in commercial activities.

The organic portion of coal is heterogeneous, but under microscopic examination certain regions can be grouped by their physical form and reflectivity. Depending on appearance, these are usually placed into one of three general maceral groups: vitrinite, exinite (liptinite), or inertinite. Vitrinite is typically the most prevalent maceral in bituminous coals and originates largely from the metamorphosis of woody tissue. Exinite is composed of resinous and other hydrogen rich materials. Inertinite is the most aromatic maceral with properties similar to charcoal and has the highest carbon and lowest hydrogen content.

The properties of coals vary with the degree of geological maturity, a concept embodied in the use of the term coal rank. The rank of a coal is a qualitative indication of the geological maturity or degree of coalification. Low-rank coals usually date from the Tertiary or Mesozoic age. The bituminous coals and anthracite are usually from the Carboniferous age. Although it is convenient to view the coal maturation process as progressing in the orderly sequence of peat to brown coal or lignite, to bituminous coal, and ultimately to anthracite, in actuality this is somewhat of an oversimplification and the ultimate rank depends on the severity of the transformation conditions. The low-rank coals tend to be soft and friable with a dull, earthy appearance and a high oxygen and moisture content. Higher rank coals are usually harder and stronger with a black vitreous luster. An increase in rank is attended by an increase in carbon and energy contents and a decrease in moisture and oxygen functionalities.

2. *COAL SAMPLING AND ANALYSIS*

Standardized laboratory procedures for coal characterization and evaluation have been developed by the American Society for Testing and Materials (ASTM), the International Organization for Standardization (ISO), and the standardization bodies of other nations. Owing to the heterogeneity of coal, these include specific procedures for collecting representative samples and sample preparation. Two procedures, the proximate analysis (ASTM D3172; ISO/CD 17246) and the ultimate analysis (ASTM D3176, ISO/CD 17247), deserve special mention because of their wide use and applicability. A sulfur forms analysis (ASTM D2492, ISO 157), which determines pyritic and sulfate sulfur and calculates organic sulfur by difference, is often performed in conjunction with the ultimate analysis.

The proximate analysis is the simplest and most common form of coal evaluation and constitutes the basis of many coal purchasing and performance indices. It describes coal in terms of moisture, volatile matter, ash and fixed carbon, which is calculated by difference. The moisture may be determined by establishing the loss in weight when heated under rigidly controlled conditions on the as received sample and on an air-dried sample. For low-rank coals, particularly for classification purposes, an equilibrium moisture may be determined by equilibrating the moisture in a sample under standard temperature and humidity conditions. The volatile matter represents the loss of weight, corrected for moisture, when the coal sample is heated to 950°C in specified apparatus under standard conditions. Ash is the residue remaining after complete combustion of the coal organic material and oxidation of the mineral matter at a specified heating rate to a specified maximum temperature (700°C to 750°C). Results are conventionally presented on an as-received or moisture-free basis. Typical results for a suite of U.S. coals of differing rank are shown in Table I.

The ultimate analysis expresses the composition of coal in terms of the percentage of carbon, hydrogen, nitrogen, sulfur, oxygen, and ash, regardless of their origin. Standard procedures are used for the determination of carbon, hydrogen, nitrogen, sulfur, and ash, and oxygen is calculated by difference. Results can be expressed on several bases, depending on the manner in which moisture, sulfur, and ash are treated. It is common to report the results on an as-received basis, a moisture-free basis, and on a moisture- and ash-free basis. One alternative is

TABLE I

Proximate Analyses of Selected Coal Samples U.S. Department of Energy Coal Sample Bank, Coal and Organic Petrology Laboratories, The Pennsylvania State University.

	Percentage moisture		Percentage ash	Percentage volatile matter	Percentage fixed carbon
Coal	As received	Equilibrium	Moisture-free basis		
Lykens #3 (DECS-21) anthracite	3.99	2.10	11.15	4.51	84.34
Pocahontas #3 (DECS-19) low volatile bituminous	1.04	4.87	4.60	18.31	77.09
Pittsburgh (DECS-23) high volatile C bituminous	2.00	2.50	9.44	39.42	51.14
Illinois #6 (DECS-24) high volatile C bituminous	13.20	11.59	13.39	40.83	45.78
Rosebud (DECS-10) subbituminous	21.58	23.98	12.56	41.67	45.77
Beulah (DECS-11) lignite	33.88	35.22	9.56	56.08	34.36

TABLE II

Ultimate and Sulfur Forms Analyses of Selected Coal Sample U.S. Department of Energy Coal Sample Bank, Coal and Organic Petrology Laboratories, The Pennsylvania State University.

Coal	Percentage carbon	Percentage hydrogen	Percentage nitrogen	Percentage oxygen	Percentage pyritic S	Percentage sulfate S	Percentage organic S
	Moisture and mineral-matter free basis				Moisture-free basis		
Lykens #3 (DECS-21) anthracite	91.53	4.06	0.81	3.60	0.06	0.00	0.43
Pocahontas #3 (DECS-23) 19 low volatile bituminous	90.61	4.94	1.14	3.30	0.21	0.00	0.52
Pittsburgh (DECS-23) high volatile A bituminous	84.64	5.82	1.54	8.00	2.23	0.01	1.63
Illinois #6 (DECS-24) high volatile C bituminous	80.06	5.56	1.38	12.99	2.64	0.25	2.64
Rosebud (DECS-10) subbituminous	79.69	4.30	1.07	14.94	0.68	0.02	0.47
Beulah (DECS-11) lignite	74.10	4.51	1.01	20.37	0.31	0.03	0.39

illustrated by Table II, which presents the total sulfur and sulfur forms analysis separately on a moisture-free basis. Carbon, hydrogen, nitrogen, and oxygen may then be presented on a moisture- and mineral matter-free basis.

3. COAL CLASSIFICATION

Coals are classified to systematically represent and order their properties and to facilitate commercial use. Many systems of classification exist as each nation with domestic coal resources tended to develop its own scheme for its particular applications. However, all coal classification schemes use a measure of rank as one of the parameters to classify coal. No single measured property or empirically determined characteristic is usually sufficient to classify coals.

The ASTM coal classification system (ASTM D388) in Table III is used extensively in North America and many other parts of the world. This classification is applicable to coals that are composed mainly of vitrinite. It divides coals into four classes—lignite, subbituminous, bituminous, and anthracite—and differentiates groups within these classes. The primary parameters used to classify coals are the fixed carbon (dry, mineral matter-free basis) for the higher rank coals and gross calorific value (moist, mineral matter-free basis) for the lower rank coals. The agglomerating character is used to differentiate between adjacent groups in certain classes.

The National Coal Board (NCB) coal classification, last revised in 1964, is still used in the United Kingdom and has many of the features of classification systems developed outside North America. Bituminous and anthracite coals are classified primarily on the basis of volatile matter (dry, mineral matter-free basis) and coking properties. The different classes of coal are designated by a three-figure numerical code. Lignites and brown coals are outside the range of the NCB classification scheme.

An attempt by the Coal Committee of the Economic Commission for Europe to develop an international standard resulted in a rather elaborate scheme for the International Classification of Hard Coals by Type. Hard coal is defined as coal having a calorific value of more than 23.86 MJ/kg on a moisture- and ash-free basis. This system classifies hard coals into nine classes (akin to rank) based on dry, ash-free volatile matter content and moisture, ash-free calorific value. These classes are further divided into groups according to their caking properties. The groups are divided into subgroups according to their coking properties. A three-figure code number is then used to express the coal classification. The first figure indicates the class of the coal, the second figure the group, and the third figure the subgroup. A new international coal classification standard is now being developed under the sponsorship of the International Organization for Standardization.

An International Standard for Classification of Brown Coals and Lignites (ISO 2950) was published

TABLE III

ASTM Classification of Coal by Rank (ASTM D388)[a]

Class/group	Fixed carbon limits (dry, mineral-matter-free basis) percentage		Volatile matter limits (dry, mineral-matter-free basis), percentage		Gross calorific value limits (moist,[b] MJ/kg)		Agglomerating character
	Equal or greater than	Less than	Greater than	Equal or less than	Equal or greater than	Less than	
Anthracitic							
Meta-anthracite	98	—	—	2	—	—	Nonagglomerating
Anthracite	92	98	2	8	—	—	Nonagglomerating
Semianthracite[c]	86	92	8	14	—	—	Nonagglomerating
Bituminous							
Low-volatile bituminous coal	78	86	14	22	—	—	Commonly agglomerating[e]
Medium-volatile bituminous coal	69	78	22	31	—	—	Commonly agglomerating[e]
High-volatile A bituminous coal		69	31	—	32.6[d]	—	Commonly agglomerating[e]
High-volatile B bituminous coal	—	—	—	—	30.2[d]	32.6	Commonly agglomerating[e]
High-volatile C bituminous coal	—	—	—	—	26.7	30.2	Commonly agglomerating[e]
					24.4	26.7	Agglomerating[e]
Subbituminous							
Subbituminous A coal	—	—	—	—	24.4	26.7	Nonagglomerating
Subbituminous B coal	—	—	—	—	22.1	24.4	Nonagglomerating
Subbituminous C coal	—	—	—	—	19.3	22.1	Nonagglomerating
Lignitic							
Lignite A	—	—	—	—	14.7	19.3	Nonagglomerating
Lignite B	—	—	—	—	—	14.7	Nonagglomerating

[a] This classification dose not apply to certain coals, discussed in Section 1 of D388.

[b] Moist refers to coal containing its natural inherent moisture but not including visible water on the surface of the coal.

[c] If agglomerating, classify in low-volatile group of the bituminous class.

[d] Coals having 69% or more fixed carbon on the dry, mineral-matter-free basis shall be classified according to fixed carbon, regardless of gross calorific value.

[e] It is recognized that there may be nonagglomerating varieties in these groups of the bituminous class, and that there are notable exceptions in the high-volatile C bituminous group.

in 1974. This applies to coal having a calorific value of less than 23.86 MJ/kg on a moisture- and ash-free basis. This system defines six classes based on the percentage of total moisture (ash-free run-of-mine coal) and further divided into five groups according to tar yield (dry, ash-free basis). A two-digit code is then applied, the first digit indicating the class and the last digit the group. This system is similar to one published in 1957 by the Coal Committee of the Economic Commission of Europe that used a four-digit code.

4. COMPOSITION

4.1 Coal Structure

The organic portion of coal consists primarily of carbon, hydrogen, and oxygen with lesser amounts of nitrogen and sulfur. Owing to the heterogeneous character of coals, it is not possible to attribute a single unique molecular structure to a coal. Based on solubility, swelling behavior, viscoelastic properties, and evidence from other physical and chemical analyses, a prevailing view is that the coal matrix is basically a three-dimensionally cross-linked structure. However, coals are not composed of regularly repeating monomeric units, but a complex mixture of structural entities. Adsorbed or embedded but not chemically bound in this matrix is potentially extractable or volatile material of a wide range of molecular structures and molecular weights. Further complications are introduced by the admixture of mineral matter to various degrees with the other coal components.

Coals have significant aromatic character, with the fraction of aromatic carbon increasing with coal rank from 0.4 to 0.5 for low-rank coals to 0.6 to 0.7 for middle rank bituminous coals to over 0.9 for anthracite. However, contrary to some evidence from early studies, more recent studies indicate the average aromatic ring size in coals may be rather modest until the anthracite stage is approached. This information suggests the condensed units in bituminous coals are often no more than four rings with two- and three-ring structures predominating. The aromatic units in low-rank coals tend to be even smaller. These structures may be joined to hydroaromatic groups and cross-linked by methylene, ether, or similar groups. Aliphatic substituents on these structures are usually of short carbon length, predominately methyl or ethyl.

Low-rank coals are rich in oxygen functionalities, particularly carboxylate, phenolic, carbonyl, and furan groups or saturated ether groups alpha or beta to aromatic moieties. Much of the ion-exchangeable hydrogen in the oxygen functionalities of low-rank coals is typically replaced by alkali or alkaline earth cations, particularly calcium. With increasing coal rank there is a general loss of organic oxygen content with the remaining oxygen tending to be of the phenolic or furan type.

Definitive information on the chemical forms of nitrogen in coal is rather meager. Available evidence indicates the nitrogen exists primarily in pyrrolic or pyridinic structures with pyrrolic nitrogen being the most abundant. The fraction of pyridinic nitrogen appears to increase with increasing rank.

The organic sulfur, particularly in higher rank coals, is largely in heterocyclic structures. It is predominately of the thiophenic type. There is indirect evidence for the presence of disulfides and aliphatic sulfides in some low-rank coals based largely on the analysis of select samples of unusual coals of very high sulfur content.

4.2 Extractable Material

Coals contain a substantial amount of material that is not tightly bound to the insoluble organic matrix and extractable by appropriate solvents at mild temperatures ($<200°C$) below the onset of significant thermal decomposition. Coal extracts vary considerably in yield and composition, depending on the coal and its preparation, the solvent, and the extraction conditions. Specific solvents may extract as much as 10 to 20% of low- and middle-rank coals; primary amines (pyridine) are particularly good extraction agents for bituminous coals. Extract yields fall rapidly with the high-rank coals; only a few percent of a low volatile bituminous coal may be extractable and even less of anthracite.

Pyridine extracts are a complex mixture of organic components with a broad distribution of molecular weights from a few hundred to more than a thousand irrespective of the parent coal. These extracts are much more amenable to many analysis techniques than the whole coal. Since they appear to have many of the chemical compositional features of the solid organic phase, extracts have often been studied to provide insight into the possible nature of the chemical entities and functionalities that make up the insoluble coal matrix. The material extracted initially or at low temperatures with poorer extraction agents is usually hydrogen rich and, particularly for brown coals and lignites, often contains material derived from resins and waxes that maybe atypical of the constitution of the coal as a whole.

4.3 Moisture

The moisture content of coal is difficult to precisely specify and control because it exists in several forms in coal, yet it is a parameter that is of great interest in determining commercial value and suitability for many applications. The total (as-received) moisture of a coal sample is usually considered to be all of the water that is not chemically combined (e.g., as water of hydration in minerals). This includes moisture on the external surfaces of coal particles as well as water (inherent moisture) condensed in the pore structure of the coal. Standard coal analysis procedures are often referenced to a moist coal basis (i.e., including the moisture remaining in a coal sample after bringing the sample to a standard condition before analysis). Neither the total nor the air-dried moisture contents from the proximate analysis of coal include water that is chemically combined or water associated with minerals. The bed moisture is the moisture that exists as an integral part of the virgin coal seam and is approximately the amount of water a coal will hold when fully saturated at nearly 100% relative humidity.

The natural bed moisture content of coals generally decreases with increasing coal rank. It can vary from less than 5% in higher rank anthracite and bituminous coals to more that 40% in low-rank lignite and brown coals. In contrast to higher rank coals, brown coals and lignites can lose water very rapidly upon exposure to the atmosphere.

4.4 Mineral Matter

The mineral matter in coal originates from several sources including sedimentary material deposited with the coal, minerals transported by water, and inherent material in the plant precursors. The majority of the mineral matter consists of minerals from one of four groups: aluminosilicates (clays), carbonates, sulfides, and silica (quartz). The most common clays are kaolinite, illite, and mixed-layer illite-montmorillonite. Pyrite is the dominant sulfide mineral in most coals. Sulfates are found in varying amounts depending on the extent of weathering. Likewise, small amounts of elemental sulfur detected in some coals also appear to be the result of pyrite weathering. There is a wide range of mineral compositions for the carbonate minerals in coals depending on the origin. Commonly found carbonate minerals are calcite, siderite, dolomite, and ankerite. The amounts of quartz are also dependent on the origin of the coals. A multitude of accessory minerals can be identified at low concentrations indicative of the particular geological environment of the coal seam.

Owing to the difficulty of directly determining the mineral matter content of coal, it is customary practice to estimate it by calculation using empirical relationships involving more readily determined properties. Often used is the Parr formula, or some variation thereof, that relates the mineral matter to the ash content determined by standard procedures with a correction for pyritic sulfur.

4.5 Trace and Minor Elements

From the geochemical standpoint, the makeup of coal mineral matter is similar to that of the earth's crust, and it contains most of the elements on the periodic table. Although minor and trace element concentrations vary considerably between coals and from sample to sample, the elemental survey analysis in Table IV illustrates the general level of most of these elements found in one small sample of a bituminous coal from West Virginia. The most prevalent inorganically bound minor elements are those associated with the dominant minerals. Interest in the trace element content of coal stems largely from their potential release, either in volatile form or as fine particulate from coal use, and the resulting environmental impact. However, there is nothing particularly toxicologically unique about the trace element composition of most coals.

Actual trace element emissions are somewhat variable depending on coal source, combustion or utilization process, and pollution abatement equipment. Among the volatile elements considered most hazardous are Be, F, As, Se, Cd, Pb, and Hg. Power plants equipped with modern gas cleanup systems efficiently prevent most of these elements from reaching the atmosphere. An exception is mercury (Hg), where the majority may bypass electrostatic emission controls. Although present at only levels of the order of 0.1 μg/g in coals, owing to the huge tonnages of coal used by the power industry, coal use is a significant factor in the total Hg being introduced into the environment by human activity. Development of practical control technology to capture and prevent the release of mercury is a high R&D priority. In developing countries where coal is burned without emission controls or where briquettes may still be used for domestic heating, certain elements such as As and F may present localized problems.

Although considerable information exists pertaining to the concentrations of trace elements in representative coals, it is much more difficult to obtain direct information on the modes of occurrence of trace elements. On a rather rudimentary level, elements are often classified by degree of organic or inorganic association based on the analysis of fractions from physical separation procedures. Such classifications may be highly misleading and may be more a reflection of particle size rather than chemical speciation. However, in most coals many of the elements of potential environmental concern including Co, Ni, Cu, Zn, As, Se, Mo, Cd, Hg, and Pb tend to be primarily associated with the sulfide (chalcophile) minerals.

5. *PHYSICAL PROPERTIES*

5.1 Porosity

The porosity of coal may be characterized by its pore volume and density, surface area, and pore size

TABLE IV

Semi-Quantitative Trace and Minor Element Survey of West Virginia (Ireland Mine) hvAb Coal

Concentrations (μg/g)[a]

Fe =1.74% (atomic absorption)

H																
Li 1.5	Be INT											B 230	C	N	O	F ≤60
Na 300[b]	Mg 900[b]											Al 9500[b]	Si 22000[b]	P 80	S	Cl 520[b]
K 1000[b]	Ca 800[b]	Sc INT	Ti 600[b]	V 12	Cr 12	Mn 24	Fe REF	Co 3.2	Ni 22	Cu 7.6	Zn 21	Ga 4.6	Ge 5	As 7	Se 0.4	Br 18[b]
Rb 14	Sr 90	Y 3.9	Zr 15	Nb 5.1	Mo 1.1	Tc	Ru ≤0.4	Rh ≤0.2	Pd ≤0.4	Ag ≤0.1	Cd ≤0.5	In REF	Sn 1.2	Sb 0.7	Te ≤0.2	I 0.8
Cs 1.1	Ba 38	La 4.4	Hf 0.7	Ta INT	W 1.3	Re REF	Os	Ir	Pt	Au	Hg	Tl 2.3	Pb 14	Bi 0.6	Po	At
Fr	Ra	Ac	Ce 11	Pr 1.4	Nd 5.4	Pm	Sm 1.2	Eu 0.2	Gd 1	Tb 0.2	Dy 1.2	Ho 0.2	Er 0.6	Tm 0.3	Yb 0.4	Lu 0.05
			Th 1.9	Pa	U 0.6											

[a]Determined by spark source mass spectrometry on low-temperature ash renormalized to whole coal basis unless otherwise indicated. Volatile elements such as halogens, S, and Sb are largely lost in ashing procedure.

[b]Determined by X-ray fluorescence on whole coal.

Note. REF, reference; Fe is used as an internal reference; In and Re are added as additional low-concentration references; INT, intereference.

Source: National Energy Technology Laboratory unpublished data.

distribution. These can be determined by a variety of experimental techniques, however, the value obtained is procedure and technique dependent. The porosity of coal has great significance in coal mining, preparation, and utilization. The inherent moisture resides in these pores. In its natural state, the pore structure in coal contains significant amounts of adsorbed methane that is both a mining hazard and in some cases a potentially recoverable resource. The ease of diffusion of methane is determined by the pore volume and pore size distribution. A proposed strategy for mitigating greenhouse gas emissions by sequestration of CO_2 in unmineable coal seams is also dependent on the porosity and adsorption capacity of the coal.

5.1.1 Pore Size

Coals are highly porous solids with pores that vary in size from cracks of micron dimensions to small voids, which are even closed to helium at room temperature. The open pore volume of a coal, calculated as the difference between the particle density (mass of a unit volume of the solid including pores and cracks) and helium density, may comprise as much as 10 to 20% of the coal volume. Total pore volumes derived using CO_2 densities tend to be even higher.

It is common practice to classify pores greater than 50 nm diameter as macropores, pores with diameters in the range 2 nm to 50 nm as mesopores (transitional pores), and pores less than 2 nm as micropores. Conventionally, the physical adsorption of gases or fluids is used to provide measurements of the pore size distribution of coals as well as a total pore volume. The pore size distribution of the macro- and mesopores is often determined by mercury porosimetry techniques. The micropore volume is then often estimated as the difference between the pore volume calculated from helium or CO_2 adsorption data and the pore volume from mercury porosimetry. The porosity in low-rank coals (carbon content <75%) is predominately due to macropores, the porosity of high volatile bituminous coals (carbon content 75 to 84%) is largely associated with meso- and micropores, while the porosity of higher rank coals such as low volatile bituminous and anthracite is predominately due to micropores.

5.1.2 *Surface Area*

The internal surface area of coals is rather high and primarily located in the micropores. The most commonly used determination methods are based on gas adsorption; CO_2 adsorption at $-78°C$ has been used extensively. Generally CO_2 surface areas of coals are between $100\,m^2/g$ to $300\,m^2/g$, with values near $200\,m^2/g$ being typical.

The measured surface area of coals, particularly lignites and brown coals, can be significantly affected by sampling and sample preparation procedures. Implementation of gas adsorption procedures requires removal of the water present in micropores without changing the coal structure. Unless suitable procedures are employed, the structure of low-rank coals can shrink and be irreversibly altered upon drying.

5.1.3 *Density*

The density of coal is usually determined by fluid displacement. Owing to the porous nature of coal and interactions with the coal matrix, the measured density depends upon the fluid employed. The true density is usually determined by helium displacement with the assumption (not entirely valid) that helium will penetrate all of the pores without interacting chemically. The density of the organic phase usually lies between $1.3\,g/cm^3$ to $1.4\,g/cm^3$ for low-rank coals, has a shallow minimum at about $1.3\,g/cm^3$ for mid-rank bituminous coals, and then increases rapidly with rank and carbon content to as high as $1.7\,g/cm^3$ for anthracite. Other fluids often used to determine apparent densities are methanol and water. The methanol density is consistently higher than the helium density. The water density is similar to the helium density for many coals, but interpretation is complicated by the fact that water interacts with oxygen functionalities present. It may also induce swelling, particularly in low-rank coals, opening up porosity that may be closed to helium. The particle density is conventionally determined by mercury displacement.

Most of the mineral constituents in coal have densities about twice that of the organic phase. A notable exception is pyrite which has a density of about $5\,g/cm^3$. These density differences are exploited in physical coal cleaning procedures.

5.2 Optical and Spectral Properties

5.2.1 *Reflectance*

The reflectance is the proportion of directly incident light, conventionally expressed as a percentage, that is reflected from a plane polished surface, usually under oil-immersion, and under specified conditions of illumination. The reflectance of coal is rather low and is usually dominated by the vitrinite reflectance. For coals of the same rank, the liptinite maceral group shows the lowest reflectance and the inertinite group the highest reflectance. There is a continuous increase in vitrinite reflectance with carbon content and coal rank from less than 1% for low-rank coals to more than 4% for very high-rank coals. This is usually ascribed to the growth of aromatic clusters and greater structural ordering. Consequently, the reflectance is widely used as an index of coal rank. Procedures for polished specimens have been automated and standardized (ASTM D2798, ISO 7404/1-5) using polarized green light of wavelength 546 nm.

From reflectance measurements in two different media, two additional optical constants, the refractive index (n) and absorption (extinction) coefficient (k) can be derived. For vitrinites, refractive indexes measured range from 1.68 for vitrinite of 58% carbon content to 2.02 for vitrinite of 96% carbon content.

5.2.2 *Absorption Spectra*

The ultraviolet and visible absorption is unspecific and increases regularly from longer to shorter wavelengths. This is consistent with the ultraviolet and visible absorption being mainly due to electronic transitions in condensed aromatic systems. In contrast, in the infrared region, several major diagnostic absorption bands can be identified in the spectra of properly prepared coal samples despite a relatively high diffuse background. These bands can be associated with vibrations involving various structural groups in the coal.

5.3 Thermal Properties

5.3.1 *Calorific Value*

The calorific value (heating value) is a direct indication of the energy available for the production of steam and probably the most important parameter for determining the commercial usefulness of coal. It is also used as a rank classification parameter, particularly for low-rank coals. It is usually determined by the complete combustion of a specified quantity of coal in standardized calorimetric procedures (ASTM D5865; ISO 1928). The calorific value of coal varies with rank in a manner that can be predicted from the changes in chemical composition. A typical value on a moist, mineral matter free basis for bituminous coal or anthracite is about 30 MJ/kg. Values for low-rank coals tend to be less, even on a moisture free basis, because of their higher oxygen content.

5.3.2 Heat Capacity/Specific Heat

The heat capacity of coal and its products is important in the design and development of any process that includes the thermal processing of coal and is also important for the complete understanding of the fundamental properties of coal. It is usually tabulated as the specific heat, that is, the dimensionless ratio of the heat capacity of the material of interest to the heat capacity of water at 15°C. The specific heat of coal usually increases with its moisture content, carbon content, and volatile matter content. The values for specific heats of various coals typically fall into the general range of 0.25 to 0.35. They can be estimated using a semi-empirical relationship between the specific heat and the elemental analysis.

5.3.3 Thermal Conductivity

The thermal conductivity is useful in estimating the thermal gradients in coal combustion furnaces and coal conversion reactors. It is difficult to assign a single value to the thermal conductivity of in place coal in a coal bed because in many cases the thermal conductivity parallel to the bedding plane appears to be higher than the thermal conductivity perpendicular to the bedding plane. The thermal conductivity of anthracite is of the order of 0.2 W/(m. K) to 0.4 W/(m. K), and the thermal conductivity of bituminous coal is typically in the range 0.17 W/(m. K) to 0.29 W/(m. K). The thermal conductivity of coal generally increases with an increase in apparent density as well as with volatile matter content, ash content, and temperature.

However, it is pulverized coal that is of interest for many applications. The thermal conductivity of pulverized coal is lower than that of the corresponding in place coal. For example, the thermal conductivity of pulverized bituminous coal usually falls in the range 0.11 W/(m. K) to 0.15 W/(m. K).

5.4 Mechanical Properties

5.4.1 Strength and Elasticity

At room temperature, bituminous and higher rank coals behave as brittle solids, whereas brown coals and lignites often display brittleness only when the rates of stressing are extremely high. The strength of coal specimens is commonly studied by means of a compression test, because the results have direct application to estimation of the load bearing capacities of pillars in coal mines. Measurement results are dependent on the orientation of the measurement to the bedding plane of the coal. In common with many other rocks, bituminous and higher rank coals exhibit a stress-strain curve in compression with an initial nonlinear portion associated with closure of cracks followed by an essentially linear elastic range. The breaking stress decreases with increasing specimen size. A dependence of both compression and impact strength on rank has been noted with a minimum for coals with 88 to 90% carbon on a moisture- and ash-free basis.

One of the most used parameters associated with the mechanical properties of coal is Young's modulus, a measure of the compressive or tensile strength of the coal relative to the strain or deformation before fracture. Measurements of Young's modulus in compression and in bending have been made for bituminous coal and anthracite, the mean values being about 3.4×10^9 Pa to 4.3×10^9 Pa, respectively. Measurement of the elastic properties of bulk coals are closely concerned with the presence of cracks or flaws and these must be taken into account in interpreting the results of stress-strain measurements and derived elastic constants.

5.4.2 Grindability

The grindability of coal, or the ease with which it may be ground fine enough for use as a pulverized fuel, is of considerable practical importance. It is a composite physical property usually evaluated by empirical tests; the Hardgrove grindability procedure and index (ASTM D409, ISO 5074) is commonly used. In general, the easier-to-grind coals are in the medium- and low-volatile bituminous ranks. They are easier to grind than coal of the high volatile bituminous, subbituminous, and anthracite ranks. High moisture contents, such as may be encountered in low-rank coals, lead to difficulty in grinding.

5.4.3 Friability

The friability is another empirically evaluated composite coal property of great significance in coal mining and beneficiation. Standard tests (ASTM D440, ASTM D441) provide a measure of the tendency of coal toward breakage on handling. Low-rank coals with high moisture content tend to be the least friable. The friability typically increases with coal rank, reaching a maximum for low volatile bituminous coals and then decreasing for anthracite.

5.5 Electrical Properties

5.5.1 Electrical Conductivity (Resistivity)

The electrical conductivity of most coals is rather low. This parameter is usually expressed as its reciprocal, the resistivity or specific resistance. The resistivity decreases with increase in both rank and temperature. Dry bituminous coals have a specific resistance between 10^{10} and 10^{14} ohm cm at room temperature and can be classed as semiconductors. There is a rapid decrease in resistivity as the carbon content of coals becomes greater than 87% owing to an increase in the three-dimensional ordering of the carbon atoms that facilitates freer movement of electrons from one region to another. Anthracites have a specific resistance between 1 and 10^5 ohm cm, which is orientation dependent. The moisture content of coals strongly affects measurement results, since the specific resistance of water is about 10^6 ohm cm.

5.5.2 Dielectric Constant

The dielectric constant is a measure of the electrostatic polarizability of the electrons in coal. It is strongly dependent on the water content and indeed has been used to estimate the water content of low-rank coal. For carefully dried coal, the dielectric constant varies with coal rank. The value is approximately 5 for lignite, drops to a minimum of approximately 3.5 for bituminous coal, and then rises abruptly with increasing carbon content to a value of 5 for anthracite. Maxwell's relation that equates the dielectric constant of a non-polar material to the square of the refractive index only holds for coal at the minimum dielectric constant. The decrease in dielectric constant from lignite to bituminous coal may be attributed to the loss of polar functional groups. The increase in the dielectric constant as the carbon content increases toward anthracite has been attributed to the presence of polarizable electrons in condensed aromatic systems that increase in size with coal rank.

5.6 Magnetic Properties

5.6.1 Magnetic Susceptibility

The organic phase of coal is slightly diamagnetic—that is, there is a net repulsion when placed in a magnetic field. There are also traces of paramagnetism, at least partially due to stable free radicals. Some of the associated mineral matter may be ferromagnetic necessitating chemical removal before measurements. Correction for the paramagnetic contribution is possible by establishing the effect of temperature, then extrapolating to infinite temperature where paramagnetism essentially vanishes. Magnetized susceptibilities of many demineralized coals have been found to lie between $-4 \times 10^{-5}\, m^3/kg$ and $-5 \times 10^{-5}\, m^3/kg$. The susceptability of one anthracite has been found to be $-9.15 \times 10^{-5}\, m^3/kg$ and exhibited anisotropy parallel and perpendicular to the bedding layers.

Only a slight dependence of the true diamagnetic susceptibility of coals on carbon content has been demonstrated. The susceptibility can be related to the calculated sums of the atomic susceptibilities of the component elements, with modification for carbon in various ring structures. From these results, estimates of the average number of aromatic rings per structural unit can be made.

5.6.2 Magnetic Resonance

Stable free radicals or other paramagnetic species containing unpaired electrons are present in all coals. These free radicals will interact with applied magnetic fields and can be studied by electron spin resonance spectroscopy which measures the absorption of electromagnetic radiation, usually in the microwave region, by a material in an appropriately strong external magnetic field. The conventional electron spin resonance spectrum of coal consists of a single broad line without resolvable hyperfine structure, but information regarding the overall concentration of free radicals and their general nature can still be derived. The free radical concentrations in coal increase approximately exponentially with increasing carbon content up to a carbon content of about 94%, after which a rapid decrease in the number of free spins is observed. Typical free spin concentrations range from 1 to $6 \times 10^{18}\, g^{-1}$ for lignites to $5.3 \times 10^{19}\, g^{-1}$ for an anthracite. These free radicals appear to be primarily associated with the aromatic structures in the coal.

Coals also are composed of many atomic nuclei (e.g., ^1H and ^{13}C) that have a nonzero nuclear spin and the resulting nuclear magnetic moments will also interact with applied magnetic fields and can be studied by nuclear magnetic resonance techniques employing radio frequency radiation. In particular, ^{13}C nuclear magnetic resonance has provided a great deal of useful information on the chemical types and environment of carbon atoms in solid coals.

6. CHEMICAL PROPERTIES

6.1 Thermal Changes

Coal undergoes a variety of physical and chemical changes when it is heated to elevated temperatures. Adsorbed water and adsorbed gases such as methane and carbon dioxide will appear as a product of thermal treatment below 100°C. Low-rank coals, such as lignites, begin to evolve carbon dioxide by thermal decarboxylation as temperatures are raised above 100°C.

Important internal molecular rearrangements (e.g., condensation and other chemical reactions) begin to occur at temperatures as low as 175°C to 200°C. These result in alteration of the original molecular structures without generating major amounts of volatile reaction products.

With the exception of anthracite coals, massive weight losses begin to occur at temperatures between 350°C to 400°C from the evolution of volatile matter in the form of tars and light oils. These result from extensive thermally induced molecular fragmentation and concurrent stabilization and recombination reactions. A solid residue of higher carbon and aromatic content than the original coal remains. As temperatures are raised above ~550°C only light gases, primarily H_2 and CO accompanied by smaller amounts of CH_4 and CO_2, are emitted.

Of special commercial importance are medium-rank caking coals that soften in the temperature range 400°C to 500°C, go through a transient plastic stage, and agglomerate to form a solid semicoke residue. Upon heating to even higher temperatures, the semicoke shrinks, hardens, and evolves additional light gases, and eventually forms strong metallurgical coke at 950°C to 1050°C. Prime coking coals are most typically vitrinite-rich bituminous coals with volatile contents of 20 to 32%. In addition to the standard coal assays, a number of empirical laboratory tests have been developed to assess the commercial coking potential of a coal. These tests usually involve determination of the swelling, plasticity, or agglomeration characteristics upon heating at a fixed rate to a specified temperature and correlation with coke quality.

6.2 Oxidation

6.2.1 Weathering

With the exception of anthracite, freshly mined coals are susceptible to oxidation at even ambient temperatures. The oxidation of coal is an exothermic process and the rate of reaction increases markedly with temperature. It is also a function of the surface area and promoted by moisture. Under appropriate conditions, spontaneous ignition of coal can occur. Care must be exercised in the mining, storage, and shipment of many coals, such as lower rank coals from the western United States, to maintain conditions that will prevent the occurrence of spontaneous ignition. Extensive weathering may also negatively impact the heating value and caking characteristics of a coal.

6.2.2 Combustion

Most mined coal is burned to produce steam for electric power generation. An obvious property of interest is the heating value, which determines the amount of steam that can be generated and is determined by a standardized calorimetric test procedure. Numerous other properties determined by standard tests are of interest in selecting a coal for a particular combustion system. The moisture content of the coal affects handling characteristics and freight costs as well as tending to lower the flame temperature. The amount of volatile matter affects the ease of combustion and flame length. In general, high-rank coals are more difficult to ignite than low-rank coals. The amount of ash governs the loss of sensible heat in ash rejection and affects ash disposal and transport costs. The sulfur content of the feed coal is of importance in meeting environmental emission requirements.

6.2.3 Gasification (Partial Oxidation)

If supplied with less than the stoichiometric amount of oxygen for complete combustion at elevated temperatures, usually above 700°C, coals will yield a gaseous product rich in CO and H_2 that is suitable for use either as a source of energy or as a raw material for synthesis of chemicals, liquid fuels, or other gaseous fuels. Industrial gasifiers commonly utilize a mixture of steam and air or oxygen with the amount of oxygen being generally one-fifth to one-third the amount theoretically required for complete combustion. The distribution and chemical composition of the products will depend not only on the coal feedstock and supplied atmosphere, but on actual gasifier conditions such as temperature, pressure, heating rate, and residence time.

Gasification technology has been adapted to a wide range of coals from brown coals to anthracites. However, reactivity tends to decrease with increase in coal rank. Alkaline elements such as Ca may also have catalytic properties that increase the char gasification reactivity. Depending on the design of the gasifier system, ease of grindability and a low ash content are usually desirable. Characteristics of the resulting coal ash are very important, particularly the ash fusion temperature. The composition of the gasifier slag impacts gasifier refractory life. Coals with low sulfur and halogen content may be preferred to minimize downstream corrosion.

Gasification technologies, particularly Integrated Gasification Combined Cycles (IGCC), have become increasingly attractive as an option to improve power generation efficiency while decreasing atmospheric emissions and providing the possibility of cogeneration of hydrogen or other ultra clean fuels and chemicals with electricity.

6.3 Hydrogenation/Hydrolysis

Although the organic portion of coal will undergo hydrogenation and hydrolysis reactions at relatively low temperatures ($\sim$100°C) in certain strong chemical reducing systems (e.g., metal/amines), most large-scale practical applications utilize reaction with H_2. The latter requires conditions of some severity, that is, elevated temperature and pressure to proceed at reasonable rates. Compared to other hydrocarbons, coal may be considered a hydrogen deficient natural product with an H/C ratio of about 0.8 versus 1.4 to 1.8 for most hydrocarbon oils and a ratio of 4.0 for methane. The addition of hydrogen is a means of not only saturating the hydrogen deficient products of the thermal breakdown of the coal structure, but also terminating side reactions that would lead to higher molecular weight products and solid residues. It also promotes the removal of nitrogen, oxygen, and sulfur as NH_3, H_2O, and H_2S.

The direct hydrogenation of coal has been extensively investigated as a means of producing liquid fuels or petroleum substitute liquids. Modern multistage process concepts involve subjecting coal slurried with recycled process derived vehicle under a hydrogen-rich atmosphere to elevated temperature and pressure. First stage temperatures are often 400°C to 450°C and pressures $\sim$14 MPa to promote coal dissolution. Subsequent stages are typically catalytic and at slightly lower temperatures to promote hydrogenation and upgrading of the first stage product. Depending on the coal, these processes can convert one ton of coal into as much as 4.5 barrels of refinery feedstock. However, although technically feasible, direct hydrogenation of coal is not at the moment practiced on a large commercial scale owing to the availability of economic and proven alternatives.

6.4 Miscellaneous Reactions

Other than the direct reaction with oxygen and hydrogen, the organic material in coal will react with a variety of chemical reagents. Owing to the complexity and heterogeneity of the coal structure, such reactions tend to be somewhat nonselective and yield a complex mixture of products that complicates detailed analysis and interpretation. Reagent accessibility and the extent of reaction are also potential problems. These reactions are typically carried out on a laboratory scale and are now of most interest from the standpoint of gaining insight into the fundamental structure of coal or for analytical applications.

The organic coal structure will react in a rather nonselective manner with many common oxidants such as nitric acid, peroxides, performic acid, permanganate, dichromate, and hypochlorites. Mild oxidation tends to result in an increase in high molecular weight alkali soluble material (humic acids), particularly in the case of low-rank coals. Upon further oxidation, these degrade to lower molecular weight oxygenated components and ultimately to carbon dioxide.

A number of acid or base catalyzed reactions have been used to partially depolymerize or increase the solubility of coals to provide products more amenable to analysis. Typical examples are acid catalyzed reactions such as treatment with BF_3/phenol, reductive alkylation using an alkali metal/tetrahydrofuran, Friedel-Crafts alkylation or acylation, and base hydrolysis. Although these reactions have provided some insight into the chemical structures in coals, the exact relationship of the ultimate reaction products to their precursors in coal is not easily established.

The organic structure of coal will undergo addition and substitution reactions with halogens producing a complex mixture of products that depend on the coal type and extent of halogenation.

Certain chemical functional groups, particularly those containing oxygen such as phenolic OH and carboxyl, can be quantified by procedures based upon derivatization reactions with selective chemical reagents.

7. COAL DATABASES

Extensive tabulations of the properties of coals are maintained by various government agencies and other organizations in the major coal producing nations. In the mid-1970s, the U.S. Geological Survey (USGS) initiated a project to create a comprehensive national coal information database. The original goal was to obtain and characterize at least one sample per coal bed from every geographic quadrangle (approximately 50 square miles) underlain by coal in the United States. The database, now known as the National Coal Resources Data System, is the largest publicly available database of its kind. Readily accessible are comprehensive analyses of a subset of more than 13,000 samples of coal and associated rocks from every major coal-bearing basin and coal bed in the United States. Also available from the USGS is coal quality information on selected coals of the Former Soviet Union compiled in cooperation with the Vernadsky State Geologic Museum and the Russian Academy of Sciences.

The Premium Coal Sample Program implemented at Argonne National Laboratory resulted in the availability of samples of eight highly uniform, premium (unexposed to oxygen) coals for laboratory researchers investigating coal structure, properties, and behavior. The suite of coals include a lignite, subbituminous, high-volatile, medium-volatile, and low-volatile bituminous, as well as a liptinite-rich, an inertinite-rich, and a coking coal. Associated with this program is a comprehensive database of information developed from these representative coals and an extensive bibliography of published research carried out using them.

The Energy Institute at the Pennsylvania State University maintains the Penn State Coal Sample Bank and Database, which still has more than 1100 coal samples available for distribution. Particularly useful is the subset maintained as the Department of Energy Coal Sample Bank and Database. This comprises a suite of 56 carefully sampled coals stored in argon in foil/polyethylene laminate bags under refrigeration and an associated computer accessible database of analyses.

The British Coal Utilization Research Association (BCURA) maintains a coal sample bank containing 36 well-characterized coal samples ranging from lignite to anthracite. These are primarily coals from the United Kingdom, but also now include a number of internationally traded, primarily power station coals.

The Center for Coal in Sustainable Development (CCSD) also maintains a coal bank of 14 representative Australian coals that have been analyzed by standard procedures and are available for research purposes.

SEE ALSO THE FOLLOWING ARTICLES

Clean Coal Technology • Coal Conversion • Coal, Fuel and Non-Fuel Uses • Coal Industry, Energy Policy in • Coal Industry, History of • Coal Mine Reclamation and Remediation • Coal Mining, Design and Methods of • Coal Mining in Appalachia, History of • Coal Preparation • Coal Resources, Formation of • Coal Storage and Transportation • Markets for Coal

Further Reading

Argonne National Laboratory. "Users Handbook for the Argonne Premium Coal Sample Program." Found at www.anl.gov.

ASTM Committee D-5 on Coal and Coke (2003). Gaseous fuels, coal and coke."Annual Book of ASTM Standards," Vol. 05.06. American Society for Testing and Materials. West Conshohocken, PA.

Berkowitz, N. (1985). "The chemistry of coal. Coal Science and Technology, 7." Elsevier, New York.

The British Coal Utilization Research Association (BCURA) Coal Bank. Found at www.bcura.org.

International Organization for Standardization (ISO). Technical Committee 27, Subcommittee 5 (Solid Mineral Fuels). Found at www.iso.ch.

NAS-NRC Committee on Chemistry of Coal (1963). "Chemistry of Coal Utilization, Supplementary Volume" (H. H. Lowry, Ed.), Chapters 1–6. John Wiley & Sons, New York.

NAS-NRC Committee on Chemistry of Coal (1981). "Chemistry of Coal Utilization, Second Supplementary Volume" (M. A. Elliott, Ed.), Chapters 1–8. John Wiley & Sons, New York.

National Research Council Committee (1945). "Chemistry of Coal Utilization" (H. H. Lowry, Ed.), Vol. 1. John Wiley & Sons, New York.

The Pennsylvania State University Energy Institute, Coal and Organic Petrology Laboratories, Penn State Coal Sample Bank and Database. Found at www.ems.psu.edu.

U.S. Geological Survey, National Resources Data System, U.S. Coal Quality Database. Found at www.energy.er.usgs.gov.

van Krevelen, D. W. (1961). "Coal." Reprinted 1981 as "Coal Science and Technology," Vol. 3. Elsevier, Amsterdam, The Netherlands.

Coal Conversion

MICHAEL A. NOWAK
U.S. Department of Energy, National Energy Technology Laboratory
Pittsburgh, Pennsylvania, United States

ANTON DILO PAUL and RAMESHWAR D. SRIVASTAVA
Science Applications International Corporation, National Energy Technology Laboratory Pittsburgh, Pennsylvania, United States

ADRIAN RADZIWON
Parsons, National Energy Technology Laboratory
South Park, Pennsylvania, United States

1. Combustion
2. Pyrolysis and Coking
3. Gasification
4. Liquefaction

Glossary

gasification Partial oxidation of coal to produce a mixture consisting primarily of carbon monoxide and hydrogen.

hydrogenation Chemical addition of hydrogen to an unsaturated molecule.

integrated gasification combined cycle A process that integrates a combustion turbine fed by a coal gasifier and waste heat recovery to produce steam for a steam turbine or process heat.

NO_x Chemical shorthand for a mixture of oxides of nitrogen; the components of NO_x are acid rain precursors and also participate in atmospheric ozone chemistry.

SO_2 Chemical shorthand for sulfur dioxide; SO_2 is an acid rain precursor.

Chemical conversion of coal can be divided into broad categories: combustion to provide process heat and steam and chemical conversion to liquid or gaseous fuels and chemicals. This article discusses the combustion of coal for the purposes of making electricity and the conversion of coal into solid, gaseous, or liquid products that can be used in metallurgical processing or converted into finished fuels and chemicals.

1. COMBUSTION

There are several technologies currently in use to burn coal. Whichever technology is used, the basic chemistry of combustion is the same. Although coal is used in a wide range of processes, the vast majority is simply burned to obtain heat. The majority of that heat is used to produce steam for electric power production. The balance is used to provide process steam and steam for space heating. This is the case worldwide, more so in the United States. The following are the basic important reactions that summarize the combustion process:

$$C + O_2 \rightarrow CO_2$$
$$H_2 + 1/2O_2 \rightarrow H_2O$$
$$S + O_2 \rightarrow SO_2$$
$$N\,(\text{fuel-bound}) + O_2 \rightarrow NO_x$$
$$N_2\,(\text{atmospheric}) + O_2 \rightarrow NO_x$$

The first two reactions are important because they account for nearly all the heat released in the combustion process. The last three are important since they produce pollutants that must be controlled if coal is to be burned in an environmentally acceptable manner. In addition to SO_2 and NO_x, particulate matter is also a pollutant that needs to be controlled, with the current emphasis being on particles that are less than 2.5 μm in diameter. Particulate matter is readily controlled with electrostatic precipitators or fabric filters. Development work continues to improve the performance of these devices and to allow them to operate at higher temperatures. Work is also being carried out on high-temperature ceramic filters. A few coal-fired plants also use Venturi scrubbers to remove particulates, but this technology is little used today. SO_2 is typically controlled with calcium- or sodium-based sorbents. It can be injected into the boiler or ductwork, wet or dry, and removed in the particulate control system. The sorbent can also be used in the form of a slurry or solution and used to contact the flue gas in a system installed downstream of the boiler. These wet systems typically show the greatest removal rates. There are also several technologies (NOXSO, copper oxide, and SNOX) that can remove the SO_2 and recover elemental sulfur or sulfuric acid, which can be sold. These processes also provide NO_x control. Some wet SO_2 removal systems that use a calcium-based reagent produce a marketable gypsum by-product. NO_x is controlled by several techniques or combinations of techniques. One set of NO_x control technologies involves carrying out the initial, hottest part of the combustion process in an air-deficient zone and then introducing additional air to complete the combustion process after the reaction products have cooled to temperatures that are less conducive to NO_x formation. This can be accomplished by specially designed burners or operating burners in an air-deficient mode and injecting overfire air into the cooler region. Another technique, reburning, entails operating the burners with normal coal:air ratios and then injecting additional fuel downstream of the burners, thus creating a reducing zone in which NO_x is destroyed. Overfire air is then injected to complete the combustion reactions at lower temperature. NO_x is also lowered by using selective noncatalytic reduction (SNCR) or selective catalytic reduction (SCR). In both cases, a reducing agent, such as ammonia or urea, is injected into the flue gas to react with the NO_x to form N_2 and water vapor. In the SNCR process, the reagent is injected into the boiler, where conditions are optimal for the reaction to proceed. The SCR process uses a catalyst bed at lower temperatures to carry out the reaction. SCR is both the most expensive and the most effective of the NO_x control strategies. Although there are no requirements to control mercury and control technologies are still under development, it is anticipated that controls may become necessary.

As stated previously, there are several technologies in use to burn coal. The first step toward modern coal combustion (Table I) systems was the stokers. Of those commonly in use today, the underfed stoker is the oldest. It uses a fixed grate and coal is forced into the bottom of the furnace by a mechanical ram.

TABLE I

Time Line of Coal Combustion Technology

300 BC	Greeks record use of coal combustion
1700s	First steam engines
1820s	Stoker boiler
1881	First steam electric plant
1920s	Pulverized coal boiler
1920–present	Atmospheric fluidized bed combustor
	Pulverized coal boiler to 38% efficiency
	Supercritical boilers
	Cyclone combustors (1950s)
	Magnetohydrodynamics
	Pressurized fluidized bed combustor (1980s)
	Environmental controls for No_x and SO_2

As additional coal is fed into the furnace, the coal is forced up and onto the grate. As burning proceeds, the ash is forced off the grate and removed. These units are inefficient, limited in size, and have relatively high maintenance costs. These have been largely replaced by traveling grate and spreader stokers. Both have continuous grates that slowly move through the furnace. In the traveling grate, coal is fed at the front end, and the grate moves front to back, where the ash is discharged. The feed system maintains a supply of coal at the end of the grate outside the combustion zone. The coal is carried into the furnace as the grate moves and the quantity is controlled by a fixed plate a few inches above the bed. Ash is discharged from the grate at the back of the furnace. In the spreader stoker, the grate moves back to front and the ash is discharged below the feed mechanism. The feed mechanism consists of a small conveyor that feeds the coal to a device (reciprocating or rotary) that essentially tosses the coal outward over the grate. The spreader and traveling grate stokers can be made larger than the underfed but are still limited in size compared to more modern boilers. This is especially important when environmental controls are needed since reverse economies of scale make such investments more costly on a unit capacity basis.

The workhorse of today's coal-fired generation is the pulverized coal boiler (Fig. 1). The coal is typically pulverized so that 70% will pass through a 200-mesh (74-μm opening) screen. These units are usually supplied as wall-fired or tangentially fired units. A few are also built with the burners mounted on top with the flame directed downward. The burners of a wall-fired boiler are mounted on one or

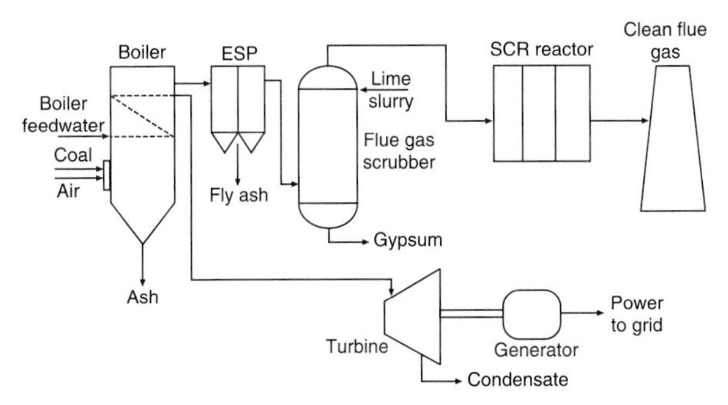

FIGURE 1 Block diagram of a typical pulverized coal-fired power plant.

two (opposing) walls and direct the coal and air horizontally into the boiler. The burners of a tangentially fired boiler are normally mounted on the corners of the boiler and the flame is directed to a point near, but not at, the center of the boiler to impart a swirling motion to the gases that promotes good mixing of the air and fuel. All burners are designed to provide the desired degree of mixing the air and coal. Although the burners of wall- and tangentially fired boilers are basically horizontal, they can be inclined to allow better control of the overall combustion operation. In low-NO_x burners, the design is such that there is some delay in mixing all the air with the fuel to inhibit NO_x formation. These boilers can either be wet- or dry-bottom depending on whether the ash is removed as a molten slag or as dry solid.

Cyclone boilers, which came into use after pulverized coal-fired boilers, have combustors mounted on the outside of the boiler. They inject the air tangentially into the combustor to create the cyclonic motion. This cyclonic motion promotes intense combustion and causes the ash to accumulate on the walls of the combustor, from which it is removed as a molten slag. These burners use a coarser coal than the pulverized coal boilers. Cyclone burners are mounted so that they are sloped toward the boiler. Cyclone boilers are seldom considered for new installations since the intense combustion also promotes NO_x formation.

In recent years, fluidized bed boilers have become more widely used. Fluidized bed systems offer several advantages over pulverized coal and cyclone boilers, including higher thermal efficiency, better environmental performance, the ability to burn a much wider range of coals and other fuels (including waste coal), and a relatively low combustion temperature that inhibits NO_x formation. If necessary, they can be equipped with SCR or SNCR to achieve extremely low NO_x levels. In these boilers, combustion takes place in a fluidized bed of sand or ash. The combustion air is injected through a distribution plate or bubble caps at the bottom of the boiler to maintain fluidization in the bed. Coal and limestone (for SO_2 removal) are injected into the bed at a point above the distribution plate. With no airflow (slumped bed) the bed is simply a bed of solid particles. As airflow increases, the bed becomes fluidized. If airflow increases, the air moves upward as air bubbles. This is referred to as a bubbling fluidized bed. If airflow is further increased, more solids are entrained, removed in an internal cyclone, and returned to the bed, which is now a circulating fluidized bed. Steam is generated in tubes mounted on the walls and tube banks immersed in the bed, with additional heat recovery taking place downstream. This arrangement results in excellent heat transfer characteristics. The first fluidized bed combustors to enter the market operated at atmospheric pressure. Next came the pressurized fluid bed combustors, which can operate at pressures up to 20 atm. In addition to generating power from the steam, additional power is generated by passing the hot, pressurized combustion products (after particulate removal) through an expansion turbine to drive the air compressor and generate additional electric power. The next version of fluidized bed is a pressurized fluidized bed coupled with a partial gasifier that produces a fuel gas that is used in a gas turbine for power generation. The char from the partial gasifier fuels the fluidized bed combustor, which operates in the same manner as when coal is fed directly to the combustor. By careful integration of the subsystems, very high thermal efficiencies can be achieved.

2. PYROLYSIS AND COKING

Although combustion is an exothermic reaction involving stoichiometric amounts of oxygen and achieving flame temperatures as high as 1650°C, coal pyrolysis is an endothermic process. When coal is heated, it undergoes thermal decomposition that results in the evolution of gases, liquids, and tars, leaving behind a residue known as char. Coal pyrolysis is an important process for making metallurgical coke. Coal pyrolysis is a very old technique based on relatively simple technology that dates back to the 18th century. Most pyrolysis systems used in the late 1800s to the early 1900s were located in Europe, where the objective was the production of a smokeless fuel (char) for domestic heating and cooking purposes. It was soon realized in the manufacture of char for fuel that the coal tar fraction contained valuable chemical products that could be separated through refining. However, as inexpensive petroleum appeared in the mid-1900s, interest in coal pyrolysis and its by-products faded.

Coal pyrolysis processes are generally classified as low temperature (<700°C), medium temperature (700–900°C), or high temperature (>900°C). Coal undergoes many physical and chemical changes when heated gradually from ambient temperature to approximately 1000°C. Low-temperature changes include loss of physically sorbed water, loss of oxygen, and carbon–carbon bond scission. At higher temperatures (375–700°C), thermal destruction of the coal structure occurs, as reflected by the formation of a variety of hydrocarbons, including methane, other alkanes, polycyclic aromatics, phenols, and nitrogen-containing compounds. In this temperature range, bituminous coals often soften and become plastic (thermoplastic) to varying degrees. At temperatures between 600 and 800°C, the plastic mass undergoes repolymerization and forms semicoke (solid coke containing significant volatile matter). At temperatures exceeding 600°C, semicoke hardens to form coke with the evolution of methane, hydrogen, and traces of carbon oxides. Pyrolysis of coal is essentially complete at approximately 1000°C.

During pyrolysis, the yield of gaseous and liquid products can vary from 25 to 70% by weight, depending on a number of factors, including coal type, pyrolysis atmosphere, heating rate, final pyrolysis temperature, and pressure. Although certain operating conditions may increase product yield, achieving these conditions may result in increased costs. Coal rank is the predominant factor in determining pyrolysis behavior. Higher rank coal, particularly high volatile A bituminous coals, produce the highest yield of tar. However, higher rank coals give products that tend to be more aromatic (i.e., they have a lower hydrogen:carbon ratio).

Other factors that can improve pyrolysis yields are lower pyrolysis temperatures, utilization of smaller particle size coal, and reduced pressure or employing a reducing (i.e., hydrogen) atmosphere. A reducing atmosphere also improves the yield of liquid and lighter products. Heating rate also affects yield distribution, with rapid heating providing a higher liquid:gas ratio. Pyrolysis in the presence of other gases, particularly steam, has been investigated and is reported to improve liquid and gas yields, but little information is available.

Process conditions also affect the product char. At temperatures higher than 1300°C, the inorganic components of coal can be separated from the products as slag. Rapid heating, conducted in a reactive atmosphere, produces a char with higher porosity and reactivity. The use of finer coal particles, lower temperatures resulting in longer process times, and the complexities of using vacuum, pressure, or hydrogen all add to the cost of coal pyrolysis.

The uses of coal pyrolysis liquids can be divided into two broad categories: (i) direct combustion, requiring little or no upgrading, and (ii) transportation fuels and chemical, requiring extensive upgrading. Much attention with regard to low-temperature tar processing has been devoted to hydroprocessing techniques, such as hydrotreating and hydrocracking, with the primary objectives of reducing viscosity, reducing polynuclear aromatics, and removing heteroatoms (sulfur, nitrogen, and oxygen) to produce usable fuels and chemicals. The cost of hydrogen is the primary impediment to tar upgrading. The tar fraction can be used as a source for chemicals, such as phenolics, road tars, preservatives, and carbon binders, but these uses do not constitute a large enough market to support a major coal pyrolysis industry.

Probably the most common application of coal pyrolysis is in the production of coke. The production of metals frequently requires the reduction of oxide-containing ores, the most important being production of iron from various iron oxide ores. Carbon in the form of coke is often used as the reducing agent in a blast furnace in the manufacture of pig iron and steel. The blast furnace is basically a vertical tubular vessel in which alternate layers of iron ore, coke, and limestone are fed from the top. Coal cannot be fed directly at the top of a blast furnace because it does not have the structural strength to support the column of iron ore and limestone in the furnace while maintaining sufficient porosity for the hot air blast to pass upward through the furnace.

Not all coals can produce coke that is suitable for use in a blast furnace. The property that distinguishes coking coals is their caking ability. Various tests are performed to identify suitable coal for conversion to coke through pyrolysis. Frequently, several coals are blended together to achieve the necessary coal properties to produce a suitable coke. Commercial coke-making processes can be divided into two categories: nonrecovery coke making and by-product coke making.

In nonrecovery coke plants, the volatile components released during coke making are not recovered but, rather, are burned to produce heat for the coke oven and for auxiliary power production. One of the earliest nonrecovery units was the beehive oven, which for many years produced most of the coke used by the iron and steel industry. With these ovens, none of the by-products produced during coking were recovered. Because of their low efficiency and pollution problems, beehive ovens are no longer in use in the United States.

Nonrecovery coking takes place in large, rectangular chambers that are heated from the top by radiant heat transfer and from the bottom by conduction through the floor. Primary air for the combustion of evolved volatiles is controlled and introduced through several ports located above the charge level. Combustion gases exit the chamber through down comers in the oven walls and enter the floor flue, thereby heating the floor of the oven. Combustion gases from all the chambers collect in a common tunnel and exit via a stack that creates a natural draft for the oven. To improve efficiency, a waste heat boiler is often added before the stack to recover waste heat and generate steam for power production.

At the completion of the coking process, the doors of the chamber are opened, and a ram pushes the hot coke (approximately 2000°F) into a quench car, where it is typically cooled by spraying it with water. The coke is then screened and transported to the blast furnace.

The majority of coke produced in the United States comes from by-product coke oven batteries. By-product coke making consists of the following operations: (i) Selected coals are blended, pulverized, and oiled for bulk density control; (ii) the blended coal is charged to a number of slot type ovens, each oven sharing a common heating flue with the adjacent oven; (iii) the coal is carbonized in a reducing atmosphere while evolved gases are collected and sent to the by-product plant for by-product recovery; and (iv) the hot coke is discharged, quenched, and shipped to the blast furnace.

After the coke oven is charged with coal, heat is transferred from the heated brick walls to the coal charge. In the temperature range of 375–475°C the coal decomposes to form a plastic layer near the walls. From 475 to 600°C, there is marked evolution of aromatic hydrocarbons and tar, followed by resolidification into semicoke. At 600–1100°C, coke stabilization occurs, characterized by contraction of the coke mass, structural development of coke, and final hydrogen evolution. As time progresses, the plastic phase moves from the walls to the center of the oven. Some gas is trapped in the plastic mass, giving the coke its porous character. When coking is complete, the incandescent coke mass is pushed from the oven and wet or dry quenched prior to being sent to the blast furnace. Modern coke ovens trap the emissions released during coke pushing and quenching so that air pollution is at a minimum.

Gases evolving during coking are commonly termed coke oven gas. In addition to hydrogen and methane, which constitute approximately 80 volume percent of typical coke oven gas, raw coke oven gas also contains nitrogen and oxides of carbon and various contaminants, such as tar vapors, light oil vapors (mainly benzene, toluene, and xylene), naphthalene, ammonia, hydrogen sulfide, and hydrogen cyanide. The by-product plant removes these contaminants so that the gas can be used as fuel. The volatiles emitted during the coking process are recovered as four major by-products: clean coke oven gas, coal tar, ammonium sulfate, and light oil. Several processes are available to clean coke oven gas and separate its constituents.

In the past, many products valuable to industry and agriculture were produced as by-products of coke production, but today most of these materials can be made more economically by other techniques. Therefore, the main emphasis of modern coke by-product plants is to treat the coke oven gas sufficiently so that it can be used as a clean, environmentally friendly fuel. Coke oven gas is generally used in the coke plant or a nearby steel plant as a fuel.

3. GASIFICATION

Coal was first gasified in England by William Murdock in 1792, and the world's first coal gas company was chartered in England in 1812. Coal gas was first produced in the United States in 1816 in Baltimore, and by 1850 more than 55 commercial coal gasification plants in the United States were generating gas for lighting and heating. During the late 1800s and early 1900s, a large number of coal gasifiers operated commercially in the United States and Europe to produce industrial and residential fuel gas.

Most of the early gasifiers were moving bed units, charged with sized coal and blown with steam and air to generate "producer gas" (150 Btu/scf). Operating these moving bed gasifiers in a cyclic mode (blowing first with air to heat the coal, followed by contact with steam to produce "water gas") increased the heating value of the product gas to 300 Btu/scf. The heating value was further increased to approximately 500 Btu/scf by cofeeding oi1 with steam to produce "carbureted water gas," which contained hydrocarbons in addition to H_2 and CO. An early gasification process, still in use today, was that developed by Lurgi.

Extensive process development was carried out in the United States in the late 1940s to mid-1950s, prompted by a concern that natural gas reserves were limited. Recent interest in coal gasification has been driven by the potential of integrated gasification combined cycle (IGCC) facilities to increase the efficiency of power production while reducing emissions.

The initial step in coal gasification involves grinding and/or pretreatment of the coal to put it into a form suitable for injection into the gasifier. In the gasifier, the coal is heated in the presence of a reactive gas (air, oxygen, and steam), the nature of which depends on the product desired. Reactions occurring during gasification of coal can be divided into three groups: pyrolysis reactions (thermal decomposition of the coal), gasification reactions (gas–solid reactions), and gas–gas reactions. The following are major reactions:

Pyrolysis reaction

$$C_xH_y \rightarrow (x - y/4)C + y/4CH_4$$

Gas−solid reactions

$$C + 1/2O_2 \rightarrow CO$$

$$C + O_2 \rightarrow CO_2$$

$$C + H_2O \rightarrow CO + H_2$$

$$C + CO_2 \rightarrow 2CO$$

$$C + 2H_2 \rightarrow CH_4$$

Gas−gas reactions

$$CO + H_2O \rightarrow H_2 + CO_2$$

$$CO + 3H_2 \rightarrow CH_4 + H_2O$$

The first reaction that occurs is pyrolysis or devolatilization. Depending on the type of gasifier, condensable hydrocarbons may be collected as a by-product or may be completely destroyed. Combustion reactions are the fastest gasification reactions and are highly exothermic. Oxidation reactions occur rapidly with essentially complete oxygen utilization. The primary combustion products are CO and CO_2. The gas–solid reactions illustrate the gasification of char by reaction with various gases. In addition to the carbon–steam reaction, steam undergoes a side reaction called the water gas shift reaction. Because coal chars are highly microporous, most of the gasification reactions take place inside the char particles.

The reducing nature of the product gas causes most heteroatoms (sulfur, nitrogen, and chlorine) to appear in reduced form as hydrogen sulfide, ammonia, and hydrogen chloride. In most cases, these materials are scrubbed from the product gas before it is burned or reacted. Ammonia and HCl are very water soluble and easily removed by a water wash, and several processes are available for removal of H_2S. Typical gasifier raw gas has a composition of 25–30% H_2, 30–60% CO, 5–15% CO_2, 2–30% H_2O, and 0–5% CH_4. Other components include H_2S, COS, NH_3, N_2, Ar, and HCN, depending on the composition of the coal, the purity of the oxygen used, and the operating conditions.

Gasification processes can be classified into three major types: moving bed (countercurrent flow), fluidized bed (back-mixed); and entrained-flow (not backmixed). Moving bed gasifiers consist of a downward-moving bed of coal in contact with a countercurrent flow of gas moving upward through the bed. As it moves down the bed, the coal undergoes the following processes: drying, devolatilization, gasification, combustion, and ash cooling. Moving bed gasifiers can be operated at atmospheric or higher pressure with either air or oxygen as the oxidant, with ash removed either dry or molten, and with or without stirrers to prevent agglomeration. At the top of the bed, the hot upward-flowing gases dry the coal. As the coal moves down, its temperature increases until at approximately 315–480°C pyrolysis occurs, generating gases, oils, and tars. As the resulting char moves further down the bed, it is gasified by reaction with steam, carbon dioxide, and hydrogen to produce a mixture of carbon monoxide, hydrogen, methane, carbon dioxide, unreacted steam, and a small amount of other gases. Below this zone, char is combusted by reaction with oxygen.

The ratio of steam to oxygen in the gasifier controls the temperature in the combustion zone. If dry-bottom operation is desired, sufficient steam is added to offset the exothermic oxidation reactions with endothermic steam–carbon reactions to stay below the ash fusion temperature. Slagging gasifiers operate at a higher temperature, and ash is removed in a molten state and then quenched in a water bath. Moving bed gasifiers require sized coal for proper operation, typically in a range 0.25–2 in.

In a fluidized bed gasifier, reactant gases are introduced through a distributor at the bottom of the bed at a velocity sufficient to suspend the feed particles. The result is a bed of highly mixed solids in intimate contact with the gas phase. The high degree of mixing leads to a uniform temperature throughout the bed and results in reaction rates that are typically higher than those experienced in moving bed gasifiers.

Exit gas temperature for a fluidized bed gasifier is higher than that for a moving bed gasifier, resulting in further pyrolysis of intermediate hydrocarbon products so that the product gas contains a much lower concentration of tar/oil. Unconverted char and ash are removed as dry solids. If strongly caking coals are used, pretreatment is required. Fluidized bed gasifiers can be operated at atmospheric or higher pressure. The fluidizing gas is a mixture of steam with either air or oxygen.

In an entrained-flow gasifier (Fig. 2), a mixture of finely ground coal entrained in a reactant gas flows through the reactor, with little or no backmixing.

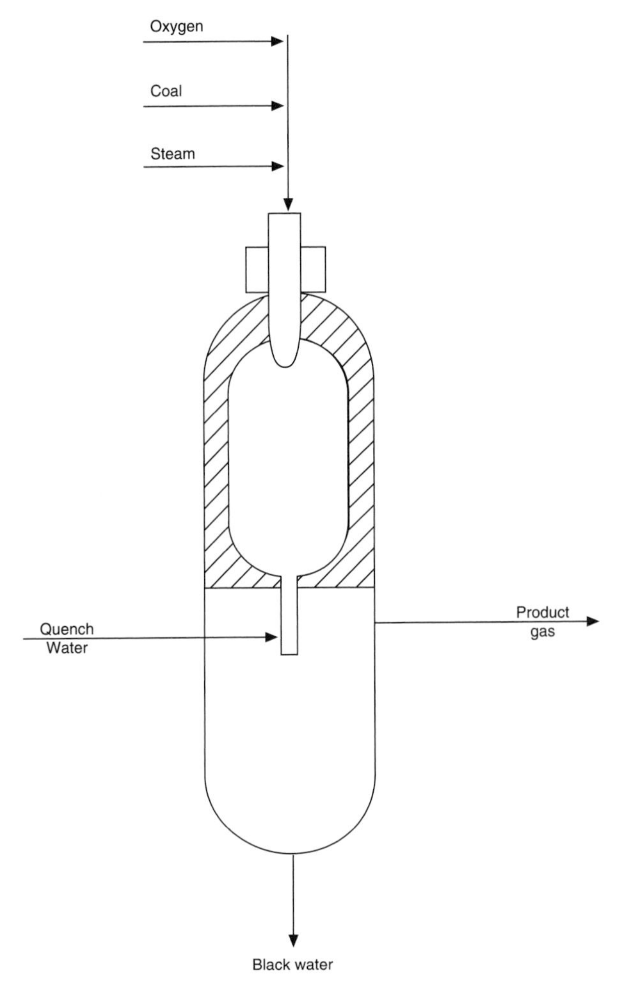

FIGURE 2 Configuration for Texaco coal gasification process.

This type of gasifier may be either single stage or two stage. In general, high temperatures (650–960°C) are used to achieve complete gasification of the coal in a mixture with steam and oxygen or air.

Entrained-flow gasifiers can handle all coals, including strongly caking coals, without pretreatment. The high temperature of operation produces a gas free of methane, oils, and tar. In two-stage gasifiers, the incoming coal is first entrained with reactant gases to produce syngas and char; the resultant char is gasified further in a second stage, which may or may not be entrained. Because the more reactive incoming coal can be gasified at a lower temperature than the less reactive char, staged operation achieves better overall thermal efficiency without sacrificing higher throughput. Entrained-flow gasifiers can be operated at atmospheric or higher pressure, and ash may be removed either dry or molten.

Many different gasifiers have been developed, at least through the demonstration stage, by a variety of organizations, both public and private. However, not all these gasifiers have achieved commercial success, and improved processes now supercede some technologies that were widely used in the past. In 1999, the Texaco, Shell, and Lurgi (dry ash) processes accounted for more than 75% of the installed and planned coal gasification capacity, with Texaco being the leader with almost 40% of installed capacity.

High reliability, acceptable capital and operating costs, and minimal environmental impact make coal gasifiers attractive for utility applications. Numerous studies confirm that gasifiers coupled with a gas turbine/steam turbine combined cycle represent one of the most promising technologies for future coal-based power generation. IGCC technologies offer the potential for high efficiencies with low pollutant emissions. High efficiencies are achieved in IGCC operation by combining efficient combustion turbines with a steam turbine bottoming cycle.

A major goal of power production is minimal environmental impact. Since the product gas from IGCC systems is purified before combustion, burning this clean fuel results in low pollutant emissions. Ash leaving the system is usually in the form of molten slag that is water quenched to form benign vitreous material suitable for a variety of uses or for disposal in a nonhazardous landfill. On balance, coal gasification systems are environmentally superior to alternative coal utilization technologies and can meet rigorous environmental standards for SO_2, NO_x, and particulates. Also, because of their increased efficiency, IGCC plants emit less CO_2, the major greenhouse gas, per unit of electricity generated.

4. LIQUEFACTION

Coal liquefaction is the process of making a liquid fuel from coal. The fundamental difference between coal, a solid, and liquid fuels is that the liquid fuels have a higher hydrogen:carbon ratio. Liquid fuels have lower ash contents and are easier to upgrade (i.e., to remove unwanted impurities such as nitrogen and sulfur). There are three major methods of producing liquid fuels from coal: pyrolysis, direct liquefaction, and indirect liquefaction.

Pyrolysis is a thermal process in which coal is heated to produce gases, liquids, and tars, leaving behind a hydrogen-depleted char. The gases, liquids, and tars are usually further treated (refined) to produce conventional liquid fuels. In direct coal liquefaction, finely divided coal is mixed with a solvent and heated under pressure in a hydrogen atmosphere and in the presence of a catalyst. After separation from ash and unconverted coal, the resulting liquids are usually further refined to produce finished products. Indirect liquefaction is a two-step process. The first step involves gasification of coal to produce a mixture of hydrogen and carbon monoxide, called synthesis gas. In the second step, the synthesis gas is heated in the presence of a catalyst to make hydrocarbons, with water being the major by-product. The second step most frequently considered is called Fischer–Tropsch synthesis. The hydrocarbon product is then refined into the desired finished products.

The conversion of coal to liquid fuels is not practiced commercially in the United States. Both direct and indirect liquefaction were practiced in Germany during World War II and were used to fuel the German war effort. Research and development efforts have been ongoing, at various levels of effort, since the 1940s to make coal liquefaction a commercially viable enterprise. The only commercial liquefaction process for manufacturing fuel is conducted in South Africa by Sasol using the Fischer–Tropsch process. Sasol's primary products are high-value chemicals in addition to fuels. Eastman Chemical operates a coal conversion plant in Kingsport, Tennessee, where each day coal is converted into 260 tons of methanol to supply Eastman's acetate synthesis plants. In 2002, China announced plans to construct a direct coal liquefaction plant.

Direct coal liquefaction can be traced back to the work of the German chemist Friedrich Bergius, who, in approximately 1913, began studying the chemistry of coal. In his studies, he attempted the hydrogenation of coal using temperatures exceeding 350°C and hydrogen under high pressure. The first commercial-scale direct liquefaction plant was constructed by I. G. Farben in Leuna in 1927. The Germans constructed several more plants during the 1930s. Some fundamental but no large-scale research was conducted in the 1940s–1960s. It was not until the 1970s that interest in direct coal liquefaction was revived.

As coal is heated in a suitable solvent, it is rapidly converted to soluble entities that have high molecular weight and an average elemental composition very similar to that of the starting feed coal. The primary chemical reaction of direct coal liquefaction is hydrogenation of the coal to break the coal into smaller molecules and increase the hydrogen: carbon ratio. Addition of hydrogen also removes the heteroatoms nitrogen, sulfur, and oxygen by converting them to ammonia, hydrogen sulfide, and water, respectively. Some combination of iron, nickel, cobalt, and molybdenum catalysts can be used to lower the required temperature and pressure and improve the rate and extent of hydrogenation.

The direct liquefaction process (Fig. 3) employs a coal ground to a fine powder, solvent, hydrogen, and catalyst employed at elevated pressures (1000–2500 psig) and temperatures (400–450°C). The solvent mixed with the coal forms a pumpable paste.

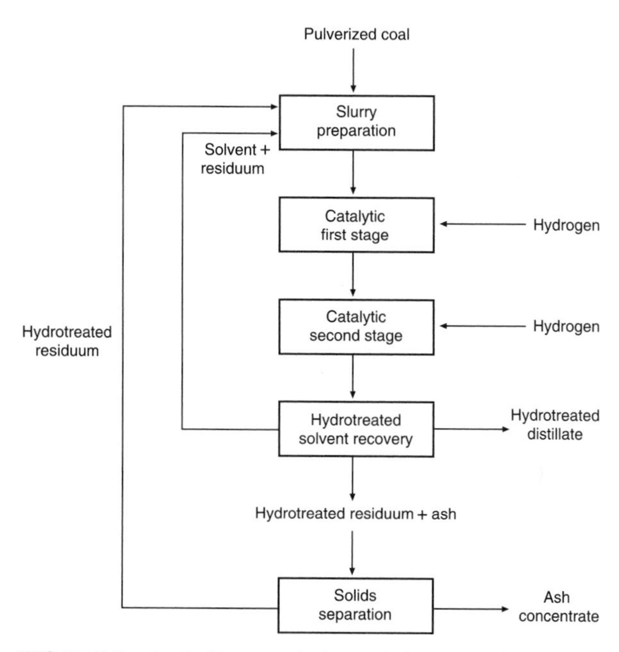

FIGURE 3 Bock diagram of advanced direct coal liquefaction process.

Intimate contact between solid coal and solvent is easier to achieve than that between coal and gaseous hydrogen; therefore, in addition to serving as vehicle and solvent, the coal liquefaction solvent also serves as a hydrogen shuttle or donor solvent. After transferring hydrogen to the coal, the hydrogen-depleted donor solvent reacts with the gaseous hydrogen, regenerating itself. Processes can be designed so that they rely primarily on dissolution, use one-stage or two-stage hydrogenation with or without catalyst addition, and use various schemes for recovery and regeneration of solvent.

If dissolution is the primary mechanism for converting coal, the product is really an ash-free solid that, if desired, can undergo further processing. When hydrogenation schemes are employed, the resulting liquid products are rich in aromatic and olefinic materials, and after separation from mineral matter and unconverted coal they need to be upgraded to finished products through conventional refinery operations such as distillation and hydrotreating. The exact product slate will depend on the type of coal used, solvent employed, and specific operating conditions.

Indirect liquefaction is the production of fuels or chemicals from synthesis gas derived from coal. Synthesis gas is a mixture of hydrogen and carbon monoxide. A hydrogen:carbon monoxide ratio of approximately 2:1 is required for hydrocarbon production. The by-product is water. In coal liquefaction, the synthesis gas is produced by coal gasification.

Franz Fischer and Hans Tropsch first reported the hydrogenation of carbon monoxide in 1925. By-products to the straight-chain hydrocarbons were a mixture of alcohols, aldehydes, and acids. Nine Fischer–Tropsch plants operated in Germany during World War II. Japan also had Fischer–Tropsch plants during World War II. In 1953, South Africa, well endowed with coal resources but lacking significant petroleum sources, began construction of a Fischer–Tropsch plant at Sasolburg, near Johannesburg. The plant is operated by the South African Coal Oil and Gas Corporation, commonly known as Sasol. In 1973, Sasol began construction of a second plant, called Sasol-2, at Secunda. Eventually, a third plant, Sasol-3, came online in 1989. No commercial Fischer–Tropsch plants have been built since, although there is considerable interest in manufacturing fuels from synthesis gas obtained by the steam reforming of methane.

The Fischer–Tropsch reaction occurs over a catalyst, typically iron or cobalt based. The Fischer–Tropsch reaction is basically a chain polymerization reaction followed by hydrogenation. The reactions are conducted at pressures ranging from 1 to 30 atm and at 200–350°C. The growth of the linear chains follows an Anderson–Schultz–Flory (normal) distribution and can be described by the term α. When $\alpha = 0$, methane is the product. As α approaches 1, the product becomes predominantly high-molecular wax. Iron catalysts may be employed when the gasification step produces a synthesis gas with insufficient hydrogen because iron is known to promote the water gas shift reaction, thus synthesizing additional hydrogen *in situ*.

The Fischer–Tropsch reaction is highly exothermic. The first commercial units used fixed bed shell and tube reactors. Because of the excess heat produced, the diameter of the tubes is limited. Tube packing is also cumbersome. By using fluidized beds, tubes could be made larger and temperature control still maintained. However, the fluidized bed requires a higher operating temperature (340°C) and deposition of products on the catalysts disrupts fluid bed operations. A newer concept, the slurry bubble contact reactor (SBCR), uses finely divided catalyst dispersed in inert liquid while reactive gases are introduced from the bottom of the reactor. Heat exchanger tubes provide temperature control. The SBCR offers high throughput and simplicity of design. Sasol commissioned a commercial SBCR (2500 bbl/day) in 1993.

The resulting hydrocarbon products need to be separated from oxygenated by-products and further refined into finished fuels or petrochemical feedstocks. When wax is a predominant product, the wax can be cracked to produce high-quality diesel fuels and jet fuels. Because the gasification process allows for ready removal of contaminants from the synthesis gas product, the resulting Fischer–Tropsch products have sulfur levels below the detection limits. Olefinic products can be used in place of petrochemical feedstocks.

As suggested by the discussion on Fischer–Tropsch synthesis, synthesis gas can also be converted to oxygenated liquids, including methanol, higher alcohols, and ketones. In the presence of a catalyst, often a zinc–chromium or copper–zinc catalyst, synthesis gas can be converted to methanol. The coal conversion plant at Kingsport, Tennessee, combines coal gasification with methanol synthesis in a SBCR. Methanol can be used as a fuel or as a chemical feedstock. It can also be dehydrated and polymerized over a zeolite catalyst to produce gasoline-like products.

SEE ALSO THE FOLLOWING ARTICLES

Clean Coal Technology • Coal, Chemical and Physical Properties • Coal, Fuel and Non-Fuel Uses • Coal Industry, Energy Policy in • Coal Industry, History of • Coal Mine Reclamation and Remediation • Coal Mining, Design and Methods of • Coal Mining in Appalachia, History of • Coal Preparation • Coal Resources, Formation of • Coal Storage and Transportation • Markets for Coal

Further Reading

Alpert, S. B., and Wolk, R. H. (1981). Liquefaction processes. *In* "Chemistry of Coal Utilization" (M. A. Elliott, Ed.), 2nd Suppl. Vol., pp. 1919–1990. Wiley, New York.

Bartok, W., and Sarofim, A. F. (eds.). (1993). "Fossil Fuel Combustion—A Source Book." Wiley, New York.

Bartok, W., Lyon, R. K., McIntyre, A. D., Ruth, L. A., and Sommerlad, R. E. (1988). Combustors: Applications and design considerations. *Chem. Eng. Prog.* **84**, 54–71.

Berkowitz, N. (1979). "An Introduction to Coal Technology." Academic Press, New York.

Brown, T., Smith, D., Hargis, R., and O'Dowd, W. (1999). Mercury measurement and its control: What we know, have learned, and need to further investigate. *J. Air Waste Management Assoc.* **49**, 628–640.

Derbyshire, F. (1986). Coal liquefaction. In "Ullmann's Encyclopedia of Industrial Chemistry," Vol. A7. VCH, New York.

El Sawy, A., Gray, D., Talib, A., and Tomlinson, G. (1986). A techno-economic assessment of recent advances in direct coal liquefaction, Sandia Contractor Report No. SAND 86–7103. Sandia National Laboratory, Albuquerque, NM.

Elliott, M. A. (ed.). (1981). "Chemistry of Coal Utilization." Wiley–Interscience, New York.

Gavalas, G. R. (1982). Coal pyrolysis. *In* "Coal Science and Technology," Vol. 4. Elsevier, Amsterdam.

Howard, J. B. (1981). Fundamentals of coal pyrolysis and hydropyrolysis. *In* "Chemistry of Coal Utilization" (M. A. Elliott, Ed.), 2nd Suppl. Vol., pp. 665–784. Wiley, New York.

Juntgen, H., Klein, J., Knoblauch, K., Schroter, H., and Schulze, J. (1981). Liquefaction processes. *In* "Chemistry of Coal Utilization" (M. A. Elliott, Ed.), 2nd Suppl. Vol., pp. 2071–2158. Wiley, New York.

Khan, M. R., and Kurata, T. (1985). "The feasibility of mild gasification of coal: Research needs, Report No. DOE/METC-85/4019, NTRS/DE85013625." Department of Energy, Washington, DC.

Penner, S. S. (1987). Coal gasification: Direct applications and synthesis of chemicals and fuels: A research needs assessment, Report No. NTIS-PR-360. Office of Energy Research, Washington, DC.

Probstein, R. F., and Hicks, R. E. (1982). "Synthetic Fuels." McGraw-Hill, New York.

Richardson, F. W. (1975). "Oil from Coal." Noyes Data, Park Ridge, NJ.

Schobert, H. H. (1987). "Coal: The Energy Source of the Past and Future." American Chemical Society, Washington, DC.

Seglin, L., and Bresler, S. A. (1981). Low-temperature pyrolysis technology. *In* "Chemistry of Coal Utilization" (M. A. Elliott, Ed.), 2nd Suppl. Vol., pp. 785–846. Wiley, New York.

SFA Pacific (1993). Coal gasification guidebook: Status, applications and technologies, Research Project No. 2221-39. EPRI, Palo Alto, CA.

Singer, J. G. (ed.). (1991). "Combustion Fossil Power—A Reference Book on Burning and Steam Generation." Combustion Engineering, Windsor, Ontario, Canada.

Solomon, P.R., and Serio, M. A. (1987). Evaluation of coal pyrolysis kinetics. *In* "Fundamentals of Physical Chemistry of Pulverized Coal Combustion" (J. Lahaye and G. Prado, Eds.). Nijhoff, Zoetermeer.

Stultz, S. C., and Kitto, J. B. (1992). "Steam—Its Generation and Use." Babcock & Wilcox, Barberton, OH.

U.S. Department of Energy (1989). "Coal liquefaction: A research and development needs assessment, DOE Coal Liquefaction Research Needs (COLIRN) Panel Assessment Final Report, Vol 2, DOE/ER-0400." U.S. Department of Energy, Washington, DC.

U.S. Department of Energy, Morgantown Energy Technology Center (1991). "Atmospheric Fluidized Bed Combustion—A Technical Source Book," Publication No. DOE/MC/1435-2544. U.S. Department of Energy, Morgantown Energy Technology Center, Morgantown, WV.

U.S. Department of Energy, National Energy Technology Laboratory (1987). "Clean Coal Technology Compendium." U.S. Department of Energy, National Energy Technology Laboratory, Pittsburgh, PA.

Coal, Fuel and Non-Fuel Uses

ANTON DILO PAUL
Science Applications International Corporation
Pittsburgh, Pennsylvania, United States

1. Coal Use History
2. Current Fuel Use of Coal
3. Environmental Issues Associated with Coal Use
4. Current Non-Fuel Uses of Coal
5. Coal Combustion By-Products
6. Modern Coal Plants
7. The Future of Coal Use

Glossary

acid rain Precipitation containing small amounts of sulfuric acid and nitric acid, which derive from sulfur and nitrogen oxides emitted during coal combustion.

boiler A device that utilizes the heat from a furnace to convert water into high-pressure steam. When steam production is the sole objective, the boiler is usually integrated with the furnace and the two terms are used synonymously in industry parlance.

coal combustion emissions Gaseous pollutants released during the combustion of coal.

coal combustion products (CCPs) Residual material from the combustion of coal. These materials were disposed of as waste in the past, but are now used in other industries.

cogeneration The use of coal (or other fuel) to produce electricity, whereby the residual heat is made available for external use.

furnace/hearth A device that burns coal efficiently using forced air circulation inside a closed container.

greenhouse gases (GHGs) Various naturally occurring atmospheric gases (particularly carbon dioxide) that are also emitted during coal combustion. Excess quantities are suspected to cause changes in weather patterns and an overall warming of the earth.

steam engine/steam turbine A device that converts the heat energy in steam into mechanical energy.

The use of coal ranges from simple combustion for heat to complex partial oxidation to produce heat, gaseous/liquid fuels, and chemical feedstocks. Heat produced from coal combustion was traditionally used for home cooking and heating. From the dawn of the Industrial Revolution until about the 1950s, coal was the primary fuel to supply process heat for industrial processes and transportation engines. Apart from its use as a fuel, coal provided chemical raw materials that were the starting blocks for many household and industrial chemical products. Unrestrained coal use during this period, however, soon led to severe air, land, and water pollution. Environmental legislation commenced in the 1970s in the United States to limit the emissions of air pollutants from coal. Prior to this, oil began to replace coal as a transportation fuel, and natural gas competed with coal as a heating fuel. These factors significantly reduced coal use, but the world still consumes about 5 billion tons/year, largely for the generation of baseload electric power. With new coal combustion technologies, coal use today is environmentally clean and shows promise of becoming even cleaner in the future.

1. COAL USE HISTORY

1.1 Home Heating and Cooking

Documented coal use dates back to the 1700s, when coal was used as a heating fuel, but historical evidence indicates that coal was used in 100–200 AD by the Romans during their occupation of England. Until about the 18th century, cooking took place in an open hearth with firewood as fuel. Hearth cooking was hard work and messy; trees had to be chopped into firewood and brought indoors and large amounts of fine-sized ash had to be hauled outdoors. Coal was a welcome relief, because its high heating value per unit mass (7000–13,000 Btu/lb) compared to wood meant that less fuel was needed. However, coal is hard to ignite and, without sufficient air supply, produces a smoky flame, but it can be burned inside a closed container with a controlled airflow to produce a much hotter flame than is produced by wood fuel. Using this principle, the coal-fired pot-bellied stove made its appearance in almost every home and building during the 1800s and early 1900s. Coal stoves not only provided heat for cooking, but also warmed the room in which they were situated. They were often hooked up to a hot-water tank to provide running hot water as well as steam for steam baths.

1.2 Coal Use during the Industrial Revolution

The Industrial Revolution of the 18th and 19th centuries greatly expanded the use of coal as a fuel to run the new industries, particularly for the manufacture of iron products. Prior to the Industrial Revolution, wood charcoal, or char, was the source of heat in industrial hearths. Wood char provided more heat compared to wood, but even so, char hearths rarely produced enough heat to melt iron. Instead, these hearths could only heat iron until it became sufficiently malleable to be hammered into shapes, yielding wrought-iron products. On the other hand, coal hearths provided enough heat to melt iron into its liquid form. Liquid iron could then be poured into special molds to manufacture cast-iron products—a faster way of producing iron goods. Coal hearths speeded operations in other heat-intensive industries, such as the smelting of iron ore in a blast furnace or the production of steel from pig iron. Regions where coal and iron ore were present together, for example, South Wales in England, flourished during the Industrial Revolution because of the abundance of both raw materials in one location.

1.3 Coke and Gas Production

Another historic use of coal was for street lighting. Coal was heated in ovens with minimal air to produce "town gas" or "manufactured gas." The main components of town gas are carbon monoxide and hydrogen. The gas was piped throughout a town in underground pipes and burned in street lamps for illumination. Town gas was common in many European cities, notably London, even up to the 1950s. The residue from the production of town gas is coke, which is nearly pure carbon (except for the ash). Coke and the by-products of coke manufacture established an important role in the iron/steel and chemical manufacturing industry from 1850 to 1950. (For further discussion on this topic, see Section 4.1)

Early coke-making processes used "beehive" ovens that vented unburnt volatile matter in coal into the atmosphere. Such coke-making practices produce severe air pollution and are nonexistent today in the United States and other developed countries. The more common practice is to recover all emissions from a coke oven and distill them into different fractions. These fractions contain many chemicals (including naphthalene, ammonia, methanol, and benzene) that form the basis or feedstock for many consumer products. Aspirin (acetylsalicylic acid), road tar, creosote, fertilizers, and disinfectants are a few examples of products traditionally manufactured from coal. A comprehensive depiction of the many products that can be produced from coal is seen in the "coal tree," which may be accessed on the Appalachian Blacksmiths' Association web site at http://www.appaltree.net/aba/coalspecs.htm.

1.4 The Rise and Fall of "King" Coal

The invention of the steam engine by James Watt in the late 1700s greatly expanded the market for coal, because it provided a method for converting the heat energy in coal into mechanical energy. The golden era of the railroads dawned in the early 1900s as coal-fired steam locomotives and steam wagons quickly replaced horse-drawn wagons and carriages. In the locomotive's boiler, a coal fire heated water and converted it to steam. The energy in the steam moved back and forth a piston that was attached to the locomotive's wheels, providing motive power. Coal fueled many forms of transport, from giant locomotives that hauled goods and passengers hundreds of miles across cities, to steam wagons and streetcars that transported the same cargo across a few city blocks. Compared with traveling on horseback or stagecoach, train travel was comfortable and fast—passengers relaxed in Pullman cars heated by pot-bellied stoves, dined on meals cooked over coal-fired stoves, and traveled in fast trains pulled by steam locomotives.

Stationary coal-fired steam engines provided mechanical power to drive factory machinery through an elaborate system of pulleys. The availability of mechanical power led to the development of machine tools that rapidly increased industrial productivity. The construction industry boomed as steam shovels moved tons of earth for new buildings and steamrollers smoothed bumpy roads into highways. The Panama Canal, the Hoover Dam, the Empire States Building, and the first few U.S. highways could not have been built during their time without these coal-fired machines.

From the reciprocating steam engine evolved the steam turbine, in which steam impinges on the vanes on a shaft to produce rotary motion instead of moving a piston back and forth. The steam turbine ran smoother than the reciprocating engine and was more suited for marine propulsion and electric power generation. Shipbuilders replaced masts and sails with smoke funnels and built larger and faster ships geared toward transatlantic travel. The first crossing of the Atlantic Ocean solely under steam power occurred in 1838 when two British steamships, Great Western and Sirius, made the trip in 15 days, compared to 30–45 days in a sail ship. This started the era of the luxurious ocean liners, including the famed Titanic, which were powered by coal carried in gigantic bunkers within the ship's hull.

Coal also fueled large furnaces for glass making, to process wood pulp, to produce iron and steel, and to heat lime and cement kilns. During the late 1800s to the 1950s, coal was the world's primary heating and transportation fuel and the backbone of several interdependent industries. Steam shovels helped mine coal and mineral ores that were transported by steam trains to blast furnaces and coke ovens. Coal-fired furnaces processed the ores into metals and finished metal goods. These goods were transported to their final destination by coal-powered trains and steam wagons.

Coal's reign as a premium fuel diminished with the discovery of crude oil. By refining crude oil, liquid fuels such as gasoline and diesel became available to power internal combustion engines. Diesel–electric locomotives, automobiles, trucks, and buses soon became the common mode of transportation, replacing steam-driven vehicles and eventually the use of coal as a direct transportation fuel. Oil and natural gas became the preferred fuels for heating and cooking in residential and commercial buildings. By the late 1970s, coal's use as a transportation fuel had diminished significantly in most developed countries, but it remained as a fuel for generating electric power, particularly in countries with large minable coal reserves. Fig. 1 shows the historic change in coal use for transportation and residential and industrial heating in the United States since 1949.

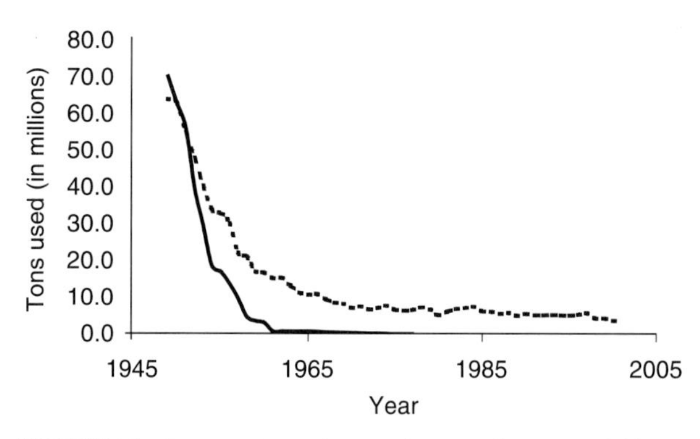

FIGURE 1 Decrease in coal use in the United States. Solid lines, transportation use; dashed lines, residential and commercial use. Data from Energy Information Agency, Annual Energy Review (2001), Section 7—Coal.

2. CURRENT FUEL USE OF COAL

2.1 Generation of Electric Power

The largest use of coal today is in the generation of electric power by electric utilities. According to the U.S. Energy Information Administration, about 65% of coal mined in the world is used for power generation. The next largest use of coal is in the production of coke for the iron and steel industries. Coal is still used for industrial heating and even commercial and residential heating in certain countries.

When coal is used to generate electric power, it is combusted in large furnaces. The heat generated is transferred to tubes inside the furnace. Water in these tubes is converted to high-pressure steam, which is piped to steam turbines that convert the thermal energy in the steam into mechanical (rotational) energy. An electric power generator connected to the turbine converts the mechanical energy into electric power, which is transmitted over high-voltage lines to a power network and is sold to industrial, commercial, and residential users. The spent (low-pressure) steam from the turbines is condensed into water and is returned to the boiler. Fig. 2 is a simplified schematic depicting the use of coal for electric power generation.

The conversion of coal to electric power can be viewed as a convenient use of coal. It is easier to burn coal continuously in a single large furnace than it is to burn it intermittently in many small home stoves. Electric power obtained from converting the heat energy derived from a large coal furnace provides a method for distributing coal energy to multiple users. In homes that depend on coal for heating and cooking, electricity generated from coal provides the same energy, but without the dust, soot, and physical labor associated with firing and stoking a coal fire. For transport systems, electricity generated from coal can move electric trains in a far more efficient and less polluting manner compared to burning coal in several steam locomotives. Hence, coal still plays a role in the transportation sector today, but in an indirect manner through the generation of electricity.

2.2 Coproduction of Heat and Electricity

After utility company power production, the next largest use of coal is for the production of heat and electricity for commercial buildings and industrial processes. Rather than purchase electricity from a central grid, an industrial plant or building owner may choose to operate a boiler that supplies both heat and electric power. The concept of cogeneration, or combined heat and power (CHP) production, is similar to the use of coal for electric power production, except that the spent (low-pressure) steam from the turbines is piped to an industrial plant or building before being condensed and returned to the boiler. Residual heat in the low-pressure steam supplies process heat for industrial application or temperature control—for example, to maintain a greenhouse at constant temperature so that vegetables and flowers can be grown during cold seasons. Cogeneration plants are common in Europe, where they are used for district heating and generating electricity. In the United States, cogeneration plants are common on industrial sites, hospitals, and college and university campuses. Table I illustrates coal consumption in the United States by end-use application.

2.3 Home Heating and Cooking

Coal is still used to a small extent for home heating and cooking. In the homes of more affluent nations,

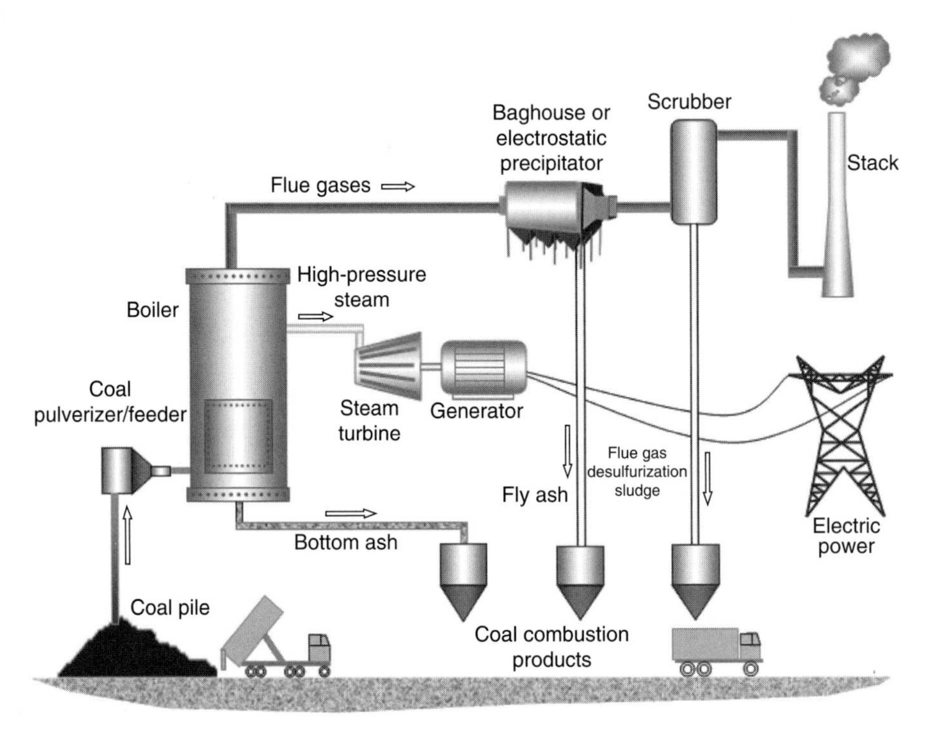

FIGURE 2 Schematic diagram illustrating coal use for electric power generation.

TABLE I

U.S. Coal Consumption by End-Use Sector, 1996–2002[a]

Year	Electric utilities	Coke plants	Other industrial	Residential and commercial	Other power products	Total
1996	874,681	31,706	71,689	6006	22,224	1,006,306
1997	900,361	30,203	71,515	6463	21,603	1,030,114
1998	910,867	28,189	67,439	4856	26,941	1,038,292
1999	894,120	28,108	64,738	4879	52,691	1,044,536
2000	859,335	28,939	65,208	4127	12,3285	1,080,894
2001	806,269	26,075	63,631	4127	15,0637	1,050,470

[a]In thousands of short tons. Data adapted from the Energy Information Agency Review (2001), Coal Consumption Statistics.

coal is used for recreational or nostalgic reasons, rather than for routine heating and cooking. Coal stoves designed for this purpose are aesthetically attractive and are engineered for efficient operation with near-zero indoor air pollution. In contrast, coal used for cooking and heating in developing countries is burned in crude stoves with inadequate ventilation. With no chimneys or hoods to vent the smoke, emissions inside homes can quickly reach levels that cause respiratory diseases. Even when stoves are vented, smoke emissions significantly increase local air pollution. In rural South Africa, for example, improperly designed or damaged stoves, coupled with poor ventilation, produce high risks to health and safety, including high levels of carbon monoxide. Robert Finkelman from the U.S. Geological Survey has reported that over 3000 people in the Guizhou Province of southwest China suffer from severe arsenic poisoning. Coal from this region contains high levels of toxins, including arsenic, which accumulates in chili peppers and other foodstuffs that are dried over unvented coal stoves. Polycyclic aromatic hydrocarbons, released during unvented coal combustion have also been cited as a primary cause for lung cancer. The World Bank, the World Health Organization, and the U.S. Geological Survey are among the many organizations that are educating developing and less developed nations about the dangers of unvented coal combustion.

3. ENVIRONMENTAL ISSUES ASSOCIATED WITH COAL USE

3.1 Emissions from Coal Combustion

As discussed in the previous section, coal emissions from poorly designed stoves are hazards to human health. However, coal emissions from large furnaces are equally hazardous. From the time of the industrial revolution until about the 1970s, little attention was paid to the air pollution created by coal combustion. Meanwhile, coal-fired power plants and industrial boilers spewed out tons of gaseous and particulate pollutants into the atmosphere. During combustion, the small amounts of sulfur and nitrogen in coal combine with oxygen to form sulfur dioxide (SO_2), sulfur trioxide (SO_3), and the oxides of nitrogen (NO_x). If these gases are not captured, they are released into the atmosphere through the power plant's stack. They combine with water vapor in the upper atmosphere to form droplets of sulfuric and nitric acids, which return to the earth's surface as acid rain. Acid rain is blamed for killing fish in streams and lakes, damaging buildings, and upsetting the ecological balance.

Flue gas from a coal plant's stack contains tiny particles of mineral matter (fly ash) and unburned carbon (soot). When fog and smoke are trapped beneath a cold air mass, they form smog, which can sometimes be deadly. For example, two killer smogs occurred in London, England in December of 1892 and 1952, lasting 3–5 days and resulting in over 5000 deaths. Menon and co-workers studied climate changes caused by coal fires; they found that soot from these fires absorbs sunlight and heats the surrounding air, but blocks sunlight from reaching the ground. Heated air over a cool surface makes the atmosphere unstable, creating heavy rainfall in some regions and droughts in others.

Smog and air pollution from coal-fired plants are responsible for many health problems observed in people who live in highly industrialized areas. Examples of highly industrialized areas include regions of Poland, the Czech Republic, and Germany, comprising the Black Triangle, one of Europe's most heavily industrialized and polluted areas; England's Black Country, a 19th century industrialized area west and north of Birmingham; and in the United States, Pittsburgh, Pennsylvania, which supplied most of the steel needed for World War II. Choking coal dust and toxic gas emissions from coke plants, blast furnaces, and coal-fired electric power and district heating plants once plagued the residents of these areas with respiratory problems, heart disease, and even hearing losses. Fortunately, most of the coal-based heavy industries that polluted these areas have been cleaned up using modern coal-burning and emission control technologies.

3.2 Mitigation of Coal Combustion Emissions

A combination of legislation and technology has helped clean up many of the world's coal-burning plants. Both developed and developing countries have adopted increasingly stringent environmental regulations to govern emissions from coal-fired power plants. In the United States, all coal-fired power plants built after 1978 must be equipped with postcombustion cleanup devices to capture pollutants before they escape into the atmosphere. Cyclones, baghouses, and electrostatic precipitators filter out nearly 99% of the particulates. Flue gas scrubbers use a slurry of crushed limestone and water to absorb sulfur oxides from flue gas. The limestone reacts with the sulfur dioxide to form calcium sulfate, which may be used to produce wallboard. Staged combustion and low-NO_x burners are used to burn coal to minimize NO_x formation. Another strategy, selective catalytic reduction, reacts ammonia with NO_x over a catalyst to produce nonpolluting nitrogen and water vapor.

Conventional coal-fired power plants capture pollutants from the flue gas after it leaves the boiler. Circulating fluidized bed (CFB) combustors capture most of the pollutants before they leave the furnace. Crushed coal particles and limestone circulate inside the CFB combustor, suspended by an upward flow of hot air. Sulfur oxides released during combustion are absorbed by the limestone, forming calcium sulfate, which drops to the bottom of the boiler. The CFB combustor operates at a lower temperature (1400°F) compared to pulverized coal (PC) boilers (2700°F), which also helps reduce the formation of NO_x.

Precombustion coal cleaning is another strategy to reduce sulfur emissions by cleaning the coal before it arrives at the power plant. Sulfur in coal is present as pyrite (FeS_2), which is physically bound to the coal as tiny mineral inclusions, and as "organic sulfur," which is chemically bound to the carbon and other atoms in coal. Pyrite is removed in a coal preparation plant, where coal is crushed into particles less than 2 inches in size and is washed in a variety of devices that perform gravity-based separations. Clean coal floats to the surface, whereas pyrite and other mineral impurities sink. Additional cleaning may be performed with flotation cells, which separate coal from its impurities based on differences in surface properties. Precombustion removal of organic sulfur can be accomplished only by chemical cleaning. So far, chemical cleaning has proved to be too costly, thus flue gas scrubbers are often required to achieve near-complete removal of sulfur pollutants.

The tightening of environmental regulations is likely to continue throughout the world. In the United States, for example, by December 2008, it is anticipated that coal-fired power plants will have to comply with maximum emission levels for mercury. Emissions of mercury and other trace metals, such as selenium, are under increasing scrutiny because of suspected adverse effects on public health.

3.3 Cofuel Applications

Coal is sometimes combusted with waste material as a combined waste reduction/electricity production strategy. The disposal of waste from agriculture and forestry (biomass), municipalities, and hospitals becomes costly when landfill space is limited. Some wastes, particularly biomass, are combustible, but their low energy density (compared with coal) limits their use as an electricity production fuel. Blending coal with these fuels provides an economical method to produce electric power, reduce waste, and decrease emissions. Most wood wastes, compared to coal, contain less fuel nitrogen and burn at lower temperatures. These characteristics lead to lower NO_x formation. In addition, wood contains minimal sulfur ($<0.1\%$ by weight) and thus reduces the load on scrubbers and decreases scrubber waste.

Numerous electric utilities have demonstrated that 1–8% of woody biomass can be blended with coal with no operational problems. Higher blends may also be used, but require burner and feed intake modifications as well as a separate feed system for the waste fuel. Cofiring in fluidized bed boilers may avoid some of these drawbacks, but the economics of cofiring are not yet sufficiently attractive to make it a widespread practice.

4. *CURRENT NON-FUEL USES OF COAL*

4.1 Coke Manufacture

Select coals are used to manufacture coke; the high pure carbon content, porous structure, and high resistance to crushing of coke have made it a necessary material in the production of molten iron (or hot metal, in iron/steel industry parlance). Coke also serves as a fuel in blast furnaces, as well as serving several nonfuel functions. In a typical blast furnace, iron ore, coke, and other materials, including sinter and fluxes, are loaded in layers. Air, preheated to 900°C, is blasted through the bottom of the furnace through nozzles called tuyeres. Oxygen in the air reacts with coke to form carbon monoxide and generate heat. Carbon monoxide reduces iron ore to hot metal, which flows downward and is tapped off at the bottom of the blast furnace. Coke provides physical support for the materials in a blast furnace by maintaining sufficient permeability for the upward flow of gases and the downward flow of hot metal.

During the past century, the coke industry grew concurrently with the growth of the iron and steel industry. Large integrated steel mills operated coke-making plants, blast furnaces, and basic oxygen furnaces that converted iron ore into steel. However, coke manufacture is a highly polluting operation, and coke cost is one of the major expenses of operating a blast furnace. Consequently, methods of producing steel without coke are under development and some are in operation today. The most straightforward approach is the recycling of steel using electric arc furnaces (also known as minimills) that melt and process scrap steel for reuse, thereby dispensing with the need for fresh steel. Direct reduced iron (DRI) is another process that reduces coke need by directly reducing iron ore without the use of coke. Another process is the direct injection of granulated coal into a blast furnace. Such a system is in operation at Bethlehem Steel's Burns Harbor Plant in the United States as well as in other locations throughout the world.

4.2 Coal-Derived Solid Products

The most common solid product derived from coal is activated carbon. This product is used as a filter medium for the purification of a wide range of products, such as bottled water, prepared foods, and pharmaceuticals. Activated carbon plays an important role in the treatment of contaminated ground water by absorbing toxic elements. Impregnated activated carbons are used for more complex separations, such as the removal of radioactive isotopes, poisonous gases, and chemical warfare agents.

Activated carbon is produced from premium coals, or a form of coal known as anthracite, but other materials, including coconut shells, palm-kernel shells, and corncobs, may also be used. Coal, or other feed material, is first heated in a controlled atmosphere to produce a char, which is then thermally or chemically activated to yield a highly porous final product. Surface areas of activated carbons range from 500 to 1400 m^2/g, with macropores ($>0.1\,\mu m$ in diameter) and micropores (0.001–$0.1\,\mu m$ in diameter). Macropores provide a passageway into the particle's interior, whereas micropores provide the large surface area for adsorption. The starting material and the activation process greatly influence the characteristics and performance of the finished activated carbon. Coal, for example, produces highly macroporous activated carbons and is the preferred base material for many impregnated activated carbons.

4.3 Coal-Derived Liquid Products

Coal-derived liquid products have been traditionally obtained as by-products of coke manufacture. These products contain chemicals that form the basis or feedstock for many consumer products. In a coke plant, the volatile components emanating from coal are collected as unpurified "foul" gas, which contains water vapor, tar, light oils, coal dust, heavy hydrocarbons, and complex carbon compounds. Condensable materials, such as tar, light oils, ammonia, and naphthalene, are removed, recovered, and processed as chemical feedstocks. However, both the availability and the market for coke by-products are declining. The highly polluting nature of coke manufacture requires costly environmental controls to comply with tightening environmental regulations, and some steelmakers have found it more economical to close down their coke plants and import coke from countries with less stringent environmental regulations. Consequently, the availability of coking by-products has diminished significantly in the United States and in Europe, thereby limiting the production of liquids from coal. More importantly, products from petroleum refining are a more cost-effective method to obtain the various chemical feedstocks that were formerly obtained from coal.

Another liquid product from coal is obtained via the direct or indirect liquefaction of coal. In direct liquefaction, coal is heated under pressure with a solvent and a catalyst to produce a liquid, which can then be refined into other liquids, including transportation fuels (gasoline, diesel fuel, and jet fuel), oxygenated fuel additives (methanol, ethers, and esters), and chemical feedstocks (such as olefins and paraffin wax). With indirect liquefaction, coal is first gasified in the presence of oxygen and steam to generate a gas containing mostly carbon monoxide and hydrogen (i.e., syngas). Next, syngas is cleaned of its impurities and converted into liquid fuels and chemical feedstocks.

The conversion of coal into liquid products is not widely practiced on a commercial scale. For most countries, it is more economical at present to use crude oil to produce liquid fuels than to produce them from coal; hence, coal liquefaction does not constitute a large part of coal use today. The exceptions are two commercial coal liquefaction operations, namely, the Sasol Chemical Industries plant in South Africa and the Eastman Chemical plant in the United States, that use coal liquefaction to produce high-value chemicals.

4.4 Carbon Products

Coal is being tested to produce carbon products, using both coal and the liquid products derived from coal liquefaction. Carbon products are becoming increasingly common over a wide spectrum of applications, ranging from sports equipment to engineered materials for fighter aircraft and space shuttles. In addition, carbon materials such as carbon fibers and carbon foams are finding new applications and are becoming cost-effective alternatives to steel in automotive and aircraft components as well as in many other industries.

Starting feedstocks for carbon products are presently obtained as by-products from the oil-refining industry and the manufacture of coke. These include impregnating pitches, mesophase pitches, needle cokes, and isotropic cokes. However, several processes are under investigation to produce such materials directly from coal. Chemical solvents, including *N*-methylpyrrolidone and dimethyl sulfoxide, are used to dissolve the carbon in coal. The resulting gel-like substance is filtered and separated from the mineral matter present in coal, after which it is hydrogenated and processed to produce binder and impregnation pitches, anode and needle cokes, fibers, foams, specialty graphites, and other carbon products. The Consortium for Premium Carbon Products from Coal (CPCPC) is one organization that promotes work on the production of high-value carbon products from coal.

5. COAL COMBUSTION BY-PRODUCTS

Coal used in an electric power-producing application generates several by-products, including fly ash, bottom ash, and flue gas desulfurization sludge. Coal combustion products (CCPs) were historically considered waste and were dumped in landfills. However, with the growing scarcity of landfill space, CCPs have found applications as low-cost additives in the manufacture of cement, roofing tiles, structural fills, and fertilizers.

CCPs contain mineral elements similar to those in the earth's crust, including silicon dioxide, aluminum oxide, iron oxide, calcium oxide, and trace amounts of sulfur trioxide, sodium oxide, and potassium oxide. These elements can act as replacement for natural materials and offer the environmental benefit of avoiding the need to mine additional natural resources. The three most widely used CCPs today are the residue from flue gas desulfurization (FGD), commonly known as FGD sludge, fly ash, and bottom ash.

Flue gas desulfurization residue is produced in flue gas scrubbers by the reaction of lime with sulfur oxides. It is chemically similar, but not identical in its physical properties, to natural gypsum. Because of this, it has limited application as a raw material in the manufacture of wallboard. It has also found application as a soil conditioner and fertilizer. It has been stabilized with fly ash and used as roadbase material in highway construction.

Fly ash refers to the fine particles suspended in the flue gas that are captured by electrostatic precipitators and baghouses. Fly ash can be used as a direct replacement for Portland cement or as an additive in the manufacture of cement. A challenge in the use of fly ash in cement manufacture is the presence of unburned carbon, which is present in the fly ash of most boilers that use staged combustion. Carbon makes fly ash unusable as a cement additive because it creates color problems and reduces concrete durability under freeze and thaw conditions. Research efforts are underway to remedy this problem.

Bottom ash is the coarse, hard, mineral residue that falls to the bottom of the furnace. Bottom ash and fly ash have similar chemical compositions, but bottom ash is of larger size and lacks cementitious properties. It is used as an aggregate or filler material in the construction industry. Another form of bottom ash, produced by older, slagging combustors, is hard and glassy and has found a market as a low-silica blasting grit.

6. MODERN COAL PLANTS

Coal use today is no longer evocative of dirty power plants with polluting black smoke billowing from their smokestacks. Many of these plants have been transformed through technology to operate more efficiently and with significantly lower emissions. Some fire coal with other waste materials and others produce both electric power and heat. Cases of plant retrofits and their new performance statistics are documented by various institutions, including the Energy Information Administration (http://www.eia.doe.gov) and the World Coal Institute (http://www.wci-coal.com). The following examples highlight clean coal use throughout the world:

- The Belle Vue power plant in Mauritius cofires coal with bagasse to generate electricity in an environmentally clean fashion. Bagasse, the residue after extracting juice from sugar cane, is available in large quantity in Mauritius. Instead of being a waste product incurring disposal costs, bagasse is now cofired with coal to produce electric power.

• The Turów power plant located in Europe's formerly polluted Black Triangle now operates as a clean power plant using flue gas scrubbers, low-NO_x burners, and electrostatic precipitators.

• A coal-fired cogeneration plant in Northern Bohemia now provides clean electric power to the Skoda automobile production plant and steam for district heating of the town of Mladá Boleslav.

• A 1970 coal plant in Ulan Bator, Mongolia was refurbished in 1995 with higher efficiency boilers and improved emission controls. The plant now provides electric power and heat for district heating with significantly lower pollution emissions and failure rates compared to its pre-1995 operation.

• The 35-year-old Northside Generating Station in Jacksonville, Florida in the United States was renovated under the U.S. Clean Coal Technology Program and is now ranked as the cleanest burning coal power plant in the world. Using circulating fluidized bed combustors and additional pollution controls, the plant cuts down on emissions by nearly 98% and generates 2.5 times more electric power.

• The Mirant-Birchwood Power Plant Facility in Virginia in the United States produces 238 MW of electric power as well as process steam, which is used to heat local greenhouses. In addition, it converts 115,000 tons of flue gas scrubber ash into 167,000 tons of lightweight aggregate for use in the manufacture of lightweight masonry blocks or lightweight concrete.

7. *THE FUTURE OF COAL USE*

7.1 Coal Use Trends

According to the U.S. Energy Information Administration, the world consumes around 5 billion short tons of coal annually, but coal use has been in a period of slow growth since the 1980s. This is largely because of growing environmental concerns, particularly in the developed countries. Coal used for coke manufacture is expected to decline due to advances in steel production technology, steel reuse via electric arc furnaces, and the replacement of steel by plastics and carbon composites.

Coal use is expected to decline in Europe and the former Soviet Union over the next 20 years, largely due to the growing use of natural gas. Other factors contributing to reductions in coal use include growing environmental concerns and continuing pressure on member countries by the European Union to reduce domestic coal production subsidies. In contrast, coal use is expected to grow in Asia, especially in China and India. These countries have vast reserves of coal with limited access to other sources of energy, creating a situation for increased coal use to fuel their rapidly growing industrial sector. China has also shown interest in coal liquefaction and is planning to build their first direct coal liquefaction plant in the coal-rich province of Shanxi.

7.2 New Environmental Challenges

In the quest for a cleaner, healthier world, there is increasing pressure for a cleaner operation of coal-using operations. Tighter environmental regulations that require further reductions in the emissions of SO_2, NO_x, air toxics, and fine particulates are likely to be enacted over the next several years. Compliance with these increasingly stringent restrictions on emissions could be costly for coal users and lead to a reduced coal use. The most significant emerging issue against coal use is climate change associated with the production of greenhouse gases (GHGs). Although still hotly debated, many scientists believe GHGs are responsible for observed changes in the world's climate. Coal-fired plants, along with oil and natural gas-fired power plants, will require significant reductions in CO_2 emissions to mitigate the impact of GHGs. The challenge to continued coal use is the development of technologies that can operate efficiently and economically within tighter environmental regulations.

Coal gasification is a growing technology to use coal with near-zero emission of pollutants. It also provides the flexibility to generate a wide range of products from coal, including electricity, gaseous and liquid fuels, chemicals, hydrogen, and steam. Coal (or any carbon-based feedstock) is reacted with steam at high temperatures and pressures under carefully controlled conditions to partially oxidize the coal. The heat and pressure break apart chemical bonds in coal's molecular structure and form a gaseous mixture, not unlike the gas mixture generated by the indirect liquefaction of coal (discussed earlier). The hot gases are sent to a special filter that removes nearly 99% of sulfur pollutants and GHGs. The clean, coal-derived gas is similar to natural gas and can be used for electric power generation or converted to clean liquid fuels and chemical feedstocks. Pollutants containing nitrogen and sulfur are processed into chemicals and fertilizers and residual solids are separated and sold as coal combustion products.

Coal gasification is a more efficient way to generate electricity compared to conventional coal-fired power plants. Instead of burning coal in a boiler to produce steam that drives a steam turbine/generator, coal gasification uses a special cycle, an integrated gasification combined cycle (IGCC), whereby the hot, high-pressure coal-derived gas is first used to power a gas turbine/generator. The exhaust from the turbine still contains sufficient heat to be sent to a boiler to produce steam for a steam turbine/generator. This dual generation of electric power represents an increased extraction of heat from coal. An IGCC plant operates with an efficiency of 45–55%, compared to 33–35% for most conventional coal-fired power plants. Higher efficiencies translate into better economics and inherent reductions in GHGs. Pressurized fluidized bed combustion (PFBC) boilers and IGCC will likely be used in future coal-fired power plants.

7.3 The Vision 21/FutureGen Plan

Vision 21/FutureGen represents a new approach to coal use that is being developed by the U.S. Department of Energy and private businesses. Unlike the single-fuel, single-product (usually electricity) plant, a Vision 21/FutureGen plant would produce a range of products—electric power, liquid fuels, chemical feedstocks, hydrogen, and industrial process heat. It also would not be restricted to a single fuel, but could process a wide variety of fuels such as coal, natural gas, biomass, petroleum coke, municipal waste, or mixtures of these fuels. The Vision 21/FutureGen concept envisions a suite of highly advanced technologies in modular form that can be interconnected and customized to meet different market requirements. Such modules include clean combustion/gasification systems, advanced gas turbines, fuel cells, and chemical synthesis technologies, coupled with the latest environmental control and carbon dioxide sequestration technologies.

The Vision 21/FutureGen plant may use a fuel cell (instead of a turbine/generator) to convert coal into electricity. The syngas produced from coal gasification is processed via an advanced water–gas shift reactor to separate hydrogen, and the carbon monoxide is converted to carbon dioxide and sequestered. The hydrogen is supplied to a fuel cell to produce electric power in a clean and efficient manner.

A Vision 21/FutureGen plant will be nearly emissions free, and any waste generated would be sequestered and recycled into products. Plant operating efficiencies are expected to be greater than 60% for coal-based systems and 75% for natural gas-based systems. Further details on the Vision 21/FutureGen plant are available by browsing through the U.S. Department of Energy's Web site at http://www.fossil.energy.gov/.

SEE ALSO THE FOLLOWING ARTICLES

Acid Deposition and Energy Use • Clean Coal Technology • Coal, Chemical and Physical Properties • Coal Conversion • Coal Industry, Energy Policy in • Coal Industry, History of • Coal Mine Reclamation and Remediation • Coal Mining, Design and Methods of • Coal Preparation • Coal Storage and Transportation • Cogeneration • Greenhouse Gas Emissions from Energy Systems, Comparison and Overview • Markets for Coal

Further Reading

Ashton, T. S. (1999). "The Industrial Revolution, 1760–1830." Oxford University Press, Oxford.

Bell, M.L., Davis, D.L. (2001). Reassessment of the lethal London fog of 1952: Novel indicators of acute and chronic consequences of acute exposure to air pollution. *Environ. Health Perspect.* **109** (Suppl. 3), 389–394.

Finkelman, R.E., Skinner, H.C., Plumlee, G.S., Bunnell, J.E., (2001). Medical geology. Geotimes (Nov. 2001), pp. 20–23 Available at www.geotimes.org.

Menon, S., Hansen, J., Nazarenko, L., and Luo, Y. (2002). Climate effects of black carbon aerosols in China and India. *Science* **297**, 2250–2253.

Paul, A.D., Maronde, C.P. (2000). The present state of the U.S. coke industry. Proc. 8th Int. Energy Forum (ENERGEX 2000), Las Vegas, pp. 88–93. Technomic Publ. Co., Lancaster, Pennsylvania.

Plasynski, S., Hughes, E., Costello, R., and Tillman, D. (1999). "Biomass Co-firing: A New Look at Old Fuels for a Future Mission." Electric Power '99, Baltimore, Maryland.

Wyzga, R.E., Rohr, A.C. (2002). Identifying the components of air pollution/PM associated with health effects. In "Air Quality III, Proceedings." Energy and Environmental Research Center, University of North Dakota.

Coal Industry, Energy Policy in

RICHARD L. GORDON
The Pennsylvania State University
University Park, Pennsylvania, United States

1. Land Disturbance from Coal Mines
2. Health and Safety Practice
3. Air Pollution
4. Rent Seeking and the Coal Industry
5. The Search for Appropriate Charges for Land
6. The Case of Coal Leasing: An Overview
7. Coal Protectionism
8. Promoting Coal Utilization Technology
9. Electricity Regulation and Coal
10. Notes on the Economics of Regulation
11. Regulation and Public Land
12. Conclusion

Glossary

air pollution The harmful discharges from combustion and other human activities.
due diligence A legal requirement for adequate performances such as time limits on implementing a mineral lease.
economic rents Income in excess of (economic) costs, often due to access to superior-quality resources.
externalities The effects of economic activities on bystanders.
fair market value An estimate of the probable selling price of an asset being purchased or sold by a government.
lease bonus A fixed advanced payment for mineral rights.
public utilities Firms regulated by the government because of concern regarding lack of competition.
publicness The simultaneous provision of commodities to everyone.
royalties Payments associated with production levels; usually a percentage of selling price but can be set as an amount per unit of production.
severance taxes Tax payments associated with production levels; usually a percentage of selling price but can be set as an amount per unit of production.
subsidy Government aid payments.
surface mining Extraction by uncovering surface material.
underground mining Extraction by digging under the surface cover.

Throughout the world, coal is profoundly affected by public policies. In almost every case, policies are directed at the consequences of coal production and use. Rarely is the promotion or even the limitation of coal use the primary objective. The best known policies affecting coal are those seeking to control air pollution from coal burning and water pollution and land disturbance from coal mining. Special, separate regulations govern efforts to protect the health and safety of coal. United States government public land policy includes complex rules for leasing coal resources on public lands. Since coal use is increasingly concentrated in electricity generation, regulatory activities in the electric power sector profoundly affect coal. This article reviews these policies.

1. LAND DISTURBANCE FROM COAL MINES

Mining disturbs the surface to some extent whether or not it occurs on the surface. This may involve damages with widespread effects. For example, abandoned mines are considered eyesores and may contain material that causes water pollution. Some sites may comprise ecosystems valuable for contributions beyond the marketable goods that can be produced. Socially beneficial biota may be present.

Although coal mining states had imposed controls, Congress decided that federal action was imperative. Thus, a national law, the U.S. Surface Mining Control and Reclamation Act of 1977, was passed. The law sets specific reclamation goals and requires the enforcing agency to create even more specific implementation rules. The basic goals are to restore the land to the "approximate original contour," restore the land to support its prior use or better ones, segregate topsoil so it can be replaced, and designate some land off limits to surface mining.

The law established an Office of Surface Mining (OSM) in the Department of the Interior (DOI) and created principles for guiding reclamation. OSM was to supervise the development of state programs to implement the law or, if a state chose not to develop a program, regulate mines in that state. The law covered reclamation of both active and abandoned coal mines. Taxes were levied on surface and underground coal mines to finance reclamation of abandoned mines. For operating mines, the law required the establishment of permitting systems under which the right to mine was granted only when extensive data, including the mining and reclamation plans and information on the impacts of mining, were submitted. Mine inspections were also mandatory. The law had complex provisions prohibiting or discouraging surface mining on certain types of land, such as "prime agricultural land."

Since this law produces benefits that are largely difficult to quantify, impact studies are limited. Neither the impacts on the industry nor the benefits of control are accurately known. Eastern surface mining started declining at approximately the time that the act was enacted, but surface mining growth continued in the western United States, albeit at a markedly lower rate. The act was not the only force at work and its effects are difficult to isolate.

2. HEALTH AND SAFETY PRACTICE

The traditional concern on the production side is worker safety. After a severe mine disaster in 1968, regulation was revised in the Coal Mine Health and Safety Act of 1969. Safety regulation was made more stringent and regulations of health effects were added. Moreover, administration was transferred from the U.S. Bureau of Mines of the DOI to a newly created Mine Safety and Health and Safety Administration, which was initially a part of the DOI but was transferred in 1977 to the Department of Labor.

Regulation has involved both setting rules for operating mines and federal inspection of the mines. The 1969 act both tightened the rules and increased the frequency of inspection. New rules required that no mine work occur under unsupported roofs; improved ventilation; use of curtains and watering to ensure reduction of "respirable dust" levels and increase the flow of air to the "face" (i.e., where mining was occurring); and increases in the monitoring of mines to determine levels of methane, dust, and other dangerous material. The number of inspections increased greatly. In the early years, many mines had daily inspections because of special problems or because they were so large that only daily visits sufficed to ensure that the required tasks could be completed.

Simultaneously with the passage of the Act, underground coal mining productivity began declining from a peak of 1.95 short tons per worker hour in 1969 to a low of 1.04 short tons in 1978. Then a reversal occurred so that by 2000, the level was up to 4.17 short tons.

Much effort was devoted to determining the role of the Act in the initial productivity declines. Evaluation was hindered by the standard problems of multiple, interrelated influences on which data were unavailable. In particular, the passage of the Act coincided with (and perhaps was a major cause of) an influx of inexperienced miners. As a result of the Act, more labor use in mines appears to have been necessary. The need to increase inspections contributed to the need for new workers because experienced miners were recruited as inspectors. The effects that this had on mine productivity are unclear, but whatever they were, they have been substantially reversed.

3. AIR POLLUTION

Concerns about air pollution from coal began with widespread coal use. The initial and still relevant focus was on the visible effects of spewing soot (particulates) from coal. Coal has a substantial but variable particle content. At best, the content is much greater than that of oil or gas.

Coal contains a variety of other contaminants. The most notable is sulfur. Sulfur contents of coals are also widely variable, and the lowest sulfur coals may have considerably less sulfur than many crude oils. During combustion, nitrogen oxides (NO_x), carbon dioxide (CO_2), and carbon monoxide form. In the 20th century, the problem of largely invisible pollution from SO_x and NO_x emerged, as did concern regarding smaller, invisible particulates. Recently, attention has turned to CO_2. It is the main greenhouse gas suspected of contributing to increased temperatures in the atmosphere (global warming).

Pollution control can be performed in several ways. Naturally less polluting fuels can be used, precombustion cleanup can be performed, and the pollutants can be captured after combustion. However, precombustion treatment is not an economically feasible option with coal. Available coal washing and sorting technologies can only modestly reduce particle and sulfur content.

More radical transformation is technically possible but economically unattractive. Prior to the development of a natural gas distribution network, coal gasification provided the input for local gas distribution systems. Other technologically established possibilities are to produce a higher quality gas than obtained from the earlier gasifiers, synthesize petroleum products, or simply produce a liquid that would compete successfully with oil and gas. All these processes would have the further benefit of facilitating the removal of pollutants. To date, the costs have proven too great to permit competition with crude oil and natural gas. Thus, emphasis is on changing the type of fuel and postcombustion cleanup. Changing to oil and gas lessens particulate and SO_x emissions. Another approach to SO_x is to use lower sulfur coal.

An alternative is to trap the particles and sulfur before discharge into the air and deposit the trapped wastes on nearby lands. The latter approach is widely used. Particulate control is invariably effected by postcombustion methods. Two standard methods are available—a mechanical filter known as a bag house or devices called electrostatic precipitators that trap the particles by electrical charges. Particle control techniques are long extant and have low control costs. Thus, visible discharges are largely eliminated. Little coal burning occurs in facilities too small to control particulates. Even the best particulate control devices, however, cannot capture all the smallest particles, and many pre-1960 regulations did not require the greatest control that was physically possible.

Sulfur oxide control proved more expensive and more difficult to perfect than expected when control rules were imposed. The wastes from these cleanup facilities can cause disposal problems. The reduction in sulfur and particulate emission associated with a switch to oil or gas is associated with increases in nitrogen oxides and unburnt hydrocarbons.

Similarly, the lowest cost, low-sulfur coal in the United States is subbituminous from Wyoming, which has a higher ash content than eastern coal. Thus, lowering sulfur by changing fuel may result in increased particulate emissions. Some observers argue that the close association between particulate and SO_x emissions precludes definitive assignment of impacts. SO_x may be blamed for effects that are actually caused by the accompanying particulates, and coal-source shifting to fuel lower in SO_x may be harmful.

Complex air pollution control programs have been developed. United States air pollution laws have resulted in multitiered regulations that are periodically tightened. The practice involves increasing the number of regulated pollutants. In the major cases, a two-phased approach prevails. One set of rules determines air-quality objectives, and another controls emissions from different sources.

The overall goals specify the allowable concentrations in the atmosphere of the pollutants with which the law is concerned. States must establish implementations plans (SIPs) to ensure that they meet these goals. Failure to do so results in a region being labeled a nonattainment area.

In addition, a federal court ordered that the preamble to the law that called for maintaining air quality require "prevention of significant deterioration" (PSD) in areas in compliance with the rules. Widespread failure to meet SIP goals and the vagueness of the PSD mandate were treated in the 1977 Clean Air Act amendments. The country was separated into nonattainment areas in which actions must be taken to comply and PSD regions in which increases in pollution were to be limited to amounts specified by the law. Continued nonattainment resulted in further remedial legislation in 1990. This effectively divided the country into nonattainment areas legally obligated to come into compliance with the goals and PSD areas. Attaining compliance is difficult, and the major amendments in 1977 and 1990 to the Clean Air Act contained extensive provisions to stimulate compliance further.

Similarly, ever more complex rules have been developed to determine how major polluters should limit pollution. Newer facilities have long been subjected to more severe new source performance standards. The new source rules originally simply imposed stricter emission controls on new facilities. This was justified by the fact that the cost of incorporating controls into a newly designed plant was less than that of adding controls to an old plant. The disincentives to adding and operating new plants were ignored. (An additional factor that became controversial was precisely defining how much refurbishment of an old facility was consistent with maintaining its status as an existing plant.)

The initial rules resulted in considerable changes in the type of fuel used. Western coal was heavily used in such producing states as Illinois and Indiana and in states such as Minnesota that previously relied on Illinois coal. Mainly in response to this change, the rules were changed in the 1977 Clean Air Act amendments to restrict how emissions were reduced.

The amendments required use of best available control technology for preventing the discharge of pollution. The available option was employing scrubbers, a cleanup technique that emphasizes capture of pollutants between burning and discharge of waste up the smokestacks. Since some cleanup was necessary, the advantage of shifting to a cleaner fuel was diminished.

Under the 1990 Clean Air Act amendments, a new program targeted the reduction of emissions in the most heavily polluting existing electric power plants. Reflecting a long-standing debate about "acid rain," the 1990 amendments instituted a complex program for reducing sulfur and nitrogen oxide emissions from existing sources.

This two-phased program initially imposed roll-backs on the units (specifically named in the law) at 111 power plants that exceeded both a critical size and level of sulfur oxide emissions. By January 1, 2000, electric power plants were required to undertake a 10 million ton reduction (from 1980 levels) of sulfur dioxide and a 2 million ton reduction in nitrogen oxide emissions. The named plants had to undertake the reductions in the first phase of the program. The law encouraged cheaper abatement by allowing the polluters opportunities to buy offsets to their activities. The provisions thus involve a paradox. Congress first extended the command and control approach by specifying limits on specific plants. Then a concession to market forces was added.

The compliance choice for existing plants, in principle, is limited by the constraints on the plants that must make changes and the supply of lower sulfur coals. Thus, boiler design may prevent the use of some types of coal and result in loss of capacity when other coals are used. At the time of enactment, those modeling the impacts debated the availability and usability of low-sulfur coal from different states. Factors such as proximity to coal fields, railroads, and waterways affect the ease of procurement from alternative suppliers. The nature of available receiving facilities is another influence. Some plants were built near coal mines to which they are connected by conveyer belts. Others are served by private rail lines and in one case a slurry pipeline. Some lack land to add large new facilities.

In practice, implementation involved fuel changes rather than increased scrubber use. Wyoming coal was shipped farther than expected (e.g., Alabama and Georgia). Emissions trading proved effective in lowering compliance costs.

Global warming, an increase in temperature produced by discharge of carbon dioxide from fossil fuel burning, became prominent in the late 1980s. It is contended that emission of carbon dioxide into the atmosphere will inevitably and undesirably raise atmospheric temperatures.

There is an enormous amount of literature on the effect of global warning and what it might mean. First, questions arise about the exact extent of warming that will occur. Second, uncertainties prevail about the physical effects of this warming. Third, numerous estimates have been made about the economic effects. Should the United States commit to massive reductions in greenhouse gas emissions, it may result in severe problems for the coal industry. Coal burning produces more greenhouse gas than the use of other fuels.

4. RENT SEEKING AND THE COAL INDUSTRY

Another profound influence on the coal industry is government policy toward the ownership and taxation of coal resources. Ownership and taxation are not inherently linked but have become so in practice. Ownership and taxation are alternative ways to extract income. The sovereign can and has used its ownership, taxing, and regulatory authorities to affect ownership, utilization, and disposition of the resulting income. Disputes occur perennially.

An important consideration is that private and public owners can and do divide and separately sell mineral and surface rights. Some governments reserve all mineral rights for the state. The U.S. government adopted a practice of retaining mineral rights to some lands it sold to private owners. Separate ownership of the land's subsurface and the surface separates an indivisible property and guarantees conflicts. The subsurface cannot be exploited without affecting the rights of the surface property owner. Defining and, more critically, implementing an acceptable system of surface owner compensation are a major problem of government or private retention of mining rights. In particular, regulations strong enough to protect surface owners from misuse of sovereign power may constitute de facto privatization of the mineral rights to surface owners.

Starting with the 1920 Mineral Leasing Act, Congress moved toward what was to prove a standard policy of extending access on a rental rather than ownership basis. The 1920 act established leasing as the means of access to fossil fuels (oil, gas, and coal) and fertilizer minerals. Periodic revisions of the leasing act have tightened rules of access.

The increase in federal environmental legislation has also greatly influenced public land administration. One pervasive influence was the requirement in the 1969 National Environmental Policy Act that the federal government examine the environmental impacts of its major actions. The result was a thriving adversarial process in the 1970s. Those opposed to federal actions successfully delayed actions by challenges to the federal evaluation process. However, this tactic subsequently became irrelevant with regard to energy but apparently still affects may other government decisions.

5. THE SEARCH FOR APPROPRIATE CHARGES FOR LAND

Policymakers and economists have long debated the method of charging for rights of access to those who extract minerals. The concerns focus on economic rents—the extra profits of a superior resource. Many alternative terms exist for this extra income, such as excess profits, surplus, and windfalls. Whatever the name, it consists of the differences between revenues and the social costs, including the necessity to earn the required rate of return on investment.

For a mineral deposit, there are rent-generating advantages with a better location, higher quality material, easier mining conditions, and similar attractive features. This applies more broadly to all natural resources. Thus, a charge on these rents allows "the public" to share in the bounty of naturally superior resources without distorting resource allocation. Therefore, policies that transfer rents to the public are often deemed desirable. However, a paradox prevails. The charges can be imposed using a classic textbook example of a tax with no effects but, in practice, the charges employ a method that follows an equally classic textbook approach that clearly discourages desirable activity.

Economic analyses of rents concentrate on securing as much government revenue as possible without inefficiently reducing production. A transfer of rent has no effect on effort if the terms are independent of any action of the payer. The relevant classic textbook case is a lump-sum charge, a payment that is a fixed amount that is not alterable by the buyer's actions. The preferable approach is to impose a charge only at the time of transfer of rights to avoid the temptation subsequently to impose distortionary charges. The arrangement may be an outright sale or a lease in perpetuity with neither control over end use nor any ability to alter charges.

A lease bonus, a single payment made when the lease is implemented, is the quintessence of such a preset total payment. The bonus, by definition, cannot be altered by any action of the firm, and no subsequent decisions can be affected by having paid a bonus. In economic jargon, the bonus is a sunk cost. That fixed unalterable decision at the time of sale is administratively feasible is demonstrated by the history of U.S. public land sales and by the millions of outright sales that occur in the private sector.

In the case of land grants, competitive bidding to pay a fixed sum before production begins would capture rents. The maximum anyone would pay is the (present-discounted) value of the expected rents. Vigorous competition would force payments of the maximum.

Rent transfers thus seem ideal for tax collectors. The rents are excess profits that can be taxed without affecting output, and the transfers are seen as fair because the rents, in principle, provide rewards unnecessary to produce the effort.

In practice, deliberate decisions are universally made not to realize the potential for efficient transfer of income. As economics textbooks routinely warn, basing payments on output or its sale creates a disincentive to produce. Since the severance taxes used on minerals are sales taxes by another name and royalties have the same impacts, the analysis applies to royalties and severance taxes. Funds that consumers wanted to give producers are diverted to the imposer of the tax. This revenue transfer discourages producers from acting. The amount ultimately purchased by consumers declines because of these disincentives. This is a violation of the central economic principle that every expansion of output that costs less than its value to consumers should occur. The technical language is that all outputs with marginal costs (the cost of expanding output) less than price should be undertaken. Therefore, the charges imposed by federal land agencies, other governments throughout the world, and private landlords differ greatly in practice from ideal rent taxes and take a form that eliminates the differences from reliance on established tax collection agencies.

Moreover, royalties and severance taxes also create difficult collection problems and thus increased administrative cost. Requiring more payments means more compliance efforts by government and land users. DOI administration can and has encountered efforts to conceal production. A percentage of revenue royalty can be difficult to administer because of lack of satisfactory data on market prices. Lump-sum payments eliminate the expenses of monitoring.

The imposition of output-related charges is justified by claiming that because of imperfect competition, the defects of linking the payments to activity are outweighed by the income gains.

A further complication is that U.S. fossil fuel leasing policy involves both bonus payments and royalties. The drain of royalties will lower the amount of an initial payment. At a minimum, this hinders the evaluation of the adequacy of bonus bids. The residual after payment of royalties is necessarily less than the present value of untaxed rent. It is a residual of a residual—namely, the rent.

6. THE CASE OF COAL LEASING: AN OVERVIEW

The U.S. government owns many of the nation's most economically attractive coal resources. In particular, it is the dominant owner of coal in such western states as Wyoming, Montana, Colorado, Utah, North Dakota, and New Mexico. Most U.S. coal output growth since 1970 has occurred in these states. The coal leasing law and its administration changed little from 1920 to 1971. Leases were freely granted under the provisions of the 1920 Mineral Leasing Act.

These policies allowed satisfaction of most desires for leases. Coal leasing proceeded modestly until the late 1960s, when it grew rapidly. At least retrospectively, the 1960s leasing surge is explicable by anticipation of the expansion of western output that actually occurred.

Starting in the late 1960s, both the legislative and administrative processes changed radically. Congress passed a series of laws directly or indirectly affecting coal leasing. Federal coal policy became so polarized that a hiatus in leasing has prevailed for most of the period from 1971 to today. In 1971, the DOI became concerned about an acceleration of leasing without a concomitant increase in output. The result is that coal leasing ceased from 1971 to 1981, resumed briefly in 1981, was stopped in 1983, and its resumption remained indefinite as of 2003. The resumption in 1981 was the first test of 1976 revisions in the laws governing coal leasing. Enforcement proved problematic.

The Coal Leasing Amendment Act of 1976 radically altered coal leasing policy. President Ford vetoed the act, but the veto was overridden. The central components of the Act were that it (i) required competitive bidding on all leases; (ii) required the receipt of fair market value on all leases; and (iii) required a royalty of at least 12.5% of the selling price on all surface-mined coal, with the Secretary of the Interior given discretion to set a lower rate on underground-mined coal.

Many ancillary provisions were provided, but a complete review is not critical here. The provision that proved to have the greatest practical relevance was that requiring forfeiture of any lease that was not developed within 10 years (the due diligence requirement). Such requirements are a response to the output-retarding effects of heavy royalties. This accepts the belief in lease limits criticized later in this article. Such diligence requirements are either redundant or harmful. In the most favorable circumstances, the requirement will leave the decision about whether and when to operate unchanged. Cases can arise in which it is more profitable to start too soon than to lose the lease. When surrender is preferable to operating and occurs, no guarantee exists that reoffer will occur before the optimal starting time.

Diligence requirements are also disincentives to arbitrage in bidding and to prevention of mining when superior private uses exist. If lease lengths were unlimited, a leaseholder who believed that other uses were more profitable than mining could refrain forever from mining the land.

The 1983 moratorium was inspired by charges that some lease auctions had failed to secure fair market value. The Commission on Fair Market Value for Federal Coal Leasing was created to evaluate the charges and proposed solutions. Given that the problem was inadequate data to support the decisions, the commission proposed devising more elaborate evaluation procedures. (The DOI could find only one public record of a sale of coal reserves, a defect that later critics, including the commission, could not remedy.) These were created, but large-scale leasing never resumed. The reluctance to resume leasing occurred because the amount of coal already leased appeared sufficient to meet predicted consumption levels for several more decades.

As noted previously, coal leasing ceased despite all the safeguards imposed. This suggests that land management has become so controversy bound that even when stringent safeguards are in place, the critics remain dissatisfied. Critics of the experience, such as Nelson, Tarlock, and Gordon, argue that unrealistic expectations governed the adopted processes.

7. COAL PROTECTIONISM

Efforts to promote coal prevail but are problematic. Two are prominent. Given the drawbacks of coal use, persistent although not extensive efforts are made to develop better ways to use coal. The term clean coal technology characterizes the desired results. Governments in many countries, most notably in Western Europe and Japan, have sought to prevent or limit the decline of uneconomic coal industries.

The United States also protected coal from imported oil but on a much smaller scale. Until the oil price increase of the 1970s, the United States maintained oil import quotas. These were imposed largely to support a domestic oil price program that state agencies had established in the 1930s. However, proponents also recognized the benefits of such controls to the coal industry. Nevertheless, one of the first breaks in the policy was the 1969 change that effectively removed the restrictions on imports of residual fuel oil into the eastern United States, which was the main competition for coal.

On a much more modest scale, the anthracite-producing region of Pennsylvania was represented in the U.S. Congress by two well-placed representatives who regularly secured government support for assistance programs. The most effective was the requirement of export of anthracite for use by the U.S. military in Germany. Otherwise, the effort was limited to studies. This culminated at the start of the Carter administration. A commission was created with the stated goal of significantly stimulating anthracite production.

However, the available options were unattractive. Charles Manula, a professor of mining engineering at The Pennsylvania State University, indicated that the best prospect for increasing production was to expose and remove the coal left behind in abandoned mines. Although the commission nominally supported this proposal, it received no further support. The drawback was that the approach created much land disturbance to produce little more coal than the output of one western strip mine.

Much more extensive and sustained programs prevailed in Western Europe and Japan. For more than 45 years, resistance to rapid rationalization of the coal industry has distorted European energy decision making. The difficulties started long before. Among the problems caused by World War I was a disruption of European coal industries that worsened due to the effects of the Great Depression and World War II. Europeans agonized about how to revive the industry.

After World War II, initially it was widely believed that coal was a mismanaged industry that could thrive with the right management. It was believed that a vigorous rebuilding program would restore the coal industry to its former might. Britain and France nationalized the coal industries. They operated on a break-even basis, which means rents subsidized outputs. The Germans and the Belgians preferred reliance on a heavily regulated and subsidized, largely privately owned coal industry.

The Dutch had developed their coal industry early in the century as a nationalized venture. The Belgians recognized that the large number of small old mines in the south were beyond saving but seemingly had to be protected for political reasons.

A turning point occurred in 1957. In the aftermath of the 1956 Suez Canal crisis, optimism about European coal was eliminated. Middle East oil could no longer be dismissed as a transitory energy source. Conversely, the contention that sufficient effort would make coal more competitive was demonstrated to be false by the pervasive failure of massive revival efforts. Oil could be sold for less than coal for many decades. Following typical patterns of aid to distressed industries, the responses were programs to slow the retreat. Consolidation was proposed and slowly effected.

Even the oil shocks of 1973–1974 and 1979–1980 failed to help. The situation for European coal was so bad that competition with higher oil prices was still not feasible. Moreover, ample supplies of cheaper coal were available elsewhere. Even, as in coking coal, when coal use was efficient, given the ample supplies available in the United States, it was cheaper to import. Thus, a policy allegedly designed to limit dependence on unreliable supplies from the Middle East actually mostly harmed reliable suppliers from first the United States and then, as the world coal industry developed, Australia, South Africa, and Colombia.

Major national differences caused each of the coal-producing countries to adopt largely independent policies. The countries had important differences regarding both goals and programs. The fervor of German devotion to coal and the willingness to impose regulations were particularly intense. The Germans erected an elaborate structure of defenses and were the most reluctant to contract. The critical factor in Britain was the ability of the mineworkers union successfully to make resistance politically untenable until the union overreached in the Thatcher years.

The impact of the differences among countries is most clearly seen by comparing production trends.

The Dutch discovered a gas field that could displace coal and produce for export. In 1974, the Dutch were the first to exit the coal industry. The main adjustment was that the coal mining company was made a participant in gas production.

The French and the Belgians moved toward substantial but slow contraction. The French contraction continues; Belgian production ended in 1992. The preservation programs in France, Britain, Spain, and Belgium were largely ones of direct subsidies to coal mines. (Japan took a similar approach.)

British aid was largely direct but reinforced with a contract between the coal industry (originally the National Coal Board and then renamed British Coal before privatization) and the former Central Electricity Generating Board. This contract yielded higher prices than those of competing imports. Similarly, before its privatization, British Steel apparently also had a cozy relationship with British Coal.

During the Thatcher years, there was an impetus for radical reform. First, in 1984, the leader of the mineworkers union chose to implement a strike without an authorizing vote. The combination of incomplete support from the members and strong resistance by the Central Electricity Generating Board resulted in defeat for the union.

After the failure of the strike, a restrained move was made toward reducing assistance. British Steel shifted to imported coal after it privatized in 1988. Electric power restructuring in 1990 led to radical reductions in utility purchase of British coal. The splitting of the Central Electric Generating Board into three companies, two of which are privatized fossil fuel-using organizations, created pressures to alter procurement practices. The new companies were more aggressive about fuel procurement. The increase in gas-using independent power producers added competition for coal. In this changing climate, the Major government effected privatization of coal. All these factors accelerated the industry's decline. Output declined from 116 million tonnes in 1983, the year before the strike, to 30 million tonnes in 2002.

The Germans had the most complex and intractable of the systems. Previously, support was financed by funds from general federal and state revenues, revenues from a special tax on electricity consumers, and forcing consumers to pay more than market prices. Aid was dispensed in several forms: direct subsidies of specific current activities of the coal industry, programs to promote the use of German hard coal in coking and electricity generation, and assumption of the cost burdens (mostly miner pensions) from past mining. (The last policy is standard among Western European coal-assistance programs.) The electricity tax paid for direct and indirect costs of coal use but was ultimately banned by a German court. A long-term contract committing domestic electric utilities to buy preset, modestly increasing tonnages in Germany at unfavorable prices prevailed from the mid-1980s to 1995. A move to direct subsidies similar to those elsewhere was made in response to the court ruling and resistance to renewing the coal purchase contract.

Since the coal crisis, the Germans have periodically set output goals presumably for many years. As costs mount, a reduction is undertaken. The goals set in 1973 called for attainment in 1978 of 83 million tonnes. The 1974 revision at the height of concern over the oil shocks called for 90 million in 1980; actual 1980 output was 94 million. A 1981 report seemed to advocate maintaining output at approximately 90 million tonnes. A 1987 review called for a 15 million tonne decrease to 65 million by 1995. A 1991 revision set goals of reducing output from approximately 80 million tonnes in 1989 to 50 million tonnes by 2001 and 43 million tonnes by 2010. Dissenters on the advisory committee called for a cut to 35–40 million tonnes. Output in 2002 was 27 million tonnes.

German aid averaged more than $100 per tonne starting in the 1990s. In contrast, nominal dollar prices of coal delivered to Europe were approximately $45 per tonne for coal for power plants and less than $60 per tonne for coal used to make coke meeting iron-making requirements.

8. PROMOTING COAL UTILIZATION TECHNOLOGY

As noted previously, in principle there are numerous ways to improve coal utilization, ranging from better ways to employ solid coal to transforming the coal into a different fuel. Programs to develop these technologies have prevailed with variable vigor for many years. However, neither transformation technologies nor new ways to use coal have proved to be commercial successes.

The U.S. government has developed a series of programs in this area. In the early 1970s, the Office of Coal Research was combined into the Energy Research and Development Administration largely composed of the research arm of the former Atomic Energy Commission. This, in turn, became the main component of the Department of Energy. In 1979, the Carter administration decided to expand its energy initiatives and in particular to encourage synthetic fuel development from coal. This policy was proposed even though research up to that point suggested that synthetic fuels from coal were not economical. A government-owned Synthetic Fuels Corporation was established, funded a few projects including a large lignite gasification facility in North Dakota, and was abolished in the 1980s. What was left was a much smaller clean coal technology program in the Department of Energy.

9. ELECTRICITY REGULATION AND COAL

Extensive, convoluted government involvement in the activities of electricity generation, transmission, and distribution existed long before the 1970s. However, the most profound effects on coal occurred with 1970s energy turmoil. The 1970s witnessed increasing prices in energy markets and policy changes. State regulators independently adopted measures to increase control over utilities and responded to federal mandates to alter practice.

The initial response to frequent requests for rate increases necessitated by the changing economic climate was multifaceted resistance. Denial of requests was supplemented by numerous forms of increased examination of realized and anticipated actions. Both major cost elements—construction and fuel purchases—were scrutinized. Management audits were also initiated.

The mass of federal energy regulations during the period from 1973 to 1992 had many elements that at least potentially had significant direct or indirect impacts on electric power. Two proved central. The 1978 Public Utility Regulatory Policy Act required states to consider many kinds of changes in regulation. However, the only major response was to the requirement to establish a mechanism to encourage utilities to purchase power from nonutility generators (NUGs) deemed worthy, mainly those using renewable resources (other than traditional large hydroelectric facilities) and "cogeneration" (electricity production by industrial companies in the facilities providing energy to their operations). Utilities were to be required to purchase from these sources when they were cheaper or, in the jargon that was developed, when the avoided cost was lower. However, the states were granted considerable discretion about how these costs would be estimated.

Given that the process involved choosing among investment options, estimates of lifetime costs had to be made. Given the climate of the time, estimates in states partial to NUGs incorrectly assumed substantial, steady increases in the cost of conventional sources of power. Utilities became committed to contracts at higher prices than were justified by prevailing market conditions. Again imitating prior experience (with fuel procurement contracts), the move to the economically appropriate task of contract renegotiation is only slowly emerging.

Small plants had several features that made them economically attractive. Their short lead time and low capital investment characteristics made them more flexible in an era of low and uncertain growth than the traditional large plants, with their high unit and total capital costs and long lead times. To the extent that suppliers would assume the risk of these ventures, they were even more attractive. NUGs provided an increasing portion of capacity additions. At least part of the industry saw independent generation and perhaps even the adoption of the proposal for totally independent generation as desirable.

Under prior law, almost all subsidiary–parent relationships were treated as public utility holding companies and subject to stringent controls. Revisions enacted in 1992 exempted operations that were incidental to the main activities of the firm. Thus, the power-producing subsidiary of a gas pipeline, an engineering firm, or even an electric company located elsewhere in the country no longer would be subject to stringent regulation.

The evidence suggests that from fairly modest beginnings in the mid-1980s, additions of NUG capacity increased while utilities lowered their own additions. These relied heavily on gas turbines often linked to a conventional boiler that uses the waste heat this is called a combined cycle. Simultaneously, there was cessation of building of conventional large coal-fired plants.

In the 1990s, several states tried to deal with these problems by altering how they regulated electric utilities. This generally involved deregulating wholesale transactions and lessening controls on rates to final consumers. The details differed greatly among states. One major difference related to ownership of generation. At one extreme (New York, Maine, and Massachusetts), divesting of all generation was required. At the other extreme, some states (e.g., Texas and Pennsylvania) made the sales voluntary.

These requirements and voluntary actions increased the role of at least nominally independent producers. Both established electric utilities from other states and specialists in independent power plants purchased the divested plants. Two forms of nominal independence were allowed. Several independent power producers purchased integrated electric utilities and were allowed to treat the acquired generating plants as independent production. Conversely, regulators agreed to treat new generating subsidiaries of some established integrated utilities as independent producers.

The ability to sustain this pattern is doubtful. Although gas-supply potential has been underestimated, it is unlikely that the gas industry can economically serve its traditional markets plus a large part of electricity generation and transportation. In 2002, the electric power industry consumed 29 quadrillion BTUs of coal and nuclear energy and 5.7 quadrillion BTUs of natural gas. Total U.S. natural gas consumption was only 23 quadrillion BTUs.

10. NOTES ON THE ECONOMICS OF REGULATION

From an economic perspective, regulation is marred by a persistent tendency to adopt clumsy ways to attain goals. Often, the goals are disputed. The concerns are based on the limits that economic analysis places on the role of government. Standard economic theory employs the concept of market failure to delineate the reasons why government action might be desirable.

The critical concern is what economists call the public-good-supply problem, provision of services whose existence makes them available to everyone. Examples include national defense, police protection, and pollution control. As a vast literature indicates, enormous problems arise in determining and efficiently satisfying the demands for such goods.

Prior to 1960, it was widely believed that causing an "externality," an impact on others than the decisionmaker, also inherently required intervention. However, Ronald Coase published an enormously influential article on the subject. He showed that side effects could be and were privately corrected when the number of affected parties was sufficiently small to make negotiation feasible. It was only when the number of participants became too large that the private solution broke down. Implicitly, impacts became a public good and their control an example of the problem of public good provision.

As is rarely noted, Coase qualified his conclusions. Throughout his writings on economic policy, Coase stressed that governments have limited capacity for action. Therefore, it is unclear when it is preferable for the government to be the financer. In a follow-up, he illustrated his point by showing that a classical example of the need for government action, provision of lighthouses, was invalid. In England, lighthouses were originally successfully privately provided. The government stepped in only after the private success. Coase's implicit point is that too much stress is given problems of ensuring that all beneficiaries participate in private programs. The resulting inadequacy is still clearly preferable to failing to secure any government aid and might even be superior to a halfhearted government effort.

A more critical issue is why governments so often intervene when no clear publicness prevails. Key examples relate to the mass of regulations affecting labor relations, such as coal mine health and safety laws. The stated rationale is that the bargaining is unfair and governments must intervene to offset this defect. This may in some cases be a form of publicness. An alternative view is that such policies are simply examples of favoritism toward the politically influential. The issue is why a supposedly effective labor union could not have secured the benefits in the national negotiations it conducted.

The resulting problem is establishing the desirability of specific policies. First, the reality of a problem must be established. Second, it must be ascertained whether the impacts justify intervention. Finally, the best way to attain the regulatory goals must be determined. The first area involves expertise in many disciplines because environmental impacts may affect human health, damage property, cause unsightliness, and alter climate. People outside these disciplines must somehow evaluate the credibility of the often tentative and conflicting claims that are made.

The next essential step of appraisal of the desirability of action involves the economic approach of cost–benefit analysis. The problem is standard in applied economics. The methodology is well established; the data for implementation are severely deficient. An analyst must associate values to the imperfect impact estimates developed in other disciplines and relate them to flawed estimates of compliance costs. Thus, the actual estimates are uncertain.

Economics also produces a preference for decentralized decision making. Economists have long disparaged the presumption of politicians to decide what is preferable for others. Thus, among economists, the tendency toward detailed instructions that characterizes environmental regulation is censured as command and control regulation.

The starting point of the argument is that politicians have considerable difficulty determining the proper overall goals of public policy and great problems determining the degree and nature of response to the goals. Therefore, if regulation is to occur, governments should adopt systems in which broad goals are set and the maximum possible flexibility is given with regard to implementation.

In particular, economists suggest two basic approaches: charges for damages or permits to damage. The underlying principle is that economically pollution control is a commodity whose benefits and cost vary with the amount of effort exerted. Thus, the standard economic technique of supply and demand analysis applies. An optimal level of control exists. At that level, the price that the public is willing to pay for that quantity of control attained has a value equal to the marginal cost of the control. Thus, as with all commodities, a market-clearing price is associated with the optimum level of control.

This viewpoint simply formalizes realities behind actual regulation. Total control is never attempted because of the high cost of perfection. Indeed, perfection may be technologically impossible (an infinite cost). Thus, denunciations of economic analyses as immoral are apparent denials of the realities. What really seems to be involved is concern that the limits of knowledge will cause less stringent controls than are justified.

In any case, the key is the association of a price with the market-clearing level of control. With sufficient information, both the market-clearing level of control and the associated price are knowable. This forms the basis for the tax and marketable permits basis for regulation. Under the tax system, every polluter is charged a tax equal to the value of efficient abatement. The permit system instead sets a total limit on pollution and allocates shares in that limit. In both systems, the affected parties decide how much and in what way to abate.

The previously noted problems of inadequate information imply that charges, transferable rights, and command and control all will fail to attain the unknowable efficient level of control. To make matters worse, the magnitude of the resulting policy defects depends critically on the magnitude of the error in the policy goals. Therefore, no *a priori* presumptions are possible about which policy fails least.

11. REGULATION AND PUBLIC LAND

Government land ownership is politically popular. However, the economists working on the subject are skeptical about this acquiescence. The standard economic arguments for intervention have little or no relevance. Briefly, most U.S. federal land use is by private individuals for ordinary commercial activities, such as grazing, forestry, and mineral extraction. Thus, the usual reasons for preferring private ownership prevail.

Another basic problem is that, as noted previously, while access charges in principle can be levied without affecting production, the standard practice is to adopt charge systems that discourage production. The view that not to charge is to cause irreparable losses to the Treasury is overstated. Any business established on public lands, its suppliers, and its employees will pay their regular taxes. The greater the efficiency of the firm, the more regular taxes will be paid. Thus, the normal tax system is at least a partial corrective to the failure directly to charge. It is unclear that this approach produces less tax revenue than a royalty system.

Moreover, the concept of rent transfer to government must coexist with a rival policy approach to rents—transfer to consumers. Although this objective has not arisen for U.S. coal, it was a major consideration with regard to European coal (and with regard to 1970s U.S. policy for oil and natural gas).

Another consideration is that the concept of optimal lease timing is invalid. More precisely, you can sell too late but never too soon. Market value is determined by expected profits from optimal development. The development schedule is determined by the trends in the markets for the output of the resource. The value is determined by the existence of the mineral, not its ownership. The land sale or lease transfers control without altering mineral existence and value.

Any private firm making a lease or purchase before it is efficient to use the property will wait until the socially most desirable exploitation date. Failure to realize this basic principle results in invalid concerns about adequate payments and calls to limit leasing. It is possible only to lease too late. Resources may not be made available to a qualified operator until after the most desirable time for exploitation. Moreover, the case does not depend on perfect markets. Delaying leasing cures no market failure. At best, time is gained that might be used to eliminate the failures.

Another problem is that politicians and critics of leasing impose unattainable standards of proof on the acceptance of offers of initial payments. Fears of giveaways prevail. The intrinsic drawbacks of rent taxation are aggravated by the insistence on detailed estimates of whether selling prices are sufficient.

Concerns also exist about the feasibility of administering the system and ensuring public faith in the integrity of the system.

In public land policy debates, leases or sales of public land are conditional on payment of "fair market value." Fair market value more generally is the legal term used to define the appropriate price to be paid in the purchase and sale of property by government agencies. Its receipt is a perennial issue. The term seems simply to mean what the asset would sell for in a competitive market. The U.S. government policy manual on valuation, "Uniform Appraisal Standards for Federal Land Acquisitions," implies such an interpretation, as does at least one Supreme Court decision.

Economic principles indicate that vigorous competition for leases will guarantee receipt of fair market value. For skeptics, some assurance such as price data may be needed. Although oil and gas leasing continues because its long history produced extensive information, the effort to establish competitive bidding for coal leases was undone by the absence of data.

12. CONCLUSION

The prior discussion suggests the diverse and profound ways that regulation affects coal. Clearly, many questions arise about these actions and the available information is inadequate to resolve the concerns. What is clear is the underlying belief that coal is such an unattractive fuel that its production and use must be tightly controlled. Thus, although the 1970s were years of advocacy of increased coal use, the 1990s saw advocacy of discouraging coal consumption.

SEE ALSO THE FOLLOWING ARTICLES

Air Pollution from Energy Production and Use • Clean Coal Technology • Coal, Chemical and Physical Properties • Coal, Fuel and Non-Fuel Uses • Coal Industry, History of • Equity and Distribution in Energy Policy • Markets for Coal • Natural Gas Industry, Energy Policy in • Transportation and Energy Policy

Further Reading

Coase, R. H. (1988). "The Firm, the Market and the Law." Univ. of Chicago Press, Chicago.

Ellerman, A. D., Joskow, P. L., Schmalensee, R., Montero, J-P., and Bailey, E. M. (2000). "Markets for Clean Air: The U.S. Acid Rain Program." Cambridge Univ. Press, Cambridge, UK.

Gordon, R. L. (1988). Federal coal leasing: An analysis of the economic issues, Discussion Paper No. EM88-01. Resources for the Future, Energy and Materials Division, Washington, DC.

Gordon, R. L. (2001). Don't restructure electricity: Deregulate. *Cato J.* **20**(3), 327–358.

McDonald, S. (1979). "The Leasing of Federal Lands for Fossil Fuels Production." Johns Hopkins Univ. Press for Resources for the Future, Baltimore.

Mueller, D. C. (2003). "Public Choice III." Cambridge Univ. Press, Cambridge, UK.

Nelson, R. H. (1983). "The Making of Federal Coal Policy." Duke Univ. Press, Durham, NC.

Parker, M. J. (2000). "Thatcherism and the Fall of Coal." Oxford Univ. Press for the Oxford Institute for Energy Studies, Oxford.

Samuelson, P. A. (1954). The pure theory of public expenditure. *Rev. Econ. Statistics* **36**(4), 387–389. [Reprinted in Stiglitz, J. E. (Ed.). (1966). "The Collected Papers of Paul A. Samuelson," Vol. 2, pp. 1223–1225. MIT Press, Cambridge; and Arrow, K. J., and Scitovsky, T. (Eds.) (1969). "Readings in Welfare Economics," pp. 179–182. Irwin, Homewood, IL]

Samuelson, P. A. (1969). Pure theory of public expenditures and taxation. *In* "Public Economics" (Margolis, J., and Guitton, H., Eds.), pp. 98–123. St. Martins, New York. [Reprinted in Merton, R. C. (1972). "Collected Scientific Papers of Paul A. Samuelson," Vol. 3, pp. 492–517. MIT Press, Cambridge, MA]

Tarlock, A. D. (1985). The making of federal coal policy: Lessons for public lands from a failed program, an essay and review. *Natural Resources J.* **25**, 349–371.

U.S. Commission on Fair Market Value Policy for Federal Coal Leasing (1984). "Report of the Commission." U.S. Government Printing Office, Washington, DC.

U.S. Congress, Office of Technology Assessment (1984). "Acid Rain and Transported Air Pollutants: Implications for Public Policy." U.S. Government Printing Office, Washington, DC.

U.S. National Acid Precipitation Assessment Program (1991). "Acidic Deposition: State of Science and Technology." U.S. Government Printing Office, Washington, DC.

Coal Industry, History of

JAAK J. K. DAEMEN
University of Nevada
Reno, Nevada, United States

1. Introduction
2. Pre- and Early History
3. Middle Age and Renaissance
4. Precursors to the Industrial Revolution
5. Industrial Revolution
6. 19th Century
7. 20th Century
8. The Future
9. Coal and Culture

Glossary

coal A black or brown rock that burns; a solid combustible rock formed by the lithification of plant remains; a metamorphosed sedimentary rock consisting of organic components; a solid fossil hydrocarbon with a H/C ratio of less than 1, most commonly around 0.7 (for bituminous coal); a solid fossil fuel; fossilized lithified solar energy.

Coal Age The historical period when coal was the dominant fuel, from late 18th through middle 20th century; name of a trade journal devoted to the coal industry (ceased publication summer 2003).

Coalbrookdale A town in England's Black Country, symbol of the Industrial Revolution; site of first iron production using coke produced from coal; center of early steam engine and railroad development; celebrated in paintings by Williams, de Loutherbourg, Turner, and others, damned in poems by Anna Seward and others; part of the Ironbridge Gorge UNESCO World Heritage Site; a British national monument.

coal field Region in which coal deposits occur.

coal gas Fuel rich gas produced by partial oxidation (burning) of coal (also producer gas).

coal liquefaction Method for producing liquid fuels (e.g., gasoline, diesel) from coal.

coal preparation Treatment of coal to prepare it for any particular use; improve the quality of coal to make it suitable for a particular use by removing impurities, sizing (crushing, screening), and special treatments (e.g., dedusting); upgrading (beneficiation) of coal to a more uniform and consistent fuel (or chemical feedstock) of a particular size range, with specified ash, sulfur, and moisture content.

Industrial Revolution Controversial term to design the period when modern industrial society developed, initially applied primarily to Britain, later to other countries as well. From about 1760 to about 1850.

International Energy Agency (IEA) Autonomous body within OECD (Organisation for Economic Co-operation and Development) to implement an international energy program. Publishes reports on the coal industry, many of which are posted on its Web page.

water gas Gas produced by the interaction of steam on hot coke, used for lighting (primarily during the 19th through early 20th century) and as fuel (well into 20th century).

In a narrow sense, the coal industry can be considered as the coal mining industry, including coal preparation. More broadly it includes major users, such as steam generators for electric power, coking for steel production, and, historically, coal as a feedstock for chemicals and as transportation fuel, especially steam locomotives and ships, as well as the wide range of applications of steam engines. The transportation of coal, given its bulk and low value, can be a major cost of its use. Coal exploration and geology are part of the front end of the coal mining cycle.

1. INTRODUCTION

Coal has been mined for centuries. Until shortly before the Industrial Revolution, its use was local, and it did not make a significant contribution to the overall energy consumption of the world. Coal use increased rapidly in the two centuries before the Industrial Revolution and has continued to grow ever since, with occasional temporary down dips. Coal was the dominant fuel during the 19th century and the first half of the 20th century. The rise of modern society is intimately intertwined with the growth of the coal industry. Developments driven by coal with a major impact on technological progress include the steam engine, the railroad, and the steamship, dominant influences on the formation of modern society. For over a century, coal was the major energy source for the world. Its relative contribution declined over the second half of the 20th century. In absolute terms its contribution continues to increase.

The environmental disadvantages of coal have been recognized for centuries. Efforts to reduce these disadvantages accelerated over the last third of the 20th century. Because coal remains the largest and most readily available energy source, its use is likely to continue, if not increase, but will have to be supported by improved environmental control.

2. PRE- AND EARLY HISTORY

Where coal was mined and used initially is not clear. Secondary sources vary as to when coal use may have started in China, from as early as 1500 BC to as late as well after 1000 AD. It has been stated that a Chinese coal industry existed by 300 AD, that coal then was used to heat buildings and to smelt metals, and that coal had become the leading fuel in China by the 1000s. Marco Polo reported its widespread use in China in the 13th century. Coal may have been used by mammoth hunters in eastern central Europe.

The Greeks knew coal, but as a geological curiosity rather than as a useful mineral. Aristotle mentions coal, and the context implies he refers to "mineral" or "earth" coal. Some mine historians suggest he referred to brown coal, known in Greece and nearby areas. Theophrastus, Aristotle's pupil and collaborator, used the term *ανθραζ*(anthrax), root of anthracite. Although Theophrastus reported its use by smiths, the very brief note devoted to coal, compared to the many pages dealing with charcoal, suggest, as does archaeological evidence, that it was a minor fuel. The term is used with ambiguous meanings, but at least one was that of a solid fossil fuel. Theophrastus describes spontaneous combustion, still a problem for coal storage and transportation.

Coal was used in South Wales during the Bronze Age. The Romans used coal in Britain, in multiple locations, and in significant quantities. After the Romans left, no coal was used until well into the second millennium. Lignite and especially peat, geological precursors to coal, were used earlier and on a larger scale in northern and western Europe. Pliny, in the first century AD, mentions the use of earth fuel by inhabitants of Gaul near the Rhine mouth, to heat their food and themselves. All indications are that he describes the use of peat in an area currently part of The Netherlands, where peat still was used as a fuel nearly twenty centuries later. Romans observed coal burning near St. Étienne, later a major French coal mining center. The Romans carved jet, a hard black highly polishable brown coal, into jewelry. Jet was used as a gemstone in Central Europe no later than 10000 BC.

The recorded history of coal mining in India dates from the late 18th century. Place and river names in the Bengal-Bihar region suggest that coal may have been used, or at least that its presence was recognized, in ancient times.

Coal use was rare, even in most parts of the world where it was readily accessible in surface outcrops, until well into the second millennium, at which time its use became widespread, be it on a small scale.

The Hopis mined coal at the southern edge of Black Mesa, in northern Arizona, from about the 13th through the 17th century AD. Most was used for house fuel, some for firing pottery. Coal was mined by a primitive form of strip mining, conceptually remarkably similar to modern strip mining. At least a few underground outcrop mines were pursued beyond the edge of the last (deepest) strip, a practice reminiscent of modern auger mining. Gob stowing was used to prevent or control overburden collapse.

3. MIDDLE AGE AND RENAISSANCE

Coal and iron ore formed the basis of the steel industry in Liège in present Belgium. Coal mining rights were granted in Liège in 1195 (or 1198?) by the Prince Bishop. At about the same time, the Holyrood Abbey in Scotland was granted coal mining rights. A typical medieval situation involved abbeys or cloisters driving technology. In France, a 13th-century real estate document established property limits defined by a coal quarry. The earliest reliable written documentation in Germany appears to be a 1302 real estate transaction that included rights to mine coal. The transaction covered land near Dortmund in the heart of the Ruhr.

Coal use started before it was documented in writing. Religious orders tended to keep and preserve written materials more reliably than others. It seems reasonable to postulate that coal use began no later than the 12th century in several West European countries. Documented evidence remains anecdotal for several more centuries but indicates that coal use increased steadily from the 12th century on. England led in coal production until late in the 19th century. Contributing to this sustained leadership were that wood (and hence charcoal) shortages developed earlier and more acutely there than in other countries, there was ready access to shipping by water (sea and navigable rivers), and large coal deposits were present close to the surface.

By the 13th century, London imported significant amounts of coal, primarily sea-coal, shipped from Newcastle-upon-Tyne. The use of coal grew steadily, even though it was controversial, for what now would be called environmental impact reasons. Smoke, soot, sulfurous odors, and health concerns made it undesirable. From the 13th century on, ordinances were passed to control, reduce, or prevent its use. None of these succeeded, presumably because the only alternatives, wood and charcoal, had become too expensive, if available at all. Critical for the acceptance of coal was the development of chimneys and improved fire places. By the early 1600s, sea-coal was the general fuel in the city. The population of the city grew from 50,000 in 1500 to more than 500,000 by 1700, the coal imported from less than 10,000 tons per year to more than 500,000 tons.

Early coal use was stimulated by industrial applications: salt production (by brine evaporation), lime burning (for cement for building construction), metal working, brewing, and lesser uses. By the late Middle Ages, Newcastle exported coal to Flanders, Holland, France, and Scotland.

By the early 16th century, coal was mined in several regions in France. In Saint Étienne, long the dominant coal producer, coal was the common household fuel, and the city was surrounded by metals and weapons manufacturing based on its coal. Legal and transportation constraints were major factors in the slower development of coal on the European continent compared to England. In the latter, surface ownership also gave subsurface coal ownership. In the former, the state owned subsurface minerals. Notwithstanding royal incentives, France found it difficult to promulgate coal development on typically small real estate properties under complex and frequently changing government regulations. Only late in the 17th century did domestic coal become competitive with English coal in Paris, as a result both of the digging of new canals and of the imposition of stiff tariffs on imported coal. Coal mining remained artisanal, with primitive exploitation "technology" of shallow outcrops, notably in comparison with the by this time highly developed underground metal mining.

The earliest coal mining proceeded by simple strip mining: the overburden was stripped off the coal and the coal dug out. As the thickness of the overburden increased, underground mines were dug into the sides of hills, the development of drift mines. Mining uphill, strongly preferred, allowed free water drainage and facilitated coal haulage. Some workings in down-dipping seams were drained by driving excavations below the mined seams.

Shafts were sunk to deeper coal formations. When coal was reached, it was mined out around the shaft, resulting in typical bell pits. Coal was carried out in baskets, on ladders, or pulled out on a rope. Where necessary, shafts were lined with timber.

In larger mines, once the coal was intersected by the shaft, "headings" were driven in several directions. From these, small coal faces were developed, typically about 10 ft wide, usually at right angles from the heading, and pillars, blocks of coal, were left in between these so-called bords, the mined-out sections. Pillar widths were selected to prevent roof, pillar, and overburden failure, which could endanger people and could lead to loss or abandonment of coal. Each working face was assigned to a hewer, who mined the coal, primarily with a pick. The hewer first undercut the coal face: he cut a groove along the floor as deep as possible. He then broke out the overhanging coal with picks, wedges, and hammers. Lump coal was hand loaded into baskets, usually by the hewer's helper. The baskets were pushed, dragged, or carried to and sometimes up the shaft. This haulage commonly was performed by women and children, usually the family of the hewer. The hewer operated as an independent contractor. He was paid according to the amount of coal he and his family delivered to the surface. Once established as a hewer, after multiple years of apprenticeship, the hewer was among the elite of workers in his community.

In deeper or better equipped mines, the coal baskets or corves were hoisted up the shaft using a windlass, later horse-driven gins. These also were used for hoisting water filled baskets or buckets, to dewater wet mines. By the end of the 17th century, shaft depths reached 90 ft, occasionally deeper. Encountering water to the extent of having to abandon pits became frequent. Improving methods to cope with water was a major preoccupation for mine operators.

Ventilation also posed major challenges. Larger mines sank at least two shafts, primarily to facilitate operations. It was recognized that this greatly improved air flow. On occasion shafts were sunk specifically to improve ventilation. During the 17th century, the use of fires, and sometimes chimneys, was introduced to enhance air updraft by heating the air at the bottom of one shaft. In "fiery" mines, firemen, crawling along the floor, heavily dressed in wetted down cloths, ignited and exploded gas accumulations with a lighted pole. This practice was well established in the Liège basin by the middle of the 16th century. As mines grew deeper, and production increased, explosions became more frequent, on occasion killing tens of people.

In North America, coal was reported on Cape Breton in 1672. Some coal was mined for the French garrison on the island, some was shipped to New England before 1700. Coal was found by Joliet, Marquette, and Father Hennepin, in 1673, along the Illinois river. It is possible that some was used at that time by the Algonquins in this area.

4. PRECURSORS TO THE INDUSTRIAL REVOLUTION

The best sun we have is made of Newcastle coal.
—Horace Walpole, 1768

Whether or not one subscribes to the (controversial) concept of a first industrial revolution, the use of coal increased significantly from about the middle of the 16th century. A prime cause was the shortage of firewood, acute in England, noticeable in France. In Britain coal production increased from less than 3 million tons in 1700 to more than 15 million tons in 1800 (and then doubled again to more than 30 million tons per year by 1830). Driving were the growing industrial uses of coal, in particular the technology that made possible the smelting and forging of iron with coal. Steam power, developed to dewater mines, created a voracious demand for coal.

Land positions and legal statutes facilitated the development of coal. The frequent closeness of iron ore, limestone, and coal deposits stimulated the iron industry. Canals provided the essential low cost water transportation for inland coal fields. Tramways and turnpikes fed canal transport and led to railroads and highways.

Major technical advances were made in mining technologies: dewatering, haulage, and ventilation, challenges to mining that continue to this day and presumably always will.

Thomas Savery's *Miners Friend*, patented in 1698, promised the use of steam for a mine dewatering pump. Thomas Newcomen's steam-driven pump fulfilled the promise. The first Newcomen steam pump, or "fire-engine" or "Invention to Raise Water by Fire," was installed at a coal mine in 1712. Although expensive and extraordinarily inefficient by modern standards or even by comparison with Watt's steam engine of 60 years later, 78 Newcomen engines operated by 1733. The first Boulton and Watt engine, based on Watt's patent, was installed at a coal mine in 1776. By the end of the century, Boulton and Watt engines also found increasing use for shaft hoisting. For most of the century, horse power had been the dominant shaft hoisting method.

Underground haulage improved. Wheels were mounted on the baskets previously dragged as or on sledges. Planks were placed on the floor and were replaced by wooden and eventually iron rails—forerunners of the railroad. Horses replaced boys for pulling, at least in those mines where haulageways were large enough.

Dewatering became practical to great depths. Haulage improved greatly. Coping with explosive gases proved difficult. Fires at shaft bottoms remained the dominant method to induce airflow. Numerous ingenious devices were invented to course air along working faces, as air coursing was pursued vigorously. But explosions remained a major and highly visible hazard, resulting in an increasing cost in lives.

The physical coal winning—hewer with pick—remained unchanged. Improvements in production were associated with better lay-outs of the workings. Most mines continued to operate the room and pillar method, leaving over half the coal behind. An exception was the development of the longwall method. Here, a single long face was mined out entirely. As the coal was mined, the roof was supported with wooden props. The gob, the mined out area, was back-filled with small waste coal. This method allowed a nominal 100% recovery or extraction but resulted in surface subsidence. It now has become, in a highly mechanized form, the dominant underground coal mining method in the United States, Australia, and Europe.

The European continent lagged far behind Britain in all aspects of coal industry development: production, technology, transportation, use, and legal framework. By 1789, France produced somewhat more than 1 million tons per year, about one-tenth of that produced in Britain. The Anzin Coal Company, formed after considerable government prodding to try to increase coal production, mined 30% of all French coal. France now recognized the growing threat posed by British industrialization.

Although by 1800 some 158 coal mines operated in the Ruhr, production that year barely exceeded 250,000 tons. The first steam engine at a coal mine in the region became operational only in 1801.

In North America, wood remained widely available at low cost, as well as water power in the more developed areas (e.g., New England). There was no need to use coal, and little was used. Major deposits were discovered and would prove extremely significant later on.

Coal was known to exist near Richmond, Virginia, by 1701, and some was mined by 1750. Some was sold to Philadelphia, New York and Boston by 1789. In 1742, coal was discovered in the Kanawha Valley, West Virginia. The Pocahontas seam, a major coal source for many decades, was mapped in Kentucky and West Virginia in 1750. An extensive description of coal along the Ohio River was made in 1751. George Washington visited a coal mine on the Ohio River in 1770. Thomas Jefferson recorded coal occurrences in the Appalachian mountains.

5. INDUSTRIAL REVOLUTION

We may well call it black diamonds. Every basket is power and civilization. For coal is a portable climate. It carries the heat of the tropics to Labrador and the polar circle; and it is the means of transporting itself whithersoever it is wanted.
—Emerson, ***Conduct of Life****, 1860*

Coal fueled the Industrial Revolution, arguably the most important development in history. It transformed the world. While the term "Industrial Revolution" is controversial, it remains helpful to identify that period when fundamental changes developed in the basic economic structure, initially in Great Britain, shortly thereafter on the European and North American continents, and eventually throughout much of the world. The transition to the predominant use of fossil fuels, initially coal, was a major aspect and driving force of industrialization.

In Britain, the Industrial Revolution usually is considered the period from about 1760 through about 1840. In Belgium it started about 1830; in France, the United States, and Germany it started within the following decades. Other West European countries followed well before the end of the 19th century. Russia and Japan initiated their industrial revolutions by the end of the century.

The reasons why Britain took the lead are complex and involve all aspects of society. Readily accessible coal and iron ore in close proximity to each other was a major factor. Mine dewatering needs led to the development of the steam engine and mine haulage to railroads, two core technologies of the Industrial Revolution.

Belgium had coal and iron ore of good quality in close proximity and the necessary navigable river system. In France and Germany, coal and iron ore were far apart, and in France they were not of a quality that could be used for steel production until late in the century, when its metallurgy was understood. France and Germany lacked the water transportation necessary for the development of a heavy industry (Fig. 1). Steam locomotives and steam ships permanently altered travel and perceptions of space and time. Both were major users of coal and required that coal be provided at refueling points all around the world, starting worldwide coal trade and transportation. For many decades, Britain dominated this coal shipping and trading. Coal bunkering ports were major way stations toward building the British Empire.

Steam converted the Mississippi into the commercial artery it still is, although it is now diesel fueled. Railroads opened up the Midwest, the West, and later Canada. As the railroads expanded, so did the coal mines feeding them. Anthracite built eastern Pennsylvania, bituminous coal built Pittsburgh.

Coal gas and water gas dramatically, often traumatically, increased available lighting and changed living habits. Initially used in manufacturing plants, to allow around the clock work, gas light slowly conquered streets and homes and turned Paris into the "city of light."

Social, political, and cultural revolutions accompanied the Industrial Revolution. Working and living conditions of the lower classes became a societal issue and concern and industrial societies accumulated the wealth needed to address such concerns. Dickens, Marx, Engels, and many others identified social problems and proposed solutions either within or outside the existing political structure.

Coal mining was marked by difficult labor relations from the beginning and continues to be so in most parts of the world where it is a significant economic activity. Responsibility of the government in the social arena found its expression in multiple studies and reports and numerous labor laws. A focal point was child and female labor. The age at which children were allowed to start working underground was gradually raised. In Belgium, women miners were

FIGURE 1 *The Coal Wagon.* Théôdore Géricault (1821). Transportation cost always has been a major item in coal use. Road haulage was and is extremely expensive. A significant advantage for early British industrial development was its ready access to sea, river, and canal haulage of coal. Courtesy of the Fogg Art Museum, Harvard University Art Museums Bequest of Grenville L. Winthrop. Gray Collection of Engravings Fund. © President and Fellows of Harvard College. Used with permission.

more adamant and more successful in their opposition to being banned from underground and worked there legally until late in the 19th century (Fig. 2).

Although science and the scientific approach were well developed by this time, they had little influence on the early technological developments. The lack of geological understanding made the search for deeper coal uncertain and expensive. Geological sciences grew rapidly toward the end of this period and would soon make major contributions to finding coal and understanding the complexities of coal—even though its origin and formation remained a matter of controversy for decades. While the official geological community had little interaction with coal mining, William Smith, author of the first geological map of Britain, a masterpiece of geological mapping, was involved in coal mines, although his prime engineering work was for canals, mainly coal shipping canals. "Strata Smith" was a founder of stratigraphy because of his recognition of the possibility to use fossils to correlate strata. The first paper published by James Hutton, the founder of modern geology, was a coal classification. Hutton devised a method to allow customs agents and shippers to agree on a coal classification to set customs fees. While Hutton recognized the usefulness of the vast amount of geological information disclosed by coal mining, his later interest was strictly scientific. He did present a theory for coal formation based on metamorphosis of plant materials. While coal mining influenced the early development of geology, its impact was less than that of metal mining.

During the 1840s, the leading British geologist Charles Lyell visited several coal mining areas in North America. He proposed igneous activity combined with structural disturbance as the mechanism of formation of the anthracites in northeastern Pennsylvania, then the booming heart of the U.S. Industrial Revolution. He used the creep of mine roof and floor as an analog for the slow large deformations of rock formations.

As in Britain, much early U.S. geological mapping was for surveys for canals built during the beginning of the U.S. Industrial Revolution. Because much of the initial geological mapping was utilitarian, including looking for building stone and metal ore, at the time when the coal industry was developing, finding and mapping coal was a major objective of early geologists. This included the First Geological Survey of Pennsylvania, and the mapping of the upper Midwest by David Dale Owen, first state geologist of Indiana.

Although the connection between coal and geology may seem obvious, and it was far less so then than now, even less obvious might be relations between Industrial Revolution technology and fundamental geology. Yet Hutton, good friend of James Watt, was influenced by the *modus operandi* of the steam engine in his understanding of the uplifting of mountains.

FIGURE 2 *The Bearers of Burden*. Vincent Van Gogh (1881). The first bearer carries a safety lamp, suggesting she came from underground, which still was legal in the Belgian Borinage coal mining region when Van Gogh worked there in the 1870s. Collection Kröller-Müller Museum, Otterlo, The Netherlands. Photograph and reproduction permission courtesy of the Kröller-Müller Foundation.

Among the members of the Oysters Club in Edinburgh that included James Watt and James Hutton was Adam Smith. *The Wealth of Nations* addresses many strengths and weaknesses of the coal industry, while laying the theoretical economic basis for the free market economy that allowed it and the Industrial Revolution to thrive. Adam Smith identified the need for a coal deposit to be large enough to make its development economically feasible, the need for it to be located in the right place (accessible to markets, i.e., on good roads or water-carriage), and the need for it to provide fuel cheaper than wood (i.e., to be price competitive). Also, as Smith stated, "the most fertile coal-mine regulates the price of coals at all other mines in its neighbourhood." He discussed the appropriate purchase price and income of a coal mine and the economic differences between coal and metal mines—in sum, the basics of mineral and fuel economics.

The French engineer-scientist Coulomb reported on steam engines based on his observations during a trip to Britain and pushed French industrialists toward the use of coal. It was recognized that steam engines remained extremely inefficient. Scientific investigations of the performance of the engines evolved toward the science of thermodynamics. In 1824, Carnot published a theoretical analysis of the heat engine, a founding publication of thermodynamics, although the term was introduced only 25 years later by Lord Kelvin in his *Account of Carnot's Theory*. From 1850 on there was considerable interaction between the development of the steam engine and of thermodynamics, as demonstrated by the involvement in both of Rankine, one of the Scottish engineers-scientists who drove the technological developments of the Industrial Revolution.

6. 19TH CENTURY

The 19th century was the century of King Coal. Coal production grew dramatically in most countries that became industrialized, and coal use grew in all of them. While Britain continued to lead in coal production, the growth rate in the United States was so fast that by the end of the century its production exceeded Britain's (Table I). In 1900, coal exports accounted for 13.3% of total British export value. Coal trading was greatly liberated during the first half of the century. Direct sales between coal producers and consumers (e.g., gas producers) became legal. In this free and open domestic trading environment coal flourished. Internationally, the Empire encouraged and protected its domestic industry, supporting domestic production for export.

TABLE I

Annual Coal Production for Some Countries, in Millions of Tonnes[a] (Mt) per Year

Year	World	United States	United Kingdom	Belgium	Germany	France	Japan	China	Australia	South Africa	Indonesia	Poland[c]	Russia/USSR/FSU[d]	Colombia	India
1700			2.5			<0.1									
1800		0.1	>10 (15?)		1	1									
1815			22.6		<1.3	0.9									
1830		0.9	22.8 (30.9)	2.3	1.8	1.9									
1850		6.4	69.5	5.8	4.2	5									
1860	135	13.2	80.0	9.6	12.3	8.3									
1870									0.9			8	0.4		0.4
1871		42.5	117.4	13.8	24	13.2									
1880	332	64.9	147.0	16.9	47	19.4	0.9		1.6	<0.1		14	2		1
1890	512	143.2	181.6	20.4	70.2	26.1	2.6								
1900	700	244.2	225.2	23.5	109.3	33.4	7.4	<1	6.5	0.9	0.2	31	10		6.2
1910	1160	454.6	264.4	25.5	151.1	38.3	15.7	4.2							
1913	1341	517.2	287.4	22.9	277.4[b]	40.9	21.3	14.0	12.6	8.9	0.5	36	32.2	<0.1	16.5
1920	1320	597.1	233.2	22.4	131.4	24.3	30.5	19.5	13.2	10.4	1.1	32	7.6		18.2
1930	1414	487.1	262.1	27.4	142.7	53.9	33.4	26.4	11.5	12.2	1.7	34.0	36.2		21.9
1932	1124	326.1	212.1	21.4	104.7	46.3	28.1	28.0	11.3	9.9		28.8	53.7		22.0
1940	1497	462.0	227.9	25.6	240.1	39.3	57.3	46.5	11.8	17.2	2.0	77.1	148.7	0.5	29.9
1950	1508	509.4	220	27.3	113.8	52.5	38.5	41.1	16.5	26.5	0.8	78.0	185.2	1.0	32.8
1952	1496	484.2	228.5	30.4	143.7	52.4	43.4	66.6	19.4	28.1	1.0	84.4	215.0	1.0	36.9
1960	1991	391.5	197.8	22.5	148.0	56.0	57.5	397.2	21.9	38.2	0.7	104.4	374.9	2.6	52.6
1970	2208	550.4	147.1	11.4	118.0	37.8	40.9	354.0	45.4	54.6	0.2	140.1	206.7	2.75	73.7
1980	2810	710.2	130.1	8.0	94.5	20.2	18.0	620.2	72.4	115.1	0.3	193.1	553.0	4.2	113.9
1990	3566	853.6	94.4	2.4	76.6	11.2	8.3	1050.7	158.8	174.8	10.5	147.7	527.7	21.4	211.7
2000	3639	899.1	32.0	0.4	37.4	4.4	3.1	1171.1	238.1	225.3	78.6	102.2	321.6	37.1	309.9

[a] 1 tonne = 1000 kg = 2204.6 lb = 1.102 (short) tons

[b] Production for pre–World War I territory, sum of hard coal and brown coal (lignite).

[c] There are considerable differences in data, especially before 1920, depending on territory considered—that is, on how changed boundaries (one of which has intersected, in different ways, Upper Silesia, the major coal producing area) affected coal production. Given is an exceedingly simplified summary of very approximate production data for the area currently (2003) Poland.

[d] FSU = Former Soviet Union (data from 1980 through 2000). Data from Coal Information (2001), "International Energy Agency," Paris, France (2001); "Energy Information Adminstration," U.S. Department of Energy, Washington, D.C.; "Minerals Yearbook, U.S. Department of the Interior," Washington, D.C., (1918), (1923), (1934), (1950); "Historical Statistics of the United States to 1957: A Statistical Abstract Supplement," Washington, D.C. (1960); B. R. Mitchell, "International Historical Statistics, Europe (1750–1988)," 3rd ed. Stockton Press, New York, NY (1992); B. R. Mitchell, "International Historical Statistics, Africa, Asia & Oceania," 2nd Rev. ed., Stockton Press, New York, NY (1995); K. Takahashi (1969), "The Rise and Development of Japan's Modern Economy," The Jiji Tsushinsha (The Jiji Press, Ltd.), Tokyo; A. L. Dunham (1955), "The Industrial Revolution in France (1815–1848)," Exposition Press, New York; B. R. Mitchell (1962), "Abstract of British Historical Statistics," Cambridge at the University Press; W. W. Lockwood (1954), "The Economic Development of Japan," Princeton University Press, Princeton, NJ; A. S. Milward and S. B. Saul (1973), "The Economic Development of Continental Europe," Rowman and Littlefield, Totowa, NJ; W. Ashworth (1975), "A Short History of the International Economy Since 1850," Longman, London; M. Gillet (1973), "Les Charbonnages du Nord de la France au XIXe Siècle," Mouton, Paris; R. Church (1986), "The History of the British Coal Industry, Vol. 3 (1830–1913): Victorian Pre-eminence," Clarendon Press, Oxford; V. Muthesius (1943), "Ruhrkohle (1893–1943)," Essener Verlagsanstalt, Essen, Germany; B. R. V. Mitchell (1980), "European Historical Statistics (1750–1975)," *Facts on File*, New York; M. W. Flinn (1984), "The History of the British Coal Industry, Vol. 2 (1700–1830): The Industrial Revolution," Clarendon Press, Oxford; J. A. Hodgkins (1961), "Soviet Power," Prentice-Hall, Englewood Cliffs, NJ; H. N. Eavenson (1935), "Coal Through the Ages," AIME, New York; S. H. Schurr and B. C. Netschert (1960), "Energy in the American Economy (1850–1975)," The Johns Hopkins Press, Baltimore; "A History of Technology," T. I. Williams, Ed. (1978), Clarendon Press, Oxford; "COAL, British Mining in Art (1680–1980)," Arts Council of Great Britain, London; T. Wright, Growth of the modern chinese coal industry, "Modern China," Vol. 7, No. 3, July 1981, 317–350; R. L. Gordon (1970), "The Evolution of Energy Policy in Western Europe," Praeger, New York; J. S. Furnivall (1939) (1967), "Netherlands, India, A Study of Plural Economy," Cambridge at the University Press; D. Kumar, Ed. (1983) "The Cambridge Economic History of India," Cambridge University Press; Z. Kalix, L. M. Fraser, and R. I. Rawson (1966), "Australian Mineral Industry: Production and Trade (1842–1964)," Bulletin No. 81, Bureau of Mineral Resources, Geology and Geophysics, Commonwealth of Australia Canberra; P. Mathias and M. M. Postan, Eds., "The Cambridge Economic History of Europe," Vol. VII, Part I, Cambridge University Press, Cambridge (1978); N. J. G. Pounds, "The Spread of Mining in the Coal Basin of Upper Silesia and Southern Moravia," *Annals of the Association of American Geographers*, Vol. 48, No. 2, 149–163 (1958). Most data prior to 1900 must be considered as subject to large uncertainties and to significant differences between different sources.

Capital was needed to develop the mines—for example, to sink shafts, construct surface and underground operating plant (e.g., pumps, hoisting engines, fans), build loading, processing, and transportation facilities. Coal mine ownership slowly shifted from private and partnerships to stock-issuing public corporations. Employment in British coal mining grew from about 60,000 to 80,000 in 1800 to nearly 800,000 in 1900.

In 1800, the United States produced barely over a 100,000 tons of soft coal and almost certainly much less anthracite. Although many Pennsylvania anthracite outcrops were well known by then and were within hauling distance from Philadelphia, Boston, and New York City, no satisfactory method for burning anthracite had been developed.

The American Industrial Revolution started around 1820, in parallel with the growth of the anthracite industry in northeastern Pennsylvania. Canals were built to provide transportation but were soon superseded by railroads. Anthracite had been promoted and used somewhat as a domestic fuel late in the 18th century. It did not become accepted until efficient fire grates became available, well into the 19th century, and until the price of wood fuel had risen significantly, particularly in the large cities. Following intense development of burning technologies, anthracite became the fuel of choice for public buildings and was used in steam engines (stationary, e.g., for coal mine pumping, manufacturing, and mobile: locomotives and boats) and iron production. The commercialization of improved and easier to use stoves made anthracite the home heating fuel of choice for well over the next century.

Early mining of anthracite was easy, as it was exposed in many outcrops. Many small operators could start mining it with little or no capital, manpower, or knowledge required. Once underground, the geological complexity of the intensely folded and faulted deposits required operating knowledge that led to consolidation in the industry. The drift mines and shallow shaft mines (rarely deeper than 30 ft) started in the 1810s were followed in the 1830s by slope mines. Breast and pillar mining was common: coal pillars were left in between the breasts (faces, rooms) that were mined out. The pillars carried the weight of the overburden. Horses and mules pulled cars running on rails. Black powder was used to shoot the coal (break out the coal). Wooden props (timber posts) provided safety by holding up the immediate roof.

Contract mining was standard. Each miner worked a breast, assisted by one or two helpers who loaded the coal and installed timber. The miner was paid on a piece rate and paid his helpers. The miner worked as an independent, with minimal supervision. Death and injury rates were high, predominantly from roof falls, not from the explosions that occurred all too often. By the end of the century the fatality rate in U.S. coal mining approached 1000 per year.

Over the course of the 19th century, numerous safety lamps were developed, typically designed with a protective wire mesh screen and glass to prevent the flame from igniting an explosion. The Davy lamp is the most famous, although it was not the one most widely used. Whether these lamps improved safety, (i.e., reduced accidents) or whether they simply allowed miners to work in more gaseous (i.e., more dangerous) conditions remains controversial.

Bituminous or soft coal mines operated with room and pillar mining, similar to the anthracite mines, but predominantly in flat or nearly flat beds—not the steeply dipping beds of the anthracite region. In 1871, virtually all coal was still undercut by hand and shot with black powder. Coal preparation was almost nonexistent. Animal haulage was universal.

Major efforts were started to mechanize coal mining. Cutter machines, designed to replace the manual undercutting, were patented, but it took several more decades before they performed well and became accepted. Late in the century, electric locomotives were introduced. They quickly gained widespread acceptance, replacing animal haulage. In 1871, most mines that used artificial ventilation—and many did not—used furnaces. By the end of the century, mechanical fans dominated. They were driven by steam engines, as were the pumps that dewatered wet mines.

By the end of the century, a pattern of difficult labor relations was well established in coal fields around the world. The first strike in the Pennsylvania anthracite region took place in 1849. Labor actions and organizations started in Scotland, England, and Wales in the 18th century. Even though by the early 19th century miners' wages were substantially higher than those in manufacturing, the relentless demand for labor, driven by the rapidly increasing demand for coal, facilitated the growth of labor movements, notwithstanding the dominant political power of the coal producers. The high fatality and injury rate gave impetus to a labor emphasis on improving safety.

The 1892 first national miners strike in Britain stopped work from July through November. Two men were killed in riots. In the 1890s the UMWA (United Mine Workers of America) was trying to organize. General strikes were called in 1894 and 1897. In 1898, an agreement was reached between the UMWA and the main operators in Illinois, Ohio, Indiana, and western Pennsylvania. Missing from this list are the anthracite region of northeastern Pennsylvania and West Virginia.

By the end of the century, coal had broadened its consumption base. In 1900, the world produced 28 million tons of steel, the United States, 11.4 million tons. Coking coal for steel had become a major coal consumer. To improve steel quality, steel producers tightened specifications on coke and thereby on the source coal.

Coal gas was introduced early in the century, and had become the major light source in both large and small cities by midcentury. Electric light was introduced by the end of the century. Electric power generation became a major coal user only by the middle of the 20th century, however. Both the conversion to gas light (from candles) and the later one to electric light (from gas) required adjustments on the part of the users. For both changes a major complaint was the excessive brightness of the new lights. (Initially electric light bulbs for domestic use were about 25 W.)

The heavy chemical industry received a major boost when it was discovered that coal tar, a waste by-product of coke and gas production, was an excellent feedstock for chemicals, notably organic dyes. Discovered in Britain, shortly after mid-century, Germany dominated the industry by the end of the century and did so until the first world war. The industry became a textbook example of the application of research and development (R&D) for the advancement of science, technology, and industry.

An important step during the 19th century was the development of coal classifications. Coal is complex and variable. It can be classified from many points of view. Geologically, coal classification is desirable to bring order in a chaotic confusion of materials (macerals) with little resemblance to the mineralogy and petrography of other rocks. Chemical classification is complicated by the fact that the material is a highly variable, heterogeneous mixture of complex hydrocarbons. From the users and the producers point of view, the buyer and the seller, some agreement needs to be reached as to what is the quality of the delivered and the received product. Depending on the use, "quality" refers to many different factors. It includes calorific value (i.e., how much heat it generates), moisture content, and chemical composition (e.g., how much carbon or hydrogen it contains). Impurities are particularly important (e.g., ash and sulfur content and the behavior of the coal during and after burning or coking).

From 1800 to 1889, world production of coal increased from 11.6 to 485 million tons. In 1899, coal provided 90.3% of the primary energy in the United States. It was indeed the century of King Coal.

7. 20TH CENTURY

Well into the second half of the 20th century, coal remained the world's major primary energy source. Throughout most of the century, the relative importance of coal declined—that is, as a fraction of total energy production coal decreased. Worldwide coal use increased steadily, but the major production centers shifted. The coal industry changed fundamentally.

Two world wars changed the world and the coal industry. During both wars most industrial countries pushed their steel production, and hence their coal production, as high as possible. Due to the wartime destructions, both wars were followed by severe coal shortages. In response, coal-producing nations pushed hard to increase coal production. In both cases, the demand peak was reached quickly and subsided quickly. A large overcapacity developed on both occasions, resulting in steep drops in coal prices, in major production cutbacks, and in severe social and business dislocations. After the second world war, two major changes impacted coal: railroads converted from steam to diesel and heating of houses and buildings switched to fuel oil and natural gas. (Diesel replaced coal in shipping after World War I.) In the United States, railroads burned 110 million tons of coal in 1946, 2 million tons in 1960. Retail deliveries, primarily for domestic and commercial building heating, dropped from 99 million tons in 1946 to less than 9 million tons in 1972. One hundred fifty years of anthracite mining in northeastern Pennsylvania was ending.

A remarkable aspect of the coal user industry is the growth in efficiency in using coal. Railroads reduced coal consumption per 1000 gross ton-miles from 178 lbs in 1917 to 117 lbs in 1937. One kWh of electrical power consumed 3.5 lbs of coal in 1917, 1.4 lbs in 1937. A ton of pig iron required 2,900 lbs of coking coal in 1936, down from 3500 lbs in 1917. Improvements in use efficiency continued until late into the century. Coal consumption per kWh of electric power dropped from 3.2 lbs in 1920 to 1.2 lbs in 1950, and to 0.8 lbs in the 1960s. After that efficiency decreased somewhat due to inefficiencies required to comply with environmental regulations. Super efficient steam generating and using technologies in the 1990s again improved efficiencies. Even more dramatic was the efficiency improvement in steel production (i.e., the reduction in coal needed to produce steel). Concurrent with the loss of traditional markets came the growth in the use of coal for generating electrical power, the basis for the steadily increasing demand for coal over the later decades of the century and for the foreseeable future.

Major shifts took place in worldwide production patterns. Britain dropped from first place to a minor producer. Early in the century, the United States became the largest coal producer in the world and maintained that position except for a few years near the very end of the century when the People's Republic of China became the largest producer. Most West European countries and Japan reached their peak production in the 1950s, after which their production declined steeply. In the much later industrialized eastern European countries, in Russia (then the Soviet Union), and in South Korea, the peak was reached much later, typically in the late 1980s.

In Australia, India, and South Africa, coal production increased over most of the century, with major growth in the last few decades. The large production growth in China shows a complex past, with major disruptions during the 1940s (World War II) and the 1960s (cultural revolution). The recent entries among the top coal producers, Colombia and Indonesia, grew primarily during the 1980s and 1990s.

Worldwide production patterns have changed in response to major transportation developments. Large bulk carrier ships reduced the cost of shipping coal across oceans. While some international coal trading existed for centuries (e.g., from Newcastle to Flanders, Paris, and Berlin and later from Britain to Singapore, Cape Horn, and Peru), only during the second half of the 20th century did a competitive market develop in which overseas imports affect domestic production worldwide. Imports from the United States contributed to coal mine closures in Western Europe and Japan during the 1950s. Imports from Australia, South Africa, Canada, Poland, and the United States contributed to the demise of coal mining in Japan, South Korea, and most of Western Europe.

Inland, unit trains haul coal at competitive cost over large distances: Wyoming coal competes in Midwestern and even East Coast utility markets. It became feasible to haul Utah, Colorado, and Alberta coking coal to West Coast ports and ship it to Japan and South Korea.

Coal mining reinvented itself over the 20th century. A coal hewer from 1800 would readily recognize a coal production face of 1900. A coal miner from 1900 would not have a clue as to what was going on at a coal face in 2000.

The most obvious and highly visible change is the move from underground to surface mining (Fig. 3). Large earthmoving equipment makes it possible to expose deeper coal seams by removing the overburden. Although large-scale surface mining of coal started early in the century, by 1940 only 50 million tons per year was surface mined, barely over 10% of the total U.S. production. Not until the 1970s did surface production surpass underground production.

FIGURE 3 Early mechanized surface coal mining. A 1920s vintage P&H shovel loading in a coal wagon pulled by three horses. Photograph courtesy of P&H Mining Equipment, A Joy Global Inc. Company, Milwaukee, WI. Used with permission.

By 2000, two-thirds of U.S. coal production was surface mined.

Room and pillar mining dominated underground U.S. coal mining until very late in the century. Early mechanization included mechanical undercutting and loading. Conventional mining, in which the coal is drilled and blasted with explosives, decreased steadily over the second half of the century and was negligible by the end of the century. Continuous mining grew, from its introduction in the late 1940s (Fig. 4), until very late in the century, when it was overtaken by longwall mining (Fig. 5). Modern mechanized longwall mining, in which the coal is broken out mechanically over the entire face, was developed in Germany and Britain by the middle of the century. Geological conditions made room and pillar mining impractical or even impossible in many European deposits. In the last two decades of the century, American (and Australian) underground coal mines adopted longwalling, and greatly increased its productivity. In conjunction with the increased production arose a serious safety problem: coal is being mined so fast that methane gas is liberated at a rate difficult to control safely with ventilation systems.

Technological advances depend on equipment manufacturers. Surface mining equipment size peaked in the 1960s and 1970s. The largest mobile land-based machine ever built was the *Captain*, a Marion 6360 stripping shovel that weighed 15,000 tons. *Big Muskie*, the largest dragline ever built, swung a 220 cu yd bucket on a 310 ft boom. The demise of the stripping shovel came about because coal seams sufficiently close to the surface yet deep enough to warrant a stripping shovel were mined out. While the stripping shovel was exceedingly productive and efficient, its large cost required that it operate in a deposit that could guarantee a mine life of at least 10 to 20 years. Capital cost for a stripping shovel was markedly higher than for a dragline.

Large draglines shipped during the 1980s were mostly in the 60 to 80 cu yard bucket size range. A few larger machines (120 to 140 cu yd) were build in the 1990s. By the end of the century, the conventional mine shovel reached a bucket size approaching that of all but the largest stripping shovels ever built.

Worldwide research was conducted in support of the coal industry. The U.S. Bureau of Mines was established in 1910 and abolished in 1996. Its mission changed, but it always conducted health

FIGURE 4 Early attempt at mechanized underground coal mining. The Jeffrey 34F Coal Cutter, or Konnerth miner, introduced in the early 1950s. The machine was designed to mechanize in a combined unit the most demanding tasks of manual coal mining: breaking and loading coal. Photograph and reproduction courtesy of Ohio Historical Society, Columbus, OH. Used with permission.

FIGURE 5 A major, highly successful advance in mechanizing underground coal mining: replacing manual undercutting by mechanical cutting. Jeffrey 24-B Longwall Cutter. Photograph and reproduction courtesy of Ohio Historical Society, Columbus, OH. Used with permission.

and safety research. The Bureau tested electrical equipment for underground coal mines, permissible explosives, designed to minimize the chances of initiating a gas or dust explosion, and improved ground control. The Bureau produced educational materials for health and safety training. In 1941, Congress authorized Bureau inspectors to enter mines. In 1947, approval was granted for a federal mine health and safety code. The 1969 Coal Mine Health and Safety Act removed the regulatory authority from the Bureau, and transferred it to the Mine Safety and Health Administration (MSHA).

Organizations similar to the Bureau were established in most countries that produced coal. In Britain, the Safety in Mines Research and Testing Board focused on explosions, electrical equipment, and health, the latter particularly with regard to dust control. In West Germany, the Steinkohlenbergbauverein was known for its authoritative work in ground control, especially for longwalls. CERCHAR in France, INICHAR in Belgium, and CANMET in Canada studied coal mine health and safety. In the Soviet Union and the People's Republic of China, highly regarded institutes ran under the auspices of their respective National Academy of Science.

Over the course of the 20th century, the classification of coal took on ever more importance, resulting in a proliferation of classification methods. Early in the century, when international coal trade was not common and user quality specifications less comprehensive, national and regional classifications were developed. As international coal trade grew, over the second half of the century the need arose for classification schemes that could be applied worldwide.

In situ coal gasification has been demonstrated and could be developed if economics made it attractive. Conceptually simple, a controlled burning is started in a coal seam to produce gas containing CO, H_2, CH_4, and higher order hydrocarbons. The complexity of the fuel, the variability of the deposits, and the potential environmental impacts complicate implementation.

You can't dig coal with bayonets.
—John L. Lewis, president, UMWA, 1956

You can't mine coal without machine guns.
—Richard B. Mellon, American industrialist

Difficult labor relations plagued coal mining through much of the century in most free economy countries that mined coal. In many parts of the world, coal miners formed the most militant labor unions. The West Virginia mine wars, lasting for most of the first three decades of the century, were among the most prolonged, violent, and bitter labor disputes in U.S. history. In Britain, the number of labor days lost to coal mine strikes far exceeded comparative numbers for other industries. Intense violent labor actions, frequently involving political objectives, have recurred throughout much of the century in Britain, France, Germany, Belgium, Australia, Canada, and Japan. During the last few decades of the century, strikes in Western Europe, Poland, Japan, and Canada were driven largely by mine closure issues. Strikes in Russia, the Ukraine, and Australia dealt primarily with living and working conditions. Coal miners in Poland, Serbia, and Rumania were leaders, or at least followers, in strikes with primarily political objectives. The last major strikes in Britain also had a strong political component, although pit closure concerns were the root cause. In the United States, the last two decades of the century were remarkably quiet on the labor front, especially compared to the 1970s.

The structure of the coal mining industry changed significantly over the course of the 20th century. In the United States during the 1930s, many family-owned coal mining businesses were taken over by corporations. Even so, the historical pattern of coal mining by a large number of small producers continued until late in the century. Production concentration remained low compared to other industries and showed an erratic pattern until late in the century. In 1950, the largest producer mined 4.8% of the total, in 1970 11.4%, in 1980 7.2%. In 1950, the largest eight producers mined 19.4% of the total; in 1970, 41%; in 1980, 29.5%. High prices during the energy crisis of the 1970s facilitated entry of small independents. The top 50 companies produced 45.2% of the total in 1950, 68.3% in 1970, 66.3% in 1980, confirming the significant reduction of the small producers.

During the 1970s, oil and chemical companies took over a significant number of coal companies because they believed widely made claims during that decade of an impending depletion of oil and gas reserves. As the hydrocarbon glut of the 1980s and 1990s progressed, most of these coal subsidiaries were spun off and operated again as independent coal producers.

Toward the end of the century, there was significant consolidation of large coal producers, domestically and internationally. Even so, the industry remained characterized by a relatively large number of major producers. In 2000, the 10 largest private companies controlled barely over 23% of the world production. In the United States the largest producer, Peabody, mined 16% of U.S. coal, the second largest one, Arch Coal, 11%. Coal remained highly competitive, domestically and internationally.

8. THE FUTURE

Coal resources are the largest known primary energy resource. Supplies will last for centuries, even if use grows at a moderate rate. Reserves are widely distributed in politically stable areas. The many uses of coal, from electrical power generation to the production of gaseous or liquid fuels and the use as petrochemical feedstock, make it likely that this versatile hydrocarbon will remain a major raw material for the foreseeable future.

The mining and especially the use of coal will become more complicated and hence the energy produced more expensive. Coal mining, coal transportation, and coal burning have been subjected to ever more stringent regulations. This trend toward tighter regulations will continue. A major environmental factor that will affect the future of coal is the growing concern about global warming. While technologies such as CO_2 capture and sequestration are being researched, restrictions on CO_2 releases will add significantly to the cost of producing energy, in particular electricity, from coal.

Predictions for the near future suggest a modest, steady increase in coal production. The main competition in the next few decades will be from natural gas. Natural gas reserves have risen steadily over the past 30 years, in parallel with the increased demand and use, and hence the increased interest in exploration for gas. Natural gas is preferred because it is richer in hydrogen, poorer in carbon, and hence the combustion products contain more steam rather than CO_2. If, in the somewhat more distant future, the predictions of a reduction in supply of natural gas and oil were to come through—and for over a century such predictions have proved "premature"—coal might once again become the dominant fossil fuel. The rise in demand for electric power seems likely to continue in most of the world to reach reasonable living standards and in the developed world for such needs as electric and fuel cell–driven vehicles and continued growth in computers and electronics in general.

To make coal acceptable in the future, steps need to be taken at all phases of the coal life cycle, from production through end use. A major focus in the production cycle is minimizing methane release associated with mining. Methane (CH_4) is a greenhouse gas. It also is a main cause of coal mine explosions. Great strides have been made in capturing methane prior to and during mining. In gassy seams, it now is collected as a fuel. In less gassy seams, especially in less technologically sophisticated mines, it remains uneconomical and impractical to control methane releases. Extensive research is in progress to reduce methane releases caused by coal mining.

Other environmental problems associated with mining coal include acid mine drainage, burning of abandoned mines and waste piles, subsidence, spoil pile stability issues, and mine site restoration and reclamation. Technological remedies exist, but their implementation may need societal decisions for regulatory requirements.

Coal preparation is critical for improving environmental acceptability of coal. Super clean coal preparation is feasible. Technically, virtually any impurity can be removed from coal, including mercury, which has drawn a great deal of attention over the last few years. Coal transportation, particularly in ocean going vessels, has modest environmental impacts, certainly compared to oil.

Coal users carry the heaviest burden to assure that coal remains an acceptable fuel. Enormous progress has been made in reducing sulfur and nitrogen oxide emissions. Capturing and sequestering CO_2 will pose a major challenge to the producers of electric power.

More efficient coal utilization contributes to the reduction in power plant emissions. Modern power plants run at efficiencies of about 37%. During the 1990s, power plants have come on stream that run at over 40%. It is likely that 50% can be achieved by 2010. Increasing efficiency from 37 to 50% reduces by one-third the coal burned to generate electricity and reduces by one-third gas (and other) emissions.

Coal has been attacked for environmental reasons for over seven centuries. With ups and downs, its use has grown over those seven centuries because of its desirable characteristics: low cost, wide availability, ease of transport and use. It will be interesting to see whether it can maintain its position as a major energy source for another seven centuries or whether more desirable alternatives will indeed be developed.

9. COAL AND CULTURE

Given the pervasive influence of coal and in particular of its uses on the fundamental transformations of societies, especially during the 19th century, it is not surprising to see coal reflected in cultural contexts. Pervasive was the sense of progress associated with industrial development, the perception of dreadful social problems associated with the industrial progress, and the disintegration of an older world.

Zola's *Germinal* remains the major novel rooted in coal mining, a classic in which have been read different, contradictory meanings from revolutionary to bourgeois conservative. D. H. Lawrence grew up in a coal mining town, and it and its collieries pervade several of his masterpieces. Again, these incorporate deep ambiguities with respect to coal mining: admiration for the male camaraderie in the pits, the daily dealing with a hostile dangerous environment, the solidarity in the mining community, and the stifling constraints of it all. Similar ambiguities are found in Orwell's *The Road to Wigan Pier* and in the reactions to it. It was published by the Left Book Club. These publishers inserted an apologetic introduction, justifying why they published this book, frequently so critical of the left. The book includes a widely quoted description of underground mining practices in Britain in the 1930s, as well as sympathetic descriptions of life in the mining communities—not deemed sufficiently negative by many on the left.

Richard Llewellyn's popular *How Green Was My Valley* introduces a small coal mining community in South Wales through a voluptuous description of rich miner's meals. (Orwell, as described in his personal diary, published long after *Wigan Pier*, similarly was struck by the rich miners meals.)

Lawrence's descriptions of coal mine towns in the early 20th century gained wide distribution through films made of his works, several of which were successful commercially. The critically acclaimed and commercially successful movie adaptation by John Ford of *How Green Was My Valley* introduced a wide audience to a coal mining community in South Wales. While justifiably criticized as sentimental, the book and film offer a sympathetic view of a community that often has felt looked down upon, examples of nostalgic flashback descriptions of mining communities often found in the regional literature of mining districts.

Joseph Conrad, in his 1902 story "Youth," fictionalized, although barely, the loss at sea, to spontaneous combustion, of a barque hauling coal from Newcastle to Bangkok. The self-ignition of coal, particularly during long hauls, was a major problem during the late 19th century and continues to pose headaches for those who haul or store coal over extended periods of time. "Youth" shows that

Conrad, an experienced coal hauling sailor, was thoroughly familiar with spontaneous combustion.

Upton Sinclair, a leading American muckraker of the early 20th century, based *King Coal* on the long and bitter 1913–1914 Colorado coal strikes. Neither a critical nor a commercial success (the sequel *The Coal War* was not published until over half a century later and then decidedly for academic purposes), it describes the recurring problem that has plagued coal forever: difficult labor relations.

Van Gogh sketched and painted female coal bearers during his stay in the Borinage, the heart of early coal mining in Belgium. Also originating in the Borinage were the sculptures by Constantin Meunier, including many of miners. The tour of his work through Eastern and Midwestern cities (1913–1914) brought to the American public a visual art that recognized the contribution, hardships, and strengths of blue collar workers.

More visible than coal mining, visual celebrations of the Industrial Revolution focused on the offspring of coal mining and its customers, railroads and the steam locomotive. Among the better known ones is *Rain, Steam and Speed*, by J. W. Turner, the painter of the British industrial landscape, immortalized in *Keelmen Heaving in Coals by Night, Coalbrookdale by Night*, and his paintings of steamships.

The opening up of the American landscape and space, or the intrusion into the landscape, was depicted by Cole's 1843 *River in the Catskills*, his *The Oxbow*, Melrose's *Westward the Star of Empire Takes Its Way*, Bierstadt's *Donner Lake*, and especially George Inness' *The Lackawanna Valley*, Henry's *The First Railway on the Mohawk and Hudson Road*, and Durand's *Progress*. During the 19th century, Currier & Ives' prints popularized progress and penetration of the wilderness thanks to the railroads and the steam engine. Walt Whitman's *Ode to a Locomotive* summarized the widely held view that a continent was being conquered and opened up thanks to the railroad and its most visible emblem, the locomotive's steam.

Nikolai Kasatkin, a leading Russian realist and one of the Itinerants (peredvizhniki—wanderers, travelers), lived several months in the Donetsk coal basin at a time (late 1800s) when Russian coal production was increasing rapidly. One of his most famous paintings, *Miners Changing Shift* (Fig. 6), is often referenced but rarely reproduced. His *Poor People Collecting Coal in an Abandoned Pit* and *Woman Miner*, both far more lyrical than the subject might suggest, also are far more readily accessible. Closer to socialist realism is Deineka's *Before Descending into the Mine*. One of the better known Soviet railroad celebrations is *Transport Returns to Normal* by Boris Yakovlev.

Charles Sheeler and Edward Hopper continue the American tradition of painting railroads in land- and cityscapes. American primitive John Kane, born in Scotland, includes coal barges on the river in his

FIGURE 6 *Coal Miners: Shift Change*. Nikolai A. Kasatkin (1895). Photograph and reproduction courtesy of Tretyakov State Gallery, Moscow.

exuberant *Monongahela River Valley* and *Industry's Increase*. Probably the only one among these artists who worked as a coal miner ("the best work I knew and enjoyed"), Kane published his autobiography *Sky Hooks*, one of the sunnier recollections among the many written by people who grew up in mining communities. More representative of the bleak conditions during the 1930s may be Ben Shahn's sad and haunting *Miners' Wives*.

In France, Théodore Géricault painted the horse haulage of coal, common well into the 19th century, in *Coal Wagon*. Manet, Monet, Caillebotte, Seurat, and Pissarro recorded their impressions of railroads, railroad stations, steamboats, and their impact on landscape and society. The responses ranged from reluctant acceptance, at best, by Pissarro, to the enthusiastic celebrations by Monet of railroads, steam engines, and in particular the Gare St. Lazare, the first railroad station in France and Paris. Monet's *Men Unloading Coal*, a rather darker picture than most Monets, illustrates the manual unloading of coal from river barges late in the 19th century.

SEE ALSO THE FOLLOWING ARTICLES

Coal, Chemical and Physical Properties • Coal, Fuel and Non-Fuel Uses • Coal Industry, Energy Policy in • Coal Mining, Design and Methods of • Coal Mining in Appalachia, History of • Coal Resources, Formation of • Electricity Use, History of • Manufactured Gas, History of • Markets for Coal • Natural Gas, History of • Nuclear Power, History of • Oil Industry, History of

Further Reading

Berkowitz, N. (1997). "Fossil Hydrocarbons: Chemistry and Technology." Academic Press, San Diego, CA.

Bryan, A. M., Sir. (1975). "The Evolution of Health and Safety in Mines." Ashire, London.

Gregory, C. E. (2001). "A Concise History of Mining." A. A. Balkema, Lisse, The Netherlands.

Jones, A. V., and Tarkenter, R. P. (1992). "Electrical Technology in Mining: The Dawn of a New Age." P. Peregrinus, London, in association with the Science Museum.

Peirce, W. S. (1996). "Economics of the Energy Industries." 2nd ed. Praeger, Westport, CT, London.

Pietrobono, J. T. (ed.). (1985). "Coal Mining: A PETEX Primer." Petroleum Extension Service, The University of Texas at Austin, Austin, TX, in cooperation with National Coal Association, Washington, DC.

Shepherd, R. (1993). "Ancient Mining." Elsevier Applied Science, London and New York.

Stefanko, R. (1983). "Coal Mining Technology." Society of Mining Engineers of AIME, New York.

Thesing, W. B. (ed.). (2000). "Caverns of Night: Coal Mines in Art, Literature, and Film." University of South Carolina Press, Columbia, SC.

Trinder, B. (ed.). (1992). "The Blackwell Encyclopedia of Industrial Archaeology." Blackwell, Oxford, UK.

Coal Mine Reclamation and Remediation

ROBERT L. P. KLEINMANN
National Energy Technology Laboratory
Pittsburgh, Pennsylvania, United States

1. Introduction
2. Potential Environmental Problems Associated with Coal Mining
3. Land Reclamation
4. Mine Water Remediation
5. Other Environmental Aspects

Glossary

acid mine drainage (AMD) Acidic water that drains or discharges from a mine; the acidity is caused by the oxidation of iron sulfide (FeS_2), or pyrite, which also causes the water to have elevated concentrations of sulfate and dissolved iron; also known as acid rock drainage (ARD), because the condition can result from non-mining-related pyrite exposure.

approximate original contour (AOC) A requirement that the final topographic configuration of backfilled and reclaimed surface-mined land approximate premining slopes and the general aspect of the premine topography.

bonding Mining companies are often required to post a bond to ensure that mined land will be properly reclaimed. The bond is typically released in phases, as remediation proceeds, either by recognizing successfully completed reclamation for part of the permitted area or by reclamation stages. In the latter case, most of the bond money is released when backfilling and regrading is completed, but some is kept for up to 5 years to ensure that revegetation remains adequate and self-sustaining and that water quality requirements are met. In the past, bond monies, when forfeited by companies that went bankrupt, were never adequate to reclaim a mine site and to treat the contaminated water, so government agencies are now requiring higher bonding levels.

coal refuse The waste left when raw coal is prepared or "cleaned" for market; usually contains a mixture of rock types, but is typically elevated in pyrite, causing it to be a source of acid mine drainage if not handled carefully. In the past, this material was simply piled up; nowadays, disposal requires compaction and stabilization measures.

fugitive dust The silt, dust, and sand that become airborne, carried away from a mine property by the wind.

mountaintop removal An often profitable but controversial method of surface mining in mountainous terrain where the coal seams lie beneath the ridge tops. All of the overburden is removed, exposing 100% of the coal for recovery. The controversy is due to the fact that the initial spoil must be placed in valley or head-of-hollow fills, though most of the subsequent spoil is placed on the level land left after mining proceeds. The final surface configuration, compared to the original landscape, has less relief.

Office of Surface Mining (OSM) A United States federal agency created by the Surface Mine Control and Reclamation Act of 1977, with responsibilities that include overseeing state enforcement of reclamation regulations, collecting fees per ton of coal mined, disbursing those funds (within the limits provided annually by Congress) to the states for reclamation of abandoned mined lands, and providing technical assistance and guidance to the states.

remining Reclamation of previously mined land by private companies, which extract the coal that was left behind by earlier mine operators; water treatment liabilities are not acquired by the new operator unless the new activity makes the water quality worse.

subsidence When fracturing and collapse of strata extend to the land surface; often associated with underground excavations.

Modern coal mines often cause environmental problems, but it must be recognized that these problems are much less extensive and less severe than those associated with mines that were abandoned before regulatory controls went into effect. The progress that has occurred over time is due in large part to the efforts of the environmental activists and the environmental regulations that they inspired, but it is also due, in part, to advances in technology that have made cost-effective environmental progress possible. Such technology can be used to avoid and prevent, or to reclaim and remediate, environmental problems.

1. INTRODUCTION

Any intensive use of the earth's resources carries with it potential environmental consequences. Mining is no exception. Historically, mining companies have extracted the earth's resources wherever economics made it feasible, secure in their knowledge that their products were essential to society. Prior to the passage of the Surface Mining Control and Reclamation Act in 1977, 1.1 million acres (445,000 ha) of coal-mined land in the United States was left unreclaimed and over 10,500 miles (16,900 km) of streams and rivers were adversely affected in the Appalachian region alone. At about the same time, 55,000 ha of mined land was left derelict in Great Britain, though not all of that was due to coal extraction. Although numbers are not readily available for other countries, it can safely be said that leaving land unreclaimed was common around the world during the time period prior to environmental legislation in the United States (Fig. 1).

However, over the past 25 years, a new generation of miners has taken over the industry. Most of these miners grew up with an environmental awareness, and have bought into the philosophy of environmentally responsible mining. Coal mining operations today are designed to comply with environmental regulations and to minimize adverse environmental impacts; exceptions exist, however, and serve to inspire environmental activists and conservation groups to resist and reject mining wherever it is proposed. This societal tension serves as a backdrop to the dynamics of issuing or rejecting mining permit applications.

FIGURE 1 Unreclaimed mine site; vegetation is largely volunteer growth.

2. POTENTIAL ENVIRONMENTAL PROBLEMS ASSOCIATED WITH COAL MINING

2.1 Surface Mines

The most obvious environmental effect of surface mining is the disruption of the land surface; surface mining is generally more visually obtrusive compared to underground mining. In most cases, coal occurs at depth and is covered by soil and rock that must be removed to allow access to the coal, leaving large holes and piles of removed material. Even when the coal outcrops on the land surface, removing the coal leaves a hole in the ground. At prelegislation mines, the land was sometimes left in a disrupted state, but mining companies are now required to separate topsoil from the overburden material, to fill in the excavation as mining proceeds, to cover the regraded land with topsoil, and then to revegetate the land surface. Moreover, in most countries, the regraded and reclaimed land has to be left in a final condition, the approximate original contour (AOC), that approximates the topography that existed before mining. Also, in the United States, bond money must be posted before mining is permitted; before the bonds can be released, the land must pass certain sustainable revegetation requirements, based on the premining soil conditions. For example, if the land was originally prime farmland, the reclaimed land must be productive enough to qualify as prime farmland (Fig. 2). Finally, the reclaimed land surface must be stable; measures must be taken to avoid erosion.

Given the fact that coal has been extracted, it might seem problematic to refill the excavated pit completely. In practice, the problem is typically the reverse; once the consolidated overburden rock is disrupted and broken, it cannot be placed in the same volume of space that it occupied before mining. Excess material (often called "spoil" or "fill") must be placed so as to be stable, either on level ground created during the mining process or in nearby valleys. At some sites, a pit is intentionally left and allowed to flood, providing a pond or small lake.

FIGURE 2 Properly regraded and reclaimed mined land.

This avoids the expense of hauling back spoil material already placed elsewhere, but the issue of water quality in such ponds, and indeed at all mine sites, is a key concern.

At many coal mines, iron sulfide (pyrite) associated with the coal and the overburden strata oxidizes on exposure; this produces water with elevated concentrations of sulfate, acidity, and dissolved metals. If there is sufficient alkalinity present (typically in the form of limestone), the alkalinity will neutralize the acidity. In such cases, the environmental impact on water quality may be relatively minor. However, if there is insufficient alkalinity, the water becomes acidic. This type of water is known both as acid mine drainage (AMD), in most coal-mining regions and especially in the eastern United States, or as acid rock drainage (ARD), in most metal ore-mining regions. The term ARD is also preferred by those who like to point out that the same water quality results from such rock exposure in road cuts and construction projects (e.g., the Halifax Airport). ARD can typically be recognized, wherever it occurs, by the color of the water, i.e., red or yellow-orange, which is caused by dissolved and suspended iron (Fig. 3).

The prediction of postmining water quality is a key component in obtaining a mining permit. This is typically determined by analyzing rock cores for pyrite and alkalinity-producing rock strata. Then, using one of several procedures, a prediction is made about whether the water will be acidic. If it appears that it will probably generate AMD, or have other adverse hydrologic consequences, the regulatory agency may deny permission to mine or require the mining company to undertake special measures to decrease the likelihood of AMD. For example, the mining company may be required to handle pyritic strata selectively, and place it in such a manner that it will be less exposed to air and/or water, or to mix it with the more alkaline rock strata, or to import additional alkalinity to the site.

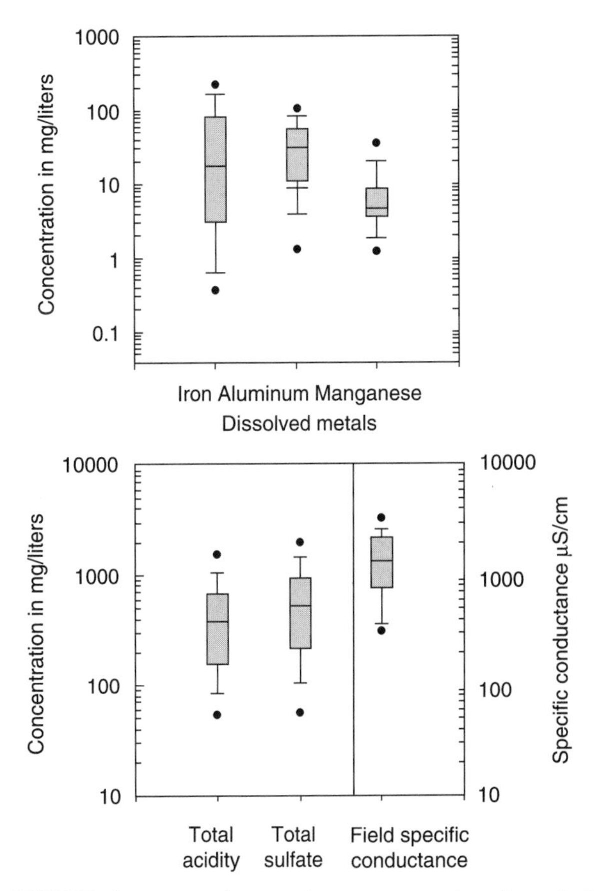

FIGURE 3 Range of contaminant concentrations in typical acidic coal mine drainage.

AMD can contaminate many miles of stream down-gradient from a mine, sometimes rendering it toxic to aquatic life. By regulation, any site that generates water not meeting regulatory standards (typically pH 6–9, no net acidity, iron less than 3 mg/liter, and manganese less than 2 mg/liter) must treat the water before it can be discharged. The water treatment must continue for as long as the untreated water fails to meet these discharge criteria. However, it should be noted that at surface mines, after the pyritic rock is buried, AMD typically moderates. Acid salts formed during exposure to the atmosphere continue to dissolve, but once these are dissipated, acid generation begins to decrease. It may take decades, but water quality, at even the worst of these sites, does improve.

In the United States, AMD is a major problem in Pennsylvania, northern West Virginia, western Maryland, and eastern Ohio, but also occurs at many mine sites in other states. Elsewhere in the world, acid drainage is often associated with coal mining, though the extent of the problem varies with local geology, site conditions, mining methods, etc. In the western United States, AMD is less likely to be a problem, due in part to the lower concentrations of pyrite associated with the strata there and in part due to the fact that the climate is drier. In fact, the lack of adequate precipitation can make reclamation difficult, and also introduces the problem of sodic spoils. Such materials are exposed during mining and contain elevated concentrations of sodium salts, which dissolve and add to the salinity of the soil and the downstream waterways.

Other environmental problems associated with surface mines include fugitive dust (airborne particles that blow away in the wind), ground vibrations and noise associated with blasting, loss or conversion of wildlife habitat, and loss of aesthetics (visual resources). The first two are temporary problems, but the latter two can be temporary or permanent, depending on the eventual land use. However, if an area is viewed as important to the life cycle of an endangered species, permit denial is almost certain. In contrast, mining companies sometimes take advantage of land disturbance and reclamation to enhance wildlife activity; for example, in the United States, an exception was granted to the AOC requirement to allow a mining company to establish an appropriate habitat for a bird that was native to the area but declining in population.

The lack of aesthetics generally attributed to mined and reclaimed land, though it varies with the individual site, generally refers to the fact that mined land is typically reclaimed in a bland and uniform manner. The lack of visual contrast, rather than the actual disturbance, is what is noticed. This is one, of many, objections that citizens often make about mountaintop removal, which is currently the most controversial form of permitted surface mining, and deserves to be specifically mentioned here. Mountaintop removal, or mountaintop mining, is used in mountainous areas such as southern West Virginia to extract multiple seams of coal over large areas. The excess overburden is placed in what were previously valleys, with French drains constructed to handle the intermittent stream flow that might have been there previously, though major drainage patterns are preserved. The original rugged ridge and valley appearance of the land is converted to a more sedate topography, creating usable flat land where before there was none, and typically reducing the hazard of storm-related flooding; local inhabitants, although they may appreciate the jobs brought to the area, find their neighborhood forever changed (some would say ruined). Local biota is of course affected as well. There have been numerous court fights attempting to end the practice of mountaintop removal, and it is not yet clear how the issue will finally be resolved, but an interesting and so far unexplained finding that elevated levels of selenium have been found downstream of West Virginia mountaintop removal operations may turn out to be significant.

2.2 Underground Mines

Although underground mine operations are not as visible as surface mining, their overall environmental impact can be greater than that of the typical surface mine. A key environmental problem is subsidence. Underground mines are large cavities in the rock, and depending on the strength of the intervening strata, the depth of the mine, and the type of mining and roof support, the rock walls can fail, causing cracks and land collapse at the surface. Typically, coal seams at depths greater than about 200 feet are extracted by underground mining methods rather than by surface mining, with the exact depth principally based on the relative amount of coal and overburden. However, before improved technology made surface mining so affordable, the trade-off occurred at much shallower depths; some abandoned underground mines are only 35 feet below the land surface.

In longwall mines, the subsidence is induced as mining proceeds. Most of the subsidence occurs within a few months after mining. Changes in surface elevation can be significant, affecting highways, waterways, etc., though mining beneath such features may be restricted. The extensive fractures and settling also disrupt aquifers, though deeper aquifers typically recover once the void spaces fill with water. In many European countries, where mining is centrally planned and directed, subsidence above longwall mines is typically anticipated and planned for (for example, in how buildings are constructed) years ahead of time. Elsewhere, where mining proceeds based on the decisions of mining companies and local and regional regulatory agencies, subsidence may be anticipated, but is rarely coordinated with other regional development activities. In contrast, more traditional room-and-pillar mining may resist subsidence for decades, failing only as pillars, left to support the overburden rock, erode and then collapse. The lateral extent of the subsidence area may not be as great as above longwall mines, but because the collapse is more localized, there is a greater risk of differential subsidence (for example, one corner of a house subsides more than the rest of the house, severely damaging or destroying it (Fig. 4). The ability to predict the extent of subsidence has improved over the past decade, but this is still a very inexact science.

The effect of subsidence on streams and waterways can be subtle or profound. In the worst cases, subsidence fractures reach the streambed and partially or totally drain the stream, diverting the water underground. Such fractures can be detected using terrain conductivity, and inexpensively sealed using an appropriate grout, but until that is done, the stream may go dry, and mining may be impeded or even temporarily halted. Even where waterways are not affected by subsidence fractures, the changes in slope will cause profiles to change, causing erosion in some areas and sediment deposition in others, and locally affecting stream biota.

Depending on the coal seam and location, AMD generation in underground mines can exceed that at surface mines. Because pyrite forms in a swampy environment, coal seams often contain more pyrite than do the overlying strata. At a surface mine, virtually all of the coal is removed, but in an underground mine, significant amounts of coal remain behind to provide roof support; if the coal is pyritic, it is problematic. In addition, alkalinity that may be present in the overburden strata is not as exposed to dissolution as it is when it is disrupted by surface mining. Finally, an underground mine is essentially a void, and behaves like a well; water flows through the surrounding rock into the mine void. When mining ceases and the water is no longer being pumped, the mine begins to flood and the water table rises. Thus, the volume of AMD that eventually discharges to the surface can be very great, completely overwhelming the buffering capacity of the receiving waterways. If the coal seam is completely inundated, the large mine pool will gradually improve in quality once all of the acid salts are washed away, because inundated pyrite does not continue to oxidize to a significant extent. However, coal seams typically slope (or "dip"); this may mean that only part of the seam is underwater, and the rest is continually exposed to the atmosphere, creating an ideal acid-generating environment that can continue to produce AMD for centuries.

FIGURE 4 An abandoned coal refuse pile that leaches acidic drainage; many of these waste piles can be converted into a resource by burning the material in a fluidized bed combustion unit.

2.3 Refuse Piles

Mined coal often contains other rock types (e.g., shale and clay associated with the coal or lying immediately above or below the coal seam) and impurities (such as pyrite). Such coal is considered "high ash," because the impurities remain behind in the ash after combustion. To improve the coal value, such coal is typically crushed and "cleaned" or "washed" in a preparation plant. This process separates the coal from the waste material, which is transported to a disposal area. This coal refuse is typically very high in pyrite, low in alkalinity, and quite reactive because it has been crushed to a relatively fine grain size. Today, there are strict procedures on how the material must be compacted and covered; even so, AMD is a common problem. Abandoned piles of prelegislation coal refuse, which frequently lie in or near waterways, are problematic (Fig. 5). Such piles can generate extremely acidic mine drainage and sometimes catch on fire. However, many of these old waste piles contain quite a bit of coal, given the fact that coal cleaning procedures were not always as efficient as they are nowadays. Such piles can be converted into a resource, because the material can be burned in a fluidized bed combustion (FBC) unit.

2.4 The Environmental Legacy Associated with Abandoned and Orphaned Mines

Currently, site reclamation is planned for during the permitting process and is incorporated into the mining operation. However, this was not always the case. Many mine sites were legally abandoned in an unreclaimed or poorly reclaimed condition because mining was completed before environmental

FIGURE 5 A house damaged by subsidence caused by the collapse of pillars in an abandoned underground mine.

FIGURE 6 Excavation and extinguishment of a burning coal refuse pile.

regulations went into effect. These abandoned mines are scars on the landscape and cause most of the water pollution attributed to mining. These old mines are considered abandoned because, in most countries, no one is legally required to reclaim the land or to treat the water. A similar problem occurs at mine sites that are or were operated by companies that have gone bankrupt. In theory, "bonds" (money held in escrow or guaranteed by a third party, required in many countries) posted by the companies will pay for reclamation and water treatment; however, at most sites, the amount of money required to reclaim the site exceeds the required bonds. In the United States, forfeiting a bond means that the company can no longer mine, so it is serious, but sometimes the costs of environmental compliance exceed the resources that a company has available. Several orphan underground mines are located near the Pennsylvania–West Virginia borders. These mines are gradually flooding with AMD, and are projected to begin discharging contaminated water to tributaries of the Monongahela River within the next 2 years. The anticipated contaminant load (contaminant concentration x flow) is such that it would have a major impact on water quality for many miles of the Monongahela River downstream; state and federal agencies are scrambling, trying to figure out what to do.

Regardless of whether a mine is abandoned or orphaned, the land remains inadequately reclaimed and the water that discharges from the site is typically not treated. In addition to the potential environmental problems associated with active mines, and already discussed, abandoned and orphan mines often have additional problems that can affect the health and safety of people who live in the area. For example, open pits with steep cliffs inevitably attract children, as do unsealed, unsafe underground mines. Unreclaimed waste rock piles and mine dumps pose a risk of slope failure and landslides. Another potential problem (discussed in detail in the next section) is mine fires, which are more typically viewed as a health and safety problem. However, at abandoned mines, such fires can smolder for decades, generating toxic gases that can flow through cracks into basements, and can accelerate subsidence events (Fig. 6).

So how are these problems dealt with? In the United States, the Surface Mining and Reclamation Act of 1977 (as extended and amended by the Abandoned Mine Reclamation Act of 1990 and the Energy Policy Act of 1992) did more than require mine operators to maintain environmental standards during mining and reclamation. It also imposed a fee of 35 cents/ton of coal mined by surface methods, 15 cents/ton of coal mined underground, and 10 cents/ton of lignite. These fees go into a fund managed by the Office of Surface Mining (OSM) in the U.S. Department of the Interior. This money can be used by the states and the OSM to address problems at abandoned mine sites. Most of these funds have gone to remediate sites that are hazardous to people and property, but increasingly funds are also being used to remediate environmental problems. These funds cannot be used to resolve problems at orphan mines.

In addition, regional and national governments have often provided additional funds for environmental remediation of abandoned sites. In particular, Pennsylvania should be singled out for its unique approach; it encourages the formation of watershed associations, which can apply for funds to remediate abandoned mines. Through the efforts of these enthusiastic volunteers, Pennsylvania accomplishes much more remediation than would be otherwise possible.

3. LAND RECLAMATION

Under natural conditions, a landscape represents a balance of geomorphic processes; this dynamic stability is disrupted by mining. In this context, the goal of reclamation is the reestablishment of the steady state. In the United States, regulations tightly control the reclamation process at active operations, dictating the slope (AOC), the extent of revegetation, and the rate at which reclamation and revegetation must proceed. Failure to comply results in loss of bond money and/or fines. Other countries (e.g., Canada) have more flexibility built into their regulatory structure. They consider the imposition of AOC inappropriate at many sites—for example, in areas where flat land would be desirable; enhanced flexibility allows such aspects to be negotiated. In the United States, exceptions can be granted by the OSM, but are atypical. Instead, operators have learned to operate within the limits of the regulation. For example, they may choose to reduce erosion potential on steep slopes by installing terraces and manipulating infiltration rates.

When reclamation agencies are attempting to reclaim derelict or abandoned operations, no attempt is typically made to restore the land to AOC. Instead, the primary intent is to remove potential hazards (e.g., extinguish potentially dangerous mine fires, seal mine openings); secondarily, the intent is to bury acid-forming material, and, finally, to create a stable surface that will resist erosion with minimal maintenance. Funds for reclamation of abandoned sites are quite limited and represent a small fraction of what would be necessary to reclaim all abandoned mine sites.

Many abandoned sites have developed a natural, if sometimes sparse, vegetative cover of volunteer species. In addition, an unusual incentive has recently developed that may encourage companies to reclaim abandoned sites that have not naturally revegetated. As pressure gradually grows on companies to lower carbon emissions, proposals have been made that credit should be given for revegetating barren lands, because this would sequester carbon from the atmosphere. Arid deserts and barren abandoned mine sites are the two most likely types of places for such an effort, if and when it becomes codified.

Another aspect of mined land remediation is subsidence control. Because subsidence events above old room-and-pillar operations commonly cluster, when subsidence begins in an area, state and OSM emergency personnel are quickly mobilized. Typically, they attempt to backfill, inject, or stow, hydraulically or pneumatically, as much solid material into the mine void as possible, figuring that reducing the void volume with fine solids (sand, fly ash, etc.) will both decrease the amount of subsidence that will occur and reinforce pillars that might otherwise fail. However, because the mine voids that they are injecting the material into are often flooded, it is sometimes difficult to ascertain if these efforts make much of a difference. An alternative option pursued by many landowners in undermined areas is government-subsidized subsidence insurance. This insurance fund pays property damage after subsidence has occurred.

4. MINE WATER REMEDIATION

Before the passage of regulations dictating mined land reclamation and mine water discharge standards, streams and rivers down-gradient of mine sites were often contaminated with high levels of suspended and dissolved solids. In the eastern United States, AMD was also a major problem. Nowadays, streams and rivers near active mine sites have much less of an impact. Sediment ponds are constructed to collect suspended solids and if the mine water does not meet regulations, chemicals [typically lime, $Ca(OH)_3$] are added to neutralize acidity and precipitate dissolved metals. Consequently, the remaining sources of contaminated water discharging to streams and rivers in mined areas are generally from abandoned or orphan mines. However, before moving on to water remediation at such sites, the techniques used to prevent, ameliorate, or remediate mine water problems at active operations need to be addressed. As alluded to previously, a powerful tool in preventing AMD is the permitting process. Variation exists in how much certainty of AMD generation a regulatory agency requires before a permit is denied. In the United States, Pennsylvania is the most conservative, and can legitimately boast that it has improved the accuracy of its permitting decisions from about 50% in the 1980s to 98%; however, it should be noted that this definition of success is based on AMD generation at permitted operations (following reclamation) and does not include mine sites that had permits denied and might not have produced AMD if allowed to operate.

In contrast, adjacent states (and many areas outside of the United States) permit a higher percentage of mines to operate, but generally require that measures be taken to reduce the risk of acid generation. In addition to the selective handling of overburden and the importation of alkalinity from off-site mentioned earlier, minimizing exposure of the pyrite to either oxygen or water can decrease the amount of acidity generated. Restricting exposure to oxygen generally means that the pyritic material must be placed beneath the eventual water table, which is difficult to do at many surface mines. Alternatively, the operator can attempt to place the pyritic material in an environment where it is dry most of the time, well above the potential water table and capped by material of low permeability. Compaction can also be used to reduce permeability, though not on the final soil surface, where compaction makes it difficult to establish a vegetative cover. Water can be diverted around the mine site by intercepting overland flow with ditches, and by constructing drainage ways along the final highwall to intercept groundwater and prevent it from flowing through the overburden material.

At many sites, mining companies have been given special permission to remine old abandoned mines to improve water quality. The mining companies harvest coal left behind by the old operations. For example, old room-and-pillar mines sometimes left as much as 50% of the coal behind to support the roof rock. Some of these old mines are relatively shallow and can be inexpensively surface mined using modern machinery. Because the coal pillars are sources of acid generation, removing them generally improves water quality. Similarly, old surface mines could not economically remove as much overburden as is now possible. The exposed highwall can now be economically mined with modern machinery. The mining companies are required to reclaim the land to current standards, but their water discharge requirements typically only require them to at least meet the water quality that existed before the remining operation. In most cases, the water quality improves and the land is reclaimed, at no cost to the public.

A more exotic approach of controlling AMD involves the inhibition of iron-oxidizing bacteria. These ubiquitous bacteria normally catalyze pyrite oxidation; inhibiting them reduces acid generation significantly. In practice, however, it is difficult to do this. The only cost-effective approach that has been developed involves the use of anionic surfactants (the cleansing agents in most laundry detergents, shampoos, and toothpaste); their use selectively inhibits the iron-oxidizing bacteria by allowing the acidity that they generate to penetrate through their cell walls. This approach has been used effectively to treat pyritic coal refuse, reducing acid generation 50–95%. Slow-release formulations have been developed for application to the top of the pile before topsoil is replaced and the site is revegetated.

FIGURE 7 Mine water can be directed through a specially constructed wetland to passively improve the water quality.

Discharge criteria must be met, and if the at-source control measures are not completely effective, some form of water treatment is required. Sometimes water treatment is required only during the mining and reclamation operation, and water quality improves soon after reclamation is completed. Sometimes the water quality remains poor. Chemical treatment is simple and straightforward, but is expensive to maintain for long periods.

Passive and semipassive water treatment technologies, though by no means universally applicable, are an option at many sites. These techniques developed as a result of observations at natural and volunteer wetlands, which were observed to improve the quality of mine drainage. The simplest of these techniques, appropriate for near-neutral mine water that is contaminated with iron, involves the construction of shallow ponds, planted with plants that will tolerate the water, such as *Typha* (commonly called cattails in North America) (Fig. 7). The iron in the water oxidizes and precipitates in this constructed wetland instead of in the stream, allowing natural stream biota to repopulate. These constructed wetlands cannot be viewed as equivalent to natural wetlands; they have been designed to optimize water treatment, and any associated ecological benefits, though they may be significant, have to be viewed as secondary.

Water that is acidic can be neutralized by the inexpensive addition of alkalinity. Several passive methods have been developed, generally using limestone and/or sulfate-reducing bacteria. Alkaline waste products (e.g., steel slag) have also been used. Water quality, site considerations, and flow dictate which approach is most cost-effective at a given location.

The realization that passive techniques can be used to treat low to moderate flows of mine water has made it possible for state reclamation agencies and local watershed associations to remediate mine water at abandoned and orphan mines without a long-term financial commitment for chemicals. However, highly contaminated AMD or high flows still cannot be cost-effectively treated passively. But the technology is continuing to evolve. Semipassive techniques, such as windmills and various water-driven devices, are being used to add air or chemical agents to mine water, extending and supplementing the capabilities of the passive treatment technologies.

Contaminated water in abandoned or orphan underground mines represents the ultimate challenge because of the large volume of water that must be treated. One option currently being explored in the United States is whether or such mine pools can be used as cooling water for a power plant. Locating new power plants is becoming more difficult because of the water needed for cooling purposes. A large mine pool could be a good resource for such an operation. The power plant would have to treat the mine water chemically, but this cost could be partially subsidized by the government, which would otherwise have to treat the water chemically or allow the mine discharge to flow untreated into the local streams and rivers.

5. OTHER ENVIRONMENTAL ASPECTS

The issue of aesthetics has already been introduced. In many ways, it can be the most challenging obstacle to popular acceptance of mining in an area, because, to many, anything less than complete restoration of the land is unacceptable. However, mine operators have learned to minimize the aesthetic impact of mining by avoiding extremely visible sites, by using screens (e.g., trees left in place as a visual barrier or revegetated topsoil stockpiles placed in locations where they hide the mine pit from view), by minimizing the duration of impact, and by ensuring that regrading and revegetation proceeds as quickly as possible.

Finally, the issue of ultimate land use of the reclaimed land must be considered. Nowadays, this is actually addressed as part of the permitting process. Some mine operators have successfully created planned industrial sites, some have created highly productive farmland, but much of the mined land is reclaimed for wildlife use. Costs for seed and plantings are relatively low, and plans are often coordinated with local conservation groups to reclaim the land in an appropriate manner for wildlife use. To be effective in this regard, these reclaimed lands must be physically connected to other areas where such wildlife currently exists; isolated fragments of habitat will not serve the desired purpose of restoring wildlife use disrupted by mining activity and road construction. With regulatory approval, ponds can be left in locations where they will serve migratory bird populations or provide watering areas for permanent wildlife populations.

SEE ALSO THE FOLLOWING ARTICLES

Coal, Chemical and Physical Properties • Coal Industry, Energy Policy in • Coal Mining, Design and Methods of • Coal Mining in Appalachia, History of • Coal Preparation • Coal Storage and Transportation • Nuclear Power Plants, Decommissioning of • Uranium Mining: Environmental Impact

Further Reading

Berger, J. J. (1990). "Environmental Restoration." Island Press, Washington, D.C.

Brady, K. B. C., Smith, M. W., and Schueck, J. (eds.). (1998). "Coal Mine Drainage Prediction and Pollution Prevention in Pennsylvania." Pennsylvania Dept. of Environmental Protection, Harrisburg, Pennsylvania.

Brown, M., Barley, B., and Wood, H. (2002). "Minewater Treatment." IWA Publ., London, England.

Kleinmann, R. L. P. (ed.). (2000). "Prediction of Water Quality at Surface Coal Mines National Mine Land Reclamation Center." West Virginia University, Morgantown, WV.

Marcus, J. J. (ed.). (1997). "Mining Environmental Handbook." Imperial College Press, London, England.

PIRAMID Consortium. (2003). Engineering Guidelines for the Passive Remediation of Acidic and/or Metalliferous Mine Drainage and Similar Wastewaters, 151 pp., accessible at http://www.piramid.info.

Skousen, J., Rose, A., Geidel, G., Foreman, J., Evans, R., and Hellier, W. (1998). "Handbook of Technologies for Avoidance and Remediation of Acid Mine Drainage." National Mine Land

Reclamation Center, West Virginia University, Morgantown, West Virginia.
Watzlaf, G. W., Schroeder, K. T., Kleinmann, R. L. P., Kairies, C. L., and Nairn, R. W. (2003). "The Passive Treatment of Coal Mine Drainage." U.S. Department of Energy (CD available on request), Pittsburgh, Pennsylvania.
Younger, P. L., Banwart, S. S., and Hedin, R. S. (2002). "Mine Water." Kluwer Academic Publ., London, England.

Coal Mining, Design and Methods of

ANDREW P. SCHISSLER
The Pennsylvania State University
University Park, Pennsylvania United States

1. Introduction
2. Surface Mining Methods and Design
3. Underground Methods and Design

Glossary

area mining A surface system that removes coal in areas.

contour mining A surface system that removes coal following the contours of hillsides.

extraction ratio The ratio of coal mined divided by the coal originally in place before mining commenced, expressed as a percentage.

full extraction mining An underground system that removes the coal in large areas causing the void to be filled with caved rock.

interburden The rock that occurs in between coal seams.

overburden The rock that occurs over coal seams.

partial extraction mining An underground mining system that removes coal in small areas in which the roof rock remains stable.

raw in place tons The total weight of coal contained within coal seams of a reserve, including in-seam impurities.

raw recoverable tons The total weight of coal, including in-seam impurities that would be mined by a specific underground or surface mine plan applied to a reserve.

salable tons The total weight of coal mined and sold. The impurities are removed by surface coal processing methods applied at the mine.

strip mining A surface system of mining that removes coal in rectangular and parallel pits.

The design and methods of coal mining is made up of the use of land, labor, and capital to remove, mine, or extract coal from a reserve, utilizing a planned and systematic approach called a mine plan. A reserve of coal is represented by a seam of coal, which is a layered deposit of the mineral. A coal reserve is usually described by the quantity of weight, expressed in tons contained within a specified boundary in plan. Tonnage for a reserve can be quantified several ways that include raw in-place tons, raw recoverable tons, and salable tons.

1. *INTRODUCTION*

1.1 Coal Geology

Coal seams originate and are formed by the consolidation of biota, minerals, and natural chemicals through geologic time. Compression, heat, sedimentation, erosion, and chemical energy are agents of the coal formation process.

Multiple coal seams can exist in a reserve, with each coal seam separated by sedimentary rock interburden. Coal, and the rock above and below the coal seam, is primarily layered sedimentary rock in origin, as opposed to igneous or metamorphic rock. The noncoal layers are usually sandstone, limestone, shale, siltstone, and mudstone. The layers can be very symmetric and consistent in thickness, or they can appear in the reserve in nonregular patterns such as sandstone channels, which can exist above the coal seam or replace the coal seam by scouring during geologic formation. Sandstone channels can also create vertical downward displacement of the coal seam by differential compaction. Chemical and heat processes during mineral formation can also produce small pockets of nonsedimentary rock within the coal seam and surrounding rock such as iron pyrite, mica, and other minerals. These minerals, depending on their hardness, can interrupt the mining process as most coal mining methods are designed to extract soft mineral having uniaxial compressive strength less than 6000 psi to 8000 psi.

The processes of the geologic environment in which the coal was formed determines the end characteristics of the coal seam, such as thickness, seam pitch, energy value (generally quantified as BTU per lb), nonvolatile content, moisture content, depth below the surface, structure (faults, folds, rock fractures), the presence of an igneous intrusives called dikes, and characteristics of the rock immediately above and below the coal seam. All of these characteristics affect the mine design and methods. Layered coal seams and surrounding rock thickness range from less than 1 ft to over 150 ft. Figure 1 shows a typical cross section of coal seams interspersed in layers of sandstone and shale.

1.2 Mining Methods

Coal is mined by three methods: surface, underground, or *in situ*. The percentage of coal produced by each mining method can be identified from production statistics kept by the Energy Information Administration (EIA), a U.S. agency. According to the EIA, in 2002 745 surface mines produced 736 million tons (67% of total U.S. production) and 654 underground mines produced 357 million tons (33% of total U.S. production). For practical purposes, coal produced from *in situ* mining methods is nil.

Coal mine design and methods are chosen based on the reserve characteristics. Where the coal seam is near the ground surface, surface mining methods prevail over underground methods. Generally, surface mining methods are less expensive than underground methods when the strip ratio is less than an economic limit. Strip ratio is the quantity of overburden (expressed in cubic yards) that has to be removed to uncover 1 ton of coal. Companies change mining method from surface to underground when the strip ratio from an existing surface mine increases due to seam pitch, which in turn increases production cost. At the economic limit, the cost of production from surface methods equals underground methods.

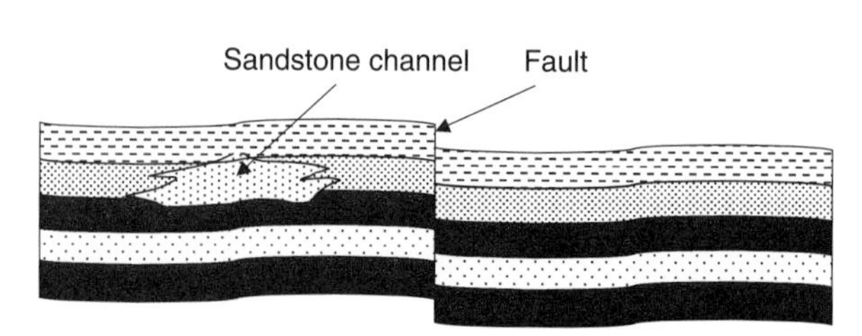

FIGURE 1 Typical cross section of shale, siltstone, coal, sandstone, and coal reserve.

2. *SURFACE MINING METHODS AND DESIGN*

The surface mining method of coal involves three ordered steps: overburden removal, coal seam removal, and reclamation. The first step in surface mining is to remove the rock over the coal seam called the overburden (Fig. 2).

Overburden is first drilled and fractured with explosives. The overburden is then removed and displaced to a previously mined pit, where the coal has already been removed, or to other disposal areas. Other disposal areas can include adjacent hollows, or topographic lows. After the overburden is removed, the coal seam that is now exposed is removed by digging methods and hauled or conveyed to surface facilities that process the coal for shipment. Downstream transportation such as railroads or overland trucks hauls the coal to markets. The third step in surface mining is reclamation. Reclamation in surface mining consists of placing the overburden rock into the hole, pit, or hollow, and recovering the rock with topsoil that was removed and stockpiled before the overburden removal began. Vegetation is replanted in the topsoil to complete the reclamation process.

Overburden removal, coal seam removal, and reclamation occur in order, and simultaneously within the life of a mine or reserve. In other words, a mine does not remove all of the overburden in a deposit before removing the coal. Surface mine design involves the planning of a series of pits *within* the overall deposit boundary that allow overburden removal, followed by coal seam removal, which is then followed by reclamation. The planning process of these sequential operations is called operations planning or sequencing.

The primary surface mine design parameter and driver of unit production cost is the strip ratio. Strip ratios for minable coal reserves range from 1:1 to over 35:1. A decrease in the strip ratio will decrease

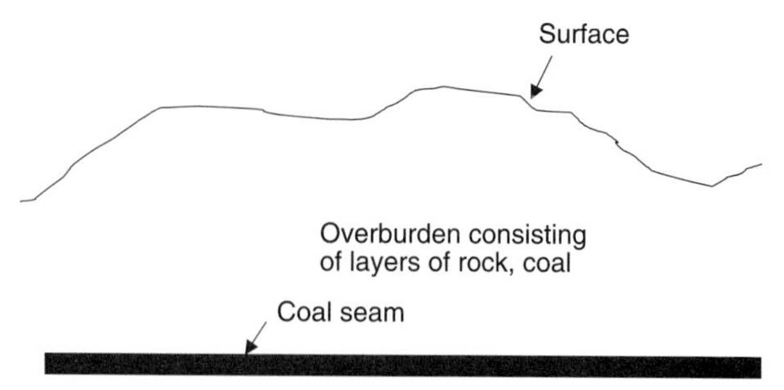

FIGURE 2 Step 1 in the surface mining process, overburden removal (in cross section).

the cost of producing the coal. Underground mining methods are employed when the amount of overburden that would have to be removed by surface mining methods is cost prohibitive (i.e., the strip ratio is beyond an economic limit).

Surface mine design is determined by the stripping ratio, the overall shape of the reserve from a plan view, and the type of rock. Three main surface methods are utilized, with each having the same common steps of overburden removal, coal seam removal, and reclamation. The three surface mining methods are contour stripping, area stripping, and strip mining.

2.1 Contour Stripping

Coal seams and associated overburden can outcrop on the edges of hills and mountains. Figure 3 shows a cross-section of a typical contour mine plan. Contour surface mining takes advantage of the low strip ratio coal reserves adjacent to the outcrop. Contour mining equipment includes a drill, loader, bulldozers, and trucks. The steps in contour mining begin with drilling and blasting a triangle shaped portion of overburden (in cross section) above the coal to be mined. The fractured overburden is then loaded into rubber-tired trucks. The loading equipment utilizes buckets such as front-end loaders, front-loading shovels, and backhoes. The coal seam is now uncovered and is loaded with the same type of bucket-equipped machinery as loaded the overburden. Some coals require drilling and blasting as the breakout force required to remove the coal without fragmentation by explosives is beyond the capabilities of the bucket-equipped machinery. Dumping the overburden into the pit begins reclamation. The overburden is usually hauled around the active pit where coal is being removed. The overburden is dumped and regraded to replicate as much as possible the original contour or slope of the hillside. Bulldozers grade and smooth the overburden in preparation for topsoil application. Topsoil is placed on the smoothed overburden and replanted with vegetation. Contour stripping gets its name because the mining operation follows the contour of where the coal outcrop meanders along the hillside (Fig. 4).

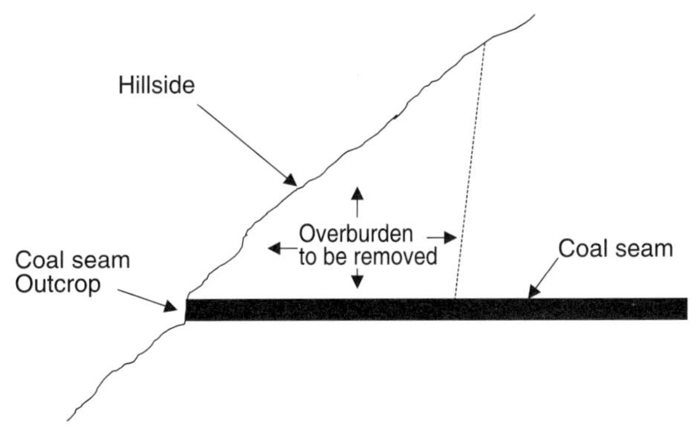

FIGURE 3 Contour mine design, surface mining method (in cross section).

The advantages of contour mining include the ability to control strip ratio during the mining process. The strip ratio is controlled by the height of the triangle portion of overburden removed. Controlling the ratio allows the unit cost of production to be largely controlled. The disadvantages of contour mining include the difficulty of maintaining the stability of steep, reclaimed hillsides and the limited width for effective operations on the coal seam bench.

2.2 Area Stripping

Area stripping is a surface mining method that is usually employed in combination with strip mining. Area stripping utilizes truck and shovels to remove the overburden down to a predetermined depth in the layers of rock and coal. The strip mining method continues to mine the overburden below the depth at which the area was area-stripped. Draglines or continuous excavators, which are used to strip the overburden from the lower seams, are large pieces of equipment and require large, level areas from which to work. Draglines and continuous excavators have dig depth limitations. Area stripping allows the draglines and continuous excavators to operate within their depth limitations. When area stripping is used in combination with draglines and continuous excavators, area stripping is called prestripping. The purpose of prestripping is to provide a flat surface for the dragline or excavator on which to set, move, and operate. The depth of overburden rock removed by area stripping is governed by dig depth capacity of the equipment utilized for subsequent strip mining.

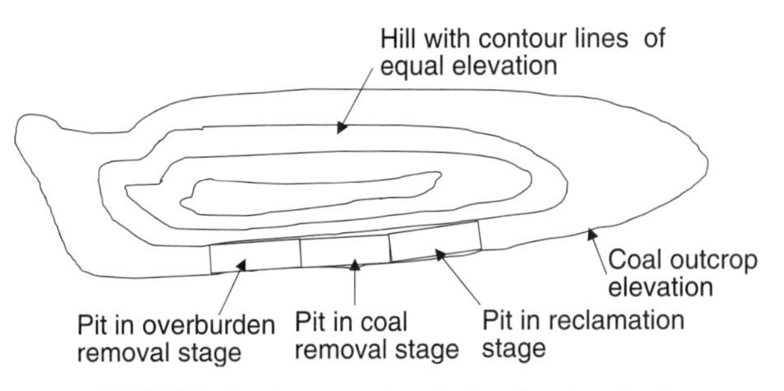

FIGURE 4 Contour mine design (in plan view).

Usually, the prestripping mine plan is arranged in rectangular blocks of overburden to be removed that coincide with pits to be mined by stripping methods underneath the prestripping area (Fig. 5).

Area stripping application includes the removal of overburden in reserves that are not conducive to strip mining because the lack of rectangular areas (in plan). The machinery used for area stripping such as shovels, front-end loaders, and trucks allow for flexibility in the mine design layout.

2.3 Strip Mining

Strip mining is employed in coal reserves where the overburden is removed in rectangular blocks in plan view called pits or strips. The pits are parallel and adjacent to each other. Strip mining is fundamentally different from contour or area mining on how the overburden is displaced, called spoil handling. In contour or area stripping, the overburden is hauled with different equipment than what digs or removes the overburden. In strip mining, the overburden is mined and moved by the same equipment: draglines or continuous excavators. The movement of overburden in strip mining is called the casting process.

The operating sequence for each pit includes drilling and blasting, followed by overburden casting, then coal removal. Some overlap exists in operational steps between pits. Draglines and continuous excavators move or displace the overburden from the active pit to the previous pit that has had the coal removed.

The primary planning mechanism used in strip mining is the range diagram, which is a cross-sectional plan of the shape of the pit in various stages of mining. The range diagram allows the dragline or continuous excavator equipment characteristics of dig depth, reach, and physical size to be placed on the geologic dimensions of depth to seams (overburden), and depth between seams (interburden). By comparing machinery specifications with dimensional characteristics of the geology, the mine designer can plan the pit width and dig depth (Fig. 6).

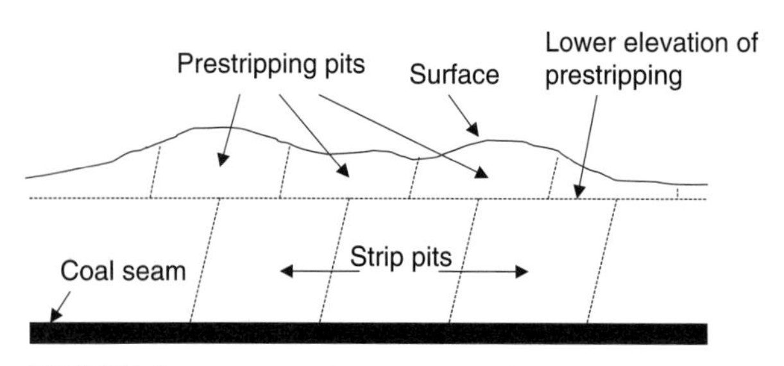

FIGURE 5 Prestrip surface mining method (in cross section).

As the dragline or continuous excavator moves the overburden to the adjacent empty pit where the coal has been removed, the rock swells in volume. Earth or rock increases in volume, called the swell factor, when the material is removed from its *in situ* or in-ground state and placed into a pit or on the surface. The range diagram allows the mine planner to identify the equipment dump height required to keep the displaced overburden (spoil) from crowding the machinery and mining operations. In certain cases of mining multiple coal seams from one pit, a coal seam can provide the boundary between the prestrip and strip elevations.

In a relatively new technique that originated in 1970s to early 1980s, explosives are used to move or throw the overburden into the previous pit in a process called cast blasting. The difference in the quantity of explosives required to fragment rock in place versus fragment and cast or throw the rock across the active pit and into the previous pit is cost-effective. Many surface strip mines use explosives to move overburden in addition to the primary swing equipment (dragline or continuous excavator), displacing up to 35% of the overburden by cast blasting. When cast blasting is used, the dragline may excavate from the spoil side of the pit, sitting on the leveled, blasted overburden.

Surface mine design principles emanate from the operational characteristics of surface mining, which are drilling and blasting, spoil handling, coal removal, and haulage. Except in a few circumstances, overburden in surface mining requires the rock to be fractured by explosives to allow it to be excavated. The goal of drill and blast design is to optimize rock fracturing, which optimizes digging productivity. Fracturing is optimized by using the correct amount of explosive per cubic yard of overburden employed in the drill hole spacing in plan view. The amount of explosive in weight per cubic yard of overburden is called the powder factor. Drill and blast design is

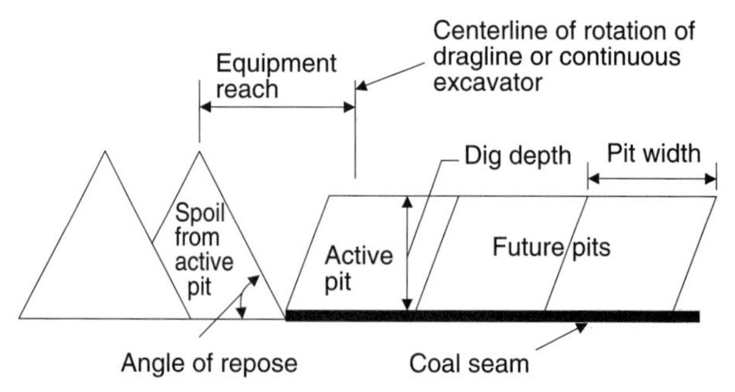

FIGURE 6 Range diagram for the strip mining method (in cross section).

accomplished by empirical methods and by experience. The drill hole layout and powder factor change when cast blasting is utilized.

Spoil handling design is of critical importance, as this function is usually the most expensive cost element in surface mining. When the surface mining method utilizes trucks, spoil handling is designed to minimize the overall haul distance for logical units of spoil volume, which may be driven by pit layout, topography, or area stripping requirements. Mine plan alternatives are evaluated to minimize the distance that spoil volumes are moved from the beginning centroid of mass to the ending centroid of mass. Spoil handling design goals for strip mining surface methods that utilize draglines and continuous excavators also include the minimization of spoil haulage distance. For the dragline, the average swing angle is identified by evaluating alternative mine plan layouts. The goal is to minimize the swing angle, which maximizes productivity.

The goals of coal removal and haulage design in surface mining include minimizing the distance coal is hauled from pits to surface processing and loadout facilities in near term years, locating haul road ramps out of the pits to minimize interference with overburden removal, and engaging excavation practices and equipment that minimize coal dilution by mining noncoal rock floor.

Surface mining has two design parameters that affect mine cost, which are minimizing rehandle and maximizing pit recovery. Rehandle occurs when overburden is handled twice and sometimes multiple times during excavation and spoil placement. Having 0% rehandle of the original inplace overburden is not achievable because of inherent design requirements of surface mining such as ramps into the pit and mining conditions such as sloughing ground that covers the coal. Simulating alternative mine plans and anticipating where overburden will be placed can minimize rehandle. Rehandle can more than double the cost of mining portions of the overburden.

The goal of coal pit recovery is to obtain as close to 100% as possible. One method to maximize pit recovery is to minimize drill and blast damage to the top of the coal. Drill and blast damage is reduced by stopping the drill holes from touching the coal seam or by placing nonexplosive material in each drill hole, called stemming. Pit recovery is also maximized by matching the pit width with the characteristics of the machinery used to extract the coal. Again, the range diagram as a planning tool is used in this evaluation.

3. UNDERGROUND METHODS AND DESIGN

Underground coal mining methods fall in two categories: full extraction and partial extraction. Full-extraction underground mining creates a sufficiently large void by coal extraction to cause the roof rock to cave into the void in a predictable and systematic manner. Partial-extraction systems disallow caving by mining reduced areas which sustains the roof rock above the tunnel in a stable or uncaved *in situ* state. When coal seams are completely below the ground surface, vertical shafts and tunnels driven on angles to the seam below called slopes are mined through the overburden rock to access the seam. When the edges of coal seams are exposed to the surface, called outcrops, access to the main reserve can be obtained by the creation of tunnels in the coal seam. Coal outcrops occur when sedimentary rock and coal is layered within a hill or mountain and the rock and coal outcrops above the surface ground level of drainages that define the topography. The main coal reserve is divided up into areas or panels (in plan view) that allow orderly extraction of the coal by the chosen equipment and panel mining method.

Tunnels excavated in coal are called entries. Entries are mined to gain access to the reserve and the production panels. Entries are excavated adjacent to one another in sets of entries. The sets of entries have additional tunnels, called crosscuts that are normally excavated at right angles to the entries. The entries and crosscuts form pillars of coal. The entire set of entries, crosscuts, and pillars, are referred to as mains, submains, or gateroads, depending on the degree of proximity to production panels. Entries excavated in production panels are sometimes called rooms.

3.1 Panel-Mining Methods

The methods to mine panels of coal in an underground mine layout are partial extraction using continuous miners to mine rooms, partial extraction using conventional mining methods to mine rooms, full extraction using continuous miners to mine rooms and pillars, full extraction using longwalls to extract a block of coal outlined by entries around the panel perimeter, and other underground methods.

3.2 Partial Extraction with Continuous Miners

Continuous miners are machines that mechanically mine or extract the coal without the use of explosives. Continuous miners consist of a rotating drum having carbide-tipped steel bits that strike the coalface. As the bits strike the face, tension cracks are created on the coal surface. These cracks separate blocks of coal. The blocks of coal are propelled out of the face by the bits and drop into a portion of the continuous miner called the gathering pan. The action of the bits extracting the coal by the creation of tension cracks is called mechanical mining. Coal is conveyed through the continuous miner and is loaded onto vehicles that haul the coal to be further transloaded onto conveyors, which convey the coal out of the mine to surface processing facilities.

Within panels, continuous miners and associated equipment mine entries and crosscuts leaving the pillars of coal. Panels are arranged adjacent to one another, usually with the long axis of the panel dimension 90° to the main or submain entries (Fig. 7). Panel size is determined by the size and shape of the overall reserve in plan view.

3.3 Partial Extraction with Conventional Methods

In partial extraction with conventional methods, explosives fragment the coal at the face in contrast to continuous miners where bits or tools mechanically mine the coal. In conventional mining, holes are drilled in the coalface allowing explosives to be loaded into the holes. The explosives fracture and loosen the coal allowing loading and removal of the coal into the haulage system of the mine. Rubber-tired scoops load the coal. Thus, entries and crosscuts are created by the conventional method. Panel layout for conventional methods is the same as for continuous mining.

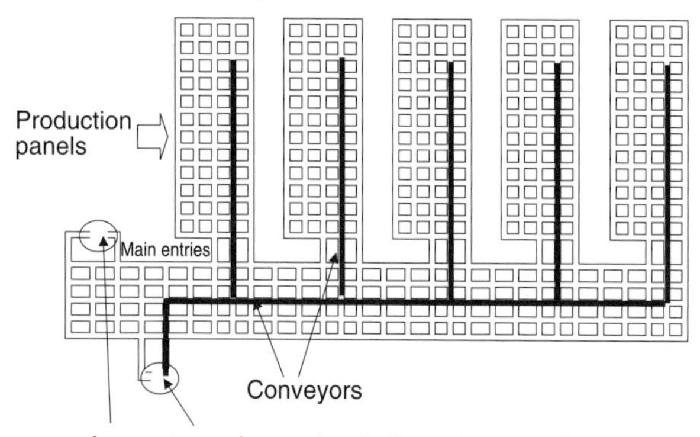

FIGURE 7 Partial extraction with continuous miners (plan view).

3.4 Full Extraction Using Continuous Miners

Full extraction using continuous miners is the same as partial extraction except that the pillars that are surrounded by entries and crosscuts in the panels are mined and removed. By removing the pillars, a sufficiently large area is created that causes the roof rock to become unstable and fall into the void. The caving process in full extraction techniques is an integral part of the mining method. Figure 8 shows the general method of individual pillar extraction.

The stumps shown in Fig. 8 are small coal pillars that are left after the extraction process. The stumps are necessary to provide roof support for the pillar during mining, but are not sufficiently large to inhibit the caving process.

3.5 Full Extraction Using Longwalls

The longwall method utilizes a machine group that consists of shields, face conveyor, stage loader, and shearer. This group of equipment is installed along one entry that bounds the short dimension of a rectangular block of coal. The block of coal is entirely surrounded by entries that outline the perimeter, and this block area is called a longwall panel. The longwall equipment extracts the coal from this rectangular panel. Each longwall panel is separated from an adjacent panel by a group of entries called the gateroad system. These entries, as

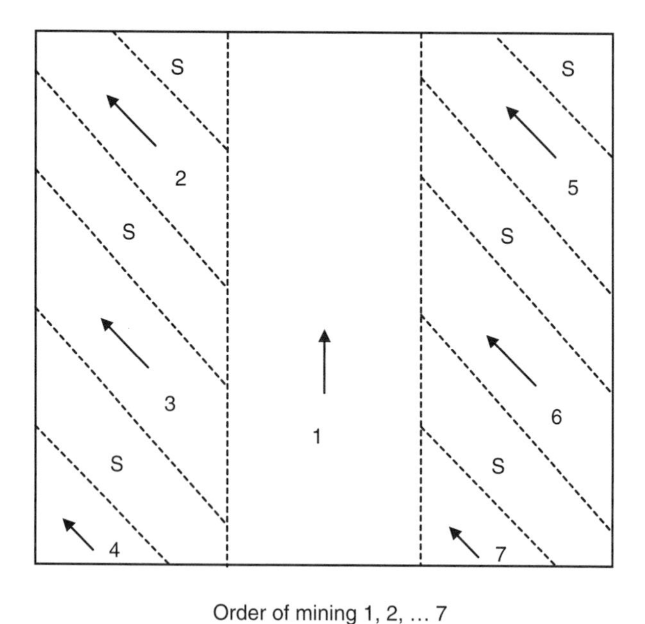

FIGURE 8 Typical pillar extraction sequence (plan view).

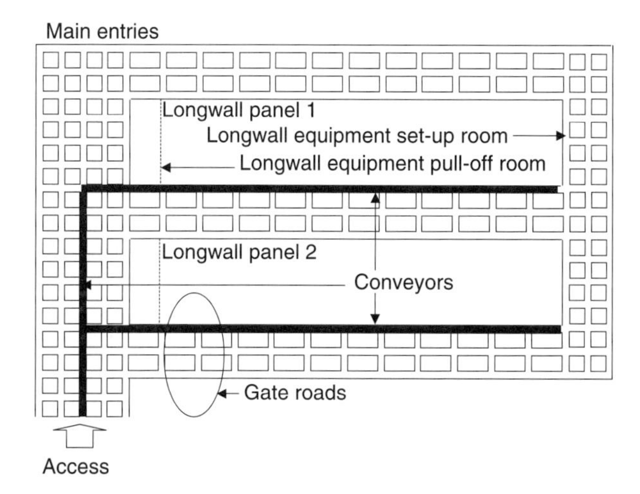

FIGURE 9 Longwall mine plan (plan view).

with the mains and submains, are mined with continuous miners. Once the longwall extracts the coal from the rectangular block, the equipment is disassembled at the pull-off room and moved and reassembled at the end of the next panel of coal in the startup room (Fig. 9).

Access to a coal seam for the longwall mining method is accomplished by outcrop access, shafts, or slopes. Gateroad systems can consist of one to five entries. Figure 9 portrays an example mine plan consisting of two longwall panels. The number of longwall panels contained within a mine plan is determined by the shape of the reserve and by safety considerations. The larger the reserve, the more longwall panels can be planned parallel to each other. Longwall panels are sometimes grouped together into subsets called districts. Longwall districts within an overall mine plan are separated by large blocks of coal called barrier pillars. Barrier pillars compartmentalize the reserve. A compartmentalized reserve allows worked out areas to be sealed, which increases safety and efficiency with respect to handling water, gas, and material.

Longwall panels range in size (in plan view), from 500 ft wide by 2000 ft long to over 1200 ft wide by 20,000 ft long. Size is determined by the reserve and the availability of capital to buy equipment. Some large underground mines have as much as 50 longwall panels separated into six or more districts.

3.6 Other Underground Methods

Other underground methods still incorporate entries, crosscuts, and pillars, but differ from previous methods by the method to extract the coal from the working face. They are hand loading and hydraulic mining. Hand loading is using human-powered pick and shovels to remove and load coal from a face into haulage equipment. This method is not used in the United States.

Hydraulic mining utilizes the force of water sprayed under high pressure to extract coal. Hydraulic mining utilizes continuous miners to mine entries to gain access to production panels. Once a production panel is established, high-pressure water cannons are used to break the coal off the face. The coal-water mixture falls into a metal sluice trough that carries the coal to further haulage equipment, usually conveyors. Production panels in a hydraulic mine are oriented to utilize gravity to carry the coal-water mixture away from the face and in the troughs. Hand loading and hydraulic mining can be either partial or full extraction depending on the extent of pillar removal in the panels.

3.7 *In Situ* Methods

In situ methods extract the coal with no human contact inside the coal seam, such as miners going underground to attend machinery. These methods include augering, highwall mining, and leaching or dissolving the coal by heat, pressure, or chemical means followed by capture of the media.

Augering consists of extracting coal from an outcrop by the repeated and parallel boring of holes into the seam. The machine, called an auger, is literally a large drill that bores 1.5 ft to 5 ft diameter holes up to 600 ft deep into the coal seam. Augering sometimes occurs in tandem with the surface contour mining method. Contour surface mining creates a flat bench on which to operate the auger.

Highwall mining is similar to augering. Highwall mining bores rectangular holes into a coal seam from an outcrop using a remote-controlled continuous miner connected to an extensible conveyor, or haulage machine. Highwall mining also can work in tandem with contour mining.

3.8 Underground Mine Design Parameters

As introduced, underground mining methods fall into two groups: full extraction and partial extraction. Key mine design considerations for underground mining include roof control, ventilation, seam pitch, extraction ratio, and industrial engineering as it pertains to maximizing safety and productivity.

3.9 Roof Control

Roof control is defined as the systematic design and installation of coal pillars and artificial structures to support the roof, floor, and sides of entries to protect the miner from falling and propelled rock. Roof, floor, and entry sides are subjected to three main forces that cause instability and require that support be installed in underground coal mines. These forces are gravity, mining-induced forces, or forces derived tectonics and mountain building. Gravity primarily loads underground workings with vertical stress. The general rule for U.S. underground mining is that gravity stress equals 1.1 psi per ft of depth of overburden. This value is derived from an assumed weight of sedimentary rock that overlays coal seams of 158.4 lb per ft^3.

Mining-induced forces are caused by the transference of overburden load from the void to pillars adjacent to the void's perimeter. As the void increases in size during full extraction, the weight transferred to the abutment pillars increases until the rock caves. In partial extraction mine design, the void area is limited to the spans across the entry roof. The roof rock in the entry remains stable. In full extraction mining, the caved rock is called gob or goaf, which is broken rock. As the gob consolidates and forms the ability to support weight, some load that was originally transferred to the pillars is reaccepted and transferred back to the gob. This process is called gob formation.

Tectonic forces in mining are caused by plate tectonics, regional mountain building forces, stress that resides in structural geology such as faults and slips, and other natural geologic processes. The result of tectonic force and stress in underground coal mining is increased stress in all directions, particularly in the horizontal plane, or the plane parallel to the bedding (layers of rock). Coal mining can reactivate stresses along structural geologic features. The primary coal mine design criteria for high horizontal stress is that the long axis of a production panel rectangle (in plan view) be oriented 15° to 45° off the maximum horizontal stress. Other than high horizontal stress, other forces from tectonic origin are difficult to incorporate into design because of the uncertainty of the force's magnitude, direction, and dynamics.

The primary design criteria in roof control in underground coal mining is to control the roof overhead where miners work and to provide abutment support in the form of pillars and blocks of coal on the edges of the excavation or void area. Pillars are designed by empirical methods and by mathematical stress modeling. Empirical formulas have been derived from actual mine case histories in all coal producing regions of the world. Most of these formulas have three primary variables to determine pillar strength: *in situ* coal strength, pillar height, and pillar width.

For pillar design in the United States, the vertical stress, assumed to be 1.1 psi per foot of depth, plus other mining induced stresses caused by load transfer would be compared against pillar strength. A pillar size would be chosen that has a satisfactory safety factor. The safety factor is strength divided by load. Each empirical method and mathematical stress modeling approach has recommended design pillar safety factors. Pillar design needs to recognize the variability of geology and the nonhomogeneous nature of coal. Rock mechanics design for underground coal mines requires a thorough design analysis of the local site geology and the comparison of mining conditions at neighboring mines. Civil engineering structures have more uniform design methodology because materials used in buildings, bridges, roads, and other structures are composed of homogeneous materials and the forces demanded on a structure are more predictable.

The main fixture used for primary support in underground coal mining is the roof bolt. A roof bolt is a steel rod from 0.5 in. to more than 1.5 in. diameter and from 2 ft to over 10 ft long. Roof bolts are installed on cycle, meaning that the bolts are installed after the entry or tunnel is excavated a specific distance. This distance the entry is advanced is called the depth-of-cut. The depth-of-cut of an entry, which ranges from 2 ft to over 40 ft, is determined by the ability of the roof rock to remain in place before being bolted; that is, to be stable. Roof bolts are anchored in a hole drilled into the roof by resin grout or by mechanical systems. Roof bolts systems bind the layered roof rock together by forming a beam or by suspending the rock layers from a zone of competent anchorage in the geologic cross section.

Roof is also supported by what is called secondary support. Secondary support is installed after the entry is advanced and after roof bolts are installed. Secondary support is generically called cribbing, which is named after wooden blocks stacked Lincoln-log style to form a vertical, free-standing structure that is wedged between the roof and floor of the entry. Other forms of cribbing or secondary support, in addition to wood blocks, include columns of cementaceous grout, steel structures, wooden posts, and wooden block structures (other than Lincoln-log stacked). Secondary support is used to fortify the entries around a full-extraction area to assist in roof support when ground load is transferred from the void to abutment pillars. The distance the load is transferred is a function of the strength and thickness characteristics of the rock over the void. The presence of strong and massive sandstones in the overburden can resist caving over long spans. The cantilevering nature of the sandstone beam can create large stress increases in the abutment areas, which requires special mine designs be implemented.

3.10 Ventilation

Ventilation, along with roof control, is critical to the safe and efficient extract of coal in underground coal mining. Ventilation is necessary to provide air to all areas of an underground coal mine, except those areas that have been intentionally sealed, to dilute mine gases, dilute airborne dust, and provide fresh air for miners. A coal mine is ventilated by main mine fans located on the surface, which either push air through the mine called a blowing system or pull air through the mine called an exhausting system. Ventilated air moves through the mine in the entries that comprise independent circuits of air called splits. The main fan that pumps air though the mine is designed to deliver a minimum quantity of air to each area of the mine where miners and equipment extract the coal. The fan size and horsepower are designed to overcome friction losses that accumulate because of turbulent airflow against the rubbing surface of the entry roof, floor, and sides. The fan creates a pressure differential that causes air to flow from areas of high pressure to areas of low pressure. Air pressure differential can also be created by atmospheric variations caused by weather patterns and seasonal and daily temperature fluctuations. These pressure variations can move air through mines, particularly in abandoned mines or idled mines that are missing mechanical ventilation devices (fans). This form of ventilation is called natural ventilation. Ventilation design for mines with regard to the size of main fans must recognize natural ventilation.

Ventilation circuits within underground coal mines can be a single split or a double split of air. Double split is often called fishtail ventilation (Fig. 10).

As shown in Fig. 10, air movement is directed to the working faces in the entries. The stoppings shown are called ventilation control devices. Stoppings are walls, usually constructed of cinder block and mortar, built in the crosscuts to separate air circuits or splits of air. Without the stoppings, air flow would short circuit and mix with other air

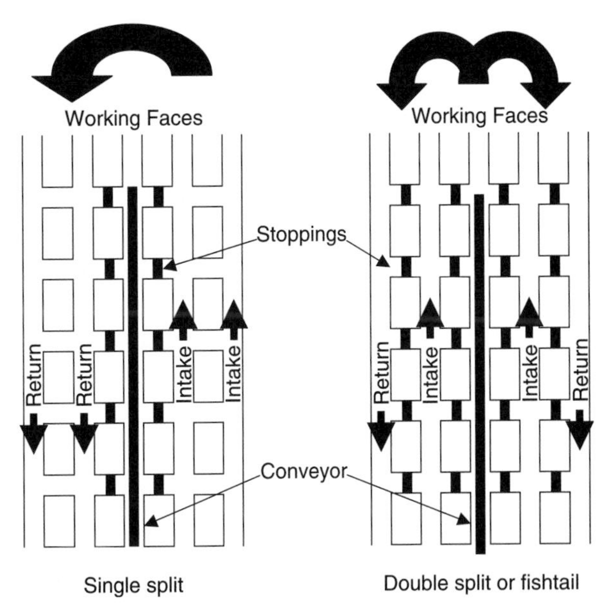

FIGURE 10 Single-and double-split ventilation showing air flow (plan view).

before reaching the working faces. Intake air is fresh air delivered to the working face that has not been used to ventilate other working faces where coal is produced. Return air is air that has been used to ventilate the working face. Return air contains coal dust and mine gases, particularly methane that has been liberated from the coal during production. Sometimes two different airways cross at an entry and a crosscut. A ventilation control device called overcast or undercast separates the two different types of air (i.e., intake and return).

Figures 7, 9, and 10 illustrate the entries containing the conveyor. The entry containing the conveyor usually contains a minimum of airflow and is not used to ventilate the working face. The airflow velocity is lessened in this entry to keep coal dust from becoming airborne.

Some mines utilize conveyor air to ventilate the working face. This practice occurs in longwall mines with long panels. Long panels require parallel intake or return air splits to increase air quantity to the working face and minimize resistance.

Ventilation design for a coal mine is based choosing a mine fan or fans, as in the case of a mine with multiple access openings, that can overcome the resistance to airflow and maintain the required air quantity at the working faces. Mine resistance varies with air quantity and obeys basic empirical laws that have friction factors based on measured data. Increasing the number of parallel entries of the spilt reduces resistance in that split.

3.11 Seam Pitch

Mines are designed to incorporate seam pitch. Rarely are coal seams completely flat. Geologic processes of coal formation, deposition, and structure cause seams to have pitch, which can vary from 0 degrees to vertical. Partial- and-full extraction mining methods can efficiently handle up to approximately 10 degrees of pitch. An underground mine plan is designed to take advantage of pitch when possible for water handling, pumping, material haulage, and gas migration.

3.12 Extraction Ratio

The extraction ratio is defined as the volume of coal mined divided by the total volume of coal within a reserve. Coal is a nonrenewable resource. Once it is combusted to make steam for electrical power generation, steel production, or other uses, all that remains is ash and gaseous substances. Maximizing the volume of coal mined from a given reserve is important for the overall energy cycle and to minimize the cost of production. With the exception of in situ mining systems, partial extraction underground coal mining methods range in extraction ratio from 20% to 50%, and full-extraction mining methods have extraction ratios that range from 50% to over 85%. Achieving a 100% volumetric extraction ratio is difficult for three reasons. First, pillars that are necessary for roof support in the plan cannot be mined. Second, coal is sometimes left in place on the floor or roof of an entry to improve the quality of the coal extracted. The immediate top or bottom layer of a coal seam can have a greater content of noncombustible material than the central portion of the coal seam. Finally, underground coal mining methods and machinery have a maximum practical mining height of approximately 14.5 ft.

Surface mining methods cannot achieve 100% extraction. Coal is damaged from drilling and blasting and cannot be recovered. Also, overburden dilution in pit recovery operations reduces recovery.

3.13 Industrial Engineering

Underground coal mining, as in any producing enterprise, requires an infusion of industrial engineering to maximize safety and productivity. Coal mine design practice has shown that safety and productivity are mutually inclusive goals.

Underground mining methods are a mix of continuous and batch processes. For example, while the continuous miner is extracting the coal from the face of the entry, the process is in continuous mode. The continuous miner has to stop mining frequently to allow loaded shuttle cars to switch out with empty shuttle cars, to advance the face ventilation system, and to allow roof bolts to be installed. These stoppages cause the face operations to be an overall batch process. Once the shuttle cars dump the coal onto a conveyor belt, the material is continuously transported outside to a stockpile. The coal is then fed through a processing facility that crushes the coal and removes impurities to customer specifications in continuous processing. Underground coal mine design is optimized when the batch processes that occur in underground mining are engineered to become continuous processes. An example of an effort in this area has been the introduction of continuous haulage equipment, which in certain conditions replace the shuttle car: a batch process.

Mine surveying plays a key role in both surface and underground coal mining. Surveying is used to direct the mining advancement as planned, to monitor performance giving feedback on issues of underground extraction ratio and surface pit recovery, to define overburden rehandle, and to perform checks against weighed coal production for inventory control.

SEE ALSO THE FOLLOWING ARTICLES

Clean Coal Technology • Coal, Chemical and Physical Properties • Coal Conversion • Coal, Fuel and Non-Fuel Uses • Coal Industry, Energy Policy in • Coal Industry, History of • Coal Mine Reclamation and Remediation • Coal Mining in Appalachia, History of • Coal Preparation • Coal Resources, Formation of • Coal Storage and Transportation • Markets for Coal • Oil and Natural Gas Drilling

Further Reading

Hartman, H. L. (ed.). (1992). "SME Mining Engineering Handbook". 2nd ed. pp. 1298–1984. Society of Mining, Metallurgy, and Exploration, Littleton, CO.

Hartman, H., and Mutmansky, J. (2002). "Introductory Mining Engineering." 2nd ed. John Wiley & Sons, Hoboken, NJ.

Mark, C. (1990). "Pillar Design Methods for Longwall Mining." U S Bureau of Mines Information Circular 9247, Washington, DC. pp. 1–53.

Peng, S. S. (1978). "Coal Mine Ground Control." John Wiley & Sons, New York.

Pfleider, E. P. (1972). "Surface Mining." American Institute of Mining Metallurgical and Petroleum Engineers, New York.

Woodruff, S. D. (1966). "Methods of Working Coal And Metal Mines," Volume 3. Pergamom Press, Oxford.

Coal Mining in Appalachia, History of

GEOFFREY L. BUCKLEY
Ohio University
Athens, Ohio, United States

1. Introduction
2. Historical Overview
3. Breaking Ground
4. Life in the Company Town
5. Working in a Coal Mine

Glossary

Appalachian Regional Commission Federal agency established in 1965 "to serve the needs of 21,000,000 people residing in one of the most economically distressed regions of the country."

black lung The common name for coal workers' pneumoconiosis, a respiratory disease caused by the inhalation of coal dust.

longwall mining A method of mining in which long sections of coal—up to 1000 feet across—are mined and deposited directly onto a conveyor system with the help of shields that support the roof and advance the longwall rock sections.

mountaintop removal A surface mining technique in which the top of a mountain is blasted away to expose a seam of coal; debris is then deposited in a valley fill.

rock dusting White, powdered limestone is sprayed on the roofs, bottoms, and ribs of a mine to mitigate the explosive qualities of coal dust.

roof bolting A process in which holes are drilled into the roof of a mine; long bolts coated with glue are then screwed into the holes to support the roof.

scrip Nonlegal tender issued to workers in place of cash; generally, redeemable only at the company-owned store.

timbering The use of wood for constructing a roof support in an underground mine.

tipple An elevated structure, located near a mine entrance, that receives coal from mine cars or conveyors and from which coal is dumped or "tipped," washed, screened, and then loaded into railroad cars or trucks.

welfare capitalism An industrial strategy in which company flexibility and benevolence are exercised in an attempt to maximize production and promote social control of the workforce.

yellow-dog contracts Agreements that prohibit employees from joining or supporting labor unions.

Appalachia—as defined by the United States Appalachian Regional Commission (ARC)—is a sprawling region encompassing approximately 400 counties in 13 eastern states: New York, Pennsylvania, Ohio, Maryland, West Virginia, Virginia, Kentucky, Tennessee, North Carolina, South Carolina, Georgia, Alabama, and Mississippi. Rich in natural resources, such as timber, coal, oil, iron ore, gold, and copper, and located within relatively easy reach of major industrial centers along the Atlantic seaboard, this expansive area supplied the raw materials that fueled America's Industrial Revolution. With vast deposits of bituminous ("soft") coal underlying approximately 72,000 square miles from Pennsylvania to Alabama, and an additional 484 square miles of anthracite ("hard") coal concentrated in northeastern Pennsylvania, exploitation of this one resource uniquely shaped the region.

1. INTRODUCTION

Opening the Appalachian coalfields was no easy feat. Nor was it achieved overnight. For aspiring "captains of industry," there were great financial risks involved. Substantial capital investment was required to survey and purchase coal lands, to construct mines and provide for the needs of a large workforce, and to establish a reliable and cost-effective means of delivering this bulky and relatively low-value good to market. For many, the road to success led instead to financial ruin. For those

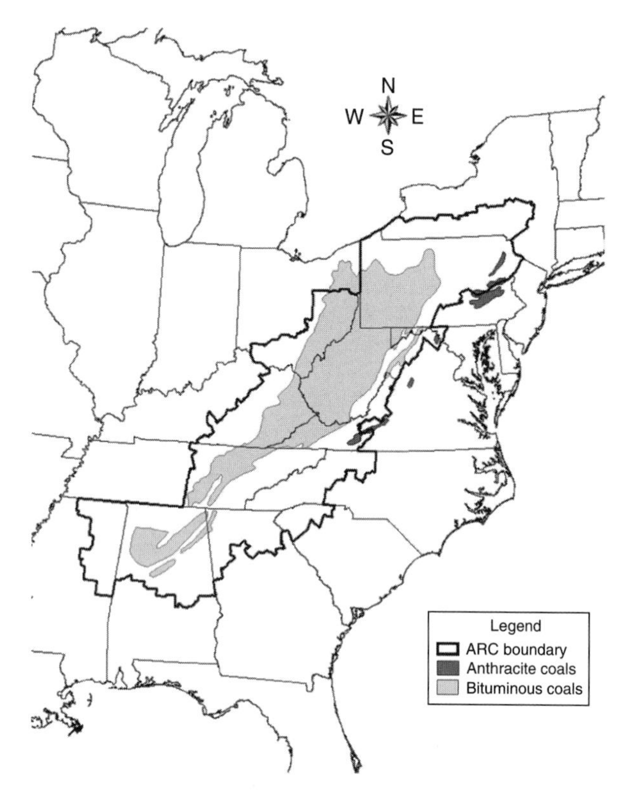

FIGURE 1 Location of mineral resources in Appalachia. ARC, Appalachian Regional Commission. Adapted from Drake (2000).

charged with the task of extracting "black diamonds" from the earth, there were great personal risks involved. Mining is and always has been hazardous work. Historically, periodic layoffs and the threat of black lung added to the worries of the miner and his family. In the final analysis, it must be concluded that coal mining in Appalachia has had as profound an impact on the region's inhabitants as it has had on the region's forest and water resources. Although much diminished, King Coal's influence is still discernible today—both in the region's historic mining districts and in those areas where underground and surface mining are still being carried out (Fig. 1).

2. HISTORICAL OVERVIEW

In his opening address to the White House Conference on Conservation in 1908, President Theodore Roosevelt reminded those in attendance that coal's ascendancy was a relatively recent occurrence: "In [George] Washington's time anthracite coal was known only as a useless black stone; and the great fields of bituminous coal were undiscovered. As steam was unknown, the use of coal for power production was undreamed of. Water was practically the only source of power, save the labor of men and animals; and this power was used only in the most primitive fashion." Over the course of the next century, however, coal's utility would be demonstrated and its value to industry affirmed. By 1900, the year before Roosevelt took office, coal accounted for nearly 70% of the national energy market in the United States. In short, it had become America's fuel of choice and Appalachia was the nation's primary producer.

2.1 Fueling the Industrial Revolution

During the early years of the 19th century, coal was used primarily to fire blacksmith forges and as a domestic heating fuel. In ensuing years, it was valued as a fuel for salt and iron furnaces, brick and pottery kilns, and steam engines. Although annual coal production increased significantly between 1800 and 1860—from a mere 100,000 tons to 20,000,000 tons—it was the post-Civil War years that first witnessed truly explosive growth in coal production and consumption in the United States. In 1885, the annual production figure had climbed to 110,000,000 tons. Stimulated by growth in domestic consumption, as well as increased use as a fuel for steamship travel, coal was now being used to stoke the furnaces of an emerging steel industry and to provide the motive power for steam locomotives. By 1900, production topped 243,000,000 tons, making the United States the world's leading producer. With respect to this tremendous growth, historian Duane A. Smith notes: "Without mining—from coal to iron to gold—the United States could not have emerged as a world power by the turn of the century, nor could it have successfully launched its international career of the twentieth century."

It would be difficult to overstate the importance of Appalachian coal in propelling America into the industrial age. Prior to World War II, virtually all coal mined in the United States came from the Appalachian fields. In 1925, for example, the Appalachian region produced 92% of the total amount of coal mined in the United States, with Pennsylvania alone accounting for approximately one-third of this output. Between 1880 and 1930, Pennsylvania consistently ranked as the leading producer of both bituminous and anthracite coal. By the mid-1920s, however, it had become apparent that all was not well with the coal industry in Appalachia.

2.2 The Decline of King Coal

Several years in advance of the stock market crash of 1929, the coal industry in general and miners in particular felt the first tremors of the Great Depression. In Appalachia, overexpansion stimulated by World War I, interstate competition with midwestern states, and competition between "union" and "non-union" mines led to market gluts, falling prices, spiraling unemployment, and labor unrest. Although mining had weathered economic downturns in the past, most notably in the mid-1890s, nothing compared with the depression that beset the industry during the 1920s and 1930s. West Virginia, which lost 33,000 coal jobs as production fell from 146 million tons in 1927 to 83.3 million tons in 1932, was hit particularly hard.

In subsequent years, the rising popularity of new energy sources—petroleum, natural gas, and electricity—challenged coal's dominance of the domestic heating market. Although World War II provided a boost to the region's coal-based economy, it was only temporary. When railroads completed the conversion from coal to diesel after the war, the industry received another blow, contributing further to the industry's recent history of boom and bust. The coal operators' response was to accelerate the process of mechanization begun in earnest during the 1930s and to opt for surface mining over underground mining wherever feasible. Economic instability coupled with widespread utilization of new labor-saving machinery, typified most noticeably by the introduction of enormous electric-powered shovels to surface operations, forced many miners and their families to leave the region altogether in the post-World War II era.

Since the 1960s, the only major growth market for coal has been to supply electric power producers and for export abroad. With respect to the former, coal remains an important player. Today 52% of U.S. electricity is generated by coal-fired plants. However, increased production from western states starting in the late 1950s, but gaining momentum especially during the 1980s and 1990s, has further eroded Appalachia's share of coal markets. Compliance with environmental laws since the 1970s, including the Clean Air Act and its amendments, the Clean Water Act, and the Surface Mining Control and Reclamation Act (SMCRA), not to mention various state-level regulations, has also contributed to a rise in coal production costs. By 1999, more than half of the coal mined in the United States was coming from states located west of the Mississippi River, and far and away the largest single producer was Wyoming, not Pennsylvania. Nevertheless, coal mining has managed to remain viable in some portions of the Appalachian region due in large part to controversial mining techniques such as mountaintop removal.

3. *BREAKING GROUND*

Given the increase in demand for coal spurred on by industrialization, the existence of coal so near to the Atlantic seaboard aroused early interest among local boosters, land speculators, foreign investors, and others wishing to purchase mineral lands and mine them for profit. Before large-scale coal-mining operations could commence, however, certain obstacles had to be surmounted. First, coal lands had to be surveyed and evaluated, and the rights to the coal purchased or leased. Equally important, a reliable and cost-effective means of transporting the coal to eastern and Great Lakes markets, as well as to the steel mills of the North and South, had to be financed and developed. As historian Crandall A. Shifflett has pointed out: "All of the land and coal was worthless... if it could not be gotten out of its mountain redoubts. There was no use to open mines if the coal could not be transported to distant markets."

3.1 Railroads Usher in the Industrial Era

As early as the 1820s, coal mining had become an important, albeit seasonal, activity in some remote locations. During the first decade of the 19th century, for example, small quantities of coal, along with lumber and a variety of agricultural goods, were sent down the Potomac River via flatboat to Georgetown. On arrival, the large rafts were dismantled and their crews walked home. Such trade could only take place during the spring season when floodwaters raised the level of the river. Without significant investment in transportation improvements, it was clear that coal could not be mined profitably on a large scale.

Although coal and coke (fuel derived from coal) were slow to replace wood and charcoal in the manufacture of iron prior to the Civil War, coal found an early use in blacksmith shops and as a domestic heating fuel. After about 1840, coal gradually began to replace charcoal in the furnaces of the burgeoning iron industry. By this time, it was also being recognized as an important source of power for steam engines. After the Civil War, it would play an indispensable role in the manufacture of steel.

Initially, it was the anthracite fields of northeastern Pennsylvania that attracted great attention. Once problems associated with the burning of hard coal were resolved, production rose sharply. Between 1830 and 1860, production of Pennsylvania anthracite, much of it destined for the Northeast, increased more than 40-fold. As Andrew Roy put it in 1903, "the 'black rocks' of the mountains, heretofore deemed useless, at once rose in public estimation." Miners' villages sprouted wherever new mines were opened as English, Welsh, Irish, Germans, and, in later years, eastern and southern Europeans entered the region in search of high wages and a new beginning in America.

Before America's Industrial Revolution could "take off," another revolution—this one in the transportation sector—was needed. As the country grew in geographic extent, government officials became increasingly aware that improvements in communication and trade depended largely on the development of new, more efficient modes of travel. In 1808, Albert Gallatin, Secretary of the Treasury under Thomas Jefferson, envisioned a system of roads and canals crisscrossing the country and financed by the federal government. Although the Gallatin Plan was never carried out, great changes were in the works. The spectacular success of New York's Erie Canal served as a catalyst. Indeed, the completion of the Erie Canal in 1825 ushered in a period of canal building in the United States as eastern cities competed with one another in a race to expand their hinterlands and capture market share. Government officials and businessmen in Pennsylvania, Maryland, Ohio, and Virginia soon unveiled plans for canal systems they hoped would allow them to remain competitive with New York. In the end, none of the canal systems to the south could match the financial success of the Erie Canal and its branch connections. Difficult terrain and competition from a new form of transportation, the railroad, ensured that the reign of the canal would be relatively short (Fig. 2).

FIGURE 2 Rodgers locomotive on turntable at Corning, Ohio, ca. 1910. Reprinted from the Buhla Collection, Ohio University, Archives and Special Collections, with permission.

On the eve of the Civil War, large parts of northern Appalachia had established outlets for coal via rivers, canals, and railroads. In addition to the anthracite fields of eastern Pennsylvania, bituminous coal from the great "Pittsburgh" seam in western Pennsylvania, western Maryland, northern West Virginia, and Ohio was now being tapped. In contrast, central and southern Appalachia remained relatively unaffected by the transportation revolution that had seized the north. After the Civil War, an already impressive network of railroads in the Appalachian north was expanded. As these railroads, and to a lesser extent canals, penetrated the mountains, operations were expanded in western Maryland, western Pennsylvania, northern West Virginia, and southeastern Ohio, contributing to the growth of the iron industry and, especially after the introduction of the Bessemer hot-blast furnace in the 1860s, the nascent steel industry. Following the steady progress of the railroads, new fields were opened up in southern West Virginia, northeastern Tennessee, and northern Alabama, and finally, by the 1890s, southwestern Virginia and eastern Kentucky. As was the case with Pennsylvania's anthracite fields, the opening of mines in these areas was accompanied by a flurry of coal town construction and an influx of immigrant labor. As historian Richard Drake reminds us, in addition to carrying a great deal of coal, rail lines such as the Baltimore and Ohio, Chesapeake and Ohio, Norfolk and Western, Louisville and Nashville, and the Southern Railway, to name but a few, figured prominently in the growth of major industrial centers concentrated in the northern and southern extremities of the region. Of particular importance was the development of Pittsburgh and its industrial satellites—Johnstown, Altoona, Morgantown, and Fairmont—and the "A.B.C." triangle of the Deep South—Atlanta, Birmingham, and Chattanooga.

3.2 Clearing the Way

Long before the new transportation lines reached their destinations, however, land speculators were clearing the way for this initial wave of industrialization. During the first half of the 19th century, corporations interested in mining coal and iron ore and cutting timber, and backed by capital from centers of commerce such as New York, Baltimore, Boston, and London, dispatched geologists and engineers into the mountains to measure the size and evaluate the quality of coal and iron ore deposits. When favorable reports were returned, they began accumulating property rights, a process that was often mediated by land speculators. Some of these speculators and businessmen were outsiders who entered the region as representatives of mining interests headquartered in the east. Others were local residents who recognized the value of coal and accumulated property rights based on the assumption that the arrival of the railroad would cause real estate values to soar. Still others were opportunistic politicians—men like Henry Gassaway Davis, Johnson Camden, and Clarence Watson—who used their power and influence in the chambers of the U.S. Senate and in various state capitols to finance railroads, purchase coal lands, and amass personal fortunes. Invariably the result was the same. By the time the railroad arrived, the coalfields were controlled for the most part by outside interests eager to begin the mining process.

One of the best known of these early entrepreneurs was Pike County, Kentucky native John C. C. Mayo. During the 1880s and 1890s, Mayo traveled throughout eastern Kentucky purchasing land or mineral rights from local farmers. Using the infamous "Broadform Deed," which permitted the mountain farmer to retain rights to the surface while authorizing the owner of the mineral rights to use any means necessary to extract the minerals, Mayo acquired thousands of acres of land that he eventually sold to companies such as the giant Consolidation Coal Company. As late as the 1980s, coal companies in Kentucky could count on the courts to accept the binding nature of Broadform Deeds signed in the late 19th and early 20th centuries. However, rather than extract the coal using underground mining techniques, as was the practice during the 19th and early 20th centuries, companies were now employing surface mining techniques. Essentially, the Broadform Deeds permitted coal operators to destroy the surface to reach the mineral deposits. The stability of many rural communities was sorely tested. In 1988, the citizens of Kentucky passed an amendment to the state constitution requiring coal companies to obtain permission from surface owners before strip mining land. Five years later, the Kentucky Court of Appeals accepted the constitutionality of this amendment.

4. LIFE IN THE COMPANY TOWN

A vexing problem that confronted coal operators in the early stages of mine development was a shortage of labor. Because exploitable seams of coal were often found in sparsely populated areas, operators were forced to recruit and import labor. Although some coal operators used prison labor, as in northern Alabama, or drew from the pool of displaced mountain farmers, by far the greater number chose to lure white labor from Europe or, in the years leading up to World War I, to enlist black migrants from the South. Thus, in addition to mine development and railroad construction, coal companies had to provide housing and other amenities for newly arrived miners and their families.

These company towns, as they came to be known, played an especially important role in central and southern Appalachia during the 19th and early 20th centuries. In eastern Kentucky, southern West Virginia, and southwestern Virginia during the 1920s, for example, an estimated two-thirds to three-fourths of all miners and their families resided in such communities. In southern West Virginia, the figure approached 98%. Beginning in the 1930s, the company town's usefulness began to fade. A number of factors contributed to this decline, including the diffusion of the automobile, which greatly enhanced worker mobility; the introduction of labor-saving devices, which made workers redundant; and, later, the adoption of surface-mining techniques, which further reduced the need for a large labor force. No longer profitable, by the 1950s and 1960s most mining firms had sold or otherwise disposed of company housing.

4.1 A Distinctive Feature of the Mining Landscape

There can be little doubt that the company town is one of the most distinctive features associated with Appalachia's historic mining districts. Although some scholars have argued that company towns were a necessity, at least at the outset, and that many companies treated their workers with respect and fairness, others contend that these carefully planned and controlled settlements allowed companies to exert an objectionable level of power over miners and other residents. Considering the extent to which some companies dominated local law enforcement, monitored educational and religious institutions, curtailed personal liberties, and influenced worker behavior, their power was pervasive. As Shifflett reminds us, however, some miners' families, especially the first generation to leave hardscrabble farms in rural Appalachia, may have favored life in the company town, especially the more durably constructed "model" towns: "Contrary to the views that the coal mines and company controlled towns led to social fragmentation, disaffection, and alienation of the workforce, many miners and their families found life in them to be a great improvement over the past. Miners viewed the company town, not in comparison to some idyllic world of freedom, independence and harmony, but against the backdrop of small farms on rocky hillsides of Tennessee, Kentucky, and Virginia, or a sharecropper's life in the deep South states."

Company towns in Appalachia generally shared several distinguishing features (Fig. 3). First, the town bore the stamp of the company in every conceivable way. Everything in the town, from the houses and the company store to the schools, churches, hospitals, and the recreational facilities, if such "frills" even existed, was provided by the company. Second, housing generally conformed to a standard design. In general, one worker's house was identical to the next with respect to style and construction. A third trait was economy of construction. Companies typically limited their investment in towns because they knew they were not likely to evolve into permanent settlements. Renting houses to workers ensured that even a substantial investment in housing would be recovered quickly. Because companies generally preferred to hire married men with families over single men, believing they were less inclined to quit their jobs, single- and two-family dwellings were far more common in the coalfields than were large boarding houses. A fourth feature was that the company houses and other structures were usually laid out in a standard gridlike pattern, with long, narrow lots lined up along the railroad tracks. Finally, there was an outward manifestation of socioeconomic stratification and segregation according to racial and ethnic group expressed in the landscape. With respect to the former, sharp distinctions could be found in the size and quality of housing occupied by miners and company officials. Regarding the latter, historian Ronald Lewis notes that some companies sought a "judicious mix" of whites, blacks, and immigrants as a means of maintaining order and fending off the advances of union organizers. Thus it was not unusual to find distinct "neighborhoods"—an immigrant town, a Negro town, and an "American" town—existing side by side in the same town, each with its own separate facilities and social and cultural life.

FIGURE 3 Railroad station, Fleming, Kentucky, ca. November 1915. Reprinted from the Smithsonian Institution, National Museum of American History, with permission.

One of the most controversial features of the typical coal mining town was the company store. Indeed, it represented both the best and worst of company town life. Often doubling as a meeting hall, lodge, and recreation facility, while also housing the company offices, the store carried a great variety of goods, including food and clothing, mining supplies, tools, and durable household goods. It also served as an important informal gathering place for both miners and managers alike. Given the remote location of many towns, the company store was often the only commercial retail establishment to which the mining population had regular access. In some cases, coal companies prohibited the establishment of other stores; in other cases, they penalized miners for making purchases in neighboring towns. Customer loyalty was achieved through the issuance of company scrip. Paying wages at least partially in scrip ensured that a portion of a worker's paycheck would be funneled back to the company. Although company officials claimed that such an arrangement protected families, critics have argued that companies charged higher prices in an attempt to keep the cost of coal production down. It is not surprising, then, that charges of debt peonage, coercion, price gouging, and differential pricing were leveled at the company store.

Some companies encouraged the planting of vegetable and flower gardens to improve the appearance of towns and to supplement food supplies. With the men at work in the mines, primary responsibility for the planting and care of the gardens rested on the shoulders of women and children. While they contributed significantly to the miners' larder, the gardens may have served another purpose. By encouraging miners to grow much of their own food, company officials could justify paying lower wages. Such a strategy enabled companies located at a considerable distance from major coal markets to keep the price of coal low and remain competitive with companies located nearer to centers of industry and population.

4.2 Culturally Diverse Communities

Coal towns in Appalachia often exhibited a high degree of cultural diversity. During the first half of the 19th century, immigrants to the minefields typically came from England, Wales, Scotland, Ireland, and Germany. By the end of the century, large numbers were being recruited from southern and eastern Europe, a trend that continued well into the first two decades of the 20th century (Fig. 4). When immigration from Europe plummeted during the war years, the door was opened wide for large numbers of African-Americans to enter the mines. Discouraged by low wages in the mines of northern Alabama or in search of seasonal work when there was a lull in activities on the family farm, this group made up a significant proportion of the workforce in places such as southern West Virginia and eastern Kentucky from approximately 1910 to the 1930s.

Cultural heterogeneity in the company town manifested itself in many and varied ways. It was not uncommon to find safety instructions posted at the mine opening written in several languages. Nor was it uncommon to find a diversity of religions represented in a single company town. Although some have argued that building churches and paying ministers' salaries allowed company officials to exercise even greater control over residents of the company town, others, citing low attendance, have downplayed the importance of religion in these communities.

FIGURE 4 Miners in front of Oakdale or Jumbo Mine No. 311, Ohio, 1887–1907. Reprinted from the Buhla Collection, Ohio University, Archives and Special Collections, with permission.

Although some coal operators looked on their creations with a certain measure of pride, believing them superior to anything one might find in outlying areas, living and working conditions, in fact, varied considerably from one location to the next. The time and money an operator was willing to invest in the development of a town depended on several factors, including the projected life of the mine, the number of houses that needed to be built, and the amount of capital the company had at its disposal. Towns constructed of durable materials and offering a range of amenities were clearly meant to last. Ramshackle housing, on the other hand, was a sure sign that a company's investment in a mine site was limited. According to contemporary theories of corporate paternalism and welfare capitalism, well-built towns equipped with electricity and indoor plumbing, and offering a wide range of amenities, such as churches, recreation buildings, meeting halls, ballparks, and stores, attracted a more dependable and loyal breed of miner—one who, in the end, would reject the temptation to join a union. Because the company maintained ownership of housing, the threat of eviction was never far from the mind of the miner. In the years following the First World War, companies fought off the advances of the unions by equating union membership with communist sympathy or radicalism. Companies employing a high number of foreign-born workers kept their workforce in line by equating loyalty to the company with patriotism and "Americanism." Eager to prove they were good American citizens and avoid persecution, recent immigrants were often reluctant to participate in union activities.

5. WORKING IN A COAL MINE

For the most part, coal miners during the 19th and early 20th centuries worked side by side in a dark, damp, and often dangerous environment, regardless of their cultural or ethnic background (Fig. 5). Given the stressful nature of their work and the dangers they faced, it was absolutely essential that safety and teamwork take precedence over all other matters in the mines. Dust and gas accumulations, roof falls,

FIGURE 5 Miner loading car at Mine No. 26, West Virginia. Reprinted from the Smithsonian Institution, National Museum of American History, with permission.

and heavy equipment malfunctions and accidents were just a few of the hazards miners dealt with on a daily basis. Under such conditions, even the most safety-conscious miner could be caught off guard. If a miner was fortunate enough to avoid a serious lost-time injury over the course of his career, black lung, a condition caused by the inhalation of coal dust, could cut his life short. Aboveground, miners went their separate ways. As Robert Armstead recently detailed in an autobiographical account, segregation and Jim Crow were a fact of life when the shift was over.

The coal mining industry in Appalachia has witnessed a great many changes since the days when miners toiled underground with picks and shovels and hand-loaded coal into wagons for transport to the surface. Although many of these changes have had a positive effect on the lives of miners and their families, e.g., improved safety, wage increases, and retirement benefits, to name just a few, others, such as the introduction of labor-saving equipment, have had the effect of putting miners out of work. Indeed, the effect that mechanization had on miners and mining communities during the middle years of the 20th century was far reaching.

5.1 Social Relations among Miners

Although relations between whites, blacks, and European immigrants were sometimes strained, especially when blacks or immigrants were brought in as strikebreakers, antagonism appears to have been the exception rather than the rule, at least on the job. According to Ronald Eller, "[a] relatively high degree of harmony existed between the races at a personal level. Working side by side in the mines, the men came to depend upon each other for their own safety, and the lack of major differences in housing, pay, and living conditions mitigated caste feelings and gave rise to a common consciousness of class." If integration typified relations in the mines, segregation characterized relations on the surface. When the workday was over, blacks and whites showered in separate facilities (if such facilities were, in fact, available) and then walked or drove home to their own neighborhoods. Foreign-born miners and their families were often subjected to similar, albeit more subtle, forms of discrimination.

Although small numbers of women worked in small "country bank" mines, they were generally discouraged from engaging in such work. In some states, such as Ohio, they were prohibited by law from coal mining work unless they actually owned the mine. Their absence is frequently attributed to the superstitious belief that a woman's presence in the mines was sure to bring bad luck. Starting in the early 1970s, small numbers of women were hired to join the ranks of male coal miners. Hiring peaked in 1978 and then tailed off again during the 1980s. The historical record seems to support geographer Richard Francaviglia's statement on the matter: "mining has traditionally been 'man's work', and no amount of neutralized language can…conceal that fact."

5.2 Mechanization

Coal mining was one of the last major industries in America to mechanize production. From the beginning of the 19th century through the first two decades of the 20th century, the manner in which coal was extracted from underground mines and brought to the surface changed very little. Before the advent of the mechanical loading machine in the 1920s, miners typically broke coal from the mine face using picks and wedges and shoveled it by hand into coal cars. Loaded cars were then pushed to the surface by miners or pulled by horses or other animals. By the end of the 19th century, miners were "undercutting" the seam and using drills and explosives to increase productivity. Considering that a miner's pay was directly linked to the amount of coal he mined and the number of cars he filled, new methods of mining that saved labor and increased productivity were quickly adopted.

As demand for coal increased over time, companies sought to overcome bottlenecks in the production process. When overexpansion and competition began to cut into profits, and unionization forced the cost of coal production to go up, the drive to mechanize underground mining gained momentum. The logical starting point was to replace the hand-loading system with one that emphasized the use of mechanical loading machines. From the 1930s to the 1970s, the introduction of mechanical loaders, cutting machines, continuous miners, and longwall mining changed forever the way coal would be mined.

Another aspect of mining transformed by mechanization was in the area of haulage. During the pick and hand-loading era, miners shoveled coal into wagons and pushed them to the surface. Drift mines were generally constructed so that they pierced the hillside at a slight upward angle to allow for drainage and to facilitate the transport of loaded cars to the tipple. As mines grew in size and the distance to the tipple increased, companies turned to animal power as the principal means of haulage. When larger coal cars came on the scene, mechanical and electrical haulage, including conveyors, became increasingly necessary.

From the vantage point of the coal operator, mechanization offered several advantages. It allowed for more easily loaded coal, reduced the need for explosives, and lowered timbering costs. Most important, mechanization permitted greater amounts of coal to be mined. Mechanization had its drawbacks, however, at least from the perspective of the miner, for the introduction of laborsaving equipment greatly reduced the need for a large workforce. The introduction of the mechanical coal-loading machine, for instance, reduced the need for coal miners by approximately 30% industrywide between 1930 and 1950. According to Ronald Lewis, black miners bore the brunt of layoffs during this period because they were disproportionately employed as hand loaders. The continuous miner, which combined cutting, drilling, blasting, and loading functions in one machine, had a similar impact. After the introduction of the continuous miner, the number of miners working in West Virginia was cut by more than half between 1950 and 1960. The impact of mechanization also shows up clearly in production and employment figures for Ohio. Between 1950 and 1970, coal production in this state climbed steadily despite a 90% decrease in the number of underground mines and an 83% decline in the size of the labor force. Starting in the 1970s, longwall mining cut even further into employee rolls.

Mechanization presented miners with other problems as well. In mining's early days, miners walked to and from the mine face and relied on natural ventilation to prevent the buildup of mine gas. As mines expanded in size, miners had to walk greater distances to get to work, and problems with ventilation and illumination arose. To facilitate the movement of men, electric trolleys were introduced to the mines. Dust and gas problems were alleviated somewhat by the installation of mechanical fans. Meanwhile, illumination was improved when safety lamps and electric lights were substituted for candles. While solving some problems, these and other solutions contributed to others. In the words of one authority on the matter, now workers could be "crushed, run over, squeezed between cars, or electrocuted on contact with bare trolley wires." In addition, some of the new machinery generated sparks and produced tremendous amounts of ignitable coal dust.

5.3 Health and Safety

Coal mining has always been and continues to be a hazardous occupation. Indeed, fatalities and serious nonfatal injuries occur still. The risks miners face today, however, pale in comparison to the perils miners faced prior to World War II. Accidents resulting from roof falls, ignition of mine gas and coal dust, the handling of mechanical haulage equipment, and the operation of electric-powered equipment of all types contributed to a high accident rate in the United States. Increased demand, fierce competition, advanced technology, negligence on the part of miners, and a poorly trained immigrant workforce have all been blamed for the high accident and fatality rates. Coal operators must share a portion of the blame as well. Opposed to unions and wary of burdensome safety regulations, operators often stood in the way of meaningful safety reform.

Initially, enactment of mine safety legislation rested with the states. In 1870, Pennsylvania passed the first substantive mine safety law. Over the next 25 years, several states passed mine inspection laws of their own. By the mid-20th century, 29 states had mine safety laws on the books. Unfortunately, these laws were often poorly enforced. Although the U.S. Bureau of Mines had been created in 1910, it was primarily an "information-gathering" agency. Authority to inspect underground mines was not conferred until 1941. Enforcement powers were not granted until 1953. Continued high fatality rates eventually sparked an effort to pass comprehensive federal legislation. Proponents of federal legislation had to wait until 1969 before such measures were enacted.

By far, the greatest number of deaths was caused by roof falls. Nevertheless, it was the ignition of methane gas and coal dust that captured the attention of the news media. Wherever miners used candles or open-flame safety lamps for illumination and black powder to break coal from the face, and where the mine was deep, "gassy," and poorly ventilated, conditions were ripe for an explosion. During the 10-year period from 1891 to 1900, 38 major explosions rocked American mines, resulting in 1006 deaths. Some of the worst explosions occurred during the first decade of the 20th century, including the worst one in American history, the dreadful Monongah mine disaster. On December 6, 1907, 362 miners employed by the Fairmont Coal Company of West Virginia were killed when the company's No. 6 and No. 8 mines blew up. Previously, Andrew Roy had noted that "more miners are annually killed by explosions in West Virginia, man for man employed, or ton for ton mined, than in any coal producing State in the Union or any nation in the world." It was clear that much needed to be done to improve mine safety in the United States.

Discovering the cause of mine explosions was of paramount importance to officials at the fledgling U.S. Bureau of Mines. Several causes were identified. First, as mines grew deeper, providing fresh air to miners and dispersing mine gas became more of a challenge. Given that miners typically used candles or open-flame safety lamps to light their way, this was a particularly serious problem. There was also a preference on the part of the miners to use greater amounts of explosives to "shoot off the solid," that is, to blast without first undermining the coal. As with the open flame, explosives were another source of ignition. Finally, there was the widely held belief that coal dust alone could not set off an explosion. Disproving this belief was particularly important (one that Europeans had confirmed by a much earlier date) given the increasing amounts of dust being produced by new mining machinery. The adoption of new technologies eventually reduced the risk of explosion. Improved ventilation, electric cap lamps, rock dusting, safer mining machinery, first-aid and mine rescue training, and passage of legislation permitting federal inspection of mines combined to improve conditions considerably. Although the frequency of mine explosions diminished considerably after the 1930s, the Farmington, West Virginia disaster of 1968, in which 78 miners were killed, served as a reminder that mine explosions were still a potential threat. With respect to roof falls, more effective supervision, systematic timbering, and roof bolting would, in time, reduce the frequency of these accidents.

5.4 Coal Mine Unionism

No history of coal mining in Appalachia would be complete without mentioning organized labor. Given the conditions that coal miners often had to work under, it is not surprising that they were among the first workers in the United States to organize. Given the power and control coal companies wielded throughout the better part of the 19th and early 20th centuries, it is also not surprising that coal operators fought the unions at every turn. Proponents of corporate paternalism and welfare capitalism believed that company benevolence could effectively stave off the unions, but others believed that coercion and intimidation would more effectively produce the desired results.

Although coal mine unionism in Appalachia can be traced to the early 1840s, the modern era of unionism began in 1890 when two groups, the National Progressive Union of Miners and Mine Laborers and the National Assembly Number 135 of Miners of the Knights of Labor, joined forces in Columbus, Ohio to form the United Mine Workers of America (UMWA). After winning recognition from Central Competitive Field Operators in 1897, the UMWA set out to stake a middle ground between ineffectual "company" unions and more radical unions, such as the National Miners Union. Under the able leadership of John Mitchell, the UMWA won strikes and concessions from the coal operators and built membership. Withdrawing from what they saw as a temporary truce with union organizers during the war years, coal operators sought to roll back the gains the UMWA had made during the first two decades of the 20th century. Citing a 1917 U.S. Supreme Court decision in which yellow-dog contracts forbidding union membership were ruled legal, coal operators set out to reclaim lost ground. An attempt by new union leader, John L. Lewis, to organize the southern coalfields during the middle of the decade was crushed. With demand for coal down and nonunion mines undercutting the cost of production in union mines, UMWA membership plummeted during the 1920s.

The 1930s, and more specifically President Franklin D. Roosevelt's New Deal legislation, breathed new life into the beleaguered union. Passage of the National Industrial Recovery Act (1933), the Bituminous Coal Code, or "Appalachian Agreement" (1933), the National Labor Relations, or "Wagner,"

Act (1935), and the first and second Bituminous Coal Conservation, or "Guffey," acts (1936 and 1937) shifted the advantage to the unions once again. Particularly important was the fact that the union had finally won the right to be recognized as a collective bargaining agent. Among the union's "victories" at this time was approval of an 8-hour workday, a 40-hour workweek, and higher wages. A minimum work age of 17 was also set, ending, in theory at least, the industry's long-standing practice of utilizing child labor. In addition, miners were no longer forced to use company scrip, to shop only at the company store, or to rent a company house, all mechanisms by which some companies attempted to keep up with the rising cost of production.

The gains of the 1930s and 1940s were replaced by the uncertainties of the 1950s and 1960s. The terms of the Love–Lewis agreement, signed in 1950 by Bituminous Coal Operators Association president George Love and UMWA president Lewis, reflect the ever-changing fortunes of the miners and their union. Although the agreement provided a high wage and improved health benefits for miners, it prevented the union from opposing any form of mechanization. The adoption of new technologies, including auguring and stripping, resulted in layoffs and a decline in membership. Although certain gains were made, e.g., portal-to-portal pay, improved insurance and pension benefits, and passage of black lung legislation, the 1970s and 1980s saw many workers return to nonunion status.

Sometimes, relations between management and labor took a violent turn. Students of Appalachia know well the price that was paid in human life in places such as Matewan, West Virginia, "bloody" Harlan County, and at Blair Mountain, where 3000 UMWA marchers clashed on the battlefield with an army of West Virginia State Police, company mine guards, and assorted others representing the interests of the coal operators. In more recent times, violence has flared up in places such as Brookside, Kentucky. In 1969, charges of fraud and the murder of a UMWA presidential candidate tarnished the reputation of the union.

Although unions, the UMWA in particular, can take justifiable pride in improving the lot of miners, some scholars have argued that the higher labor costs associated with these victories ended up increasing the cost of coal production and accelerating the move toward mechanization. Thus, an impressive and hard-earned package of benefits was eventually passed down to an increasingly smaller pool of beneficiaries.

SEE ALSO THE FOLLOWING ARTICLES

Clean Coal Technology • Coal, Chemical and Physical Properties • Coal, Fuel and Non-Fuel Uses • Coal Industry, History of • Coal Mine Reclamation and Remediation • Coal Mining, Design and Methods of • Coal Storage and Transportation

Further Reading

Armstead, R. (2002). "Black Days, Black Dust: The Memories of an African American Coal Miner." The University of Tennessee Press, Knoxville.

Corbin, D. (1981). "Life, Work, and Rebellion in the Coal Fields: The Southern West Virginia Miners, 1880–1922." University of Illinois Press, Urbana.

Crowell, D. (1995). "History of the Coal-Mining Industry in Ohio." Department of Natural Resources, Division of Geological Survey, Columbus, Ohio.

Drake, R. (2000). "A History of Appalachia." The University Press of Kentucky, Lexington.

Eller, R. (1982). "Miners, Millhands, and Mountaineers: Industrialization of the Appalachian South, 1880–1930." The University of Tennessee Press, Knoxville.

Francaviglia, R. (1991). "Hard Places: Reading the Landscape of America's Historic Mining Districts." University of Iowa Press, Iowa City.

Gaventa, J. (1980). "Power and Powerlessness, Quiescence and Rebellion in an Appalachian Valley." University of Illinois Press, Urbana.

Harvey, K. (1969). "The Best-Dressed Miners: Life and Labor in the Maryland Coal Region, 1835–1910." Cornell University, Ithaca, New York.

Hennen, J. (1996). "The Americanization of West Virginia: Creating a Modern Industrial State, 1916–1925." University Press of Kentucky, Lexington.

Lewis, R. (1987). "Black Coal Miners in America: Race, Class, and Community Conflict, 1780–1980." University Press of Kentucky, Lexington.

Roy, A. (1903). "A History of the Coal Miners of the United States: From the Development of the Mines to the Close of the Anthracite Strike of 1902 Including a Brief Sketch of British Miners." J. L. Trauger Printing Co., Columbus, Ohio.

Salstrom, P. (1994). "Appalachia's Path to Dependency: Rethinking a Region's Economic History, 1730–1940." University Press of Kentucky, Lexington.

Shifflett, C. (1991). "Coal Towns: Life, Work, and Culture in Company Towns of Southern Appalachia, 1880–1960." The University of Tennessee Press, Knoxville.

Thomas, J. (1998). "An Appalachian New Deal: West Virginia in the Great Depression." University Press of Kentucky, Lexington.

Williams, J. (1976). "West Virginia and the Captains of Industry." West Virginia University Library, Morgantown.

Coal Preparation

PETER J. BETHELL
Bethell Processing Solutions LLC
Charleston, West Virginia, United States

GERALD H. LUTTRELL
Virginia Polytechnic Institute & State University
Blacksburg, Virginia, United States

1. Introduction
2. Coal Sizing
3. Coal Cleaning
4. Solid-Liquid Separation
5. Miscellaneous

Glossary

ash A measure of coal purity that consists of the noncombustible residue that remains after coal is completely burned.

coking coal A coal product having properties that make is suitable for consumption by the metallurgical market for the manufacture of steel.

cut point The density corresponding to a particle that has an equal probability of reporting to either the clean coal or reject product of a density-based separator.

cut size The particle size that has an equal probability of reporting to either the oversize or undersize product of a sizing device such as a screen or hydraulic classifier.

heavy media An aqueous suspension of micron-sized particles (usually magnetite) that is used to create an artificial, high-density fluid that is passed through various coal washing processes to separate coal from rock based on differences in density.

middling A composite particle containing both organic and inorganic matter.

open area The percentage of the total cross-sectional area of a screening surface that is open for particles to pass. A larger open area improves screen capacity and efficiency.

organic efficiency An indicator of separator performance (usually expressed as a percentage) that is calculated by dividing the actual clean coal yield by the theoretical maximum yield that could be achieved at the same ash content according to washability analysis.

steam coal A coal product having properties that make it suitable for consumption by the power industry for the production of electricity.

throw The maximum displacement created by the rotary vibrator of a screening mechanism.

washability analysis A laboratory procedure used to assess the potential cleanability of coal that uses dense liquids (usually organic) to partition a particulate sample into various density fractions (also called float-sink tests).

washing A generic term used to describe the process by which mined material is treated by various particulate separation processes to remove unwanted impurities from the valuable organic matter (also called *cleaning*).

yield An indicator of separation performance (usually reported as a percentage) that is calculated by dividing the clean coal tonnage by the feed coal tonnage.

Coal preparation is the process in which mined coal is upgraded to satisfy size and purity specifications that are dictated by a given market. The upgrading, which occurs after mining and before shipping to market, is achieved using low-cost unit operations that include screening, cleaning, and dewatering. The process is typically required for economic reasons that are driven by a desire to improve utilization properties, reduce transportation costs, or minimize environmental impacts.

1. INTRODUCTION

1.1 Why Process Coal?

Mined coal is a heterogeneous mixture of both organic (carbonaceous) and inorganic (mineral) matter. The inorganic matter is comprised of a wide range of mineral components including shale, slate, and clay. During combustion, these impurities reduce the heating value of the coal and leave behind an ash residue as waste. The impurities also influence the suitability of the coal for high-end uses such as the manufacture of metallurgical coke or generation of petrochemicals. Some of the mineral impurities, such as pyrite, form gaseous pollutants when burned that are potentially harmful to the environment. The impurities may represent a significant portion of the mined coal weight and, as a result, need to be removed to minimize the cost of transporting the coal to the customer. For these reasons, a tremendous incentive exists for the removal of inorganic impurities prior to the downstream utilization of the coal. Coal purchasing agreements commonly include limitations on ash (mineral matter), sulfur, and moisture.

1.2 Brief History

Little attention was given to coal preparation in the early industrial years. Hand sorting of rock and sizing by screening were the major methods of preparing mined coals for market. Most of the fine (<0.5 mm) coal was either discarded or sold for very low prices. More recently, increased emphasis has been placed on coal preparation because of the depletion of higher quality reserves and increased levels of out-of-seam dilution resulting from mechanized mining. Consumers in the steam (utility) and metallurgical (coking) coal markets have also placed new demands on coal producers to provide consistent high-quality products. The combined influence of these factors has made coal preparation far more important than it was in the past. There are estimated to be more than 2000 plants operating worldwide with capacities ranging from as little as 100 ton per hour to more than several thousand ton per hour. In terms of annual production, coal preparation facilities produced just over 1.3 billion tons of the 3.5 billion tons of coal sold by the top 10 coal-producing countries in 1998. The United States produced more than one-third of this tonnage (35%), followed by China (20%), South Africa (14%), and Australia (13%). The average plant capacity in North America is approximately 750 ton per hour.

1.3 Typical Systems

Despite the perception of coal preparation as a simple operation, it has become far more complicated than most people realize. Modern coal preparation plants can be as complex and sophisticated as processing facilities used in the chemical industries. However, for simplicity, plant flowsheets can be generically represented by a series of sequential unit operations for sizing, cleaning, and dewatering (Fig. 1). This sequence of operations, which is commonly called a circuit, will be repeated several times since the processes employed in modern plants each have a limited range of applicability in terms of particle size. Modern plants may include as many as four separate processing circuits for treating the coarse (plus 50 mm), intermediate (50×1 mm), fine (1×0.15 mm), and ultrafine (minus 0.15 mm) material. Although some commonalities exist, the selection of what number of circuits to use, which types of unit operations to employ, and how they should be configured are highly subjective and largely dependent on the inherent characteristics of the coal feedstock.

2. *COAL SIZING*

2.1 Background

Run-of-mine coal produced by mechanized mining operations contains particles as small as fine powder and as large as several hundred millimeters. Particles too large to pass into the plant are crushed to an appropriate upper size or rejected where insufficient recoverable coal is present in the coarse size fractions. Rotary breakers, jaw crushers, roll crushers, or sizers are used to achieve particle size reduction. Crushing may also improve the cleanability of the coal by liberating impurities locked within composite particles (called middlings) containing both organic and inorganic matter. The crushed material is then segregated into groups having well-defined maximum and minimum sizes. The sizing is achieved using various types of equipment including screens, sieves, and classifying cyclones. Screens are typically employed for sizing coarser particles, while various combinations of fine coal sieves and classifying cyclones are used for sizing finer particles. Figure 2 shows the typical sizes of particles that can be produced by common types of industrial sizing equipment.

2.2 Screening Principles

Screens are mechanical sizing devices that use a mesh or perforated plate to sort particles into an undersize product (particles that pass through the openings in the screen) and an oversize product (particles that are retained on the screen surface). In coal preparation, screens are used for a wide range of purposes including the extraction of large rock and trash prior to treatment and the production of optimum size

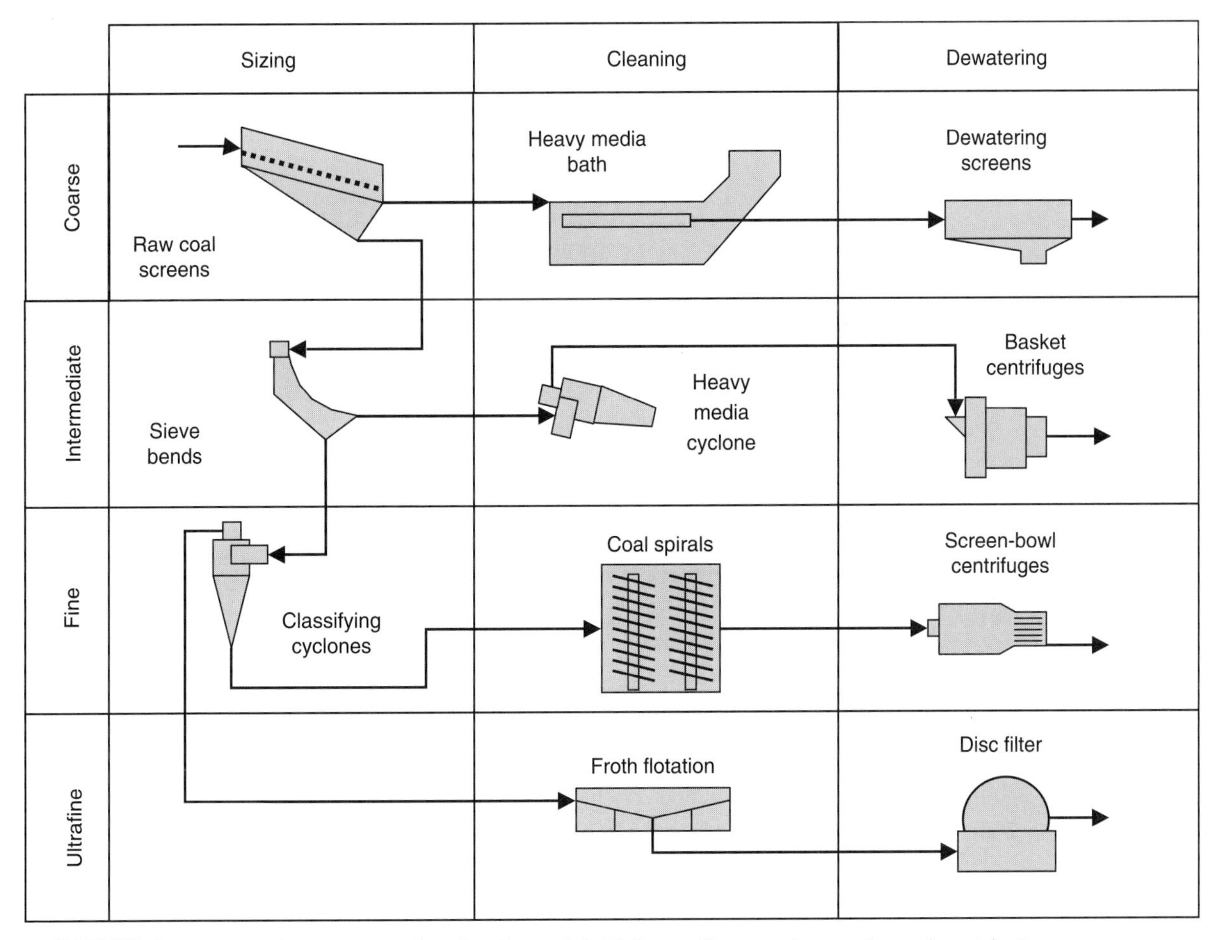

FIGURE 1 Conceptual preparation plant flowsheet subdivided according to unit operation and particle size.

fractions for separation devices. In addition, screens are commonly employed for the recovery of heavy media suspensions and the separation of water from solid particles. The efficiency of the screening operation is influenced by the material properties of the feed and the operating and design characteristics of the machine, particularly its size.

2.3 Types of Screens

Several types of screens are commonly used in coal preparation. These include the grizzly, vibrating screens and high-frequency screens. A grizzly is a series of evenly spaced parallel bars that are positioned in the direction of material flow. This simple device is used only for gross sizing of very coarse particles (>50 mm) and is generally applied only for the protection of equipment from large oversized material.

Vibrating screens are the most common equipment for sizing particles and are used in almost every aspect of coal preparation. These units utilize an out-of-balance rotating mechanism to create a vibrating motion to sort particles and to move material along the screen surface (called a deck). The three common versions of vibrating screens are the incline screen, horizontal screen, and banana screen. The operating characteristics of these units are compared in Table I. An incline screen uses a combination of gravity flow and machine movement to convey material along the screen by mounting the screen deck 15 to 20 degrees from horizontal. Incline screens are available in single-, double-, and triple-deck arrangements with dimensions up to 3 m (10 ft) wide and 7.3 m (24 ft) long. For a horizontal screen, material is conveyed without the aid of gravity, thus a higher speed and throw are required. These units may be as large as 3 m (10 ft) wide and 7.3 m (24 ft) long and are available in single-, double-, and triple-deck arrangements. Finally, banana screens are designed so that the slope of the screen surface changes along the length of the unit (Fig. 3). The slope at the feed end will typically be 30

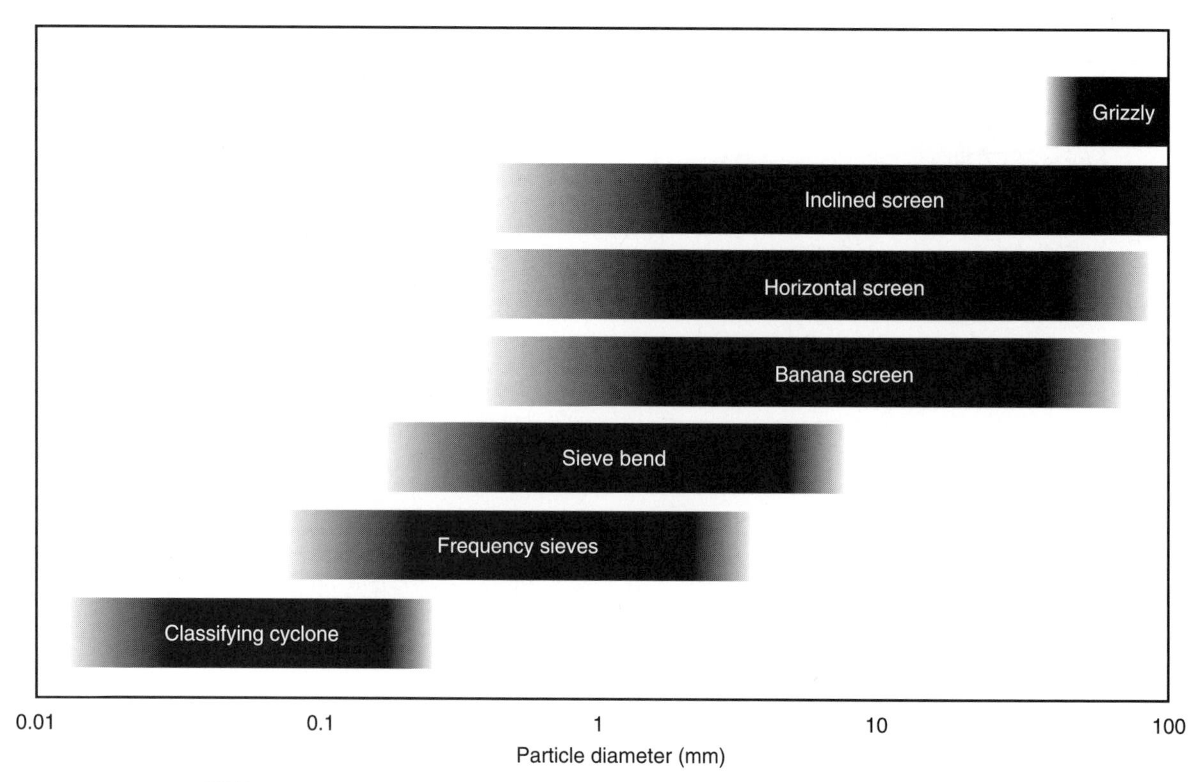

FIGURE 2 Typical ranges of applicability for various sizing equipment.

TABLE I

Typical Operating Characteristics of Vibrating Screens

	Incline	Horizontal	Banana	High frequency
Application	Raw coal sizing scalping	Raw coal sizing prewet drain and rinse dewatering	Raw coal sizing deslime drain and rinse dewatering	Dewatering fines recovery
Incline (degrees)	15–25	Horizontal	Variable	Opposing
Motion	Circular	Linear	Linear	Linear
Speed (rpm)	750–800	930	930	1200
Throw	9.6–11.2 mm (0.38–0.44 inch)	11.2 mm (0.44 inch)	11.2 mm (0.44 inch)	6.4 mm (0.25 inch)
Travel rate	0.5–0.6 m/s (100–120 fpm)	0.2–0.23 m/s (40–45 fpm)	0.9–1.0 m/s (180–200 fpm)	0.13–0.18 m/s (25–35 fpm)

to 35 degrees, while the slope at the discharge end can be 0 to 15 degrees. The steep initial slope enhances the ability of fines to pass since the higher rate of travel reduces the depth of the particle bed. The shallower slope at the discharge end of the screen increases screening efficiency by slowing the travel rate and providing more opportunities for near-size particles to pass. Commercial units are available with dimensions exceeding 3.7 m (12 ft) wide and 7.3 m (24 ft) long in single- and double-deck arrangements. Banana screens are becoming increasingly popular since they offer increased capacity and lower capital and operating costs compared to conventional horizontal and inclined screens.

High-frequency screens are normally used for dewatering fine coal or refuse (Fig. 4). These screens are designed to retain feed slurry between two opposing inclines. The screen motion transports dewatered cake up a gradual slope to the discharge end of the screen. The screen vibrator, which may be an unbalanced rotor or linear exciter, operates at a higher speed and with a shorter thrown than vibrating

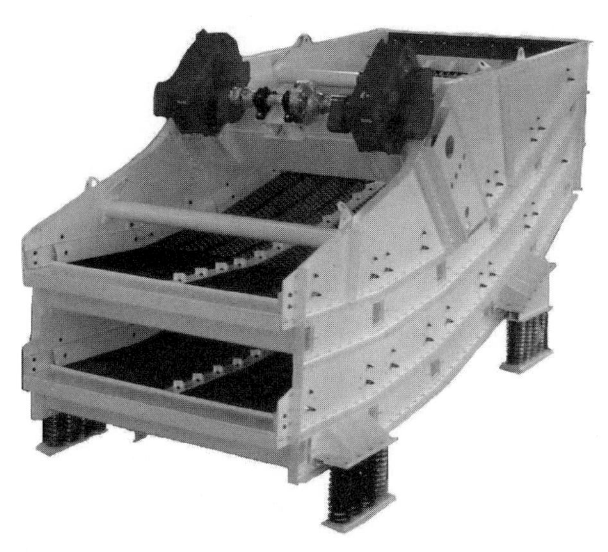

FIGURE 3 Production-scale banana (multislope) screen. Courtesy of Innovative Screen Technologies.

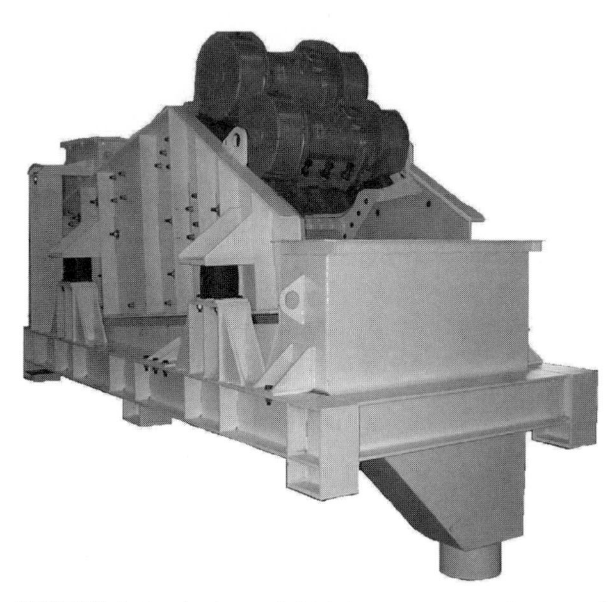

FIGURE 4 Production-scale high frequency screen. Courtesy of Innovative Screen Technologies.

screens. The operating characteristics are beneficial for the efficient dewatering of fine solids. Commercial units are now available in single-deck configurations that exceed 2.4 m (8 ft) wide and 3.7 m (12 ft) long.

2.4 Screen Surfaces

A wide variety of screen surfaces are commercially available for coal sizing. The most common types of media include punch plate, woven wire, rubber, urethane, and profile wire (Fig. 5). The selection of the most appropriate media is dictated by type of

A

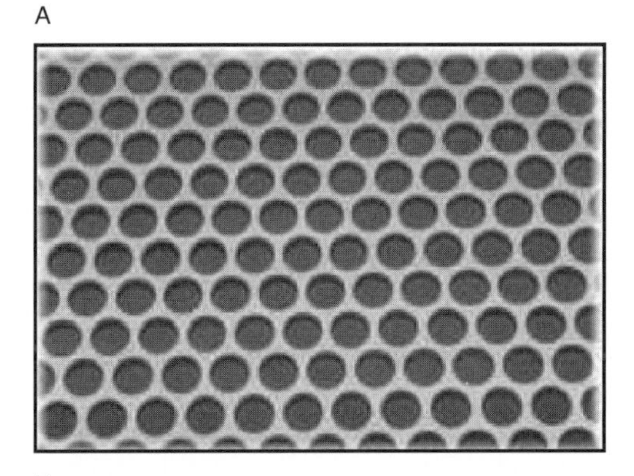

B

C

D

FIGURE 5 Common types of screening media. (A) punch plate, (B) woven wire, (C) polyurethane, and (D) profile wire.

application and considerations associated with wear and maintenance. Punch plate typically consists of steel sheets that are perforated with square, round, or rectangular holes. This type of media is used for primary sizing. Woven wire media consists of fine wires that are woven together to form a mesh pattern with the desired hole size. Also, the woven mesh must be strong and wear resistant (which requires larger wires) and still possess sufficient open area for the efficient passage of particles (which requires smaller wires). The final selection is usually based on an economic tradeoff between replacement frequency and sizing efficiency. Both rubber and urethane media can be manufactured in modular panels that can be easily installed and replaced. These materials offer a longer wear life and are less likely to become plugged than other types of screening media. Urethane panels can be manufactured with openings ranging from 0.15 to 100 mm. The major concern with this type of media is the loss of open area that normally occurs when they are used to replace woven wire media. In cases where fine (3×0.5 mm) sizing is performed, profile wire may be used to reduce problems with plugging and to increase the open area of the screen. The wires in this type of media are constructed with tapered cross sections (usually an inverted triangle or trapezoid) that allow particles to freely pass once they have cleared the top surface of the screen media. This type of screening media improves open area and screen capacity but typically has a shorter wear life.

2.5 Fine Coal Sieves

Sizing of fine coal is difficult due to the increased likelihood of plugging and the large number of particles that must pass through the screen surface. Fine sizing is typically performed either utilizing banana screens or sieve bends in conjunction with a horizontal screens. A sieve bend (Fig. 6) consists of a curved panel constructed from inverted trapezoidal bars (profile or wedge wire) placed at right angles to fluid flow. The capacity of the sieve is relatively high since the leading edge of the trapezoidal bars "slice" material from the flowing stream. Although the screen has no moving parts, passage of solids is also greatly enhanced by the centrifugal force created by the fluid flow along the curved sieve surface. Because of the unique geometry, the size retained by the sieve is approximately three-fourths the bar spacing (e.g., 0.5 mm spacing retains 0.38 mm particles). Sieve bends are commonly used for fine particle dewatering and are often used ahead of vibrating screens. A variation on the sieve bend theme incorporates the use of high-frequency vibration.

FIGURE 6 Production-scale sieve bend for fine coal sizing. Courtesy of Conn-Weld Industries.

2.6 Classification

Classification is used in coal preparation for fine sizing where conventional screening becomes impractical or too expensive. Classification is usually applied to dilute suspensions of <1 mm diameter particles. This technique is generally less efficient than screening or sieving since the separation can be influenced by particle shape and density, as well as particle size. The most common type of classification device used in the coal industry is the classifying cyclone (Fig. 7). In this device, dilute slurry is pumped under pressure to multiple parallel units (banks) of cyclones. The rotating flow field within the cyclone creates a centrifugal force that increases the settling rate of the particles. Larger particles are forced to the wall of the cyclone where they are carried to the discharge port (apex) at the bottom of the cyclone. Finer particles exit out the top of the cyclone (vortex finder). Classifying cyclones are commonly applied to size (cut) at 0.10 to 0.15 mm and represent the only practical option for sizing ultrafine particles (at a cut of 0.045 mm). Smaller

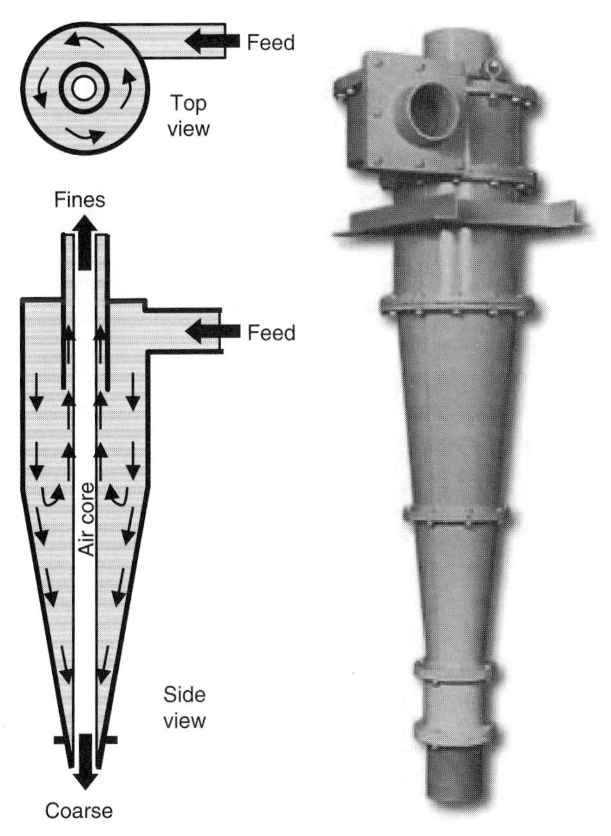

FIGURE 7 Illustration of a classifying cyclone. Courtesy of Krebs Engineers.

cyclones are typically used to size finer particles since these units provide a higher centrifugal force. Classifying cyclones have become very popular in the coal industry since they offer a high unit capacity, operational flexibility, and low maintenance requirements.

3. COAL CLEANING

3.1 Background

One of the primary goals of coal preparation is to separate the valuable carbonaceous material (organic matter) from contaminants in the surrounding host rock (mineral matter). The separation is typically accomplished using low-cost processes that exploit differences in the physical properties that vary according to mineralogical composition. Some of the common properties that are used to separate coal and rock include size, density, and surface wettability. The effectiveness of the separation depends largely on the characteristic properties of the feed material and the operational efficiency of the separator. The effectiveness of most separators is limited to a relatively narrow range of particles sizes (Fig. 8) due to limitations associated with processing efficiency and equipment geometry. Several different types of cleaning processes are incorporated into the plant flowsheet to treat the wide range of particle sizes present in the run-of-mine coal. Density separators, such as heavy media baths and cyclones, are commonly used to treat coarse and intermediate sized particles. Froth flotation, which separates particles based on differences in surface wettability, is the only practical method for treating very fine coal (<0.15 mm). Particles in the intermediate size range between 0.15 mm and 2 to 3 mm may be treated by a variety of processes including water-only cyclones, spirals, teeter bed separators, or combinations of these units. Each of these units is described in greater detail in the following sections.

3.2 Coal Washability

Most of the physical separation processes used in coal preparation take advantage of the differences in the density of coal and rock. Coal preparation terminology commonly interchanges density and specific gravity (a dimensionless parameter defined as the density of a substance divided by the density of water). The specific gravity (SG) for pure coal is approximately 1.3, while the SG for common coal contaminants such as shale and sandstone are in the 2.3 to 2.7 SG range. The density of pyrite ($SG = 4.8$) is significantly higher. Unfortunately, mined coal does not consist of pure coal and pure rock. Instead, middlings material consisting of composite particles of varying composition are also present to a greater or lesser extent. A typical distribution of the composite material is shown in Table II. There is an approximate linear relationship between the density of the composite particle and its ash (mineral matter) content. The SG at which preparation plants operate is adjusted to provide operating levels which satisfy customer quality constraints.

3.3 Coarse Coal Cleaning (Plus 12.5 mm)

One of the oldest forms of coarse coal cleaning is jigging. Jigging is the process of particle stratification due to alternate expansion and compaction of a bed of particles by a vertical pulsating fluid flow. A single jigging vessel can treat large particles (up to 200 mm) at capacities approaching 680 tonnes (750 tons) per hour. Common versions include the baum and batac designs. The baum jig, which is most common, uses a

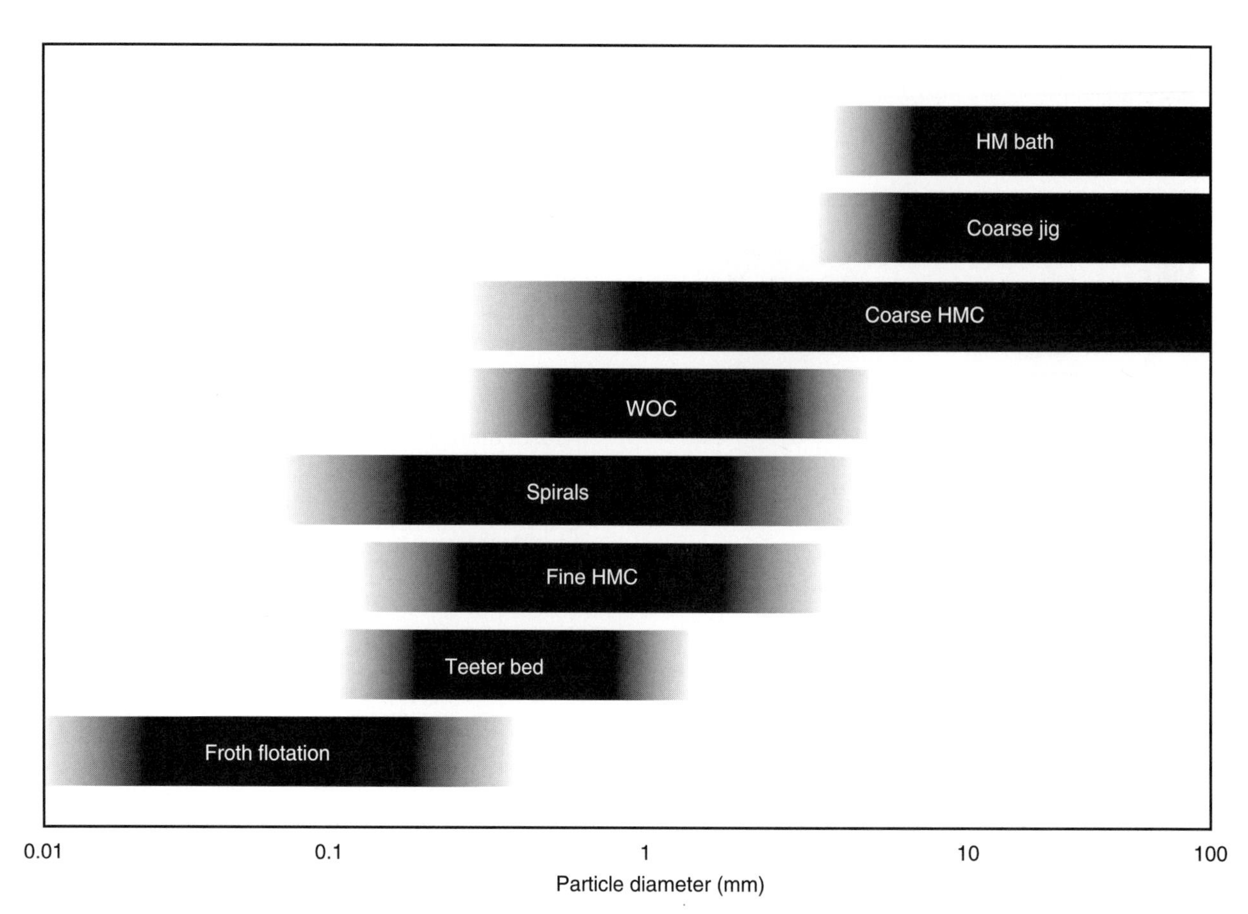

FIGURE 8 Typical ranges of applicability for various cleaning equipment.

U-shaped vessel that is open on one size and closed on the other (Fig. 9). Water is pulsed through the particle bed atop the perforated plate by sequentially injecting and releasing air in and out of the enclosed chamber. A batac jig is a variation of this design that uses an air chamber below the perforated plant to trap air and create the pulsing action. Jig installations require specific conditions in order to be effective and are significantly less efficient than competitive heavy media processes. As a result, the use of jigs for coal cleaning has declined rapidly. However, jigs can treat a wider size range than single heavy media processes and provide an obvious advantage in applications where capital is limited.

The most common process for upgrading coarse coal (>12.5 mm) is the heavy media bath. A common type of heavy media bath is the Daniels vessel (Fig. 10). This unit consists of a large open vessel through which an aqueous suspension of finely pulverized magnetite (SG = 5.2) is circulated. The magnetite creates a suspension with an artificial density that is between that of pure coal and pure rock. Lighter coal particles introduced into media float to the surface of the vessel where they are transported by the overflowing media to a collection screen. Waste rock, which is much denser, sinks to the bottom of the vessel where it is collected by a series of mechanical scrapers (called flights). The flights travel across the bottom of the vessel and eventually drag the waste rock over a discharge lip at one end of the separator. The unit is highly flexible, since the quality of the clean coal product can be readily adjusted by varying the density of the media from 1.3 to 1.7 SG by controlling the amount of magnetite introduced into the suspension. Sharp separations are possible even in the presence of large amounts of middlings, since SG can be controlled very precisely (as close as 0.005 SG units in some cases) using online instrumentation. A wide variety of different designs of heavy media baths exist, with the major differences being the mechanisms used to discharge the clean coal and waste rock products. Common examples include heavy media cones, drums, and several variations of deep and shallow baths. Drums, usually Wemco or Teska, are widely

TABLE II
Example of Coal Washability Data

Specific gravity		Individual		Cumulative float		Cumulative sink	
Sink	Float	Weight (%)	Ash (%)	Weight (%)	Ash (%)	Weight (%)	Ash (%)
	1.30	47.80	4.20	47.80	4.20	100.00	28.38
1.30	1.40	15.60	14.50	63.40	6.73	52.20	50.52
1.40	1.50	6.60	22.80	70.00	8.25	36.60	65.88
1.50	1.60	2.20	31.20	72.20	8.95	30.00	75.35
1.60	1.70	2.10	39.60	74.30	9.81	27.80	78.85
1.70	1.90	5.60	62.50	79.90	13.51	25.70	82.05
1.90		20.10	87.50	100.00	28.38	20.10	87.50
		100.00	28.38				

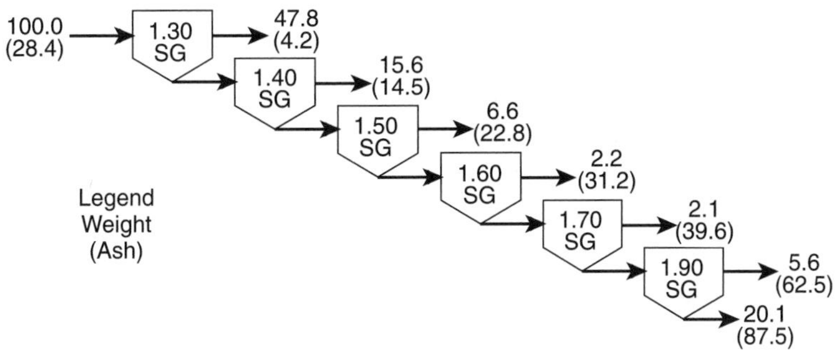

100.0 (28.4) → 1.30 SG → 47.8 (4.2)
1.40 SG → 15.6 (14.5)
1.50 SG → 6.6 (22.8)
1.60 SG → 2.2 (31.2)
1.70 SG → 2.1 (39.6)
1.90 SG → 5.6 (62.5); 20.1 (87.5)
Legend
Weight
(Ash)

used in South Africa and to a certain extent in Australia. This is where coals are difficult to wash and very high efficiency levels are required.

The primary disadvantage of a heavy media process is the ancillary equipment that must be installed to support the operation of the separator (Fig. 11). A typical circuit requires that the clean coal and refuse products pass over drain-and-rinse screens to wash the media from the surfaces of the products and to dewater the particles. The media that is recovered from the drain section is circulated back to the separator by means of a pump, while media from the rinse section is diluted by spray water and must be concentrated before being fed back to the separator. Fortunately, magnetite is magnetic and can be readily recovered from a dilute suspension using magnetic separators. A rotating drum magnetic separator is commonly used for this purpose (Fig. 12). These devices are highly efficient and generally recover >99.9% of the magnetite to the magnetic product. For cost reasons, losses of magnetite must be kept as low as possible. Typical losses fall in the range of about 0.45 kg of magnetite loss per tonne (1 lb per ton) of feed coal treated. A portion of the circulating media from the drain section may also be diverted (bled) to the magnetic separator so that the buildup of contaminants such as ultrafine clay can be prevented. Such a build increases the viscosity of the circulating media and adversely impacts the separation efficiency.

Heavy media processes make use of a variety of instrumentation to monitor and control the performance of the separator. In most cases, this consists of a gamma (nuclear) density gauge or other device that continuously monitors the specific gravity of the circulating media. Automated controls are used to add water or magnetite so as to maintain a constant density in the circulating media.

3.4 Intermediate Coal Cleaning (12.5 × 1 mm)

Heavy media cyclones, which were originally developed by the Dutch State Mines, are the most effective method for treating coal particles in the size range between 1.0 and 12.5 mm. These high-capacity devices (Fig. 13) make use of the same basic principle as heavy media baths—that is, an artificial media is used to separate low-density coal from high-density rock. In this case, however, the rate of separation is greatly increased by the gravitational effect created by passing media and coal through

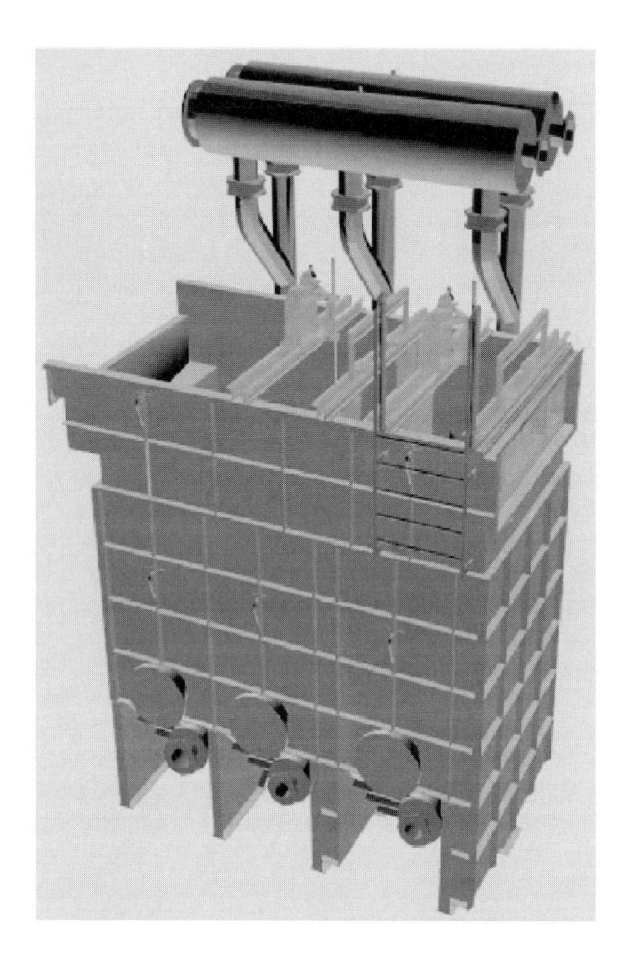

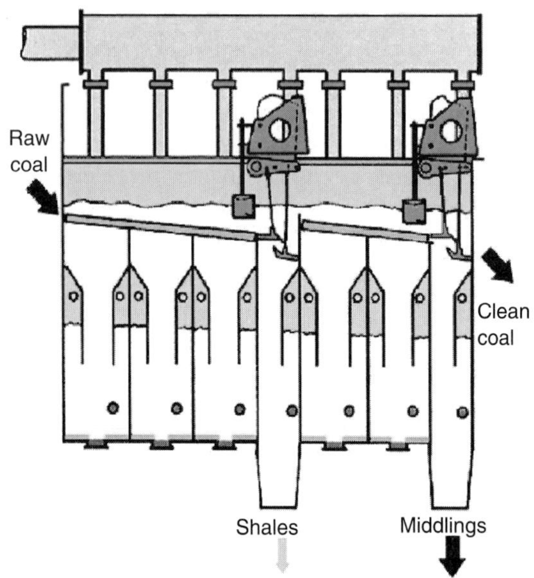

FIGURE 9 Illustration of a baum jig. Courtesy of Bateman.

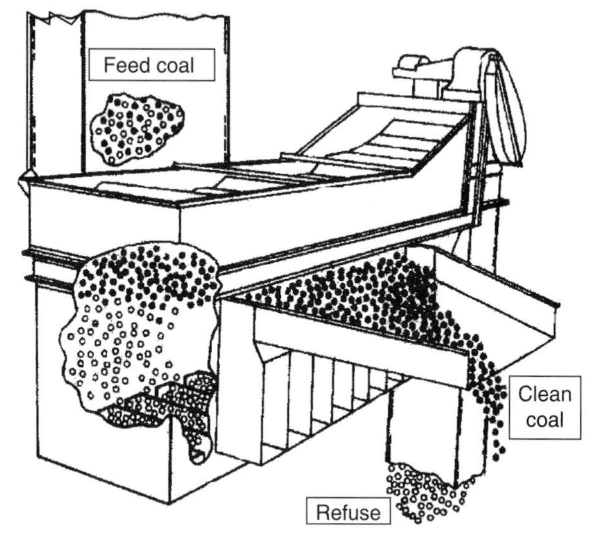

FIGURE 10 Illustration of a heavy media bath. Courtesy of Peters Equipment.

one or more cyclones. Heavier particles of rock sink in the rotating media and are pushed toward the wall of the cyclone where they are eventually rejected out the bottom. Lighter particles of coal rise to the surface of the rotating media and are carried with the bulk of the media flow out the top of the cyclone. Major advantages of heavy media cyclones include a high separation efficiency, high unit capacity, and ease of control. As with baths, heavy media cyclones have the disadvantage of higher costs associated with the ancillary operations required to support the separator. Several different configurations are used to feed heavy media cyclones. These include gravity flow, pump fed, and wing tanks. Of these, the pump fed configuration (Fig. 14) offers the greatest operational flexibility. The circuit is similar in function to that employed by a heavy media bath.

Advances in cyclone technology and improved wear-resistant materials (ceramic linings) have allowed the upper particle size limit to increase in many applications to 60 to 75 mm. Massive

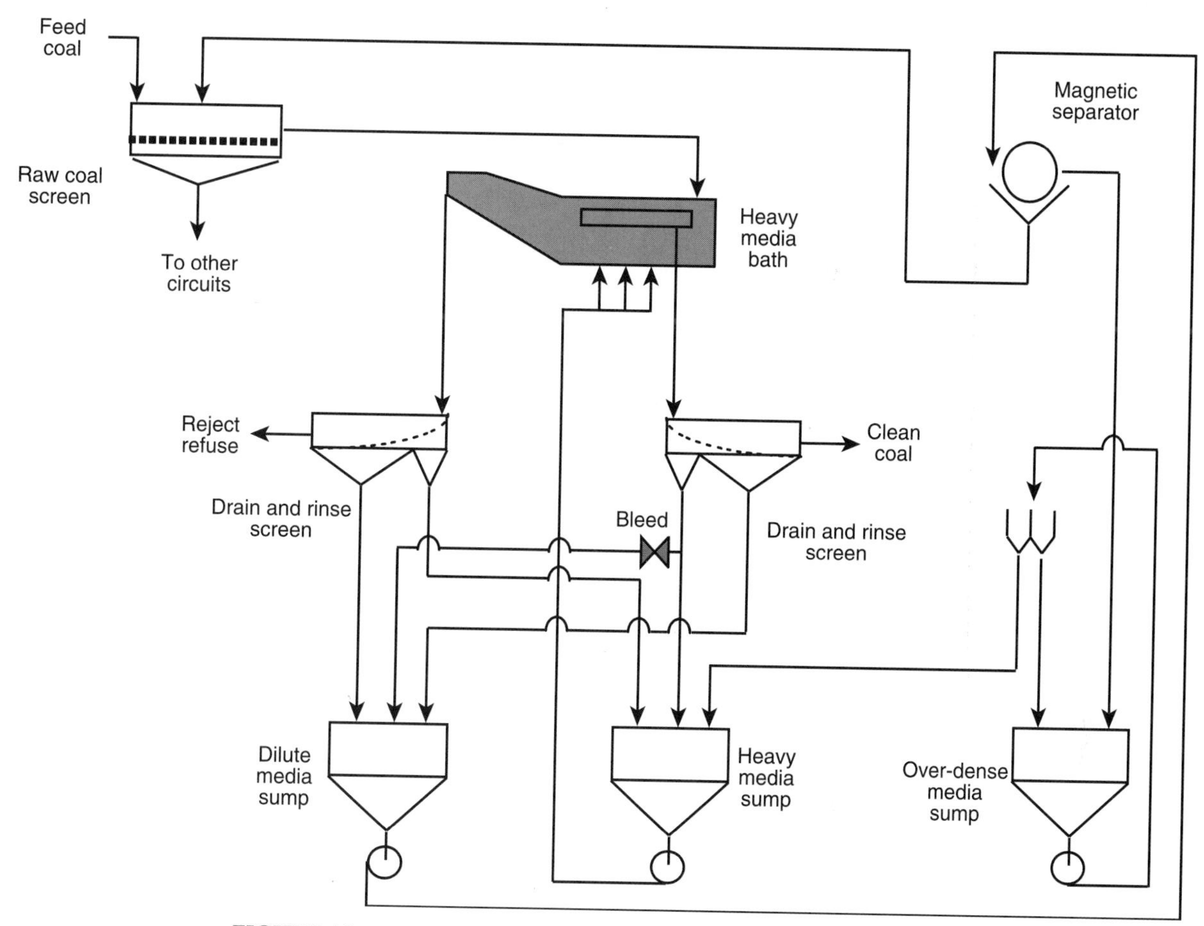

FIGURE 11 Simplified schematic of a heavy media bath circuit.

increases in throughput capacity have also been realized through the use of larger cyclones having diameters of 1.2 m and more. A single unit of this size is capable of treating more than 360 to 410 tonnes per hour (400 to 450 tons per hour) of feed coal. Unfortunately, increased diameter is accompanied with poorer efficiency, particularly in the finer size fractions. The ability to increase both throughput and the upper particle size for cyclones makes it possible to eliminate the heavy media bath in many new plants. This capability is particularly useful for those plant treating very friable coals containing minimal quantities of larger (>50 mm) particles. The simplified circuitry also has significant benefits in terms of capital cost and ease of maintenance. For high reject or hard coals, significant advantages are still present for the bath/cyclone circuit. This results from the reduced necessity for crushing, reduced wear in the plant, and improved total product moisture content. If low density separations are required (<1.4 SG), then the two circuit plant incorporating both baths and cyclones may be necessary since low density separations are easy in baths but difficult in cyclones.

Another option for cleaning coarse coal is a large coal dense media separator, known as a larcodem. This unit was originally developed to handle coal feeds in the size range between 0.6 and 100 mm. The device has the major advantage of being able to treat in one process what was traditionally processed in two separate heavy media devices (heavy medium bath and heavy medium cyclone). Larcodems can also clean far more efficiently what was previously fed to jigging operations. Capital and operating costs for circuits incorporating Larcodems are considerably less that would be the case for traditional heavy medium designs. Despite these advantages, Larcodems have not been widely used in the United States due to lower fine coal cleaning efficiency and the ability of the new generation of cyclones to handle similar top-size material.

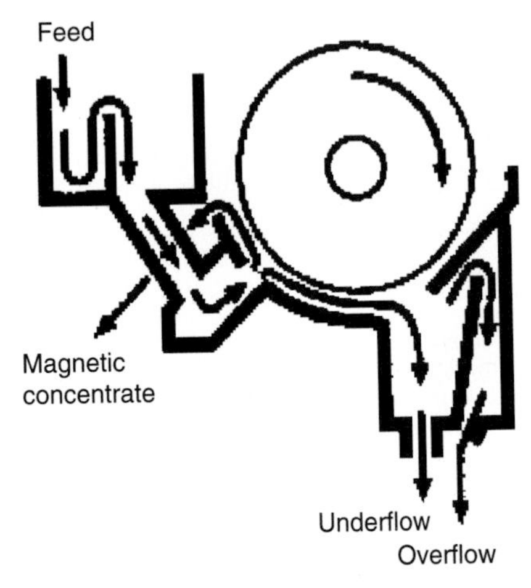

FIGURE 12 Illustration of a drum-type magnetic separator. Courtesy of Eriez Manufacturing.

3.5 Fine Coal Cleaning (1 × 0.15 mm)

A wide variety of options are available for upgrading fine coal in the size range between 0.15 and 1 mm. Common methods include water-only cyclones, spirals, and teeter-bed separators. A water-only cyclone (WOC) is a water-based process that employs no extraneous media to separate rock from coal. Water-only cyclones were introduced in the early 1950s for the treatment of fine coal (<5 mm). Water-only cyclones are similar to classifying cyclones, but typically have a somewhat stubby wide-angled conical bottom and are usually equipped with a long axially mounted vortex finder. Separations occur due to differences in the settling rates of coal and rock in the centrifugal field within the cyclone body. The separation is also enhanced by the formation of autogenous media created by the natural fines already present in the feed slurry. These units are often employed in two stages or in combination with other water-based separators to improve performance. Water-only cyclones do offer some advantages for selected applications including high throughput capacity, operational simplicity, and low operating and maintenance costs.

A separator that has grown tremendously in popularity is the coal spiral (Fig. 15). A spiral separator consists of a corkscrew shaped conduit (normally four turns) with a modified semicircular cross section. The most common units are constructed from polyurethane-coated fiberglass.

FIGURE 13 Illustration of a heavy media cyclone bank. Courtesy of Krebs Engineers.

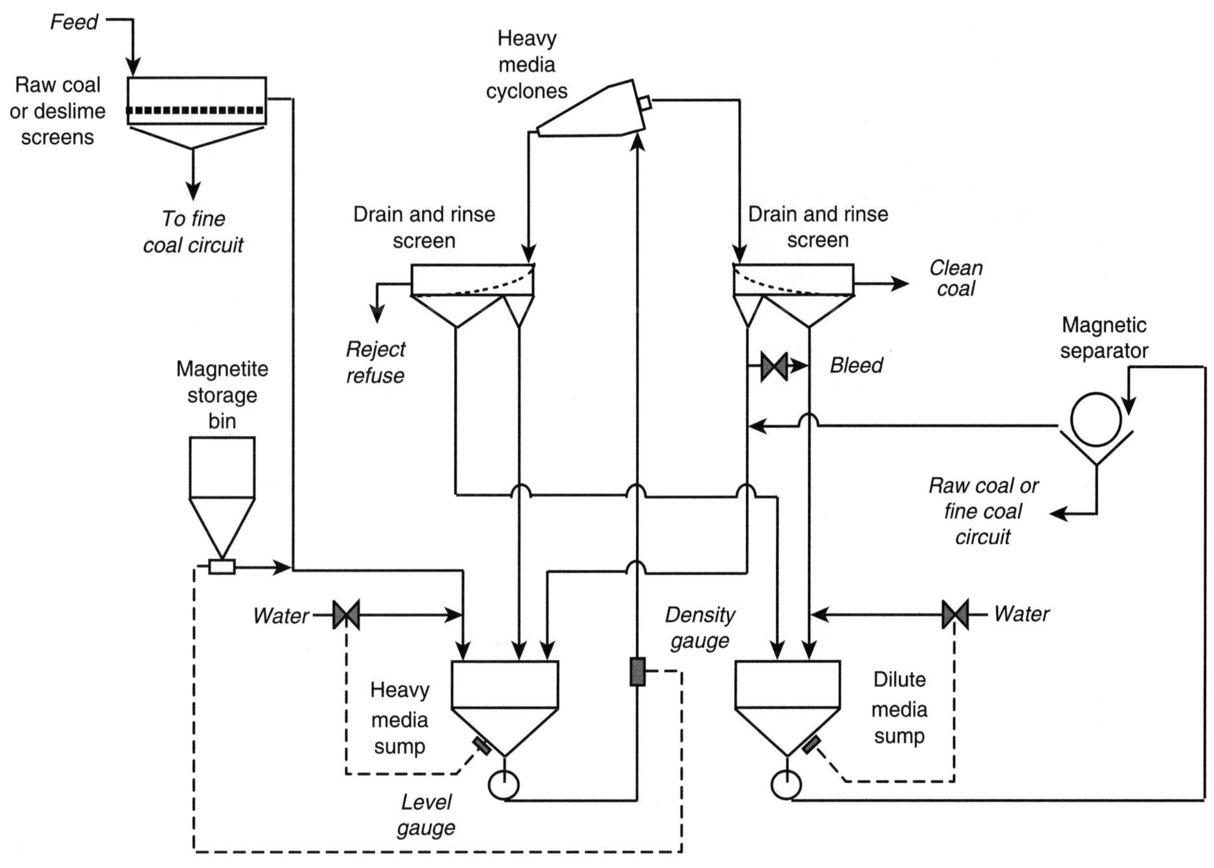

FIGURE 14 Simplified schematic of a pump fed heavy media cyclone circuit.

During operation, feed slurry is introduced to the top of the spiral and is permitted to flow by gravity along the helical path to the bottom of the spiral. Particles in the flowing film are stratified due to the combined effects of differential settling rates, centrifugal forces, and particle-particle contacts. These complex mechanisms force lighter coal particles to the outer wall of the spiral, whereas heavier particles are forced inward to the center of the spiral. The segregated bands of heavy to light materials are collected at the bottom of the spiral. Adjustable diverters (called splitters) are used to control the proportions of particles that report to the various products. A three-product split is usually produced, giving rise to three primary products containing clean coal product, middlings, and refuse. Because of the low unit capacity 1.8 to 3.6 tonnes per hour (2 to 4 tons per hour), spirals are usually arranged in groups fed by an overhead radial distributor. To save space, several spirals (two or three) may be intertwined along a single central axis. In practice, the middlings product should be retreated using a secondary set of spirals to maximize efficiency. The secondary circuit, which is often 10 to 20% of the size of the primary circuit, enables any misplaced material from the primary circuit to be captured. Circuits have been designed to use spirals in treating wider size ranges (e.g., 0.1×3.0 mm), but performance has been poor and efficient spiral cleaning at present would appear to be limited to the 0.15×1.5 mm size range. The main advantages of the spiral circuits are comparatively low capital and operating costs coupled with ease of operation. A major drawback of spiral circuitry is the inability to make a low gravity separation since spirals normally cut in the 1.7 to 2.0 SG range. A consistent volumetric feed rate and minimum oversize (>2 mm) and undersize (<0.1 mm) in spiral feed are also essential requisites to good operation. Spirals have been successfully utilized in combination with water-only cyclones to improve the efficiency of separating fine coal. Latest spiral technologies incorporate primary and secondary cleaning in a single unit. Clean coal and middlings being produced in the upper section are reprocessed in an extended spiral totaling seven turns.

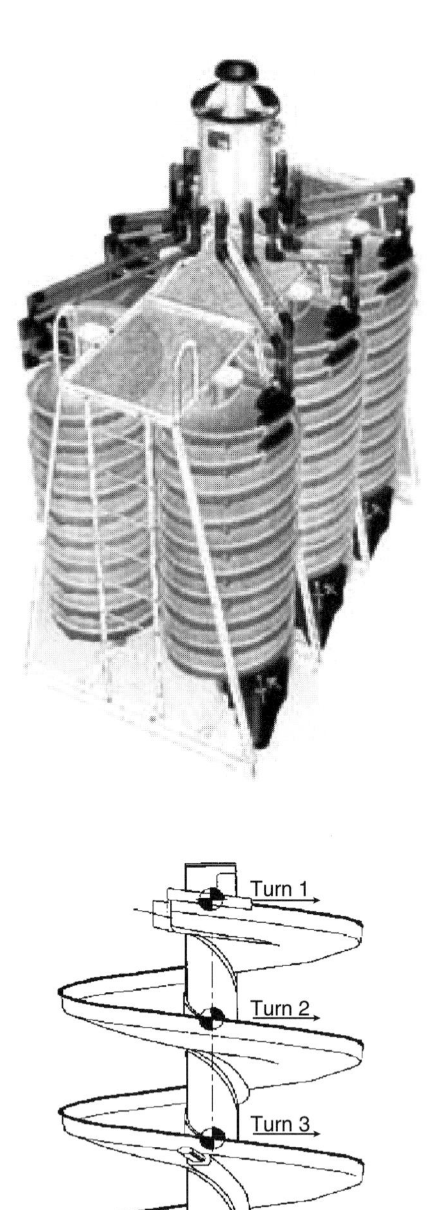

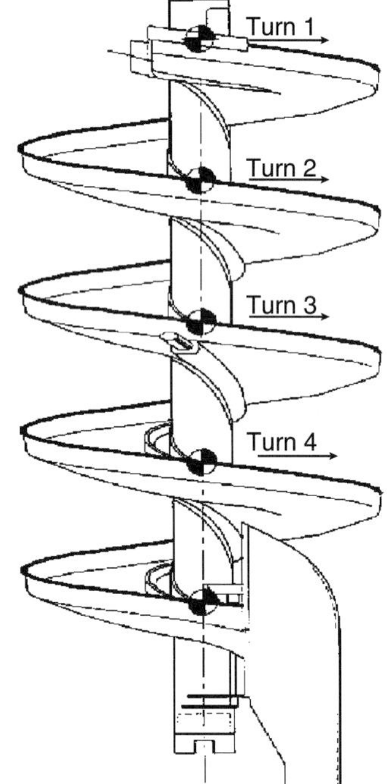

FIGURE 15 Illustration of a coal spiral. Courtesy of Multotec.

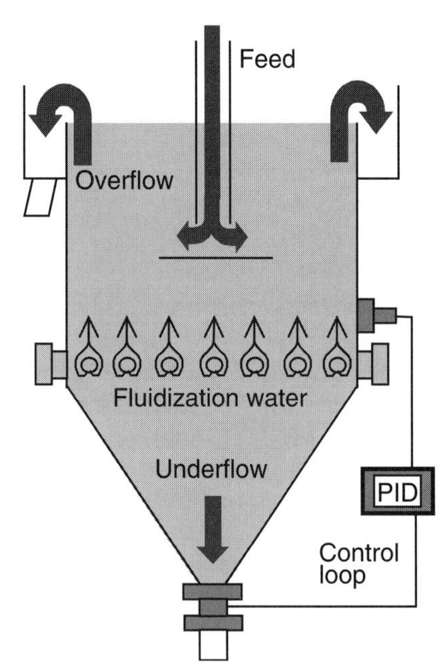

FIGURE 16 Illustration of a teeter-bed separator. Courtesy of Eriez Manufacturing.

Heavy media cycloning of fine coal is practiced, but usually only where a low SG of separation is required. High efficiencies and low impurity levels (ash) are possible. Operating and capital costs are high.

Teeter bed separators (Fig. 16) are gaining popularity, particularly in Australia. Separation is achieved as a result of the different setting rates of coal and rock in water currents. These units have high capacity and can make lower SG cut points than spirals. The major unit disadvantage is the limited size range which can be efficiently treated (4:1 top size to bottom size).

3.6 Ultrafine Coal Cleaning (<0.15 mm)

The most widely accepted method for upgrading ultrafine coal is froth flotation. Froth flotation is a physicochemical process that separates particles based on differences in surface wettability. Flotation takes place by passing finely dispersed air bubbles through an aqueous suspension of particles (Fig. 17). A chemical reagent, called a frother, is normally added to promote the formation of small bubbles. Typical addition rates are in the order of 0.05 to 0.25 kg of reagent per tonne (0.1 to 0.5 lb per ton) of coal feed. Coal particles, which are naturally hydrophobic (dislike water), become selectively attached to air bubbles and are carried to the surface of the pulp. These particles are collected from a coal-laden froth bed that forms atop the cell due to the frother addition. Most of the impurities that associate with coal are naturally hydrophilic (like water) and remain suspended until they are discharged as dilute slurry waste. Another chemical additive, called a collector, may be added to improve the adhesion between air bubbles and coal particles. Collectors are commonly hydrocarbon liquids such as diesel fuel or fuel oil. In the United States, flotation is typically performed on only the finest fractions of coal (<0.1 mm), although coarse particle flotation (<0.6 mm) is practiced in Australia where coal floatability is high and the contaminants in the process feed low. The removal of clay slimes (<0.03 mm) is carried out ahead of some flotation circuits to minimize the carryover of this high ash material into the froth product. The ultrafine clay slimes are typically removed using large numbers of small diameter (15 mm or 6 inch) classifying cyclones (Fig. 18).

Most of the industrial installations of flotation make use of mechanical (conventional) flotation machines. These machines consist of a series of agitated tanks (four to six cells) through which fine coal slurry is passed. The agitators are used to ensure that larger particles are kept in suspension and to disperse air that enters down through the rotating shaft assembly. The air is either injected into the cell using a blower or drawn into the cell by the negative pressure created by the rotating impeller. Most commercial units are very similar, although some variations exist in terms of cell geometry and impeller design. Industrial flotation machines are now available with individual cell volumes of 28.3 m^3 (1000 ft^3) or more. Coal flotation is typically performed in a single stage with no attempt to reprocess the reject or concentrate streams. In some cases, particularly where high clay concentrations are present, advanced flotation processes such as column cells have been used with great success. Conventional flotation cells allow a small amount of clay slimes to be recovered with the water that reports to the froth product. A column cell (Fig. 19) virtually eliminates this problem by washing the clay slimes from the froth using a countercurrent flow of wash water. This feature allows columns to produce higher quality concentrate coals at the same coal recovery.

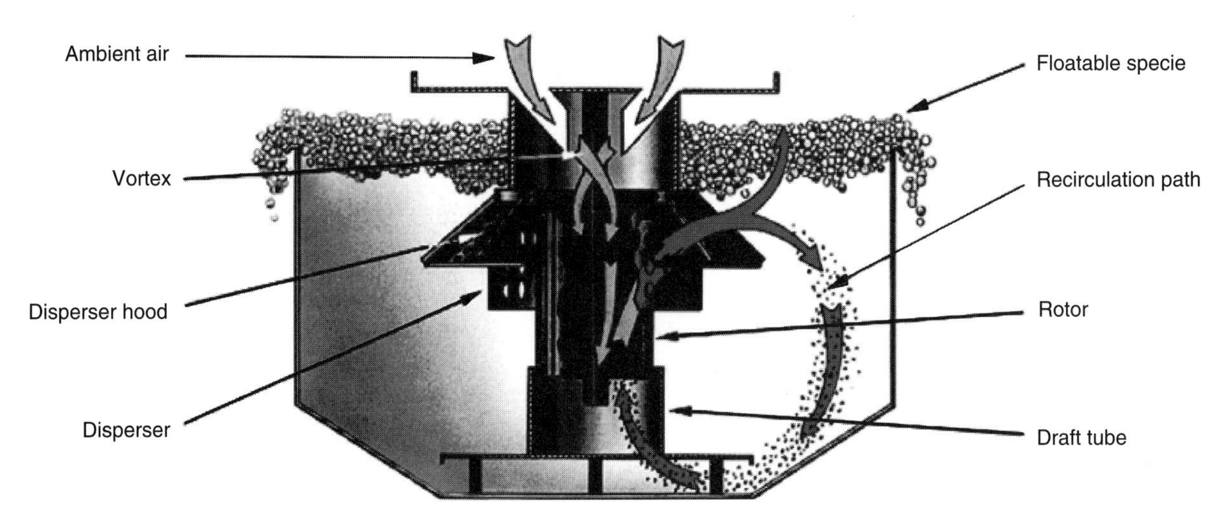

FIGURE 17 Illustration of a conventional flotation cell. Courtesy of Wemco.

FIGURE 18 Bank of small diameter classifying cyclones. Courtsey of Multotec.

4. SOLID-LIQUID SEPARATION

4.1 Background

Practically all of the processes employed in coal preparation use water as the medium. After treatment, the coal must be dewatered. The purpose of dewatering is to remove water from the coal to provide a relatively dry product. A coal product with an excessively high moisture content possesses a lower heating value, is subject to freezing, and is more costly to transport.

4.2 Mechanical Dewatering

A variety of methods are used to mechanically dewater the different sizes of particles produced by coal cleaning operations (Fig. 20). Dewatering of coarse coal is predominantly carried out using screens or centrifugal dryers. For screens, the prime force involved in dewatering is gravity. This is a relatively simple process for particles larger than approximately 5 mm. Dewatering screens have the advantage of being simple, effective, and inexpensive. Disadvantages include blinding and poor performance for smaller particles. Finer particles, which have a much higher surface area, tend to have correspondingly higher moisture content and are typically dewatered by centrifugal methods.

Centrifugal dryers are comprised of a rotating screen or basket, an entrance for feed slurry, a power drive for rotation, and a mechanism for solids removal. Centrifugal dewatering equipment utilizes sedimentation and filtration principles that are amplified by centrifugal forces. Centrifuges can be used to dewater particles from 50 mm to a few microns. However, no single unit can effectively treat this entire size range (Fig. 20). For coarser fractions, two popular units are the vibratory centrifuge and the screen-scroll centrifuge. Both are available in horizontal or vertical configurations. The primary difference in these units is the mode by which solids are transported. The vibrating centrifuge uses a reciprocating motion to induce the flow of solids across the surface of the dewatering basket (Fig. 21a). The screen-scroll centrifuge uses a screw-shaped scroll that rotates at a slightly different speed than the basket to positively transport the solids through the unit (Fig. 21b). The vibratory centrifuge works best for coarser feeds up to 30 to 50 mm. Both the screen-scroll and the vibratory centrifuges are appropriate for intermediate sizes, although the screen-scroll centrifuge typically provides lower product moistures. For finer material (<1 mm), both the screen-scroll centrifuge and another popular design, called a screen-bowl centrifuge, may be used. The screen-bowl centrifuge is a horizontal unit that consists of a solid bowl section and a screen section

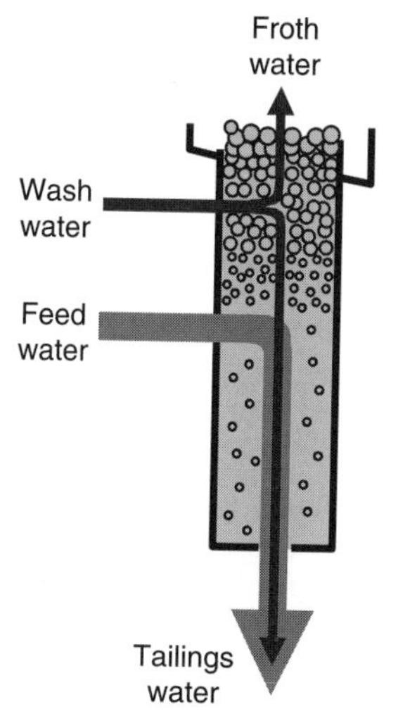

FIGURE 19 Illustration of column flotation technology.

(Fig. 22). Feed slurry is introduced into the bowl section where most of the solids settle out under the influence of the centrifugal field. A rotating scroll transports the settled solids up a beach and across the screen section where additional dewatering occurs before the solids are discharged. Solids that pass through the screen section are typically added back to the machine with the fresh feed. This unit is capable of providing low product moistures, although some ultrafine solids are lost with the main effluent that is discarded from the solid bowl section. This process has become dominant in the United States for drying combined fine and ultrafine coal due to low operating cost and ease of maintenance.

If high coal recovery is desirable, then higher-cost filtration processes may be used to dewater fine coal. Filtration involves the entrapment of fine solids as a cake against a porous filtering media. Traditionally, flotation concentrates have been dewatered using some form of vacuum filtration. These units are capable of maintaining high coal recoveries (>97%) while generating product moisture contents in the 20 to 30% range. One of the most popular types of vacuum filters is the disc filter (Fig. 23). This device consists of a series of circular filtration discs that are constructed from independent wedge shaped segments. At the beginning of a cycle, a segment rotates down into the filter tub filled with fine coal slurry. Vacuum is applied to the segment so that water is drawn through the filter media and a particle cake is formed. The resultant cake continues to dry as the segment rotates out of the slurry and over the top of the machine. The drying cycle ends when the vacuum is removed and the cake is discharged using scrapers and a reverse flow of compressed air.

More recently, horizontal belt vacuum filters have become increasingly popular for dewatering fine coal (particularly in Australia). This device consists of a continuous rubber carrier belt supported between two end rollers (Fig. 24). Filter fabric is placed along the top surface of the belt to serve as filter media. Coal slurry is introduced at one end of the moving belt and is dewatered by the vacuum that is applied along the entire length of the underside. As the belt reaches the end roller, the vacuum is removed and the dewatered cake is discharged. A significant advantage of this design is the ability to clean the filter cloth on the return side using high-pressure sprays. Other advantages include the ability to control the cake thickness (by varying belt speed) and the independent control of cake formation and drying cycle times.

4.3 Thermal Drying

Thermal drying is used when mechanical dewatering is incapable of producing a coal moisture content that is acceptable for market. Although many types of thermal dryers have been tried including indirect heating dryers, fluidized bed dryers have become prominent in drying coals of bituminous rank, often producing product moistures in the 5 to 6% range. Fuel used is most often coal either in a pulverized form or as a stoker product. As far as lower rank coal drying is concerned, numerous thermal drying

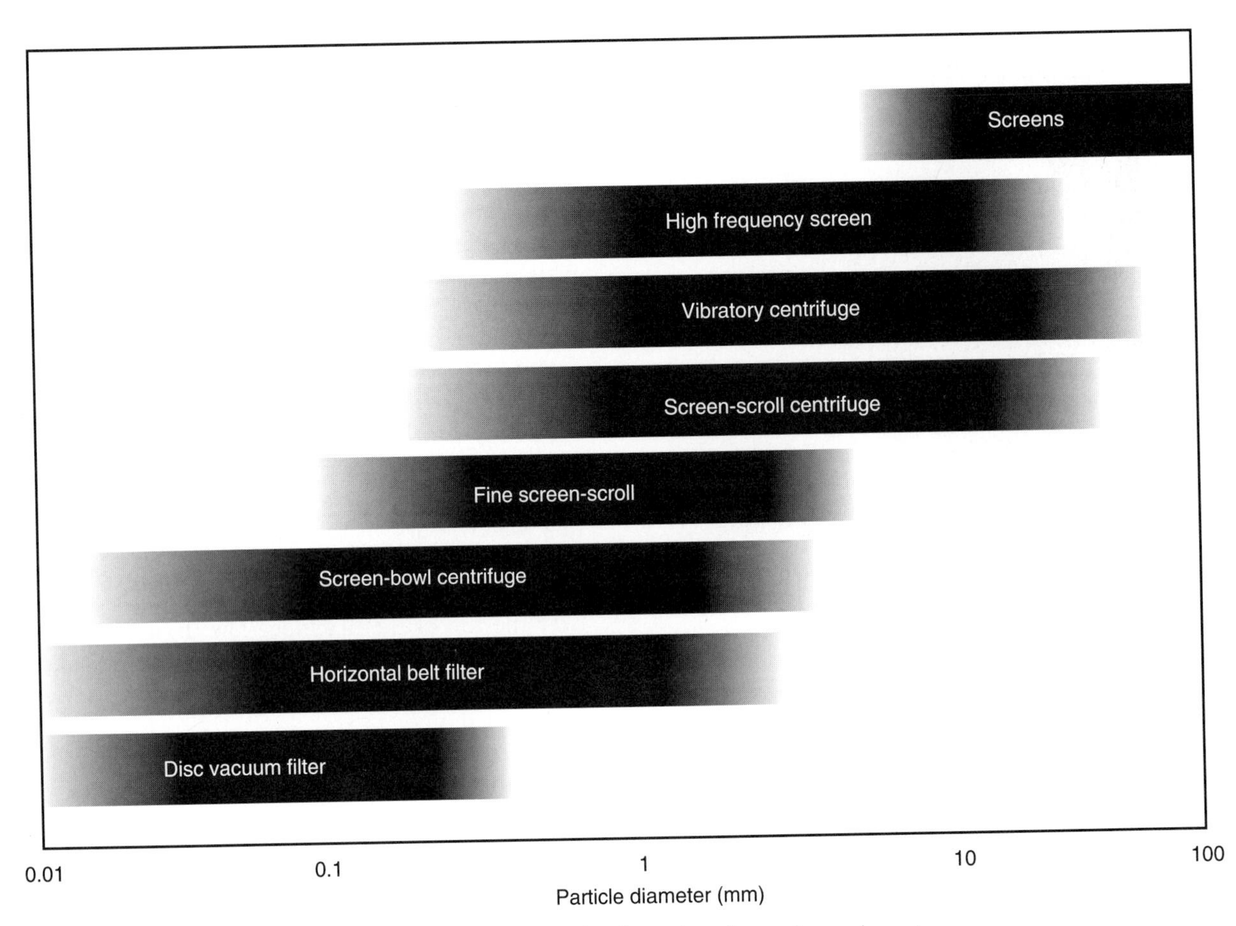

FIGURE 20 Typical ranges of applicability for various dewatering equipment.

processes have been developed, several of which include some degree of surface pyrolysis to reduce the tendency of the coal to reabsorb water or to be subject to spontaneous combustion. Briquetting is also utilized in several processes to deal with the fines inevitably generated in the drying process. Advantages of thermal drying include the ability to achieve very low moistures and the potential to improve other coal properties (e.g., coking). Disadvantages include high operating and capital costs, fugitive dust emissions, and safety hazards associated with potential fires and explosions.

4.4 Thickening

The process of cleaning coal is a wet one, using large quantities of water (i.e., thousands of gallons per minute). As a result, vast quantities of dirty (fine rock-laden) water are produced. Beneficiation processes require clean water. Consequently, dirty water is processed, solids removed, and clean water recycled for further use in the plant. This process is called thickening. A thickener is simply a large tank in which particles (usually flocs) are allowed to settle, thereby producing a clean overflow and a thickened underflow (Fig. 25). Thickeners are typically fed dilute slurry (<5 to 10% solids) that is rejected from the fine coal cleaning circuits. The clarified water that overflows along the circumference of the thickener is recovered and reused as process water. The thickened underflow, which typically contains 20 to 35% solids, is transported to the center of the thickener by a series of rotating rakes. The thickened sludge is typically pumped to an appropriate disposal area or is further dewatered prior to disposal. Chemical additives are usually introduced ahead of the thickener to promote the aggregation of ultrafine particles so that settling rates are increased. The pH levels are also monitored and controlled.

Conventional thickeners, which are typically 15 to 60 m (50 to 200 ft) diameter, require constant monitoring to ensure that overflow clarity and underflow density is maintained and the rake mechanism is not overloaded. A torque sensor on

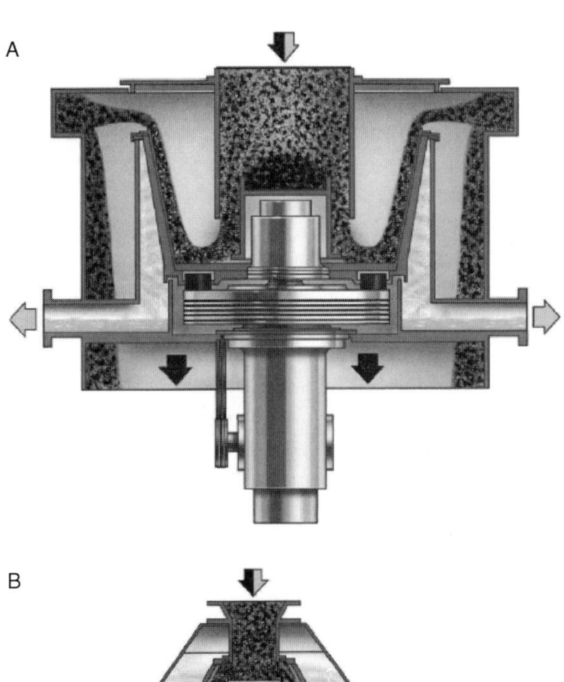

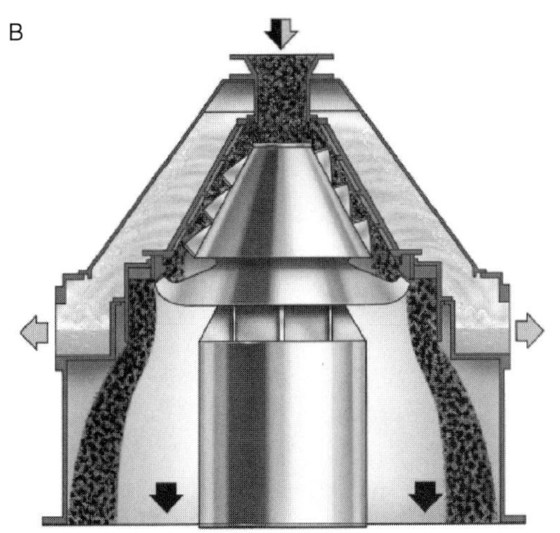

FIGURE 21 Illustrations of (A) vibratory and (B) screen-scroll centrifugal dryers. Courtesy of Centrifugal Machines, Inc.

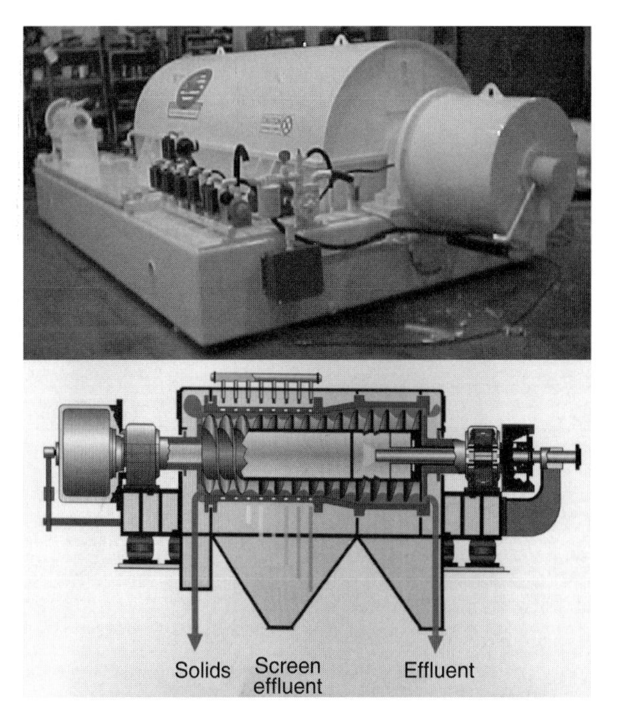

FIGURE 22 Illustration of a screen-bowl centrifuge. Courtesy of Decanter Machine, Inc.

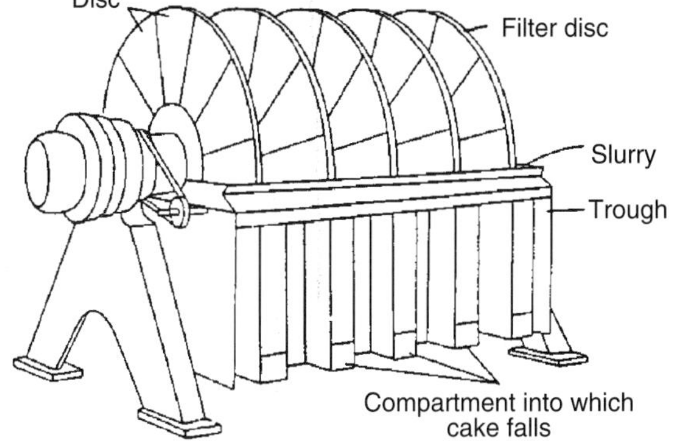

FIGURE 23 Illustration of a conventional disc filter. Courtesy of Peterson Filters.

the central drive is typically used to monitor rake loading, while a nuclear density gauge is used to monitor underflow density.

5. *MISCELLANEOUS*

5.1 Process Control

A state-of-the-art coal processing facility typically uses a programmable logic controller (PLC) as the heart of the plant control system. The PLC is a computer based controller that acquires data (input) representing the status of all the units and control loops in the plant, makes decisions based on the data input and programmed instructions, and sends data (output) to the plant units reflecting the decisions.

FIGURE 24 Production-scale horizontal belt filter. Courtesy of Delkor Pty, Ltd.

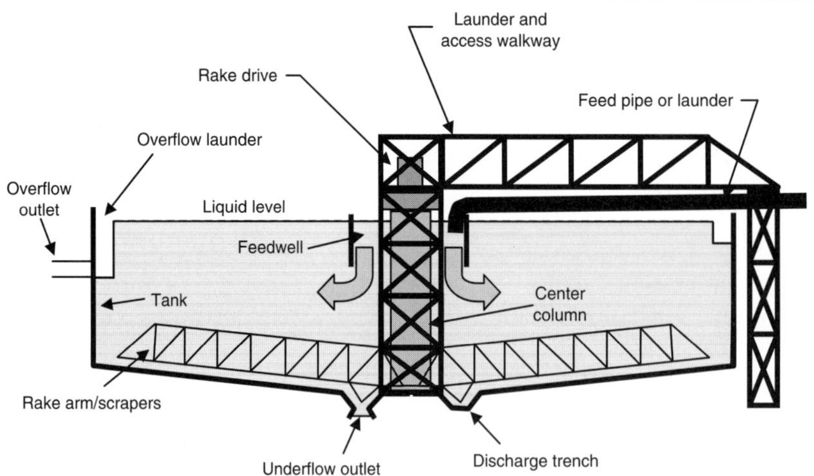

FIGURE 25 Illustration of a refuse thickener.

Inputs include a wide variety of data such as the electrical status of motor control circuits, operating/position status of moving equipment, electrical and mechanical faults, and operating status of control loops (levels, temperatures, flow rates, pressure, etc.). Field devices that provide input to the PLC include electronic transmitters for levels, temperature, pressure, position, and flow rates. Nuclear devices are used for determining online measurements for specific gravity and solids content of slurry streams, for mineral and elemental analysis of process streams, and for flow rates. Digital (microprocessor-based) equipment is becoming available that provides faster communication, less hard wiring, and more reliable data. Digital equipment is also being used to replace standard motor control circuits, which provides better monitoring for those circuits. Outputs include various instructions to field devices such a start/stop signals to motors, open/close or position signals to process equipment (valves, gates, etc.), and signals to the control loops (valves, diverters, feeders, gates, etc.). Field devices that translate PLC output instructions into control actions include standard relay based or digital motor control circuits; electric or electro-pneumatic actuators for valves, gates, and so on; and electronic or digital variable speed controllers for feeders, pumps, conveyors, and so on.

The PLC in conjunction with a full color graphical interface (typically based on a personal computer) provides the operator with all relevant data concerning the operation of the facility. The capabilities of the system includes real time monitoring and control. A major advantage of the PLC system in is the ability to modify the program quickly to change interlocks, startup/showdown procedures. Two other techniques being used to enhance monitoring and control are fiber optics communications and radio frequency telemetry systems. The use of fiber optics for data transmission over long distances provides faster communications and immunity to electrical (lightning) interference. Radio frequency telemetry (direct and satellite relay) provides communications where it is impractical to install long runs of wires and conduits for remote monitoring and control.

During the past decade major enhancements in online analysis of coals have occurred. A variety of ash analyzers are now commercially available for monitoring the quality of the clean coal products produced by a preparation plant. These analyzers are typically mounted on a conveyor belt for the through-the-belt analysis of process streams or at the coal loadouts where the product streams are intercepted. These units can be used to monitor product quality, but more usually in the blending of coals of different qualities. This is done to meet customer quality specifications. They can use the principle of measuring the attenuation of a beam of nuclear radiation as it passes through a process stream. There are also slurry ash analyzers available, including one that uses submersible probes and another than is similar to a nuclear density gauge.

Several commercial monitors are also available of online determination of product moisture content. The principle of measuring the attenuation of a beam of nuclear or microwave radiation as it passes through a process stream is utilized in moisture determination. Accuracy for the microwave analyzers decreases as the variability in the size consists of the process stream increase.

5.2 Waste Disposal

Coal washing processes produce large volumes of waste rock that must be discarded into refuse piles or impoundments. Refuse piles are designed to receive coarse particles of waste rock that can be easily dewatered. This material is relatively easy to handle and can be transported by truck or belt haulage systems to the disposal area. In contrast, impoundments are designed to retain fine waste rock that is difficult to dewater. The waste is normally pumped from a thickener to the impoundment as slurry containing water, coal fines, silt, clay, and other fine mineral particulates. In most cases, the slurry is retained behind a embankment (earthen dam) constructed from compacted refuse material. The impoundment is designed to have a volume that is sufficiently large to ensure that fine particles settle by gravity before the clear water at the surface is recycled back to the plant for reuse. In some cases, chemical additives may be used to promote settling and to control pH. Impoundments, like any body of water contained behind a dam, can pose a serious environmental and safety risk if not properly constructed, monitored, and maintained. Potential problems include structural failures, seepage/piping, overtopping, acid drainage, and black water discharges. In the United States, strict engineering standards are mandated by government agencies to regulate the design and operation of impoundments. In some countries, combined refuse systems are employed in which both coarse and fine refuse are disposed together as a solid product or alternatively as a pumped slurry.

SEE ALSO THE FOLLOWING ARTICLES

Clean Coal Technology • Coal, Chemical and Physical Properties • Coal Conversion • Coal, Fuel and Non-Fuel Uses • Coal Industry, Energy Policy in • Coal Industry, History of • Coal Mine Reclamation and Remediation • Coal Mining, Design and Methods of • Coal Mining in Appalachia, History of • Coal Resources, Formation of • Coal Storage and Transportation • Markets for Coal • Oil Refining and Products

Further Reading

Bethell, P. J. (1998). "Industrial Practice of Fine Coal Processing" (Klimpel and Luckie, Eds.). Society for Mining, Metallurgy, and Exploration (SME), Littleton, CO, September, pp. 317–329.

Coal Preparation in South Africa. (2002). "South African Coal Processing Society." Natal Witness Commercial Printers (Pty) Ltd., Pietermaritzburg, South Africa.

Laskowski, J. S. (2001). "Coal Flotation and Fine Coal Utilization: Developments in Mineral Processing." Elsevier Science Publishing, Amsterdam, The Netherlands.

Luttrell, G. H., Kohmuench, J. N., Stanley, F. L., and Davis, V. L. (1999). "Technical and Economic Considerations in the Design of Column Flotation Circuits for the Coal Industry." SME Annual Meeting and Exhibit, Denver, Colorado, March, Preprint No. 99-166.

Wood, C. J. (1997). "Coal Preparation Expertise in Australia: In-Plant Issues and the Potential Impact of Broader Applications," Proceedings, Coal Prep '97, Lexington, KY, pp. 179–198.

Coal Resources, Formation of

SHREEKANT B. MALVADKAR, SARAH FORBES, and GILBERT V. MCGURL
U.S. Department of Energy, National Energy Technology Laboratory
Morgantown, West Virginia, United States

1. **Origin of Coal**
2. **Formation of Coal**
3. **Describing and Classifying Coal**
4. **Constituents of Coal and Their Impacts**
5. **Compositional Analysis and Properties of Coal**
6. **Frontiers of Coal Research**

Glossary

ASTM Abbreviation for American Society for Testing and Materials.
coalification Geochemical process by which buried peat is converted into increasingly higher ranking coal.
coal rank American Society for Testing and Materials-adopted coal classification system, accepted in the United States, that includes brown coal, lignite, subbituminous, bituminous, and anthracite as increasingly higher coal ranks.
lithology Characterization of the structural aspects of the coal bed and the coal texture obtained from the macro- and microscopic observation of the coal.
macerals The altered plant remains, classified into three groups: vitrinite, liptinite, and inertinite.
peatification Biochemical conversion of plant material by anaerobic bacterial action.
petrography Systematic quantification of different bands of material in coal and their characteristics by microscopic study.
petrology Study of the origin, composition, and technological application of different bands of material in coal.

Coal is defined by the American Society for Testing and Materials (ASTM) as "brown to black combustible sedimentary rock composed principally of consolidated and chemically altered plant remains." Coal, formed by biological and geological processes over millions of years, is a heterogeneous and nonuniform material containing macromolecular organic compounds, mineral matters, elements, moisture, gases, and oily and tarry material. Its properties are the result of types of plants, geological age, and biological and geological processes to which plants are subjected.

1. ORIGIN OF COAL

An understanding of coal origin and factors influencing coal formation is important because the demand for electricity, a large fraction of which comes from coal-fired power plants, is increasing. Locating coal sources, assessing their extent and quality, and using coal in an environmentally sound manner are aided by such an understanding.

1.1 Influence of Age

The first plants to grow on land made their appearance toward the end of the Silurian period. These were the earliest vascular plants, leafless shoots that transported nutrients and water throughout the plant. Extensive forests existed by the middle of the Devonian period from which the oldest coal comes. The environment was warm and humid, allowing club mosses, scouring rushes, and horsetails to flourish.

1.1.1 Carboniferous Period

During this period, under a worldwide tropical/subtropical climate with long growing seasons and moderate to heavy rainfall, plants extended their geographic coverage and increased in size. A forest could cover 20,000 to 25,000 square miles, and more plant species existed than do today. Ground pines and club mosses were 30 to 75 feet tall. Giant forerunners of today's horsetail and scouring rushes were abundant. Forests were in swampy areas located on low-lying flat land near shallow seas or rivers. Periodic flooding or drought conditions altered the nature of the forests and led to the formation of anthracite or bituminous coal.

1.1.2 Permian Period

A cooler climate and seasonal changes characterized this period. Except in China, deposition leading to formation of coal in the North declined. In the South, important deposits leading to bituminous coal were laid down in South Africa and Australia.

1.1.3 Cretaceous Period

During this period, the climate was so mild that vegetation grew even in the polar regions, leading to the formation of coalfields there. A major shift in vegetation occurred, namely, plants emerged with seeds having protective coverings. Because of its relatively young age, the coal of this period is of low rank.

1.1.4 Tertiary Period

Deposition of plant materials continued during this period. Ashes, willows, and maples were among the common plants. The younger age of these deposits led to lignite formation.

A strong relationship is found between the rank of coal and the geological age of plant deposition. For example, Texas lignite deposits are from the Tertiary Ecocene period, whereas subbituminous coal in the Big Horn basin and eastern Wyoming was formed during the Paleocene epoch and bituminous coal in western Pennsylvania and eastern Ohio came from the Carboniferous period. However, age alone does not determine the rank. For example, brown coal and lignite found near Moscow, Russia, were formed from plant deposits of the Carboniferous period. However, the plants were not buried deeply. Thus, the temperatures and pressures to which the plant materials were subjected were not high enough to form high-ranking coals.

1.2 Influence of Geology

In spite of seeming abundance in various regions of the world, coal beds form only a very small percentage of the total thickness of the overall sedimentary deposits termed "coal measures" in coal-bearing areas. In a simplified description of coal formation, there are two sequential processes: peat formation (peatification) by biological processes and conversion of peat to coal (coalification) by geological processes. Geological activities also play an important role during the peatification phase. The composition, thickness, hardness, and regularity of the strata both below and above coal beds are created by geological processes over millions of years. These characteristics must be considered in assessing coal mineability and its economics and in selecting mining approaches. For example, in underground mining, the weight and strength of the overlying material ("roof") have a significant bearing on the underground support structure needed for safe operation. Similarly, the "seat" rock on which a coal bed lies may become unduly soft or sticky when wet or excessively plastic when under pressure, causing problems in mining operation.

Under some geological conditions, peat-forming material accumulated at or close to the place of plant growth (autochthonous origin). Under different conditions such as along beaches, along river streams, or when flooding occurs, plant material was moved to a different location, where it was deposited (allochthonous origin) possibly on top of existing plants. Coals formed by the allochthonous mechanism are relatively rare and have less commercial value because of their higher mineral contents.

During peatification, lasting 10,000 to 15,000 years, biochemical activities predominated. Plants and partially decomposed plants were subjected to varying environments (e.g., flooding, drought), addition of plants from other locations, and so on. The extent to which plants were buried under water had a large effect on the type of biochemical activity (e.g., anaerobic vs aerobic fermentation). Some geological changes destroyed swamps and halted the formation and accumulation of peat. Long periods of drought promoted aerobic action, leading to inferior coalfields.

The aerobic process, if it proceeds unimpeded, destroys plant material completely. The anaerobic process leads to peat, and eventually to various ranks of coal, depending on the prevailing geological conditions. If lightning strikes during peatification, fires set by it lead to the formation of charcoal. The charcoal may be incorporated in the coal to a lithotype form of coal called fusain, also termed mineral charcoal or mother-of-coal. An alternate theory for fusain formation also has a geological basis: a drier environment is claimed to lead to the formation of fusain. Cannel coal is theorized to be formed in aerated water, which decomposes all but the spores and pollen.

In stagnant water, the concentrations of acids increase as microbial activities proceed. These acids tend to kill bacteria and short-circuit peatification. In rushing water, this mechanism is less important because acids formed by microbial activities are removed or diluted. Water itself acts as a shield against attack by atmospheric oxygen. Thus, a freshwater environment is most favorable for coal formation.

Coalification primarily involves the cumulative effects of temperature and pressure over millions of years. The sediment covering the peat provides the pressure and insulation, and heat comes from the earth's interior. (Temperature increases ~4–8°C each 100 m.) The change from bituminous coal to anthracite coal involves significantly higher pressure (as in mountain building activity under tectonic plate movements) and higher temperatures than does the formation of bituminous coal. Pressure affects the physical properties of coal (e.g., hardness, strength, porosity). Higher temperatures can be the result of several factors: (1) relative proximity of the earth's magma to formative coal seams, (2) close approach, or contact with, igneous intrusions/extrusions, (3) volcanic activity, and (4) depth of burial. Tectonic plate movement itself creates additional heat. Volcanic activity near the peat or partially formed coal can have a negative impact such as sulfur, mercury, and other environmentally undesirable mineral matters being incorporated into the coal bed. Rapid coalification may lead to parts of coalfields to have natural coke.

The kinds of sediments surrounding the coals are also important. For example, clays do not resist pressure well and deform relatively easily. Thus, a peat deposit in contact with clays reflects the results of pressure more. Although rare, geological processes introduce in some coal beds relatively narrow, vertical, or steeply slanting veins of clay, sometimes partially pyritized.

Several geological phenomena have an impact on the coal bed placement underground:

- *Partings*. Partings are formed when plants buried in a swamp get covered with layers of mud, silt, or clay material followed by additional plant accumulation trapping nonplant materials in between two coal seams. Often, the partings are not uniform and, instead, form a wedge shape that is progressively thicker toward the source of deposited material. Partings impair the mining procedures and, especially if pyritized, decrease the quality of the mineable coal.
- *Concretions*. Formations, perhaps several feet in diameter, that impede mining operations are formed during the overlaying of nonplant material (lenticular masses composed of pyrite, calcite, or siderite or their combinations).
- *Heterogeneity*. During coal bed formation, valleys are formed by the downward erosion caused by nearby streams, occasionally leading to exposure of existing coal beds during their formative stages. The beds are then covered with additional plants or a different type of sedimentation.
- *Geological stresses*. These stresses induce folding and sharp deformation and lead to specific dislocation ("faults"). Such dislocations may range up to hundreds of feet vertically and thousands of feet laterally.

Throughout coalification, formative coal beds are subjected to the action of ground waters. Most of the mineral constituents in the coal bed, such as pyretic material and gypsum, are brought in by ground water either as suspension or by ion exchange. During the bituminous coal stage, the beds develop vertical or nearly vertical fractures that are often filled with coatings of pyrite, calcite, kaolinite, or other minerals deposited by circulating ground water. Similar deposition occurs during peatification, but with a difference. Peat tends to be highly porous and acts as a filter. Secondly, living organisms within peat interact with dissolved minerals such as sodium, potassium, and calcium to exchange ions.

1.3 Influence of Microbes

Microbial activity leads to peat formation (peatification). Microbial activity continues for many years even after the geochemical phase gets under way so long as pressure and temperature are not excessive.

Microbial activity can be of fungal, aerobic, or anaerobic type. The first two need oxygen and cause complete plant decay. If the plant is submerged in water, anaerobic activity dominates. Decomposition of plant material by microbes can be classified as shown in Table I. For the formation of a commercially viable product (coal), peatification or putrefaction is required. Attacks of microbes, oxygen, dissolved oxygen, enzymes, and water-borne chemicals cause degradation of plant material, forming new organic macromolecules so that the original plant structure is not recognizable. Methane and carbon dioxide are also generated under these conditions, hence the term "marsh gas" for methane.

Different parts of plant material have different levels of resistance to microbial activities. Also, their products have different levels of toxicity. For example, cellulose, a major constituent of cell walls, and protein are more susceptible to microbial attacks than are plant constituents such as lignin, tanins, fats, waxes, and resins. Humic acids and other humic substances are important peatification products. Woody material passes through a plastic gel-like

TABLE I

Decomposition of Plant Material via Microbial Actions

Type of action	Type of product
Disintegration (aerated)	Gases, liquids, limited resins
Mouldering (aerobic bacteria and fungi)	Inertinite
Peatification (anaerobic/aerobic bacteria and fungi)	Vitrinite
Putrefaction (anaerobic bacteria)	Exinite

Source. Mehta, A. (Project Manager) (1986). "Effects of Coal Quality on Power Plant Performance and Costs," Vol. 4: "Review of Coal Science Fundamentals," (EPRI CS-283). Electric Power Research Institute, Palo Alto, CA.

stage before solidifying to huminite and eventually vitrinite. Formation of other maceral structures, such as liptinite and inertinite, is also the result of microbial activities.

2. *FORMATION OF COAL*

2.1 Petrology, Petrography, and Lithotypes

Careful visual examination of a coal sample shows that it is layered; that is, it consists of bands of different materials with distinct entities distinguishable by optical characteristics. Coal petrology is the study of the origin, composition, and technological application of these materials, whereas coal petrography involves systematic quantification of their amounts and characteristics by microscopic study. Lithological characterization of coal pertains to (1) the structural aspects of the coal bed and (2) the coal texture obtained from the macro- and microscopic observation of the coal.

The texture of the coal is determined by the type, grain, and distribution of its macro- and microscopic components. Banded coals composed of relatively coarse, highly lustrous vitrain lenses (6 mm or more in thickness) are labeled as coarsely textured. As the thickness narrows, the texture becomes fine-banded and then microbanded. Coals with homogenous texture are rare and regarded as nonbanded or canneloid coals. The lithotype nomenclature is often not suitable for lignite due to the presence of plant pieces.

Some geologists consider coal to be a rock that has microscopically identifiable altered plant remains and minerals. The altered plant remains are called macerals, which are classified into three groups: vitrinite, liptinite, and inertinite. Vitrinite is derived from coalified plant tissues and gels. Liptinite is formed from plants' protective coatings, reproductive organs, and resinous cell fillings. Inertinite is produced from oxidation of vitrinite and liptinite precursors. Lithotype classification is partly independent of coal rank. In spite of the similarity of naming conventions, macerals are not minerals because they do not have definite chemical compositions and are not crystalline.

The quantities of various types of macerals in coal can be estimated petrographically. Bulk chemical data can be used to assess the influence of macerals on coal composition. Petrographic analysis involves the use of a microscope to measure optical properties. A widely accepted method is called vitrinite reflectance.

2.1.1 The Stopes–Heerlen System

The Stopes–Heerlen system has been codified by the International Committee for Coal Petrology. In this system, coal is classified either as hard coal or brown/lignite coal on the basis of visual appearance. The hard coals are further divided into banded/humic coals and saporpolic coal (nonbanded coal with large concentrations of algae or spores). The coal types are further divided into four lithotypes based on visual observation of features greater than 3 mm:

- *Vitrain:* bright, black, brittle
- *Clarain:* semi-bright, black, finely stratified
- *Durain:* dull, black–gray, black, hard, rough
- *Fusain:* charcoal-like, silky, black, fibrous, friable

The lithotypes are further divided into microlithotypes, which must be at least 50 microns wide and contain at least 5% of the minor components. The combination of individual macerals in a particular coal sample identifies it and determines whether it is mono-, bi-, or trimaceral. Table II provides a list of the combinations of macerals found in coal and the names of their resultant hybrids. The optical classification of brown and lignite coals has recently been standardized along lines similar to those used for hard coals.

2.1.2 Impact of Lithotype on Utilization Potential

Lithotype descriptions are primarily used to characterize coal beds in geological investigations. This includes the study of lateral variabilities in thickness and structure. They are used as the principal basis for coal resource and reserve determination, but not to predict specific coal properties such as ash content and coal rank.

A wide range of petrographic applications have been developed to assess coals for their specific use,

TABLE II
Characteristics of Hard Coal

Group maceral	Maceral	Submaceral	Maceral variety	Cryptomaceral	Occurrence (percentage)	Reflectance class
Vitrinite	Telinite	Telinite 1	Cordeltetelinite		75–98	T0–T21
			Fungotelinite			
			Xylotelinite			
		Telinite 2	Lepidophytotelinite			
			Sigillariotelinite			
	Collinite	Telocollinite		Cryptotelenite		C0–C21
		Gelocellinite		Cryptocorpocollinite		
		Desmocollinite		Cryptogelocollinite		
		Corpocollinite		Cryptovitrodetrinite		
	Vitrodetrinite					
Exinite or liptinite	Sporinite		Tenuisporinite		0–10	St0–St15
			Crassisporinite	Cryptoexosporinite		
			Microsporinite	Cryptointosporinite		
			Macrosporinite			
	Cutinite				0–1	Ct0–Ct15
	Resinite				0–5	R0–R15
	Suberinite					
	Alginite				0–1	At0–At15
	Liptodetrinite					
	Fluorinite					
	Bituminite					
	Exudatinite					
Inertinite	Micrinite				0–10	M0–M18
	Macrinite				0–10	
	Semifusinite				0–20	SF00–SF21
	Fusinite	Pyrofusinite			0.1–10	
		Degradofusinite				
	Sclerotinite			Plectenchyminite	0–1	
				Corposclerotinite		
				Pseudocorposclerotinite		
	Inertodetrinite				Common	

Sources. Mehta, A. (Project Manager), (1986). "Effects of Coal Quality on Power Plant Performance and Costs," Vol. 4: "Review of Coal Science Fundamentals," (EPRI CS-283). Electric Power Research Institute, Palo Alto, CA; Simon, J. A. Coal. *In* "Encyclopedia of Energy," 2nd ed. McGraw–Hill, New York; Vorres, K. S. Coal. *In* "Encyclopedia of Chemical Technology", (J. Kroschwitz, Ed.), 4th ed. John Wiley and Sons, New York.

predict potential problems in their application, and solve problems when they are encountered. Petrographic data are used for determination of seam correlations, geological history, and ease of coal oxidation as well as for measurements of rank, interpretation of petroleum formation from coal deposits, and prediction of coke properties. For example, constituents of seams can be measured over considerable distances apart, and these measurements can then be used to establish profiles for entire seams or coal basins. The vitrinite reflectance histogram can provide a quantitative measure of the ease of coal burning.

A vital role is played by petrographic analyses of various coals under consideration as potential feed material to form metallurgical coke. The coking behavior of the coal depends on the rank, maceral constituents, and their relative proportions. Sometimes, as many as eight different coals have to be blended, maintaining a balance between "reactive" and "inert" maceral constituents, to attain the desired coking properties. Vitrinite has good plasticity and swelling properties and produces an excellent coke, whereas inertinite is nearly inert and does not soften on heating. Exinite becomes extremely plastic and is nearly completely distilled as asphaltenes (tar). By carefully controlling the blends of coals based on their petrological compositions, behavior during carbonization can be controlled.

Petrographic analyses are also useful for the evaluation of coals for liquefaction and gasification. Bituminous and lower ranking coals that are rich in exinite and vitrinite macerals are desirable for liquefaction. Based on microscopic analysis of dispersed "coaly" materials in shales and other sedimentary rocks, geologists can identify favorable areas for oil and gas exploration.

Extensive work has been carried out to correlate elemental composition, functional forms in which elements are present, and physical properties of coal with maceral forms and their proportions in the coal.

2.2 Factors Influencing Formation of Coal

The net result of peatification and coalification is reduction in hydrogen, oxygen, and free moisture and a corresponding increase in fixed carbon content of the original plant material.

2.2.1 Components of Plants

A perspective of the biochemical degradation leads to three plant components:

a. Structural components meant for the plant's rigidity and form include cellulose (polymerized glucose), hemicellulose (polymerized xylose, or wood sugar), lignin (polymerized *n*-propyl benzene), and calcium pectate.

b. Stored food includes carbohydrates, starch, fats and oil, and proteins. Carbohydrates tend to decompose earlier during peatification.

c. Specialty materials are meant to provide various services to the plant. These materials, functioning to protect the plant, include resin, cutin (protection of leaf surfaces), waxes (moisture retention), and sporopollenin (protective outer layer of spores and pollen). They are relatively resistant to biochemical degradation.

2.2.2 Biochemical Phase (Peatification)

Although bacteria and fungi attack all parts of plants simultaneously, the first ones to succumb are carbohydrates, starch, and calcium pectate. This is followed by decomposition of hemicellulose and then decomposition of cellulose. Lignin is more resistant, specialty materials are even more so, and sporopollenin is the most resistant.

Brackish or saline water often contains enough sulfur to support bacterial activities that lead to conversion of sulfate ions to hydrogen sulfide or elemental sulfur. These two by-products, instead of escaping formative peat, may be incorporated into preexisting pyrites or into the organic structures in the peat.

As peatification proceeds, more plants may be dumped on top of the previously buried plant material, compacting the peat-like material underneath, squeezing out some of the trapped water, and increasing its density. Degradation of cellulose under bacterial attack opens up cell walls and helps to disintegrate the large pieces of plant tissue. The structure of coal itself changes drastically during this phase, increasing carbon content from 50 to 60%.

2.2.3 Geochemical Phase (Coalification)

The geochemical phase begins when sedimentation such as silt, mud, sand, and clay is placed on top of the peat or partially formed peat. The weight of the sedimentation compresses the peat and buries it deeper as more sedimentation is piled up.

Transformation of peat into brown coal, lignite, subbituminous, bituminous, and (finally) anthracite is a sequential and irreversible process. The primary factors that determine coalification progress are temperature, pressure, and duration of coalification to which peat is subjected. With a shorter time, coalification does not proceed beyond the rank of brown coal or lignite. Incomplete coalification produces coal that has recognizable plant fragments.

Higher temperatures are critical for the metamorphosis of peat to increasingly higher ranks of coal. The coalification of brown coals and lignite occurs at temperatures less than 100°C. These coals retain much water and volatile matter and more of the functional groups of the original plant. Temperatures in the range of 100 to 150°C yield bituminous coals through the loss of moisture, aliphatic and oxygenated groups, and hydrogen. In the process, there is a decrease in the pore structure and an increase in the carbon and heat content of the coal. Low- and medium-volatile coals are produced by further exposure to temperatures in the range of 150 to 180°C. Anthracites are produced at temperatures greater than 180°C, with further loss of hydrogen and organic functional groups. Loss of hydrogen during this stage actually decreases slightly the heating value of coal. Anthracite formation also increases the fine pore structure of the coal. Anomalous coal of higher ranking has been produced during relatively short coalification periods when peat is subjected to magma intruding near it.

2.3 Structure of Coal

Because coal is the result of the partial decay and deposition of vegetation, it consists mostly of organic material. In addition, inorganic matter is incorporated in the coal bed, some of which becomes an integral part of the organic material in coal. Coal also contains free water (inherent moisture). The percentage of inherent moisture ranges from 40 to 50% in lignite to less than 5% in anthracite. Coal composition and structure vary substantially as the rank changes from lignite to anthracite. As rank increases, fixed carbon increases and oxygen and inherent moisture/volatile matter decrease.

In general, rank and maceral composition are viewed as substantially fixed for a particular coalfield, certainly within a particular coal seam. The structure of the coal seam/bed is itself highly irregular, depending on its geological history. Thus, the height of the seam and its horizontal spread are often unpredictable.

2.3.1 Bonding within a Macromolecule

A variety of macromolecules with different molar masses, proportions of functional groups, and structural features exist in any coal piece. These molecules consist of planar fused ring clusters linked to nonpolar hydroaromatic structures. The overall structure is complex, irregular, and open. Entanglement between macromolecules occurs, as does cross-linking of hydrogen-bonded species. These different molecular shapes and sizes in coal lead to irregularities in packing and porosity in coal and make it amorphous. In the extracted material of the coal, the average molar mass is 1000 to 3000 for 5 to 50% fraction. Unextractable coal contains even larger macromolecules. The magnetic resonance technique was applied to coals of various ranks to generate the intramolecular functional group data given in Table III.

The structure of vitrinites, the major maceral groups in bituminous and anthracite coals, has been analyzed and found to have the following characteristics:

a. The organic molecules contain a number of small aromatic clusters, with each typically having one to four fused benzene rings. The average number of clusters depends on the rank, increasing slowly up to 90% fixed carbon and then rapidly for higher carbon content. (Lignites typically have 0.60 aromaticity, whereas anthracite has 0.95.)

b. The aromatic cluster linkage is, in part, by hydroaromatic, alicyclic, and ring structures, which also contain six carbons. On dehydrogenation, these ring structures can become part of the cluster through molecular rearrangement, increasing the average cluster size.

TABLE III

Carbon Structural Distribution: Fraction of Carbon Type

	Lignite	Subbituminous	High-volatile bituminous	Medium-volatile bituminous	Low-volatile bituminous
sp^2 hybridized carbon	0.61	0.63	0.72	0.81	0.86
sp^2 hybridized carbon in aromatic ring	0.54	0.55	0.72	0.81	0.86
Carbon in carbonyl	0.07	0.08	0.00	0.00	0.00
Protonated aromatic	0.26	0.17	0.27	0.28	0.33
Nonprotonated aromatic	0.28	0.38	0.45	0.53	0.53
Phenolic or phenolic ether carbon	0.06	0.08	0.06	0.04	0.02
Alkylated aromatic carbon	0.13	0.14	0.18	0.20	0.17
Aromatic bridgehead carbon	0.09	0.16	0.22	0.29	0.34
sp^3 hybridized carbon	0.39	0.37	0.28	0.19	0.14
CH or CH_2 aliphatic carbon	0.25	0.27	0.16	0.09	0.08
CH_3 aliphatic carbon	0.14	0.10	0.12	0.10	0.06
Oxygen-bound aliphatic carbon	0.12	0.10	0.04	0.02	0.01

Source. Vorres, K. S. Coal. *In* "Encyclopedia of Chemical Technology" (J. Kroschwitz, Ed.), 4th ed. John Wiley and Sons, New York.

c. Clusters can also form linkages employing short groups such as methylene, ethylene, and ether oxygen.

d. A large fraction of hydrogen sites on the aromatic and hydroaromatic rings are substituted by methyl or larger aliphatic groups.

e. Oxygen is usually present in phenolic and ether groups, substituting for hydrogen on the aromatic ring.

f. In a small fraction of the aromatic and hydroaromatic rings, oxygen replaces carbon in the ring, making the ring heterocyclic. (Some of these rings are five-membered.)

g. Compared to oxygen in phenolic groups, a lesser fraction of oxygen appears in the carbonyl structure (quinone form on aromatic and ketone form on hydroaromatic rings). In both cases, oxygen is apparently hydrogen bonded to adjacent hydroxyl groups.

h. Nitrogen, which is less abundant in vitrinite than is oxygen, is usually present as a heteroatom in a ring structure or as a nitrile.

i. Most of the sulfur, especially if its content exceeds 2%, is in the form of inorganic material. Organic sulfur occurs in both aliphatic and aromatic forms. Sulfide and disulfide groups are involved in linking clusters.

2.3.2 Bonding between Macromolecules

Individual macromolecules in coal structure are bonded to each other by a variety of forces and bonds to form a solid, three-dimensional coal structure. Aromatic clusters are linked by connecting bonds through oxygen, methylene or longer aliphatic groups, or disulfide bridges. The proportions of the various functional groups change with rank. Acid groups decrease early in the rank progression and other groups follow, leaving ether groups to act as cross-linking groups. Another type of linkage involves hydrogen bonds that, for example, hold hydroxy and keto groups together in solids.

3. DESCRIBING AND CLASSIFYING COAL

Coal samples from around the world show differences in appearance and properties—soft, moist, and brownish; very hard, glossy, and black solid; or numerous variations in between. Coal is primarily used for electric power generation. Other coal uses include drilling fluids, briquettes, and formation of coke for iron smelting. For more than a century, it supplied hundreds of chemical intermediates to prepare medicines, dyes, paints, and perfumes. These coal applications require different physical, chemical, and other properties of coal.

A number of classification systems have been developed to facilitate the production, sale, and use of coal and to satisfy the needs of major stakeholders. They seek a systematic way by which to assess performance and economic value of coals in particular applications and to facilitate assessing potential operational problems.

3.1 The Ranks of Coal

In the United States in 1938, the American Society for Testing and Materials (ASTM) adopted classification of coals "by rank" as a standard means of coal specifications (D388), parameters of which are provided in Table IV. Lower ranking coals are classified by their calorific values on a moisture- and mineral-free basis. Higher ranking coals are specified in terms of percentage of fixed carbon ($\geq$69%) or volatile matter ($\leq$31%) on a moisture- and mineral-free basis. ("Volatile matter" is defined by the ASTM D3175 procedure.) Many properties of coal vary in a fairly regular way with rank. This allows one to judge the properties of a particular coal on the basis of its rank alone.

Anthracite, bituminous, subbituminous, and lignite are ranks of coal in descending order. Peat itself is not considered coal. Less mature lignite is called brown coal. A further distinction is made within bituminous and subbituminous coals on the basis of volatile matters and agglomerating properties. Lignite and subbituminous are low-ranking coals, whereas bituminous and anthracite are high-ranking coals.

3.1.1 Descriptions of Coals by Their Ranks

3.1.1.1 Brown Coal Geologically least mature, brown coal has high moisture, sometimes more than 60%. On drying, it tends to disintegrate. Although susceptible to spontaneous ignition, it has a low heating value ($\sim$3000 BTU/pound as mined).

3.1.1.2 Lignite Geologically not mature, with easily recognizable plant parts present, lignite has less than 75% carbon on a moisture- and ash-free basis and moisture as high as 40 to 42%. Because of its high volatile matter, it ignites easily, but with a smoky flame. It has a low heating value and tends to disintegrate on drying.

TABLE IV

Classification of Coals by Rank

	Fixed carbon (percentage)[a]		Volatile matter (percentage)[a]		Gross calorific value (kJ/kg)[b]		
	≥	<	≥	<	≥	<	Agglomerating character
Coal rank							
Anthracite							
Meta-anthracite	98			2			Nonagglomerating
Anthracite	92	98	2	8			Nonagglomerating
Semi-anthracite[c]	86	92	8	14			Nonagglomerating
Bituminous							
Low volatile	78	86	14	22			Commonly agglomerating[e]
Medium volatile	69	78	22	31			Commonly agglomerating[e]
High volatile							
A		69	31		32,500[d]		Commonly agglomerating[e]
B					30,200[d]	32,500	Commonly agglomerating[e]
C					26,700	30,200	Commonly agglomerating[e]
					24,400	26,700	Agglomerating
Subbituminous							
A					24,400	26,700	Nonagglomerating
B					22,100	24,400	Nonagglomerating
C					19,300	22,100	Nonagglomerating
Lignite							
A					14,600	19,300	Nonagglomerating
B						14,600	Nonagglomerating

Source. Vorres, K. S. Coal. *In* "Encyclopedia of Chemical Technology" (J. Kroschwitz, Ed.), 4th ed. John Wiley and Sons, New York.

[a] Dry, mineral-matter-free basis.

[b] Moist, mineral-matter-free basis (i.e., contains inherent moisture)

[c] If agglomerating, classify in low-volatile group of the bituminous class.

[d] Coals having 69% or more fixed carbon on the dry, mineral matter-free basis are classified according to fixed carbon, regardless of gross calorific value.

[e] There may be nonagglomerating varieties in the groups of bituminous class, and there are notable exceptions in high-volatile C bituminous group.

3.1.1.3 Subbituminous Coal ("Black Lignite") Geologically more mature, as manifested by the absence of woody texture, subbituminous coal has a tendency to disintegrate and spontaneously ignite like lignite, but it burns more cleanly. Since the 1970s, environmental concerns have increased interest in low-sulfur subbituminous coal.

3.1.1.4 Bituminous Coal Bituminous coal is black and often appears to be banded with alternating layers of glossy and dull black colors. It has a higher heating value and lower moisture and volatile matters than does subbituminous coal. It has little tendency to disintegrate or ignite spontaneously.

3.1.1.5 Anthracite With very low volatile matter, sulfur, and moisture, anthracite burns with a hot clean flame and no smoke or soot. Stable in storage, it can be handled without forming dust. Hardest and densest of all the coal ranks, anthracite is jet black and has a high luster.

Although not part of the standard ASTM classification, some coals, with 25 to 50% ash by weight on a dry basis, are termed "impure coal." The mineral matter is introduced during deposition, mostly as clay or as fine-grained detrital mineral matter (bone coal) or later by secondary mineralization (mineralized coal). Other deposits in which mineral matter predominates, such as oil shale, coaly carbonaceous shale, and bituminous shale, are also not classified as coal.

3.2 The International Systems for Coal Classification

Many countries have official national organizations that develop and maintain standards for analyzing coal and classification schemes that are used in the coal trade and compilation of coal resource and reserve data. In the United Kingdom, the British Standards Organization serves this purpose, as does the ASTM in the United States.

3.2.1 National Coal Board Classification for British Coals

Proposed in 1946 by the Department of Scientific and Industrial Research, this classification is based on (1) volatile matter on a dry, mineral-free basis and (2) coking power, as measured by the Gray–King assay. Classification of lower ranking coals is based primarily on the Gray–King assay. Because this classification is applied to coals having less than 10% ash, high-ash coals need to be first cleaned by methods such as a float–sink separation.

3.2.2 International Classification (Hard Coal)

The rapid increase in the international coal commerce since 1945 necessitated an international system of coal classification. In 1956, the Coal Committee of the European Economic Community agreed on a system designated as the International Classification of Hard Coal by Type. Among the parameters considered are volatile matter and gross calorific value on a moisture- and ash-free basis. Table V provides various national classifications and their relationships to the classes of the international system.

In coal classification, most countries use volatile matter as the primary criterion, with a secondary criterion being coking properties. Criteria employed by some countries are as follows:

- Belgium: none
- Germany and The Netherlands: appearance of coal button
- France and Italy: Free Swelling Index (FSI)
- Poland: Roga Index
- United Kingdom: Gray–King assay

The International Organization for Standardization (ISO) based in Geneva, Switzerland, formed a committee to develop international standards. These standards, which incorporate both coal rank and coal facies parameter, replaced the UN Economic Commission of Europe's classification for hard coal.

A three-digit classification of the international system combines rank (class), caking (group), and coking (subgroup) parameters into a three-code system, as shown in Table VI. The first digit indicates the class or rank (higher digits for lower ranks). Coals having less than 33% volatile matter are divided into classes 1 to 5, and those with higher volatile matter are divided into classes 6 to 9. The second digit is based on the caking properties such as the Roga Index and the FSI. The third digit defines a subgroup based on coking properties.

3.2.3 International Classification (Brown Coal and Lignite)

These types of coals are defined as those having less than 23,860 kJ/kg heating values. Their classification employs a four-digit code. The first two digits (class parameter) are for total moisture content of freshly mined coal on an ash-free basis. The third and fourth digits are defined by the tar yield on a dry, ash-free basis.

4. CONSTITUENTS OF COAL AND THEIR IMPACTS

4.1 Organic Components in Coal

The functional groups within coal contain primarily elements C, H, O, N, and S. The significant oxygen-containing groups found in coals are carbonyl, hydroxyl, carboxylic acid, and methoxy. The nitrogen-containing groups include aromatic nitriles, pyridines, carbazoles, quinolines, and pyrroles. The sulfur-containing groups are thiols, dialkyl and aryl-alkyl thioethers, thiophenes, and disulfide.

The relative amounts of various functional groups depend on coal rank and maceral type. In vitrinites of mature coals, the principal oxygen-containing functional groups are phenolic and conjugated carbonyls as in quinones. In exinites, oxygen is present in functional groups such as ketones (unconjugated carbonyls).

The aromaticity of macromolecules in coal increases with coal rank. Calculations based on several coal models indicate that the number of aromatic carbons per cluster varies from 9 for lignite to 20 for low-volatile bituminous coal, and the number of attachments per cluster varies from 3 for lignite, to 4 for low-volatile bituminous coal, to 5 for subbituminous through medium-volatile bituminous coal.

4.2 Inorganic Constituents in Coal

To determine inorganic constituents in coal, a variety of instrumental techniques are available, including

TABLE V

The International System and Corresponding National Systems of Coal Classes

International system			National classifications							
Class number	Volatile matter (percentage)	Calorific value (kJ/g)[a,b]	Belgium	Germany	France	Italy	The Netherlands	Poland	United Kingdom	United States
0	0–3					Anthraciti Speciali		Meta-antracyt		Meta-anthracite
1A	3.0–6.5		Maigre	Anthrazit	Anthracite	Anthraciti Comuni	Anthraciet	Antracyt	Anthracite	Anthracite
1B	6.5–10.0							Polantracyt		
2	10–14		$\frac{1}{4}$ Gras	Magerkohle	Maigre	Carboni Magri	Mager	Chudy	Dry steam	Semianthracite
3	14–20		$\frac{1}{2}$ Gras	Esskohle	Demigras	Carboni Semigrassi	Esskool	Polkoksowy meta-koksowy	Coking steam	Low-volatile bituminous
4	20–28		$\frac{3}{4}$ Gras	Fettkohle	Gras à Courte flamme	Carboni Grassi Corta fiamaa	Vetkool	Orto-koksowy	Medium volatile coking	Medium-volatile bituminous
5	28–33		Gras	Gaskohle	Gras Proprement dit	Carboni Grassi Media fiamma		Gazowo koksowy		High-volatile bituminous A
6	>33(33–40)	32.4–35.4				Carboni da gas	Gaskool			
7	>33(32–44)	30.1–32.4			Flambant gras	Carboni grassi da vapore	Gasvlamkool	Gazowy	High volatile	High-volatile bituminous B
8	>33(34–46)	25.6–30.1		Gas flammkohle	Flambant scc	Carboni Secchi	Vlamkkol	Gazowoplomienny		High-volatile bituminous C
9	>33(36–48)	<25.6						Plomienny		Subbituminous

Source. Vorres, K. S. Coal. *In* "Encyclopedia of Chemical Technology" (J. Kroschwitz, Ed.), 4th ed. John Wiley and Sons, New York.
[a] Calculated to standard moisture content.
[b] To convert to BTU/pound from kJ/g, multiply by 430.2.

TABLE VI

International Classification of Hard Coals

	Value	Volatile matter (percentage; dry ash-free basis)	Heating value (MJ/kg; moist ash-free basis)
First digit: coal class	0	0–3	
	1A	>3–6.5	
	1B	>6.5–10	
	2	>10–14	
	3A	>14–16	
	3B	>16–20	
	4	>20–28	
	5	>28–33	
	6	>33–41	>32.45
	7	>33–44	32.45–30.15
	8	>35–50	30.15–25.55
	9	>42–50	25.55–23.78
	Value	**Free swelling index**	**Roga index**
Second digit: coal group	0	0–0.5	0–5
	1	1–2	>5–20
	2	2.5–4	>20–45
	4	>4	>45
	Value	**Gray–King coke type**	**Dilatometer, maximum dilatation (percentage)**
Third digit: coal subgroup	0	A	Nonsoftening
	1	B–D	Contracting only
	2	E–G	<0
	3	G1–G4	>0–50
	4	G5–G8	>50–140
	5	>G8	>140

Source. Damberger, H. H. Coal. *In* "Encyclopedia of Science and Technology," 7th ed. McGraw–Hill, New York.

X-ray diffraction, infrared spectroscopy, differential thermal analysis, electron microscopy, and petrographic analysis. Typically, the coal sample has to be first prepared by low-temperature "ashing" to remove the organic material. Analyses have shown the presence of nearly all of the elements in coal, at least in trace quantities. Certain elements (e.g., Ge, Be, B, Sb) are found primarily with organic matter, whereas others (e.g., S, P, Na, K, Ca, Mg) are found in both inorganic and organic components of coal. The primary elements forming inorganic constituents are Al, Si, Ca, Mg, Na, and S, with the secondary elements being Zn, Cd, Mn, As, Mo, and Fe.

The presence of inorganic constituents in coal is a result of several separate processes. These include (1) natural occurrence in plants as, for example, silica in cell walls of scouring rushes, (2) incorporation by ion exchange (cations linking with carboxylic acid groups or formation of chelating complexes), and (3) incorporation by mechanisms such as filtration. Some come from the detrital minerals deposited during peatification, whereas others come from secondary minerals that crystallize from water that percolates through the coal seams. During the early stages of coal development, minerals can be deposited in the peat either allogenically (minerals carried and deposited unaltered from outside the peat bed) or authigenically (minerals altered during the deposition in the peat bed).

4.3 Minerals Mixed with Coal

Table VII provides a list of different minerals that have been found in coal. Mineral matter in coal can be classified as inherent mineral matter, included mineral matter, and excluded mineral matter. Inherent mineral matter is bound to organic matter in the coal's structure. Included mineral matter is that which forms an inorganic entity within pulverized coal particles. Excluded mineral matter is that which is external to the coal seam, taken in accidentally.

Mineral matter occurs in coal seams as layers, nodules, fissure fillings, rock fragments, and small particles finely distributed throughout. These are classified as inherent (plant derived), detrital, syngenetic, or epigenetic. Inherent minerals contribute less than 2% to the coal ash.

Detrital minerals are those that were washed into the peat swamp by water streams or blown in by wind. The various clay minerals are the most common detrital minerals. Other common detrital minerals include quartz, feldspar, garnet, apatite, zircon, muscovite, epidote, biotite, augite, kyanite, rutile, staurolite, topaz, and tourmaline. The secondary minerals are kaolinite, calcite, and pyrite.

Syngenetic minerals are formed *in situ* from colloidal, chemical, and biochemical processes as organic matter accumulates in swamps. Pyrite is the most common type, and in some coals certain clay materials are syngenetic in origin. Epigenetic minerals are formed during later stages of coalification. These include calcite, dolomite, ankerite, siderite,

TABLE VII

Minerals Found in Coals

Mineral	Occurrence	Mineral	Occurrence
Clay minerals		**Chloride minerals**	
Montmorillonite	Common	Halite	Common
Halloysite	Rare	Sylvite	Rare
Illite-Sericite	Abundant	Bischofite	Rare
Kaolinite	Dominant	**Silicate minerals**	
Mixed layer	Common	Quartz	Dominant
Clay minerals/smectite		Tourmaline	Rare
Sulfide minerals		Staurolite	Rare
Pyrite (isometric)	Common	Sanidine	Rare
Chalcopyrite	Rare	Hornblende	Rare
Millerite	Rare	Biotite	Rare
Sphalerite	Rare	Garnet	Rare
Pyrrhotite	Rare	Epidote	Rare
Arsenopyrite	Rare	Orthoclase	Rare
Galena	Rare	Topaz	Rare
Marcasite (orthorhombic)	Rare	Zircon	Rare
Carbonate minerals		Kyanite	Rare
Calcite	Common	Albite	Rare
Siderite	Rare	Augite	Rare
Dolomite	Common	Feldspar	Common
Witherite	Rare	**Oxide and hydroxide minerals**	
Ankerite (ferroan dolomite)	Common	Hematite	Rare
Sulfate minerals		Limonite	Rare
Barite	Rare	Lepidocrocite	Rare
Bassanite	Rare	Magnetite	Rare
Rozenite	Rare	Goethite	Rare
Roemerite	Rare	Rutile/Anatase	Rare
Sideronatrite	Rare	Diaspore	Rare
Gypsum	Common	**Phosphate minerals**	
Jarosite	Rare	Apatite	Rare
Melanterite	Rare		
Mirabilite	Rare		
Anhydrite	Rare		
Szomolnokite	Rare		
Coquimbite	Rare		
Kieserite	Rare		

Source. Mehta, A. (Project Manager). (1986). "Effects of Coal Quality on Power Plant Performance and Costs," Vol. 4: "Review of Coal Science Fundamentals," (EPRI CS-283). Electric Power Research Institute, Palo Alto, CA.

pyrite, marcasite, sphalerite, barite, certain clay minerals (especially kaolinite), gypsum, carbonates, and chlorides. They commonly precipitate from solutions as they percolate through the peat or along joints and pores in the coal. Sometimes, they are coarsely crystalline or in the shape of nodules.

4.3.1 Impact of Mineral Matter on Coal Use

Mineral matters in coal cause major problems in coal-fired power plants. Hard minerals cause erosion and wear of equipment during handling, crushing, and grinding. Ash, produced by mineral matters, needs to be disposed of in an environmentally safe manner, adding to the power generation cost. High-temperature ash discharge as a molten slag involves substantial loss of sensible heat. Slagging and fouling on boiler and superheater walls reduce heat transfer efficiency, increase fan power requirements, and in extreme cases block passages in boilers. Major air pollution is caused by mineral matters such as sulfur and mercury. Minerals in coal can behave as poisons for a variety of coal conversion processes. For example, titanium compounds in coal act as poisons in coal gasification and liquefaction.

Mineral matters do not always have a negative impact on coal use. A high sulfur content of feed coal in the synthetic fuels process is deemed to be best for catalysis. A high calcium carbonate content in combustion promotes self-cleaning of emissions. The presence of sodium in flue gas enhances the performance of electrostatic precipitators.

To assess the potential adverse effects of minerals in the coals, it is insufficient to depend on the "ultimate and trace element" analyses. The forms in which various elements are present in coal are also important. For example, sulfur in coal can have 30 different forms. Organically bound sulfur exists in a variety of functional forms. Selenium occurs in pyrite and sphalerite or as organically bound selenium, lead selenide, and ion-exchangeable, water-soluble selenium.

In general, one needs to know four factors about mineral matters in the coal to assess their effects on coal use:

- Chemical composition of the minerals
- Particle size distribution of mineral matters
- Percentage compositions of minerals
- Extent of crystallanity of the minerals

4.3.2 Metallurgical Coke

The quality of metallurgical coke depends on ash yield and composition and the sulfur content of the coal in addition to its rank, maceral composition, and density. Phosphorus and vanadium in coal are undesirable in the coke-forming process. Depending on the maceral and other properties, cokes of various types and strength are generated. Therefore, to produce cokes of a specific strength to support the iron ore in a blast furnace, coal blending is necessary.

4.3.3 Fouling

The buildup of sintered ash deposits ("fouling") on the heat exchange surfaces of coal-fired boilers is one of the most costly problems. Fouling drastically reduces boiler efficiency and also promotes corrosion and erosion. The size and strength of these deposits depend on the configuration and operation of the boiler and the mineral contents of the coal. Although sodium is often cited as the primary cause of fouling, the formation of sintered ash deposits is the result of the interaction of many inorganic constituents. Research indicates that ash yield and sodium, calcium, and magnesium concentrations in ash are the most important factors influencing the fouling behavior of low-rank coal.

4.3.4 Slagging

Slag is a molten ash deposited on boiler walls in zones subjected to radiant heat transfer. Minerals that have low melting points tend to stick to the boiler walls and accumulate to form massive fused deposits. Many slags reach a point at which the viscosity begins to rise rapidly with very small temperature changes. Above this "temperature of critical viscosity," slag becomes too viscous to flow and drain from the equipment.

The iron content of the coal ash has a major influence on the ash fusion temperature and the slagging potential of a coal. Iron sulfides tend to promote slagging more than do other iron-bearing minerals. Illite, a clay mineral, also may play a role in the slagging process by acting as an initiator.

4.3.5 Corrosion

Corrosion is enhanced by nonsilicate impurities in coal that volatilize and collect on the metal walls. Low-temperature corrosion can be caused by acidic gases (chlorine and sulfur oxides) or condensing acids (sulfuric acid), leading to most severe damage of the boiler's coolest parts. Therefore, the flue gas temperature needs to be above the acid dew point to minimize sulfur corrosion effects. High-temperature corrosion is caused by oxidation, sulfidation, or chlorination of the metal. Sodium and potassium sulfates play an important role in the corrosion process. High levels of chlorine (>1000 ppm) and (to a lesser extent) fluorine are considered to be potentially corrosive. At high levels ($>0.6\%$), chlorine, normally associated with sodium and complex sulfate deposits, initiates deposition on superheater tubes and also initiates stress corrosion cracking of tubes.

Although the elements arsenic, lead, boron, and (perhaps) zinc are typically present in trace quantities, they also enhance corrosion under certain conditions. Under high-firing temperature conditions, phosphorus at levels greater than 0.03% leads to phosphate deposits.

4.3.6 Erosion and Abrasion

Erosion and abrasion lead to wear. They cause wear in boilers and in mining and grinding equipment. Abrasive minerals in coal include quartz and pyrite. Particle size and shape affect the extent of erosion and abrasion. Coal technologists use the silicon/aluminum ratio as an indicator of abrasiveness, based on the assumption that the higher the ratio, the greater the quartz content. Ratios greater than 3 are deemed abrasive.

4.3.7 Environmental Impact

There are many ways by which coal use can negatively affect the environment. Coal mining involves blasting, dredging, and excavation that leads to air pollution by particulate matter being discharged into the atmosphere. Waste material dumped in local streams contaminates them. Coal dust explosions, mining accidents, and black lung disease among miners are some of the major health, safety, and hazard issues of mining. During coal transportation in open railroad cars, some coal particles fly off and spread over the land.

The 1990 Amendments to the Clean Air Act (CAAA) cited 12 elements in the Hazardous Air Pollutant List (P, Sb, As, Be, Cd, Cr, Co, Pb, Mn, Hg, Ni, and Se), several of which are present in coal. Many of the same elements (As, Be, Cd, Cr, Pb, Hg, and Se) appear in the U.S. Environmental Protection Agency's (EPA) list of potential pollutants in drinking water.

The EPA found that most coal wastes are not hazardous as defined by the Resources Conservation and Recovery Act. However, it reported the presence of groundwater contamination in the vicinity of waste disposal sites. At one site, nearby drinking water wells were contaminated with vanadium and selenium. The release of selenium from coal combustion wastes has caused extensive fish kills at two sites in North Carolina and at one site in Texas. Small amounts of selenium can be potentially toxic to livestock. Although coals in the Powder River Basin contain less than 1 ppm selenium, coal mining there releases selenium in sufficient quantity to exceed the permissible groundwater selenium concentration limit for livestock drinking water.

In power plants, mineral matters are collected as bottom or fly ash. Sometimes, they are sluiced up for environmentally sound disposal. To collect fly ash, electrostatic precipitators, bag houses, or cyclones are employed. Roughly 20 million tons of ash, or 25% of all particulate matter discharged into the air from all sources, is generated each year. Much of the ash is discarded as waste, is returned to the mines, or is dumped in the ocean. Ash has been used to make concrete. It is also a potential raw material to make aluminum. Some ashes contain valuable quantities of gallium, gold, and silver, making them attractive revenue sources.

4.3.7.1 Sulfur Pollution Sulfates in coal, constituting approximately 15% of the sulfur, are retained as ash. Pyritic and organically bound sulfur is combusted to form sulfur oxides, which in the absence of emission controls escape to the atmosphere. Sulfur oxides and sulfurous and sulfuric acids are corrosive and harmful to the health of humans, animals, and plants. Acid rains have caused deforestation and damaged buildings and monuments.

To comply with increasingly stringent government regulations, power plants have had to resort to a variety of actions. They include (1) switching from coal to alternate fuels, (2) switching from high-sulfur to low-sulfur coal, (3) blending coals, (4) beneficiating coal to lower its sulfur content, and (5) desulfurizing flue gas. Low-sulfur subbituminous coal from eastern Wyoming has come into high demand to lower sulfur emissions.

4.3.7.2 Mercury Pollution Mercury in flue gases can be present simultaneously as elemental mercury, oxidized mercury, and/or "particulate mercury," the form of which has not been determined by current methods. No reliable methods that can control mercury regardless of its form are yet available. The leading candidate is injection of activated carbon. Currently, U.S. power plants are exempt from having to control mercury pollution resulting from their operations. With the 2002 "Clear Skies" initiative, the U.S. government proposed to reduce mercury emissions from power plants by approximately 70% by 2018. The EPA is also separately considering mercury regulations under the CAAA. Thus, some mercury emission regulations can be expected, but their timing and stringency are not known.

4.3.7.3 Radionuclides Pollution Less known is the fact that coal contains trace quantities of radionuclides such as uranium, polonium, thorium, and radium. They are removed by wet scrubbers and electrostatic precipitators, and they tend to concentrate in the ash that is disposed. The use of ash in building materials has led to radon problems in residential and commercial buildings in some cases. Based on a U.S. Department of Energy's (DOE) comprehensive study showing emission values at or below the crustal values of the earth, the EPA concluded that levels of radionuclide emissions are not harmful to the ecosystem or humans and exempted power plants from reporting them.

5. COMPOSITIONAL ANALYSIS AND PROPERTIES OF COAL

5.1 Sampling Issues

Assessment of coal resources and their efficient use require knowing the coal properties in sufficient detail. The sample must be representative and large enough for all planned testing procedures and their replications. To ensure integrity of the analyses, the sample must be properly handled. The most widely accepted authority on sampling methods is U.S. Geological Survey (USGS) bulletin 1111-B. The procedures for taking a sample, reducing the particle size of the sample, and separating a smaller portion for later analysis are in ASTM procedures D2234, D2013, and BS 1017.

5.2 Proximate Analysis (ASTM D3172)

In this analysis, a coal sample is gently heated at 105°C in an inert atmosphere, and loss of weight is attributed to "moisture" in the coal (D3173). However, this loss is also due to the evolution of gaseous matter trapped in coal.

When loss of weight at 105°C stabilizes, coal is further heated at 750°C, again under inert atmosphere. The loss in weight under these conditions is assigned to volatile matter consisting of gases, organic oil, and tar, which normally is not further analyzed (D3175).

In the last step of proximate analysis, the remaining char is burned off, leaving behind ash (D3175, D388). By subtraction of moisture, volatile matter, and ash masses from the original sample mass, the fixed carbon mass is obtained (D3174).

Although not part of the standard proximate analysis, low-temperature oxidation of fixed carbon (and organic combustible material) at 120 to 150°C using oxygen-rich plasma is employed to ensure that the original minerals remain essentially unaltered. The resulting ash is then subjected to a wide range of analyses such as X-ray diffraction.

Proximate analysis is easy and quick to perform and is of great practical value in making empirical predictions of the coal combustion behavior. The volatile matter/fixed carbon ratio predicts how the coal will burn, that is, the flame characteristics and ease of ignition. Because high-rank coals contain less volatile matter than do low-rank coals, the former are more difficult to ignite and have a shorter flame than do the latter. This makes low-rank coals more suited for cyclone burners in which rapid intense combustion leads to maximum carbon use and minimum smoke emissions. Ash data generated by the proximate analysis allow engineers to predict the frequency and quantity of ash disposal. The proximate analysis data can be used to (1) establish the ranks of coals, (2) aid in buying and selling decisions, and (3) evaluate the need for beneficiation and the like.

5.3 Ultimate Analysis (ASTM D3176)

This is an elemental analysis limited to carbon, hydrogen, sulfur, nitrogen, and oxygen. Carbon and hydrogen are measured by burning the coal and collecting the resulting carbon dioxide and water (D3178).

Because of increasingly stringent environmental restrictions on SO_x emissions, particular interest has developed in the total sulfur part of the ultimate analysis. Sulfur determination (the Eschka method) (D3177) is based on conversion of all types of sulfur in the coal (e.g., pyrite, sulfates, sulfides, organically bound sulfur) to sodium sulfate. A coal sample is ignited with magnesium oxide and sodium carbonate. The resulting sodium sulfate is then converted to water-insoluble barium sulfate, which is filtered, dried, and weighed. There are two additional sulfur determination methods. In the bomb-washing method, sulfur is precipitated as barium sulfate from the oxygen bomb calorimeter washings. In the high-temperature combustion method, the sample is burned in a tube furnace and sulfur oxides are collected and quantified by titration. Forms of sulfur are determined by the D2492 method.

Nitrogen is usually determined by first liberating elemental nitrogen from the sample by the digestion method (Kjeldahl–Gunning), converting the elemental nitrogen catalytically to ammonia, and (finally) titrating the resulting ammonia with acid (D3179).

There are no relatively simple analytical methods for oxygen determination. The most commonly used method requires a neutron source such as a nuclear reactor. Neutrons are bombarded on the sample to convert oxygen into nitrogen-16, a very unstable isotope of nitrogen. Nitrogen-16 decays rapidly by emitting beta and gamma rays that are counted. Because most laboratories do not have access to a neutron source, oxygen content is calculated by subtracting the masses of other elements from the mass of the sample (D3176).

Chorine may also be included in the ultimate analysis. It is determined by either the bomb calorimeter or the Eschka method (D2361, D4208).

5.4 Heating Value or Calorific Value

The heating value of coal is measured by an adiabatic calorimeter (D2015). The procedure gives the "gross heating value." The "net heating value" is based on the assumption that all of the water resulting from combustion remains in vapor form (D407). The heating value of a coal determines the amount of coal a power plant needs.

5.5 Other Chemical Analysis Methods

Since the 1970s, environmental concerns have led to increasingly restrictive levels of permissible emissions of some of the trace elements in coal. A procedure has been adopted by USGS, outlined in Table VIII, to analyze trace mineral matters. MAS, ICP/AES, and ICP/MS analyses are performed on ash produced at 525°C.

5.6 Physical, Mechanical, and Other Properties of Coal

5.6.1 Physical Properties

Most of the physical properties of coal discussed here depend on the coal orientation and history of the piece of coal. Properties may vary from one coal piece to another coming from the same seam or mine.

The specific electrical conductivity of dry coal is in the semiconductor zone, 10^{10} to 10^{14} ohm.cm, increasing with rank and temperature. Moisture in coal also enhances conductivity.

The dielectric constant also varies with rank, with a minimum at about 88% carbon. Polar groups in low-rank coal and higher electrical conductivity in high-rank coal contribute to their dielectric constants.

Magnetic susceptibility measurements indicate that the organic part of coal is diamagnetic, with traces of paramagnetic behavior resulting from the presence of free radicals or unpaired electrons.

Thermal conductivity and diffusivity of coal depend on pore and crack structure. Conductivity ranges from 0.23 to 0.35 W/(cm · K), and diffusivity ranges from 1×10^{-3} to 2×10^{-3} cm^2/s. At 800°C, these ranges increase to 1 to 2 W/(cm · K) and (1–5) $\times 10^{-2}$ cm^2/s, respectively. The increases are due to radiation across pores and cracks at higher temperatures. The thermal conductivity of coal is influenced by moisture content, the chemical and physical nature of the coal during reaction, and convective and radiant heat transfer within the particle. Below 400°C, thermal conductivity of coal has been measured in the range of (5–8) $\times 10^{-4}$ cal/(cm · s · K).

The specific heat of coal is estimated to be in the range of 1 to 2 J/(g · K) at 20°C to 0.4 J/(g · K) at 800°C. The specific heat is affected by the oxidation of the coal.

5.6.2 Mechanical and Other Properties of Coal

These properties have a much more direct bearing on the operations of power plants or other coal processing units.

5.6.2.1 Density The true density of coal, which increases with carbon content or rank, is most accurately obtained by measuring helium displacement. Values of 1.4 to 1.6 g/ml have been reported for greater than 85% carbon coal. The bulk density (D291), typically about 0.8 g/ml, depends on the void fraction. It affects coal's storage, hauling requirements, and sizing for its conveyance.

5.6.2.2 Porosity Fine porosity of coal restricts the access of oxygen to the internal structure. This retards combustion rates and depresses burnout in power plant boilers. Porosity is the result of the random arrangement of organic macromolecules within coal. Total porosity of coal ranges from 0.02 to 0.2%, with no apparent trends with carbon

TABLE VIII

Methods Employed by U.S. Geological Survey

Elements	Methods employed
Hg, F	Wet chemical analysis
Cl, P	X-ray diffraction
Se	Atomic absorption spectroscopy hydride ASS
Na, Be, Cu, Li, Th, Zn, Co, Mn, Ni, V, Cr, Sc, Sr, Y	Induced coupled plasma/atomic emission spectroscopy (acid digestion)
Si, K, Ba, Al, B, Ca, Ti, Zr, Mg, P, and Iron oxides	Induced coupled plasma/atomic emission spectroscopy (sinter)
Ag, As, Au, Bi, Cd, Cs, Ga, Ge, Mo, Nb, Pb, Sb, Sn, Te, Tl, U	Induced coupled plasma/mass spectroscopy (acid digestion)
Ce, Dy, Er, Eu, Gd, Hf, Ho, La, Nd, Pr, Sm, Tb, Tm, W, Yb, Ta	Induced coupled plasma/mass spectroscopy (sinter)
S	MAS

Source. Stanton, R. W., and Finkelman, R. B. Coal quality. *In* "Energy Technology and the Environment" (A. Bisio and S. Boots, Eds.), Vol. 1. John Wiley and Sons, New York.

content. Porosity is often subdivided into macropore ($d = 300A^\circ$), intermediate pores ($d = 12–300A^\circ$), and micropores ($d < 12A^\circ$). Micropore volume appears to be low for low-rank coals and high for coals with carbon contents between 74 and 80% and greater than 90%. This is one reason why low-rank coals burn more easily.

5.6.2.3 Equilibrium Moisture Content Equilibrium or inherent moisture (D1412) is the moisture-holding capacity of coal at 30°C in an atmosphere of 97% relative humidity. The organic reflux moisture test (ISO R348) has been used to determine the residual moisture in the prepared sample.

Total moisture or "as received" moisture is determined by a two-step procedure (D3302, D3173) involving surface moisture measurement, reduction of gross sample, and determination of residual moisture.

Higher total moisture leads to a lower heating value of the coal and lower flame temperature, and it lowers the plant efficiency. It also causes higher haulage costs per million BTUs and causes the power plant to purchase more coal. A number of processes have been patented to dry coal at the mine site before its shipment, but none of them has been commercialized, due partly to excessive capital requirements.

5.6.2.4 Ash Fusibility (D1857) The ash fusibility test provides the "melting point" of ash. Although empirical, the procedure requires strict adherence to ensure reproducible results. Ash fusibility data are critical to power plants because they are used to assess operational difficulties encountered in using the coal (e.g., slagging, loss of heat transfer rate, frequency of boiler shutdown).

Ash fusibility under oxidizing conditions tends to be higher than that under reducing or neutral conditions. In the procedure, the condition of the starting cone of ash is reported as it undergoes several physical changes as temperature increases. These include (1) initial (cone tip) deformation, (2) softening (height/base = 1.0), (3) hemispherical shape formation (height/base = 0.5), and (4) fluid (puddle) formation.

A related criterion, namely T_{250}, the temperature at which slag has a viscosity of 250 poise, depends on base/acid and silica/alumina ratios. The lower the base/acid ratio and the higher the silica/alumina ratio, the higher the T_{250} of the ash.

5.6.2.5 Ash Analysis This term is applied to the analysis of the major elements commonly found in coal and coal ash (D2795). Oxides of Si, Al, Fe, Ti, Ca, Mg, Na, K, and P are measured. Phosphorus oxide has importance in the coke used in steelmaking processes. Two other ASTM methods for ash composition based on atomic absorption are D3682 and D3683.

The ASTM ash analysis is widely used to predict slagging and fouling. However, the way by which ash is produced in the boiler is considerably different from the way by which it is produced by the ASTM procedure.

5.6.2.6 Free Swelling Index This index is determined by quickly heating a coal in a nonrestraining crucible (D720). The values are assigned from 0 to 9, with noncaking, nonswelling coals being assigned the value 0. The higher the FSI value, the greater the swelling property.

The FSI is not recommended for predicting coal behavior in a coke oven. The index is not deemed important for pulverized coal-fired boilers. Some units (e.g., retort stokers) form coke in their normal operation. For certain coals, the FSI values reduce while in storage, indicating coal deterioration and partial oxidation.

Gray–King and ROGA tests, employed in Europe, serve a function similar to that of the FSI test in the United States.

5.6.2.7 Washability A large part of mined coal is cleaned to reduce its mineral contents before it is sold. Washability tests are performed to predict the effectiveness of coal cleaning. This is a float–sink test in heavy liquids that generates various gravity samples that are analyzed for their heating values, ash and sulfur contents, and so on. Sometimes, efforts to lower ash content by washing are uneconomical.

5.6.2.8 Strength and Hardness These are important coal properties in mine stability and comminution. Various tests are available to estimate coal's compressive and shearing strengths and hardness. On the Mohs hardness scale, lignites range from 1 to 3, bituminous from 2.5 to 3.0, and anthracite from 3 to 4.

5.6.2.9 Grindability The Hardgrove Grindability Index (HGI) method is used to determine the relative ease of coal pulverization in comparison with a series of standard coals (D409). The data can be used to determine power and sizing of grinding equipment and its expected wear rate. For low-rank coals, subbituminous and lignite, the HGI is found to depend on coal moisture, which decreases as grinding proceeds.

5.6.2.10 Friability The breaking of coal into dust during handling, transporting, and pulverizing causes explosion hazards, control of which requires special sensors and control equipment. The most friable coal contains approximately 90% carbon. This point also corresponds to the maximum HGI and minimum Vickers microhardness values. Charcoal-like fusain tends to be soft and friable.

5.6.2.11 Slaking Slaking causes coal disintegration during storage due to coal drying. This is an important phenomenon for lignite and subbituminous coals but not for high-rank coals, as lignite and subbituminous coals have high moisture content.

5.6.2.12 Spontaneous Ignition Temperature of a coal pile builds up slowly if it is not turned over. This triggers spontaneous ignition, which is a serious hazard requiring monitoring and preventive measures. Coal particle size distribution and its oxygen content have a great influence on spontaneous ignition. Other important factors include pyrite content, oxygen circulation in the pile, moisture, and rank of coal.

5.6.2.13 Coal Petrography/Vitrite Reflectance Test Extensive use has been made of vitrite reflectance measurements (D2797, D2798) and reflectograms along with the physical microscopic components (D2797, D2799) to assess coking coals. Vitrinite reflectance limits data for various coal rank classes are given in Table IX.

5.6.2.14 Gieseler Plasticity Plasticity measurement (D2639) gives the coal's tendency to fuse or soften and reach a fluid state on heating during simulated coking stages.

5.6.2.15 Audibert Arni Dilatometer/Sole Heated Oven Determination The dilatometer measures the contraction and expansion of coal during carbonization (ISO R349). The corresponding ASTM test is called Sole Heated Oven Determination (D2014). These data are important in predicting whether coke can be removed during production without damaging the equipment.

5.6.2.16 Other ASTM Tests for Coal Table X lists other ASTM tests that are relevant for coal characterization.

6. *FRONTIERS OF COAL RESEARCH*

6.1 Advanced Instrumental Methods of Measurements

The advanced measurements listed here overcome some of the limitations of the standard methods by providing more detailed and accurate characterization of specific coal. These methods can be grouped into (1) composition, (2) combustion characteristics, and (3) tests for improved correlations.

6.1.1 Composition

Table XI lists some of the advanced methods that have been used with some success, but much work still needs to be done to ensure their reliability and to assess the limits of their applicability. This is a

TABLE IX

Vitrinite Reflectance Limits Taken in Oil versus Coal Rank Classes

Coal rank	Maximum reflectance (percentage)
Subbituminous	<0.47
High-volatile bituminous	
C	0.47–0.57
B	0.57–0.71
A	0.71–1.10
Medium-volatile bituminous	1.10–1.50
Low-volatile bituminous	1.50–2.05
Semianthracite	2.05–3.00 (estimated)
Anthracite	> 3.00 (estimated)

Source. Vorres, K. S. Coal. *In* "Encyclopedia of Chemical Technology" (J. Kroschwitz, Ed.), 4th ed. John Wiley and Sons, New York.

TABLE X

Other ASTM Tests for Coal

Test	ASTM method	Parameters	Plant performance
Calorific value	D3286, D2015	—	General combustion
Fineness	D197	75–1180 micron sieve	Burner
Sieve analysis	D410	3/8- to 8-inch sieve	Coal handling/ storage
Drop–Shatter test	D440	Friability	Pulverization
Tumbler test	D441	Friability	Pulverization
Dustiness	D547	Dustiness indices	Handling
Sulfur in ash	D1757	Sulfur	SO_x pollution

TABLE XI

Advanced Coal Composition Measurements

Method	Parameters
Electron microscopy	Ultra-fine granular structure
	Pore structure
X-ray diffraction	Size distribution of aromatic ring systems
	Diameter and thickness of lamellae
	Mean bond length
	Mineral phase evaluation
Mass spectroscopy	Molar mass and formulae
	Analysis to determine types of hydrocarbons
	Carbon number distribution
	Trace elements
Nuclear magnetic resonance	H_{ar}/H and C_{ar}/C ratios
	Distribution of isotopic tracers
Electron spin resonance	Stable free radicals
	Other species with unpaired electrons
Infrared absorption	Functional groups (e.g., OH, CH_{ar})
	Minerals
Ultraviolet–visible absorption	Difficult to apply to whole coal samples but can be used to identify components in extracts
Electron spectroscopy for chemical analysis	Elemental distribution and chemical state of surface
Optical emission spectroscopy	Elemental analysis
	Single-element and survey analysis
	Major, minor, and trace element analysis for metals and semi-metals
	Applies to coal ash only
X-ray fluorescence	Inorganic element analysis
Atomic absorption	Precise quantitative determinations of metals and semi-metals
	Single element analysis for minor and trace constituents
Mossbauer spectroscopy	Iron compounds and association with sulfur
Scanning electron microscope automatic image analysis	Size, shape, and composition of minerals in coal
Neutron activation analysis	Elemental analysis
Spark source mass spectroscopy	Semiquantitative elemental analysis of coal ash
Inductively coupled plasma spectroscopy	Quantitative multi-element analysis of coal ash
Electron microprobe analysis	Elemental analysis of micrometer-size areas
Thermometric techniques	Chemical characterization of coal ash
Optical microscopy	Identification of minerals in coal or coal ash
Ion microprobe mass analysis	Isotopic characterization of minerals or macerals
Extended X-ray analysis fine structure	Chemical environment of an atom

Sources. Mehta, A. (Project Manager). (1986). "Effects of Coal Quality on Power Plant Performance and Costs," Vol. 3: "Review of Coal Quality and Performance Diagnostics" (EPRI CS-4283). Electric Power Research Institute, Palo Alto, CA; Stanton, R. W., and Finkelman, R. B. Coal quality. *In* "Energy Technology and the Environment" (A. Bisio and S. Boots, Eds.), Vol. 1. John Wiley and Sons, New York.

dynamic area because new methods and applications are being developed continually. The methods are used to analyze microstructure, trace elements, organic material structure, and mineral matter composition and structure. They make it possible to characterize coal with more details and better accuracy. However, they require expensive instrumentation and highly skilled personnel to operate.

6.1.2 Combustion Characteristics

Although proximate analysis provides key coal data, it provides no data on devolatalization and combustion characteristics. These data are provided by thermogravimetric analyses. Four important methods needed to design boilers and assess performance of existing boilers are the following:

- Burning profile (Babcock & Wilcox)
- Volatile release profile (Babcock & Wilcox)
- Thermogravimetric analysis of char (Combustion Engineering)
- Drop tube furnace system (Combustion Engineering)

6.1.3 Tests for Improved Correlations

Procedures have been developed to improve the correlation of coal properties with coal's performance in power plant systems. These include analysis of physical separation of pulverized coal (pulverizer wear and slagging), analysis of alkalies in coal ash (fouling), sintering strength of fly ash (fouling), free quartz index (pulverizer wear), and continuous coal grindability (pulverizer capacity).

6.2 Active Areas of Coal Research

Active coal research areas worldwide include the following:

- Advances in pollution controls for NO_x, SO_x, hazardous air pollutants (HAPs), and particulate matter
- Development of economically viable mercury emissions control technologies
- Advances in coal liquefaction and gasification
- Advances in power generation systems
- Studies in greenhouse gases sequestration
- Advanced studies to correlate physical structure, petrography, and properties
- Advanced processing studies (e.g., management of coal use by-products, coal chemicals, pyrolysis, and coal–oil–water interaction)
- Studies of coal gasification fundamentals

6.3 Carbon Sequestration

Major activities are under way worldwide to find economically viable means of sequestration of carbon dioxide, a major greenhouse gas. Although its emissions from fossil fuel power plants are only approximately 39% of its total emissions in the United States (32% from coal-fired plants), sequestration of these power plant emissions is the focus of research and development and analysis because they are localized. CO_2 use for enhanced resource recovery is being considered along with the use of underground saline formations, possibly oceans, and mineral carbonation for its sequestration. The United States has not signed the International Climate Change agreement (Kyoto Protocol). Nevertheless, its government has undertaken a 10-year demonstration project, "FutureGen," to create a coal-based zero-emission power plant.

6.4 Computational Methods

In support of experimental research programs, much work continues to be carried out in coal-related computational methods, including the following:

- Constitutive laws of multiphase flow
- Process simulation and modeling (e.g., black liquor gasification, circulating and spouted beds, spontaneous combustion, oxidative pyrolysis, coal–biomass co-firing, char gasification)
- Computational fluid dynamics in support of process design and optimization
- Three-dimensional modeling of subsurface conditions (e.g., simulation of coal bed for methane recovery)
- Engineering design models (e.g., hydroclone, circulating fluidized bed, transport reactors)
- Modeling for pollution control (e.g., NO_x formation)

SEE ALSO THE FOLLOWING ARTICLES

Clean Coal Technology • Coal, Chemical and Physical Properties • Coal Conversion • Coal, Fuel and Non-Fuel Uses • Coal Industry, Energy Policy in • Coal Industry, History of • Coal Mine Reclamation and Remediation • Coal Mining, Design and Methods of • Coal Mining in Appalachia, History of • Coal Preparation • Coal Storage and Transportation • Markets for Coal • Oil and Natural Gal Liquids: Global Magnitude and Distribution • Peat Resources

Further Reading

Damberger, H. H. (1992). Coal. *In* "Encyclopedia of Science and Technology," 7th ed. McGraw–Hill, New York. [The following subsections of this article have been excerpted with permission from McGraw–Hill: 1.2 (Influence of Geology) and 2.1 (Petrology, Petrography, and Lithotypes).]

Mehta, A. (Project Manager) (1986). "Effects of Coal Quality on Power Plant Performance and Costs," Vol. 3: "Review of Coal Quality and Performance Diagnostics" (EPRI CS-4283). Electric Power Research Institute, Palo Alto, CA. [The following subsections of this article have been excerpted with permission from EPRI: 5.6.2.4 (Ash Fusibility [D1857]), 5.6.2.16 (Other ASTM Tests for Coal), and 6.1.2 (Combustion Characteristics).]

Mehta, A. (Project Manager) (1986). "Effects of Coal Quality on Power Plant Performance and Costs," Vol. 4: "Review of Coal Science Fundamentals" (EPRI CS-283). Electric Power Research Institute, Palo Alto, CA. [The following subsections of this article have been excerpted with permission from EPRI: 2.1 (Petrology, Petrography, and Lithotypes), 2.2.3 (Geochemical Phase [coalification]), and 4.3 (Minerals Mixed with Coal).]

Perry, M. B. (Conference Chairman), and Ekmann, J. M. (Program Chairman). (2001). "Proceedings of the 11th International Conference on Coal Science." San Francisco, CA.

Savage, K. I. (1977). Coal: Properties and Statistics. *In* "Energy Technology Handbook" (D. M. Considine, Ed.). McGraw–Hill, New York. [The following subsections of this article have been excerpted with permission from McGraw–Hill: 1.2 (Influence of Geology), 5.1 (Sampling Issues), 5.6.2.3 (Equilibrium Moisture Content), and 5.6.2.6 (Free Swelling Index).]

Schobert, H. H. (1987). "Coal: The Energy Source of the Past and Future." American Chemical Society, Columbus, OH. [The following subsections of this article have been excerpted with permission from the American Chemical Society: 1.1 (Influence of Age), 2.2.1 (Components of Plants), 3.1.1 (Descriptions of Coals by Their Ranks), 4.2 (Inorganic Constituents in Coal), 4.3 (Minerals Mixed with Coal), 5.2 (Proximate Analysis [ASTM D3172]), and 5.3 (Ultimate Analysis [ASTM D3176]).]

Simon, J. A. (1981). Coal. *In* "Encyclopedia of Energy," 2nd ed. McGraw–Hill, New York. [The following subsections of this article have been excerpted with permission from McGraw–Hill: 1.2 (Influence of Geology) and 4.3 (Minerals Mixed with Coal).]

Stanton, R. W., and Finkelman, R. B. (1995). Coal quality. *In* "Energy Technology and the Environment" (A. Bisio and S. Boots, Eds.), vol. 1. John Wiley, New York. [The following subsections of this article have been excerpted with permission from John Wiley and Sons: 4.3.1 (Impact of Mineral Matter on Coal Use) and 4.3.7 (Environmental Impact).]

Vorres, K. S. (1992). Coal. *In* "Encyclopedia of Chemical Technology" (J. Kroschwitz, Ed.), 4th ed. John Wiley, New York. [The following subsections of this article have been excerpted with permission from John Wiley and Sons: 1.2 (Influence of Geology), 2.1 (Petrology, Petrography, and Lithotypes), 2.3.1 (Bonding within a Macromolecule), 3.2 (The International Systems for Coal Classification), and 5.6.1 (Physical Properties).]

Coal Storage and Transportation

JAMES M. EKMANN and PATRICK H. LE
U.S. Department of Energy
Pittsburgh, Pennsylvania, United States

1. Coal Transport
2. Coal Preparation
3. Stockpiles
4. Handling Coal at the End-Use Site
5. Storing Coal at the End-Use Site

Glossary

coal preparation A process that includes not only crushing and breaking methods (coal pretreatment) but also all the handling and treatment methods necessary to remove impurities and prepare coal for commercial use.

coal pretreatment A process consisting of breaking, crushing, and screening run-of-mine coal to produce coal with a predetermined top size.

coal screening A method to separate coal particles by passing them through a series of screens, of decreasing size.

coal sizing A process to separate raw coal into various particle sizes, either by screening or by other classification methods.

coal spontaneous combustion The process whereby coal ignites on its own, because the rate of heat generation by the oxidation reaction exceeds the rate of heat dissipation.

coal storage The storage process to accomplish various objectives: (1) to reclaim coal promptly and economically, (2) to carry-out blending/homogenization in order to achieve coal with uniform or desirable characteristics, and (3) to store various coals whose price and market demand vary seasonally, thus maximizing profits.

dead storage of coal The concept that coal is not immediately accessible because it is in a compacted pile to prevent weatherization and some means is needed to reclaim the coal.

live storage of coal The concept whereby stored coal is immediately accessible by the principal means of reclaiming in coal piles or by gravity flow in silos.

run-of-mine (ROM) coal A heterogeneous material that is often wet and contaminated with rock and/or clay and is unsuitable for commercial use without treatment.

unit train A railroad technique to efficiently transport coal on a predetermined schedule, by using a dedicated train and the associated equipment for loading and unloading.

This article deals with the coal management system that encompasses coal preparation, storage, and transportation. Coal must first be transported from the mines to the point of consumption, often an electric power plant. Railroads are the predominant form of transportation, sometimes in cost-effective "unit trains." Barges and ships are the other major mode of coal transportation, with trucks seeing limited use due to higher costs. At the preparation plants, the coal undergoes preparation processes that remove rocks, noncombustibles, and other impurities from run-of-mine coal and also crush and size the coal for its end use. After preparation, coal is usually stored. Since even prepared coal is not homogenous, various methods are employed to ensure that the coal properties remain fairly uniform when burned. Also, coal will weather and oxidize, changing its properties, and in certain cases, coal can self-ignite, unless precautions are taken.

1. COAL TRANSPORT

Coal has relatively low energy content per unit weight and competes on the market with other fuels as a low-cost boiler fuel. Consequently, a low-cost and efficient delivery system based on many alternatives is essential to maintain its competitiveness.

Based on statistics from Consol Energy, the modes of coal transportation in the United States are distributed as follows: 62% by rail, 12% by truck, 16% by barge and/or vessel, and 10% by conveyor belt or slurry pipeline; these numbers were presented at a coal and power training course in July 2000.

1.1 Rail Transport

Railroads are very well positioned to transport U.S. coals from the mines to end-use destinations, such as utility and industrial plants. Their dominance as the major carrier of coal is expected to continue. The U.S. railway system has 126,682 miles of Class 1 track, 47,214 miles of Class 2 and 3 track, and 1.2 million freight cars to carry all commodities, according to data from Consol Energy presented by Robert J. Schneid at the above-mentioned coal and power training course.

The trend in the industry is toward a consolidation through state ownership or a merger of privately owned systems. Dominant companies include the following: in the West, Union Pacific and Burlington Northern Santa Fe, and in the East, CSX and Norfork Southern. Table I gives an insight into the railroad consolidation of U.S. companies, with Stan Kaplan of PA Consulting Group as the source of information in year 2002.

Several strategies have been considered to improve railroad services and profitability: (1) cost cutting and productivity improvements; (2) merger; (3) differential pricing; (4) diversification; (5) new services; and (6) investment.

Strategies (1), (2), and (4) have exhausted their potential. Diversification is being abandoned. Cost cutting may be reaching a point of diminishing returns. Mergers have not proved to be a financial success. As a result, the focus of the railroad industry is on improved intermodal and other truck-competitive services, but not for coal as a high-priority service. The tension between demands for shipment at low rates with high-quality services and objectives to improve profitability will persist.

TABLE I

Consolidation of U.S. Railroad Companies

Old companies	New company
Seaboard Coast line	CSX
Chessie System	
Conrail	
Conrail	Norfolk Southern (NS)
Norfolk & Western	
Southern Railroad	
Missouri Kansas Texas	Union Pacific (UP)
Denver & Rio Grande Western	
Southern Pacific	
Union Pacific	
Missouri Pacific	
Chicago North Western	
Santa Fe	Burlington
Burlington Northern	Northern Santa Fe
Kansas City Southern	Kansas City Southern
Illinois Central Gulf	Canadian Pacific

Source. Adapted from Kaplan (2002).

In general, railroads have offered three major types of service to transport coal: (1) multiple-carload, (2) trainload, and (3) unit trains. Another less important type of service is single-carload rates for low volumes of smaller mines. Multiple-carload rates ranging from as few as 5 or 10 cars to as many as 100 cars are usually significantly lower than single-carload rates. Trainload rates are characterized by large minimum-volume or car requirements, since they move from origin to destination as one unit. Not meeting the annual volume requirements could lead to a higher penalty rate. Unit train rates usually have the lowest rates possible; the difference between them and trainload rates is that unit train rates use dedicated equipment and operate on a regular schedule.

The unit train (approximately 100 to 150 cars) came into existence in the early 1960s in response to other competitive alternatives in the transportation industry. Whereas most unit trains consist of 100 cars, they can also operate with as few as 20 cars. For a train unit of 100 to 150 cars, the payload would range from 10,000 to 15,000 tons and require four to six locomotives per train rated at 4000 to 5000 hp each. If lightweight aluminum cars are used, a unit train of 100 to 120 cars, carrying 121 tons (110 tonnes) each, could move more than 14,000 tons (12,700 tonnes) in total.

In terms of motive power for unit trains, five or six earlier diesel locomotive models (3000 hp) would be required to move a train of 100 to 125 cars. In the 1990s, distributed power became the most efficient method of transporting coal by rail. In this system, a remote-controlled engine is placed in the middle of the train. In the early 2000s, two or three modern diesel electric locomotives (6000 hp) based on microprocessor and alternating current motor technology could do the work of five earlier models. In addition, advances in control technology, called microwave centralized traffic control systems, have facilitated the safe movement of trains with minimal between-train clearance, therefore leading to significant increases in moving capacity.

1.1.1 Loading and Unloading Railcars

Typical coal railcars can carry 80 to 125 tons per car and the average load is 100 to 110 tons. Two types of railcars are in service: the solid bottom gondola and the hopper car.

A rotary dump system can handle gondola and hopper cars, whereas the undertrack pit can handle only hopper cars with bottom-opening doors.

The selection of the unloading method requires an evaluation of several trade-offs, such as the following: (1) gondola cars are the least expensive, need minimal maintenance, and can be completely unloaded with minimum coal hang-up but do require an expensive rotary dumper with the capability to accurately position the cars; and (2) hoppers (or bottom-dump cars) are more expensive and require more maintenance than the gondolas, due to the gate mechanism, but they can be unloaded more quickly on a relatively inexpensive track hopper. In addition, hopper cars are more difficult to unload under freezing conditions.

In addition to loading and unloading the coal, the issue of windage losses is frequently overlooked. Windage losses, i.e., coal blown out by wind from open-top cars, cause economic loss and environmental concern. One solution is a lid system, such as a flip-top lid, to control windblown loss, reduce dust, and prevent additional moisture from penetrating the coal, thus reducing the potential for freezing. Frozen coal is a troublesome problem related to unloading coal because the exposed top surface, the size plates of the car, and the surfaces of the sloped hoppers could be frozen. Several methods to thaw the cars are available, ranging from bonfires and oil-burning pans to sophisticated sheds. Treating the car sides and bottoms of the cars with oil can minimize freezing and ease the coal unloading process.

In terms of loading systems, three types of trackage are commonly used for unit trains. These are (1) the loop or balloon track, (2) the single track that is situated above and below the loading station, and (3) the parallel tracks that allow loading on both tracks. The loading system consists of locomotive, car haul, tripper conveyor, and hydraulic ram. The locomotive moves the cars in one pass under the loading station. The car-haul system moves two strings of cars in opposite directions on parallel tracks under two loading stations. The tripper conveyor system moves two-way loading chutes over two stationary parallel strings of cars. The hydraulic ram system moves the cars into loading position, after the locomotive has been removed.

In terms of unloading systems, there are two major types: the rotary dump and the undertrack pit. The rotary dump system handles the rollover type of car (hopper or gondola design) and the undertrack pit handles bottom-up cars.

1.2 Transport by Barge and Ship—Water Transport

The U.S. Great Lakes and inland waterway systems are a very valuable asset for moving coal and other commodities, as shown in Table II. In addition, the U.S. river network is connected to the Gulf Intracoastal Waterway, allowing highly efficient barge and lake carrier services. The Great Lakes facilitate shipments to the export market but have seasonal limitations. The inland waterway system is characterized by open top barges and the size of the tow. The tow is a function of the width and depth of the respective river and the configurations of its locks and dams.

1.2.1 Barge Transportation—Loading and Unloading

Barge transportation offers a lower cost alternative to railways for the majority of power plants and coke ovens that are located on inland waterways. In this waterway system, 25,000 miles of navigable rivers, lakes, and coastlines and a system of 192 locks and dams are ideal means for moving coal over long distances. Consequently, barge transportation could be considered competitive because it requires less fuel, the infrastructure is already in place or needs only to be updated, and barges carry large loads. A jumbo barge can carry an amount of coal equivalent to 15 railcars or 58 truck loads.

In addition, barge transportation is a good example of an intermodal carrier. With mines near a navigable waterway, several combinations are possible: (1) for short distances, trucks or conveyors

TABLE II

U.S. Riverways

Characteristics
• River network with portion highly navigable (depth of 9 ft or more)
• Connects with the Gulf Intracoastal Waterway
• Spans the country from Pennsylvania and West Virginia to the east and Minnesota and Wisconsin to the north
• Alabama and Florida on the Gulf Coast to the southeast
• Louisiana and Texas on the Gulf coast to the southwest
• Great Lakes allows heavy freight distribution
— From Illinois, Wisconsin, and Minnesota to the west and to Ohio, Pennsylvania, and New York to the east
— Access for exports to Canada through the St. Lawrence Seaway to overseas customers (Europe and Mediterranean region)

Source. Adapted from Lema (1990).

are used to transport coal to a barge terminal (for instance, up to 10 miles for conveyors and 30 or more miles for trucks, away from the mine); and (2) for longer distances, railroads can be used to transport coal to the river terminal (for instance, over 100 miles away from the mine).

However, the economics of barge transportation is affected by the circuitry of the waterways, their nature, the weather, and market conditions. Barge size is subject to the size, number, and condition of the locks. Low water levels and frozen rivers can halt shipment. In shipment, coal competes with other commodities to be transported and backhaul is a good solution to reduce coal rates.

To increase productivity, barge size has increased from a standard 1000-ton capacity to the jumbo-sized barge (9 ft draft) capable of carrying 1400 to 1800 tons. Usually, the jumbo-sized barge would carry 1800 tons, which enables the tow to move approximately 72,000 tons of coal. This is comparable to the loading of five unit trains. Towboats have pushed up to 48 jumbo barges on the Lower Mississippi River where there are no locks, but 40 barges are more common. In other instances, 15 jumbo barges (as maximum) are more common for a single lock, such as the 1200 ft lock on the Ohio River.

In the United States, five categories of loading plants are used, depending on the loading requirements. These categories are as follows:

1. The simple dock, which is used only to a small extent, mostly by small operators and for temporary facilities. It depends on a relatively stable water level. The trucks dump directly into the barges or into a fixed chute.
2. The stationary-chute system has declined in popularity; it is relatively economical and has the same disadvantages as the simple dock system (sensitive to water level fluctuations).
3. The elevating-boom system is commonly used; it is very flexible and adjustable to variations in water levels. It can handle high capacities. The common features of this system found in most plants are the hinged boom, the pantograph trim chute, and the extensive dock area.
4. The spar-barge system is suitable for rivers with fluctuating levels. It has two essential components: the floating spar barge and the shuttle. The floating spar-barge (as floating dock) contains the loading boom and the barge-positioning equipment. The shuttle or retracting loading belt leads to the spar-barge.
5. The traveling-tripper system is the method used to stockpile coal at large power plants; it is applied to load barges. The barges are held stationary by a barge-mooring system. The traveling tripper (as a movable loading point), with a reversible traverse shuttle belt and telescoping chutes, discharges coal in the barges. This barge loading method can load eight barges at a time, four on each side.

Regarding barge unloading, coal operators usually receive barge sizes and types that will accommodate their barge unloading systems. Earlier, barge unloading was performed using whirler-type cranes with clamshell buckets. Most barge coal is still unloaded by clamshell buckets that are mounted on either stationary straight-line-type towers or on moving tower barge unloaders; in the latter case, the tower is mounted on rails, it straddles the dock belt, and the barges remain stationary.

1.2.2 Ocean Transport

The U.S. export coal market has primarily been exporting coking coal for use in steel production overseas, but the market for steam coal has developed at a fast rate. Consequently, modern vessels and modern or upgraded docking facilities (e.g., larger and faster processing capabilities) are needed.

In addition, several possible schemes to transport steam coal from the mine to overseas markets have been considered. These schemes are based on coupling ocean transport with pipeline or railroad transport and moving the coal in either dry or slurry form. The first scheme involves transportation from the mine to the preparation plant (size reduction, storage, and pumping), to the pipeline, and to the ship, and from the ship to the pump station/pipeline and to the dewatering plant/thermal drying and to the end-use. The second scheme is similar to the first scheme, except that the coal is dewaterized before being loaded into the ship and on the other shore the coal is turned into slurry again before being shipped through a pipeline. In the third scheme, the coal is reduced in size and washed before it is loaded on the railway. The railroad brings the coal to a maritime terminal, where on the other shore it is again transported by rail to a grinding plant and to the end-use destination.

Coal is exported from the United States through several major coal loading facilities located in Baltimore (Maryland), Philadelphia (Pennsylvania), Mobile (Alabama), Jersey City (New Jersey), Charleston (South Carolina), New Orleans (Louisiana), Newport News (Virginia), and Norfolk (Virginia).

The last two facilities have undergone extensive modernization and new construction to satisfy the increasing demands for export coal. Coal is now shipped in modern vessels, such as Panamax vessels (60,000 dwt [deadweight tonnage]) and Capesize carriers (200,000 dwt or greater).

1.3 Truck Transport

For short distances, trucks play an important role in transporting the coal (or coke to foundries) to a stop station before the coal is shipped to its final destination or to mine-mouth power plants. Due to routing flexibility and low capital investments, trucks can economically move coal approximately 100 miles (at most) one way in relatively small batches. Trucks are used predominantly for short-distance hauls of 50 miles or less. The size of the payload varies according to state gross vehicle weight limitations; for instance, the payload is 20 to 40 tons per truck east of the Mississippi River and 40 to 60 tons per truck west of the Mississippi River.

Basically, there are two types of off-highway trucks: the rear-dump truck and the bottom-dump truck. In a conventional rear-dump truck, the body is mounted on the chassis and lifted up by means of a hydraulic hoist system. The bottom-dump truck, as indicated by its name, dumps material through the bottom and is not well suited to hilly or rough-terrain operation.

Rear-dump trucks have been tested in sizes up to 350 tons; however, 200 tons is considered a large truck and 100 tons is viewed as a standard off-highway truck. Typically, rear-dump trucks carry approximately 25 to 50 tons. Bottom-dump trucks are normally in the range of 35 to 180 tons.

Trucks of up to 85 tons are usually equipped with diesel engines having power shift transmission and torque converters. In the range of 85 to 100 tons, the diesel engine of these trucks could transmit power to the drive wheels by either mechanical or electrical means. At 100 tons and over, the diesel engine drives a generator that produces electricity for electric traction motors that move the wheels.

1.4 Transport by Slurry Pipelines

Pipelines carrying solids such as coal, iron concentrate, and copper concentrate follow the same process: (1) the solids are ground to a fine size, (2) mixed with a liquid vehicle such as water, and (3) pumped through a buried steel pipeline.

Although advantages related to economics, the environment, and simplicity of operation for gas and liquid pipelines are applicable to solid-carrying pipelines, only a few coal pipelines have been built: two in the United States (108 and 272 miles long) and one in the former Soviet Union (38 miles long).

Few coal pipelines have been built because several barriers exist, including long distances, high volumes required, inadequate rail service, and long-term commitment (20 years).

Slurry pipe technology is not new:

1. There are conventional coal water slurries (CWS) and coal water fuels (CWF). In the CWS case, pipelines ship coal at a ratio of approximately 50:50 coal and water by weight in a turbulent flow. In addition to water, other types of fluids can be considered vehicle fluids, such as oil, liquid carbon dioxide, or methanol. The carbon dioxide can be used for enhanced oil recovery, after separation from coal. In the CWF case, fuels are a mixture of coal and water, flowing in a laminar regime, and dry coal weight varies from 60 to 75% depending on the coal type and on the stabilizing additives.
2. In the slurry preparation process, coal slurry suitable for transportation is produced by performing size reduction (crushing and grinding) and chemically treating (corrosion inhibition and thinning) it, if needed, to improve the final characteristics of the product. Two parameters are important for controlling the slurry: density and the product top size. The slurry concentration affects the friction loss of the pipeline and the critical velocity of the flow. The top size is directly related to potential pipeline plugs caused by large settling particles (safety screens are used as countermeasures).
3. To be suitable for transportation, a compromise is made between pumpability and dewatering requirements. If particle size is fine, pumpability is enhanced, but if particle size is too coarse, dewatering at the destination is affected because the slurry must be pumped at higher flow rates to maintain a suspension.
4. To inhibit corrosion in the pipeline, chemical agents, such as sodium dichromate, are added to form a film on the wall, or oxygen scavengers, such as sodium sulfite, are added to prevent attack by dissolved oxygen.
5. A test loop is normally included in the system as the final control. All slurry must pass through it before entering the pipeline. Any increase above the specified limits of pressure drop would indicate the formation of a bed of fast-settling particles in the test loop.
6. For dewatering slurries, continuous vacuum filters and centrifuges are used. For the first U.S.

pipeline, dewatering was performed by thickening, vacuum filtration, and thermal drying.

7. Although the water requirements for coal slurry pipelines are significant, they normally are at approximately 15% of the required power plant makeup water; therefore, the slurry water is commonly used as part of the makeup water in conventional practice.

8. Some advanced transport modes have been considered based on pneumatic and container concepts, e.g., capsules in pipelines. The capsules may be rigid (cylinders or spheres of the cast material), semirigid (flexible container with the material inside), or paste plugs (material mixed with a liquid not miscible with the carrier liquid). These concepts have not yet been developed for commercial application.

1.4.1 *Commercial Projects of Coal Slurry Pipelines*

Consolidation Coal Company built the first major U.S. coal slurry pipeline from Cadiz, Ohio, to Cleveland, Ohio. This 108-mile pipeline was successful but was mothballed in 1963 after 6 years of operation. The second successful U.S. pipeline is the 273-mile-long Black Mesa pipeline built in 1970 in Arizona. The pipeline has a 5 million ton/year capacity and transports pumped coal slurry (47% solids by weight) through an 18 in. pipe. The Black Mesa pipeline is available 99% of the time. After being "dewatered" by centrifuges, the coal cake from the pipeline is used as fuel in two 750 MW power plants in southern Nevada.

1.5 Economics of Coal Transport Alternatives

Cost comparisons of coal transport alternatives on identical bases are not available. Assuming the best conditions is not appropriate because transport costs are strongly influenced by several factors, such as terrain conditions, sunk cost elements (such as existing railway infrastructure), and inflation.

In the past, several studies have provided some trends:

1. A study performed by Ebasco has indicated that pipelines are most economical for long-distance and high-quantity applications. In this study, the different forms of transporting coal energy have been considered, e.g., by rail, pipeline, rail–barge combination, and extra high voltage power lines, the assumptions being transportation distances of 500–1500 miles and power plant sizes of 1600–9000 MW. If water transport is available, the combination of rail–water transport is competitive, assuming that existing tracks are utilized.

2. The U.S. Department of the Interior conducted a study comparing costs based on the premise that 25 million tons of coal slurry was transported for 1000 miles. It led to the following generalizations:

- Water transport is almost always the most inexpensive way to move coal; it usually involves coupling it with another transport mode depending on the distance between mine and riverway—rail, pipeline, and trucks are possible.
- Truck transport is not competitive with rail or pipeline for high tonnage and long distances (e.g., more than 50 miles and 1 million tons per year). Trucks are the preferable mode for short distances from mine mouth to a terminal (preparation plant or en route to a final destination).

3. A U.S. Department of Energy-sponsored study by Argonne National Laboratory, involving Williams Technologies Inc., identified the potential for U.S. coal exports to the Asian Pacific Basin. High inland transportation costs combined with a relatively low heating value of western coal had resulted in coal prices exceeding world market levels. In a portion of this study, Williams considered how to lower the transportation costs of western coal to the Pacific Basin. The first target is a price of less than $2.05 per MBtu and the second target is $2.9 to $3.25 per MBtu. Four U.S. coal regions/states considered as sources were Four Corners, Utah, Powder River Basin, and Alaska. For Four Corners and Utah, the coal water slurry systems and the coal water fuel systems would go to Stockton, California, and Los Angeles, California; for Powder River Basin, pipeline delivery will go to Puget Sound. For Alaska, two routes were evaluated: (1) from Nenana to Cook Inlet and (2) from Beluga to Tyonek Dock. The findings of the study indicated that Four Corners and Utah, being close to existing ports, provide opportunities for both conventional coal slurries and coal–water fuels, whereas Powder River Basin offers potential for coal–water fuels.

In summary, railroads and slurry pipelines are usually the choices for high tonnage and long distances. Normally, railroad distances are longer than pipeline distances; railroads outperform pipelines when no track must be built, but if new trackage is needed, pipelines are more economical. If inflation is considered, the percentage of transport tariffs subject to inflation is higher for railroad than for pipeline.

2. COAL PREPARATION

Coal preparation is defined as the process of removing the undesirable elements from run-of-mine (ROM) coal by employing separation processes to achieve a relatively pure, uniform product. These separation processes are based on the differences between the physical and surface properties of the coal and the impurities. Coal preparation would include not only sizing (crushing and breaking), but also all the handling and treating techniques necessary to prepare the coal for market. Table III gives a general overview of the characteristics of different coal types offered in the market.

ROM coal is usually a heterogeneous material with no size definition and can consist of pieces ranging from fine dust to lumps; it is often wet and contains rock and/or clay, making it unsuitable for commercial use, such as in combustion. The objective of coal preparation, in terms of combustibility, is to improve the quality of the coal by removing these extraneous noncombustible materials leading to:

1. Ash reduction (lower particulates, reduced ash handling at power stations)
2. Sulfur reduction (lower SO_x emission at power plants)
3. Carbon consumption and NO_x emission reduction (at power plants)
4. Reduced transportation costs per unit of heat
5. A guaranteed consistent heat value (improved plant efficiency) to meet customers' requirements
6. For optimum calorific value (heating value) consistent grindability, minimal moisture, and ash variability, such as in the area of power generation
7. Economical design and operation of steam-generating equipment

TABLE III

Characteristics of Coal Types

Coal type	Characteristics
Lignite	4500 to 7000 Btu/lb (2500 to 3890 kcal/kg)
	High moisture (25%) with ash
Low-grade bituminous	5000 to 9000 Btu/lb (2780 to 5000 kcal/kg)
	High ash (+20%) with moisture (10%)
Thermal bituminous	8000 to 12,000 Btu/lb (4450 to 6675 kcal/kg)
	Medium ash (10 to 12%) with moisture (9%)
High-grade bituminous	+12,000 Btu/lb (+7000 kcal/kg)
	Low ash (<7%) with low moisture (<8%)

Source. Adapted from Sharpe (2002).

2.1 Size Reduction

Pretreatment, or size reduction, is a common term used to include the processes of breaking, crushing, and screening ROM coal to provide a uniform coal feed with a certain predetermined top size. The number of stages in the size reduction process will depend on the end use of the coal. For instance, in power generation, the number of stages is more than the number needed for coking coal. In breaking the coal, four types of equipment are available: rotary breakers; roll crushers; hammer mills (ring mills), and impactors.

Before describing these four systems, the following terms frequently used in pretreatment should be defined:

1. The top size (or upper limit) of the crushed material is the smallest sieve opening that will retain a total of less than 5% of the sample (ASTM D431);
2. The nominal top size is the smallest sieve opening that will retain a total of 5 to 20% of the sample;
3. The reduction ratio is defined as the ratio between feed top size and product top size;
4. The Hardgrove grindability index is the index determined after passing 50 g of 16 × 30-mesh dried coal in a standardized ball-and-race mill and sieve this sample through a 200-mesh sieve (ASTM D409) to determine the amount of material going through it.

The four types of equipment for size reduction of run-of-mine coal ([2] and [3]) are described as follows:

1. The rotary breaker causes the coal to break by lifting it to a given height and dropping it against a hard surface. It is a cylindrical unit (operating at low speed) that subjects the coal particles on the liner to undergo centrifugal motion; the liners of the rotary breakers are screen plates that allow peripheral discharge of the properly sized material along the length of the rotary breaker.

2. The roll crusher is the workhorse of coal size reduction to shear and/or compress the material. It is compactly engineered in the form of a rotating roll and a stationary anvil (single roll crusher) or two equal-speed rolls (double roll crusher) rotating in opposite directions.

3. The hammer mill is the most commonly used coal pulverizer. In this device, feed coal is impacted

by rotating hammers and further by grid breaker plates. Usually a high portion of fines is produced. A ring-type hammer mill (rings instead of hammers) would minimize the amount of fines in the product.

4. The impactor impacts the coal, which is then projected against a hard surface or against other coal particles. The rotor-type impact mill uses rotors to reduce the size of the material; the shattered material rebounds in the rotor path repetitively until the product is discharged from the bottom.

The sizing process usually follows the crushing process. The three major reasons for sizing the coal are to separate coal into various sizes for marketing, to feed washing and dewatering units, and to recover the solids used to control the specific gravity in these washing devices. Sizing of the coal is performed by screening or classification.

The most common screening method includes bringing each particle against an opening or aperture where the particle will either pass through or be retained. Several screen configurations have been developed, such as the following: (1) the Dutch sieve bend is a fixed screen with no moving parts and is situated at a level lower than the feed, (2) the Vor-Siv, developed in Poland, and (3) the loose-rod deck, developed in the United States by Inland Steel Corp., which consists of steel rods set at a suitable pitch to create the screening aperture.

Although most sizing of coal is performed with screening devices, classifiers are also in use. Classification refers to achieving size separation by exploiting different flow rates through a fluid. It includes dry classification and wet classification. Dry classification uses air as the ambient fluid and involves higher fluid volumes and higher velocities than those found in wet classification.

2.2 Cleaning Methods

Coal preparation has been evolving for more than a century and the core of most of the process development work is in the area of separating the impurities from the coal, utilizing the physical difference in specific gravity between the coal and the rock impurities. An overview of the different coal cleaning systems is provided in Table IV.

2.3 Coal Dewatering and Drying Systems

After the cleaning process, the coal retains a significant amount of moisture, which can have negative effects on its transport and handling and its calorific value. Therefore, there is a need to reduce the moisture content by either dewatering or drying. Dewatering means mechanically separating water from the coal and drying means separation by thermally evaporating water.

Table V outlines the different technologies used in the dewatering process or in the drying process of coal.

2.4 Arrangement of a Coal Preparation Facility

Given all the above-mentioned techniques for preparing the coal, it is useful to elaborate on the flow of coal within a coal preparation facility through the following stages:

1. Size reduction of coal
2. Sizing of the coal to match the proper feed sizes with the various types of coal-washing equipment
3. Washing the coal
4. Dewatering the products
5. Recycling the process water (treatment/reuse)
6. Blending the product streams into storage
7. Loading out the marketable products

Table VI gives an example of the flow of coal through a coal preparation facility and the equipment involved in the subprocesses.

3. *STOCKPILES*

3.1 Stacking and Reclaiming of Stockpiles

Coal pile management, an integral and vital element in the coal chain, is carried out at coal mines, coal preparation plants, the transshipment sites for export/import of coal, and end-user sites; end users usually include power plants, iron and steel plants, coking plants, and cement plants. Since these "stockyards" tie up capital with no return on investment, the owners of these plants are interested in optimizing coal inventories to minimize this capital and addressing the issues of stockpiling of coals. The significant issues are stock size, stock pile turnover period, and availability of cheaper coals on the international market. All of these issues involve the best strategy for blending coals with the accuracy required by the customer and just-in-time delivery. Coal pile management includes examining why stockyards are used and how they are designed/erected and operated in compliance with environmental regulations.

TABLE IV

Coal Cleaning Systems

Beneficiation method		Systems
Jigs		• Predominant washing technique throughout the world, as pulse water separators • Two jigs known as Tacub jig (with uniform air distribution underneath the screen) and Batac jig (with improved air valves and control mechanism in addition)
Cyclones	Dense-medium cyclonic separator	• First patent in 1958[a] using metal chloride salts, as first dense-medium cyclone • Dense-medium cyclones effect sharper separation between coal and impurity over a wider specific gravity range than jigs • Several variations (Japanese Swirl cyclone, three Russian cyclone models, British Vorsyl cyclone, and the Dynawhirlpool cyclone)
	Water-only cyclones	Not regarded as dense-medium cyclone that has a much greater included angle in the cone—up to 120°—and a longer vortex finder • Constant velocity separators • Do not achieve a separation as sharp as dense-medium cyclones; nevertheless, provide clean coal products free of high-ash materials in all but the finest sizes • Simplicity/small space requirements/economical operation attributes lead to adoption of water-only cyclones as a rougher device for decreasing the load on downstream cleaning units
Concentrating tables		• One of the oldest most widely used cleaning devices for washing coals • Configurations of two-deck and four-deck systems • Potential to remove pyritic sulfur down to a certain size
Froth flotation (aeration) cells[a]	Flotation air	• Coal particles adhere to air bubbles and form a froth a top the pulp • Froth skimmed off and separated from the refuse that remains in the pulp
Oil agglomeration		• A process of removing fine particles from suspension by selective wetting and agglomeration with oil • Preferential wetting of the coal by the oil; if air admitted, bubbles attach to the oil–wet coal surfaces, leading to increased buoyancy and rapid separation of heavier, unoiled mineral particles
Sand suspension process	Chance process[a]	• Patented in 1917, first applied to bituminous coal in 1925 (specific gravity separations of 1.45 to 1.60 commonly used) • One of the most successful heavy-medium cleaning processes; involves a large inverted conical vessel with sand in suspension in an upward current of water; increasing or decreasing the amount of sand in suspension causes a change in density of the fluid
Magnetite process	Conklin process[a]	• Used powdered magnetite in 1922 • After modification, as Tromp process in 1938 for magnetic medium
Pneumatic process	Air process	• Use air as separating medium

Source. Adapted from Sharpe (2002) and from Deurbrouck and Hucko (1981).
[a]Studies on flotation reagents reported in 1973 and 1976.

With the mounting pressure imposed by economic constraints associated with using smaller stockyards, blending various coals to the accuracy demanded by the customers, and addressing the just-in-time concept, the main functions of the coal pile are very clearly defined: (1) to serve as a buffer capacity to accommodate the differences between supply and demand, thereby acting as a strategic stock against short- and long-term interruptions, and (2) to provide a uniform feedstock of the required quality by accurate blending of coals available on the market.

In the case of a power plant or a process industry, stockpiling involves two important types of storage, namely, live and dead storage. Live storage is defined as "coal within reach" that can be retrieved by using available equipment or by gravity flow. In contrast, dead storage would require using

TABLE V

Technologies for Dewatering of Drying Coal

Method	Technology	Characteristics of technology
Dewatering	Screens	• Dutch State Mines sieve bend is a stationary wedge bar screen with the bars oriented at right angles and across the flow stream; advantageous feature is that the screen generally does not blind • Polish Vor-Siv is a stationary device to handle high volumes of solid in-water slurry; it combines certain characteristics of cyclones, sieve bends, and cross-flow screens • Conventional screens handle less flow volume than Vor-Siv per unit surface area
	Cyclones	• Normally processes a pulp with a low percentage of solids (5 to 20%) and produces a relatively thick suspension by removing a portion of water; solids discharged at the bottom, and water and slimes exit at the top of the unit
	Centrifuges	• Two main types available: solid bowl centrifuges and screen bowl centrifuges; there are many variations of these two types
	Filters	• Most often used is the vacuum disk filter; for best performance, solids in the feed should be coarse • Occasionally used is the drum-type filter; able to operate under varying load resulting from changes in feed rate • Also done is heating the feed slurry (steam filtration) by covering a conventional vacuum filter with a hood and applying steam under the hood to facilitate additional water drainage from the filter cake
	Emerging techniques	• Shoe rotary press developed by Centre de Recherche Industrielle du Quebec • Electro-acoustic apparatus developed by Battelle Columbus Laboratories • Tube filter press developed by English China Clays Ltd. • Membrane pressure filter developed by National Coal Board (England)
Thermal drying	Direct drying	• Before 1958, screen, vertical tray cascade, multilouvers, and suspension dryers dominated the market of thermal drying • After 1958, fluidized bed dryers gained rapid favor over the previous dryer techniques • After 1967, there was a decline in the use of fluidized bed dryer due to increased costs because of stricter emission regulations
	Indirect drying	• Eliminates the potential problem of excessive fines carryover of direct drying systems • Based on contacting the coal with a heated surface; two designs: Torus (or Thermal Disk) and Joy Holo-Flite (variation of a helicoid screw) with oil as heating medium • Another concept (Bearce Dryer) is to heat metallic balls and mix the balls with the wet coal as it travels down a sloping/rotating cylinder

Source. Adapted from Deurbrouck and Hucko (1981) and from Parek and Matoney (1991).

a secondary or an auxiliary means to reclaim this coal. When a coal arrives at a power plant, a decision is made either to add it to live storage or send it to long-term storage in the reserve stockpile; reserve stockpiles ensure against disruptions of deliveries or are opportunities for taking advantage of low prices on the spot market. Buffer stockpiles are sometimes available at power plants and act as a buffer between live and reserve stockpiles. Coals of a similar type can be stacked in the pile, but different coal types/grades are normally stored separately.

3.1.1 Features of Coal Stockpiles

Live and dead storage of large quantities of coal is usually carried out by storing the coal in open stockpiles on the ground, unless stringent environmental considerations require a roof over the stockpile, or enclosures. In this case, silos and bunkers are the alternatives to open piles, but these methods would be more expensive and capacity-limited due to the feasible size of the bunkers or silos.

For an efficient management of the stockpiles, the stockyard layout would satisfy the following requirements:

1. Optimize the land use (tons/acre) consistent with storage and environment constraints;
2. Maintain easy access to stored coal;
3. Maximize the load distance efficiency factor;
4. Position coal such that stacking and reclaiming rates can be achieved with the least effort;

TABLE VI

An Example of a Coal Preparation Plant

Coal preparation facility		
Step	**Interim products**	**Equipment involved**
Run-of-mine storage		• Ground piles • Bunkers • Raw coal blending
Size reduction		• Crushers and rotary breakers
Sizing/dedusting/desliming		• Cylindrical screens • Grizzlies • Vibrating screens
Cleaning method for various types of coal	Option (1) Pneumatic cleaning	• Refuse stream • Product as dry product
	Option (2) Coarse coal washers (+1/4 inch)	• Heavy media vessels • Jigs • Refuse stream
	Option (3) Fine coal washers (1/4 inch × 28 mesh)	• Heavy media cyclones • Fine coal jigs • Tables • Water-only cyclones • Refuse stream
	Option (4) Ultrafine coal washers (−28 mesh)	• Water-only cyclones • Flotation cells • Cleaned with coarse coal • Oil agglomeration
Dewatering	Option 2	• Vibrating screens • Centrifuges • Stream going to water clarification • Magnetite recovery circuit (with dryers)
	Option 3	• Sieve bends • Vibrating screens centrifuges, vibrating basket-solid screen bowls • Magnetite recovery circuit (with dryers)
	Option 4	• Raped sieve bend • High-speed screens • Filters • Solid bowl centrifuge
Water clarification (return to circuit use)	Options 2 and 4	• Thickeners and cyclones for option 4 • Refuse dewatering for option 4 with filters, solid bowl centrifuges, filter presses, and ponds
Thermal drying	Options 3 and 4	• Fluid bed dryer • Fine coal dryer
Dry product		• From Option 1
Product load-out		Coming from all the options

Source. Adapted from Speight (1994).
Note. See Fig. 6.1. of Speight (1994), for further details.

5. Attain the desired level of remote control of the equipment and the desired degree of automation of the plant;
6. Meet the homogenization and blending requirements if accuracy is requested;
7. Maintain/improve the uniformity, integrity, and quality of the stored coal;
8. Maximize equipment capability and reliability and minimize manpower requirements;
9. Provide a safe and dependable system consistent with environmental regulations, such as potential stock fires, dust emission, noise levels, and water drainage;

10. Minimize the overall costs (dollars per ton of coal handled) in terms of operation and capital costs.

Characteristics of the land location (in terms of size, shape, and load-bearing capacity) play an important role in defining the choice for the overall storage system of an open pile. These will determine the pile and height of the coal that can be stored.

Several measures must be considered in preparing the ground for a new coal pile:

1. The topsoil must be removed and a stabilizing layer of crushed rock or similar material be laid down;
2. Adequate site drainage must be in place;
3. Utility pipes and conduits should not be run under the pipes;
4. Long stockpiles should have their narrow side facing the prevailing winds to reduce dusting and oxidation.

In addition to the kidney-shaped stockpiles, the most commonly found shapes are the following:

1. Longitudinal beds (or linear beds): the beds are located either side-by-side or end-on. For instance, in the case of two beds, one pile is stacked and the other is reclaimed.

2. Circular beds: they are used for environmental reasons, requiring covering of the stockpiles; many small circular piles are preferred over a single large one for reducing the risk of a breakdown of the reclaimer system and for convenience of maintenance.

In today's business climate, the end user has no plausible reason to invest in excessive coal piles. The drawback of a circular bed is that it has only a limited storage capacity and cannot be increased once built. However, the circular beds eliminate the nonstandard end cones that are formed in the linear beds; these end cones are not representative of all layers in the stockpile; not reclaiming these end cones could cause short-term grade variations when these ends are processed. But the linear beds can store more coal, provided that space is available. The circular beds, which have limited pile capacity and short layer length, will have relatively smaller input lot size but will have many different lots. Capital operating and maintenance costs of circular beds are less than those of linear beds, but the costs for civil engineering work are higher for circular beds.

3.1.2 Methods of Stacking and Reclaiming Coal Piles

3.1.2.1 Stacking Coal During stacking, a stockpile is built with layers of coal, creating a pile with a triangular cross section; the layers may come from the same coal type or from different coal types. In this stacking process, several issues need to be addressed adequately, such as the following:

1. Minimize coal size segregation; the segregation could lead to spontaneous combustion and could contribute to flowslides;
2. Avoid dropping coal from excessive heights, thereby lessening the formation of dust;
3. Take into consideration coals that are more friable, that is, could lead to more formation of fines that, when wet, could cause pluggage of chutes.

Taking these issues into consideration, there are several main methods for stacking coal based on different components and arrangements. These methods are further based on the capability of the different types of stackers. The stacker runs on rails laid along the total length of the bed. The capability of the stacker will partly determine the ability to switch from one stacking method to another. The stacking equipment for longitudinal stockpiles is designed to reflect the different amounts of flexibility, the varying ability to blend/process the coal types, and the range of investments. To address these features, there are several types of stackers: (1) stacker with rigid boom; (2) stacker with luffing boom; (3) stacker with rigid and retractable boom; (4) stacker with luffing, rigid, and retractable boom; (5) stacker with slewing boom; (6) stacker with luffing and slewing boom.

In the cone shell method, a pile is formed by stacking coal in a single cone from a fixed position. When this cone is full, the stacking system moves to a new position to form a new cone against the shell of the previous cone. This method is not efficient for blending coals and is used in cases where homogenization is not required. The strata method stacks the coal in horizontal layers and will create alternating layers when two or more coals are to be blended; each layer will extend from one end of the bed to the other end and the whole bed should be completed before reclaiming is performed from the end of the pile. The skewed chevron is the strata method with the difference that the layers are inclined and reclaiming is carried out from the side of the bed.

The chevron, windrow, and chevron–windrow systems are all intended to have coal reclaiming carried out from one end of the bed and they can achieve a high blending efficiency. The chevron system is the simplest system since it requires only a stacker with one discharge point: the stacker moving along the central axis of the pile. The drawback is that segregation of material will lead to fine particles depositing in the central part of the pile and coarse particles depositing on the surface and at the bottom of the pile. The windrow and chevron–windrow methods will require more expensive stackers with a slewable boom allowing multiple discharge points; both of these methods minimize particle segregation. The Chevcon method (continuous chevron) is a combination of the cone shell and chevron methods and utilizes a circular rather than a linear bed. This circular bed has a round base with a ring-shaped pile where coal is stacked at one end and reclaimed from the other end.

3.1.2.2 Coal Reclaiming The technology of equipment for reclaiming coal has evolved from a discontinuous process originally to a continuous process, with the emergence of bucket-wheel, bridge-type, and portal scraper reclaimers.

In North America, the first bucket-wheel stacker reclaimer was installed in the Potomac Electric Power Plant at Chalk Point (Maryland) in 1963. Prior to this reclaiming method, coal stockpiles at large power plants relied on gravity systems that moved coal on a belt conveyor in a tunnel beneath the pile. The bucket-wheel stacker reclaimer that operates in the United States has two additional features not available in the early European version: (1) a movable counterweight instead of a fixed one (safety hazard eliminated) and (2) complete automation (labor-saving feature).

These reclaimers have become popular and make up approximately 90% of the total coal applications. The ability to perform homogenization/blending efficiently varies from one type to another. They reclaim different coal types stored in separate beds in a sequence as defined by the blending process required to meet a customer's expectations. This key feature is helpful at transshipment facilities tasked with blending imported coals.

A description of the types of reclaimers in light of their key features and functionality is given below:

1. The bucket-wheel recalimer is designed with a bucket wheel to reclaim the coal and operates from the side of the bed. This type of reclaimer is suitable only for operation in the open (due to the counterweight); it is deployed in stockyards with high throughput. Its homogenization/blending efficiency is not good. The bucket wheel is often combined with a stacker.

2. The gantry-type reclaimer may have multiple bucket wheels to improve the homogenization/blending efficiency.

3. The drum-type reclaimer uses buckets attached to a long rotating drum to reclaim the coal across the entire width of the pile, resulting in an improvement of the homogenization/blending efficiency.

4. The bridge-type scraper is designed with a scraper chain to reclaim coal across the front face of the longitudinal pile. This reclaimer has a very good blending/homogenization efficiency, but it cannot jump over stockpiles (whereas side-face reclaimers can).

5. The portal, semiportal, and cantilever scraper systems reclaim coal from the side of the bed. As side-face reclaimers, they are suitable for removing coal with signs of imminent spontaneous combustion.

6. The portal bridge scraper combines the advantages of the side- and front-face scraper reclaimers. A special version has been developed to reclaim circular beds; this special version is based on a bridge-type scraper and a cantilever scraper reclaimer, both mounted on a central column.

7. Front-end loaders and bulldozers are used for smaller stockpiles to stack and reclaim coals.

8. The gravity reclaim is utilized in some stockpiles. This system uses hoppers (located in a tunnel below the stockpile) and feeders (to control the discharge into the conveyors) to reclaim coal. Vibratory stockpile reclaimers would improve the reclaiming process. The use of multiple hopper systems (allowing intersection of flow channels) would improve the reclaiming performance. Blending efficiency within the stockpile is poor.

In addition to the reclaimers, some other types of mobile handling equipment play a role in stacking/reclaiming coal at a smaller volume than the above-mentioned reclaimers do. Front-end loaders and bulldozers are used to compact the coal and to move coal from reserve pile to live pile.

3.1.3 Bins and Silos

Coal can be stored in silos or bunkers as an alternative to open piles on the ground or in areas with enclosures. However, this method of storing coal in silos or bunkers is more expensive and constrained by the feasible size of the units.

Silos and bunkers are either steel or concrete, usually have round cross sections, and can have single or multiple outlet cones. An internal lining of stainless steel is a big advantage when storing high-sulfur coals to prevent corrosion. Regarding structural integrity of the foundations, loads should not be offset on the foundations when eccentrically charging or eccentrically withdrawing coal from these units.

A silo with a large diameter and a flat bottom will have multiple outlets. Bunkers with a height-to-diameter ratio of 2:1 to 3:1, designed to be self-cleaning, will have a cone bottom (with a suitable angle, e.g., a suitable flow factor) and a single outlet.

These coal storage devices do offer unique benefits such as: (1) controlling the dust and moisture content of the coal, (2) controlling runoff from rainwater, and (3) providing blending capacities by having several units feeding a single reclaim conveyor.

3.1.4 Trends in Management of Coal Piles

Traditionally, the coal stockyard for U.S. power plants contained approximately one-fourth of the plant's annual burn, such as the 90- to 120-day storage level at several American Electric Power plants, but that level in the United States has declined to much lower coal inventories. In the United States, stockpile levels have decreased to a level of approximately 50 days since mid-1993.

In comparison, some power plants have even kept 7- to 10-day coal inventories when their plants were located in the vicinity of their mines, such as the case of the South African Kendal mine-mouth power plant.

With a just-in-time delivery system, some power plants have coal stockpiles of only 5–8 days, such as those in the Netherlands or in U.S. centralized plants. Relative to just-in-time delivery, the decision to buy is driven more by demand than lower on-the-spot coal prices. As such, the decision could be subject to increases in coal prices due to regional coal shortages. These shortages could affect several utilities drawing on the same pool of mines. Consequently, before reducing its coal inventory, a power plant would need to carefully consider many factors such as: (1) transport options (rail, barge, truck, and ROM conveyor) in terms of reliability; (2) number of supply sources; (3) backup sources, such as other power plants; (4) reduced burn rate; and (5) delivery time for coal from supplier to the power plant (in terms of days).

Another trend in management of coal stockpiles is outsourcing; this concept is suitable for mine-mouth plants or power plants served by a single supplier.

3.2 Sampling, Analysis, and Auditing

With larger, more expensive shipments of various compositions, coal must be sampled and analyzed before storage for the following reasons:

1. Sampling is performed during loading operations at terminals or at power plants to ensure that the coal meets the contract specification (penalty and bonus clauses related to coal properties).
2. Sampling leads to knowledge about the coal consignment; this information allows the coal operators to store coal in different piles—coals of similar grade are stockpiled together and blending of different types of coal can be performed based on this knowledge.
3. Sampling during the reclamation process helps evaluate changes in properties during the storage period.

Given the above-mentioned reasons, a good sampling technique is critical for obtaining a truly representative sample of the coal consignment and of paramount importance for the analysis of the coal. The usual procedure is to divide the sample into three parts—one for the supplier, one for the customer, and one for independent analysis in case of a contract dispute.

3.2.1 International Standards for Coal Sampling Techniques

Coal is one of the most difficult materials to sample because it has the tendency to segregate by size and mass. The ideal objective of a sampling would be to have every particle size represented—ash content and heating value may be different for each particle size.

Other factors in the sampling process are the use of analytical results, the availability of sampling equipment, the quantity of the sample, and the required degree of precision according to standards (ASTM D2234).

Since biased results can be introduced into the sampling procedure, certain precautions to reduce the sources of bias are used, such as: (1) selecting the most suitable location for sampling purposes; (2) using sampling equipment that conforms to necessary specifications; and (3) taking special precautions when performing specific sampling procedures, such as sampling for total moisture content or size.

Table VII gives an overview of the international standards on sampling and sampling procedures. These standards are applicable only for sampling coal from moving streams.

In general, the standards specify the number and weight of the increments to be taken for each sampling unit. The increment is defined as a small portion of the coal lot collected in a single operation of the sampling device; the higher the number of increments, the greater the accuracy. A greater number of increments is recommended when sampling blended coal. The individual increments can be evenly spaced in time or in position, or randomly spaced, and taken out of the entire lot. The increments are then combined in a gross sample; the gross sample is then crushed and divided, according to standard procedures, into samples for the analysis. In the analysis part, standards differ in detail and often do not take into account the characteristics of the coal being sampled. These coal characteristics would include particle size distribution, density, particle shape of the coal, and the distribution of the parameter of interest. Consequently, sampling models have been developed to supplement these standards.

3.2.2 Sampling Systems and Bias Testing

Sampling in a stockpile *in situ* presents an enormous challenge in terms of obtaining a representative sample given the following:

1. Since the coal is subject to weather/oxidation and to size segregation, its quality is always different at the various locations of the pile (top, sides, or in the middle).
2. With many additions of different coals to the pile, the pile may contain coal zones that vary markedly.
3. Because sampling coal in the middle of the pile is difficult, sampling from a moving stream during the stacking or reclaiming process is preferable; however, if sampling of stockpiles *in situ* is necessary, the sample should be taken by drilling or augering into the pile.
4. Drill sampling and auger sampling have limited applications. The drill sampling method is used to penetrate the full depth of the pile at each point sampled and extract the whole column of coal. However, this technique does not collect all of the fine material at the bottom of the pile, because of the drill or auger design. Because this technique can

TABLE VII

International Standards on Sampling Coal from Moving Streams

Organizations	Standards
American Society for Testing and Materials (ASTM)	• D2013 Preparation of samples for analysis • D2234 Collection of a gross sample • D4702 Inspection of coal-sampling systems for conformance with current ASTM standards • D4916 Mechanical auger sampling
Standards Australia (AS)	• AS 4264.1 Coal and coke—Sampling; with Part 1 of sampliing procedures for higher rank coal • AS 4264.3 Coal and coke—Sampling; with Part 3 of sampling procedures for lower rank coal • AS 4264.4 Coal and coke—Sampling; with Part 4 of determination of precision and bias
British Standards Institution (BSI)	• BS 1017: Part 1 as sampling of coal and coke; methods for sampling of coal revised to adopt ISO/DIS 13909
International Organization for Standardization (ISO)	• ISO 1988 Hard coal—Sampling • ISO 5069/1 Brown coals and lignite—Principles of sampling; Part 1 of sampling for determination of moisture content and general analysis • ISO 9411-1 Solid mineral fuels—Mechanical sampling from moving streams; Part 1: Coal • ISO/DIS 13909-1 Hard Coal and coke—Mechanical sampling; Part 1: General Introduction • ISO/DIS 13909-2 Hard coal and coke—Mechanical sampling; Part 2 of coal—Mechanical sampling from moving streams • ISO/DIS 13909-3 Hard coal and coke—Mechanical sampling; Part 3 of coal—Mechanical sampling from stationary lots • ISO/DIS 13909-4 Hard coal and coke—Mechanical sampling; Part 4 of coal—preparation of test samples • ISO/DIS 13909-7 Hard coal and coke—Mechanical sampling; Part 7 of methods for determining the precision of sampling • ISO/DIS 13909-8 Hard coal and coke—Mechanical sampling; Part 8 of methods of testing for bias

Source. Adapted from Carpenter (1999).

break the particles, it is not recommended when collecting samples for size analysis and bulk density determination.

As mentioned above, standards specify the method and the number of increments to be collected. The method of collection is particularly important in terms of sampling location and tool. Standard scoop tools allow every coal particle size to have an equal chance of being sampled; spades are inappropriate and give a sample that is mainly fines. Regarding sampling location, coal can be collected while it is transported on the belt conveyors or to and from the stockpile.

The most accurate sampling method is the stopped-belt or reference method (a full cross section of the flow would be obtained). If stopping the belt is impractical or uneconomical, sampling could be done manually either from the moving belt or from a falling column of coal at a certain point. For high tonnage rates, sampling should be done by mechanical samplers. Some standards, such as BS 1017 and ISO 1988, stipulate that manual sampling should not be exercised in cases where the nominal coal top size is above 63 mm or when the coal flow rate is greater than 100 tons per hour.

Mechanical samplers provide a representative sample that is difficult to attain with manual sampling. The advantage of mechanical samplers is that statistics of precision and bias are applicable. However, these mechanical samplers cannot be applied to all situations, such as sampling within a stockpile or at facilities with small tonnages. The common characteristics of the mechanical samplers are as follows: (1) they provide a full cross-sectional representation instead of a partial cross-sectional sample, (2) their design and operation are covered in various standards, (3) their functionality is to avoid separation of the various coal densities and/or sizes by minimizing disturbances of the coal, (4) they require sufficient capacity of coal to perform sampling without any loss or spillage, (5) they require material to flow freely through all stages of the sampling system without clogging or losing material, and finally (6) the standards specify the design and the number of increments to be collected and the size of the cutter opening (typically three times the coal top size) and its speed.

In terms of categories, these mechanical samplers are divided into two types:

1. Cross-belt samplers (sweep arm or hammer samplers) sweep a cross section of the coal on the moving conveyor (or belt) into the hopper. For the sake of accuracy, these devices need to be adjusted so that no coal fines are left on the belt.
2. Cross-stream (or falling-stream or cross-cut) cutter samplers collect a cross section of a freely falling stream of coal. The location of this sample collection is typically a gap at a transfer point, such as between two conveyor belts.

All sampling systems should be checked for bias, because systematic errors may be introduced. These systematic errors generally are caused by a loss or gain in the mass of increments during collection or by cyclical variations of coal quality at time intervals coinciding with systematic sampling time. Bias testing is discussed in certain standards, such as ISO/DIS 13909 Part 8 for bias testing of mechanical samplers and in ASTM D4702 with guidelines for inspecting cross-cut, sweep arm, and auger mechanical sampling systems.

3.2.3 On-line Analyzers

Obtaining representative samples of coal from large stockpiles is burdensome and time-consuming when laboratory analysis of these samples based on standards is applied. The results do not reflect the variations in coal quality. Real-time information about these variations would help stockyard operators manage their assets much more efficiently. In response to this problem, on-line analyzers have been developed to continuously determine the coal quality and in real time.

Advantages of on-line analyzers over conventional analyzers are obvious when it comes to the following: (1) determination of coal quality in real time, thereby allowing the performance of trend and variability analyses; (2) simplification of sampling requirements in the context of large coal quantities to be analyzed; and (3) determination of coal quality in tons per hour instead of grams per day.

On the downside, on-line analyzers are expensive because they are too site-specific and application-specific. Their performance is related strongly to the initial installation, calibration, maintenance, and application environment. Analysis of coals beyond the initial calibration will not produce the same accuracy. Calibration drifts over time result in the need for recalibration of the analyzer.

International standards (such as ISO CD 15239, ASTM, and AS 1038.24) do not provide particular test methods because of the multitude of on-line analyzers and their interaction with sampling/analysis systems. On-line analyzers have gradually reached the level of precision and reliability to be used in several commercial applications:

1. Monitoring the incoming coal to determine whether it meets specified requirements, as is done at the Herdrina plant in South Africa;
2. Sorting and segregating coal into different stockpiles (as is done at the Rotowaro coal handling facility in New Zealand);
3. Blending coal from different stockpiles to meet customer requirements (the Rotowaro coal handling facility);
4. Monitoring coal during reclamation to check for the desired specifications.

On-line analyzers available on the market have been developed based on various technologies, such as: (1) electromagnetic principles, (2) X-ray fluorescence, and (3) capacitance/conductance methods. Table VIII lists the application areas for these on-line analyzers based on the principle of operation.

All of these on-line coal analysis techniques are capable of meeting ASTM measurement standards for a predetermined range of coals. Table IX gives an overview of the four main on-line measurement techniques that are in use.

3.2.4 Stockpile Audits

Coal pile management also involves coal stock audits; the intent is to adjust the book inventory. In this task, the stockpile tonnage (calculated by measurements of volume and density), as the measured inventory, is to be reconciled with the book inventory.

In the book inventory, recorded weights of coal going in and out of the stockpile are documented.

Consequently, accurate weighting systems are important to gain better control over the stockpile and eliminate overpayment; it is recommended that the systems are properly situated, installed, maintained, and regularly calibrated. By knowing more precisely how many tons of coal are in the stockpile, one can write off coal used more quickly and/or can reduce inventories that are too large and thereby achieve significant savings.

Both the book inventory and the measured inventory are subject to measurement errors. Whereas inventory is subject to errors coming from volume and density measurements, several sources of errors are associated with the book inventory: (1) weight errors associated with the stockpile; (2) unaccounted changes in coal moisture in the stockpile; (3) losses due to spontaneous combustion; (4) losses due to runoff and wind erosion; and (5) losses due to contamination of the coal at the base of the pile in contact with the underlying strata.

Because stockpiles have various shapes and sizes, measurements must be used to determine volume. Both ground and aerial surveys are used to determine volume.

In an aerial survey using a photogrammetry method, two independent photogrammetrists using the same photograph on stockpiles between 50,000 and 500,000 tons can obtain accuracies of $\pm 5\%$.

For ground surveys, an electro-optical distance ranging system has been developed. It uses a theodolite (based on a laser) for distance measurement. The theodolite applies numerous point measurements to provide three-dimensional information about the surface of the pile. Using density contour information, specialized computer software computes the mass of the stockpile.

For density determination, the methods used would include volume displacement procedures and nuclear surface and depth density determination. Among these methods, a nuclear depth density gauge is probably the most dominant method, although the procedures have not yet been standardized. The gauges take horizontal and vertical density measurements at different locations and depths within the pile. However, these gauges require calibration for

TABLE VIII

Applications Areas for On-line Analyzers

Principle of operation	Suitable for measuring	Energy source
Prompt gamma neutron activation analysis	S, Btu, Ash, H, C, Si, Fe, N, Cl, Al, Ti, Ca	Californium-252
Gamma-energy transmission	Ash	Barium-133, cesium-137, americium-241
Gamma-energy backscatter	Ash	Americium-241
Microwave energy attenuation	Moisture	Microwave generator

Source. Adapted from Connor (1987).

TABLE IX

Overview of Four Main On-line Measurement Techniques

On-line analyzer technique	Characteristics
Natural gamma systems	• Requires no radioactive source • Measures the gamma emission from the conveyed coal • Combines this measurement with a measurement of the weight of the load to calculate the ash content • Not the most accurate system but the least expensive
Dual-energy gamma-ray transmission systems	• Combines measurements of the intensity of two narrow beams of high-and low-intensity gamma rays passing vertically on the conveyor belt to determine the ash content; coal must be mixed well • Instrumentation to split the beam and with multiple detectors has been developed to determine ash content across the full belt width • Varying chemical composition, especially iron content, can cause inaccuracies; accuracy is better for low-ash coals • Triple-energy gamma transmission systems developed for improving accuracy
Prompt gamma neutron activation analysis (PGNAA)	• Provides an elemental analysis of coal through measurement of the gamma radiation emitted when coal is exposed to a neutron source (such as californium-252) • Extra radiation shielding required for safety precaution and sophisticated signal-processing equipment needed to interpret the gamma-energy signals • Energy of the gamma ray is characteristic of the emitting element; gamma rays detected and collected in a spectrum and later correlated to coal composition • Interpretation of the spectrum is a complicated process; each manufacturer has a corresponding deconvolution process • Carbon, hydrogen, sulfur, nitrogen, and chlorine are measured directly • Ash content measured indirectly by its elements (mainly silicon, iron, calcium, aluminum potassium, and titanium) • Calculation of heating value possible if moisture analyzer is present • Instrumentation with multiple sodium iodide detectors developed to deal with coals from multiple sources • PGNAA is the most accurate analyzer but is also significantly more expensive than gamma analyzers
Microwave moisture meters	• Determine moisture content by measurement of energy attenuation and phase shift of microwaves through coal—Moisture of coal absorbs energy by a "back and forth" rotation of the water molecules in tune with the microwave frequency • Often incorporated in dual-energy gamma-ray transmission or on PGNAA analyzers to enable calculation of heating value • Measurement of coal density or inference of coal density required, usually provided by a high-energy source similar to that in gamma-energy attenuation ash analyzers • Not applicable for measuring frozen coal because ice and coal matter have similar energy-absorbing properties

Source. Adapted from Carpenter (1999) and from Connor (1987).

each type of coal and therefore encounter problems when measuring a stockpile with different types of coals. In addition, their accuracy depends on penetrating the coal without disturbing the *in situ* density and they are expensive to acquire.

3.3 Coal Deterioration

During storage in open stockpiles, most coals are subject to weathering and atmospheric oxidation; these processes alter the properties and structure of the stored coal and could render the coal less valuable for its intended use. Assessing the effects of weathering and atmospheric oxidation of the stored coal would mean examining the following key factors: (1) heating value losses, (2) handling related to coal moisture content, (3) coal cleaning (coal flotation), (4) combustion behavior, and (5) coking behavior.

3.3.1 Heating Value Losses

Oxidation can reduce the heating value of coal. Coal is affected by several factors, such as the characteristics (composition, rank, and particle size) of the stored coal, the storage conditions (temperature and moisture content in the stockpile, storage time, location in the depth of the pile, and ventilation pattern through the pile), and the climatic conditions. The oxidation rate is coal-type-dependent. Low-rank coals are more susceptible to weathering and atmospheric oxidation than high-rank coals. Smaller particles, given their increased surface area, tend to oxidize more rapidly than larger particles. The heating value of the coal decreases with an increase in ash content. Also, the temperature and moisture content in the stockpile affect the rate of coal oxidation and hence the heating value. In a stockpile, the heating value losses decrease with depth; these losses are highest at the surface of the pile where coal comes in intimate contact with the atmosphere. Local hot spots can lower the heating value, due to the airflow through the pile. Compacted coal piles have less heating value losses than uncompacted piles.

Climatic conditions cause coal to lose or gain moisture. Climatic factors, such as solar intensity and rainfall frequency, cause heating value losses. Coal with ice formation on the surface can reduce the oxidation rate.

3.3.2 Coal Handling—Coal Moisture Content

The performance of the coal handling equipment is strongly affected by the moisture content of the coal. In open stockpiles, weather conditions (heavy rainfall or freezing) can increase the coal moisture content or freeze the coal so that equipment failures occur in belt conveyors, chutes, and bins. In hot and dry climates, through moisture loss by evaporation, the stored coal becomes more friable and dustiness increases. A high percentage of fines in combination with high moisture content can lead to a high potential for pluggage. In all these cases, weathering and oxidation increase the friability of the coals and the production of more fines.

Several practices at the stockyard should be avoided. Others are beneficial to control the coal moisture content of the pile. The practices that lead to increased moisture content and that should be avoided are the following: (1) creating large stock piles with little, if any, runoff; (2) allowing "ponding" on the surface of the pile; (3) moving coal through low areas with standing water during the reclaiming process; and (4) having pile levels lower than groundwater levels.

The practices that are beneficial to controlling the coal moisture content include the following: (1) stacking the pile with a maximum height over the smallest area, consistent with stable side slopes; (2) avoiding flat tops for the piles; (3) compacting the surface and the side slopes of the stockpile to enhance runoff; (4) developing a good strategy for efficient drainage; and (5) reducing the moisture content of the incoming coal to the maximum level practical at coal preparation plants.

Other practices include adjusting the equipment to handle wet coal by inserting plastic liners in the chutes and by using intermittent air blasts and vibrators. Chemicals should not be added since they rarely can be applied in sufficient quantities and mixed well enough to be effective. In addition, these chemical treatments affect the end-use behavior of the coal.

3.3.3 Coal Cleaning—Flotation

Coal cleaning processes, such as flotation, are negatively impacted by weatherized or oxidized coal. No method for efficiently cleaning the coal has been found.

Acidic oxidation products on the surface of the coal have dissolved in the process water and lowered the pH. Consequently, this has led to a reduction in flotation recovery and equipment wear. In response, wear-resistant material has been used for surfaces exposed to these adverse elements and magnesium hydroxide is added to raise the pH of the process water. Blending the oxidized coal (up to 20%) with fresh coal to dilute these adverse effects has been practiced.

Another adverse effect of the oxidation process is that oxidation causes a change in the wettability of coal that consequently affects the surface-based separation processes (flotation and oil agglomeration).

The performance of these separation processes decreases with the storage time. As remedies to improve the performance of these coals, several measures have been carried out:

1. neutralizing these adverse acidic oxidation products with alkalis;
2. using surface conditioners (or collectors);
3. using cationic collectors (amines);
4. using low-concentration electrolytes (iron chlorides);
5. grinding the stored coal to −28-mesh (standard flotation size) before subjecting it to flotation;
6. washing the stored coal with water to remove the dissolved metal ions prior to the flotation process.

3.3.4 Impact on Combustion Behavior

Coal stockpiles stored in the open for a long period of time undergo changes that affect their combustion behavior and consequently affect boiler operations. The NO_x emission levels, carbon burnout, and slagging behavior could be impacted by changes in swelling index, maceral composition, and chlorine. Moisture increase in the stored coal would have significant consequences in the boiler plant in terms of the ability to handle the coal, mill capacity, and boiler efficiency.

In addition to lowering the heating value of the stored coal, atmospheric oxidation of the coal causes a delay in devolatization, leading to a change in coal reactivity. Stored coals with high free swelling index (FSI) have shown a significant loss in the FSI value and hence a loss in reactivity. A decrease in FSI and reactivity also means the appearance of oxidized vitrine, which is found to produce compact char structures requiring long combustion times.

The loss-on-ignition (LOI) directly related to char burnout increases significantly when weathered coals are fired. Any increase of unburned carbon in the fly ash, as the consequence of LOI, would affect the sale of the fly ash. To ensure a satisfactory burnout of the char, a longer residence time in the furnace burning zone is needed.

In general, to minimize the above-mentioned negative effects on the performance of the boiler (power plant), several measures have been put into practice: (1) modifying the plant operating parameters; (2) blending the oxidized coals with unoxidized coals; and (3) minimizing the storage time of sensitive coals.

3.3.5 Coking of Coal

Weathering and oxidation have several negative effects on the coking of coal: (1) reduction in coke stability; (2) issues of coal bulk density (related to particle size distribution and moisture content); (3) reduced coke production; (4) overheated charges; (5) carbon deposits; (6) oven damage, such as to coke oven walls; (7) generation of fines; (8) increased coke reactivity; and; (9) decreased coking rate.

Bulk density decreases with decreases in particle size distribution and increases with additional moisture over 8%. To produce a good metallurgical coke, two properties are important: good thermoplastic and caking properties. Deterioration of these properties due to weathering/oxidation of the coal in open stockpiles would affect coke production. Alvarez *et al.* presented the influence of natural weathering on coal in two papers in 1995 and 1996. The study was performed on a typical blend of 13 different coals in a 100-ton open stockpile in Spain during a 1-year period. Four of the component coals (in piles of 50–60 tons) were stored without grinding (unlike the blend) and were exposed to weathering for a period of 5 to 6 months. With mild weathering conditions, results based on the test parameters of the free swelling index, Gieseler maximum fluidity, and Audibert-Arnu dilation demonstrate that there was no significant change in free swelling index for the blend for each of the four coals, except for one coal after 138 days, and the Gieseler maximum fluidity factor (as the most sensitive indicator for loss of thermoplastic properties and oxidation) decreases over time for the blend and the four coals.

Another result of the study was that weathering/oxidation does not necessary lead to deterioration of coke quality. Mild weathering and oxidation can help produce a coke of improved quality in high-volatility and high-fluidity coals and in some medium-volatility coals.

Other studies indicated that the free swelling index generally is not sensitive to slight-to-moderate weathering, although there are exceptions. Studies also suggest that moderate caking coals (FSI between 3 and 5) are more prone to lose their caking properties than coals with higher FSIs.

With the increased trend to use pulverized coal injection into the tuyeres of steel blast furnaces and to blend coals, coke use has declined. Because coke strength and reactivity are especially important, steel works are very interested in having the right coke blend.

3.4 Spontaneous Coal Combustion

Spontaneous coal combustion is an important concern due to its effect on safety, the environment, economics, and coal handling. Economic losses occur because of fires in the coal pile and subsequent unsuitability of the coal for its intended use. Even if spontaneous combustion does not happen in coal piles, fires can occur later in confined spaces, such as on ships or in railroad cars. Ship fires, such as those reported on coal shipments from New Orleans, Louisiana, to the Far East in 1993, have led to the practice of not loading coal at temperatures greater than 104°F (40°C).

Although coal producers cannot always control the factors that govern spontaneous combustion, they can control management of the coal piles.

The tendency of coal to heat spontaneously in storage is related to its tendency to oxidize. This oxidation process is closely related to the rank of the coal (the higher the rank, the less potential for oxidization), the size of the coal in the pile, the stacking method, the temperature at which the coal is stored, the external heat additions, the amount/size of pyrite present in the coal, the moisture content, and the storage conditions (in terms of oxidation heat release and heat dissipation by ventilation).

In lower rank coals, the amount of carbon is low and the amount of oxygen is high. In the higher rank coals, the reverse is true. Hydrogen remains relatively constant. Sulfur in the coal is present in the form of iron-free sulfur, organic sulfur, pyrite, and sulfate in various percentages, ranging from less than 0.5 to 8%.

Spontaneous combustion is the result of oxidation of the coal when heat generated by complex chemical reactions exceeds the rate of heat dissipation. Of the coal types, subbituminous coals and lignite require special handling to prevent spontaneous combustion. The lower the rank of the coal, the more easily it can absorb oxygen. In addition, the low-rank coals are higher in moisture, oxygen content, and volatile matter, all of which contribute to rapid oxidation. Also, coal can gain or lose moisture during storage. When dried or partially dried coal absorbs moisture, heat is generated (heat of wetting) and the temperature of the coal can increase significantly if this heat is not removed. Consequently, spraying water to extinguish fires in coal piles is not the solution, but instead chemical dust suppressants could be used.

Other factors that can contribute to coal oxidation are coal porosity, particle size, segregation during stockpile construction, and mineral content and composition. A more porous coal is more susceptible to oxidation because of the larger number of exposed areas. The effects of particle size are linked to coal porosity; a decrease in the particle size provides a greater surface, and hence a higher oxidation effect.

Regarding particle size segregation, differently sized coals should not be stacked. During construction of the pile, size segregation can occur in conical piles when there is a range of particle sizes; the outer area of the pile contains large-size particles and the middle of the pile has fine particles. This will lead to a so-called "chimney effect," whereby air passes up freely from the outer to the inner areas of the pile, creating the potential for heat production. In terms of mineral content and composition of the coal, some mineral elements can accelerate the spontaneous heating process. Pyrite, sodium, and potassium (in the form of organic salts) can act as catalysts for spontaneous combustion.

3.4.1 The Five Stages of Spontaneous Combustion

Although opinion differs on the exact mechanism of oxidation, the process of oxidation occurs in five steps:

1. Coal starts to oxidize slowly until a temperature of approximately 120°F is reached.
2. The coal–oxygen reaction rate increases at an increasing rate and the coal temperature increases until the temperature reaches 212 to 280°F.
3. At approximately 280°F, carbon dioxide and most of the water are given off.
4. The liberated carbon dioxide increases rapidly until a temperature of 450°F is attained; at this stage, spontaneous combustion may occur.
5. At 660°F, the coal ignites and burns vigorously.

For pyrite, the chemical reaction of FeS_2 with water and oxygen leads to the formation of iron sulfate ($FeSO_4$) and sulfuric acid (H_2SO_4) and is an exothermic reaction. In an open pile, pyrite, through its transformation to bulkier materials, is sometimes responsible for causing slacking and forming fines. Consequently, coal operators are reluctant to store high-sulfur coals for extended periods of time, but coals with low sulfur content are no guarantee for safe storage either.

3.4.2 Safe Storage of Coal through Prevention

Coal producers and users must know the self-heating propensity of stored coals. Practices for long-term storage include the following:

1. The storage area should be dry and on solid ground or sloped to allow good drainage of any pockets of water.
2. The area should be cleared of any debris and combustible materials with low-ignition temperatures, such as wood, oil-soaked rags, paper, and chemicals.
3. The area should be void of any buried steam lines, sewer lines, hot water lines, and other heat sources—stockpiles should not be added to on hot, sunny days.
4. The formation of conical piles should be avoided because larger pieces tend to roll away, creating a chimney effect for the fines situated at the center; for conical piles, removal should occur according to the principle of "first-in, first-out."
5. The windrow method is preferred over the chevron method for stacking coals that are prone to spontaneous combustion.

6. Air must be able to circulate freely in the pile or the pile must be packed in 1 ft layers or less to ensure the absence of air circulation (either restrict the entry of oxygen into the pile or provide adequate ventilation to remove the generated heat).

7. Upright structural supports in or adjacent to the storage piles could create chimney effects.

8. Particle-size segregation should be avoided.

9. Stoker-sized coals may be treated with oil before storage to slow the absorption of oxygen and moisture. Road tar, asphalt, coal tar, and various surfactant dust-control binders have been used as a seal; among these, asphalt is unsatisfactory in some applications.

10. Seals of compacted fine coal may be used to minimize the undesirable formation of fines if the coal has a tendency to slack. Chemical inhibitors have been used to reduce the reactivity of coal; these inhibitors include phenol, aniline, ammonium chloride, sodium nitrate, sodium chloride, calcium carbonates, and borates.

11. Good access should be available around the perimeter of the pile to allow the removal of hot spots.

3.4.3 Detection and Control

Early detection of conditions that cause spontaneous combustion is critical for safe storage of the coal and elimination of conditions, such as subsurface voids, that can endanger operators. The cost and effort to monitor stockpiles are generally small compared to the value of the coal piles. Spontaneous combustion is often detected visually by smoke or steam coming out of the pile or by the melting of ice or snow at certain locations in the pile. In hot weather, a hot area is identified by the lighter color of dry coal surfaces due to the escaping heat, a situation that requires immediate action. To detect/monitor the coal self-heating process, two means are available: (1) temperature measurements, through metal rods, thermocouples, and infrared detectors; and (2) gas analysis, such as monitoring the carbon dioxide, hydrocarbon, and hydrogen concentrations emitted.

Once self-heating is under way, two basic control options are available:

1. Removing the hot coal and using it immediately; if it is too hot for the transport equipment, it should be cooled first.
2. Removing the hot coal, cooling it by spreading into thin layers, and repiling and recompacting it; the coal can be left overnight to cool or can be sprayed with water. However, wet coal is difficult to handle.

Water is generally not recommended for extinguishing fires in coal piles. Dry ice (solid carbon dioxide) has been used successfully to control spontaneous coal combustion in an enclosed or covered area. However, dry ice is not recommended in an open area because wind would disperse it.

3.5 Dust Emissions

Fugitive dust emissions are considered monetary losses and pose environmental problems in the areas surrounding a stockyard. Coal dust poses health risks and damages equipment. Environment regulations are becoming more stringent regarding emissions of particulates such as PM_{10} and $PM_{2.5}$.

Dust emissions can be caused by wind erosion in a coal pile and stacking and reclaiming operations of coal piles.

The magnitude of these coal dust emissions depends on the following: (1) the coal characteristics (moisture content, particle shape, and particle size distribution); (2) the stockpile arrangement (layout and construction); and (3) the local climatic conditions (wind velocity, sun, and rainfall).

From these above-mentioned factors, the moisture content of the coal plays an important role in defining the level of dust generated. This dust emission potential is viewed as severe when the surface moisture of the coal lies between 0 and 4%, mild if it is between 4 and 8%, and low at 8% or above. In addition to this moisture factor, the volume of fines is another factor affecting dust emission. Small-size coal piles with a high volume of fines will produce higher dust emission levels than coal piles with larger particles; coal particles greater than 1 mm in size are not likely to become airborne. The next factor affecting wind-borne dust emission is the particle size distribution (which is related to the friability of the coal). Low-rank coals are generally more friable and consequently more prone to generate dust. Very high rank coals are also susceptible to dust emission.

Regarding the stockpile arrangement and its layout, an appropriate selection of the stacking and the reclaiming method, when combined with a series of control measures, can help lower dust emissions.

3.5.1 Methods for Dust Emission Control and Monitoring

Several methods are available to address dust suppression: (1) wind control (stockpile layout and construction with fences/berms); (2) wet suppression systems (water, chemical wetting agents, and foams);

(3) chemicals binders and agglomerating agents; (4) sealants (chemicals or vegetation); and (5) stockpile enclosure (complete or partial).

Table X gives an overview of the applicability of these dust control methods.

These dust control methods are applied during stacking and reclaiming of coal and applicable to active and inactive (long-term) storage piles. In the stacking process, controlling the size and the particle distribution of the stock piles is important. Conical piles are subject to particle segregation; minimizing the drop height by fitting appropriate stacking chutes can decrease particle fragmentation and consequently reduce dust emissions. Among reclaiming systems, gravity reclamation (through bottom reclaimers underneath the pile) typically generates the least dust, but safety precautions must be exercised due to fire and explosion hazards related to dust generation.

The orientation of the stockpiles can help minimize the dust emission level. This orientation can include presenting the smallest surface area (narrowest side) of the piles to the prevailing wind (wind direction according to climatic conditions), taking advantage of any natural shelters, or setting berms and fences. In addition to acting as wind breaks, berms (earth embankments) would also reduce noise pollution and diminish the visual adverse impact of the piles on the landscape. Fences, if erected upwind, can reduce the wind speed and turbulence; if placed downwind, they can prevent dust transport but can also create dead zones where dust can be deposited before being redispersed.

3.5.2 Dust Suppressants

Several dust control and suppressing agents are available for use on coal piles, including water, wetting agents, foams, binders, agglomerating agents, and chemicals. These agents are as follows:

1. Water is the simplest, cheapest method in the short-tem, but not always the most effective. Water cannons are used to wet the coal. There are many drawbacks, such as short-lived effect, reduced heating value of the coal with excessive water content (compromising contract moisture specification), and accelerated surface weathering of the coal (the coal breaks down). Also, as a consequence, handling of wet coal and water runoff must be taken into consideration.

2. Wetting agents are added to reduce the amount of water used; they reduce the surface tension of water to enable fine water droplets and to enable the ability of water to wet the coal. The concentration of wetting agents ranges from 0.05 to 2% in spray water. A substantial number of wetting agents is offered on the market, but the appropriateness of the wetting agent to a certain type of coal must be established in laboratory or field tests and needs to satisfy environmental requirements (biodegradability and aqueous toxicity).

3. Foams are produced from water, air, and a foaming agent. If properly applied, they are better at controlling the moisture content of coal than water or a wetting agent.

4. Agglomerating agents (also called binders) bind the particles to one another or to larger particles; they include oils, oil products (such as asphalt), lignins, resins and polymer-based products. Oil and oil products are suitable when freezing problems are encountered. Safety (fire) and environmental issues (groundwater) need to be considered; polysulfides may be toxic to aquatic organisms. The binders that claim to be nontoxic are Dustcoat series and Coalgard (oil emulsion).

5. Sealants, as crusting or coating agents, are sprayed over the stockpile with a water cannon to form a water-insoluble crust. This method is suitable only for long-term stockpiles and requires regular inspection; reapplication of the crusting agent is necessary once the surface of the pile is broken for reclaiming. The crusting agents include polymer-based products (polyvinyl acetates, acrylics, and latex), resins, lignosulfonates, lignins, waste oils,

TABLE X

Applicability of Dust Control Methods

	Wind control	Wet suppression	Agglomerating agents	Sealants	Enclosures
Stacking		X	X		
Reclaiming		X	X		
Active piles	X	X	X		X
Long-term piles			X	X	

Source. Adapted from Carpenter (1999).

and bitumens. Lignins and waste oil have drawbacks in terms of water solubility and heavy metal contamination. Other novel approaches as sealant solutions are applications of grass or fly ash to the pile.

6. A complete or partial enclosure of piles is a solution to limit dust formation. Circular stockpiles are cheaper to cover than longitudinal piles. Safety measures would include ventilation to inhibit fire and explosion (dusts and methane). In addition, provision of ample openings to allow access for moveable handling equipment should be considered.

Monitoring of dust levels around a stockyard not only helps determine the effectiveness of dust control practices, but also helps to confirm that ambient air quality standards have been satisfied.

A successful monitoring of coal dust would require adequate measurement points (taking into consideration the direction of the prevailing wind), knowledge of the meteorological conditions (wind speed, temperature, humidity, and rainfall), and coal-handling operations. Although instrumentation to measure dust concentration or dust deposition is available, particulate measurements are generally known to be difficult to standardize.

3.6 Runoff and Flowslides

3.6.1 Coal Pile Runoff

In light of more stringent environmental regulations, rising costs of water management, and disposal of waste products, stockyard operators are encouraged to incorporate systems that will address runoff and flowslide of coal piles.

Coal pile runoff occurs when water seeps through the pile or runs off the ground surface. Several elements could generate this runoff: rain, snow, piles left to drain to meet moisture content requirements, and underground streams beneath the pile.

The amount of rainfall has the greatest impact on the amount of runoff; soil permeability under the pile also has an impact but to a lesser extent. Factors influencing the coal pile runoff include the following: the composition and size distribution of the coal; the drainage patterns in terms of contact time between coal and infiltrating water; and the amount of water seeping through the pile.

In general, runoff is characterized by high levels of solids in suspension, a variable pH, and heavy metal content. The composition of this effluent can vary widely. Acidic leachates show high levels of many metals. The pH of these leachates also depends on the presence of the neutralizing elements in the coal. Thus, bituminous coals produce runoff that is acidic; subbituminous coals produce neutral to alkaline runoff.

3.6.2 Runoff Collection

Site evaluation and site preparation are necessary steps to limit soil contamination caused by runoff. The potential site is evaluated in terms of its soil characteristics, bedrock structure, drainage patterns, and climatic conditions (precipitation records and potential flooding). Site preparation would involve compacting the soil, layout, and the construction of drainage ditches around the perimeter of the piles; the ground of the piles would be slightly sloped toward the drainage ditches and sloped away from the reclaiming area. The drainage system is designed to collect the runoff, collect rainwater from the entire stockyard, and handle the storm water. Several ways to prevent runoff and groundwater contamination include the installation of an impenetrable layer under the stockpile by using the following: (1) a plastic liner with compacted sand as the protected layer (Ashdod coal terminal, Israel); (2) a reinforced polypropylene membrane layer with fly ash/bottom ash mixed with an activator as the protective layer (Morgantown, West Virginia, Generating Station); and (3) an impermeable membrane protected by a cement-treated base overlain by rejected waste coal as the protective barrier (Los Angeles, California, Export Terminal).

3.6.3 Runoff Treatment

For recovery of the coal fines, coal pile runoff is directed to settling ponds located close to the corners of the stockyard and away from the stacking and reclaiming areas. Another alternative would be to install presettling ponds ahead of the settling ponds so that the bulk of the fines would be recovered in the primary ponds. This would reduce the amount of fines collected in the settling ponds, leading to a less frequent cleanup of these ponds. Different types of coal produce different types of runoff. In certain effluents (runoff), additives are used to remove the suspended solids and to modify the pH. Examples of these additives are alum (added to the runoff of the Powder River Basin [PRB] subbituminous coal at the U.S. St. Clair Power Plant) and cationic polymer flocculant (added to the runoff at Bolsover Works, United Kingdom). In addition, filter presses, other mechanical equipment, and chemical processes for wastewater treatment may be involved. Bioremediation techniques can be applied to produce biodegradable by-products and recover metals.

3.6.4 Coal Pile Flowslides

Instability in coal piles results from two types of slope failure: shallow slipping and deep-seated sliding, which is the type that causes flowslides.

In the first case, wetting of the surface layer can lead to erosion gullies and to shallow slipping with small flows depositing saturated coal at the toe of the slope. In the second case, deep-seated sliding causes a major flowslide within a short time. The stockpile shows a flat final slope and a temporary steep scarp that subsequently slumps back to the angle of repose. This slope failure results in economic loss associated with cleanup costs, loss of production, damage to equipment, and danger to personnel. Significant conditions that could lead to flowslides include the following:

1. Saturation of the stockpile base due to the infiltration of heavy rainfall, leading to the potential of a deep-seated slip;
2. Redistribution of moisture within the coal at placement (there is a threshold moisture content below which no saturated zone develops);
3. Loosely stacked coal (prone to structural instability);
4. Particle size distribution (migration of fine coal particles under the influence of water flow causes local water saturation and reduction in shear strength).

Failures can happen with relatively fresh coal (placed within the previous 2 weeks), and a significant coal pile height increases the risk of collapse.

Ideally, strategies to prevent and control flowslides would mean modifying the stacking method and the overall stockpile installation, but in reality operational and other factors may preclude major modifications to the existing operations. Consequently, a more practical approach is to adopt safety precautions and adjust the height of the stockpiles as prevention measures:

1. Excluding pedestrians and mobile equipment from stockpiles thought to be prone to flowslides or from coal piles with modified characteristics, particularly in the week after heavy rain or placement of wet coal.
2. Minimizing stockpile height during the rainy season; in case of busy shipping schedules and heavy rain, access roads should be closed to personnel; if a coal slide occurs, it should be cleaned up and restacked.
3. Reclaiming the coal from the toe of the pile by front-end loaders should be avoided because of the hazard; as coal dries out and adhesion between particles becomes weaker, the unstable sides of the pile may collapse.

To control flowslides, experience has shown that:

1. Controlling the moisture content of the pile can decrease the risk of flowsliding.
2. Compacting the entire coal pile, or selected areas of it, can reduce the tendency for the saturated coal to suddenly lose strength and to flow.
3. Building drainage slopes can facilitate surface runoff.
4. Constructing the pile with an appropriate profile and top facilitates runoff (with convex longitudinal profile and no flat-topped center, or slightly concave cross-sectional profile, or slightly crowned top for compacted piles).
5. Sealing the coal pile is useful in heavy rain areas but is not practical for working piles.
6. Enclosing the pile keeps the coal dry but is an expensive option and not applicable at coal blending terminals.
7. Draining the stockpile toe is beneficial for preventing minor instabilities that could lead to an overall failure.

4. HANDLING COAL AT THE END-USE SITE

4.1 Pneumatic Transport of Solids

Pneumatic transport of particles occurs in almost all industrial applications that involve powder and granular materials. The purpose of pneumatic transport is to protect the products from the environment and protect the environment from the products. Although this is not an energy-efficient method of transport because it requires power to provide the motive air or gas, it is easy and convenient to put into operation.

Five components are included in a pneumatic system: conveying line, air/gas mover, feeder, collector, and controls.

Pneumatic systems are broken down into three classifications: pressure system, vacuum system, and pressure/vacuum system. Their modes of transport are categorized as dilute, strand (two-phase), or dense phase and take into account the characteristics of the particles in terms of the following:

1. Material size and size distribution;
2. Particle shape;
3. Overall force balance of the particle/gas system, taking into account the acceleration, the drag forces, the gravity, and the electrostatic forces; terminal velocity is usually the characterizing term for particles—the higher the terminal velocity value, the greater the size and/or density of the particle;
4. The pressure drop and the energy loss for operational costs; the pressure loss calculations would take into account the voidage, the particle velocity, and the friction factors;
5. Acceleration of the flow at several locations of the pneumatic system (bend or connection, outlet or inlet);
6. Saltation (deposition of particles at the bottom surface of the pipe);
7. Pickup velocity, defined as the feed point of the solids and the velocity required to pick up particles from the bottom of the pipe;
8. Compressibility, which is of paramount importance for long distances and high-pressure systems;
9. Bends, creating significant pressure loss for systems with several bends and a relatively short length (<300 ft);
10. A dense phase that includes all types of flow, except dilute-phase flow (a lightly loaded flow); and
11. Choking conditions (in a vertical direction).

4.2 Pneumatic Transport of Coal

Three different types of air movers are commonly used for pneumatic transport: fans, blowers, and compressors. The fans are the most inexpensive movers, the blowers are the workhorse of pneumatic transport, and the compressors are used for pressure ratios over 2:1.

When any of these movers are used in coal-fired steam power plants, primary air fans are used to supply air to the mills. There are many varieties of feeders, such as rotary feeders with different design variations (rotor, blades, and casing), gate lock feeders in pressurized systems, and screw feeders. In addition, filters and cyclones are available for separating, dividing, or redirecting the flow.

4.3 Transport Properties of Slurries or Wet Solids

Understanding and accurately predicting the flow behavior of solid–liquid mixtures through the slurry fuel-handling system is key to the successful operation of these handling systems. Slurries are solid–liquid mixtures. By definition, slurries should contain sufficient liquid to wet the surface of every particle, small and large. Therefore, particle size and the relative density between solids and liquids are important and highlight four types of slurry flow regimes: homogenous; heterogeneous; saltation; and moving bed.

These flow regimes are investigated in laboratory-scale and pilot-scale rheological tests to elucidate slurry behavior, e.g., its viscosity under certain flow conditions. In these tests, various geometries and instruments can be used to measure the rheological properties of the slurry. These properties are expressed in terms of shear stress (force/area), shear rate, or strain rate (velocity/distance). Equations or charts are developed to distinguish between Newtonian and non-Newtonian fluids. Virtually all slurries display non-Newtonian behavior and are represented commonly by shear stress characteristics, such as velocity gradient curves (pseudo-plastic, plastic, Newtonian, and dilatant curves). The Bingham plastic is the most common curve; it is actually characterized by a straight line with an intercept on the shear stress axis. This helps in designing the handling systems and the operating parameter ranges. For heterogeneous slurries, the design of this type of transport is based almost entirely on correlations developed from experience.

Pipelines are used to transport slurries over long distances and these slurries must be stable enough to withstand the demands of transport, readily separable after reaching their destination, and cost-competitive with other forms of transport.

For pipelines, operating at lower velocities would result in less pressure loss per length of pipe (smaller pumps) and greater pipeline lifetime (less erosion). The minimum operating velocity is the lowest velocity at which all the particles stay in suspension.

As expected, for a given particle size distribution, the greater the solids loading, the more difficult it is to pump the slurry. However, adding wood pulp fibers to the slurry has considerably reduced the drag because of changes in the turbulent flow regimes. For instance, in aqueous slurries of 20% coal with and without wood pulp fibers flowing at 15 and 60 ft/s, the drag has been reduced up to 30% over this range of velocities by the addition of 0.8% fibers.

In addition to the type of drag-reducing additives that help increase the solid loading in the slurry (to 70% solids with additives, compared to 55% solids without additives), there are other additives that will help to retard particle settling. These can increase the shelf-life of the slurry from a few days to several months.

4.4 Slurry Transport of Coal

A particle size must be created to transport coal slurry, requiring a balance between pumping and dewatering characteristics. If the slurry is too fine, the pumpability is good but the process of separating water from solids could be difficult. If the slurry is too coarse, it is not homogenous and higher pumping power is needed. In short, two parameters need to be controlled: the density of the slurry and the top particle size. The first parameter, in terms of slurry concentration, has an important impact on friction loss and the critical velocity of the slurry. The second parameter prevents pluggage of pipelines by controlling the coal particles. In addition, corrosion inhibitors, such as sodium dichromate, will protect the pipe wall, and other additives, such as sodium sulfite, will act as oxygen scavengers.

In commercial pipelines, a so-called critical concentration range exists where the slurry provides maximum support for the particles without an untenably high critical velocity (for the pumping stations). At the destination, the objective is to remove the water and recover the solids with minimum moisture content by using centrifuges or filters.

Coal slurries find applications in power generation. The IEA Coal Research report of 1993 gives an insight into the impact of the coal on the power station performance. In conventional steam power plants, coal slurry technology has been commercialized. One example is the 273-mile Black Mesa coal pipeline that provides all the fuel requirements (approximately 5 million tons/year) to two 750 MW generating units in southern Nevada.

In pressurized fluidized bed combustion, coal and sorbent (usually lime or limestone) are injected into a fluidized bed as a water-mixed paste using concrete pumps or pneumatically as a dry suspension via lock hoppers. In the first case, the coal could be crushed by roll crushers to a certain size distribution, such as a top size of 6 mm, and the sorbent could be crushed through hammer mills to a certain top size, such as 3 mm. After that, the coal and sorbent are mixed to form a pumpable slurry. Optimizing the paste moisture content at 20–30% is an acceptable compromise relative to boiler efficiency. At the Tidd Plant in Brilliant, Ohio, for example, the coal paste had a nominal value of 25% weight water.

Regarding atmospheric circulating fluidized bed combustion, coal slurry represents a preferable solution when using low-grade coals and coal wastes. Three options are available: a dilute slurry ($>$40% water), a dense slurry ($<$40% water), or a nominally dry material ($\approx$12% water). Coal slurry has also been used in integrated gasification combined cycles, such as at the Wabash power station and at the Tampa Electric Company's Polk power station.

TABLE XI

Example of Coal Storage Requirements at a Coal-Fired Power Plant

	Coal			
	Lignite	Subbituminous	Midwestern bituminous	Eastern bituminous
Heat to turbine	4591×10^6 kJ/h	4591×10^6 kJ/h	4591×10^6 kJ/h	4591×10^6 kJ/h
Boiler efficiency	83.5%	86.2%	88.5%	90.5%
Coal heat input	5499×10^6 kJ/h	5326×10^6 kJ/h	5185×10^6 kJ/h	5073×10^6 kJ/h
Coal HHV	14.135 MJ/kg	19.771 MJ/kg	26.284 MJ/kg	33.285 MJ/kg
Coal flow rate	429 tons/h	297 tons/h	217 tons/h	168 tons/h
		Design storage requirements, tons		
Bunker (12 h)	5148	3564	2604	2016
Live (10 days)	102,960	71,280	52,080	40,320
Dead (90 days)	926,640	641,520	468,720	362,880
		Storage time for eastern bituminous plant design[a]		
Bunker	4.7 h	6.8 h	9.3 h	12.0 h
Live	3.9 days	5.7 days	7.7 days	10.0 days
Dead	35 days	51 days	69 days	902 days

Source. Adapted from Folsom *et al.* (1986).

[a] Equivalent storage time for a plant designed for eastern bituminous coal but fired with the coals listed.

5. STORING COAL AT THE END-USE SITE

5.1 Safe Storage of Crushed Coal

Crushed coal storage can be divided into two types: live storage (e.g., active storage with a short residence time) and dead storage (e.g., reserve storage). A common practice is to directly transfer part of a shipment to live storage and divert the remainder to reserve storage.

Usually, the storage components are sized to a certain capacity equivalent to a fixed period of firing at full load. The length of time for each storage type is as follows: 12 h for in-plant storage in bunkers, 10 days for live storage, and more than 90 days for reserve storage. These time periods are usually set by the operating procedures of the plant (power station capacity, heat rate, and coal heating value) and by other constraints (e.g., stocking strategies). If the actual heating value of the coal changes and if the steam generator is expected to operate at a certain fixed heat input, the storage capacity and the quantities delivered to the plant must increase. As a result, silos or bunkers represent an important link to provide an interrupted coal flow to the pulverizers. Typically at coal-fired power plants, a number of round silos with conical bottoms are stationed on each side of the boiler (steam generator).

Silos and bunkers are constructed of either concrete or steel and are generally round. Loading and unloading of coal from the silo are carefully controlled to prevent dangerous offset loading on the foundations. The coal is supplied to the mills and pulverized continuously and a mixture of pulverized coal/air is pneumatically transported to the burners. A single mill can supply several burners (tangentially fired systems and wall-fired systems). Table XI gives an example of coal storage requirements at a coal-fired power plant in the 1980s.

TABLE XII

Preferred Range of Coal Properties for Mills

Properties	Mill type: Low speed	Medium speed	High speed
Maximum capacity	100 tons/h	100 tons/h	30 tons/h
Turndown	4:1	4:1	5:1
Coal feed top size	25 mm	35 mm	32 mm
Coal moisture	0–10%	0–20%	0–15%
Coal mineral matter	1–50%	1–30%	1–15%
Coal quartz content	0–10%	0–3%	0–1%
Coal fibre content	0–1%	0–10%	0–15%
Hardgrove grindability index	30–50, 80	40–60	60–100
Coal reactivity	Low	Medium	Medium

Source. Adapted from Sligar (1985).

5.2 Milling of Pulverized Coal

At the mill, fine grinding of coal involves producing coal particle sizes such that 70% or more are finer than 75 μm (200-mesh) to ensure complete

TABLE XIII

Impact of Coal Properties on Mill Performance

Mill performance	Low-speed and medium-speed mills
Drying	• Moisture affects primary air requirements and power consumption of mills • Volatile matter affects the susceptibility to mill fires • Total moisture causes a −3% throughput for a 1% moisture increase for low-speed mill; a −1.5% throughput for 1% moisture increase above 12% moisture for medium-speed mill
Mill throughput	• Total moisture (see above) • −1% throughput for 1-unit reduction in HGI[a] for both mill types (caution when using HGI for coal blend behavior) • Raw coal top size: −3% throughput for 5 mm increase in top size for low-speed mill; no loss in throughput below 60 mm top size for medium-speed mill • Pulverized fuel size distribution: reduction of fraction going through <75 μm mesh screen by 0.35% for a 1% increase in throughput for low-speed mill; reduction by 0.9% for a 1% increase in throughput for medium-speed mill
Wear	• Mineral matter and size distribution influence the operation and maintenance of both types of mills

Source. Adapted from Lowe (1987).
[a] *Note.* HGI, Hardgrove grindability index.

combustion and minimization of ash and carbon on the heat exchanger surfaces.

Three types of mills, according to speed, are available: low-speed mills of ball/tube design; medium-speed mills of vertical spindle design; and high-speed mills with a high-speed rotor.

Table XII gives the preferred range of coal properties applicable to the types of mills and Table XIII shows the impact of coal properties on the performance of the low-speed and medium-speed mills.

SEE ALSO THE FOLLOWING ARTICLES

Clean Coal Technology • Coal, Chemical and Physical Properties • Coal Conversion • Coal, Fuel and Non-Fuel Uses • Coal Industry, Energy Policy in • Coal Industry, History of • Coal Mine Reclamation and Remediation • Coal Mining, Design and Methods of • Coal Preparation • Coal Resources, Formation of • Hydrogen Storage and Transportation • Markets for Coal • Natural Gas Transportation and Storage

Further Reading

Alvarez, R., Barriocanal, C., Casal, M. D., Diez, M. A., González, A. I., Pis, J. J., and Canga, C. S. (1996). Coal weathering studies. *In* "Ironmaking Conference Proceedings, Pittsburgh, PA, March 24–27, 1996." Vol. 55.

Alvarez, R., Casal, M. D., Canga, C. S., Diez, M. A., González, A. I., Lázaro, M., and Pis, J. J. (1995). Influence of natural weathering of two coking coals of similar rank on coke quality. *In* "Coal Science, Proceedings of the 8th International Conference on Coal Science, Oviedo, Spain, September 10–15, 1995." Elsevier Science, Amsterdam/New York.

Brolick, H. J., and Tennant, J. D. (1990). Innovative transport modes: Coal slurry pipelines. *In* "Fuel Strategies—Coal Supply, Dust Control, and By-product Utilization, The 1990 International Joint Power Generation Conference, Boston, MA, October 21–25, 1990." National Coal Association, Washington, DC.

Carpenter, A. M. (1999). "Management of Coal Stockpiles." IEA Coal Research, October 1999. International Energy Agency, London, UK.

Chakraborti, S. K. (1995). American Electric Power's coal pile management program. *Bulk Solids Handling* **15**, 421–428.

Chakraborti, S. K. (1999). An overview of AEP's coal handling systems. *Bulk Solids Handling* **19**, 81–88.

Craven, N. F. (1990). Coal stockpile bulk density measurement. *In* "Coal Handling and Utilisation Conference, Sydney, NSW, Australia, June 19–21, 1990."

Department of Energy (1990). "Innovative Alternative Transport Modes for Movement of U.S. Coal Exports to the Asian Pacific Basin," DOE Report DOE/FE-61819-H1, Prime Contractor, Argonne National Laboratories; Subcontractors, William Technologies Inc., The Fieldston Co., and TransTech Marine Company, March 1, 1990.

Deurbrouck, A. W., and Hucko, R. E. (1981). Coal preparation. *In* "Chemistry of Coal Utilization, Secondary Volume," Chap. 10 (M. Elliott, Ed.). John Wiley & Sons.

Dong, X., and Drysdale, D. (1995). Retardation of the spontaneous ignition of bituminous coal. *In* "Coal Science, Proceedings of the 8th International Conference on Coal Science, Oviedo, Spain, September 10–15, 1995", Vol. 1, pp. 501–504. Elsevier Science, Amsterdam, the Netherlands/New York.

Folsom, B. A., Heap, M. P., Pohl, J. H., Smith, J. L., and Corio, M. R. (1986). Effect of coal quality on power plant performance and costs. *In* "Review of Coal Quality Impact Evaluation Procedures," EPRI-CS-4283-Vol. 2 (February 1986). EPRI Technical Information Services.

Kaplan, S. (2002). "Coal Transportation by Rail—Outlook for Rates and Service, Energy Information Administration, March 12, 2002," PA Consulting Group, Washington, D.C.

Klinzing, G. E. (1995). Pneumatic conveying. *In* "NSF Short Course on Fluid/Particle Processing," June 5–9, 1995, University of Pittsburgh, Pittsburgh, PA.

Lema, J. E. (1990). Attributes of selected transportation modes. *In* "Fuel Strategies—Coal Supply, Dust Control, and By-product Utilization, The 1990 International Joint Power Generation Conference, Boston, MA, October 21–25, 1990." National Coal Association, Washington, DC.

Leonard, J. W., III, and Hardinge, B. C. (eds.). (1991). "Coal Preparation," 5th ed. Society for Mining, Metallurgy and Exploration, Inc., Littleton, CO.

Lowe, A. (1987). Important coal properties for power generation. *In* "Geology and Coal Mining Conference," pp. 188–196. Sydney, Australia, October 13–5, 1987.

Meikle, P. G., Bucklen, O. B., Goode, C. A., and Matoney, J. T. (1991). Post preparation/storage and loading. *In* "Coal Preparation" (J. W. Leonard, III and B. C. Hardinge, Eds.), 5th ed., Chap. 9.

O'Connor, D. C. (1987). On-line coal analysis: A new challenge. *In* "Coal Sampling, Fundamentals and New Applications–Belt Conveyor Systems." The 1987 Joint Power Generation Conference, October 4–8, 1987.

Parek, B. K., and Matoney, J. P. (1991). Dewatering. *In* "Coal Preparation," 5th ed., Chap. 8 (Leonard, J. W., III, and Hardinge, B. C., Eds.). Society for Mining, Metallurgy, and Exploration Inc., Littleton, CO.

Schneid, R. J. (2000). "Coal Transportation, Coal and Power Industry Training Course," Presented to U.S. Department of State Foreign Service Officers and Other U.S. Government Employees, July 25, 2000. Consol Energy Inc.

Scott, D. H., and Carpenter, A. M. (1996). "Advanced Power Systems and Coal Quality, IEA Coal Research, May 1996." International Energy Agency, London, UK.

Sharpe, M. A. (2001). "Coal Preparation, Coal and Power Industry Training Course, Presented to the U. S. Department of State Foreign Service Officers, July 25, 2001." CLI Corp, Canonsburg, PA.

Skorupska, N. M. (1993). "Coal Specifications—Impact on Power Station Performance, IEA Coal Research, January 1993." International Energy Agency, London, UK.

Sligar, J. (1985). Coal pulverizing mill section. *In* "Course on the Characterization of Steaming Coals," pp. 10.1–10.23. University of Newcastle, Institute of Coal Research, Newcastle, NSW, Australia.

Speight, J. G. (1994). Preparation and transportation. *In* "The Chemistry and Technology of Coal," 2nd ed., Chap. 6, p. 121. Dekker, New York.

Sujanti, W., and Zhang, D.-K. (1999). Low-temperature oxidation of coal studied using wire-mesh reactors with both steady-state and transient methods. *Combustion Flame* **117**, 646–651.

Thomson, T. L., and Raymer, F. B. (1981). Transportation, storage and handling of coal. *In* "Chemistry of Coal Utilization" (M. A. Elliott, Ed.), Chap. 9, pp. 523–570. Wiley, New York.

Walker, S. (1999). "Uncontrolled Fires in Coal and Coal Wastes, CCC/16, IEA Coal Research, April 1999." International Energy Agency, London, UK.

Wildman, D., Ekmann, J., and Mathur, M. (1995). Slurry transport and handling. *In* "NSF Short Course on Fluid/Particle Processing," June 5–9, 1995, Pittsburgh Energy Technology Center, Pittsburgh, PA.

Zumerchik, J. (2001). Coal transportation and storage. *In* "Macmillan Encyclopedia of Energy," Vol. 1, (J. Zumerchik, Ed.). Macmillan Co., New York.

Cogeneration

DOUG HINRICHS
Clean Energy Consultant
Silver Springs, Maryland, United States

1. Time Line
2. Technical Characterization
3. Applications and Markets
4. Regulatory and Rate Structure Issues
5. U.S. Federal Government Support

Glossary

absorption chiller An absorption chiller transfers thermal energy from the heat source to the heat sink through an absorbent fluid and a refrigerant. The absorption chiller accomplishes its refrigerative effect by absorbing and then releasing water vapor into and out of a lithium bromide solution.

cogeneration The simultaneous production of electric and thermal energy in on-site, distributed energy systems; typically, waste heat from a prime mover is recovered and used to heat/cool/dehumidify building space. Technically, if cooling is an end result, the process is called trigeneration.

combined heat and power Synonymous with cogeneration, combined heat and power (CHP) is more commonly used in the United States than in Europe or China, where cogeneration is more common. The term CHP applies more to the industrial sector, in which "waste heat" may be used in applications other than heating or cooling building space.

desiccant A solid or liquid material with an affinity for absorbing water molecules.

desiccant dehumidifier Desiccant dehumidifiers use recovered heat from the prime mover to dry building space using a desiccant material. The two general types are based on the desiccant used: solid (dry desiccant) and liquid (liquid desiccant).

distributed generation In contrast to centralized utility electricity generation, distributed generation involves small-scale prime movers that can be located at or near the building or facility where the energy is used to provide greater power reliability and reduced emissions.

district energy District energy systems produce steam, hot water, or chilled water at a central plant and then pipe that energy out to buildings in the district for space heating, domestic hot water heating, and air-conditioning. A district energy system can obviate the need for individual buildings to have their own boilers or furnaces, chillers, or air conditioners. A district energy system can serves many customers from one location; use a variety of conventional fuels, such as coal, oil, and natural gas; and transition to use renewable fuels, such as biomass, geothermal, and combined heat and power.

"first-generation" packaged cooling, heating, and power systems On August 8, 2001, Energy Secretary Abraham announced awards to seven industry teams for research, development, and testing of first-generation packaged cooling, heating, and power systems. In doing so, the U.S. Department of Energy is helping manufacturers work together to integrate their individual cogeneration components into easier to use packages and to make "plug-and-play" packages more readily available in the commercial and industrial marketplace; objectives include reduction of engineering time/cost at the installation site.

fuel cell Device for producing electricity using a chemical process that combines hydrogen and oxygen rather than conventional combustion processes with electric generators.

prime mover The electric and thermal energy generator; typically an on-site distributed generator. For cogeneration purposes, microturbines, reciprocating engines, and industrial turbines are most commonly used; other prime movers include fuel cells, concentrating solar power systems, and Sterling engines.

Public Utilities Regulatory Policy Act (PURPA) Congress passed PURPA in 1978. The nation had recently survived several energy crises resulting from the escalation of crude oil prices coupled with curtailments of production. Policymakers belatedly realized that fuel was dependent on the forces of international interests and beyond the control of governmental officials and industries in the United States. Public utility regulators and Congress determined that steps to avoid future energy crises should include both focused conservation programs and the development of alternative sources for producing electricity. PURPA was a major response to these crises and started the power industry on the road to competition by ending promotional rate structures, establishing unconventional qualifying facilities, and encouraging conservation. Unintentionally, PURPA started the competitive process because it enabled nonutility generators to produce power for use by customers attached to a utility's grid. PURPA broke the stranglehold on power companies' previous monopoly in the generation function because now any unregulated cogenerator or renewable energy producer could sell electricity into the power grid, and regulated utilities were unable to dictate terms. Of particular note regarding the cogeneration industry, PURPA promoted cogeneration as an energy-efficient technology.

thermally activated technologies Thermally activated technologies (TATs) represent a diverse portfolio of equipment that transforms heat for a useful purpose, such as energy recovery, heating, cooling, humidity control, thermal storage, or bottoming cycles. TATs are the essential building blocks for cogeneration systems.

trigeneration The simultaneous production of electric and thermal energy in on-site distributed energy systems; typically, waste heat from a prime mover is recovered and used to heat and cool/dehumidify building space. Technically, if cooling is an end result, the process is called trigeneration.

waste heat The portion of the energy input to a mechanical process that is rejected to the environment.

The efficiency of central electricity generation has not improved much in the past 40 years, with the exception of an increase in natural gas utilization and the use of combined cycle plants by central plants. Currently, power plants are approximately 33% efficient, meaning that only one-third of the energy is used to generate new power. The remaining two-thirds is lost, vented as "waste heat" to the environment that, along with carbon/nitrogen/sulfur-based emissions from coal or oil, causes significant ecosystem degradation. Cogeneration, or combined heat and power, is a clear alternative to central electricity generation: Using smaller scale generation technologies that primarily use natural gas at or near a building, campus, or industrial facility, cogeneration systems take advantage of waste heat by using it to heat, cool, or dehumidify building space, with overall system efficiencies in the 70–90% range. (Technically, an energy system that makes heat, power, and cooling is considered trigeneration. In this article, cogeneration includes trigeneration.) This doubling or tripling of efficiency from cogeneration or trigeneration has far-reaching implications for the possibility of a more sustainable energy paradigm in commercial and industrial sectors in industrialized countries.

1. *TIME LINE*

1.1 Historical

Cogeneration is not new. It first appeared in the late 1880s in Europe and in the early 20th century in the United States, when most industrial plants generated their own electricity using coal-fired boilers and steam-turbine generators. Many of these plants used the exhaust steam for industrial processes. According to COGEN Europe, an advocacy group based in Belgium, cogeneration produced as much as 58% of the total power from on-site industrial power plants in the United States in the early 1900s.

As a result of improvements in the cost and reliability of electricity generated by the separate electric power industry as well as increasing regulation, the electric generation capacity at most cogeneration facilities was abandoned in favor of more convenient purchased electricity. A few industries, such as pulp and paper and petroleum refining, continued to operate their cogeneration facilities, in part driven by high steam loads and the availability of by-product fuels.

In the mid-1900s, the pendulum began to swing in the opposite direction with the advent of central electric power plants and reliable utility grids, which allowed industrial plants to begin purchasing lower cost electricity. As a result, on-site industrial cogeneration declined significantly to only 15% of total U.S. electrical generation capacity by 1950, and it then declined to approximately 4% by 1974.

The first dramatic rise in fuel costs and uncertainty in the fuel supply in the 1970s triggered an upswing in cogeneration, particularly in large industrial applications that required vast quantities of steam. In recent years, smaller cogeneration systems have begun to make inroads in the food, pharmaceutical, and light manufacturing industries, commercial buildings, and on university campuses.

In the late 1970s interest in cogeneration in the United States was renewed in response to the Public Utilities Regulatory Policy Act of 1978 (PURPA), which included measures to promote cogeneration as an energy-efficient technology. With the passage of PURPA, multiple megawatt projects at large pulp and paper, steel, chemical, and refining plants were undertaken. Although PURPA created an enormous incentive for these large-scale projects that helped increase cogeneration capacity significantly, the act may have caused utilities to discourage beneficial projects because of the regulatory mandate. However, these attitudes are changing: Forward-thinking utilities are actively pursuing new cogeneration projects as part of their business strategy. PURPA has played a critical role in moving cogeneration into the marketplace by addressing many barriers that were present in the early 1980s.

During the 1990s, approximately 9500 miles of new high-voltage transmission lines were constructed (an approximately 7% increase)—short of the need—which represented a new opportunity for efficient distributed generation (DG) and cogeneration use. The U.S. Department of Energy (DOE) estimates that more than 390 GW of new generating capacity will be needed by 2020 to meet growing U.S. energy demand and offset power lost from retired power plants. However, obtaining approval for construction of new transmission lines and acquiring right-of-ways is very difficult.

According to Tom Casten, founder of Trigen and currently chairman and CEO of Private Power, "The public simply detests new transmission lines, and no one wants them in their back yard. Most recent U.S. power problems were caused by lack of adequate transmission and distribution (T&D), and nobody close to the industry believes enough new transmission can be built." Distributed cogeneration systems require no new construction of T&D lines. Tom Casten cites industry forecasts that estimate that the United States will need 137,000 MW of new capacity by 2010. According to Casten, meeting this demand will require $84 billion for new power plants and $220 billion for new T&D, for a total of $304 billion. Meeting the same demand with DG will require $168 billion for new plants but $0 for T&D. Casten notes, "A distributed energy future would save $136 billion of capital investment and reduce the cost of new power by about 3¢/kW."

1.2 Present

The Energy Information Administration (EIA) reports that as of 2000, cogeneration accounted for approximately 7.5% of electricity generation capacity and almost 9% of electricity generated in the United States. Private Power studied data from an unpublished EIA survey and other EIA data and determined that total DG was 10.7% of U.S. totals in 2000.

In stark contrast, in The Netherlands, more than 40% of electricity is obtained from combined heat and power (CHP) systems, and in Denmark, approximately 60% is obtained from these systems. In the United Kingdom, cogeneration's share of electric power production has doubled in the past decade, with additional growth targeted by the government.

2. TECHNICAL CHARACTERIZATION

2.1 Efficiency Gains

According to the U.S. Combined Heat and Power Association, cogeneration technologies produce both electricity and steam from a single fuel at a facility located near the consumer. These efficient systems recover heat that normally would be wasted in an electricity generator, and they save the fuel that would otherwise be used to produce heat or steam in a separate unit.

Cogeneration offers dramatic advantages in efficiency and much lower air pollution than conventional technologies (Fig. 1). A wide variety of cogeneration technologies generate electricity and meet thermal energy needs (direct heat, hot water, steam, and process heating and/or cooling) simultaneously at the point of use. Cogeneration can result in system efficiencies of 70–90%, compared with the national average of 30% efficiency for conventional electricity plants and perhaps 50% for conventional thermal applications.

The environmental benefits of cogeneration are apparent. Cogeneration systems in the United States:

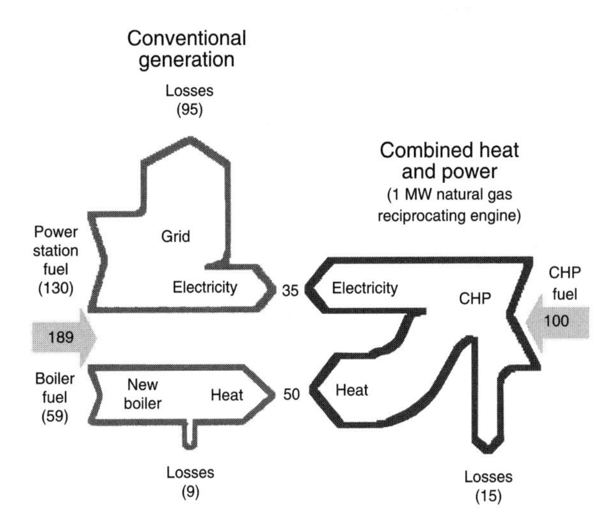

FIGURE 1 This diagram compares the typical fuel input needed to produce 35 units of electricity and 50 units of heat using conventional separate heat and power. For typical electric and thermal efficiencies, cogeneration is nearly twice as efficient.

- Decrease energy use by approximately 1.3 trillion Btus/year.
- Reduce NO_x emissions by 0.4 million tons/year.
- Reduce sulfur dioxide (SO_2) emissions by more than 0.9 million tons/year.
- Prevent the release of more than 35 million metric tons of carbon equivalent into the atmosphere.

2.2 Technology Advances

Nearly all U.S. buildings and industries use thermal energy from boilers for space heating, hot water, steam systems, and direct process heating applications. Most cogeneration systems installed in the United States are used for industrial applications, but there is increasing usage by U.S. commercial and institutional building owners and district energy systems worldwide. Newer thermally activated technology applications include recycling waste heat for cooling and humidity control, which has significant energy and economic potential.

Recent advances in technology have resulted in the development of a range of efficient and versatile systems for industrial and other applications. Improvements in electricity generation technologies, particularly advanced combustion turbines and engines, have allowed for new configurations that reduce size but increase output. Distributed, on-site power generation technologies, such as microturbines, reciprocating engines, industrial turbines, and fuel cells, are becoming more effective at generating electricity and producing recoverable thermal energy.

Recovered heat from these generation technologies can be used to power thermally activated technologies, control humidity with desiccant dehumidifiers, and heat buildings with steam or hot water in cogeneration systems.

2.3 Prime Movers

Prime movers generate electricity and heat, or electric and thermal energy. They include several technologies. First, microturbine generator units are composed of a compressor, combustor, turbine, alternator, recuperator, and generator. In a simple cycle turbine (without a recuperator), compressed air is mixed with fuel and burned under constant pressure conditions. The resulting hot gas is allowed to expand through a turbine to perform work.

Recuperated units use a heat exchanger (recuperator or regenerator) that recovers some of the heat from the turbine exhaust and transfers it to the incoming air stream for combustion in the turbine. By using recuperators that capture and return waste exhaust heat, existing microturbine systems can reach 25–30% cycle efficiency. The incorporation of advanced materials, such as ceramics and thermal barrier coatings, could further improve their efficiency by enabling a significant increase in engine operating temperature.

Microturbines offer many advantages over other technologies for small-scale power generation, including the ability to provide reliable backup power, power for remote locations, and "peak shave." Other advantages include less maintenance and longer lifetimes because of a small number of moving parts, compact size, lighter weight, greater efficiency, lower emissions, and quicker starting. Microturbines also offer opportunities to use waste fuels such as landfill gas.

Second, gas-fired reciprocating engines are the fastest selling, least expensive distributed generation technology in the world. Commercially available natural gas versions of the engine produce power from 0.5 kW to 10 MW, have efficiencies between 37 and 40%, and can operate down to NO_x levels of 1 g per horsepower-hour. When properly treated, these rugged engines can run on fuel generated by waste treatment (methane) and other biofuels. By using recuperators that capture and return waste exhaust heat, reciprocating engines can also be used in cogeneration systems in buildings to achieve energy-efficiency levels approaching 80%.

Gas-fired reciprocating engines offer many advantages over other technologies for small-scale power generation, including the ability to provide highly reliable, inexpensive backup power; provide power for remote locations; and generate on-site power during peak periods when utility charges are highest.

Reciprocating engines require fuel, air, compression, and a combustion source to function. Depending on the ignition source, they generally are classified into two categories: spark-ignited engines, typically fueled by gasoline or natural gas, and compression-ignited engines, typically fueled by diesel oil.

Third, tremendous efficiency strides with regard to industrial gas turbines have been made in recent years. Combustion turbines are a class of electricity-generation devices that produce high-temperature, high-pressure gas to induce shaft rotation by impingement of the gas on a series of specially designed blades. Industrial gas turbines are used in many industrial and commercial applications ranging from 1 to 20 MW.

Because gas turbines are compact, lightweight, quick starting, and simple to operate, they are used widely in industry, universities and colleges, hospitals, and commercial buildings to produce electricity, heat, or steam. In such cases, simple cycle gas turbines convert a portion of input energy from the fuel to electricity and use the remaining energy, normally rejected to the atmosphere, to produce heat. This waste heat may be used to power a separate turbine by creating steam. The attached steam turbine may generate electricity or power a mechanical load. This is referred to as a combined cycle combustion turbine since two separate processes or cycles are derived from one fuel input to the primary turbine.

These reliable distributed generation technologies are becoming the mainstay of peak shaving equipment. When needed, this equipment can be started immediately and can quickly generate full capacity.

The waste heat generated by the turbine can be used to generate steam in a heat recovery steam generator (HRSG), which can power a steam turbine, heat living space, or generate cooling using steam-driven chillers. The advantages of these types of systems are inexpensive electrical power and better reliability since the user may be independent from the grid. These systems can be started even if the grid has failed.

Advanced materials, such as ceramics, composites, and thermal barrier coatings, are some of the key enabling technologies under development to improve the efficiency of distributed generation technologies. Efficiency gains can be achieved with materials such as ceramics, which allow a significant increase in engine operating temperature. The increased operating temperature also lowers greenhouse gas and NO_x emissions.

Two successful applications of HRSG systems are at Princeton University and the Massachusetts Institute of Technology (MIT). Each school deploys a gas turbine for electrical generation to meet the majority of the campus's needs. The exhaust heat from these turbines drives an HRSG. The steam from the HRSG provides heating for campus living spaces and cooling through steam-driven chillers. Frequently, summer peak-time operation of electrical chillers is expensive. Producing steam on-site enables the use of steam-driven chillers. Operating steam-driven chillers acts to peak shave the user's high demand operations, thus saving money. The electricity generated by the gas turbine may also be used to operate conventional electric chillers. Since the user generates this electricity, the price may be lower than that of the local utility and does not have a peak or demand charge associated with it.

Fourth, fuel cells were discovered in 1839 when scientist William Grove accidentally let a hydrolysis experiment run backwards. He was surprised to learn that electron flows were created when hydrogen and oxygen combined. NASA made this technology well-known with its use in space shuttles.

Fuel cells consist of two electrodes (cathode and anode) around an electrolyte. The generation process involves introducing hydrogen to one side of the fuel cell (the anode), where it breaks apart into protons and electrons. The electrode conducts protons but not electrons. The protons then flow through while the electrons travel through the external circuit and provide electrical power. The electrons and protons are reunited at the other end of the fuel cell (the cathode). When combined with oxygen from the air, fuel cells produce water and heat in a process that is practically silent, nearly emission free, and involves no moving parts.

Fuel cells are categorized by the kind of electrolyte they use. Electrolyte types used in the building sector include phosphoric acid, molten carbonate, solid oxide, and proton exchange membrane. High-temperature molten carbonate and solid oxide fuel cells are undergoing full-scale demonstration with some existing applications. Manufacturers are actively pursuing this technology for widespread, general use in commercial buildings, homes, and transportation.

Although price and performance data on reciprocating engines and turbines are fairly well established, data for fuel cells are based on a limited number of demonstration projects. As a result, comparisons of price and performance should be interpreted with caution.

The price and performance of engines, turbines, and fuel cells are summarized in Table I.

Finally, concentrating solar electric generators concentrate solar power during the high solar radiation periods onto tubes of water (or oil) to make usable steam, which can then be used to make electricity in a steam generator or to power a chiller. Some concentrating solar electric generators use a hybrid renewable/fossil fuel system to diversify their energy source.

California's Mojave Desert is home to the world's largest concentrating solar power facility. The Solar Electric Generating System plants have a combined capacity of 354 MW and are configured as hybrids. The nine hybrid plants generate more than 75% of the total power output from the sun, with natural gas providing the balance.

TABLE I

Cost and Performance of Distributed Generation Technologies for Cogeneration Systems[a]

Technology	Reciprocating engine	Turbine and microturbine	Fuel cell
Size	30 kW–8 MW	30 kW–20 + MW	100–3000 kW
Installed cost ($/kW)[b]	300–1500	350–1500	2000–5000
Electricity efficiency (LHV)	28–42%	14–40%	40–57%
Overall efficiency[c]	~80–85%	~85–90%	~80–85%
Variable O&M ($/kWh)	0.0075–0.02	0.004–0.01	0.002–0.05
Footprint (ft^2/kW)	0.22–0.31	0.15–0.35	0.9
Emissions (lb/kWh unless otherwise noted)			
Diesel			
NO_x	0.22–0.025	3–50 ppm	<0.00005
CO	0.001–0.002	3–50 ppm	<0.00002
Natural gas			
NO_x	0.0015–0.037		
CO	0.004–0.006		
Fuels	Diesel, natural gas, gasoline, digester gas, biomass and landfill gas; larger units can use dual-fuel (natural gas-diesel) or heavy fuels	Natural gas, diesel, kerosene, naphtha, methanol, ethanol, alcohol, flare gas, digester gas, biomass and landfill gas	Natural gas, propane, digester gas, biomass and landfill gas (potentially)

[a] *Source.* Resource Dynamics Corporation (2002).
[b] Cost varies significantly based on siting and interconnection requirements as well as unit size and configuration.
[c] Assuming cogeneration.

2.4 Thermally Activated Technologies

Thermally activated technologies (TATs) take advantage of the waste heat from prime movers. There are several types of TATs.

First, absorption chillers provide cooling to buildings by using heat. This seemingly paradoxical but highly efficient technology is most cost-effective in large facilities with significant heat loads. Absorption chillers not only use less energy than conventional equipment but also cool buildings without the use of ozone-depleting chlorofluorocarbons.

Unlike conventional electric chillers, which use mechanical energy in a vapor compression process to provide refrigeration, absorption chillers primarily use heat energy with limited mechanical energy for pumping. These chillers can be powered by natural gas, steam, or waste heat.

An absorption chiller transfers thermal energy from the heat source to the heat sink through an absorbent fluid and a refrigerant. The absorption chiller accomplishes its refrigerative effect by absorbing and then releasing water vapor into and out of a lithium bromide solution. The process begins as heat is applied at the generator and water vapor is driven off to a condenser. The cooled water vapor then passes through an expansion valve, where the pressure reduces. The low-pressure water vapor then enters an evaporator, where ambient heat is added from a load and the actual cooling takes place. The heated, low-pressure vapor returns to the absorber, where it recombines with the lithium bromide and becomes a low-pressure liquid. This low-pressure solution is pumped to a higher pressure and into the generator to repeat the process.

Absorption chiller systems are classified by single-, double-, or triple-stage effects, which indicate the number of generators in the given system. The greater the number of stages, the higher the overall efficiency. Double-effect absorption chillers typically have a higher initial cost, but a significantly lower energy cost, than single-effect chillers, resulting in a lower net present worth. Triple-effect systems are underdevelopment.

Chillers employ cogeneration thermal output during cooling periods when heating uses are limited to domestic hot water loads or zonal heating, which may be small in many building types. These units involve a complex cycle of absorbing heat from the cogeneration system to create chilled water. The waste heat from the cogeneration system is used to boil a solution of refrigerant/absorbent. Most systems use water and lithium bromide for the working solution. The absorption chiller then captures the refrigerant vapor from the boiling process and uses the energy in this fluid to chill water after a series of condensing, evaporating, and absorbing steps are performed. This process is essentially a thermal compressor, which replaces the electrical compressor in a conventional electric chiller. In so doing, the electrical requirements are significantly reduced, requiring electricity only to drive the pumps that circulate the solution by single-effect chillers (Table II).

Second, engine-driven chillers (EDCs) are essentially conventional chillers driven by an engine instead of an electric motor. They employ the same thermodynamic cycle and compressor technology as electric chillers but use a gas-fired reciprocating engine to drive the compressor. As a result, EDCs can provide cooling economically where gas rates are relatively low and electric rates are high. EDCs also offer better variable-speed performance, which yields improved partial load efficiencies (Table III).

Third, desiccant dehumidifiers use thermal energy to dehumidify and to some extent control the indoor air quality of building space. With the recent push toward building humidity control, manufacturers of desiccant dehumidification systems have developed new packages designed to treat building ventilating air efficiently. A desiccant dehumidifier uses a drying agent to remove water from the air being used to condition building space. Desiccants can work in concert with chillers or conventional air-conditioning systems to significantly increase energy system efficiency by allowing chillers to cool low-moisture air. Desiccants can run off the waste heat from distributed generation technologies, with system efficiency approaching 80% in cogeneration mode.

The desiccant process involves exposing the desiccant material (e.g., silica gel, activated alumina, lithium chloride salt, and molecular sieves) to a high relative humidity air stream—allowing it to attract and retain some of the water vapor—and then to a lower relative humidity air stream, which draws the retained moisture from the desiccant. In some applications, desiccants can reduce cooling loads and peak demand by as much as 50%, with significant cost savings.

In stand-alone operations, a solid (dry) desiccant removes moisture from the air; a heat exchanger and evaporative cooler then chill the air. A desiccant dehumidifier wheel, composed of lightweight honeycomb or corrugated matrix material, dries the air as the wheel rotates through the supply air before reaching the building. Once the desiccant material reaches its saturation point, heat must be added or the material must be replaced to regenerate the moisture-absorbing capability. Natural gas, waste heat, or solar energy can be used to dry moisture that

TABLE II

Cost and Performance of Single-Effect, Indirect-Fired Absorption Chillers[a]

Tons	Cost ($/ton)	Electric use (kW/ton)	Thermal input (Mbtu/ton)	Maintenance cost ($/ton annual)
10–100	700–1200	0.02–0.04	17–19	30–80
100–500	400–700	0.02–0.04	17–18	20–50
500–2000	300–500	0.02–0.05	17–18	10–30

[a] *Source.* Resource Dynamics Corporation (2002).

TABLE III

Cost and Performance of Engine-Driven Chillers[a]

Tons	Cost ($/ton)	Electric use (kW/ton)	Thermal input (Mbtu/ton)	Maintenance cost ($/ton annual)
10–100	800–1050	0.05–0.07	9–12	45–100
100–500	650–950	0.01–0.05	8–11	35–75
500–2000	450–750	0.003–0.01	7–8	25–60

[a] *Source.* Resource Dynamics Corporation (2002).

is either absorbed or collected on the surface to regenerate the material.

Liquid absorption dehumidifiers spray the air with a desiccant solution that has a lower vapor pressure than that of the entering air stream. The liquid has a vapor pressure lower than water at the same temperature, and the air passing over the solution approaches this reduced vapor pressure. The conditioner can be adjusted so that it delivers air at the desired relative humidity. Liquid desiccant materials include lithium chloride, lithium bromide, calcium, chloride, and triethylene glycol solutions (Table IV).

Finally, hybrid power systems combine different power generation devices or two or more fuels for the same device. When integrated, these systems overcome limitations inherent in either one. Hybrid power systems may feature lower fossil fuel emissions as well as continuous power generation when intermittent renewable resources, such as wind and solar, are unavailable.

Solar concentrators/chiller systems use sunlight, which is plentiful in many regions of the country and is a load follower for afternoon cooling demand, to power an absorption cooling in low- to medium-temperature-driven systems, with either nontracking or tracking solar energy collectors and a two-stage absorption chiller. The resultant hybrid systems can produce an average cooling season solar fraction of at least 60% (40% of the cooling load supplied by natural gas) so that the net primary source fuel COP for the fossil fuel used is 3.0. Therefore, solar/gas absorption cooling systems will consume one-third the average primary source energy used for electric air-conditioning in the United States.

The widespread use of solar/thermal systems would have a significant impact on reducing the demand for electric power and for extending natural gas reserves. Every 1000 tons of installed solar/natural gas absorption air-conditioning equipment relieves the electric power grid of an average of 1 MW of demand.

3. APPLICATIONS AND MARKETS

New cogeneration project growth throughout the United States has been curtailed because of natural gas and oil prices and price volatility, making spark spread an unreliable cogeneration potential indicator. Natural gas price volatility and historic high prices have dampened new cogeneration natural gas-fueled projects in general. Market factors causing these price excursions are complex and not fully understood. Concurrently, there has been a similar increase in oil prices.

Because of the uncertainty in fuel prices and availability, spark spread is a difficult dimension to quantify or qualify, and industry growth should take precedence over other standard measures of cogeneration economic feasibility such as spark spread—the cost differential between electricity generated from central plants vs distributed generation—according to the American Council for an Energy-Efficient Economy (ACEEE).

Cogeneration systems can be especially useful in areas of the country in which development is constrained due to poor air quality. This is particularly significant in older, industrial cities such as Chicago. In air emission zones in California, the Northeast, and the Midwest, new development can create the need for emissions offsets. Cogeneration systems can meet this requirement, especially when using output-based emission standards that capture these system's high fuel efficiencies.

3.1 Utilities

Grid reliability could be a strong driver for cogeneration growth; for example, New York City's power system needs an additional 7100 MW to enhance reliability, according to the Center for Business Intelligence. With utility restructuring, access to downtown generation capacity from a cogeneration

TABLE IV

Cost and Performance of Desiccant Dehumidification Systems[a]

SCFM	Cost ($/SCFM)	Thermal input (hourly Btu/SCFM)	Maximum latent removal (hourly Btu/SCFM)
1500–5000	8–18	30–100	30–60
5000–10,000	6–11	30–100	30–60
10,000+	6–9	30–100	30–60

[a] *Source.* Resource Dynamics Corporation (2002).

facility may be more attractive and less problematic to the local electric utility in terms of projected load loss.

Addressing the need for distributed energy systems to interconnect to the electric utility grid, and to obtain supplemental and backup power, is critical to the success of cogeneration projects throughout the United States. Including multiple, dispersed generating and cogeneration units throughout the grid is a relatively new concept—one that has to be incorporated into the existing technical, regulatory, and institutional framework.

Some utilities are considering investing in the cogeneration value proposition to not only help them meet air quality objectives in their service territories but also to respond to customer service demands. Some utilities understand that cogeneration systems can significantly lower power plant, offer affordable incremental power costs, help optimize natural gas resources, hold gas costs down, and may open the door to new thermal energy business.

Also, if regulators do not permit, or if utilities disincent, cogeneration competition on an interconnected basis, end users will be forced to install the on-site systems on an "islanded" basis, leaving electric utilities out of the energy business. Now is an opportune time for utilities to embrace cogeneration as part of a national energy security campaign and as part of their own strategic business plans.

3.2 Industrial

Cogeneration has been most successful in large industrial applications that require large amounts of steam, but it is also well suited to use in hospitals, laundries, and health clubs. Since 1980, approximately 50,000 MW of cogeneration capacity has been built in the United States.

Cogeneration in the large to medium industrial market is typically found in the process industries, such as petroleum refining, pulp and paper, and chemicals. These systems have installed electricity capacities >25 MW (often hundreds of megawatts) and steam generation rates measured in hundreds of thousands of pounds of steam per hour. This sector represents the largest share of the current installed cogeneration capacity in the United States and is the segment with the greatest potential for short-term growth. Some facilities of this type are merchant power plants using combined cycle configurations. They are owned by an independent power producer that seeks an industrial customer for its waste stream and sells excess electricity on the wholesale market.

Thousands of boilers provide process steam to a broad range of small industrial U.S. manufacturing plants. These boilers offer a large potential for adding new electricity generation between 50 kW and 25 MW by either modifying boiler systems to add electricity generation (e.g., repowering existing boilers with a combustion turbine) or replacing the existing boiler with a new cogeneration system. Small manufacturers represent an important growth segment during the next decade.

Various industrial markets, including petrochemical, food products, bioproducts, and biotech, rank as the highest priority markets for future industry growth. The petrochemical market is the largest cogeneration market sector nationally, representing approximately 40% of existing cogeneration capacity. Bulk chemicals are in economic decline domestically. Some sectors of the chemicals industry are continuing to experience significant growth, including pharmaceuticals, specialty chemicals, and bioproducts including ethanol and biofeedstocks.

Food products manufacturing is a fast growing and stable market with demonstrated opportunity for cogeneration. The ACEEE has identified food products as one of the most geographically dispersed industry groups, with a significant presence in almost all states. In contrast to wood products and bulk chemicals industries, which are in economic decline domestically, the food products industry is among the fastest growing industry groups.

The growing biotech industry has overtaken the computer and semiconductor industry as a leader in projected economic growth. Both sectors have similar power reliability and thermal management needs. The biotech market has not experienced the same slowdown that has been seen in the computer-related markets.

3.3 District Energy

District energy systems are a growing market for cogeneration because these systems significantly expand the amount of thermal loads potentially served by cogeneration. District energy also has a major added benefit of reducing the requirement for size and capital investment in production equipment due to the diversity of consumer loads. These systems also tend to use larger and more efficient equipment and can take advantage of such things as thermal energy storage that are not economically effective on a small scale. Additionally, district energy systems aggregate thermal loads, enabling more cost-effective cogeneration. District energy systems may be installed at large, multibuilding sites, such as universities, hospitals, and government complexes. District energy systems can also serve as merchant thermal systems providing heating (and often cooling) to multiple buildings in urban areas.

According to the International District Energy Association (IDEA), there are three primary markets for district energy: colleges and universities, downtowns, and airports. The college and university market seems to be the most promising due to several factors. Colleges and universities are installing cogeneration in response to campus load growth, asset replacement of aging boiler capacity, and the favorable economics from fuel efficiency improvements with cogeneration.

Frequently, large campuses centralize heating and cooling and have large electrical loads. Many options are available for satisfying this electrical, heating, and cooling demand. A common way is to size a boiler for the entire campus's heating load. Large facilities may be able to negotiate attractive fuel contracts that lower heating costs and may choose to oversize the boiler, enabling the use of excess steam to power a steam turbine for electrical generation or a steam-driven chiller. This cheap, on-site secondary power source (steam) may make the generation of electricity or cooling less expensive than conventional methods such as purchasing from the grid and using conventional electric chillers.

The University of North Carolina at Chapel Hill, the University of Texas at Austin, and Cornell University represent examples of this situation. In each case, the universities have large boilers making steam for a variety of uses. Each campus ties the boilers to a steam turbine and the campus's heating system. The steam turbine generates electricity for campus use, reducing the dependency on the local grid and saving operating dollars for the campus. This cheap electricity may also be used to operate electric chillers for cooling, further reducing cooling costs. The remainder of the steam is used for heating of domestic water and all the campus's office, classroom, and dormitory spaces. Steam generated by the boilers may also be used by steam-driven chillers for cooling campus living space.

IDEA reports that college and university markets have significant pressures for "greening the campus," especially in the Northeast. Princeton, MIT, Rutgers, and the New Jersey Higher Education Partnership for Sustainability are exemplary, but there are other promising college and university examples throughout the country.

IDEA also notes that many colleges and universities conduct critical research and therefore need reliable power to avoid losing months or years of research and to assure students and the community that test animals are being properly cared for. Cogeneration could provide both power reliability and space conditioning.

Major urban centers are also a very promising market for adding cogeneration to existing district energy systems. Many district energy steam plants were originally cogeneration facilities that generated both power and steam when owned by the local electric utility. There is a growing need for local grid support in light of utility divestiture of generating capacity coupled with solid market growth in downtown district energy systems.

Airports represent another promising opportunity; these facilities are often in NO_x nonattainment areas and face significant emissions pressure from both regulators and the community. With large space conditioning and electrical load with long hours of operation, airports are often well suited to add cogeneration to their district energy systems.

3.4 Federal Government

The federal government is the largest energy consumer in the United States, with new mandates to meet increased demand, reduce peak operating costs, enhance energy security, and improve the reliability of electric power generation through DG and cogeneration. The Federal Energy Management Program was created to reduce government costs by advancing energy efficiency, water conservation, and the use of solar and other renewable energies. Executive Order 13123, "Greening the Government through Efficient Energy Management,: specifies that federal facilities shall use combined cooling, heat, and power systems when lifecycle costs indicate that energy reduction goals will be achieved. This market sector resembles the district energy sector, although the two sectors have different drivers and budgetary pressures.

According to the ACEEE, the largest cogeneration potential at federal facilities has been identified at military facilities, many of which are managed by various agencies within the Department of Defense.

3.5 Commercial/Institutional

The United States consumed more than 94 quadrillion Btus of energy in 2000. Of this total, commercial buildings accounted for more than 16 quadrillion Btus—equivalent to the amount of energy of gasoline consumed in the United States in 1 year. The growth of the economy, as well as the nation's increasing population, is leading to greater numbers of larger and more energy-intensive homes and commercial buildings, resulting in increased energy consumption in this sector.

Several reports indicate CHP potential and interest in the large commercial office buildings, hospital/health care, supermarket, hotel/motel, restaurant, and large retail markets. Some markets appear to have strong drivers to adopt CHP; for example, grocery stores have significant dehumidification loads to keep water from condensing on freezer displays, and hospitals and health care facilities have significant reliability needs. Schools appear to be a prime market for desiccant dehumidifiers, which control humidity and therefore reduce the likelihood of mold and bacteria growth.

In 2002, the Federal Communications Commission (FCC) adopted a plan that will give consumers access to digital programming over television by requiring off-air digital television (DTV) tuners on nearly all new TV sets by 2007. By enacting a 5-year rollout schedule that starts with larger, more expensive TV sets, the FCC is minimizing the costs for equipment manufacturers and consumers but creating new demands on power reliability for TV stations.

Commercial brownfield sites, or "real property, the expansion, redevelopment, or reuse of which may be complicated by the presence or potential presence of a hazardous substance, pollutant, or contaminant," are an appealing market for cogeneration. Redeveloping contaminated brownfield sites removes serious environmental hazards. Cogeneration can play an important role in the effort. Suzanne Watson of the Northeast-Midwest Institute notes that "with the right new technology fixes, there is the potential to create developments that improve the natural environment. Cogeneration is one of these technology solutions."

The DOE is helping manufacturers work together to integrate their individual CHP components into easier to use packages and to make "plug-and-play" packages more readily available in the commercial marketplace. On August 8, 2001, Energy Secretary Abraham announced awards to seven industry teams for research, development, and testing of "first-generation packaged cooling, heating, and power systems.

3.6 Residential

The DOE and the U.S. Department of Housing and Urban Development (HUD) are currently encouraging residential cogeneration units for individual or aggregated loads, which could be achieved in the multifamily markets under HUD jurisdiction.

There is significant potential for cogeneration in the privately owned assisted and public housing multifamily markets, where electric and thermal loads are already aggregated. According to HUD, the assisted housing market has a greater number of properties with less than 50 units, as opposed to the public housing market (20 units was used as a reasonable cogeneration entry point for both markets). In the assisted housing market, there are approximately 109,000 HUD properties with 20–49 units and 53,000 properties with more than 50 units. In the public housing market, there are approximately 4535 HUD properties with 20–49 units and 7154 properties with more than 50 units.

4. REGULATORY AND RATE STRUCTURE ISSUES

4.1 Standardizing Interconnect

Addressing the need for distributed energy systems to interconnect to the electric utility grid, and to obtain supplemental and backup power, is critical to the success of cogeneration projects. Until recently, the U.S. electric system was based on central station generation, tied to customers through the use of transmission lines. Including multiple, dispersed generating, and cogeneration units throughout the grid is a relatively new concept—one that has to be incorporated into the existing technical, regulatory, and institutional framework.

The DOE has been working for several years with industry through the Institute of Electrical and Electronics Engineers (IEEE) to develop a uniform national standard for interconnection. The IEEE P1547, "Standard for Distributed Resources Interconnected with Electric Power Systems," will provide interconnection guidelines for performance, operation, testing, safety, and maintenance. It is hoped that many utilities and state utility regulators will adopt it, giving their customers more power generation choices.

The Federal Energy Regulatory Commission has a rule making under way to establish standards and practices for interconnection of DG systems. The comment period for the proposed rule closed in early 2002.

In some cases, cogeneration projects would be able to proceed if the interconnection procedures for generators were standardized through a process of law making or rule making with the state simplifying, or at least setting, a recognized standard for the process. Both Texas and California have developed state-level interconnection standards (CEC 2001 and Texas 2000). These standards represent a compromise between utilities and generators. Although the guidelines differ in some specifics and main motivation, both create a relatively rapid and standard procedure for all generators seeking to connect to the grid.

The Texas Public Utility Regulatory Act of 1999 granted all utility customers access to DG. On February 4, 1999, the Texas Public Utility Commission (PUCT) adopted interconnection standards (PUCT 1999). PUCT continued to investigate DG and in May 2002 published a guidebook to interconnection in the state.

California's decision to expedite the streamlining of interconnection procedures resulted from their well-publicized energy problems in 2000 and 2001. Although there are many reasons why California had to resort to rolling blackouts, increasing DG is a way to remove pressure from the grid and reduce the likelihood that the event will be repeated.

The California Energy Commission (CEC) completed a state DG plan (CEC 2002), which included a plan for creating standardized guidelines for interconnection to the grid. In 2002, the CEC's Rule 21 Interconnection Working Group completed this guideline. Soon thereafter, the California Public Utilities Commission adopted the standards. Individual utility versions of the interconnection standards can be found at www.energy.ca.gov.

4.2 Output-Based Standards

Implementation of output-based standards, pollution allowances per kilowatt-hour, or useful electrical and thermal energy recognizes that cogeneration systems achieve significant reductions in environmental emissions due to their much higher efficiencies of fuel conversion relative to conventional separate heat and power systems. These standards would measure environmental emissions, or the rate at which emissions are discharged by the technology to the environment per unit of total usable energy produced by the technology as opposed to emissions per unit of fuel consumed. These standards should value electric and nonelectric output equally and be independent of fuel source. They should not discriminate against different ownership situations.

A promising development that the U.S. Environmental Protection Agency (EPA) is sponsoring is "demand response" run by regional transmission organizations. End users can bid back to the grid pool to reduce their electricity use by using on-site power or energy efficiency at set rates.

4.3 Time-of-Use Pricing

Time-of-use pricing is a rate structure intended to encourage customers to shift their energy use away from peak hours when wholesale electricity prices are highest. Real-time pricing can be readily used in large commercial and industrial facilities to manage thermal loads, making cogeneration more cost-effective. Chilled water and ice thermal energy storage, for example, are effective means to shift load to off-peak periods.

5. U.S. FEDERAL GOVERNMENT SUPPORT

The U.S. federal government is supporting cogeneration at the highest administration levels as well as through various agency programs and association support.

5.1 Administration

Released in May 2001, the Bush administration's National Energy Plan recognized the important role that cogeneration can play in meeting national energy objectives and maintaining comfort and safety of commercial and office buildings. Section 3–5 states:

> *A family of technologies known as combined heat and power (CHP) can achieve efficiencies of 80% or more. In addition to environmental benefits, CHP projects offer efficiency and cost savings in a variety of settings, including industrial boilers, energy systems, and small, building scale applications. At industrial facilities alone, there is potential for an additional 124,000 MW of efficient power from gas-fired CHP, which could result in annual emissions reductions of 614,000 tons of NO_x emissions and 44 million tons of carbon equivalent. CHP is also one of a group of clean, highly reliable distributed energy technologies that reduce the amount of electricity lost in transmission while eliminating the need to construct expensive power lines to transmit power from large central power plants.*

5.2 U.S. Environmental Protection Agency

Currently, power plants are responsible for two-thirds of the nation's annual SO_2 emissions, one-fourth of the NO_x emissions, one-third of the mercury emissions, and one-third of the CO_2 emissions, a leading greenhouse gas. These emissions contribute to serious environmental problems, including global climate change, acid rain, haze, acidification of waterways, and eutrophication of critical estuaries. They also contribute to numerous health problems, such as chronic bronchitis and aggravation of asthma, particularly in children.

The EPA understands that resolving these problems must start with pollution prevention, which equates to using fewer energy resources to produce goods and services. The National Energy Plan includes four specific recommendations to promote efficient cogeneration, three of which were directed to the EPA for action:

1. Promote cogeneration through flexible environmental permitting.
2. Issue guidance to encourage development of highly efficient and low-emitting cogeneration through shortened lead times and greater certainty.
3. Promote the use of cogeneration at abandoned industrial or commercial sites known as brownfields.

As a follow-up to these recommendations, the EPA joined with 18 *Fortune* 500 companies, city and state governments, and nonprofits in February 2002 in Washington, DC, to announce the EPA Combined Heat and Power Partnership. Cogeneration advances cogeneration as a more efficient, clean, and reliable alternative to conventional electricity generation.

As another follow-up to the National Energy Plan, the EPA is continuing work on a guidance document that clarifies how source determinations should be made for district energy and cogeneration projects under the New Source Review and Title V permitting regulations. The EPA intends for this guidance to help permitting officials and project developers streamline the development and permitting of these projects nationwide.

Through multiple programs, the DOE and its network of national laboratories have been working with manufacturers, end users, and other government offices to support increased use of cogeneration technologies.

5.3 U.S. Department of Energy

The DOE's Office of Distributed Energy Resources coordinates the CHP Initiative to raise awareness of the energy, economic, and environmental benefits of cogeneration and highlight barriers that limit its increased implementation. The initiative supports a range of activities, including regional, national, and international meetings; industry dialogues; and development of educational materials.

The DOE is also helping manufacturers work together to integrate their individual cogeneration components into easier to use packages and to make plug-and-play packages more readily available in the commercial and industrial marketplace. On August 8, 2001, Energy Secretary Abraham announced awards to seven industry teams for research, development, and testing of first-generation packaged cooling, heating, and power systems.

Integrated energy systems (IES) is a new approach to integrating all types of energy technologies into a building's energy system, including DG, cogeneration, HVAC, doors, windows, distribution systems, controls, insulation, building materials, lighting, and other building equipment. The link between building design and energy use is key to IES. As a DOE National User Facility, Oak Ridge National Laboratory (ORNL) provides manufacturers access to the user facility to help them evaluate and optimize the performance of IES and develop IES certification protocols.

The IES test center at the University of Maryland is designed to show building owners and managers how to save energy without sacrificing comfort in their buildings. The center is testing various cogeneration systems in an office building on campus. Phil Fairchild, program manager for Cooling, Heating, and Power for ORNL, explains,

> *What we're trying to do at the test center is to put together different pieces of equipment, use recoverable energy, and in doing so, understand the science of integration so that we can assist or better advise manufacturers about how to integrate the equipment in the future. If we reach our goals, manufacturers will have a better incentive, in terms of efficiency gains, to offer packaged or modular systems to the commercial building sector.*

The DOE is supporting a series of regional CHP application centers to provide application assistance, technology information, and education to architects and engineering companies, energy services companies, and building owners in eight states. The Midwest CHP application center was the first, established in March 2001 as a partnership between the DOE, the University of Illinois at Chicago's Energy Resources Center, the Gas Technology Institute, and ORNL. The center's educational material includes a Web site, accredited training classes, engineering tool kits, case studies, and application screening tools.

The DOE's Industrial Technologies Program (ITP) coordinates a combustion program with the industrial combustion community. The program promotes research and development of combustion systems used to generate steam and heat for vital manufacturing processes, heat-processing materials from metals to chemical feedstocks, and change the mechanical and chemical properties of materials and products.

The Steam Challenge is a public–private initiative sponsored by ITP. The goal of Steam Challenge is to help industrial customers retrofit, maintain, and operate their steam systems more efficiently and profitably. Steam Challenge provides steam plant operators with tools and technical assistance while promoting greater awareness of the energy and environmental benefits of efficient steam systems.

SEE ALSO THE FOLLOWING ARTICLES

Coal, Fuel and Non-Fuel Uses • Conservation of Energy, Overview • District Heating and Cooling • Energy Efficiency, Taxonomic Overview • Fuel Cells • Industrial Energy Efficiency • Solar Thermal Power Generation • Thermal Energy Storage

Further Reading

Garland, P., Hinrichs, D., and Cowie, M. (2003). Integrating CHP into commercial buildings. *Cogeneration On-Site Power Production*, May/June.

Hinrichs, D. (2002). CHP in the United States: Gaining momentum. *Cogeneration On-Site Power Production*, July/August.

Jimison, J., Thornton, R., Elliott, N., and Hinrichs, D. (2003). Emerging drivers for CHP market development in the U.S. *Cogeneration On-Site Power Production*, July/August.

Oak Ridge National Laboratory (2002). "Analysis of CHP Potential at Federal Sites." Oak Ridge National Laboratory, Oak Ridge, TN.

Onsite Sycom Energy (2000). "The Market and Technical Potential for Combined Heat and Power in the Commercial/Institutional Sector." Onsite Sycom Energy.

Onsite Sycom Energy (2000). "The Market and Technical Potential for Combined Heat and Power in the Industrial Sector." Onsite Sycom Energy.

Resource Dynamics Corporation (2002). "Integrated Energy Systems (IES) for Buildings: A Market Assessment–Final Report." Resource Dynamics.

U.S. Combined Heat and Power Association, U.S. Department of Energy, and U.S. Environmental Protection Agency (2001). "National CHP Roadmap: Doubling Combined Heat and Power Capacity in the United States by 2010." U.S. Combined Heat and Power Association, U.S. Department of Energy, and U.S. Environmental Protection Agency.

Combustion and Thermochemistry

SHEN-LIN CHANG[†]
Argonne National Laboratory
Argonne, Illinois, United States
CHENN QIAN ZHOU
Purdue University Calumet
Hammond, Indiana, United States

1. Introduction
2. Combustion Elements
3. Thermochemistry
4. Hydrodynamics and Mixing
5. Chemical Reactions
6. Radiation Heat Transfer
7. Tools for Combustion Analysis
8. Pollution Control Techniques
9. Concluding Remarks

Glossary

adiabatic flame temperature The highest flame temperature that is achievable under the condition of no external heat transfer.

combustion The burning of fuel and oxidant to produce heat and/or work.

computational fluid dynamics (CFD) Computer program for calculating fluid dynamics properties of a flow.

diffusive flame Separate fuel and oxidant inlets before burning takes place.

flammability Combustion reaction criteria.

gasification Incomplete fuel-rich combustion that produces CO and/or H_2.

ignition Initial conditions for sustained combustion.

Much solar energy is stored in hydrocarbons. Early humans, seeking protection and comfort, learned to convert hydrocarbon energy to fire to survive. Combustion, the most important form of energy conversion, is a chemical reaction between a hydrocarbon fuel and an oxidant. Combustion generates significant amounts of light, heat, and/or mechanical work. Furnaces and combustors have been developed to generate electricity and manufacture tools and equipment. Combustion engines have enabled transportation in every region of Earth and even into space. In recent years, concerns over energy availability and the environmental impacts of combustion processes have attracted more and more attention. To use energy resources efficiently and responsibly has become a main goal of many societies. To achieve this goal, knowledge of combustion science has to be learned and applied.

[†]Deceased.

1. INTRODUCTION

Combustion includes thermal, hydrodynamic, and chemical processes. Combustion starts with the mixing of fuel and oxidant, and sometimes in the presence of additional compounds or catalysts. The fuel/oxidant mixture can be ignited with a heat source; chemical reactions between fuel and oxidant then take place and the heat released from the reactions makes the process self-sustaining. Instability may occur, inhibiting or strengthening combustion. Combustion products include heat, light, chemical species, pollutants, mechanical work, and plasma. Sometimes, a low-grade fuel, e.g., coal, biomass, and coke, can be partially burned to produce a higher grade fuel, e.g., methane. The partial burning process is called gasification. Various combustion systems, e.g., furnaces, combustors, boilers, reactors, and engines, have been developed to utilize combustion heat, chemical species, and

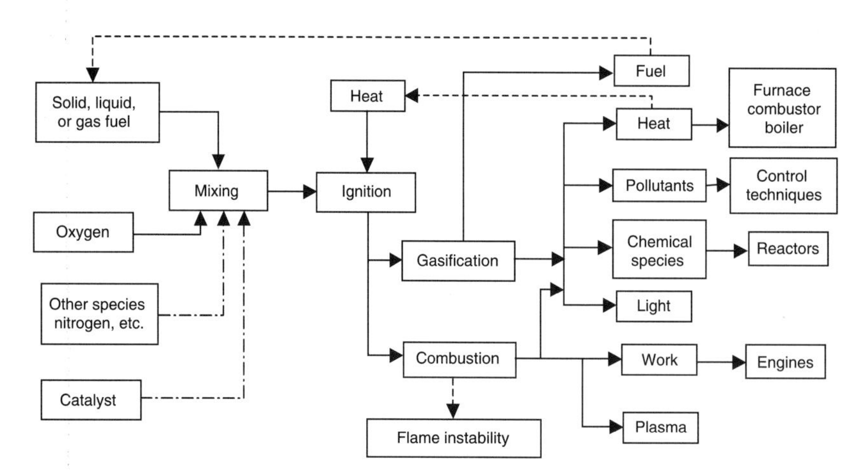

FIGURE 1 Typical combustion processes.

work. Figure 1 shows a flow diagram of typical combustion processes.

Advanced measurement/control devices have been traditionally used as tools for analysis of combustion systems. The analysis can lead to lower energy use and lower pollutant emissions by a system. In recent years, the computational analyses, including computer models of kinetics and fluid dynamics, have been used as an added tool for combustion analysis. The focus here is to assess the overall process of analyzing a combustion system.

A computational combustion analysis starts with the identification of the participating elements and their thermochemical properties and relations. This is followed by derivation of the governing equations and development of the required phenomenological models, including turbulent mixing, droplet evaporation, particle devolatization, radiation heat transfer, and interfacial interactions. Then, the governing equations are solved numerically on a computer. The results are validated with experimental data. Thus, the validated computer code can be used to evaluate the combustion system and to control combustion products.

2. COMBUSTION ELEMENTS

The elements (or species) involved in a combustion process include reactants, inert and intermediate species, products, and catalysts. Reactants generally include a fuel and an oxidant. Fuel can be in solid, liquid, or gas phase. Solid fuels include coal and biomass. Liquid fuels include hydrogen, gasoline, diesel, kerosene, and jet fuels. Gas fuels include natural gas, methane, and propane. The oxidant is mostly oxygen but may include F_2O, F_2O_2, NF_3, ClF_3, ClF_3, ClO_3F, O_3, and N_2F_4 in some rocket applications. The fuel and the oxidant can also be formed as a monopropellant in rocket applications. Catalysts are sometimes used to enhance combustion.

Inert species are not directly involved in the combustion process, but at high temperatures these species can be dissociated or ionized. These species include the nitrogen in the air and the ash in coal. Major combustion products include CO_2 and H_2O. Because these species can emit and absolve radiation energy, they are regarded as greenhouse gases that are related to the global warming issue. Many intermediate species, such as OH, CO, CH, are present during combustion. Combustion products also include some minor species, i.e., NO_x, SO_x, unburnt hydrocarbons (UHCs), and soot. These species are regarded as pollutants. NO_x and SO_x are believed to cause acid rain, UHCs play a role in smog, and soot chemicals, including polynuclear aromatics, may be carcinogenic.

In a combustion reaction, there can be so many combustion elements that it is impossible to analyze them all. Therefore, many combustion elements are lumped into one category for the convenience of analysis. For example, coal consists of thousands of hydrocarbons, but in most coal combustion analyses, the hydrocarbons are grouped into four categories: moisture, volatiles, char, and ash. The moisture is water, the volatiles represent all of the hydrocarbons that can be vaporized during combustion, the char includes all hydrocarbons that cannot be vaporized, and the ash is inert material, mostly metals.

3. THERMOCHEMISTRY

Assume that Reaction (1) represents a combustion reaction of I lumped reactants and J lumped products:

$$\sum_{i=1}^{I} n_i \text{ Reactants} \rightarrow \sum_{j=1}^{J} n_j \text{ Products.} \qquad (1)$$

In the reaction, the n_i values on the reactant side and n_j values on the product side are called the stoichiometric coefficients. The coefficients can be on a molar or a mass basis. On a molar basis, the unit is moles; on a mass basis, the unit is kilograms.

A computational combustion analysis calculates the heat, work, and/or species generated from a combustion process, which includes the combustion reaction, fluid dynamics, and heat transfer. Thermodynamic and chemical (thermochemical) properties of the combustion elements are needed to calculate the heat, work, and species. The thermochemical properties most often used in a combustion analysis include temperature, pressure, density, enthalpy, specific heat, viscosity, diffusivity, thermal conductivity, Gibbs free energy, molecular weight, species concentration, heat of combustion, latent heat, and equilibrium constant.

3.1 Equations of State

The thermochemical properties of a combustion species can be obtained by direct measurements or theoretical calculation. These properties are not all independent. Thermodynamic properties of a combustion species are generally functions of temperature and pressure. In most practical applications, these properties can be expressed in terms of temperature only. Sometimes, even constant property values are used. Most properties of a combustion species have been determined and tabulated in tables and other references produced by the joint interagency (Army, Navy, NASA, and Air Force) committee, JANNAF (formerly JANAF, before NASA was included). Properties of a mixture are determined by adding the properties of its components according to their concentrations. The partial pressure of a component is defined as the total pressure times its molar concentration.

Other than direct measurement, the thermochemical properties of members of a grouped ("lumped") combustion element (or species) can be obtained by adding the properties of its major constituents according to their proportion. For example, a fuel composed of 90% methane (CH_4) and 10% ethane (C_2H_6) becomes a lumped species $C_{1.1}H_{4.2}$. Its molecular weight is 17.1 kg/mol, calculated by adding 90% of 16 kg/mol (molecular mass of methane) and 10% of 30 kg/mol (molecular mass of ethane). Similarly, other properties of the lumped species can be obtained from those of the chemical species.

3.2 Chemical Equilibrium

At constant temperature and pressure, a combustion reaction such as Reaction (1) will reach an equilibrium state. The equilibrium constant K_p is defined as

$$K_p = \prod_{j=p} P_j^{n_j} / \prod_{i=r} P_i^{n_{ij}} \qquad (2)$$

in which the P_i values are partial pressures of the reactants (r) and products (p). The equilibrium constants of many combustion reactions have been determined and are tabulated in most combustion textbooks.

A combustion process generally consists of many reactions. The equilibrium species concentrations can be calculated from the equilibrium constants and the initial species concentrations. Bittker and Scullin developed a general chemical kinetics computer (GCKP) program to perform equilibrium calculations of hydrogen/air combustion and many other combustion processes.

3.3 Adiabatic Flame Temperature

Combustion of a fuel produces heat. When all of the heat is used to heat the products, without heat transfer from external sources, the final temperature is called the adiabatic flame temperature. The temperature can be determined by balancing the heat of combustion, the heat of dissociation, and the total enthalpy of the combustion products. At a combustion temperature higher than 2000 K, a small portion of the combustion products CO_2 and H_2O dissociates into intermediate species, such as CO, O, and OH, and the dissociation uses up a significant amount of the heat of combustion. Thus, the energy balance equation needs to be solved in conjunction with the equilibrium calculation to obtain an adiabatic flame temperature. Assuming the initial temperature and pressure are under standard conditions, the adiabatic flame temperature of a methane/air mixture is about 2200 K. Chang and Rhee developed a set of empirical formulations that can be used to estimate adiabatic flame temperature of lean fuel/air mixtures. The adiabatic flame temperature is an important thermochemical property in a combustion analysis because it is the upper bound of the temperature.

4. HYDRODYNAMICS AND MIXING

A fuel and an oxidant must mix to react. The mixing process is a part of flow dynamics. Gas phase mixing is mainly due to turbulence; liquid injection in droplet sprays is commonly used to enhance the mixing between liquid droplets and gas, and some solid fuels are pulverized to increase the mixing between the particles and gas.

4.1 Governing Equations

Combustion generates heat, work, and product species. In a computational combustion analysis, heat is calculated from the temperature and velocity; work is calculated from pressure and density; and the species are calculated from concentrations and velocities. The governing equations of these properties are derived from fundamental thermodynamic, hydrodynamic, and chemistry principles. The derivation can be on a fixed mass basis or a control volume basis. In the following discussion, a control volume approach is used.

The principles of mass, species, and momentum conservation are used to determine pressure, concentration, and velocity. The first law of thermodynamics (or conservation of energy) is used to determine temperature. The equation of state is used to determine density. In deriving the governing equations, many derived properties are created. For example, enthalpy is used in the energy equation and specific heat is used in the caloric equations of state. Equation (3) is a typical energy equation derived on a Cartesian coordinate system:

$$\sum_{i=1}^{3} \frac{\partial}{\partial x_i}\left(\rho u_i h + \frac{\mu_e \partial h}{\sigma_h \partial x_i}\right) = S_{\text{comb}} + S_{\text{rad}} + S_{\text{int}}, \quad (3)$$

in which x_i is a coordinate, ρ is density, u_i is velocity components, h is enthalpy, μ_e is effective viscosity, σ_h is a turbulent scale factor for the enthalpy, S_{comb} is the combustion heat release rate, S_{rad} is net radiation heat flux, and S_{int} is heat transfer from the liquid or solid phase. Effective diffusivity is used in the momentum, energy, and species equations. It is the sum of both laminar and turbulent diffusivities. The source terms of the governing equations need to be determined from various models. In the energy equation (3), there are three source terms. The combustion heat release rate needs to be calculated from combustion kinetics, radiation heat flux needs to be solved from a radiative transport equation, and the last term needs to be determined from an interfacial interactions model.

4.2 Turbulent Mixing

Diffusive flames, commonly used in a furnace or a combustor, are mostly turbulent. Turbulence is understood to be the small eddies that enhance mixing and combustion, but understanding the physics of the turbulence is still far from complete. An enclosure type of the $k-\varepsilon$ turbulence model is commonly used for combustion analysis. The model expresses the turbulent viscosity μ_t in terms of two turbulent parameters, kinetic energy k and dissipation rate ε,

$$\mu_e = \mu + \mu_t = \mu + C_\mu \rho k^2/\varepsilon,$$

in which C_μ is an empirical constant.

Lauder and Spalding proposed two additional transport equations to solve for these two turbulent parameters. Turbulent diffusivities for the enthalpy and species equations are also expressed in terms of the turbulent viscosity with a scaling factor. In a multiphase flow, solid particles or liquid droplets tend to inhibit eddy motion and reduce turbulent diffusion and the adjustment of the turbulent viscosity is needed. In addition, statistical turbulence models are used to describe the turbulence mixing effects. In this approach, probability density functions are chosen to represent the fluctuation of the flow properties. Large eddy and direct numerical simulations have been introduced with the hope of improving the description of turbulent flow.

4.3 Liquid Spray and Droplet Vaporization

Liquid fuel can be burned in pools or sprays. The burning processes start with the vaporization of the liquid fuel. Next, the fuel vapor is mixed with oxygen. Finally, the combustion reaction takes place. The pool burning is mostly fire. Because of limited contact area, the mixing of fuel vapor and oxygen is poor. Therefore, combustion is mostly fuel rich and a lot of smoke is produced. A fuel injector breaks up the liquid into tiny droplets and the high pressure makes droplets penetrate deep into the gas flow. Empirical models have been developed to describe the droplet size and number distributions off the tip of an injector. The penetration enhances the mixing of the fuel and the oxygen and the high ratio of surface area to volume speeds up the evaporation process.

Liquid fuel vaporizes in two modes. At low temperature, vaporization occurs due to the diffusion of mass. At a temperature higher than the boiling point, the liquid boils and the vaporization rate is controlled by the heat transfer rate from the surroundings to the liquid. In most combustion systems, the heat transfer is convective. Empirical correlations have been developed to calculate the interfacial heat transfer rate from other flow properties, such as, velocity, conductivity, and viscosity.

Liquid droplets are generally treated in a Lagrangian or an Eulerian approach. The Lagrangian approach takes one droplet at a time. It traces the motion of a single droplet in the flow and takes the average effect of all the droplets. Zhou and Yao also included the group effects on the spray dynamics. The Eulerian approach treats a group of droplets as a continuum. Governing equations are derived for each droplet group for solving the flow field.

4.4 Devolatization of Solid Fuel

Pulverized coal is used in many utility boilers. Due to the small particle size, pulverized coal can be transported pneumatically. Coal particles contain four major components: moisture, volatiles, char, and ash. When heated, the moisture is dried and the volatiles are devolatized. The volatile vapor mixes with oxygen and the mixture burns. The char left on the particles is burned with oxygen when contacted. The drying and devolatilization processes are controlled by the heat transfer from the gas to the particles. Similar to the liquid vaporization, the heat transfer is convective. Empirical correlations are used to determine the interfacial heat transfer rate. Similar Lagrangian and Eulerian approaches can be applied for particles.

5. CHEMICAL REACTIONS

The mixture of fuel and oxidant is burned in chemical reactions. A combustion reaction is the effective collisions of reactants (and catalyst), breaking the chemical bonds of the reactants, forming the products, and releasing heat. Three types of reactions are significant in a combustion analysis: ignition, heat release, and pollutant formation.

5.1 Ignition and Flammability

Fuel and oxidant can be self-ignited (autoignition) or by an external heat source. For example, diesel engine combustion is autoignition and gasoline engine combustion is ignited by a spark. At an ignition temperature, the combustion reactions generate heat and the heat is lost to the surroundings by conduction, convection, and radiation heat transfer. If the heat generation is larger than the heat loss, the combustion reactions are sustained.

There are mixture ratios of fuel and oxidant that the combustion will not sustain after the ignition source is removed. For premixed laminar flame, there exist flammability limits that are frequently defined as percentage fuel by volume in the mixture. The lower limit is the leanest mixture that will allow steady flame, and the upper limit represents the richest mixture. Flammability limits in both air and oxygen are found in the literature. For example, the lower flammability limit of methane is about 5% by volume.

5.2 Combustion Reactions

Assume that fuel is a hydrocarbon C_mH_{4n}. The combustion of the fuel and oxygen is shown in Eq. (4):

$$C_mH_{4n} + (m+n)O_2 \xrightarrow{k} mCO_2 + 2nH_2O. \quad (4)$$

The extent of reaction, ξ, is defined as the molar concentration of the consumed fuel. For each mole of ξ consumed, the sum $(m+n)$ moles of oxygen is consumed and m moles of CO_2 and $2n$ moles of H_2O are produced. The reaction rate $d\xi/dt$ of the combustion reaction can be expressed in an Arrhenius formulation:

$$\frac{d\xi}{dt} = -\frac{d}{dt}[C_mH_{4n}] = -\frac{1}{m+n}\frac{d}{dt}[O_2] = \frac{1}{m}\frac{d}{dt}[CO_2] = \frac{1}{2n}\frac{d}{dt}[H_2O], \quad (5a)$$

$$\frac{d\xi}{dt} = k_{30}\exp(-E/RT)[C_mH_{4n}][O_2]^{(m+n)}, \quad (5b)$$

in which [ ] represents molar concentration of a species, t is reaction time, k_0 is a preexponential constant, E is an activation energy, and R is the universal gas constant. The preexponential constant and the activation energy are empirical constants extracted from experimental data. Thus, the species consumption/generation rates can be obtained from Eqs. (5a) and (5b) and the heat release rate becomes

$$dq_{comb}/dt = (d\xi/dt)H_o, \quad (6)$$

in which H_o is the heat of combustion.

5.2.1 Gasification Reactions

Gasification is an incomplete combustion of the fuel. In a fuel-rich combustion, the intermediate species such as CO becomes a major product species. If water is added to the reaction, hydrogen is also produced. Because both CO and H_2 can be burnt later with oxygen and can release heat, the gasification has been used to convert low-grade fuels, such as, coal and biomass, to a higher grade gaseous fuel, CO and H_2. The gasification process is now also used to convert diesel or gasoline fuel to hydrogen for fuel-cell applications.

5.2.2 Heterogeneous Reactions

Heterogeneous reactions include char burning with oxygen and gaseous reactions on catalyst surfaces. Char burning is regarded as similar to a combustion reaction. Empirical constants are extracted from experimental data. The catalytic reaction is generally treated as a third-body reaction and an Arrhenius formula is used for analysis.

5.3 Pollution Reactions

In addition to the major heat release reaction, many intermediate species are produced in numerous combustion reactions. Pitz and Wesbrook have compiled hundreds of combustion reactions and the list is still growing. Some of the intermediate species attract more public attention because they are pollutants. Pollution is defined as the contamination of the environment. All matter emitted from combustion chambers that alters or disturbs the natural equilibrium of the environment must be considered as a pollutant. Pollutants of concern include particular matter, the sulfur oxides, unburned and partially burned hydrocarbons, nitrogen oxides, carbon monoxide, and carbon dioxide.

5.3.1 Soot Kinetics

Soot is formed in the fuel-rich region of a combustion flow. Because soot is a strong contributor to the radiation heat transfer, a soot model is needed to calculate soot concentration for the radiation heat transfer calculation. The following simplified soot model assumes that soot (st) is generated in the fuel-rich flame by the cracking of the fuel and is burnt up by oxygen. The soot formation/oxidation rate is expressed as

$$df_{st}/dt = A_{O2} f_{fu} \exp(-E_2/RT) - A_{O3} f_{st} f_{ox} T^{1/2} \exp(-E_3/RT). \quad (7)$$

5.3.2 NO_x Kinetics

NO_x is formed by combustion in the presence of nitrogen. There are three types of NO_x formation: thermal, prompt, and fuel. Thermal NO_x is formed in high-temperature reactions, prompt NO_x comes from reactions with intermediate species, and fuel NO_x comes from the nitrogen in the fuel, e.g., coal. The Gas Research Institute (GRI) has compiled a long list (over 300) of reactions related to NO_x formation. However, lumped reduce mechanisms with many less reactions are used in practice.

5.3.3 SO_x Formation

Many coals contain significant amount of sulfur. The sulfur reacts with oxygen during combustion and produces SO_x. Sulfur is also contained in many lubricants. In diesel combustion, SO is formed and becomes a major source of soot particle nucleation.

6. RADIATION HEAT TRANSFER

At combustion temperature, radiation is an important if not dominating mode of heat transfer. In a combustion process, radiatively participating species include soot and gaseous species such as CO_2 and H_2O. Heat is emitted, absorbed, and scattered in waves of various wavelengths. Assuming the scattering effect in the combustion flow is negligible, the radiative heat transfer becomes the balance of emissive and absorption powers. For each wavelength, local net radiation heat power is obtained by integrating the absorption of the incoming radiation from all other locations and subtracting the emitted power. A radiative transport equation is obtained by integrating the net radiation heat power over all wavelengths, Eq. (8):

$$q_r(x, y, z) = \int_0^{\infty} \left[\oint \kappa'_\lambda e'_{b\lambda}(T') e^{-\int \kappa'' dl''} dv' - \kappa_\lambda e_{b\lambda}(T) \right] d\lambda, \quad (8)$$

in which λ is wavelength, κ is volume absorptivity, l is optical length, v is a control volume, and $e_{b\lambda}(T)$ is the blackbody radiation function. The spectral volumetric absorptivities of these media must be determined from gas and soot radiation models.

6.1 Gas Radiation Model

The H_2O species has five strong radiation bands, centered at wavelengths of 1.38, 1.87, 2.7, 6.3, and

20 μm, and the CO_2 species has six bands, at 2.0, 2.7, 4.3, 9.4, 10.4, and 15 μm. A wideband model was introduced to calculate total band absorptance of these species. For each band, species concentrations, pressure, and temperature are used to determine a set of semiempirical parameters: the integrated band intensity, the bandwidth parameter, and the line width parameter. A semiempirical correlation is then used to calculate the total band absorptance from these parameters.

6.2 Soot Radiation Model

If the scattering is negligible, the Rayleigh-limit expression of the soot volume absorptance $\kappa_{s\lambda}$ can be used:

$$\kappa_{s\lambda} = \frac{36n^2k(\pi/\lambda)f_v}{[n^2(1+k^2)+2]^2+4n^4k^2}. \qquad (9)$$

Soot volume absorptance is proportional to the soot volume fraction f_v and inversely proportional to the wavelength. The optical refraction indices, n and k, are weak functions of wavelength that can be derived from classical electromagnetic theory.

7. *TOOLS FOR COMBUSTION ANALYSIS*

Measurement and testing have long been the ways to analyze and evaluate combustion systems. However, computational software has now become an additional tool for combustion analysis.

7.1 Computational Fluid Dynamics

Combustion flow calculations adopt a control volume approach to convert the governing equations to algebraic equations on a staggered, discretized grid system. The algebraic equations are solved iteratively with proper boundary conditions. Exit flow conditions have a strong impact on the convergence of the calculation. Because most flows are not fully developed, the exit flow conditions that satisfy global mass balance appear to improve numerical convergence the most.

The major species flow calculation uses Patankar's SIMPLER computational scheme to solve the pressure-linked momentum equations. A major species flow calculation is considered to have converged if the local and global mass balances are smaller than a set of predetermined criteria. The subspecies flow is solved using the flow properties calculated from the major species flow calculation. Free from the interactions of the pressure and velocity fluctuations, the calculation of the partially decoupled subspecies transport equations becomes very stable numerically.

Chang and colleagues developed an advanced methodology to solve the radiative transfer equation (RTE) in a Cartesian coordinate system. The methodology guarantees the integrity of the energy balance between emitting and absorbing powers. The RTE solution routine divides the radiation heat flux into many bands of wavelength. The blackbody radiation function can be discretized in the wavelength domain by using a closed form solution. For each wavelength band, a calculation is performed to balance the emitted and absorbed energy. For each node in the computational fluid dynamics (CFD) grid system, energy emitted is calculated. This emitted energy is then traced along the optical paths of all angles. The energy absorbed in every node (including the wall surface node) is calculated and added to the absorption energy of the initial node.

7.2 Kinetics Calculations

Many schemes have been proposed for the number of equations and reaction rates that may be used to simulate combustion reactions. One of examples is the GCKP program developed at the National Aeronautics and Space Administration (NASA). Another example is the CHEMKIN program, a widely used general-purpose package for solving chemical kinetics problems. CHEMKIN was developed at Sandia National Labs and provides a flexible and powerful tool for incorporating complex chemical kinetics into simulations of fluid dynamics. The latest version includes capabilities for treating multifluid plasma systems that are not in thermal equilibrium. These features allow researchers to describe chemistry systems that are characterized by more than one temperature, in which reactions may depend on temperatures associated with different species; i.e., reactions may be driven by collisions with electrons, ions, or charge-neutral species.

Both thermodynamic and kinetic databases are needed for the kinetics calculations. The GRI-Mech combustion model is a widely used database of detailed chemical kinetic data. The latest version is an optimized mechanism designed to model natural gas combustion, including NO formation and reburn chemistry, containing 325 reactions and 53 species. It provides sound basic kinetics and also furnishes the best combined modeling predictability of basic combustion properties.

7.3 Measurements

Measurements of flame properties can be made by the insertion of probes into a flame, by optical methods applied to emissions from the flames, or by laser probing. The techniques for probing in flames include temperature measurement by thermocouples, velocity measurement by pressure impact probes and particle-tracking photography, and species concentration measurements using suction probes followed by gas analysis.

The optical methods for making measurements in flames include interferometry, schlieren, and photographic methods. All of the probing methods may cause some form of disturbance to the fluid flow, heat transfer, and species concentrations. Optical methods provide information that is integrated over the optical length of the measurement and the spatial resolution is poor. Lasers provide noninvasive probing of flames and high spatial resolution of measurements. The measurement of temperature and species concentration in flames can be carried out by laser spectroscopy. Laser anemometers are widely used to measure velocity components, turbulent characteristics, and particle size in flames. Various advanced combustion sensors have been developed with the potential to enable direct measurement of key combustion parameters for improved combustion control.

8. *POLLUTION CONTROL TECHNIQUES*

Control of pollutant emissions is a major factor in the design of modern combustion systems. Pollution control techniques can be mainly classified as two types. One involves process modification to prevent or reduce the formation of pollutant emissions. The other is postcombustion treatment to remove or destroy pollutants from combustion before being emitted into the ambient air.

8.1 Process Modification

The techniques to modify combustion for reducing pollutant emissions include fuel changes and combustion modifications. Switching vehicles from gasoline to compressed natural gas, propane, or ethanol can reduce vehicular air pollutant emissions. Adding oxygenated compounds to motor fuels can change the combustion reaction to lower CO emissions significantly. Using low-sulfur fuels can reduce sulfur dioxides emissions.

The principle of combustion modification is widely used in NO_x emission control. Modifying burner configuration or operation conditions can change distributions of temperature, oxygen, and nitrogen to reduce the formation of NO_x. Such techniques include using low excess air, staged combustion, low NO_x burners, water injection, oxyfuel combustion, exhaust gas recirculation (EGR), and reburn. These techniques can be economical and effective. However, in some applications, combustion system modification alone is unable to reduce pollutant emission levels below legislated standards, and postcombustion treatment may be required.

To reduce SO_2 emissions, fluidized bed combustion, in which limestone particles are used as bed materials, is an alternative way to burn coal. Another combustion modification alternative for SO_2 reduction that is in the demonstration stage is the combined-cycle power plant.

8.2 Postcombustion Treatment

Postcombustion treatment involves installation of pollution control devices downstream of the combustor. For example, selective noncatalytic reduction (SNCR) and selective catalytic reduction (SCR) processes are widely used for NO_x reduction. These techniques involve a nitrogen-containing additive, either ammonia, urea, or cyanuric acid, that is injected and mixed with flue gases to effect chemical reduction of NO to N_2, without or with the aid of a catalyst. Other postcombustion treatment techniques utilize wet or dry limestone scrubbers and spray dryers to reduce SO_2 emissions; cyclones, electrostatic precipitators (ESPs), filters, and scrubbers for particle removal; or three-way catalytic converters for the control of NO_x, CO, and hydrocarbon (HC) emissions from automobiles.

To stabilize and ultimately reduce concentrations of the greenhouse gas CO_2, carbon sequestration (carbon capture, separation, and storage or reuse) is a major potential effective tool for reducing carbon emissions from fossil fuel combustion.

9. *CONCLUDING REMARKS*

This article is intended to guide a reader through the overall process of analyzing a combustion system, especially using the newly added computational tools, and to advocate the good use of combustion analyses to convert combustion processes for cleaner and more efficient energy production.

SEE ALSO THE FOLLOWING ARTICLES

Acid Deposition and Energy Use • Greenhouse Gas Emissions from Energy Systems, Comparison and Overview • Heat Transfer • Temperature and Its Measurement

Further Reading

Bittker, D. A., and Scullin, V. J. (1972). "General Chemical Kinetics Computer Program for Static and Flow Reactions, with Application to Combustion and Shock-tube Kinetics," NASA TN D-6586. NASA, Washington, D.C.

Chang, S. L., and Rhee, K. T. (1983). Adiabatic flame temperature estimates of lean fuel/air mixtures. *Combust. Sci. Technol.* **35**, 203–206.

Chang, S. L., Golchert, B., and Petrick, M. (2000). Numerical analysis of CFD-coupled radiation heat transfer in a glass furnace, No. 12084. *Proc. 34th Natl. Heat Transfer Conf.*, Pittsburgh, Pennsylvania.

Glassman, I. (1977). "Combustion." Academic Press, San Diego.

Kee, R. J., Rupley, F. M., Meeks, E., and Miller, J. A. (1996). "CHEMKIN-III: A Fortran Chemical Kinetics Package for the Analysis of Gasphase Chemical and Plasma Kinetics," Rep. SAND96-8216. Sandia National Laboratories, Livermore, CA.

Lauder, B. E., and Spalding, D. B. (1974). The numerical computation of turbulent flows. *Comput. Meth. Appl. Mech. Eng.* **3**, 269–289.

Patankar, S. V. (1980). "Numerical Heat Transfer and Fluid Flow." Hemisphere, Washington, D.C.

Pitz, W. J., and Westbrokk, C. K. (1986). Chemical kinetics of the high pressure oxidation of *n*-butane and its relation to engine knock. *Combust. Flame* **63**, 113–133.

Smith, G. P., Golden, D. M., Frenklach, M., Moriarty, N. W., Eiteneer, B., Goldenberg, M., Bowman, C. T., Hanson, R. K., Song, S., Gardiner, W. C., Jr., Lissianski, V. V., and Qin, Z. (2002). GRI-Mech 3.0. Available on the Internet at http://www.me.berkeley.edu/.

Stull, D. R., and Prophet, H. (1971). "JANAF Thermochemical Tables," 2nd ed., NSRDS-NBS 37. U.S. Dept. of Commerce, National Reference Standard Data System-National Bureau of Standards, Washington, D.C.

Turns, S. R. (2000). "An Introduction to Combustion: Concepts and Applications." McGraw Hill, New York.

Zhou, Q., and Yao, S. C. (1992). Group modeling of impacting spray dynamics. *Int. J. Heat Mass Transfer* **35**, 121–129.

Commercial Sector and Energy Use

J. MICHAEL MACDONALD
Oak Ridge National Laboratory
Oak Ridge, Tennessee, United States

1. Definition and Extent of Commercial Sector
2. Magnitude and Significance of Commercial Energy Use
3. Measuring Energy Performance
4. Performance Rating Systems
5. Energy Efficiency Variation and Improvement
6. Sectoral Data and Modeling
7. Sectoral Views Formed by Models
8. Driving Toward Energy Efficiency
9. Future Energy Performance and Use

Glossary

annual total energy The sum of all energy used from all sources in a year.

British thermal unit (Btu) Generically, the amount of energy or heat required to raise the temperature of 1 lb of water (about 0.5 quart or 0.5 liter) 1 degree Fahrenheit (equals about 1055 joules).

commercial sector The portion of buildings in a nation or the world including all buildings that are not residential, industrial, or agricultural.

electricity losses Energy lost in generation, transmission, and distribution of electricity.

empirical Obtained through physical measurement or observation.

energy intensity Annual total energy divided by some normalizing factor, usually gross floor area of a building.

energy performance An empirical value indicating the energy efficiency of one commercial building compared to other, usually similar, commercial buildings.

Energy Star rating system Energy performance rating systems developed by the United States Environmental Protection Agency for specific U.S. commercial building types, which as of June 2003 included offices, hospitals, primary and secondary schools, grocery stores, and hotels and motels, with adjustment capabilities for special spaces such as computer centers and parking garages.

exajoule (EJ) 10^{18} joules.

joule Watt-second, the energy to maintain 1 W for 1 sec.

kBtu 1000 British thermal units.

microdata Detailed survey data arranged in an electronic file for computer analysis.

normalization A method to adjust a quantity, in this case energy use, to account for allowable differences in operational or other factors, such as worker density, hours of operation, and personal computer density.

quads A quadrillion British thermal units; 1 quadrillion is 10 raised to the fifteenth power.

rating scale A numerical scale matched to a range of values of interest, such as normalized annual energy use.

regression An analysis method for statistically determining functional relationships between quantities that are correlated, often using empirical data.

standards Authoritative or legally fixed bases for comparison, valuation, or compliance.

subsectors Subsets of a sector; office buildings, for example, represent a subsector of the commercial sector.

Energy use in commercial buildings is complicated to understand, due to the wide range of building uses and ownership, variations in the size and complexity of energy systems, differences in energy system operation, and other factors. Due to this complexity, as national economies grow toward more specialized services and enterprise management, the share of national energy use held by the commercial sector also tends to grow relative to other sectors. Increased understanding of commercial building energy performance, probably through performance ratings and certifications, is needed to help reduce commercial sector energy growth relative to other sectors in advancing economies.

1. DEFINITION AND EXTENT OF COMMERCIAL SECTOR

The commercial sector is defined typically in terms of including everything else that other sectors do not include. The energy use breakout of national economies related to buildings typically covers residential and commercial buildings; where commercial buildings are typically all buildings that are not residential, industrial, or agricultural. Sometimes this broader grouping of commercial buildings is separated further into institutional and commercial, or governmental and commercial. Thus the commercial sector consists of buildings used by businesses or other organizations to provide workspace or gathering space or offer services. The sector includes service businesses, such as shops and stores, hotels and motels, restaurants, and hospitals, as well as a wide range of facilities that would not be considered commercial in a traditional economic sense, such as public schools, specialized governmental facilities, and religious organizations. Many other types of buildings are also included.

The wide range of building uses is one factor contributing to the complexity of the commercial sector. In the United States, a major national survey of commercial buildings, examining an extensive set of their characteristics and their energy use, is conducted every 4 years. The latest available survey, for 1999, includes in the detailed microdata over 40 different types of building uses that can categorize commercial buildings (Table I). Commercial buildings in the United States, where perhaps the most specialized services and enterprise management exist, are estimated to number 4–5 million, with a total gross floor area of over 6 billion m^2 (over 65 billion ft^2). The approximate breakout of commercial buildings (Table I), showing the range of building uses, the number of buildings, and total floor area estimates, provides an informative starting point for understanding the commercial sector in the United States and elsewhere. Although comparison of these estimates with other, more detailed estimates for subsectors, such as schools, would show some discrepancy in estimates, the scope and size of the commercial sector are well illustrated. For even the least developed countries, the same general range of facilities that comprise the commercial sector will exist, although the aggregate size relative to other sectors typically is smaller.

2. MAGNITUDE AND SIGNIFICANCE OF COMMERCIAL ENERGY USE

The amount of energy consumed in the commercial sector often must be estimated as a fraction of energy use in the combined residential and commercial sectors; national energy use in buildings is often tracked within the major sectors, categorized as industrial, transportation, and "other," with residential and commercial buildings aggregated and accounting for most of the energy use in this "other" sector. Thus some quick checks on world total energy consumption are useful. The units used to sum world energy use are not easily comprehended by most people, so the important knowledge to retain is the relative values. Total world energy consumption in the year 2000 was about 395 quads (1×10^{15} Btu), or about 420 EJ (exajoules). Fuel processing, non-energy use of fuels, and other losses reduce the total final energy consumption in the major energy-using sectors to about 270 quads (280 EJ).

Because the importance of energy use in the industrial and building sectors can be misunderstood if losses associated with generation and distribution of electricity are not included, comparisons that show both totals are useful. The estimated sectoral breakouts (Table II), without accounting for electricity losses, are 87 quads for industry, 71 quads for transport, and 109 quads for "other," which is primarily residential and commercial buildings. Adding approximate electricity losses (Table II) brings the totals to 122 quads for industry, 73 quads for transport, and 156 quads for "other," for a total of about 350 quads, or 370 EJ.

For the world overall in the year 2000, commercial sector energy use is approximately 30% of the "other" energy use, which amounts to a little over 30 quads (35 EJ) when electricity losses are not included. This energy use represents about 12% of the approximately 270 quads of total final energy consumption for the world. When electricity losses are included, commercial sector energy use is about 45 quads (50 EJ).

Although commercial sector energy use is only 12% of the world total, as economies develop, energy use in this sector tends to rise relative to other sectors and is one of the most difficult to reduce, due to complexities of systems, building ownership, and building uses. The rise in energy use relative to other sectors appears to result from the need for increasingly sophisticated facilities to handle activities in this sector as national economies advance, as well as from a concurrent rise in income within the sector relative to the cost of facilities.

The history of the commercial sector relative to the residential sector in the United States provides an example of this pattern of change. Specialization in services and enterprise management grew significantly in the United States throughout the last

TABLE I

U.S. Commercial Buildings, 1999; Approximate Breakout

Building use	No. of buildings (thousands)	Floor area (billions) (m^2)	Floor area (billions) (ft^2)
Administrative/professional office	503	0.83	8.92
Auto dealership/showroom	56	0.04	0.41
Auto service/auto repair	210	0.13	1.42
Bank/financial	128	0.09	0.97
Clinic/outpatient health	37	0.03	0.34
College/university	25	0.11	1.18
Courthouse/probation office	3	0.02	0.23
Doctor/dentist office	111	0.10	1.03
Dormitory/fraternity/sorority	35	0.17	0.80
Dry cleaner/laundromat	58	0.03	0.30
Elementary/middle/high school	230	0.61	6.52
Enclosed mall	3	0.15	1.64
Entertainment (theater/sports arena/nightclub)	35	0.07	0.78
Fire station/police station	52	0.04	0.48
Government office	33	0.11	1.21
Grocery store/food market	49	0.06	0.62
Hospital/inpatient health	7	0.15	1.65
Hotel	28	0.17	1.81
Jail/reformatory/penitentiary	6	0.02	0.26
Laboratory	27	0.04	0.42
Library/museum	28	0.05	0.52
Motel/inn/resort	61	0.10	1.04
Nonrefrigerated warehouse	589	0.89	9.56
Nursing home/assisted living	25	0.06	0.68
Other education	40	0.04	0.40
Other food sales or service	129	0.04	0.39
Other health care	22	0.03	0.31
Other lodging	4	0.02	0.17
Other office	25	0.03	0.37
Other public assembly	21	0.06	0.63
Other public order and safety	11	0.02	0.20
Other retail	88	0.04	0.42
Other service	118	0.08	0.82
Post office/postal center	25	0.03	0.33
Preschool/daycare	32	0.05	0.51
Recreation (gymnasium/bowling alley/health club)	88	0.11	1.22
Refrigerated warehouse	14	0.08	0.87
Religious worship	307	0.32	3.40
Repair shop	68	0.05	0.51
Restaurant (bar/fast food/cafeteria)	345	0.17	1.84
Social meeting center/convention center	132	0.11	1.22
Store	389	0.36	3.92
Strip shopping center	131	0.37	3.94
Vacant	253	0.18	1.90
Other	75	0.07	0.80
Total	4654	6.22	67.00

TABLE II

World Sectoral Energy Consumption in 2000[a]

	Without electric losses		With electric losses	
Sector	Quads	EJ	Quads	EJ
Industry	87	92	122	129
Transport	71	75	73	77
Other	109	115	156	165
Total	267	282	351	370

[a]Values are approximate and can also vary depending on the detail in accounting for different types of losses or fuel processing. Sectors may not add to totals due to rounding.

half of the 20th century. Allowing some time to pass after World War II, so that wartime effects and rationing-induced behaviors can be discounted, energy use in the commercial sector was about 50% of residential energy use in the year 1960. This ratio grew to about two-thirds by 1980, and in the year 2000 was over 83% (Fig. 1). Residential sector energy use in the United States was 20% of total national energy use in both 1960 and 2000, whereas commercial sector use increased from 10 to 17% of the total. Thus, commercial sector energy use was about 30% of sectoral energy use in the building or "other" category in the United States in 1960, but by 2000 had grown to 45% of the building sector total.

Variations among countries and location are dramatic. Although the commercial sector in the United States accounts for about 17% of all energy use, in China it accounts for about 5%. Commercial buildings in rural areas typically are less complicated

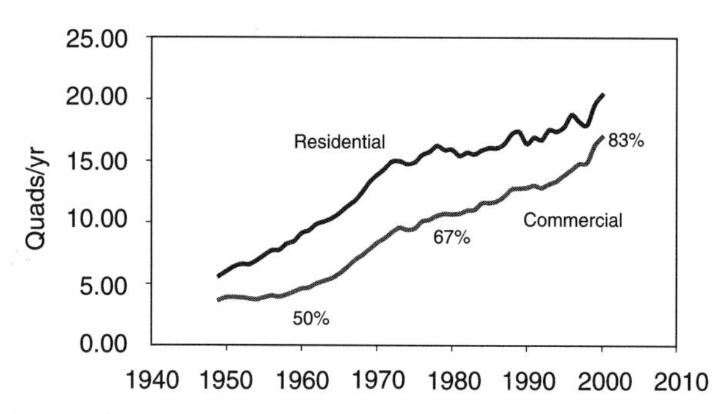

FIGURE 1 Comparison of commercial sector and residential energy use in the United States from 1960 to 2000; the ratio of commercial to residential energy use grew from 50 to 83%. Electricity losses are included.

and use less energy, as compared to those in metropolitan areas; over 80% of commercial sector energy use is in metropolitan areas.

3. MEASURING ENERGY PERFORMANCE

Interest in rating the real-life energy performance of buildings has increased in recent years, and the real-life efficiency performance rating of buildings is important for any sustainable energy future. The ability to compare the energy performance of one commercial building with that of another is important for determination of national and international energy efficiency because comparison allows meaningful measurements of potential relative improvements. This ability also may allow different classes of buildings to be analyzed together (e.g., offices and hospitals).

The European Union has been examining requirements for improving the energy performance of residential and commercial buildings, because a large potential improvement in energy performance has been determined to exist. Among the requirements examined is the establishment of a general framework for a common methodology for calculating the integrated energy performance of buildings.

The United States Environmental Protection Agency (EPA) has established an empirical energy performance rating system for some commercial building types, the Energy Star rating system, whereby a normalized energy performance rating scale is developed. The energy use of a specific building is normalized based on the factors in the method, and the normalized energy is compared to the performance rating scale. Buildings scoring in the top 25% on the scale have energy performance level that makes them eligible for consideration of award of an Energy Star label.

Commercial building energy performance, or energy efficiency, is often measured to a certain degree by building energy experts, and even many nonexperts, without using any real standards. To judge how well a specific building is doing, however, energy performance measurement should involve a comparison of building energy use to some type of standard, which in the past has typically been the energy use of other, similar buildings. The challenge over the years has been to determine a true standard for comparison and to determine what a "similar" building is. Because the historical methods of comparison had known limitations, building energy experts developed their own sense of what constitutes an energy-efficient building. This expert sense is based on experience with similar buildings, the types of activities within specific buildings, and any history of achieving reductions in energy use in comparable buildings. This expert knowledge has gaps and is not easily transferable, because it is usually based on several years of experience concerning expected patterns of energy use for different buildings and impacts of schedules, uses, geographic location, and system configurations. This expert knowledge is used to "adjust" the measure of the performance of a commercial building to provide a more informed measure of performance. However, this knowledge is ad hoc, with multiple practitioners probably arriving at differing assessments of the same building. In the end, the result is essentially a subjective expert opinion, albeit possibly a very good one, but also possibly not.

Five generic classes of building energy data analysis methods have been identified as useful in measuring the energy performance of commercial buildings:

1. Annual total energy and energy intensity comparisons.
2. Linear regression and end-use component models.
3. Multiple regression models.
4. Building simulation programs.
5. Dynamic thermal performance models.

All of these analytical approaches can be used to develop building energy performance measurement methods, but the most effective current approach in use today, based on results achieved, involves the third approach, multiple regression models. When calculating commercial building energy performance using multiple regression models, the effects of many factors can be modeled, potentially factoring out influences such as the number of people in a building or occupant density. The Energy Star rating system develops its performance rating scales using multiple regression models.

The limitations of the other methods include their inability to cover wide ranges of buildings without an inordinate amount of data. Some of the other methods require large volumes of data to develop empirical results. In the following discussion of performance rating systems, both simple annual total energy intensity comparisons (Method 1 above) and multiple regression method information will be addressed; the first is useful both as an example and as a well-understood quantity, whereas multiple regression analysis illustrates the current state-of-the-art methodology.

4. PERFORMANCE RATING SYSTEMS

Many people confuse building simulation and energy audits or energy assessments with energy performance ratings. Energy performance ratings are less detailed and provide much less information regarding potential causes of specific energy performance. Instead, what is provided is a true indication of overall energy performance relative to similar buildings. Highly technical assessments, including calibrated simulations, are a tremendous tool for diagnosing root causes of specific building energy performance. But these approaches typically provide only very limited information about performance relative to other buildings or relative to any ranking scale based on performance of similar buildings, and their complexity makes them impractical for extensive use in rating performance.

Performance rating can be done many ways. The EPA Energy Star rating system for buildings uses a percentile rating scale of 1 to 100 for a particular building type, with a rating of 75 or greater required to qualify for an Energy Star label. The energy performance rating of 1 to 100 can be obtained simply by using the rating tool. The rating scale is developed from a regression analysis of energy use versus key characteristics of the class of buildings against which energy use is to be normalized. A simple and straightforward way of quantifying and comparing building energy performance is accomplished by using the annual total energy and energy intensity data. Annual total energy is the sum of the energy content of all fuel used by the building in one year. Energy intensity (also called energy use intensity or index, EUI) is the total energy used divided by the total floor area. It would also be possible to examine annual energy or energy intensities for individual fuels.

The strength of the total energy and energy intensity comparisons is their ease of use and widespread familiarity. However, knowledge is lacking regarding causes of variation that have been observed and the relative impacts of factors such as schedules, functional uses, and density of use on the energy performance. This general approach to rating commercial building energy performance is of interest for quick comparison of one building's energy use from one year to another or quick comparisons of many buildings, but information to adjust for at least some of the wide variation typically observed across a data set with many buildings is lacking.

In cases in which a performance rating system is desired for a specific type of building in a relatively homogeneous climate region, rating scales based on total energy or energy intensity by floor area have some practicality. In the Czech Republic, "labels" of actual, measured energy performance (energy intensity) have been studied, tested, and are now required for apartment buildings. The European Union is likely to require energy performance certificates for buildings by the year 2010. These certificates may have to be renewed every 5 years. With reasonably small ranges of climatic differences, certificates for specific types of buildings based on simple energy intensity values can be a moderately reasonable approach to an energy performance rating system. However, even with a common building type and common climate, there are other variations in key factors that should typically be considered. The basic annual energy intensity accounts for floor area, which has been found to be the most important factor to use in normalizing energy use. If multiple climates must be considered, adjustments for climatic variation, such as normalization for heating degree-days and cooling degree-days, should be included in an energy performance rating system.

Beyond floor area and climate, there will typically be variations in other important factors based on building use, e.g., hospitals and offices will have different normalization factors. Decisions on such factors can be, and at times are, arbitrary. Also, differing policy perspectives can strongly influence consideration of what parameters should be evaluated for normalization of energy performance. Rating systems of the more sophisticated, multifactor type typically consist of parameters for normalization of energy use, a normalization equation or calculation method, and a normalized distribution of energy use. The normalized distribution is typically matched to some scale, often a percentile scale of 1 to 100, to allow a simplified rating result to be obtained. After the energy use of a building is normalized for the factors in the method, the normalized result is compared to the rating scale to determine a rating.

A generic rating system developed for the entire commercial sector in the United States included normalization factors (Table III) that adjusted for floor area, climate, amount of building cooled and heated, worker density, personal computer density, extent of food service and education/training facilities, hours open, and average adjustments for specific types of building space uses. These factors were found to account for over 70% of the variation in energy use in the entire combined U.S. commercial sector, with its wide range of building types. The percentile scale for this particular rating system (Fig. 2) was based on the building energy use index that included electricity losses.

TABLE III

Normalization Factors in a Generic Energy Performance Rating System for All U.S. Commercial Buildings

Floor area (ft^2)
Annual cooling degree days (base 65°F)
Annual heating degree days (base 65°F)
Floor area cooled is less than 25% (yes or no)
Floor area cooled is 50 to 75% (yes or no)
Floor area cooled is 75 to 100% (yes or no)
Floor area heated is 75 to 100% (yes or no)
Personal computers per 1000 ft^2
Workers per 1000 ft^2
Food seats per 1000 ft^2
Educational seats per 1000 ft^2
Hours per week open
Fraction of area that is laboratory
Fraction of area that is nonrefrigerated warehouse
Fraction of area that is food sales (grocery)
Fraction of area that is outpatient health care
Fraction of area that is refrigerated ware house
Fraction of area that is worship space
Fraction of area that is public assembly
Fraction of area that is educational space
Fraction of area that is restaurant
Fraction of area that is inpatient health care
Fraction of area that is strip shopping mall
Fraction of area that is larger shopping mall
Fraction of area that is service

Rating system analyses for specific types of buildings in the United States have found that about 90% or more of the variation in energy use in a specific building type can be normalized based on reasonable factors. As an example, for offices, in addition to floor area, climate, fraction of building heated or cooled, worker density, personal computer density, and hours of operation shown in the generic model (Table III), the number of floors was also a factor found to be important for normalization. Additional examples of the extensive information on the Energy Star rating system for specific building types can be found on the Energy Star Web site (www.energystar.gov).

Another factor that might be considered important for energy performance rating systems is the unit price of energy in a particular location. The energy unit price is significant; analyses comparing all-electric buildings with those that are not indicate an important statistical difference between the energy use distributions of these two categories, suggesting that they should not be treated as equivalent unless some adjustment is made for the difference. Study has shown that this disparity between all-electric buildings and other buildings can be accounted for fairly well by including electricity losses in total energy use as a surrogate for the typical energy unit price differential and other factors related to remote efficiency losses. But the average unit price of energy also appears to adjust fairly well for these differences between all-electric and other buildings, as well as introducing an adjustment for the local economic incentive to be efficient.

Building energy performance rating systems are important tools that offer reasonably quick building energy performance assessment without rigorous

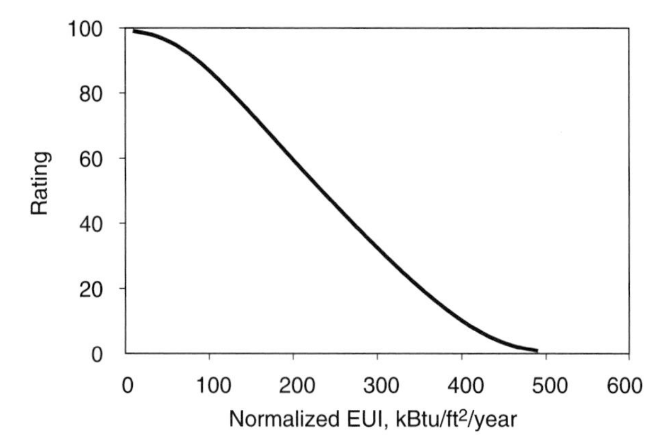

FIGURE 2 Rating scale for the entire U.S. commercial sector. Normalized annual energy use intensity (EUI) matched to percentile scale. Electricity losses are included.

evaluation. In addition, energy performance ratings provide an empirical statement of energy performance not available with other methods, even those that are more rigorous and complicated. Because documenting building energy performance has been determined to be important for many nations, understanding and improving systems for performing such ratings appear to be important for continued progress in commercial building energy efficiency.

5. ENERGY EFFICIENCY VARIATION AND IMPROVEMENT

Energy performance ratings tell what the energy performance of a building is, but if the energy performance of a building is to be improved, the causes of lower than desired performance must be understood, and methods of achieving improved performance must be determined. Causes of variation in energy performance among commercial buildings are understood to a degree, but much remains to be learned.

Estimates of potential improvements in energy efficiency of commercial buildings, i.e., the potential to reduce energy use, have, over many years, indicated that a lot has been and still could be accomplished. Issues of economic incentive and resource allocation influence the estimated savings values, but reductions of 20–40%, on average for the entire sector, appear reasonably achievable if a great enough need exists. In the European Union, a potential savings of 40% has been presented as possible. A savings of 20% of the worldwide annual commercial sector energy use of 30 quads (35 EJ) is 6 quads/yr (7 EJ/yr), which is about 2% of total world energy use. Unfortunately, energy costs are often not large enough, relative to other costs of running a business or organization, to receive major attention, so efficiency improvements have a lower priority.

Research has indicated many reasons why energy efficiency varies so much in commercial buildings. The causes of variation in efficiency can be categorized as variations in: efficiency of operation, efficiency of systems, and efficiency of equipment. Of these three, about half of the potential improvement for the sector would result from operational improvements, with the remainder from equipment and system improvements.

Many studies have shown the importance of operational improvements, with typical savings of 10–20% possible in a wide range of buildings. Other

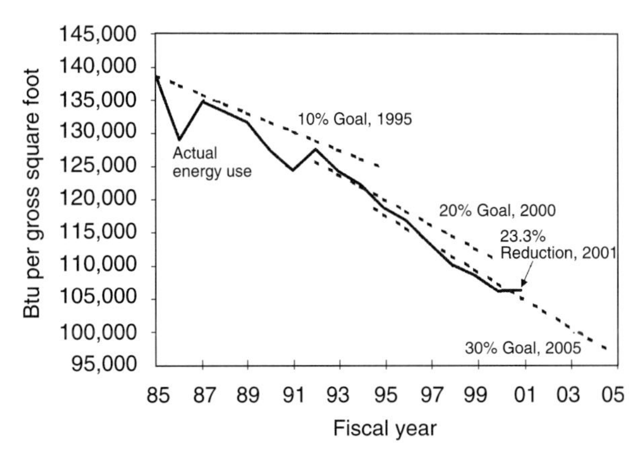

FIGURE 3 Energy efficiency progress for government buildings in the United States. Electricity losses are not included in the annual energy intensity.

studies show significant savings from equipment and systems improvements. The United States Federal Energy Management Program is responsible for achieving reductions in annual energy intensity for most U.S. government buildings. Significant reductions have been achieved through attention to all three areas of efficiency improvement: operation, equipment, and systems. From 1975 to 1985, a reduction in annual energy intensity of about 20% was achieved. After 1985, additional goals were required by Executive Order and other means, and the annual energy intensity of U.S. government buildings is on track to reduce annual energy intensity an additional 30% by the year 2005 relative to the year 1985 (Fig. 3).

The potential for improvement in energy efficiency in the commercial sector is large, if the desire to improve is there, as witnessed by the progress in U.S. government buildings. European estimates also show a large savings potential. If the sources of energy inefficiency in commercial buildings can be reduced through attention to the three major areas of efficiency improvement, about a 2% reduction in world energy use appears achievable.

6. SECTORAL DATA AND MODELING

Extensive data are collected on energy use and sectoral characteristics throughout the world. Sectoral data covering energy use of major economic energy sectors are available for many countries, and world data are available from the International

Energy Agency and from some major nations. The commercial sector, as defined here, is often called by other names, such as "general" or "tertiary," indicating the "everything else" nature of the sector, including all buildings other than residential, industrial, or agricultural.

As an example of data for a nation with one of the most specialized commercial sectors in the world, the United States collects extensive energy data for this sector, including consumption by fuel type, prices by fuel type, and expenditures by fuel type for the overall sector. The special sampling survey of 5000 to 6000 buildings that is conducted every 4 years provides detail on the annual consumption of each fuel and extensive characteristics data, allowing additional extensive analyses to be performed and supporting national modeling of energy use in this sector.

Energy data provide a historical record and are used to forecast trends into the future. Several organizations model world energy use, although often not with the commercial sector separated. Many individual nations also track historical energy use and forecast future energy use. The National Energy Modeling System (NEMS), used in the United States for energy use forecasts and other analyses, provides an example of an extensive sectoral and national modeling system. The NEMS Commercial Sector Demand Module is a simulation tool based on economic and engineering relationships; it models commercial sector energy demands, with breakout detail at the geographic level of nine census divisions, using 11 distinct categories of commercial buildings.

Projections of future energy use involve selections of equipment for the major fuels of electricity, natural gas, and distillate fuel, and for the major services of space heating, space cooling, water heating, ventilation, cooking, refrigeration, and lighting. The equipment choices are made based on an algorithm that uses life-cycle cost minimization constrained by factors related to commercial sector consumer behavior and time preference premiums. The algorithm also models demand for the minor fuels of residual oil, liquefied petroleum gas, coal, motor gasoline, and kerosene. The use of renewable fuel sources (wood, municipal solid waste, and solar energy) is also modeled. Decisions regarding the use of distributed generation and cogeneration technologies are performed using a separate cash-flow algorithm.

The NEMS Commercial Module generates mid-term (20- to 30-year) forecasts of commercial sector energy demand. The model facilitates policy analysis of energy markets, technological development, environmental issues, and regulatory development as they interrelate with commercial sector energy demand. Input to this model is quite extensive, and in addition to the sectoral energy data includes building lifetime estimates, economic information such as demand elasticities for building services and market forecasts for certain types of equipment, distributed electricity generation/cogeneration system data, energy equipment market data, historical energy use data, and short-term energy use projections from another model. The primary output of the modeling process is a forecast of commercial sector energy consumption by fuel type, end use, building type, census division, and year. The module also provides forecasts of the following parameters for each of the forecast years:

- Construction of new commercial floor space by building type and census division.
- Surviving commercial floor space by building type, year of construction, and census division.
- Equipment market shares by technology, end use, fuel, building type, and census division.
- Distributed generation and cogeneration of electricity and fuels used.
- Consumption of fuels to provide district services.
- Nonbuilding consumption of fuels in the commercial sector.
- Average efficiency of equipment mix by end use and fuel type.

The NEMS Commercial Module interacts with and requires input from other modules in NEMS. This relationship of the Commercial Module to other components of NEMS is depicted schematically in Fig. 4. Not shown are the other sectoral modules and the central controlling module in NEMS that integrates and reconciles national level results for all sectors.

In contrast to such extensive sectoral models that are used to provide forecasts of energy use, the energy performance rating systems presented previously provide another type of model for the commercial sector; they allow additional understanding of current energy use and the influence of certain operating parameters of buildings. The type of information and the understanding of the commercial sector generated by different modeling approaches differ in perspective and ability to understand potential for improvements in energy efficiency. Effects of these differing perspectives are presented next.

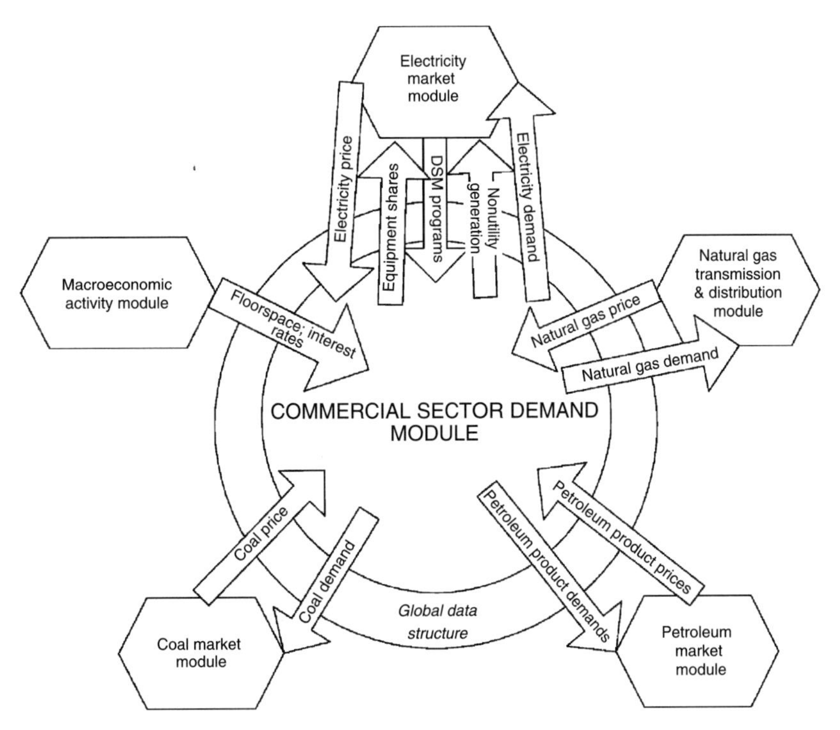

FIGURE 4 Relationship of the commercial sector and other National Energy Modeling System modules. DSM, demand-side management.

7. SECTORAL VIEWS FORMED BY MODELS

Energy models such as NEMS can be called economic-engineering models. Such models use engineering data and analysis results to feed into and partially interact with an economic model of commercial sector changes and demand for energy. Because changes in the efficiency of buildings and the energy use patterns of buildings tend to take many years to evolve, such economic-engineering models often do a good job of representing the energy use of a large group of buildings, including the entire commercial sector in a country. These models can also forecast the end uses of energy in buildings (Table IV) and are usually good at doing so.

Although NEMS forecasts about a 50% increase in U.S. commercial sector energy use between the years 2001 and 2025, the energy intensity of these buildings remains almost constant over this forecast period. This information indicates that growth in commercial sector energy use is attributed almost exclusively to growth in floor area, and that any efficiency improvements are modeled as offset by growth in end uses of energy not affected by the efficiency improvements. Clearly, with little change in annual energy intensity over a 25-year period, a tendency to limit certain types of change can be seen in this modeling approach. In addition, although the model provides a breakout of energy according to end uses such as heating, cooling, lighting, and seven other uses (Table IV), data on the impacts of the density of workers or occupants, schedule of operations, and density of personal computers in buildings are not modeled and are not known. To forecast total energy use, this type of normalization of energy is not required, because it represents a different way of looking at the sector, and normalized energy is not the desired output.

Economic-engineering models are also capable of forecasting impacts of new energy technologies and more efficient operations on energy use, but new energy technologies and impacts of those technologies on new buildings are more capably modeled in NEMS than are improvements in operations, which must typically be treated as changes in annual energy intensity. Unfortunately, this characteristic means that impacts of improvements in energy system operations are not understood and cannot be estimated well with this modeling approach.

Detailed engineering simulation models such as DOE-2, Energy Plus, and BLAST, which have the capabilities to model energy use in commercial buildings, also have difficulty modeling improvements

TABLE IV

NEMS Commercial Sector Energy End-Use Baseline and Forecast[a]

	Year						
Use	2000	2001	2005	2010	2015	2020	2025
Space heating	2.13	1.95	2.22	2.26	2.3	2.37	2.42
Space cooling	1.35	1.39	1.35	1.37	1.39	1.41	1.43
Water heating	1.15	1.12	1.21	1.26	1.3	1.35	1.39
Ventilation	0.56	0.55	0.55	0.56	0.56	0.56	0.57
Cooking	0.38	0.37	0.39	0.41	0.42	0.44	0.45
Lighting	3.34	3.31	3.54	3.73	3.81	3.87	3.89
Refrigeration	0.69	0.69	0.71	0.73	0.75	0.77	0.78
Office equipment (PC)[b]	0.5	0.52	0.6	0.74	0.85	0.95	1.05
Office equipment (non-PC)	0.98	0.99	1.12	1.44	1.8	2.21	2.69
Other uses	6.13	6.56	6.89	7.65	8.54	9.6	10.6
Total	17.2	17.44	18.59	20.15	21.72	23.52	25.33

[a] In quads/year; electricity losses are included.
[b] PC, personal computer.

in energy system operations, because the simulation routines are all set up to model systems and components that work correctly. Simulating improvements in operations often requires knowing the answer first, and then tricking the simulation program into calculating the correct results. Again, the limitations of the models influence decisions about appropriate energy efficiency improvements for buildings. Limitations in ability to model impacts of operational improvements on energy efficiency lead to a view of the commercial sector that essentially ignores the potential of such improvements. A problem with this situation is that policy officials typically also do not receive information on the potential of operational improvements and lack an understanding of the importance of improving operations in commercial buildings. Because improvements in operations represent about half of the potential energy savings that could be achieved in the commercial sector, acceptable means of modeling impacts and potential efficiency benefits would be helpful. Until acceptable means of modeling are developed, the benefits of operational improvements must continue to be determined empirically.

Improved understanding of commercial sector energy use, and the potential of operational improvements for saving energy in commercial buildings, would result from the ability to model the effects of operational improvements on forecasts of energy use in the commercial sector. Improved understanding of the potential for energy efficiency improvements appears possible through modeling of normalized energy use for the sector, whereby adjustments for a few key factors known to cause variation in energy use in commercial buildings, together with performance rating values, could allow analysis and pursuit of scenarios of improvement toward higher efficiency ratings. Advances in modeling such as these could be important for the future, if the percentage increase in commercial sector energy use in advancing economies becomes a challenge for world energy efficiency and climate impacts.

8. DRIVING TOWARD ENERGY EFFICIENCY

As the need for energy efficiency becomes more pronounced, the drive toward efficiency in the commercial sector will be impeded by its complicated mix of building sizes and uses, the complicated systems often used in commercial buildings, and the relative lack of understanding of operations factors impacting energy use and how to achieve efficiency.

In the United States, commercial energy use has increased from 10 to 17% or more of national energy use between the years 1960 and 2000. A significant reason for this increase is the low cost of energy relative to the other costs of conducting business, but the difficulty in understanding energy systems and energy use in commercial buildings is also an important contributing factor. Policy officials often have difficulty understanding discussions of the needs for improvements in commercial buildings.

As economies advance and commercial sector energy use begins to grow relative to other sectors, an improved understanding of methods of measuring commercial energy performance, and the means of achieving efficiency improvements in this sector, will be important in any drive toward efficiency.

One warning sign of the need to increase understanding of energy performance is an increase in the use of air conditioning in commercial buildings. When air conditioning use increases, energy system complexity and indoor space quality issues also increase significantly. If air conditioning use is increasing, any proposal for increased efficiency that relies heavily on thermal insulation should be treated warily, because insulation optimization becomes more difficult, and other system complexities tend to become much more important.

A warning should be given overall for energy standards for buildings, as they currently exist around the world, because, despite the existence and use of these standards for many years, the effect of standards has been only moderate in most cases. The shortcoming of existing standards is that they rely too heavily on simulation of expected performance, without conducting true empirical studies to verify the effectiveness of what the standards achieve. One major reason such empirical studies have been conducted in only limited and mostly ineffective fashion is that an empirical method of measuring energy performance of commercial buildings, although still adjusting for legitimate building use differences, has only recently been established in concept, only for certain building types, and not with the stated intent of being a performance standard (Energy Star label for buildings). Interestingly, however, if such energy performance standards existed, energy standards currently in use, with their typically complicated requirements, would not necessarily be needed any longer as standards.

Use of energy performance certificates may be necessary to overcome the difficulty users, occupants, code and policy officials, and owners have in understanding commercial energy use and performance. Without certification of energy performance, the complexity of systems and uses makes understanding energy performance by anyone other than an expert, and even by some experts, difficult. However, without increased understanding of the most appropriate means to normalize energy use for legitimate differences in building function and use, energy performance certificates may offer an unsatisfactory solution, due to inequities that will be obvious to many, if reasonable normalizations are not applied.

Increased energy efficiency in the commercial sector is an important piece of the national efficiency strategy in advanced economies, in which the priority for efficiency must be increased. Complexities in commercial buildings have made progress in energy efficiency for this sector less than desirable in notable cases. Fortunately, methods and knowledge needed to increase success in this sector have begun to be developed, and better solutions can be offered in the near future to help increase energy efficiency in the commercial sectors of countries where the need is pressing.

9. FUTURE ENERGY PERFORMANCE AND USE

Current forecasts call for solid growth in world energy use over the next 20 years, potentially increasing 60% above current use. With the forces in place to keep energy use patterns the same, a safe, conservative assumption would be that the commercial sector will contribute about 12% to final total energy consumption in the year 2020. If world energy use grows to 600 quads, or 630 EJ, by the year 2020, and total final consumption in the energy-using sectors is about 400 quads, or 420 EJ, final consumption in the commercial sector, at 12%, would be about 50 quads/yr (or about 50 EJ), without electricity losses included. This energy use would be two-thirds more than the energy use in the year 2000.

Without significant changes in energy performance of commercial buildings, this scenario of 50 quads of commercial energy use in the year 2020 is likely to occur, absent major world upheaval. If important progress on improving the energy performance of commercial buildings in the advanced economies can be made, potentially a reduction of 2–4 quads or more in commercial sector worldwide use in 2020 (a 4–8% reduction) could be achieved. Such an achievement would require that energy performance certifications become the norm, that operational standards increase significantly, and that 25–40% of buildings in the advanced economies see significant improvements in their performance.

Methods for certifying the energy performance of commercial buildings have been developed to the conceptual and practical applications stages. However, these methods are in their infancy. Energy policy and research attention to the commercial sector have been lacking relative to the growth observed. Without increased attention to improving commercial sector energy efficiency, energy use growth relative to other sectors will continue to make this sector a challenge when decreased energy use and emissions are sought.

SEE ALSO THE FOLLOWING ARTICLES

District Heating and Cooling • Industrial Ecology • Industrial Energy Efficiency • Industrial Energy Use, Status and Trends • International Comparisons of Energy End Use: Benefits and Risks • Obstacles to Energy Efficiency • Thermal Comfort

Further Reading

Energy Information Administration. (2001). Model Documentation Report: Commercial Sector Demand Module of the National Energy Modeling System. DOE/EIA-M066(2002). Department of Energy, Washington, D.C.

Energy Information Administration. (2002). International Energy Outlook 2002. DOE/EIA-0484(2002). Department of Energy, Washington, D.C.

Energy Information Administration. (2003). Annual Energy Outlook 2003. DOE/EIA-0383(2003). Department of Energy, Washington, D.C.

Energy Star Web site. (2003). Available at http://www.energystar.gov/.

European Commission. (2001). Proposal for a directive of the European Parliament and of the Council on the energy performance of buildings. COM(2001) 226 final, 2001/0098 (COD). EC, Brussels.

International Energy Agency. (2000). Key World Energy Statistics Pamphlet. IEA, Paris.

Complex Systems and Energy

MARIO GIAMPIETRO
Istituto Nazionale di Ricerca per gli Alimenti e la Nutrizione
Rome, Italy

KOZO MAYUMI
University of Tokushima
Tokushima, Japan

1. Introduction
2. The Energetics of Human Labor: An Epistemological Analysis of the Failure of Conventional Energy Analysis
3. The Problematics Related to Formal Definitions of Energy, Work, and Power in Physics and Their Applicability to Energy Analysis
4. The Metaphor of Prigogine Scheme to Characterize Autocatalytic Loops of Energy Forms in Hierarchically Organized Adaptive Systems
5. Conclusion

Glossary

complex system A system that allows one to discern many subsystems, depending entirely on how one chooses to interact with the system.

dissipative systems All natural systems of interest for sustainability (e.g., complex biogeochemical cycles on this planet, ecological systems, human systems when analyzed at different levels of organization and scales beyond the molecular one); self-organizing open systems, away from thermodynamic equilibrium.

epistemological complexity Complexity that is in play every time the interests of the observer (the goal of the mapping) are affecting what the observer sees (the formalization of a scientific problem and the resulting model).

hierarchical systems Systems that are analyzable into successive sets of subsystems; when alternative methods of description exist for the same system.

hierarchy theory A theory of the observer's role in any formal study of complex systems.

scale The relation between the perception of a given entity and its representation.

The purpose of this article is to examine key epistemological issues that must be addressed by those willing to apply energy analysis to complex systems and particularly to the sustainability of adaptive systems organized in nested hierarchies. After a general introduction, sections 2 and 3 provide a critical appraisal of conventional energy analysis. Such a criticism is based on concepts developed in the field of hierarchy theory. Section 4 suggests an alternative view of the metaphorical message given by Prigogine's scheme in nonequilibrium thermodynamics—a message that can be used to develop innovative and alternative approaches for energy analysis. Section 5 presents the conclusions.

1. INTRODUCTION

The revolution entailed by complex systems theory is finally forcing many scientists to acknowledge that it is impossible to have a unique substantive representation of the reality. Different observers with various "scale-dependent" measuring schemes and goals (e.g., humans using either a microscope, the naked eye, or a telescope for different purposes) will perceive different aspects of a particular ontological entity. These different views do reflect the special relation established between the observer and the components of the entity observed. This predicament gets worse when dealing with hierarchically organized adaptive systems. These systems are, in fact, organized on different levels (e.g., cells, organs, individual humans, households, communities, countries). This implies that their identity is associated with the ability to express various patterns simultaneously on different scales. This implies that when doing energy analysis on these systems, wholes, parts, and their contexts require the parallel use of nonequivalent descriptive domains. That is, the definitions of parts, wholes, and contexts referring to different levels are not reducible to each other. To make things more difficult for modelers, dissipative systems are becoming in time, meaning that their identities change in time. Their hierarchical nature entails that the pace of evolution of the identities of the parts will be different from that of the identity of the whole.

The disciplinary field of energy analysis seems to have ignored the epistemological challenge posed to the analysts by the fact that energy assessments must be a result of different perceptions and representations of the reality that cannot be reduced to each other. This problem is particularly serious considering that hierarchically organized adaptive systems, by definition, are operating across multiple scales. Not only do the energetic descriptions have to be based on the use of nonequivalent descriptive domains, but they also have to be updated in time. Thus, scientists dealing with the sustainability of complex dissipative systems adapting in time must be able to (1) individuate relevant causal relations and relevant scales in their modeling efforts, (2) use different selections of variables for representing changes in relevant attributes, and (3) continuously check the validity of these selections.

The main epistemological implication of studying systems organized across hierarchical levels and scales is that there is a virtual infinite universe of possible ways of perceiving and representing them. For example, there is virtually an infinite number of possible combinations of views of a human that can be generated when looking at him or her using microscopes, ultrasound scanners, X-ray machines, blood tests, conventional cameras, radars, and telescopes simultaneously. Any formal representation—the information space used to make models in terms of variables and formal systems of inference—must be finite and closed. This represents a huge challenge for scientists.

The main message of the new scientific paradigm entailed by complexity can be expressed using two crucial points. When dealing with the sustainability of complex adaptive systems, first, scientists must accept dealing with an unavoidable nonreducibility of models defined on different scales and reflecting different selections of relevant attributes. This implies acknowledging the impossibility of reaching an uncontested agreement on a substantive and unique description of the reality that can be assumed to be the "right" one. Second, scientists must accept dealing with incommensurability among the priorities used by nonequivalent observers when deciding how to perceive and represent nested hierarchical systems operating across scales. Nonequivalent observers with different goals will provide logically independent definitions of the relevant attributes to be considered in the model. This implies the impossibility of having a substantive and agreed-on definition of what should be considered the "most useful" narrative to be adopted when constructing the model.

This new paradigm associated with complexity, which is rocking the reductionist building, is the offspring of a big epistemological revolution started during the first half of 19th century by classical thermodynamics (e.g., by the works of Carnot and Clausius) and continued during the second half of the 20th century by nonequilibrium thermodynamics (e.g., by the works of Schroedinger and Prigogine). Both revolutions used the concept of "entropy" as a banner. The equilibrium thermodynamics represented a first bifurcation from mechanistic epistemology by introducing new concepts such as irreversibility and symmetry breaking when describing real-world processes. Then, the introduction of the new nonequilibrium paradigm implied a final departure from reductionist epistemology in that it implies the uncomfortable acknowledgment that scientists can work only with system-dependent and context-dependent definitions of entities. Put differently, food is a "high-quality" energy input for humans but not for cars. Saying that 1 kg of rice has an "energy content" of a given amount of megajoules can be misleading because this "energy input" is not an available energy input for driving a car that is out of gasoline. In the same way, the definition of what should be considered "useful energy" depends on the goals of the system that is expected to operate within a given associative context. For example, fish are expected to operate under the water to get their food, whereas birds cannot fly to catch prey below the ground.

Classical thermodynamics first, and nonequilibrium thermodynamics later on, gave a fatal blow to the mechanist epistemology of Newtonian times, introducing brand-new concepts such as disorder, information, entropy, and negentropy. However, we cannot replace the hole left by the collapse of Newtonian mechanistic epistemology just by using these concepts, as if they were "substantive" concepts that are definable in a strictly physical sense as well as context independent. For example, the concept of "negative entropy" is not a substantive concept. Rather, it is a "construction" (an artifact) associated with the given identity of a dissipative system that is perceived and represented as operating at a given point in space and time. Therefore, the concept of negative entropy reflects the perception and representation of the "quality" of both energy inputs and energy transformations that are associated with a particular pattern of dissipation (or typology of metabolism). But this quality assessment is valid only for the specific dissipative system considered that is assumed to operate within a given associative context or an admissible environment. These can seem to be trivial observations, but they are directly linked to a crucial statement we want to make in this article. It is not possible to characterize, using a "substantive formalism," qualitative aspects of energy forms, and this is applicable to all conceivable dissipative systems operating in the reality. That is, it is impossible to define a "quality index" that is valid in relation to all of the conceivable scales and when considering all of the conceivable attributes that could be relevant for energy analysis. Conventional variables used in mechanics can be relevant attributes to define the stability of a dissipative process, but there are situations in which unconventional attributes, such as smell and color, can become crucial observable qualities (e.g., in ecology in relation to the admissibility of the expected associative context). This is the norm rather than the exception for complex adaptive systems. In this case, it is not always easy to quantify the relevance of these attributes in terms of an energy quality index.

To make things more challenging, we must bear in mind the fact that not only the observed system but also the observer is becoming something else in time. That is, an observer with different goals, experiences, fears, and knowledge will never perceive and represent energy forms, energy transformations, and a relative set of "indexes of quality" in the same way as did a previous one.

The consequences of this fact deserve attention. This entails that, on the one hand, various concepts derived from thermodynamic analysis used to generate energy indexes of quality (e.g., output/input energy ratios, exergy-based indexes, entropy-related concepts, embodied assessments associated with energy flows) do carry a powerful metaphorical message. On the other hand, like all metaphors, they always require a semantic check before their use. That is, concepts such as efficiency, efficacy, and other quality indexes are very useful in a discussion about sustainability issues. The problem is that they cannot and should not be used to provide normative indications about how to deal with a sustainability predicament in an algorithmic way. That is, policies cannot be chosen or evaluated on the basis of the maximization of a single parameter (e.g., efficiency, power, rate of supply of negentropy) or on the basis of the application of a set of given rules written in protocols to a given simplified model of the reality.

The wisdom of Carnot, well before the complexity revolution, led him to similar conclusions in 1824 that can be found in the closing paragraph of his *Reflections on the Motive Power of Fire, and on Machines Fitted to Develop That Power*:

> *We should not expect ever to utilize in practice all the motive power of combustibles. The attempts made to attain this result would be far more harmful than useful if they caused other important considerations to be neglected. The economy of the combustible is only one of the conditions to be fulfilled in heat-engines. In many cases it is only secondary. It should often give precedence to safety, to strength, to the durability of the engine, to the small space which it must occupy, to small cost of installation, etc. To know how to appreciate in each case, at their true value, the considerations of convenience and economy which may present themselves; to know how to discern the more important of those which are only secondary; to balance them properly against each other in order to attain the best results by the simplest means; such should be the leading characteristics of the man called to direct, to co-ordinate the labours of his fellow men, to make them co-operate towards a useful end, whatsoever it may be.*

2. THE ENERGETICS OF HUMAN LABOR: AN EPISTEMOLOGICAL ANALYSIS OF THE FAILURE OF CONVENTIONAL ENERGY ANALYSIS

Attempts to apply energy analysis to human systems have a long history. Pioneering work was done by, among others, Podolinsky, Jevons, Ostwald, Lotka, White, and Cottrel. However, it was not until the 1970s that energy analysis became a fashionable scientific exercise, probably due to the oil crisis surging during that period. During the 1970s, energy input/output analysis was widely applied to farming systems and national economies and was applied more generally to describe the interaction of humans with their environment by Odum, Georgescu-Roegen, and Pimentel, among others. At the IFISA workshop of 1974, the term "energy analysis" (as opposed to "energy accounting") was officially coined. The second energy crisis during the 1980s was echoed by the appearance of a new wave of interesting work by biophysical researchers and a second elaboration of their own original work by the "old guard." However, quite remarkably, after less than a decade or so, the interest in energy analysis declined outside the original circle. Indeed, even the scientists of this field soon realized that using energy as a numeraire to describe and analyze changes in the characteristics of agricultural and socioeconomic systems proved to be more complicated than one had anticipated.

We start our critical appraisal of the epistemological foundations of conventional energy analysis using one of the most well-known case studies of this field: the attempt to develop a standardized tool kit for dealing with the energetics of human labor. This has probably been the largest fiasco of energy analysis given the huge effort dedicated to this by the community of energy analysts.

Looking over the vast literature on the energetics of human labor, one can find agreement about the need to know at least three distinct pieces of information, which are required simultaneously, to characterize indexes of "efficiency" or "efficacy":

1. *The requirement (and/or availability) of an adequate energy input needed to obtain the conversion of interest.* In the case of human labor, a flow of energy contained in food is energy carriers compatible with human metabolism.
2. *The ability of the considered converter to transform the energy input into a flow of useful energy to fulfill a given set of tasks.* In this case, a system made up of humans has to be able to convert available food energy input into useful energy at a certain rate, depending on the assigned task.
3. *The achievement obtained by the work done, that is, the results associated with the application of useful energy to a given set of tasks.* In this case, this has to do with the usefulness of the work done by human labor in the interaction with the context.

If we want to use indexes based on energy analysis to formalize the concept of performance, we have to link these three pieces of information to numerical assessments based on "observable" qualities (Fig. 1). For this operation, we need at least four nonequivalent numerical assessments related to the following:

1. *Flow of a given energy input that is required and consumed by the converter.* In the case of the study of human labor, it is easy to define what should be considered food (energy carriers) for humans: something that can be digested and then transformed into an input for muscles. If the converter were a diesel engine, food would not be considered an energy input. This implies that the energy input can

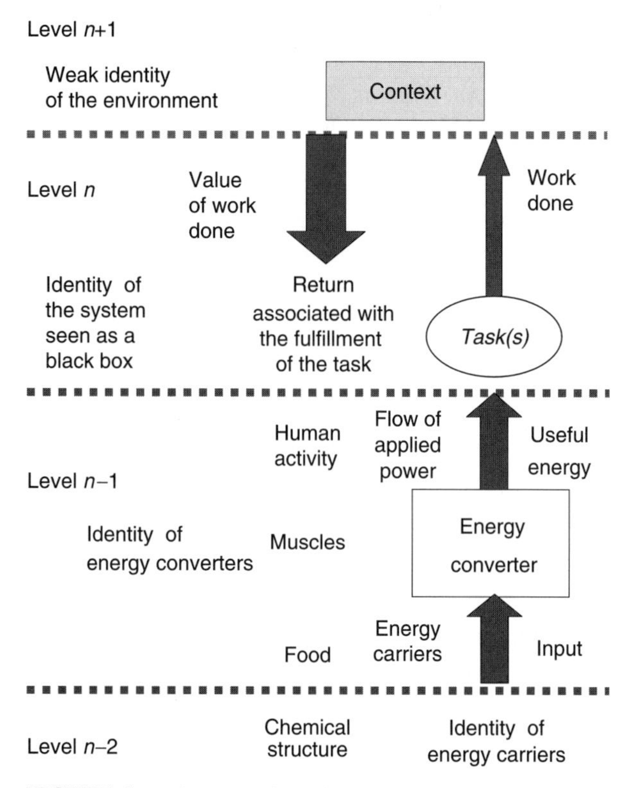

FIGURE 1 Relevant qualities for characterizing the behavior of an energy converter across hierarchical levels. From Giampietro, M. (2003). "Multi-Scale Integrated Analysis of Agroecosystems." CRC Press, Boca Raton, FL. Used with permission.

be characterized and defined only in relation to the given identity of the converter using it.

2. *The power level at which useful energy is generated by the converter.* This is a more elusive "observable quality" of the converter. Still, this information is crucial. As stated by Odum and Pinkerton in 1955, when dealing with the characterization of energy converters, we always have to consider both the pace of the throughput (the power) and the output/input ratio. A higher power level tends to be associated with a lower output/input ratio (e.g., the faster one drives, the lower the mileage of one's car). On the other hand, it does not make any sense to compare two output/input ratios if they refer to different throughput rates. In fact, we cannot say that a truck is less efficient than a small motorbike on the basis of the information given by a single indicator, that is, by the simple assessment that the truck uses more gas per mile than does the motorbike. However, after admitting that the power level is another piece of information that is required to assess the performance of an energy converter, it becomes very difficult to find a standard definition of "power level" applicable to "complex

energy converters" (e.g., to compare human workers to light bulbs or economic sectors). This is especially true when these systems are operating on multiple tasks. Obviously, this is reflected in an impossibility to define a standard experimental setting to measure such a value for nonequivalent systems. Moreover, power level (e.g., 1000 horsepower of a tractor vs 0.1 horsepower of a human worker) does not map onto either energy input flows (how much energy is consumed by the converter over a given period of time used as a reference [e.g., 1 year]) or how much applied power is delivered (how much useful energy has been generated by a converter over a given period of time used as reference [e.g., 1 year]). In fact, these two different pieces of information depend on how many hours the tractor or human worker has worked during the reference period and how wisely the tractor or human worker has been operating. This is reflected by the need to introduce a third piece of information to the analysis.

3. *Flow of applied power generated by the conversion.* The numerical mapping of this quality clearly depends heavily on the previous choices about how to define and measure power levels and power supply. In fact, an assessment of the flow of applied power represents the formalization of the semantic concept of "useful energy." Therefore, this is the measured flow of energy generated by a given converter (e.g., an engine) that is used to fulfill a specified task (e.g., water pumped out of the well). To make things more difficult, the definition of usefulness of such a task can be given only when considering the hierarchical level of the whole system to which the converter belongs (e.g., if the water is needed to do something by the owner of the pump). Put differently, such a task must be defined as "useful" by an observer operating at a hierarchical level higher than the level at which the converter is transforming energy input into useful energy. What is produced by the work of a tractor has a value that is generated by the interaction of the farm with a larger context (e.g., the selling of products on the market). That is, the definition of "usefulness" refers to the interaction of the whole system—the farm, which is seen as a black box (to which the converter belongs as a component) with its context. Therefore, the assessment of the quality of "usefulness" requires a descriptive domain different from that used to represent the conversion at the level of the converter (gasoline into motive power in the engine). This introduces a first major epistemological complication: the definition of usefulness of a given task (based on the return that this task implies for the whole system) refers to a representation of events on a given hierarchical level (the interface $n/n+1$) that is different from the hierarchical level used to describe and represent (assess indexes of efficiency of) the conversion of the energy input into useful energy (the interface $n-1/n$).

4. *Work done by the flow of applied power, that is, what is achieved by the physical effort generated by the converter.* Work is another very elusive quality that requires a lot of assumptions to be measured and quantified in biophysical terms. The only relevant issue here is that this represents a big problem regarding energy analysis. Even if assessments 3 and 4 use the same measurement unit (e.g., megajoules), they are different in terms of the relevant observable qualities of the system. That is, assessment 3 (applied power) and assessment 4 (work done) not only do not coincide in numerical terms but also require different definitions of descriptive domain. In fact, the same amount of applied power can imply differences in achievement due to differences in design of technology and know-how when using it. The problem is that it is impossible to find a "context-independent quality factor" that can be used to explain these differences in substantive terms.

The overview provided in Fig. 1 should already make clear to those familiar with epistemological implications of hierarchy theory that practical procedures used to generate numerical assessments within a linear input/output framework cannot escape the unavoidable ambiguity and arbitrariness implied by the hierarchical nature of complex systems. A linear characterization of input/output using the four assessments discussed so far requires the simultaneous use of at least two nonequivalent and nonreducible descriptive domains. Therefore, a few of these four assessments are logically independent, and this opens the door to an unavoidable degree of arbitrariness in the problem structuring (root definitions of the system). Any definition or assessment of energy flows, as input and/or output, will in fact depend on an arbitrary choice made by the analyst about what should be considered the focal level n, that is, what should be considered as a converter, what should be considered as an energy carrier, what should be considered as the whole system to which the converter belongs, and what has to be included and excluded in the characterization of the environment when checking the admissibility of boundary conditions and the usefulness of work done. A linear representation in energy analysis forces the analyst to decide "from scratch," in a situation of total arbitrariness, on the set of formal identities adopted in the model to represent energy flows and conversions using the finite set of variables used to describe changes in the various elements of interest. This choice of a set of formal identities for representing energy carriers, converters, and the black box—the finite set of characteristics related to the definition of an admissible environment—is then translated into the selection of the set of epistemic categories (variables) that will be used in the formal model. It is at this point that the capital sin of the assumption of linear representation becomes evident. No matter how smart the analyst may be, any assessment of energy input (the embodied energy of the input) or energy output (what has been achieved by the work done) will be unavoidably biased by the arbitrary preanalytical choice of root definitions. In terms of hierarchy theory, we can describe this fact as follows.

The unavoidable preliminary triadic filtering needed to obtain a meaningful representation of the reality is at the basis of this impasse. For this filtering, we have to select (1) the interface between the focal level n and the lower level $n-1$ to represent the structural organization of the system and (2) the interface between the focal level n and higher level $n+1$ to represent the relational functions of the system. Ignoring this fact simply leads to a list of disorganized nonequivalent and nonreducible assessments of the same concepts. A self-explanatory example of this standard impasse in the field of energy analysis is given in Table I, which reports an example of several nonequivalent assessments of the energetic equivalent of 1 h of labor found in the literature.

That is, every time we choose a particular hierarchical level of analysis for assessing an energy flow (e.g., an individual worker over a day), we are also selecting a space–time scale at which we will describe the process of energy conversion (e.g., over a day, over a year, over the entire life). This, in turn, implies a nonequivalent definition of the context or environment and of the lower level where structural components are defined. Human workers can be seen as individuals operating over a 1-h time horizon where muscles are the converters, or they can be seen as citizens of developed countries where machines are the converters. The definitions of identities of these elements must be compatible with the identity of energy carriers such that individuals eat food, whereas developed countries eat fossil energy. This implies that, whatever we choose as a model, the various identities of energy carriers, parts, whole, and environment have to make sense in their reciprocal conditioning. Obviously, a different choice of hierarchical level considered as the focal level (e.g., the worker's household) requires adopting a different system of accounting for input and output.

3. *THE PROBLEMATICS RELATED TO FORMAL DEFINITIONS OF ENERGY, WORK, AND POWER IN PHYSICS AND THEIR APPLICABILITY TO ENERGY ANALYSIS*

In the previous case study, we have actually dealt the problematics related to formal definitions of energy, power, and work when dealing with an energetic assessment of human labor. Because this is a point carrying quite heavy epistemological implications, we would like to elaborate a bit more on it. To be able to calculate substantive indexes of performance based on concepts such as efficiency (maximization of an output/input ratio) and efficacy (maximization of an indicator of achievement in relation to an input) within energy analysis, we should be able to define, first, three basic concepts—"energy," "work," and "power"—in substantive terms. That is, we should be able to agree on definitions that are independent of the special context and settings in which these three concepts are used. However, if we try to do that, we are getting into an even more difficult situation.

3.1 Energy

Much of the innate indeterminacy of energy analysis, especially when applied to complex systems, has its roots in the problematic definition of energy in physics. As Feynman and colleagues pointed out in 1963, we have no knowledge of what energy is, and energy is an abstract thing in that it does not tell us the mechanism or the reasons for the various formulas. In practice, energy is perceived and described in a large number of different forms: gravitational energy, kinetic energy, heat energy, elastic energy, electrical energy, chemical energy, radiant energy, nuclear energy, mass energy, and so on. A general definition of energy, without getting into specific context and space–time scale-dependent settings, is necessarily limited to a vague expression such as "the potential to induce physical transformations." Note that the classical definition of energy in conventional physics textbooks, "the potential to do

TABLE I

Nonequivalent Assessments of Energy Requirement for 1 h of Human Labor

Method[a]	Worker system boundaries[b]	Energy input per hour of labor		
		Food energy input[c] (MJ/h)	Exosomatic energy input[d] (MJ/h)	Other energy forms across scales[e]
1	Man	0.5[f]	Ignored	Ignored
1	Woman	0.3[g]	Ignored	Ignored
1	Adult	0.4[h]	Ignored	Ignored
2	Man	0.8[f]	Ignored	Ignored
2	Woman	0.5[g]	Ignored	Ignored
2	Adult	0.6[h]	Ignored	Ignored
3	Man	1.6[f]	Ignored	Ignored
3	Woman	1.2[g]	Ignored	Ignored
3	Adult	1.3[h]	Ignored	Ignored
4	Man	2.5[f]	Ignored	Ignored
4	Woman	1.8[g]	Ignored	Ignored
4	Adult	2.1[h]	Ignored	Ignored
5	Household	3.9[i]	Ignored	Ignored
6	Society	4.2[j]	Ignored	Ignored
7a	Household	3.9[i]	39 (food system)[k]	Ignored
7b	Society	4.2[j]	42 (food system)[k]	Ignored
8	Society	Ignored	400 (society)[l]	Ignored
9	Society	Ignored	400 (society)[l]	2×10^{10} MJ[m]

Source. From Giampietro, M. (2003). "Multi-Scale Integrated Analysis of Agroecosystems." CRC Press, Boca Raton, FL. Used with permission.

[a]The nine methods considered are (1) only extra metabolic energy due to the actual work (total energy consumption minus metabolic rate) in 1 h; (2) total metabolic energy spent during actual work (including metabolic rate) in 1 h; (3) metabolic energy spent in a typical work day divided by the hours worked in that day; (4) metabolic energy spent in 1 year divided by the hours worked in that year. (5) Same as method 4 but applied at hierarchical level of household; (6) same as method 4 but applied at the level of society; (7) besides food energy, also including exosomatic energy spent in food system per year divided by the hours of work in that year (7a at household level, 7b at society level); (8) total exosomatic energy consumed by society in 1 year divided by the hours of work delivered in that society in that year; and (9) assessing the megajoules of solar energy equivalent of the amount of fossil energy assessed with method 8.

[b]The "systems delivering work" considered are typical adult man (Man), typical adult woman (Woman), average adult (Adult), typical household, and an entire society. Definitions of "typical" are arbitrary and serve only to exemplify methods of calculation.

[c]Food energy input is approximated by the metabolic energy requirement. Given the nature of the diet and food losses during consumption, this flow can be translated into food energy consumption. Considering also postharvest food losses and preharvest crop losses, it can then be translated into different requirements of food (energy) production.

[d]We report here an assessment referring only to fossil energy input.

[e]Other energy forms, acting now or in the past, that are (were) relevant for the current stabilization of the described system even if operating on space–time scales not detected by the actual definition of identity for the system.

[f]Based on a basal metabolic rate for adult men of BMR = 0.0485 W + 3.67 MJ/day (W = weight = 70 kg) = 7.065 MJ/day = 0.294 MJ/h. Physical activity factor for moderate occupational work (classified as moderate) 2.7 × BMR. Average daily physical activity factor 1.78 × BMR (moderate occupational work). Occupational workload: 8 h/day considering work days only; 5 h/day average workload over the entire year, including weekends, holidays, and absence.

[g]Based on a basal metabolic rate for adult women of BMR = 0.0364 W + 3.47 (W = weight = 55 kg) = 5.472 MJ/day = 0.228 MJ/h. Physical activity factor for moderate occupational work 2.2 × BMR. Average daily physical activity factor (based on moderate occupational work) 1.64 × BMR. Occupational workload: 8 h/day considering work days only; 5 h/days average work load over the entire year, including weekends, holidays, and absence.

[h]Assuming a 50% gender ratio.

[i]A typical household is arbitrarily assumed to consist of one adult male (70 kg, moderate occupational activity), one adult female (55 kg, moderate occupational activity), and two children (male of 12 years and female of 9 years).

[j]This assessment refers to the United States. In 1993, the food energy requirement was 910,000 TJ/year and the work supply was 215 billion h.

[k]Assuming 10 MJ of fossil energy spent in the food system per 1 MJ of food consumed.

[l]Assuming a total primary energy supply in 1993 in the United States (including the energy sector) of 85 million TJ divided by a work supply of 215 billion h.

[m]Assuming a transformity ratio for fossil energy of 50 million EMJ/joule.

work," refers to the concept of "free energy" or "exergy," and this is another potential source of confusion. Both of these concepts require a previous formal definition of work and a clear definition of operational settings to be applied.

3.2 Work

The similar ambiguous definition of energy is also found when dealing with a general definition of work, a definition applicable to any specific space–time scale-dependent settings. The classical definition found in dictionaries refers to the ideal world of elementary mechanics: "work is performed only when a force is exerted on a body while the body moves at the same time in such a way that the force has a component in the direction of the motion." Others express work in terms of "equivalent to heat," as is done in thermodynamics: "the work performed by a system during a cyclic transformation is equal to the heat absorbed by the system." This calorimetric equivalence offers the possibility to express assessments of work in the same unit as energy, that is, in joules. However, the elaborate description of work in elementary mechanics and the calorimetric equivalence derived from classical thermodynamics are of little use in real-life situations. Very often, to characterize the performance of various types of work, we need inherently qualitative characteristics that are impossible to quantify in terms of "heat equivalent." For example, the work of a director of an orchestra standing on a podium cannot be described using the previously given definition of elementary mechanics, nor can it be measured in terms of heat. An accounting of the joules related to how much the director sweats during the execution of a musical score will not tell anything about the quality of his or her performance. "Sweating" and "directing well" do not map onto each other. Indeed, it is virtually impossible to provide a physical formula or general model defining the quality (or value) of a given task (what has been achieved after having fulfilled a useful task) in numerical terms from within a descriptive domain useful in representing energy transformations based on a definition of identity for energy carriers and energy converters. In fact, formal measurements leading to energetic assessments refer to transformations occurring at a lower level (e.g., how much heat has been generated by muscles at the interface level $n/n-1$), whereas an assessment of performance (e.g., how much the direction has been appreciated by the public) refers to interactions and feedback occurring at the interface level $n/n+1$). This standard predicament applies to all real complex systems and especially to human societies.

3.3 Power

The concept of power is related to "the time rate of an energy transfer" or, when adopting the theoretical formalization based on elementary mechanics, to "the time rate at which work is being done." Obviously, this definition cannot be applied without a previous valid and formal mapping of the "transfer of energy" in the form of a numerical indicator or of "doing the work." At this point, it should be obvious that the introduction of this "new" concept, based on the previous ones, does not get us out of the predicament experienced so far. Any formal definition of power will run into the same epistemological impasse discussed for the previous two concepts given that it depends on the availability of a valid definition of them in the first place.

It should be noted, however, that the concept of power introduces an important new qualitative aspect—a new attribute required to characterize the energy transformation—not present in the previous two concepts. Whereas the concepts of energy and work, as defined in physics, refer to quantitative assessment of energy without taking into account the time required for the conversion process under analysis, the concept of power is, by definition, related to the rate at which events happen. This introduces a qualitative dimension that can be related either to degree of organization of the dissipative system or to the size of the system performing the conversion of energy in relation to the processes that guarantee the stability of boundary conditions in the environment. To deliver power at a certain rate, a system must have two complementing but distinct relevant features. First, it must have an "adequate organized structure" to match the given task. In other words, it must have the capability of doing work at a given rate. For example, an individual human cannot process and convert into power 100,000 kcal of food in a day. Second, it must have the capability of securing a sufficient supply of energy input for doing the task. For example, to gain an advantage from the power of 100 soldiers for 1 year, we must be able to supply them with enough food. In short, "gasoline without a car" (the first feature) or "a car without gasoline" (the second feature) is of no use.

This is the point that exposes the inadequacy of the conventional output/input approach. In hierarchically organized adaptive systems, it is not thinkable to have an assessment of energy flows without properly addressing the set of relevant characteristics of the process of transformation that are level and scale dependent in their perception and representation. The analyst must address the existence of expected differences in the perception and representation of concepts such as energy, power, and work done on different hierarchical levels and associated with the multiscale identity of the dissipative system in the first place. This means that the three theoretical concepts—energy, work, and power—must be simultaneously defined and applied within such a representation in relation to the particular typology of dissipative system (engines of cars are different from muscles in humans).

At this point, if we accept the option of using nonequivalent descriptive domains in parallel, we can establish a reciprocal entailment on the definitions of identity of the various elements (converter, whole system, energy carriers, and environment) to characterize the set of energy transformations required to stabilize the metabolism of a dissipative system.

In this case, we have two couples of self-entailment among identities:

Self-Entailment 1: The identities adopted for the various converters (on interface level n and level $n-1$) referring to the power level of the converter *define/are defined by* the identities adopted for the energy carriers (on interface level $n-1$ and level $n-2$).

Self-Entailment 2: The identities of the set of energy forms considered to be useful energy on the focal level, that is, characterization of the autocatalytic loop from within (on interface between level $n-1$ and level n) in relation to their ability, validated in the past, to fulfill a given set of tasks, *define/are defined by* the compatibility of the identities of the whole system in relation to its interaction with the larger context, that is, characterization of the autocatalytic loop from outside (on the interface between level n and level $n+1$). This compatibility can be associated with the determination of the set of useful tasks in relation to the availability in the environment of the required flow of input and sink capacity for wastes.

When using these two set of relations and identities, which depend on each other for their definitions, simultaneously, we can link nonequivalent characterizations of energy transformations across scales (Fig. 2).

Bridge 1: conversion rates represented on different levels must be compatible with each other. This implies a constraint of compatibility between the definition of identity of the set of converters defined at level $n-1$ (triadic reading: $n-2/n-1/n$) and the definition of the set of tasks for the whole defined at level $n+1$ (triadic reading: $n-1/n/n+1$). This constraint addresses the ability of the various converters to generate "useful energy" (the right energy form applied in the specified setting) at a given rate that must be admissible for the various tasks. This bridge deals with qualitative aspects of energy conversions.

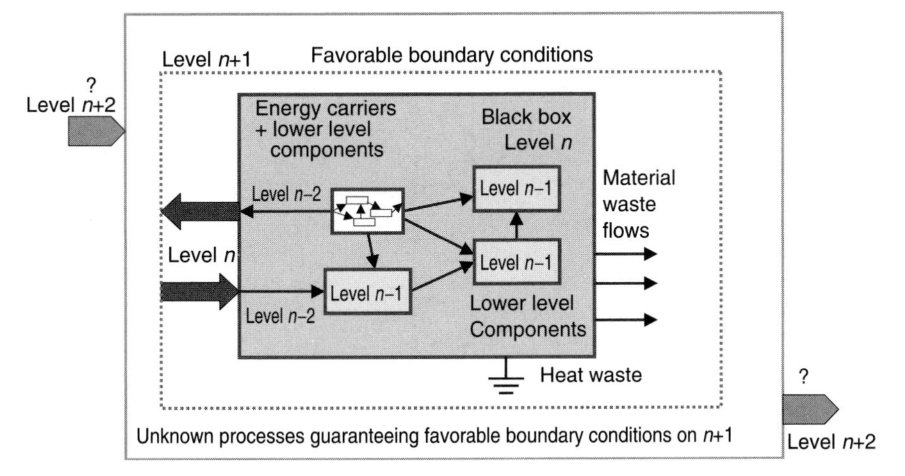

FIGURE 2 Hierarchical levels that should be considered when studying autocatalytic loops of energy forms. From Giampietro, M. (2003). "Multi-Scale Integrated Analysis of Agroecosystems." CRC Press, Boca Raton, FL. Used with permission.

Bridge 2: The flow of energy input from the environment and the sink capacity of the environment must be enough to cope with the rate of metabolism implied by the identity of the black box. This implies a constraint of compatibility between the size of the converters defined at level $n-1$ (triadic reading: $n-2/n-1/n$) and the relative supply of energy carriers and sink capacity related to processes occurring at level $n+1$. The set of identities of the input (from the environment to the converters) and output (from the converters to the environment) is referring to a representation of events belonging to level $n-2$. In fact, these energy carriers will interact with internal elements of the converters to generate the flow of useful energy and will be turned out into waste by the process of conversions. However, the availability of an adequate supply of energy carriers and of an adequate sink capacity is related to the existence of processes occurring in the environment that are needed to maintain favorable conditions at level $n+1$. Differently put, the ability to maintain favorable conditions in the face of a given level of dissipation can be defined only by considering level $n+1$ as the focal one (triadic reading: $n/n+1/n+2$).

This means that when dealing with quantitative and qualitative aspects of energy transformations over an autocatalytic loop of energy forms, we have to bridge at least five hierarchical levels (from level $n-2$ to level $n+2$).

Unfortunately, by definition, the environment (processes determining the interface between levels $n+1$ and $n+2$) is something we do not know enough about. Otherwise, it will become part of the modeled system. This means that when dealing with the stability of "favorable boundary conditions," we can only hope that they will remain the way they for as long as possible. On the other hand, the existence of favorable boundary conditions is a must for dissipative systems. That is, the environment is and must generally be assumed to be an "admissible environment" in all technical assessments of energy transformations.

Therefore, the existence of favorable boundary conditions" (interface level $n+1/n+2$) is an assumption that is not directly related to a definition of usefulness of the tasks (interface level $n/n+1$) in relation to the issue of sustainability. The existing definition of the set of useful tasks at level n simply reflects the fact that these tasks were perceived as useful in the past by those living inside the system. That is, the established set of useful tasks was able to sustain a network of activities compatible with boundary conditions (*ceteris paribus* at work). However, this definition of usefulness for these tasks (what is perceived as good at level n according to favorable boundary conditions at level $n+1$) has nothing to do with the ability or effect of these tasks in relation to the stabilization of boundary conditions in the future. For example, producing a given crop that provided an abundant profit last year does not necessarily imply that the same activity will remain so useful next year. Existing favorable boundary conditions at level $n+1$ require the stability of the processes occurring at level $n+2$ (e.g., the demand for that crop remains high in the face of a limited supply, natural resources such as nutrients, water, soil, and pollinating bees will remain available for the next year). This is information we do not know, and cannot know enough about, in advance. This implies that analyses of "efficiency" and "efficacy" are based on data referring to characterizations and representations relative to identities defined only on the four levels $n-2$, $n-1$, n, and $n+1$ (on the *ceteris paribus* hypothesis and reflecting what has been validated in the past). Because of this, they are not very useful in studying coevolutionary trajectory of dissipative systems. In fact, they (1) do not address the full set of relevant processes determining the stability of favorable boundary conditions (they miss a certain number of relevant processes occurring at level $n+2$) and (2) deal only with qualitative aspects (intensive variables referring to an old set of identities) but not quantitative aspects such as the relative size of components (how big is the requirement of the whole dissipative system—an extensive variable assessing the size of the box from the inside—in relation to the unknown processes that stabilize the identity of its environment at level $n+2$). As observed earlier, the processes that are stabilizing the identity of the environment are not known by definition. This is why, when dealing with coevolution, we have to address the issue of "emergence." That is, we should expect the appearance of new relevant attributes (requiring the introduction of new epistemic categories in the model, that is, new relevant qualities of the system so far ignored) to be considered as soon as the dissipative system (e.g., human society) discovers or learns new relevant information about those processes occurring at level $n+2$ that were not known before.

4. THE METAPHOR OF PRIGOGINE SCHEME TO CHARACTERIZE AUTOCATALYTIC LOOPS OF ENERGY FORMS IN HIERARCHICALLY ORGANIZED ADAPTIVE SYSTEMS

This section applies the rationale of hierarchical reading of energy transformations to the famous scheme proposed by Prigogine to explain the biophysical feasibility of dissipative systems in entropic terms. As Schroedinger showed, living systems can escape the curse of the second law thanks to their ability to feed on "negentropy." Differently put, living systems are, and must be, open systems that can preserve their identity thanks to a metabolic process.

This original idea has been further developed by the work of the school of Prigogine in nonequilibrium thermodynamic. These researchers introduced the class of dissipative systems in which the concept of entropy is associated with that of self-organization and emergence. In this way, it becomes possible to better characterize the concept of metabolism of dissipative systems. The ability to generate and preserve in time a given pattern of organization depends on two self-entailing abilities: (1) the ability to generate the entropy associated with the energy transformations occurring within the system and (2) the ability to discharge this entropy into the environment at a rate that is at least equal to that of internal generation. As observed by Schneider and Kay in a famous article in 1994, the possibility to have life, self-organization, and autocatalytic loops of energy forms is strictly linked to the ability of open dissipative systems to generate and discharge entropy into an admissible environment. Actually, the more such systems can generate and discharge entropy, the higher the complexity of the patterns they can sustain.

Dissipative systems have to maintain a level of entropy generation and discharge that is admissible in relation to their identity of metabolic systems. Differently put, the pattern of energy dissipation associated with their expected identity in terms of an ordered structure of material flows across hierarchical levels must be compatible with a set of favorable boundary conditions within the environment. Using the vocabulary developed by Schroedinger and Prigogine, this requires compensating the unavoidable generation of entropy associated with internal irreversibility (dS_i) by importing an adequate amount of negentropy (dS_e) from their environment. The famous relation proposed by Prigogine to represent this idea can be related as

$$dS_T \Leftrightarrow dS_i + dS_e \quad (1)$$

Relation (1) indicates that the identity of the system defined at interface level $n/n-1$ must be congruent or compatible with the identity of the larger dissipative system defined at interface level $n+1/n$. In other words, the two flows of entropy internally generated (dS_i) and "imported/exported" (dS_e) must be harmonized with change in entropy of the system (dS_T) in which the identity of the first is embedded.

Let us now try to interpret, in a metaphorical sense, the scheme proposed by Prigogine to characterize autocatalytic loops of energy forms within nested metabolic systems.

First, dS_i refers to the representation of the mechanism of internal entropy production that is associated with the irreversibility generated to preserve system identity. This assessment necessarily must be obtained using a representation of events related to a perception obtained within the black box on interface level $n-2/n-1/n$. This can be done by using a mapping of an energy form that makes it possible to represent how the total input used by the black box is then invested among the various parts (the internal working of the metabolism).

Second, dS_e refers to the representation of imported negative entropy. The term "negentropy" entails the adoption of a mechanism of mapping that has to be related to the previous choices made when representing dS_i. This term requires establishing a bridge between the assessment of two forms of energy: (1) one used to describe events within the black box in terms of dS_i (the mapping used to assess the effect of internal mechanisms associated with irreversibility) and (2) another used to describe how the metabolism of the larger system (the environment) providing favorable boundary conditions is affected by the activity of the black box. This implies using the mapping of a second energy form. This second energy form, reflecting the interference of the black box over the metabolism of the environment, has to be effectively linkable to the first energy form.

The first energy form is used to represent relevant mechanisms inside the black box. The second energy form is used to represent the effects of the interaction of the black box with its associative context. It represents the view of the metabolism of the black box from the outside.

Third, dS_T translates into a required compatibility, in quantitative and qualitative terms, of the overall interaction of the black box with the environment according to the particular selection of mapping of the two energy forms. Whenever $dS_T < 0$, the system can expand its domain of activity, either by growing in size (generating more of the same pattern, so to speak) – or by moving to another dissipative pattern (developing emergent properties). On the contrary, whenever $dS_T > 0$, the system has to reduce its activity. It can remain alive, at least for a while, by sacrificing a part of its activities. In this case, a certain level of redundancy can be burned, providing a temporary buffer, to get through critical moments and cyclic perturbations (disappearance of expected favorable conditions).

As noted earlier, it is not possible to obtain a simultaneous check in formal terms—and in a substantive way—of the relation among the three dSs defined on different descriptive domains. On the other hand, an autocatalytic loop of different energy forms can be used to study the forced congruence across levels of some of the characteristics of the autocatalytic loop defined on different descriptive domains (Fig. 3).

In Fig. 3A, we resume the key features of the energetic analysis of dissipative systems. In particular, we provide a different perspective on the discussion related to the representation of hierarchically organized dissipative systems (Fig. 1). That is, the network represented in Fig. 2 can be represented by dividing the components described at level $n-1$ into two compartments: (1) those that do not interact directly with the environment (aggregated in the compartment labeled "indirect" in Fig. 3A) and (2) those that interact directly with the environment by gathering environmental inputs and providing those useful tasks aimed at such a gathering (aggregated in the compartment labeled "direct" in Fig. 3A). Thanks to the existence of

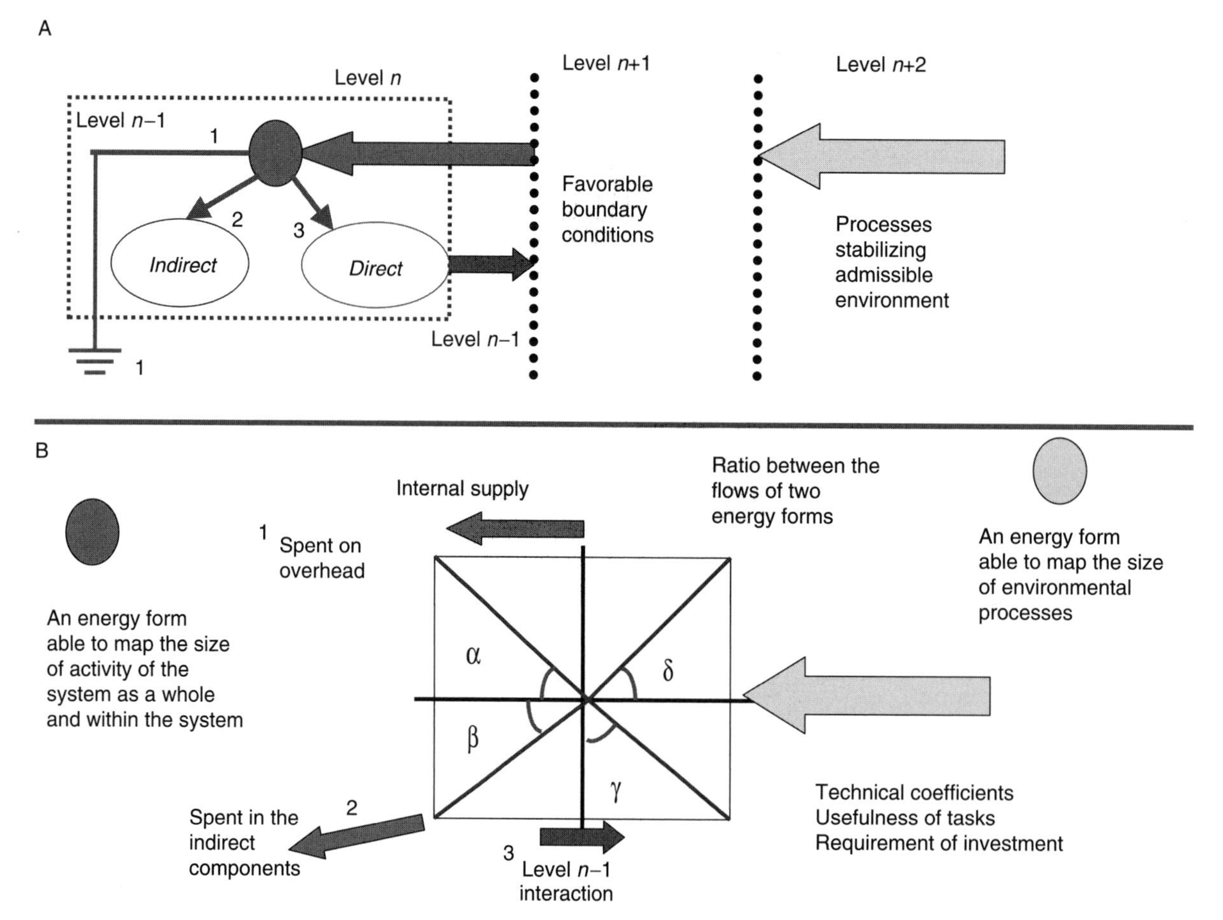

FIGURE 3 Congruence across levels of characteristics of the autocatalytic loop defined over nonequivalent descriptive domains. From Giampietro, M. (2003). "Multi-Scale Integrated Analysis of Agroecosystems." CRC Press, Boca Raton, FL. Used with permission.

favorable conditions (perceived at level $n+1$), the whole system (at level n) can receive an adequate supply of required input. This input can then be expressed in an energy form that is used as reference to assess how this total input is then invested over the lower level compartment. Therefore, we can expect that this total input will be dissipated within the black box in three flows (indicated by the three arrows in Fig. 3A): a given overhead (1) for the functioning of the whole, (2) for the operation of the compartment labeled "indirect," and (3) for the operation of the compartment labeled "direct." The favorable conditions perceived at level $n+1$, which make possible the stability of the environmental input consumed by the whole system, in turn are stabilized due to the existence of some favorable gradient generated elsewhere (level $n+2$), and this is not accounted for in the analysis. This available gradient can be exploited thanks to the tasks performed by the components belonging to the direct compartment (energy transformations occurring at level $n-1$). The return between the energy input made available to the whole system (at level n) per unit of useful energy invested by the direct compartment into the interaction with the environment will determine the strength of the autocatalytic loop of energy generated at level $n-1$ by the exploitation of a given energy resource. This integrated use of representations of energy transformation across levels is at the basis of a new approach that can be used to describe the sustainability of complex adaptive systems that base their identity on metabolic processes. In particular, three very useful concepts are (1) a mosaic effect across hierarchical levels, (2) impredicative loop analysis, and (3) narratives capable of surfing in complex time. These concepts are required to provide coherent representation of identities of energy forms generating an autocatalytic loop over five contiguous hierarchical levels (from level $n-2$ to level $n+2$). These three concepts can be used to structure the information space in a way that makes it possible to perform a series of congruence checks on the selected representation of the autocatalytic loop of energy forms.

An example of this congruence check is shown in Fig. 3. It is possible to establish a relation among the representation of parts, whole, and interaction with the environment across scales using a four-angle figure that combines intensive and extensive variables used to represent and bridge the characterization of metabolic processes across levels. The two angles on the left side (α and β) refer to the distribution of the total available supply of energy carriers (indicated on the upper part of the vertical axis) among the three flows of internal consumption. The angle α refers to the fraction of the total supply that is invested in overhead. The angle β refers to the fraction of the total supply that is spent on the operation of internal components. The value of the lower part of the vertical axis at this point refers to the amount of primary energy (the mapping related to the internal perception of energy form) that is invested in the direct interaction with the environment. The two angles on the right (γ and δ) refer to a characterization of the interaction of the system with the environment.

The selection of mapping (what set of variables have to be used in such a representation) must fulfill the double task of making it possible to relate the perception and representation of relevant characteristics of what is going on inside the black box to characteristics that are relevant to studying the stability of the environment. That is, this choice does not have the goal of establishing a direct link between a model studying the dynamics inside the black box and models studying relevant dynamics in the environment. To reiterate, this is simply not possible. This analytical tool simply makes it possible to establish bridges among nonequivalent representations of the identity of parts and wholes.

An example of three impredicative loop analyses is presented in Fig. 4. These formalizations reflect three logically independent ways of looking at an autocatalytic loop of energy forms according to the scheme presented in Fig. 3. It is very important to note that these three formalizations cannot be directly linked to each other because they are constructed using logically independent perceptions and characterizations. However, they share a meta-model used for the semantic problem structuring that made it possible to tailor in the formalization (when putting numerical assessment in it) according to the system considered.

5. CONCLUSION

The existence of both multiple levels of organization in the investigated system and different useful perspectives for looking at the reality entails the use of multiscale analysis when dealing with the sustainability of dissipative systems organized in nested hierarchies. This translates into a severe epistemological challenge for energy analysis. Innovative

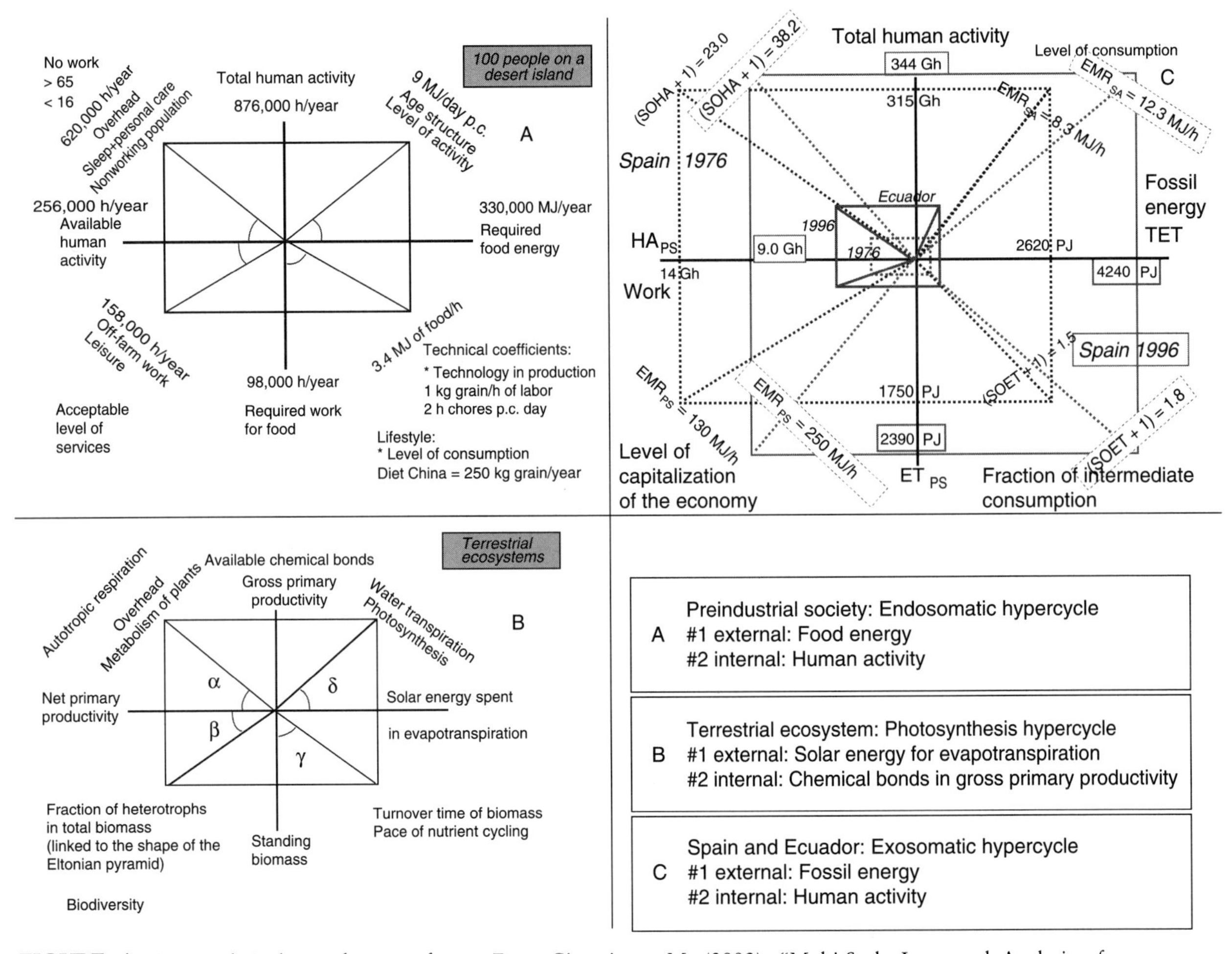

FIGURE 4 Autocatalytic loop of energy forms. From Giampietro, M. (2003). "Multi-Scale Integrated Analysis of Agroecosystems." CRC Press, Boca Raton, FL. Used with permission.

analytical methodologies can be developed taking advantage of principles derived from complex system theory and nonequilibrium thermodynamics. However, this innovative approach requires accepting two negative side effects.

First, the same autocatalytic loop can be represented in different legitimate ways (by choosing different combinations of formal identities). That is, there is no such thing as the "right model" when dealing with an analysis of the sustainability of an autocatalytic loop of energy. Rather, all models are incomplete. Analysts have the role of helping in the societal discussion about how to select narratives and models that are useful.

Second, there is an unavoidable degree of ignorance associated with any formal representation of an autocatalytic loop of energy forms. In fact, the very set of assumptions that make it possible to represent the system as viable (e.g., about the future stability of the relation between investment and return, about the future admissibility of the environment on level $n+2$, about the relevance of the selected set of attributes) guarantees that such a representation is affected by uncertainty. Both the observer and the observed system are becoming "*ceteris* are never *paribus*" when coming to the representation of the evolution in time of autocatalytic loops. The concept of entropy, in this case, translates into a sort of "yin yang" predicament for the analysis of complex adaptive systems. The more a dissipative system is successful in defending and imposing its identity on the environment, the sooner it will be forced to change it.

In short, any representation in energy terms of the predicament of sustainability for a complex dissipative system is (1) just one of many possible alternatives and (2) unavoidably affected by uncertainty, ignorance, and certain obsolescence.

SEE ALSO THE FOLLOWING ARTICLES

Human Energetics • Input–Output Analysis • Multi-criteria Analysis of Energy • Thermodynamics, Laws of • Work, Power, and Energy

Further Reading

Allen, T. F. H., and Starr, T. B. (1982). "Hierarchy." University of Chicago Press, Chicago.

Giampietro, M. (2003). "Multi-Scale Integrated Analysis of Agroecosystems." CRC Press, Boca Raton, FL.

Hall, C. A. S., Cleveland, C. J., and Kaufmann, R. (1986). "Energy and Resource Quality." John Wiley, New York.

Herendeen, R. A. (1998). "Ecological Numeracy: Quantitative Analysis of Environmental Issues." John Wiley, New York.

Odum, H. T. (1996). "Environmental Accounting: Energy and Decision Making." John Wiley, New York.

O'Neill, R. V. (1989). Perspectives in hierarchy and scale. *In* "Perspectives in Ecological Theory" (J. Roughgarden, R. M. May, and S. Levin, Eds.), pp. 140–156. Princeton University Press, Princeton, NJ.

Pimentel, D., and Pimentel, M. (eds.). (1996). "Food, Energy, and Society," 2nd ed. University of Colorado Press, Niwot.

Prigogine, I. (1978). "From Being to Becoming." W. H. Freeman, San Francisco.

Prigogine, I., and Stengers, I. (1981). "Order out of Chaos." Bantam Books, New York.

Schneider, E. D., and Kay, J. J. (1994). Life as a manifestation of the second law of thermodynamics. *Math. Comp. Modelling* **19,** 25–48.

Smil, V. (1991). "General Energetics: Energy in the Biosphere." John Wiley, New York.

Rosen, R. (2000). "Essays on Life Itself." Columbia University Press, New York.

Rotmans, J., and Rothman, D. S. (eds.). (2003). "Scaling Issues in Integrated Assessment." Swets & Zeitlinger, Lissen, Netherlands.

Salthe, S. N. (1985). "Evolving Hierarchical Systems: Their Structure and Representation." Columbia University Press, New York.

Schrödinger, E. (1967). "What Is Life?" Cambridge University Press, Cambridge, UK.

Computer Modeling of Renewable Power Systems

PETER LILIENTHAL
U.S. Department of Energy, National Renewable Energy Laboratory
Golden, Colorado, United States

THOMAS LAMBERT
Mistaya Engineering, Inc.
Calgary, Alberta, Canada

PAUL GILMAN
U.S. Department of Energy, National Renewable Energy Laboratory
Golden, Colorado, United States

1. Introduction
2. Renewable Resource Modeling
3. Component Modeling
4. System Modeling
5. Models for Renewable Power System Design
6. Conclusions

Glossary

beam radiation Solar radiation that travels from the sun to Earth's surface without any scattering by the atmosphere. Beam radiation is also called direct radiation.

clearness index The fraction of solar radiation striking the top of Earth's atmosphere that is transmitted through the atmosphere to strike Earth's surface.

declination The latitude at which the sun's rays are perpendicular to the surface of Earth at solar noon.

diffuse radiation Solar radiation that has been redirected by atmospheric scattering.

extraterrestrial radiation Solar radiation striking the top of Earth's atmosphere.

global radiation The sum of beam radiation and diffuse radiation; sometimes called total radiation.

life-cycle cost The summation of all appropriately discounted costs, both recurring and nonrecurring, associated with a system or component of a system during its life span.

sensitivity analysis An investigation into the extent to which changes in certain inputs affect a model's outputs.

spinning reserve Surplus electrical generating capacity that is operating and able to respond instantly to a sudden increase in the electric load or to a sudden decrease in the renewable power.

Renewable power systems produce electricity from one or more renewable resources, and often incorporate nonrenewable power sources or energy storage devices for backup. Computer models of such systems are mathematical equations or algorithms programmed into a computer that can simulate system behavior or costs. Renewable power system designers use computer modeling to predict performance and explore different design options.

1. INTRODUCTION

Recent improvements in the economics of renewable power sources have made them competitive with conventional power sources in many situations. But the design of renewable power systems is made difficult by the novelty of the technologies, the intermittency and geographic diversity of the renewable resources on which they depend, and the large number of possible system configurations. Computer models, which are simplified mathematical representations of real systems, can help overcome these challenges. Models allow the designer to simulate the long-term performance and economics of renewable power systems, explore different design options, and test the effects of input assumptions.

This article deals with the modeling of solar, wind, small hydro, and biomass power systems suitable for small-scale (less than 1 MW) electric power generation, both stand-alone and grid-connected. The focus will be on long-term performance and economic modeling rather than dynamic modeling, which is concerned with short-term system stability and response. The material is structured to first address the modeling of each of the renewable resources, then the renewable technologies, then the power systems in which resources and technologies may be combined with each other and with more conventional power sources. In the final section, there is a description of the types of models currently available, and examples of each.

2. RENEWABLE RESOURCE MODELING

Characteristics of the renewable resources influence the behavior and economics of renewable power systems. Resource modeling is therefore a critical element of system modeling. In this section, the focus is on those aspects of each renewable resource that are important for modeling renewable power systems.

2.1 Solar Resource

Solar power technologies convert solar radiation into electricity. The goal of solar resource modeling is to estimate the amount of solar radiation available for conversion at some location, and how that amount varies over time. Four factors determine the amount of radiation striking a plane on Earth's surface: the amount of extraterrestrial radiation, the fraction of that radiation being transmitted through the atmosphere, the orientation of the plane, and shading of the plane by nearby objects such as buildings or mountains.

Extraterrestrial radiation varies in an entirely predictable fashion as Earth rotates on its axis and revolves around the sun. For any time between sunrise and sunset, the following equation gives the solar radiation incident on a horizontal plane at the top of the atmosphere:

$$G_o = G_{sc}\left(1 + 0.033\cos\frac{360n}{365}\right) \times (\cos\phi\cos\delta\cos\omega + \sin\phi\sin\delta)$$

where G_{sc} is the solar constant (normally taken as 1367 W/m^2), n is the day of the year, ϕ is the latitude (north positive), ω is the hour angle (zero at solar noon, increasing at 15° per hour throughout the day), and δ is the declination, given by

$$\delta = 23.45\sin\left(360\frac{284+n}{365}\right)$$

Integrating the extraterrestrial radiation over 1 hour gives the energy per unit area received over that hour. Integrating from sunrise to sunset gives the energy per unit area received daily. Summing this daily total for each day of the month gives the monthly total.

Atmospheric scattering and absorption reduce the amount of solar radiation striking Earth's surface to some fraction of the extraterrestrial radiation. This fraction, called the clearness index, varies stochastically in time, due principally to the fluctuating amount of water vapor and water droplets contained in the atmosphere. The equation for the clearness index is as follows:

$$k_t = G/G_o$$

where G is the radiation incident on a horizontal plane on the surface of Earth and G_o is the radiation incident on a horizontal plane at the top of the atmosphere. Atmospheric scattering affects the character as well as the amount of surface radiation. Under very clear conditions, the global radiation may be as much as 80% beam radiation and only 20% diffuse radiation. Under very cloudy conditions, the global radiation may be almost 100% diffuse radiation. Duffie and Beckman provide correlations that relate this diffuse fraction to the clearness index. The diffuse fraction is of interest to modelers for two reasons. First, the effect of surface orientation differs for beam and diffuse radiation, because beam radiation emanates only from the direction of the sun whereas diffuse radiation emanates from all parts of the sky. Second, some solar conversion devices (those that focus the sun's rays) can only make use of beam radiation.

Researchers have developed several algorithms for calculating the beam and diffuse radiation striking a surface with a particular slope (the angle between the surface and the horizontal) and azimuth (the direction toward which the surface faces), as well as means of accounting for shading from nearby objects (details can be found in the book by Duffie and Beckman, *Solar Engineering of Thermal Processes*). Most measured solar resource data are reported as global radiation on a horizontal surface, averaged for each hour, day, or month of the year. Models can use the above-mentioned algorithms to calculate the corresponding clearness indices, resolve the global radiation into its beam and diffuse components, and calculate the radiation striking a particular surface. The National Aeronautics and Space Administration

(NASA) Surface Solar Energy Web site provides satellite-measured monthly average radiation values for any location worldwide at a 1° resolution. Numerous other sources provide monthly or hourly data for more limited areas at higher spatial resolution. Several authors, including Mora-López and Sidrach-de-Cardona, have proposed algorithms for the generation of synthetic hourly radiation data from monthly averages.

2.2 Wind Resource

Wind turbines convert the kinetic energy of the wind into electricity. The goal of wind resource modeling is to describe the statistical properties of the wind at a particular location in order to estimate the amount of power a wind turbine would produce over time at that location. The wind power density, or power per unit area, depends on the air density ρ and the wind speed v according to the following equation:

$$P/A = (1/2)\rho v^3$$

The air density depends on pressure and temperature according to the ideal gas law. Air pressure varies with elevation above sea level, but does not vary significantly in time. Temperature does vary significantly with time at many locations, but because wind power density varies with the cube of the wind speed, the most significant factor in the variation of wind power with time is the fluctuating wind speed.

Driven by complex large-scale atmospheric circulation patterns and local topographic influences, wind speeds at a particular location typically exhibit seasonal, daily, and short-term fluctuations. Short-term fluctuations (on the order of minutes or seconds) can affect a wind turbine's structural loads and power quality, but they do not significantly affect long-term production or economics. The seasonal and daily patterns of wind speed are more difficult to predict than are those of solar radiation, because the wind speed is not driven by any well-defined phenomenon comparable to the solar extraterrestrial radiation. Accurate determination of seasonal and daily patterns typically requires anemometer measurements spanning months or years.

A defining characteristic of a wind regime is its long-term distribution of wind speeds. Long-term wind speed observations typically conform well to the two-parameter Weibull distribution, the probability distribution function of which is given by

$$f(v) = \frac{k}{c}\left(\frac{v}{c}\right)^{k-1} \exp\left[-\left(\frac{v}{c}\right)^{k}\right]$$

where v is the wind speed, c is the scale parameter (in the same units as v), and k is the dimensionless shape factor. The following equation relates the two parameters c and k to the mean wind speed:

$$\bar{v} = c\Gamma\left(\frac{1}{k}+1\right)$$

where Γ is the gamma function. The Rayleigh distribution is a special case of the Weibull distribution, with $k=2$. The Weibull distribution provides a convenient way to characterize the wind resource, because a mean wind speed and a Weibull k factor are sufficient to quantify the energy potential. Figure 1 shows the Weibull probability distribution function for a mean wind speed of 6 m/s and three values of the shape factor k. Lower values of k correspond to broader distributions. Weibull distributions fitted to measured wind data typically have k values between 1.0 and 3.0. Stevens and Smulders have determined several methods for fitting a Weibull distribution to measured data (see Fig. 1).

The wind speed tends to increase with height above ground, an important factor when using wind measurements taken at one height to estimate the production of a wind turbine at a different height. Modelers typically assume that the wind speed's increase with height follows either a logarithmic profile or a power law profile. The logarithmic profile is given by

$$\frac{v(z_1)}{v(z_2)} = \frac{\ln(z_1/z_0)}{\ln(z_2/z_0)}$$

where $v(z_1)$ and $v(z_2)$ are the wind speeds at heights z_1 and z_2, and the parameter z_0 is the surface roughness length. The power law profile is given by

$$\frac{v(z_1)}{v(z_2)} = \left(\frac{z_1}{z_2}\right)^{\alpha}$$

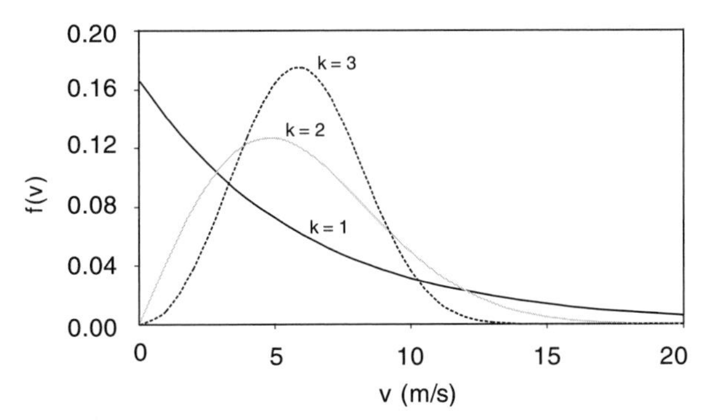

FIGURE 1 The Weibull distribution with a mean wind speed of 6 m/s and three values of k.

where α is the power law exponent. Manwell, McGowan, and Rogers have provided guidance on the selection of surface roughness length and the power law exponent, as well as a more complete examination of wind resource modeling.

2.3 Hydro Resource

Hydropower systems convert the energy of flowing water into electricity, usually by diverting a portion of a watercourse through a penstock (also called a pressure pipe) to a turbine. Hydro resource modeling aims to estimate the amount of power that a hydropower system could extract from a particular watercourse.

The ability of falling water to produce power depends on the flow rate and the head (the vertical distance through which it falls) according to the following equation:

$$P = \rho g h \dot{Q}$$

where ρ is the density of water, g is the gravitational acceleration, h is the head, and $\dot{Q}$ is the flow rate. Therefore the head (which is determined by local topography) and the available flow rate define the hydro resource at any particular location. The flow rate often varies seasonally and may in some cases vary according to the time of day, although it generally does not display the intermittency inherent to the solar and wind resources. Modelers sometimes characterize a watercourse's flow rate by computing its flow duration curve, an example of which appears in Fig. 2. This example shows that the flow rate exceeds 5 m^3/s about 37% of the time, and exceeds 10 m^3/s about 13% of the time. Flow duration curves can be created using long-term flow measurements or synthesized based on precipitation data within the catchment basin. Because most small hydro systems do not incorporate storage, the critical value is the firm flow, usually defined as the flow rate available 95 or 100% of the time. Designers typically size hydro systems based on the firm flow, or some portion thereof. The Web site www.microhydropower.net contains information on many aspects of small hydro systems, including guidance on the measurement of head and flow.

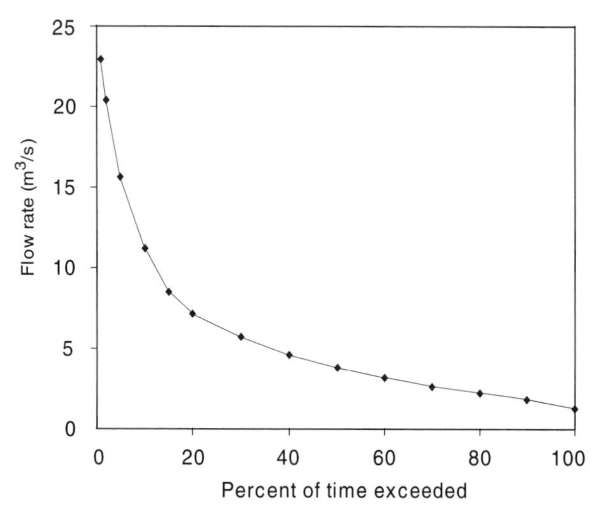

FIGURE 2 Sample flow duration curve.

2.4 Biomass Resource

Biomass power systems produce electricity from the chemical energy contained in organic matter. Many different types of biomass are suitable for power production, including wood waste, agricultural residue, animal waste, and energy crops. The goal of biomass resource modeling is to quantify the availability, price, and physical properties of the feedstock.

Three factors distinguish the modeling of the biomass resource from that of the other renewable resources. First, the amount of available biomass depends largely on human action, because the feedstock must be harvested, transported, and, in the case of dedicated energy crops, even sowed. So, although the gross amount of biomass depends on natural factors such as growing season and rainfall, the intermittency and unpredictability that characterize the other renewable resources do not apply to the biomass resource. Second, the biomass feedstock is often easily stored, further reducing intermittency and increasing flexibility. Third, the feedstock is often converted to a liquid or gaseous fuel, which can be stored and transported. As a result of these factors, biomass resource modeling typically focuses not on availability, but rather on the properties of the feedstock (such as moisture content) and the costs of delivering it in a usable form. McKendry has provided a thorough overview of the biomass resource.

3. COMPONENT MODELING

The characteristics of renewable technologies relevant to their modeling are briefly outlined here. Because renewable power systems often include some form of backup power to compensate for the intermittent nature of the renewable resources, fossil-fueled generators, energy storage devices, and grid connection are also covered here.

3.1 Photovoltaic Modules

Photovoltaic (PV) modules convert solar radiation directly to direct current (DC) electricity, with sizes ranging from a few watts to hundreds of kilowatts. The output current of a photovoltaic module increases linearly with increasing incident global radiation, and decreases linearly with increasing module temperature. At any fixed temperature and radiation value, current (and hence power) varies nonlinearly with voltage, as shown in Fig. 3.

The peak of the power curve in Fig. 3 is called the maximum power point. The voltage corresponding to the maximum power point varies with radiation and temperature. If the photovoltaic module is directly connected to a DC load (or a battery bank), then the voltage it experiences will often not correspond to the maximum power point, and performance will suffer. A maximum power point tracker is a solid-state device that decouples the module from the load and controls the module voltage so that it always corresponds to the maximum power point, thereby improving performance. The complexity of modeling the power output from photovoltaic modules depends on the presence of the maximum power point tracker. In the absence of a maximum power point tracker, the modeler must account for the effects of radiation, temperature, and voltage. Duffie and Beckman have provided algorithms to do so. A maximum power point tracker eliminates the nonlinear dependence on voltage and allows the modeler to use the following equation for module efficiency:

$$\eta = \eta_{\text{ref}} + \mu(T_c - T_{\text{ref}})$$

where T_c is the module temperature, T_{ref} is the reference temperature, η_{ref} is the module efficiency at the reference temperature, and μ is the efficiency temperature coefficient. The efficiency temperature coefficient is a property of the module and is typically a negative number. From the efficiency, a modeler can calculate the power output of a photovoltaic module using the following equation:

$$P = \eta G_T A$$

where η is the efficiency, G_T is the global (total) solar radiation incident on the surface of the module, and A is the area of the module.

The life-cycle cost of a photovoltaic subsystem is dominated by initial capital cost. The lack of moving parts (with the rare exception of tracking systems) results in long lifetimes and near-zero operation and maintenance costs. Economic models of photovoltaic modules commonly assume 20- to 30-year lifetimes and neglect operation and maintenance costs altogether.

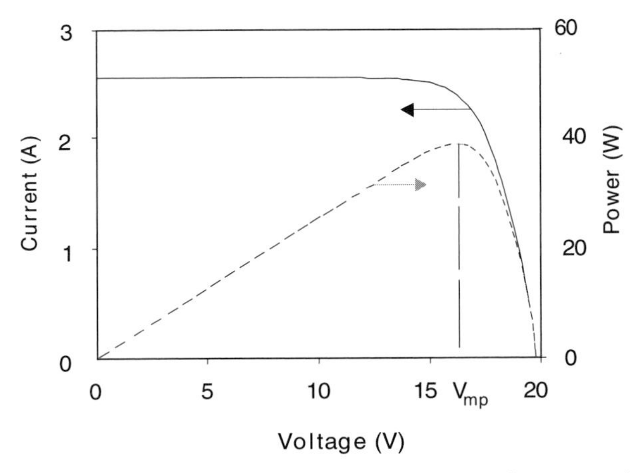

FIGURE 3 Current and power versus voltage for a typical photovoltaic module.

3.2 Wind Turbines

Wind turbines generate alternating or direct current (AC or DC) power, in sizes ranging from a few watts to several megawatts. Each wind turbine has a characteristic power curve that describes its power output as a function of the wind speed at its hub height. Figure 4 shows an example. Power curves typically display a cut-in wind speed below which the output is zero, then a section wherein the output varies roughly with the cube of the wind speed, then a section wherein the output peaks and declines due

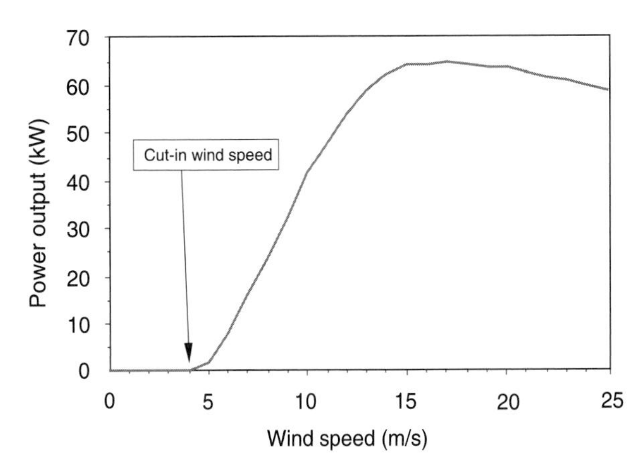

FIGURE 4 A wind turbine power curve.

to limitations designed into the turbine. The power curve describes turbine performance at a reference air density, typically corresponding to standard temperature and pressure. For a turbine operating at a different air density, the modeler can calculate its power output using the following equation:

$$P = P_{pc}(\rho/\rho_{ref})$$

where P_{pc} is the power output predicted by the power curve, ρ is the actual air density, and ρ_{ref} is the reference air density. Because the wind turbine output is so nonlinear with wind speed, and the wind speed is so variable, a modeler must exercise care when estimating the output of a wind turbine over long periods. The direct use of annual, monthly, or even daily average wind speeds causes unacceptably large inaccuracies. Time series data can be used directly only if its averaging interval is 1 hour or less. For longer term averages, the modeler must consider the frequency distribution of wind speeds.

Initial capital costs (including installation) dominate the life-cycle cost of wind turbines, but maintenance costs do occur. Economic models of wind turbines often assume a 15- to 25-year lifetime and an annual operation and maintenance cost of 1–3% of capital cost.

3.3 Small Hydropower Systems

Small hydropower systems generate AC or DC electricity from flowing water. The following equation gives the output power of a hydropower system:

$$P = \eta_{turb}\eta_{gen}\rho g h_{eff}\dot{Q}$$

where P is the output power in watts, η_{turb} is the efficiency with which the hydro turbine converts water power into shaft power, η_{gen} is the efficiency with which the generator converts shaft power into electrical power, ρ is the density of water in kg/m^3, g is the gravitational acceleration in m/s^2, h_{eff} is the effective head (the available head minus the head loss due to turbulence and friction) in meters, and $\dot{Q}$ is the volumetric flow rate through the turbine in m^3/s.

The turbine efficiency depends on the type of turbine and the flow rate. Hydro turbines fall into two broad categories, impulse turbines and reaction turbines. In an impulse turbine, water flows through a nozzle and emerges as a high-velocity jet, which strikes the turbine blades and causes them to rotate. In a reaction turbine, the flowing water surrounds the turbine blades, which are shaped such that the flowing water creates lift forces, causing rotation. Most small hydro systems use impulse turbines such as the Pelton, Turgo, or crossflow, but some low-head applications use reaction turbines such as the Francis, propeller, or Kaplan. Figure 5 shows typical efficiency curves. Although the impulse turbines exhibit near-constant efficiencies throughout their operating range, the reaction turbines (excluding the Kaplan) are much less efficient at lower flow rates. The Kaplan turbine, an adjustable-blade propeller turbine, exhibits an efficiency curve similar to that of the impulse turbines (see Fig. 5).

Friction and turbulence in the penstock cause a reduction in the water pressure upstream of the turbine. This pressure loss can be expressed as a head loss. The effective head can then be calculated by subtracting this head loss from the available head (the vertical drop from the penstock inlet to the turbine). Modelers can calculate the head loss using the standard techniques of fluid mechanics (Darcy friction factor, Moody diagram) or estimate it using rules of thumb.

Small hydro systems rarely incorporate storage reservoirs, because the economic and ecological costs of reservoirs often exceed their value in compensating for variation in the flow rate and adding a measure of control. Most run-of-river systems (ones that do not incorporate storage) have no means of restricting the flow rate though the turbine, so the flow rate in the preceding equation is simply equal to the flow through the penstock, which cannot exceed the stream flow.

Like photovoltaic modules and wind turbines, small hydro systems have high initial costs and low operating costs. The capital cost of a hydro system, which includes the cost of the required civil works plus that of the turbine and related machinery,

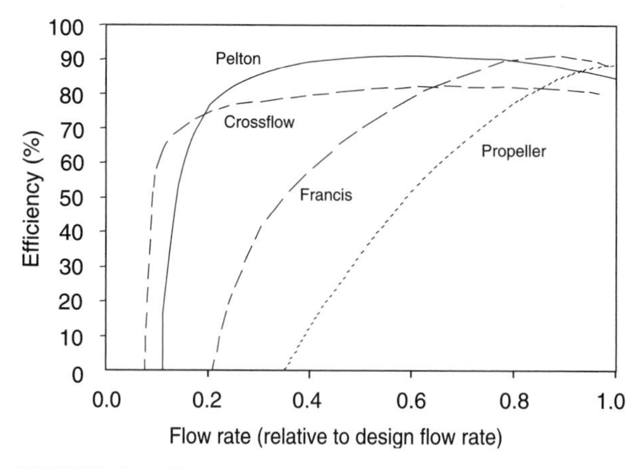

FIGURE 5 Efficiencies curves of common small hydro turbines. The Pelton and crossflow are impulse turbines; the Francis and propeller are reaction turbines. From Paish (2002), with permission.

depends very strongly on the local topography. Sites with high head and low flow generally have more favorable economics than do those with low head and high flow, because the required size of the machinery depends most strongly on the flow rate. Hydro systems are very long-lived, and lifetimes of 30 to 50 years are not uncommon.

3.4 Biomass Power Systems

The conversion of biomass into electricity typically makes use of either a thermochemical process (gasification, pyrolysis, or direct combustion), or a biochemical process (fermentation or anaerobic digestion). With the exception of direct combustion, each of these processes results in a gaseous or liquid fuel that can be used like any conventional fuel (sometimes in combination with a conventional fuel) in an internal combustion engine, microturbine, or fuel cell. In a direct combustion process, heat generated by burning biomass feedstock drives a thermal cycle. From a modeling perspective, these biomass conversion technologies are more akin to conventional forms of electrical production than to the other renewable power technologies. In particular, their ability to produce power on demand and their relatively high operating costs distinguish them from solar, wind, and hydropower technologies. The differences between biomass power and conventional power lie in technical issues (such as fuel handling and combustion characteristics) and the possibly seasonal availability of the biomass-derived fuel.

3.5 Fuel-Fired Generators

The diesel generator, an electrical generator driven by a reciprocating diesel engine, is the most common source of electric power in remote locations. Sizes range from a few kilowatts to 1 MW. Gasoline- and propane-fueled generators are also common below about 25 kW. Microturbines have become commercially available in recent years in the 25- to 250-kW size range, and fuel cells may become widespread in years to come.

Their ability to produce power on demand and their fast response to a fluctuating electric load make conventional generators valuable in stand-alone systems. When used in combination with renewable power sources, a generator can provide backup power during times of insufficient renewable output. For constant-speed reciprocating engine generators as well as microturbines, the rate of fuel consumption is well approximated by the following first-order

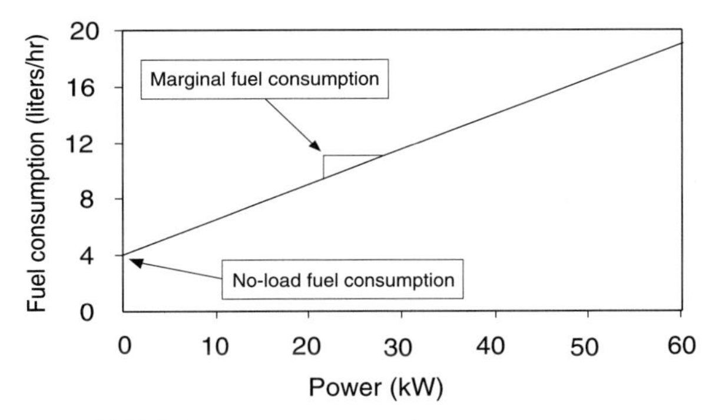

FIGURE 6 Typical generator fuel consumption curve.

equation:

$$F = F_0 + F_1 \cdot P$$

where F is the rate of fuel consumption, F_0 and F_1 are constants (the no-load fuel consumption and the marginal fuel consumption, respectively), and P is the output power of the generator. This relation is shown in Fig. 6. Skarstein and Uhlen found that, for diesel generators, the value of F_0 scales linearly with generator size. The fuel consumption of variable-speed reciprocating engine generators is better approximated with a higher order function, although they still exhibit a no-load fuel consumption.

Most models assume that a generator will need replacement (or a major overhaul) after a certain number of operating hours. To avoid engine damage, constant-speed diesel generators typically do not operate below about 30% of their rated capacity, and most other generator types have similar restrictions. Compared to renewable power sources, conventional generators have a low capital cost per kilowatt capacity, and a high operating cost per kilowatt-hour of electricity produced. Importantly, the operating cost per kilowatt-hour decreases with increasing output power. This operating cost has two components: maintenance and fuel. The maintenance cost per hour of operation is essentially independent of the output power; so on a per-kilowatt-hour basis it is inversely proportional to output power. Due to the aforementioned shape of the fuel consumption curve, the fuel cost per kilowatt-hour also decreases with output power. Generators are therefore most economically efficient when running at full capacity, a fact that has implications for power system design and operation.

3.6 Energy Storage Technologies

Energy storage devices can play an important role in stand-alone renewable power systems. In systems without generators, they can provide backup power for intermittent renewable power sources. In systems with generators, they can save fuel and help avoid inefficient generator operation by serving the load during times of low electric demand, where the generator is least efficient. By buffering against the fluctuations in renewable power output, energy storage can also reduce the frequency of generator starts. Beyer, Degner, and Gabler found that in high-penetration wind/diesel systems (where the installed wind capacity exceeds the average load), even a very small amount of storage dramatically reduces the frequency of diesel starts.

Batteries, in particular lead-acid batteries, remain the predominant energy storage device. Many competing battery types (nickel–cadmium, nickel–metal hydride, lithium ion, sodium–sulfur, metal–air, flow batteries) surpass the lead-acid battery in one or more aspect of performance such as cycle life, round-trip efficiency, energy density, charge and discharge rate, cold-weather performance, or required amount of maintenance. In most applications, however, their lower cost per kilowatt-hour of capacity makes lead-acid batteries the optimal choice. Alternatives such as flywheels, ultracapacitors, or hydrogen storage may become commercially successful in the future, but are rare presently. In this section, the focus is exclusively on batteries.

For technoeconomic modeling, the three most important aspects of battery performance are efficiency (or losses), charge and discharge capacity, and lifetime. The simplest way to model the efficiency is to assume a constant energetic efficiency, meaning that some fixed percentage of the energy absorbed by the battery can later be retrieved. A more accurate alternative is to model the battery as a voltage source in series with a resistance. This "internal resistance" causes a difference in voltage between charging and discharging, and losses can be calculated either based on that voltage difference (charge is conserved, so 1 amp-hour of energy enters the battery at a high voltage and leaves at a lower voltage) or, equivalently, by calculating the I^2R losses directly.

The charge and discharge capacities of a battery are a measure of how quickly the battery can absorb or release energy. The simplest battery models assume unlimited charge and discharge capacities. More sophisticated models account for the fact that both the battery's state of charge and its recent charge and discharge history affect its charge and discharge capacities. The kinetic battery model, proposed by Manwell and McGowan and illustrated in Fig. 7, does this by modeling the battery as a two-tank system. The energy in one tank is available for immediate use, but the energy in the other is chemically bound. A conductance separates the two tanks and limits the rate that energy can move from one to the other. Some models also account for self-discharge that occurs within batteries, although the rate of self-discharge is typically small compared to the rates of discharge typically experienced by batteries in power systems. Regardless of the method chosen to calculate charge and discharge capacity, performance models should account for the fact that most batteries do not tolerate being discharged below a certain critical state of charge. For lead-acid batteries, this minimum state of charge ranges from about 30 or 40% for deep-discharge varieties to as much as 80% for common car batteries.

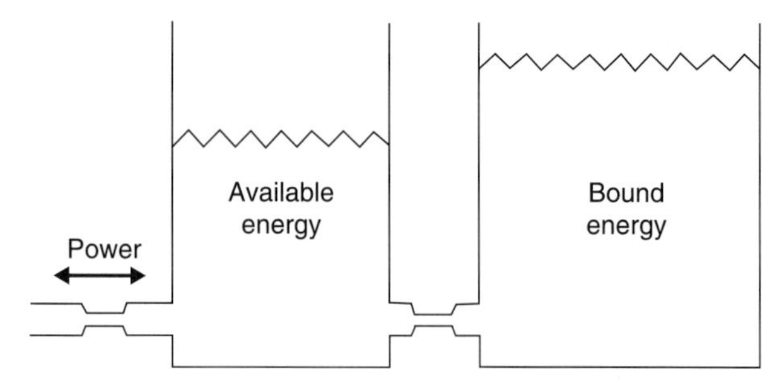

FIGURE 7 Kinetic battery model.

Spiers and Rasinkoski proposed three types of factors limiting the lifetime of a battery. Primary factors include the cycle life (governed by the amount of energy cycled through the battery) and the float life (governed by internal corrosion). Secondary factors include effects such as sulfation of lead-acid batteries caused by extended periods at low states of charge. Catastrophic factors include manufacturing defects, misuse, and freezing. Because secondary factors normally have little effect and catastrophic failures can be prevented, simple battery lifetime models typically account only for the primary factors. Spiers and Rasinkoski suggest calculating the float life and the cycle life, and assuming that the battery lifetime will be equal to the lesser of the two. The float life can be either assumed constant or calculated based on the average ambient temperature. Spiers and Rasinkoski suggest a simple formula for this calculation. The cycle life can be modeled most simply by assuming that a fixed amount of energy can cycle through the battery before it needs replacement, a good assumption for most lead-acid batteries. Alternatively, one could determine the number of cycles of different depths experienced by the battery per year, and then refer to the battery lifetime curve (cycles to failure versus depth of discharge) to calculate the resulting years of lifetime. Downing and Socie have proposed two algorithms for counting cycles. The effect of temperature on the storage capacity, charge/discharge capacity, and internal resistance may be significant for certain modeling applications. Perez has offered insight into this and other aspects of battery performance.

3.7 AC/DC Converters

Renewable power systems commonly comprise a mixture of AC and DC loads and components, and hence require AC/DC conversion. Solid-state inverters and rectifiers typically perform this conversion, although some systems use rotary converters. For technoeconomic modeling, the important performance characteristics of a conversion device are its maximum capacity and its efficiency. Another characteristic important for modeling an inverter is whether it can operate in parallel with an AC generator; those that cannot are sometimes called "switched inverters."

Sukamongkol *et al.* have proposed that the power flow through solid-state inverters may be modeled by the following equation:

$$P_{\mathrm{O}} = \alpha(P_{\mathrm{I}} - \beta)$$

where P_{O} is the output power, P_{I} is the input power, and α and β are constants. The value of β, often called the "standing losses," is small compared to the capacity of the inverter. Many models therefore ignore standing losses and assume a constant inverter efficiency. The same treatment applies to rectifiers. Solid-state inverters typically exhibit efficiencies near or above 90%, with rectifier efficiencies slightly lower. Rotary converters typically have slightly lower efficiencies, compared to solid-state converters.

3.8 Grid Connection

Distributed generation systems supply electricity to grid-connected energy consumers and either store the excess or sell it back to the utility. Such systems may also supply heat, in which case they are called cogeneration or combined heat and power (CHP) systems. The utility rate structure (or tariff structure) often determines the economic viability of distributed generation systems. This rate structure typically consists of an energy charge (on the number of kilowatt-hours purchased per billing period) and a demand charge (on the peak demand within the billing period). Both may vary according to season, time of day, quantity, or other factors. Some utilities allow the sale of excess electricity back to the grid, usually at the utility's avoided cost or according to a net metering scheme whereby the customer pays for net monthly or annual consumption. Utilities may also assess an interconnection charge (a one-time fee for connecting a power system to the grid) or standby charges (monthly or annual fees for providing backup power).

4. *SYSTEM MODELING*

Several characteristics of renewable power technologies combine to complicate the process of designing renewable power systems. The principal virtue of renewable power sources, that they do not require fossil fuel, makes them suitable for many remote and off-grid applications. Such "stand-alone" systems require careful design because of the need to balance electrical supply and demand. Models can help by simulating the performance of a system to determine its viability, robustness, and stability.

The intermittent and seasonal nature of renewable resources often makes it necessary (or at least advantageous) to combine renewable power sources with each other, with conventional power sources such as diesel generators, and with energy storage technologies such as batteries. Hence renewable power systems frequently comprise several dissimilar components, often both AC and DC. Models can help both in simulating the complex behavior of a composite system and in evaluating the many different possible combinations of component types and sizes.

Renewable power sources typically have high capital costs and low operating costs, in contrast to most conventional (fossil-fueled) power sources. Any fair comparison of the relative economics of conventional and renewable sources of energy must therefore consider operating costs in addition to initial capital costs. This requires more sophisticated economic modeling than may be necessary when simply choosing between conventional power sources.

The design of renewable power systems involves many variables, some inherently uncertain. Computer models can help the system designer take this uncertainty into consideration. A model may, for example, allow a designer to assess the technical and economic implications of an increase in fuel price, growth in electric demand, or an inaccurate estimate of wind speed.

4.1 Temporal Variability

Renewable power systems experience two kinds of temporal variability. Seasonal or long-term variability results from the cyclic annual pattern of the renewable resources. Short-term variability, which applies in particular to systems comprising solar or wind power, results from the hour-to-hour, even second-to-second, intermittency in the solar radiation, wind speed, and electric load. Both types of variability influence power system design.

The seasonal variability of resources often makes it impossible to rely on a single renewable power source year-round without backup power from a fuel-fired generator or utility grid. If stream flow declines sharply during the winter, for example, a diesel generator may be needed to supplement the output of the hydro turbine for part of the year. The degree to which different resources complement each other also affects the optimal combination of power sources. For example, a combination of wind and solar power will more likely be optimal if wind speeds peak during the winter (so that increased wind power output makes up for decreased solar power output) than if they peak during the summer.

Short-term variability has wide-ranging implications for system design and operation. It is the principal factor in determining the optimal amount of energy storage, and an important consideration when sizing dump loads (which manage excess renewable power) and AC/DC converters. Short-term variability also plays a role in the control of backup generators, which must modulate their output in response to fluctuations in the renewable output and electric demand.

4.2 Component Selection and Sizing

A power system designer must choose which power sources the system should comprise, decide whether to include energy storage, and select an appropriate size for each system component. This process of selecting the best system configuration from a long list of possibilities is an optimization problem well suited to computer modeling. The objective is typically to minimize the cost of energy or the total system cost, subject to any number of constraints. For stand-alone systems, the principal constraint is normally that the system must supply some fraction of the annual electric demand, often 100%. Other constraints may limit such parameters as fuel usage, emissions, or generator run time.

The optimal system design varies widely, depending on the magnitude and character of the electric demand, the quality of the renewable resources, and several economic factors. Relevant economic factors may include component costs, fuel price, interest rate, and, in the case of grid-connected systems, the utility rate structure. Models allow the designer to explore the many different possible system configurations in search of the optimum. Some models perform the optimization procedure by simulating multiple different system configurations and comparing them based on user-specified criteria. Others simply simulate one system at a time, in which case the designer can perform the optimization process manually.

4.3 System Operation and Control

For some renewable power systems, the designer must also consider the method of operation. A small photovoltaic/battery or wind/battery system needs no complicated control logic, other than appropriate constraints on battery charging and discharging. The system simply charges the battery with excess renewable power and discharges the battery when necessary. Similarly straightforward control logic would apply to a grid-connected photovoltaic system or a low-penetration wind/diesel system in which the diesel always operates and the wind power simply saves fuel. But a high-penetration wind/diesel system, in which the wind power may occasionally exceed the electric load, needs control logic to determine if and when the diesel can safely shut down without risking a loss of load. Spinning reserve is an important consideration in such decision-making. The manner in which the controller makes dispatch decisions can affect both the economics and the optimal design of the power system.

Control strategies are also important for any system comprising more than one dispatchable power source. A stand-alone power system that includes both a generator and a battery bank is one example. Two common control strategies for such a system are load following and cycle charging. Under the load-following strategy, whenever the generator operates, it produces only enough power to serve the load (surplus renewable power charges the batteries). Under the cycle-charging strategy, whenever the generator operates, it produces excess power to charge the battery bank. Barley and Winn have presented a thorough analysis of these two strategies and their variations, and proposed a design chart to help select the better of the two, depending on the configuration and economics of the system.

Other examples of systems comprising more than one dispatchable power source include a stand-alone system containing multiple diesel generators, or a grid-connected biomass-fueled generator. In either case, the system must depend on some control strategy to decide how to operate various power sources that may differ in cost, fuel source, efficiency, thermal output, noise, or emissions. Modeling can help the system designer explore the consequences of different control strategies.

4.4 Economic Modeling

Economics frequently figure prominently in system design. Even when nonmonetary factors are involved, economic figures of merit can help in selecting among design options. To account properly for both initial capital costs and recurring operating costs, economic models should consider the time value of money. One approach is to calculate the present value of all cash flows (annual fuel payments, maintenance costs, grid power costs, component replacement costs, etc.) that occur within the project lifetime. This yields the system's total life-cycle cost. An equivalent approach is to annualize all one-time costs and add them to the recurring annual costs to calculate the system's total annualized cost. This number divided by the total energy delivered annually to the load yields the levelized cost of energy. For system retrofits, when an investment in renewable energy results in future cost savings, the payback period and internal rate of return are also useful figures of merit.

When calculating life-cycle cost it is helpful to make use of two economic concepts, the present value of a future cash flow and the present value of a sequence of equal annual cash flows. The present value of a cash flow occurring N years in the future is found by multiplying by the present worth factor (PWF), given by

$$PWF = 1/(1+i)^N$$

where i is the annual discount rate. Inflation can be factored out by using the real annual discount rate, given by

$$i = (i' - f)/(1+f)$$

where i' is the nominal interest rate and f is the inflation rate.

The present value of a sequence of equal annual cash flows is equal to the annual cash flow divided by the capital recovery factor (CRF). This factor is given by

$$CRF = i(1+i)^N/[(1+i)^N - 1]$$

where N is the number of years over which the annual cash flows occur. The capital recovery factor is also used to calculate the equivalent annualized value of a one-time cash flow (for a more complete discussion of these factors and economic modeling in general, see DeGarmo *et al.* or any introductory economics text).

4.5 Sensitivity Analysis

Regardless of how carefully a designer models the cost and performance of a power system, uncertainty in key inputs invariably causes actual and modeled results to differ. In a wind/diesel system for example, if the actual average wind speed turns out to be lower than the estimate used during modeling, the system will likely consume more fuel than predicted, hence the life-cycle cost will be greater than predicted. Similarly, an underestimate of the diesel fuel price over the project lifetime will also result in an underestimate of life-cycle cost.

A modeler can take this uncertainty into consideration by performing multiple simulations under different input assumptions. For example, the modeler could analyze the system cost and performance using several different values of fuel price to establish the sensitivity of key outputs to that input. The results of such a sensitivity analysis are often plotted as a spider graph such as the one shown in Fig. 8. This example shows how the cost of energy varies with changes in particular input variables. The modeler calculated the cost of energy assuming five different values of fuel price: a best estimate, plus two values above and two values below that best estimate. The modeler did the same for two more input variables, the average wind speed and the average solar radiation. The graph shows that the cost of energy increases with increasing fuel price, and decreases with increasing wind speed or solar radiation. Changes in wind speed have a more dramatic effect on the cost of energy than do changes in solar radiation.

For a more formal treatment of uncertainty, a modeler can make use of the techniques of decision analysis, a field devoted to decision-making in the presence of uncertainty. The canonical decision analysis approach is to assign probabilities to multiple

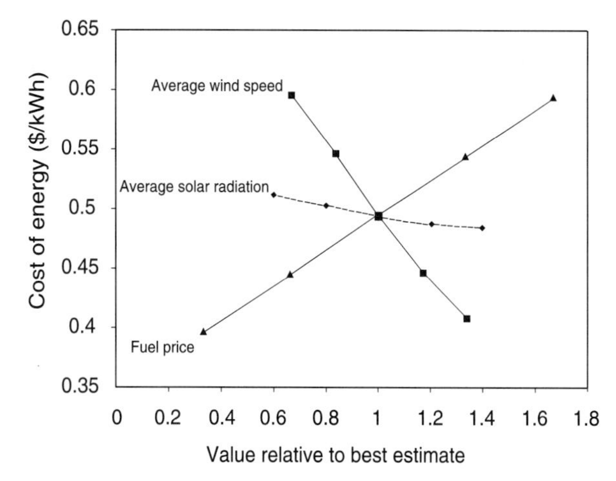

FIGURE 8 Spider graph showing results of a sensitivity analysis for a small photovoltaic/wind/diesel system.

input scenarios, simulate each scenario, and then calculate the resulting probability-weighted average value of any output variable. This "expected value" represents the best estimate of that output variable given the uncertainty in the inputs. When applied to the power system design problem, decision analysis can help select a system that is optimal not just for a fixed (and often dubious) set of input assumptions, but rather with consideration of a wide range of possible scenarios and their attendant risks. None of the models presented in the next section has a built-in decision analysis capability, but any one of them could help inform the decision analysis process by simulating input scenarios.

5. MODELS FOR RENEWABLE POWER SYSTEM DESIGN

Many available computer models simulate the performance and economics of renewable power systems. Each model uses representative load and resource data to predict the performance of the system, typically over a 1-year period. Performance parameters of interest may include the amount of energy produced by a renewable power source, the amount of fuel consumed by a generator, the amount of energy cycled through the batteries, or the amount of energy purchased from the grid. It is common to extrapolate these performance data over many years to calculate economic figures of merit such as life-cycle cost or cost of energy.

Models can be categorized as statistical models or time series models according to the technique used to predict system performance under highly variable conditions of load and renewable resources. Statistical models consider large time intervals (monthly or annual) and rely on statistical techniques to predict the outcome of the short-term variability occurring within each time interval. Time series models approach the issue of variability by dividing the year into time steps of 1 hour or less and simulating the behavior of the system at each time step.

5.1 Statistical Models

Statistical models usually account for seasonal variability by simulating system performance separately for each month of the year. But to account for short-term variability, they depend on some form of statistical manipulation. For example, a common technique when simulating a stand-alone PV/battery system is to calculate the loss-of-load probability each month. This is the probability that the system will experience a supply shortage at some time during the month, a scenario that may occur after the load exceeds the PV output long enough to drain the battery bank completely. A model can calculate the loss-of-load probability using information about the variability of the solar resource and its correlation with electric load. If the loss-of-load probability exceeds some user-specified critical value, the system is unacceptable.

Compared to time series models, statistical models tend to be simpler to use and quicker to provide results. Their simplicity stems partly from their limited input requirements; they require monthly or annual average load and resource data, rather than the hourly data required by most time series models. They also tend to simulate system performance and cost using less detailed algorithms that require fewer, more easily obtainable inputs. This input simplicity makes statistical models relatively easy to learn. It also means that a designer can quickly assemble the necessary load, resource, and component data to analyze a particular system. A further advantage of statistical models over time series models is their generally faster computation speed; most produce outputs virtually instantaneously. The combination of faster assembling of inputs and faster production of outputs makes statistical models ideal for quick system comparisons and cost estimates.

The simplicity of statistical models, however, comes at the cost of reduced accuracy. Because they rely on simplifying assumptions, statistical models sometimes overlook important aspects of real system behavior. For example, when modeling a system comprising diesel generators, a statistical model may assume that the generator's fuel consumption is linear with power output, meaning the generator's efficiency is constant. Although greatly simplifying system modeling, this assumption neglects the fact that generators are much less efficient at low loads, a fact that has important implications for the control and economics of the system. Their reduced accuracy means that statistical models often cannot reliably model the behavior of complex systems such as high-penetration wind/diesel systems. In such systems, details for which a simplified model cannot account may have a substantial impact on both performance and economics.

Statistical models also tend to allow less flexibility in the configuration of the system. Because of their inability to simulate complicated systems, for example, statistical models typically allow a designer to simulate at most one renewable power source and one dispatchable generator. They typically cannot analyze systems comprising multiple renewable power sources, multiple generators, or other advanced features such as cogeneration. An additional drawback of statistical models is that despite using fewer and generally simpler inputs, a small number of their inputs can be difficult to understand and to estimate. A statistical model may, for example, require the user to specify the degree to which the electric load correlates with the solar radiation, or the fraction of the wind turbine output that will have to be dumped as excess. Time series models require neither of these inputs; the solar and load correlation is unnecessary because the model has access to the hourly solar and load data, and a time series simulation can determine the wind turbine's excess energy fraction.

In summary, statistical models offer simplicity and speed at the cost of accuracy and flexibility. They are suitable for prefeasibility analyses of many types of renewable power systems. In fact, certain uncomplicated systems (such as PV/battery or grid-connected PV) may require no further modeling. However, in the design of more complex systems, particularly those requiring sophisticated control strategies or careful balancing of electrical supply and demand, statistical modeling should be supplemented with time series modeling.

Several statistical models for renewable power system design are currently available and may be accessed on the Internet. PVSYST, created by the University of Geneva, can simulate PV/battery and grid-connected PV systems. Orion Energy's NSol! can simulate these systems as well as PV/generator/battery systems. RETScreen, created by Natural Resources Canada, can simulate these systems as well as grid-connected wind, low-penetration wind/diesel, and small hydro systems.

5.2 Time Series Models

The premise of time series modeling is that by simulating system performance at a sufficiently fine temporal resolution, the effects of short-term variability can be modeled directly. Time series models typically perform annual simulations using a 1-hour time step, and assume that the time-dependent variables (electric load and renewable resources) remain constant within that time step. Shorter time steps lead to improved accuracy but increased computation time. That, along with the difficulty of obtaining higher resolution resource and load data, explains why most models use a 1-hour time step.

At the core of time series modeling is the energy balance performed each time step. In this energy balance, the model calculates the available renewable power, compares it to the electric load, and simulates how the system would deal with any surplus or deficit. In the case of insufficient renewable power, the model may draw power from the grid, dispatch one or more generators, or discharge the battery bank to avoid unmet load. In the case of an oversupply of renewable power, the model may sell power to the grid, charge the battery bank, serve an optional load, or dissipate the excess power in a dump load. When making such dispatch decisions, time series models typically refer to some preset operating strategy, and many models allow the comparison of several different strategies so that the user can determine which is optimal for a particular application.

The division of the year into many short intervals allows time series models to simulate even complex systems accurately. As a result, time series models allow the modeler to simulate a wide variety of system types, sometimes comprising multiple renewable power sources, multiple generators, battery storage, and grid connection. With this flexibility, a modeler can evaluate many more design options than is possible with a statistical model.

Hourly simulations require hourly load and resource data. The limited availability of such data can hinder the use of time series models, although several algorithms exist for synthesizing hourly load, solar radiation, and wind speed data from more readily available monthly or annual average data. Virtually all time series models, for example, allow the user to construct hourly load data from typical daily load profiles. Algorithms that synthesize hourly solar and wind data typically require monthly averages plus a few parameters. Some models integrate such algorithms directly into their interfaces, but external resource data generators can produce data suitable for any time series model.

The computation time required to simulate annual system performance with a time series model depends strongly on whether the model tracks the voltage of the DC components. This DC bus voltage can fluctuate widely over time, in contrast to the AC bus voltage, which is fixed by the system controller. Because of the feedback effect between the DC bus voltage and the performance of batteries, photovoltaic modules, and DC wind turbines, calculating the DC bus voltage requires an iterative solution procedure within each time step. Models that track the DC bus voltage therefore typically require many seconds or a few minutes of computation time per simulation. Models that simply track power flows without calculating voltage do not require this iteration, and so can simulate systems much faster.

Several time series models for renewable power system design are available and may be accessed on the Internet. PV-DesignPro, a product of Maui Solar Energy Software Corporation, can model stand-alone or grid-connected systems comprising PV modules, wind turbines, batteries, and a generator. PV*SOL, created by Valentin EnergieSoftware, can model stand-alone and grid-connected systems comprising PV modules, batteries, and a generator. PV*SOL uses a statistical modeling approach to perform a quick initial simulation, then a time series modeling approach for a more accurate final simulation. HOMER, a product of the National Renewable Energy Laboratory, can model systems comprising PV modules, wind turbines, small hydro, biomass power, batteries, hydrogen storage, grid connection, and as many as three generators, serving any combination of AC, DC, and thermal loads. HOMER also performs optimization and sensitivity analysis. Hybrid2, created by the National Renewable Energy Laboratory and the University of Massachusetts, can model stand-alone systems comprising PV modules, wind turbines, batteries, and as many as eight generators, with a time step as short as 1 minute. Hybrid2 also models a wide variety of different system control and load management strategies. Though it is a time series model, Hybrid2 incorporates some aspects of statistical modeling to account for the short-term variability occurring within each time step.

6. CONCLUSIONS

In the design of renewable power systems, significant complexity arises due to the intermittent nature of renewable power sources, the large number of possible system configurations, the uncertainty in key input parameters, and the need to compare different design options on the basis of life-cycle cost. Computer modeling can help the system designer to account for the many factors affecting the performance and economics of renewable power systems, to evaluate diverse design options, and to explore multiple input scenarios.

Various computer models are currently available for use in renewable power system design. Statistical models, offering speed and ease of use, are well suited to the type of prefeasibility analyses commonly performed early in the design process. Time series models, generally simulating renewable power systems of a wider variety and with a higher accuracy than statistical models, are suitable for more precise modeling later in the design process.

SEE ALSO THE FOLLOWING ARTICLES

Biomass for Renewable Energy and Fuels • National Energy Modeling Systems • Neural Network Modeling of Energy Systems • Renewable Energy and the City • Renewable Energy Policies and Barriers • Renewable Energy, Taxonomic Overview • System Dynamics and the Energy Industry

Further Reading

Barley, C. D., and Winn, C. B. (1996). Optimal dispatch strategy in remote hybrid power systems. *Solar Energy* **58**, 165–179.

Beyer, H. G., Degner, T., and Gabler, H. (1995). Operational behaviour of wind diesel systems incorporating short-term storage: an analysis via simulation calculations. *Solar Energy* **54**, 429–439.

Clemen, R. T. (1991). "Making Hard Decisions: An Introduction to Decision Analysis." PWS-Kent, Boston.

DeGarmo, E. P., Sullivan, W. G., and Bontadelli, J. A. (1988). "Engineering Economy." 8th ed. Macmillan, New York.

Downing, S. D., and Socie, D. F. (1982). Simple rainflow counting algorithms. *Int. J. Fatigue* **4**, 31–39.

Duffie, J. A., and Beckman, W. A. (1991). "Solar Engineering of Thermal Processes." 2nd ed. Wiley, New York.

Hunter, R., and Elliot, G. (1994). "Wind–Diesel Systems." Cambridge Univ. Press, Cambridge, Massachusetts.

Manwell, J. F., and McGowan, J. G. (1993). Lead acid battery storage model for hybrid energy systems. *Solar Energy* **50**, 399–405.

Manwell, J. F., McGowan, J. G., and Rogers, A. L. (2002). "Wind Energy Explained." Wiley, New York.

McKendry, P. (2002). Energy production from biomass (Part 1): Overview of biomass. *Bioresource Technol.* **83**, 37–46.

McKendry, P. (2002). Energy production from biomass (Part 2): Conversion technologies. *Bioresource Technol.* **83**, 47–54.

Mora-López, L. L., and Sidrach-de-Cardona, M. (1998). Multiplicative ARMA models to generate hourly series of global irradiation. *Solar Energy* **63**, 283–291.

Paish, O. (2002). Small hydro power: Technology and current status. *Renew. Sustain. Energy Rev.* **6**, 537–556.

Perez, R. (1993). Lead-acid battery state of charge vs. voltage. *Home Power* **36**, 66–69.

Skarstein, O., and Uhlen, K. (1989). Diesel considerations with respect to long-term diesel saving in wind/diesel plants. *Wind Engineer.* **13**, 72–87.

Spiers, D. J., and Rasinkoski, A. A. (1996). Limits to battery lifetime in photovoltaic applications. *Solar Energy* **58**, 147–154.

Stevens, M. J. M., and Smulders, P. T. (1979). The estimation of the parameters of the Weibull wind speed distribution for wind energy utilization purposes. *Wind Engineer.* **3**, 132–145.

Sukamongkol, Y., Chungpaibulpatana, S., and Ongsakul, W. (2002). A simulation model for predicting the performance of a solar photovoltaic system with alternating current loads. *Renewable Energy* **27**, 237–258.

Conservation Measures for Energy, History of

JOHN H. GIBBONS and HOLLY L. GWIN
Resource Strategies
The Plains, Virginia, United States

1. Introduction
2. Decade of Crisis: 1970s
3. Decade of No- and Low-Cost Options: 1980s
4. Decade of Globalization: 1990s
5. Decades Ahead

Glossary

end-use efficiency Substituting technological sophistication for energy consumption through (1) obtaining higher efficiency in energy production and utilization and (2) accommodating behavior to maximize personal welfare in response to changing prices of competing goods and services.

energy conservation The wise and thoughtful use of energy; changing technology and policy to reduce the demand for energy without corresponding reductions in living standards.

energy intensity The amount of energy consumed to produce a given economic product or service; often measured as the ratio of the energy consumption (E) of a society to its economic output (gross domestic product, GDP), measured in dollars of constant purchasing power (the E/GDP ratio).

energy services The "ends," or amenities, to which energy is the "means," e.g., space conditioning, lighting, transportation, communication, and industrial processes.

externalities The environmental, national security, human health, and other social costs of providing energy services.

least-cost strategy A strategy for providing individual and institutional energy consumers with all of the energy services, or amenities, they require or want at the least possible cost. It includes the internal economic costs of energy services (fuel, capital, and other operating costs) as well as external costs.

The history of energy conservation reflects the influence of technology and policy on moderating the growth of demand for energy. It traces the evolution of concepts of conservation from curtailment of energy use to least-cost strategies through end-use efficiency. Through understanding the past efforts to address the need for energy conservation, the future potential of the role for energy conservation in the global economy is better understood.

1. INTRODUCTION

Just as beauty is in the eye of the beholder, energy conservation means different things to different people. Concerned about deforestation in Pennsylvania, Benjamin Franklin took heed of his own advice ("a penny saved is a penny earned") by inventing the Franklin stove. In his *Account of the New Invented Fireplaces*, published in 1744, Franklin said "As therefore so much of the comfort and conveniency of our lives for great a part of the year depend on this article of fire; since fuel is become so expensive, and (as the country is more clear'd and settle'd) will of course grow scarcer and dearer; any new proposals for saving the wood, and for lessening the charge and augmenting the benefit of fire, by some particular method of making and managing it, may at least be thought worth consideration." The Franklin stove dramatically improved the efficiency of space heating. In Franklin's time, however, others viewed energy consumption as a signal of economic progress and activity. This view was epitomized by company letterheads picturing a factory with chimneys proudly billowing clouds of smoke!

Viewed as a matter of how much energy it takes to run an economy—often referred to as a society's energy intensity—energy conservation has been with us since the middle of the 20th century. Between 1949 and 2000, the amount of energy required to produce a (1996 constant) dollar of economic output fell 49%. Much of this improvement occurred by the combined effects of improved technology (including increased use of higher quality fuels in new technologies, such as high-efficiency gas turbines) and perceptions of higher energy costs after the oil shocks of the 1970s, when energy intensity began to fall at an average rate of 2% per year. (Fig. 1).

Although largely oblivious to long-term improvements in energy intensity, consumers undoubtedly felt the oil shocks, and came to equate energy conservation with waiting in long lines to purchase gasoline. For many people, even to the present day, energy conservation connotes curtailment or denial of the services energy provides, e.g., turning down the heat on a cold winter's day, turning off the lights, or driving lighter cars at slower speeds. This viewpoint came into sharp focus one night during the 1973 oil embargo, when one author (Gibbons), director of the newly established Federal Office of Energy Conservation, fielded a call from an irate Texan who proclaimed that "America didn't *conserve* its way to greatness; it *produced* its way to greatness!"

For purposes of this article, we accept and adapt the term "wise and thoughtful use" for energy conservation. This term implies that, when technically feasible and economically sound, a society will use technology and/or adopt social policy to reduce energy demand in providing energy services. It also implies that society's effort to bring energy supply and energy demand into balance will take into full account the total costs involved. Market prices for energy are generally less than its total social cost due to various production subsidies and nonmarket costs, or externalities, such as environmental impacts and security requirements. This approach has been labeled a "least-total cost" strategy for optimizing the investment trade-off between an increment of supply versus an increment of conservation.

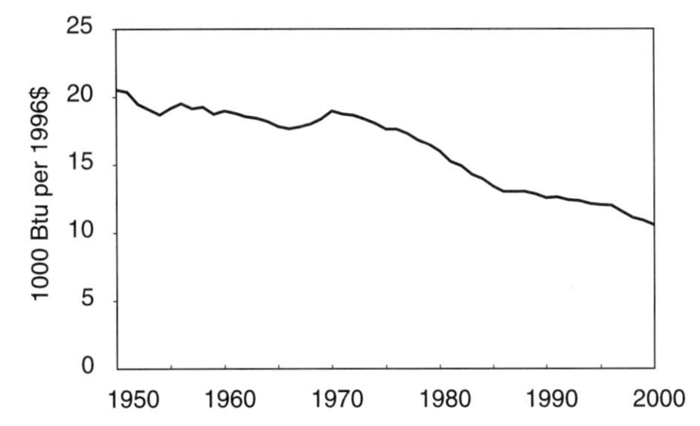

FIGURE 1 Energy use per dollar of gross domestic product. From the U.S. Energy Information Administration, *Annual Review of Energy 2000*.

2. *DECADE OF CRISIS: 1970s*

Events of the late 1960s highlighted many of the external costs of energy production and consumption. Many of the worsening problems of air and water pollution were directly attributable to energy use. The growth in the power of the Organization of the Petroleum Exporting Countries (OPEC) and increasing tensions in the Middle East contributed to a growing wariness of import dependence. High projected electric demand growth exacerbated worries not only about coal and the environment, but also about the sheer amount of capital investment that was implied. Gas curtailments began to show up along with spot shortages of heating oil and gasoline. And, surprising to many in the energy field, there were growing doubts about the nuclear option.

2.1 Rethinking Energy Demand

Those forces converged to trigger work on improved and alternative energy sources as well as on the dynamics and technologies of demand. Public support for analysis of conservation strategies came at first not from the "energy agencies" but from the National Science Foundation (NSF). In 1970, Oak Ridge National Laboratory (ORNL) researchers (who were well-sensitized to the vexing problems of coal and fission) seized on the opportunity provided by the NSF to investigate demand dynamics and technologies. The owner/operator of ORNL, the Atomic Energy Commission (AEC), was not keen on the idea but accepted the new direction for research as "work for others"!

It quickly became apparent that despite declining energy prices in previous years, energy efficiency had been slowly improving. Technology relevant to efficient use was advancing so impressively that more efficiency was economically attractive in virtually every sector of consumption. Further examination at ORNL and elsewhere showed that such improvements would accelerate more rapidly if energy prices were to rise and/or if consumers became better informed. For example, it was demonstrated that cost-effective technical improvements in refrigerators could double the energy efficiency of household refrigerators for a very modest price increment (Fig. 2). Clearly, major environmental benefits would also accrue from more conservation.

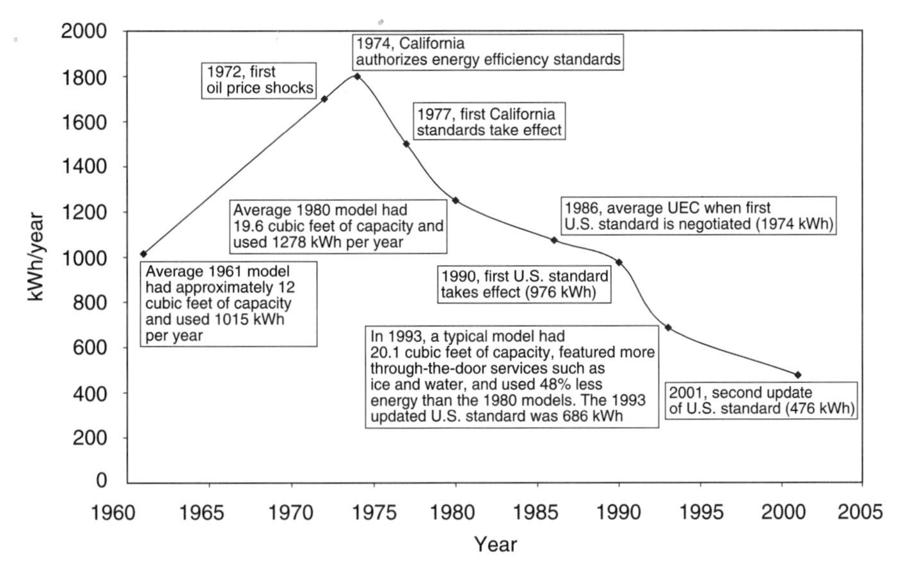

FIGURE 2 Changes in energy demand of refrigerators in the United States since 1961. UEC, unit energy consumption (kWh/year). From the Lawrence Berkeley National Laboratory.

The environmental concerns of the 1960s prompted the first serious efforts to project future energy demand. Scenario analyses, prepared by many different authors throughout the 1970s, produced a wide range of estimates, helping to shape a diverse range of views on the importance of conservation. Eleven forecasts of consumption made during the 1970s turned out to be remarkably consistent but ludicrously high, even for 1985 (ranging from 88 to 116 quads—the actual number was 75 quads). Prior to the oil embargo of 1973, most projections of U.S. energy consumption for the year 2000 fell in the range of 130–175 quads. The actual number turned out to be about 100 quads. Some of the difference can be attributed to lower than anticipated economic growth, but the dominant difference is due to the unexpected role of conservation.

Limits to Growth, published in 1972, was the first analysis to capture the public's attention. This computerized extrapolation of then-current consumption trends described a disastrous disjuncture of the growth curves of population and food supply and of resource use, including energy, and pollution. One scenario climaxed in a collapse mode, the sudden and drastic decline in the human population as a consequence of famine and environmental poisoning. This apocryphal report correctly made the point that exponential growth in finite systems is unsustainable, but its macroeconomic model failed because it did not consider responses to resource prices or advances in technologies.

A Time to Choose: America's Energy Future, a group of scenarios commissioned by the Ford Foundation and published in 1974, became notable, and controversial, for its "zero energy growth" (ZEG) scenario. This ZEG scenario clearly was meant to show that energy growth could technically be decoupled from economic growth. ZEG called for a leveling off in energy demand at about 100 quads per year shortly after 1985. In addition to those economically feasible measures taken by 1985, reduced demand would be effected by means of energy consumption taxes and other constraints on use. Taxes would increase the price of fuels so that substitution of energy-saving capital investments could be made economical while simultaneously encouraging shifts to less energy-intensive services. *A Time to Choose* also included a "technical fix" scenario that estimated 1985 consumption at 91 quads by coupling a microeconomic model with analysis of the engineering potential for conservation as a response to increased energy price and government policies. In 1976, Amory Lovins published a far more controversial and attention-grabbing scenario called "Energy Strategy: The Road Not Taken" in the journal *Foreign Affairs*. Lovins advocated a "soft energy path" based on three components: (1) greatly increased efficiency in the use of energy, (2) rapid deployment of "soft" technologies (diverse, decentralized, renewable energy sources), and (3) the transitional use of fossil fuels. His assertion that the "road less traveled" could bring America to greatly reduced energy demand (less than 50 quads per year in 2025) is both venerated and despised to this day.

These private sector analyses competed with scenarios developed or commissioned by government agencies for the attention of energy policy makers throughout the 1970s. *Guidelines for Growth of the Electric Power Industry*, produced by the Federal Power Commission (FPC) in 1970, was a classic study of energy demand based largely on an extrapolation of past trends. The FPC had credibility because of its 1964 success in predicting 1970 electricity demand, but its forecast of a doubling of energy demand from 1970 to 1990 was dramatically far from the mark. The Atomic Energy Commission made more accurate estimates in *The Nation's Energy Future*, published in 1973, but it mistakenly forecast U.S. independence from imported oil by 1980—no doubt caused or at least influenced by President Nixon, who was determined to achieve energy independence before the end of his term.

In 1975, the U.S. National Academy of Sciences established a Committee on Nuclear and Alternative Energy Strategies (CONAES). The CONAES Panel on Demand and Conservation published a path-breaking paper in the journal *Science* in 1978: "U.S. Energy Demand: Some Low Energy Futures." The CONAES scenarios not only incorporated different assumptions for gross domestic product (GDP) and population growth, but also, for the first time, incorporated varied assumptions about energy prices. The lowest CONAES estimate, based on a 4% annual real price increase coupled with strong nonmarket conservation incentives and regulations, came in at only 58 quads for 2010. The highest estimate, based on a 2% annual real price increase coupled with a rapidly growing economy, came in at 136 quads for 2010. This study clearly illustrated the importance in scenario analysis of assumptions about everything from income, labor force, and the rate of household formation, to worker productivity, energy productivity, energy price, income, and energy substitutability in scenario analysis. It also highlighted the importance of capital stock turnover in energy futures, in that the largest technical opportunities for increased efficiency derive from new stock, where state-of-the-art technology can be most effectively employed. The CONAES project had a major influence on all of the energy modeling work that followed.

2.2 Oil Shocks

In October 1973, the United States helped Israel beat back the Yom Kippur invasion by Egypt and Syria. OPEC (dominated by Arab oil producers) retaliated with an oil embargo on the United States and a 70% oil price increase on West European allies of the United States. Homeowners were asked to turn down their thermostats and gas stations were requested to limit sales to 10 gallons per customer. Illumination and speed limits were lowered, and the traffic law allowing a right turn on a red light came into use. Gasoline lines, conspiracy theories, and recession followed in rapid succession. As Adlai Stevenson once observed, "Americans never seem to see the handwriting on the wall until their backs are up against it." We perceived the wall behind our backs and responded accordingly—for a brief spell. American humor also surfaced. Some wonderful cartoons depicted innovative ways to save energy—for example, by sleeping with a chicken, which has a normal body temperature of 107°F, and public interest highway signs appeared proclaiming "Don't be fuelish!"

OPEC lifted the embargo in 1974, but political unrest in the Middle East again precipitated a major oil price increase in 1979. When anti-Western Islamic fundamentalists gained control of Iran, oil production in that nation dropped off dramatically and U.S. crude oil prices tripled. These price increases also led to skyrocketing inflation and recession, and rekindled interest in energy efficiency.

2.3 Policy Responses

The need to curtail energy consumption in response to U.S. vulnerability to oil supply disruptions dominated energy policy in the 1970s. In the summer of 1973, President Nixon, anticipating possible shortages of heating oil in the coming winter, and summer gasoline shortfalls, established the Office of Energy Conservation to coordinate efforts to cut energy consumption in federal agencies by about 5%, raise public attention to the need for increased efficiency, encourage private industry to give greater attention to saving energy, and devote more federal research and development to efficiency. After the first oil shock, President Nixon urged voluntary rationing and pushed for a commitment to U.S. energy independence by 1985. President Carter emphasized the importance of conservation in the effort to achieve energy independence, but appeared to equate conservation with wearing a sweater in a chilly White House. Early actions by our political leaders certainly contributed to a negative association of conservation with sacrifice. But they also contributed to an atmosphere conducive to serious public and private sector actions to incorporate efficient end-use technologies in the U.S. economy. Congress enacted legislation to create standards for Corporate Average Fuel Economy (CAFE) and building energy performance. They also created a cabinet-level Department of Energy (DOE) by merging the functions of preexisting agencies scattered throughout government. In response to an analysis by the congressional Office of Technology Assessment, Congress explicitly included energy conservation in the DOE's new mandate.

3. *DECADE OF NO- AND LOW-COST OPTIONS: 1980s*

The combination of the oil price shocks with increasing public concern about the environmental consequences of energy use prompted policy makers to take a serious look at policy tools for decreasing energy demand growth during the 1980s. Faced with higher oil prices, industrial and commercial oil consumers also sought out cost-effective investments in energy conservation during this decade. A plethora of no- and low-cost options were ripe for the picking, and progressive industries did just that, with great success at the bottom line.

3.1 Policy Tools

At the beginning of the 1980s, energy analysts converged on a set of principles for a comprehensive energy policy that had conservation at its core. These principles included the following ideas:

1. Consider the production and use of energy as means to certain ends, not as goals in and of themselves. Remember always that, given time and the capital for adjustment, energy is a largely substitutable input in the provision of most goods and services.
2. Application of technical ingenuity and institutional innovation can greatly facilitate energy options.
3. Energy decisions, like other investment decisions, should be made using clear signals of comparative total long-run costs, marginal costs, and cost trends.
4. It is important to correct distorted or inadequate market signals with policy instruments; otherwise, external costs can be ignored and resources can be squandered. This correction includes internalizing in energy price and/or regulation, to the extent possible, the national security, human health, and environmental costs attributable to energy.
5. Investment in both energy supply and utilization research and development is an appropriate activity for both the public and private sectors, because costs and benefits accrue to both sectors.
6. There are other, generally more productive, ways (for example, assistance with insulation or lighting retrofits, fuel funds, or even refundable tax credits) to assist underprivileged citizens with their energy needs, rather than subsidizing energy's price to them.
7. In a world characterized by tightly integrated economies, we need to increase our cognizance of world energy resource conditions and needs with special regard for international security as well as concern for the special needs of poor nations.

By the mid-1980s, the national energy situation had begun to reflect some of these ideas. CAFE standards reached their peak (27.5 miles per gallon) in 1985. Congress created the Strategic Petroleum Reserve (SPR) to respond to energy supply emergencies. Oil and gas price deregulation was virtually complete, but that was only the first step in internalizing the costs of energy consumption. Energy prices still did not reflect the manifold environmental costs of production and use, the costs of defending Middle East oil production and shipping lanes, the costs of purchasing and storing oil to meet emergencies, or the impacts of U.S. competition for scarce oil resources on developing nations. As the price of imported oil decreased, U.S. policy makers were lulled into complacency about the nation's energy strategy. A dearth of energy research and development expenditures symbolized this declining interest. Before the end of the decade, however, Congress passed the National Appliance Energy Conservation Act, which authorized higher appliance efficiency standards and efficiency labeling that have significantly impacted the market for energy-efficient appliances.

3.2 Private Sector Activities

Immediately following the oil shocks, private sector efforts to save energy focused on changing patterns of energy use within the existing infrastructure, such as lowering thermostats. Most actions involved investments in technology, however—either retrofits of existing technology, such as insulating existing homes, or new investments in technology, such as energy-efficient new construction or autos with improved mileage. Later in the 1980s, energy efficiency gains accrued as incidental benefits to other, much larger investments aimed at improving competitiveness of U.S. products in world markets.

All in all, energy efficiency investments in the 1970s and the 1980s turned out to be generally easier and much more cost-effective than finding new sources of energy. Nongovernmental institutions helped to keep the conservation ball of progress rolling, especially in the residential and commercial sectors, with information to increase awareness of cost-effective conservation opportunities, rebates, and prizes. In the first decade after the oil shocks, industry cut energy requirements per unit of output by over 30%. Households cut energy use by 20% and owners and operators of commercial buildings cut energy use per square foot by more than 10%. Transportation energy efficiency steadily improved as CAFE requirements slowly increased fleet average mileage. However, political maneuvering by car companies resulted in excepting "light trucks," including sport utility vehicles (SUVs), from the mileage requirements for passenger cars. This has resulted in a major slow down of improvement in fleet performance.

Some surprising partnerships formed to promote energy efficiency. For example, the Natural Resources Defense Council designed a Golden Carrot Award to encourage production of a superefficient refrigerator. In response, 24 major utilities pooled their resources to create the $30 million prize. Whirlpool won the competition and collected its winnings after manufacturing and selling at least 250,000 units that were at least 25% more efficient than the 1993 efficiency standard required. By the end of the 1980s, energy efficiency improvements lost momentum. One reason was the decline in oil prices: corrected for inflation, the average price of gasoline in 1987 was half that in 1980. Natural gas and electricity prices also fell, and the price-driven impetus for consumers and businesses to conserve was diminished.

3.3 Lessons from Experience

By the end of the decade, conservation advocates were urging policy makers to take several lessons from the experiences of 1970 and 1980s:

1. First and foremost, that energy conservation worked. Nothing contributed more to the improved American energy situation than energy efficiency. By the late 1980s, the United States used little more energy than in 1973, yet it produced 40% more goods and services. According to one estimate, efficiency cut the nation's annual energy bill by $160 billion.

2. Oil and gas price controls are counterproductive. Price increases of the 1970s and 1980s sparked operating changes, technology improvements, and other conservation actions throughout the United States.

3. Technological ingenuity can substitute for energy. A quiet revolution in technology transformed energy use. Numerous efficient products and processes—appliances, lighting products, building techniques, automobiles, motor drives, and manufacturing techniques—were developed and commercialized by private companies. Energy productivity rose as more efficient equipment and processes were incorporated into buildings, vehicles, and factory stock.

4. Complementary policies work best. The most effective policies and programs helped overcome barriers to greater investment in conservation—barriers such as the lack of awareness and low priority among consumers; lack of investment capital; reluctance among some manufacturers to conduct research and to innovate; subsidies for energy production; and the problem of split incentives, exemplified by the building owner or landlord who does not pay the utility bills but does make decisions on insulation and appliance efficiency.

5. The private sector is the primary vehicle for delivering efficiency improvements, but government–industry cooperation is essential.

Many activists advocated adoption of a new national goal for energy intensity in order to regain momentum toward a secure energy future. The United States achieved a 2.7% annual rate of reduction in energy intensity between 1976 and 1986. The DOE was predicting, however, that U.S. energy intensity would decline at less than 1% per year through the rest of the 20th century. This is the rate of improvement now widely assumed as part of "business as usual." Recognizing that economic structural change had also influenced this rapid improvement and that many of the least expensive investments in efficiency had already been made, the present authors and others, including the American Council for an Energy Efficient Economy, nonetheless recommended a goal of reducing the energy intensity of the U.S. economy by at least 2.5% per year into the 21st century. For the United States, even with an apparently disinterested Administration, it is now proposed that energy intensity over the next decade be improved by 18% (roughly the average yearly gain over the past two decades).

4. DECADE OF GLOBALIZATION: 1990s

The United States went to war for oil at the beginning of the 1990s. Oil prices initially spiraled upward, temporarily reinvigorating interest in conservation investments. That interest dwindled as quickly as prices fell. Still, after an inauspicious beginning, during the economic boom of the late 1990s, with major investments in new capital stocks, energy intensity dropped at an average rate of over 2% per year.

The role of carbon emissions in global climate change began to dominate the debate over energy in the 1990s. Although U.S. total energy consumption far outstripped any other country's, consumption was projected to grow at much higher rates in the developing countries, and U.S. policy makers shifted their attention to U.S. participation in international cooperation on energy innovation. In 1993, the U.S. government and the U.S. automobile industry forged an unprecedented alliance under the leadership of President Clinton and Vice President Gore. The partnership included seven federal agencies, 19 federal laboratories, and more than 300 automotive suppliers and universities and the United States Council for Automotive Research, the precompetitive research arm of Ford, DaimlerChrysler, and General Motors. The Partnership for a New Generation of Vehicles (PNGV) supported research and development of technologies to achieve the program's three research goals: (1) to significantly improve international competitiveness in automotive manufacturing by upgrading manufacturing technology; (2) to apply commercially viable innovations resulting from ongoing research to conventional vehicles, especially technologies that improve fuel efficiency and reduce emissions; and (3) to develop advanced technologies for midsized vehicles that deliver up to triple the fuel efficiency of today's cars (equivalent to 80 miles per gallon), without sacrificing affordability, performance, or safety. The research plan and the program's progress were peer-reviewed annually by the National Research Council. PNGV made extraordinary progress toward achieving its aggressive technical goals. In March 2000, PNGV unveiled three concept cars demonstrating the technical feasibility of creating cars capable of getting 80 miles per gallon. All three cars employ some form of hybrid technology that combines a gasoline- or diesel-powered engine with an electric motor to increase fuel economy. The three major automakers also confirmed their commitment to move PNGV technology out of the lab and onto the road by putting vehicles with significant improvements in fuel economy into volume production and into dealers' showrooms. Work continues on technologies that might contribute to the full achievement of goals for the 2004 prototype.

In 1999, the President's Committee of Advisors on Science and Technology (PCAST) issued recommendations for improving international cooperation on energy conservation. Reflecting a growing national consensus, they reported that efficient energy use helps satisfy basic human needs and powers economic development. The industrialized world depends on massive energy flows to power factories, fuel transport, and heat, cool, and light homes, and must grapple with the environmental and security dilemmas these uses cause. These energy services are fundamental to a modern economy. In contrast, many developing-country households are not yet heated or cooled. Some 2 billion people do not yet have access to electric lighting or refrigeration. The Chinese enjoy less than one-tenth as much commercial energy per person as do Americans, and Indians use less than one-thirtieth as much. Raising energy use of the world population to only half that of today's average American would nearly triple world energy demand. Such growth in per capita energy use coupled with an increase by over 50% in world population over the next half-century and no improvement in energy efficiency would together increase global energy use more than four times. Using conventional technologies, energy use of this magnitude would generate almost unimaginable demands on energy resources, capital, and environmental resources—air, water, and land.

Energy-efficient technologies can cost-effectively moderate those energy supply demands. For example, investments in currently available technologies for efficient electric power use could reduce initial capital costs by 10%, life-cycle costs by 24%, and electricity use by nearly 50%, compared to the current mix of technologies in use. Modest investments in efficient end-use technologies lead to larger reductions in the need for capital-intensive electricity generation plants. Conversely, when unnecessary power plants and mines are built, less money is available for factories, schools, and health care.

In the language of economics, there is a large opportunity cost associated with energy inefficiency. It is both an economic and environmental imperative that energy needs be satisfied effectively. Incorporating energy efficiency measures as economies develop can help hold energy demand growth to manageable levels, while reducing total costs, dependence on foreign sources of energy, and impacts on the environment. Ironically, energy is most often wasted where it is most precious. Developing and transition economies, such as in China, the world's second largest energy consumer, use much more energy to produce a ton of industrial material compared to modern market economies. Developing-nation inefficiency reflects both market distortions and underdevelopment. Decades of energy subsidies and market distortions in developing and transition economies have exacerbated energy waste. Energy efficiency requires a technically sophisticated society. Every society in the world, including the United States, has at one time encouraged energy inefficiency by controlling energy prices, erecting utility monopolies, subsidizing loans for power plant development, and ignoring environmental pollution. Many nations, including the United States, continue these wasteful practices. These subsidies and market distortions can seriously delay technological advances, even by decades, compared to best practice. The worldwide shift to more open, competitive markets may reduce some of the distortions on the supply side, but will do little to change the inherent market barriers on the demand side.

China, India, Brazil, Russia, and other developing and transition economies are building homes and factories with outdated technology, which will be used for many decades. In cold northern China and Western Siberia, for example, apartments are built with low thermal integrity and leaky windows and are equipped with appliances half as efficient as those available in the United States. Developing nations thus tend to build in excessive energy costs, lock out environmental protection, and diminish their own development potential. There are important exceptions. Progress can be both rapid and significant. Over recent decades, China has, through energy-sector reform and pursuit of least-cost paths via energy efficiency opportunities, made unprecedented progress in energy intensity reduction, faster than any developing nation in history. China has held energy demand growth to half of its rate of overall economic growth over the past two decades (Fig. 3). This example demonstrates that economic development can proceed while restraining energy demand growth; indeed, by employing cost-effective energy-efficient technologies, funds are freed for investment in other critical development needs. Energy efficiency can thus be a significant contributor to social and economic development.

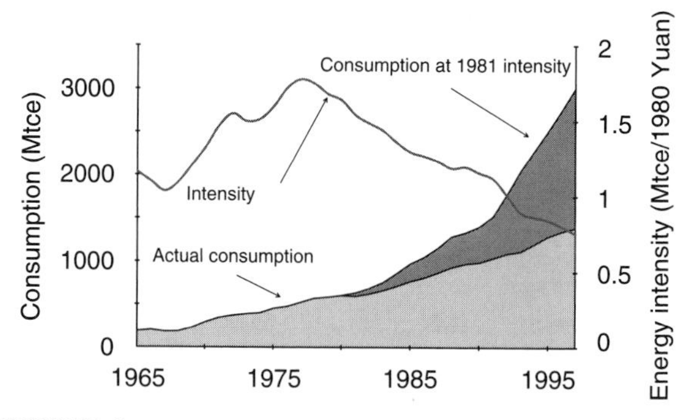

FIGURE 3 Energy intensity and consumption in China. Mtce, Million tons of coal equivalent. From Advanced International Studies Unit, Battelle Pacific Northwest National Laboratory.

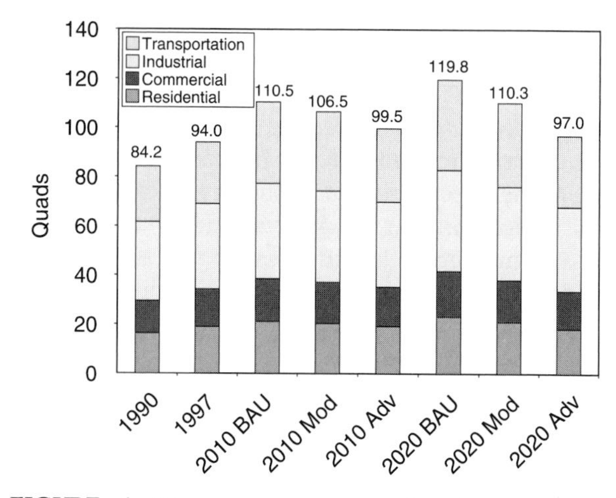

FIGURE 4 Primary energy consumption by sector. BAU, Business-as-usual model, compared to moderate (Mod) and advanced (Adv) conservation models. From Interlaboratory Working Group (2000).

5. DECADES AHEAD

Energy projections indicate differences in U.S. energy demand in 2020 between the high and low scenarios of approximately 25 quads (Fig. 4). The difference between these scenarios is substantially due to differing assumptions about development and deployment of energy-efficient technologies. The differences also reflect the plausible range of key consumption-driving factors that, operating over 20 years, can have a profound influence. These factors include economic growth rate and the (rising) urgency to cut emissions of greenhouse gases in order to slow global climate change.

5.1 Cultural Barriers to Conservation

Sadly, there remains a persistent struggle between advocates of energy supply and conservation advocates. On the supply side, the energy industry is massive and concentrated, wielding great political influence (e.g., obtaining tax subsidies, allowing unhealthy levels of environmental emissions). On the conservation side, the relevant industry is diffusely spread over the entire economy, and energy efficiency is seldom the primary concern. Furthermore, decisions regarding the energy efficiency of appliances (e.g., heating, ventilation and air-conditioning systems and refrigerators) are frequently made (e.g., by developers, not consumers) on the basis of minimum first cost rather than life-cycle cost, thus generally favoring lower efficiency.

Two streams of progress over the past 50 years have combined to accentuate the attention given to energy conservation. First, steady and ubiquitous advances in technology (such as computers and various special materials) have enabled remarkable improvements (lower cost, higher efficiency) in productivity, including energy. Second, there has been a growing awareness of major direct and indirect externalities associated with energy production and conversion. Although such externalities (e.g., health, ecological stress, climate change, and air and water pollution) are now widely recognized, they still are not generally incorporated into national economic accounts. Instead, various standards, regulations, and policies have been developed or proposed to reflect the external costs by acting as a "shadow price," in order to level the playing field of price signals.

Despite its profound impacts, the conservation revolution has resulted in only a fraction of its ultimate potential. Here are some of the reasons:

1. Lack of leadership. Whereas an energy consumer can make short-term adjustments to save energy (driving slower, driving fewer miles and traveling less, adjusting the thermostat, or turning off lights), these mostly have the unpopular effect (other than saving money) of curtailing the very energy services that we seek. Sadly, such heroic short-term curtailment measures are too often posed as the essence of conservation rather than emergency measures. On national television, President Carter advocated energy conservation as a moral imperative, but delivering his speech while wearing a sweater and sitting by an open fire connoted sacrifice and denial of comfort. Worse yet, President Reagan referred to conservation as being hot in summer and cold in winter! In the short run, curtailment is about all anyone can do, but if we stop there, we lose by far the most impressive opportunities. After the shocks of the 1970s subsided, attention shifted to the more substantial but also more challenging task of capturing longer term opportunities.

2. Time lags. The largest opportunities for energy conservation derive from incorporating more efficient technologies into new capital investments (e.g., cars, buildings, appliances, and industrial plants). These logically accrue at a rate that reflects depreciation of existing capital stock, ranging from years to many decades. Just as for the case of major transitions in energy systems, the major opportunities for conservation inherently require decades to be fully achieved.

3. Discount rates. With a few exceptions, capital purchase decisions tend to favor minimum first cost, rather than minimum life-cycle cost. Generally, the minimum first-cost model corresponds to the least energy-efficient model. Thus the "least-cost" solution is typically not taken by most consumers and this slows the rate at which efficiency is gained.

5.2 Remaining Potential for Conservation of Energy Resources

Liquid fuels, so attractive because they can be stored and transported with relative ease, loom as a global challenge because the main supply (petroleum) is very unevenly distributed geographically and is often concentrated in insecure locations. Technologies for petroleum conservation are steadily improving, from elegant, new methods of exploration and production to highly efficient combustion in motors and turbines. Thus the petroleum resource base, although headed for depletion in this century, can be stretched (conserved) by a variety of technologies. Advances in high-temperature and low-corrosion materials now enable a variety of combustion turbines that can be employed to generate electricity with 60% or better thermodynamic efficiency—double the best-practice several decades ago.

Probably the greatest challenge for liquid fuel conservation for the future is in the transportation sector. Technology, the auto industry, public policies, and consumer preference all play important roles. The industry is most profitable in making large, heavy and powerful cars and SUVs; consumers in the United States presently have a love affair with the image of power, ruggedness, and convenience with little regard for efficiency. And the record shows that U.S. public policy is to keep fuel price below total cost. Advances in automotive efficiency such as hybrid-electric/advanced diesel and fuel cells could help alleviate the situation. For example, over the past decade, work on streamlining, electrical power control systems, high-compression/direct-injection engines, electric drive trains, etc. has resulted in the recent market entry of hybrid cars that are very popular and have about 50% better mileage; in addition, rapid advances in fuel cell technology promise within a decade or so more conservation as well as lower environmental emissions. However, there seems to be little reason to hope for major gains in efficiency without major increases in fuel price or its equivalent shadow price (performance regulation such as CAFE).

Electricity generation is getting very efficient and the technologies that use electricity are also advancing. The world is still electrifying at a rapid pace. Electricity can now be generated from diverse energy sources and with diverse sizes of generators. Source diversity means inherent resilience of the system and capability for disaggregated generation means that "waste heat" from generators can be more readily utilized (cogeneration). Electricity is such a high-quality and versatile energy form that it is likely it will continue to increase in market share. Thus further advances in the efficiency of conversion and use of electricity should find ready markets.

The historic succession of energy sources over the past 150 years has led to progressively higher hydrogen-to-carbon (H/C) ratios for the fuels (e.g., coal, followed by oil, followed by natural gas). Global climate protection argues that we need to move with dispatch toward even higher H/C ratios in the coming decades. This implies a combination of removal (sequestering) of carbon from combustion releases, increased carbon recycle (e.g., renewable fuels from biomass), direct use of solar energy, and increased use of nuclear energy. Under this scenario, hydrogen becomes the most likely candidate for the derived fuel. This underscores the promise of high-efficiency fuel cells for energy conversion to electricity.

The amount of conservation achieved through advancing technology has been outstanding over the past half-century. Many more gains are possible, but we can also discern limits dictated by limits imposed by natural law. Carnot's thermodynamic rule for efficiency of heat engines is one limit. On the other hand, fuel cells give an opportunity to bypass Carnot. Success in this area could effectively enable a doubling of the useful work out of a unit of energy compared to an internal combustion engine, and with basically no environmental emissions. Success beckons but will require many technical advances in special materials and also in providing an acceptable primary energy source—probably hydrogen—all at a much lower production cost than now. The artificially low current price of gasoline constitutes a serious barrier to market entry of the new technologies.

Human ingenuity in the form of technological innovation will continue to provide new options to society. Energy conservation, which derives from sophisticated and elegant technologies, has every chance to continue paying big dividends to technologically capable societies. Conservation enables humanity to receive more "goods" along with fewer "bads." But most things have limits. Ultimately, we must recognize the folly of trying to follow an exponential path of population growth and resource consumption in its present form. We must equip ourselves with new tools, including conservation, and move with dispatch on a journey toward sustainability in the 21st century. There is little time to spare.

SEE ALSO THE FOLLOWING ARTICLES

Alternative Transportation Fuels: Contemporary Case Studies • Economics of Energy Efficiency • Energy Efficiency, Taxonomic Overview • Internal Combustion Engine Vehicles • International Comparisons of Energy End Use: Benefits and Risks • Obstacles to Energy Efficiency • Oil Crises, Historical Perspective • OPEC Market Behavior: 1973–2003 • Passenger Demand for Travel and Energy Use • Transportation and Energy, Overview

Further Reading

Chandler, W. (2000). "Energy and Environment in the Transition Economies." Westview Press, Boulder and Oxford.

Chandler, W., Geller, H., and Ledbetter, M. (1988). "Energy Efficiency: A New Agenda." GW Press, Springfield, Virginia.

Demand and Conservation Panel of the Committee on Nuclear and Alternative Energy Systems. (1978). U.S. Energy Demand: Some Low Energy Futures(14 April 1978). *Science* **200**, 142–152.

Gibbons, J. (1997). "This Gifted Age: Science and Technology at the Millennium." AIP Press, Woodbury, New York.

Gibbons, J., and Chandler, W. (1981). "Energy: The Conservation Revolution." Plenum Press, New York and London.
Gibbons, J., Blair, P., and Gwin, H. (1989). Strategies for energy use. *Sci. Am.* **261,** 136–143.
Interlaboratory Working Group. (2000). "Scenarios for a Clean Energy Future." Oak Ridge National Laboratory and Lawrence National Laboratory, Oak Ridge, Tennessee, and Berkeley, California.
President's Committee of Advisors on Science and Technology. (1997). "Federal Energy Research and Development for the Challenges of the 21st Century." Office of Science and Technology Policy, Washington, D.C.
President's Committee of Advisors on Science and Technology. (1999). "Powerful Partnerships: The Federal Role in International Cooperation on Energy Innovation." Office of Science and Technology Policy, Washington, D.C.
Sant, R. (1979). "The Least-Cost Energy Strategy: Minimizing Consumer Costs through Competition." Carnegie-Mellon University Press, Pittsburgh, Pennsylvania.
U.S. Congress, Office of Technology Assessment. (2004). "OTA Legacy" (includes reports on Conservation). U.S. Govt. Printing Office, Washington, D.C. See also http://www.wws.princeton.edu/ota/.

Conservation of Energy Concept, History of

ELIZABETH GARBER
State University of New York at Stony Brook
Stony Brook, New York, United States

1. Motion and Mechanics before 1800
2. Caloric Theory and Heat Engines
3. Equivalence of Heat and Work
4. Conservation of Energy
5. Energy Physics
6. Energy in 20th Century Physics
7. Energy Conservation beyond Physics

Glossary

caloric theory The theory that heat is a substance that flows into or out of a body as it is heated or cooled; this theory was replaced by the notion of the conservation of energy around 1850.

conservation of energy The notion that in a closed system, energy cannot be created or destroyed but can only be converted into other forms.

energy The capacity of a body to do work; it comes in various forms that can be converted into each other.

ether A medium that was assumed to fill space to explain the propagation of light, heat, and electromagnetic waves; it was discarded during the early 20th century.

heat engine Any device that takes heat and converts it into mechanical work.

mechanical equivalent of heat A constant that expresses the number of units of heat in terms of a unit of work.

mechanical theory of heat The theory that heat consists of the motions of the particles that make up a substance.

perpetual motion The fallacious theory that a machine, producing work, can be operated without any external source of power; it was discussed by philosophers and sold to kings but has been ignored by engineers since ancient times.

vitalism The theory that life depends on a unique force and cannot be reduced to chemical and physical explanations; this theory has been discarded.

work done Measured by force multiplied by the distance through which its point of application moves.

The modern principle of the conservation of energy emerged during the middle of the 19th century from a complex of problems across the sciences, from medicine to engineering. It was quickly accepted as a principle of nature and was applied to problems in mechanics and electromagnetism, and then in other sciences, even when it was not the most useful approach. It survived challenges even into the 20th century.

1. MOTION AND MECHANICS BEFORE 1800

Some ancient philosophers and most engineers assumed that perpetual motion was impossible, but nobody examined the efficiency of machines. The latter did not become central to engineering until the 18th century, after exhaustive metaphysical and methodological debates about motion and mathematics as necessary components of a comprehensive philosophy of nature. Natural philosophers and mathematicians argued over the conditions under which particular conservation laws were valid, and through these debates motion and collisions drew more attention from engineers and mathematicians.

In Rene Descartes's natural philosophy, the universe consisted of matter, whose quantity was measured by size, and motion. Collisions between particles was the mechanism for change, and the total amount of motion in the universe remained constant, although motion could be lost or gained in each collision. Descartes also assumed that the natural motion of bodies was linear—not circular—and was a vector. In collisions, the quantity of motion, size multiplied by velocity, could be transferred from one particle to another. The force of a particle was its size multiplied by the square of its velocity. Although Descartes modeled his philosophical method on geometry, very little mathematics was contained within it. It was left to later generations to put his philosophy of nature into mathematical form.

One of the most successful was Isaac Newton, who began as a follower of Descartes and ended as his scientific and philosophical rival. In his explorations of Descartes's system, Newton became dissatisfied with Cartesian explanations of cohesion and other phenomena. He eventually accepted the reality of the concept of "force," banned by Descartes as "occult," as acting on micro and macro levels. Newton's universe was as empty as Descartes's was full, and God was necessary to keep His vast creation going. In such a universe, conservation laws and collisions were not of central importance. Yet Newton introduced a measure of force, momentum, that became part of the debates on conservation laws in mechanics.

One of the first entries in this debate was that of the Cartesian Christian Huyghens, who examined circular motion and the collisions of perfectly "hard" (perfectly elastic) bodies. He demonstrated that in such collisions, the sum of the magnitude of the bodies multiplied by the square of the velocity was a constant. However, for Huyghens, this expression was mathematically significant but not physically significant.

This pattern was repeated as 18th century continental mathematicians debated the measure of force and the quantities conserved in collisions. Gottfried Leibniz argued that Descartes's measure of force was insufficient. According to Leibniz, a body dropped a certain distance acquired just enough force to return to its original position, and this was a truer measure of force. Descartes had confused motive force with quantity of motion. Force should be measured by the effect it can produce, not by the velocity that a body can impress on another. In perfectly hard body collisions "vis viva," living force ($\frac{1}{2}mv^2$) was conserved. Central to subsequent debates were the experimental results of Giovanni Poleni and Wilhelm s'Gravesande, who measured the depths of the indentations left by metal balls dropped at different speeds into clay targets. If balls of different masses but the same size were dropped, the indentations remained the same. Lighter balls left the same indentations if they were dropped from heights proportional to 1/masses. s'Gravesande's experiments also showed that the force of a body in motion was proportional to v^2, that is, to its vis viva. He also did some elegant experiments in which hollow metal balls were filled with clay to investigate collisions of "soft" bodies. The results only fueled debates about the conditions over which measure of force was appropriate, mv or $\frac{1}{2}mv^2$.

s'Gravesande began his experiment to prove Newton's theories were correct, but the experiments seemed to Cartesians to support their point of view. Despite this evidence, s'Gravesande remained a Newtonian; for him, $\frac{1}{2}mv^2$ remained a number of no physical significance. Mathematicians in these debates over force included Madame de Chatelet, who published a French translation of Newton's *Principia* just 2 years after she had published *Institutiones* to spread Leibniz's ideas among French mathematicians. The subject of vis viva became a part of ongoing debates in the salons and Paris Academie des Sciences over Newtonian or Cartesian natural philosophy and the accompanying politics of each of those institutions.

Nothing was resolved, and mathematicians eventually moved on to other problems. They were inclined to use whatever physical principles seemed the most appropriate in their quest to develop the calculus using mechanical problems of increasing sophistication. Mechanics, or problems within it, became the instrument to display mathematical expertise and brilliance rather than explorations into physical problems. A perfect illustration of this is the work of Jean Louis Lagrange. By considering a four-dimensional space and using calculus, Lagrange reduced solid and fluid dynamics to algebra. He began with the fundamental axiom of virtual velocity for mass particles,

$$\sum m\frac{d\vec{v}}{dt}\cdot\delta\vec{r} = \sum \vec{F}\cdot\delta\vec{r},$$

where m is mass, v is velocity, $\delta\vec{r}$ is virtual displacement, and $\vec{F}$ is the applied force acting on a particle. Lagrange then introduced generalized coordinates, $q_1 \ldots q_n$, and generated the equations,

$$\frac{\partial T}{\partial q_i} - \frac{d}{dt}\left(\frac{\partial T}{\partial \dot{q}_i}\right) - \frac{\partial \Phi}{\partial q_i} = 0,$$

where T is the vis viva and Φ is a function. Thus, he succeeded in making rational mechanics as abstract, general, and algebraic as possible. No diagrams, geometry, or mechanical reasoning graced the pages of his text. It consisted "only of algebraic operations."

2. CALORIC THEORY AND HEAT ENGINES

During the 18th century, French engineers developed entirely different, yet mathematical, approaches to mechanics. Mathematicians had used problems on the construction of bridges, the construction of pillars, and the design of ships to explore the calculus. The education of French engineers began to include some calculus to solve engineering problems, especially in those schools of engineering within branches of the military. Here they developed an independent approach to the understanding of structures and the operations of machines.

Out of this tradition emerged Lazare Carnot, whose 1783 treatise on the efficiency of machines also began with the principle of virtual velocities. From this, he developed a relationship among the moment of activity, force multiplied by distance (work), and vis viva. He concluded that to get the maximum "effect" from the cycles of a machine, it was necessary to avoid inelastic impact in the working of its parts. During the 1820s, his son Sadi turned the same analytical approach toward the new "machine" that was revolutionizing the economy of Great Britain, the steam engine. As Carnot noted, removing this source of power would "ruin her prosperity" and "annihilate that colossal power." By this time, steam engines had been used in Britain for more than a century as pumps for mines. More recently, through the development by James Watt of the separable condenser and a governor, the steam engine became a new driving force for machinery. By the 1830s, the use of steam power and associated machinery was integrated into American and British cultures as a measure of the "civilized" states of societies.

Carnot wanted to establish the conditions under which maximum work could be produced by a heat engine. He stripped the heat engine down to its essential elements and operations. Through a series of examples, he demonstrated that successive expansions and contractions from heating and cooling—not heat alone—produced the motive power. Steam was not unique; any substance that expanded and contracted could be a source of power. Gases generally afforded the most expansion and contraction on heating and cooling. Carnot thought of heating and then cooling as the "reestablishment of equilibrium" in caloric and the substance of heat as the source of motive power. Therefore, he began to search for the general conditions that would produce the maximum motive power, and in principle, this could be applied to any heat engine.

Sadi Carnot used the same approach as his father and other French engineers in analyzing machines by looking at infinitesimal changes. In this case, the temperatures of the heat source and sink differed only slightly from that of the working substance. Carnot examined the changes in temperature and pressure of the working substance in a complete cycle of the engine, again a familiar approach in French engineering. Initially, the body is in contact with the heat source, and its piston rises and its volume expands at a constant temperature. It is then removed from this source, and the expansion continues without any heat input and the temperature of the working substance decreases to that of the heat sink. Placed in contact with the cooler body, the gas is compressed at constant temperature and then is removed and further compressed until the cycle is closed and the working substance attains the temperature of the heat source. The Carnot cycle is then complete. However, Carnot's analysis of its operation was in terms of caloric theory. In the first operation, the caloric is removed from the heat source to the working substance, and motive power is produced as the caloric attains the temperature of the heat sink. Equilibrium is restored as the caloric again reaches the temperature of the heat source. Using the incomplete image of a waterfall used to drive machinery, the caloric is conserved. It has fallen from one temperature to a lower one, and in this fall we can extract motive power just as we do from the gravitational fall of water.

The caloric theory of heat was accepted across Europe by the latter part of the 18th century. Experimental research into heat occupied many chemists across Europe and had an important place in the development of the reformed chemistry of Antoine Lavoisier. "Airs" (gases) were also the object of avid research throughout Europe by many chemists from before the flight of the Montgolfier brothers in 1784 through the first third of the 19th century. Caloric theory drew these two strands of research together. During the 1780s, Lavoisier had measured the specific heats of a number of substances with Pierre Simon de Laplace. These experiments were aspects of Lavoisier's research on airs and heat. In another series of such experiments, Lavoisier studied digestion and animal heat. Using an ice calorimeter, he demonstrated that respiration was a slow form of combustion and predicted heat outputs that agreed with his experimental results. Both Lavoisier and Laplace believed that heat was a material substance that manifested itself in two forms: free caloric (detected by thermometers) and latent caloric. Laplace and Simon Poisson developed caloric theory mathematically by assuming that the caloric, Q, in a gas was a function $Q = f(P, \rho, T)$, where P is the pressure, T is the temperature, and ρ is the density of the gas. Manipulating the partial differential equations in a theory of adiabatic compression, this general equation led to relationships in accord with known experimental results such as the ratio of the specific heats of gases. Caloric was connected to the notion of work through Laplace's theory. Free heat was vis viva, combined (latent) heat was the loss of vis viva, and heat released was an increase in vis viva. Caloric theory was sophisticated and was taught at the Ecole Polytechnique when Sadi Carnot was a student, although toward the end of his short life he rejected it to explore the idea that heat was the motion of particles.

Carnot used the latest experimental results on the properties of gases to demonstrate his contentions and then discussed the limitations of heat engine design at the time. He concluded that because of gases' great expansion and contraction with heating and cooling, and because of the need for large differences in temperature to gain the most motive power, gases were the only feasible working substances for heat engines. He used current research on gases to demonstrate that any gas might be used.

Although Carnot's small treatise, *Reflections on the Motive Power of Heat,* received a good review from the *Revue Encyclopedique,* only one engineering professor at the Ecole Polytechnique developed Carnot's ideas. Emile Clapyeron gave geometrical and analytical expression to Carnot's analysis, creating the first pressure–volume (PV) diagram of a Carnot cycle. In his analysis, this cycle was reduced to an infinitesimally small parallelogram for the case pursued by Carnot. Again using the results from work on the properties of gases, Clapyeron obtained an expression for the "maximum quantity of action" developed in a cycle from the area within the parallelogram of the heat cycle for a perfect gas heat engine. Again, the analytical expressions were in terms of general expressions for heat, $Q = Q(P, V, T)$, $P = P(V, T)$, and $V = V(P, T)$. In the concluding paragraphs of his memoir, Clapyeron described the mechanical work available from his heat engine in terms of the vis viva the caloric developed in its passage from the boiler to the condenser of the engine. Many of his expressions were independent of any particular theory of the nature of heat and remained valid even as their physical explanation changed during the 1850s.

The works of both Carnot and Clapyeron sank into oblivion. Carnot died young and tragically in an asylum, but his oblivion was partially political. His father, Lazare, had enthusiastically embraced the French Revolution, was important in its early wars, and rose to prominence in Napoleon's regime. He remained loyal to Napoleon until the end and died in exile. The Carnot name (except that of his brother, a prominent politician), was politically tainted even into the 1830s. And however useful caloric might be for engineers, among scientists the caloric theory held less interest during this decade. They extended the wave theory of light to heat, and through it they explained other scientifically pressing problems such as the conduction and radiation of heat.

So deep was Carnot's oblivion that in 1845 William Thomson, a young ambitious Scottish academic, could find no copies of his treatise in Paris. Thomson had recently graduated Second Wrangler and first Smith's Prizeman from Cambridge University and was actively seeking the empty chair in natural philosophy at Glasgow University. The Glasgow that Thomson grew up in was Presbyterian and was the center of heavy industry (especially shipbuilding) and chemical engineering in Great Britain. Although his father was a distinguished professor of mathematics at Glasgow, the younger Thomson needed the support of the rest of the faculty and of the town representatives to ensure his election. He had to cultivate the image of the practical experimenter rather than flaunt his mathematical abilities, hence Thomson's sojourn at the laboratory of Henri Victor Regnault, well known for his empirical studies of steam engines. These were the circumstances that led Thomson to search out Carnot's work. Undaunted, Thomson made full use of Clapyeron's publications in a series of papers (published in 1847 and 1850) extending Clapyeron's analysis. In his first paper he postulated an absolute zero of temperature implied by Clapyeron's analysis and put forward the notion of an absolute scale of temperature. Both emerged from the mathematical forms independent of their physical interpretation. Thomson's brother, James, noted that one of the results of Clapyeron's work was that the freezing point of water decreased with an increase in pressure. William Thomson confirmed this through experiments done at Glasgow. The position of caloric theory seemed to be becoming firmer once again. However, in the middle of this theoretical and experimental triumph, Thomson went to the 1847 annual meeting of the British Association for the Advancement of Science (BAAS), where he heard a paper given by James Joule on experiments whose results contradicted caloric theory.

3. EQUIVALENCE OF HEAT AND WORK

Joule was the son of a very successful brewer in Manchester, England. He was educated privately and with ambitions to spend his life in science. His first papers dealt with "electromagnetic engines," and in the course of his experiments he focused on what we now call Joulean heating in wires carrying electric currents. This drew attention to the production of heat in both physical and chemical processes. From his experiments and a developing theory, he concluded that an electric generator enables one to convert mechanical power into heat through the electric currents induced into it. Joule also implied that an electromagnetic "engine" producing mechanical power would also convert heat into mechanical power. By 1843, he presented his first paper to the British Association for the Advancement of Science (BAAS) on the "mechanical value of heat." Joule believed that only the Creator could destroy or create and that the great agents of nature, such as electricity, magnetism, and heat, were indestructible yet transformable into each other. As his experiments progressed, he elaborated his mechanical theory of nature. Important to observers, and to Thomson, was Joule's idea that mechanical work and electricity could be converted into heat and that the ratio between mechanical work and heat could be measured. In a series of papers on the mechanical equivalent of heat presented annually to BAAS and politely received, Joule labored alone until the stark reactions of Thomson in 1847. Some scientists could not believe Joule's contentions because they were based on such small temperature changes. Presumably, their thermometers were not that accurate. However, Thomson understood that Joule was challenging caloric theory head on. Until 1850, Thomson continued publishing papers within a caloric framework and an account of Carnot's theory. He had finally located a copy of the treatise. Clearly, he began to doubt its validity, noting that more work on the experimental foundations of the theory were "urgent," especially claims that heat was a substance. It took Joule 3 years to convince Thomson of the correctness of his insight. Only in 1850 did Thomson concede that heat was convertible into work after he read Rudoph Julius Emmanuel Clausius's papers and those of William McQuorne Rankine on heat. He then reinterpreted some parts of his caloric papers that depended on heat as a substance in terms of the interconversion of heat and work. Also, he rederived and reinterpreted his expression for the maximum amount of work that could be extracted from the cycle of a perfect heat engine and other results and then gave his version of the second law of thermodynamics.

4. CONSERVATION OF ENERGY

Others had come to conclusions similar to those of Joule before Thomson. One of the most notable was Hermann von Helmholtz. Trained as a physician while studying as much mathematics and physics as possible, his doctoral dissertation was a thorough microscopic study of the nervous systems of invertebrates. His mentor was Hermann Mueller, who (along with Helmholtz's fellow students) was determined to erase vitalism from physiology. Their goal was to make physiology an exact experimental science and to base its explanatory structures in physics and chemistry. Helmholtz managed to continue his physiological research while a physician in the Prussian army, where he was dependent on his army postings and his early research on fermentation and putrefaction was cut short by the lack of facilities. He turned to a study of animal heat and the heat generated by muscular action and found that during contraction chemical changes occurred in working muscles that also generated heat. In the midst of this research, Helmholtz published "On the Conservation of Force." This paper displayed the extent of his mechanical vision of life and his commitment to a Kantian foundation for his physics. Central forces bound together the mass points of his theory of matter. These commitments resurfaced during the 1880s with his work on the principle of least action.

With considerable pressure brought to bear on the ministry of education and the Prussian military, Helmholtz was released from his army obligations and entered German academic life. He never practiced medicine again, yet his research over the subsequent two decades began with questions in physiology that expanded into studies on sound, hearing, and vision. His work on nerve impulses in 1848 developed into a study of electrical pulses and then into a study of the problem of induction before growing into a critical overview of the state of the theory of electricity during the 1870s.

The closeness of Helmholtz's work in physics and physiology is evident in his paper on the conservation of force. He had actually stated the principle earlier (in 1845) in a paper reviewing the work of Humphry Davy and Antoine Lavoisier on animal heat. He concluded that if one accepts the "motion theory of heat," mechanical, electrical, and chemical forces not only were equivalent but also could be transformed one into each other. This assumption bound together results already known from chemistry, electricity, and electromagnetism. The structure of this review formed the outline of Helmholtz's 1847 paper. He added a theory of matter that he believed demonstrated the universality of the principle of conservation. Helmholtz began in metaphysical principles from which he extracted specific physical results. In his model, force was either attractive or repulsive. Change in a closed system was measured by changes in vis viva expressed as changes in the "intensity of the force," that is, the potential of the forces that acted between the particles, $\frac{1}{2}mv_1^2 - \frac{1}{2}mv_2^2 = \int_r^R \varphi dr$, where m is the mass of the particle, whose velocity changes from v_1 to v_2, and ϕ is the intensity of the force, whose measure is constructed by looking at changes in v^2. Because both velocities and forces were functions of the coordinates, Helmholtz expressed these last changes as $\frac{1}{2}md(v^2) = Xdx + Ydy + Zdz$, where $X = (x/r)\phi$. He demonstrated that for central forces, vis viva was conserved and, thus, so was the intensity of the force. He then expanded the argument to a closed system of such forces. The unifying concept in the paper was "tensional force" and its intensity ϕ. This allowed Helmholtz to extend his argument to electricity to obtain the "force equivalent" of electrical processes. In this case, the change in vis viva of two charges moving from distance r to R apart was $\int_1^2 \frac{1}{2}md(v^2) = -\int_r^R \varphi dr$, where he identified ϕ with Gauss's potential function, to which Helmholtz gave physical meaning. However, he did not give this entity a new name. He was reinterpreting physically a well-known approach used when mathematizing a physical problem. Unlike mathematicians, Helmholtz drew important physical conclusions from his equations, including the Joulean heating effect. He also included Franz Neumann's work on electrical induction in his conceptual net. Helmholtz used examples from chemistry and electricity to argue that heat was a measure of the vis viva of thermal motions, not a substance. He used a broad range of experimental results across the sciences to illustrate the plausibility of his contentions.

Helmholtz had difficulty in getting these speculations into print. They eventually appeared as a pamphlet. The thrust of the speculations went against prevailing standards in the emerging field of physics. German physicists had rejected grand speculative visions of nature and its operations for precision work in the laboratory, from which they assumed theoretical images would emerge. The model of this new approach was Wilhelm Weber. In Weber's case, precise, painstaking numerical results, analyzed in detail for possible sources of error, led to a center-of-force model that explained the range of electrical phenomena he had researched experimentally. Neumann was known for his careful experiments on crystals and their optical properties. Helmholtz presented his mechanical model before his analysis of the experimental work of others that ranged across disciplinary boundaries. This approach also dissolved the recently hard-won academic and specialist designation of physics in German academia.

Helmholtz was not the only German claimant to the "discovery" of the principle of conservation of energy. After his graduation in 1840, Julius Robert Mayer, a physician, became a doctor on a ship bound for the East Indies. Mayer noted the deeper color of venous blood in the tropics compared with that in Northern Europe. He interpreted this as more oxygen in the blood, with the body using less oxygen in the tropics because it required less heat. He speculated that muscular force, heat, and the chemical oxidation of food in the body were interconvertible forces. They were not permanent but were transformable into each other. On his return, Mayer learned physics to demonstrate the truth of this conviction and then set about to calculate the mechanical value of a unit of heat. To do this, he looked at the works on the thermal properties of gases and took the difference between the two principal specific heats as the amount of heat required to expand the gas against atmospheric pressure. This heat was transformed into work, and from this he computed the conversion factor between heat and work. Further memoirs followed, extending his use of available experimental results from the chemistry and physics of gases. Mayer even included the solar system. He argued that meteorites falling under gravity built up enormous amounts of mechanical force that yielded considerable amounts of heat on impact. This could account for the sun's seemingly inexhaustible supply of heat.

Mayer also encountered difficulties in publishing his work. His first paper of 1842, "Remarks on the

Forces of Inorganic Nature," was published in a chemistry journal rather than a physics journal. As with Helmholtz, the combination of metaphysics and calculations derived from other people's experiments was not acceptable in the physics community in the German states. In addition, Mayer did not have the academic credentials or connections of Helmholtz. In fact, Mayer had to underwrite the publication of his later papers. At the end of the decade, Mayer suffered personal tragedy and in 1850 tried to take his own life. He was confined to a series of mental institutions. However, from 1858 onward, he began to gain recognition for his work from his colleagues, including Helmholtz, and he received the Copley Medal from the Royal Society in 1871.

So far, there were several approaches leading to the statement of the first—and even the second—law of thermodynamics. None of these claimants replaced Carnot's and Clapeyron's work mathematically from a thermodynamic point of view except the work of Clausius. In his 1850 paper, Clausius began from the observation that the conduction of heat is possible only from a hotter to a colder body. This led him logically to replace Carnot's principle, that heat is conserved in the cycle of a perfect heat engine, to that of the equivalence of heat and work. As a consequence, when work is done, heat is consumed in any heat engine. Although he kept many of the mathematical forms developed by Clapeyron, Clausius assumed that the mathematical characteristics of the functions that entered into his equations determined the physical characteristics of his system. In analyzing the Carnot cycle, Clausius chose the perfect gas as his physical system and looked at the state of the gas as it traversed an infinitely small cycle. Following Clapeyron, he constructed the heat added or expended for each leg of the cycle and constructed an expression for the inverse of the mechanical equivalent of heat, A, as the ratio of the heat expended over the work produced. To construct an expression for the work done in completing the cycle, Clausius calculated the expressions for the heat added at each leg in the cycle and constructed

$$\frac{d}{dt}\left(\frac{dQ}{dV}\right) - \frac{d}{dV}\left(\frac{dQ}{dt}\right) = \frac{A \cdot R}{V},$$

where dQ is the heat added, V is the volume of the gas, and R is the gas constant. Arguing mathematically that Q could not be a function of A and t so long as they were independent of each other, the right-hand side of the equation must be zero. This brought the equation into the form of a complete differential,

$$dQ = dU + A \cdot R\left(\frac{a+t}{V}\right)dV,$$

where $a+t=$ the absolute temperature and $U=U(V, t)$, is the heat necessary for internal work, and depends only on the initial and final states of the gas. The second term on the right-hand side, the work done, depended on the path taken between the initial and final states of the gas. Clausius then introduced $1/T$, where T is the temperature that, as a multiplier of dQ, makes a complete differential of the form $Xdx + Ydy$. He applied this combination of physics and mathematics to the behavior of gases and saturated vapors, with the latter being of great interest due to the engineering importance of steam engines during the mid-19th century. In this analysis, he restated the results, some mathematically equivalent to those of Clapeyron and Thomson, but interpreted them in terms of the equivalence of heat and work. At the end of this first paper, he gave an initial statement of the second law of thermodynamics.

Therefore, there are several different paths and claimants to some statements equivalent to the first law of thermodynamics. During the decades immediately following 1850, many others were championed as "discoverers" of the principle of the conservation of energy among the engineers and scientists working on related problems. These claims came in the wake of a change in language from the equivalence of heat and work to the conservation of energy. Some of the assertions gave rise to bitter complaints, especially the contentions of Peter Guthrie Tait's, whose *Thermodynamics* trumpeted Joule as the "discoverer" of the equivalence of heat and work yet readily recognized Helmholtz's priority in the matter of the first law. Mayer's work was recognized belatedly by Helmholtz and Rankine against Joule's claims. Mayer published an estimate of the mechanical equivalent of heat a year before the appearance of Joule's first value. And so it went. A great deal was at stake in these disagreements, which reflected on industrial prowess as well as on intellectual authority in the growing political and industrial rivalry between Germany and Great Britain. Using hindsight, later historians also tried to claim some of the credit for other Europeans.

5. ENERGY PHYSICS

A decade before the development of the mechanical meanings of the conservation of energy, Thomson,

Clausius, and Helmholtz had extended its reach into electrostatics, thermoelectricity, the theory of electromagnetism, and current electricity. During the 1860s, one of the more important applications of this new approach was the establishment of an absolute system of electrical units. Again this involved national prestige given that the Germans already had their own system based on the measurements of Weber. The BAAS committee included Thomson, James Clerk Maxwell, and the young electrical engineer Fleming Jenkin. Maxwell and Jenkin measured resistance in absolute (mechanical) units and deduced the others from there.

So far, what existed were disparate statements and uses for energy, none of which shared a common language or set of meanings. Thomson and Tait created that language for energy in their 1868 textbook on mechanics, *A Treatise on Natural Philosophy.* The text was directed to the needs of their many engineering students. During the 19th century, numerous new technical terms were coined, many of which derived from ancient Greek, and energy was no exception. Derived from the Greek word *ενεργεια*, energy was commonly used for people who demonstrated a high level of mental or physical activity. Also, it expressed activity or working, a vigorous mode of expression in writing or speech, a natural power, or the capacity to display some power. More technically, Thomson and Tait used energy as the foundation of their 1868 text. They claimed that dynamics was contained in the "law of energy" that included a term for the energy lost through friction. They introduced and defined kinetic and potential energy and even claimed that conservation of energy was hidden in the work of Newton. William Whewell grumbled about the book's philosophical foundations, but it was translated into German as well as being widely used in Great Britain.

Conservation of energy was becoming the foundation for teaching physics as well as for new approaches in research in the physical sciences. Thomson considered that conservation of energy justified a purely mechanical vision of nature and tried to reduce all phenomena, including electromagnetic phenomena, to purely mechanical terms. In his early papers in electromagnetism, Maxwell used conservation of energy in constructing and analyzing mechanical models as mechanical analogues to electrical and magnetic phenomena. He then abandoned this mechanical scaffolding and established his results on general energetic grounds, including his electromagnetic theory of light. In defense of his approach to electromagnetism, Maxwell criticized Weber's action-at-a-distance approach as being incompatible with the conservation of energy. In 1870, Helmholtz began a systematic critique of Weber's electrical theory and repeated Maxwell's judgment that he located in Weber's use of velocity-dependent forces. Helmholtz had first stated this criticism in his 1847 paper on the conservation of force, and it became a starting point for alternative approaches to electromagnetism among German physicists. These included Neumann and Bernard Riemann, who developed the notion of a retarded potential. By 1870, this criticism undermined the validity of his work and Weber was forced to answer. He demonstrated that, indeed, his theory was compatible with conservation of energy.

Clausius was the first to try to develop an expression for the laws of thermodynamics in terms of the motions of the molecules of gases. His models were mechanical, whereas the later ones of Maxwell and Ludwig Boltzmann were statistical. Maxwell became more interested in the properties of gases, such as viscosity, heat conduction, and diffusion, relating them to statistical expressions of energy and other molecular properties. Boltzmann developed a statistical expression for the first and second laws of thermodynamics. During the late 19th century, thermodynamics itself was subject to great mathematical manipulation, which in some hands became so remote from physics that it once again became mathematics.

By the time of Maxwell's death in 1879, the conservation of energy had become a principle by which physical theories were judged and developed. This was particularly true in Great Britain, where "energy physics" propelled the theoretical work of most physicists. Much of this work, such as that of Oliver Lodge, was directed toward the development of Maxwell's electromagnetism, although the work was expressed as mechanical models. Lodge also drew up a classification of energy in its varied physical forms, from the level of atoms to that of the ether. He also used an economic image for energy, equating it with "capital." He extolled John Henry Poynting's reworking of Maxwell's *Treatise* in 1884 and saw it as a new form of the law of conservation of energy. In his treatment of the energy of the electromagnetic field, Poynting studied the paths energy took through the ether. He investigated the rate at which energy entered into a region of space, where it was stored in the field or dissipated as heat, and found that they depended only on the value of the electric and magnetic forces at the boundaries of the space. This work was duplicated by Oliver Heaviside, who then set about reworking Maxwell's equations into their still familiar form. The focus was now firmly on the "energetics" of the electromagnetic field.

The emphasis on energy stored in the ether separated British interpretations of Maxwell's work from those on the continent, including that of Heinrich Hertz. However, energy in the German context became the foundation for a new philosophy of and approach to the physical sciences, energetics. Initiated by Ernst Mach, energetics found its most ardent scientific advocate in Wilhelm Ostwald. The scientific outcome of Ostwald's work was the development of physical chemistry. Yet his dismissal of atoms led to clashes with Boltzmann, and during the 1890s his pronouncements drew the criticism of Max Planck. Planck had, since his dissertation during the 1860s, explored the meaning of entropy, carefully avoiding atoms and molecules. He was forced into Boltzmann's camp in his research on black body radiation, essentially a problem of the thermodynamics of electromagnetic radiation. Planck was not alone in understanding the importance of the problem; however, other theorists were tracking the energy, whereas Planck followed the entropy changes in the radiation.

6. ENERGY IN 20TH CENTURY PHYSICS

As an editor of the prestigious journal *Annalen der Physik,* Planck encouraged the early work of Albert Einstein, not only his papers in statistical mechanics but also those on the theory of special relativity. Planck then published his own papers on relativistic thermodynamics, even before Einstein's 1907 paper on the loss of energy of a radiating body. Two years before this, Einstein had investigated the dependence of the inertia of body on its energy content. In this case, the mass of the body in the moving system decreased by E/c^2, where E is the energy emitted as a plane wave and c is the velocity of light. Einstein concluded that "the mass of a body is a measure of its energy content." Einstein constructed his relativistic thermodynamics of 1907 largely on the foundation of Planck's work and reached the conclusion that $dQ = dQ_0\sqrt{(1 - v^2/c^2)}$, where dQ is the heat supplied to the moving system, dQ_0 is that for the system at rest, and v is the velocity of the moving system to that at rest.

Thus, the law of conservation of energy was integrated into relativity theory. However, its passage through the quantum revolution was not as smooth. In 1923, Niels Bohr expressed doubts about its validity in the cases of the interaction of radiation and matter. This was after Arthur Holly Compton explained the increase in the wavelengths of scattered X rays using the conservation of energy and momentum but treating the X rays as particles. The focus of Compton's research, the interaction of radiation and atoms, was shared with many groups, including that around Bohr in Copenhagen, Denmark. Bohr did not accept Einstein's particle theory of light, even with the evidence of the Compton effect. By using the idea of "virtual oscillators" that carried no momentum or energy, Bohr, H. A. Kramers, and John Slater (BKS) hoped to maintain a wave theory of the interaction of radiation and matter. To do so, they had to abandon the idea of causality, the connection between events in distant atoms, and the conservation of energy and momentum in individual interactions. Struggling with the meanings of the "BKS theory" became crucial in the development of Werner Karl Heisenberg's formulation of quantum mechanics. However, the BKS theory was bitterly opposed by Einstein as well as by Wolfgang Pauli, who, after a brief flirtation with the theory, demonstrated its impossibility using relativistic arguments.

Conservation laws were reinstated at the center of physical theory. Conservation of energy became the criterion for abandoning otherwise promising theoretical approaches during the late 1920s and early 1930s in quantum field and nuclear theory. In the latter case, Pauli posited the existence of a mass-less particle, the neutrino, to explain beta decay. The neutrino remained elusive until the experiments of Ray Cowan and Fred Reines during the 1950s. Conservation of energy also governed the analysis of the collisions of nuclear particles and the identification of new ones, whether from cosmic rays, scattering, or particle decay experiments. Relativistic considerations were important in understanding the mass defects of nuclei throughout the atomic table. The difference between the measured masses and those calculated from the sum of the masses of their constituent particles was explained as the mass equivalence of the energy necessary to bind those components together.

7. ENERGY CONSERVATION BEYOND PHYSICS

Beyond physics, conservation of energy became a foundation for other sciences, including chemistry and meteorology, although in this case the second law of thermodynamics was more powerful in explaining vertical meteorological phenomena. In the natural sciences, the most dramatic demonstration of the conservation of energy's usefulness was in the study of animal heat and human digestion. The production of heat as synonymous with life can be seen in the aphorisms of Hippocrates during the fifth century BCE. In Galen's physiology, developed during the second century AD, the source of bodily heat was the heart. Heat was carried in the arterial system throughout the body, where it was dissipated through the lungs and the extremities. Aspects of Galen's physiology remained in William Harvey's explanation of the function of the heart. Yet Harvey's experiments, published in 1628 as *Anatomical Experiments on the Motion of the Heart and Blood in Animals,* demonstrated the circulation of the blood and the pumping action of the heart. This mechanical function of the heart became part of Descartes's mechanical philosophy that changed debates over the source of heat in the body. Was it generated in the heart or from friction as the blood flowed through the vessels of the body? Harvey's argument that the lungs dissipated the heat from the body fitted into this mechanical vision of the body.

During the 18th century, Descartes's mechanistic vision of the body was replaced by vitalism based on chemical experiments. This was the vitalism that Helmholtz sought to root out from medicine during the 1840s. The experiments were those performed on "airs" consumed and expended by plants that vitiated or revived the atmosphere. Combined with these experiments were others showing that combustion and respiration made the atmosphere less able to sustain life, usually of birds and small mammals. After the introduction of Lavoisier's reform of chemical nomenclature, which also carried with it a new theory of chemistry and chemical reactions, experiments on animal chemistry became more quantitative and lay within his new theoretical framework. Treadmill studies by Edward Smith in 1857 demonstrated that muscles derived energy, measured by heat output, from foods not containing nitrogen. However, Justus Leibig's idea that bodily energy derived from muscle degradation was not easily deposed. It took the mountain-climbing experiments of Adolph Fick and Johannes Wislicenus in 1866 and the analysis of them by Edward Frankland to finally displace Leibig's theory. Frankland pointed out that Fick and Wislicenus's results could be interpreted by looking at the energy involved in combustion in a unit of muscle tissue and considering the work output the equivalent of this energy. Conservation of energy was necessary in understanding the dynamics of the human body. Frankland also showed that the energy the body derived from protein equaled the difference between the total heat of combustion and the heat of combustion of the urea expelled, with the urea being the pathway for the body to rid itself of the nitrogen in proteins. Thus, conservation of energy entered the realm of nutrition, biophysics, and biochemistry. In later experimental studies, W. O Atwater measured, by a whole body calorimeter, that the heat given off by the human body equaled the heat released by the metabolism of the food within it. Even when the complexities of the pathways of amino acids and other sources of nitrogen were understood, along with their routes through the digestive system, conservation of energy remained the foundation for interpreting increasingly complex results.

Thus, conservation of energy has proved to be crucial for analyzing the results of experiments across—and in some cases beyond—the scientific spectrum. There have even been experiments—so far unsuccessful—aimed at measuring the mass of the soul on its departure from the body.

SEE ALSO THE FOLLOWING ARTICLES

Heat Transfer • *Mechanical Energy* • *Storage of Energy, Overview* • *Thermodynamic Sciences, History of* • *Thermodynamics, Laws of* • *Work, Power, and Energy*

Further Reading

Adas, M. (1989). "Machines as the Measure of Man." Cornell University Press, Ithaca, NY.

Caneva, K. L. (1993). "Robert Mayer and the Conservation of Energy." Princeton University Press, Princeton, NJ.

Cardwell, D. (1971). "From Watt to Clausius." Cornell University Press, Ithaca, NY.

Carpenter, K. J. (1994). "Protein and Energy: The Study of Changing Ideas in Nutrition." Cambridge University Press, Cambridge, UK.

Cassidy, D. (1991). "Uncertainty: The Life and Times of Werner Heisenberg." W. H. Freeman, San Francisco.

Cooper, N. G., and Geoffrey, W. B. (eds.). (1988). "Particle Physics: A Los Alamos Primer." Cambridge University Press, Cambridge, UK.

Einstein, A. (1905). Does the inertia of a body depend upon its energy content? *In* "The Collected Papers of Albert Einstein" (A. Beck Trans, Ed.), vol. 2, Princeton University Press, Princeton, NJ.

Einstein, A. (1907). On the relativity principle and the conclusions drawn from it. *In* "The Collected Papers of Albert Einstein" (A. Beck Trans, Ed.), vol. 2, Princeton University Press, Princeton, NJ.
Everitt, C. W. F. (1975). "James Clerk Maxwell: Physicist and Natural Philosopher." Scribner, New York.
Helmholtz, H. (1853). On the conservation of force. *In* "Scientific Memoirs" (J. Tyndall and W. Francis, Eds.). Taylor & Francis, London. (Original work published 1847.)
Smith, C. (1998). "The Science of Energy: A Cultural History of Energy Physics in Victorian Britain." University of Chicago Press, Chicago.
Smith, C., and Norton Wise, M. (1989). "Energy and Empire: A Biographical Study of Lord Kelvin." Cambridge University Press, Cambridge, UK.

Conservation of Energy, Overview

GORDON J. AUBRECHT II
The Ohio State University
Columbus, Ohio, United States

1. Energy Preliminaries
2. The Physical Principle of Energy Conservation
3. Thermal Systems and Energy
4. Energy Conservation as Preservation of Resources

Glossary

conservative force The force that allows a potential energy to be constructed. A conservative force that acts along a closed path does no work. See potential energy.

district heating The use of waste heat from energy generation for space heating within a specified distance of the generating plant.

first law of thermodynamics The statement that the total internal energy of a system is changed both by heat transferred into or out of a system and the work done by or on a system. The principle of conservation of energy was at first applied only to mechanical systems (conservation of mechanical energy). When it was realized that thermal energy was another form of energy, the reason for the "disappearance" of energy from frictional work was realized.

heat Thermal energy that is transferred between reservoirs at different temperatures.

hybrid An automobile that has both a gasoline motor and an electric motor. It is more fuel efficient than a pure gasoline engine because braking recharges the battery and because the gasoline motor can be shut off when it is not accelerating the car.

industrial ecology An approach to design of industrial systems that uses analogies to biological systems to minimize energy and resource demand. The focus of industrial ecological design is thus on the physical parameters such as minimization of resource flows and efficiency of energy budgets.

potential energy A potential energy $U(\mathbf{r})$ can be constructed as

$$U(\mathbf{r}_2) - U(\mathbf{r}_1) = -\int_{\mathbf{r}_1}^{\mathbf{r}_2} \mathbf{F} \cdot d\mathbf{r}.$$

The conservative force F does work that is independent of the path taken. In this case, only the end points of the motion matter, and because only the difference in potential energy is defined physically, the potential energy $U(\mathbf{r})$ may only be defined up to an overall constant. In many cases, this constant is fixed by the choice $U(\mathbf{f}) = 0$, as is usually the case for the electrical potential energy between two charges q_1 and q_2:

$$U(\mathbf{r}) = \frac{1}{4pe0}\frac{q_1 q_2}{\mathbf{r}}.$$

principle of conservation of energy The idea that the total energy of the universe is conserved (it is neither created nor destroyed). The homogeneity of the universe with respect to time translations is a symmetry of nature. As a consequence of this symmetry, the universe's energy is conserved. For example, it is axiomatic for science that an experiment, prepared identically, will return the same result no matter when the experiment is performed.

second-law efficiency The actual work performed divided by the available work. The available work is the maximum amount of work along any path that a system (or a fuel) can attain as it proceeds to a specified final state that is in thermodynamic equilibrium with the atmosphere.

Second Law of Thermodynamics There are many formulations of the second law. A useful one is that there is no way to get work from a single temperature reservoir (one at a fixed temperature). Clausius proved that systems that are isolated will move toward equilibrium with their surroundings.

thermal systems A system that depends on the temperature. Heat transfer is a central feature of thermal systems.

waste heat Thermal energy transferred to the low-temperature reservoir in a thermal engine.

The physical principle of conservation of energy always holds. It reflects the homogeneity of physical laws with respect to time. In the context of thermal systems, this principle is known as the First Law of Thermodynamics. Energy obtained through thermal transformations is subject to a maximal efficiency due to the second law of thermodynamics, which severely constrains the utility of thermal energy as it is transformed to other kinds of energy. Not all sorts of energy have the same ease of use, even though it is always possible to transform one into another. High-quality energy is the most versatile, and thermal energy the least versatile. Energy conservation in ordinary speech refers to preservation of resources the provision of energy resources more efficiently or at lower cost. Many opportunities still exist for reducing costs by raising efficiencies, by encouraging cooperation so that waste is minimized, and by rethinking traditional designs. The transportation and agricultural systems may have the best prospects for increasing energy efficiency.

1. *ENERGY PRELIMINARIES*

Energy is defined as the capacity to do work, but as a practical matter, energy is difficult to measure with a machine (but it is easy to calculate). Work is defined by the differential relation

$$dW = \mathbf{F} \bullet d\mathbf{r},$$

where $d\mathbf{r}$ is an infinitesimal displacement of the massive object subject to the force $\mathbf{F}$. If the force is constant, we find that the work done in displacing the object by $\Delta\mathbf{r}$ is

$$\Delta W = \mathbf{F} \bullet \Delta\mathbf{r} = F|\Delta\mathbf{r}|\cos\theta,$$

where θ is the angle between $\mathbf{F}$ and $\Delta\mathbf{r}$. In many texts, this relation is simplified to read

$$\Delta W = F_{||}d = Fd_{||},$$

where $F_{||}$ is the component of the force along the displacement direction, d is the total distance traveled, and $d_{||}$ is the displacement component along the direction of the force. A key to understanding energy is to recognize that different forms of energy may be transformed into other forms of energy. In some cases, the work that has to be done against a force to move an object from one particular point A to another particular point B is the same, no matter what way the work is done. The work does, of course, depend on which points are used as end points, but not on the path traversed between them. In such a case, the force is called conservative. The gravitational force is an example of a conservative force. For such a conservative force, work may be done against the force to move a massive object from one position to another. This work represents the change in energy of the massive object. For example, a person may pick up an object that is on the floor (such as a fork) and put it on a table. If the external force on the object is $\mathbf{F}$ (in this case, $\mathbf{F}$ would be the gravitational force on the fork, otherwise known as the fork's weight), the work the person does opposing $\mathbf{F}$ to pick the fork up is

$$-\int_{\mathbf{r}_1}^{\mathbf{r}_2} \mathbf{F} \bullet d\mathbf{r}.$$

If the fork were to fall off the table onto the floor, we would expect that work to reappear somehow (as some other form of energy, perhaps). It has the potential to reappear if the fork is put into a position to fall, and so this energy is known as gravitational potential energy.

In general, there are many types of potential energy; the general potential energy of a particle is defined as

$$U(\mathbf{r}_2) - U(\mathbf{r}_1) = -\int_{\mathbf{r}_1}^{\mathbf{r}_2} \mathbf{F} \bullet d\mathbf{r}.$$

Potential energy may only be defined for a conservative force. For a conservative force $\mathbf{F}$, this integral is independent of the path and can be expressed in terms of the end points only; here these are the starting and ending positions $\mathbf{r}_1$ and $\mathbf{r}_2$. It should be clear from the definition of potential energy that only differences in potential energy matter.

Energy in science is measured in the International System of Units as the product of the units of force and distance (newtons times meters); this energy unit is given the special name joule (1 J = 1 N m), but many other energy units exist. The most common are the British thermal unit (1 Btu = 1055 J), the kilowatt-hour (3.6×10^6 J = 3.6 MJ), and the kilocalorie [food calorie] (1 kcal = 4186 J). The quadrillion Btu ("quad") is often used to describe American energy use. In 2001, this was 98.5 quads (1.04×10^{20} J, or 104 EJ).

We may recognize two uses of the phrase energy conservation. One use is scientific and refers to the physical principle taught in school and essential to contemporary understanding of nature; this is discussed in Section 2. A background in thermal physics, which bridges the two uses of the word, is sketched in Section 3. The other use of the word refers to the preservation or wise use of energy resources and is discussed in detail in Section 4.

2. *THE PHYSICAL PRINCIPLE OF ENERGY CONSERVATION*

The impetus for the principle of energy conservation originally is the observation that in simple systems on Earth, one may define an energy of position (the gravitational potential energy) and connect it to another form of energy, kinetic energy. For a simple mechanical system with no friction, using Newton's second law ($\mathbf{F} = m\mathbf{a}$), we recast the work as

$$\int_{r_1}^{r_2} \mathbf{F} \bullet d\mathbf{r} = \int_{r_1}^{r_2} m\mathbf{a} \bullet d\mathbf{r} = m \int_{r_1}^{r_2} \mathbf{a} \bullet d\mathbf{r}$$
$$= m \int_{t_1}^{t_2} \mathbf{a} \bullet \frac{d\mathbf{r}}{dt} dt = m \int_{t_1}^{t_2} \mathbf{a} \bullet \mathbf{v} dt$$
$$= m \int_{v_1}^{v_2} v \bullet dv = \frac{1}{2} m v_2^2 - \frac{1}{2} m v_2^1.$$

The quantity $T = \frac{1}{2} mv^2$ is called the kinetic energy (it represents energy associated with the motion of a massive object). With this nomenclature, we find from the definition of potential energy that

$$U(\mathbf{r}_2) - U(\mathbf{r}_1) = -[T_2 - T_1],$$

and rewriting slightly, we obtain

$$T_2 + U(\mathbf{r}_2) = T_1 + U(\mathbf{r}_1).$$

Note that the quantity on the left refers only to position 2, while that on the right refers only to position 1. Since 1 and 2 were completely unspecified, this equation states a general result: in such a simple system, the quantity $T + U$ must be a constant (or conserved). This is referred to as conservation of mechanical energy (and of course, "conservative force" is a retrodiction that follows from the observation of conservation of mechanical energy).

More generally, the total potential energy is the sum of all the potential energies possible in a given situation. There is no new physics in the principle of conservation of mechanical energy. Everything follows from Newton's laws (refer back to the introduction of kinetic energy and potential energy).

It was natural to physicists to expand this idea by including every possible form of energy and to believe that the total amount of energy in a closed system is conserved. As a consequence of work by the German mathematical physicist Emmy Noether, it became clear that each conservation law (such as that of energy or momentum) is connected to a symmetry of the universe. The law of conservation of momentum is connected to translation invariance, the law of conservation of energy to temporal invariance. This means, respectively, that moving an experiment in a reference frame and repeating it when prepared identically has no effect of the observed relations among physical quantities, and that doing an experiment prepared identically at different times should produce identical results.

From these observations it is clear that relations among reference frames are essential to consider if discussing conservation laws. In the early 20th century, Albert Einstein worked to understand such relations in the context of electromagnetism and extended it beyond electromagnetism to more general physical situations. Einstein created two theories, the special theory of relativity (1905) regarding transformation among inertial reference frames, and the general theory of relativity (1915), which allowed extension of special relativity to accelerated (noninertial) reference frames.

In special relativity, in an inertial frame of reference an energy-momentum four-vector p^μ consisting of a time component (E/c) and spatial momentum components p_x, p_y, and p_z, may be defined that satisfies $p^\mu p_\mu = m^2c^2$, which is an expression of energy conservation in special relativity (written in more conventional notation as $E^2 - p^2c^2 = m^2c^4$). That is, the fact that the vector product of two four-vectors is invariant leads to conservation of energy. In special relativity, there is a special frame of reference for massive objects known as the rest frame; in that frame $p = 0$, so the relativistic energy equation is $E = mc^2$, a rather widely known result. All massive objects have energy by virtue of possessing mass, known as mass-energy.

The relativistic energy relation $E^2 = p^2c^2 + m^2c^4$ may be recast as $E = \gamma mc^2$ for massive objects (and, of course, as $E = pc$ for massless objects), where the relativistic momentum is $\mathbf{p} = \gamma m\mathbf{v}$ and $\gamma = \left[1 - \left(\frac{v}{c}\right)^2\right]^{-1/2}$. In this case, the kinetic energy is $T = (\gamma - 1)mc^2$ because T plus mass-energy (mc^2) must be equal to total energy E.

In general relativity, the connection to the Noether theorem is clearer, as one defines a so-called stress-energy tensor, $T^{\mu\nu}$. The stress-energy tensor describes momentum flow across space-time boundaries. The flow (technically, flux) of the stress-energy tensor across space-time boundaries may be calculated in a covariant reference frame and the covariant derivative of the stress-energy tensor turns out to be zero. This means that energy is conserved in general relativity.

There is no proof of the principle of conservation of energy, but there has never so far been disproof, and so it is accepted. It is a bedrock of modern physics. In physics, as in other sciences, there is no possibility of proof, only of disproof. The cycle of model (hypothesis), prediction, experimental investigation, and evaluation leads to no new information if the experimental result is in accord with the prediction. However, if there is disagreement between prediction and experiment, the model must be replaced. Over the historical course of science, faulty models are thrown away and the surviving models must be regarded as having a special status (until shown to be false as described earlier). Special relativity, general relativity, and conservation of energy are theories that have withstood the test of time. (In fact, by superseding Newton's law of universal gravitation, Einstein's theory of General Relativity shows where Newton's successful theory's boundaries of applicability are.)

3. THERMAL SYSTEMS AND ENERGY

As mentioned in Section 1, energy comes in many forms, all of which may be transformed into other forms of energy. Someone who knew only the physics principle of conservation of energy could be forgiven for wondering why people were speaking of energy conservation and meaning something else entirely. Why was generalization of mechanical energy conservation so difficult? The reason lies with the characteristics of thermal systems.

3.1 Thermal Energy

To help clarify the issues involved, let us consider a simple system: a box resting on flat ground to which a rope is attached. The rope is pulled by a person so that the force on the box is constant. Anyone who has actually done such an experiment (a class that contains virtually the whole human population of Earth) knows that, contrary to the assertion that the kinetic energy increases, it is often the case that the speed of the box is constant (the kinetic energy is constant).

What is missing is consideration of the friction between the box and the ground. Figure 1 shows the situation. In Fig. 1A, the rope's force on the box (caused by the person pulling the rope) is shown and the box is moving along the ground. However, many forces that act on the box are not shown. Figure 1B shows the free body diagram for the box, including all the forces that act. The box has weight, **W**, which must be included. The ground pushes directly upward to prevent the box from falling through the surface with a force $\mathbf{F}_{\text{ground on box}}$ and exerts a frictional force **f** on the box; and, of course, a force is exerted by the person on the rope, and hence the rope exerts a force on the box, **F**.

The presence of the frictional force is what is responsible for the lack of change in the kinetic energy. That is, the speed can be constant because the horizontal components of the two forces add to zero. So where has all the work done gone? It has apparently disappeared from the mechanical energy. In fact, the work is still there, but thermal energy, the random motion of the atoms in the box has increased at the expense of the work done. In Fig. 2, an energy bar chart representation shows what has happened: The thermal energy of the system has increased.

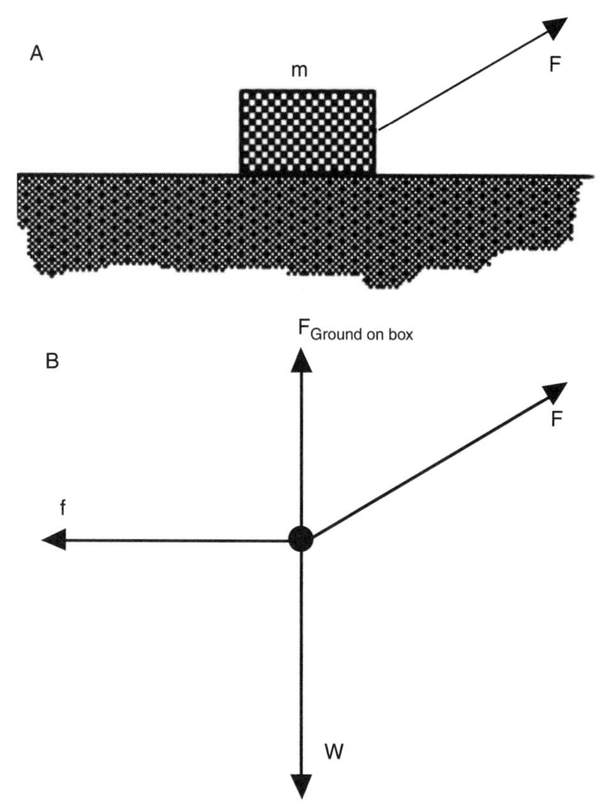

FIGURE 1 (A) A box being pulled along the ground. (B) A free body diagram showing all forces acting.

Thermal energy is a new form of energy that must be considered in working with conservation of energy. In fact, the consideration of thermal energy was first codified by the physicist Sadi Carnot in the early part of the 19th century. We might call the 19th century the century of thermal energy because of the work of other giants of thermal physics such as Boltzmann, Clausius, Joule, Kelvin, Maxwell, Planck, and many others who created the science of thermodynamics.

3.2 Thermodynamics: The Zeroth Law

Systems involving measurements of temperature or exchanges of thermal energy are thermal systems.

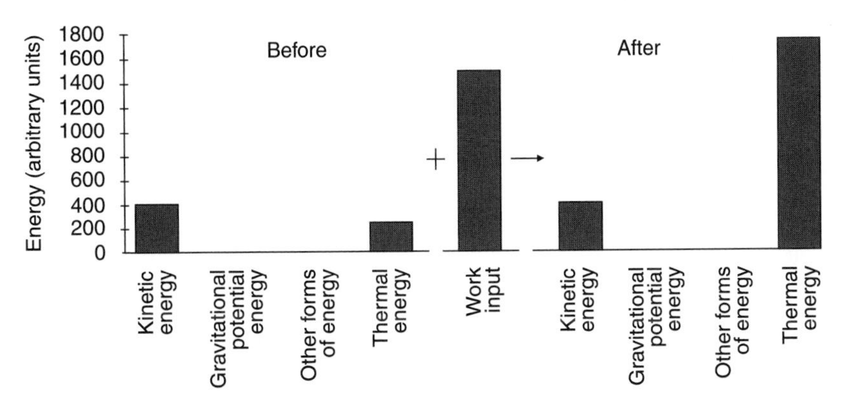

FIGURE 2 An energy bar chart representation of the change in energy of the box because energy is brought into the box system from outside by the rope.

Thermal systems are characterized by their temperatures. Temperature is determined by the mean of the random motions of the atoms making up a material. A device that measures temperature is known as a thermometer. Temperatures are measured by assigning numbers to systems we designate as cold or hot and by using the idea of thermal equilibrium.

If a spoon from the drawer is put into a pot of boiling water, thermal energy will flow from hot water to cooler spoon. Thermal energy is exchanged between the water and the spoon, but eventually the spoon and water do not exchange any more energy. Thermal equilibrium is the state in which there is no net interchange of thermal energy between bodies. The spoon comes to thermal equilibrium with the boiling water.

Temperatures are measured by allowing the thermometer and the system to come to thermal equilibrium. Two bodies are said to be in a state of thermal equilibrium if there is no net interchange of thermal energy between those bodies when they are brought into thermal contact. The measurement of the temperature of the thermometer therefore also gives a measurement of the temperature of the system.

The most common temperature scale used on Earth, and often used in science, is the Celsius scale. It sets the zero temperature at the freezing point of water and sets 100°C at the boiling point of water at sea level under standard conditions. The choice of the boiling end point is arbitrary scientifically. As virtually everyone living in the temperate regions of Earth knows from personal experience, there are temperatures below zero using this scale. A brisk day might have a temperature around 10°C, and a very hot day a temperature around 40°C. Human body temperature is 37°C. Comfortable room temperature is about 20°C.

In science, the arbitrariness of the Celsius scale makes it less useful in communication. Lord Kelvin devised a temperature scale, known as the absolute temperature or the Kelvin temperature scale, that begins with the lowest temperature possible in the universe as 0 K (absolute zero) and has a temperature of 273.15 K at the triple point of water (the temperature at which ice, water, and water vapor are in thermal equilibrium). With this definition, the kelvin interval is the same as the degree Celsius (1 K = 1°C).

3.3 Thermodynamics: The First Law

In the 18th century, steam engines began to contribute mechanical energy to replace human labor. Carnot began to study various engines that operate, as steam engines do, between two different temperatures. In a steam engine, burning fuel boils water to make steam, which pushes a piston, turning the steam's thermal energy to mechanical energy. That mechanical energy can be used in a machine. In early steam engines, the spent steam was vented to the air, the piston retracted, and the cycle began again (modern steam engines work slightly differently). The spent steam's thermal energy was released from the system.

The word "heat" refers to energy in transit between bodies at two different temperatures. It is one of the most misused words in physics. Heat is not a separate form of energy, but an artifact of the transfer of thermal energy to a body. There is no heat content to a body, as was first shown in 1798 in an experiment carried out by the American inventor Benjamin Thompson (1753–1814), also known as Count Rumford. As the person in charge of making cannon for the king of Bavaria, he was able to bore channels for the shot in cannon; there was need for cooling water as long as the boring proceeded. There was no end to the amount of heat being transferred.

Any thermal engine operates between two temperatures. It is useful to think of the original steam at the higher temperature and the air at the lower temperature as temperature reservoirs, or large "containers" of thermal energy, so large that taking some or adding some leaves the system unchanged. A thermal engine runs through a cycle in which thermal energy is taken in at the high temperature reservoir, work is done, and thermal energy is exhausted at the low temperature reservoir.

Of course, energy is conserved in this cycle. The energy the system takes in from the high temperature is equal to the sum of the work done by the system and the energy the system exhausts to the low temperature reservoir. As applied to thermodynamics, the principle of conservation of energy is known as the first law of thermodynamics. From the first law of thermodynamics, no machine is possible that could create energy from nothing.

3.4 Thermodynamics: The Second Law

We define the efficiency of any energy transformation as

$$\text{Efficiency} = \frac{\text{energy output}}{\text{energy input}};$$

for a machine that produces work from other forms of energy, this relation would be

$$\text{Efficiency} = \frac{\text{work output}}{\text{energy input}}.$$

In the case of a heat (or thermal) engine, the first law of thermodynamics reads

$$\text{Energy input at } T_H = \text{Work output} + \text{Energy output at } T_L.$$

As a result, the efficiency is

$$\text{Efficiency} = \frac{\text{energy input at } T_H - \text{energy output at } T_L}{\text{energy input at } T_H},$$

clearly less than 100%. Note that the necessity of exhausting thermal energy as heat to the low-temperature reservoir necessitates a severe restriction on the efficiency of a heat engine. This also has deleterious consequences, because this so-called waste heat must be absorbed by the local environment (usually local water is used as the low temperature reservoir) and can adversely affect aquatic life.

Carnot studied a special cycle (Fig. 3) in which the thermal energy is absorbed at the high temperature, the system expands, then the system contracts as thermal energy is exhausted at the low temperature. This cycle is known as the Carnot cycle. By studying this special Carnot cycle, which was the most efficient cycle that could possibly exist between the two temperatures, Carnot could find the maximum possible efficiency in terms of the absolute temperatures of the reservoirs for a heat engine operating between these two temperatures.

The second law of thermodynamics, originally conceived by Rudolf Clausius, is the statement that any real thermal engine is less efficient than the Carnot cycle. (There are many other formulations of the second law.) The Carnot efficiency is found to be

$$\text{Maximum possible efficiency of a heat engine} = \frac{T_H - T_L}{T_H}.$$

Again, because of the need to exhaust thermal energy at the low temperature, the efficiency of even the ideal Carnot engine is restricted. For example, suppose that a heat engine (an electric generator) operates between a high temperature of 500°C and a low temperature of 20°C (room temperature). The maximum possible efficiency of operation is

$$\text{Maximum possible efficiency of a heat engine} = \frac{773.15\text{ K} - 293.15\text{ K}}{773.15\text{ K}} = 0.621(62.1\%).$$

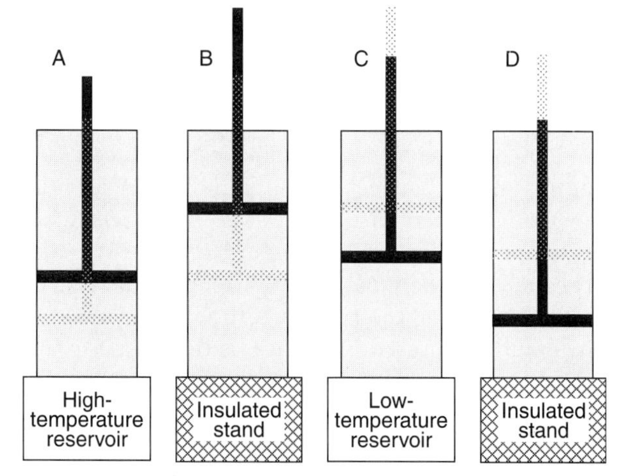

FIGURE 3 The four parts of the Carnot cycle. (A) The system expands, drawing in thermal energy, at absolute temperature T_H while it is attached to a high-temperature reservoir. (B) The system is isolated, so no energy enters or leaves, and the system expands further, lowering the absolute temperature from T_H to T_L. (C) At absolute temperature T_L, the system contracts as the piston pushes inward, transferring thermal energy to the low-temperature reservoir. (D) The system is isolated, so no energy enters or leaves, and the system contracts back to the original volume.

Real electric generators average about 33% efficiency, much lower than the maximum possible (Carnot) efficiency.

3.5 Quality and Second-Law Efficiency

It may be seen from the above that when transforming thermal energy into other forms of energy, the amount available is not the total energy available, but is smaller. However, if work is done by an agent, *all* the work is available; if electrical energy is available, *all* of it may be transformed directly into other, usable, forms of energy. Thermal energy is said to be low-quality energy because of its limitations in the conversion process, while work and electrical energy are high-quality energy.

Thermal energy may be used directly in space heating and in cooking, but if we want to run a motor with it, not all the thermal energy would be available to us. It would make sense to use forms of energy (if available) that could deliver all their energy to the motor. Conversely, if we want to increase the temperature of a room, it is silly to use high-quality energy to do it.

While the equation for efficiency is found using the first law of thermodynamics, the American Physical Society study "Efficient Uses of Energy" points out that the second law of thermodynamics, which shows that the Carnot engine operating between two temperatures is the most efficient, allows us to compare the actual efficiency to the theoretical maximum efficiency for producing a given amount of work; this is called the second-law efficiency for the system. For real thermal systems, the second-law efficiency measures how much room for improvement remains.

4. ENERGY CONSERVATION AS PRESERVATION OF RESOURCES

Often exhortations to conserve energy or other resources such as land or water are heard. Many people view the probable effect of the requested conservation measures in their daily lives as having to "do without" something desirable. Conservation might represent "personal virtue" through sacrifice; it involves paying a personal price. It may represent the religious idea of stewardship. Alternatively, they may see these measures as making their lives more complicated in a world that already seems full of intimidating tasks. They are interested in living their lives, in having life be as comfortable and convenient as possible.

Energy conservation here refers to the provision of the same level of goods or services with a smaller drain than at present on stored energy resources. In their common perceptions about conservation, those people who are wary of conservation efforts are both wrong and right. They are wrong because people who conserve energy are not sacrificing or doing without but are benefiting from increased efficiency or reduced costs. However, they are right because this reorientation may have a cost in personal convenience since it may involve additional time, thought, or change in habits.

In the following, we discuss possibilities for "energy conservation" with the understanding that, physically, energy is always conserved as discussed in Section 2; we are discussing here reduction of waste and increase of efficiency of the economic and physical system. The reliance on thermal systems for transforming energy makes the second-law efficiency a good measure of the efficacy of a proposed conservation measure. The following examples show how it has been or may in the future be possible to provide the same services at reduced monetary and energy cost.

4.1 Space Conditioning

Space heating and cooling are responsible for a considerable amount of energy use. In the United States in 1993, 4.3×10^{17} J of electricity, representing 1.3×10^{18} J of thermal energy, 3.9×10^{18} J of thermal energy from natural gas, and 1.3×10^{18} J of thermal energy from kerosene, fuel oil, and liquefied petroleum gas (a total of 6.5×10^{18} J = 6.5 EJ of thermal energy) was used for residential space heating. Another 1.50 EJ was used for residential air conditioning, of which 18% (0.26 EJ) was used for room air conditioners. This represents over 8% of national energy use. Commercial buildings in the United States in 1995 used 8.3 EJ of thermal energy as electricity, 2.1 EJ of thermal energy from natural gas, and 0.25 EJ of thermal energy from fuel oil, a total of 10.6 EJ of thermal energy, roughly 10% of national energy use. The great majority of energy use is for space conditioning (4.1 EJ) and lighting (3.3 EJ).

In the aftermath of the 1973 energy crisis, many buildings were built with more attention paid to energy-conserving features, resulting in a lowered energy use per area, but this impetus was lost with a return to cheap oil in the 1980s and 1990s, and in early 1990s commercial buildings were built to use one-third more energy per square meter than average 1980s buildings.

Clearly, with over 10% of United States energy use committed to space conditioning, systematic savings could have a great impact. Savings are available from installation of replacement glazing, increased insulation, and replacement of antiquated heating and cooling systems.

Active and passive solar energy systems have been used for millennia to achieve space conditioning at little or no cost, and these methods continue to hold promise of reduced reliance on fossil fuels and uranium. Passive systems are characterized by construction methods that allow people to cool rooms during the day and heat by night through thermal storage—for instance, in masonry walls—of energy from the sun. Use of surrounding vegetation to help maintain house temperature has been common since humans began building shelters. Natural ventilation systems are also passive in nature. Active solar systems use some fossil energy to pump fluids, but the main input is free energy from the sun.

Heat pumps, especially groundwater-based heat pumps, use a heat engine in a clever way. Groundwater is about the same temperature the year around, and in the winter it is much warmer than the ambient air in colder climates, while in summer it is much cooler than the ambient air. In winter, the heat pump is run by taking in thermal energy at the groundwater low-temperature reservoir, doing work on it, and exhausting the thermal energy plus the work at the high-temperature reservoir (the room). Only the extra energy (the work input) must be supplied by fossil fuels. The advantage of this system is that the user pays only for the work done, and the extra energy used to heat the space is the "free" thermal energy in groundwater or the atmosphere. In summer, the heat pump works backward, taking thermal energy from the high-temperature reservoir (the room), doing work, and exhausting thermal energy to the low-temperature reservoir (the groundwater). Commercial freestanding heat pumps use the atmosphere as the reservoir and can be substantially less efficient than groundwater-based heat pumps.

In cold weather, much ambient thermal energy is transferred out of buildings, where it is desirable to keep it, to the outside through walls and windows. In warm weather, ambient thermal energy is transferred from outside to inside, where it is undesriable, again through the walls and windows. Advanced technology can contribute to reduced fossil energy use through design of various types of windows—low emissivity coatings, double- or triple-glazed windows with evacuated spaces, or smart window coatings that can act as an insulating shade. Aerogel-based insulating materials may also contribute to future energy savings in buildings, as they already have for appliances.

Air conditioners have become more efficient over the years since the 1970s energy crisis. The state of California has led the way by mandating increasingly efficient room and home air conditioning units, and the federal government has followed. The rollback of Clinton administration seasonal energy efficiency ratio (SEER) regulations by the incoming Bush administration (which may have been illegal, since the rules had been published in the *Federal Register*) was a setback to this trend. The payback time for the increased cost of a SEER-13 unit over a SEER-12 unit is a little over 1 year. Fortunately, because it has such a large population, California's energy-efficiency regulations tend to drag the rest of the nation along despite federal foot-dragging, and even the rolled-back SEER targets are far above the average SEER of 5 to 6 in 1994. The efficiency regulation might actually be looked at as an antipoverty regulation, because the least-efficient units (having the lowest purchase price) are bought by landlords, while the typically lower-income tenants must pay the electric bills. Wealthier Americans tend to buy the higher-SEER units because they themselves pay their electricity bills.

4.2 District Heating

In many European cities and some older American cities, electric utilities use the waste heat from the boilers or heat exchangers to keep buildings warm in the cooler months of the year. Hot water is circulated through pipes through nearby buildings, and cooler water returns to the utility. This is known as district heating because an entire region can be served by this system. In the United States, 5.6×10^{17} J (0.56 EJ) of energy was distributed as district heating in 1995.

Savings are possible because of economy of scale (one central heating system instead of many individual systems), the reduced maintenance per unit of output characteristic of a larger system, and because the waste heat would have to be disposed of by other means were it not used for this useful purpose. District heating is only feasible when there is a large but compact region having a relatively large population density through at least part of the day. This is more characteristic of European and Asian conditions than American or African ones.

4.3 Illumination

As noted earlier, commercial buildings use a substantial amount of energy for lighting, roughly 3% of total American energy use. Lighting is also an important part of household and commercial energy consumption. There are many ways of saving fossil energy. For example, commercial buildings may be built with daylighting features, in which the architect designs the space and the windows such as clerestories to provide light to workspaces during the day (clearly lamps are needed when it is not light enough outside, but in most locations days are not severely overcast most of the time).

Incandescent lighting, in which a filament is heated to a white-hot temperature and the resulting light is used, converts only about 5% of the electric energy into radiant energy. Ordinary fluorescent lighting is four times as efficient, converting about 20% of the electric energy into radiant energy. The development of electronic ballasts for fluorescent bulbs in the 1970s after the first energy crisis contributed significantly to fluorescents' energy efficiency advantage.

In the 1990s, compact fluorescent bulbs became available. Now these bulbs, while more expensive than incandescent bulbs to buy, deliver light for a longer time and use much less electricity and so are much cheaper to use. Consider a 100 W incandescent bulb that costs $0.25 in a package of four bulbs for $1, and a 23 W compact fluorescent that delivers about the same amount of light but costs $4.95. The cheap incandescent bulb last on average 500 hours, while the compact fluorescent lasts 10,000 hours. That means it takes 20 incandescent bulbs to replace the compact fluorescent, at a cost of $5 (the same as the price of the compact fluorescent bulb). However, over 10,000 hours of operation, compact fluorescent use costs (at a rather typical 8 cents per kilowatt-hour)

$$\begin{aligned}&23\,\mathrm{W} \times 10,000\ \text{hours} \times \$0.08/\mathrm{kWh}\\&\quad = 230 \times \$0.08 = \$18.40,\end{aligned}$$

while incandescents' use cost

$$\begin{aligned}&100\,\mathrm{W} \times 10,000\ \text{hours} \times \$0.08/\mathrm{kWh}\\&\quad = 1000 \times \$0.08 = \$80.00.\end{aligned}$$

Obviously, it is much more rational for a homeowner or businessperson to use the compact fluorescent. However, inertia—the resistance to change—and the high initial cost has limited compact fluorescents' market penetration, only about 10% of lighting sales. The Department of Energy has estimated that a complete switch to compact fluorescents would save the United States about 32 billion kilowatt-hours of electric energy annually, or 0.36 EJ of fossil fuel energy—the equivalent of the output of almost four 1000 MW generating plants.

4.4 Appliances

In the aftermath of the energy crises of the 1970 s and 1980 s, attention turned to energy use in appliances as a means of energy savings. Newer compressors, more effective insulation, and better design of refrigerators has decreased continuing expenses substantially. In 1973, the average refrigerator used over 2000 kWh/year. By 2000, this average was reduced to 1140 kWh/year, and the most energy-efficient units (volume $20\,\mathrm{ft}^3 \sim 0.6\,\mathrm{m}^3$) used just 440 kWh/year. Several factors contributed to increases in efficiency. In the aftermath of the energy crises, legislation mandated the posting of energy guides on appliances (including refrigerators) so consumers could make rational decisions on purchases. In addition, California introduced efficiency regulations for refrigerators (and the California market is so large, it made no sense for manufacturers to make a separate California model). Finally, a $30 million prize was offered for development of the most energy-efficient refrigerator (the prize was won by Whirlpool).

Europeans have long paid substantially more for energy than their American counterparts, and appliance manufacturers have developed many cost savings that had not been adopted in the United States. With the consolidation of appliance manufacturers worldwide, more European energy-efficient technology has made its way west. Whirlpool, for example, is now manufacturing virtually identical energy-efficient washing machines in Europe and the United States.

A uniquely American idea is the EnergyStar rating system, which is an extension of the energy guides to appliances not covered by the legislation voluntarily entered into by the appliance manufacturers. Appliances meeting EnergyStar guidelines display the EnergyStar sticker and consumers can know that they are among the most efficient available. The criteria change with time; for example, for dishwashers to be eligible for EnergyStar designation prior to January 2001, their efficiency had to be greater than 13% above federal standards, while to earn the EnergyStar after January 2001, the dishwasher had to be over 25% more energy efficient. The efficiencies are achieved by locating heater boosters inside the dishwashers and by use of less water per cycle.

Computers have been increasingly used in many households and businesses. The use of flat screens for televisions and computer monitors saves considerable energy compared to the CRTs typically used, which is why they were originally introduced for the notebook computers, which must often run from batteries. The sleep mode of computers and computer peripherals uses much less energy than would be used if the computers were fully "awake."

Some years ago, appliances had to "warm up" before they came fully on. Consumers disliked this wait time, and manufacturers of televisions, radios, and other appliances began to manufacture instant response units. Consumers liked the feature, but didn't realize the hidden costs. The units are turned on all the time and collectively use between 10% and 25% of the electricity used in households. Such devices have been called energy vampires. The problem may be limited to developed countries, but the cost is not a necessary one. Electronic transformer use would substantially decrease this drain, and they already are used in countries where mandated by law (they cost only 0 to 20 cents more per unit, but are not installed voluntarily by cost-conscious manufacturers). Computers use much more than the 10 W or so used by VCRs and TVs in their sleep mode, though, of course, the power is much lower than when the computer is awake.

4.5 Transportation

Automobile engines have a Carnot efficiency of around 60% but are about 25% efficient in practice. As long as the internal combustion (Otto) engine continues to be used, large increases in efficiency are unlikely, but sizable incremental savings are still available. Alternatives include electric vehicles and fuel cell vehicles. The transportation sector is important because transportation accounts for 27% of total United States energy use. Most of the energy used for transportation is petroleum based. That means that importing countries depend on the exporting countries' good faith and on their continued willingness to supply oil at a relatively cheap price. While the United States pays some of the highest prices for oil, both because domestic oil is more expensive than imported oil and because the average transportation cost is large, the price of gasoline at the pump is among the lowest in the world. This is mainly due to the difference in tax structures among countries.

In Europe, automobiles are generally smaller than in the United States because of the higher energy prices consumers pay for gasoline and diesel fuel (because of the high local taxes). Smaller cars use less material to manufacture and, because the weight is smaller, cost less to run. Many drivers in the United States are reluctant to purchase small cars because they are concerned about safety (this is apparently partly responsible for the boom in sport utility vehicle, or SUV, sales). Ironically, though people in SUVs are less likely to be killed or injured in a direct front, back, or side collision, the overall death rate is higher in SUVs than in smaller cars because of the large number of fatalities in rollover accidents, which have a relatively high probability of occurrence.

American cars are much lighter than in the 1970s, a direct result of the first energy crisis. A conventional suburban car gets a mileage around 30 mi/gal. It is clear that the weight of most cars could be reduced further. A countervailing trend is for more and more SUVs and pickup trucks to be sold. These vehicles weigh more than normal cars and as a result get worse mileage. By 2000, fewer than half of all vehicles purchased for personal transportation in the United States were ordinary automobiles. A large SUV might get a mileage of just 17 mi/gal, while a small pickup gets 25 mi/gal. The reason for the better car mileage is legislative—Congress adopted Corporate Average Fuel Economy (CAFE) standards after the first energy crisis to force manufacturers to raise mileage, and it worked. The rise in SUVs and pickups have pleased the manufacturers because these vehicles were not covered by the CAFE standards. In addition, the low average energy prices of the late 1990s lulled drivers into expecting cheap imported oil would continue to be available. Political pressure for increasing CAFE standards abated, and car mileage stagnated. As a result, the fleet average for American vehicles experienced a decrease in mileage. The lowest fleet average mileage was achieved about 1989. Even imported cars' mileage has slipped. On the basis of history, only legislative mandate will be successful in increasing mileage in the absence of external threats such as giant leaps in world oil prices.

Before the 1970s, most American cars were rear-wheel drive; now most are front-wheel drive, which is more energy efficient. Most modern automobile engines are more efficient overhead cam engines. However, much more energy could be saved by continuing to pursue ways to decrease drag, and the installation of advanced automatic transmissions that are controlled electronically. Other measures, some partly introduced include improved fuel injection systems, improved tires, improved lubricants, and reduced engine friction. All these measures are good conservation measures because they come at no or low cost (low cost means short payback times).

California's quest for less-polluting cars led to designation of quotas for lower-emission vehicles. Many ideas for automobiles that run on electricity or energy stored in high-tech epoxy-based flywheels have been proposed. The high hopes California had for so-called zero-emission vehicles were apparently unwarranted, and few total electric vehicles have been designed or produced. There have been difficulties with battery design (conventional lead-acid automobile batteries weigh quite a lot). The nickel-metal hydride battery is the apparent technology of choice for electric vehicles. In addition, battery-powered cars do not perform well at low ambient temperatures, so these cars are not common outside the southwest. Another strike against battery-powered cars is the lack of charging infrastructure. And such vehicles are not really *zero* emission, because gas, oil, or coal must be burned to produce the electricity used to recharge the batteries.

California's quest has had the effect of boosting prospects for low-emission vehicles that are hybrid (combined gasoline engines and electric systems that work together). Hybrid features that save fuel include automatic shutdown of the engine at stops and regenerative braking (converting the work done in braking into energy stored in the battery). In hybrids, the gasoline engine runs only at or near peak efficiency, and excess energy recharges the batteries. Since the fuel is conventional and the electric system recharges in use, there is no need for a special infrastructure as is the case with pure electrics. Honda introduced the first commercial hybrid, the Insight, followed quickly by Toyota with the Prius. Honda has begun to sell Civic hybrid models.

Fuel cells, which are devices that combine fuel (ideally hydrogen) with oxygen from the air without combustion, are expected to supply automobiles with energy in the near future. If the hydrogen were produced by electrolysis using solar energy, such a vehicle really would involve zero emission (outside the water made from combining hydrogen and oxygen). Fuel cells have been used in the past for transportation, but usually liquid metal fuel cells in buses because of the relatively high temperature of operation (300 to 1200°C). The development of low-temperature proton-exchange membrane fuel cells have led the auto manufacturers to engineering development. These are typically 50% efficient at transforming fuel to energy, about twice as efficient as Otto engines. (Very small fuel cells are being used as an alternative to battery packs in some electronic devices.)

The fuel cell option suffers from a lack of an infrastructure to supply the hydrogen to vehicles and the problem of how to carry sufficient hydrogen. Hydrogen is a dilute gas, and it has a very much smaller amount of energy per volume than liquid fuels. Possible solutions to the storage problem range from mixing hydrogen into borides to pressured tanks to storage as hydrides in metal containers. However, many questions still surround the storage question. Alternatively, methanol (a liquid) can be chemically transformed to produce hydrogen, and this may be a more probable choice for the short term. The disadvantage is the presence of carbon dioxide in the exhaust stream.

4.6 Industry

Together, commercial and industrial energy uses total 54.3 EJ, of which 41 EJ is industrial energy use. Industry was hardest hit by rising energy prices in the aftermath of the energy crises and had the largest incentive to reform. Indeed, the industrial sector has saved a considerable amount of energy since then (Fig. 4 shows how the industrial sector energy use breaks down). Work has been done on design of variable-speed motors and control mechanisms for machines used in industrial processes. Motors are much more efficient than those of a generation ago. Compressors are especially important to consider, as many heavy industries use compressors for many different purposes.

Many industries have the same opportunity as commercial business for reduction in energy use in offices—installation of newer, more efficient HVAC systems, replacement of ballasts for fluorescent lighting, or installation of compact fluorescents in place of incandescents, and so on. Industries often buy their electricity on a "interruptible" basis, meaning that the utility can cut them off if demand from other users is high; the industry gets concessional electricity rates in return. The processes can be designed to result in "load leveling" for the utility's demand. For example, the industry can use high demand when the utility's other customers are near the low point of demand; this helps the utility by increasing base load slightly (base load is cheapest). Many energy management control systems have been designed that automatically shuffle load on and off in response to external conditions.

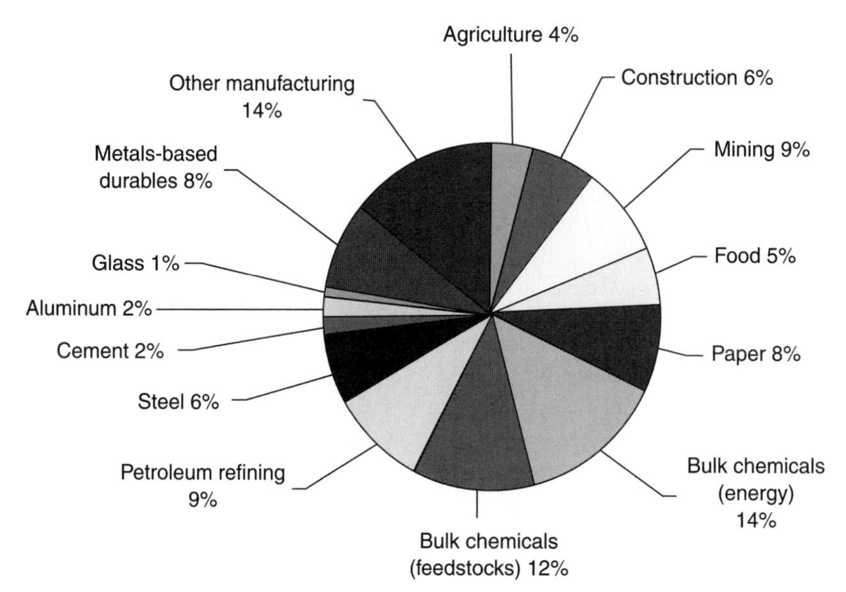

FIGURE 4 Breakdown of the energy use by industrial subsector 1992. From Interlaboratory Working Group, "Scenarios for a Clean Energy Future," Figure 5.1.

While much work has been done, opportunities remain for reducing industrial process energy use. In a 1992 study concerning reduction of carbon dioxide emissions by the United States, a National Academy expert panel identified billions of dollars in savings that would result from changing processes and materials with the aim of lowering carbon emissions (i.e., negative real cost), even in as mature an economy as was current in the United States at that time. Few of its recommendations have been adopted, mostly, it seems because of the attitude that we have not done it that way in the past (in other words, another case of inertia holding back rational change).

Opportunities remain, especially in mining technology, metals refining and working, and chemicals (large subsectors from Fig. 4). In steel processing, hot rolling of metals is being replaced by direct casting and rerolling in small hot strip mills as needed. Scrap to be melted can be preheated using furnace waste heat, reducing overall energy use. More than 50 such energy-saving measures have been identified in the steel industry alone.

Dow Chemical has been a leader in making waste streams into resource streams for other parts of the company or for other companies. This benefits society both by reducing waste emissions and reducing the impact of obtaining the resource. Dow has experienced economic returns over 100% on some programs. If such a well-managed company can find such opportunities, it is clear that practically all other companies can benefit. Experiences such as this and the success of the independent companies colocated in Kalundborg, Denmark, in exchanging materials to mutual benefit have led people to speak of a new field of study they call industrial ecology. There is a hope that bringing attention to these possibilities will result in greater savings and efficiencies in the future and that companies can learn from the best practices of other companies.

The science of chemical catalysis, originally strongest in the petroleum industry, has spread to other industrial sectors. Catalysts are chemicals that increase reaction rates without being themselves consumed in the process. Increases in reaction rates at low temperatures saves the application of thermal energy to raise the temperature (as reaction rates generally increase with temperature).

Glassmaking is an industry that uses a lot of energy per kilogram of output because of the necessity of melting the constituents of glass (or melting glass when it is recycled). New refactory materials have reduced energy use in the glass industry at no cost.

The agricultural industry is one area where penetration of ideas is slow; more energy conservation measures need to be implemented. Individual farmers are rugged individualists and apt to think in many instances that any change is for the worse. One promising avenue is for a combination of chemical and agricultural industry to grow chemical feedstocks and to achieve ways to grow plastics in the field. For this to be entirely successful, negative attitudes held by the general public toward genetic manipulation may have to be abandoned.

4.7 The Utility Industry

Utilities are large-scale users of fossil fuels. In 2000, utilities used 42.6 EJ of mostly fossil energy, 41% of total U.S. energy use. Because they use so much energy, there is a built-in incentive to adopt the most energy-efficient equipment possible. While the average 33% efficiency in transforming thermal energy into electrical energy seems small, it is far improved from the original generators in the industry a century ago. Progress has generally been incremental, but the advent of integrated gas combined cycle power plants (known as IGCC plants) in the late 1980s and 1990s has boosted output efficiency substantially. Some combined cycle plants are operating near 60% efficiency, not far from the Carnot cycle limit. Gas turbine generators now use aircraft turbines almost unmodified and are much higher in efficiency than previous gas-fired turbines.

One way for the utility to save money is to even out its demand, as referred to briefly earlier. It is to the utility's benefit to assure that as much energy sold as possible is base load, the cheapest energy the utility can generate. The smaller the demand for high-cost peak energy, the better off the utility is. Therefore, utilities offer preferential rates to industrial users who are willing to have their electricity shut off or radically decreased when demand spikes from other users. This is known as load leveling.

Even individual users can sign up for interruptible rates if they are willing to allow the utilities to shut them (or their largest appliances such as air conditioners or water heaters) down as needed. Electronic devices can be installed that give the utility control over whether the individual interruptible appliances are on or off. The homeowner or business owner agrees to allow the utility to shut them off as needed when demand for electricity is high.

4.8 Cogeneration and Net Metering

By cogeneration, we mean generation of electricity as an incidental outcome of industrial production. For example, the paper industry uses a great deal of hot water, and produces the hot water in boilers. It could use the steam produced to run turbines and generate electricity. In the early days of electricity distribution, this occurred regularly, but with the formation of utility monopolies in the late 1800s, industrial producers were not allowed to sell their energy offsite. Given the obvious advantage (two gains for the price of one), restrictions were first loosened after the energy crises of the 1970s and 1980s, and then vanished altogether. Cogeneration has become more widespread since the 1980s. The utility typically signs a purchase agreement with the cogenerator with a contracted rate schedule. The sale of electricity contributes to the seller's bottom line.

The same dynamic is at work on the much smaller scale of the individual electric utility customer in the question of whether states have "net metering" laws. Cogeneration can occur at the scale of an individual home. Customers who install solar energy generators but remain connected to the grid for backup in states without net metering can decrease their electricity costs by using "home-generated" electricity in place of buying, but they cannot sell any excess back (the utility gets it for free). In states with net metering, the excess electricity is bought by the utility (albeit at a lower rate than the customer usually pays for electricity), in some cases leading to reverse bills—utility payments to customers for energy sold back that exceed the value of energy sold to the customer.

SEE ALSO THE FOLLOWING ARTICLES

Cogeneration • Conservation Measures for Energy, History of • District Heating and Cooling • Mechanical Energy • Storage of Energy, Overview • Thermodynamic Sciences, History of • Thermodynamics, Laws of • Work, Power, and Energy

Further Reading

Carnahan, W., Ford, K. W., Prosperetti, A., Rochlin, G. I., Rosenfeld, A., Ross, M., Rothberg, J., Seidel, G., and Socolow, R. H. (eds.). (1975). "Efficient Use of Energy: The APS studies on the Technical Aspects of the More Efficient Use of Energy." American Institute of Physics, New York.

Department of Energy (1995). "Buildings and Energy in the 1980s." U.S. Government Printing Office, Washington, DC.

Department of Energy (1995). "Household Energy Consumption and Expenditures, 1993." U.S. Government Printing Office, Washington, DC.

Department of Energy (1997). "1995 Commercial Buildings Energy Consumption Survey." U.S. Government Printing Office, Washington, DC.

Department of Energy (2001). "Annual Energy Review, 2000." U.S. Government Printing Office, Washington, DC.

Interlaboratory Working Group (2000). "Scenarios for a Clean Energy Future." Oak Ridge National Laboratory, Oak Ridge,

TN, ORNL/CON-476, Lawrence Berkeley National Laboratory, Berkeley, CA, LBNL-44029, and National Renewable Energy Laboratory, Golden, CO, NREL/TP-620-29379.
International Energy Agency (2001). "Things That Go Blip in the Night: Standby Power and How to Limit It." Organization for Economic Cooperation and Development, Paris.
National Energy Policy Development Group (Cheney, R., Powell, C. L., O'Neill, P., Norton, G., Veneman, A. M., Evans, D. L., Mineta, N. Y., Abraham, S., Allbaugh, J. M., Whitman, C. T., Bolten, J. B., Daniels, M. E., Lindsey, L. B., and Barrales, R.) (2001). "National Energy Policy: Report of the National Energy Policy Development Group." U.S. Government Printing Office, Washington, DC.
Panel on Policy Implications of Greenhouse Warming (Evans, D. J., Adams, R. M., Carrer, G. F., Cooper, R. N., Frosch, R. A., Lee, T. H., Mathews, J. T., Nordhaus, W. D., Orians, G. H., Schneider, S. H., Strong, M., Tickell, C., Tschinkel, V. J., and Waggoner, P. E.), National Academy of Sciences (1992). "Policy Implications of Greenhouse Warming: Mitigation, Adaptation, and the Science Base." National Academy Press, Washington, DC.
Van Heuvelen, A., and Zou, X. (2001). Multiple representations of work-energy processes. *Am. J. Phys.* **69**, 184–194.

Consumption, Energy, and the Environment

SIMON GUY
University of Newcastle
Newcastle Upon Tyne, United Kingdom

1. Introduction
2. Theories of Technical Change
3. Barriers to Energy Efficiency
4. Leaping the Barriers
5. Contexts of Consumption
6. Sociotechnical Theories of Change
7. Conclusion: Rethinking Energy Consumption

Glossary

nontechnical barriers The idea that the "attitudes" of decision makers toward technology are the key determinants of innovation and that apathy, ignorance, and lack of financial interest are the major obstacles.

socially relevant actors The broad networks of individuals and organizations that may have an influence on technical change.

sociotechnical The idea that science is a sociocultural phenomenon and that the technical is always in relationship with wider social, economic, and political processes.

technical frames The sets of assumptions and culturally conditioned ways of seeing that underpin individual or organizational perspectives on technical change.

technoeconomic The idea that if technical knowledge is rigorously tested and demonstrably proved, then consumption choices will be made rationally.

universe of choice The multiple rationalities shaping technical innovation and how these relate to the contrasting contexts in which design and development actors operate.

The contribution of building science to our stock of energy efficiency knowledge is substantial. Energy-saving technologies and materials have been successfully developed and manufactured, and energy-efficient building designs have been constructed, tested, and widely publicized. In fact, for some time, designers have possessed the technical knowledge and identified the best practice design techniques necessary to construct zero-energy buildings. Moreover, extensive monitoring of local, national, and international building stocks means that more is known than ever before about the precise potential for improved energy performance. Having perfected energy efficiency technologies, the next task is to explain why apparently proved technical knowledge is often ignored in design practices, or why building occupants consistently fail to adopt energy-saving techniques. Facing the "inadequate diffusion" of apparently cost-effective energy-conserving technologies, the practice of building science is extended to consider the nontechnical processes of technology transfer and arrives at the complex question of consumption.

1. *INTRODUCTION*

> *Those who call ourselves energy analysts have made a mistake...we have analysed energy. We should have analysed human behaviour.*
>
> —*Lee Schipper, 1987*

> *A renewal of the social theory which informs energy consumption and conservation is called for in the face of environmental challenges.*
>
> —*Hal Wilhite, 2003*

Many researchers, like Schipper and Wilhite, are now questioning the notion that the challenge of improving the energy performance of buildings is simply a task for "building science," a term now widely used to describe the growing body of knowledge about the relevant physical science and its application to buildings. Across international boundaries and research cultures, the common aim of building science is to produce more and better knowledge about buildings and how they perform in order to identify promising solutions to the problem of energy inefficiency. The critical issue here is the definition of "better knowledge" about building performance and the subsequent identification of problems and solutions within a scientific, or techno-economic, paradigm. As with other branches of physical science, the reactions of materials and other inanimate objects are the legitimate interests of the building scientist. Like problems of rain penetration, fire safety, or structural soundness, the issue of energy efficiency in buildings becomes defined as a technical problem amenable to scientific methods and solutions. The belief is that proved technical solutions are transferable and readily applicable to other technically similar situations. Consequently, the research agenda of building science is geared toward the scientific resolution of what are taken to be physical problems. Energy is no different. As Loren Lutzenhiser points out, to date, energy management strategies have "focused almost entirely on the physical characteristics of buildings and appliances." Vast sums of money have been, and continue to be, invested in scientific research to identify and fulfill (at least experimentally) the technical potential for energy savings in buildings.

2. THEORIES OF TECHNICAL CHANGE

Acknowledgment that an actual improvement in buildings has not always matched growth in knowledge has led to the realization that building scientists must appreciate the contribution of the life sciences. Here, the most natural fit has proved to be between building science and economic theory. Economists reinterpret social processes in terms of a market arena that neatly divides the world into separate, but interlinked, domains of knowledge and action. A world of perfect information, on the one hand, and a definable logic of (utility-maximizing) rationality, on the other hand, are posited. This confidence in the capacity of individual decision makers to quantify the benefits of energy efficiency economically is central to the technoeconomic view of technology transfer. The ability of building users literally to count the costs of energy consumption will ultimately lead to more rational energy use. This approach has found eloquent expression in the field of energy economics, and the use of energy audits that enable the new user to cope with the complex conversion equations and calculation of energy costs per standard unit. As Chadderton makes clear, the link to economics is explicit:

> *An energy audit of an existing building or a new development is carried out in a similar manner to a financial audit but it is not money that is accounted. All energy use is monitored and regular statements are prepared showing final uses, costs and energy quantities consumed per unit of production or per square metre of floor area as appropriate. Weather data are used to assess the performance of heating systems. Monthly intervals between audits are most practical for building use, and in addition an annual statement can be incorporated into a company's accounts.*
>
> —*D. V. Chadderton, 1991*

This model of rational action suggests that financial and energy consumption-related decisions are comparable, and that business and domestic users alike are self-interested, knowledgeable, and economically calculative when considering energy measures. Lutzenhiser illustrates this in regard to business users in particular who, skilled at marginal cost and future calculations, would "only seem to need to see the potential competitive economic advantage of innovation to move towards energy efficiency." Following a similar logic, the development of a visible market for energy efficiency would, in turn, encourage domestic consumers to minimize their costs by substituting more efficient ways of satisfying needs for energy services such as heating, cooling, or water heating. Here, the methodological presumption of building science, that with careful monitoring and scientific control it is possible to reproduce the technical achievements of the laboratory universally, is mirrored by an economic logic that promises that technological potential will be fulfilled in any market situation that demonstrates a "static, intertemporal, and intergenerational Pareto optimality." Put simply, the view is that economically rational actors, replete with the necessary technical and economic information, will consistently put science into practice. As Hinchliffe puts it, "a rational, profit maximising man is visualised at work, at home and at play," whereas "the engineering model tends to picture humans as optimally utilising technologies after the fashion of their creators." This epistemological coalition of building science and economics has led, internationally, to a prescriptive view of technological diffusion based on the twin technical and economic logics of proved, replicable, science and idealized consumer behavior. As Lutzenhiser suggests,

> *As a result, a physical–technical–economic model (PTEM) of consumption dominates energy analysis, particularly in energy demand forecasting and policy planning. The behaviour of the human 'occupants' of buildings is seen*

as secondary to building thermodynamics and technology efficiencies in the PTEM, which assumes 'typical' consumer patterns of hardware ownership and use.

—*L. Lutzenhiser, 1993*

The technoeconomic view of energy efficiency suggests that if technical knowledge is rigorously tested and demonstrably proved, and if "market forces" are not "disturbed," then consumption choices should be taken rationally, with the "right decisions being taken by millions of individual consumers, both at home and in their place of work." The role of policymakers is clear, "to set the background conditions and prices such that consumers will take decisions which are both in their own and the national interest."

3. BARRIERS TO ENERGY EFFICIENCY

This way of seeing technical change is not merely a matter of the attitudes or perceptions of individual policymakers. Rather, it provides an organizational logic or operating principle on which energy research and policy proceeds. This epistemic view represents a specific social organization of technical development and diffusion. The outcome of this view in policy terms is that human and financial resources are, almost exclusively, committed to demonstrating continually the technical efficacy of energy-efficient innovation through regulation, the provision of information about technical means to reduce energy consumption, and the development of best practice demonstration schemes. That policymakers, industry, and nongovernmental organizations all find consensus in this model of technological diffusion is perhaps not surprising, given its strikingly familiarity in the wider world of environmental policy debates.

Stephen Trudgill has, approvingly, formalized the technoeconomic way of seeing innovation in his formulation of the Acknowledgment, Knowledge, Technology, Economic, Social, Political (AKTESP) barriers to the resolution of environmental problems. Trudgill presents an image of technological innovation as a path from ignorance to enlightenment, with the evils of social, political, and economic reality cast as unpredictable obstacles to an otherwise assured technical utopia. Critically, the key to overcoming AKTESP barriers is almost always located within individual motivations. As Trudgill puts it, "motivation for tackling a problem comes from our moral obligation and our self-interest in enhancing the resource base and its life—thus enhancing, rather than destroying, planetary ecosystems and plant and animal species, including ourselves." When such motivation arises, we are in a position to solve the remaining barriers to the solution and to the implementation of environmental problems. The vocabulary of solution and implementation contains similarly individualized terms: "inadequacy of knowledge," "technological complacency," "economic denial or complacency," "social morality/resistance/leadership," and "political cynicism/ideology." Within this model of technical change, individuals are consistently the linchpins of effective energy-saving action. Changes in the level of the energy demand of the built environment is seen as a process involving "thousands" of individual judgments by property owners and other decision makers. The technical, organizational, and commercial complexity of energy-related decision-making is here replaced by an image of autonomous actors, each free to commit themselves to a more sustainable urban future. As Nigel Howard has argued, "there are lots of decision-takers. There are lots of people who have to be influenced, right from government to local authorities, developers, designers, material producers, professional and trade bodies. They all have a role to play. And what we have to do is try and influence all of them." This way of seeing energy views more or less "rational" individuals as both the solution to and the cause of energy-related environmental problems. Bewildered by the apparent irrationality of the social world, government, commercial, and voluntary interests in energy efficiency have all committed to a technoeconomic view of technology transfer, whereby established, proved, scientific knowledge is impeded by lack of information and/or by other nontechnical barriers. As Kersty Hobson puts it, "the drive to provide individuals with information, either to 'create' responsibility or to affect consumption, underpins the choice of national policy mechanisms used to forward sustainable consumption issues."

We are faced here with a self-sustaining, mutually reinforcing package of beliefs spinning between the realms of technology and policy without really belonging to either. Each element of the pervasive bundle, the transferability of technical knowledge, the individualistic theory of technical change, the sequential logic of research and development, and the implicit distinction between the social and the technical, feeds into the next, creating a web of belief strong enough to encapsulate technical researchers and their project officers and elastic enough to span countries and continents.

4. LEAPING THE BARRIERS

By defining what is acceptable as evidence, certain privileged methods also act to exclude other sorts of data. It is in this way that certain questions remain unmasked, and certain types of evidence are ignored or dismissed as invalid.

—*M. Leach and R. Mearns, 1995*

Leach and Mearns' observations highlight how wide acceptance of the technoeconomic view of technology transfer has hitherto marginalized explanations of technical innovation concerned with the social shaping of technical change. As Thomas Hilmo suggests, the notion of barriers suggests that "problems are absolute" and that "some actors [scientists] know the truth about a problem," whereas "other actors [nonscientists] do not and obstruct the solutions in different ways." Hence, the role of social scientist is in turn reduced to that of market researchers, typically undertaking attitudinal surveys designed to identify the human barriers to good energy practice that the promotional campaigns are designed to overcome. This research has typically been described as exploring the behavioral or human dimensions of energy efficiency. While drawing on diverse intellectual sources from economic sociology to environmental psychology, such research shares a view of energy efficiency as the product of a slow cascade of more or less rational individual choices. Although Stern and Aronson usefully identify five archetypes of energy user (the "investor," "consumer," "member of a social group," "expressor of personal values," and "avoider of problems"), their narrow focus on the attitudes and motivations of individual energy consumers tends to isolate and atomize the decision-making process. Little is heard in this research about the consumption practices of organizations, or the role of the organizational actors and groups that actually design, finance, and develop buildings. As Janda points out, although this approach "usefully delineates differences in individuals' attitudes, human dimensions research does not reveal where these differences originate, how they develop, or if they can be changed." Attitudes and decisions are always shaped and framed within wider social processes. Abstraction of the opinions and outlook of energy consumers, and of design and development actors from the contexts of production and consumption, tends to isolate and freeze what are always contingent practices. This narrow, behavioral view of technical change fails to recognize the routine complexities of energy-related decision making. In particular, there is little room in this model to view technical innovations and social processes as interrelated, or to assess how energy-efficient choices may be embedded in the mundane routines of domestic life and commercial practice.

The technoeconomic way of seeing energy efficiency can be contrasted with an alternative view, the sociotechnical, which has questioned what Michael Mulkay has termed the "standard view of science," a view that exists, already formed, outside society. In its place, a broad church of sociologists have presented a revised image of science as a "sociocultural phenomenon" that questions the "authority of science," locates "knowledge-claims in their social context," and identifies the "relationship between such contexts and wider economic and political processes." This approach to understanding science and technical change can be characterized by a series of questions. Volti asks, for example, "What accounts for the emergence of particular technologies? Why do they appear when they do? What sort of forces generate them? How is the choice of technology exercised?" Rather than analyze the process of technical change internally, in terms of developments within the technology, or by reference to the ideas of famous scientists, inventors, and entrepreneurs, such research is interested in discovering to what extent, and how, does the kind of society we live in affect the kind of technology we produce. Or, as McKenzie and Wajcman ask, "What role does society play in how the refrigerator got its hum, in why the light bulb is the way it is, in why nuclear missiles are designed the way they are?"

Questioning the notion that technological change has its own logic, this way of seeing technical change recognizes the wider social contexts within which design solutions emerge and patterns of consumption evolve. Rather than viewing science and technology as asocial, nonpolitical, expert, and progressive, innovation is viewed by Andrew Webster as a "contested terrain, an arena where differences of opinion and division appear." In relation to analysis of energy problems, such research avoids individualist explanations of technological innovation (the rational energy consumer), rejects any form of technological determinism (technical innovation as handmaiden to an energy-efficient economy), and, critically, refuses to distinguish prematurely between technical, social, economic, and political aspects of energy use. As Thomas Hughes has graphically illustrated, in exploring the development of the American electricity system, "sociological, techno-scientific and economic analyses are permanently woven together in seamless web." In this world without seams, social groups and institutions are considered, alongside technological artifacts, as actors who actively fashion their world according to their own particular logic of social action. This sociotechnical analysis of energy use replaces technoeconomic descriptions of universal barriers to energy-efficient innovation (apathy, ignorance, lack of financial interest), with analysis of the ways in which the changing social organization of energy-related choices structures opportunities for more efficient energy use.

There are complementary methodological issues at stake here. Bridging the theoretical gap between the social and technical features of energy use demands a more qualitative research agenda. Rather than solely relying on positivist research tools such as surveys, opinion polls, and statistical analysis, undertaking sociotechnical research means attempting to peer over the shoulder of the actors making energy-related decisions by following actors through their professional and personal routines. In this way, research into energy efficiency finds itself on what Michel Callon has termed a new terrain: that of "society in the making." This refocuses analysis away from pure energy questions to a wider set of debates about design conventions, investment analysis, development costs, space utilization, and market value. Idealized notions of a rational energy user or best technical practice make little sense here. Instead, the social and the technical form a network of associations that serve to frame the meaning of energy efficiency, thereby encouraging or delimiting opportunities for innovation over time and space, and between organizational settings.

5. CONTEXTS OF CONSUMPTION

A significant amount of academic work has been undertaken in an attempt to put energy use into its social context. Social psychologists and sociologists have emphasized the importance of seeing energy problems and solutions in terms of "social systems rather than single causes," the need to design energy systems for "adaptability as an alternative to detailed planning," and, as Stern and Aronson argue, to treat energy policies and programs as forms of "social experiment." Reviewing this work, Lutzenhiser found "a consensus in the literature" that to understand the sociotechnical complexity of energy-saving action, policymakers must be concerned more directly with "the social contexts of individual action." As Lutzenhiser points out,

> *While the physical–technical–economic model assumes consumption to be relatively homogenous and efficiency to be driven by price, the empirical evidence points towards variation, non-economic motives, and the social contexts of consumption. Economics can supply normative guides regarding when investments would be economically desirable, but it tells us little about how persons actually make economic decisions.*
>
> —*L. Lutzenhiser, 1993*

Both Lutzenhiser and Wilhite draw attention to the social nature of energy consumption and the social dynamics of change as a form of "social load." These social loads have peaks that drive the dimensioning of peak energy loads. Wilhite provides examples from his anthropological research into lighting use in Norway. Here, home lighting is designed not simply to provide a sufficient degree of brightness to support basic human activity, but rather to provide a "cozy aesthetic." A dark house was described by Wilhite's respondents as a "sad house." Extra lighting fittings are always fitted and utilized to make sure guests feel cozy, and for social visits on a winter evening, lights are left on in every room to provide a welcoming glow as guests arrive. This, of course, means the use of extra energy. In his study sample, Wilhite discovered an average of 11.5 lights per living room. This social peak "drives the dimensioning of the material and energy system behind lighting, which based on a straightforward provision of lumens would be far smaller." As Wilhite argues elsewhere, "the things we use energy to achieve—a comfortable home, suitable lighting, clean clothes, tasty food—have also been assumed in models of consumption to be generic and physically determined." Wilhite goes on to argue that there is an "urgent need for the development of a more robust theory of consumption, one which incorporates social relations and cultural context, as well as perspectives on individual agency and social change."

6. SOCIOTECHNICAL THEORIES OF CHANGE

In Table I, the two broad positions outlined so far are contrasted. The effect of these analytical assumptions is cumulative, with a technoeconomic or sociotechnical view of buildings as an artifact leading correspondingly to a particular way of viewing energy-efficient design, energy-saving action, technical innovation, or market failure. Commitment to

TABLE I

Ways of Viewing Energy Efficiency in Buildings

Aspect	Perspective	
	Technoeconomic	Sociotechnical
Buildings	Materially similar, physical structures	Material product of competing social practices
Energy-efficient design	Replicable technical solutions	Outcome of conflicting sociocommercial priorities
Energy-saving action	Individual, rational decision-making in a social vacuum	Socially structured, collective choices
Technological innovation	Series of isolated technical choices by "key" decision makers	Technical change embedded within wider sociotechnical processes
Market failure	Existence of social barriers	Lack of perceived sociocommercial viability
Image of energy consumers	More or less rational	Creative, multirational, and strategic
Role of social science research	Evaluation of technical potential and the detection of environmental attitudes and nontechnical barriers	Identification of context-specific opportunities for technological innovation
Energy policy	Provision of information, granting of subsidies, and setting of regulations	Forging of context-specific communities of interest and promotion of socially viable pathways of innovation

either a predictable linear or socially shaped diffusion pathway for energy-efficient technologies frames the contribution of social science research to understanding innovation and the direction of energy policy-making.

These competing ways of seeing energy consumption raise a number of research dilemmas. Commitment to a technoeconomic or sociotechnical perspective on innovation focuses attention on different actors (key decision makers vs. relevant social groups), different practices (technical design vs. development strategies), and different processes (technological diffusion vs. social and commercial change). We are seemingly faced with a series of choices over how we conceptualize the problem of energy consumption in buildings. As Groak suggests, we can view buildings in terms of their physical attributes, as "essentially static objects" formed in a relatively standardized manner from an assembly of interconnecting construction materials. Seen this way, buildings appear remarkably similar. Irrespective of geographical location, ownership patterns, or operational function, the technical character of building form appears comparatively homogeneous. Alternatively, we could view buildings as material products of competing social practices. This would suggest a different analytical approach. For sociologists such as Bruno Latour, understanding "what machines are" is the same task as understanding "who the people are" that shape their use. Seen this way, technologies and technological practices are as Bijker and Law have illustrated, "built in a process of social construction and negotiation," a process driven by the shifting social, political, and commercial interests of those actors linked to technological artifacts of design and use. Although the complexity of buildings differs from the individual technologies often studied within sociological studies of science and technology, we can nevertheless develop a similar analytical approach. Thus, to understand buildings we must trace the characteristics of the "actor world" that "shapes and supports" their production. Adopting this perspective would mean relating the form, design, and specification of buildings to the social processes that underpin their development. So, although two identical buildings, standing side by side, may well appear physically and materially similar, investigation of their respective modes of production and consumption may reveal profoundly different design rationales, which in turn might help explain variations in energy performance.

This stress on the social organization of design is at odds with the technoeconomic perspective that emphasizes how a repertory of well-tried technical solutions provides reliable precedents for designers. Here, new technical challenges are seen as solvable by shifts in design emphasis, mirroring the march of scientific progress. Although the form and specification of buildings may well vary spatially with climate and culture, the objective is always viewed as the same, i.e., the provision of the universal needs, shelter and comfort. Technical design is viewed as a process of adaptation and modification to suit changing physical circumstances. Emphasizing the social logic of design raises a different set of questions. Rather than supporting a linear model of innovation, studies in the sociology of technical change have revealed the multidirectionality of the technical design process. Rather than one preordained process of change, we are faced with competing pathways of innovation. Typically, Pinch and Bijker describe the development process of a technological artifact as "an alternation of variation and selection." For example, Wiebe Bijker's study of the development of fluorescent lighting points to a range of innovation pathways that could not be resolved by appeals to technical superiority. Instead, a standoff between lighting manufacturers (who supported a high-efficiency daylight fluorescent lamp) and electric utility companies (who, worried about the effect on electricity sales, supported an energy-intensive tint-lighting fluorescent) led to the introduction of a third design alternative, the high-intensity daylight fluorescent lamp, which combined efficiency with a high light output, thereby maintaining electricity demand. Bijker's study illustrates how a socially optimal design was established from a range of technically feasible possibilities through a process of compromise between competing social interests. In doing so, he highlights the "interpretative flexibility" of design. We might similarly ask how rival energy-efficient designs are valued by different members of the design and development process and how this process of contestation and compromise frames the resulting design strategies.

In approaching these questions, we might begin to draw attention to the idealism surrounding the technoeconomic image of enlightened, rational individuals motivated by a growing stock of technical knowledge. Although these more, or sometimes less, knowledgeable individuals are placed in a hierarchy of influence—from the key decision makers, designers, and top managers, to technicians and lower management, to domestic consumers—their social, spatial, or temporal situation appears of marginal importance. From a sociotechnical perspective, the relationship between individual and context is emphasized and made fluid. Particular technical choices are viewed as expressive of the prevailing social, political, and commercial pressures operating within spatially and temporally contingent contexts. Here, technical choices are not considered to be solely determined by knowledge or motivation, but are shaped by the existence of a more or less socially favorable context. For a variety of reasons, consumers may be unable or prefer not to use particular technologies, or may even use technologies in unpredictable ways not envisaged in the original design. To understand this social structuring of technical choice, Ruth Schwartz Cowan, in her history of home heating and cooking systems in America, treats the consumer "as a person embedded in a network of social relations that limits and controls the technological choices that she or he is capable of making." For Cowan, consumers come in "many different shapes and sizes," and operate in a variety of social contexts. Her analysis of the introduction of cast iron stoves to replace open hearth fires for cooking traces the interconnections between stove producers and merchants, fuels suppliers and merchants, and their networks of influence through production, wholesale, retail, and household domains over both urban and rural consumers. This emphasis on the embedding of the decision maker within wider social networks focuses analytical attention on the "place and time at which the consumer makes choices between competing technologies." This allows Cowan to unpack the "elements" more "determinant of choices" and the technical pathways that "seemed wise to pursue" or that appeared "too dangerous to contemplate." As Cowan points out, "today's 'mistake' may have been yesterday's 'rational' choice." Drawing on this approach, a sociotechnical perspective on energy efficiency might identify the multiple rationalities shaping innovation and how these relate to the contrasting "universe of choice" in which design and development actors operate.

If we begin to accept that the nature and direction of technical change are subject to interpretative flexibility, underpinned by multiple rationalities of context-specific choice, then we would also have to begin to alter our understanding of the process of technological innovation. As we have seen, the technoeconomic perspective views technical change as following an almost preordained pattern of design, development, and diffusion. By contrast, a sociotechnical analysis would explore why particular technical solutions emerged at a certain time and in a particular place. For example, in studying the electrification of cities, David Nye illustrates how the emergence of street lighting was less connected to the rational technical ordering of urban space, and more intimately linked to the need of utility companies to increase load and to the commercial instincts of shopkeepers keen to attract business. Nye suggests that "shopkeepers understood lighting as a weapon in the struggle to define the business center of the city dramatizing one sector at the expense of others." Framed this way, "electric lighting could easily be sold as a commercial investment to increase the competitiveness of a business." As a result, the subsequent spread of street lighting was accelerated as "electrification of one street quickly forced other commercial areas to follow suit or else lose most of their evening customers." Following this lead, a sociotechnical analysis of energy efficiency would ask what roles were played by architects, developers, governments, investors, manufacturers, retailers, and consumers in fashioning innovation in building design and use. This approach would mean widening the nature of analytical inquiry away from explicit decisions about energy efficiency studied in isolation, to an examination of the embedding of energy-related choices in the manufacture, distribution, and retailing of the technologies, and the commercial processes framing building design and development.

Finally, asking different questions about the process of technical innovation may provide a different understanding of the market success or failure of proved energy-saving technologies. Instead of characterizing nontechnical barriers as both universal and timeless in nature, a sociotechnical approach would explore the degree to which the marketability of technical innovation can be identified as a socially, and temporally, contingent process. For example, Gail Cooper highlights how the seemingly pervasive nature of air-conditioning systems in the United States masks a more contested story about the emergence of heating, cooling, and ventilating technologies. She shows how attempts by engineers to promote the most "technically rational design," which demanded sealed buildings and passive use, was resisted by what engineers saw as "irrational users" who preferred mobile, plug-in air-conditioning systems that, although less efficient, provided greater flexibility and active control. As Cooper suggests, "the engineering culture that characterised the custom-design industry did not produce the best technology, neither did the market forces that dominated the mass production industry." The result is localized compromises that reflect the "seesawing power relations surrounding the development of air-conditioning." In particular, Cooper's study underlines the contrasting image of the energy consumer underpinning the technoeconomic and sociotechnical views. Put simply, the technoeconomic "irrational user" is translated in the sociotechnical literature into a "guerrilla fighter of those disenfranchised from the design process." Herein lies a key distinction. Rather than assume the intrinsic marketability of technically proved innovations, a sociotechnical approach would assess the sociocommercial viability of the artifact in varying social contexts. Instead of explaining market failure in terms of ubiquitous and timeless barriers, a sociotechnical analysis would seek to explain market success or failure in situationally specific situations. Some contexts may favor innovation, others may not. Mapping what Bijker terms the "technical frames" of "socially relevant actors" is vital here, for, as Cowan points out, any one of those "groups or individuals acting within the context of their group identity…may be responsible for the success or failure of a given artefact."

7. CONCLUSION: RETHINKING ENERGY CONSUMPTION

> *The understanding of present and future energy use depends upon understanding how conventions and practices evolve over time and within and between cultures.*
>
> —*Hal Wilhite, 2003*

The aim of this new agenda is not to produce abstract social theory. Instead, a growing number of social scientists are striving to articulate a new role for social science in research and policy debates about energy consumption and environmental change. These researchers are exploring the heterogeneous and contested nature of consumption practices that shape energy use. Rather than simply assuming that people use energy, they are analyzing how energy intersects with everyday life through diverse and cultural inscribed practices such as heating and cooling, cooking, lighting, washing, working, and entertaining. To achieve this, they are drawing on methods and theories beyond science and economics, including sociology, anthropology, geography, psychology, and cultural studies, and are learning lessons from consumption debates beyond energy and the environment, including fashion, food, and shopping.

The scope of this research agenda takes us far beyond the technoeconomic analysis of energy consumption, which tends to be dedicated to identifying the potential scope and scale of energy performance improvements in different building types and in different building sectors, and to the setting of technically feasible CO_2 abatement targets. It also suggests a role for social scientists much deeper than the forms of market research, evaluating the attitudes of what are taken to be key decision makers toward energy that typify research on human behavior and attitudes. In developing a sociotechnical approach to energy consumption, the concern is more with identifying context-specific opportunities for inserting technically proved technologies into appropriate social practices. But rather than being led by calculations of technical potential, the task is to seek to identify how specific social, spatial, and temporal configurations of energy consumption encourage, or mitigate against, effective energy-saving action. The result is stories about the circumstances in which energy efficiency and conservation practices do or do not flourish. This focus takes us far from the world of building science and the paradigmatic certainties of the technoeconomic perspective, and instead reveals the construction of energy knowledge in varying social worlds and reflects the contested nature of energy consumption practices.

SEE ALSO THE FOLLOWING ARTICLES

Conservation of Energy, Overview • Economics of Energy Efficiency • Energy Efficiency and Climate Change • Energy Efficiency, Taxonomic Overview • Obstacles to Energy Efficiency • Technology Innovation and Energy • Vehicles and Their Powerplants: Energy Use and Efficiency

Further Reading

Bijker, W. E. (1995). Sociohistorical technology studies. *In* "Handbook of Science and Technology Studies" (S. Jasanoff *et al.*, Eds.), pp. 229–256. MIT Press, Cambridge, Massachusetts.

Bijker, W. E., and Law, J. (eds.). (1992). "Shaping Technology/Building Society." MIT Press, Cambridge, Massachusetts.

Bijker, W. E., Hughes, T. P., and Pinch, T. (eds.). (1987). "The Social Construction of Technological Systems." MIT Press, Cambridge, Massachusetts.

Callon, M. (1987). Society in the making: The study of technology as a tool for sociological analysis. *In* "The Social Construction of Technological Systems" (W. E. Bijker, T. P. Hughes, and T. Pinch, Eds.), pp. 83–103. MIT Press, Cambridge, Massachusetts.

Chadderton, D. V. (1991). "Building Services Engineering." E & FN Spon, London.

Cooper, G. (1998). "Air Conditioning America: Engineers and the Controlled Environment 1900–1960." Johns Hopkins Univ. Press, Baltimore.

Cowan, R. S. (1987). How the refrigerator got its hum. *In* "The Social Shaping of Technology" (D. McKenzie and J. Wajcman, Eds.), pp. 202–218. Milton Keynes: Open Univ. Press, Philadelphia.

Groak, S. (1992). "The Idea of Building: Thought and Action in the Design and Production of Buildings." E & FN Spon, London.

Guy, S., Marvin, S., and Moss, T. (2001). "Urban Infrastructure in Transition: Networks, Buildings, Plans." Earthscan, London.

Hilmo, T. (1990). Review of barriers to a better environment. *Geografiska* **72B**, 124.

Hinchliffe, S. (1995). Missing culture: Energy efficiency and lost causes. *Energy Policy* **23**(1), 93–95.

Hobson, K. (2002). Competing discourses of sustainable consumption: Does the 'rationalisation of lifestyles' make sense? *Environ. Politics* **11**(2), 95–120.

Howard, N. (1994). Materials and energy flows. *In* "Cities, Sustainability and the Construction Industry," pp. 11–14. Engineering and Physical Science Research Council, Swindon.

Hughes, T. P. (1983). "Networks of Power: Electrification in Western Society, 1880–1930." Johns Hopkins Univ. Press, Baltimore.

Janda, K. (1998). "Building Change: Effects of Professional Culture and Organisational Context on Adopting Energy Efficiency in Buildings." Unpublished Ph.D. thesis, University of California, Berkeley.

Latour, B. (1987). "Science in Action." Milton Keynes: Open University Press, Philadelphia.

Leach, M., and Mearns, R. (1995). "The Lie of the Land: Challenging Received Wisdom in African Environmental Change and Policy." James Currey, London.

Lutzenhiser, L. (1993). Social and behavioural aspects of energy use. *Annu. Rev. Energy Environ.* **18**, 247–289.

Lutzenhiser, L. (1994). Innovation and organizational networks: Barriers to energy efficiency in the US housing industry. *Energy Policy* **22**(10), 867–876.

McKenzie, D., and Wajcman, J. (1985). "The Social Shaping of Technology." Milton Keynes: Open Univ. Press, Philadelphia.

Mulkay, M. (1979). "Science and the Sociology of Knowledge." George Allen and Unwin, London.

Nye, D. E. (1998). "Consuming Power: A Social History of American Energies." MIT Press, Cambridge, Massachusetts.

Pinch, T., and Bijker, W. T. (1989). The social construction of facts and artefacts: Or how the sociology of science and the sociology of technology might benefit each other. *In* "The Social Construction of Technological Systems" (W. E. Bijker, T. P. Hughes, and T. Pinch, Eds.), pp. 17–50. MIT Press, Cambridge, Massachusetts.

Schipper, L. (1987). Energy conservation policies in the OECD: Did they make a difference? *Energy Policy*, December, 538–548.

Stern, P. C., and Aronson, E. (eds.). (1984). "Energy Use: The Human Dimension." W. H. Freeman, New York.

Trudgill, S. (1990). "Barriers to a Better Environment." Belhaven Press, London.

Volti, R. (1992). "Society and Technological Change." St. Martins Press, New York.

Webster, A. (1991). "Science, Technology and Society." Macmillan, London.

Wilhite, H. (1997). "Cultural Aspects of Consumption." Paper (unpublished manuscript) presented at the ESF-TERM workshop on Consumption, Everyday Life and Sustainability, Lancaster University, Lancaster, U.K.

Wilhite, H. (2003). What can energy efficiency policy learn from thinking about sex? *In* "Proceedings of European Council for an Energy-Efficient Economy (ECEEE)," pp. 331–341. ECEEE, Paris.

Wilhite, H., and Lutzenhiser, L. (1998). Social loading and sustainable consumption. *Adv. Consumer Res.* **26,** 281–287.

Wilhite, H., Shove, E., Lutzenhiser, L., and Kempton, W. (2000). "Twenty Years of Energy Demand Management: We Know More about Individual Behaviour but How Much Do We Really Know about Demand?" Proceedings of the American Council for an Energy-Efficient Economy (ACEEE) 2000 Summer School on Energy Efficiency in Buildings. ACEEE, Washington, D.C.

Conversion of Energy: People and Animals

VACLAV SMIL
University of Manitoba
Winnipeg, Manitoba, Canada

1. Mammalian Metabolism
2. Human Energetics
3. Sustained Output and Extreme Exertions
4. Working Animals
5. Bovines
6. Horses

Glossary

basal metabolic rate A measure of nutritional energy needed by an organism during a fasting and resting state.

bovines Large ungulate mammals belonging to genus *Bos* (cattle) and *Bubalus* (water buffalo) that were domesticated for dairy production and draft labor.

draft animals Domesticated large ungulate mammals of several families (bovines, horses, and camelids) harnessed for fieldwork or transport duties.

horses Large ungulates of the genus *Equus* that were domesticated for heavy agricultural labor and transport draft as well as for riding and warfare.

metabolic scope Multiple of the basal metabolic rate achieved during periods of maximum exertion.

The power of human and animal muscles, limited by metabolic rates and mechanical properties of working bodies, restricted the productive capacities, and hence the typical quality of life, of all preindustrial civilizations. Only the widespread adoption of inanimate energy converters replaced these animate (or somatic) energies as the most important prime movers of human society. This epochal energy transition began in parts of Western Europe and North America during the early 19th century and it was basically completed during the first half of the 20th century. In contrast, in scores of low-income countries both the subsistence agricultural production and many industrial tasks are still energized by human and animal muscles.

Human labor was the only prime mover in all subsistence foraging (gathering and hunting) societies. Simple wooden, stone, and leather tools (including digging sticks, bows and arrows, spears and knives, and scrapers) were gradually invented and adopted in order to increase and extend the inherently limited muscle power. The only extrasomatic energy conversion mastered by foraging societies was the use of fire for warmth and cooking. Even those early agricultural societies that domesticated large animals continued to rely largely on human muscles, but they increased their effectiveness by multiplying individual inputs through concentrated applications of labor and by using simple but ingenious mechanical devices.

The first approach had logistic problems. For example, only a limited number of people can grasp a small but heavy object to carry it, and merely multiplying the number of available hands makes no difference if the object is to be raised far above the people's heads. The second approach relied largely on the three simplest aids, levers, inclined planes, and pulleys, and their common variations and combinations, including wedges, screws, wheels, windlasses, treadwheels, and gearwheels. These tools and machines were used by virtually all old high cultures. They enlarged the scope of human action but their use had obvious physical limits (e.g., the maximum practical length of levers or pulleys). However, some societies that commanded just the muscle power were able to complete, aided by careful organization of complex tasks, such astounding projects as the megalithic structures at Stonehenge or on the Easter Island, the pyramids of Egypt and Mesoamerica, and massive stone temples, walls, and fortresses on every continent except Australia.

Harnessing of large domesticated animals made the greatest difference in agricultural productivity. Depending on the soil type, a peasant working with a hoe would take 100–180 h to prepare 1 ha of land for planting cereals. A plowman guiding a single medium-sized ox pulling a primitive wooden plough could finish this task in just over 30 h. Use of a pair of horses and a steel plough reduced the time to less than 10 h and in light soil to just 3 h. Hoe-dependent farming could have never reached the scale of cultivation made possible by draft animals, and many tasks, including deep plowing of heavy soils, would have been entirely impossible without draft animals.

Draft animals harnessed to wheeled vehicles also made major contributions to land transport, construction, and industry. Wheelbarrows provided the most efficient way for an individual to transport goods, with an adult man covering approximately 10–15 km a day and moving usually less than 50 kg. Oxen were not much faster but could pull wagons up to 10 times that weight. Heavy horse-drawn wagons covered 30–40 km a day, and passenger horse carts on good roads could travel 50–70 km (galloping messengers on Roman roads could ride up to 380 km per day). Oxen and horses were deployed to transport heavy materials to building sites, such as medieval cathedrals or Renaissance villas. Horses were also used for many tedious tasks in the early stages of expanding coal extraction and incipient industrialization of the 17th and 18th centuries. Here, too, there were obvious logistic limits, no matter if the animals worked alone or in groups. Large and heavy animals could not be deployed in low and narrow mine tunnels, and smaller animals were often inadequate to perform tasks requiring hours of nonstop draft.

Harnessing of more than 20 animals to be guided as a single force is difficult, and harnessing more than 100 of them to get 100 kW of sustained power is impossible. Stabling of horses placed enormous demand on urban space, and during the late 19th century dense horse-drawn traffic often created scenes of chaos on the streets of the largest cities (Fig. 1). Even heavy animals could not demonstrate their powerful draft because of the poor preindustrial roads. Late Roman specifications restricted the loads to 326 kg for horse-drawn and up to 490 kg for slower ox-drawn carriages. By 1850, the maximum allowable load on French roads was approximately 1.4 tons, still a small fraction of today's truckloads (commonly more than 15 tons).

FIGURE 1 High density of horse-drawn traffic on a London bridge in 1872.

Maintaining larger numbers of more powerful animals also required more feed (even during periods when the animals were idle), and this was often impossible given the limited amount of land that could be taken away from food cultivation. Combustion of fossil fuels and generation of primary electricity displaced virtually all hard labor in affluent nations, but reliance on animal and human muscles remains high in most low-income countries. Unfortunately, the latter toll still includes such debilitating tasks as producing gravel by shattering rocks with hand hammers or transporting outsize loads on bent backs of porters. Indeed, this dependence on animate prime movers is one of the most obvious physical distinctions between the rich and the poor worlds.

1. MAMMALIAN METABOLISM

Basal metabolic rate (BMR) is the amount of energy needed to maintain critical body functions of any heterotrophic organism. The rate varies with sex, body size, and age, and its measurements in adults show large individual variations from expected means. To determine BMR in humans, the person must be at complete rest in a thermoneutral environment at least 12 h after consuming a meal. Neutral temperature (approximately 25°C) is required to preclude any shivering and sweating, and the wait eliminates the effects of thermogenesis caused by food digestion, which peaks approximately 1 h after the meal and can add up to 10% to the BMR. Resting energy expenditure (REE) is higher than BMR because it includes the thermic effect of meals. Mammals thermoregulate at 34–40°C (humans at approximately 37°C) and heart, kidneys, liver, and brain are the metabolically most active organs; in humans, these four organs account for slightly more than one-half of adult BMR.

Max Kleiber pioneered the systematic study of mammalian BMR during the 1930s, and in 1961 he concluded that dependence of this rate (expressed in watts) on total body weight (mass in kilograms) is best expressed as $3.4\,m^{0.75}$ (the exponent would be 0.67 if BMR were a direct function of body surface area). Kleiber's famous mouse-to-elephant line became one of the most important and best known generalizations in bioenergetics (Fig. 2). Closer examination of available BMRs shows slightly different slopes for several mammalian orders, and individual outliers illustrate genetic peculiarities and environmental adaptation. For example, cattle have a relatively high BMR, requiring more feed per unit of body weight gain than would be expected based on their mass, whereas the relatively low BMR of pigs makes them efficient converters of feed to meat and fat. Various explanations of the three-fourths slopes have been offered in the biological literature, but the definitive solution remains elusive.

In order to obtain total energy requirements, BMRs must be increased by energy needs for growth and activity. Conversion efficiency of growth depends on the share of newly stored proteins (which are always more costly to synthesize than lipids) and on age (efficiencies range from more than 40% during postnatal growth to well below 10% for mature adults). In humans, the average food energy need is approximately 21 kJ/g of new body mass in infants, equal to a conversion efficiency of approximately 50%; adult growth efficiencies are up to 35% lower. Regarding the conversion of chemical energy of food and feed to kinetic energy of working muscles, it has been known since the late 19th century that its peak efficiency is slightly more than 20%.

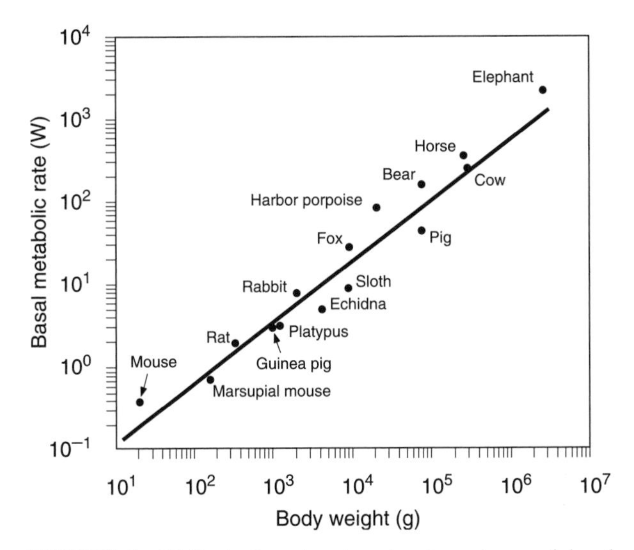

FIGURE 2 Kleiber's line showing the dependence of basal metabolic rate on body mass.

The first fairly accurate estimates of animate power were made in 1699 when Guillaume Amontons studied the effort of French glass polishers. In 1782, James Watt calculated that a mill horse works at a rate of 32,400 foot-pounds per minute, and the next year he rounded this figure to 33,000 foot-pounds, making 1 horsepower equivalent to 745.7 W. By 1800, it was known, correctly, that the power of useful human labor ranges from less than 70 to slightly more than 100 W for most steadily working adults. This means that when working at 75 W, 10 men were needed to equal the power of a good horse for tasks in which either of these animate prime movers could be used.

2. HUMAN ENERGETICS

Healthy people fed balanced diets can convert chemical energy in the three basic classes of nutrients with high efficiency: 99% for carbohydrates, 95% for lipids, and 92% for proteins (however, more than 20% of dietary protein is lost daily through urine). With the exception of people engaged in very heavy work or in strenuous and prolonged exercise, most of the digested energy is used to satisfy the basic metabolic needs. Much BMR and REE data have been acquired mostly by measuring oxygen consumption, which can be easily converted into energy equivalents: One liter of O_2 equals 21.1 kJ/g of starch, 19.6 kJ/g of lipids, and 19.3 kJ/g of protein.

The relationship between BMR and body size can be expressed by simple linear equations (Fig. 3), but these are excellent predictors of individual BMRs only for children and adolescents. More important, using these equations may lead to exaggerated estimates of energy requirements for populations outside the affluent world in general and outside of Europe in particular. For example, BMR equations favored by the Food and Agriculture Organization (FAO) overpredict BMRs of people living in tropical regions by 1.5–22.4%, with the difference being largest for adults older than age 30.

Adult BMRs vary substantially not only among individuals of the same population but also among different population groups. There are no indisputable, comprehensive explanations for these disparities.

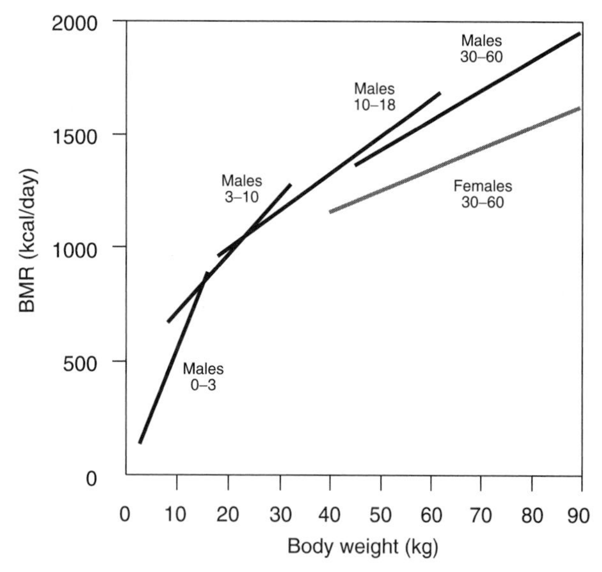

FIGURE 3 Linear predictions of basal metabolic rate (BMR) based on equations recommended by FAO/WHO/UNU.

Disparities among different populations may be due simply to different shares of metabolizing tissues (i.e., muscles and especially internal organs) in the overall body mass. Although individual BMRs vary widely, their lifetime course follows a universal trend. After peaking between 3 and 6 months of age, they decline to approximately half the peak value by late adolescence; this reduction is followed by an extended plateau. Metabolic decline, due to a steady loss of metabolizing lean tissue and increased fat deposition, resumes during the sixth decade and continues for the rest of life. By the seventh decade, fat comprises approximately one-fourth of typical male and slightly more than one-third of female body mass.

As for other mammals, human energy requirements for growth become marginal after reaching adulthood, and food intakes are accounted for almost completely by the combination of BMR and energy needed for various activities. Determination of energy costs of labor and leisure activities has a century-long tradition based on respirometry that requires masks, hoods, or mouthpieces and hoses to measure gas exchange. This is not a problem when testing a cyclist on a stationary bicycle, but it is a challenge when done in restricted workspaces and during aerobic sport activities. Nevertheless, many diverse and interesting values, including those for playing cricket in India and climbing Alpine peaks, have been obtained during decades of such research. Portable systems that can be used for continuous measurements of oxygen uptake for long periods are in widespread use, and two convenient methods allow nonintrusive measurement of energy expenditures.

The first, developed in the mid-1950s, has been used in humans since the mid-1980s: It monitors exponential losses of the doubly labeled water (marked with stable heavy isotopes of ^{2}H and ^{18}O) drunk by subjects during a period of 1 to 2 weeks. When the two heavy isotopes are washed out of the body and replaced with dominant ^{1}H and ^{16}O, loss of deuterium is a measure of water flux and the elimination of ^{18}O indicates not only the water loss but also the flux of CO_2, a key final product of metabolism, in the expired air. The difference between the washout rates of the two isotopes can be used to determine energy expenditure over a period of time by using standard indirect calorimetric calculations. The other technique involves continuous monitoring of heart rate and its conversion to energy expenditure by using calibrations predetermined by a laboratory respirometry.

3. SUSTAINED OUTPUT AND EXTREME EXERTIONS

Energy costs of individual activities, or so-called physical activity levels (PALs), are best expressed as multiples of the BMR or the REE, that is, energy flows equal to approximately 70–90 W for most adult males and 55–75 W for females. Typical averages, corresponding to the reference man of 65 kg and the reference woman of 55 kg, are 78 and 64 W, respectively. Contrary to a common belief, the high basal metabolism of the brain, which always requires approximately one-fifth of the adult food energy intake, increases only marginally when a person is engaged in the most challenging mental tasks: Typical energy increase is only 5% higher than the BMR.

Minimal survival requirements, including metabolic response to food and energy needed to maintain basic personal hygiene, require PALs between 1.15 and 1.2. Standing requires the deployment of large leg muscles, resulting in a PAL 1.3–1.5 times the BMR. Light exertions typical of numerous service and manufacturing jobs that now dominate the modern economies, such as secretarial work, truck driving, assembly line work, or car repair, require no more than 2.5 times the REE. Frequent carrying of loads, lifting of heavy burdens, deep digging, and cleaning of irrigation canals increase the exertion averages into the moderate or heavy expenditure range.

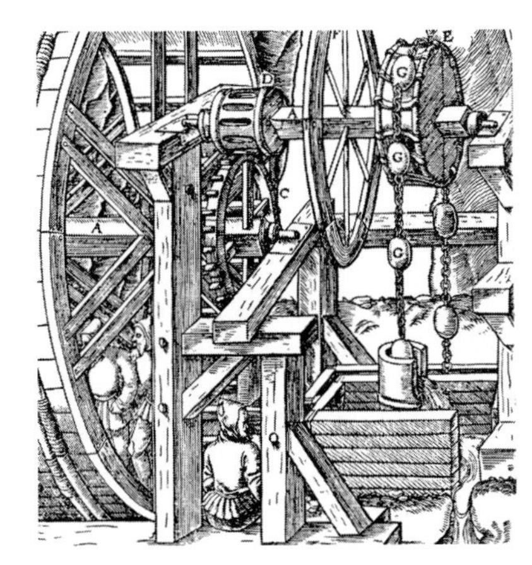

FIGURE 4 Men powering large treadwheels depicted in Agricola's famous "De re Metallica" (1556).

Activities requiring moderate (up to five times the REE), and sometimes even heavy (up to seven times the REE), exertion are now common only in traditional farming (ploughing with animals, hand weeding, and cleaning of irrigation canals are particularly taxing), forestry, and fishing. From the consensus meeting organized by the FAO, the World Health Organization, and the United Nations University PALs of 1.55 for males and 1.56 for females were determined to be the average daily energy requirements of adults whose occupations require light exertions; PALs of 1.64 for females and 1.78 for males engaged in moderately demanding work and 1.82 for females and 2.1 for males with jobs demanding heavy exertions were chosen.

A great deal of experimental work has been performed to determine the limits of human performance. Tiny amount of ATP stored in muscles (5 mmol/kg of wet tissue) can be converted to just 4.2 kJ, enough to energize contractions for less than 1 s of maximal effort. Replenishment through the breakdown of phosphocreatine can add a total of 17–38 kJ, enough to support a few seconds of intensive exertion. Maximum metabolic power achievable in such ephemeral outbursts is large—3.5–8.5 kW for a typical Western man and as much as 12.5 kW for a trained individual. Longer exertions, lasting 30 s to 3 min, can be energized by muscular glycogen. Maximum power of this anaerobic glycolysis is 1.8–3.3 kW for average men. All prolonged efforts are powered primarily by aerobic metabolism, with higher energy demands requiring linear increases in pulmonary ventilation.

Healthy adults tolerate many hours of work at 40–50% of their maximum aerobic capacity. This translates into 400–500 W for most adult men, and with typical kinetic efficiencies of approximately 20% it amounts to 80–100 W of useful work. Untrained individuals have a metabolic scope (the multiple of their BMR) no higher than 10; for active individuals and laborers habituated to hard work, metabolic scope is higher than 15, and for elite endurance athletes (none being better than Nordic skiers and African long-distance runners) the peak aerobic power can be 25 times their BMR or more than 2 kW. Among all mammals, only canids have higher metabolic scopes. Limits of human performance are variable. In total energy terms, peak aerobic capacities are l.5–3.5 MJ for healthy adults, higher than 10 MJ for good athletes, and peak at an astonishing 45 MJ. Both glycogen and fat are the substrates of oxidative metabolism, with lipids contributing up to 70% during prolonged activities.

Overall energy conversion efficiency of aerobic metabolism is 16–20%, whereas anaerobic conversion is less efficient (10–13%). The most efficient, albeit clearly demeaning and stressful, way of using people as prime movers in preindustrial societies was to put them inside (or outside of) treadwheels (Fig. 4). Their operation deployed the body's largest back and leg muscles to do tasks ranging from lifting loads at construction sites and harbors to operating textile machinery. The most efficient use of humans as modern prime movers is in cycling. Bursts of 1 kW are possible for a few seconds while pedaling, and rates of 300–400 W can be sustained for up to 10 min.

4. *WORKING ANIMALS*

Domesticated animals, mainly bovines (cattle and water buffalo) and equines (horses, ponies, mules, and donkeys), have been the most powerful and highly versatile providers of draft power since the beginning of Old World sedentary farming. Horses and camels also provided the fastest means of land transport in preindustrial societies, and these and other species (including donkeys, yaks, and llamas) were also used as pack animals, whose caravans were particularly useful in mountainous or desert terrain.

Some Asian societies used working elephants (particularly in forestry operations), several Eurasian circumpolar cultures relied on reindeer as pack animals, and in preindustrial Europe dogs turned spits over kitchen fires or pulled small carts or wheelbarrows. In the Western world, the use of draft animals continued for several decades after the introduction of internal combustion engines and electric motors at the end of the 19th century, and tens of millions of them are still indispensable for fieldwork and transportation in low-income societies of Asia and Africa and in poorer regions of Latin America.

The power of working animals is approximately proportional to their weight (less than 200 kg for small donkeys and approximately 1 ton for heavy horses), but their actual field performance is also determined by a number of external variables. By far the most important are the animal's sex, age, health, and experience; soil and terrain conditions; and the kind of harness used. Only heavier animals could handle such tasks as deep plowing or extracting deep tree roots, but lighter animals were superior in muddy and dandy soils or on steep paths. Moreover, mechanical considerations favored smaller draft animals: Their line of pull is lower, and a more acute angle between this line and the direction of traction will produce a more elongated parallelogram of vectors and hence yield a greater efficiency of work.

Sustainable pulls are equal to approximately 15% of body weight for horses and only 10% for other species. Working speeds range mostly from 0.6–0.8 m/s for oxen to 0.9–1.1 m/s for horses. Thus, depending on their body mass, these draft animals can work at sustained rates of 300–500 and 500–800 W, respectively (Fig. 5). Assuming sustained human exertion at 75 W, the power of a working equine or bovine animal is typically eight times, and usually not less than six times, greater. Actual draft needed for specific fieldwork varies mainly with the soil type.

Deep plowing requires drafts of 120–170 kg, with a range of 80–120 kg for shallow plowing, heavy harrowing, and mowing. Cereal harvesting with a mechanical reaper and binder requires approximately 200 kg. Small horse (350 kg) or a good ox

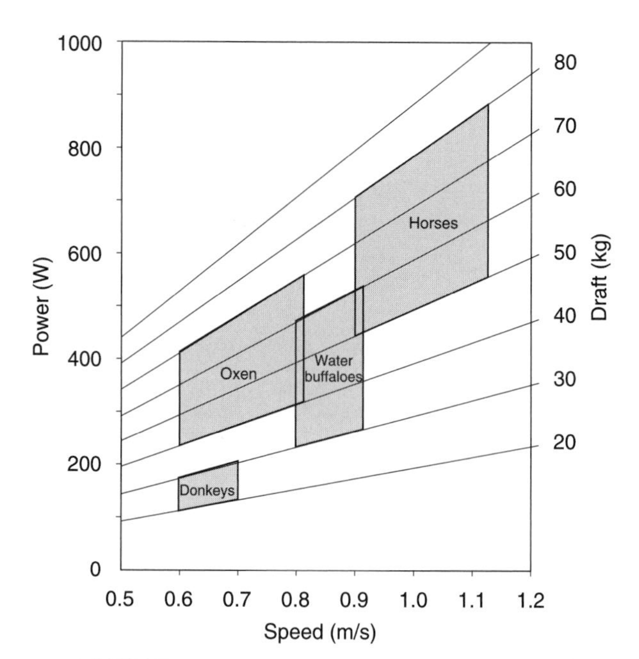

FIGURE 5 Typical performances of draft animals.

(500 kg), capable of sustained draft of approximately 50 kg, will thus not be able to adequately perform most field tasks alone. Teams of two, and later four, draft animals were common in European and American fieldwork, and efficient harnessing eventually resulted in more than 30 horses pulling the world's first combines during the 1880s.

Although the numbers of draft animals continued to increase in every country well into the 20th century, their combined power was rapidly overshadowed by new mechanical prime movers. In the United States, draft animals accounted for more than half of all prime mover power as late as 1870, but by 1950 their share was less than 0.2%. By 2000, the combined power of the world's half a billion draft animals was approximately 200 GW, less than 0.5% of the power installed in internal combustion engines.

5. BOVINES

Currently, bovines account for approximately two-thirds of the remaining draft power. More common reliance on bovines rather than on draft horses everywhere during the earlier stages of agricultural intensification, and in poorer countries and regions today, is easy to explain. Oxen (castrated males) may be difficult to train, less powerful than equally massive horses, and slower paced, but these drawbacks are compensated by their stolidity, inexpensive harness, the lack of need for shoeing, and especially the fact that they are easy to feed. The efficient digestive system of these ruminants, with microbes (bacteria, fungi, and protists) in their rumen that are able to decompose cellulose of plant roughages that other mammals are unable to digest or can use only very inefficiently, means that during slack periods they can be kept fit only on grasses (fresh or dried) and crop residues (mostly cereal straws) that other animals find difficult or impossible to consume.

Consequently, ruminants require supplementary feeding by highly digestible concentrates (grains or milling residues) only during periods of hard work. Cattle, except for their susceptibility to trypanosomiasis, also perform better in the tropics than do horses. However, water buffaloes are unmatched in wet tropics. Their body mass ranges from 250 to 700 kg, but even the most massive buffaloes are quite nimble, moving easily on narrow dividers between rice fields. Their large hoofs and flexible pastern and fetlock joints made it easier to walk in the slippery, deep mud of rice fields. They eat many grasses, can graze on aquatic plants while completely submerged, require approximately 40% less feed energy per unit of gain than do cattle, can work reliably for at least 10–15 years, and children can take care of these docile beasts.

The energetic advantages of working bovines did not end with their draft power. After the end of their working lives, the animals became a valuable source of meat and leather, and during their lifetime they provided a great deal of organic wastes that were (usually after fermentation) returned to fields in all traditional agricultures that practiced intensive cropping. However, there were some drawbacks when working with bovines. Both the neck and head yokes used to harness draft cattle had a high traction point and required a pair of equally sized animals even when one beast would suffice for lighter work (Fig. 6). The slow pace of bovine work was another major limitation, particularly where multicropping was practiced, in which speed was essential for timely harvesting and planting of a new crop. For example, a pair of oxen working in a dry soil could plow 0.2 or 0.3 ha per day, whereas a good pair of horses could plow 0.5–0.8 ha. European experience during the 19th century was that a pair of good horses easily did 25–30% more fieldwork in a day than a team of four oxen.

In addition, oxen have neither the endurance nor the longevity of horses. The working day for many oxen was 5 h, whereas horses commonly worked twice as long. Also, although both oxen and horses started working at 3 or 4 years of age, oxen lasted usually less than 10 years, whereas horses could work for 15–20 years. Poor peasants were often forced to use their only head of cattle, a family cow, as a draft animal, compromising her milk production and getting only weak traction. More powerful oxen were, and remain, the most commonly used draft bovines: At least 250 million of them were working worldwide in 2000, mostly in Asia's two most populous countries.

FIGURE 6 Head yoke commonly used for harnessing working oxen.

6. HORSES

There are five major reasons why horses are better draft animals than cattle. Unlike in cattle, frontal parts of horses' bodies are heavier than their rears (the ratio is approximately 3:2) and this gives them an advantage in inertial motion. Their unique arrangement of the suspensory ligament and a pair of tendons makes it possible to "lock" their legs without engaging any muscles and hence without any additional energy cost incurred during standing. Horses also grow generally larger and live longer than cattle, and they have greater work endurance. However, their superior strength was efficiently harnessed only with the widespread adoption of the collar harness and iron horseshoes, and their inherent power and endurance became readily available only with better feeding.

Fitted, and later also padded, collar harnesses provided a desirably low traction angle and allowed for the deployment of powerful breast and shoulder muscles without any restriction on the animal's breathing (Fig. 7). Its precursor was first documented in China in the 5th century of the CE, and an improved version spread to Europe just before the end of the first millennium. Another important innovation in harnessing was swingletrees attached to traces in order to equalize the strain resulting from uneven pulling: Their use made it possible to harness an even or odd number of animals. Iron horseshoes prevented excessive wear of hooves and they also improved traction.

However, collars and horseshoes alone could not guarantee the widespread use of horses: Only larger and better fed horses proved to be superior draft animals. The body mass, and hence power, of European draft horses began to increase only after the working stock benefited from several centuries of breeding heavy war animals needed to carry armored knights. However, these more powerful horses needed at least some cereal or legume grains, not just grasses fed to weaker animals and cattle. Production of concentrate feed required intensification of farming in order to provide yields high enough for both people and animals. Agricultural intensification was a slow process, and it first began in northwestern Europe during the late 18th century.

Even larger horses were taxed when pulling wooden ploughs, whose heavy soles, wheels, and moldboards generated enormous friction, particularly in wet soils. Moreover, the absence of a smooth, curved fitting between the share and the flat moldboard caused constant clogging by compressed soil and weeds. Iron moldboard ploughs were introduced to Europe from China during the 17th century, and cast iron shares were replaced with smooth, curved steel ploughshares by the mid-19th century during the rise of the modern steel industry. Horses then became the principal energizers of the world's largest extension of arable land that took place on the plains and prairies of North America, pampas of Argentina, and grasslands of Australia and southern Russia during the latter half of the 19th and the first half of the 20th century.

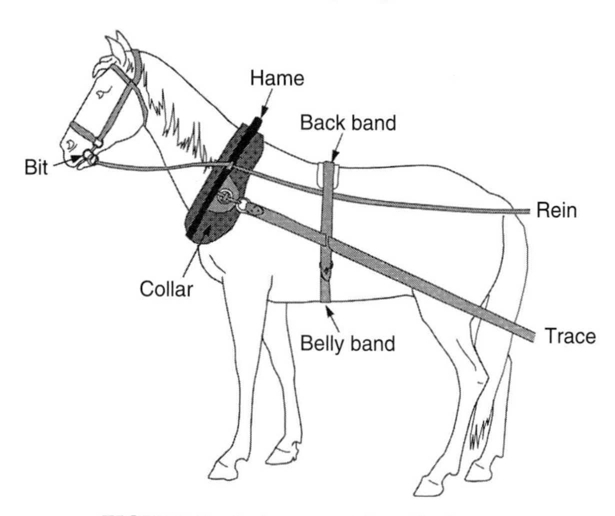

FIGURE 7 Basic parts of the collar harness.

The heaviest horse breeds—French Percherons, English Shires, and German Rheinlanders—could work briefly at rates of more than 2 kW (approximately 3 horsepower), and they could steadily deliver 800–1000 W. Feed requirements of these animals were high, but their net energy benefit was indisputable: A horse eating 4 kg of oats per day preempted cultivation of food grain that would have fed approximately 6 adults, but its power could supplant that of at least 10 strong men. Larger and better fed horses also provided essential traction during the initial stages of industrialization, powering road and canal transport, turning whims in mining, and performing tasks in food processing and in numerous manufactures. In most of these tasks, they were displaced by steam engines before the mid-19th century, but even as the railways were taking over long-distance transport, horse-drawn carts, trucks, and streetcars continued to move people and goods in all rapidly growing cities of the late 19th century. Only the diffusion of internal combustion engines and electric motors ended the use of horses as urban prime movers.

However, in fieldwork, horses remained important well into the 20th century. Obviously, their largest numbers were supported where abundant farmland made it easy to produce the requisite feed. The total number of farm horses (and mules) peaked at 21 million animals in 1919 in the United States, when at least 20% of the country's farmland was needed to cultivate their feed. Subsequent mass adoption of tractors and self-propelled field machines was inevitable: Even small tractor engines could replace at least 10 horses and needed no farmland for support. The U.S. Department of Agriculture discontinued its count of draft animals in 1960, when approximately 3 million working horses were still left on American farms. However, in China, the total number of draft horses continued to increase even after 1980, when the country began its economic modernization. Horses, as well as water buffaloes, yaks, camels, and donkeys, remain important draft and pack animals in parts of Asia and Africa.

Virtually complete substitution of animate prime movers by engines and motors is one of the key attributes of modern high-energy society, in which humans act as designers and controllers of increasingly more powerful energy flows rather than as weak prime movers, and in which a rare working animal is an object of curiosity. Nothing illustrates the wide gap between the affluent and the subsistence areas of the world better than the fact that heavy, repetitive, and dangerous human exertions (including those of millions of children) and the draft of hundreds of millions of animals continue to be indispensable prime movers in large areas of the three continents on which modernity is still a promise for tomorrow.

SEE ALSO THE FOLLOWING ARTICLES

Conservation of Energy, Overview • Cultural Evolution and Energy • Human Energetics • Transitions in Energy Use • Work, Power, and Energy • World History and Energy

Further Reading

Aiello, L. C., and Wheeler, P. (1995). The expensive-tissue hypothesis. *Curr. Anthropol.* **36**, 199–221.

Alexander, R. M. (1992). “The Human Machine.” Columbia Univ. Press, New York.

Brown, J. H., and West, G. B. (eds.). (2000). “Scaling in Biology.” Oxford Univ. Press, Oxford.

Burstall, A. F. (1968). “Simple Working Models of Historic Machines.” MIT Press, Cambridge, MA.

Durnin, J. V. G. A., and Passmore, R. (1967). “Energy, Work and Leisure.” Heinemann, London.

Food and Agriculture Organization/World Health Organization/ United Nations University (1985). “Energy and Protein Requirements.” World Health Organization, Geneva.

Hochachka, P. W. (1994). “Muscles as Molecular and Metabolic Machines.” CRC Press, Boca Raton, FL.

Kleiber, M. (1961). “The Fire of Life.” Wiley, New York.

Lacey, J. M. (1935). “A Comprehensive Treatise on Practical Mechanics.” Technical Press, London.

Landels, J. G. (1980). “Engineering in the Ancient World.” Chatto & Windus, London.

Needham, J., *et al.* (1965). “Science and Civilisation in China. Vol. 4, Part II: Physics and Physical Technology.” Cambridge Univ. Press, Cambridge, UK.

Smil, V. (1991). “General Energetics: Energy in the Biosphere and Civilization.” Wiley, New York.

Smil, V. (1994). “Energy in World History.” Westview, Boulder, CO.

Whitt, F. R., and Wilson, D. G. (1993). “Bicycling Science.” MIT Press, Cambridge, MA.

Corporate Environmental Strategy

BRUCE PIASECKI
American Hazard Control Group
Saratoga, New York, United States

1. Corporate Environmental Strategy
2. Beyond Compliance Initiatives
3. Making the Business Case for Sustainable Development
4. Relating Corporate Environmental Strategy to Corporate Change Initiatives
5. How CES Fits into a Normal Corporate Structure
6. Relating CES to Product Innovation and Strategy
7. The New Urgency for CES after Climate Change
8. Social Response Product Development
9. Case Example of CES in Practice
10. Summary

Glossary

corporate environmental strategy (CES) A functional business slogan represented by journals by the same name, owned by Elsevier, and a set of mimic journals and books utilizing the same title and concept. In addition, it is a measured business practice, starting after the tragic accident at Bhopal, India, the Exxon *Valdez* oil spill, and other environmental catastrophes at the end of the 20th century, that allows a firm to lessen its environmental liabilities, compliance costs, and product development costs as well as pursue competitive advantage in an organized and reportable fashion.

governance and management systems The means by which the board and key advisors of a firm manage its direction and performance. After the Bhopal and *Valdez* disasters, these systems increasingly involved environmental and public liability reduction efforts.

leading change, corporate change A studied area of management whereby systems are put in place to manage change in products, leadership, and corporate direction.

market-based environmentalism A term designed to suggest that leading firms based their search for environmental performance improvements on market opportunities, not just on environmental regulation and law.

social response product development Companies restructure their operations to actively shape market desires by creating new products that bridge the gap between traditional expectations of performance, safety, and environmental responsibility.

sustainable value creation A term developed after the demise of Enron and WorldCom in the early parts of the 21st century and designed to capture in a focus phrase the need for reform in how companies position themselves and report to the investment community on the paths they are taking to make their firms more sustainable.

Corporate environmental strategy (CES) involves the tools, management programs, processes, and product development choices that allow a firm to pursue competitive advantage through environmental management strategies.

1. CORPORATE ENVIRONMENTAL STRATEGY

Management scholars such as Deming and Juran spent several decades after World War II making sure that quantity and quality processes entered the plans of corporate strategy, along with the classical concerns of price, technical quality, and distribution matters. In a similar but often more diffuse manner, the proponents of corporate environmental strategy began, in the 1970s through the 1990s, to alter the standard decision models of corporate strategy to include externalities that challenged the future growth of corporations, such as new environmental regulations or irregularities in energy markets and pricing.

This massive set of changes in corporate behavior was first motivated by the need to extinguish both legal and public liabilities of the major multinationals, after the tragic accident at Bhopal and the highly visible oil spill of the Exxon *Valdez*. Later in the 21st century, CES took on a more "proactive" or "sustainable value creation" thrust, as described in this article.

Classic post-Bhopal examples of CES include the following:

1. Honda's and Toyota's development of the hybrid powertrain automobile that uses an electric battery recharged by braking and the conventional combustion engine.
2. British Petroleum's (BP's) and Shell's investments in solar power and their "beyond petroleum" campaigns of the late 20th century.
3. A range of high-efficiency appliances from Whirlpool and Electrolux that go beyond regulatory requirements.

By 2003, CES was a visible and stated corporate strategy and management goal for many multinational corporations. Some suggest that CES became a stated mantra of the Fortune 500; others believe it was spotty in the top companies across the globe. In either case, CES began to have clout in leadership councils and began to tie together energy conservation, materials savings in product design, and the classical environmental regulatory elements.

The core innovations in CES in the last decade of the 20th century occurred in the heavy industries, such as petroleum refining, where firms such as Shell and BP competed on lessening their environmental footprint to attract new partners and new investment capital.

CES also spilled rapidly into the chemical industry, where firms such as DuPont, Celanese AG, and Dow offered new products, such as water-based emulsions.

2. BEYOND COMPLIANCE INITIATIVES

Competition in domestic and international markets and legal requirements led to the rapid development of CES in the late 1990s. The approaches signified by this concept are not academic in nature and were not, on the whole, developed by academia, which in the 1990s spent much of its intellectual energy developing concepts in sustainable development, industrial ecology, and alternative energy.

Instead, CES is an evolving practitioner concept that is defined by corporate practice and function and has been implemented directly by thousands of companies since Bhopal.

There are at least three competing forms of CES. These are as follows: grudging compliance by companies with regulatory requirements while seeking to turn a profit; going beyond regulations and standards while developing products that have market value and demand but are also environmentally sound; a mix of the first two—namely, the combination of willing compliance with innovation in the market.

Whereas most leadership councils felt in practice that the third mix was the best, some firms pursued the purity of the first two strategies into the new century. Then, market pressures, after the Enron and WorldCom scandals, encouraged the attention of corporate leaders on governance reforms and Sarbanes Oxley laws upgrading management systems, corporate reporting, and the overall transparency of modern corporate reporting.

During the first decade of the 21st century, these corporate reforms became attached to CES in complex and still to be studied ways.

Many attempts were made in the 1990s and the new century to differentiate the essential or shared ingredients of CES, but most firms found that "one size does not fit all," as CES applications range widely from the mining and extractive industries (where the cost of energy is the focus), to automotive and petrochemical firms (where efficiency is often the key), to consumer-based firms (where firms such as Avon, Coca Cola, and Nike began to compete on elements of corporate social responsibility that are based on environmental strategies). As a result, what is shared is the approach, some tools and metrics, and general direction. Yet there is tremendous variance in results.

For example, CES has become a rallying term that often appears all too inclusive. In fact, sometimes CES seems, in sloppy parlance, a surrogate for classic business strategy and simply another way to make money by serving changing customer expectations. Some view it as a tool to measure emerging issues, where Intel or Hewlett Packard may use CES tools to track how their core customers are looking at new environmental risks in the metals and plastics used in computer and information technology systems.

Some corporations view CES as a tool for new product positioning, so that Rohm and Haas and Celanese AG are building emulsions plants in Asia that bypass the usual environmental legacy questions from traditional manufacturing of adhesives. Others see CES as a pathway whereby a firm can achieve near-term shareholder value through corporate responsibility.

It is precisely because of this significant variance of results that the history and the readiness of firms to use the term CES have been at best checkered since Bhopal.

3. MAKING THE BUSINESS CASE FOR SUSTAINABLE DEVELOPMENT

By 2003, a number of advantages to CES were being investigated and documented by the management consulting profession and by groups such as the Global Environmental Management Initiative (GEMI) and the Conference Board. These advantages serve to explain why a firm would invest in efforts to move it beyond basic legal compliance with all laws.

Competitive advantages gained by environmental strategies include the following:

1. Margin improvement—seeking cost savings at every stage of the product life cycle through more efficient use of labor, energy, and material resources.
2. Rapid cycle time—reducing time to market by considering environmental issues as part of the concurrent engineering process during the early stages of design.
3. Market access—developing global products that are environmentally "preferable" and meet international ecolabeling standards in Europe, Japan, and other regions.
4. Product differentiation—introducing distinctive environmental benefits, such as energy efficiency or ease of disassembly, that may sway a purchase decision.

CES is the blend of classical environmental regulatory compliance matters with these new strategic dimensions and elements. CES, at its best, is an integral input in the creation of business strategy, product selection, and a corporation's public affairs and reporting under Sarbanes Oxley, and related laws of disclosure both before the U.S. Securities and Exchange Commission and before the investment community.

From 1985 to 2003, different firms excelled in these four highlighted attributes: margin improvement, rapid cycle time, product differentiation, and market access. Intel was motivated to pursue CES to enhance its needs for rapid cycle time reductions, and marketing-based firms such as Nike and Proctor & Gamble set up CES goals for reasons of basic product differentiation. In the end, CES began as a means of improving corporate bottom lines through cost containment and new product opportunity programs.

If one were to estimate the degree of penetration of CES into corporate boardrooms, it kept its stigma as being an added cost, or what some have referred to as a "bag on the side of corporate strategy," for many years after Bhopal. Although the diligent and deliberate attempt to lessen a firm's environmental liabilities is a prime force behind CES, the relentless thirst for profit has proven a more reliable and less distracting guide in the new thinking regarding CES. The list of four key elements and goals of CES can be thought of as the turning point from the old century thinking into the new century, where the pressures for sustainable product mixes and policies became far more palpable.

A firm can achieve CES by a sequence of related efforts, including the following:

1. Make environmental management a business issue that complements the overall business strategy.
2. Change environmental communication within the company to use traditional business terms that reflect business logic and priorities.
3. Adopt metrics to measure the real costs and business benefits of the environmental management program.
4. Embed environmental management into operations, similar to the successful design-for-environmental models (here, environmental footprint reduction elements are entered into the equation at the product design stage).
5. Radically change the job descriptions and compensation of environmental managers—and line managers—to reflect the realities of doing business.

Others answered the question of how a firm would begin to move beyond compliance by pointing to the cost and administrative inconvenience of penalties and fines. The classic and more vivid example is the demise of Union Carbide, and its eventual purchase by Dow, after Bhopal, a tragic accident that killed over 2700 people in Bhopal, India. Lessons from Union Carbide for practitioners of improving environmental management systems, under the rubric of CES, include this summary of Union Carbide's major changes after Bhopal:

1. An intricate classification scheme of standards, regulations, and audits, which allows executives more

insight into the firm's liabilities, priorities, and high-risk areas. Technical staff functions of regulatory compliance can then be linked to the legal department's liability containment measures and strategies.

2. A large database of audit findings statistically amenable to analysis. From this corporate base, executives from different divisions compete for capital resources on more exact terms. Its size and consistency make it a management tool of considerable consequence.

3. A systematic monitoring and control program, more reliable than the traditional environmental, health, and safety systems. For instance, Union Carbide's worldwide computer program prioritizes risk at its 1200 facilities worldwide and assists senior management with its growth or phase-down plans.

4. New management tools, based on the audit system, provide quality assurance and executive reassurance that the company's environmental practices do not violate the law or threaten human health or the environment. Nationwide, more than 200 executives have served jail time for environmental crimes. This executive assurance program—developed for Union Carbide by A. D. Little—helps senior management feel "in touch" with the firm's liabilities, compliance issues, and long-term viability.

Other recognized drivers in corporate behavior for corporate environmental strategy included Superfund Authorization and Reform Act (SARA) Title III, which put public pressure on companies by requiring disclosures on chemical stockpiles, warehouses, and such, and the Chemical Manufacturers Association's responsible care initiative, which outlined a code of behavior.

The disruption caused by the Union Carbide accident spurred the above regulations and conduct codes throughout the industry. The public lost faith in the industry and the negative opinion polls increased. Corporate environmental strategy is, in part, a response to the negative public image and public concerns about environmental, health, and safety performance.

4. RELATING CORPORATE ENVIRONMENTAL STRATEGY TO CORPORATE CHANGE INITIATIVES

There are several options that, ultimately, must be integrated when attempting to realize a change in corporate culture: the legal remedy, which is necessary to spur on the slow movers; the money remedy, which does not mean blindly throwing money at an environmental problem, but instead taking a more strategic approach; and finally, the regulatory remedy, which moves beyond simple regulatory compliance to a level of strategy that allows a company to stay ahead.

Some feel that without law there would be no corporate environmentalism, and yet there are those that stood at the forefront of an issue, pushing it to become law. These leaders from industry, academia, government, and the environmental movement are leading CES more than the law is, and for a company to find success through CES, it must stay well ahead with those leaders, not lagging behind with the laws. The law is functional and offers a glimpse at a company's next step, but this is not strategy.

The money remedy does not translate, as some might believe, to throwing money away, the way some companies have done. It is about information management, including audit information, environmental liability and insurance information, product selection, announcement, and positioning information, stockholder and stakeholder information, new and international market information, regulatory and government liaison information, technical information, legal information, and strategic and financial information. Without the full integration of these nine types of information, the amount of money invested in them will not matter.

Finally, there is the regulatory remedy, which can be seen in companies who have begun mass disclosure measures to create corporate transparency, well beyond simple regulatory compliance. Strategy here allows leaders to stay ahead of environmental issues and avoid the old way of constantly cleaning up past mistakes; both company executives and the public are able to see the company's liabilities and assets more easily. This also creates an environment of corporate trust for the public.

These three remedies together can create a strategic advantage for companies willing to make the effort of changing "business as usual." Table I outlines a list of five elements of success in shifting corporate culture.

Table I was based on a survey by the author's firm of over 150 companies of the Fortune 500. Whereas the five corporate change elements noted above are quite inclusive, and extend well beyond what lawyers or engineers in a firm would normally consider environmental management, they do capture the "strategic" or management and leadership council feel of CES in practice.

Without a doubt, the actual benefits of CES are most often kept quite confidential, as they are close enough to the heart of a corporation's business strategy, when they are functional and not based on public relations, that they are kept close to the chest of the leading executives in a firm. This makes it difficult for scholars in the field to do much in their diagnostics except case studies and retrospective summaries of these elements of success.

5. HOW CES FITS INTO A NORMAL CORPORATE STRUCTURE

After Bhopal and the Exxon *Valdez* spill, it became clear that environmental management was not sufficiently integrated into normal business functions, reviews, and organizations. A 1991 survey, seen in Fig. 1, revealed some interested gaps. What the study indicated, as it was replicated for different firms over subsequent years, is that there is no natural or ultimate home for CES in an organization. Depending on the firm, it might be centered in the legal, risk reduction, or regulatory divisions of a firm. But consistently, most firms feel there is a wall between these CES functions and the financial disciplines that run a firm. As a result, CES functions are often perceived as being of secondary or tertiary importance in many firms.

TABLE I

Elements of Success in Shifting Corporate Culture

1. Role for strategic recognition:
Including mission statements and performance reviews
2. Role for changes in staff:
Title and function
3. Adaptations in conventional management tools:
Such as audits or return-on-investment equations
4. Increasing role for external strategies:
Such as community relations and voluntary disclosure
5. The unexpected role for strategic alliances

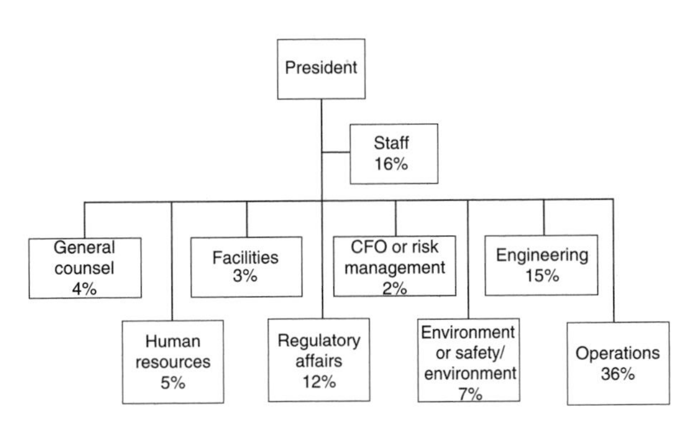

FIGURE 1 Reporting requirements in environmental management. From Coopers and Lybrant (1991), "Environmental Management in the 1990s: A Snapshot of the Profession." Reprinted from Piasecki (1995), with permission from John Wiley & Sons, NY.

6. RELATING CES TO PRODUCT INNOVATION AND STRATEGY

In the pursuit of superior cars, electronic products, and computing, several leading multinational corporations began in the last quarter of the 20th century to tie CES to their product strategy. Leaders such as Toyota and Honda are classic examples in the automotive industry, as are Shell and BP in the petroleum sector.

Those firms that had successfully integrated CES into their normal business functions by the new century shared a common set of attributes. These attributes are described below as "generic" elements of allowing CES to be elevated within a corporate setting, as they often depend on organizational dynamics or a unique set of executive interests and needs. These organizational elements of CES involve "action plans" or goals, such as the following:

1. A role for strategic recognition including mission statements and performance reviews.
2. A role for changes in staff, title, and functions where chief environmental officers and their staff become more visible and respected in corporate organizations.
3. Adaptations in conventional management tools, such as audits or return-on-investment equations.
4. An increasing role for external strategies, such as community relations and voluntary disclosure efforts.
5. In some of the more curious and interesting developments, CES also involved the unexpected role of strategic alliances, where past enemies meet.

Some individuals, such as Steve Percy, the former CEO and Chairman of BP America, have claimed that BP's purchase of Amaco was motivated by CES, since BP saw and appreciated the large natural gas holdings of Amaco and their solar research portfolio as an opportunity to advance BP's need for progress in CES as a reputation enhancer.

Other examples of these strategic alliances were found in the new century in utilities such as Exelon, chemical giants such as Dow, Celanese, and DuPont, and select automakers such as Toyota and Honda.

CES includes knowing when to play by the conventional corporate and financial rules and when to change the game. In a small set of successful corporations, this involved the integration of "new rules" with the ever-changing game of strategy, product positioning, and product reformulation. Evidence of such "radical" or "incremental" changes became hotly debated with GEMI companies, for example. In the new century, GEMI extended its sunset clause and allowed its continued existence in order to provide a range of companies with these stated requirements in their membership rules.

GEMI companies must reach toward corporate environmental strategy:

- By sharpening existing tools and measurement systems to help managers refine and enhance their environmental performance;
- By popularizing to the press and the public the belief that corporations have a role in the search for environmental excellence, both domestically and internationally;
- By explaining how knowledge from certain business sectors can be shared, coordinated, and then adapted across different industrial sectors and nations;
- By asserting boldly that environmental strategy involves more than regulatory compliance.

7. THE NEW URGENCY FOR CES AFTER CLIMATE CHANGE

Although it is significant that many of the above four imperatives are qualitative, not quantitative, such new efforts have begun to add an elevated level of urgency to corporate environmental strategy.

Whereas the Bhopal and *Valdez* tragedies may have precipitated the birth of CES, global environmental concerns, such as water shortages in China, the energy crisis and electricity grid problems in the northeastern United States, and especially the complexities surrounding the economic impacts and costs of climate change, have begun to add a new turn-of-the-century urgency to the debates over CES.

By 2003, CES began to help firms link price, technical performance, and social response product development. In other words, conventional business tools focusing on competitive price and product strategies, in some instances, began to notice the appeal of shaping or reformulating products relative to external energy and environmental threats, such as limitations in water supply, spiking energy costs, or climate change and other carbon-related challenges. Figure 2 sums up these evolving links between CES and product development.

When multinational firms such as Toyota, Honda, Electrolux, and Whirlpool began making claims before the investment community that their efforts in corporate environmental strategy were paying off, the new emphasis on linking CES to product development and differentiation began to be in vogue.

8. SOCIAL RESPONSE PRODUCT DEVELOPMENT

Historians of business know that vogue is an element of consequence, as evidenced by the passion for Six Sigma tools that General Electric brought to firms as different as 3M, Dow, and Celanese AG. This new product-based vogue for CES enabled a more rapid embrace of CES in select companies, as illustrated in Fig. 3.

"Social response" (i.e., incorporating the sustainability dimension) product development means that instead of manufacturing products solely in response

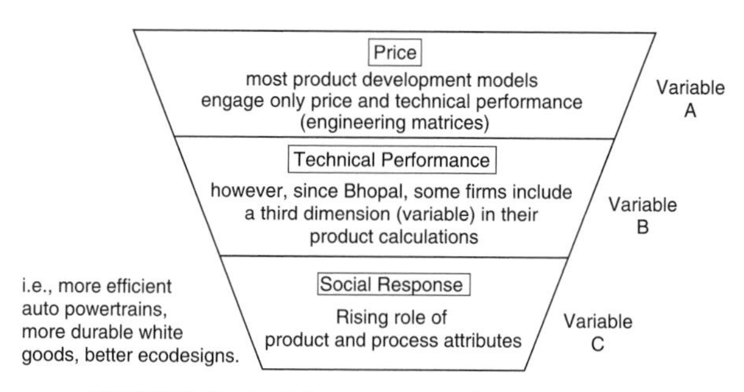

FIGURE 2 Social response product development.

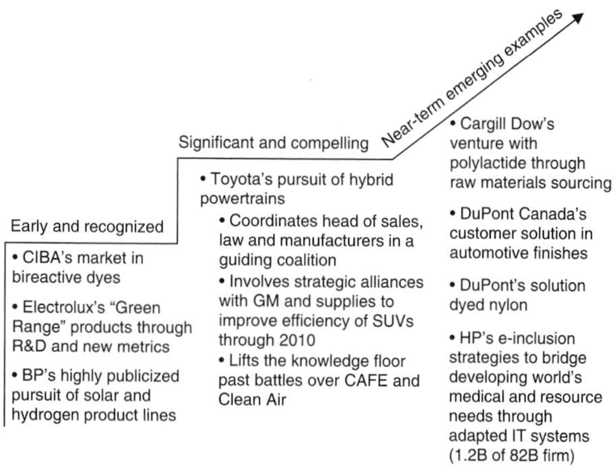

FIGURE 3 Defining social response product development through select examples.

to consumer preferences as discerned and measured by price and technical performance through market research methodologies, companies restructure their operations to actively shape market desires by creating new products such as Toyota's and Honda's hybrid cars.

Although such efforts are logically supported by new manufacturing, quality control, and resource management techniques, what is new is that they bridge the gap between traditional expectations of performance, safety, and environmental responsibility. Such expanded product development models allow a more sustainable value creation at present and through the coming decades. Social response product development also allows product differentiation in mature product lines in a fashion that attracts new capital and new forms of partnering. Examples here include Toyota, BP, Shell, and DuPont. A sustainable value approach, based on this kind of product differentiation and development, supports all other actions in the company. Examples of social response approaches include the following:

- From 1996 to 2001, Anheuser-Busch established a multiyear annual "Pledge and Promise" staff and stakeholder retreat to award eco-innovation teams. The retreat includes presidents and chief executive officers of key Anheuser-Busch stakeholder groups as external judges to select top five sustainable value leaders each year.
- From 1998 to 2002, British Petroleum established a multiyear technical task force to identify, select, and finance sustainable value product lines and plans.

Social response product development shapes how executives approach their task of accomplishing seemingly disparate goals. In chemicals, in high-performance plastics, and in new markets for cars and computers, social response product development delivers both resource efficiency and customer-demanded performance. As an example of social response product development being used as a CES strategy, Toyota's search for a superior automobile is explored.

9. *CASE EXAMPLE OF CES IN PRACTICE*

In this new century, there is considerable pressure on the top six automakers to reduce their environmental footprint. The automaker that wins the race to build and sell the superior car will shape consumer preferences, thereby boosting sales and profits. The winning firm will fashion a corporate strategy that drives automobile emissions to near zero while simultaneously providing high levels of performance, safety, and comfort.

The model of social response product development as a CES strategy is built on four dimensions, like the structure of a house: knowledge depth—the ever-growing basement; knowledge durability—the expanding sides; knowledge dependence—the ceiling of use; knowledge floors—the ground base. The breakthrough at Toyota was recognizing that new technologies mean that a new knowledge floor can be laid, one that makes it possible to meet all consumer expectations of performance, safety, and environment (fuel efficiency, emissions, resource efficiency, alternative fuels) without the tradeoffs previously required.

Toyota and Honda have used a new form of CES to change the rules of the game by leveraging their knowledge depth to introduce superior new cars, such as the four-door, five-passenger Prius. The Prius achieves fuel efficiency rates of 55 miles per gallon (mpg), twice the regulated corporate average fleet requirement. The move to hybrid cars—those that can run on multiple fuels, most commonly, gasoline and electricity—will be understood as truly exceptional when the realization sinks in how absolutely ordinary it will make the hundreds of millions of cars now littering the current knowledge floor. This floor—containing everything from gas-guzzling sport utility vehicles and commercial trucks to Europe's newest smart cars that are still dependent on the 100-year-old combustion engine—has been permanently superseded.

Some firsts of the Toyota Prius include the following:

1. The Prius has been designed to be recycled. Although this is most relevant to Japanese and European consumers, it does highlight how forward-thinking Toyota has become. The amount of waste avoided through this redesign is staggering, separating Toyota from General Motors, Ford, and DaimlerChrysler.
2. At 2003 gas prices, the average North American would save at least $1/day, or a total exceeding $360, at the pump by driving a Prius rather than a typical compact sedan.
3. Consumers will enjoy the convenience of driving 600 miles between fill-ups.
4. The American version of the Prius has a battery pack with 20% more power than the Japanese equivalent, yet weighs 20% less.

5. The Prius has fundamentally changed the debate over clean cars, almost eliminating the shift to all-electric cars.

6. The Prius has already become legend. It can be found at the New York Museum of Modern Art's exhibit "Different Roads." In June 1999, Toyota became the first automaker in the world to receive the United Nations Global 500 Award, not only because of the Prius, but also because of all of Toyota's environmental commitments to reduce dependence on landfills.

What is noteworthy about this case example is that much of this innovation occurred in anticipation of changes in corporate average fuel economy standards and related air pollution legislation across the globe. This represents a new form of CES as it involves aggressive product change before the regulations take final shape, as a new means of competitive advantage.

10. SUMMARY

Although it would be historically inaccurate to claim that the above-described efforts of BP, Toyota, and DuPont represent a norm, they do signify a way in which corporate environmental strategists are succeeding in the early part of the 21st century.

More and more corporate boards, after the Bhopal and *Valdez* accidents and following a second wave of reforms starting after Sarbanes Oxley required a more accurate disclosure of a firm's lasting liabilities and product assets, are taking this series of steps in linking CES to their larger corporate and social responsibility obligations and plans. Their eventual success will prove to be based on linking their mission and goals, originally compelled by a desire to meet regulations and limit a firm's liability, with products.

As in most corporations, where the product is king, this move may prove the means by which CES reinvigorates itself in the new century.

Acknowledgments

Celeste Richie, of American Hazard Control Group, is gratefully acknowledged for performing the research for this article.

SEE ALSO THE FOLLOWING ARTICLES

Business Cycles and Energy Prices • Clean Air Markets • Consumption, Energy, and the Environment • Ecological Risk Assessment Applied to Energy Development • Energy Services Industry • Risk Analysis Applied to Energy Systems • Service and Commerce Sector, Energy Use in • Subsidies to Energy Industries • Trade in Energy and Energy Services

Further Reading

American Hazard Control Group. (2002–2003). "Corporate Strategy Today," Executive Monograph Series. Lexis Nexis, Saratoga Springs, NY.

Baram, M. (1991). "Managing Chemical Risks." Tufts Center for Environmental Management, Bedford, MA.

Elsevier. (1994–2003). *Corp. Environ. Strategy.*

Epstein, M. (1995). "Measuring Corporate Environmental Performance." The IMA Foundation for Applied Research, New York.

Friedman, F. (1997). "Practical Guide to Environmental Management," 7th ed., pp. 68–70. Environmental Law Institute, Washington, DC.

Hoffman, A. (2000). "Competitive Environmental Strategy." Island Press, Washington, DC.

Kotter, J. (1996). "Leading Change." Harvard Business School Press, Boston, MA.

Laszlo, C. (2003). "The Sustainable Company: How to Create Lasting Value through Social and Environmental Performance." Island Press, Washington, DC.

Piasecki, B. (1995). "Corporate Environmental Strategy: The Avalanche of Change since Bhopal." Wiley, New York.

Piasecki, B., and Asmus, P. (1990). "In Search of Environmental Excellence: Moving beyond Blame." Simon & Schuster, New York.

Renfro, W. (1993). "Issues Management in Strategic Planning." Quorum Books, Westport, CT.

Cost–Benefit Analysis Applied to Energy

J. PETER CLINCH
University College Dublin
Dublin, Ireland

1. Definition of Cost–Benefit Analysis Applied to Energy
2. History of Cost–Benefit Analysis
3. Rationale for Cost–Benefit Analysis Applied to Energy
4. Cost–Benefit Methodology
5. Valuation of Energy Externalities
6. Further Issues in the Application of Cost–Benefit Analysis to Energy

Glossary

benefits transfer A process of taking environmental valuation estimates from existing studies and applying them outside of the site context in which they were originally undertaken.

discounting Adjusting a cost or benefit in the future so that it may be compared to costs and benefits in the present.

energy externality A cost or benefit that arises from an energy project but is not borne or appropriated by those involved in the energy project (e.g., pollution from energy generation).

environmental valuation The process of placing monetary values on environmental effects.

revealed preference methods Environmental valuation methods that use data from actual behavior to derive values for environmental assets.

sensitivity analysis A presentation of various possible outcomes of a future project or policy under conditions of uncertainty.

shadow pricing The process of adjusting a distorted market price to better reflect the true opportunity cost to society of using that resource.

stated preference methods Direct environmental valuation methods that ask people in a survey to place a value on an environmental asset.

total economic value A categorization of the various elements of value of a resource, project, or policy.

value of statistical life The translation of the value of a small reduction in the risk of death into a value for a project that would involve the equivalent of one life saved or lost.

welfare criterion The criterion in a cost–benefit analysis to assess whether a project or policy benefits society.

welfare measure A measurement tool to assess the strength of consumers' preferences for or against a project or policy in order to assess its costs and benefits.

Cost–benefit analysis applied to energy is the appraisal of all the costs and all the benefits of an energy project, policy, or activity, whether marketed or not, to whomsoever accruing, both present and future, insofar as possible in a common unit of account.

1. DEFINITION OF COST–BENEFIT ANALYSIS APPLIED TO ENERGY

Cost–benefit analysis (CBA), also known as benefit–cost analysis, is rooted in applied welfare economics. It is a way of organizing and analyzing data as an aid to thinking. It provides a set of procedures for comparing benefits and costs and is traditionally associated with government intervention and with the evaluation of government action and government projects. The underlying rationale for CBA is rational choice; that is, a rational agent will weigh the costs and benefits of any proposed activity and will only undertake the activity if the benefits exceed the costs.

A confusion that sometimes exists relates to the difference between social cost–benefit analysis and cost–benefit analysis. However, to an economist, social impacts and environmental impacts are economic impacts, so any analysis purporting to be a CBA must address social and environmental impacts.

In this way, CBA differs from financial analysis in that it assesses the costs and benefits to the whole of society rather than to a portion of society. For example, a financial analysis of a proposed new power plant by a private energy supplier will not take into account the costs imposed on wider society (such as environmental impacts), whereas a CBA, by definition, will do so. In addition, there are various other subtle but important differences between financial analysis and CBA related to the fact that market prices do not always reflect the true value of goods and services to society. This provides the basis for a technique known as shadow pricing, which adjusts distorted market prices such that they may be used to value costs and benefits appropriately.

As the previous definition implies, a CBA must include all the costs and all the benefits to whomsoever accruing. Thus, the definition instructs that whether the costs or benefits are marketed or not makes no difference. In other words, even if some of the costs or benefits are not exchanged in markets at all and so do not have a price, they may still be of value. This provides the basis for the inclusion of environmental and wider social impacts in CBAs of energy projects and policies. The common unit of account in CBA is money. Therefore, insofar as is possible and reasonable, all impacts, including environmental and social, should be evaluated in monetary terms so that they can be compared with one another. This provides the basis for a technique known as nonmarket or environmental valuation—that is, placing monetary values on goods and services not exchanged in markets, such as environmental improvements and damages.

In many cases, the costs and benefits of a project or policy will arise at different times. This is the case in relation to most energy activities, where the consequences of actions may not be known until far into the future, particularly in the case of environmental impacts such as global warming. This provides the basis for a technique known as discounting, which allows future costs and benefits to be compared with those in the present.

CBA has been used *inter alia* to estimate the environmental costs of power generation, the net social benefit of different power supply options, the net benefit of the use of renewable energy sources (e.g., solar, wind, and hydroelectric power), the net benefits of alterative modes of transport, the benefits of energy efficiency programs in the commercial and domestic sectors, as well as for policy analysis of energy efficiency regulations such as energy rating and labeling.

2. *HISTORY OF COST–BENEFIT ANALYSIS*

Detailed histories of CBA can be found elsewhere, but the essence is as follows. In 1808, U.S. Secretary of the Treasury Albert Gallatin gave what is believed to be the first recommendation for the employment of CBA in public decision making when he suggested that the costs and benefits of water-related projects should be compared. In 1936, the U.S. Flood Control Act deemed flood control projects desirable if "the benefits to whomsoever they may accrue are in excess of the costs." In 1950, after a debate on the foundations of CBA in welfare economics, the U.S. Federal Inter-Agency River Basin Committee published a guide to CBA known as the "Green Book." From then on, there was considerable interest in CBA and the development of cost–benefit rules, with seminal contributions by a number of authors, particularly during the 1970s. Although the distortions created by government intervention in the 1970s provided much room for the application of CBA, the technique went through a lull in the early 1980s when structural reform replaced piecemeal intervention. However, CBA has made a comeback since then, mainly as a result of increased public awareness of environmental degradation and the associated demand that this cost be accounted for in analyses of projects and policies. However, for a number of years, the economist's view that environmental impacts could be assessed in monetary terms had little impact on policy, with many environmentalists having a particular problem with this concept. For example, in the U.S. Clean Air Act (1970), it was forbidden to use CBA, and the U.S. Clean Water Act (1972) set the objective of the elimination of all discharges into navigable waters. However, in the 1980s, it became a requirement that the U.S. Environmental Protection Agency perform CBAs on all environmental directives to assess their economic efficiency and significant resources were devoted to researching cost–benefit techniques and methods of placing monetary values on the environment. Although it was not until the Single European Act was introduced in 1987 that a legal basis for European Union (EU) environmental policy was provided, the nature of environmental policy reflected the origins of the EU being an economic one. EU policy that "the potential benefits and costs of action or lack of action must be taken into account in the preparation of policy on the environment" provides an obvious platform for the use of CBA.

3. RATIONALE FOR COST–BENEFIT ANALYSIS APPLIED TO ENERGY

The need for CBA applied to energy results from the existence of market failure. If there were no market failure and all energy suppliers were privately owned, there would be no need for CBAs of energy projects and policies. If market signals were appropriate, those activities and projects that are in the best interests of society would be implemented automatically and financial analyses by firms would simply replace any need for CBAs. However, the market does not work perfectly. Market failure necessitates CBA of energy projects, policies, and activities as a result of the existence of externalities, the associated need for environmental policy instruments, state ownership of energy generation and supply, assessment of sustainability, and "green" national accounting.

3.1 Energy Externalities

One of the most significant areas of environmental concern today relates to the environmental impacts of energy generation from fossil fuels, particularly acidification precursors, greenhouse gases, and localized air quality impacts. Much of these emissions result from power generation and the transport sector. These environmental impacts occur at local, national, regional, and global levels.

The environment can be defined as those parts of our physical and psychological endowment that we somehow share, which are "open access." An economic analysis shows that it is precisely this shared nature of many environmental endowments that threatens their quality and character. In a market system, goods are allocated by the price mechanism. In a free market, the price of a good is determined by the demand for, and supply of, the good. Price is therefore a reflection of society's willingness to pay for, or the valuation it places on, the good in question. However, the shared nature of many environmental assets, such as air, the atmosphere, and water resources, means that they are not owned and so do not have a price. When goods are seen as free they are overused. Thus, the market fails to protect environmental assets adequately.

This failure of the market can be best demonstrated with a simple example: If a power plant produces SO_2 emissions resulting in acid rain, which causes forest damage and damage to buildings, in a free market these costs will not be reflected in the cost of power generation. Thus, the price of the energy will not reflect the true cost to society of generating it. In this case, more energy will be generated and more of the environmental asset depleted than is in the best interest of society. When a cost is imposed on those other than the person or business that produces the cost, this is known as an external cost. Of course, many energy projects will reduce such externalities, such as through renewable energy sources and improved heating and insulation systems.

When assessing the social efficiency of energy activities using CBA, it is necessary to include these externalities and other wider societal impacts. However, because by definition such costs and benefits are not traded in markets, there is no price to reflect their value. This provides the basis for the tools of environmental valuation.

3.2 Calibration of Environmental Policy Instruments

Unless some mechanism is found to internalize these externalities, they will cause a misuse of society's scarce resources. This provides the basis for government intervention in the market. The notion of externalities was addressed by economists as far back as the 1930s. Pollution was seen as the consequence of an absence of prices for scarce environmental resources and so the prescription for solving the problem was to introduce a set of surrogate prices using environmental (Pigouvian) taxes and charges. A large literature has developed on these so-called market-based instruments. One of the more controversial concepts in environmental economics is that there is an optimal level of pollution. Pollution is usually the by-product of a useful activity, such as the production of goods and services; in other words, it is an input into the production process. Reducing pollution usually involves a cost in terms of output forgone and/or the cost of the technical process of abatement. Therefore, there is a trade-off between the environmental improvement and the costs of pollution reduction. The optimal level of pollution occurs where the benefit of another unit of abatement (the marginal benefit) equals the (marginal) cost, with the (marginal) cost being higher after that point. In the case of an extremely hazardous pollutant, the optimal level of pollution will be zero or close to it, but in most cases the optimal level of pollution will be positive. CBA provides the tools to assess these trade-offs. In reality, few economic instruments are actually based on such valuations. Research on so-called target setting using environmental valuation is still in a preliminary stage. However, it could be used to assess the optimal level of an energy tax.

3.3 State Ownership of Energy Generation and Supply/State Support of Energy Efficiency Initiatives

In some countries, energy supply companies are owned by the state. In some cases, the original rationale for such ownership was that no private company would bear the large fixed costs of developing an energy supply system. In addition, it is sometimes suggested that state companies can perform as natural monopolies and thereby exploit economies of scale while producing at minimum average cost and subsidizing energy for certain sections of society. CBA has a role to play in assessing the merits of such state initiatives and is particularly relevant when assessing the merits of government intervention to improve energy efficiency, for example, by subsidizing wind power or a domestic energy efficiency program.

3.4 Sustainability Assessment and Green National Accounting

CBA is useful in assessing sustainability; that is, what is the value of our resource stock, how is it being depleted, and is man-made capital that is replacing it of equal value? Green accounting adjusts traditional measures of economic performance such as gross national product to reflect environmental damage and the depletion of resources. In relation to energy, CBA may be used to assess the social costs and benefits of depleting energy resources such as fossil fuels.

4. *COST–BENEFIT METHODOLOGY*

4.1 Welfare Criterion

In using CBA to assess the costs and benefits of a project or policy to society, it is first necessary to define the social efficiency criterion. The most frequently used criterion in microeconomic theory is Pareto optimality. Using this criterion, a project would be deemed to pass a CBA if it made at least one person better off without making anybody else worse off. For an investment to pass the Pareto optimality test, compensation must be paid to those bearing the costs so as to leave them indifferent between their welfare *ex post* and *ex ante*. Most policies and investments leave somebody worse off and thus most would fail to meet the Pareto criterion. Frequently in CBA, the Kaldor–Hicks test is used. This test embodies the potential compensation principle. Kaldor categorized an action as efficient if those who gain from the action could compensate the losers and still be better off. Hicks framed the question slightly differently and labeled the action efficient if the potential losers from the action could compensate the potential beneficiaries of the action for not initiating the action. In both cases, the action is an improvement regardless of whether the compensation is actually paid. Thus, if the CBA shows the project to result in a welfare improvement under the Kaldor–Hicks test, this is essentially a potential Pareto improvement since there is the potential, in theory, to pay compensation to the losers such that nobody would be worse off. This will be the case as long as the benefits of the project outweigh the costs (i.e., the benefit:cost ratio is greater than 1).

Whether or not this potential payment is enough to show that a project is good for society has been the matter of much debate. Some authors have suggested that despite the shortcomings of the Kaldor–Hicks criterion, it is often useful to measure the effects of a change in the total value of output independently of the distribution of output; however, in judging the social efficiency of a project, equity aspects should also be considered. A two-part test may be used in which a project or policy is put to the Kaldor–Hicks test and then tested to determine if it improves (or at least does not disimprove) the distribution of income. Approaches that consider questions of equity therefore allow the possibility of acceptance of a project or policy that returns a negative figure for the sum of individual welfare changes if the gains from income distribution outweigh these losses. However, when considering equity the difficulty is in the assignment of weights to individual welfare changes. Consideration of equity issues and distributional consequences are particularly relevant in the case of projects that affect the price of energy and/or health. Energy price increases may be regressive unless poverty-proofing measures are put in place because energy is a necessity and therefore requires a larger proportion of the budget of less well-off people. On the other hand, improvements in domestic energy efficiency may bring significant health and comfort benefits to the elderly, who in some countries have been shown to be more likely to live in energy-inefficient homes and therefore more likely to be subject to winter morbidity and mortality.

4.2 Welfare Measures

In order to assess the costs and benefits of a project or policy, it is necessary to examine the strength of consumers' preferences for or against the project. The standard approach is to measure the benefit an individual obtains from a project or policy by the maximum willingness to pay on his or her part for such a benefit or the maximum willingness to accept compensation for not receiving such a benefit. Similarly, a cost can be measured by the maximum willingness to pay on the part of an individual to avoid such a cost or the maximum willingness to accept compensation for such a cost being imposed on the individual. Such welfare measures may be approximated using shadow pricing and/or direct valuation methods from surveys.

4.3 Temporal Considerations

Account must be taken of the fact that costs and benefits arise at different times. Thus, future costs and benefits must be discounted so that they can be compared with present costs and benefits. The rate of return for an investment can be strongly affected by the discount rate if there is a long time span involved. Therefore, the choice of discount rate is very important. In some estimates of damages from global warming resulting from the burning of fossil fuels, a small change in the assumed discount rate will have significant implications for climate change policy. However, despite the enormous literature on discounting, it is a difficult task to calculate the social rate of discounting. The discount rate is measured by some combination of a social rate of time preference and a measure of the opportunity cost of capital. Economists have been unable to provide a definitive figure for the social rate of discount, and so a rate tends to be chosen rather than measured.

In an application of CBA to any project or policy, a range of discount rates should be used. Often, for a government project, a published government test discount rate is used for making policy recommendations. These vary by jurisdiction. For example, UK treasury guidelines for the opportunity cost of public investment take the rate of return to marginal private sector investment as 6%. For pricing output sold commercially, public sector agencies are expected to achieve a required average rate of return of 8% to be consistent with the private sector. In the United States, recommended rates for investment projects vary. The Office of Management and Budget recommends a rate of 10%, whereas the General Accounting Office prefers a rate of 2.5% based on government bond yields. The World Bank has traditionally used an 8% rate in project analysis.

4.4 Uncertainty and Risk

In *ex ante* CBAs, investments with long time horizons are subject to uncertainty, particularly in relation to future prices. In addition, when evaluating costs and benefits *ex ante*, there may be considerable uncertainty regarding the magnitude of various parameters. This is particularly the case in relation to potential environmental impacts and their valuation. In cases in which there is a high degree of uncertainty in relation to the coefficients being used to give the monetary value of outputs of a project, lower bound estimates are sometimes used because, if the results are to be the basis for the recommendation of a change in policy, it is thought by some authors that it is better for the benefits to be underestimated rather than overestimated. It is important to note that lower bound results do not necessarily imply the lowest possible results. Rather, they denote the lower limits of a range of reasonable estimates.

Sensitivity analysis, whereby results are calculated for various scenarios, should be used in all CBAs. In addition, confidence intervals should be presented where possible. A general rule is that the more results the better while giving guidance and reasons for the particular estimates chosen by the author(s) for the purposes of recommendations based on the CBA results. Risk assessment may be incorporated by including probability estimates for various scenarios.

4.5 Shadow Prices and Wages and Community Stability

Shadow prices must be used when there are distortions within an economy that render certain prices invalid as measures of the social opportunity costs and benefits of using resources in the project in question. In the real world, market prices may not reflect the social value of goods or services because of distortions in the market, such as monopoly power (which inflates price), indirect taxes and subsidies (unless they are correcting some other distortion), and unemployment. Thus, when calculating each value of a project, it is necessary to assess the appropriateness of using the market value and, where necessary, these values should be adjusted for market distortions. This can be a problem in energy projects for which a particular piece of technology may be supplied by one company and/or estimates must be made regarding how price would decline should such a piece of technology be widely adopted, for example, in a national domestic energy efficiency retrofit.

The approach to valuing labor is worth detailing. In reports purporting to be CBAs, labor or "employment" is sometimes treated as a benefit rather than a cost. Labor is an input into a project and, therefore, a cost even though it may have some additional benefits such as community stability, as discussed later. However, that cost may be less than the market wage would suggest. If labor markets clear, the shadow price of labor will equal the market wage. Any increase in employment in one sector of the economy will merely reduce the availability of labor to another sector. This is likely to be the case when the jobs are highly skilled and/or there is a shortage of such skills. However, for a country with a high unemployment rate, it could be argued that an increase in the demand for labor in one sector of the economy will not necessarily displace a job in another sector; that is, there may be employment additionality whereby if the new job is filled by a person who was previously unemployed, a very low cost (not zero, unless it is considered that the unemployed's spare time is totally unproductive) in terms of output forgone is imposed on society. In this case, the shadow price of labor would be close to zero. It has been the practice of some government CBAs to assume a shadow price of labor of zero. However, this is against the international practice of setting the shadow price of labor at most a fraction below the market wage. For example, in Canada, the shadow wage is usually 95% of the market wage, and in the United Kingdom they are usually considered equal. In many energy projects, specialized skills are required and so a shadow wage close to the market wage will usually be appropriate. However, in some household energy efficiency improvement schemes, long-term unemployed people have been trained how to fit various sorts of insulation, for example. In this example, a case could be made for using a lower shadow wage rate.

A related value that should be considered is whether an energy project may contribute to community stability. For example, would the provision of a new power plant in a rural area provide jobs such that depopulation of that area would halt? This value is not the value of the job per se. Rather, it is the added value of a job being in a particular place. These values are difficult to assess but are real. For example, it is often observed that employees will be willing to accept lower salaries to live in their hometown.

4.6 Valuing Changes in Energy Use

When calculating the benefits of reductions in energy use, the price of energy may be used to value such savings. However, it may be necessary to shadow price such savings. If the state supplies energy, an assessment must be made of whether there is any form of subsidy involved; if so, this should be removed. Where there is monopoly power in the energy market, either locally or, for example, through the monopolistic behavior of the Organization of Petroleum Exporting Countries, the price of energy is unlikely to reflect the true cost to society of its generation.

Energy supply is considered so essential that a premium may be placed on its supply from domestic sources if this reduces the risk of supply interruption. Most countries have assessed the importance of such supply, and this information may be used to assess any premium that should be applied. If the increase in supply that would be generated by the energy project under consideration is marginal to the total supply in a country, it is unlikely that such a premium would be significant.

In an *ex ante* analysis, a sensitivity analysis should be undertaken for alternative scenarios regarding the possible future changes in energy prices. These are difficult to predict because there are a number of offsetting effects. Possible deflationary pressures include the development of more energy-efficient machinery and technology, improved availability (and therefore reduced prices) of non-fossil fuel energy sources and systems, and any change in the regulation of the energy sector. Possible inflationary pressures include rapid economic growth and a consequent increased demand for energy, the exercising of market power in oil production, future scarcity of fossil fuels, and the introduction of carbon taxes and/or tradable permits in light of the Kyoto Protocol. However, if the price of energy is inflated by a carbon tax/emissions-trading system set to reflect the external costs of energy use, it is important to avoid summing external costs estimated by the techniques described later with the energy savings using this price because if the price already reflects the external effects, it would be inappropriate to include them a second time.

4.7 Externalities

Shadow prices adjust for one form of market failure. However, a CBA must also take into account the potential existence of externalities and public goods, as explained previously. The consideration of such costs and benefits is extremely important in CBAs because all of the costs and all of the benefits to the whole of society are being considered. The difficulty is that these external costs and benefits are not exchanged within markets, so there is no price to reflect their value. For this reason, it is necessary to use nonmarket or environmental valuation approaches to place monetary values on these externalities.

5. *VALUATION OF ENERGY EXTERNALITIES*

Most energy projects have some impact on the environment that must be factored into a CBA. For example, a new coal-fired power plant will have implications for climate change, localized air pollution, and the production of acidification precursors. A cost–benefit analyst assessing the merits of a renewable energy project or an energy efficiency program will wish to factor in the value of reductions in such environmental effects.

5.1 Economic Value Concepts

A useful framework for assessing the effects of environmental impacts on humans is known as the total economic value approach. This expands the traditional concepts of economic value to include option value and existence value. In essence, an environmental asset is considered to have both use and nonuse value.

Use value can be broken down into actual use value and option value. Suppose we are estimating the value of damage caused to woodland from acid rain resulting from power generation. Actual use value is derived from the actual use of the asset and is composed of direct use value, such as timber from the woodland, and indirect use value, such as the value that would be yielded to inhabitants downstream should a forest reduce flooding.

Option value can be broken down into a number of components: the value individuals place on the preservation of an asset so that they may have the option of using it in the future (value of future options for the individual) for recreational purposes, for example; the value they attach to preserving the asset because it can be used by others (vicarious value); and the value they attach to preserving the asset so that future generations have the option of using it (bequest value). Thus, total use value is defined as

$$\begin{aligned}\text{Total use value} &= \text{direct use value}\\ &+ \text{indirect use value}\\ &+ \text{option value},\end{aligned}$$

where

$$\begin{aligned}\text{Option value} &= \text{value of future options for the individual}\\ &+ \text{vicarious value}\\ &+ \text{bequest value}.\end{aligned}$$

Nonuse values are existence or passive-use values. Existence values reflect the benefit to individuals of the existence of an environmental asset, although the individuals do not actually use the asset. These values may exist due to sympathy for animals and plants and/or some human approximation of the intrinsic value of nature (i.e., this value reflects a sense of stewardship on the part of humans toward the environment). Existence value is classed as a nonuse value such that total economic value (TEV) is defined as

$$\begin{aligned}\text{TEV} &= \text{actual use value} + \text{option value}\\ &+ \text{existence value}.\end{aligned}$$

It is important to note that the TEV approach attempts to encapsulate all aspects of a resource that enter into human's utility functions (i.e., it is anthropocentric). Thus, it does not attempt to find an intrinsic value for the environment, where intrinsic means the value that resides in something that is unrelated to humans. However, it does attempt to include the intrinsic value that humans bestow on nature. There is no agreement on the exact breakdown of the values in the TEV framework; however, the important point in CBA is to account for all these values as best as possible.

5.2 Environmental Valuation Methods

There are two broad categories of nonmarket valuation methods: revealed preference methods and stated preference methods.

5.2.1 Revealed Preference Methods

Revealed preference studies use data from actual behavior to derive values for environmental assets. Suppose the value of a marketed good depends in part on an environmental good. These methods try to isolate the value of the environmental good from the overall value of the marketed good. Such techniques include the following:

Hedonic pricing method: The value of a house depends, in part, on its surrounding environment. This method examines the determinants of the house's value and then isolates the effect that the environmental feature (e.g., a nearby park) has on the house. This is used as an estimate of the value of the environmental asset in question.

Travel cost method: The value of the park can be estimated by the willingness to pay to attend the park in terms of costs of travel and entrance to the park.

Production function approaches: These value a decrease (increase) in environmental quality as a result of some activity by the loss of (addition to) the value of production due to that activity. For example, the cost of acid rain resulting from pollution from fossil fuel-based power plants might be measured by the lost output of timber from forests (dose–response functions), the cost of restoring the damage to buildings (replacement cost), or the cost of defensive expenditures.

The production function approaches may seem more applicable to estimating externalities of energy generation, but the hedonic pricing method and travel cost method may be used to estimate the value of an amenity such as a forest park. This value could then be used to calculate the cost of damage to the park from air pollution.

5.2.2 Stated Preference Methods

Direct valuation methods ask people in a survey to place a value on the environmental asset in question by rating (contingent valuation), ranking (contingent ranking), or choosing trade-offs between various policy alternatives (choice procedures). These direct methods via questionnaires are the only approaches that can estimate existence value. The best known and most controversial of these methods is contingent valuation.

The contingent valuation method (CVM) collects information on preferences by asking households how much they are willing to pay for some change in the provision of an environmental good or the minimum compensation they would require if the change were not carried out. Since the mid- to late 1980 s, this method has become the most widely used approach for valuing public goods and there has been an explosion in the number of CV studies.

The CVM approach is to choose a representative sample, choose an interview technique (face-to-face, telephone, or mail), and then construct a questionnaire containing:

- A detailed description of the good being valued and the hypothetical circumstance in which it is made available to the respondent.
- Questions that elicit the respondents' willingness to pay for the good being valued: open ended (How much would you pay?), bidding (Will you pay x, if not, will you pay $x-y$), payment card (Which of these amounts would you choose?), and binary choice (If it costs x, would you pay it? Yes/no).
- Questions about the respondents' characteristics (e.g., age, education, and income), their preferences relevant to the good being valued (e.g., how concerned or interested are they in the environment), and their use of the good.

The mean or median willingness to pay from the sample of households surveyed can be multiplied up to the population level to give a total valuation.

5.3 Valuing Environmental Damage and Improvements from Energy Projects

The most significant environmental effects of energy projects that are included in CBAs tend to be the contribution of such projects to global warming from changes in CO_2 emissions and acid rain resulting from the production of NO_x and SO_2. In addition, many energy projects have implications for the production of particulate matter that may have consequences for human health and visibility.

The impacts of climate change are as follows: storm and flood damage (increased in some regions, such as the Caribbean and Pacific, but not others), increased mortality and morbidity, effects on small islands (sea levels rise due to heating water and melting ice caps), increasing salinity of estuaries, damage to freshwater aquifers, and effects on agriculture and forestry. It is estimated that a substantial portion of the existing forested area of the world will undergo major changes.

There are three approaches to placing a monetary value on the benefits of greenhouse gas reductions. The damage-avoided approach values 1 ton of carbon not emitted by the cost of the damage that would have been done by global warming in the event that it had been emitted. The offset approach measures the value of not emitting 1 ton of carbon using one method compared to the next cheapest alternative method. The avoided cost of compliance approach measures the 1 ton of saved carbon by the avoided cost of compliance with a global/regional CO_2 emissions reduction agreement. A variety of estimates have been produced regarding the benefits of reducing CO_2, including studies by the United Nations Intergovernmental Panel for Climate Change.

Several major research projects have endeavored to place monetary values on all the environmental effects related to energy. Two EU-funded projects, one on the external costs of power generation (ExternE) and the other on European environmental priorities, are very useful sources of information on such values.

The ExternE project used an "impact pathway" methodology. Under this methodology, the fuel cycle, or life cycle of the fossil fuel used for generation, is first divided into stages, such as extraction, transport, combustion, and waste disposal. The environmental burdens imposed at each stage and the likely impact of these burdens on the environment and the public are isolated. The impacts are then valued using a variety of economic valuation methods, such as contingent valuation, direct cost of damage, or cost of mitigation. These costs are then normalized to give a cost per unit energy output and a value per ton of a pollutant reduced.

The "priorities" project endeavors to derive values for various environmental externalities suitable for use in a European context. The impacts of SO_2 on acute mortality, morbidity, materials, and the fertilization effect are all included; however, some impacts are omitted, such as forest damage, adverse effects on ecosystems and cultural heritage, and the dampening effect that SO_2 has on global warming. The main areas of uncertainty relate to the value of mortality. NO_x emissions are dealt with under two headings, acidification and tropospheric ozone. The fertilization effect of NO_x is examined under acidification. For damage to health and crops, the effects of NO_x on tropospheric ozone are used. Again, the bias is toward underestimation because damage caused to forests, biodiversity, non-crop vegetation, and materials is excluded.

Care should be taken when using unit damage values such as those in the studies mentioned previously. First, the confidence intervals are large (and should always be quoted in a CBA). In addition, there is confusion about which emissions cause which adverse effects. For this reason, it is very easy to double count when using estimates derived from different studies. For example, in some studies negative health effects are included in SO_2 damage estimates, whereas in others the most significant health effects result from particulate matter emissions. Finally, the damage caused by 1 ton of most air pollutants (excluding CO_2) will vary depending on the geographical origin of the emission. However, most estimates are derived from an exercise known as benefits transfer.

5.4 Benefits Transfer

To reduce costs and where there is no alternative procedure, it is standard practice in CBAs to apply existing monetary valuation studies outside of the site context in which they were originally undertaken. This is known as benefits transfer. There are a number of rules when using this method, the most obvious being that the contexts of the studies should be very similar. In addition, appropriate statistical techniques should be used, including meta-analyses of best practice studies; the studies should have similar population characteristics; and property rights should be similarly distributed. The method is most accurate when the impacts are not affected by local characteristics (e.g., income and population size). The most comprehensive studies of energy externalities referred to previously involve significant benefits transfer.

5.5 Valuing Mortality and Morbidity

Placing monetary values on life and death is perhaps the most controversial area of economics. There are two major approaches for carrying out this task. The first values life directly through livelihood. This is known as the human capital method and is the oldest and most easily applied. It equates the value of life of an individual as the present value of future lost output of the individual, which is usually estimated using income. However, it is important to note the substantial drawbacks of this approach. Most obvious, it values livelihood rather than life per se. It is quite improbable that there would be a 1:1 relationship between individuals' earnings and their demand for risk reduction. In other words, the human capital approach makes no reference to how humans value risks to their own lives. One of the most glaring flaws of this approach, however, is the fact that it places a zero value on the retired population and those who do not "produce" marketed output (e.g., those who work in the home).

The appropriate way of valuing reductions in mortality is not to value a life per se but rather to value people's willingness to pay for a reduction in the risk of mortality. These data derive primarily from one of three study formats: the wage risk model (also known as the hedonic wage model), the avertive behavior method, and the contingent valuation method. Although space does not permit a detailed explanation of these methods, the general approach is to value a small reduction in risk and then translate this into the equivalent of a life saved. This is the value of a statistical life (VSL).

There has been controversy regarding this approach in relation to the estimation of the costs of power generation. The VSL estimates would be expected to be considerably lower for a developing country because the willingness of people to pay to reduce risk of death is partially dependent on income. Thus, the VSL used for a developing country (e.g., for deaths from coal mining) would be expected to be lower than the VSL used for a developed area (e.g., for deaths from air emissions in Europe by the burning of such coal). It is notable that the ExternE project steered clear of controversy by using the same VSL for each area.

The value of morbidity may be assessed using a cost of illness approach. In the case of air pollution from energy generation, this would be the sum of the cost, both public and private, of hospital stays and drugs for the treatment of associated illnesses. More positively, the health benefits of improving domestic energy efficiency would be the associated avoided costs of illnesses. Other costs avoided by wider society include the loss of productivity from those who are ill and unable to work. The avoided cost to the individual beyond drugs and hospitalization direct costs (if any) is more difficult to calculate. One approach is to estimate individuals' willingness to pay to avoid "restricted-activity days." The priorities study referred to previously contains such estimates.

6. *FURTHER ISSUES IN THE APPLICATION OF COST–BENEFIT ANALYSIS TO ENERGY*

When undertaking or evaluating a CBA of an energy project, there are a number of additional issues that must be considered. The tool is not without its controversies. These arise because of its use of money as the common unit of measurement. First, assessing welfare changes using willingness to pay and/or willingness to accept is perfectly valid in theory if one believes that income is distributed optimally. However, care must be taken when the population(s) whose welfare is being assessed has a high income variance. Poor people will be willing to pay less, all else being equal, simply because of their budget constraint. In addition, because the values and market prices used in a CBA of an energy project or policy are in part a product of the existing income distribution, if that distribution changes, prices will change, and so will market and nonmarket values. Second, distributional effects tend to be ignored in CBAs and yet energy policies often have very different consequences for different socioeconomic groups. Because energy is a necessity, energy efficiency schemes may benefit those on lower incomes more than average. However, without poverty-proofing measures, energy taxes will comprise a much higher proportion of their income than average. Third, not all support attempts to place monetary values on environmental assets, and the value of the statistical life concept is particularly controversial. Some find it unpalatable to consider that the environment might be traded with other goods and services let alone "lives." In addition, as in all public policy or human actions, the environmental values used are by definition anthropocentric.

Finally, the environmental valuation methods are limited in their ability to assess values. Many environmental outcomes assessed by natural scientists are subject to large confidence intervals even before economists endeavor to value them. Also, the valuation methods vary in their ability to assess environmental values. Economists have identified a number of criteria by which environmental valuation methods should be evaluated: validity (the degree to which it measures the total economic value of a change in the environment), reliability (the extent to which the variability in nonmarket values is due to random sources), applicability (assesses where the techniques are best applied), and cost (the practicality and viability of carrying out the valuation study). In general, revealed preference methods (e.g., productivity loss and replacement cost) are widely accepted valuation methods that provide valid and often reliable estimates of use values. However, they are limited in that they only provide estimates of use values, and it is often difficult to find suitable and reliable links between market goods and environmental amenities. The estimated values are often sensitive with respect to modeling assumptions, reducing their reliability in such cases. In general, however, revealed preference methods are applicable for valuation of policies that primarily affect use values and can provide very reliable results. Stated preference methods such as contingent valuation are capable of measuring both use and nonuse (existence/passive-use) values. They are applicable to a larger array of environmental issues than are revealed preference methods, and they are often the only applicable methods for benefit estimation. However, they do have a number of limitations and may be subject to a number of biases.

There are some simple rules for using cost–benefit techniques. Simpler environmental valuation methods should be used where possible (e.g., production function approaches) while recognizing these are primarily of interest when the policy changes will mainly influence use values. A high degree of reliability in stated preference methods may be achieved by adhering to sound principles for survey and questionnaire design and if the method is applied to suitable environmental goods. The methods work best for familiar environmental goods and for salient environmental issues. The methods are likely to be least reliable where the environmental good being valued is complex (e.g., in the case of biodiversity).

Finally, in CBAs of energy initiatives, it is important to note that the estimates of values are made at the margin. The plausibility of the techniques described depends in part on the fact that most things remain the same, with the implicit assumption that there is no paradigm shift in values, power, income, and the like during the time period of assessment.

SEE ALSO THE FOLLOWING ARTICLES

Decomposition Analysis Applied to Energy • Ecological Risk Assessment Applied to Energy Development • Energy Development on Public Land in the United States • Environmental Change and Energy • Equity and Distribution in Energy Policy • External Costs of Energy • Green Accounting and Energy • Value Theory and Energy

Further Reading

Clinch, J. P., and Healy, J. D. (2001). Cost–benefit analysis of domestic energy efficiency. *Energy Policy* **29**(2), 113–124.

Commission of the European Communities—DG Environment (2003). European environmental priorities Web site: An integrated economic and environmental assessment. Available at europa.eu.int/comm/environment. ExternE Web site: http://externe.jrc.es.

Hanley, N., and Spash, C. (1993). "Cost–Benefit Analysis and the Environment." Elgar, Aldershot, UK.

Layard, R., and Glaister, S. (eds.). "Cost–Benefit Analysis," 2nd ed. Cambridge Univ. Press, Cambridge, UK.

Zerbe, R., and Dively, D. (1994). "Benefit Cost Analysis in Theory and Practice." HarperCollins, New York.

Crude Oil Releases to the Environment: Natural Fate and Remediation Options

ROGER C. PRINCE
ExxonMobil Research and Engineering Company
Annandale, New Jersey, United States

RICHARD R. LESSARD
ExxonMobil Research and Engineering Company
Fairfax, Virginia, United States

1. Introduction
2. Composition of Crude Oils and Refined Products
3. Releases to the Environment
4. Natural Fate of Oil in the Environment
5. Responses to Spills

Glossary

API (American Petroleum Institute) gravity A scale describing the density of an oil. Oils with gravities >40° are known as light oils, and those with gravities <17° are said to be heavy.

asphaltenes Large complex molecules in crude oils and heavy fuels, often with molecular weights in the many thousands, composed of carbon, hydrogen, oxygen, and other heteroatoms. They are responsible for most of the color of crude and heavy oils and typically are not "oily" to the touch.

bioremediation Stimulating the natural process of oil biodegradation. To date, this has been restricted to the addition of fertilizers to alleviate nutrient limitations for these organisms. Adding bacteria has not been shown to be necessary.

booms Floating devices for corralling, or sometimes absorbing, oil on water. There are many different designs for different situations, including stationary booms in rivers, towed booms for collecting oil, and fireproof booms for holding oil for combustion.

Bunker C Fuel oil used to power ships' engines. It is a very heavy, viscous fuel derived from the heaviest hydrocarbon compounds that remain after processing whole crude oil.

crude oil Natural product that is the remains of biomass that was typically alive 100 million years ago. It is a complex mixture of hydrocarbons, resins, and asphaltenes.

dispersants Concentrated solutions of surfactants that can be sprayed on floating oil to cause it to disperse into the water column, where it is naturally biodegraded.

hydrocarbons Compounds composed of only carbon and hydrocarbon; these are the most abundant molecules in most crude oils and refined fuels. They are a diverse mixture of linear, branched, and cyclic saturated compounds with 1–80 carbon atoms, together with aromatic compounds, most commonly with more than one ring and with multiple alkyl substituents.

natural attenuation The natural processes that eventually remove hydrocarbons from the atmosphere.

resins Moderately large complex molecules in crude oils and heavy fuels, often with molecular weights in the thousands, composed of carbon, hydrogen, oxygen, and other heteroatoms.

skimmers Mechanical devices to remove floating oil from water. There are many designs, including moving belts, weirs, mops, and vacuum pumps.

Crude oils and other hydrocarbons have been a part of the biosphere for millennia, so it is no surprise that a ubiquitous, diverse population of hydrocarbon-consuming microorganisms are able to exploit this resource. These organisms can be recruited to help respond to human-associated oil spills, although sometimes only after physical treatments to remove bulk oil, such as mechanical pick-up or combustion:

- If oil is floating at sea, the addition of dispersants to dramatically increase the surface area available for microbial attack will stimulate biodegradation and also inhibit the formation of troublesome tar balls and minimize the amount of oil that reaches the shore.

- If oil is on a shoreline, the addition of fertilizers to provide necessary ancillary nutrients for the degrading microorganisms will stimulate oil biodegradation.
- If oil is in surface soil, adding fertilizers and tilling the soil will provide both the nutrients and the air needed to stimulate biodegradation.
- If soluble hydrocarbons are in an underground water plume, adding a source of oxygen, whether as a gas or a soluble or slow-release peroxides, will stimulate aerobic biodegradation.

Together, these approaches exemplify modern environmentally responsible remediation activities: working with natural processes to stimulate them without causing any additional adverse environmental impact.

1. INTRODUCTION

Crude oils have been a part of the biosphere for millennia, and human use extends back thousands of years. Early uses included hafting stone axes to handles, use as an embalming agent, and use as a medical nostrum. In the Bible, Genesis implies that bitumen was used as the mortar for the Tower of Babel. It also seems likely that several religions started near natural gas seeps, either as eternal flames or as sources of hallucinogenic vapors. However, these were very minor uses, and it is only in the past century and a half that oil has come to play a truly central role in our lives. Terrestrial seeps were the first locations to be drilled when oil production began in earnest in the 19th century, such as the 1859 Drake well in Pennsylvania. Today, oil is recovered from all over the world, except the Antarctic, and from great depths, including coastal margins such as offshore West Africa and South America. The scale of oil production is staggering; global use is on the order of 10^{12} U.S. gallons per year (3.25×10^9 tonnes or 3.8×10^{12} liters/year), and much of it is transported thousands of miles before it is used. For example, the United States imported 350,000 tonnes of oil per day from the Middle East alone in 1999. Unfortunately, despite the best efforts of the oil industry, some (a relatively small fraction) of this material has been spilled at production and refinery facilities and from pipelines and vessels. Such spills generate enormous public pressure for speedy clean-up, but the obvious approach of simply sucking up the oil from the environment is rarely simple or complete. This article addresses the natural fate of oil in the environment and how understanding the underlying processes allows intervention to safely stimulate them and thereby minimize the environmental impact of spills.

2. COMPOSITION OF CRUDE OILS AND REFINED PRODUCTS

2.1 Origins

Crude oils are natural products, the result of the burial and alteration of biomass from the distant past. The alteration processes are known as diagenesis and catagenesis. The initial process of diagenesis typically occurs at temperatures at which microbes partially degrade the biomass and results in dehydration, condensation, cyclization, and polymerization of the biomass. Subsequent burial under more sediments, and thus at higher temperatures and pressures, allows catagenesis to complete the transformation of the biomass to fossil fuel by thermal cracking and decarboxylation.

It is generally accepted that most petroleum reserves are derived from aquatic algae, albeit sometimes with some terrestrial material, whereas the great coal reserves of the world arose from deposits of higher plants, typically in nonmarine environments. The generation of oil and gas invariably occurs at significant depth and under great pressure, which forcibly expels the oil and gas from the initial source rock. Commercially valuable reservoirs are formed if the migrating oil becomes trapped in a geological structure so that the oil and gas can accumulate. Most petroleum reservoirs are found in sandstones, siltstones, and carbonates with porosities of 5–30%. The traps are never completely full of oil, and there is always some water, usually containing substantial amounts of inorganic salts (brine), underlying the oil. Typical commercial oils originated from biomass that was produced on the order of 100 million years ago, although the oldest commercially valuable oils are from Ordovician biomass (486 million years ago), whereas others are as young as the late Tertiary (a few million years ago).

2.2 Hydrocarbons

Crude oils and refined products are very complex mixtures of molecules, both hydrocarbons and compounds containing other elements, such as sulfur, oxygen, and nitrogen. The hydrocarbons, containing only hydrogen and carbon, are the best studied and understood, and they can be divided into broad classes based on their overall chemical properties. The hydrogen to carbon ratio of crude oils is between 1.5 and 2.0; the organic molecules are thus principally saturated molecules (i.e., the predominant form of carbon is $-CH_2-$). The convention in the oil industry is to call linear alkanes paraffins and cyclic alkanes naphthenes. There are also significant amounts of aromatic carbon in all crude oils and polar molecules containing the heteroatoms oxygen, nitrogen, and sulfur. These latter molecules can be fractionated by column chromatography, and the fractions have a variety of names, including resins and asphaltenes. In their classic 1984 book "Petroleum Formation and Occurrence," Tissot and Welte state that the average composition of 527 crude oils is 58.2% saturates, 28.6% aromatics, and 14.2% polar compounds, although the absolute values vary widely in different oils. On average, there is rough parity between paraffins, naphthenes, and aromatics.

The paraffins include the linear alkanes from methane to waxes with up to 80 carbons. Linear alkanes typically make up 10–20% of a crude oil, although their content can range from essentially undetectable to as high as 35% depending on source and reservoir conditions. There are also branched alkanes, usually most abundant in the C_6–C_8 range. Two branched alkanes found in almost all crude oils, pristane ($C_{19}H_{40}$) and phytane ($C_{20}H_{42}$), are molecular relics of the phytol chains of chlorophylls and perhaps other biomolecules in the original biomass that gave rise to the oil.

The naphthenes include simple compounds such as cyclopentane, cyclohexane, and decalin, together with their alkylated congeners. Tissot and Welte state that the average composition of the naphthene fraction of 299 crude oils is 54.9% one- and two-ring naphthenes, 20.4% tricyclic naphthenes, and 24.0% tetra- and pentacyclic naphthenes. These latter molecules are molecular relics of membrane components of the original biomass, and they are very useful fingerprints of individual oils.

The aromatic molecules in crude oils range from benzene to three- and four-ring structures, with traces of even larger ring systems. Because of the separation procedures used in the characterization of crude oils, any molecule containing at least one aromatic ring is included in the "aromatic" fraction, regardless of the presence of saturated rings and alkyl substituents. Crude oils typically contain aromatic molecules with up to four aromatic rings; one series contains just six-membered rings and their alkylated derivatives—benzene (one ring), naphthalene (two rings), phenanthrene (three rings), chrysene (four rings), etc. Another series includes one five-membered ring in addition to the six-membered ones—fluorene (three rings), fluoranthene (four rings), etc. Because of the operational definition based on chromatography, sulfur aromatic heterocycles, such as thiophenes, benzothiophenes, and dibenzothiophenes, are included in the aromatic category. Indoles and carbazoles, usually the most abundant nitrogen-containing species, and the less abundant basic nitrogen species such as quinolines are also included in the aromatic category. Alkylated aromatic species are usually more abundant than their parent compounds, with mono-, di-, and trimethyl derivatives usually being most abundant. Nevertheless, the median aromatic structure probably has one or two methyl substituents together with a long-chain alkyl substituent.

2.3 Nonhydrocarbons

The polar molecules of crude oils are the most difficult to characterize because they are often unamenable to gas chromatography, the usual method of choice for the molecular characterization of petroleum. All are thought to contain heteroatoms, such as nitrogen, oxygen, and/or sulfur, and the category includes the porphyrins (usually nickel or vanadium species) and naphthenic acids. Some of these molecules have molecular weights in the thousands or higher, and many are suspended in the oil rather than dissolved in it. The polar fraction of the oil contains the majority of the color centers in crude oil, and in isolation these materials are difficult to distinguish from recent biological residues, such as humic and fulvic acids.

2.4 Oil Classification

Oils are classified by several criteria, but among the most important is the specific gravity. The oil industry uses a unit known as API (American Petroleum Institute) gravity, which is defined as $[141.5/(\text{specific gravity})] - 131.5$ and expressed in degrees. Thus, water has an API gravity of 10°, and denser fluids have lower API gravities. Less dense fluids (e.g., most hydrocarbons) have API gravities $>10°$. For convenience, oils with API gravities $>40°$ are said to be light oils, whereas those with API gravities $<17°$ are said to be heavy (Table I). Light

TABLE I

Conversion Table for a Medium Crude Oil

1 tonne	308 U.S. gallons 7.33 barrels 858 liters

oils have higher proportions of small molecules; heavy oils are rich in larger molecules. Viscosity is inversely proportional to API gravity, but it is also dependent on the physical state of the polar compounds and longer alkanes in the oil, and it is highly dependent on temperature.

2.5 Refined Products

Crude oil is transported in the largest volumes, both in pipelines and in tankers, but refined products are also transported long distances, particularly in pipelines. Refining starts with distillation, and the simplest distinction of the various refined products can be related to this process. The most volatile liquid product is aviation gasoline, followed by automobile gasoline, jet fuels, diesel and heating oils, and the heavy oils that are used for fueling ships and some electrical generation. All ships contain large volumes of heavy fuel oil, often called Bunker C, that is barely liquid at ambient temperatures and must be kept warm to be pumped into engines.

Most of the molecules in gasoline have 4–10 carbons (Table II), most in diesel fuel have 9–20, and heavy fuel oils typically have very few molecules with less than 15 carbon atoms except for some added as a diluent to achieve the appropriate viscosity. It is important to recognize that fuels are not sold by their composition but by their properties, such as octane, cetane, or viscosity, and there are many chemical mixtures that meet these requirements. The chemical composition of fuels with the same name can thus be very different, even though all meet their specifications. The lightest products (i.e., those with the lowest boiling points, such as gasolines and diesels) are almost entirely hydrocarbons, whereas the heavy fuel oils are enriched in the polar constituents such as asphaltenes; this is reflected in the color of the products.

TABLE II

Number of Carbon Atoms in the Majority of Compounds in Refined Products

Product	Approximate carbon range
Aviation gasoline	C_4–C_8
Regular gasoline	C_4–C_{10}
JP4 jet fuel	C_5–C_{11}
JP8 jet fuel	C_8–C_{15}
Kerosene	C_9–C_{16}
Diesel No. 1	C_9–C_{17}
Diesel No. 2	C_9–C_{20}
Bunker C	C_8–C_{10}, C_{15}–C_{50}
Hydraulic oil	C_{18}–C_{36}
Automatic transmission fluid	C_{19}–C_{36}

TABLE III

National Research Council's Estimates of Annual Input of Oil to the Sea

Source	North America, tonnes	World, tonnes
Seeps	160,000	600,000
Consumers	90,000	480,000
Spills	12,100	160,000
Extraction	3,000	38,000
Total	260,000	1,300,000

3. RELEASES TO THE ENVIRONMENT

Crude oil has been released to the natural environment from seeps for millennia and, as mentioned previously, these have proved useful sources for humankind. Some seeps are relatively small, and the oil affects only a very small area. This was true for the seeps that gave rise to "perpetual flames" in Iraq and bitumen reserves in eastern North America. However, some are vast and have given rise to huge asphalt lakes and tarpits. The La Brea tarpits in Los Angeles are well-known, not least for the impressive bones from tigers that unwittingly became entrapped, and the large asphalt lakes in Trinidad (140 acres) were once exploited to deliver asphalt for paving in Newark, New Jersey. Clearly, a major determinant of the environmental impact is the scale of the release. Table III shows 2002 estimates of the scale of oil releases into the sea both in the United States and throughout the world.

3.1 Natural Seeps

Natural seeps are found all over the world, and the scale of their releases is quite large. In 2002, the National Research Council estimated that the total annual input of petroleum into the sea from all sources is approximately 1.3 million tonnes, with almost 50% coming from natural seeps. For example, natural seeps in the Gulf of Mexico release an estimated 140,000 tonnes of oil per year into the sea, those offshore Southern California release 20,000 tonnes per year, and those offshore Alaska release 40,000 tonnes.

3.2 Municipal Runoff

The National Research Council estimates that releases due to the consumption of crude oil, principally municipal runoff but also including spills from nontanker ships, recreational uses, and atmospheric deposition, are the second largest input of oil into the world's oceans, far dwarfing that from catastrophic tanker spills. Again, the sheer scale of oil use is staggering; it is estimated that just the input of lubricating oil from two-stroke engine use in the U.S. coastal seas is 5300 tonnes per year.

3.3 Tanker Spills

Almost half of all oil moves by seagoing tanker at some stage between production and use, and spills from these vessels attract by far the largest amount of attention from the media. As discussed previously, their total input is relatively small compared to other inputs, and a recent estimate is that more than 99.98% of the oil in transit reaches its destination safely. The issue, of course, is that when they do occur, tanker spills are very localized, and they can have substantial localized environmental impacts that call for a cleanup response. The industry has responded to public concerns, and tanker spills are fortunately becoming rarer; for the period 1990–1999, spillage from vessels in U.S. waters was less than one-third that released in 1980–1989. Similar improvements have occurred worldwide, although there is always need for improvement.

3.4 Pipeline Spills

The Minerals Management Service estimates that there are 21,000 miles of undersea oil pipelines installed on the U.S. continental shelf delivering oil from wellheads to collection points and refineries. Spills are rare but are estimated to be approximately 1,700 tonnes per year, mainly in the Gulf of Mexico. To put this in perspective, the largest spill in the Gulf occurred in 1979, when the IXTOC-1 well "blew out" approximately 80 km off Ciudad del Carmen on the Mexican coast. The blowout lasted 9 months and spilled 476,000 tonnes of crude oil before it could be controlled.

Total U.S. liquid petroleum pipelines include a total of 114,000 miles of crude oil pipelines and 86,500 miles of pipeline for refined products, which together transport two-thirds of all U.S. petroleum. Pipeline leaks, often caused by excavation errors, lead to annual spills of approximately 19,500 tonnes, and again there has been a substantial improvement in safety during the past few decades. Because of the physical barriers around most pipeline spills, they are usually contained, and most spilled oil is collected with pumps. Unfortunately, this is not always the case; in 1997, an underwater pipeline released 940 tonnes of south Louisiana crude oil into Lake Barre in Louisiana. The spill affected 1750 ha with a light sheen, but only 0.1 ha of marsh died back. Much of the oil evaporated, some pooled oil at the edges of some marshes was recovered, and the lightly oiled areas appeared to be clean within 2 months.

3.5 Wars, Environmental Terrorism, and Vandalism

The past decade has seen a resurgence of oil spills associated with warfare, terrorism, and vandalism. Many ships were sunk in World War II, and some still contain substantial amounts of oil; an example is the USS Mississinewa, which was sunk in Ulithi Lagoon in Micronesia in 1944 and recently located. It is estimated to contain 3300–10,000 tonnes of heavy fuel oil, and this is beginning to leak from the decaying vessel. By far the largest oils spills in history were those deliberately released by Iraqi forces retreating from Kuwait in 1991. At least 10 million barrels of oil were deliberately poured into the Arabian Gulf, and 700 oil wells were ignited or otherwise destroyed. Vast lakes of oil covering approximately 49 km^2 were formed, and most remain after a decade. The volumes are so huge that the totals remain unknown.

Smaller wars also cause the release of substantial volumes of oil, although again the totals are difficult to estimate. Colombia's 780-km Cano Limon pipeline, which ships 110,000 barrels of oil per day, was bombed at least 170 times in 2001, and estimates of total oil spilled are far higher than those of well-known tanker spills. The French supertanker Limberg was apparently attacked by terrorists offshore Yemen in 2002. Even the Trans Alaska Pipeline is not immune to vandalism; 925 tonnes were spilled when a high-velocity rifle slug penetrated the pipeline in 2001.

4. NATURAL FATE OF OIL IN THE ENVIRONMENT

The description of the various inputs of oil into our environment, mainly from natural seeps and municipal runoff, makes it clear that oil has been part of the biosphere for millennia. However, we are not overwhelmed by it. This is because a variety of natural processes contribute to its removal, the most important of which is biodegradation. Crude oils and refined products provide an excellent source of carbon and energy to organisms, principally microorganisms, able to exploit it. These organisms are responsible for the consumption of most of the hydrocarbons that get into the biosphere and that are not collected or burnt. Nevertheless, this process is relatively slow, and several physical processes typically act on spilled oil before it is eventually biodegraded.

4.1 Flotation and Spreading

Almost all oils in commerce float; thus, they spread rapidly if the spill is on water. This has the effect of providing a dramatically larger surface area for evaporation and dissolution, and eventually the floating layer becomes subject to physical disruption by waves. A few very heavy oils, such as that spilled in the Delaware River from the Presidente Rivera in 1989, are sufficiently dense that they form almost solid lumps that submerge in fresh water. They do not usually sink, however, and can be collected by subsurface nets.

4.2 Evaporation

Small hydrocarbons are very volatile, and molecules smaller than those with approximately 15 carbons typically evaporate quite readily if a spill is on the surface. This can be effectively complete in a matter of days, although underground spills do not typically evaporate very rapidly. Spills are not the only source of volatile hydrocarbons in the atmosphere; trees liberate vast quantities of isoprene and other terpenes. Whatever the source, some interact with reactive nitrogen oxides in the presence of light and in high enough concentrations can contribute to photochemical smogs, The Blue Mountains got their name from this natural phenomenon long before the onset of the petroleum era. Eventually, most volatile compounds, and the reaction products of photochemical oxidations, are "rained out" and deposited on the ground, where they are biodegraded. Some biodegradation may even take place on water droplets in the atmosphere because there are many bacteria in the air.

4.3 Dissolution

A few hydrocarbons are sufficiently soluble in water that they "wash out" of floating surface or underground spills into the water phase. This phenomenon is usually only significant for the BTEX compounds (benzene, toluene, ethylbenzene, and the xylenes) and oxygenated compounds such as methyl-tertiarybutylether. Such compounds have the tendency to wash out from floating spills, whether on surface water or on the water table underground. This phenomenon is of particular concern underground because the soluble compounds have the potential to migrate and contaminate wells and surface water. Fortunately, they are biodegradable under both aerobic and anaerobic conditions.

4.4 Dispersion

Although oils do not dissolve in water, they can disperse. Wave action breaks up floating slicks and may disperse the oil so finely that it is readily biodegraded. This is what happened to most of the 85,000 tonnes of oil lost in the 1993 spill off the Shetland Islands from the Braer. Adding chemical dispersants to encourage this process is an important oil-spill response tool. If the oil is merely broken into large globs, however, these can coalesce and form tar balls that can drift for thousands of miles before eventually beaching. It is not unusual for tar balls to accumulate other debris during their travels or even to be colonized by barnacles.

4.5 Emulsification

Many oils will form very stable water-in-oil emulsions, particularly after they have weathered sufficiently to form a critical concentration of asphaltenes that can stabilize such emulsions. This can multiply the mass of the original oil by up to five times, increasing the recovery and disposal burden. Water-in-oil emulsions can achieve extremely high viscosities, making relatively light oils behave more like heavy fuel oils. This can slow the evaporative and dissolution processes so that if the emulsions are broken, the released oil's properties are more similar to those of the original material than if emulsification had not occurred.

4.6 Physical Interactions with Sediments

The formation of long-lived "pavements" is a poorly understood phenomenon that can have a significant effect on the persistence of oil in the environment. At high enough concentrations, crude oils and heavy refined products such as bunker fuels interact with sand and gravel so that the oil becomes completely saturated with sediment and vice versa. Even quite liquid oils and fine sediments can form pavements that mimic commercial asphalt, and when this occurs the inner oil is essentially entombed and protected from biodegradation. It can then last for years without significant weathering, as happened in Tierra del Fuego following the 1974 spill of 50,000 tonnes of oil and fuel from the Metula. Spill response efforts aim to minimize and prevent the formation of these pavements.

At the microscale, however, the interaction of oil with fine sediments is proving to be an important natural process that removes oil from oiled shorelines. Fine particles become associated with droplets of oil and effectively disperse into the water column, where biodegradation occurs quite rapidly. The scale and importance of this phenomenon have only been recognized during approximately the past decade.

4.7 Photooxidation

Photooxidation by sunlight is very effective at oxidizing three-, four-, and five-ring aromatics in floating slicks and stranded oil. The molecules are not completely destroyed, but they seem to be polymerized and incorporated into the resin and polar fractions of the oil, where they are probably no longer bioavailable. Photooxidation of the surface of pavements and tar balls may contribute to their longevity. Photooxidation seems to have far less of an effect on the saturated components of oil.

Photooxidation is an important phenomenon not because it removes large volumes of oil from the environment but because it removes the molecules of most toxicological concern. Crude oils and heavy fuels contain trace amounts of potentially carcinogenic polycyclic aromatic hydrocarbons such as benzo[*a*]pyrene. Fortunately, these are among the most susceptible to photooxidation, and they are readily destroyed if an oil is exposed to sunlight.

4.8 Combustion

The natural ignition of gaseous seeps is still a source of wonder whenever it occurs. All hydrocarbons are combustible, and sometimes shipwrecks catch on fire (e.g., the 144,000-tonne spill from the Haven near Genoa in 1991) or are deliberately ignited (e.g., the 1360 tonnes of fuel oil on board the New Carissa offshore Oregon in 1999). A major problem is that the oil has to maintain a high temperature to continue burning, and heat transfer to water underneath a slick usually extinguishes any fire as the slick thins. Burning spilled oil thus usually requires containment within fireproof booms, but if successful it can be very effective; combustion of $>90\%$ has been reported in several trials. Burning may be particularly useful in ice-infested waters, in which skimming is very difficult and the ice can act as an effective boom. The residue, although viscous, is not very different from oil that weathers without heat, and there is only a small increase in pyrogenic compounds in the residue. Burning oil, however, can generate a large amount of black smoke. The smoke is very low in residual hydrocarbon, but may, in certain circumstances, present a potential health concern due to fine particulate matter. If burning occurs away from population centers, the potential for environmental and health impacts of the smoke is minimal.

4.9 Biodegradation

4.1.1 Aerobic

Biodegradation is the ultimate fate of hydrocarbons released into the environment, although it is sometimes a slow process. Biodegradation under aerobic conditions has been studied for at least a century, and the variety of organisms able to consume hydrocarbons and the diverse pathways and enzymes used in the process are well understood. More than 60 genera of bacteria and 95 genera of fungi have been described that catalyze this process, along with a few species of archaea and algae. These organisms share the ability to insert one or two of the oxygen atoms of diatomic O_2 into a hydrocarbon molecule, thereby activating it and making it accessible to the central metabolism of the organism. The oxygenated hydrocarbon then serves as a source of reductant for the growth of the organism, and much more oxygen is required as the terminal oxidant.

Almost all hydrocarbons that have been studied have been shown to be biodegraded in the presence of oxygen. The small aromatics (benzene and the substituted benzenes) and alkanes are most readily consumed, followed by the larger alkanes and the two- and three-ring aromatics and their alkylated derivatives. The branched alkanes are slightly more resistant, but they are still readily degraded. The biodegradation of alkylated three- and four-ring aromatics is slower, and some of the four- and five-ring saturated compounds, the hopanes and steranes, are among the most resistant. Indeed, they are so resistant that they can serve as conserved internal markers in the oil, and their concentration in the residual oil increases as biodegradation proceeds. They remain as "fingerprints" of the initial composition of the oil until the very last stages of biodegradation.

Estimates for the total biodegradability of the hydrocarbon fraction of different crude oils range from 70 to 97%. Less is known about the biodegradability of the resins, asphaltenes, and other polar species in crude oils, and the larger ones are not thought to be very susceptible to the process. Fortunately, these compounds lack "oiliness," are friable in small quantities, and are essentially biologically inert. Indeed, they are essentially indistinguishable from other relatively inert biomass in the environment, such as the humic and fulvic acids that make up the organic fraction of soils and sediments and are so essential for plant growth.

Aerobic oil-degrading microorganisms are ubiquitous, having been found in all natural environments in which they have been diligently pursued. Indeed, aerobic hydrocarbon-degrading organisms are the foundations of the extensive ecosystems that exploit marine oil seeps, which often include large multicellular organisms that feed on the hydrocarbon-oxidizing microbes. Oil seeps may even serve as the primary food sources for some fisheries, such as those of Atlantic Canada.

4.1.2 Anaerobic

For some time, it has been known that hydrocarbons are also biodegraded under anaerobic conditions. The hydrocarbon still acts as a reductant, and in the absence of oxygen something else must serve as a terminal electron acceptor. Sulfate (reduced to sulfide), nitrate (reduced to nitrogen), chlorate (reduced to chloride), ferric and manganic ions (reduced to ferrous and manganous ions), and carbon dioxide (reduced to methane) have all been shown to be effective oxidants for hydrocarbon degradation, at least in mixed cultures of organisms. Only a few genera of anaerobic bacteria that can catalyze these reactions have been identified, but this undoubtedly is just the tip of the iceberg and probably more an indication of the difficulty of isolating these organisms than of their scarcity in anaerobic environments. In the absence of oxygen, something else must be added to the hydrocarbon to initiate its metabolism, and at least two pathways have been identified. One adds a molecule of fumarate to the hydrocarbon, whereas the other seems to add a molecule of carbon dioxide. The anaerobic biodegradation of small hydrocarbons, especially the BTEXs, which are of most concern if they leak from storage tanks, is becoming quite well understood.

An important feature of crude oils and refined products is that although they are a rich source of carbon and energy for microbes able to consume them, they are almost uniquely unbalanced foods. Microbes typically have a carbon:nitrogen:phosphorus ratio of approximately 100:10:1, but oil contains no significant amounts of either nitrogen or phosphorus. Oil-degrading microbes must therefore obtain these elements in useful forms from elsewhere in the environment, and if there is a significant amount of hydrocarbon it is likely that the growth of the degrading organisms will be limited by the supply of these nutrients.

5. RESPONSES TO SPILLS

The first response to a spill is to stop any further release and to contain and pick up as much as possible. On land, this is relatively simple, albeit time-consuming and expensive. Indeed, most of the "oil lakes" left in Kuwait from the Gulf War are still awaiting treatment. At sea and on lakes and rivers, however, any collection requires corralling the oil with booms to stop it from spreading.

5.1 Booms, Skimming, and Combustion

Many different types of boom are available; some are inflatable, and others are rigid. Some are designed to withstand ocean waves, and others are designed to work in sheltered harbors. Often, booms are used to keep oil out of particularly sensitive areas, such as fish hatcheries and marshes. Others are designed to be towed to "sweep up" oil, and still others are fire resistant and designed to allow the oil within the boom to be ignited.

Once the oil has been corralled, it has to be collected. This is done with skimmers, which remove oil while accumulating as little water as possible. There are many different designs; mops, moving belts, and floating suction devices are all used. Each is best suited for a particular type of oil and may be ineffective against other oils, but all attempt to collect the oil without collecting much water or debris.

Igniting oil in fireproof booms is not a trivial matter, and special igniters are usually required.

5.2 Physical Removal

If oil reaches a sandy shoreline, it can often be collected with scrapers and bulldozers. This was very effective following the 1996 spill of 72,000 tonnes of crude oil from the Sea Empress off the south coast of Wales. Sometimes, manual removal by teams of workers with shovels is appropriate, as in the case of the 2002 spill from the Prestige off the coast of Spain. If the shoreline is inappropriate for heavy equipment—such as the shorelines of Prince William Sound, Alaska, affected by the spill from the Exxon Valdez—the oil can be flushed back into the water, corralled within booms, and collected with skimmers. Recent developments include special surfactant products that help lift oil from the shoreline while allowing its retention within booms for subsequent skimming, thus allowing more oil to be released with cooler flushing. In the United Kingdom, it is permissible to use dispersants for this purpose in certain circumstances, provided that they have passed the required approval protocols. In North America, use of dispersants on shores is not permitted, and so a new generation of products have been developed that loosen the oils for removal by water stream without the unwanted effect of oil being dispersed into the near-shore water column. These products are specially formulated with amounts of surfactants sufficient to wet the surface but insufficient to foster dispersion. Some of these products have been successfully used on spills in Canada and the United States. Their main objective is to allow weathered oils to be removed by washing without having to use heated water, which could have detrimental side effects on biological communities.

5.3 Bioremediation

Although the physical removal of small near-shore spills can usually be carried out quite effectively and completely, this is rarely possible with large spills at sea. Even when physical collection or combustion are fairly effective, there is almost always some oil that escapes capture. As discussed previously, the ultimate fate of this residual oil is biodegradation, and there are several environmentally responsible ways of stimulating this process. Colloquially known as bioremediation, these can be divided into two complementary strategies. While oil is afloat at sea, the surface area available for microbial attack limits its biodegradation. Adding dispersants to disperse the oil into the water, and dramatically increase the surface area for microbial colonization, is an important option. In contrast, the biodegradation of oil remaining on a shoreline is likely limited by the supply of other necessary microbial nutrients, such as biologically available nitrogen and phosphorus, and carefully adding fertilizers is a very useful option.

5.3.1 Dispersants

Dispersants are concentrated solutions of surfactants that can be applied from the air or from ships equipped with spray booms or fire hoses. Typical application rates are approximately 5% of the oil volume, so quite large stockpiles are required, and indeed are available, in central depots. The dispersants need to be added fairly soon after a spill, but there is usually a window of several days during which they are effective. Even mild wave energy disperses the oil very effectively into the water column, and it has been demonstrated that this substantially increases the rate of biodegradation. Some dispersants also include molecules that actively stimulate oil biodegradation either by providing limiting nutrients or by adding preferred substrates. Dispersion also minimizes the beaching of spilled oils. Dispersants were used very effectively following the 1996 spill of 72,000 tonnes of a relatively light crude oil from the Sea Empress off the south coast of Wales.

5.3.2 Fertilizers

Heavily oiled sites that have been washed or otherwise physically treated to remove the bulk of the oil always retain some residual oil, and this is an ideal target for bioremediation through the addition of fertilizers. Lightly oiled beaches may also provide excellent opportunities for use of this technology without pretreatment. Aerobic oil-degrading microbes are ubiquitous, but their abundance is usually limited by the availability of oil. An oil spill removes this limitation, and there is invariably a "bloom" of oil-degrading microbes, both in terrestrial and in aquatic environments. Their growth becomes limited by the availability of something else. In marine environments, this is usually nitrogen, and although a few oil-degrading microorganisms are able to fix atmospheric nitrogen, most have to await the arrival of low levels of nitrogen with every tide. Carefully applying fertilizers to partially overcome this limitation can substantially increase the rate of biodegradation with no adverse environmental impact. Slow-release and oleophilic fertilizers (designed to adhere to the oil) seem most effective, and they were used with significant success following the 1989 spill from the Exxon Valdez in Alaska. The oleophilic fertilizer was applied with airless sprayers, and the granular slow-release fertilizer was applied with agricultural "whirligigs." Careful monitoring revealed no adverse environmental impacts from this approach and a stimulation of the rate of oil biodegradation of between two- and fivefold. In other words, although the technique did not make the oil disappear overnight, it sufficiently stimulated a process predicted to take a decade so that it occurred in a couple of years.

Since then, bioremediation has been used on a limited site as part of the cleanup of the 1996 Sea Empress spill and has been demonstrated on experimental spills in marine or brackish environments on the Delaware Bay, a Texas wetland, a fine-sand beach in England, mangroves in Australia, and an Arctic shoreline in Spitsbergen. Experience suggests that fertilizer should be added to maintain soluble nitrogen at approximately 100 μM in the interstitial water of the oiled sediment, and this may require reapplication every few weeks. Such a level of fertilizer nutrients does not seem to cause any adverse environmental impact but stimulates the rate of oil biodegradation severalfold.

Fertilizers have also been used with success at many terrestrial spill sites after the bulk of the spilled oil has been collected. Often, biodegradation at terrestrial sites is also limited by the availability of oxygen, and this can be ameliorated by tilling of surface soils or by the addition of oxygen as air or dissolved or slow-release peroxides for subsurface plumes. One approach suspends slow-release peroxides in a series of wells that intercept flowing groundwater plumes; others sparge air into similar wells.

5.3.3 Natural Attenuation

Biodegradation is such an effective environmental process that in some cases it can be relied on to clean an oil spill without further human intervention. Sometimes this is necessary because human intervention may cause more environmental impact than the spill, such as attempts to remove small amounts of oil from marshes. Other situations in which such an approach can be appropriate include underground spills, in which soluble hydrocarbons enter the groundwater. Usually, these are degraded *in situ*, and if the source is plugged the contaminants may be degraded before the plume intercepts any sensitive receptor. In such a case, it may be best to rely on the natural biodegradation process rather than attempt to stimulate it. Clearly, it is important to have a realistic understanding of the natural processes that will remediate the spill before natural attenuation is accepted as the appropriate response.

Together, these approaches exemplify modern environmentally responsible remediation activities—working with natural processes to stimulate them without causing any additional adverse environmental impact.

SEE ALSO THE FOLLOWING ARTICLES

Acid Deposition and Energy Use • Crude Oil Spills, Environmental Impact of • Ecosystems and Energy: History and Overview • Hazardous Waste from Fossil Fuels • Occupational Health Risks in Crude Oil and Natural Gas Extraction • Oil Pipelines • Petroleum System: Natures's Distribution System for Oil and Gas • Public Reaction to Offshore Oil • Tanker Transportation

Further Reading

Canadian Coast Guard (1995). "Oil Spill Response Field Guide." Canadian Coast Guard, Ottawa.

Fingas, M. (2000). "The Basics of Oil Spill Cleanup." 2nd ed. CRC Press, Boca Raton, FL.

National Research Council (1989). "Using Oil Spill Dispersants in the Sea." National Academy Press, Washington, DC.

National Research Council (1993). "In Situ Bioremediation. When Does It Work?" National Academy Press, Washington, DC.

National Research Council (2002). "Oil in the Sea III: Inputs, Fates and Effects." National Academy Press, Washington, DC.

Ornitz, B. E., and Champ, M. A. (2002). "Oil Spills First Principles." Elsevier, New York.

Tissot, B. P., and Welte, D. H. (1984). "Petroleum Formation and Occurrence." Springer-Verlag, Berlin.

Crude Oil Spills, Environmental Impact of

STANISLAV PATIN
Russian Federal Research Institute of Fisheries and Oceanography
Moscow, Russia

1. Crude Oil Spills: Characteristics, Causes, and Statistics
2. Ecotoxicology of Oil in Marine Environments
3. Environmental Impacts of Oil Spills
4. Impacts on Living Resources and Fisheries
5. Oil Spill Response and Prevention Strategies

Glossary

assessment An orderly process of gathering information about a system and determining the significance and causes of any observed changes following a stress on the system.

benthos The region at the sea bottom and the complex of living organisms and their specific assemblages that inhabit the bottom substrates.

biota The totality of all forms and species of living organisms within a given area or habitat.

disturbance A chemical or physical process (including stresses), often caused by humans, that may or may not lead to a response in a biological system, at the organismic, population, or community level.

ecotoxicology A discipline aimed at studying adverse effects of chemicals on biological and ecological systems.

hazards Combinations of properties and characteristics of a material, process, object, or situation that are able to harm and cause environmental, economic, or any other kind of damage.

littoral A nearshore ecological zone influenced by periodic tidal processes.

pelagial An offshore ecological zone located in open waters and inhabited by water-column organisms.

plankton The microscopic and near-microscopic plants (phytoplankton) and animals (zooplankton) that passively drift in the water column or near the sea surface.

response to stress The physiological, biochemical, reproductive, and other adaptive reactions, induced by stress, of a single organism or a whole assemblage (environmental impact).

risk The probability of realization of a particular hazard under specific conditions and within a certain period of time.

stress A chemical or physical process that leads to a response from a single organism, or at the level of a whole population or community. Acute and chronic stresses are distinguished, depending on duration of impact in situations of short-term and long-term exposure.

The largest percentage of accidental oil input into the sea is associated with oil transportation by tankers and pipelines (about 70%), whereas the contribution of drilling and production activities is minimal (less than 1%). Large and catastrophic spills belong to the category of relatively rare events and their frequency in recent decades has decreased perceptibly. Yet, such episodes can pose serious ecological risk and result in long-term environmental disturbances. Because oil spills are under particularly close scrutiny by environmentalists and strongly shape public opinion toward offshore oil development, they deserve special consideration.

1. CRUDE OIL SPILLS: CHARACTERISTICS, CAUSES, AND STATISTICS

Accidents and associated oil spills have been and continue to be inevitable events at all phases of oil and gas field development. On land, oil spills are usually localized and thus their impact can be eliminated relatively easily. In contrast, marine oil spills may result in oil pollution over large areas and present serious environmental hazards. A wealth of statistical data on accidental oil spills has been accumulated over several decades of hydrocarbon production on the continental shelves of many countries. Some of these data are presented in Fig. 1 and Tables I and II. These data as well as other corresponding materials demonstrate a number of points:

- Overall (with the exception of catastrophic spills that occurred during the Persian Gulf War in 1991), there is a trend toward lower volumes of oil spills during all maritime operations, as compared to the late 1970s.
- Particularly large oil losses result from accidental spills during tanker shipments, whereas the contribution of offshore drilling and production activities is minimal and the losses associated with oil storage and pipeline transportation occupy an intermediate position.
- Small and quickly eliminated oil spills are the most usual types of spills; because they occur frequently, they have the potential to create "stable" (long-term) oil contamination in areas of intense offshore oil production and transportation.
- Catastrophic spills (releasing more than 30,000 tons of oil) occur at a rate of zero to several incidents per year, yet these spills are responsible for the highly uneven distribution of anthropogenic input of crude oil into the sea and for its overall volume.

At least half of the oil produced offshore is transported by more than 6000 tankers. In spite of the clear tendency toward accident-free tanker operations (see Fig. 1), tanker-related accidents remain one of the major sources of ecological risk. The list of very large oil spills (tens of thousands of tons each), long as it is, may become even longer at any time. The probability of an accident involving a crude oil spill from an oil tanker is usually estimated as a function of the covered transport distance. According to one such estimate, based on many years of statistical data, the probability of an accident with more than 160 tons of oil spilled is 5.3×10^{-7} per kilometer of the tanker's route. Similar estimates can also be derived based on the total amount of transported oil.

According to a classification by the International Tanker Owners Pollution Federation (ITOPF), oil spills are categorized into three size groups: small (less than 7 tons), intermediate (7–700 tons), and large (over 700 tons). Information is now available on more than 10,000 spills, the vast majority of which (over 80%) are less than 7 tons. Analysis of causes of intermediate spills from 1974 to 2000 indicates that 35% of such spills occurred during routine operations (mainly oil loading/unloading). Accidents such as grounding and collisions gave rise to 42% of spills in this size category. The major cause of large spills relates to accidents (83%), with grounding and collisions alone making up 64%. The probability of catastrophic blowouts and large spills (more than 1000 tons) during drilling and production operations is rather small (one accident per 10,000 wells).

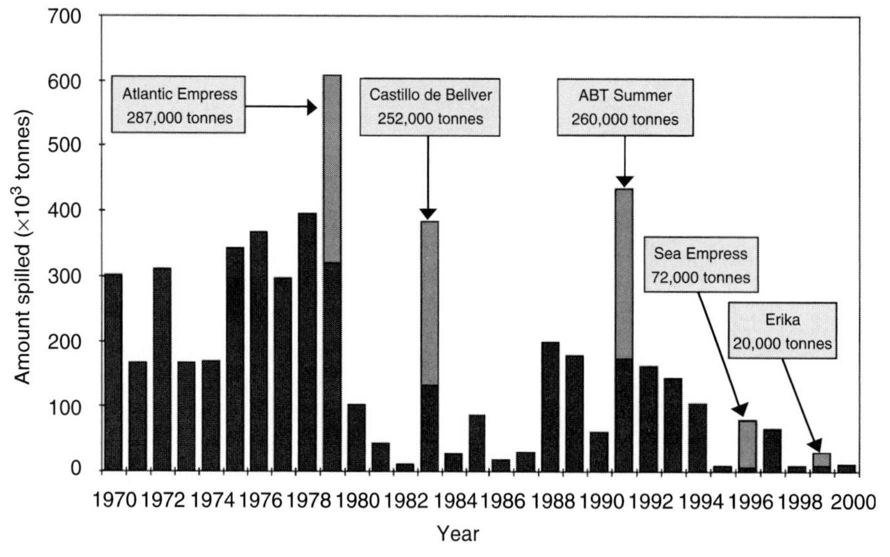

FIGURE 1 Quantities of oil spilled as a result of tanker incidents. Data from the International Tanker Owners Pollution Federation.

TABLE I

Oil Spills Associated with Different Types of Offshore Activities in 1997[a]

Activity	Number of spills	Average spill amount (tons)	Total spill amount	
			Tons	Percentage[b]
Drilling and production	4	191	764	0.5
Oil storage and handling	12	2579	30,948	19.4
Pipeline transport	63	788	49,644	31.2
Tanker transport	23	3391	77,993	48.9
Total	102	1562	159,349	100

[a] Summary from the International Oil Spill Database.
[b] Percentage of total amount spilled in 1997.

TABLE II

Selected Major Oil Spills and Accidental Releases

Source of accident and spill	Year	Location	Oil lost (tons)
Tankers, terminals, installations, Persian Gulf War events	1991	Kuwait; off coast of Persian Gulf and in Saudi Arabia	Up to 1,000,000
Exploratory well *Ixtoc I*	1979	Mexico; Gulf of Mexico, Bahia del Campeche	480,000
Tanker *Atlantic Empress*	1979	Off Tobago, West Indies	287,000
Platform No. 3 well (Nowruz)	1983	Iran; Persian Gulf, Nowruz Field	270,000
Tanker *ABT Summer*	1991	Open water; 1300 km off Angola	260,000
Tanker *Castillo de Bellver*	1983	South Africa; 64 km off Saldanha Bay	252,000
Tanker *Amoco Cadiz*	1978	France; off Brittany	223,000
Tanker *Haven*	1991	Italy; port of Genoa	144,000
Tanker *Odyssey*	1988	Canada; 1175 km off Nova Scotia	132,000
Tanker *Torrey Canyon*	1967	United Kingdom; Scilly Isles	130,000
Pipeline, Kharyaga-Usinsk	1994	Russia; Usinsk	104,000
Tanker *Braer*	1993	United Kingdom; Shetland Islands	85,000
Tanker *Aegean Sea*	1992	Spain; La Coruna harbor port	74,000
Tanker *Sea Empress*	1996	United Kingdom; Mill Bay, Milford Haven harbor port	72,000
Tanker *Prestige*	2002	Spain; over 200 km off Spanish coast	70,000
Tanker *Exxon Valdez*	1989	United States; Alaska, Prince William Sound	37,000

Over 30 years of offshore oil operations, only a few such episodes have occurred, primarily in the early 1970s.

To date, more than 100,000 km of subsea pipelines have been laid worldwide for transporting oil and other hydrocarbons. According to different estimates, the probability of a pipeline accident accompanied by a spill is anywhere between 10^{-4} and 10^{-3} spills/km/year. The data summarized in Table I suggest that the probability of pipeline accidents can be estimated at about 6.3×10^{-4} spills/km/year, with an average spill size of about 800 tons and worldwide oil losses of about 50,000 tons/year. Regional estimates may, of course, vary widely, depending on the local conditions. Accidental spills may be caused by many factors, including pipe corrosion, mechanical damage due to tectonic shifts, and impact by ship anchors. Pipelines become less prone to accidents as their diameter increases, but in all cases the probability of damage and ecological risk increases with aging of seabed pipelines. In general, pipelines are considered to be the safest way to accomplish hydrocarbon transport at sea and on land.

2. ECOTOXICOLOGY OF OIL IN MARINE ENVIRONMENTS

Crude oil is a naturally occurring material; its influx from seabed seepage alone amounts to more than 600,000 tons/year. Additionally, more than 10 million tons/year of aliphatic and other hydrocarbons are produced by marine organisms (mainly bacteria and phytoplankton). Unlike permanent natural sources, accidental oil spills usually occur within a short period of time and produce local zones of abnormally high hydrocarbon concentrations that are hazardous to marine life.

Chemically, crude oil (petroleum) is a complex of natural (mainly organic) substances. The basic component (up to 98%) consists of saturated and unsaturated aliphatic and cyclic hydrocarbons. In seawater, the original petroleum substrate rapidly (within a few hours or days) disappears as it breaks down into fractions (forms) in different states of aggregation. The most common aggregates include surface films (slicks), dissolved and suspended forms, emulsions (oil-in-water, water-in-oil), solid and viscous fractions precipitated to the bottom, and compounds accumulated in marine organisms. Under conditions of long-term (chronic) pollution, the dominant form is usually emulsified and dissolved oil. The fate (behavior, distribution, migration) of oil substances in the marine environment is controlled by a variety of interrelated processes. The major processes, usually known as weathering, include physical transport, evaporation, dissolution, emulsification, sedimentation, biodegradation (decomposition by microorganisms), and oxidation (Fig. 2).

The background contamination levels of petroleum hydrocarbons in the marine environment vary rather widely, from 10^{-5} to 10 mg/liter in seawater

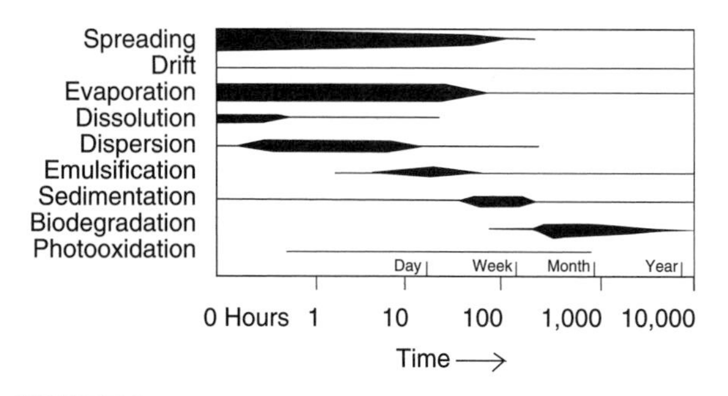

FIGURE 2 Processes acting on an oil slick as a function of time following a spill. The line thickness indicates the relative magnitude of each process. Reproduced from Swan *et al.* (1994) after Wheeler (1978).

and from 10^{-1} to 10^4 mg/kg in bottom sediments, depending on a spectrum of natural and technogenic factors. Maximum concentrations tend to occur in coastal and littoral waters as a result of many types of offshore, coastal, and land-based activities (developing hydrocarbon resources, transportation, industrial discharges, wastes dumping, etc.). Bioaccumulation (i.e., intake and assimilation by living organisms) of oil substances strongly depends on their hydrophilic and lipophilic properties. The concentration of polycyclic aromatic hydrocarbons (PAHs) in marine organisms is often up to three orders of magnitude above their content in seawater. The most pronounced capacity to accumulate PAHs without noticeable metabolic degradation in tissues is exhibited by filter-feeding bivalves (e.g., mussels).

From the ecotoxicological standpoint, oil is a complex multicomponent toxicant with nonspecific effects. Its basic component is a group of the most soluble and toxic aromatic hydrocarbons. High-molecular-weight PAHs with five and more benzene rings (e.g., benzo[*a*]pyrene) are able to produce carcinogenic effects, especially in benthic organisms, which inhabit heavily polluted bottom sediments. The primary responses of marine biota exposed to background levels of PAHs can be reliably detected by specific biochemical methods (e.g., measuring the induced activity of enzyme systems in individual marine organism tissues and organs).

For different groups of marine biota, toxic concentrations of oil (primarily aromatic hydrocarbons) in seawater and bottom sediments vary within several orders of magnitude. These differences are most pronounced in benthic communities of coastal zones polluted by oil, including those impacted by spills. Most species of the marine fauna exposed to oil are particularly vulnerable at early life stages (embryos and larvae). A simplified sequence of biological responses in marine biota, depending on oil concentrations in seawater and bottom sediments, is shown in Fig. 3.

A summary of experimental (toxicological) data and actual (measured) levels of oil content in seawater and bottom sediments of different areas of the worldwide oceans is given in Fig. 4. The upper limits of the zone in which there are "no observed effect" concentrations (NOECs), in which biological effects are either absent or manifest as primary (mostly reversible) physiological and biochemical responses, lie within the ranges of 10^{-3} to 10^{-2} mg/liter in seawater and 10–100 mg/kg in

Lethal effects
Sublethal effects
Impaired reproduction
Reduced growth rate
Reduced feeding activity
Behavior reactions
Tolerance zone
NOEC
0 1 10 100 1,000 10,000
Oil concentration in seawater, μg/liter
NOEC
0 1 10 100 1,000 10,000
Oil concentration in sediments, mg/kg

FIGURE 3 Developing characteristic effects and responses in marine biota, depending on the concentrations of oil in seawater and bottom sediments. NOEC, "No observed effect" concentration; see text for definition.

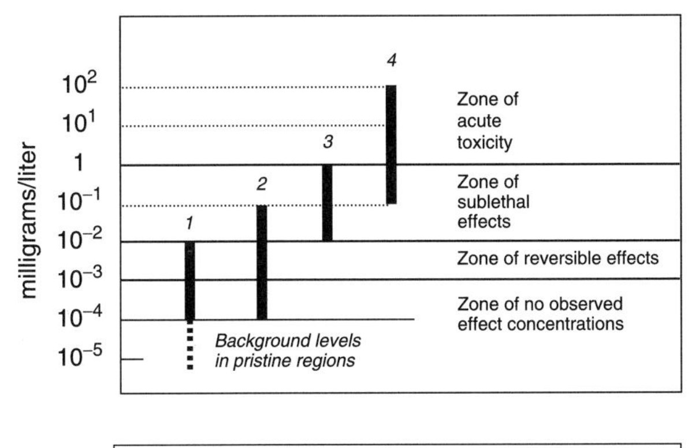

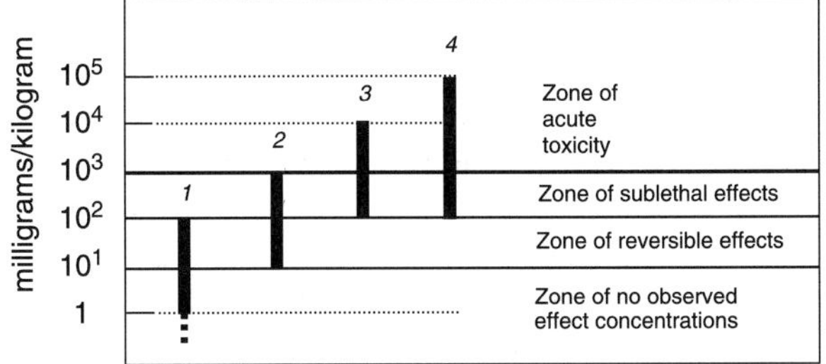

FIGURE 4 Approximate zones of biological effects and ranges of typical concentrations of oil hydrocarbons (mainly aromatic) in seawater (top) and bottom sediments (bottom). Zones: 1, pelagic areas (open waters); 2, coastal and littoral areas; 3, estuaries, bays, and other shallow semiclosed areas; 4, areas of local pollution, including oil spills.

bottom sediments. These ranges can be roughly considered as the limits of maximum permissible (safe) concentrations of oil hydrocarbons dissolved in seawater and accumulated in bottom sediments, respectively.

3. ENVIRONMENTAL IMPACTS OF OIL SPILLS

3.1 Methodology of Environmental Impact Assessment

Oil spills show one of the most complex and dynamic patterns of pollutant distribution and impact in the marine environment. In a sense, each spill is unique and unrepeatable because of the practically unlimited possible combination of natural and anthropogenic factors at a given location and at a given time. This poses a challenge that complicates the modeling of crude oil distribution, not to mention making quantitative predictions of the ecological effects of spills. At the same time, over the past several decades, a great number of special studies and field observations on oil behavior and the environmental hazards of oil spills have been conducted in a variety of ocean regions worldwide. These studies provide a good base for realistic assessment of the negative impact of oil spills on the marine environment and on living resources.

It should be noted that, to this day, no standard and widely recognized methodology of environmental impact assessment (EIA), including EIAs with respect to oil spills, actually exists. The most common methods used at different stages of EIAs include expert estimations, matrix analysis, simulations (modeling), hazards and risks theory, ecotoxicological studies, and field and laboratory experiments. Apparently, such a great diversity of methods and approaches extremely complicates standardization of the EIA procedure. At the same time, the general outlines of this procedure, including the contents and sequence of its stages, have already taken final shape in many countries.

Any system for assessing the state of environmental and human impacts has to be accompanied by a number of selected scales (classifications) for ranking spatial and temporal parameters of impact, as well as for specific evaluation of environmental hazards and risks. One of such classification used here is shown in Table III. Certainly, any scales and relative estimates of this kind will inevitably be circumstantial, but there is no other way to render the results of EIA more specific or objective. Otherwise, we are doomed to ineffective attempts to describe highly complex multifactorial natural systems with an endless number of direct and feedback relations (most often the nonlinear ones). This would leave plenty of room for subjective interpretations and uncertain estimates without any

TABLE III

Impact Scales and Gradation of Oil Spill Ecological Hazards and Consequences in Marine Environments

Impact	Definition
	Spatial scale
Point	Area under impact is less than 100 m^2
Local	Area under impact ranges from 100 m^2 to 1 km^2
Confined	Area under impact ranges within 1–100 km^2
Subregional	Area under impact is more than 100 km^2
Regional	Area under impact spreads over shelf region
	Temporal scale
Short term	From several minutes to several days
Temporary	From several days to one season
Long term	From one season to 1 year
Chronic	More than 1 year
	Reversibility of changes
Reversible (acute stress)	Disturbances in the state of environment and stresses in biota that can be eliminated naturally or artificially within the time span of several days to one season
Slightly reversible	Disturbances in the state of environment and stresses in biota that can be eliminated naturally or artificially within the time span of one season to 3 years
Irreversible (chronic stress)	Disturbances in the state of environment and stresses in biota that exist longer than 3 years
	General assessments
Insignificant	Changes in the state of environment and biota are absent or not discernible against the background of natural variability
Slight	Observed disturbances in the environment and short-term reversible stresses in biota (below minimum reaction threshold–0.1% of natural population reactions) are possible
Moderate	Disturbances in the environment and stresses in biota are observed without signs of degradation and loss of a system capacity for recovery; changes up to 1% of natural population reactions are possible
Severe	Stable structural and functional disturbances in biota communities (up to 10% of natural population and community parameters) are observed
Catastrophic	Irreversible and stable structural and functional degradation of a system is evident; disturbances in ecosystem parameters are more than 50% of statistical norm

assurance that the ultimate objectives of the EIA will be attained at all.

3.2 Oil Spill Types and Scenarios

From the environmental point of view, distinction should be made between two basic types (scenarios) of oil spills. One scenario includes spills that occur in open waters, without having the oil contact the shoreline and/or bottom sediments (pelagic scenarios). In this case, under the influence of wind, currents, and other hydrodynamic processes, the oil slicks are quickly (within a few hours or days) transformed and dispersed atop and under the sea surface. The dispersed and emulsified oil undergoes active biodegradation, its toxicity significantly decreases, and biological systems display attributes of acute (short-term) stress. The other type of spill, unfortunately more common and potentially more hazardous, involves influx of the oil onto the shore, its accumulation in sediments along the coast, and subsequent long-term ecological disturbances both in the littoral (intertidal) areas and on the shore. Such situations should certainly be treated as chronic stresses. The two scenarios often develop simultaneously, especially when an accidental oil spill occurs within a marine coastal zone not far from the shoreline.

According to available statistics, a majority of accidents and accompanying oil spills take place in coastal waters. In such cases, the likelihood of oil contact with the shoreline depends on many factors, including oil spill characteristics (volume and type of oil spilled, distance from the coast, etc.) and local oceanographic and meteorological conditions, particularly velocity and direction of winds and currents. In most accidents, the probability of an oil slick migration toward the coast and contact with the

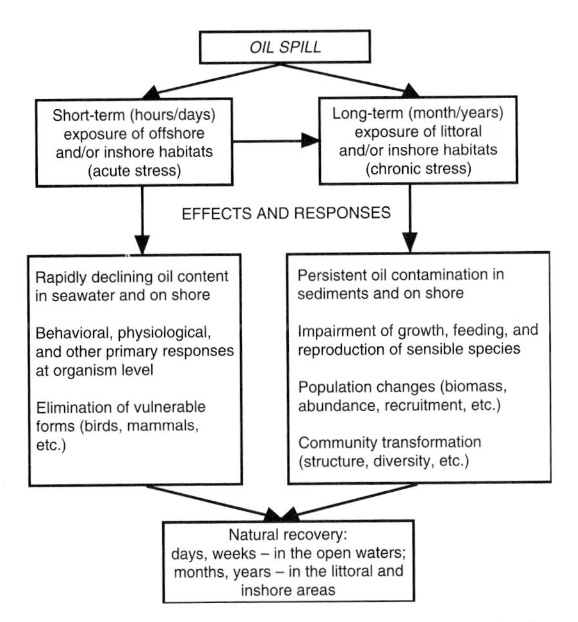

FIGURE 5 Conceptual scheme of characteristic biological effects and environmental hazards of oil spills in the sea.

shoreline is quite high (up to 50%), and the length of the coastal line impacted by oil can reach tens and even hundreds of kilometers. The conceptual scheme of developing biological effects and consequences of an oil spill under acute and chronic stresses is presented in Fig. 5. Depending on the duration, level, and scale of oil contamination, a wide range of effects may occur both in the water column and in littoral sediments—from behavioral responses of organisms at the initial phases of the spill up to long-term populational disturbances under chronic impact in the littoral zone.

3.3 Pelagic Scenarios (Oil Does Not Contact Shoreline and Bottom Sediments)

Ecological effects of oil spills in the open sea are mainly associated with short-term oil pollution of the surface water layers, where organisms such as plankton, nekton (mainly fishes), marine birds, and mammals may be affected when the situation is one of acute oil stress. Pelagic scenarios involve the formation of oil slicks in which total oil concentration ranges from 1 to 10 mg/liter in the surface seawater layers, at a depth of several meters. At horizons deeper than 10 m, oil contamination usually does not exceed background levels. After several days (sometimes weeks), surface slicks on open water are usually dispersed and eliminated under the influence of the processes and factors reflected in Fig. 2.

3.3.1 *Impact on Plankton*

Among all ecological forms, the hyponeuston, species that live within the most polluted surface microlayer (several centimeters deep at the atmosphere/sea interface), might be assumed to experience the heaviest toxic impact from oil spills. Nevertheless, in open waters, both hyponeuston and planktonic biota as a whole do not demonstrate any persistent structural and functional changes during pelagic oil spills. This is the result of a number of factors and processes, including a rapid decrease in the concentration of spilled oil to harmless levels, due to its dispersion and degradation in seawater; the rapid recovery of phytoplankton and zooplankton populations (within a few hours and days) due to high reproduction rates; and the influx of planktonic organisms from adjoining marine areas. There is not a single known publication in which oil spills have been shown to produce irreversible long-term impact on the planktonic populations of open waters.

3.3.2 *Impact on Fish*

Conclusions about the absence of any persistent long-term changes at the population level in planktonic biota after pelagic oil spills are fully valid in relation to fish populations. Due to oil localization within the surface layer of seawater, the mortality of water column organisms, including pelagic fish, is practically excluded. Fish mass mortality has never been observed even after catastrophic oil spills in the open sea. Available publications on the subject suggest that adult fish are capable of detecting and avoiding zones of heavy oil pollution.

An adverse impact of oil spills on fish is most likely to be observed in the shallow coastal areas of the sea where the water dynamics are slow. Fish in early life stages are known to be more vulnerable to oil, compared to adults, and, therefore, some younger fish may be killed by exposure to high concentrations of toxic components of crude oil. However, calculations and direct observations indicate that such losses could not be reliably detected against the natural background of high and variable mortality at embryonic and postembryonic stages of fish development.

3.3.3 *Impact on Birds and Mammals*

Sea birds and mammals are among the most vulnerable components of marine ecosystems in relation to oil pollution. Even short-term contact with spilled oil affects the insulating functions of hairy or feathery coats and results in quick death.

Studies of mass mortality (by tens of thousands of individuals) of birds and mammals after oil spills in coastal waters (even without oil contacting the shoreline) have been described in numerous publications.

The exact consequences of such spills depend primarily on the population traits of different species. Abundant populations with a high reproductive potential are the least susceptible to population stresses. At the same time, adverse impacts for the less abundant species with longer life spans are usually more serious and protracted. Of particular importance are the distribution and congregation patterns of birds and mammals. Even a small oil spill may severely affect a large number of birds or sea mammals that form compact colonies within a small area. In assessing such impacts, one must take into consideration the effect of natural factors, particularly natural mortality caused by extreme weather conditions, lack of food, epizootic diseases, etc. Most data suggest that populations of marine birds and mammals are able to return to their natural level (optimal for given conditions) within several years (seasons) after the oil spill impact.

3.3.4 General Assessment

Using the scales and ratings in Table III, typical oil spill impacts in pelagic scenarios can be assessed as varying from local to confined and from short term to temporary. Potential ecological hazards and biological effects in the form of acute stress for the marine pelagic biota can be assessed as predominantly reversible and varying from insignificant to moderate.

3.4 Littoral Scenarios (Oil Contacts Shoreline and Bottom Sediments)

When spilled oil reaches the shoreline, the major processes of oil accumulation, transport, and transformation occur in the littoral zone exposed to wind-induced waves, storms, surf, currents, and tides. The width of this zone may vary within tens of meters. Its potential self-cleaning capacity depends, first of all, on the geomorphology of the shoreline and characteristics of near-shore sediments and shore deposits (composition, fineness, etc.), as well as on the energy of wave and tidal processes. Numerous observations in different parts of the world indicate that oil persistence and, consequently, its adverse impact sharply increase from open rocky shores to sheltered gravel and pebble coasts (Table IV).

Some examples and characteristics of potential ecological impacts of oil spills in the littoral zone and adjoining shallow (several meters deep) sublittoral habitats are given in Table V. These assessments are based on the analysis of nearshore spill situations and describe biological effects resulting from oil pollution of both littoral sediments and seawater. Typically in such cases, local transformations of species population structures in littoral communities result in the predominance of more resistant species (e.g., some polychaetes) and rapid elimination of sensitive ones (especially crustaceans, such as amphipods, and marine birds and mammals). The time of return to environmental quality and community structure may vary widely, from a month to several years, depending on the exact combination of different natural and anthropogenic factors.

TABLE IV

Classification of Shorelines Based on Their Potential Oil Spill Recovery Capabilities

Self-cleaning capacity	Type of shoreline	Process characteristics
High (type I)	Open rocky shoreline	As a result of wave and surf action, oil is removed within several weeks; there is no need to clean inshore habitats
Medium (type II)	Accumulative coasts with fine-sand beaches	Oil does not usually penetrate deeply into fine, compact sand and can be removed by mechanical means; oil contamination persists for several months
Low (type III)	Abrasive coasts with coarse sand, pebble, and gravel stretches	Oil rapidly penetrates into deeper layers (up to 1 m) and can persist for years
Very low (type IV)	Sheltered coasts with boulder, gravel, and pebble material; wetlands (marshes, mangroves)	Favorable conditions for rapid accumulation and slow degradation of oil; tars and asphalt crusts can be formed

TABLE V

Potential Biological Effects of Oil Spills in the Littoral (Intertidal) Zone

Shoreline type[a]	Recovery capacity	Typical oil content levels in: Water[b] (mg/liter)	Bottom (mg/kg)	Characteristic biological effects
I	High	<0.1	$<10^2$	Mortality among the more sensitive species within the first days of contact; sublethal effects; primary structural changes in communities; recovery time, up to a month
II	Medium	0.1–1.0	10^2–10^3	Elimination of crustaceans (especially amphipods), predominance of polychaetes; decreasing biomass and structural changes in communities; recovery time, up to half a year
III	Low	1–10	10^3–10^4	Mass mortality among the more sensitive species (crustaceans and bivalves); steadily decreasing biomass and poorer species diversity; recovery time, up to a year
IV	Very low	>10	$>10^4$	Mass mortality among most species; pronounced declined biomass, abundance, and species diversity; recovery time, more than a year

[a] See classification in Table IV.
[b] Dissolved and emulsified forms of oil.

There are numerous records of oil spills in coastal and littoral areas in many marine regions, as well as hundreds of publications with detailed descriptions of associated environmental impacts. Besides the well-known catastrophic accident involving the tanker *Exxon Valdez* in the nearshore area of Alaska in 1989, a number of similar events have been documented, including oil spills in the coastal waters of the United States, France, United Kingdom, and Spain, among other countries (see Table II). One of the largest oil spills occurred in February 1996 when the tanker *Sea Empress* ran aground within a kilometer of the South Wales shore and 72,000 tons of crude oil were released. Despite favorable weather conditions (only 5–7% of the spilled oil reached the shore) and effective clean-up operations, the accident resulted in heavy pollution of 200 km long the coastline and littoral zone. About 7000 oily birds were collected on the shore alone. At the same time, no mass mortality of commercial fish, crustaceans, or shellfish was recorded. The time of environmental recovery in different locations was quite variable (from several months to a year and more), depending on adopted criteria and specific conditions within the affected littoral zone.

The list of this kind of accident in coastal areas is rather long. All these situations, unlike the purely pelagic spills (oil does not contact shoreline and bottom sediments), have two major phases, interpretable as acute and chronic stresses. Acute stress usually occurs within the first few days (sometimes weeks) and may result in mortality of the oil-impacted organisms (especially birds and mammals). Chronic stress lasts for several months, seasons, or even years and manifests mainly as adaptive modifications of benthic communities in response to the long-term oil contamination of littoral and coastal (inshore) sediments. In a majority of cases (excluding heavy oil pollution during very large catastrophic spills), these modifications do not exceed critical population levels and do not result in irreversible changes in structure and function of benthic populations.

3.4.1 General Assessment

In terms of ratings and definitions in Table III, typical oil spill impacts in littoral scenarios can be assessed as varying from local to subregional and from temporary to chronic. Environmental hazards and biological effects in the form of chronic stress for the marine biota (mainly for benthic organisms) can be assessed as predominantly slightly reversible and varying from slight to severe.

4. IMPACTS ON LIVING RESOURCES AND FISHERIES

Until recently, there has been no direct evidence of any detectable impact of oil spills on the stock and biomass of commercial species. This conclusion is supported by a number of special studies, including those conducted in the situations of catastrophic accidents. Results of modeling and calculations of potential losses of commercial fish and invertebrates during oil spills lead to similar conclusion. Most estimates indicate that losses of commercial species during spills, even in the most pessimistic scenarios, usually do not exceed hundreds/thousands tons of biomass and cannot be reliably distinguished against the background of extremely high population variability (mainly due to environmental changes, natural mortality, and fishing).

Of particular concern is the possibility of the negative impact of oil spills on the anadromous fish (primarily, salmons) during their spawning migration shoreward. In this regard, it should be remembered that in the open sea, the oil concentration beneath the oil slick decreases sharply with depth, approaching background values at horizons of 5 to 10 m. Detailed studies conducted during a number of years after oil spills near the shores of Alaska and in other salmon breeding regions have shown that all basic parameters of commercial populations (abundance, structure, reproduction, migration, stock, and catches) remain within the natural ranges. Thus, there is a reasonable consensus that negative effects of oil spills on marine living resources in terms of their population parameters are negligible and could not be practically detected due to the enormous natural variability and dynamics of marine populations and communities. This conclusion does not mean, however, that oil spills do not exert any adverse impact on fisheries and the entire fishing industry. Numerous studies have shown that the actual economic losses to fisheries occur as a result of temporary restrictions on fishing activities, interference with aquaculture (i.e., growing and reproduction of marine organisms in marine farms) routines, loss of marketable values of commercial species due to petroleum taste (tainting), and oil contamination of fishing gear. The most serious economic losses for fisheries result from the limitations (bans, closures) imposed on fishing following oil spills in impacted coastal areas. Commercial fishermen are deprived of their livelihood unless they are able to exploit alternative stocks. Naturally, the exact nature and extent of such losses may vary widely, depending on a combination of diverse factors, such as oil spill size, time of year, and weather conditions. In some cases, during the catastrophic accident involving the tanker *Exxon Valdez* near Alaskan shores in the spring of 1989, for example, the losses to local fisheries were estimated at hundreds millions of dollars. This amount was eventually paid by court decisions in the United States as compensation for the restriction on fishing because of heavy oil pollution. Significant losses are known to have been sustained by aquaculturists, as was the case with many farms specializing in growing fish and shellfish during oil spills near British shores in 1993 and 1996. Similar events occurred in the winter of 2002 due to catastrophic oil pollution of the Atlantic coasts of Spain and Portugal after an incident involving the tanker *Prestige* caused release of 70,000 tons of heavy fuel oil.

Another rather common yet poorly controlled type of damage to fisheries inflicted by oil spills is associated with accumulation of petroleum hydrocarbons (especially light aromatic compounds) in marine organisms. This does not usually lead to intoxication of the organisms but results in an oily odor and taste, which inevitably means deterioration of marketable qualities of marine products and corresponding economic losses.

Legal and regulatory mechanisms of compensation for losses sustained by the fishing sector as a result of oil spills are yet to be fully elaborated, although two special international conventions are already in effect: the International Convention on Civil Liability for Oil Pollution Damage (Brussels, 1969) and the International Convention on the Establishment of an International Fund for Compensation for Oil Pollution Damage (London, 1971; amended by the 1992 Protocol). These conventions, administered by member states of the United Nations, set forth a number of measures and provisions for paying compensations to countries in cases of accidental oil spills within the bounds of their continental shelf. In the context of the conventions, fishery closures are expected to minimize or prevent economic damage that might otherwise occur.

5. OIL SPILL RESPONSE AND PREVENTION STRATEGIES

Public concern and anxiety about oil spills in the sea have been clearly augmented since the *Torry Canyon* tanker accident off the southwest coast of the United Kingdom in March 1967, when 130,000 tons of spilled oil caused heavy pollution of the French and British shores, with serious ecological and fisheries consequences. This accident was followed by a number of similar catastrophic episodes (see Table II). These events produced an acute public reaction and clearly elucidated a need for developing effective prevention and response strategies for appropriate reactions in such situations. Since the time of these spills, an impressive technical, political, and legal experience in managing the problem has been acquired in many countries and on the international level, mainly through the International Maritime Organization. In 1990, the International Convention on Oil Pollution Preparedness, Response, and Co-operation was adopted by a number of countries. The convention provided the framework for coordination of the national and international efforts to create special funds, centers, technical means, and methods for dealing with the consequences of oil spills in coastal areas. In practically all nearshore countries, specialized oil spill response centers and services have been established. Additionally, any offshore oil project is required to develop "action emergency plans." In many countries, national plans on preparedness and responses to oil spill accidents have been developed and appropriate legal rules, norms, and regulations are enforced.

An idea about the scope of economical losses and costs of oil spills may be derived from the *Exxon Valdez* incident in 1989; the clean-up operations were conducted during 2 years with about 10,000 people participating, and response costs totaled approximately $2.8 billion (total costs exceeded $9 billion). In the aftermath of this incident, the U.S. Congress adopted the 1990 Oil Pollution Act, which expanded the oil spill prevention, preparedness, and response responsibilities of both the federal government and industry.

The wide spectrum of contemporary techniques for oil spill response and clean-up operations may be divided into four major groups:

1. Mechanical means (booms, skimmers, etc.) for collection and removal of oil from both the sea surface and inshore habitats.
2. Burning of spilled oil at the site.
3. Chemical methods for dispersing surface oil slicks, aimed at accelerating oil degradation by natural factors.
4. Microbiological methods for enhancing oil degradation, usually applied in combination with mechanical and chemical methods.

The great variety of means and methods are quite relevant, taking into account the enormous variation in emergency situations as well as the extremely complex behavior of crude oil in marine environments. In the field, a combination of diverse methods is commonly used, with preference for mechanical means of oil containment and removal. Chemical and microbiological methods are applied predominantly in addition to the mechanical operations.

There are two situations when oil spill response and clean-up operations are not justified. First, oil spills in the open sea in areas with active surface water mass hydrodynamics, at great depths, and far from any shoreline need not be mitigated. As already noted, in such situations, oil slicks are relatively rapidly (usually within several days) dispersed without any significant damage to water column organisms and pelagic ecosystems. Second, shorelines are classified according to their vulnerability to oil spills. As was shown in Table IV, along open rocky shorelines, oil can be quickly dispersed and eliminated as a result of favorable natural processes (i.e., strong wave, tide, and wind activities). It is quite evident that in these situations human intervention may be senseless. Unfortunately, such situations are the exception rather than the rule. The littoral scenarios, with oil entering and accumulating in shallow coastal zones with low-energy hydrodynamic processes, are more usual. In these cases, the necessity of cleaning operations is quite evident.

Clearly, the techniques and means for alleviating the impacts of oil spills describe herein represent only a few of the armamentarium of contemporary oil spill response and prevention strategies. These strategies have been implemented worldwide and represent the development of regional systems of accident monitoring and reporting and coastal sensitivity analysis and mapping, involving adoption of detailed plans for dealing with emergencies and production of databanks and information systems for assisting spill response management. In some countries (e.g., the United Kingdom), integrated booming strategy plans (employing anchored oil-blocking booms) are being developed for protection of environmentally sensitive areas.

In conclusion, it should be emphasized that accidental oil spills are only one of a multitude of various factors of human impact on marine environments. The protection of oceans has always been and shall always be a priority task of international significance.

SEE ALSO THE FOLLOWING ARTICLES

Crude Oil Releases to the Environment: Natural Fate and Remediation Options • Ecological Risk Assessment Applied to Energy Development • Gulf War, Environmental Impact of • Hazardous Waste from Fossil Fuels • Occupational Health Risks in Crude Oil and Natural Gas Extraction • Oil and Natural Gas: Offshore Operations • Oil Pipelines • Tanker Transportation • Wetlands: Impacts of Energy Development in the Mississippi Delta

Further Reading

Cairns, W. J. (ed.). (1992). "North Sea Oil and the Environment. Developing Oil and Gas Resources, Environmental Impacts and Responses." Elsevier Applied Science, London and New York.

Davies, J. M., and Topping, G. (eds.). (1997). "The Impact of an Oil Spill in Turbulent Waters: The Braer." The Stationery Office, Edinburgh.

Etkin, D. S. (1999). Historical Overview of Oil Spills from All Sources (1960–1998). *In* "Proceedings of the 1999 International Oil Spill Conference." American Petroleum Institute, Washington, D.C.

Joint Group of Experts on the Scientific Aspects of Marine Pollution (GESAMP). (1993). "Impact of Oil and Related Chemicals and Wastes on the Marine Environment." GESAMP Reports and Studies, No. 50. International Maritime Organization, London

Patin, S. A. (1999). "Environmental Impact of the Offshore Oil and Gas Industry." EcoMonitor Publ., New York.

Swan, J. M., Neff, J. M., and Young, P. C. (eds.). (1994). "Environmental Implications of Offshore Oil and Gas Development in Australia." Australian Petroleum Exploration Association, Sydney.

Cultural Evolution and Energy

RICHARD N. ADAMS
University of Texas at Austin
Austin, Texas, United States

1. Introduction
2. Energy as a Conceptual Tool for Social Analysis
3. Dynamics of Energy Flow
4. Energy in Evolution of Human Society
5. Concluding Note

Glossary

circumscription theory As formulated by Robert L. Carneiro, the process whereby threats to the survival of a society stemming usually from population pressure on resources, but also from competition from other societies, and environmental constraints, lead to the increasing complexity of the sociopolitical organization.

cultural evolution (1) The process whereby human beings adapt their social and cultural behavior to deal with challenges to their survival; (2) the process whereby human societies become more complex as they adapt to deal with challenges to their survival; such challenges may be internal or external.

dissipative structure An open energy system with a form created by the constant incorporation of energy and matter as inputs and dissipation as output.

energy maximization principle As formulated by Alfred Lotka, the principle that, among living organisms, "natural selection will so operate as to increase the total flux through the system, so long as there is presented an unutilized residue of matter and available energy."

energy triggers The process whereby the dissipation of one energy flow releases dissipation from another energy source.

levels of sociocultural integration See **stages of cultural evolution.**

minimum dissipation principle As formulated by Ilya Prigogine and J. M. Waime, the principle that "when given boundary conditions prevent the system from reaching thermodynamic equilibrium (i.e., zero entropy production) the system settles down in the state of 'least dissipation'."

stages of cultural evolution Successive series of configurations of sociocultural organization that emerge as societies become increasingly complex; these various configurations are also known as levels of sociocultural integration.

Human societies, like all other living phenomena, exist by virtue of extracting energy from the environment and dissipating it in specific ways. The laws, principles, and measures applying to energy dissipation are as applicable to human society and culture as they are to the rest of nature. Theories of social and cultural evolution have developed with limited attention to this fact. The bulk of the present account is divided into three parts. The first two concern using energy dynamics in social analysis and are directed at social scientists interested in the question. The second is concerned with cultural evolution, and are directed more at those unfamiliar with the development of theory on this topic.

1. INTRODUCTION

Sociocultural evolution was first seriously explored in the 19th century by Herbert Spencer, Edward B. Tylor, and Lewis Henry Morgan, but fell under the cloud of cultural relativism in the early 20th century. The concept of sociocultural evolution was subsequently revived by Leslie White, V. Gordon Childe, and Julian Steward, and was further developed by Elman R. Service, Marvin Harris, and Robert L. Carneiro and by many of their students. Sociocultural evolution has proved to be of interest in archeology for modeling sociocultural change over extended eras of time and comparatively between different cultural traditions. It has been taken up in sociology by Gerhard Lenski and Stephen K. Sanderson and in a more Darwinian form by Donald Campbell. Also, under the influence of sociobiology, a distinctive branch of coevolutionary thinking (which will not concern us here) has been explored by Robert Boyd and Peter J. Richarson, Charles J. Lumsden and Edward O. Wilson, and William H. Durham.

The notion that energy was relevant to understanding the social process was early described by Herbert Spencer in his *First Principles* (1863): "Based as the life of a society is on animal and vegetal products, and dependent as these are on the light and heat of the Sun, it follows that the changes wrought by men as socially organized, are effects of forces having a common origin...Whatever takes place in a society results either from the undirected physical energies around, from these energies as directed by men, or from the energies of the men themselves." Following this assessment, the connectedness of energy and cultural evolution was primarily pursued by scholars with interests other than cultural evolution. By the early 20th century, it had been seriously discussed by Wilhelm Ostwald, Patrick Geddes, and Frederick Soddy, to be followed by Alfred Lotka, T. N. Carver, and Lewis Mumford. In 1943, anthropologist Leslie White proposed it as a tool for social analysis. It was then explored in more detail by sociologist Fred Cottrell, ecologist Howard Odum, economists Nicolas Georgescu-Roegen and Herman Daly, anthropologist Richard N. Adams, and, recently, sociologists Eugene A. Rosa and Gary E. Machlis. In the latter part of 20th century, world events produced an oil crisis and with it an explosion of interest in the political economy of energy in society.

2. ENERGY AS A CONCEPTUAL TOOL FOR SOCIAL ANALYSIS

Because energy is dissipated when anything happens, it is a substantive process that provides a common denominator for all activities, human and nonhuman. The second law of thermodynamics defines the irreversibility (and therefore in an important sense, directionality) of the model and thus is a measure for everything that happens. Energy can be thought of as a material cause of everything, but in an instrumental, or agential, sense energy causes nothing; it merely requires that all events evolve so as to reduce the capacity to do work that may be available in the system. This provides a measure for the analysis of human history, one that is culture free and value free and uncolored by political or moral concerns. This permits comparison among all historically known societies of any culture.

Although not always explicit, two perspectives—the agential and holistic—have been used to describe the relationship between energy and human society. The first treats energy as a resource that is controlled by human beings to secure their survival and adaptation, i.e., to sustain their life. Seen in this way, the environment is full of energy resources—biota (including human beings), fossil fuels, water and solar radiation, subnuclear sources, etc.—that human beings harness and control to work for them. Somewhat problematic in this picture is human labor, because individuals control their own labor and may control the labor of others. Thus, when referring to the energy per capita expended by a society in a contemporary industrial nation, we are usually talking about nonhuman expenditure, but when dealing with preindustrial societies and particularly in hunting and gathering societies, human labor is the major energy source. Two phases emerge from this situation: (1) energy (including one's own energy) as controlled by individual human beings and (2) energy as controlled by society, usually by some agent for the collectivity. In complex societies (i.e., with two or more levels of social integration), leaders or authorities decide how energy resources, both nonhuman and human collectivities, are to be used. This perspective readily separates human energy expenditure as an agential process that controls and is impacted by on-going energy-based events.

In the holistic perspective, human beings and the society and the environment relevant to them are a continuing and fluctuating field of multiple energy flows. Within this field, there are islands or sectors of potential energy (materials, fuels, stored goods) held in equilibrium, the energy of which may be released for human use when an appropriate technology is applied. There are also dissipative structures (see later) wherein human labor is seen as merely one part of the flow, as simply one more energy resource. In this perspective, the triggers (see later) that release potential energy are energy flows. This perspective allows us to see social processes in terms of a series of principles, laws, and working hypotheses derived from the physical and biological sciences. If sociocultural evolution is merely one facet of general biological evolution, then these concepts and principles should be equally applicable to social evolution.

3. DYNAMICS OF ENERGY FLOW

There are a number of energetic processes useful for understanding sociocultural evolution.

3.1 Laws of Thermodynamics

The basic dynamics of energy are described in the first and second laws of thermodynamics, which require no elaboration here. The fact that energy is dissipated makes it unidirectional and measurable. What has happened can never be repeated—no matter how much new events may appear to duplicate history. The presence of historical similarities is a challenge to find out why and how they came to be.

3.2 Equilibrium and Dissipative Structures

Although it was recognized that energy was central to the social process, Ilya Prigogine's 1947 formulation of the thermodynamics of nonequilibrium structures provided a key concept—the dissipative structure—for treating human beings and human societies as dissipative energy forms within a field of both equilibrium and dissipative structures. Equilibrium structures are energy forms in relative equilibrium, in which no exchange of matter or energy is, for the moment, taking place with the environment. Dissipative structures are open energy systems that constantly incorporate energy and matter as inputs and dissipate them as output. The form they take is dissipation. They are characterized by fluctuations that are essentially indeterministic when seen in a microscopic field, but, with the operation of natural selection, may contribute to directionality over longer courses of time.

As adapted from Prigogine's model the dissipative structure provides a remarkably accurate representation of how living bodies and societies operate. It combines the inputs of matter and energy but the output also involves the discarding of waste materials that may impact and change other energy flows. Particularly significant are impacts that release potential energy of equilibrium structures and that divert or modify other dissipative processes. To use dissipative structure with sociological models requires only that we find the energy values associated with the reality being represented.

One criticism of dissipative structures as an analytical model for human societies is that it was "a nice analogy, yet while it can explain certain qualitative changes in a system by showing how order emerges from fluctuations, it cannot account for complex cultural systems with their interdependent subsystems, which influence self-organization. And it cannot explain the new structure after a phase of instability." The point is not well taken because such models are not intended to explain the specifics of historical cases. Rather, they provide a model of structural dynamics—in this case, the presence of fluctuations inherent in dissipative systems, which, when energy flow is increased, may cause the structure to become more complex. This happens at critical points where the expansion forces new structures to emerge, or the system will fragment or collapse. Such dynamics characterize Carneiro's 1970 theory of the emergence of the state. The model cannot predict the cultural details of how it occurs—things that depend on specific historical antecedents.

3.3 Energy Maximization

Treating society as a dissipative process allows us to map sociological cases on the energy model and permits the measurement of the energy dynamics at work. With this, there are some energy principles that become useful. One of the earliest of these was Alfred Lotka's 1922 principle of energy maximization: "In every instance considered, natural selection will so operate as to increase the total flux through the system, so long as there is presented an unutilized residue of matter and available energy." It is not difficult to find illustrations of this, but wars are an obvious case. Cavalry dominates over foot soldiers, firearms dominate over the bow, nuclear bombs dominate over gunpowder. In more general terms, one indication that natural selection favors societies with the greatest dissipation of energy may be seen in the reality that societies that have survived and flourished into the present day, dominating competitors, are precisely those that have dissipated the most energy. The society or species that monopolizes resources leaves none for others. The second clause of Lotka's statement, however, states that when energy is not available, maximization cannot take place. This may result from the exhaustion of resources (the path to the "the limits of growth") but may also be due to the presence of boundary conditions or technological inadequacies that prevent access or expansion.

3.4 Minimum Dissipation

Boundaries to energy systems are critical because they may be imposed from the outside, or they may be created by society. This points to a second proposition, the Prigogine–Waime principle of minimum dissipation: "When given boundary conditions prevent the system from reaching thermodynamic equilibrium (i.e., zero entropy production) the system settles down in the state of 'least dissipation'." This states the conditions under which maximization will not take place. Adult human life is a fluctuating steady state, a homeostatic system, in which internal controls maintain a level of energy use that does not exceed the available and anticipated rate or amount of input. Boundaries may be imposed by disastrous droughts, floods, and plagues, or gradually but consistently, such as by endemic diseases or endemic warfare. Boundaries are often imposed by society through controlling life, such as infanticide, careful birth spacing practices, and putting the elderly to death. Steady-state societies are those that have found ways of controlling the consumption of energy so as to survive on the energy available.

Another way that the principle operates may be seen in the splitting off of part of a society when it is growing too large for comfort. Thus, growing hunting bands and agricultural villages, reaching a certain population size, will hive off, forming new bands or settlements. New boundaries are created because the old boundaries prove inappropriate for survival. In this way, societies follow the minimum dissipation principle because the amount of energy flow is kept to some minimum within the society; but they may succeed in doing this only by creating other societies with boundaries to separate them from the old.

Classic cases have been the peasant philosophy of the "limited good," a pattern of social control whereby those who try to surpass others in some endeavor—and thereby stretch the exploitation of available resources—are condemned and punished by their fellows and restrained from excelling. Some highland New Guinea societies sponsor periodic pig feasts, as described by Roy Rappaport: "The operation of ritual...helps to maintain an undegraded environment, limits fighting to frequencies which do not endanger the existence of the regional population, adjusts man–land ratios, facilitates trade, [and] distributes local surpluses of pig throughout the regional population...." Socially controlling for minimum dissipation is usually achieved in societies only after hard experiences with the consequences of exceeding energy resources.

The advent of capitalism and industrial exploitation of the fossil energy resources, when coupled with an era of rapid population expansion, has generally obscured the implications of the Prigogine–Waime principle. Contemporary industrial societies have sought to expand energy use rather than to limit it. Although societies differ in the degree to which they attempt to shift to more dependable sources—e.g., energy derived directly from solar radiation, river or tidal waters, wind, or hydrogen—none has as yet made a marked shift in that direction. In most industrial countries, such concerns are marginalized by commercial and political interests that promote a policy of continual expansion of dependency on traditional sources. Everywhere, however, the Prigogine–Waime principle operates as available energy resources decline.

3.5 Trigger-Flow Ratio

Energy enters society principally as solar energy but also as moving water or wind and in equilibrium or quasiequilibrium forms, as potential energy—foodstuffs, minerals, forest products, etc. The release of potential energy is literally triggered and then, depending on the technology, controlled as it dissipates into lower grades and entropy, or leaves as a residue of material artifacts and waste. Because triggers and subsequent controls require energy to operate, it is generally expected that the ratio of their energy cost to the amount of energy flow dissipated should be low. Consistent failure to achieve this can lead to the trigger energy exhausting the energy available to the system.

Energy triggers are parts of a larger process, sometimes seen as an investment of energy. It has been proposed that this energy investment is a useful analytical tool, and as such has been referred to variously as embedded energy or, as by Howard Odem, as emergy. The relationship of this investment to the energy it causes to be released is seen as a net energy gain. Odum argued that an increase in the ratio of embedded energy to the energy released constitutes an increase in energy quality. The utility of this concept of energy quality is problematic because, as Joseph A. Tainter and colleagues have noted, "Energy quality in a human system varies with context, as it may also in other living systems." Yasar Demirel prefers to see quality as "the ability of energy to be converted into mechanical work under conditions determined by the natural environment." Emergy and embedded energy are not easy to apply to real situations because they imply an infinite series of antecedent inputs; in fossil fuels alone they include "the energy that previously went into the organism from which these fuels originate as well as the energy from the chemical and geological processes that transformed dead organisms into coal, gas, or petroleum." The importance of net energy lies in economic analyses of the more immediate energy costs.

Because food is everywhere required by human beings, changes in its dynamics reveal much about the course of cultural evolution. History has seen four major kinds of trigger technologies to produce food, all of which require human input: (1) human power in hunting and gathering, (2) cultivation based solely on human energy, (3) cultivation using work animals, and (4) industrialized agriculture. The ratio of trigger energy costs to resulting dissipation changes markedly in these different phases.

Minimally, trigger costs must involve human beings, but other energy sources are also required (tools, fuel, etc.). Table I orders data comparing societies at various levels of energy consumption in terms of the relative amounts of human and nonhuman energy used in food procurement. Early human beings, like contemporary hunters and gatherers, depended almost wholly on their own human energy, as illustrated by the !Kung bushmen (Table I, column C). !Kung gatherers' energy output was almost four times as much as their food energy input (Table I, column F). The development of cultivation increased this gain from 11 to 24 times as much (column F). However, because other energy resources were also used, the output/input gain was somewhat less (column E). With the introduction of work animals, the input of nonhuman energy increases (column C). In general, however, the use of draft animals adds to the trigger costs and provides relatively less yield than can be gained by human beings alone (compare the Mexican production in Table I); the advantage lies in the saving in time it allows. Industrialization brings a major increase in yield over human energy inputs (Table I, column F). With respect to total trigger costs, however, industrialization reduces the net energy gain to a level below that of the !Kung, and much below that of any all the subsistence agriculturalists (column E). In some cases, when the caloric yield is not the object of the cultivation—as with spinach and brussels sprouts—there is a clear net energy loss in the production (column E). Human activity continues to be important in industrial situations. Comparing Japanese and U.S. rice production shows that human activity yields greater efficiency in industrial situations. The Japanese farm requires much more human labor (column C), but

TABLE I

Trigger Energy Cost and Released Flow under Different Food Production Regimes[a]

	A	B	C	D	E	F
Food production effort	Human energy input (kcal)	Total energy input (kcal)	Total energy input/human energy input	Energy output (kcal)	Energy output/ total energy input	Energy output/ human energy input
Human energy gathering						
!Kung bushmen nuts	2680	2680	1.0	10,500	3.92	3.92
Human energy cultivation						
African cassava	821,760	838,260	1.0	19,219,200	22.93	23.39
New Guinea swidden	686,300	739,160	1.1	11,384,462	15.40	16.59
Mexican corn	589,160	642,338	1.1	6,901,338	10.74	11.71
Iban Borneo rice	625,615	1,034,225	1.7	7,318,080	7.08	11.70
Work animal energy						
Thailand peanuts (buffalo)	585,040	1,923,410	3.3	4,992,000	2.60	8.53
Mexican corn (oxen)	197,245	770,253	3.9	3,340,550	4.34	16.94
India wheat (bullocks)	324,413	2,827,813	8.7	2,709,300	0.96	8.35
Industrialized agriculture						
Japanese rice	297,600	8,221,040	27.6	22,977,900	2.80	77.21
U.S. brussels sprouts	27,900	8,060,328	288.9	5,544,000	0.69	198.71
U.S. spinach	26,040	12,759,849	490.0	2,912,000	0.23	111.83
U.S. rice	11,000	11,017,000	1001.5	23,642,190	2.15	2149.29
U.S. corn	4650	10,535,650	2265.7	26,625,000	2.53	5725.81

[a] From Pimentel and Pimentel (1996).

has a higher total net yield compared to the much more energy-expensive U.S. production (column E).

Although high-energy production is costly in kilocalories, it saves time. Pimentel and Pimentel calculate that the energy required to till a hectare of land by human power alone requires 200,000 kcal in 400 hr; with oxen, energy input increases to $\approx$300,000 kcal in 65 hr; with a 6-hp tractor, it increases to $\approx$440,000 in 25 hr; with a 50-hp tractor, it increases to $\approx$550,000 in 4 hr. Time, of course, may be critical to survival.

Comparable changes took place in the other areas in the Industrial Revolution. Whereas human labor in fossil-fueled industry at first expanded, the fossil fuels gradually displaced human labor in most of the production process. Adams showed that in Great Britain, for example, the human population in agriculture and industry grew to over 30% of the total in 1860, then declined by 1990 to well below 20%. In Japan, between 1920 and 1990, it also declined, from 36% to, again, below 20%. All countries that have become significantly industrialized have experienced this decline. However, although human energy used in triggers has declined, total energy costs have increased profoundly, drawing on what was thought to be an unlimited availability of fossil and gravitational energy. In industrial society, the energy cost of triggers grew not only through the accessibility of fossil fuels, but through the emergence of a vast complex system of linked triggers, comprising commercial, industrial, governmental, and societal organizations. Even where consumption is relatively low, the ever-increasing need for energy is a serious challenge to the future of industrial societies.

3.6 Entropy and Energy

Some scholars of energy and society have found entropy to be useful for societal analyses. Particularly influential was Nicholas Georgescu-Roegen's work in economics, directly arguing the implications of the second law of thermodynamics against more classical views of the inevitably progressive dynamics of the market. The focus on entropy in the biological and social sciences, however, took wings when Claude Shannon and Warren Weaver used the term to characterize the statistical patterns in information processes, because they conformed to Ludwig Boltzmann's algebraic formula for thermodynamic entropy.

Scholars differ in their acceptance of the appropriateness of identifying the informational process with thermodynamics. Jeffrey Wicken argues that the former does not refer to changes in physical space but rather in complexity of structure. He argues that it is really an "entropy analogue." It is not a process for which reversibility is highly improbable because complex states may have a low or high probability of becoming more or less complex. The identity has, nevertheless, been found useful by various scholars. Daniel R. Brooks and E. O. Wiley propose a formulation of biological evolution whereby "the strictly thermodynamic entropy of the organism... increases along with the informational entropy, since increasing structural complexity requires increasing energy to maintain the steady state." Sociologist Kenneth D. Bailey tends to put aside thermodynamic questions and elaborates a macrotheory of society that focuses on entropy in terms of the probabilities of order and variation.

Students of cultural evolution have not generally taken up the challenge of the entropy formulations, although Jeremy Rifkin, in a popular account, traced the increasing use of energy as being inextricably interrelated with the creation of disorder, chaos, and waste over the course of human history. The apparent preference of scholars for the analytic utility of energy over entropy may be because the materials of social evolution and history are more easily measurable in terms of the former than the latter.

4. *ENERGY IN EVOLUTION OF HUMAN SOCIETY*

4.1 Culture and Evolution

Sociocultural evolution is a part of the evolution of life, a process that has been best formulated in Darwinian terms. Although some discussions differentiate biological and social evolution, in fact, the individual and the social setting are necessarily complementary parts of a single process. Human life is generated in the biology of individuals, but it survives and reproduces by virtue of the aggregate activities and interactions that take place as a part of societies. Whereas all living forms are in some way complex, social complexity is best understood as the compounding of successive levels of integration, of which the individual is one. Thus, no matter the complexity of a human cell, an individual, a family, a band, or a nation, this complexity is compounded by requiring additional interrelationships—and hence organization—among the aggregated components, thereby creating new levels of integration.

Human society grows by increasing energy flow. This is done first by biological expansion of the population. Natural selection has favored those that organized in order to reproduce, thus creating societies—what Spencer called the superorganism. It is the evolution of human superorganisms that we call social or cultural evolution. The Darwinian processes of reproduction, variation, and natural selection concern individuals, but societies exist to propagate individuals and continue to survive while individuals die. Human society is a facet of natural selection in that it organizes people to reproduce the human species.

To a degree unmatched by other species, human beings control a vast range of extrasomatic energy forms that are captured, gathered, cultivated, and mined from the environment and other societies. This extraction of energy is a cost to the environment caused by the emergence and evolution of human society. It may also be seen as a coevolution in which the human component differentially promotes and inhibits the evolution of many other species. Culture consists of the mental models people make to deal with the external world, and the behavioral and material forms that correspond to these representations. People use these mental models to manipulate the external world, to control external energy forms. In actual historical fact, the models may be the basis for creating controls, or the model may be created to describe a control that has been discovered to work; in any event, the external controls and their models develop in an interactive process. Mental models trigger energy flows. When we speak of the evolution of culture, we are speaking of both the change in the mental models and the changes in energy forms that correspond to these models. Given that both are involved, the term "cultural evolution" has been used to refer to two phases of changing human behavior: (1) the process whereby human beings adapt their mental models and social and cultural behavior to deal with challenges to their survival, and (2) the process whereby mental models, societies, and cultures become more complex as they adapt to deal with challenges to their survival.

Some treatments of cultural evolution see the first as adaptation and prefer to see the increasing complexity in terms of more complicated or compounded social organizations, or in more complicated technology that allows for the control of greater sequences of energy. The relationship between energy and cultural evolution lies in two phases of the process. One is the construction of ever longer and more complex linked trigger-flow systems. The other is the harnessing of additional energy from the environment, allowing more opportunities to tap the flow through its successive stages of dissipation. The Australian aborigines set savanna fires that release an immense amount of energy but exercise no further control over it. There is no successive system of triggers. In contrast, the fire in a steam locomotive converts water to steam that pushes a piston, that directly turns the wheels, that travels along tracks and pulls a number of railway cars, etc. The detonation of a nuclear bomb requires the careful arrangement of a chain of trigger actions before the explosion emerges.

4.2 Population Growth

The most important dynamic process in sociocultural evolution is population expansion, a thing that, in itself, entails a proportionally greater extraction of nutritional energy from the environment. That energy, however, is then further consumed by human activity that further promotes human life.

Population pressure is central to Robert L. Carneiro's circumscription theory and to Marvin Harris' depletion theory of cultural evolution; it is also at the core of Esther Boserup's concern with technological change. As such, it risks being tautological; but it need not be. Circumscription theory proposes that an expanding population pressing against an unyielding environment produces changes in social organization. Harris's depletion theory holds that societies inevitably deplete their environment and must, therefore, intensify their production to retain their living standard. The problem is that the evidence for the population pressure and for depletion lies in the fact that the social organization changes. To avoid the tautology, the problems entailed by this pressure may be specified. The amount of energy flow is central to both processes. In both hunter-gatherer and agricultural societies, changes in population density or in level of food production threaten the availability of energy inputs to the society. Land:human ratio reduction and/or environmental depletion will reduce food availability to the point that social survival is threatened. The same argument applies in defense of Boserup's thesis that farming intensified because of declining production. Specifying the connectivity avoids the tautological by showing how social organization changes due a demonstrable need to improve the food supply.

A response to such pressure in simple societies is that the society fragments, allowing the separate segments to hive off in the search for other lands. Another response is that the society reaches out for other resources by warfare or trade, using local resources, such as mineral or aquatic resources, to trade for the inputs needed. Where no other possibility is available—or at least known within the society's cultural repertory—members may turn to warfare. This in itself, especially in complex societies, requires additional expenditures of energy. Wars may fail or may succeed in eliminating the enemy and in taking their land and resources without producing major changes in social complexity beyond those necessary for war. However, if they capture the other society and control its production, then the emergence of political subordination, slavery, etc. requires a new level of integration, i.e., greater complexity. It is only in the last case that new external sources of energy are incorporated, and it is there that the society becomes more complex.

Although there is good reason to believe that population pressure and density led to social expansion and intensification in the preindustrial eras of human history, the emergence of states decisively changed the dynamics. States often want to hold a territory, regardless of whether it has any immediate productive use, and also to hold territories that may be important for production, but may be sparsely inhabited. States therefore may occupy vast territories that have little relevance to the immediate needs of their population. Given high levels of economic control and the relative ease of international migration, relatively small states may feel population pressures that are continually eased by emigration, both temporary and permanent. Urbanization, begun with the appearance of advanced agricultural systems, continued to be a central part of the solution under the state. The consequence of these differences is that even though hunting bands combine low energy consumption with low population density, there are some states that have low energy consumption—such as Haiti, Rwanda, and Bangladesh—that have very high population density; and there are states with high-level energy consumption—Australia, Canada and the Iceland—with very low densities. Such differences would not be possible in early (i.e., low energy consumption) phases of sociocultural evolution.

R. B. Graeber has proposed a mathematical theory of cultural evolution, relating population, territorial area, societal numbers, density, and mean societal size. He concludes that density and mean societal size are "indicators of the cultural-evolutionary state attained by a population." Although suggestive for earlier phases, the argument fails to deal with the contradictory cases among contemporary states.

4.3 Stages of Sociocultural Evolution

Anthropologists have long used stages of sociocultural evolution to characterize classes of societies in evolution. Although various sequences and scales of complexity have been proposed, no systematic effort has been made to relate them to the level of energy use. In the 19th century, Morgan, Tylor, Spencer, and Karl Marx all proposed sequences that were generally progressive and argued to constitute some kind of an "advance." In the past century, Childe, Cottrell, Service, and Morton Fried have made such propositions. The formulation of stages, however, has consistently reflected an author's particular interests, and some of the proposals were not made with energy in mind. Karl Marx proposed stages concerning modes of production (tribal, ancient, feudal, and capitalist) but he evinced no interest in thermodynamic issues. His follower, Friedrich Engels, was more explicit; for him to consider energy in economics was "nothing but nonsense."

The concern for technology early appeared in Morgan's sequence of savagery, barbarism, and civilization, whereby the transition between the first two was specifically based on domestication, and therefore with energy consumption and technology. The criteria for a civilization referred not to energy changes but to the appearance of cities and the population densities that were thereby implied. Energy conversion technology implicitly came to include control of human energy by the state. On this basis, Childe proposed two major revolutions in prehistory. The first was the Neolithic Revolution that specifically increased food production. The second was the Urban Revolution that was based on the invention of the plow, but its major manifestation was the emergence of cities based on the large economic surpluses. The sequence suggested in 1971 by Service—bands, tribes, chiefdoms, and state—shifted the concern from a technology to the relative complexity of social organization.

Evolutionary stages have been useful as sequences of kinds of societies reflecting differences in, among other things, energy consumption. Otherwise, the interest shown by social anthropologists in energy has been related to interests in ecology from an ethnographic perspective. Actual measurements have been forthcoming in a few field studies, such as those by Roy A. Rappaport in New Guinea, W. B. Kemp among Eskimo, Pertti Pelto with the Skolt Lapps, and Richard Lee and L. J. Marshall with the !Kung Bushmen.

Contemporary treatments of the stages of cultural evolution have dispensed with the notions of progress or advance that were inherent in 19th and early 20th century versions. However, there have been few major formulations that take into account the potential exhaustion of energy resources for the species as a whole and the consequent potential decline that it implies. Tainter has explored the process of decline and collapse of specific societies and civilizations, but, given the dominant role of historical contingencies, there is little basis for predicting just how cultural devolution will emerge. There is reason to suspect, however, that although the societies and cultures became more complex with increases in energy, the internal dynamics of complex organizations will lead to decreasing returns in the future. Moreover, the social life of the species has evolved as innumerable separate societies, and these will each continue to trace their own course.

4.4 Energy and Complexity

No matter how heuristically useful stages of growth have been in ordering materials, they tend to imply an unwarranted punctuated quality to the course of cultural evolution. In fact, the process has been much more gradual, fluctuating over time and space, and the central quality that reflects change in energy flow is the change in complexity.

In 1943, Leslie White formulated the first specific proposition relating energy consumption to cultural evolution:

"Culture develops when the amount of energy harnessed by man per capita per year is increased; or as the efficiency of the technological means of putting this energy to work is increased; or, as both factors are simultaneously increased." Although it has recently been claimed that "the thermodynamic conception of culture has suffused cultural evolutionism," there is little work in the anthropological literature to support this. Whereas the domestication of plants and animals was accompanied by increases in population, fluctuating increases in energy consumption, and imperial expansions over the past 5000 years, the exponential per capita growth of both population and energy flow most markedly began with the Industrial Revolution. Boserup calculates that in 1500 AD the population density in Europe was still under 8 persons/km^2, and by 1750 it was still below 16 persons/km^2; by 1900 it had reached averages as high as 128 persons/km^2.

In defining sociocultural complexity, it helps to differentiate the social from the cultural. It is useful to see social complexity as the creation or differentiation of social relationships in new configurations. Most importantly, sociocultural complexity refers to the creation of new levels of social integration, hence to stratification, centralization, inequality, and emergence of hierarchy. Service's four stages outlined a succession of compounding organizations. Cultural complexity, however, refers to the heterogeneity and multiplicity of culture, i.e., meaning systems and the corresponding material and behavioral forms—as in technology (communication, engineering), social organization (voluntary associations, networking), and mental models (intellectual and symbolic systems) allowing for mathematics and astronomy, to suggest just a few. A number of scales of complexity have been proposed for simpler societies. One by Carneiro proposed the relevance of over 600 traits in 72 societies and another by George Peter Murdock and Catarina Provost was based on 10 traits in 180 societies; the two scales correlate highly. Henri J. M. Claessen and Pieter van de Velde have proposed a specific model of evolution as a complex interaction.

The salience of comparative energy consumption has been clearly related to social complexity in the study of contemporary societies at the national level. Somewhat contrary concerns for energy shortages on the one hand, and environmental damage on the other, have stimulated a concern over the past four decades for the relation between levels of energy consumption and various aspects of living. The interests here differ from those concerning the stages of cultural evolution, not merely because they are contemporary and deal with nation states, but because they are motivated by the concern for the survival of society in the short and long terms. At this level, there is little doubt that there is a relationship between various measures of quality of living and energy consumption. In 1970, Boserup divided 130 nation-states evenly into five groups based on technology and demonstrated the consistent difference in some indices of development and welfare. The first group had an average per capita energy consumption of 50, the third had 445, and the fifth had 4910. The corresponding life expectancies were 41, 55, and 72 years. The percentage of the population that was literate was 12, 46, and 98%, respectively. However, in a much-cited 1965 study, Ali Bulent Cambel reported that in countries with over $1500 per capita, the energy consumption varied widely from below 50 to over 175 Btu per capita.

In relating the amount of energy used by a society to its complexity, there are a number of conditioning factors. First, in both the agential and holistic perspectives, the control of energy must distinguish between controls that merely release energy but do not further control it; and controls that not only release but can limit or amplify the flow, its direction, and its impact, i.e., modulate the flow. In the first case, the energy is totally an output, a degrading that removes it from the society, such as setting an uncontrolled fire, detonating a bomb, or pushing a person off a cliff; the impacts are randomized. In the second, the available energy is kept within control of the society, and the impacts are controlled, such as the adjustment of the level of electric current, the ability of a court to vary a prison sentence, or of a farmer to decide when to plant a crop. The Australian aborigines who set extensive range fires to clear out savanna litter have no control over its dissipation. According to one estimate, such fires may release up to 500 million kcal of biomass energy a year—six times that used by the average American on a per capita basis—but the fires and the dissipation cannot be controlled.

Second, the increase of societal complexity depends not merely on the total energy released, nor even that for which there is control over the dissipation, but whether the controlled energy is effective in the self-reproduction of the society. In terms of comparative societal evaluation, this is technically more difficult to ascertain. One has to judge whether ceremonial structures and performances dedicated to appeasing a rain god are in fact more cost-effective than the construction of an irrigation system, or at what point the energetic costs of imperial expansion cease to be productive for the society's reproduction. Returning to White's proposition, a revised version might be as follows: the evolution of sociocultural complexity varies directly with the amount of energy controlled per capita per year and with the efficiency of the technological means of putting this energy to work in behalf of societal reproduction.

Although there are cases in which culture appears to be devolving and to be returning to some earlier form, in fact such a return simply does not happen. If there were a vast and gradual decline of energy available for consumption, it would require cultural simplification and reduction in complexity, but this would require new adaptations. But there would be no strict reversion to the social and cultural forms that emerged during the expansive phases of the species over the past few million years, or of the human civilizations of the past 50,000 years. The so-called devolution would, among other things, be the evolution of cultures that would be formally different from their predecessors, but necessarily simpler in the diversity of traits and relationships.

4.5 Partitioning of Energy

A facet of energy flow analysis that can help clarify the nature of society is how it is differentially partitioned among different internal functions and external relations. The way that energy is channeled and how this changes throw light on what is taking place in the structure of the society. For example, the proportion of the total energy flow involved in feeding the society seems remarkably similar in both simple and complex industrial societies. John A. Whitehead compared two Australian aboriginal bands with the population in the United States. Whereas the aboriginal bands used 20–30% of their energy resources for food gathering, in the United States, household plus commercial (which included more than food) consumption was estimated at 27%. Joel Darmstadter and colleagues found in eight Euro-American countries that the household–commercial sector constituted between 24 and 32% of the output, and 19% in Japan. This suggests that there is a consistent relationship between the nutrition requirements of the human beings and the economic process that surrounds them, irrespective of the energy level of the society.

In contrast, some differences appear in human energy partitioning as the total energy flow increases. In comparing a series of nations in the past 150 years, Adams found that the proportion of the population engaged in productive activities—i.e., agriculture and industry—was as high as 40% in the 19th century, but then dropped, often below 20% by the end of the 20th century. This was true of both highly industrialized countries as well as much of the Third World that increasingly received the industrial produce of advanced countries. In contrast, the population engaged in administration and commerce was everywhere between 2 and 4% in the early censuses, but then rose to as high as 20%, varying directly with the increase in nonhuman energy. The expansion of this regulatory sector of the population was systematically more marked in the countries with high levels of energy consumption.

Changes in the partitioning of energy consumption were fundamental in the evolution of the capacity for culture. Leslie C. Aiello and Jonathan C. K. Wells write that "the emergence of *Homo* is characterized by...an absolute increase, to greater body size [and] a shift in relative requirements of different organs, with increased energy diverted to brain metabolism at the expense of gut tissue...." The nutrition patterns changed profoundly as the range of the human diet allowed greater territorial range, and digestive efficiency increased and the relative amount of energy required decreased. This decreasing proportion of energy dedicated to food processing was accompanied by an increase in the size of the brain and the relative amount of energy contributing to it. The increased efficiency of food processing was thus related to an increase in diet quality, which is consistent with the observation that primates with higher diet quality have higher levels of social activity.

These changes occurred relatively rapidly as the hominid brain from *Australopithecus* to *Homo sapiens* grew at a rate almost three times that which characterized ungulates and carnivores in general. Part of this growth was increase in human brain size, and more particularly the appearance of fissures that Harry J. Jerison argued allowed "the evolutionary building of integrated neural circuits through which diverse sensory information could be used to create increasing complex mental images." This suggests that the ability to create models of reality—an important component in culture—existed 50 millions years ago. The energetic dimension of this change was reflected in the changing partitioning of energy use between the brain and the gut. Today, according to Marcus E. Raichle and Debra A. Gusnard, the brain represents 2% of body weight, but accounts for 20% of calorie consumption by the body.

4.6 The Limits of Growth

Although the conveniences made possible by increasing energy use and intensifying consumption were viewed as human progress by some, the vulnerability it implied was also recognized. Soddy early devoted a number of works to this aspect of contemporary societies, arguing that energy could be a limiting factor in social growth. He recognized that the shift from renewable agricultural resources to an increasing dependence on coal and, later, petroleum posed serious problems for advanced societies. Lotka's proposition on the maximization of energy in living systems carefully specified that the dynamics of greater energy flows exercising dominance over lesser flow could take place only "so long as there is presented an unutilized residue of matter and available energy."

For some, the question of energy shortages was less critical than a preoccupation with the damage done to the environment by the impacts of increasing energy expenditure. Concern for this was evinced in the 19th century by George Perkins Marsh and Alexander Ivanovich Voikof; more recently, Rachel Carson's *Silent Spring* (1962) helped trigger the environmental movement in the United States. The remainder of the 20th century saw the expansion of environmental politics and studies concerned with the destruction of natural resources over much of the world. Besides the exhaustion of nonrenewable energy resources, another major concern has been the effect of excessive combustion of fossil fuels on the atmosphere, global warming, and related deterioration of the ecology of the planet. The chemical and material pollution of oceans, lakes, and the air was coupled with the deposition of millions of tons of waste over the earth's surface. In all this, the role of increasing energy use was fundamental, both in the despoiling of resources and in polluting the environment. The central question is shifting from simply how to get more energy to how to reduce consumption.

The concern with the availability of energy resources in the nations consuming high levels of energy, especially the United States, replays the problems of circumscription and depletion in earlier human societies. The concern has dominated both economic and political policy and action to a degree that is seldom explored to the fullest. Availability has directly affected the development of environmental concerns in areas of potential oil exploitation, pitting those concerned for the long-term environmental and human sustainability issues against those interested in the monetary yield of exploiting as many resources as fast as possible and controlling resources in other parts of the world. The response of the United States to the terrorist attacks of 2001 was not unlike those postulated for early societies facing threats to their resources created by population growth and competitive societies.

5. CONCLUDING NOTE

Spencer recognized the importance of energetic processes to the evolution of culture and society in his formulation of the modern concept of evolution in the 19th century. It was formally articulated in 1943 by Leslie White, and interest since that time has been limited mainly to archaeologists, who have found in it a useful basis for comparing societies and for modeling the activities of prehistoric societies, and to a few ethnologists working in technologically limited societies. Independently an understanding of nonequilibrium thermodynamic systems was opened in mid-20th century by the work of Prigogine, yielding the model of the dissipative structure. This has provided the basis for the mapping of social processes onto energetic dynamics, a potential as yet little explored. The importance of energy to the organization of society and to its future has become a matter of major concern to contemporary nations because of their increasing demand for energy, and by environmentalists and globally conscious politicians and economists in their concern for ecological survival and the planet's ability to support human life. What may seem to be an arcane interest in the role of energy in cultural evolution is directly related to the imperative concern for the survival of human society.

SEE ALSO THE FOLLOWING ARTICLES

Development and Energy, Overview • Early Industrial World, Energy Flow in • Earth's Energy Balance • Global Energy Use: Status and Trends • Oil Crises, Historical Perspective • Population Growth and Energy • Sociopolitical Collapse, Energy and • Sustainable Development: Basic Concepts and Application to Energy • War and Energy

Further Reading

Adams, R. N. (1988). "The Eighth Day: Social Evolution as the Self Organization of Energy." University of Texas Press, Austin.

Boserup, E. (1981). "Population and Technological Change." University of Chicago Press, Chicago.

Carneiro, R. L. (2003). "Evolutionism in Cultural Anthropology." Westview Press, Boulder, Colorado.

Claessen, H. J. M., van de Velde, P., and Smith, M. E. (1985). "Development and Decline, The Evolution of Sociopolitical Organization." Bergin & Garvey Publ., South Hadley, Massachusetts.

Darmstadter, J., Dunkerley, J., and Alterman, J. (1977). "How Industrial Societies Use Energy." Johns Hopkins University Press, Baltimore.

Khalil, E. L., and Boulding, K. E. (eds.). (1996). "Evolution, Order and Complexity." Routledge, London and New York.

Lenski, G., and Lenski, J. (1996). "Human Societies." Fifth Ed. McGraw-Hill, New York.

Pimentel, D., and Pimentel, M. (eds.). (1996). "Food, Energy and Society," Rev Ed. Niwot, University of Colorado Press, Colorado.

Prigogine, I., and Stengers, I. (1984). "Order out of Chaos." Bantam Books, New York.

Rosa, E. A., and Machlis, G. E. (1988). Energetic theories of society: An evaluative review. *Sociol. Inquiry* 53, 152–178.

Sanderson, S. K. (1990). "Social Evolutionism." Blackwell, Cambridge and Oxford.

White, L. (1959). "The Evolution of Culture." McGraw-Hill, New York.

Decomposition Analysis Applied to Energy

B. W. ANG
National University of Singapore
Singapore

1. Introduction
2. Basics of Decomposition Analysis
3. Methodology and Related Issues
4. Application and Related Issues
5. Conclusion

Glossary

aggregate energy intensity The ratio of total energy use to total output measured at an aggregate level, such as industrywide and economywide.

decomposition technique The technique of decomposing variations in an aggregate over time or between two countries into components associated with some predefined factors.

Divisia index A weighted sum of growth rates, where the weights are the components' shares in total value, given in the form of a line integral.

energy intensity The amount of energy use per unit of output or activity, measured at the sectoral or activity level.

gross domestic product (GDP) A measure of the total flow of goods and services produced by the economy during a particular time period, normally 1 year.

index number A single number that gives the average value of a set of related items, expressed as a percentage of their value at some base period.

Laspeyres index An index number that measures the change in some aspect of a group of items over time, using weights based on values in some base year.

structure change A change in the shares or composition of some attribute, such as sector output in industrial production and fuel share in energy use.

This article discusses the decomposition techniques formulated using concepts similar to index numbers in economics and statistics. They have been called the index decomposition analysis. Approximately 200 energy and energy-related studies have been reported since 1978. Of these, slightly less than 20% deal with methodological or related issues and the remaining are application studies. Several other decomposition analysis methodologies have also been adopted in energy analysis, including the input–output structural decomposition analysis, which uses the input–output approach in economics to carry out decomposition. In terms of the number of studies reported, these techniques are relatively less popular compared to index decomposition analysis. The basics of index energy decomposition analysis are presented here. This is followed by a review of the methodological and application issues.

1. INTRODUCTION

Accounting for the largest share of primary energy demand in most countries, energy use in industry attracted considerable attention among policymakers, researchers, and analysts in the aftermath of the world oil crisis in 1973 and 1974. Effort was made by researchers and analysts to understand the mechanisms of change in industrial energy use through engineering process studies, energy audits, and industrywide analyses. At the industrywide level, a new research area began to emerge in approximately 1980, namely the study of the impact of changes in industry product mix, or structural change, on energy demand. It was realized that structural change could greatly influence energy demand trends, and there was a need to identify and quantify its impact to assist in policymaking. This line of research has since expanded substantially in terms of both methodology and application. It is now called decomposition analysis in the energy literature.

An important feature of decomposition analysis in the 1980s was its close linkages with the study of the aggregate energy intensity of industry. As an aggregate for the entire spectrum of industrial sectors/activities, this indicator is defined as the total industrial energy use in a country divided by its total industrial output. A decrease in the aggregate energy intensity is generally preferred since it means that a lower level of energy requirements is needed to produce a unit of industrial output. At approximately the same time, several cross-country comparative studies of industrialized countries were conducted, and countries with comparatively low aggregate energy intensity were considered more effective in energy utilization.

Researchers and analysts then began to examine factors contributing to changes in the aggregate energy intensity of countries over time (and variations in the aggregate energy intensity between countries). In the simplest form, two contributing factors were identified, namely structural change and energy intensity change. Industry is an aggregate comprising a wide range of sectors (or activities), such as food, textiles, petrochemical, transport equipment, and electronics. Structural change is associated with the varying growth rates among these sectors, which lead to a change in the product mix. Energy intensity change is determined by changes in the energy intensities of these industrial sectors. The sectoral energy intensity, or simply energy intensity, is the ratio of energy use to output for a specific sector defined at the finest level of sector disaggregation in the data set. To determine the impacts of these two factors, the observed change in the aggregate energy intensity is decomposed using an appropriate decomposition technique. Since 1980, numerous decomposition techniques have been proposed.

Information obtained through decomposition analysis has direct policy implications. The composition of industrial activity is effectively beyond the control of the energy policymaker except through indirect measures such as energy pricing. On the other hand, many measures can be taken to influence sectoral energy intensity change, such as taxation, regulatory standards, financial incentives, and information programs. The ability to accurately quantify the relative contributions of structural change and energy intensity change provides insight into trends in industrial use of energy. The success or failure of national energy conservation programs may be evaluated. The decomposition results also provide a basis for forecasting future energy requirements and for evaluating appropriate courses of action.

2. BASICS OF DECOMPOSITION ANALYSIS

Since index decomposition analysis was first applied to analyze energy use in industry, it is appropriate to explain the underlying concept with reference to the decomposition of a change in the aggregate energy intensity of industry. To keep the discussion simple, we consider the two-factor case in which a change in the aggregate intensity is decomposed to give the impacts of structural change and sectoral energy intensity change.

2.1 Notations and Basic Formulae

Assume that total energy consumption is the sum of consumption in n different industrial sectors and define the following variables for time t. Energy consumption is measured in an energy unit and industrial output in a monetary unit.

E_t = Total energy consumption in industry

$E_{i,t}$ = Energy consumption in industrial sector i

Y_t = Total industrial production

$Y_{i,t}$ = Production of industrial sector i

$S_{i,t}$ = Production share of sector i $(= Y_{i,t}/Y_t)$

I_t = Aggregate energy intensity $(= E_t/Y_t)$

$I_{i,t}$ = Energy intensity of sector i $(= E_{i,t}/Y_{i,t})$.

Express the aggregate energy intensity as a summation of the sectoral data:

$$I_t = \sum_i S_{i,t} I_{i,t},$$

where the summation is taken over the n sectors. The aggregate energy intensity is expressed in terms of production share and sectoral energy intensity. Suppose the aggregate energy intensity varies from I_0 in Year 0 to I_T in Year T. Such a change may be expressed in two ways, $D_{tot} = I_T/I_0$ and $\Delta I_{tot} = I_T - I_0$, respectively a ratio and a difference. Accordingly, decomposition may be conducted either multiplicatively or additively, respectively expressed in the forms

$$D_{tot} = I_T/I_0 = D_{str} D_{int}$$
$$\Delta I_{tot} = I_T - I_0 = \Delta I_{str} + \Delta I_{int},$$

where D_{str} and ΔI_{str} denote the impact of structural change, and D_{int} and ΔI_{int} denote the impact of sectoral intensity change. In the multiplicative case, all the terms are given in indices, whereas in the additive case all the terms, including the aggregate being decomposed, have the same unit of measurement.

To quantify the impacts, two approaches—the Laspeyres index approach and the Divisia index approach—have been widely adopted by researchers and analysts.

2.2 Laspeyres Index Approach

The Laspeyres index approach follows the Laspeyres price and quantity indices by isolating the impact of a variable by letting that specific variable change while holding the other variables at their respective base year values (in this case, Year 0). The formulae for multiplicative decomposition are

$$D_{str} = \sum_i S_{i,T}I_{i,0}/\sum_i S_{i,0}I_{i,0}$$

$$D_{int} = \sum_i S_{i,0}I_{i,T}/\sum_i S_{i,0}I_{i,0}$$

$$D_{rsd} = D_{tot}/(D_{str}D_{int}).$$

In additive decomposition, the formulae are

$$\Delta I_{str} = \sum_i S_{i,T}I_{i,0} - \sum_i S_{i,0}I_{i,0}$$

$$\Delta I_{int} = \sum_i S_{i,0}I_{i,T} - \sum_i S_{i,0}I_{i,0}$$

$$\Delta I_{rsd} = \Delta I_{tot} - \Delta I_{str} - \Delta I_{int}.$$

The Laspeyres index approach does not give perfect decomposition. The residual terms D_{rsd} and ΔI_{str} denote the part that is left unexplained. Decomposition is considered perfect if $D_{rsd} = 1$ and $\Delta I_{str} = 0$. These are also target values of the residual terms for evaluating the performance of a decomposition technique.

2.3 Divisia Index Approach

The Divisia index is an integral index number. In this approach, first the theorem of instantaneous growth rate is applied to $I_t = \sum_i S_{i,t}I_{i,t}$, which gives

$$d\ln(I_T/dt) = \sum_i \omega_i\left[d\ln(S_{i,t})/dt + d\ln(I_{i,t})/dt\right],$$

where $\omega_i = E_{i,t}/E_t$ is the sector share of energy consumption and is known as the weight for sector i in the summation. Integrating over time from 0 to T,

$$\ln(I_T/I_0) = \int_0^T \sum_i \omega_i\left[d\ln(S_{i,t})/dt\right] + \int_0^T \sum_i \omega_i\left[d\ln(I_{i,t})/dt\right],$$

Taking the exponential, the previous equation can be expressed in the multiplicative form $D_{tot} = D_{str}D_{int}$, where

$$D_{str} = \exp\left\{\int_0^T \sum_i \omega_i\left[d\ln(S_{i,t})/dt\right]\right\}$$

$$D_{int} = \exp\left\{\int_0^T \sum_i \omega_i\left[d\ln(I_{i,t})/dt\right]\right\}.$$

Since in practice only discrete data are available, the weight function has to be explicitly defined in actual application. Two of the weight functions that have been proposed in energy decomposition analysis are described next.

The first is the arithmetic mean of the weights for Year 0 and Year T, where

$$D_{str} = \exp\left\{\sum_i (\omega_{i,T} + \omega_{i,0})/2\ln(S_{i,T}/S_{i,0})\right\}$$

$$D_{int} = \exp\left\{\sum_i (\omega_{i,T} + \omega_{i,0})/2\ln(I_{i,T}/I_{i,0})\right\}.$$

Like the Laspeyres index approach, the approach does not give perfect decomposition, and we can write $D_{tot} = D_{str}D_{int}D_{rsd}$. The additive version can be derived in the same manner, and the relevant formulae are

$$\Delta I_{str} = \sum_i (E_{i,T}/Y_T + E_{i,0}/Y_0)/2\ln(S_{i,T}/S_{i,0})$$

$$\Delta I_{int} = \sum_i (E_{i,T}/Y_T + E_{i,0}/Y_0)/2\ln(I_{i,T}/I_{i,0}).$$

The decomposition is also not perfect, and we have $\Delta I_{tot} = \Delta I_{str} + \Delta I_{int} + \Delta I_{rsd}$.

The second and more refined weight function is given by the logarithmic mean of the weights for Year 0 and Year T. The corresponding formulae are

$$D_{str} = \exp\left\{\sum_i \frac{L(E_{i,T}/Y_T, E_{i,0}/Y_0)}{L(I_T, I_0)}\ln(S_{i,T}/S_{i,0})\right\}$$

$$D_{int} = \exp\left\{\sum_i \frac{L(E_{i,T}/Y_T, E_{i,0}/Y_0)}{L(I_T, I_0)}\ln(I_{i,T}/I_{i,0})\right\}$$

$$\Delta I_{str} = \sum_i L(E_{i,T}/Y_T, E_{i,0}/Y_0)\ln(S_{i,T}/S_{i,0})$$

$$\Delta I_{int} = \sum_i L(E_{i,T}/Y_T, E_{i,0}/Y_0)\ln(S_{i,T}/S_{i,0}),$$

where the function $L(a,b)$ is the logarithmic mean of two positive numbers a and b given by

$$L(a,b) = (a-b)/(\ln a - \ln b),\ \textit{for}\ a \neq b$$
$$= a,\ \textit{for}\ a = b$$

The advantage of the logarithmic mean weight function is that perfect decomposition is ensured. It can be proven that the residual term does not exist (i.e., $D_{rsd} = 1$ and $\Delta I_{str} = 0$).

2.4 An Example

Table I shows a hypothetical case in which industry comprises two sectors. The sectoral energy intensity decreases from Year 0 to Year T for both sectors. In percentage terms, it decreases by 33% in sector 1 and by 20% in sector 2. By examining the sectoral data, one would conclude that at the given sectoral level, there are significant improvements in energy efficiency. At the industrywide level, however, the aggregate energy intensity increases by 20%. This apparent contradiction arises because of structural change. Sector 1, the more energy intensive of the two sectors, expands substantially in production level and its production share increases from 20 to 50%. The achievements in energy conservation, assuming that conservation measures have been successfully implemented within sectors 1 and 2, are therefore not reflected in the industrywide aggregate indicator.

Application of appropriate decomposition techniques to give the impacts of structural change and energy intensity change helps to resolve the previous apparent contradiction. The decomposition results obtained using the Laspeyres index method and the two Divisia index methods are summarized in Table II. Based on the results for the logarithmic mean Divisia method, we conclude that in multiplicative decomposition, the contribution of structural change increases by 70%, whereas that of sectoral energy change decreases by 29%, resulting in a net 20% increase in the aggregate energy intensity as observed. In the case of additive decomposition, structural change accounts for an increase by 0.58 units, whereas sectoral energy intensity change accounts for a decrease by 0.38 units, leading to a net increase in the aggregate energy intensity of 0.20 units.

It may be seen that the impacts of structural change and energy intensity change can be conveniently estimated and the relative contributions identified. In actual application, where very fine data are available, the number of sectors can be as many as a few hundred. The decomposition results obtained are method dependent, as shown in Table II, and the choice of an appropriate technique is often an application issue. The methodology has been applied to study problems with more than two factors and in many energy-related areas.

TABLE I

An Illustrative Example (Arbitrary Units)

	Year 0				Year T			
	E_0	Y_0	S_0	I_0	E_T	Y_T	S_T	I_T
Sector 1	30	10	0.2	3.0	80	40	0.5	2.0
Sector 2	20	40	0.8	0.5	16	40	0.5	0.4
Industry	50	50	1.0	1.0	96	80	1.0	1.2

TABLE II

Decomposition Results Obtained Using the Data in Table I

	Laspeyres approach	Divisia approach: Arithmetic mean	Divisia approach: Logarithmic mean
Multiplicative			
D_{tot}	1.2000	1.2000	1.2000
D_{str}	1.7500	1.6879	1.6996
D_{int}	0.7200	0.7020	0.7060
D_{rsd}	0.9524	1.0127	1[a]
Additive			
ΔI_{tot}	0.2000	0.2000	0.2000
ΔI_{str}	0.7500	0.5920	0.5819
ΔI_{int}	−0.2800	−0.3913	−0.3819
ΔI_{rsd}	−0.2700	−0.0007	0[a]

[a] Perfect decomposition.

3. METHODOLOGY AND RELATED ISSUES

The methodological development of energy decomposition analysis since the late 1970s may be divided into three phases: the introduction phase (prior to 1985), the consolidation phase (1985–1995), and the further refinement phase (after 1995). It is a multifaceted subject, and the advances have been made due to the collective effort of researchers and analysts with backgrounds varying from science to mathematics, engineering, and economics.

3.1 The Introduction Phase (Prior to 1985)

The techniques proposed prior to the mid-1980s are intuitive and straightforward. Typically, the impact of structural change was singled out by computing the hypothetical aggregate energy intensity that would have been in a target year had sectoral energy intensities remained unchanged at their respective values in a base year. The difference between this hypothetical target year aggregate energy intensity and the observed base year aggregate energy intensity is taken as the impact of structural change. The difference between the observed aggregate energy intensity and the hypothetical aggregate energy intensity, both of the target year, is taken as the impact of sectoral energy change. These techniques were first proposed by researchers to study trends of industrial use of energy in the United Kingdom and the United States in approximately 1980. No reference was made to index numbers in these studies, which means that the techniques were developed independent of index number theory. However, it was later found that these techniques were very similar to the Laspeyres index in concept. Methodologically, it is appropriate to refer to them as the Laspeyres index approach or the Laspeyres index-related decomposition approach.

3.2 The Consolidation Phase (1985–1995)

In this phase, several new techniques were proposed, with the realization that the Laspeyres index concept need not be the only one that could be adopted. The arithmetic mean Divisia index method was proposed in 1987 as an alternative to the Laspeyres index method. At approximately the same time, as an extension to the Laspeyres index formula that assigns the total weight to the base year, methods with equal weight assigned to Year 0 and Year T were formulated. This modification leads to formulae that possess the symmetry property.

In the early 1990s, attempts were made to consolidate the fast growing number of methods into a unified decomposition framework. A notable development was the introduction in 1992 of two general parametric Divisia index methods that allow an infinite number of decomposition methods to be specified. In these two methods, the weights for Year 0 and Year T are treated as variables whose values are explicitly defined. By choosing the weights appropriately, many earlier methods, including those based on the Laspeyres index approach and the Divisia index approach, can be shown to be special cases of these two general methods. Alternatively, the weights can be determined in an adaptive manner based on energy consumption and output growth patterns in the data set. In chaining decomposition, the weights would be updated automatically over time. By the mid-1990s, this technique, called the adaptive weighting parametric Divisia index method, was the most sophisticated decomposition technique. It was later adopted by some researchers and analysts.

3.3 Further Refinement Phase (after 1995)

Until the mid-1990s, all the proposed energy decomposition techniques left an unexplained residual term. Studies found that in some situations the residual term could be large, especially for the Laspeyres index approach. This issue became a concern because if a large part of the change in the aggregate is left unexplained, the interpretation of decomposition results would become difficult if not impossible, and the objective of decomposition analysis would be defeated.

Researchers began to search for more refined decomposition techniques, and the first perfect decomposition technique for energy decomposition analysis was proposed in 1997. This technique is similar to the logarithmic mean Divisia index method. In addition to having the property of perfect decomposition, the logarithmic mean Divisia index methods are also superior to the arithmetic mean Divisia method in that they can handle data with zero values. After the application of decomposition analysis was extended to include the effect of fuel mix in the early 1990s, the existence of zero values in the data set became a problem for some decomposition techniques. Zero values will appear in the data set when an energy source, such as nuclear energy or natural gas, begins or ceases to be used in a sector in the study period.

In the Laspeyres index approach, the residual terms, such as those shown in Table II, are known to researchers to consist of interactions among the factors considered in decomposition. A scheme to distribute the interaction terms, whereby perfect decomposition is achieved, was proposed in 1998. It has been referred to as the refined Laspeyres index method because it may be considered a refinement to the Laspeyres approach. In 2002, researchers introduced the Shapley decomposition, a perfect decomposition technique that has long been used in cost-allocation problems, to energy decomposition analysis. It has been determined that the refined Laspeyres index technique introduced in 1998 and the Shapley decomposition technique are exactly the same. Several other techniques that give perfect decomposition have been reported, including one that uses the concept of the Fisher ideal index.

3.4 Other Methodological Issues

In addition to method formulation, the following methodological issues are relevant to the application of decomposition analysis: multiplicative decomposition versus additive decomposition, disaggregation level of the data, and decomposition on a chaining basis versus a nonchaining basis. Each of these issues is briefly described.

For a given data set, the analyst may use either the multiplicative scheme or the additive decomposition scheme, and generally only one will be chosen by the analyst. The decomposition formulae for the two schemes are different, as are the numerical results obtained. However, the main conclusions are generally the same. As previously mentioned, the numerical results of the estimated effects are given in indices in multiplicative decomposition, but they are given in a physical unit in additive decomposition. Preference of the analyst and ease of result interpretation are the main considerations in the choice between the two decomposition schemes. Methodologically, the logarithmic mean Divisia index method has an advantage over other methods because the results obtained from the multiplicative and the additive schemes are linked by a very simple and straightforward formula. This unique property allows the estimate of an effect given by one of the two schemes to be readily converted to that of the other, and the choice between the multiplicative and additive schemes made by the analyst before decomposition becomes inconsequential.

The disaggregation level of the data defines the level or degree of structural change to be estimated. In the case of energy use in industry, past decomposition studies used levels varying from 2 sectors, which is the minimum, to approximately 500 sectors. The choice of disaggregation level dictates the results since structural change and energy efficiency change are dependent on the level chosen. Hence, result interpretation must be done with reference to the level defined and generalization should be avoided. A fine level of sector disaggregation is normally preferred. For instance, energy efficiency change is often taken as the inverse of the estimated energy intensity change. Derived in this manner, a good estimate of the real energy efficiency change requires that industry be broken down into a sufficiently large number of sectors. The quality of the estimate improves as the number of sectors increase.

When year 0 and year T are not consecutive years but two points in time separated by 1 year or more, nonchaining decomposition simply uses the data of the two years in the analysis without considering the data of all the intervening years. In contrast, decomposition on a chaining basis involves yearly decomposition using the data of every two consecutive years starting from year 0 and ending at year T, with the decomposition results computed on a cumulative basis over time. Chaining decomposition has also been referred to as time-series decomposition and rolling base year decomposition. In nonchaining decomposition, the year-to-year evolution of factors over time between year 0 and year T is not considered. It has been shown that there is less variation between the results given by different decomposition techniques when decomposition is performed on a chaining basis. This consistency, which arises because decomposition is "path dependent," is a distinct advantage. As such, the chaining procedure should be preferred when time-series data are available. However, nonchaining decomposition is the only choice in some situations, such as in studies that involve a very large number of sectors for which energy and production data are not collected annually.

3.5 Further Remarks

The methodological development of energy decomposition analysis has also been driven by the need of energy researchers to study an increasingly wider range of problems. In the simplest form and with only two factors, many decomposition techniques have a strong affinity with index numbers (i.e., product mix and energy intensity as equivalent to commodity quantity and price on an aggregate commodity value in economics). However, the problems studied after 1990 seldom have two factors. A typical decomposition problem on energy-related national carbon dioxide (CO_2) emissions, for instance, involves five factors—total energy demand per unit of gross domestic product (GDP), per capita GDP, population, energy demand by fuel type, and the CO_2 emission coefficient by fuel type.

There is a good collection of decomposition techniques that can cater to almost every possible decomposition situation or suit the varied needs of the analyst. Each of these techniques has its strengths and weaknesses. Perfect decomposition has been widely accepted as a very desirable property of a decomposition technique in energy analysis. In some situations, especially in cross-country decomposition, some of the techniques that do not give perfect decomposition could have a residual term many times larger than the estimated main effects. In addition, there are techniques that fail when the data set contains zero values or negative values. Some techniques, such as those based on the Divisia index, have the advantage of ease of formulation and use, whereas others, including those based on the Laspeyres index, have complex formulae when more than three factors are considered.

In general, whether a specific technique is applicable or the "best" among all depends on the following: (i) the properties of the technique, which are related to the index number with which the technique is associated with; (ii) ease of understanding and use; (iii) the problem being analyzed, such as whether it is a temporal or cross-country problem; and (iv) the data structure, such as whether zero or negative values exist in the data set. A good understanding of the strengths and weaknesses of the various decomposition techniques would allow analysts to make sound judgments as to which technique to use. These issues are addressed in some publications.

4. APPLICATION AND RELATED ISSUES

The origin of energy decomposition analysis is closely linked to the study of energy use in industry. Not surprisingly, this application area accounted for most of the publications on the subject until 1990. Since then, the application of decomposition analysis has expanded greatly in scope. The main reasons are the growing concern about the environment and the increased emphasis on sustainable development globally. Moreover, decomposition analysis can be conveniently adopted as long as structural change can be meaningfully defined and data are available at the chosen disaggregation level. Indeed, its simplicity and flexibility opens up many possible application areas. Some of these, such as the development of national energy efficiency indicators to monitor national energy efficiency trends, are unique to index decomposition analysis as a decomposition methodology. In terms of application by country, decomposition analysis has been adopted to study energy and energy-related issues in a large number of countries, including all major industrialized countries, most Eastern European countries including Russia, and many developing countries. The following sections focus on developments in several key application areas.

4.1 Energy

With the successful application to study trends of energy use in industry, extensions to decomposition analysis were made to study energy use in the economy as a whole, in energy-consuming sectors other than industry, and in the energy supply sector. In the case of national energy demand, the aggregate to be decomposed is generally energy use per unit of GDP, commonly known as the energy-output ratio in the energy literature, where energy use covers all sectors of the economy and structural change refers to changes in GDP product mix. However, such studies often have difficulty matching energy consumption with economic output. This is due to the fact that energy uses in some sectors, such as private transportation and residential, have no direct match with output sectors/activities in the GDP, and as such some adjustments to sector definition and data are often needed.

Regarding energy consumption in major sectors other than industry, studies have been performed on energy use in passenger transportation and in freight transportation. For passenger transportation, aggregate energy use per passenger-kilometer has been decomposed to give the impacts of transportation modal mix and modal energy intensity given by modal energy requirement per passenger-kilometer. Similarly, for freight transportation, aggregate energy use per tonne-kilometer of freight moved has been decomposed. For residential energy use, decomposition has been conducted to quantify the impact of changes in the mix of residential energy services consumed. For electricity production, aggregate generation efficiency has been decomposed to give the impacts of generation mix and the generation efficiency of each fuel type.

4.2 Energy-Related Gas Emissions

Since the early 1990s, due to the growing concern about global warming and air pollution, an increasing number of studies on the decomposition of energy-induced emissions of CO_2 have been performed. The extension of the methodology presented in Section 2 to decompose an aggregate CO_2 indicator is fairly straightforward. Assume that the ratio of industrial energy-related CO_2 emissions to industrial output is the aggregate of interest. A declining trend for the indicator is generally preferred, and the motivation is to study the underlying factors contributing to changes in the aggregate. This would normally be formulated as a four-factor problem with the following factors: industrial output structure change, sectoral energy intensity change, sectoral fuel share change, and fuel CO_2 emission coefficient change. Similar decomposition procedures have been applied to study energy-related CO_2

emissions for the economy as a whole, other major energy-consuming sectors, and the energy supply sector. Although the focus among researchers and analysts has been on CO_2 emissions, some studies have dealt with emissions of other gases, such as sulfur dioxide and nitrogen oxides. The number of decomposition studies dealing with energy-related gas emissions equaled that of studies on energy demand by the late 1990s and exceeded the latter in 2000.

4.3 Material Flows and Dematerialization

Another extension of decomposition analysis is the study of the use of materials in an economy. The materials of interest to researchers include a wide range of metals and nonmetallic minerals, timber, paper, water, cotton, etc. The list also includes oil, coal, and natural gas, which are treated as materials rather than energy sources. The aggregate of interest would be the consumption of a material divided by GDP or some other meaningful output measure. In material flows studies, dematerialization refers to the absolute or relative reduction in the quantity of material to serve economic functions, and its concept is similar to reduction of aggregate energy intensity. The technique of energy decomposition analysis can be easily extended to material decomposition analysis to study trends of material use. Indeed, a number of such studies have been performed, and it has been found that decomposition analysis is a useful means of analyzing the development of the materials basis of an economy.

4.4 National Energy Efficiency Trend Monitoring

Increasingly more countries see the need to have appropriate national energy efficiency indicators or indices to measure national energy efficiency trends and progress toward national energy efficiency targets. The energy efficiency indicators may be defined at the national level or at the sectoral level of energy consumption, such as industrial, transportation, commercial, and residential. Here, the primary focus of decomposition analysis is not to study the impact of structural change. Rather, it is the elimination of the impacts of structural change and other factors from changes in an aggregate such as the ratio of national energy consumption to GDP, whereby the impact of energy intensity change computed is sufficiently refined to be taken as inversely proportional to real energy efficiency change. Advances in the decomposition analysis methodology, including the use of physical indicators to measure output or activity level, have allowed the development and use of more meaningful and robust energy efficiency indicators or indices. A related methodological framework that would more accurately track trends in energy efficiency has been developed by many countries and international organizations, including the Energy Efficiency and Conservation Authority of New Zealand, the U.S. Department of Energy, the Office of Energy Efficiency of Canada, the International Energy Agency, and the European Project and Database on Energy Efficiency Indicators.

4.5 Cross-Country Comparisons

Cross-country comparisons involve the quantification of factors contributing to differences in an aggregate, such as aggregate energy intensity or the ratio of energy-related CO_2 emissions to GDP, between two countries. The factors considered are similar to those in the study of changes in the indicator over time in a country. In decomposition analysis, the data for the two countries are taken to be the same as those for two different years in a single-country study. Such studies have been carried out for a number of countries. Studies have also compared CO_2 emissions between world regions, such as between North America and the developing world. For the factors considered, because the variations tend to be larger between two countries/regions than between two years for a country, the decomposition tends to give a large residual term if nonperfect decomposition techniques are applied. In general, the use of a perfect decomposition technique is recommended for cross-country decomposition analysis.

5. CONCLUSION

The methodology development and application of index decomposition analysis have been driven by energy researchers since the 1970s, immediately after the 1973 and 1974 world oil crisis. In the case of the basic two-factor decomposition problems, the tool has a strong affinity with index number problems. Nevertheless, over the years, energy researchers and analysts have continuously refined and extended the methodology for the unique situations and requirements in the energy field. As a result, decomposition analysis has reached a reasonable level of sophistication and its usefulness in energy studies has been firmly established.

A well-defined set of decomposition approaches/methods is now available to accommodate an arbitrary number of factors to be decomposed. Decomposition can be conducted multiplicatively or additively, on a chaining or nonchaining basis, and on a temporal or cross-sectional basis. Due to its simplicity and flexibility, the index decomposition analysis has been adopted in a wide range of application areas. The number of application studies has increased rapidly over time, particularly since the early 1990s, and interest among energy researchers in the methodology has been growing.

The previously mentioned developments were driven by the concern about world energy supply that began in the 1970s and were reinforced by concern about global warming and the need to reduce the growth or cut emissions of CO_2 and other greenhouse gases since the late 1980s. Because energy and the environment will remain important global issues and sustainable development will become increasingly more important in the following years, it is expected that index decomposition analysis will continue to play a useful role in providing information for policymakers to address energy, environmental, and resource depletion problems that mankind will continue to face.

SEE ALSO THE FOLLOWING ARTICLES

Aggregation of Energy • Commercial Sector and Energy Use • Industrial Ecology • Industrial Energy Efficiency • Industrial Energy Use, Status and Trends • International Comparisons of Energy End Use: Benefits and Risks • Modeling Energy Supply and Demand: A Comparison of Approaches • National Energy Modeling Systems

Further Reading

Ang, B. W., and Choi, K. H. (1997). Decomposition of aggregate energy and gas emission intensities for industry: a refined Divisia index method. *Energy J.* **18**(3), 59–73.

Ang, B. W., and Zhang, F. Q. (2000). A survey of index decomposition analysis in energy and environmental studies. *Energy* **25**, 1149–1176.

Ang, B.W., Liu, F.L.,Chew, E. P. (2003). Perfect decomposition techniques in energy and environmental analysis. *Energy Policy* **31**, 1561–1566.

Boyd, G., McDonald, J. F., Ross, M., and Hanson, D. A. (1987). Separating the changing composition of U.S. manufacturing production from energy efficiency improvements: a Divisia index approach. *Energy J.* **8**(2), 77–96.

Farla, J. C. M., and Blok, K. (2000). Energy efficiency and structural change in The Netherlands. 1980–1995. *J. Ind. Ecol.* **4**(1), 93–117.

Jenne, J., and Cattell, R. (1983). Structural change and energy efficiency in industry. *Energy Econ.* **5**, 114–123.

Lermit, J., and Jollands, N. (2001). "Monitoring Energy Efficiency Performance in New Zealand: A Conceptual and Methodological Framework." National Energy Efficiency and Conservation Authority, Wellington, New Zealand.

Liu, X. Q., Ang, B. W., and Ong, H. L. (1992). The application of the Divisia index to the decomposition of changes in industrial energy consumption. *Energy J.* **13**(4), 161–177.

Marlay, R. (1984). Trends in industrial use of energy. *Science* **226**, 1277–1283.

Schipper, L., Unander, F., Murtishaw, S., and Mike, T. (2001). Indicators of energy use and carbon emissions: Explaining the energy economy link. *Annu. Rev. Energy Environ.* **26**, 49–81.

Sun, J. W. (1998). Changes in energy consumption and energy intensity: a complete decomposition model. *Energy Econ.* **20**, 85–100.

Wade, S. H. (2002). "Measuring Changes in Energy Efficiency for the Annual Energy Outlook 2002." U.S. Department of Energy, Energy Information Administration, Washington, DC.

Depletion and Valuation of Energy Resources

JOHN HARTWICK
Queen's University
Kingston, Ontario, Canada

1. Depletion Pressure
2. Depletion Economics
3. The Scarcity of Clean Energy Supplies
4. Summary

Glossary

backstop A substitute supply that is expected to become active when current supplies are exhausted. This is taken as a "renewable" supply available indefinitely at a constant unit cost. The "price of the backstop" would be this "high" unit cost. Commercially viable fusion reactors have been taken as the backstop energy source.

hotelling rent The rent per unit extracted that exists because of the finiteness (exhaustibility) of the stock being extracted from.

marginal cost of extraction Given an equilibrium, the cost of extracting one more unit

Nordhaus (Herfindahl) model A linked sequence of "Hotelling models," each with a higher unit cost of extraction. The demand schedule is unchanging in the Herfindahl version.

rent The positive gap between market price per unit and marginal cost for supplying an additional unit.

Ricardian rent The rent per unit that exists because of a quality advantage of the unit in question.

royalty A payment to the owner of a deposit by the extractor, who in effect is renting or borrowing the deposit.

In the early years of the 21st century, industrialized nations exist in a world of cheap energy. Direct outlays by households for electricity, household heating, and fuel for transportation occupy small fractions of almost all family budgets. Canada and the United States, with 5% of the world's population, consume 25% of global energy. In per capita terms, the United States consumes 354 million Btu/year, Western Europe consumes 170 million Btu/year, and India consumes 12 million Btu/year. A Btu is the amount of heat required to raise the temperature of 1 lb of water 1°F. Energy use in North America grew 31% between 1972 and 1997. Is an era of high energy costs just around the corner? Is there a signal of a higher priced era to be read out there today? In the period following oil price hikes in 1973–1974, the U.S. Department of Energy predicted $150 a barrel for oil in the year 2000 in year 2000 prices. Exxon predicted a price of $100. What has abundant energy accomplished for us? Since approximately 1800, the four most revolutionary changes in a person's life in industrialized countries are (1) the substitution of machine power for animal and manpower, (2) the more than doubling of life expectancy, (3) the arrival of relatively low-cost transportation for persons and commodities, and (4) the arrival of global and low-cost communication. Clearly (1) and (3) are outcomes of new forms of energy use. The dramatic increase in longevity is probably mostly a consequence of improvements in the processing of water, food, and wastes—that is, improvements in the public health milieu, broadly conceived rather than a result of low-cost power, directly. But the impact of reduced demands on humans and animals for power on a person's life expectancy should not be dismissed. Improved, low-cost communication is related to the arrival of electricity and electronics, not power per se. The words labor, from Latin, and work, from Greek, both capture the idea of an expenditure of painful physical or mental effort, of energy expenditure. People who do no labor or work then are those expending little energy. Revolutionary in human history is the substitution of machine power for human and animal power. Hoists (pulleys) and levers were early simple machines that magnified human and animal power considerably. Watermills and windmills arrived and showed what could be accomplished with new energy sources captured by machines and new types of power output—namely, steady and continuous rather than episodic. But it was the arrival of steam engines (heat machines) that revolutionized the concept of usable power. The concept of transporting goods and people changed dramatically and powering a factory was freed from ties to watermills and windmills.

1. DEPLETION PRESSURE

It is easy to make the case that depletion pressures induced invention and innovation in energy use. Since at least the time of Great Britain's Elizabeth I, say 1600, conservation of forests for future fuel and timber supply was prominent on the public agenda. In the early 1700s, English iron production was using up approximately 1100 km^2 of forest annually. The notion that forest depletion drove the development of British coal fields is a popular one. But causality is tricky here — did the arrival of cheap coal lead to a liquidation of remaining forests or did the depletion of forests induce innovation in coal supply? This really has no ready answer. Fuel scarcity apparently led to the importation of iron by England in the mid-18th century. Around 1775, coal became viable in smelting iron and England's importing ceased. Coinciding with the coalification of the British economy was the development of the steam engine, first as a water pump and later as a source of power in factories, such as cotton mills. Jevons (1865, p. 113) opined that "there can be no doubt that an urgent need was felt at the beginning of the 17th century for a more powerful means of draining mines." Savery's steam pump was operating in 1698 and Smeaton made Newcomen's engine, with a piston, sufficiently efficient to be commercially viable. In 1775, the new steam engine of James Watt was developed.

Given the documented concern in Great Britain about timber supply, say, before 1700, one should observe a rise in the price of wood for fuel, house-building, and ship-building at some point. The year 1700 seems like a reasonable year to select as having recorded an increase in wood price. Firewood prices grew three times faster than the general price level in England between 1500 and 1630 and the 17th century was one of "energy crisis." Over one-half of the total tonnage entering British ports in the 1750s was timber, and fir imports grew 700% from 1752 to 1792. France also experienced rapid increases in fuel wood prices over this latter period. The prevailing "high" price in 1700 should, of course, spur the search for substitute energy sources. Importation is one form of substitution and coal development would be another. Given flooding problems with new coal deposits, an investment in reasonable-quality pumps was a low-risk strategy. The deposits were known to exist and the demand for coal seemed certain. The Saverys and Newcomens rose to the occasion. This is a possible explanation for the invention and introduction of steam pumps after 1700. It shifts the emphasis from spontaneous invention to profit-driven invention and innovation. A wood supply crisis is the ultimate impetus in this scenario. Current or anticipated fuel scarcity raises fuel costs and induces a search for substitutes. Coal is the obvious substitute but mine-draining looms up as a serious constraint on coal production—hence, the pressure to invent and to make use of steam pumps. Cheap coal transformed the British economy. In 1969, Landes estimated coal consumption in Great Britain at 11 million tons in 1800, 22 million tons in 1830, and 100 million tons in 1870. The British appear to have been sensitive to energy use over the long run. The Domesday Book, published in 1086, records over 5000 water-wheels in use in England—one mill for every 50 households.

Attractive in this scenario is the low-risk nature of investment in pumping activity. The coal was there, the demand was there, and one simply had to keep the water level in the mines low. Recall that a Newcomen engine was part of a multistory structure and thus represented a serious outlay. However, the technology had been proven in small-scale engines and the minerals were there for the taking. The second watershed moment in the history of steam power, associated with James Watt, was its development for powering whole factories. It can be argued that it was the depletion of good watermill sites, well documented for the Birmingham region by Pelham in 1963, that spurred invention. Richard Arkwright's invention of the cotton-spinning machine and mill and James Watt's invention of the improved steam engine compete in the literature on the industrial revolution for the label, "the key events." Spinning mills required a power source and water-wheels were the standard at the time. But convenient sites for water-wheels had mostly been taken up by the 1770s. The price of those sites would have exhibited a significant rise in, say, 1775, the year Watt was granted his special patent for his engine with a separate condenser, an engine vastly more efficient than the Newcomen type. The climate was right for invention in the power supply field. Depletion of good water-wheel sites drove up their prices and the resulting umbrella of profitability made invention in this area attractive. At approximately this time, serious research of water-wheels was undertaken by Smeaton with a view, of course, to obtain more power from the waterflow. Obviously, a profit-driven view of events is not the only plausible one, but there is no doubt that profits accrued to Watt and his partner, Boulton.

Coal and oil compete as bulk fuels in many parts of the world for powering electricity-generating plants but steam engines for motive power have been displaced by gasoline, diesel, and electric engines in most areas of the world. Steam locomotives were common in industrialized countries as late as the 1950s. This is the classic case of a low-cost, clean energy source (e.g., a diesel engine for a locomotive) being substituted for a dirtier and, presumably, higher cost source. The conspicuous aspect of energy use is the capital goods that come with the energy flow. For example, wind is the energy source and the mill transforms the energy flow into usable power in, for example, grinding wheat. Similarly, the rushing stream is the energy source and the wheel and mount are the capital goods that allow for the production of usable power. What Savery, Newcomen, Watt, and others did was to invent a capital good that turned a heat flow into usable power. The role of the inventor is to envisage a capital good, a machine, that allows for the transformation of an energy flow into a power flow. From this perspective, it is difficult to conceive of an energy supply crisis. One looks forward to larger and better machines harnessing old energy flows better and new flows well. An oil supply crisis in the early years of the 21st century would translate into a transportation cost crisis but not an end-of-civilization crisis. Approximately one-fifth of electricity in the United States was generated with petroleum in 1970. In 2002, that figure is one one-hundredth. Coal burning produces approximately one-third more carbon dioxide than oil burning and approximately twice as much as the burning of natural gas, per unit of heat. Transportation fuel can be synthesized from coal and methane, but only at relatively high costs, given today's technology. The Germans used the Fischer-Tropsch process to produce gasoline from coal during World War II and the search for a catalyst to produce gasoline from methane is under way.

The appropriate aggregate production function to consider for an economy takes the form of output flow, $Q = F[N, K, E(K^E, R)]$, where $E(\cdot)$ is power derived from an energy flow R, say, oil, and capital equipment, K^E, K is other capital, and N is labor services. Before 1800, K^E was windmills and water-wheels for the most part and R was wind and water flow. In the 21st century, food is cooked in and on electric stoves rather than in hearths, light is provided by electric units rather than candles or lanterns, and heat is provided by furnaces and distribution devices rather than by fires, fireplaces, and stoves. $E(K^E, R)$ is dramatically different even though basic houses are similar in concept and shape and eating styles and habits are not hugely different. What would the world be like with steam power but no electricity? There was an interval of approximately 75 years during which this state existed. Steam power was decisive in liberating people from much heavy labor. Steam power could drive hoists, diggers, and pumps. It could also power tractors, locomotives, and even automobiles. Factories could be well mechanized with steam engines. In 1969, Landes suggested that the 4 million hp of installed steam engines in Britain in 1870 was equivalent to 80 million men, allowing for rest, or 12 million horses. A worker with steam power was like a boss with, say, 16 healthy workers at his bidding. In 1994, Smil estimated that a worker can put out 800 kJ net per hour. From this, he inferred that the Roman empire's 85,000 km of trunk roads required 20,000 workers, full time, working 600 years for construction and maintenance. The gain in bulk power production from the use of electric motors rather than steam motors seems marginal when history is viewed on a large time-scale. Electricity rather created new activities such as listening to the radio, watching television, and talking on the telephone. These are large gains in the range of consumer goods, essentially new attractive commodities, but they are on a different order from innovations relieving humans of great and steady physical exertion. Electric stoves and washers have undeniably made housework easier, and air-conditioning (room cooling) has the lessened the discomfort of labor for many as well as altered the geographical distribution of settlement.

With regard to personal travel, steam power might have been revolutionary enough. Steamships were quite different from sailing ships and trains were quite different from stagecoaches. Steam-powered automobiles and trucks are technically feasible. The internal combustion engine, drawing on gasoline and diesel fuel, is not a crucial technology, though steam power for vehicles was quickly displaced. Steam power revolutionized the concept of daily work and of personal travel. Huge numbers of people ceased to be beasts of burden. Huge numbers of people were able to visit relatively distant places quite rapidly. Diets changed significantly when new foods could be transported from distant growing places. Crucial to modernization, however, was the substitution of machine power for human and animal power, which steam engines effected, and the spread of reasonable living standards to low-income families. Steam power brought about this latter effect only indirectly, by making mass production techniques widespread. Mechanization brought down the prices of many goods (e.g., cotton) but displaced workers in the handicraft, labor-intensive sectors. Wages rose in Great Britain some 50 years after the industrial revolution was well rooted. Machine-intensive production relieves workers of much grinding physical exertion but usually eliminates many traditional jobs as well. A large virtue of market-oriented (capitalistic) development is that entrepreneurs create and often invent new jobs for the workers displaced by mechanization. Wages and living standards have risen, particularly with a decline in family size.

2. DEPLETION ECONOMICS

Though in 1914 Gray first formalized the idea that impending natural stock (say, oil) scarcity would show up in current price as a rent, a "surplus" above extraction cost, Hotelling presented the idea most clearly in 1931. A royalty charged by an owner of a deposit to a user of the deposit is, in fact, "Hotelling rent." Thus, the idea of "depletion" being a component of current mineral price is ancient. Competing, however, with the Gray–Hotelling theory is a theory that has current oil price rising up the market demand schedule because each subsequent ton is becoming more expensive to extract. Deeper is more costly. This approach was adopted by the great W. S. Jevons in his monograph on coal use in Great Britain: "...the growing difficulties of management and extraction of coal in a very deep mine must greatly enhance its price. It is by this rise of price that gradual exhaustion will be manifested ..." (Jevons, 1865, p. 8) Jevons' son, also a professor of economics, reflected on his father's work, pointing out that the price of coal had indeed risen over the 50 years since his father's book was written, by, however, just 12% relative to an index of prices. And reflecting on his father's analogy of high energy prices being like a tax, Jevons recommended accumulating alternative forms of capital to compensate future generations for their shrunken coal capital. A standard modern view has the current rent in oil price, a reflection of the profit or rent derivable from currently discovered new deposits. Hotelling theory sees current oil price as a reflection of (1) how long current stocks will carry society until users are forced to switch to substitute sources and (2) what price those substitute sources will "come in at." It was then the value of remaining stocks that held current price up, rather than pressure from currently steadily worsening qualities. As Nordhaus has made clear, in Hotelling's theory of depletion or "resource rent," current price turns crucially on the cost of supply from the future substitute source (the backstop) and on the interval that current stock will continue to provide for demand. See Fig. 1 for an illustration of Hotelling theory.

r is the interest rate in the market and Hotelling's approach is often referred to as the r percent theory of resource rent because profit-maximizing extraction by competitive suppliers implies that rent rises over time at r percent. Hence, the quantity sequence in an equilibrium extraction program is the one displaying rent rising at r percent per period. Current rent (price minus current unit extraction cost) turns out to be simply discounted terminal rent, with this latter term being defined by the unit cost of energy from the backstop or substitute source. To say that the price of oil is high today could mean that remaining stock is not abundant and substitutes will

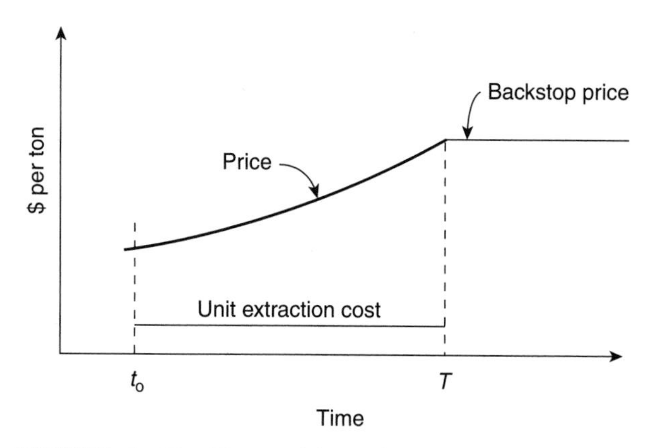

FIGURE 1 Hotelling model. In this competitive case, rent (price less unit extraction cost) rises r% until the stock is exhausted, at date T. Price "fed into" a market demand schedule yields current quantity supplied at each date. Unit cost, backstop price, stock size, and the demand schedule are parameters. Interval (t_0, T) is endogenous.

be available at high prices (Hotelling theory) or that future supplies will come from lower quality, high-cost deposits (Jevons theory). These two theories have been brought together in one model in a simple version by Herfindahl (see Fig. 2) and in a more complicated version in 1977 by Levhari and Leviatan and many others. The Herfindahl version has scarcity or depletion rent rising at r percent and, although unit extraction costs are constant, then jumping down as the next deposit or source is switched to. For the case of many deposits, Fig. 2 depicts a combination of intervals with rising rent linked by jumps down in rent. Quality decline causes rent to jump down. The Hotelling or depletion effect moves counter to the quality-decline effect. Thus, even in this simple world of no uncertainty or oligopoly in supply, one cannot be accurate in saying that exhaustibility is showing up in current price movement as an increasing rent or depletion effect (Fig. 2).

Hotelling provided a systematic theory of how depletion would show up in current price and how finiteness of the stock would show up in the pace of stock drawdown. In the simplest terms, exhaustibility is represented by a wedge between current unit cost and price. Thus, if world oil stocks were dwindling rapidly, one would expect oil prices to be high, well above unit extraction costs. And in fact, episodes of high world oil prices have been observed in recent decades, times when average world oil prices were well above unit extraction costs. Such wedges are labeled "rents" in standard economics and, with something like oil, the wedge might be referred to as scarcity rent or Hotelling rent. The world demand schedule does not shift in these episodes but current extracted output does shift back

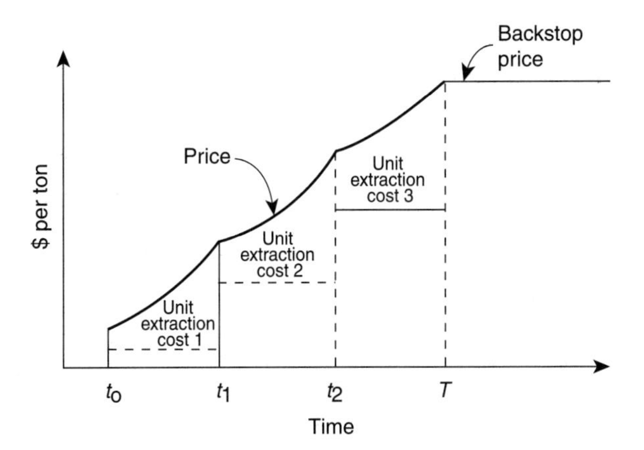

FIGURE 2 Nordhaus (Herfindahl) model. A three-source, linked Hotelling model. Low-cost supplies are exhausted first. Rent rises r% except at source switch. Rent jumps down across source switch. Intervals of source use are endogenous.

for some reason. It is these fairly sudden contractions in current production that cause current prices to jump up and "create" extra rent. The first large jump was in the fall of 1973. Another occurred in 1979, with the collapse of the Palhevi monarchy in Iran, and a large drop in price occurred over mid-1985 to mid-1986. Here the argument is that discipline among a cartel of oil suppliers broke down and world supply increased fairly quickly.

Price changes are largely driven by changes in current production flow or supply and current production is a complicated function of parameters, including estimated remaining stock or cumulative supply. Hotelling theory is an intertemporal theory of discounted profit maximization by competitive producers that links current production to remaining stock, extraction costs, demand conditions, and backstop price. It makes precise the exhaustibility effect or depletion in current resource price. Hotelling theory obtains a "depletion effect" in current price relatively simply in part because it makes use of the traditional assumption of perfect foresight by profit-maximizing agents. Economic theory relies on this assumption because it becomes an essential component of intertemporal profit maximization. The real world is, of course, replete with events that are very difficult to anticipate, such as wars, government regime switches, and technical changes. For example, research suggests that oil may be generated deep in the earth by unexpected geothermal processes. Uncertainty makes simple calculation of a depletion effect in current price a dubious course.

In its original form, Hotelling theory abstracted from uncertainty completely. And Hotelling did not take up quality variation in current deposits or stocks. This has, fact, led many observers to argue that Hotelling theory is fundamentally flawed as an explanation of rent or price paths for, say, oil. Also, Hotelling did not address oligopoly in supply and such market structures have been considered to be crucial in explaining oil price movements in recent decades. In the face of those well-known difficulties in arriving at a theory of oil price determination, in 1985 Miller and Upton tested "Hotelling's valuation principle" for a sample of oil-producing companies whose shares were being traded on the New York Stock Exchange. Employing heroic simplifying assumptions, they derived an equation to estimate in a sequence of straightforward steps and fitted their equation to the data, with positive results. It seems undeniable that extraction to maximize discounted profit is the reasonable maintained hypothesis and in a sense Miller and Upton confirm this position. But profit maximization by an extractive firm in an oligopoly setting, for instance, yields a quite different extraction program than that for a firm in a competitive setting. Miller and Upton provide weak evidence in favor of competitive extractive behavior by oil extraction companies. Implicit in their work is an argument that Hotelling rent is positive and present in the price of oil. However, the plausible "crowding" of such rent by stock quality decline cannot be evaluated from their analysis. In 1990, Adelman argued that current oil rent reflects the marginal cost of discovering new supply and most rent earned by producers is a Ricardian rent, one attributable to a quality advantage possessed by current suppliers lucky enough to be currently working a low-cost deposit.

Consider the reality of a backstop source of energy. Suppose the world does have a 20-year supply of "cheap" oil available, as Deffeyes suggested in 2001. (The Edison Electric Institute estimates that proven reserves of oil will last 37 years, those of natural gas will last 61 years, and those of coal will last 211 years.) In 2001, Kaufmann and Cleveland presented a careful, empirically based critique of Hubbert's bell-shaped supply schedule. Why is a foreseeable energy supply crisis not showing up in markets today? The answer appears to be that backstop supplies are fairly abundant, in the form of large deposits of coal, natural gas, tar sands oil, shale oil, and uranium. None of these supplies is as attractive and as low-cost as oil but it would not be complicated to power a modern economy with a combination of these energy feedstocks. Directly difficult would be fuel for transportation, given current engine technology and driving habits. Whatever transpires with long-term oil supplies, the era of low-cost individualistic transportation will likely end. The century 1920–2020 may well go down in history as the curious low-cost, automobile period, an era of individuals dashing about the surface of continents in big cocoons of steel on expensive roadways, while seriously polluting the atmosphere.

In 1974, Nordhaus brought his combination of Hotelling and backstop economics together in an ingenious empirical exercise. Given a backstop supply price and a sequence of world demands for energy into the distant future, and given world stocks of oil, coal, and uranium and unit extraction costs, he let his computer solve for the Hotelling transition path from his base period, 1973, to the era of constant cost fusion power. Naturally, stocks with lower unit extraction costs provide energy early on. See Fig. 2. Thus, one observes an oil supply era of endogenous length, a coal era, a uranium era, and then the fusion power era with constant unit cost into the indefinite future. A critic can easily dismiss this investigation as computer-aided, crystal-ball gazing. However, the Nordhaus model was subjected to sensitivity testing at low cost, and the empirical output appeared to be fairly robust. Central to this is, of course, discounting. Events beyond 30 years register little today, given discounting. Hence, fusion power costs largely fade out in the current price of energy, as do many other parameters of the model. It remains, nonetheless, a very good framework for organizing one's thinking about the future of energy. When the oil runs out, modern life does not end, just the curious era of abundant, low-cost energy, especially for personal transportation.

Nordhaus's "simulation" yielded a current price (1973) of energy somewhat below the observed market price. He inferred that this occurred because his model was "competitive," *a priori*, whereas the real world contained much oligopoly in energy production—hence, an obvious need to refine the computer formulation. However, the introduction of oligopoly to extraction models is tricky, if only for reasons of incorporating the dynamic consistency of agents' strategies. There must be no obvious possibilities for profit-taking simply by reneging on one's "original" strategy selected at time zero. To rule out such possibilities for profits, problems must be analyzed and solved backward, from tail to head, and this turns out to be much more complicated than, for example, Nordhaus's essentially competitive formulation. There is an important lesson here. If one believes that current energy prices reflect noncompetitive behavior among suppliers, which seems most reasonable, then quantifying such capitalizations is known to be difficult. One can rule out simple explanations for current energy price levels. The combination of Hotelling theory and backstop theory indicates that traditional supply and demand analysis is an inadequate way to think about current energy price formation. The addition of oligopoly considerations complicates an already complicated process of current price determination.

In the simplest case of current price shift, as for, say, October 1973, suppliers revise estimates of remaining stock downward and current price jumps up. "Mediating" this price shift is a process of agents discounting back from "the terminal date" to the present, discounting along a Hotelling extraction path. Current quantity extracted jumps down and current price jumps up. If demand is inelastic in the region, aggregate revenue will jump up. Thus, the motivation to raise current price could be for reasons of revenue. With some coordination of quantities in, say, a cartel, the current price jump may be an oligopoly or cartel phenomenon rather than an outcome of stock size revision. Somewhat analogous is a revision by suppliers of the anticipated backstop supply price. A revision of expectations upward would lead to a current price jump upward and vice versa. A price drop occurred briefly when University of Utah researchers announced "cold fusion." And the popular interpretation of the large drop in world oil prices during 1985–1986 is that cartel discipline broke down Organization of Petroleum Exporting Countries (OPEC) and key members raised their outputs above their quota levels.

It is undeniable that the current interaction of the world supply curve for oil and the world demand curve yield the current price. At issue is how depletion rent and oligopoly rent are capitalized in the current costs of production, costs that determine the characteristics of the effective supply schedule. In the mid-1970s, reasonable observers debated whether the 1973 jump in world oil price was the result of production restraint by OPEC, an oligopoly effect, or a reckoning that future oil stocks would be rather smaller than had been anticipated. In any case, current production was cut back and world prices leapt up. A popular view of the poor performance of the U.S. economy in the later 1970s was that it had to accommodate unanticipated supply-side shocks in the energy sector. The notion of cost-push inflation became popular and an extensive literature emerged on this issue.

Since impending exhaustibility is predicted to manifest itself in currently high price, which is reflecting a large scarcity rent, a standard view is that the current and anticipated high prices should induce a search for substitute supply. How tight is this link between, say, rising current oil price and a spurt of invention to develop an alternative supply? It was argued above that this link is relevant for explaining two of the most significant events in human history—the effect of timber scarcity in Great Britain on the timing and intensity of the invention of mine-draining machinery (the steam engine) and the effect of watermill site scarcity on the timing and intensity of invention in steam engine efficiency (Watt's separate condenser). It may be the anticipated high price rather than the current actual high price per se that induces inventors to perform, to come up with substitute supply. Simply anticipating that, say, oil will be exhausted soon while current price is not yet high may be sufficient to get invention going. Inventors can have foresight. They need a process or invention that brings in a competitive supply of energy when the oil runs out. In 2003, Mann argued that recent low oil prices have slowed innovation in the alternative energy sector.

Interesting models in which the substitute supply is, in fact, invented or perfected today in the public sector in order to force current oil suppliers to lower current prices have been developed. The invention is annouced and mothballed until the oil is exhausted. In such a scenario, it is not high prices today that speed up invention, rather it is the anticipation of the high prices in the future that induces inventors to perfect the substitute today.

3. *THE SCARCITY OF CLEAN ENERGY SUPPLIES*

Attention has shifted from energy sources with low cost, per se, to sources that can provide energy cleanly, at low costs. Smog in Los Angeles seems to have spurred policymakers to move toward designs for a future with cleaner air and with low-emission vehicles, homes, and factories. Eighty-five percent of world energy supply is currently derived from fossil fuels and 1% is renewable (excluding hydropower). The recent DICE model of global warming of Nordhaus and Boyer has unlimited energy produced at constant unit extraction cost from hydrocarbons but energy production contributes directly to an atmospheric temperature increase, via carbon dioxide emissions. The temperature increase shows up in explicit current economic damage. This is a large departure from the Hotelling view because Nordhaus and Boyer have no stock size limitation for, say, oil. Price, inclusive of an atmospheric degradation charge (a Pigovian tax on energy use), rises along the industry demand schedules because, essentially, of the increasing scarcity of moderate atmospheric temperatures. It is "moderate temperature" that is being "depleted" in this view and a "depletion charge" in the form of a marginal damage term (a carbon tax) is required for full cost pricing of energy. A world without carbon taxes is, of course, one of "low" energy prices, one in which the atmosphere is a carbon sink, with carbon dioxide dumped in at zero charge. User fees for dumping? "If U.S. consumers pay $1 billion a year, would that be enough to cover the problem? How about $10 billion? The level of economic damage that might be inflicted by greenhouse gas abatement is so uncertain that even the Kyoto treaty on global warming ... says not a word about what the "right" level of emissions should be." (Mann, 2002, p. 38) In 2002, Sarmiento and Gruber reported that atmospheric carbon dioxide levels have increased by more than 30% since the industrial revolution (the mid-18th century).

A formal statement of the world energy industry along Nordhaus–Boyer lines looks much like the Hotelling variant of Levhari and Leviatan amended to incorporate a continuous quality decline of the oil stock. Aggregate benefits (the area under the industry demand schedule) from current energy use, q, are $B(q)$. Temperature interval, T, above "normal," increases with q in $dT/dt = \alpha q$. Energy users (everyone) suffer current economic damage, $D(T)$. Thus, the only way to stop the full cost price of energy from rising is to switch to a source that does not drive up global temperatures. Here, the current price is $[dB(q)]/(dt)$ and the carbon tax in price is the discounted sum of marginal damages from temperature increments. (A constant unit extraction cost changes the formulation very little.) This price needs the right Pigovian tax on carbon emissions, a dumping fee, to be the full market price for energy.

Will the current, dirty, energy-generating technologies be invented around? Would they be invented around if current energy prices were higher? Mann thinks so. How abundant will fuel cells, solar generation installations, "pebble bed" nuclear reactors, windmill farms, synthetic fuels, and more sophisticated electric power transmission grids be 20 years from now? Are carbon taxes and attendant high current energy prices the best prescription for a cleaner energy future? Many policymakers have not espoused this approach, particularly in the United States and Canada.

4. SUMMARY

Depletion theory is the attempt to explain price formation for commodities such as oil, those with current flows being drawn from finite stocks. Price rises above unit extraction cost because of a depletion effect and one analyzes the nature of this effect in order to try to predict future price paths. Until recently, attention was focused on how the progressive scarcity of an oil stock was translating into the current high price for oil in use. In the past decade, attention has shifted from how the limitation on future supply is showing up in, say, oil price to how the effects of pollution from fossil fuel burning should be incorporated into current price. The depletion effect on the clean environment becomes a component (a pollution tax) in current full price. The hope is that full pricing of energy will signal to users how best to use energy currently and will signal to suppliers the potential profitability of innovating in the energy supply process.

SEE ALSO THE FOLLOWING ARTICLES

Coal Industry, History of • Economics of Energy Supply • Innovation and Energy Prices • Modeling Energy Supply and Demand: A Comparison of Approaches • Oil and Natural Gas Liquids: Global Magnitude and Distribution • Petroleum Property Valuation • Prices of Energy, History of • Transitions in Energy Use • Value Theory and Energy • Wood Energy, History of

Further Reading

Adelman, M. A. (1990). Mineral depletion with special reference to petroleum. *Rev. Econ. Statist* **72,** 1–10.
Deffeyes, K. (2001). "The View from Hubbert's Peak." Princeton University Press, Princeton NJ.
Gray, L. C. (1914). Rent under the assumption of exhaustibility. *Q. J. Econ.* **28,** 466–489.
Hotelling, H. (1931). The economics of exhaustible resources. *J. Polit. Econ.* **39,** 135–179.
Jevons, H. S. (1915, reprint 1969). "The British Coal Trade." A. M. Kelley, New York.
Jevons, W. S. (1865, reprint 1968). "The Coal Question: An Inquiry Concerning the Progress of the Nation and the Probable Exhaustion of Our Coal-mines," 3rd ed. A. M. Kelley, New York.
Kaufmann, R. K., and Cleveland, C. J. (2001). Oil production in the lower 48 states: Economic, geological, and institutional determinants. *Energy J.* **22,** 27–49.
Landes, D. S. (1969). "The Unbound Prometheus." Cambridge University Press, Cambridge, MA.
Levhari, D., and Leviatan, D. (1977). Notes on Hotelling's economics of exhaustible resources. *Can. J. Econ.* **10,** 177–192.
Mann, C. C. (2002). Getting over oil. *Technol. Rev.* **January/February,** 33–38.
Miller, M., and Upton, C. (1985). A test of the Hotelling valuation principle. *J. Polit. Econ* **93,** 1–25.
Nordhaus, W. (1974). The allocation of energy resources. *Brookings Papers Econ. Activ.* **3,** 529–570.
Nordhaus, W. D., and Boyer, J. (2002). "Warming the World: Economic Models of Global Warming." MIT Press, Cambridge MA.
Pelham, R. A. (1963). The water-power crisis in Birmingham in the eighteenth century. *Univ. Birmingham Hist. J.*, 23–57.
Pomeranz, K. (2002). "The Great Divergence: Europe, China and the Making of the Modern World Economy." Princeton University Press, Princeton NJ.

Rae, J., and Volti, R. (2001). "The Engineer in History," Revised ed. Peter Lang, New York.
Sarmiento, J. L., and Gruber, N. (2002). Sinks for anthropogenic carbon. *Phys. Today* **55**, 30–36.
Smil, V. (1994). "Energy in World History." Westview Press, Boulder, CO.
Thomas, B. (1980). Towards an energy interpretation of the industrial revolution. *Atlantic Econ. J.* **3**, 1–15.
United Nations. (2002). "North America's Environment: A Thirty Year State of the Environment and Policy Retrospective." United Nations, New York.

Derivatives, Energy

VINCENT KAMINSKI
Reliant Resources, Inc.
Houston, Texas, United States

1. Classification of Energy Derivatives
2. Forwards and Futures
3. Energy Commodity Swaps
4. Options
5. Main Energy Derivatives Markets
6. Options Embedded in Bilateral Contracts and Physical Assets (Real Options)
7. Valuation of Energy-Related Options

Glossary

derivative The term used in finance to describe a financial product with value dependent on the value of another instrument, commodity, or product.

forward Forward contract is a transaction that provides for delivery of the underlying commodity at a future date, at the price determined at the inception of the contract, with cash being paid at or after an agreed number of days following the delivery.

NYMEX Commodity exchange located in New York City, offers a number of listed energy contracts.

option A contract that gives the holder the right, but not the obligation, to buy (sell) a financial instrument or commodity (the underlying) at a predetermined price (a strike or exercise price) by or on certain day.

real options In the energy markets, flexibilities or rigidities embedded in bilateral transactions or physical assets that can be valued using option pricing technology developed for the financial markets instruments.

swap An exchange of streams of cash flows over time between two counterparties. The level of the cash flows depends on prices of financial instruments or commodities referenced in the swap agreement. In a typical commodity swap, one side pays a fixed price; the second side pays a floating (varying) price that corresponds to the market price of the underlying commodity.

volatility The most important input to the option pricing models. Volatility is typically defined as the annualized standard deviation of price returns.

A derivative is a financial instrument with a value based on, or derived from, the value of another stock, bond, or a commodity (the underlying). In the case of energy derivatives, the underlying is an energy commodity, energy-related derivative or a physical asset used to produce or distribute energy.

1. *CLASSIFICATION OF ENERGY DERIVATIVES*

Energy derivatives can be classified based on (1) the contract definition specifying the rules by which the instrument derives its value from the price of another traded instrument, (2) the market or platform on which they are traded, or (3) the underlying commodity or commodities. Most energy derivatives can be classified as forward (futures) contracts or options. Both types of instruments can be negotiated as physical contracts that require actual delivery of the underlying commodity as financial contracts that are cash settled, based on the prices of the underlying observed in the marketplace or published by an industry newsletter or a specialized business.

Energy derivatives are traded and valued as stand-alone contracts or may embedded in bilateral transactions and physical assets. Such options existed for a long time and only recently they have become a subject of rigorous analysis directed at their valuation and risk management of the exposures they create.

2. *FORWARDS AND FUTURES*

A forward contract is a bilateral transaction that provides for delivery of the underlying commodity at a future date, at the price determined at the inception of the contract, with cash being paid at or after an agreed number of days following the delivery. In many markets, the fair price of a forward contract can be derived from the spot commodity price, interest rate, and the storage cost, using the so-called arbitrage principle. Specifically, the forward price (F) is equal to the spot price (S) adjusted for the financing and storage cost (r – annualized interest rate, s – unit storage cost), or $F = S^*(1 + r + s)^{(T-t)}$ where $T-t$ represents the life of the contract (t is today's date, T is the delivery date, and both are measured in years). The logic behind this equation is based on the arbitrage argument that defines the upper and lower bounds on the forward price; one can prove that both bounds are equal giving the unique forward price. The buyer of the forward contract can lock in the economic outcome of the transaction by borrowing funds at rate r, buying the commodity at spot (S) and storing the commodity and taking a short position in the forward contract (selling). At contract expiration, the transaction is reversed: the commodity is delivered into the forward contract at price F and the financing and storage costs are recovered. The arbitrageurs will engage in such transactions, as long as $F > S^*(1 + r + s)^{(T-t)}$. Alternatively, the lower bound can be established by recognizing that one can borrow the commodity, sell it in the spot market, and invest the funds at rate r, taking a long position in the forward contract. This argument breaks down often in the case of commodity-related forward contracts for two reasons. Some energy commodities may be unstorable (like electricity) or the supply of storage may be limited at times. In some cases, the existing market framework does not allow for shorting of the underlying physical commodity, eliminating one of the arbitrage bounds.

Given that both the storage cost and interest rate are positive, the arbitrage argument implies that the forward price is always greater than the spot price and that the forward price curve (a collection of forward prices for different contract horizons) is an increasing function of time to expiration (the curve is always in contango). In reality, forward price curves for many commodities, including natural gas and crude, are often backwardated (the forward prices in front of the curve exceed the prices in the back). This arises from the fact that for many market participants holding physical commodity has a higher value than controlling a forward contract. The stream of benefits that accrue to the owner of physical commodity is called a convenience yield. Assuming that the convenience yield can be stated per unit of commodity, δ, the full cost-of-carry formula for an energy forward contract is given by the following equation:

$$F = S^*(1 + r + s - \delta)^{(T-t)}.$$

Futures are forward contracts traded on an organized exchange that provides credit guarantees of counterparty performance, provides trading infrastructure (typically an open outcry system or screen trading), and provides the definition of a standardized contract. Credit protection is provided by margining and by the capital contributed by the exchange members and ultimately their credit. At inception, each counterparty posts the so-called initial margin that is followed by the so-called maintenance margin deposited if the original margin is eroded by losses on the contract. The practice of marking to market the contract and providing (receiving) additional margin is the most important difference between the futures and forward contracts. The most important exchanges offering energy contracts (crude, natural gas, heating oil, gasoline, and electricity) are the New York Mercantile Exchange (NYMEX), International Petroleum Exchange (IPE) in London, Nordpool in Norway. The Intercontinental Exchange offers many energy-related contracts but should be treated more as a trading platform, not as an organized exchange.

Under some fairly restrictive conditions, the futures and forwards have the same value. Potential differences arise from the practice of margining the contract that results in stochastic financing cost (one has to borrow/invest the funds paid/received) and from the differences in default risk. An organized exchange offers more layers of credit protection than bilateral arrangements. In practice, the differences between futures and forwards have been found to be rather small in most empirical studies. The energy industry typically ignores this distinction in routine transactions.

The distinction between the forward and futures contracts has become more ambiguous in many markets. The practice of requiring and posting periodically collateral to provide guarantee of performance under a bilateral forward contract, which has become a standard practice in the United States and many other countries, has obliterated one major difference between forwards and futures. Classical forwards, unlike futures, did not trigger cash flows before contract maturity.

3. *ENERGY COMMODITY SWAPS*

Swaps can be looked at as a basket of forwards that has to be executed as one contract (no cherry picking is allowed). In the general sense, a swap is the exchange of two streams of cash flows: one of them is based on a fixed (predetermined) price and the second on a floating (variable) price. At inception, the fixed price is determined by finding the fixed price that makes the fixed and floating legs of the swap equal. The fixed leg is equal to the sum of the periodically exchanged volumes of the underlying commodity multiplied by the fixed price, discounted back to the starting date. The floating leg is equal to the sum of the volumes multiplied by the floating price, discounted back to the starting date. The floating price is obtained from the forward price curve, a collection of forward prices for contracts with different maturities. In other words, the fixed price of the swap is derived from the following equality:

$$\sum_{t=1}^{n} \frac{F_f \times v_t}{(1+r_t)^t} = \sum_{t=1}^{n} \frac{F_t \times v_t}{(1+r_t)^t}, \qquad (1)$$

where F_f denotes the fixed price of the swap, F_t – the floating price, r the applicable discount rate, and v_t – periodic deliverable volumes.

In practice, a swap may be physical or financial. A sale of an energy commodity at a fixed price, with deliveries distributed over time, is referred to and priced as a swap. A swap may be a purely financial transaction, with one counterparty sending periodically to another the check for the fixed-floating price difference (multiplied by the underlying volume), with the direction of the cash flow dependent on the sign of the difference.

4. *OPTIONS*

An option is a contract that gives the holder the right, but not the obligation, to buy (sell) a financial instrument or commodity (the underlying) at a predetermined price (a strike or exercise price, usually denoted by K), by or on certain day. The option to buy is referred to as a call, the option to sell as a put. An option that can be exercised only at maturity is known as a European option, an option that can be exercised at any point during its entire lifetime is known as an American option. In the case of an energy-related option, the underlying is an energy commodity or another energy derivative (a forward, a swap, a futures contract, or even another option). Some popular energy options are defined in terms of a number of several different commodity prices (have multiple underlying commodities or instruments).

Most energy options traded in bilateral transactions have a forward contract as the underlying and this has important implications for the valuation procedures. Through the rest of this article, we are assuming that the underlying is a forward or a futures contract. Most options transacted in the energy markets are plain vanilla European calls and puts. Certain features of energy markets created the need for more specialized structures, addressing specialized needs of commercial hedgers, with complicated payoff definitions. Most popular options with complex payoff definitions and many practical applications are Asian and spread options.

Asian options are defined in terms of an average price. The payoff of an Asian option at maturity is defined as max (avg (F)-K, 0) for a call option and max (K-avg (F), 0) for a put option, where F stands for the forward price of the underlying and K for the strike price. The term avg (F) corresponds to the average price calculated over a predefined time window. The rationale for Asian options arises in most energy markets from the fear of price manipulation on or near option expiration date, given relatively low trading volumes and the relative ease to affect the price through targeted buying or selling. Averaging mitigates this risk, reducing the impact of any single day price on the option payoff. In many cases, the exposures facing commercial energy hedgers are defined in terms of average prices. For example, an electric utility buying natural gas as a fuel is exposed to an average cost over time, given its ability to pass on its average costs to the end users. Asian options tend to be cheaper than corresponding European options, as price averaging reduces price volatility, an important input to the option pricing models (discussed later).

Spread options are defined in terms of a price difference. The payoff of a call spread option is defined as max (F_1-F_2-K, 0) and the payoff of the put as max (K-F_1-F_2, 0). The price difference may correspond to the location spread (for example, the difference between Brent crude produced in the North Sea and West Texas Intermediate, WTI) or time difference (calendar spread between price of natural gas corresponding to December natural gas and July natural gas contracts with delivery at Henry Hub location in Louisiana). Quality spread corresponds to the price difference of different grades of the same commodity (for example, sweet and sour crude oil). The rationale for spread options arises from the exposures that many commercial operations face in the energy markets: exposures to price differentials as opposed to exposures to a single price level. A refinery is exposed to the spread between the composite price of the basket of refined products and the price of crude. A power plant is exposed to the difference between price of electricity and the price of fuel. An oil company buying gasoline in Europe and shipping it to the United States faces combined location and calendar spread risks when the cargo is in transit.

The importance of price differential related exposures created specialized terminology describing different types of spreads. The difference between the price of heating oil and the price of crude is referred to as a heat spread; the difference between the price of gasoline and crude in referred to as the gas crack spread. The difference between natural gas prices at one of many trading hubs in the United States and the price at Henry Hub (the delivery point for the NYMEX natural gas futures contracts) is routinely referred to as *basis* in the U.S. natural gas markets (in more general sense, basis denotes price difference). The difference between price of electricity and the price of natural gas is known as the spark spread. In the case of the spark spread, the price of natural gas, denominated typically in $/MMBTU, has to be converted into $/MWh. This is accomplished in practice by multiplying natural gas price by the heat rate (MMBTU/MWh) that corresponds to the thermal efficiency of a specific (or notional) power plant (the energy input required to produce one MWh of electricity).

Spread options are used to value many transactions in the energy markets. One example is the contract for transmission, known as a firm or fixed transmission right (FTR). The FTR protects a provider of electricity against basis risk. This type of exposure happens when a supplier offers delivery at one point but owns generation at another point in a power pool. If no transmission capacity is available between these two points (i.e., congestion develops on the transmission grid), the prices at two locations may decouple and the supplier may be unable to deliver into his contractual obligation. The supplier would have to buy electricity at the point where his or her load is located, running the risk that he will incur a loss. To hedge against this risk he may buy FTRs offered by many power pools through special auctions. An FTR contract may be structured as a swap or an option. A swap pays the difference between the two locational prices and allows lock-in the economic outcome of the original power supply transaction, irrespective of the potential congestion. An FTR contract that is structured like an option may be priced as a spread option.

A special type of energy related option was invented to offer protection against combination of volumetric and price risks. Many energy producers and buyers (especially utilities and local distribution companies) are exposed to both uncertain prices and quantities. This risk is addressed through the so-called swing option that has a number of unique characteristics. It is typically structured as a forward start option: the strike price is set equal to the market price of the underlying on some future date. In the U.S. natural gas markets, this price is typically equal to the first of the month price for base load deliveries over the entire calendar month, published by one of the industry newsletters (like, for example, Inside FERC Natural Gas Market Report). Some swing options use a strike price that is fixed in a bilateral contract. The holder of the option is entitled to buy the commodity at the strike price n times over the defined period (typically one calendar month) within certain volumetric limits, referred to as a minimum and maximum daily quantity. The option definition may be complicated further by ratchets that allow the holder to reduce or increase the purchased volumes in certain increments. The contract may provide also for minimum and maximum takes over certain time periods (for example, months, quarters, and years) and for penalties if the periodic bounds are exceeded. The swing options are popular in the U.S. natural gas and electricity markets as a form of protection against a jump in prices combined with an increase in consumption requirements. Pricing such options is quite complicated due to the need to keep track of both types of optionality (with respect to price and volume).

Basket options are defined as contracts on a composite commodity like, for example, a package of natural gas and crude oil. The price of the underlying in this case is a (weighted) sum of prices of commodities included in the basket. Such options are typically used by producers who have exposure to multiple commodity prices and want to hedge the overall exposure, as the price fluctuations of the basket components may be mutually offsetting, providing a natural hedge. The value of a basket option, like the value of a spread option, is heavily influenced by the correlations between basket components.

5. *MAIN ENERGY DERIVATIVES MARKETS*

An alternative classification of energy derivatives is based on the market at which they are transacted. The most transparent and liquid markets are those provided by the organized or electronic exchanges. An organized exchange may be based on the principle of open outcry or may be based on screen-based trading. The most important exchanges in the United States are the New York Mercantile Exchange (NYMEX) and the Intercontinental Exchange. The latter provides an electronic trading platform and other services such as netting and settlements (with London Clearing House and Merchant Exchange). NYMEX offers both futures and options on futures (options that have a futures contract as the underlying and are exercised into a futures contract that has to be taken into delivery or closed in a separate transaction). NYMEX trading is based on the principle of the open outcry but electronic trading is available as well for certain contracts or after regular trading hours. Table I provides the summary of the contract available on NYMEX (only contracts with significant volumes are listed, as opposed to all listed contracts that may have no significant activity).

Other exchanges important in the energy markets are International Petroleum Exchange of London (futures and options on futures for Brent oil and gas oil, futures contracts for natural gas), Powernext in Paris (day-ahead electricity contracts), European Energy Exchange (EEX) in Leipzig, Germany (electricity), and Nordpool in Scandinavia (electricity).

The exchange-based transactions are supplemented by the over-the-counter (OTC) market that provides the benefits of greater flexibility of contract design and allows for customization to fit exactly the needs of the buyer. OTC derivatives lack the credit support provided by an exchange and their pricing is more opaque, given that in many cases they can be quite complicated. The most important forms of the OTC energy contracts are swaps and options.

6. OPTIONS EMBEDDED IN BILATERAL CONTRACTS AND PHYSICAL ASSETS (REAL OPTIONS)

One very important class of energy related derivatives are options embedded in energy contracts. Unlike exchange traded and OTC derivatives, they are not traded as stand alone contracts but are packaged in bilateral contracts. They typically correspond to different flexibilities and rigidities embedded in the transactions that correspond to practical needs of doing business. In principle, every flexibility granted to, or accorded by, a counterparty in terms of quantity, delivery timing, and location or quality can be translated into an option, swaption, or another derivative. The challenge of pricing options embedded in the contracts is that one is dealing with many layers of interacting optionality, and it is often difficult to separate different options (they may not be additive) and price them individually using commercially available software. It is, however, important to price such options to avoid giving them away for free or at reduced prices and to manage the exposures they create over the life of the contract.

One example of a complicated bilateral contract that may extend over many years is so-called full requirements contract. Under this contract, the energy provider takes over the resources of the customer (for example, generation plants, storage, and contractual assets like transportation and transmission contracts of a municipal utility) and commits to satisfy all the energy needs of the buyer at an agreed price. Energy is delivered using the assets of the buyer, controlled and dispatched by the supplier over the duration of the contract, and the surplus (deficit) is sold (covered) in the market. Providers also have the option to shut down the generation assets they control in case

TABLE I

NYMEX Energy Futures

Contract	Size	Delivery point	Options	Settlement type
Natural gas	10,000 MMBTUs	Henry Hub, LA	Yes	Delivery, ADP, EFP
Light sweet crude oil	1000 barrels	Cushing, OK	Yes	Delivery, ADP, EFP
Heating oil	1000 barrels	New York Harbor	Yes	Delivery, ADP, EFP
Unleaded gasoline	1000 barrels	New York Harbor	Yes	Delivery, ADP, EFP
Brent oil	1000 barrels		Yes	Financial
Central Appalachian coal	1550 tons	Ohio River, Big Sandy River	No	Delivery, EFP
PJM electricity	40 MW on peak per day, 19 to 23 days a month		No	Financial

electricity may be acquired at a lower price from the market and sell in the market fuel purchased for the generation plants. In most cases, the provider assumes the volumetric risk associated with demand fluctuations (due to weather changes) and load growth due to economic and demographic changes, within certain predetermined bounds.

Derivatives embedded in physical assets such as power plants or natural gas storage facilities are the subject of the so-called real options, the term that describes the application of the technology developed for financial markets to the analysis of investment decisions. For example, a gas-fired natural gas power plant can be looked at as a strip of spark spread options. The underlying assumption is that the power plant is dispatched if the spark spread, as defined previously, exceeds variable marginal cost. This approach is very effective but should be used with caution. Valuation of a power plant with a life extending over many years requires many inputs for which market information is not directly available. For example, one has to provide information about forward price curves for natural gas and electricity for as many as 20 years, assumptions regarding price volatility and correlations between price returns for both commodities.

In many cases, there are no markets extending that far into the future, and this means that one has to substitute forecasts for what is supposed to be market derived inputs. A more serious shortcoming of this method is that a power plants operations are subject to many physical and regulatory constraints that are difficult to accommodate in the option pricing model. For example, one has to recognize constraints related to planned and forced outages, startup and shutdown costs, emission constraints that limit the total number of hours that a plant may operate in a year, and the restriction on the number of starts (due to provisions in maintenance and service contracts). Incorporation of such restrictions increases the dimensionality of the problem and greatly increases the computational resources required to come up with realistic valuation.

In spite of the shortcomings, the real options approach represents a much better alternative to the standard discounted cash flow approach, traditionally used in making investment decisions under uncertainty. The traditional discounted cash flow approach can only address uncertainty through comparisons of a limited number of scenarios that may not capture the entire range of possible future states of the world and may sometimes represent wishful thinking of the decision makers.

Other frequent uses of the real options technology are natural gas storage facilities and pipelines. A natural gas storage facility can be looked at as a portfolio of calendar spread options. A pipeline transportation contract can be valued as a portfolio of basis (locational spread) options.

The classification of options depending on the underlying commodity is quite straightforward. One possible classification is based on the physical commodity or commodities to which the option is related. One can talk about options on crude, natural gas, and so on. If the option is written on a specific energy-related instrument, the options are referred to as options on forward, futures, other options (compound options), and swaps (swaptions).

7. VALUATION OF ENERGY-RELATED OPTIONS

Valuation of energy-related options offers many challenges due to a number of complications in their definition and market behavior of the underlyings and is an area of very intensive research efforts in a number of academic centers and many energy companies. European options on forwards and futures are valued using the so-called Black formula that represents a version of the well known Black-Scholes formula for pricing options on stocks that don't pay dividends. The call price c is given by

$$c = e^{-r(T-t)}(FN(d_1) - KN(d_2)),$$

and the put p price is given by

$$p = e^{-r(T-t)}(KN(-d_2) - FN(-d_1)),$$

where

$$d_1 = \frac{\ln\left(\frac{F}{K}\right) + \frac{\sigma^2}{2}(T-t)}{\sigma\sqrt{T-t}}, \quad d_2 = d_1 - \sigma\sqrt{T-t}.$$

In these formulas, F stands for the forward price as of the time of valuation (t), K represents the strike price. The expiration date is given by T. Both t and T are measured in years (1 year is equal to 1, 1 month to 1/12, etc.) and therefore $T-t$ is equal to the remaining life of the option. The risk-free interest rate is given by r and the expression $\exp(-r(T-t))$ is a discount factor. The critical input to the formulas is σ, or the volatility coefficient. Volatility, unlike other inputs, is not directly observable and represents the most critical assumption in the option pricing process. In practice, it is based on the insights obtained from historical estimates or is inferred from market prices of traded options. In the first case, volatility is estimated as the standard deviation of the logarithmic price returns (natural logs of the price ratios, defined as price P in period t divided by price P in period $t-1$). Volatility is quoted by convention as an annualized number. Annualization is accomplished by multiplying the standard deviation of price returns by the factor that depends on the price data frequency. In the case of monthly data, the factor is the square root of 12. In the case of weekly data, the annualization factor is the square root of 52. In the case of daily prices, one uses by convention the square root of 250 (or 260)—an approximate number of trading days in a year. In some energy markets, trading continues on Saturdays (Western power markets in the United States) or over the weekend. In such cases, one should use the appropriate number of trading days in a year.

For other energy options, there are typically no closed form valuation formulas and one must use different approximation formulas or numerical solutions. The most widely used approaches are Monte Carlo simulations and binomial (or more complicated) trees (see Table II). The Monte Carlo approach is based on repetitive simulations of price trajectories and calculating option payoffs at expiration. The payoffs are then averaged and discounted back to the current date. This approach is very flexible and allows for quick implementation. The shortcoming of this approach is that it requires significant computational resources and may be time intensive. It is more appropriate for path-dependent options (for example, Asian options) where the payoff depends not only on the terminal price but also on the trajectory along which this price was reached. In general, a Monte Carlo approach is not suitable for American options valuation; though some numerical techniques address this problem.

The binomial tree technique is well suited for pricing American options and may be described as a structured simulation. The technique is based on the assumption that a price at any point in time may evolve only into two other states (this explains the use of the term binomial), up or down from the starting level, with probabilities and the magnitudes

TABLE II

Option Valuation Using a Binomial Tree

Time	1	2	3	4	5
					89.06561
					0
				79.35276	
				0	
			70.69912		70.69912
			0		0
		62.98919		62.98919	
		0.786522		0	
	56.12005		56.12005		56.12005
	2.58979		1.499718		0
50		50		50	
5.22672		4.237387		2.859618	
	44.54736		44.54736		44.54736
	7.658791		6.743557		5.452637
		39.68935		39.68935	
		10.82827		10.31065	
			35.36112		35.36112
			14.63888		14.63888
				31.50489	
				18.49511	
					28.0692
					21.9308

of changes can be calibrated from the standard option pricing assumptions (including the volatility assumption). Specifically, probability of the up move is given by

$$p = \frac{1-d}{u-d},$$

and the probability of the down move is given by $q=1-p$. The price level in the up move is given by uF, where F is the price level and the u multiplier is given by $u = e^{\sigma\sqrt{\Delta t}}$. The down move price level is given by dF, where d is given by $d=1/u$. The term Δt corresponds to the length of the time step and is chosen by the modeler depending on the computation time and required precision requirements. The time step determines how dense is the tree (how many nodes it has for a given contract duration). Parameters u, d, and p allow for constructing the tree describing possible evolution of the price from the current level till maturity. Options can be priced through the process of backward induction by calculating the payoff of the option at terminal nodes (it is either zero or a difference between the terminal price and the strike price) and moving back in tree. At each node, (t, i), where t corresponds to the time step, i to the node level counted from the bottom of the tree, the value of the option is given by the comparison of the payoff of the option if it is immediately exercised with the expected value one obtains from keeping the option alive. This expected value is equal to the probability weighted values at two nodes connecting to the given node at time $t+1$, discounted back over the length of one time step.

The example that follows shows a tree developed for a put option with a life of 5 months using five time steps. This means that the length of one time step is 1/12 or 0.08333. The starting price is 50, the strike price K is 50, and the volatility is assumed to be 0.4 or 40%. The probability p of the up move at each node is equal to 0.471165. The risk free interest rate is 10% and this translates into a 1-month discount factor of 0.991701. For example, the value of the option at the node (3, 2), where the price level is equal to 44.554736, is equal to 6.743557. This value is calculated as follows: the immediate exercise of the option gives the option holder 50−44.54736=5.452 (it's a put option). Keeping the option alive gives the holder 6.743557 that is calculated as probability weighted average of the option values at the nodes (4,3) and (4,2), discounted back one period. To be exact, $6.743557 = 0.991701 \times (0.471165 \times 2.859618 + (1-0.471165) \times 10.31065)$. Keeping the option alive is better than killing it (exercising it immediately) and therefore the last calculation gives the value of the option at the node (3,2).

The Black formula is based on the assumption that the price of the underlying commodity follows the stochastic process known as Geometric Brownian Motion. Under this process, the instantaneous price return (dF/F) is given by the following formula:

$$dF = \mu F dt + \sigma F dZ, \tag{2}$$

where μ stand for the so-called drift (instantaneous rate of return) and dz is the so-called Wiener variable ($dz = \varepsilon\sqrt{dt}$, where ε is drawn from the standard normal distribution). In case the underlying is a forward contract, the drift is equal to zero. This can be explained by the fact that entering into a forward contract has zero cost. A nonzero drift implies that the expected return is nonzero, and this implies that one would enter into a transaction with an expected loss or gain and no cost. In the first case, the market participants would bid down the price by shorting it; in the second case (gain) they would bid the price up till the potential gain disappeared.

The challenge of pricing energy derivatives is that the prices in many commodity markets do not follow this process. Energy commodity prices display behavior that is characterized by seasonality, the tendency to gap due to shocks that affect either the demand or supply side, and also they cannot be modeled without paying attention to production costs. The research in the area of energy option valuation evolves around development of new models that are based on alternative stochastic assumptions regarding price evolution. Two examples of such processes are a mean reversion process and a jump-diffusion process. Mean reversion process recognizes the fact that prices gravitate toward the levels determined by the cost of production. The simplest form of the mean reversion process is given by the so-called Ornstein-Uhlenbeck process:

$$dF = \alpha(m - F)dt + \sigma dz, \tag{3}$$

where α (>0) denotes the speed of mean reversion and the m denotes the mean price level toward which the price gravitates. If the price drops below the mean level, the difference $m-F$ is positive and given the speed reversion that is positive, the deterministic part of the Eq. (3) will be positive as well. This will help to bring the price back to the mean level. Of course, the stochastic part (σdz) may reinforce or counter the contribution of the deterministic part. If the price F is above the mean level, the mean reversion part of the process will act to bring it down. It is important to notice that the volatility parameter has a different interpretation than the volatility parameter in the Geometric Brownian Motion process. The price change on the left-hand side of Eq. (3) is measured in dollars per commodity unit. The Wiener variable dz is dimensionless and, therefore, σ is measured in the same monetary units as dP. Volatility σ in the Eq. (2) is unitless.

The propensity of energy pieces to jump reflects one of the structural features of the energy markets. The price at any location is likely to jump (gap) due to sudden changes in supply or demand conditions that cannot be addressed due to the rigidities of the physical infrastructure (inflexible transmission or transportation grid, limited deliverability from and into storage). The modeling approach to this problem is a combination of a diffusion process (Geometric Brownian Motion or Ornstein-Uhlenbeck, for example) and a jump process. One example is the Poisson process combined with the Geometric Brownian motion:

$$dF = \mu F dt + \sigma F dz + (J - 1) F dq, \qquad (4)$$

where the variable dq assumes two possible values, 1 if a jump occurs, 0 if it does not. It is assumed that the jump occurs with the probability λdt, where dt is the time step and λ is known as the intensity of the process. J is the dimension of the jump. If the jump happens and J is equal to 0.8, then the forward price F drops immediately by 20%, in addition to changes resulting from the diffusion part of the Eq. (4). When a jump occurs, it is typically assumed to follow a log-normal distribution ($ln(J) \sim N(\gamma,\delta)$), where γ denotes the mean of the natural logarithm of the jump and δ its standard deviation. The jump diffusion process parameters may be estimated from the historical price return data and used in option valuation models based on numerical techniques. The jump diffusion approach, although very popular, suffers from many shortcomings. The parameter estimates may be unstable and it takes a long time for the results of the jump to taper off, contrary to many empirical observations that show a quick return to more normal price levels following a jump.

Mean reversion and jump diffusion are just two fairly elementary examples of the efforts under way to come up with better valuation techniques for energy options that recognize the unique features of energy markets.

SEE ALSO THE FOLLOWING ARTICLES

Business Cycles and Energy Prices • Energy Futures and Options • Energy Services Industry • Investment in Fossil Fuels Industries • Oil Price Volatility • Stock Markets and Energy Prices • Trade in Energy and Energy Services

Further Reading

Eydeland, A., and Wolyniec, K. (2002). "Energy and Power Risk Management: New Developments in Modeling, Pricing and Hedging." John Wiley, New York.

Fusaro, P. C. (1998). "Energy Risk Management: Hedging Strategies and Instruments for the International Energy Markets." McGraw-Hill, New York.

"Managing Energy Price Risk" (1999). Risk Books, London.

Pilipovic, D. (1997). "Energy Risk: Valuing and Managing Energy Derivatives." McGraw-Hill, New York.

Ronn, E. (ed.). (2002). "Real Options and Energy Management: Using Options Methodology to Enhance Capital Budgeting Decisions." Risk Books, London.

Desalination and Energy Use

JOHN B. TONNER and JODIE TONNER
Water Consultants International
Mequon, Wisconsin, United States

1. Introduction
2. Minimum Energy of Separation
3. Thermal Processes

Glossary

desalination The removal of salt, especially from seawater, to make it drinkable.

membrane desalination The removal of salt from seawater by using membranes to differentiate and selectively separate salts and water.

minimum energy of separation The thermodynamic minimum amount of energy needed to separate salts from water in the process of desalination.

thermal desalination The removal of salt from seawater by condensing purified vapor to yield a product in the form of distilled water.

This article focuses on processes involving the separation of solute and solution with the objective of reducing the solute concentration in the product stream. The focus of the following section is practical economic processes that have demonstrated sustained commercial viability.

1. INTRODUCTION

All desalination processes can be generically represented by the diagram shown in Fig. 1. Desalination processes can be segregated into two broad types, as follows:

1. Thermal processes, the most common being multistage flash (MSF); multiple-effect distillation (MED) and its variants; and vapor compression, which can be subclassified into thermal vapor compression (TVC)—most commonly combined with MED—and mechanical vapor compression;

2. Membrane processes, including reverse osmosis (RO); electrodialysis (ED); and electrodialysis reversal (EDR).

The thermal processes share the common step of condensing purified vapor to yield a product in the form of distilled water. The membrane processes have two further subclassifications with significantly different modes of operation: pressure-driven membranes and charge-selective membranes.

2. MINIMUM ENERGY OF SEPARATION

The minimum energy to separate a solute from solution can be derived from first principles. The most common approach is to consider the thermodynamic properties of pure water and a saline water, such as seawater. Two physically identical vessels, at the same temperature, containing fresh water and seawater, respectively, will have different pressures. This is due to the different vapor pressures of the two chambers, which differ because of the respective purity and salinity of the waters, and is shown diagrammatically in Fig. 2.

If the two vessels were to be connected, then the pressure difference that would exist would drive a flow in the direction of the arrow, from fresh water to saline water. A flow in this direction is not desirable for a desalination process. A day-to-day result of this situation is the reason that reverse osmosis is called reverse osmosis; some energy must be added to the system to overcome the natural direction of flow.

To achieve desalination, the pure water from the vapor space above the saline water (typically seawater) must be transported to the freshwater chamber (see Fig. 3). This can be achieved by a reversible process utilizing a hypothetical vapor compressor that can raise the pressure from the

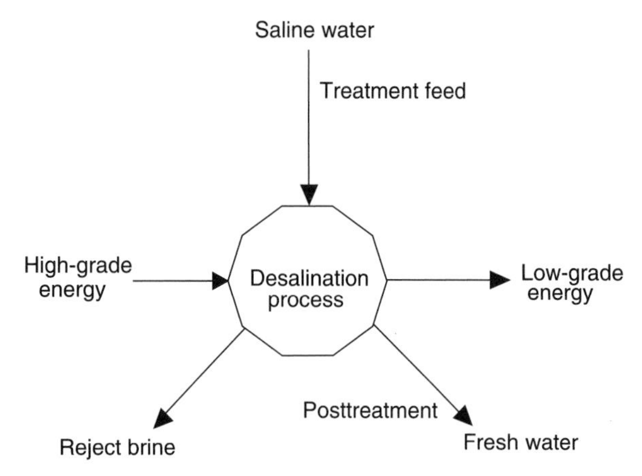

FIGURE 1 The desalination process.

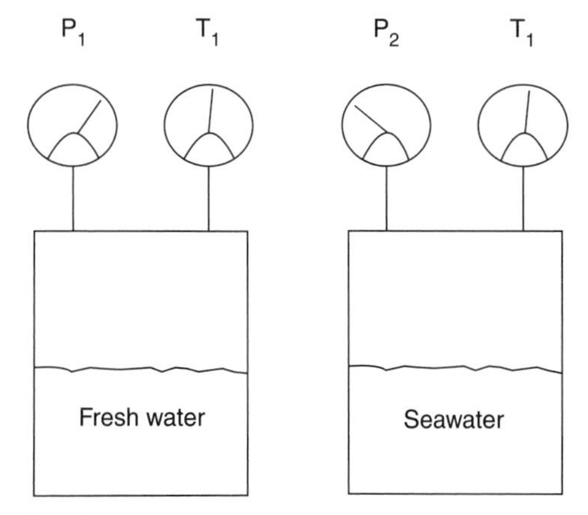

FIGURE 2 Two physically identical vessels, at the same temperature, containing fresh water and seawater, respectively, have different pressures.

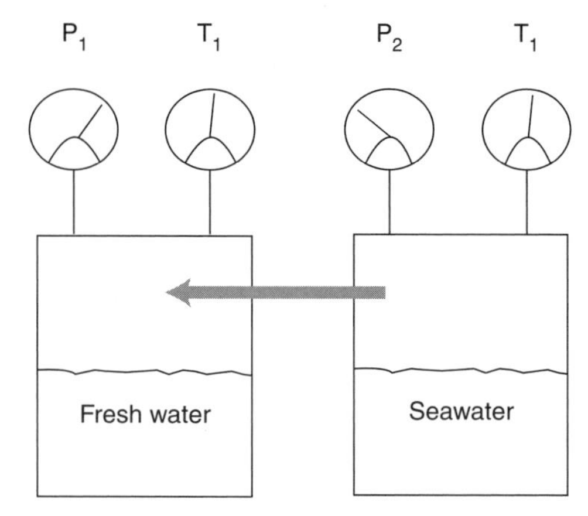

FIGURE 3 To achieve desalination, pure water from the vapor space above the seawater must be transported to the freshwater chamber.

freshwater vessel to a value just above that of the saline vessel. If the two vessels are held within a suitably large constant temperature reservoir, then it can be assumed that there is no temperature rise associated with this process.

Using this model, the minimum reversible energy consumption of a desalination process can be found by calculating the compressor power consumption that is a function of the vapor flow rate and partial pressure difference. This shows that the minimum energy of separation is a function of salinity (and also of temperature, since temperature relates directly to partial pressure). For seawater at 25°C, the vapor pressure is 3103 Pa and the specific volume is 44.17 m^3/kg; comparable figures for pure water are 3166 Pa and 43.40 m^3/kg. This yields a minimum energy consumption for the desalination of seawater of approximately 0.7 kWh per ton of water produced (0.75 kWh/ton, or 2.734 kJ/kg). This minimum energy varies from 0.65 kWh/ton at 0°C to 0.90 kWh/ton at 100°C.

This analysis was considered for vapor compression processes but is valid for all processes. Why? The analysis is based on a theoretical ideal reversible process, which means that the results are valid as minima, which cannot be bested by practical processes. All practical desalination processes have these same values as targets although each of them may be restricted by different irreversibilities.

Each desalination process has a basic thermodynamic parameter that is a primary irreversible and unavoidable loss. For example,

- Distillation must always account for ΔT_b, the boiling point elevation.
- Freezing must similarly allow for ΔT_f, the freeze point suppression.
- Reverse osmosis must overcome ΔP_o, the osmotic pressure.
- Ion exchange must overcome ΔV_p, the polarization voltage, and provide ΔV_d, the diffusion voltage.

A detailed but less intuitively obviously minimum energy analysis could be completed considering each process; however, the net result would be the same. A more practical approach is to consider the state of the art for the most commonly utilized processes.

3. THERMAL PROCESSES

MSF, MED, and TVC processes are all driven by a high-enthalpy energy source, typically steam, but hot water or oil can also be utilized. In general, all these processes incorporate a maximum operating temperature limited to between 70 and 115°C to avoid scale and corrosion associated with the solutions. This is known as the top brine temperature (TBT), which as a design parameter varies within the specified range due to slight differences in the configuration of the respective processes. This means that these processes can be heated by any energy source that is a few degrees warmer than the TBT; however, some processes utilize significantly higher enthalpy streams.

3.1 MED

MED is one of the oldest desalination methods and the simplest process with which to explain the principle of gained output ratio (GOR). GOR is a measure of how much thermal energy is consumed in a desalination process. The simplest definition is as follows:

How many kilograms of distilled water are produced per kilogram of steam consumed? Obviously, the same value for GOR is obtained when U.S. customary units of pounds are used.

Typically, the value of GOR ranges from 1 to 10 kg (H_2O)/kg (steam). Lower values are typical of applications where there is a high availability of low thermal energy. Higher values, even up to 18, have been associated with situations where local energy values are very high, when the local value or need for water is high or a combination of both.

A single-effect distiller, as shown in Fig. 4, operates with a GOR of 1 if there are no thermodynamic losses. In the absence of losses, each kilogram of steam that condenses in the effect will release sufficient heat of condensation to evaporate an equal quantity of vapor from the seawater, which then condenses in the condenser. This further assumes that the heat transfer area is infinite and there is a negligible temperature difference. Each kilogram of steam is therefore returned to the boiler as condensate and a kilogram of distilled water has been produced for other uses.

In the late 1800s, it was first noticed that this single-step process could be repeated with a gain in production, and therefore efficiency, for each effect added. This became known as Rillieux's First Principle: "In a multi-effect, one pound of steam applied to the apparatus will evaporate as many pounds of water as there are bodies in the set." This can be explained by examining the MED system depicted in Fig. 5. Once again, if losses are neglected, each kilogram of steam will release heat of condensation, which is equal to the heat of vaporization, and thus a four-effect system would produce 1 kg of condensate and 4 kg of distillate for every kilogram of heating steam utilized.

There are, however, many irreversibilities and other losses within these processes but the basic concept of Rillieux's First Principle is substantially correct and requires a 0.8–0.88 correction factor to be applied. This means that a four-effect MED system will typically yield a GOR of between 3.2 and 3.5.

MED systems have been designed and successfully operated with GORs ranging from 0.8 to approximately 10. Higher GOR designs invariably incorporate larger heat transfer areas, which then increases capital costs. The justification of higher capital costs can be made only with energy values that are appropriate or where greater demand for water suitably impacts the economics. The trade-off between capital cost (CAPEX) and operating cost (OPEX) is shown diagrammatically in Fig. 6, where it can be seen that a minimum will exist, which is project-specific, based on prevailing energy values and interest rates.

Typically, low-GOR applications include those where low-grade, low-value energy is abundant; examples include refineries and methanol facilities in arid locations where a traditional source of water is not available and a GOR of 3–4 is common. Higher GOR applications include production of distilled water for municipal supplies in areas with high-energy values such as the Caribbean Islands, where a GOR of closer to 10 is common.

Important additional points to note about the MED process are as follows:

- Each effect has an unavoidable BPE loss. Practical restrictions on TBT mean that utilizing too many effects in order to achieve high GORs is

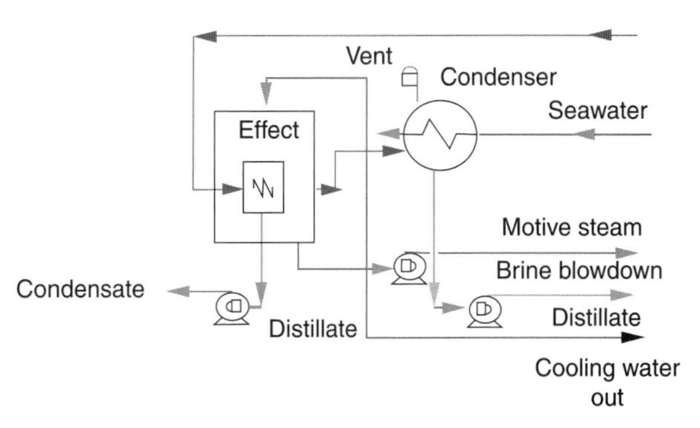

FIGURE 4 A single-effect distiller.

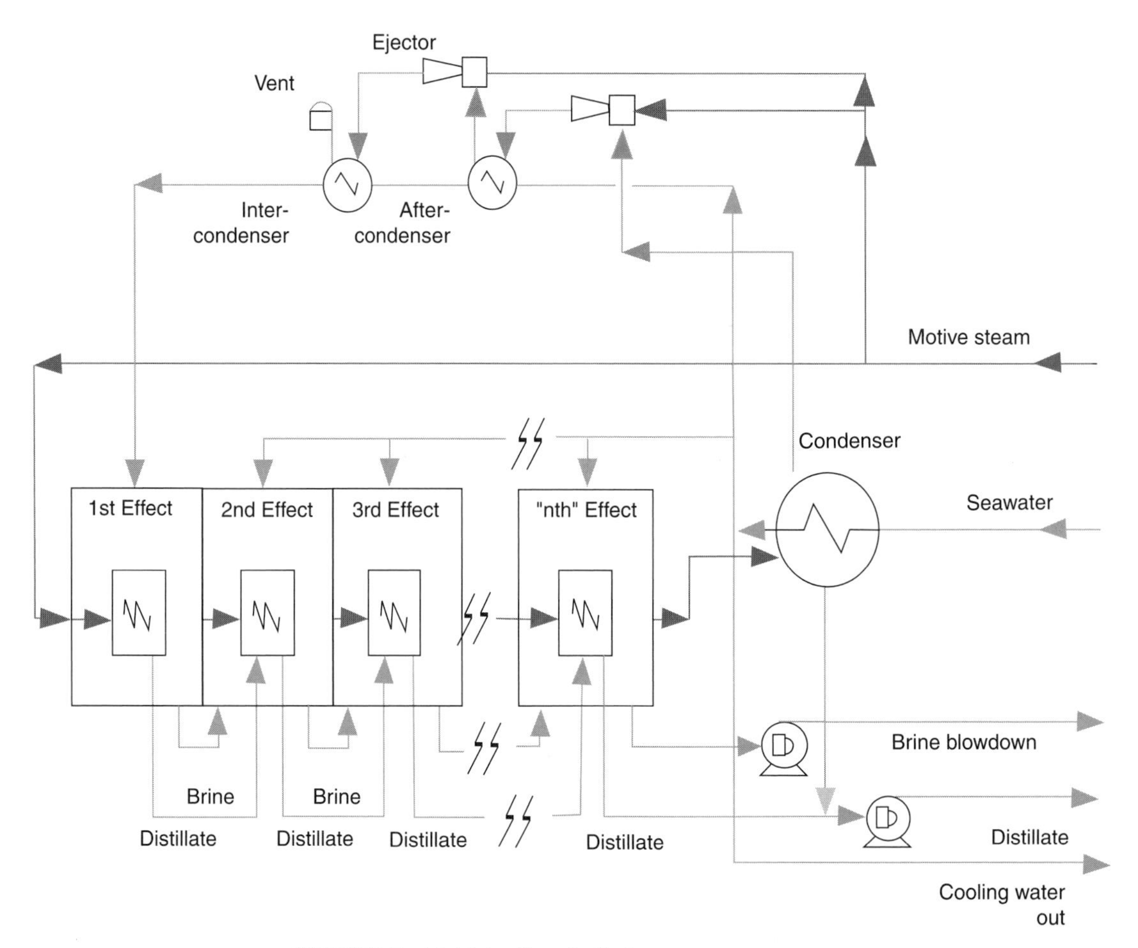

FIGURE 5 Multiple-effect distillation system.

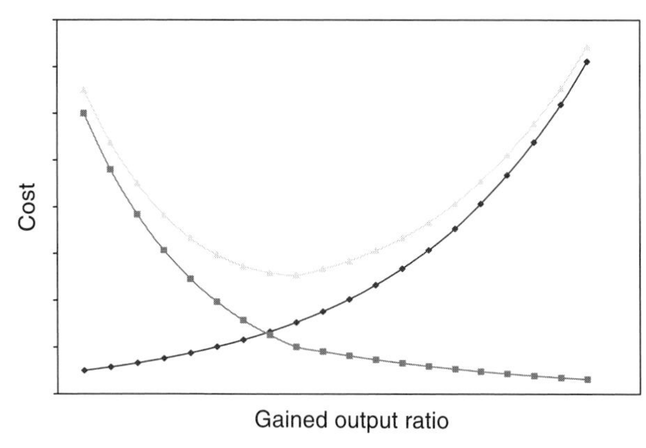

FIGURE 6 Trade-off between capital cost (CAPEX) and operating cost (OPEX). Diamonds, CAPEX; squares, OPEX; triangles, total cost.

impractical as the total BPE loss consumes a larger portion of the available ΔT.

- Pumping power is typically 0.7–1.2 kWh/ton. This includes the process pumps shown in the figures but excludes external pumps such as intake pumps. Excluding these pumps is usually valid since the cooling water that they provide is normally required even if the desalination plant is not operating as part of the facility heat balance (i.e., the steam must still be condensed). The intake pump flow rate is related to the GOR and the total dynamic head is related to site conditions, both of which also justify considering only the pumping energy within the process for comparative purposes.

3.2 MED with Thermal Vapor Compression

MED with thermal vapor compression (MED-TVC), or simply TVC, is the most rapidly growing distillation process. The process combines the basic MED system with a steam jet ejector, which enables the plant to utilize enthalpy and kinetic energy from motive steam and is shown in Fig. 7. Like the MED system, each kilogram of steam entering the first effect can yield a distillate production of $0.85N$ kg, where

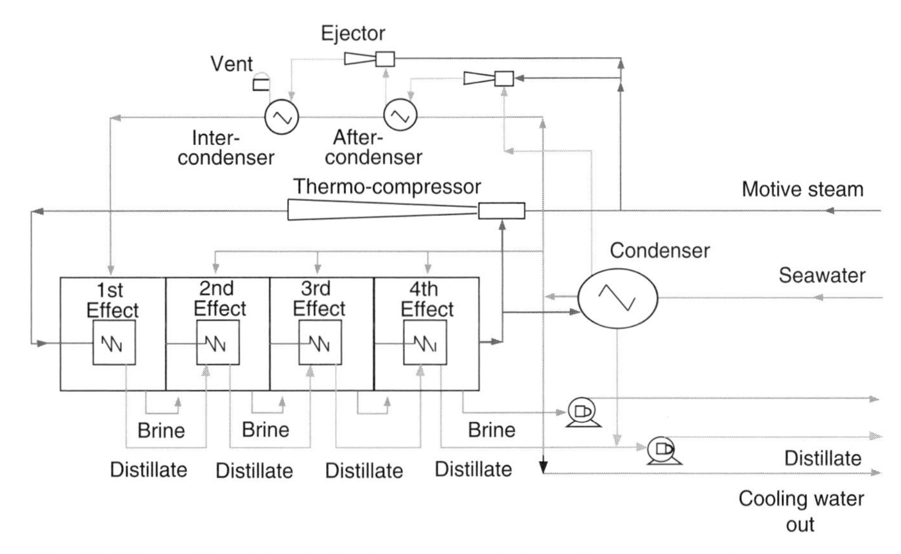

FIGURE 7 Multiple-effect distillation system with thermal vapor compression.

N is the number of effects. The important point to note is that each kilogram of steam entering the first effect is a combination of live steam and recycled low-pressure process vapor. The net result is that when utilizing medium-pressure steam, it is possible to achieve a GOR of over 8 with only four effects.

The addition of a thermocompression device adds approximately 1% to the CAPEX of a basic MED unit, yet the efficiency can be doubled. In cases where the steam is produced from waste gases, or where there is little commercial value difference between medium- and low-pressure steam, the use of TVC has grown tremendously. Using medium-pressure steam, it has been possible to achieve GORs of over 15. Such high GOR values also mean that less heat must be rejected from the system via cooling water, which can be an important consideration in many projects (for environmental or permit reasons).

The most common method for large-scale desalting of seawater had been the MSF process, which typically was combined with power production and utilized turbine extraction steam at 5 bar. Higher heat transfer coefficients, lower pumping pressures (and therefore lighter mechanical design requirements), and greater design flexibility mean that TVC designs are producing the same amount of water from this 5 bar steam but at lower CAPEX and OPEX than MSF. For distillation units in the range of 1000–25,000 ton/day capacity, the TVC process is beginning to dominate.

3.3 MSF

The MSF process is the most prevalent method of desalting seawater based on the volume of freshwater produced daily on a global basis. The process was commercialized during the 1960s based on competing Scottish and U.S. patented technology. MSF is a classic forced circulation process wherein seawater is preheated by the condensation of the vapor, which has been flashed from hot seawater. A once-through MSF process is shown in Fig. 8.

The relationship between the number of stages and the GOR is complex and also depends on the orientation of the tubes. For long tube designs where the tubes and flashing brine are always countercurrent, the number of stages, *N*, is typically 4 times the GOR. For cross-tube designs where there can be multiple tube passes within each stage and the flow through the tubes is perpendicular to the flashing brine, *N* is typically 2.5–3 times the GOR.

The once-through configuration is found mostly on marine distillation applications or similar applications that are relatively low capacity and low GOR. A typical plant of this design will have a GOR of 3 and a capacity of 1000 tons/day. Pumping power is typically 5–6 kWh per ton of distilled water. The U.S. nuclear carrier fleet utilizes this technology primarily to provide high-quality feedwater for the steam catapult launch-assist systems.

The most common MSF plants utilize cross-tube configurations in brine recycle flow schemes, shown in Fig. 9, and have unit production capacities of up to 55,000 tons/day. Many of these large systems are provided in the countries of the Gulf Cooperation Council, in Saudi Arabia, the location of the largest desalination facility at Al Jubail, and in the United Arab Emirates, the location of the largest units at Al Taweelah.

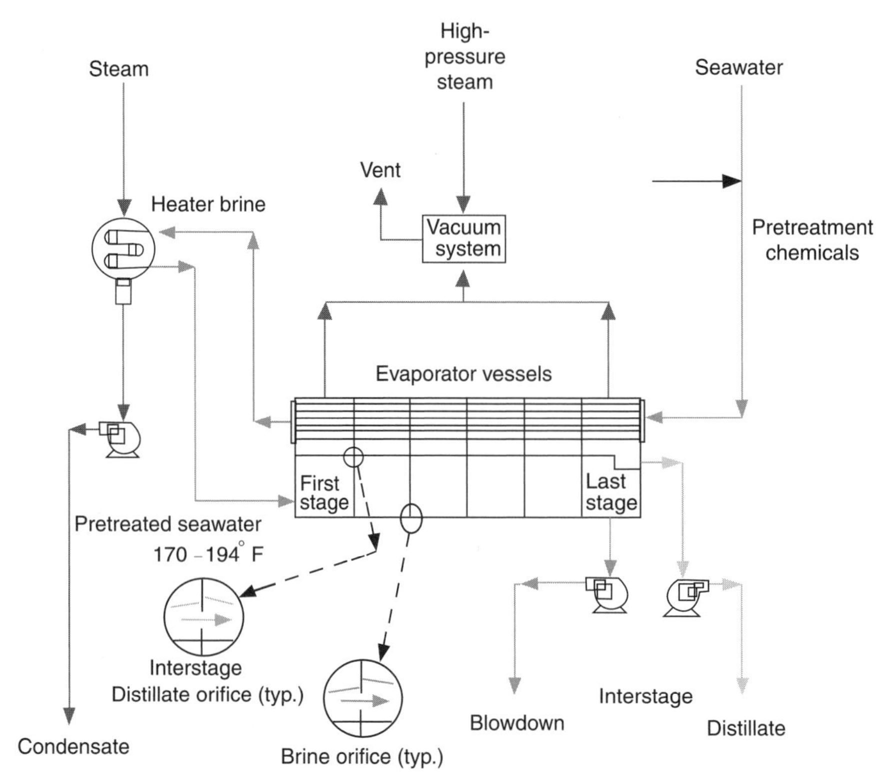

FIGURE 8 A once-through multistage flash process.

The typical GOR for these large units ranges from 7.7 to 9.0. All MSF plants are designed with electrically driven pumps but past practices included steam turbine drivers for the larger pumps. Exhaust steam from the turbines was then condensed in the brine heater, providing much of the MSF process heating. Such steam turbine drives have been replaced by electric motors not for efficiency reasons but for ease of use, simplicity, and operational flexibility. Typical pumping power for MSF units is 3.7–4.5 kWh per ton of distilled water.

3.4 RO

The RO process takes advantage of semipermeable properties of certain materials that allow solvents to pass through the material but impeding the passage of dissolved ions. The most commonly treated solvent is water. The semipermeable material can be configured in many ways but the most common is to roll flat sheets around a central pipe to create what is known as a membrane element. Even though membranes are available for performing very fine filtration, RO is not to be confused with filtration, which removes only material that is suspended within the fluid.

When a membrane is installed between saline water and fresh water, the natural flow will be toward the saline solution. The application of pressure to the saline side can stop this process, with the point of no flow being called the osmotic pressure. If a pressure above the osmotic pressure is applied to the saline side, then the flow reverses and fresh water can be produced. The basic RO flow diagram is shown in Fig. 10.

Osmotic pressure is a colligative property of the solution and generally increases as the concentration of dissolved ions increases. This means that as an RO system produces fresh water, known as the permeate, the salinity of the saline side increases, resulting in an increase in osmotic pressure. For a practical process, the RO system must be configured so that the average pressure along the membrane surface is suitably higher than the osmotic pressure, resulting in a net driving force for the process.

The osmotic pressure of typical municipal water is on the order of 0.33 bar, which is why faucet-mounted RO units can operate using the pressure of the municipal supply (typically 2 bar). The osmotic pressure of standard seawater is approximately 23 bar and seawater reverse osmosis (SWRO) systems typically operate such that the concentrated brine is

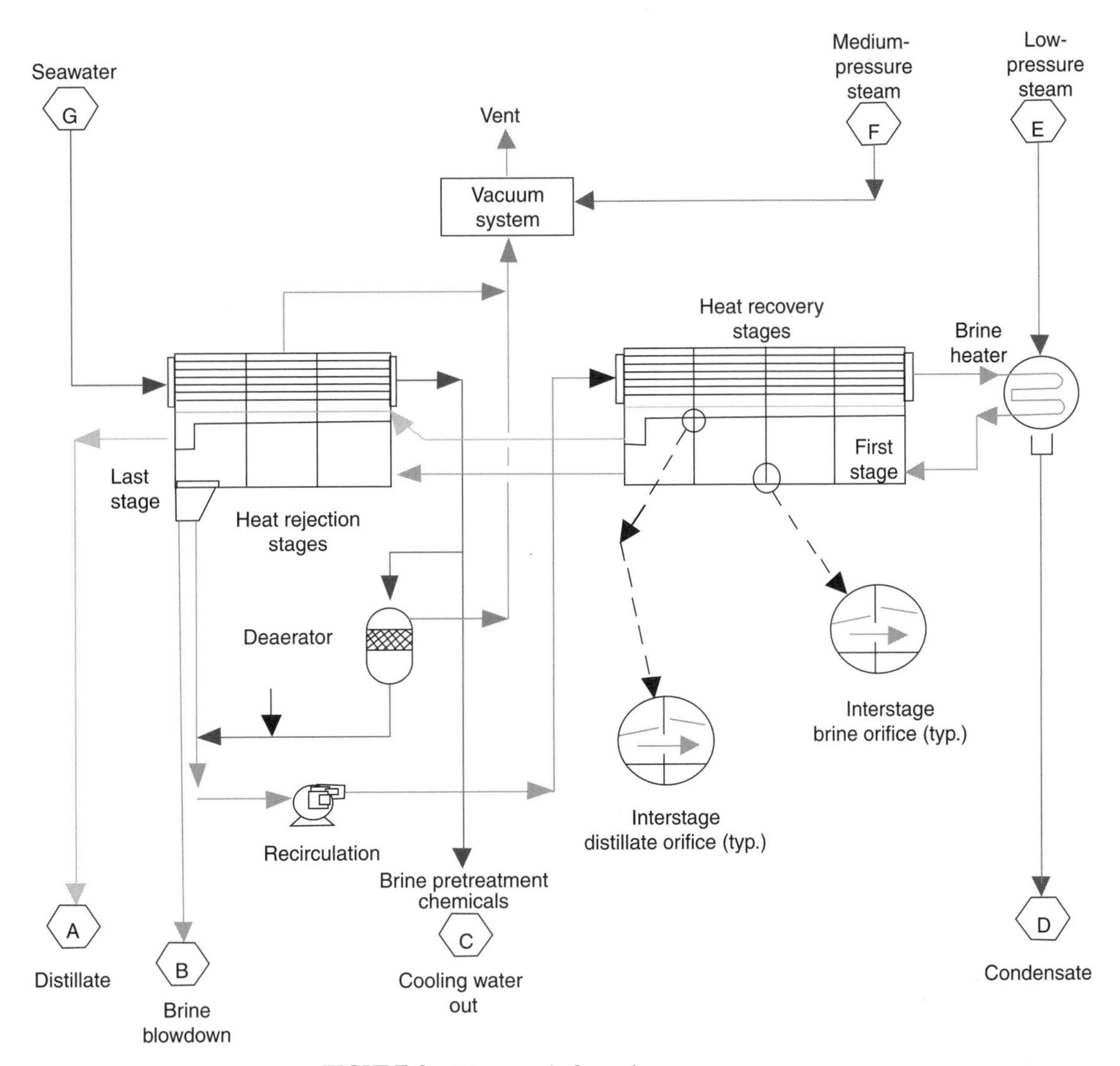

FIGURE 9 Brine recycle flow schemes.

almost two times the inlet concentration (which would be known as 50% recovery). This results in the brine having an osmotic pressure of approximately 45 bar. To achieve a practical economic driving force, the systems operate at 65–70 bar. This is close to the maximum mechanical pressure that the membranes can bear and the concentration is near the limits of solubility for certain salts. There are limited efforts in the marketplace to achieve higher concentrations, which would permit yields of more permeate than current recovery rates, which are approximately 45%.

Energy consumption for the process shown in Fig. 10 is typically 6.5 kWh per ton of permeate, which is unacceptable. The relatively large volume of high-pressure brine contains too much energy to be simply dissipated by the concentrate control valve. Energy recovery devices are installed in the brine stream to recapture the energy, as represented in Fig. 11. The most common devices are reaction turbines; often a reverse running pump; impulse turbines; normally a Pelton wheel; and pressure or work exchangers—these devices rely on proprietary methods of transferring pressure energy directly to the inlet feed stream.

The state of the art for SWRO energy consumption is approximately 2.0 kWh per ton of permeate produced; however, the average is closer to 2.75 kWh per ton of permeate produced.

Less saline water sources have much lower osmotic pressures and generally can be concentrated to 80–90% recovery before solubility limits are reached. The nature of these surface and subsurface natural waters is such that they are highly variable

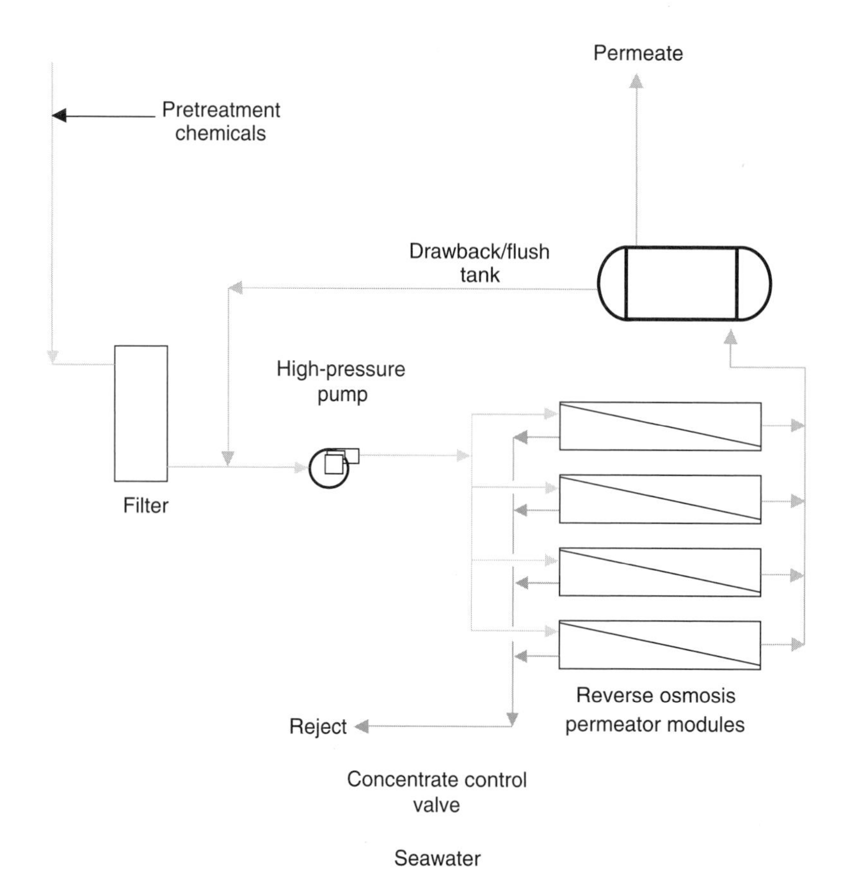

FIGURE 10 Reverse osmosis flow diagram.

from location to location and generalizations are difficult to make (unlike "standard" seawater). It is not uncommon to find waters that are already saturated in one or more ionic species and which cannot be treated (economically by standard desalination processes).

The typical brackish water RO (BWRO) operates at 85% recovery and pressures of approximately 25 bar, yielding energy consumption of 1.0 kWh/m^3. Since the pressures and brine flow volumes for BWRO are much lower than for SWRO, there has not been an economic driving force to develop energy recovery devices for BWRO. New applications of existing technology have been proposed, allowing multiple BWRO brine streams to be joined, yielding flow volumes that merit using energy recovery.

There has been work published that highlights the finding that the mechanical devices involved in RO processes (pumps, energy recovery, etc.) are reaching their practical limits and that future reductions in energy consumption must come from the membrane.

3.5 ED and EDR

Electrodialysis is quite unique among practical desalination processes since it literally desalts the feed source, whereas other processes actually remove pure water. ED is a membrane process utilizing charge-selective membranes and is pressure-based, like RO. It is possible to manufacture membranes that are charge-selective and that allow only ions with the appropriate charge to pass.

The most common configuration of ED utilizes a technique of periodically changing the polarity of the system electrodes returning in a reversed flow through the system, yielding some back-flushing of the membrane surface. This is known as electrodialysis reversal.

There are two main components of ED/EDR energy consumption: The energy of separation can be approximated as requiring 1 kWh/1000 usg per 1000 ppm of salt removed. This energy requirement varies significantly with temperature and is stated reflecting typical ambient conditions of 18–22°C.

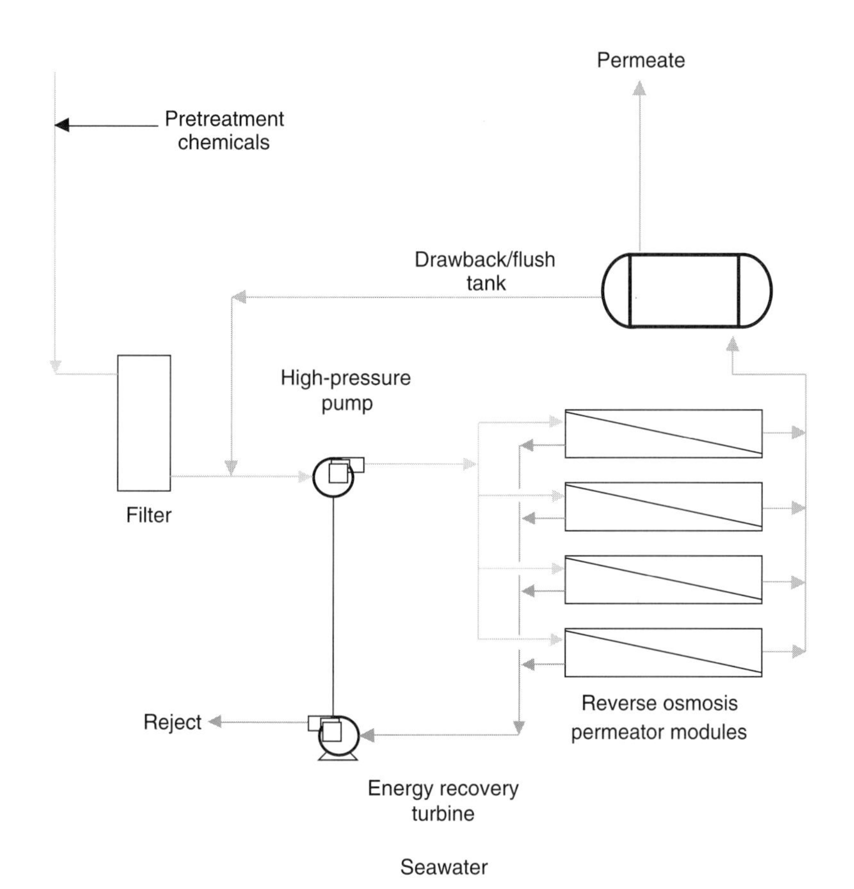

FIGURE 11 Energy recovery devices installed in the brine stream.

The energy of pumping is approximately 2 kWh/ 1000 usg of water produced.

SEE ALSO THE FOLLOWING ARTICLES

Ocean, Energy Flows in • Ocean Thermal Energy • Solar Water Desalination

Further Reading

El-Dessouky, H. T., and Ettouney, H. M. (2002). "Fundamentals of Salt Water Desalination." Elsevier, San Diego, CA.

Pankratz, T. M. (2003). "Seawater Desalination Processes: Operations and Costs." Technomic, Lancaster, PA.

Lattemann, S., and Hopner, T. (2003). "Seawater Desalination: Impacts of Brine and Chemical Discharge on the Marine Environment." Desalination Publications, L'Aquila, Italy, and DEStech Publications, Lancaster, PA.

Development and Energy, Overview

JOSÉ GOLDEMBERG
University of São Paulo
São Paulo, Brazil

1. Introduction
2. The Importance of Noncommercial Energy
3. Energy Intensity
4. Human Development Index
5. Generation of Jobs in Energy Production
6. Conclusions

Glossary

energy intensity The ratio of energy demand to gross domestic product.

gross domestic product (GDP) The total output produced within the geographical boundaries of the country, regardless of the nationality of the entities producing the output.

gross national product (GNP) Measure of the country's output of final goods and services for an accounting period (valued at market or purchaser's price).

human development index (HDI) Measures the average achievement of a country in basic human capabilities. The HDI indicates whether people lead a long and healthy life, are educated and knowledgeable, and enjoy a decent standard of living. The HDI examines the average condition of all people in a country: Distributional inequalities for various groups of society have to be calculated separately. The HDI is a composite of three basic components of human development: longevity, knowledge, and standard of living. Longevity is measured by life expectancy. Knowledge is measured by a combination of adult literacy (two-thirds weight) and mean years of schooling (one-third weight). Standard of living is measured by purchasing power, based on real GDP per capita adjusted for the local cost of living (purchasing power parity).

purchasing power parity (PPP) A rate of exchange that accounts for price differences across countries allowing international comparisons of real output and incomes. At the U.S. PPP rate, the PPP of $1 has the same purchasing power in the domestic economy as $1 has in the United States.

tonnes of oil equivalent (toe) Energy unit equivalent to 41.84 GJ or approximately 0.93 tonnes of gasoline, 0.99 tonnes of diesel oil, 0.96 tonnes of kerosene, 1.04 tonnes of fuel oil, 0.93 tonnes of liquified petroleum gas, 1.61 tonnes of coal, 6.25 tonnes of bagasse, 2.63 tonnes of fuel wood, and 1.35 tonnes of charcoal.

terawett-hour (TWh) Energy unit equivalent to 3.6 million GJ.

Although energy is a physical entity well understood and quantitatively defined, the concept of development is less well defined and there are different perceptions about its meaning. The World Bank measures development by the gross national product (GNP) and nations are classified in categories according to their GNP "per capita." This "monetization" of the concept of development is not well understood and is not accepted by many, particularly in developing countries, where income per capita varies dramatically between the poor and the rich. In 1999 U.S. dollars, high income is considered to be $9,266/year or more, whereas middle income is between $756 and $9,265 and low income is $755 or less. This is not the case in the Organization for Economic Cooperation and Development countries, where there is a large middle class and variations in income are not very large.

1. INTRODUCTION

The poor in developing countries, who represent 70% of the world's population, aspire to a "better life," meaning jobs, food, health services, housing (rural or urban), education, transportation, running water, sewage service, communication services, security of supplies of necessities, and a good environment. These things are usually measured in industrialized countries by monetary transactions, but this is not necessarily the case in many other countries. Climate and abundant and easily available natural resources can lead to a better life without great monetary resources. In some countries, cultural values are such that some items are less desirable than they are in other countries. In others, the political system privileges some solutions over others that cost much less. Thus, to compare stages of development only by gross domestic product (GDP) per capita can be quite misleading.

Energy in its various forms (mechanical, chemical, electrical, heat, and electromagnetic radiation) is essential for all the aspirations listed previously, and thus it is closely linked to a range of social issues. Conversely, the quality and quantity of energy services and how they are achieved have an effect on social issues.

The stages of the development of man, from primitive man (1 million years ago) to today's technological man, can be roughly correlated with energy consumption (Fig. 1).

- Primitive man (East Africa approximately 1 million years ago), without the use of fire, had only the energy of the food he ate (2000 kcal/day).
- Hunting man (Europe approximately 100,000 years ago) had more food and also burned wood for heat and cooking.
- Primitive agricultural man (Fertile Crescent in 5000 BC) grew crops and used animal energy.
- Advanced agricultural man (northeast Europe in 1400 AD) had coal for heating, water power, wind power, and animal transport.

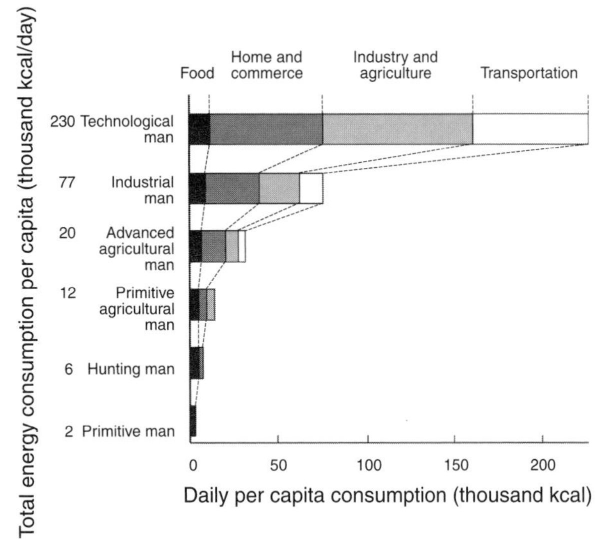

FIGURE 1 Stages of development and energy consumption. Reproduced from Cook, (1976).

- Industrial man (in England in 1875) had the steam engine.
- Technological man (in the United States in 1970) consumed 230,000 kcal/day.

From the very low energy consumption of 2000 kcal per day, which characterized primitive man, energy consumption increased in 1 million years to 230,000 kcal per day. This enormous growth of per capita energy consumption was only made possible by the increased use of coal as a source of heat and power in the 19th century; the use of internal combustion engines, which led to the massive use of petroleum and its derivatives; and Electricity generated initially in hydroelectric sites and later in thermoelectric plants. Income per capita has also increased, and one is thus tempted to determine if there is a clear correlation between energy consumption/capita and GDP/capita. Data are available for a number of countries (Fig. 2).

A linear relationship between energy consumption/capita and GDP/capita could be expected since higher income means more appliances and automobiles, larger homes, more travel, and many other activities, but empirical evidence shows that this is not the case. Clearly, there are many countries that do not fit into a linear relationship.

Moreover, some countries with the same GNP/capita can have very different energy consumption per capita, which is the case for Russia, China, and Morocco. Their GNP/capita is approximately the same, but energy consumption in Russia is four times higher than in China and 15 times higher than in Morocco.

2. THE IMPORTANCE OF NONCOMMERCIAL ENERGY

With regard to developing countries, a problem with energy data (Fig. 2), is that the vertical axis considers only "commercial energy," which is the fraction of total energy consumption that can be quantified since, by definition, it involves monetary transactions. In reality, in many areas noncommercial energy sources, such as dung, agricultural residues, and fuelwood, are used. In some countries, it represents a large percentage of total consumption, particularly for the poor. Table I shows the importance of noncommercial energy in a variety of countries.

The importance of noncommercial energy is also clearly exemplified by data that show the average energy demand by income segment in Brazil (Fig. 3).

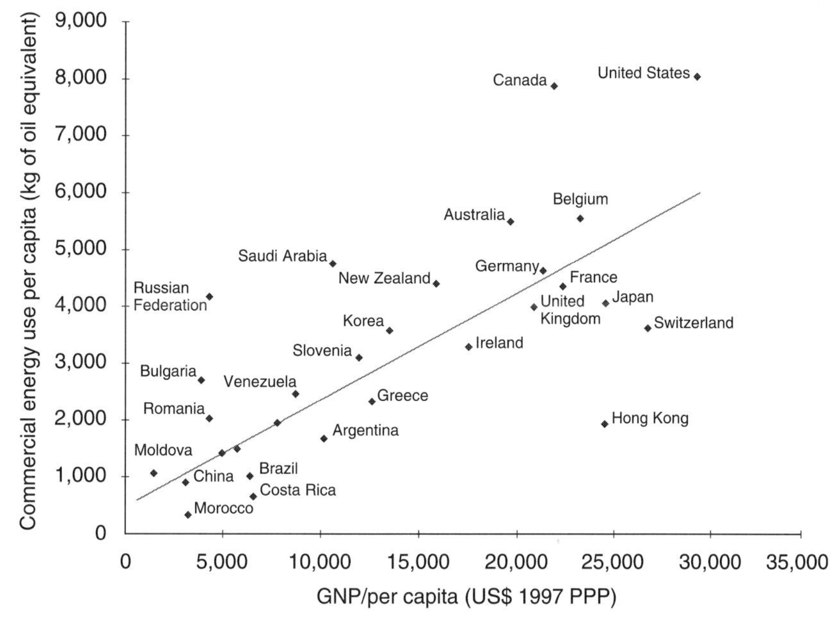

FIGURE 2 Energy use versus GNP. Reproduced from World Bank (1999).

TABLE I

Noncommercial Energy in Africa[a]

Country	1973 (%)	1985 (%)
Uganda	83	92
Malawi	87	94
Guinea-Bissau	72	67
Gambia	89	78
Guinea	69	72
Burundi	97	95
Mali	90	88
Burkina Faso	96	92

[a]Data from the World Resources Institute, United Nations Development Program, United Nations Environmental Program, and World Bank (1998).

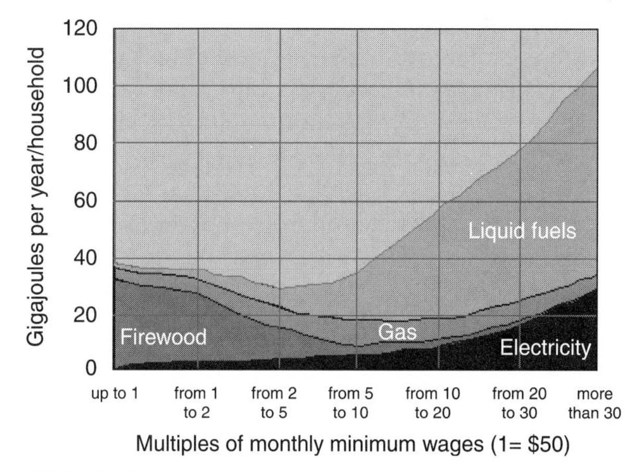

FIGURE 3 Average energy demand by income segment in Brazil, 1998. Reproduced from Almeida and Oliveira (1995).

For low-income households, firewood (usually a noncommercial energy source) is the dominant fuel. At higher income, firewood is replaced by commercial fuels and electricity, which offer much greater convenience, energy efficiency, and cleanliness. In most of Africa and India, dung and agricultural residues are used in lieu of firewood.

3. ENERGY INTENSITY

Another reason for the lack of a linear relationship between energy and GDP per capita is that the amount of additional energy required to provide energy services depends on the efficiencies with which the energy is produced, delivered, and used. Energy intensity often depends on a country's stage of development. In Organization for Economic Cooperation and Development (OECD) countries, which enjoy abundant energy services, growth in energy demand is less tightly linked to economic productivity than it was in the past (Fig. 4).

The evolution of the energy intensity ($I = E/\text{GDP}$) over time reflects the combined effects of structural changes in the economy (built into the GDP) and changes in the mix of energy sources and the

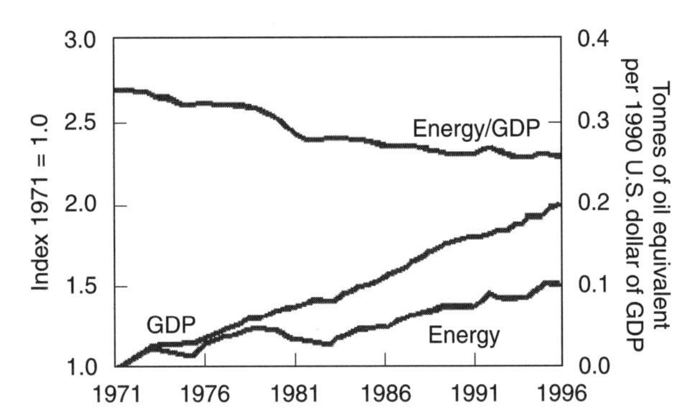

FIGURE 4 GDP and primary energy consumption in OECD countries, 1971–1996. Reproduced from the International Energy Agency (1999).

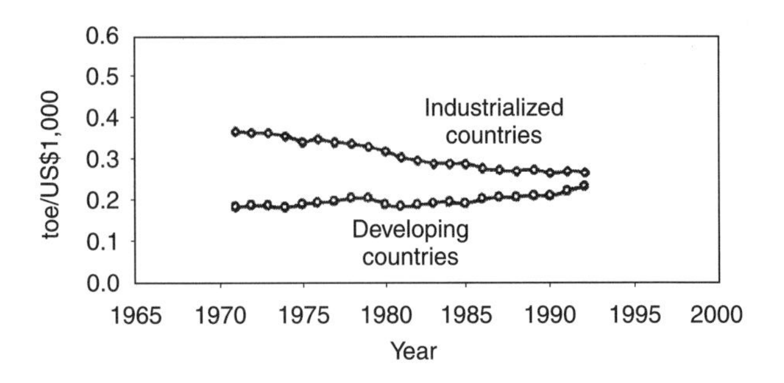

FIGURE 5 Energy intensity. Reproduced from Mielnik and Goldemberg (2000).

efficiency of energy use [built into the primary energy consumed (*E*)].

Although admittedly a very rough indicator, energy intensity has some attractive features: Whereas *E* and GDP per capita vary by more than one order of magnitude between developing and developed countries, energy intensity does not change by more than a factor of two. This is due in part to common characteristics of the energy systems of industrialized and developing countries in the "modern" sector of the economy and in part to the fact that in industrialized countries energy-intensive activities, such as jet travel, are increasingly offsetting efficiency gains in basic industries.

Energy intensity (considering only commercial energy sources) declined in OECD countries during the period 1971–1991 at a rate of approximately 1.4% per year. The main reasons for this movement were efficiency improvements, structural change, and fuel substitution. However, in the developing countries the pattern has been more varied. One study indicates that during the period 1971–1992 the energy intensity of developing and industrialized countries was converging to a common pattern of energy use. For each country, energy intensity was obtained as the ratio of commercial energy use to GDP converted in terms of purchasing power parity (PPP). The path of energy intensity of a country was given by the yearly sequence of energy intensity data during the period 1971–1994. The same procedure was followed to derive the energy intensity paths for a set of 18 industrialized countries and for 1 of 23 developing countries. The energy intensity data for each of these subsets were given by the ratio of total commercial energy use to total PPP-converted GDP for each group of countries for each year during the period 1971–1994 (Fig. 5).

4. HUMAN DEVELOPMENT INDEX

Since a clear correlation between energy and income is difficult to establish, one is tempted to search for other correlations between energy consumption and social indicators such as infant mortality, illiteracy, and fertility (Fig. 6). It is clear from Fig. 6 that energy use has many effects on major social issues, resulting in a relationship that is not necessarily casual between the parameters represented but there is strong covariance. Such behavior encouraged analysts to develop an indicator that would not only include GDP per capita but also take into account longevity and literacy. The Human Development Index (HDI), developed by the United Nations Development Program in 2001, is one way of measuring how well countries are meeting not just the economic but also the social needs of their people—that is, their quality of life. The HDI is calculated on the basis of a simple average of longevity, knowledge, and standard of living. Longevity is measured by life expectancy, knowledge is measured by a combination of adult literacy (two-thirds weight) and mean years of schooling (one-third weight), and standard of living is measured by purchasing power, based on real GDP per capita adjusted for the local cost of living (PPP).

The HDI measures performance by expressing a value between 0 (poorest performance) and 1 (ideal performance). HDI can be presented as a function of commercial energy consumption for a large number of countries in tons of oil equivalent (toe) (Fig. 7). Table II lists the characteristics of a few countries, their HDI values, and HDI rank.

An analysis of Table II indicates the importance of factors besides GDP per capita in determining the HDI rank of different countries. For example, Russia and Brazil have approximately the same GDP per capita but adult literacy in Russia (99.5%) is much

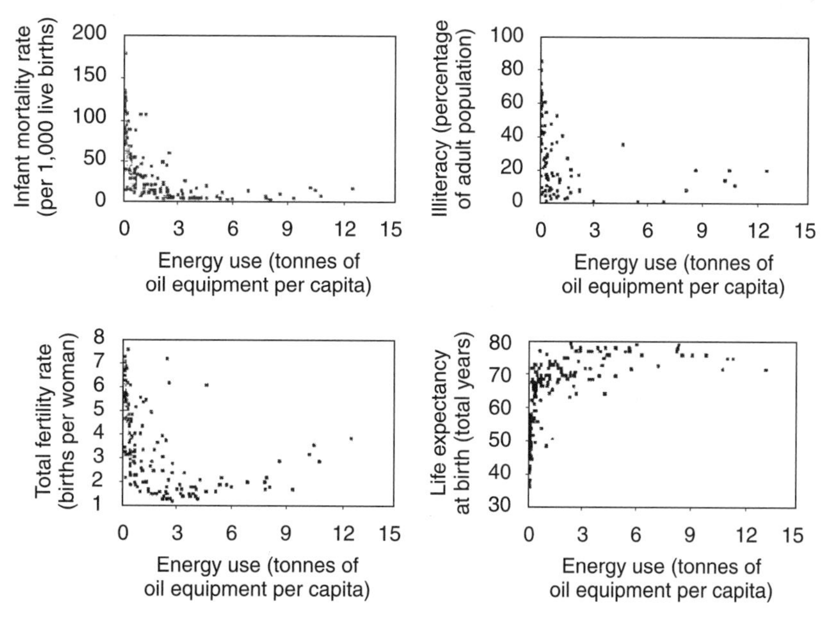

FIGURE 6 Commercial energy use and infant mortality rate, illiteracy, life expectancy at birth, and total fertility rate. Reproduced from United Nations Development Program, United Nations Department of Economic and Social Affairs, and World Energy Council (2000).

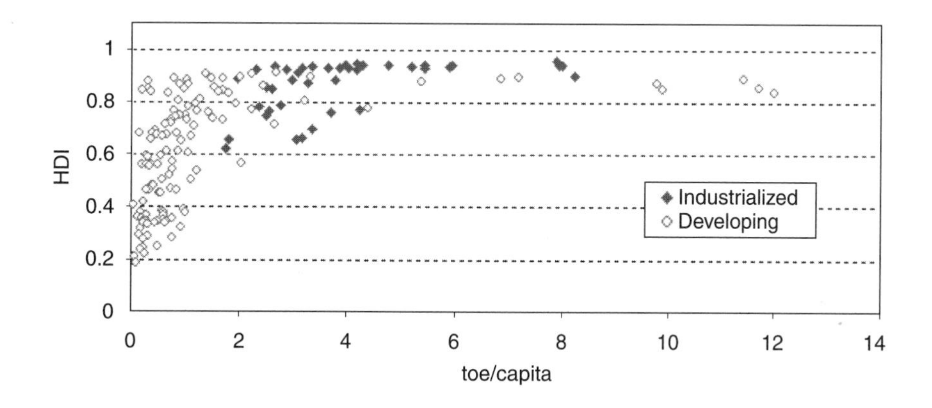

FIGURE 7 Human development index (HDI) and energy use by country. Reproduced from the World Bank (1999) and United Nations Development Program (1998).

TABLE II

***Human Development (HDI) Index for Some Countries*[a]**

Country	Life expectancy at birth (years)	Adult literacy rate (% age 15 or older)	GDP per capita (PPR U.S. $)	HDI	HDI rank
Russia	66.1	99.5	7473	0.775	55
Brazil	67.5	84.9	7037	0.75	69
Sri Lanka	71.9	91.4	3279	0.735	81
China	70.2	83.5	3617	0.718	87
Morocco	67.2	48.0	3419	0.596	112
India	62.9	56.5	2248	0.571	115

[a]Data from the United Nations Development Program (2001).

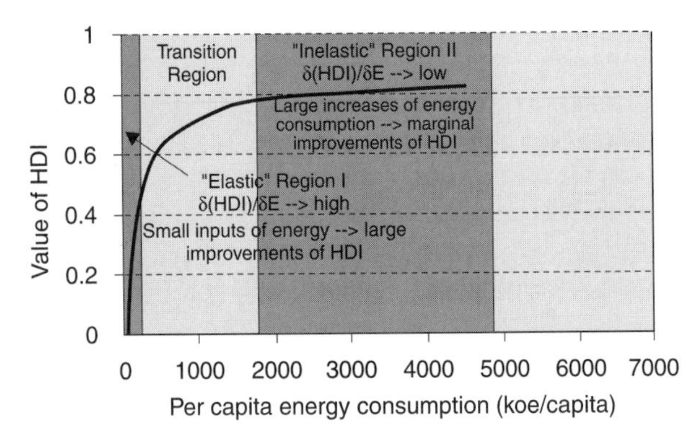

FIGURE 8 HDI versus per capita energy consumption. Reproduced from Reddy (2002).

higher than that in Brazil (84.9%), which ranks the former 55 compared to 69 for Brazil. Sri Lanka and China have a GDP per capita that is approximately the same as that of Morocco but the life expectancy and adult literacy of the former are much higher, which give Sri Lanka a ranking of 81, China 87, and Morocco 112.

It is apparent from Fig. 7 that for an energy consumption more than 1 toe/capita/year, the HDI is higher than 0.8 and essentially constant for all countries. Therefore, 1 toe/capita/year seems to be the minimum energy needed to guarantee an acceptable level of living as measured by the HDI, despite many variations in consumption patterns and lifestyles across countries.

The relationship between HDI and energy has several important implications. The relationship can be considered to consist of two regions (Fig. 8). In region I, the elastic region, – the slope δ (HDI)/δ (E) of the HDI vs E curve is high. In region II, the inelastic region, the slope δ (HDI)/δ (E) of the HDI vs E curve is low. Even large inputs of energy (large improvements of energy services) result in only marginal improvements in HDI; that is, the HDI–energy (benefit–cost) ratio is very low. In region I, enhanced energy services lead directly to improvement. Thus, the implication of the elastic and inelastic regions is that in the elastic region increased energy services guarantee direct improvement of HDI, whereas improvement of HDI via income depends on the uses of the income.

5. GENERATION OF JOBS IN ENERGY PRODUCTION

A variety of sources provide estimates of the number of jobs directly related to the production of energy.

TABLE III

Jobs in Energy Production[a]

Sector	Jobs (person-years), TWh
Petroleum	260
Offshore oil	265
Natural gas	250
Coal	370
Nuclear	75
Wood energy	1000
Hydro	250
Minihydro	120
Wind	918
Photovoltaics	7600
Ethanol (from sugarcane)	4000

[a] Data from Goldemberg (2003).

Some of these estimates are shown in Table III. These data include jobs involved in operating the generating stations, excluding the jobs involved in producing and commissioning the equipment. The very large number of jobs generated by the truly decentralized options, which are photovoltaics and ethanol from sugarcane, is striking.

Photovoltaics energy is usually generated (and used) in small modules of 100 W and the generation of 1 TWh typically requires 10 million modules to be installed and maintained. Ethanol production from sugarcane in Brazil involves agricultural production using approximately 4 million ha and industrial production in sugar–alcohol distilleries, accounting for the large number of jobs directly related to the production of energy.

There are also some rough estimates of the number of jobs generated by energy conservation. A study performed in Sacramento, California, showed that saving enough energy to operate at less than 100-MW power plant capacity requires 39 jobs compared with 15–20 jobs required to operate at the same amount of capacity at a modern coal or gas-fired power plant.

6. CONCLUSIONS

Energy has a determinant influence on the HDI, particularly in the early stages of development, in which the vast majority of the world's people, particularly women and children, are classified. The influence of per capita energy consumption on the HDI begins to decline between 1000 and 3000 kg of oil equivalent (koe) per inhabitant. Thereafter, even with a tripling of energy consumption, the HDI does not increase. Thus, from approximately 1000 koe per capita, the strong positive covariance of energy consumption with HDI starts to diminish. The efficiency of energy use is also important in influencing the relationship between energy and development.

Another aspect of the problem is the mix of supply-side resources that dominate the world's energy scene today. Fossil fuels have a dominating role (81% of supply in OECD countries and 70% in developing countries), although, as a rule, renewables are more significant for low-income populations. However, there are significant advantages to increasing the role of renewable sources since they enhance diversity in energy supply markets, secure long-term sustainable energy supplies, reduce atmospheric emissions (local, regional, and global), create new employment opportunities in rural communities offering possibilities for local manufacturing, and enhance security of supply since they do not require imports that characterize the supply of fossil fuels.

More generally, development, including the generation of jobs, depends on a number of factors in addition to GNP per capita. Furthermore, although an essential ingredient of development, energy is more important with regard to low rather than high incomes.

SEE ALSO THE FOLLOWING ARTICLES

Cultural Evolution and Energy • Economic Growth and Energy • Energy Ladder in Developing Nations • Global Energy Use: Status and Trends • International Comparisons of Energy End Use: Benefits and Risks • Population Growth and Energy • Sustainable Development: Basic Concepts and Application to Energy • Technology Innovation and Energy • Transitions in Energy Use • Women and Energy: Issues in Developing Nations

Further Reading

Almeida, E., and Oliveira, A. (1995). Brazilian life style and energy consumption. *In* "Energy Demand, Life Style Changes and Technology Development." World Energy Council, London.

Cook, E. (1976). "Man, Energy, Society." Freeman, San Francisco.

Goldemberg, J. (2003). "Jobs Provided in the Energy Sector." [Adapted from Grassi, G. (1996). Potential employment impacts of bioenergy activity on employment. *In* "Proceedings of the 9th European Bioenergy Conference" (P. Chartier *et al.*, Eds.), Vol. 1, pp. 419–423. Elsevier, Oxford, UK; Carvalho, L. C., and Szwarc, A. (2001). Understanding the impact of externalities: case studies. Paper presented at the Brazil International Development Seminar on Fuel Ethanol, December 14, Washington, DC; Perez, E. M. (2001). "Energias Renovables, Sustentabilidad y Creacion de Enpleo: Una Economía Impulsionada por el Sol (Renewable Energy, Sustainability and Job Creation: An Economy Pushed Forward by the Sun), ISBN: 84–8319 115-6; and Renner, M. (2000, September). Working for the environment: A growing source of jobs, Worldwatch Paper No. 152. Worldwatch Institute]

International Energy Agency (1999). "Energy Balances of OECD Countries." International Energy Agency, Paris.

Mielnik, O., and Goldemberg, J. (2000). Converging to a common pattern of energy use in developing and industrialized countries. *Energy Policy,* pp. 503–508.

Reddy, A. K. N. (2002). Energy technologies and policies for rural development. *In* "Energy for Sustainable Development" (T. Johansson and J. Goldemberg, Eds.). International Institute for Industrial Environmental Economics University and United Nations Development Program, New York.

United Nations Development Program (1998). "Human Development Report." United Nations Development Program, New York.

United Nations Development Program (2001). "Human Development Index." United Nations Development Program, New York.

United Nations Development Program, United Nations Department of Economic and Social Affairs, and World Energy Council (2000). "World Energy Assessment" (J. Goldemberg, Ed.). United Nations Development Program, United Nations Department of Economic and Social Affairs, and World Energy Council, New York.

World Bank (1999). "World Development Indicators for Energy Data." World Bank, Geneva.

World Resources Institute, United Nations Development Program, United Nations Environmental Program, and World Bank (1998). "World Resources 1998–1990—Environmental Change and Human Health." World Resources Institute, United Nations Development Program, United Nations Environmental Program, and World Bank, New York.

Diet, Energy, and Greenhouse Gas Emissions

ANNIKA CARLSSON-KANYAMA
Environmental Strategies Research Group at FOI
Stockholm, Sweden

1. Introduction
2. Calculations of Energy-Efficient and Low-Pollution Diets
3. Trends in Diets
4. Policy Instruments for Dietary Change
5. Conclusions

Glossary

input/output analysis (IOA) A method traditionally used to describe the financial linkages and network of input supplies and production that connect industries in a regional economy (however defined) and to predict the changes in regional output, income, and employment; also used for environmental impact analysis today.

Intergovernmental Panel on Climate Change (IPCC) - Jointly established by the World Meteorological Organization and the UN Environment Program in 1998 to assess available scientific information on climate change, assess the environmental and socioeconomic impacts on climate change, and formulate response strategies.

International Vegetarian Union (IVU) Established in 1908 as a nonprofit-making organization with membership open to any nonprofit organization whose primary purpose is to promote vegetarianism and that is governed exclusively by vegetarians; aims to promote vegetarianism throughout the world.

life cycle assessment (LCA) A method to assess the potential environmental impact of a material, product, or service throughout its entire life cycle, from the extraction of raw materials, through the production process, through the user phase, and to the final disposal.

Energy use and greenhouse gas emissions in the current food system are substantial. The food system comprises the whole chain of events resulting in edible food—production of farm inputs, farming, animal husbandry, transportation, processing, storage, retailing, and preparation. Ecological impacts are not evenly distributed among products; therefore, dietary change is a possibility for environmental improvement. From a number of studies, it is possible to detect patterns of dietary recommendations that would lower energy use and greenhouse gas emissions in the food system. They include reducing consumption of meat, exotic food, food with little nutritional value, and overprocessed food. Overall trends in dietary patterns are not conducive to higher environmental efficiency, but the trends toward vegetarianism have an environmental potential. Policy instruments for encouraging more efficient diets include information, supportive structures, and financial incentives. However, they have scarcely been used in mitigation strategies for climate change.

1. INTRODUCTION

For a long time, it has been known that the production of certain food products takes a larger toll on natural resources than that of others, and at times dietary change has been proposed as a way of conserving such resources. During the 1960s, when global concern was about overpopulation and starvation, Georg Borgström proposed a diet lower down on the food chain, implying less use of cereals for animal feed. During the same period, Francois Lappé composed a diet for a small planet based on vegetarian ingredients only to save agricultural land. During the 1970s, when the concern was about energy resources, David and Marcia Pimentel presented calculations of energy inputs over the life cycle for bread and beef. The beef was found to be more energy demanding by a factor of 29 when the two products were compared on a weight basis.

Today, when climate change has emerged as perhaps the most controversial global environmental issue, the question of dietary change as a mitigation strategy has partly resurfaced. This is because energy use and greenhouse gas emissions from the current food system are substantial. Energy use in the food system typically amounts to 12 to 20% of the total energy consumed in developed countries. Current anthropogenic emissions of carbon dioxide (CO_2) are primarily the result of the consumption of energy from fossil fuels, the main energy source in the developed parts of the world. Another 10 to 30% of the current total anthropogenic emissions of CO_2 are estimated to be caused by land-use conversion, partly related to production of cash crops. In addition, the agricultural sector is a large source of emissions of methane and nitrous oxides. It accounts for half of the global anthropogenic emissions of methane through rice farming and rearing of ruminants, whereas 70% of the global anthropogenic emissions of nitrous oxide come from agricultural soils, cattle, and feedlots.

In looking more closely at where energy is used in the food system, one finds that a surprisingly large amount is used by the consumer. For the United States, a review of several studies showed that home preparation contributed between 23 and 30% of the total, figures comparable to processing, which contributed 24 to 37%. For Sweden, a recent survey showed that household energy use for cooking and storing food was 28% of the total, whereas processing contributed 25% (Fig. 1). A different pattern may emerge when looking at emissions of carbon dioxide (Fig. 1). Here, the system for electricity production may alter the energy pattern. For Sweden and some other Nordic countries, extensive use of hydropower results in low emissions, as does nuclear power production.

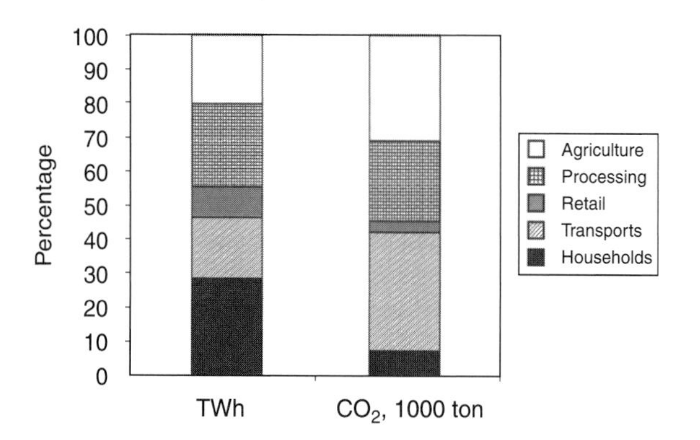

FIGURE 1 Energy use and emissions of carbon dioxide in the Swedish food system in the year 2000.

The purpose of this article is to provide an overview of current research about energy, greenhouse gas emissions, and dietary change. The approach is product oriented with a life cycle perspective, meaning that individual products are considered "from cradle to grave."

The article begins with some recent results from calculations of the environmental impacts over the life cycle for food products and diets. Several studies conducted in Europe and the United States provide sufficient basis for an informed discussion about the potential for change. In the subsequent section, some recent trends in global food consumption are reviewed for their environmental implications, and the trend toward vegetarianism is evaluated. Some lessons from research about policy instruments and dietary change are discussed with a focus on information and its efficiency. Finally, conclusions of relevance for further mitigation efforts are drawn.

2. CALCULATIONS OF ENERGY-EFFICIENT AND LOW-POLLUTION DIETS

Investigating possibilities for low-impact diets involves analyzing the resource use and emissions of the numerous individual food products that constitute a diet throughout the life cycle. Several attempts have been made to this end, and as a result broad patterns of dietary recommendations can be detected. Based on several studies, a diet with less meat and cheese, more in-season vegetables, more locally produced and fresh foods, better use of products, and more use of products from the "wild" could be recommended. The results from some of these studies are discussed in what follows, as are the methods used for computing the environmental impacts.

A team of Dutch researchers used a combination of life cycle assessment (LCA) and input/output analysis (IOA) to estimate greenhouse gas emissions from food products consumed in The Netherlands. The results were expressed as greenhouse gas emissions per Dutch florin (Dfl), and calculations were performed for more than 100 products. This method is useful in that the results can be directly combined with household expenditure profiles to estimate impacts of a whole consumption pattern but lack some detail when it comes to differentiation among various processing methods and so on. Some results from these calculations were that greenhouse gas emissions during the life cycle of rice are nearly five times the emissions during pastry life cycles and that tomatoes emit four times as much greenhouse gases as do green beans. The reasons for these differences relate to the paddy field cultivation methods for rice and the fact that tomatoes are usually cultivated in fossil fuel-heated greenhouses. Regarding the food consumption emission profile of a Dutch household, methane emissions came mainly from products originating from cattle farming and dairy production. Enteric fermentation by ruminants explains much of this. The dietary recommendations made by these authors included substituting meat and cheese with eggs, nuts, or certain vegetable products and shifting from greenhouse-cultivated vegetables to those grown in the open.

In the United States, two researchers used IOA to calculate greenhouse gas emissions for 44 food products sold in the country. Their results were expressed as amount of emissions per dollar and were combined with household expenditures to calculate total household emissions. Regarding dietary change, the authors' proposal was to lower the expenditure on meat and poultry in favor of expenditure on vegetable products such as pasta. This was because the emissions intensity of beef and pork was 166 g per U.S. dollar, whereas the same intensity for pasta was only 75 g per U.S. dollar. Lowering meat consumption was also considered to be of major importance when all activities of the household were assessed for greenhouse gas emissions and other environmental pollutants.

A Swedish researcher investigated energy use over the life cycle for food supplied in Sweden using a life cycle inventory approach. The study covered 150 food items, and results were expressed as megajoules per kilogram of product. The advantage of this approach is that results can easily be combined with dietary information and that the level of detail in the environmental calculations is high. Among the results from this study are that energy use per kilogram of ready-to-eat food can vary from 2 to 220 MJ/kg, that is, by a factor of more than 100. The highest values were found for certain kinds of fish and seafood. A multitude of factors determine the level of energy inputs, and they relate to animal or vegetable origin, degree of processing, choice of processing and preparation technology, and transportation distance. The study confirmed several earlier findings about environmentally efficient diets while adding some new perspectives. One is the role of food preparation (which can add substantial energy inputs if food is cooked in small portions), another is the remarkably high energy inputs for certain fish-harvesting techniques, and a third is the impacts from food losses on total energy inputs in food life cycles. Examples of daily total life cycle energy inputs for diets with similar dietary energy were calculated, and the result was that energy inputs could vary by a factor of 3. Table I shows meal components for two meals with high and low energy inputs but with the same amount of dietary energy.

In a study of six food products for which LCA was used, detailed calculations of greenhouse gas emissions confirmed the large differences between some products of animal origin and some of vegetable origin. In this study, emissions of greenhouse gases per kilogram of ready-to-eat beef were 135 times higher than the emissions per kilogram of ready-to-eat potato (Table II). There are several explanations for the high emissions for beef: enteric fermentation causing emission of methane, energy use during animal feed production, and transportation of food ingredients. The reasons for lower emissions for pork and chicken are partly related to the digestive systems of pigs and poultry but are also due to the fact that they are much more efficient feed converters than cattle.

TABLE I

Meal Components, Dietary Energy, and Life Cycle Energy Inputs for Two Different Dinners: High and Low

Meal component	Kilograms	Megajoules dietary energy	Megajoules life cycle inputs
Dinner: High			
Beef	0.13	0.80	9.4
Rice	0.15	0.68	1.1
Tomatoes, greenhouse	0.070	0.06	4.6
Wine	0.30	0.98	4.2
Total	0.65	2.51	19
Dinner: Low			
Chicken	0.13	0.81	4.37
Potatoes	0.20	0.61	0.91
Carrot	0.13	0.21	0.50
Water, tap	0.15	0.23	0.0
Oil	0.02	0.74	0.30
Total	0.60	2.61	6.1

Source. Carlsson-Kanyama *et al.* (2003).

TABLE II

Emission of Greenhouse Gases per Kilogram Ready-to-Eat Food for Six Food Products Supplied in Sweden

	Grams CO_2 equivalents per kilogram ready-to-eat product
Beef	13,500
Pork	4,200
Chicken	1,400
Potatoes	100
Milk	870
Bread	500
Lettuce	230

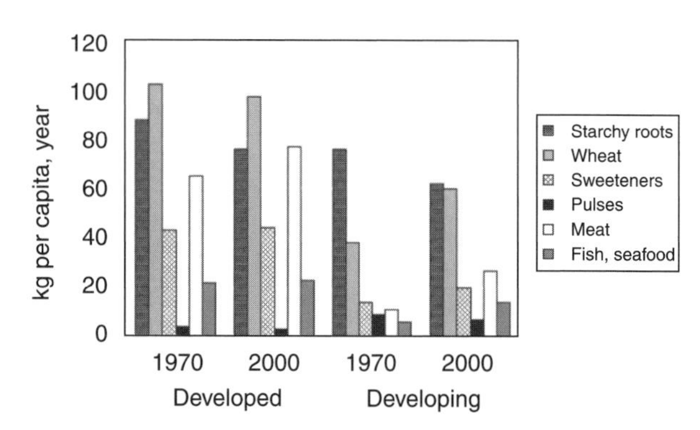

FIGURE 2 Consumption levels of some food products in 1970 and 2000 in the developed and developing worlds.

When David and Marcia Pimentel revised their book, *Food, Energy, and Society,* in 1996, they concluded,

> *One practical way to increase food supplies with minimal increase in fossil energy inputs is for the world population as a whole to consume more plant foods. This diet modification would reduce energy expenditures and reduce food supplies, because less food suitable for human consumption would be fed to livestock.*

As described previously, a number of studies have certainly shown this to be true, although detailed calculations of energy use and greenhouse gas emissions over the life cycle of different kinds of meat show that there are large differences between, for example, chicken and beef. This opens the scope for discussing the role of meat in a more environmentally efficient diet as well as for questioning the role of certain kinds of fish and vegetables and the total amount of food consumed.

How can food consumption patterns be influenced, what are the current food trends, and what is the role of information for disseminating environmental knowledge about diets? The next section contains a review of some significant elements in current food trends and their implications for energy use and greenhouse gas emissions.

3. *TRENDS IN DIETS*

From the statistics databases provided by the Food and Agricultural Organization (FAO), it is possible to provide some global figures on dietary trends of interest for the environmental impacts in the food system. Figure 2 shows some developments with respect to this using starchy roots, wheat, pulses, meat, fish, and sweeteners as examples. The rationale for choosing these products was that the life cycle-related levels of energy inputs and greenhouse gas emissions are commonly substantially different. Starchy roots and pulses are expected to have much lower impacts than do meat and fish, whereas sweeteners are interesting because they represent food products with considerable production-related energy inputs but with little or disputed nutritional value. Wheat is grown mainly in the Northern Hemisphere but is increasingly replacing traditional starch-rich crops, such as cassava and millet, in the developing parts of the world. Although wheat is not a very energy-demanding food item compared with meat and fish, it requires higher life cycle energy inputs than do some domestic starchy crops due to long transportation distances and intensive cultivation methods.

Figure 2 shows that the consumption of the selected food products in the developing parts of the world is still quite different from that in the developed world. This is particularly evident when it comes to meat consumption and sweeteners, with the 2000 levels per capita in the developed world exceeding those in the developing world by a factor of between 2 and 3. Figure 2 also shows that the most notable changes in the developed world during the past three decades were a 25% decrease in the consumption of pulses and an 18% increase in the consumption of meat. In the developing world, the changes were even more spectacular in that the consumption of meat and fish more than doubled and the consumption of wheat and sweeteners increased by approximately 50%.

The environmental implications of these dietary changes can only be interpreted as negative with respect to the prospects for lowering energy use and greenhouse gas emissions. Whether or not actual emissions have increased depends on the magnitude of efficiency improvements in the food supply system. More efficient trucks, ships, processing equipment, and refrigerators have to some extent counteracted the negative environmental impacts of dietary change.

Along with the negative dietary changes experienced on a global scale, certain trends in the developing parts of the world (e.g., vegetarianism) hold some promise for more efficient diets in an affluent environment. Vegetarianism is not a new phenomenon in Western societies but has become a mainstream feature in several countries. According to the International Vegetarian Union (IVU), there is a vegetarian subculture in nearly every European country today. It is becoming increasingly common to have vegetarian options in nonvegetarian restaurants as an alternative for those who do not adhere to a vegetarian diet on an everyday basis. No statistics exist on how many vegetarians there are globally, but estimates from several European countries show the highest percentages in the United Kingdom, Germany, and Switzerland (~8–9% of these populations) and much lower percentages in Spain, Norway, and France (~2% of these populations).

The main reasons for choosing a vegetarian diet are commonly better health and nutrition as well as animal welfare considerations. The potential for environmental efficiency has been less investigated, and when it is claimed it is mostly mentioned as a way in which to avoid excessive use of agricultural land. Recent findings about energy inputs and greenhouse gas emissions during the life cycle of food products provide a new tool for evaluating the effects of vegetarianism. A problem with such an evaluation is that the dietary recommendations for vegetarians are only exclusive. They describe what not to eat but do not prescribe any alternatives. So, meat may be replaced by a very energy-efficient alternative, such as legumes, but also by other less efficient alternatives, such as vegetables grown in greenhouses and fruits transported by airplane. However, the trend toward vegetarianism should be considered as an environmentally promising one provided that the alternatives to animal food products are chosen among those vegetable products with moderate life cycle-based resource demands. When discussing potential environmental gains from a vegetarian diet, it is important to know that the environmental impacts from meat in an ordinary diet might not be dominant. One example of this is shown in Fig. 3, where the life cycle-based energy inputs for typical diets in Sweden are displayed. The figure shows that the energy inputs from meat and fish may constitute about a third of the total. The meat and fish part of the diets in other countries may contribute differently, and if the greenhouse gas emissions profile becomes available, it could also alter this conclusion to some extent.

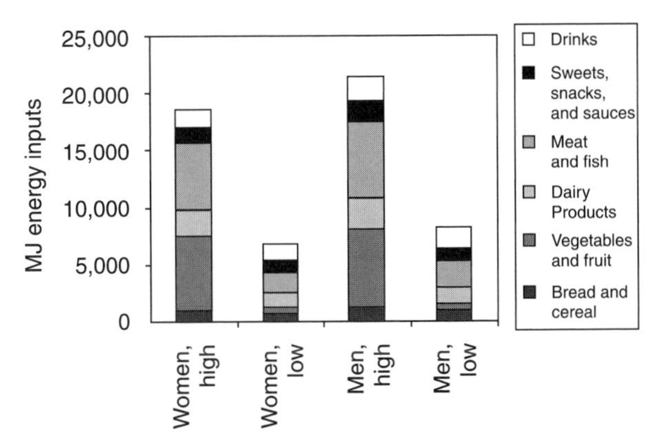

FIGURE 3 Life cycle energy inputs for the annual food consumption of average men and women in Sweden. The ranges indicate possible levels with various food choices.

Having reviewed current trends and commented on those with potential for environmental efficiency, we now turn to the issue of how policy could be used to influence diets so as to decrease energy inputs and greenhouse emissions from the food system. First, it is important to remember that experiences from large-scale campaigns for environmentally friendly diets are lacking and that the few results that can be drawn on come from isolated and relatively small research projects. In this context, it is also interesting to note that the Intergovernmental Panel on Climate Change (IPCC) does not mention dietary change among the mitigation potentials in its recent report about mitigation. Unfortunately, the same "blindness" toward dietary change as a possible environmental measure is not uncommon in policy documents. This article does not speculate on the reasons for this but rather concentrates on the results of some research in which information was used as a policy instrument.

4. POLICY INSTRUMENTS FOR DIETARY CHANGE

How to make everyday behavior more environmentally friendly has been on the agenda for nearly a half-century, and much has been learned. Both external and internal barriers must be overcome. There has been an optimistic belief in the use of information in the hope of fostering environmentally sound behavior among members of the general public. However, it has been found that there is no automatic link between knowing and doing. Norms, values, feelings, and consumption opportunities have been shown to be a vital part of the picture.

However, there are several policy tools, in addition to information, that can be used for changing everyday behavior, and they differ in influence and effect (Table III).

Four groups of policy instruments can be identified—information, economic instruments, administrative instruments, and physical improvements—and they are often used in various combinations to increase efficiency.

Information can be mediated in several ways. When conveying environmental information about products, written information in pamphlets and advertisements is often used, as is product labeling. The common denominator for all of these ways is that they should attract attention. The receiver of information is supposed to notice and benefit from the new arguments voluntarily.

Economic instruments include taxing, pricing, and refund schemes. Besides increasing the costs for pollution and providing economic benefits for environmentally friendly consumption, these instruments inspire polluters to reflect over changes in behavior or use of technology and to make plans for behavioral change to minimize the costs for polluting the environment. Economic instruments function as catalysts for changes in the future.

Administrative instruments have an immediate effect on all of the actors they are intended to affect from an announced date. The success depends largely on the negative sanctions with which those with deviant behavior are punished. Examples include fines, imprisonment, and prohibition of trade.

Physical improvements, such as construction of waste depositories and bicycle lanes, are intended to facilitate a new pattern of behavior. Such improvements are very often combined with some other instrument pushing the behavior in the same direction. For example, separate lanes for cars and bicycles are followed by traffic rules regulating where the two kinds of vehicles may be driven.

TABLE III

Policy Instruments, Influence on Actors, and Effects

Instrument	Influence	Effect
Information	Voluntary	Slow
Economic instruments	Catalytic	Short range
Administrative instruments	Immediate, forcing	Middle range
Physical improvements	Reminding, repeating	Change habits

From this knowledge, it is obvious that information will be more efficient if it is combined with other policy instruments. As an example, if households are given life cycle-based environmental information about their diets and food choices, the chances that they will lower environmental impacts of their diets increase if (1) there are tax subsidies for low-polluting food products (economic instrument), (2) the sale of food products with very high environmental impacts is prohibited (administrative instrument), and (3) a steady and diverse supply of low-polluting food products is available in nearby shops and restaurants (physical improvement).

Having reviewed the various policy instruments and how they can be integrated, this article now presents some results from research where the potential or actual impacts of environmental information given to households have been tested. In both of these examples, scientific methods were used for evaluation.

4.1 Negative versus Positive Labels

A team of environmental psychologists tested the impact of negative versus positive ecolabels on consumer behavior. The background was that all ecolabels today signal a positive outcome, for example, "Choose this product; it is better for the environment than the average product." Another strategy would be to signal a negative outcome with the purpose of making consumers avoid a product. A laboratory test was carried out on 40 students in Sweden who were asked to choose between products labeled with a three-level ecolabel inspired by traffic lights: red, green, and yellow. The red label was defined as much worse in terms of environmental consequences than the average, the green label was defined as much better than the average, and the yellow label was defined as average. All participants were asked to rate the importance attached to environmental consequences when purchasing everyday commodities. The results showed that those who attached high importance to environmental consequences when purchasing everyday commodities were most strongly affected by ecolabels of any kind. The results also showed that those who attached little importance to environmental consequences when purchasing everyday commodities were unaffected by any kind of label. However, those who had an intermediately strong interest in environmental issues were more affected by the red label than by the green one. The conclusion was that the ecolabels in use today affect mostly individuals with a strong interest in environmental issues and that a negative label influences a wider range of individuals.

4.2 Households and Researchers in Partnership When Experimenting with Alternative Diets

In a research project titled "Urban Households and Consumption-Related Resource Use," 10 households participated in an experiment where they tried to reduce life cycle energy inputs related to their food consumption. A manual for low-energy foods habits was developed, and based on this information, the households planned and implemented food habits demanding fewer life cycle energy inputs. The effect of participation in the experiment was evaluated.

Among the changes of most importance for lowering energy inputs were a reduction in consumption of meat (mostly pork); an increase in vegetables and fruits, especially vegetables grown in the open; an increase in poultry, eggs, and fish; and an increase in legumes. The households frequently bought organic food, chose products from Sweden instead of imported ones, and used electric kettles for boiling water for coffee and tea as part of their efforts to achieve more energy-efficient diets. Life cycle energy inputs decreased by at most a third. Environmental aspects other than energy inputs (e.g., animal welfare) were also important to the households. Some recommendations, such as supplementing meat with pulses, were hard to follow because the households did not know how to prepare them.

The overall conclusion from this project was that it takes time to educate households to become "green" and that forces from both inside and outside the household must support such an effort. Being involved in an experiment can affect everyday habits, as can feedback on one's own resource use. However, misconceptions and distrust are deeply rooted, and complex information is difficult to convey to consumers. All of this must be considered if we want to count on the collaboration of households for ecologically sustainable development. Table IV shows the changes in food consumption that occurred as a result of the experiment.

TABLE IV

Changes in Quantities of Food Consumed in the 10 Households Participating in the "Urban Households and Consumption-Related Resource Use" (URRU) Project

Food category	Percentage change
Legumes and seeds	132
Egg	97
Poultry	55
Vegetables	86
Fruit, berries, and jam	25
Fish	22
Pork	−30
Lamb	−22
Bread, beef, drinks, fermented milk, cheese, and yogurt	Negligible
Food composed of several food items, porridge and cereals, potatoes, pasta and rice, pastries and sweets, ice cream and desserts	Negligible

Source. Carlsson-Kanyama *et al.* (2002).

5. CONCLUSIONS

Greenhouse gas emissions and energy inputs in the current food system are substantial. Results from case studies of food products followed throughout their life cycles show important differences depending on animal or vegetable origin, degree of processing, transportation distance and mode of transport, preparation methods, and so on. This leaves room for discussing dietary change as a mitigation strategy for curbing climate change. The theoretical potential for such a strategy is large and is open mainly to affluent consumers.

A successful mitigation package will have to recognize that a combination of policy instruments will be most efficient and that any information applied by itself will have a slow change potential. However, there are many different ways in which to produce and disseminate information that will affect the outcome in terms of changes in general behavior. In general, information with feedback is more efficient than one-way communication. Moreover, involving households in experiments with alternative lifestyles has been shown to produce lasting results. Information should also be tailored to the everyday situation of the household for most impact. Although current global trends in food consumption are not promoting higher environmental efficiency, there are some trends that could develop in that direction.

Policymakers should recognize and support such trends for successful mitigation.

SEE ALSO THE FOLLOWING ARTICLES

Development and Energy, Overview • Global Material Cycles and Energy • Greenhouse Gas Abatement: Controversies in Cost Assessment • Greenhouse Gas Emissions, Alternative Scenarios of • Greenhouse Gas Emissions from Energy Systems, Comparison and Overview • Input–Output Analysis • Life Cycle Assessment and Energy Systems • Suburbanization and Energy

Further Reading

Borgström, G. (1972). "The Hungry Planet: The Modern World at the Edge of Famine." Collier Books, New York.

Brower, M., and Leon, W. (1999). "The Consumer's Guide to Effective Environmental Choices: Practical Advice from the Union of Concerned Scientists." Three Rivers Press, New York.

Carlsson-Kanyama, A., Pipping Ekström, M., and Shanahan, H. (2002). Urban Households and Consumption-Related Resource Use (URRU): Case studies of changing food habits in Sweden. *In* "Changes in the Other End of the Chain" (Butjin *et al.*, Eds.). Shaker Publishing, Maastricht, Netherlands.

Carlsson-Kanyama, A., Pipping Ekström, M., and Shanahan, H. (2003). Food and life cycle energy inputs: Consequences of diet and ways to increase efficiency. *Ecol. Econ.* **44**, 293–307.

Grankvist, G., Dahlstrand, U., and Biel, A. (2002). The impact of environmental labelling on consumer preference: Negative versus positive labels. *In* "Determinants of Choice of Eco-Labelled Products" (G. Grankvist, doctoral thesis). Department of Psychology, Göteborg University, Sweden.

Jungbluth, N., Tietje, O., and Scholz, R. W. (2000). Food purchases: Impacts from consumers' point of view investigated with a modular LCA. *Intl. J. Life Cycle Assessment* 5(3), 134–142.

Kramer, K. J., Moll, H. C., Nonhebel, S., and Wilting, H. C. (1999). Greenhouse gas emissions related to Dutch food consumption. *Energy Policy* **27**, 203–216.

Linden, A-L., and Carlsson-Kanyama, A. (2003). Environmentally friendly behavior and local support systems: Lessons from a metropolitan area. *Local Environ.* **8**, 291–301.

Pimentel, D., and Pimentel, M. (eds.). (1996). "Food, Energy, and Society," rev. ed. University Press of Colorado, Boulder.

Discount Rates and Energy Efficiency Gap

RICHARD B. HOWARTH
Dartmouth College
Hanover, New Hampshire, United States

1. Introduction
2. The Hidden Costs Argument
3. High Discount Rates?
4. Imperfect Information and Bounded Rationality

Glossary

bounded rationality The notion that people behave in a manner that is nearly but not fully optimal with respect to their goals as their available information and cognitive limitations will allow.

discount rate The rate at which someone prefers present over future benefits in a multiperiod model.

efficiency gap The difference between the actual level of investment in energy efficiency and the higher level that would be cost beneficial from the consumer's standpoint.

imperfect information This refers to a situation in which an individual makes decisions without full knowledge of costs and benefits.

Investments in enhanced energy efficiency typically involve a trade-off between short-term costs and long-term benefits. Consider, for example, a home insulation project that would generate energy savings of \$130 per year during a 30-year expected life span. If the project required an initial investment of \$900, would it be justified from a financial perspective? A naive answer to this question is that the project would provide a long-term net savings of \$3000, thereby warranting its implementation. However, although this type of reasoning seems intuitive to many people, it departs from the standard methods of financial analysis, in which costs and benefits that accrue in the future should be discounted relative to the present.

1. INTRODUCTION

The logic behind discounting is linked to people's observed behavior in financial markets. Suppose, for example, that an investor would receive an r percent rate of return on standard financial instruments such as corporate stocks. If the investor was well informed and rational, then he or she would be willing to pay no more than $B/(1+r)^t$ dollars in the present for an asset that he or she could resell to obtain B dollars t years in the future; otherwise, the investor would be better off investing his or her money at the market rate of return. In this example, economists say that r measures the discount rate or time value of money (i.e., the rate of return that people demand on each unit of investment). Based on typical rates of return in the private sector, analysts commonly use "real" (i.e., inflation-adjusted) discount rates on the order of 6% per year in evaluating projects with average degrees of risk.

More generally, consider a project that would yield a stream of costs and benefits C_t and B_t in years $t=0, 1, \ldots, T$. In this formulation, year 0 represents the present, whereas the project yields impacts that extend a total of T years into the future. Then the project is considered financially attractive if its net present value (NPV), calculated according to the formula

$$NPV = \sum_{t=0}^{T} (B_t - C_t)/(1+r)^t,$$

is positive. In this event, the project's discounted net benefits exceed the initial investment cost. This implies that the project generates a rate of return that exceeds the discount rate, thereby justifying the investment from an economic perspective.

Given a 6% discount rate, the net present value of the home insulation project described previously is

$NPV = -900 + 130/1.06 + 130/1.06^2 + \ldots + 130/1.06^{30} = \$889$. Although this example is purely illustrative, it signals an important fact about real-world investments in energy efficiency: In many cases, energy-efficient technologies are cost-effective in the sense that they generate positive net benefits when evaluated using conventional discount rates.

Empirical work on a broad range of energy use technologies, however, points to an important paradox that is known as the energy efficiency gap. Replacing incandescent light bulbs with compact fluorescents, for example, typically reduces electricity use by 75% while generating simultaneous cost savings. In a similar vein, a study by the U.S. National Academy of Sciences found that the fuel economy of passenger vehicles could be increased by as much as 33% for automobiles and 47% for light trucks (pickups, minivans, and sport utility vehicles) by implementing cost-effective design options that leave vehicle size and performance unchanged. On an economywide basis, the Intergovernmental Panel on Climate Change estimates that the energy use of advanced industrialized nations could be reduced by up to 30% through the full adoption of cost-effective technologies. However, households and businesses have shown little enthusiasm for energy-efficient technologies, so market decisions seem insufficient to realize this potential.

Three interpretations of the efficiency gap have been set forth in the literature:

1. Energy-efficient technologies have "hidden costs" that are overlooked by conventional financial analysis.
2. Households and businesses use unusually high discount rates in decisions concerning energy efficiency.
3. The adoption of cost-effective, energy-efficient technologies is impeded by issues of imperfect information, "bounded rationality," or structural anomalies in markets for energy-using equipment.

The remainder of this article discusses these themes.

2. *THE HIDDEN COSTS ARGUMENT*

Studies of the costs and benefits of energy-efficient technologies typically employ an engineering approach that accounts for purchase price, operations and maintenance costs, and anticipated energy savings. However, energy use technologies can have intangible costs that are not easily measured. The failure to include such hidden costs introduces a potential bias or source of error in financial analysis.

Consider the case of compact fluorescent light bulbs mentioned previously. Although this technology yields well-documented direct cost savings, it involves two types of hidden costs that are well recognized in the literature. First, compact fluorescent bulbs tend to be bigger and bulkier than the conventional incandescents that they replace. Although progress has been made in addressing this problem since this technology was first mass marketed in the 1990s, the fact remains that compact fluorescents do not work well with certain lighting fixtures. Second, compact fluorescents produce a different light spectrum than incandescent light bulbs. Some people find their light to be "cold" and aesthetically inferior. Although these costs are difficult to measure in financial terms, they legitimately affect people's willingness to adopt this technology.

A second example is provided by the case of automobile fuel economy. In the United States, the energy efficiency of new vehicles is regulated under the Corporate Average Fuel Economy (CAFE) standards that were introduced in response to the Arab oil embargo of 1973. Although these standards have been effective in reducing fuel consumption, some analysts argue that car manufacturers achieved increased fuel economy by reducing the size and weight of new vehicles. Since (all else being equal) the crashworthiness of vehicles is inversely proportional to their size and weight, authors such as Crandall and Graham argue a case can be made that the CAFE standards have compromised safety, increasing highway fatalities by up to 3900 persons per year. This impact, of course, is not reflected in standard engineering studies of the economic costs of improved fuel economy, although it clearly is relevant to the purchase decisions of car buyers. To be fair, it is important to note that the energy savings potential identified by the U.S. National Academy of Sciences attempted to control carefully for this effect. However, safety considerations trump fuel economy in the eyes of many consumers.

In general, examples such as compact fluorescent light bulbs and vehicle fuel economy suggest that the hidden costs issue is a real-world phenomenon with significant implications for consumer decisions. Nonetheless, it is important to consider at least two caveats that cast doubt on the generality of this reasoning. First, certain energy efficiency measures yield intangible benefits that are typically overlooked in conventional cost calculations. Weather stripping a leaky building shell, for example, yields direct cost savings through reduced energy use. In addition, it improves the comfort of building occupants by reducing drafts and hence the subjective sense that a building is cold. In a similar vein, the use of day lighting in passive solar building technologies can substantially enhance a building's aesthetic appeal, although this benefit is difficult to reduce to simple economic terms.

Second, there are many technologies for which changes in equipment characteristics yield energy savings without altering the quality of energy services provided to end users. In refrigerators, for example, the use of energy-efficient motors and compressors leads to pure energy savings through technical measures that have no observable effect on the equipment's size, functionality, or performance. Similarly, programming computers to enter a "sleep state" during periods of nonuse generates substantial energy savings without reducing the benefits of this technology.

The bottom line is that energy analysts should be aware of and sensitive to issues of comfort, convenience, and safety. In some instances, hidden costs can provide good reasons not to adopt a technology that yields apparent cost savings. On the other hand, there is no good evidence that hidden costs explain the overall empirical magnitude of the efficiency gap.

3. *HIGH DISCOUNT RATES?*

As explained previously, the discount rate reflects the rate at which people are willing to exchange present and future economic benefits. According to standard economic reasoning, discount rates are revealed by people's observed behavior in markets for savings and investment. A homeowner who takes out a loan at a 4% real (inflation-adjusted) interest rate, for example, reveals a willingness to pay back $3.24 30 years from the present to obtain just $1 dollar today. Similarly, a worker who invests $1 in a mutual fund would expect to receive $10.06 upon retiring three decades in the future given an 8% annual return.

Economists generally agree that market rates of return reveal people's discount rates. It is important to note, however, that different types of financial assets yield different rates of return. In particular, safe investments, such as money market funds, certificates of deposit, and short-term government bonds, yield real returns on the order of just 1% per year. Risky assets such as corporate stocks, in contrast, pay average real returns of 6% or more on a long-term basis. The high returns paid by stocks constitute a risk premium that rewards investors for accepting the fluctuations associated with financial markets.

If households and businesses were well informed and rational, then economic theory suggests that they would discount the benefits of energy-efficient technologies at a rate equal to the market return available on investments with similar risk. In general, the volatility of energy prices implies that the benefits of energy efficiency are uncertain. On the other hand, investments in energy efficiency reduce people's vulnerability to large swings in energy prices. In this sense, energy efficiency measures have ambiguous effects on the overall financial risk faced by households and businesses. In practical analysis, it is common to assume that investments in energy efficiency have risk characteristics similar to those associated with typical private sector investments. In theoretical terms, this assumption favors the use of a 6% discount rate.

Despite this body of theory, empirical studies have found that people behave as if they discounted benefits of energy-efficient technologies at an annual rate far in excess of 6% per year. An early study by Hausman, for example, suggested that people discounted the energy savings provided by air conditioners at a 25% annual rate. Later studies summarized by Train and Ruderman *et al.* pointed to the existence of implicit discount rates ranging from 25% to 300% for refrigerators, heating and cooling systems, building shell improvements, and a variety of other technologies. These studies employ sophisticated econometric and/or engineering–economic models to gauge the rate of time preference at which people's actual technology choices could be considered economically rational. As noted by Howarth and Sanstad, the fact that people reject efficiency technologies that yield high rates of return suggests that actual behavior may diverge from the assumption of perfect rationality.

Analysts such as Hassett and Metcalf have attempted to explain this anomaly by arguing that improvements in the state of technology over time can provide a reason to defer investing in technologies that are apparently cost-effective in the short term. The reason is that purchasing an item such as a fuel-efficient car implicitly raises the cost of buying an even more efficient vehicle a few years later. To better understand this point, consider the case of a mid-1990s computer user weighing the merits of purchasing an upgraded machine. For a cost of

$2000, the user could step up from a good system to a very good system with enhanced speed and performance. Alternatively, the user could defer the purchase for 1 or 2 years and expect to buy a far better machine for an equal or lower price. Hassett and Metcalf reasoned that this type of behavior could explain the high discount rates revealed in markets for investments in energy efficiency.

According to this perspective, the benefits of waiting imply that the early adoption of energy-efficient technologies may have a type of hidden cost (or option value) that is sometimes ignored in naive technology cost studies. An empirical analysis by Sanstad *et al.*, however, suggests that the magnitude of this effect is too small to have much impact on decisions about energy efficiency given anticipated developments in energy prices and the pace of technological change. This line of reasoning therefore appears not to explain why technologies that yield positive discounted net benefits or (equivalently) attractive rates of return are often passed up in markets for energy-using equipment.

4. IMPERFECT INFORMATION AND BOUNDED RATIONALITY

The preceding arguments attempt to explain the paradoxes surrounding discount rates and the efficiency gap in terms of the workings of what Sutherland terms normal or economically efficient markets. An alternative explanation offered by Stanstad and Howarth focuses on market barriers or market failures that impeded the adoption of cost-effective, energy-efficient technologies. In broad terms, at least two sources of market failure are emphasized in this context: imperfect information and bounded rationality.

Behavioral studies have established that households and businesses are often poorly informed about the technical characteristics of energy-using equipment. In summarizing this literature, Stern argues that people's knowledge of opportunities to save energy is "not only incomplete, but systematically incorrect. Generally speaking, people tend to overestimate the amounts of energy used by and that may be saved in technologies that are visible and that must be activated each time they are used." At a cognitive level, many people do not focus on energy trade-offs in making decisions about purchasing goods, such as appliances, new cars, or even a home. This holds true because (i) the energy requirements of most goods are not vividly observable prior to a purchase decision, and (ii) energy costs are typically only a small fraction of the total cost of equipment ownership. Hence, buyers emphasize factors such as a refrigerator's features and aesthetics, a vehicle's acceleration, and a home's location and architecture without fine-tuning their decisions to optimize energy efficiency.

The problem of imperfect information is linked to the concept of adverse selection in information economics. As Akerlof showed in his seminal "markets for lemons" article, low-quality products can crowd out high-quality products in markets in which product quality is not easily observed by consumers. In the case of the efficiency gap, the inference is that poorly informed buyers would pass up opportunities to purchase technologies that would simultaneously save energy and produce net economic benefits.

Issues of imperfect information have been partially addressed through measures such as advertising campaigns and energy labeling for appliances, vehicles, and (in recent years) new homes. According to Robinson, these efforts have typically met with limited success. One reason is that from a psychological perspective, human beings do not have the cognitive capabilities required to make perfectly informed, rational decisions in complex, real-world settings. Instead, Kempton and Layne argue that people are boundedly rational, using simple but cognitively efficient rules of thumb to make decisions in environments that are too complex to be fully understood. The literature on this subject suggests that people aim to achieve goals such as cost minimization but in practice make decisions and are flawed and (in many cases) systematically biased.

In cases in which consumer decision making is marked by imperfect information and/or bounded rationality, targeted policy interventions can often facilitate the adoption of cost-effective, energy-efficient technologies. Under the U.S. Appliance Efficiency Standards, for example, equipment manufacturers must meet minimum performance standards that are explicitly based on the criterion of cost-effectiveness. Studies by McMahon *et al.* and Geller suggest that these standards alone will yield 24 exajoules of energy and $46 billion between 1990 and 2015. These standards are set in a process of information exchange and rule making that involves a relatively small number of technology experts in government and industry. Through this process, technologies are optimized in a manner that does not occur through the engagement of buyers and sellers in the unregulated marketplace.

A second policy that has facilitated the adoption of cost-effective, energy-efficient technologies is the Green Lights program of the U.S. Environmental Protection Agency (EPA). Under Green Lights, the EPA provides technology assistance and public recognition for businesses and nonprofit organizations that voluntarily agree to implement energy-efficient lighting technologies that yield net cost savings without negative impacts on energy services. In its first 5 years of operation, EPA estimates that this program achieved $7.4 billion kWh of electricity savings that remained unexploited prior to its implementation in 1991. According to DeCanio, the typical investment carried out under Green Lights yielded a 45% annual rate of return. This example illustrates how market barriers and the efficiency gap can affect the production sector of the economy as well as how voluntary cooperation between government and the private sector can facilitate improved market performance.

Of course, programs and policies can have administrative costs that should be carefully considered in evaluating their net economic benefits. These costs are small for the U.S. Appliance Efficiency Standards and Green Lights but are quite substantial in certain situations. Joskow and Marron, for example, found that administrative costs are a major component of the total cost of reducing residential electricity use through utility-sponsored demand-side management programs. In addition, regulations that depart from the goal of cost-effectiveness can have unintended negative impacts on product characteristics and consumer choice. This issue is illustrated by the possible increase in highway fatalities caused by the CAFE standards imposed on new vehicles in the 1970s.

Nonetheless, the empirical literature on the efficiency gap suggests that there is ample scope to improve energy efficiency through well-designed policy interventions. In this context, using the net present value criterion and conventional discounting procedures provides a mechanism to identify programs and policies that would enhance consumer welfare and economic efficiency, provided that all of the relevant costs and benefits are taken into account.

SEE ALSO THE FOLLOWING ARTICLES

Cost–Benefit Analysis Applied to Energy • Economics of Energy Efficiency • Energy Efficiency, Taxonomic Overview • Industrial Energy Efficiency • Obstacles to Energy Efficiency • Vehicles and Their Powerplants: Energy Use and Efficiency

Further Reading

Akerlof, G. (1970). The market for lemons: Quality uncertainty and the market mechanism. *Q. J. Econ.* **89,** 488–500.

Crandall, R. W., and Graham, J. D. (1989). The effect of fuel economy standards on automobile safety. *J. Law Econ.* **32,** 97–118.

DeCanio, S. J. (1998). The efficiency paradox: Bureaucratic and organizational barriers to profitable energy-saving investments. *Energy Policy* **26,** 441–454.

Geller, H. (1997). National appliance efficiency standards in the USA: Cost-effective federal regulations. *Energy Buildings* **26,** 101–109.

Hassett, K. A., and Metcalf, G. (1993). Energy conservation investment: Do consumers discount the future correctly? *Energy Policy* **21,** 710–716.

Hausman, J. A. (1979). Individual discount rates and the purchase and utilization of energy-using durables. *Bell J. Econ.* **10,** 33–54.

Howarth, R. B., and Sanstad, A. H. (1995). Discount rates and energy efficiency. *Contemp. Econ. Policy* **13,** 101–109.

Intergovernmental Panel on Climate Change (1996). "Climate Change 1995: Economic and Social Dimensions of Climate Change." Cambridge Univ. Press, New York.

Interlaboratory Working Group (2000). "Scenarios for a Clean Energy Future." Oak Ridge National Laboratory, Oak Ridge, TN.

Joskow, P. L., and Marron, D. B. (1992). What does a negawatt really cost? Evidence from utility conservation programs *Energy J.* **13**(4), 41–74.

Kempton, W., and Layne, L. (1994). The consumer's energy analysis environment. *Energy Policy* **22,** 857–866.

McMahon, J. E., Berman, D., Chan, P., Chan, T., Koomey, J., Levine, M. D., and Stoft, S. (1990). Impacts of U.S. appliance energy performance standards on consumers, manufacturers, electric utilities, and the environment. *In* "ACEEE Summer Study on Energy Efficiency in Buildings," pp. 7.107–7.116. American Council for an Energy Efficient Economy, Washington, D.C.

Robinson, J. B. (1991). The proof of the pudding: Making energy efficiency work. *Energy Policy* **7,** 631–645.

Ruderman, H., Levine, M. D., and McMahon, J. E. (1987). The behavior of the market for energy efficiency in residential appliances including heating and cooling equipment. *Energy J.* **8,** 101–124.

Sanstad, A. H., and Howarth, R. B. (1994). Normal markets, market imperfections, and energy efficiency. *Energy Policy* **22,** 811–818.

Sanstad, A. H., Blumstein, C., and Stoft, S. E. (1995). How high are options values in energy-efficiency investments? *Energy Policy* **23,** 739–743.

Stern, P. C. (1986). Blind spots in policy analysis: What economics doesn't say about energy use. *J. Policy Anal. Management* **5,** 200–227.

Sutherland, R. J. (1991). Market barriers to energy-efficiency investments. *Energy J.* **12,** 15–34.

Train, K. (1985). Discount rates in consumers' energy-related decisions: A review of the literature. *Energy* **10,** 1243–1253.

U.S. Environmental Protection Agency (1997). "Building on Our Success: Green Lights and Energy Star Buildings 1996 Year in Review." U.S. Environmental Protection Agency, Office of Air and Radiation, Washington, DC.

U.S. National Academy of Sciences (2002). "Effectiveness and Impact of Corporate Average Fuel Economy (CAFE) Standards." National Academy Press, Washington, DC.

Distributed Energy, Overview

NEIL STRACHAN
Pew Center on Global Climate Change
Arlington, Virginia, United States

1. Introduction
2. Definition of DE
3. Historical Development of DE
4. DE Market Status
5. Overview of DE Technologies
6. Economic Characteristics of DE
7. Environmental Characteristics of DE
8. Other Characteristics of DE
9. Future Role of DE

Glossary

buyback tariff The price paid to the owner of the distributed energy (DE) resource for electricity sold back to the electricity grid. The price paid depends on how much and when power is exported, how regularly it is guaranteed, and where in the distribution system the DE resource is located (and hence the impact on the operation of the network).

capacity factor Also known as load factor. Electricity delivered during a time period divided by the rated electricity capacity.

combined heat and power (CHP) Also known as cogeneration. The use of waste heat to provide steam or hot water, or cooling via an absorption chiller. This can raise the overall efficiency to over 80%

efficiency Electric efficiency is the total electricity output divided by the total fuel input taking into account generation efficiency and transmission losses. Total efficiency includes the use of waste heat.

heat to power ratio (HPR) The ratio between heat use and electricity generation in a combined heat and power application. For example, if electric efficiency and total efficiency are 30 and 80%, respectively, then HPR = (80−30)/30, or 1.67.

higher heating value (HHV) Includes the heat of vaporization of water. For natural gas, HHV efficiency = lower heating value efficiency/1.1.

installation cost The cost necessary to implement a distributed energy technology or resource in an application. Includes technology cost (dollars per kilowatt), engineering studies, permitting, interconnection, and setup expenses.

interconnection The link between a distributed energy generator and the load being served by the utility electricity network

lower heating value (LHV) Assumes the heat of vaporization of water cannot be recovered. For natural gas, LHV efficiency = higher heating value efficiency × 1.1.

modularity Increments of new capacity that can be simply and flexibly added or removed to change the unit's energy output.

operation and maintenance costs (O&M Costs) These are divided into fixed costs and variable costs depending on the relation to hours of operation and capacity factor.

photovoltaic (PV) cells Also known as solar cells. A distributed energy technology that converts sunlight directly into electricity.

Distributed energy (DE) represents an alternative paradigm of generating electricity (and heat) at, or close to, the point of demand. DE includes fossil technologies (fuel cells, microturbines, internal combustion engines and Stirling engines), renewable technologies (photovoltaics and wind turbines), and energy storage options. Potential DE advantages include higher efficiency and lower cost through waste heat recovery and avoidance of transmission and distribution, reduced global air pollutants, enhanced flexibility of electricity grids, reduced investment uncertainty through modular increments of new capacity, and greater system and infrastructure reliability and security. Potential DE disadvantages include higher costs through loss of economies of scale, higher local air pollution near population areas, economic losses from stranded capital investments in generation and distribution, and increased reliance on natural gas.

1. INTRODUCTION

Distributed energy (DE) represents an alternative paradigm of generating electricity (and heat) at or close to the point of demand. DE includes fossil technologies—fuel cells, microturbines, internal combustion engines, and Stirling engines; renewable technologies—photovoltaics and wind turbines; energy storage options; and energy efficiency. Potential DE advantages include higher efficiency and lower cost through waste heat recovery and avoidance of transmission and distribution, reduced global and local air pollutants, enhanced flexibility of electricity grids, reduced investment uncertainty through modular increments of new capacity, and greater system and infrastructure reliability and security. Potential DE disadvantages include higher costs through loss of economies of scale, higher local air pollution near areas of population, economic losses from stranded capital investments in generation and distribution, and increased reliance on natural gas.

DE has generally been considered and studied in the context of niche applications, as emergency back-up power, or as limited to a small portion of grid-connected electricity supply. In many countries, the economies of scale of centralized generation, the low price of coal as a fuel for electricity generation, and regulatory barriers or disincentives to on-site generation have precluded the widespread adoption of DE. These institutional barriers have included lack of interconnection protocols, low electricity buy-back tariffs, and little consideration of the system's benefits of distributed resources. However, changes in the relative economics of centralized versus distributed energy, the increasing use of natural gas, restrictions on new electricity transmission lines, recognition of the environmental benefits of DE, and improved DE control technologies have resulted in the reconsideration of the widespread use of DE. For example, the Netherlands has seen the greatest penetration of DE technologies. By the year 2000, 6% of national electricity capacity was DE units of <1 MWe and 35% of national electricity capacity was DE units of <50 MWe.

This article gives an overview of DE. Sections 2 and 3 define DE and give an historical context. Sections 4 and 5 summarize the market status of DE and the major DE technologies. Sections 6, 7, and 8 summarize the characteristics of DE technologies including their economic, environmental, and reliability performance, as well as compatibility with existing electricity infrastructure and investment requirements. Finally, Section 9 details the opportunities and barriers to future DE deployment.

2. *DEFINITION OF DE*

DE has been defined by the size of the technology or by specific technologies. It has also been defined on the basis of connection (connected to either the electricity distribution system or, in some cases, the medium-voltage network), dispatch (not centrally dispatched), whether it is located on the customer side of the meter, and noninclusion in system production optimization studies.

However, defining DE by its relationship to the electricity network ensures that DE is a function of the characteristics of that network. For example, a country dominated by very large centralized units, such as France, may define DE as <100 MWe, but another system configuration, such as that in the Netherlands, may define DE as <1 MWe, or two orders of magnitude smaller. The size range of DE technologies thus varies from a few watts to approximately 50 MWe.

A preferable definition is as follows: Distributed energy represents an alternative paradigm of generating electricity (and heat) at, or close to, the point of demand. This definition allows discussion across national electricity infrastructures and across demand sectors, as various applications may require different characteristics of energy supply and different institutional arrangements may require different coupling to the electricity network. Second, it allows for DE to cover a range of sizes of generation for residential through industrial applications. Third, it emphasizes the opportunity for energy efficiency through the use of waste heat, which cannot be transported across distances outside the immediate area. Fourth, it allows for the consideration of a host of current and future technologies and, crucially, also energy storage and energy efficiency in a broad definition of DE. And finally, this definition does not constrain DE to specific functions within current energy delivery networks. These functions include standby power, peak load shaving, remote sites, and grid support. However, an energy infrastructure based solely on DE, bypassing the need for large-scale electricity generation, is feasible and may be desirable.

3. *HISTORICAL DEVELOPMENT OF DE*

The shift toward DE can in many ways be viewed as power generation coming full circle. In the 1880s, Thomas Edison's Pearl Street electricity system serving Wall Street and the surrounding city blocks was distributed power. This paradigm was continued over the next 20 years with schemes in the United States and across the world serving limited urban areas with small-scale direct current (DC) systems. Also popular were installations serving individual factories and supplying electricity and heat in combined heat and power (CHP) applications. One large drawback of using DC was the large losses from this low-voltage system when transmitting power over distance. An alternative system based on alternating current (AC) was promoted by a number of competitors, including George Westinghouse. AC systems had the advantage when serving larger and more spread-out service areas, as the voltage could be stepped up using a transformer to minimize losses. The superiority of the AC system was ensured when, in the late 1880s, Nikola Tesla invented the three-phase AC arrangement, which simplified the number and size of wires, and in 1893 when the universal system, a series of technical standards, allowed the interconnection of AC and DC systems and users.

With technology facilitating AC systems, a significant financial impetus for larger, more centralized systems was the requirement for consistently high-load factors on the electricity network. Large amounts of electricity cannot be stored, thus requiring that demand meets supply at all times. Economical use of the transmission and generation capacity thus requires that a diverse customer base maintain electricity demand through the days and seasons. A combination of residential, commercial, and industrial applications enables the demand profile to be smoothed out as much as possible and this generally requires a large service area.

Another stimulus of larger electricity networks was the availability of energy inputs. Hydroelectric power and coal mines for electricity generation are usually located some distance from population and industrial centers. Either the fuel source was transported to distributed power plants (an impossibility in the case of hydroelectricity sites) or electricity was transmitted at high voltages.

An additional driving force of the centralized electricity network was the rise of natural monopolies. Although the institutional structure of emerging systems was very different, ranging from private companies to state-owned enterprises, competition in electricity generation was viewed as infeasible due to the prohibitive costs of laying competing wires and the institutional difficulties (at that time) of being able to verify and charge for use of an electricity network by different firms and organizations. By the 1930s, industrialized countries had set up large electricity utilities, coalescing around the dominant generation design of the steam turbine. Smaller systems were either absorbed or shut down.

This centralized paradigm led to the developments of larger and larger generating plants as economies of scale were pursued to raise total electricity system efficiencies and reduce costs. The size of the largest generating units jumped from 80 MWe in 1920, to 600 MWe in 1960, to 1400 MWe in 1980. In order to finance these enormous capital investments in generation and related transmission, utility monopolies relied on guaranteed revenue streams from predetermined electricity prices to their "locked-in" customer base. Other advantages of this system included reliability due to the integrated nature of the electricity network and reductions of local air pollution near population centers as generation took place in remote areas. In addition, large-scale nuclear power stations were well suited to this paradigm as they required the enormous initial investment and guaranteed investment recovery promised by the centralized electricity network paradigm.

The turning point of the return to DE came with the oil shocks of the 1970s and the drive toward higher efficiencies. Despite economies of scale, overall electricity system efficiencies were restricted to 30–33% of input energy. A major reason for this was the inability to capture the large amounts of waste heat during the electricity production process and hence renewed efforts were made toward technological and regulatory arrangements to allow distributed CHP facilities with overall efficiencies of greater than 80%. The most impressive technological development was in high-efficiency combined-cycle turbines (CCGT) using natural gas. This technology became the mode of generation of choice, capturing most incremental capacity growth in most markets. Progress on smaller DE technologies (see Section 5) included significant cost, reliability, and emissions improvements of engines, microturbines, fuel cells, and photovoltaics (PV). In parallel, and especially in the past 10 years, there have been developments in control and monitoring technologies. The telecommunications revolution has enabled cheap, real-time operation and integration of small-scale power technologies.

Institutional progress has centered around the drive to restructure electricity markets, from monopolies (often publicly owned) to a competitive generation market in which market participants are free to choose technology type and their customer base. The motivations for this shift were to lower power costs and improve customer choice and also to raise revenues in the case of state-owned enterprises.

A major consequence of this market restructuring was the increase in volatility and the resultant uncertainty in recovering investment costs. Firms were less and less willing to commit huge sums in centralized plants with repayment occurring over decades. The average size of a new generating unit in the United States declined from 200 MWe in the mid-1980s, to 100 MW in 1992, and to 21 MWe in 1998. This is approximately equivalent to the average unit sizes in the years 1915–1920.

Additional incentives have stemmed from changing public pressures. It has become increasingly difficult to build additions to electricity networks, especially near population centers. In addition, the demand for very reliable and higher quality power has emerged, spurred by the rapid penetration of electronic appliances and industrial processes that are extremely sensitive to supply disruptions or degradations. Finally, other energy delivery infrastructures, especially the natural gas system, have emerged to compete with electricity transmission. Energy must be moved from source to demand, but where the conversion to electricity is carried out, and at what scale, will be part of the emerging paradigm to come. This is particularly relevant to developing countries where it may or may not be optimal to construct a capital-intensive centralized electricity network.

4. DE MARKET STATUS

It is extremely difficult, or even impossible, to provide values for the total installed capacity of DE technologies and resources. This is due to the range of technologies, their varying levels of use in different countries, and ambiguity over what constitutes a DE application as unit sizes become larger and units are sited further from demand centers. In addition, some of the most promising technologies, including fuel cells, photovoltaics, and microturbines, either are just entering commercialization or are used only in niche applications. The relative rate of technological progress and the regulatory structure of energy markets will determine to what degree and how quickly DE will penetrate the mainstream market. In addition, fuel constraints, stemming from the availability of natural gas, hydrogen, or other fuel networks, and the available renewable energy resource may limit some of the spread of DE technology.

An example of DE installed capacity from the very broadest definition is the DE database of Resources Dynamics Corporation, which as of 2003 lists DE capacity in the United States at 10.7 million DE units with an aggregate capacity of 168 GW. This constitutes 19.7% of the total U.S. installed capacity, although the definitions in this database include back-up and energy units, remote-application small-scale technologies, and a range of larger CHP units.

Better documented market penetration totals and rates can be given for some countries and DE technologies to elucidate the current and potential global DE market size. The Netherlands has the largest market percentage of DE technologies in the world. During the 1980s and 1990s, this was primarily in the form of natural gas-fired engines of less than 2 MWe each in size. By the end of the 1990s, this DE technology was found in over 5000 installations with 1500 MWe of installed capacity. This represents over 6% of the total national capacity.

The Dutch DE units are primarily operated as base-load, CHP applications. If larger gas turbine units are also considered as DE, then the total DE capacity rises to 7500 MWe or 30% of national capacity. Other DE technologies, including microturbines, fuel cells, and Stirling engines, are being introduced as relative costs, emission characteristics, reliability, and power quality considerations are traded off in different applications.

The range of typical DE applications in the Netherlands is given in Fig. 1. The largest sector was commercial horticulture, with greenhouses providing an ideal DE–CHP application with demands for heat as well as electricity for artificial lighting. The Netherlands has many DE units in various commercial building sectors, including hospitals, industry, multiresidential complexes, industrial sites, and sewage operation facilities, where they can also utilize waste gases.

Other countries, including the United Kingdom, Germany, Japan, and the United States, have also seen considerable penetration of internal combustion (IC) engines as a DE resource. These can be either diesel engines, which are often used for back-up and emergency power (as is common in the United States), or gas engines, which are used in CHP applications and can replace the need for an on-site heat boiler plant, and, depending on the regulatory structure and buy-back tariffs, can provide electricity back to the grid. Figure 2 details the numbers and capacity of total worldwide sales of engines that are 1 to 5 MWe each in size, showing both the growth in overall sales and the trend toward gas-fired base-load DE units.

Finally, regarding a nonfossil DE technology, Fig. 3 charts the shipments of PV units in the United States. The largest sectors have traditionally been in applications that are incorporated into buildings to

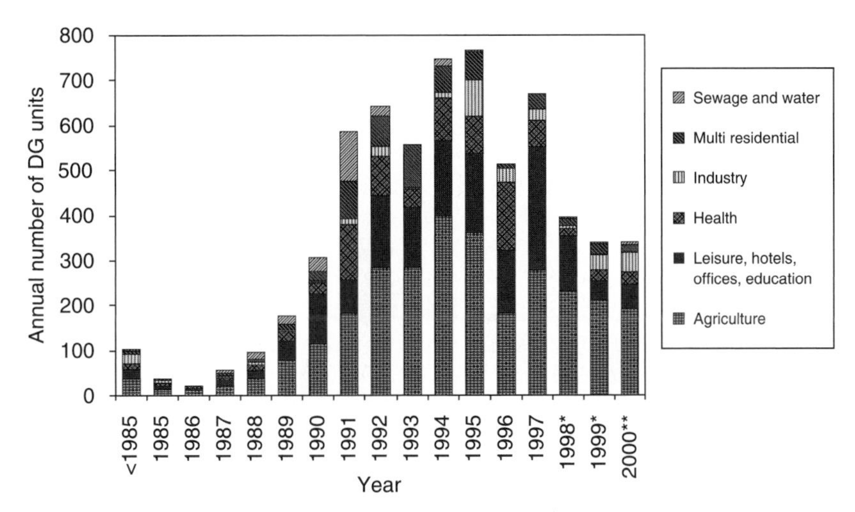

FIGURE 1 Gas IC DG (distributed generation; equivalent to distributed energy) installations by sector in the Netherlands. Years marked with an asterisk are provisional data; those marked with two asterisks are supply firms' market estimates. From Central Bureau of Statistics (1998). "Decentralized Energy Statistics of the Netherlands." Voorborg/Herleen, The Netherlands; and Cogen Nederland (1999). "Nederland Cogeneration Statistical Information." Driebergen, The Netherlands.

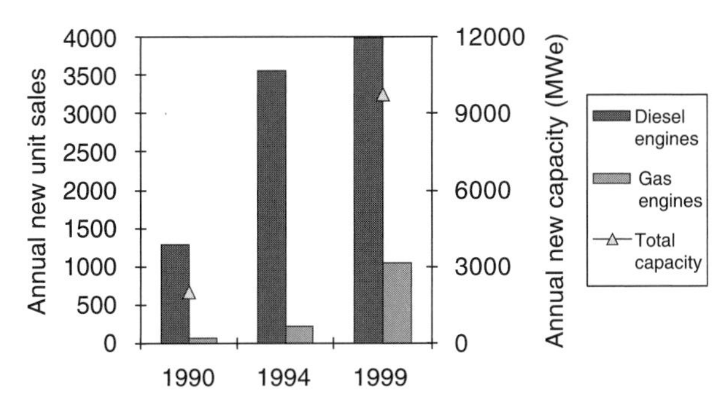

FIGURE 2 Worldwide new unit sales and capacity of DE engines. From Diesel and Gas Turbine Worldwide Power Generation Survey, available at http://www.dieselpub.com.

provide water heating and electrical requirements and in remote applications, such as communication facilities. However, grid-connected PV units have seen the fastest growth.

5. OVERVIEW OF DE TECHNOLOGIES

DE technologies have a range of technical characteristics, making different technologies more suitable in specific applications, depending on whether the determining criterion is electrical output, provision of usable heat, fuel availability, reliability, or emission levels, to name just a few criteria. However, all DE technologies have the capacity to be remotely operated and controlled. This is achieved through monitoring a range of operating parameters via a communication link and allows simplified maintenance, avoids unexpected shutdowns, and eases the grid control issues that arise from locating DE units within the electric distribution network. Another common characteristic of DE technologies is that following factory fabrication they are straightforward to install, thus minimizing implementation costs.

A useful parameter when comparing DE technologies is the heat to power ratio (HPR). This is the ratio between heat use and electricity generation in a CHP application. For example, if electric and total efficiencies are 30 and 80%, respectively, then HPR = (80−30)/30, or 1.67. Depending on the application, electrical or thermal output may be valued or required. Also, note that all electrical efficiencies in this article are quoted as higher heating value (HHV). This includes the heat of vaporization of water. For natural gas, HHV efficiency = Lower Heating Value efficiency/1.1.

This section provides a brief overview of the major DE technologies, discusses their principal strengths and weaknesses, and provides a quantitative comparison of their technical characteristics (see Tables I and II). This section is designed to be illustrative as site-specific variation, differing models, and continuing technical progress will alter these characteristics. Economic and environmental comparisons are detailed in Sections 6 and 7.

5.1 Engines

Reciprocating IC engines are an established and well-known technology. Engines for stationary power are

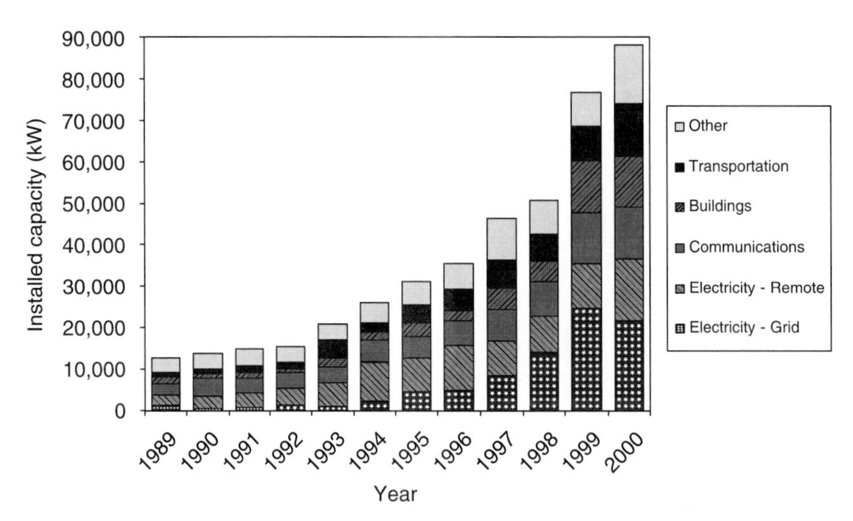

FIGURE 3 U.S. photovoltaic shipments. From Energy Information Administration (2001), Renewable Energy Annual, Table 10, U.S. Department of Energy, Washington, DC.

TABLE I

Qualitative Strengths and Weaknesses of DE Technologies

Technology	Strengths		Weaknesses	
Reciprocating engines	Low capital cost	Fuel flexibility	High air pollutant emissions	Noisy
	High reliability	Quick startup	Frequent maintenance	Thermal output at low temperatures
	Established technology			
Gas turbines	Proven reliability and availability	Low capital cost	Reduced efficiencies at part load	Sensitivity to ambient conditions (temperature, altitude)
	Established technology	High-temperature steam	Required maintenance	
	Larger sizes available			
Microturbines	Compact size	Lightweight	Low electricity efficiencies	Sensitivity to temperature, altitude
	Low emissions	Can utilize waste fuels		
Fuel cells	High electrical efficiencies	Low emissions	High capital costs	Fuel infrastructure constraint
	Quiet	Modular	Reforming natural gas	
	Operates well at part load	High reliability		
	Synergy with vehicle power trains			
Stirling engines	Low noise and vibrationless operation	Low emissions	High capital costs	Low electrical efficiency
	Low maintenance and high reliability	Multifuel capability, including solar power	Durability unresolved	
	Long life			
PV	Low maintenance	No emissions	High capital costs	Intermittency
	Free fuel		By-products of manufacture and disposal	
Wind	Low maintenance	No emissions	High capital costs	Variable power output
	Free fuel		Visual impact	

TABLE II

Quantitative Technical Comparison of DE Technologies

	Reciprocating engines		Turbines		Fuel cells						
	Gas	Diesel	Gas	Micro	MCFC	PAFC	PEMFC	SOFC	Stirling engine	PV	Wind
Commercialized	Yes	Yes	Yes	Yes	No	Yes	No	No	No	Yes	Yes
Size (kWe)	30–1000	50–2000	500–50,000	30–500	250–3000	200–500	10–250	200–2000	1–30	10–1000	10–10,000
Electrical efficiency (%)	27–30	28–35	24–38	20–30	48–55%	33–40%	33–40%	42–58%	20–30		
CHP	Yes	Yes	Yes	Yes	Yes	Yes	Yes	Yes	Yes	No	No
Thermal temperature (°F)	300–500	190–900	500–1100	400–650	170–710	140–250	140–170	350–420	150	—	—
HPR (85% total efficiency)	2.15–1.83	2.04–1.43	2.54–1.24	3.25–1.83	0.77–0.55	1.58–1.13	1.58–1.13	1.02–0.47	3.25–1.83	—	—
Dispatchable	Yes	Yes	Yes	Yes	Yes	Yes	Yes	Yes	No	No	No
Fuels	Gas	Diesel	Gas, waste fuels	Gas, waste fuels	Natural gas, H_2	Natural gas, H_2	Natural gas, H_2	Natural gas, H_2	Any	Solar	Wind
Annual maintenance (h/year)	20–250	20–250	50–700	1–20	120	50	10	40	5–10	0	0
Hours between maintenance	500–1000	1500–2000	4000–8000	750–10,000	800	2200–8700	8700	8700	8700	1–40	4000–30,000
Startup time (min)	0.02–0.2	0.2	1–100	0.5–1	1200–2400	180	60	2	0.5–1	0	0.1–10
Physical size/power output (ft^2/kW)	0.22–0.3	0.22	0.2–0.6	0.15–1.5	1–4	4	0.6–3	1.2	1–3	400–500	0.25–110
Noise (dB) (without acoustic enclosure)	60–85	80–90	60–85	50–60	60	62	50–65	60	40–50	0	55–65

Note. MCFC, molten carbonate fuel cell; PAFC, phosphoric acid fuel cell; PEMFC, proton-exchange membrane fuel cell; SDFC, solid oxide fuel cell.

derived from automobile engines. These engines are fired by either natural gas or diesel and use spark ignition or compression ignition depending on the compression ratio used. A typical energy balance is electricity, 26–39% (this range is due to type, size, and operation of engine); useful heat, 46–60%; losses (radiation from engine and exhaust, lubrication, gearbox, generator), 10–20%.

Power production engines are typically less than 1 MWe and have become the most common DE technology for both standby and peaking applications and are increasing in use as base-load combined heat and power units. Engines have low capital costs and high reliability, although they require a regular maintenance and overhaul schedule. Regular maintenance is essential for good performance and ranges from weekly oil changes to major overhauls after 25,000–40,000 h of operation. Heat recovery is from the engine jacket and cylinder head cooling water and from the hot engine exhaust, resulting in hot water at 80–120°C. A combination of combustion modifications and the use of catalytic converters has greatly reduced local air pollutant emissions, although in this regard engines are still inferior to other DE technologies.

5.2 Gas Turbines

Derived from aero-applications, power generation with gas turbines is commonplace, with an extensive installed capacity and sizes ranging from 500 kWe to $\geq$50 MWe. Low maintenance costs and high-grade heat recovery have made gas turbines a favorite in industrial DE applications.

The unit has a turbine and compressor on the same shaft and compressed combustion products at high temperature drive the turbine. Compressor losses and materials restrictions inhibit the upper temperature of the cycle and thus constrain efficiency. Hot exhaust gases are used to preheat incoming air (regeneration). The remaining thermal energy of these exhaust gases can be reclaimed as waste to heat a boiler. A typical energy balance is as follows: electricity, 30%; high-grade heat, 50%; stack losses, 14.5%; other losses, 6.5%. Turbines are classified as temperature-limited cycles and increases in efficiency continue to occur through materials developments and cooling technologies to allow the turbine to run at higher maximum temperatures.

5.3 Microturbines

Microturbines follow the same cycle as conventional gas turbines, although they are at a less developed stage of commercial development. Their development has come from the design of small, very high speed turbines (with compressors on the same shaft) rotating up to 100,000 rpm. Additional scaling down of components, particularly nozzles and burners, has been achieved. The marketed size range is 30–500 kWe.

Microturbines use regeneration, with resultant lowered exhaust gas temperatures ($\sim$300°C). The unit is air-cooled. Maintenance requirements are reduced due to the elimination of an oil-based lubricating system in favor of air bearings and because the microturbines have no gearbox. Efficiencies are a few percentage points less than those of conventional gas turbines, due primarily to lower operating temperatures.

5.4 Fuel Cells

Fuel cells convert chemical energy directly into electrochemical work without going through an intermediate thermal conversion. Thus, the second law of thermodynamics does not apply, they are not limited by Carnot's theorem, and, therefore, they offer the potential of very high electrical efficiencies. In a fuel cell, the fuel (H_2) and the oxidizer (O_2) are supplied continuously (unlike in a battery). A dynamic equilibrium is maintained, with the hydrogen being oxidized to water with electrons going around an external wire to provide useful electrical work. A major limitation of fuel cells is that direct efficient oxidation of natural gas (CH_4) is not yet possible. Therefore, a reformer is used to convert CH_4 to H_2, with resulting losses in overall efficiency and creation of CO_2 as a by-product.

There are four main types of fuel cells in development. Fuel cells can either use additional heat for the reformer or their own waste heat. They are compared in Table III.

Fuel cells have many admirable qualities for power production including a high potential efficiency, no loss in performance when operating at partial load, ultralow emissions, and quiet operation. Their high capital costs are a major barrier to development.

5.5 Stirling Engines

Stirling engines are an external combustion engine, where the fuel source is burned outside the engine cylinder. This energy source drives a sealed inert working fluid, usually either helium or hydrogen, which moves between a hot chamber and a cold chamber. Stirling engines can be very small in scale (as little as 1 kWe), which in combination with very

TABLE III

Categorization of Major Fuel Cell Types

Type	Operating temperature	Reforming	Efficiency (HHV)	Comments
Phosphoric acid (PAFC)	200°C	External heat required	33–40%	Commercialized
Proton-exchange membrane (PEMFC)	80–100°C	External heat required	33–40%	Poisoned by traces of CO
Molten carbonate (MCFC)	600–650°C	Internal heat	48–55%	Internal reforming can cause variable H_2
Solid oxide (SOFC)	900–1000°C	Internal heat	42–58%	Current R&D favorite

low maintenance, low noise, and low emissions makes them ideal for residential CHP applications. Another advantage of Stirling engines is that they can utilize almost any fuel, including direct solar energy. However, significant challenges remain in reducing the high capital costs and overcoming remaining technical barriers, primarily related to durability.

5.6 Photovoltaics

PV cells, or solar cells, convert sunlight directly into electricity. PV cells are assembled into flat plate systems that can be mounted on rooftops or other sunny areas. They generate electricity with no moving parts, operate quietly with no emissions, and require little maintenance. An individual photovoltaic cell will typically produce between 1 and 2 W. To increase the power output, several cells are interconnected to form a module. Photovoltaic systems are available in the form of small rooftop residential systems (less than 10 kWe), medium-sized systems in the range of 10 to 100 kWe, and larger systems greater than 100 kWe connected to utility distribution feeders.

Two semiconductor layers in the solar cell create the electron current. Materials, such as silicon, are suitable for making these semiconducting layers and each has benefits and drawbacks for different applications. In addition to the semiconducting materials, solar cells consist of two metallic grids or electrical contacts. One is placed above the semiconducting material and the other is placed below it. The top grid or contact collects electrons from the semiconductor and transfers them to the external load. The back contact layer is connected to complete the electrical circuit.

Commercially available PV modules convert sunlight into energy with approximately 5 to 15% efficiency. Efforts are under way to improve photovoltaic cell efficiencies as well as reduce capital costs. Considerable attention is also being given to fully building-integrated PV cells, where the PV cells are an alternative to other construction materials. A principal drawback of PV cells is their reliance on an intermittent power source.

5.7 Wind Turbines

Wind-based power generation may or may not be a DE technology, depending on whether it is located near the demand source or at a remote wind farm. Generally, wind turbines are located in areas with good winds and have annual capacity factors ranging from 20 to over 40%. The typical life span of a wind turbine is 20 years. Maintenance is required at 6-month intervals.

A wind turbine with fan blades is placed at the top of a tall tower. The tower is tall in order to harness the wind at a greater velocity and to be free of turbulence caused by interference from obstacles such as trees, hills, and buildings. As the turbine rotates in the wind, a generator produces electrical power. A single wind turbine can range in size from a few kilowatts for residential applications to greater than 5 MW.

Drawbacks of wind power include high capital costs and reliance on an intermittent resource.

5.8 Energy Storage

Energy storage technologies produce no net energy but can provide electric power over short periods of time. The principal storage options include the following.

5.8.1 Battery Storage

The standard battery used in energy storage applications is the lead–acid battery. A lead–acid battery reaction is reversible, allowing the battery to be reused. There are also some advanced sodium–sulfur, zinc–bromine, and lithium–air batteries that are nearing commercial readiness. Batteries suffer from limited storage capacity and environmental concerns over their manufacture and disposal processes.

5.8.2 Flywheel

A flywheel is an electromechanical device that couples a motor generator with a rotating mass to store energy for short durations. Conventional flywheels are "charged" and "discharged" via an integral motor/generator. The motor/generator draws power provided by the grid to spin the rotor of the flywheel. Traditional flywheel rotors are usually constructed of steel and are limited to a spin rate of a few thousand revolutions per minute and hence offer limited storage capacity.

5.8.3 Superconducting Magnetic Energy Storage

Superconducting magnetic energy storage (SMES) systems store energy in the field of a large magnetic coil with DC flowing. It can be converted back to AC electric current as needed. Low-temperature SMES cooled by liquid helium is commercially available. High-temperature SMES cooled by liquid nitrogen is still in the development stage and may become a viable commercial energy storage source in the future. SMES systems are large and generally used for short durations, such as utility switching events.

5.8.4 Supercapacitor

Supercapacitors (also known as ultracapacitors) are DC energy sources and must be interfaced to the electric grid with a static power conditioner. Small supercapacitors are commercially available to extend battery life in electronic equipment, but large supercapacitors are still in development.

5.8.5 Compressed Air Energy Storage

Compressed air energy storage uses pressurized air as the energy storage medium. An electric motor-driven compressor is used to pressurize the storage reservoir using off-peak energy and air is released from the reservoir through a turbine during peak hours to produce energy. Ideal locations for large compressed air energy storage reservoirs are aquifers, conventional mines in hard rock, and hydraulically mined salt caverns. Air can be stored in pressurized tanks for small systems.

5.9 Hybrid Systems

A combination of various DE technologies or inclusion of energy storage options (including batteries and flywheels) has been proposed to overcome specific limitations (e.g., intermittency) and to boost primary performance characteristics (e.g., electrical efficiency). Several examples of hybrid systems include the following:

- Solid oxide fuel cell (SOFC) combined with a gas turbine or microturbine;
- Stirling engine combined with a solar dish;
- Wind turbines with battery storage and diesel backup generators;
- Engines combined with energy storage devices such as flywheels.

The SOFC/gas turbine hybrid system can provide electrical conversion efficiencies of 60 to 70%, through a combined cycle where the waste heat from the fuel cell reaction drives a secondary microturbine. Stirling engine/solar dish hybrid systems can also run on other fuel sources during periods without sunlight. Wind turbines can be used in combination with energy storage and some type of backup generation (e.g., an IC engine) to provide a steady power supply to remote locations not connected to the grid. Energy storage devices such as flywheels are being combined with IC engines and microturbines to provide a reliable backup power supply. The energy storage device provides ride-through capability to enable the backup power supply to get started. In this way, electricity users can have an interruption-free backup power supply.

6. ECONOMIC CHARACTERISTICS OF DE

6.1 DE Technology Comparison

Table IV details the major cost parameters of DE technologies. These values are intended only to be illustrative due to the site-specific and application-specific factors that alter individual plant costs. Installed cost covers the capital price of the unit plus implementation costs to connect the unit to its electric load and integrate it with its fuel infrastructure, the electricity network, and, if required, the heating system of the building. Operation and maintenance (O&M) costs cover regular maintenance and scheduled overhauls. Fuel costs are based on typical U.S. prices for industrial customers and do not include any additional mark-up from distribution to commercial or residential sites.

As Table IV shows, the more mature DE technologies, engines, gas turbines, and wind turbines, have

TABLE IV

Economic Comparison of DE Technologies

	Installed cost ($/kWe)	O&M cost (cent/kWh)	Input fuel cost (cent/kWh)
Reciprocating engines			
Gas	600–700	0.8–1.4	0.7–1.7
Diesel	300–400	0.7–1.2	0.8–1.5
Gas turbines			
Simple	500–700	0.2–0.8	0.7–1.7
Micro	800	0.3–1	0.7–1.7
Fuel cells			
MCFC	2500–3000	0.3	0.7–1.7
PAFC	3000	0.3–0.5	0.7–1.7
PEMFC	3000–4000	0.1–0.4	0.7–1.7
SOFC	2500–3000	0.2–0.3	0.7–1.7
Stirling engines	2500–4000	0.1	0–1.5
PV	5000–8000	0	0
Wind	850–2500	0.1	0

Note. MCFC, molten carbonate fuel cell; PAFC, phosphoric acid fuel cell; PEMFC, proton-exchange membrane fuel cell; SOFC, solid oxide fuel cell.

the lowest costs. DE technologies utilizing renewable resources have the advantage of an essentially free fuel. Finally, a number of DE technologies with highly appealing technical and environmental characteristics (fuel cells and PV) have high capital costs that are a focus of considerable research and development efforts.

The more mature DE technologies are competitive with centralized electricity generation, especially if they are operated as CHP units, and the comparison is made with both electricity generation and on-site heat boilers. Any investment in DE technologies will depend on a number of site-specific factors including fuel availability, input fuel costs, average and peak power and heating requirements, and the negotiation of interconnection, emergency backup, safety, maintenance, and electricity export charges and tariffs. In addition, DE is likely to impose charges or produce benefits for the wider electricity network, including reduced use of existing assets, retention of stand-by capacity, or the defraying of costs of distribution upgrades.

DE technologies have additional characteristics that influence their economic attractiveness. Whereas centralized units benefit from economies of scale from larger plant size, DE technologies achieve economies of scale through mass production. This effect is expected to be enhanced by the phenomena of learning by doing and of learning by using in the production and use by early adopters. In addition, DE technologies that require significant maintenance benefit from economics of geographic scale. As a greater density of DE units is installed in a given area, the cost of maintaining each unit falls because a significant proportion of maintenance costs arises from on-site visits to maintain and repair the DE unit. Taken together, as more and more DE units penetrate the market, the economic costs of any single unit will fall, leading to higher levels of dispersal and a cycle of improving economics.

A number of factors will limit this cycle of improving costs, including costs imposed on the electricity network, less suitable sites and applications for DE units, rising costs on input fuels, and stresses on the fuel infrastructure. Unless the DE unit is powered by renewable resources at the point of use, the primary fuel delivery infrastructure will likely be natural gas. A rapid rise in natural gas demand from DE technologies will be tempered by rising natural gas prices and increasing delivery costs as the gas network is extended and reinforced.

A final general driving force of economic costs is path dependency. The lifetime of traditional large-scale energy investments (both plants and networks) is measured in years or decades. This gives rise to path-dependent technological trajectories that can "lock-in" energy supply into nonoptimal portfolios of generation and transmission infrastructure. As large amounts of DE market penetration would amount to a paradigm shift in the generation and supply of energy, closing of an older plant and bypassing of existing transmission capacity may allow a more efficient overall system, especially if DE technologies are operated as CHP plants. However, recompense for long-lived generation and transmission assets remains one of the most contentious issues as energy markets are liberalized.

6.2 Valuing DE within Electricity Networks

The benefits or costs that DE technologies bring to an electricity network are a function of the level of penetration. If DE is limited to a relatively small percentage of overall capacity, it can be incorporated into the existing paradigm of remote generation of electricity with extensive transmission and distribution networks. Within this paradigm, DE can fulfill a number of functions in addition to providing electricity and heat to individual customers. These include grid reinforcement, standby power, meeting peak demands, remote sites, and hedging against fuel volatility.

The use of DE in areas where it is difficult or expensive to upgrade the distribution network can alleviate bottlenecks in the system and enable greater system capacity and reliability. For grid reinforcement benefits to occur, utilities should know, at a minimum, when, where, and how much electricity will be exported to the grid. The use of DG for network management is much improved if utilities also have some measure of control over DG resources. DE can also be used for peak shaving to meet demands for electricity at times of highest usage. This can be particularly cost-effective using low-capital-cost DE technologies, such as engines. Other DE technologies can be used in remote areas where it is not economically feasible to extend the distribution network. Historically, this has been the application where renewable DE technologies have been employed, despite their high capital costs. Finally, as DE technologies have a range of fuel sources, including natural gas, waste gases, oil, and renewable fuels, they can contribute in a portfolio of supply technologies to hedge against volatility in fuel costs.

As DE capacity increases in the electricity sector, a range of issues related to costs imposed on the wider electricity sector become more prominent. If, as in many developing countries, the national electricity grid is only in its infancy, then a system based on DE and alternate fuel delivery options is perfectly feasible and will offer distinct characteristics in terms of environmental, economic, and reliability performance. However, in all developed nations, a national electricity network that has been designed to move large amounts of power from remote generation sites to meet a diverse set of demand loads is already in place. Incorporating DE into this system necessitates consideration of interconnection, buy-back (or export) tariffs, net-metering, stranded investment costs, and impacts of changing demands.

A general theme in the relationship between DE and the electricity network is that DE users should compensate the utility for any costs imposed on the centralized network, but also be rewarded for any benefits the use of DE gives to the network. Experience has shown that coordination between users, utilities, regulators, and technology vendors is essential for optimal use and recompense of DE to be achieved. Specific regulatory structure will dictate whether utilities can be direct partners in DE projects. Utilities constrained by a cost plus revenue arrangement, where their income is driven by the amount of electricity that flows down their wires, are most likely to view DE as a threat to their core business.

Interconnection charges are imposed on the DE user to connect a unit to the electricity network for both emergency and backup power and to allow export of DE-generated electricity back to the grid. The general small size of DE schemes makes individual assessments impractical and cost-prohibitive and a number of countries have, or are developing, protocols for simple and equitable interconnection of DE. Buy-back tariffs (or prices for electricity export) are the price paid to the DE user for electricity sold back to the grid. The fair price for this electricity will depend on where the DE unit is located in the distribution network, how much electricity is exported, when it is exported (especially in reference to peak demand times), and whether electricity exports are guaranteed. Figure 4 illustrates the range in electricity buy-back tariffs in the Netherlands and United Kingdom. Larger volumes for longer periods of time are valued more highly. The overall difference between the two countries is a result of the level of cooperation with the incumbent utilities and their willingness to use DE as an integral part of their generation and distribution portfolios. An extension of buy-back tariffs is net-metering, where a DE user can reverse the flow of electricity from the grid to offset electricity purchases at other times. A flat rate for net-metering unfairly imposes costs on the electricity utility, whereas flexible net-metering tariffs dependent on time, place, and level of guarantee have much to recommend them.

Wider cost issues for electricity networks include stranded assets, where a capital-intensive generation and distribution plant is not utilized and hence does not recover its investment. This can also occur if

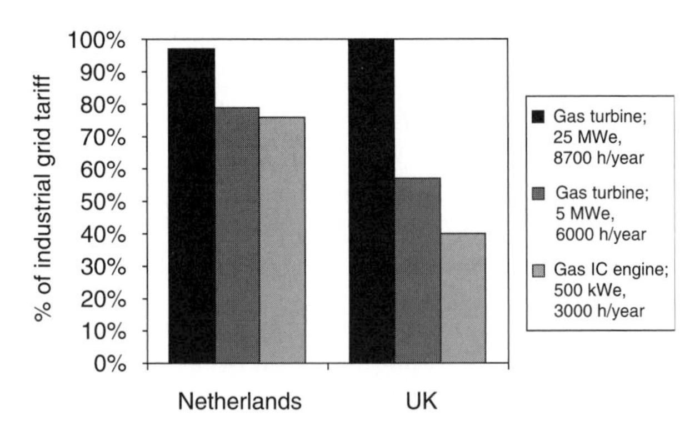

FIGURE 4 Electricity buy-back tariffs for DE technologies. From Cogen Europe (1995), The Barriers to CHP in Europe, SAVE Program of the European Commission, Brussels, Belgium.

capacity must be kept available to meet energy demands from DE unit failure. As electricity markets have been liberalized in many countries, recompense for stranded investments has been a dominant feature of the regulatory debate. Finally, as DE capacity grows, it will affect the demand profile of the area the electricity network serves. If the DE units primarily meet base-load demand, the remaining load will become more variable and hence more expensive to service with a conventional centralized generation portfolio.

7. ENVIRONMENTAL CHARACTERISTICS OF DE

DE technologies exhibit a range of emission characteristics. Carbon dioxide (CO_2) and sulfur dioxide (SO_2) emissions depend only on the carbon or sulfur content of the fuel and the efficiency of generation and supply. Local pollutants including nitrogen oxides (NO_x) also depend on the specifics of the combustion or generation process. The relative environmental performance of DE depends on what centralized electricity technology and fuel they are competing against and whether the DE technology is being used for electricity only or CHP applications. If it is a DE–CHP application, then the emissions from the heat boiler plant must also be factored in. Higher relative DE efficiencies are also aided by the avoidance of electricity distribution losses. Further issues in comparing environmental characteristics of technologies include what control technologies are assumed to be fitted and whether the comparison units are the average of the installed capacity base or only of new units.

Table V gives the latest controlled technology, efficiencies, and controlled emission rates of a range of DE technologies, as well as for a centralized coal steam turbine plant, combined cycle gas turbines (CCGT), and an on-site heat boiler plant. For calibration to alternative units, 1 lb/MWh is approximately equivalent to 0.45 g/kWh and approximately equivalent (for NO_x only) to 22 ppm.

Table V shows that some DE technologies (fuel cells, Stirling engines, PV, and wind turbines) have very low levels of global and local air pollutants. Diesel engines and coal steam turbines are the most polluting distributed and centralized technologies, respectively. Although CCGT is a much cleaner centralized technology, once emissions from heat boilers are factored in, even gas engines are equivalent in emission rates for CO_2, SO_2, NO_x, and particulate matter (PM_{10}). Engine emissions of carbon monoxide (CO) and unburned hydrocarbons (HC) remain a significant problem.

Comparing emission rates is only a first step in evaluating the environmental impact of DE technologies. It is also important to note that emissions will occur closer to population centers, and hence increase the health impacts of local air pollutants, and may also happen at specific times when environmental loading is at its peak. An example of this would be significant levels of diesel engine use in urbanized areas during hot summer afternoons when ozone (O_3) formation is at its peak. Thus, the specific health and environmental impacts of significant DE use require atmospheric

TABLE V

Emission Characteristics of DE and Power/Heat Technologies (lbs/MWh)[a]

	Control technology	CO_2	SO_2	NO_x	CO	PM_{10}	HC	Mercury (Hg)
Reciprocating gas engine	Ignition timing and SCR	1394	0.08	1.11	4.01	0.03	1.20	0
Reciprocating diesel engine	Ignition timing and SCR	1639	2.95	4.75	6.24	0.80	3.68	0
Gas turbine	Staged combustion	1406	0.08	0.67	0.68	0.10	0.98	0
Microturbine	SCR	1617	0.09	0.45	1.05	0.09	0.31	0
Fuel cell (phosphoric acid fuel cell)	None	1064	0.06	0.03	0	0	0	0
PV	None	0	0	0	0	0	0	0
Stirling engine	None	1617	0.09	0.1	0	0	0	0
Wind	None	0	0	0	0	0	0	0
Combined-cycle gas turbine	Staged combustion	980	0.06	0.47	0.17	0.10	0.12	0
Coal steam turbine	Low NO_x burners, FGD and SCR	2345	13.71	4.13	0.17	0.31	0.12	0.0038
Gas heat boiler	Low NO_x burners	449	0.03	0.49	1.05	0.09	0.11	0

[a] Divide by 2.2 to convert to g/kWh.
SCR, selective catalytic reduction; FGD, flue gas desulfurization.

modeling. As many DE technologies have thus far been "under the radar" in regulatory efforts for air pollution, such detailed analysis will be paramount.

Finally, it should be noted that there are non-emission environmental issues that ensure that all DE technologies have some environmental concerns to be addressed. Noise pollution, particularly near workplaces and residences, is a concern (see Table II for uncontrolled noise levels of DE technologies). Even renewable DE technologies with no noise or air pollution consequences need to address concerns over hazardous materials during manufacture and disposal (PV and batteries) or visual impacts (wind turbines).

8. OTHER CHARACTERISTICS OF DE

8.1 Reliability

Reliability considerations for DE can be divided into plant and system impacts. At the plant level, high levels of availability of DE units are achieved through remote monitoring and real-time control. As there can be a large number of DE units, reliability can be ensured for scheduled maintenance and unscheduled outages by transferring the electricity load to the remaining DE units. Building or process management systems can turn off nonessential loads if required. Energy storage devices can also be used to maintain supply for short periods. Finally, the dual-fuel capabilities of many DE technologies or the use of different DE units at the same location can ensure supply in the case of fuel disruptions.

In comparing DE reliability to a centralized electricity network, it should be noted that the existing electricity grid is extremely reliable, operating in excess of 99.9% availability. The majority of outage events in the centralized network are due to accidents and climatic events in the transmission and especially distribution system. DE offers a route to bypass these outages. In addition, in countries such as India, where widespread theft of electric power from the distribution network, and hence reduced reliability, is commonplace, DE can offer an alternative.

The comparison, and possible coordination, between DE and the electricity network depends on which generation source is the back-up. Traditionally, the electricity grid has been employed as emergency or stand-by generation for DE units. This imposes costs on the electricity grid (see Section 6.2) as the utility must maintain spare generation and distribution capacity to meet a potential additional load from a DE outage event. However, this disadvantage is avoided if a cluster of DE units can provide emergency power without the grid. Taking this paradigm one step further, a DE cluster can act as a source of emergency or stand-by generation to the grid, especially in times of peak demand when generation costs or distribution bottleneck issues are of utmost concern.

A final reliability issue is the comparison between the electricity network and the fuel distribution (primarily natural gas) that DE units would depend on. This is discussed below in the context of robustness.

8.2 Robustness

The robustness of an energy generation and distribution network has received considerable attention for its performance under conditions of conflict, whether in terms of war or terrorist events. A system based on DE would do little to alleviate dependence on oil imports, which is primarily related to transportation. On an individual generation unit basis, the small size and low profile of DE units make them a less important target. Although DE can be powered by a range of renewable, waste, and fossil fuels, the reliance on a natural gas infrastructure has been debated and compared to the system robustness of the electricity network. Table VI details some of the advantages and disadvantages of a natural gas DE network.

8.3 Power Quality

Evolving demands for high-quality electricity in a range of sensitive industrial processes and information technologies has refocused attention on the need for extremely high-quality power. DE technologies can help respond to this demand in a number of ways. The use of DE technologies can avoid movement of electricity where the majority of power quality problems, especially voltage sags, occur. Energy storage devices are a DE resource that is often used to provide uninterrupted power supply either as a support to the power grid during periods of very short term instability or as a bridge to allow stand-by DE technologies to begin operation and ramp up to meet the required load.

TABLE VI

Potential Robustness Advantages and Disadvantages of DE Systems

Features of natural gas DE	Conflict context advantages	Conflict context disadvantages
Increased number and smaller size of DE generators	When one generator is damaged, a much smaller proportion of the generating capacity is unavailable	Harder to physically protect individual generation sites
Greater use of natural gas	May replace nuclear facilities	More prone to risks of foreign supply interruptions (pipelines and LNG)
Decreased reliance on electricity transmission and distribution	The electricity T&D system is harder to protect than generators; having generation close to the load reduces the reliance on the vulnerable transmission system	Unless DE units are arranged in clusters to back up one another, may still have to deal with electricity grid outage effects
Increased reliance on natural gas transmission and distribution	Underground natural gas T&D systems (which would be needed anyway for on-site heat-only plants) are generally underground and therefore better protected than electrical T&D lines	Long-distance over-ground natural gas pipelines are very difficult to protect and are vulnerable to attack (with explosives or high-powered rifles)
T&D real-time operational advantages	Gas pipelines do not have the same strict real-time operational problems as electric power grids; in electric power grids, a disturbance to one part of the grid can result in cascading failures that knock out power to a much wider area than the original disturbance	
Fuel substitutability and storage	A natural gas DE system can have greater robustness through dual fuel technologies and local storage facilities for gas and/or other fuel suitable for use in ICE DE units	A centralized electricity network can also build in a diverse supply portfolio

Note: T&D, transmission and distribution; LNG, liquified natural gas.

As there are a range of customer types, each requiring, and willing to pay for, a range of power quality levels, those users requiring extremely high quality power (e.g., information technology firms) have been investigating the concept of microgrids, where a range of DE technologies, along with energy storage options, are geared to provide very high quality power in a designated geographical area. DE microgrids could be based on DC power if required for sensitive electronic equipment.

8.4 Modularity

A final characteristic of DE technologies is that they represent modular or increment investment and capacity additions, especially in comparison to large centralized power plants. This feature has been even more advantageous following the liberalization of many nations' electricity industries and the resultant increase in electricity price uncertainty. The ending of the monopoly systems where investment revenues were guaranteed to be recovered has placed a premium on projects with a shorter, less expensive construction period. Many factory-made and – assembled DE units come packaged with heat recovery, acoustic enclosure, and control interface and can be installed at a site and connected to the electricity and heating systems in 1 or 2 days.

In addition to improving investment under price uncertainty, modular capacity extensions can more easily follow increments in demand growth, alleviating shortages in generation capacity. Adding further value, the geographical flexibility of modular DE capacity allows it to alleviate bottlenecks in the distribution capacity. Finally, mobile DE units can be quickly moved to meet peak demand periods or moved on a seasonal basis to meet climate-driven high-demand periods.

9. FUTURE ROLE OF DE

The traditional policy driving forces for DE, and indeed for all energy supply and distribution options, have been the lowest economic cost and security of supply. Increasingly, motives for evolving policies will also be concerns over global and local environmental issues, reliability, robustness, and quality of electricity supply, and integration with energy infrastructures, notably electricity, natural gas, and any future hydrogen infrastructure.

Along with continuing technical developments in DE technologies, cost reductions will be largely delivered by economies of scale in production, geographical economies of scale, and the processes of learning by doing and learning by using as DE penetrates the market. Regulatory developments, especially facilitating flexible ownership and operational arrangements, will be necessary if these cost reductions are to be realized.

Any new regulatory design will have to accommodate and value the impacts of a widespread penetration of DE technologies on existing infrastructures and especially electricity generation and distribution utilities. Utilities need to be fairly compensated for costs incurred for interconnection, ensuring emergency supply, altered demand profiles, and possible stranding of investments in generation and distribution equipment. These costs need to be transparent and standardized, as the introduction of DE can be incorporated to reduce peak requirements, defray distributional capacity expansions, and meet the range of daily and seasonal demands at lowest cost. Regulatory decisions over the ownership, technical protocols, market structure, and liability aspects of the electricity industry will all affect whether and how DE is incorporated into the electricity paradigm.

A diverse range of DE technologies, including renewable DE and multifuel DE units such as microturbines and Stirling engines, can help contribute to the diversity and hence the security of energy supplies. However, if the great majority of DE units are fired on natural gas, this could exacerbate price volatility and upward price pressure on natural gas supplies, particularly in North America.

Renewable DE and CHP DE technologies have the potential to significantly reduce emissions of CO_2 and thus may be a major tool in any concerted drive to limit and reduce greenhouse gas emissions. Although most DE technologies emit very low levels of SO_2, only some (fuel cells, microturbines, Stirling engines, PV) have very low levels of a variety of local air pollutants. Given the distributional location of DE near population centers, all local air pollution impacts need to be fully assessed and further controlled if necessary. In addition, any doubts of affected local populations need to be understood and addressed.

Evolving demands for high reliability, high robustness, and high quality in liberalized electricity markets will likely result in customers specifying criteria of supply. For those customers who have very strict requirements for reliable and high-quality power, DE technologies can either supplement the grid or be used in microgrids to meet specialized demands. The use of DE can hedge against both supply and price uncertainties, particularly in view of the modularity of capital investments required for DE.

Significant expansion of DE capacity may impose greater strain, increased price levels, and greater price volatility on the natural gas network. It may also change seasonal demand profiles as more summer electricity is generated and may require rethinking of natural gas storage requirement and locations. Finally, DE may or may not be integral or compatible with the growing use of hydrogen. If hydrogen is created at central facilities (for example, using fossil fuels and carbon sequestration), the energy distribution carrier could be electricity, taking advantage of economies of scale in plants, or hydrogen, taking advantage of economies of scale in DE unit production and higher efficiencies with CHP DE units. If the transportation sector moves to a hydrogen system with a distribution pipeline to fueling stations, this is likely to support hydrogen-fueled stationary DE applications.

Finally, DE technologies and their associated infrastructure impacts and ownership arrangements potentially represent a very different paradigm than the norm of centralized electricity generation and distribution with on-site heat provision. These two paradigms have advantages and disadvantages, but in countries that already have a developed energy infrastructure, any widespread move toward DE will entail significant transitional concerns over a period of decades due to the long life of electricity-related equipment.

SEE ALSO THE FOLLOWING ARTICLES

Electricity, Environmental Impacts of • Electricity Use, History of • Electric Power Reform: Social and Environmental Issues

Further Reading

Ackermann, T., Andersson, G., and Soder, L. (2001). Distributed generation: A definition. *Electric Power Systems Res.* 57, 195–204.

Borbely, A., and Kreider, J. F. (eds.). (2001). "Distributed Generation: The Power Paradigm for the New Millennium." CRC Press, Boca Raton, FL.

Cowart, R. (2001). "Distributed Resources and Electric System Reliability." The Regulatory Assistance Project. Gardiner, ME.

Dunn, S. (2000). "Micropower: The Next Electrical Era," WorldWatch Institute Paper 51. WorldWatch Institute, Washington, DC.

Feinstein, C., Orans, R., and Chapel, S. (1997). The distributed utility: A new electric utility planning and pricing paradigm. *Annu. Rev. Energy Environ.* **22,** 155–185.

Gas Research Institute. (1999). "The Role of Distributed Generation in Competitive Energy Markets." Gas Research Institute, Chicago, IL.

Laurie, R. (2001). Distributed generation: Reaching the market just in time. *Electric. J.,* 87–94.

Meyers, E., and Hu, M. G. (2001). Clean distributed generation: Policy options to promote clean air and reliability. *Electric. J.,* 89–98.

Morgan, M. G., and Zerriffi, H. (2002). The regulatory environment for small independent micro-grid companies. *Electric. J.,* 52–57.

Patterson, W. (2000). "Transforming Electricity." Earthscan, London.

Smeers, Y., and Yatchew, A. (eds.). (1998). Distributed resources: Toward a new paradigm of the electricity business. *Energy J.,* Special Issue.

Strachan, N., and Farrell, A. (2002). "Emissions from Distributed Generation," Carnegie Mellon Electricity Industry Center Working Paper 02–04. Carnegie Mellon Electricity Industry Center, Pittsburgh, PA.

Strachan, N., and Dowlatabadi, H. (2002). Distributed generation and distribution utilities. *Energy Policy* **30,** 649–661.

Willis, H. L., and Scott, W. (2000). "Distributed Power Generation: Planning and Evaluation." Dekker, New York.

Zerriffi, H., Dowlatabadi, H., and Strachan, N. (2002). Electricity and conflict: Advantages of a distributed system. *Electric. J.* **15,** 55–65.

District Heating and Cooling

SVEN WERNER
Chalmers University of Technology
Gothenburg, Sweden

1. Five Strategic Resources
2. Market Penetration
3. History of District Heating
4. Technical Design and Construction
5. Customer Relations
6. Energy Supply
7. Environmental Impact
8. District Cooling
9. Conclusions

Glossary

cogeneration Method of simultaneous generation of electricity and heat; also known as combined heat and power (CHP).

community heating Synonym for district heating, common in Great Britain.

customer substation Installation in customer building transferring heat from the heat network to the internal heat distribution systems in the building.

district cooling Method of fulfilling many cooling demands by a seldom citywide cooling distribution network, chilled from one or many large chillers or transferring facilities.

district energy Collective term for district heating and cooling.

district heating Method of fulfilling many city heat demands by an often citywide heat distribution network, which receives heat from one or many large heat-generation or transmission facilities.

trigeneration Method of simultaneous generation of electricity, heat, and cold.

The fundamental idea of district heating is to use local fuel or heat resources that would otherwise be wasted to satisfy local customer heat demands by using a heat distribution network of pipes as a local marketplace. This idea contains the three obligatory elements of a competitive district heating system: the suitable cheap heat source, the market heat demands, and the pipes as a connection between source and demands. These three elements must all be local in order to obtain short pipes for minimizing the capital investment in the distribution network. Suitable heat demands are space heating and preparation of domestic hot water for residential, public, and commercial buildings. Low-temperature industrial heat demands are also suitable.

1. *FIVE STRATEGIC RESOURCES*

The five suitable strategic local energy resources are useful waste heat from thermal power stations (cogeneration); heat obtained from refuse incineration; useful waste heat from industrial processes; natural geothermal heat sources; and fuels difficult to manage, such as wood waste, peat, straw, or olive stones. These heat sources must be cheap in order to compensate for capital investments in the distribution network and complementing heat-generation plants for peak and backup heat demands. The latter is needed in order to meet customer heat service demands at extremely low outdoor temperatures and when the regular heat source is temporarily unavailable.

District heating is an energy service provided for immediate use directly by the customer and was commercially introduced in the 19th century as a very early example of outsourcing.

2. *MARKET PENETRATION*

Citywide district heating systems exist in Helsinki, Stockholm, Copenhagen, Berlin, Munich, Hamburg, Paris, Prague, Moscow, Kiev, Warsaw, and other large cities. Many systems supply a downtown district (such as in New York, San Francisco,

Minneapolis, St. Paul, Seattle, Philadelphia, and other cities) or a university, military base, hospital complex, or an industrial area.

Total annual heat turnover is approximately 11 EJ in several thousand district heating systems operating throughout the world. The amount of heat delivered corresponds to 3.5% of the total global final energy consumption (1999). The market penetration for district heating in various countries is presented in Table I which is based on international energy statistics. However, many countries have undeveloped or no routines for gathering statistical information from district heating systems. This is valid for the United States, Great Britain, France, China, and some other countries. An independent analysis estimated the total annual heat deliveries from all district heating systems in the United States to be more than 1000 PJ. Hence, the real-world market penetration is probably higher than that presented in Table I.

Fulfilling heat demands for space heating and domestic hot water preparation for residential, public, and commercial sectors constitutes the majority of deliveries. Among the Organization for Economic Cooperation and Development (OECD) countries, district heating has a strong, almost dominating, market position in Denmark, Finland, Sweden, Poland, and the Czech Republic. Among non-OECD countries, district heating has a strong market position in Russia, Belarus, Romania, and the three Baltic states.

3. HISTORY OF DISTRICT HEATING

Birdsill Holly, an inventor and hydraulic engineer, is often credited with being the first to use district heating on a successful commercial basis. As a result of an experiment in 1876 involving a loop of steam pipes buried in his garden, Holly developed a steam supply system in October 1877. Several district heating systems were started in North American cities in the 1880s. The fuel source was steam coal. In New York, the Manhattan steam system went into operation in 1882. It exists today as the steam division of Consolidated Edison, and it delivered 29 PJ of heat during 2000.

The oldest district heating system still in operation is located in Chaudes-Aigues, France. It is based on a geothermal heat source with a temperature of 82°C and was in operation as early as the 14th century. Old municipal documents indicate that two citizens did not pay their heat fees properly in 1332. The hot water was partly distributed using drilled tree trunks as distribution pipes.

In Europe, an early district heating system was built in Dresden in 1900, although it was not a commercial project. The main purpose was to reduce the fire risk in 11 royal and public buildings containing invaluable art treasures. A more commercial project was initiated by Fernheizwerk Hamburg Gmbh for the city of Hamburg in 1921. According to Abraham Margolis, the company chief engineer, the driving force for the project was the high cost of fuel after World War I. The Hamburg system was followed by systems in Kiel in 1922, Leipzig in 1925, and Berlin in 1927. Outside Germany, district heating systems were started in Copenhagen in 1925, Paris in 1930, Utrecht in 1927, Zürich in 1933, and Stockholm and Helsinki in 1953. Reykjavik, Iceland, started a geothermal district heating system in 1930, which today supplies almost all the 160,000 inhabitants with heat for space heating and domestic hot water (10 PJ during 2000).

In Russia, a general utilization of cogeneration and district heating was outlined in the electrification plan in 1920 in order to reduce future fuel demand. The first heat was delivered in St. Petersburg in 1924. Teploset Mosenergo was established in 1931 to manage heat distribution in Moscow, although the heat deliveries had begun in 1928. This department of Mosenergo delivered 268 PJ during 2000 and an additional 70–80 PJ was supplied by Mosteploenergo, another local distributor. Together, these companies constitute the most extensive district heating system in the world. The second largest is the St. Petersburg system.

4. TECHNICAL DESIGN AND CONSTRUCTION

Heat is transferred from the network to the building heating systems by customer substations located in the connected buildings. A building has at least two internal distribution systems that must be heated—one system for supplying heat to the radiators and one system for distribution of domestic hot water. Sometimes, a separate system also provides heat for heating the supply air in the mechanical ventilation system.

Each internal system is heated and regulated separately. The heat is often transferred by the use of heat exchangers, in which case the connection is indirect. There are also direct variants with only valves and elevator pumps. If domestic hot water is prepared by mixing district heating water and

TABLE I

Heat Flows through District Heating Systems for Various Countries and Regions in 1999[a]

Region or country	Total heat supply (PJ)	Total deliveries (PJ)	Industrial customers (PJ)	Other customers (PJ)	% of all final consumption	% of all industrial demands	% of all other demands
World	11,473	10,852	3782	6079	3.5	4.2	5.5
OECD total	2296	2120	535	1452	1.3	1.2	3.0
United States	406	345	226	83	0	2	0
Canada	33	33	33	0	0	1	0
Korea	166	161	0	161	3	0	10
Japan	25	22	0	21	0	0	0
Switzerland	16	14	6	8	2	3	2
Norway	8	6	1	5	1	0	2
Poland	371	371	57	262	12	7	21
Czech Republic	153	146	39	85	12	9	22
Hungary	71	71	19	43	9	11	11
Slovak Republic	31	29	1	23	4	0	11
European Union	1007	913	155	751	2	1	5
Germany	371	343	44	292	3	2	7
Sweden	168	158	13	145	11	2	26
Finland	126	116	27	89	11	5	26
Denmark	122	98	6	91	15	5	30
The Netherlands	105	90	39	51	4	5	6
Austria	48	43	5	38	4	2	9
France	31	31	0	31	0	0	1
Other countries in European Union	36	34	21	13	0	0	0
Other countries in OECD	9	8	0	8	0	0	0
Non-OECD total	9178	8732	3247	4627	5.9	7.2	7.4
Russia	6332	6158	2098	3607	33	36	47
China	1344	1327	719	249	3	5	2
Ukraine	576	404	258	141	11	16	8
Belarus	300	281	118	159	36	40	45
Romania	216	187	20	146	17	5	34
Uzbekistan	105	105	0	105	7	0	11
Former Yugoslavia	67	64	8	53	7	3	13
Bulgaria	55	47	11	29	10	6	22
Lithuania	52	40	5	32	19	11	37
Latvia	36	30	1	27	20	4	36
Estonia	29	24	2	21	22	7	39
Other countries in non-OECD	66	65	8	57	0	0	0

[a] According to the energy balances for OECD and non-OECD countries published by the International Energy Agency. Differences between supply and deliveries constitute distribution losses. Differences between total deliveries and sum of deliveries to customers constitute own use. All other demands do not include the transportation sector.

domestic cold water, the open connection method is used, whereas when a heat exchanger is employed, the closed connection method is used. The open connection method occurs is used in Russia, resulting in high feed water demands in the distribution network.

The most typical part of a district heating system is the citywide distribution network of pipes buried in the ground under streets, pavement, and park lawns or installed in building basements. Water is normally used as the heat carrier in heat distribution networks. Steam is completely or partly used in systems that were installed before approximately 1940 and for high-temperature industrial heat demands.

The water temperature in the forward pipes varies between 70 and 150°C, with an annual average of 80–90°C. The higher temperature is used at extreme low outdoor temperature, whereas the lower temperature is used during the summer to enable domestic hot water preparation. The temperature of the return water varies between 35 and 70°C, with an annual average of 45–60°C. However, return temperatures of 30–35°C may be obtained in the future. The current high return temperatures depend on malfunctions in substations, short-circuited radiator systems in customer buildings, and short-circuits between forward and return pipes in the distribution networks. The lower the return temperature, the more heat can be transported in the network.

The pipes are constructed so that they accommodate thermal expansion and avoid outer corrosion. Construction methods for pipes buried in the ground have varied throughout the years. In the early days of district heating, double steel pipes were insulated with mineral wool and laid in a common concrete square box duct. This early generation of distribution pipes were expensive to build but were reliable if the ducts were well ventilated and well drained. Today, the most common method is to use prefabricated steel pipes with polyethylene casing and insulated with polyurethane foam. This method has the advantages of low distribution cost, low heat losses, and high reliability. During the past 30 years, plastic pipes have been developed as heat carrier pipes. With respect to limitations in pressure and possible long-term temperatures, the new generation of district heating pipes are mostly used in secondary distribution networks separated from the primary distribution networks by heat exchangers.

The heat loss in the distribution network depends on the thermal resistance in the pipe insulation, the pipe size, the supply and return temperatures, and the linear heat density. The latter is the heat sold annually divided by the route length of double pipes in the network. Typical values for linear heat densities are 15–25 GJ/m for whole networks, more than 40 GJ/m in concentrated downtown and commercial areas, and less than 5 GJ/m in blocks with single-family houses. Typical heat loss in whole networks is 5–10% of heat generated in plants, but up to 20–30% can occur in areas with low linear heat densities. Also, the distribution cost depends on the linear heat density. The cost is low beyond 20–30 GJ/m but increases very quickly below 6–8 GJ/m.

The district heating system is operated using control equipment at four independent levels. Two of these are located in the buildings connected and two are managed by the district heating operator. The first level is the heat demand control by thermostatic valves at the radiators and mixing valves for domestic hot water. The second level is in the customer substations. The heat transfer is controlled by valves that adjust the flow of district heating water. One valve is used for each connected customer system. The third level is the control of the pressure difference between the supply and return pipes in the network, which is accomplished by adjusting the speed of the distribution pumps. This control level allows for all customer substations to receive heat since this pressure difference is the driving force for the flow to circulate through the customer substations. The final level is the control of the supply temperature by adjusting the capacity in the heat-generation facilities.

These levels of control result in an efficient demand-oriented district heating system. The district heating operator can never deliver more heat than the customers request. However, the first three levels are often not available in Russian and Eastern European systems. This situation creates a production-oriented system in which it is difficult to properly allocating the flow in the network, resulting in both low and high indoor temperatures in the buildings connected. When these systems are rehabilitated, the first step is to install at least control at the second and third levels.

5. CUSTOMER RELATIONS

The customers usually pay for the heat received according to a heat rate, which normally consists of one fixed portion related to the capacity needed and one variable portion related to the amount of heat bought. Sometimes, a water volume portion is also used in order to promote low return temperatures from the customer substations. The capacity portion is based on either the maximum heat load or the maximum water flow capacity. The amount of heat sold to the customer is recorded by integrating the product of water flow and the temperature difference between the supply and return pipes. A heat meter in each customer substation measures the district heating water flow, the supply temperature, and the return temperature. These three measurements are integrated with media constants to derive an accumulating heat volume in gigajoules or megawatt hours by the heat meter.

The point of delivery varies from system to system. If the customer owns the substation, the point of delivery is before the substation; otherwise, it is after the substation. Sometimes the customer pays a connection fee.

The pricing principle for the heat rate varies. Municipal district heating systems generally have lower prices, reflecting lower returns of capital investments. Privately owned systems often price the district heat according to the available alternative for the customer, usually the price of light fuel oil or natural gas with a discount in order to keep the customer. In some countries, the price levels in district heating systems are supervised or approved by national or regional regulatory authorities.

Typical prices for district heat deliveries were $6–13/GJ in OECD countries in 1999 and 2000. An exception was the geothermal system in Reykjavik, which sold heat for only $4/GJ. Typical prices in Eastern Europe were $3–7/GJ, except in Russia, where world market energy prices are not allowed due to a low ability to pay market prices. Heat was sold from the Moscow system for approximately $1/GJ. All these prices are lower than or equal to those from local boilers using fossil fuels with different national energy taxes, such as carbon dioxide taxes.

The lower prices in countries in Eastern Europe and the former USSR just cover the operating costs. The returns of capital investments made in the former planned economies are currently kept low by the national regulatory authorities since the average ability to pay for heating is low in the population.

Requests for deregulation of district heating systems, allowing third-party access, have been put forward from customers and competitors, for example, in New York. The requests are for the deregulation processes of telecommunication, aviation, railway, gas, and electricity markets. These markets are large, international, and mature with many market actors. The goal of deregulation of these markets is an effective price determination with many sellers and buyers in each market. However, district heating systems are normally small local markets, which are often immature, with low market penetration. One operator can easily supply a town of 150,000 inhabitants with heat from one site; thus, more operators are not needed in many district heating systems. Hence, deregulation of district heating systems cannot easily be implemented.

6. ENERGY SUPPLY

The base load heat-generation capacity in a district heating system must have a strong correlation with the five strategic resources discussed previously: cogeneration, refuse incineration, industrial waste heat, geothermal heat, and fuels difficult to manage. Internationally, the strongest driving force for district heating is the simultaneous generation of electricity and heat in cogeneration plants. Base load capacities are associated with low operating costs and high capital investment costs.

Conventional fossil fuels dominate the fuel supply for cogeneration, but the use of biomass is increasing in some countries. One example is the newly built biomass cogeneration plant in Eskilstuna, Sweden, which meets the majority of heat demand of a city with 65,000 inhabitants. Another example is the new biomass cogeneration plant for the St. Paul system. In general, cogeneration plants use more carbon lean fuels than conventional thermal power stations since natural gas, renewables, and waste are more common fuels than coal and oil. It is also possible to recover heat from nuclear power stations, but this is employed to a very limited extent due to long distances between nuclear reactors and heat demands in cities, loss of some electricity generation, fear of strong dependence on one large heat source, and lack of trust in nuclear technology in some countries.

Waste incineration is an acceptable treatment method compared to landfill for the burnable industrial and municipal waste stream. This waste remains after separation of streams for hazardous waste, recycling, and biological waste treatment. Normally, this waste stream from a city comprises only approximately 5–15% of the design capacity in a mature district heating system with a high local market penetration. However, 10% of the design capacity corresponds to approximately 30% of the annual heat generation. Within the European Union, up to 50% of municipal waste is allocated for incineration. Denmark, Switzerland, and Sweden incinerate the highest percentages of waste.

Many industrial processes release heat at a temperature higher than 100°C either in flue gases or in cooling water. This heat can be directed to a district heating system by using heat-recovery boilers or heat exchangers. In Gothenburg, the district heating system receives useful waste heat from two oil refineries, comprising 30% of the annual heat demand in the system. Revenues from the heat sale cover a major portion of staffing, operating, and maintenance costs for the refineries. This heat recovery gives them a competitive advantage over other oil refineries in Europe.

Geothermal heat is a natural energy source for heating in the same manner that wind, hydro, and solar power are natural renewable energy sources for power generation. However, this resource is not available everywhere since it is restricted to areas with the appropriate geophysical conditions. There are geothermal district heating systems in Iceland, the United States, France, China, Japan, Turkey, Poland, Hungary, and some other countries. However, the total heat delivery worldwide is low, estimated to be approximately 30 PJ/year, where one-third is supplied in the Reykjavik system.

Biomass is an example of a fuel that is normally difficult to manage for conventional heating. Regarding transport, fuel feeding into boilers, and environmental impact, these types of fuel are easier to handle in one central boiler than in several small boilers. Biomass fuels can be obtained from agricultural sources or from the forest industry. Wood waste can be delivered as raw chips or refined as small pellets or powder from forest companies or sawmills. In the early 1990s, the Cree Nation of Ouje-Bougoumou determined that a biomass district heating system would be an enormously positive tool in contributing to the development of the community's future financial base. This aboriginal community, located 1000 km north of Montreal, Canada, uses wood waste from a nearby sawmill for their two biomass boilers, which are connected to the district heating system. Importing fuel oil to this village was considered a less favorable option.

Peak and backup capacity plants complement the base load capacity plant in a district heating system. These plants are needed to meet customer heat service demands at extremely low outdoor temperatures and when the regular heat source is temporarily unavailable. These plants are normally associated with high operating costs and low capital investment costs.

The optimal heat-generation cost is obtained by choosing the base load capacity so that its marginal capacity will have the same annual total cost as peak load capacity for the same operating hours. The typical optimal base load capacity is approximately half the nominal system design capacity based on the local design outdoor temperature. The base half of the capacity represents approximately 90% of the volume of heat generated.

The frequency distribution for the outdoor temperatures determine the limit of capacity utilization in a district heating system. The annual average capacity factor is 25–50% for extreme inland and maritime climate locations. The capacity factor is defined as the actual heat generation divided by the heat generation that is possible. This situation occurs since less heating is required during autumn and spring compared to winter, and no heating is needed during summer.

7. ENVIRONMENTAL IMPACT

High contents of dust, sulfur dioxide (SO_2), and nitrogen oxides (NO_x) in urban areas were among the first pollution problems to be identified. District heating improved the urban air quality in many towns since local heat boilers were replaced and high chimneys for heat-generation plants were required.

Acidification from SO_2 and NO_x is a major problem in many regions. District heating systems provide a centralized and cheap way to eliminate these emissions by separation, use of fuels with low sulfur content, have higher conversion efficiencies, and have higher combustion quality, resulting in lower NO_x emissions.

Global warming due to the greenhouse gases is a problem more difficult to solve than the problems of urban air quality and acidification. In this respect, district heating can be considered the fifth possibility for carbon dioxide emission reduction, after deposition, lowering final energy demands, higher conversion efficiencies, and fuel substitution. District heating uses mainly waste heat from other processes, replacing the use of fossil fuels for heating buildings.

It is estimated that in 1998, all cogeneration plants and district heating systems throughout the world reduced carbon dioxide emissions by 900 Mtons. This corresponds to approximately 4% of all carbon dioxide emitted from fuel combustion. Hence, district heating systems can play an important role in the mitigation of carbon dioxide emissions.

8. DISTRICT COOLING

District cooling systems distribute chilled water to supply air-conditioning or process cooling. Cities with major downtown or commercial districts with district cooling systems include Stockholm, Hamburg, Paris, Tokyo, Chicago, Minneapolis, St. Paul, New Orleans, and Houston. The first major district cooling system was employed in Hartford, Connecticut, in 1962. Other systems were built in the 1960s and 1970s. District cooling experienced a renaissance during the 1990s, when many countries demanded the termination of the use of CFC due to ozone depletion. District cooling systems are much smaller than district heating systems. Annual cold generation is approximately 1 or 2 PJ in the largest district cooling systems.

The distribution network is similar to that of a district heating system, with one supply pipe and one return pipe. Cold water is circulated instead of warm water. The supply temperature is normally 6 or 7°C, but an ice mixture of 0°C is sometimes used. Customers heat the chilled water in the supply pipe by cooling air supplied to a building or a industrial process. The typical temperature in the return pipe is 12–17°C. The water is again chilled when it circulates through the cold-generation plant. District cooling systems are never used citywide. The low temperature difference makes district cooling distribution expensive, so district cooling networks can only be built in areas with high cooling demand densities, such as downtowns and commercial areas.

Cold generation can be accomplished by central compression or absorption chillers. Free cooling with deep, cold lake or seawater can also be used, as is the case in Toronto and Stockholm. However, this is only possible in areas with cold winters, which annually create cold reservoirs for summer use. Since the diurnal variation of the cooling demand is large at peak load, a significant portion of the peak demand can be met by cold energy storage capacity with a cold water storage tank or an ice storage facility.

Two major associations exist between district heating and district cooling systems. The evaporator of a heat pump can be used for cold generation in a district cooling system and the condenser can simultaneously generate heat in a district heating system. One part of bought electricity will then create two parts of sellable cold and three parts of sellable heat, giving a higher return of the heat pump investment. The other association is the possible use of low-cost district heat for operation of local absorption chillers in connection with customer cooling demands. In this way, citywide deliveries of cooling can be accomplished without installing a citywide district cooling system. This variant is sometimes called warm district cooling, in contrast to cold district cooling, which refers to ordinary district cooling systems. District heating systems using this warm variant are employed in Gothenburg, Mannheim, New York, and Seoul.

These two associations can be integrated into a trigeneration plant, in which power generation is integrated with the generation of both cold and heat. First, a cogeneration unit can generate electricity and heat. These products are then used for feeding a compression chiller or a absorption chiller.

There is an operation optimization benefit for district cooling. Cold generation can be optimized by using different generation technologies to meet various cooling demands. The base load demand can be met by an absorption chiller if low-cost heat is available. Next-generation units in merit order include a compression chiller with heat recovery and a similar chiller without heat recovery. Extreme peak load can be met by a cold storage.

Major advantages of district cooling systems are reduced grid power demand by supplying cooling energy through the district cooling system rather than through the local power grid, shifting power demand to off-peak periods through cold energy storage, and the use of free cooling. Advantages for customers with district cooling are the requirement for less space for cooling equipment inside the building, less operation and maintenance costs, and less electricity cost. The subscribed cooling capacity can also be adjusted to the actual measured customer cooling demand. The customer does not have to buy an expensive oversized chiller.

9. CONCLUSIONS

District heating significantly increases the overall energy system efficiency since the heat supply is coordinated with generation of electricity, refuse incineration, or industrial processes. It is also possible to use geothermal heat sources or biomass fuels such as wood waste. Demand for commercial fossil fuels for heating is reduced, resulting in lower carbon dioxide emissions.

District heating also benefits from economy of scale. This is especially valid with regard to distribution since the cost of pipes is proportional to pipe size, whereas transmission capacity increases with the square of the pipe size. Also, the investment cost for heat-generation capacity decreases with size. The variation in heat demand is also lower, especially with respect to the demand for domestic hot water preparation. However, these advantages are not sufficient to motivate centralized use of conventional fossil fuels in district heating systems. At least one of the five strategic resources must be used to develop a competitive district heating system. When using more than one of the strategic resources, an internal competition is created within the system, which will increase the competitiveness of the system.

A district heating system becomes more competitive when the advantages of higher overall system efficiency and lower environmental impact are priced by the heat market. Hence, district heating is highly competitive when fuel prices are high and when low environmental impact is appreciated. Many countries use fiscal taxes for fossil fuels and are also introducing carbon dioxide taxes for fuels. International, regional, and national carbon dioxide trading schemes have also been discussed. All these measures increase the long-term competitiveness of district heating systems.

Although district heating has many advantages, the method is still unknown in many communities, whereas it is taken for granted in other communities. Since district heating reduces the operating costs of heating buildings due to capital investment in central heat-generation facilities and distribution pipes, the local view of long-term investments in urban infrastructure is very important. Historically, district heating systems have been built in communities in which district heating has been considered part of the urban infrastructure, along with roads, bridges, water supply, and electricity distribution. The current strong market penetration of district heating in some countries in Europe can be explained by this view. The municipal electrical and technical departments became the district heating entrepreneurs in Finland, Sweden, Denmark, Germany, and Austria since by tradition they were responsible for the urban infrastructure.

SEE ALSO THE FOLLOWING ARTICLES

Biomass for Renewable Energy and Fuels • Cogeneration • Commercial Sector and Energy Use • Geothermal Direct Use • Urbanization and Energy • Waste-to-Energy Technology

Further Reading

Euroheat & Power (1999). "District Cooling Handbook," 2nd ed. Euroheat & Power, Brussels.

Euroheat & Power (2001). "District Heat in Europe, Country by Country—2001 Survey." Euroheat & Power, Brussels.

European Institute of Environmental Energy (1997). "District Heating Handbook." European Institute of Environmental Energy, Herning, Denmark. [Available in Chinese, English, Estonian, German, Latvian, Lithuanian, Polish, and Russian]

Frederiksen, S.,Werner, S. (1993). Fjärrvärme—Teori, Teknik och Funktion" ("District Heating—Theory, Technology, and Function"). Studentlitteratur, Lund. [In Swedish].

Gochenour, C. (2001). "District energy trends, issues, and opportunities, Technical Paper No 493." World Bank, Washington, DC.

Hakansson, K. (1986). "Handbuch der Fernwärmepraxis," 3rd ed. Vulkan Verlag, Essen. [In German].

International District Heating Association (currently the International District Energy Association) (1983). "District Heating Handbook," 4th ed. International District Heating Association. Westborough, MA.

Sokolov, E. J. (1982). "Teplofikazija i Teplovye Seti" ("Cogeneration and District Heating Networks"), 5th ed. Energoizdat, Moscow. [In Russian].

Werner, S. (1999). "50 Years of District Heating in Sweden." Swedish District Heating Association, Stockholm.

Westin, P., and Lagergren, F. (2002). Re-regulating district heating in Sweden. *Energy Policy* 30, 583–596.

Early Industrial World, Energy Flow in

RICHARD D. PERIMAN
Rocky Mountain Research Station, USDA Forest Service
Albuquerque, New Mexico, United States

1. Agriculture and Urbanization
2. Competition for Energy Resources
3. Wood Energy
4. Wind and Water Energy
5. The Transition to Coal Energy
6. Steam Power
7. Steam Engines and Transportation
8. Petroleum
9. Electrical Power
10. Energy flow and Industrial Momentum

Glossary

agricultural revolution The period prior to England's industrial revolution, when significant improvements in agricultural production were achieved through changes in free labor, land reform, technological innovation, and proto-industrial output.

capitalism A socioeconomic and political system in which private investors, rather than states, control trade and industry for profit.

industry Economic activity concerned with processing raw materials and producing large volumes of consistently manufactured goods, with maximum fuel exploitation.

industrial revolution The period of technological innovation and mechanized development that gathered momentum in 18th-century Britain; fuel shortages and rising population pressures, with nation-states growing in power and competing for energy, led to dramatic improvements in energy technology, manufacture, and agriculture and to massive increases in the volume of international trade.

plague Referred to as the Black Death, or Bubonic Plague, which killed one-third the population of England and Europe from the mid-14th century until around 1500; millions of people perished, and commerce, government, energy production, and industry waned as farms were abandoned and cities emptied.

proto-industry Supplemented agricultural income via the production of handcrafted work; this piecework was especially common in the creation of textiles by spinning, weaving, and sewing.

steel A strong hard metal that is an alloy of iron with carbon and other elements; it is used as a structural material and in manufacturing.

work energy Defined by natural scientist James Joule as kg m^2/s^2 or a joule (J); it enabled physicists to realize that heat and mechanical energy are convertible.

Population growth and competition for resources have continuously been an impetus for the control and use of energy. In harnessing energy, inventions evolved from previous innovations, ebbing and flowing with changes in population pressures and the availability of resources—necessity shaping history. For millennia, societies depended on resources provided by solar energy (radiation stored as vegetation), wind (a type of solar energy), animal and human energy, water, and fossil fuels. The advent of mechanized labor and its synchronized production demands transformed the lives of workers previously tuned to seasonal energy rhythms. Technological innovations during the early 18th century resulted in enormous growth in the economic wealth and sociopolitical power of nation-states. Sustained growth in Western Europe and Britain spread via exploration and trade. The competition to colonize and control the raw natural resources in the Caribbean and the Americas, in conjunction with imperialism in Indochina, the Mideast, and the Mediterranean, changed the global energy map.

1. AGRICULTURE AND URBANIZATION

Preindustrial economies in Britain and Europe were widespread in scale, dominated by agricultural production, processing, and manufacture, with populations based primarily in rural villages. A range of processed goods were essential to farms and households, although restricted incomes reduced purchasing power. Solar energy, water, regional fuels, and the ebb and flow of the seasons dictated the energy of labor expenditure in workers' lives.

There was little capital accumulation or technological progress during the widespread population destruction of the Black Death in medieval Britain and Europe. In Italy, Spain, France, and Poland, agricultural output per worker diminished, and nonagricultural employment tended to be based in rural areas, corresponding with only minor increases in urban populations. The subsequent recovery of populations after the plague led to the growth of urbanization, and local and national government structures increased their control of energy resources.

The Netherlands, Belgium, and England were distinguished from the rest of Europe by the interrelated factors of rising agricultural productivity, high wages, and urbanization. At the end of the medieval period, rural industry and urban economies expanded. Despite rates of population growth exceeding those elsewhere in Europe, economic development offset population growth, wages were increased or maintained.

By 1750, Britain had the highest portion of population growth in cities, and agricultural output per worker surpassed Belgian and Dutch levels. Urban and rural nonagricultural populations grew more rapidly in England than anywhere else in Europe. The share of the workforce in agriculture dropped by nearly half, while the trend of providing real wages rose.

The wool and cotton textile industry was the first widespread manufacturing enterprise in Britain. When farm laborers began producing piecework for cash, the social and labor organization of agrarian economies changed dramatically. Labor–market interactions and productivity growth in agriculture required a degree of labor mobility that was incompatible with previous agricultural economic models such as serfdom. England's division between landlords and tenant farmers caused reorganization to occur through reductions in farm employment, enclosure, and farm amalgamation. In Europe, where landowners/occupiers were more prevalent, farm amalgamation and enclosure were disruptive to productivity and institutional change was less profitable.

Rural industrialization was a response to population growth combined with increased urban demands for farm products, and sustained urbanization created dependency on labor based in the countryside. Substantial migrations of people to cities, and the spread of industry to the countryside, provided opportunities to escape the restrictions of guilds.

Providing enough farm produce for cities required increasing agricultural output and value per acre and introducing new crops. By the dawn of the 18th century, the industrial and commercial sectors of the British economy contributed at least one-third of the gross national product.

2. COMPETITION FOR ENERGY RESOURCES

For millennia, the Mediterranean had been the center of international trade, with trade routes among Europe, Africa, and Asia. After the plague during the 15th and 16th centuries, Mediterranean networks of international trade with Europe declined. European nation-states were not powerful enough to govern beyond their transient borders, and inadequate local resources could not support population growth. This led to each state employing protective measures to stimulate domestic industrial production, self-sufficiency, and competition for resources. During the 16th and 17th centuries, these policies were formalized, extended, and consolidated, severely limiting the growth of trade.

The high costs of inland transportation and restrictive national practices inhibited trade opportunities and led to an economic crisis. As trading opportunities ceased to expand after the plague's decimations, the need for new sources of revenue provided the impetus to search far afield for resources. With the diminishing extent of accessible markets, Europeans explored the Atlantic, Africa, the Caribbean, and the Americas, where raw resources and cheap commodity production offered immense opportunities for trade expansion. During the 15th century, the Portuguese established trading posts along the west coast region of Africa, produced sugar on slave-worked plantations, and traded gold.

The advent of the Atlantic world trading system and its abundant resources extended the production and consumption frontier of Western Europe. Because available transportation of goods was slow and expensive, unit cost of production in the Americas had to be sufficiently low for commodities to bear the cost of trans-Atlantic transportation. Large-scale production in the Americas increasingly depended on coerced labor. Indigenous peoples of the Americas were forced into slavery to mine silver in the Spanish colonies and provision Europeans. Native American land was appropriated, and populations were decimated by an unwelcome import, Old World diseases. Colonizers obtained vast quantities of acreage, becoming land rich beyond the limits of European caste and class systems. The average population density in the Americas was less than one person per square mile during the 17th century.

As indigenous peoples in the colonies suffered disease, displacement, and slavery, the production of commodities for Atlantic commerce became more dependent on slaves from Africa. Labor costs on plantations throughout the Caribbean and the Americas were below subsistence costs, enabling large levels of production and maximum exploitation of human energy resources. From Barbados to the Carolinas and the Mississippi delta, a major source of power were the thousands of enslaved Africans and Native Americans as well as indentured servants from England, Scotland, and Ireland. Plantations prospered due to the high value of cotton in the world market and by producing coffee, sugar, cocoa, tobacco, and rum. As these export commodities increased in volume, prices fell in Europe, with the commodities becoming commonly consumed goods rather than rarified luxuries. The production of raw materials for mass markets contributed greatly to the development and economic homogenization of industry and energy exploitation.

England combined naval power and commercial development to secure choice Atlantic territories and create advantageous treaties to control the colonies' resources. The majority of commodity shares, production, and trade of the slave-based economy of the Atlantic world trading system were primarily British controlled.

The estimated percentage share of these commodities produced by enslaved Africans and indentured servants in the Americas was 50 to 80%. Between 1650 and 1850, the slave-based economy of the Atlantic system transformed international trade, transformed European and British economics, and spurred technological innovation. Slavery failed as an energy source due to the socioeconomic upheavals that eventually prevented and censured its expansion as well as the fact that it could not compete with developing sources of industrial energy such as steam power. Slave labor, rather than free labor, failed as a socioeconomic source of power.

In per capita terms, the exposure of England's economy and society to the development of the Atlantic world market was greater than that of any other country, although Europe profited immensely from participation in the slave-based Atlantic world economy. Newly acquired gold and silver from the Americas bolstered weak European economies. Imported textiles from England were welcomed in the Americas, with these sustained sales creating employment in England's manufacturing regions and stimulating population growth and changes in agrarian social structures. Combined with export demands, this created a ripe environment for transforming the organization and technology of manufacturing in domestic and export industries. This international economic energy propelled the technological industrial revolution.

The transition of intensified industrialization depended on improvements in transportation systems to provide dispersion networks of fuels and raw materials. With the advent of the Victorian empire, improvements in the infrastructure of transportation (e.g., roads, canals, shipping) spread. The construction of railroad systems in Britain, Europe, Asia, Russia, Africa, and the Americas connected, amalgamated, and exploited energy resources in most regions of the world, creating an engine of economic dynamism that itself was a product of mechanized industry.

3. WOOD ENERGY

Woodlands have been an essential resource for constructing homes, vehicles, ships, furniture, tools, and machines as well as for providing heat for households and cooking. Poor management of woodlands caused low productivity, and as timber stands became more remote, logging and transporting timber became less economical. During the 13th century in England, lime kilns devoured hundreds of ancient oaks in Wellington Forest. By the 1540s, the salt industry was forced to search far afield for the wood used in its processing furnaces. British navel power judged this shortage to be a national security threat. Authorities attempted to impose stiff fines for timber poaching and forbade bakers, brewers, and tile makers from purchasing boatloads of wood.

Severe timber shortages reached national crisis proportions in Britain, impelling the import of expensive timber from the Baltic and North America. Native turf, peat, reeds, gorse, and other plant fuels simply could not provide adequate supplementation. The climate did not cooperate. Between around 1450 and 1850, winters became unusually long and harsh, the Thames river froze, and diseases and cold increased in their killing power. Populations suffered as low birth rates could not keep up with expanding death rates.

In Scotland, woodlands were carefully managed during the early medieval period, and native oak supplied the needs of developing burghs. Norse incursions pushed populations inland, allowing woodland regeneration in coastal areas. During the 15th century, the quality and quantity of native oak dwindled, and alternative sources (e.g., imported oak, conifers) were increasingly exploited. France had abundant internal supplies of timber, so there was no need to invest time, money, and politico–military resources in obtaining timber from elsewhere.

By the close of the 17th century, industry and manufacturing devoured more than a third of all the fuel burned in Britain. The textile industry's immense scale made it a leading consumer of fuels, although fuel cost was not a major constituent in manufacturing textiles, dyeing, calendaring, and bleaching. Energy consumption by industries increased in local and national terms because industry was not uniformly distributed, and this exacerbated and caused severe local and regional scarcity. Local needs concerning brewing, baking, smithing, potting, and lime processing had survived on whatever fuel was available. High temperatures were required to brew ale and beer and to process salt and sugar. The construction industry depended on kilns and furnaces for the production of bricks, tiles, glass, firing pottery, and burning lime. Smelting and working metals, and refining the base compounds of alum, copperas, saltpeter, and starch, were expensive processes.

Even when located within extensive woodlands, levels of consumption by iron forges, glass factories, and lime kilns led to shortages. Fuel needs grew voraciously as industries expanded in scale and became more centralized and urbanized. There was a fine balance between fuel needs in communities and industrial demands. Urban centers consisting of large concentrations of consumers inherently tended to outgrow the capacity of adjacent woodland fuel supplies. Transport costs commensurately inflated as supply lines were extended.

4. *WIND AND WATER ENERGY*

Windmills were developed in Persia by the 9th century BCE. This technological innovation was spurred by the need to mill corn in areas lacking consistent water supplies. Early windmills used an upright shaft, rather than a horizontal one, to hold the blades. This system was housed in a vertical adobe tunnel with flues to catch the wind (similar to a revolving door). The concept of harnessing wind energy via windmills reached Britain and Europe by 1137, where it underwent significant changes, with horizontal (rather than vertical) shafts and horizontal rotation. A form of solar energy, climate-dependent wind power has been unreliable and difficult to accumulate and store.

Roman aqueducts provided a system of fast-flowing water to watermills. By the 1st century BCE, Romans used an efficient horizontal axis where the waterwheel converted rotary motion around a horizontal axis into motion around a vertical axis; water passed under the wheel, and kinetic water energy turned the wheel. Draft animals and waterpower were harnessed to turn millstones. Rudimentary primitive watermills used vertical axes, with millstones mounted directly to waterwheel shafts. In mountainous regions, water was channeled through a chute into the wheel. Although power output was minimal ($\sim 300\,W$), this type of watermill was used in Europe until the late medieval period.

During the late 18th and early 19th centuries, waterwheel efficiency was improved. English engineer John Smeaton, expanding on the systems of early Roman waterwheels, found that more energy was generated when water was delivered to the top of the wheel via a chute. An overshot waterwheel used both kinetic and potential water energy and was capable of harnessing 63% more potential energy than was the undershot. The growing demand for energy intensified use of the waterwheel. While toiling in the gold mines of California during the 1870s, a British mining engineer, Lester Pelton, discovered that a waterwheel could be made more powerful if high-pressure jets of water were directed into hemispherical cups placed around its circumference. This waterwheel design, generating 500 horsepower (hp), was used 20 years later in the sodden mines of Alaska.

5. *THE TRANSITION TO COAL ENERGY*

Coal had been widely harvested and exploited as an energy source for thousands of years. In Asia, coal provided little illumination but plenty of heat. Romans in Britain found that stones from black outcrops were flammable, and Roman soldiers and blacksmiths used this coal to provide heat. During the 13th and 14th centuries, a multitude of industries became dependent on a regular supply of coal. In salt production, coal was used to heat enormous pans of sea water. If woodlands had not been devoted nearly exclusively to the production of timber, their yield could not have met the fuel value generated by that produced by coal mines. It took an annual yield from 2 acres of well-managed woodlands to equal the heat energy of a ton of coal. Timber became more scarce and expensive to procure during the Tudor and Stuart periods. Consequently, the processes of salt boiling, ironworking, brewing, baking, textile manufacturing, and lime burning adopted coal, rather than wood, as a readily available and economical fuel source.

Confronted by the unrelenting scarcity and high price of timber, manufacturers in fuel-intensive industries were further prompted to restrain costs and enhance profits by switching to coal. The problems caused by burning coal in processes previously dependent on wood (e.g., producing glass, bricks, tiles, pottery, alum, copperas, saltpeter, and paper; smelting lead, tin, and copper; refining soap and sugar) swelled the demand for progressive technology. Increased energy efficiency in industrial processes stimulated competition, innovation, and the emergence of fuel economies. Urbanization expanded, and the availability of inexpensive coal enabled agriculture to supplant arboriculture.

During the late 17th century, British coal mining employed 12,000 to 15,000 people, from miners to those involved in the transportation and distribution of coal to the consumer. During this period, coal prices remained virtually constant and were not dependent on revolutionary changes in structure or technology, and mine shaft depth did not increase dramatically (safe ventilation of deep pits did not occur until the 19th century). Conservatively estimated, coal production increased 370% between 1650 and 1680, creating an environment in which profound industrial advances in mining and trade occurred.

The substantial expansion of the textile industry was accommodated within preexisting systems of production, relying on traditional procedures, artisan workshops, and piecework arrangements. Advances were built on the potential that had existed in preindustrial economies, while labor and finance, marketing, transportation, and distribution evolved. Industrial enterprises incorporated factories, warehouses, forges, mills, and furnaces, and small collieries produced thousands of tons of coal per year. By 1750, the British population had nearly doubled, and urbanization and industrialization continued to amalgamate.

During the 1790s, French engineer Philippe Lebon distilled gas from heated wood and concluded that it could provide warmth, conveniently light interiors with gas in glass globes (fuel distributed via small pipes within walls), and inflate balloons. Gregory Watt, son of inventor James Watt, journeyed to France to investigate these experiments. In Cornwall, England, William Murdock lit a house with gas generated from coal. Concerned with commercial possibilities, he experimented with coal of varying qualities, different types of burners, and devices that could store enough gas to make portable light. In 1798, he returned to the foundry in Soho, England, where he illuminated the main building with coal gas using cast-iron retorts.

By 1816, London had 26 miles of gas mains, and soon there were three rival gas companies. With the availability of kerosene and gas light, the whale oil industry futilely attempted to compete. The advent of gas light enabled Victorian readers to become increasingly literate. During the 1860s, gas utilities spread in England and Europe, while the eastern United States and Canada adopted gas lighting for streets. At the turn of the 20th century, coal gas cooking and radiant heat to warm homes from gas fires became more common. The gas industry was determined to increase efficiency in all of these domestic and industrial heating functions.

Massive petroleum and gas industries resulted from the need for improved lighting during the industrial revolution. Although oil lamps had improved, the illuminating capabilities of vegetable and animal oils were limited. Gas light was superior but was restricted to urban areas. Gas oil, consisting of the fraction intermediate between kerosene and lubricants, was inexpensive and less dangerous than petroleum and also provided a convenient means of enriching the coal gas.

At the turn of the 20th century, when the automobile industry was still rudimentary, crude petroleum and rubber remained insignificant, while coal gas reached its zenith of full technological development as an energy source. Versatile and inexpensive kerosene continued to provide illuminating energy in much of the world, while gas and electricity competed in urban centers. As an illuminant, gas light began to lose momentum, while electrical power grew enormously.

The turning point for the British coal industry, and a defining moment of the industrial revolution, came with the innovation of the steam engine. It was used in processing both coal and metals, and by the close of the 18th century more than 15 million tons of coal were produced annually.

6. STEAM POWER

In the 6th century BCE, Greek philosophers described the universal elements as consisting of air, earth, fire, and water. It was established that water displaced air and that steam condensed back to liquid and could rotate wheels. During the 17th century, the physical nature of the atmosphere was explored in qualitative rather than quantitative terms; water and other liquids entering vacuous space was attributed to nature abhorring a vacuum. The persistent problem of mine drainage inspired understanding of atmospheric pressure, leading to the earliest steam engines.

When engineers failed to drain excess water from Italian mines during the 1600s, Galileo was enlisted. In 1644, his pupil, Evangelista, found that atmospheric pressure was equal to that of a column of mercury 30 inches high, corresponding to a column of water 30 feet high, and that atmospheric pressure falls with increasing altitude. Torricelli postulated that a method of repeatedly creating a vacuum would enable atmospheric pressure to be a source of power and could be used in mine drainage. Steam at 100°C equals sea-level atmospheric pressure and so can raise water 30 feet. Tremendous pressure can be generated by superheating water to create high-pressure steam; at 200°C, pressure is 15 times greater than at 100°C, lifting water 450 feet.

The first steam engines were termed atmospheric engines. In 1680, Christian Huygens, a Dutch scientist, designed a system where gunpowder exploding in a cylinder created hot gases that were expelled through relief valves. On cooling, the valves closed and a partial vacuum existed within the cylinder. The gunpowder gas occupied a smaller volume than when heated; thus, atmosphere pressure drove a piston in the cylinder. In 1690, Huygens's assistant, Denis Papin, found that 1 volume of water yields 1300 volumes of steam. A vacuum could be achieved by converting steam back into water via condensation, propelling a piston up and down in a cylinder.

Prolific inventor Thomas Savery improved on this in 1698, designing a steam pump using heat to raise water, generating approximately 1 hp. Steam from a boiler passed through a valve-fitted pipe into a vessel full of water, expelling the water up through a second pipe. When steam filled the vessel, it was condensed by pouring cold water onto the outer surface. This created a partial vacuum. When the vessel was connected with another pipe with water at a lower level, atmospheric pressure forced the water up. Savery's engine pumped water for large buildings and waterwheels, but its maximum lift was inadequate for mine drainage.

Thomas Newcomen, an inventor and blacksmith, adopted Papin's cylinder and piston and built his own version of the steam pump, replacing costly horse-drawn pump engines. Newcomen's direct experience with drainage problems in Cornish tin mines led to his construction of an efficient engine using atmospheric pressure rather than high-pressure steam. The Newcomen system had a boiler produce steam at atmospheric pressure within a cylinder, where a pump rod hung from a beam so as to apply weight to drive a piston. When the cylinder was full of steam and closed by a valve, a jet of cold water condensed the steam and atmospheric pressure forced the pump rod to push the piston down. Generating approximately 5 hp, this engine was a success in draining coal mines, lifting 10 gallons of water 153 feet through a series of pumps. Its primary users were coal mine proprietors who ran their engines on low-grade coal.

With improved control of mine flooding, the depth of mines and the coal industry grew exponentially. Use of the steam engine spread throughout Europe and to the American colonies. Newcomen's design was so reliable that even as other fuel technologies were developed, the last engine of this type worked for more than a century in Yorkshire, England, without serious breakdown (it was finally dismantled in 1934). Steam power was also used to pump water for waterwheels in cotton mills, accommodating the growing energy needs of textile manufacture as well as a multitude of other industries.

Between 1769 and 1776, James Watt discovered that energy was lost in the process of cooling and reheating the Newcomen engine. The addition of a separate condenser cylinder permitted a steady injection of steam into the first cylinder without having to continuously reheat it. In 1774, Watt and Matthew Boulton, an English manufacturer in Soho, became partners, providing Watt with skilled engineers and workers.

An impediment to steam engine design was the shortage of precise cylinders; imprecise boring allowed steam to escape between the cylinder walls and piston. John Wilkinson designed a mill for boring iron cannons that was capable of making precise cylinders for any purpose. His blast furnaces, with air from Watt's engine, produced consistent cylinders. Using this newfound precision in cylinder construction, the rotative Watt engine extracted four times as much energy, a minimum of 10 hp, from coal as the Newcomen model. By the turn of the 19th century, Watt's and Boulton's Soho factory had produced hundreds of engines powering industrial machinery.

Improvements in steam technology continued to be made during the 19th century. In 1802, a Cornish mining engineer, Richard Trevithick, built a small pumping engine with a cylinder 7 inches in diameter, a cast-iron boiler 1.5 inches thick, and steam pressure of 145 pounds per square inch—10 times atmospheric pressure. Two years later, Trevithick built 50 stationary engines and the first successful railway locomotive. By 1844, high-pressure steam engines averaged 68 million foot pounds per bushel of coal, compared with 6 million foot pounds capacity for the Newcomen engine. When Watt's engine was combined with Trevithick's design, a pressure of up to 150 pounds per square inch was achieved, referred to as "compounding." Trevithick's steam engine eliminated the huge rocking beam of Watt's design and became the new standard, suited to factories with limited space. His engines were unrivaled and remained in a variety of industrial purposes, including pumping, iron works, corn grinding, and sugar milling, until the end of the 19th century.

7. STEAM ENGINES AND TRANSPORTATION

Steam-powered engines were crucial in the development of transportation. Roads often were little more than deeply gouged tracks, slowing people, carriages, and carts in an energy-hindering quagmire. The growing demands for bulk fuel beyond what was locally available, with the high cost of transport added to the expense of energy resources, expedited innovation.

During the 1700s, British canals were improved, and tow paths allowed heavily burdened vessels to be pulled by teams of horses. Combining water and steam power, a paddlewheel steamboat ascended a river in 1783 near Lyons, France. William Symington, an English engineer, constructed an atmospheric steam engine to propel a small boat in 1788 using a horizontal double-acting cylinder with a connecting rod driving the paddlewheel's crankshaft. In the United States, Robert Fulton used a Boulton and Watt engine to operate a commercial steamer, and during the 1840s the U.S. Navy adapted this propulsion system in military craft, locating the engine below the waterline for protection.

In 1801, Trevithick built his first steam carriage (later driving one wildly through the streets of London). He combined the steam locomotive with railways, demonstrating its usefulness in conveying heavy goods and passengers. Rails had already been constructed for horses to tow carriages and coal carts more efficiently than on unreliable roads. The development of railway passenger services was a crucial formative factor in the phenomenon of the industrial revolution. Horsepower was replaced by the more powerful steam engine. By 1830, George Stephenson developed a locomotive railway line between Liverpool and Manchester without the need for stationary steam engines or horses in steep areas to supplement power. Railroad carriages were finally capable of hauling tons of coal and hundreds of passengers. Britain exported locomotives to Europe and America, where railway systems were rapidly established. Rail speeds exceeding 60 miles per hour were common, and engine efficiency increased, requiring only 23 pounds of coal per mile, although countries with abundant forests, such as the United States and Canada, still used wood fuel. Improved materials and methods of manufacture, as well as the steam turbine, resulted in higher power/weight ratios and increased fuel economy. During the 1880s, industrial steam power transformed networks of world communication, extending and unifying international economic systems.

8. PETROLEUM

Coal gas and petroleum energy share the property of having hydrocarbons as their principal constituents; coal gas is rich in the hydrocarbon methane, whereas petroleum consists of a complex mixture of liquid hydrocarbons. The use of petroleum products long antedates the use of coal gas. In Mesopotamia, concentrations of petroleum deposits provided commerce and production. Babylonians called inflammable oil "naphtha." For thousands of years, mixtures containing bitumen were used in roads, ship caulking, floor waterproofing, hard mortar, and medicines. Naphtha was militarily important as a weapon, although Romans derived their bitumen from wood pitch. Aztecs made chewing gum made with bitumen, and during the 16th century oil from seepages in

Havana was used to caulk ships. By the 15th and 16th centuries, there was an international trade in petroleum.

Distillation of crude oil revealed that its derivatives were suitable in axle grease, paints and varnishes, dressing leather, lamp fuel, and mineral turpentine. A pliable caulking for ships was made by thinning pitch thinned with turpentine, and hot asphalt pitch and powdered rock were used in flooring and steps. During the 19th century, asphalt mixed with mineral oil was used in pavements and roads, and oil cloth and linoleum became popular and inexpensive household items.

During the early 19th century, deep exploratory drilling for water and salt required sufficiently hard drills and mechanical power. Steam engines provided the energy, and the development of derricks aided in the manipulation of rapid-drilling machines. Prospecting for oil ceased to be dependent on surface seepages when hollow drills enabled drill samples to be removed, revealing the structure of underground formations. The compositions of crude oil varied; thus, methods of oil refinery were important. During the 1860s, the United States had a steady supply of 10 million barrels of petroleum annually. By 1901, Russian crude oil output reached 11 million tons, and petroleum-fueled railways connected world markets.

Kerosene was distilled from petroleum with sulfuric acid and lime, providing inexpensive and plentiful fuel for lamplight. Although gasoline remained a worthless and inflammable by-product of the industry, oil derivatives provided lubricants that became increasingly necessary with innovations in vehicles and machinery, and heavy machine oils were used on railways.

9. *ELECTRICAL POWER*

In early Greece, it was found that the *elektron* (amber), when rubbed, acquired the power of attracting low-density materials. The widespread use of electricity for heat, light, and power depended on the development of mechanical methods of generation. During the 17th and 18th centuries, static electricity was found to be distinct from electric currents. Electricity could be positive or negative as charged bodies repelled or attracted each other, distinguishing conductors from nonconductors.

In 1754, John Canton devised an instrument to measure electricity based on the repulsion of like-charged pith suspended by threads, and this was later standardized and redesigned as the gold leaf electroscope. Benjamin Franklin identified the electrical discharge of lightning, leading to the invention of the lightning conductor. In Italy, Allesandro Volta, a natural scientist, found that electricity was derived when alternate metal plates of silver or copper and zinc contacted each other within a solution. The importance of this discovery was that it provided a source of continuous electric current. The original voltaic energy battery gave stimulus to the experimental study of electricity. In a matter of months, laboratories produced electric batteries, converting the energy released in chemical reactions.

In 1820, Danish physicist H. C. Oersted described the magnetic field surrounding a conductor carrying an electric current. The relationship between the strength of a magnetic field and its electric current established that a continuous conductor in a magnetic field causes electric currents to flow. From 1834, rotating coil generators were made commercially in London. The earliest generators produced alternating currents. Converting alternating energy into direct electric currents was resolved via a mechanical commutator and rectangular coils that rotated in a magnetic field. Maximum voltage in each coil was generated in succession, alleviating the irregularities at a given speed of rotation providing constant voltage.

By 1825, electromagnets were used as an alternative to permanent magnets by the founder of the first English electrical journal, William Sturgeon. Electromagnets possessed enough residual magnetism in their iron cores to provide the necessary magnetic field to start output from an electric generator. The electric generator became a self-contained machine that needed only to be rotated to produce electrical energy. The application of a steam engine to rotate the armature generated enough electricity to supply arc lamps for lighthouses, bringing large-scale use of electricity a step closer.

An armature using a continuous winding of copper wire increased the possibilities of using arc lamps for lighting streets. Arc lamps developed after the discovery that a spark struck between two pieces of carbon creates brilliant light. Inventors Thomas A. Edison and Sir Joseph Swan developed filament lamps for domestic use during the 1870s, and an incandescent filament lamp used an electric current to heat its conductor.

The growing needs of industrial light required electric power generators to be substantial in size. In New York, Edison's electric-generating station was in operation in 1882. Electric storage batteries were used for lighting railway carriages and propelling vehicles.

At Niagara Falls during the 1890s, George Westinghouse developed the first large-scale hydroelectric installation, with a capacity of 200,000 hp. Hydroelectric generators demanded a considerable capital outlay for successful operation, so coal-fired steam engines lingered as a favored energy source at the end of the 19th century. During the late 1800s, the electricity industry expanded, output per electric power stations tripled, and coal consumption per unit of electricity fell by nearly half.

Conductors consisting of concentric copper tubes were used with alternating currents because copper had a lower electrostatic capacity and the concentric cable had no inductive effect on neighboring telegraph, telephone, or other electrical installations. The electric telegraph required a reliable battery cell to provide constant voltage that was capable of prolonged output. Attendant construction needs for telegraph poles required the production of enormous quantities of timber, porcelain insulators, and rubber insulation for cables.

Local electrical generation evolved into large centralized distribution utilities, and surplus electricity was sold to local consumers. In 1889, the London Electricity Supply Corporation's power station provided high voltage by operating four 10,000-hp steam engines to fuel 10,000-volt alternators. The economic advantages of central power stations to generate electricity at high voltages for serving large areas brought new practical and economic problems of distribution and, thus, further innovations. By the end of the 19th century, underground distribution systems of electricity had color-coded cables. Each strand was identified in this manner, a major consideration given that the electrical network beneath city streets increased in complexity.

10. ENERGY FLOW AND INDUSTRIAL MOMENTUM

During the early industrial period, increases in scale and diversity of manufacturing and trade corresponded with a shortage of readily available consistent energy, and this pressure induced developments in fuel technology industries. Competition for dwindling resources spurred the expansion of international trade and exploration, and fuel efficiency became more crucial. The fusion between mechanized growth and economic fuel use created the momentum of energy exploitation and technological innovation, and socioeconomic systems became known as the industrial revolution. Transportation systems connected a myriad of worldwide energy sources, and improved communication expedited the process of efficient global fuel exploitation and industry. Global commerce and fuel exploitation decisively affected the flow of materials via worldwide transportation and trade systems. As technological innovations were applied to harvesting energy and the subsequent momentum of labor mechanization, production uniformity and the quantity of fuel used in mass processing vast quantities of goods coalesced into an enormous worldwide energy flow.

SEE ALSO THE FOLLOWING ARTICLES

Coal Industry, History of • *Electricity Use, History of* • *Energy in the History and Philosophy of Science* • *Technology Innovation and Energy* • *Thermodynamic Sciences, History of* • *Transitions in Energy Use* • *Wind Energy, History of* • *Wood Energy, History of* • *World History and Energy*

Further Reading

Cotterell, B., and Kamminga, J. (1990). "Mechanics of Pre-Industrial Technology." Cambridge University Press, Cambridge, UK.

Crone, A., and Mills, C. M. (2002). Seeing the wood and the trees: Dendrochronological studies in Scotland. *Antiquity* **76**, 788–794.

Derry, T. K., and Williams, T. I. (1960). "A Short History of Technology from the Earliest Times to A.D. 1900." Oxford University Press, Oxford, UK.

Fields, G. (1999). City systems, urban history, and economic modernity: Urbanization and the Transition from agrarian to industrial society. *Berkeley Planning J.* **13**, 102–128.

Freese, B. (2003). "Coal: A Human History." Perseus Publishing, Cambridge, MA.

Harman, P., and Mitton, S. (eds.). (2002). "Cambridge Scientific Minds." Cambridge University Press, Cambridge, UK.

Harrison, R., Gillespie, M., and Peuramaki-Brown, M. (eds.). (2002). "Eureka: The Archaeology of Innovation and Science—Proceedings of the Twenty-Ninth Annual Conference of the Archaeological Association of the University of Calgary." Archaeological Association of the University of Calgary, Alberta, Canada.

Hatcher, J. (1993). "The History of the British Coal Industry," vol. 1: "Before 1700: Towards the Age of Coal." Clarendon, Oxford, UK.

Licht, W. (1995). "Industrializing America: The Nineteenth Century." Johns Hopkins University Press, Baltimore, MD.

Nef, J. (1964). "The Conquest of the Material World: Collected Essays." Chicago University Press, Chicago.

Theibault, J. (1998). Town and countryside, and the proto-industrialization in early modern Europe. *J. Interdisc. Hist.* **29**, 263–272.

Toynbee, A. (1956). "The Industrial Revolution." Beacon, Boston.

Wallerstein, I. (1974). "The Modern World-System I: Capitalist Agriculture and the Origins of the European World-Economy in the Sixteenth Century." Academic Press, San Diego.

Wallerstein, I. (1980). "The Modern World-System II: Mercantilism an the Consolidation of the European World-Economy 1600–1750." Academic Press, San Diego.

Wallerstein, I. (1989). "The Modern World-System III: The Second Era of Great Expansion of the Capitalist World-Economy 1730–1840s." Academic Press, San Diego.

Zell, M. (1994). "Industry in the Countryside: Wealden Society in the Sixteenth Century." Cambridge University Press, Cambridge, UK.

Earth's Energy Balance

KEVIN E. TRENBERTH
National Center for Atmospheric Research
Boulder, Colorado, United States

1. The Earth and Climate System
2. The Global Energy Balance
3. Regional Patterns
4. The Atmosphere
5. The Hydrological Cycle
6. The Oceans
7. The Land
8. Ice
9. The Role of Heat Storage
10. Atmosphere–Ocean Interaction: El Niño
11. Anthropogenic Climate Change
12. Observed and Projected Temperatures

Glossary

aerosol Microscopic particles suspended in the atmosphere, originating from either a natural source (e.g., volcanoes) or human activity (e.g., coal burning).

albedo The reflectivity of the earth.

anthropogenic climate change Climate change due to human influences.

anticyclone A high-pressure weather system. The wind rotates around anticyclones in a clockwise direction in the Northern Hemisphere and counterclockwise in the Southern Hemisphere. They usually give rise to fine, settled weather.

convection In weather, the process of warm air rising rapidly while cooler air usually subsides more gradually over broader regions elsewhere to take its place. This process often produces cumulus clouds and may result in rain.

cyclone A low-pressure weather system. The wind rotates around cyclones in a counterclockwise direction in the Northern Hemisphere and clockwise in the Southern Hemisphere. Cyclones are usually associated with rainy, unsettled weather and may include warm and cold fronts.

dry static energy The sum of the atmospheric sensible heat and potential energy.

El Niño The occasional warming of the tropical Pacific Ocean from the west coast of South America to the central Pacific that typically lasts approximately 1 year and alters weather patterns throughout the world.

El Niño–Southern Oscillation (ENSO) El Niño and the Southern Oscillation together; the warm phase is El Niño and the cold phase is La Niña.

enthalpy The heat content of a substance per unit mass. Used to refer to sensible heat in atmospheric science (as opposed to latent heat).

greenhouse effect The effect produced as certain atmospheric gases allow incoming solar radiation to pass through to the earth's surface but reduce the outgoing (infrared) radiation, which is reradiated from Earth, escaping into outer space. The effect is responsible for warming the planet.

greenhouse gas Any gas that absorbs infrared radiation in the atmosphere.

Hadley circulation The large-scale meridional overturning in the atmosphere in the tropics.

La Niña A substantial cooling of the central and eastern tropical Pacific Ocean lasting 5 months or longer.

long-wave radiation Infrared radiation in the long wave part of the electromagnetic spectrum, corresponding to wavelengths of 0.8–1000 μm. For the earth, it also corresponds to the wavelengths of thermal-emitted radiation.

short-wave radiation Radiation from the sun, most of which occurs at wavelengths shorter than the infrared.

Southern Oscillation A global-scale variation in the atmosphere associated with El Niño and La Niña events.

thermocline The region of vertical temperature gradient in the oceans lying between the deep abyssal waters and the surface mixed layer.

total solar irradiance The solar radiation received at the top of the earth's atmosphere on a surface oriented perpendicular to the incoming radiation and at the mean distance of the earth from the sun.

troposphere The part of the atmosphere in which we live, ascending to approximately 15 km above the earth's surface, and in which temperatures generally decrease with height. The atmospheric dynamics we know as "weather" take place within the troposphere.

urban heat island The region of warm air over built-up cities associated with the presence of city structures, roads, etc.

The distribution of solar radiation absorbed on Earth is very uneven and largely determined by the geometry of the sun–earth orbit and its variations. This incoming radiant energy is transformed into various forms (internal heat, potential energy, latent energy, and kinetic energy) moved around in various ways primarily by the atmosphere and oceans; stored and sequestered in the ocean, land, and ice components of the climate system; and ultimately radiated back to space as infrared radiation. The requirement for an equilibrium climate mandates a balance between the incoming and outgoing radiation and further mandates that the flows of energy are systematic. These drive the weather systems in the atmosphere and the currents in the ocean, and they fundamentally determine the climate. They can be perturbed, causing climate change. This article examines the processes involved and follows the flows, storage, and release of energy.

1. THE EARTH AND CLIMATE SYSTEM

Our planet orbits the sun at an average distance of 1.50×10^{11} m once per year. It receives from the sun an average radiation of 1368 W m^{-2} at this distance, and this value is referred to as the total solar irradiance. It used to be called the "solar constant" even though it does vary by small amounts with the sunspot cycle and related changes on the sun. The earth's shape is similar to that of an oblate spheroid, with an average radius of 6371 km. It rotates on an axis with a tilt relative to the ecliptic plane of 23.5° around the sun once per year in a slightly elliptical orbit that brings the earth closest to the sun on January 3 (called perihelion). Due to the shape of the earth, incoming solar radiation varies enormously with latitude. Moreover, tilt of the axis and the rotation of the earth around the sun give rise to the seasons, as the Northern Hemisphere points more toward the sun in June and the Southern Hemisphere points toward the sun in late December. The earth also turns on its axis once per day, resulting in the day–night cycle (Fig. 1). A consequence of the earth's roughly spherical shape and the rotation is that the average energy in the form of solar radiation received at the top of the earth's atmosphere is the total solar irradiance divided by 4, which is the ratio of the earth's surface area ($4\pi a^2$, where a is the mean radius) to that of the cross section (πa^2).

The earth system can be altered by effects or influences from outside the planet usually regarded as "externally" imposed. Most important are the sun and its output, the earth's rotation rate, the sun–earth geometry and the slowly changing orbit, the physical makeup of the earth system such as the distribution of land and ocean, the geographic features on the land, the ocean bottom topography and basin configurations, and the mass and basic composition of the atmosphere and ocean. These components affect the absorption and reflection of radiation, the storage of energy, and the movement of energy, all of which determine the mean climate, which may vary due to natural causes.

On timescales of tens of thousands of years, the earth's orbit slowly changes, the shape of the orbit is altered, the tilt changes, and the earth precesses on its axis like a rotating top, all of which combine to alter the annual distribution of solar radiation received on the earth. Similarly, a change in the average net radiation at the top of the atmosphere due to perturbations in the incident solar radiation from the changes internal to the sun or the emergent infrared radiation from changes in atmospheric composition leads to a change in heating. Changes in atmospheric composition arise from natural events such as erupting volcanoes, which can create a cloud of debris that blocks the sun. They may also arise from human activities such as the burning of fossil fuels, which creates visible particulate pollution and carbon dioxide, a greenhouse gas. The greatest variations in the composition of the atmosphere involve water in various phases, such as water vapor, clouds of liquid water, ice-crystal clouds, and hail, and these affect the radiative balance of the earth.

The climate system has several internal interactive components. The atmosphere does not have very much heat capacity but is very important—the most volatile component of the climate system with wind speeds in the jet stream often exceeding 50 m s^{-1}—in moving heat and energy around. The oceans have enormous heat capacity and, being fluid, can also move heat and energy around in important ways. Ocean currents may be >1 m s^{-1} in strong currents such as the Gulf Stream, but are more typically a few centimeters per second at the surface. Other major components of the climate system include sea ice, the land and its features [including the vegetation, *albedo* (reflective character), biomass, and ecosystems], snow cover, land ice (including the semipermanent ice sheets of Antarctica and Greenland, and glaciers),

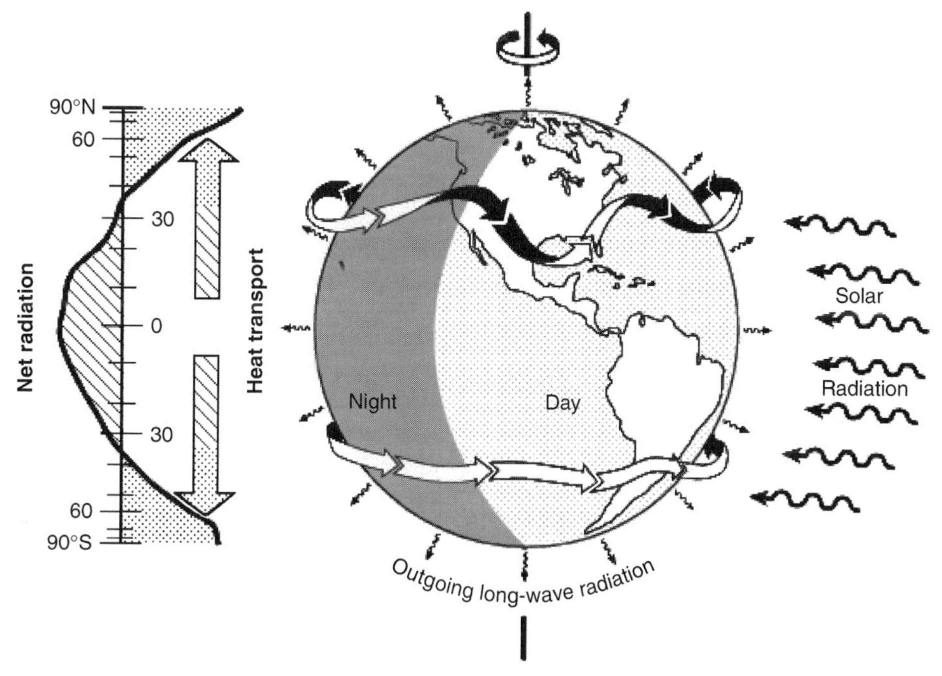

FIGURE 1 The incoming solar radiation (right) illuminates only part of the earth, whereas the outgoing long-wave radiation is distributed more evenly. On an annual mean basis, the result is an excess of absorbed solar radiation over the outgoing long-wave radiation in the tropics, whereas there is a deficit at middle to high latitudes (left) so that there is a requirement for a poleward heat transport in each hemisphere (arrows) by the atmosphere and the oceans. The radiation distribution results in warm conditions in the tropics but cold at high latitudes, and the temperature contrast results in a broad band of westerlies in the extratropics of each hemisphere in which there is an embedded jet stream (ribbon arrows) at approximately 10 km above the earth's surface. The flow of the jet stream over the different underlying surface (ocean, land, and mountains) produces waves in the atmosphere and geographic spatial structure to climate. [The excess of net radiation at the equator is $68\,W\,m^{-2}$ and the deficit peaks at $-100\,W\,m^{-2}$ at the South Pole and $-125\,W\,m^{-2}$ at the North Pole. From Trenberth *et al.* (1996).

and rivers, lakes, and surface and subsurface water. Their role in energy storage and the energy balance of the earth is discussed later.

Changes in any of the climate system components, whether internal and thus a part of the system or from external forcings, cause the climate to vary or to change. Thus, climate can vary because of alterations in the internal exchanges of energy or in the internal dynamics of the climate system. An example is El Niño–Southern Oscillation (ENSO) events, which arise from natural coupled interactions between the atmosphere and the ocean centered in the tropical Pacific. Such interactions are also discussed later from the standpoint of energy.

2. THE GLOBAL ENERGY BALANCE

The incoming energy to the earth system is in the form of solar radiation and roughly corresponds to that of a black body at the temperature of the sun of approximately 6000 K. The sun's emissions peak at a wavelength of approximately 0.6 μm and much of this energy is in the visible part of the electromagnetic spectrum, although some extends beyond the red into the infrared and some extends beyond the violet into the ultraviolet. As noted earlier, because of the roughly spherical shape of the earth, at any one time half the earth is in night (Fig. 1) and the average amount of energy incident on a level surface outside the atmosphere is one-fourth of the total solar irradiance or $342\,W\,m^{-2}$. Approximately 31% of this energy is scattered or reflected back to space by molecules, tiny airborne particles (known as aerosols), and clouds in the atmosphere or by the earth's surface, which leaves approximately $235\,W\,m^{-2}$ on average to warm the earth's surface and atmosphere (Fig. 2).

To balance the incoming energy, the earth must radiate on average the same amount of energy back to space (Fig. 2). The amount of thermal radiation

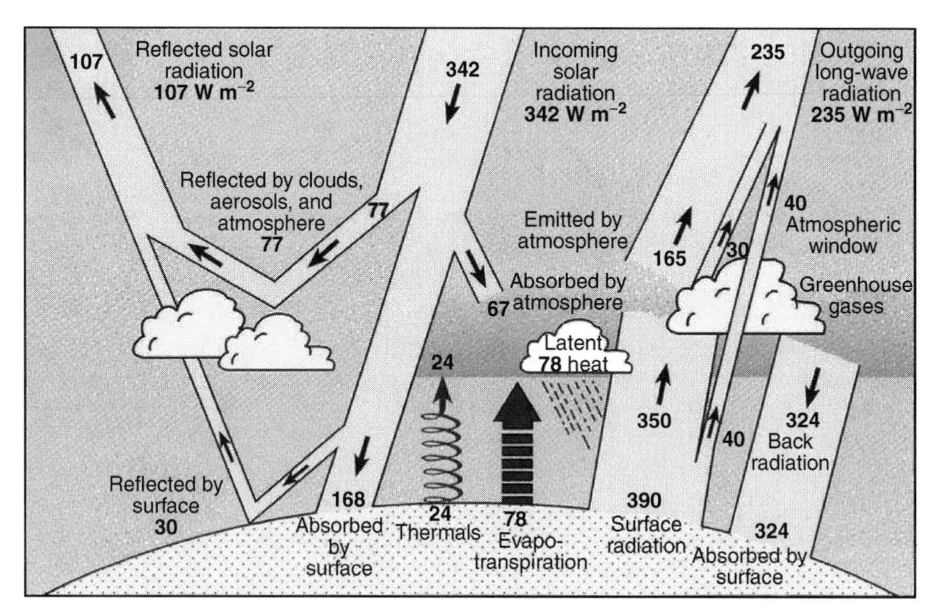

FIGURE 2 The earth's radiation balance. The net incoming solar radiation of 342 W m^{-2} is partially reflected by clouds and the atmosphere or at the surface, but 49% is absorbed by the surface. Some of this heat is returned to the atmosphere as sensible heating and most as evapotranspiration that is realized as latent heat in precipitation. The rest is radiated as thermal infrared radiation and most of this is absorbed by the atmosphere and reemitted both up and down, producing a greenhouse effect, since the radiation lost to space comes from cloud tops and parts of the atmosphere much colder than the surface. From Kiehl and Trenberth (1997).

emitted by a warm surface depends on its temperature and on how absorbing it is. For a completely absorbing surface to emit 235 W m^{-2} of thermal radiation, it would have a temperature of approximately −19°C (254 K). Therefore, the emitted thermal radiation occurs at approximately 10 μm, which is in the infrared part of the electromagnetic radiation spectrum. Near 4 μm, radiation from both the sun and the earth is very small, and hence there is a separation of wavelengths that has led to the solar radiation being referred to as short-wave radiation and the outgoing terrestrial radiation as long-wave radiation. Note that −19°C is much colder than the conditions that actually exist near the earth's surface, where the annual average global mean temperature is approximately 14°C. However, because the temperature in the lower atmosphere (troposphere) decreases quite rapidly with height, a temperature of −19°C is reached typically at an altitude of 5 km above the surface in midlatitudes. This provides a clue about the role of the atmosphere in making the surface climate hospitable.

2.1 The Greenhouse Effect

Some of the infrared radiation leaving the atmosphere originates near the earth's surface and is transmitted relatively unimpeded through the atmosphere; this is the radiation from areas where there are no clouds and which is present in the part of the spectrum known as the atmospheric "window" (Fig. 2). The bulk of the radiation, however, is intercepted and reemitted both up and down. The emissions to space occur either from the tops of clouds at different atmospheric levels (which are almost always colder than the surface) or by gases present in the atmosphere that absorb and emit infrared radiation. Most of the atmosphere consists of nitrogen and oxygen (99% of dry air), which are transparent to infrared radiation. It is the water vapor, which varies in amount from 0 to approximately 3%, carbon dioxide, and some other minor gases present in the atmosphere in much smaller quantities that absorb some of the thermal radiation leaving the surface and reemit radiation from much higher and colder levels out to space. These radiatively active gases are known as greenhouse gases because they act as a partial blanket for the thermal radiation from the surface and enable it to be substantially warmer than it would otherwise be, analogous to the effects of a greenhouse. Note that while a real greenhouse does work this way, the main heat retention in a greenhouse actually comes through protection from the wind. In the current climate, water vapor is estimated to account for approximately 60% of the greenhouse effect, carbon dioxide 26%, ozone 8%, and other gases 6% for clear skies.

2.2 Effects of Clouds

Clouds also absorb and emit thermal radiation and have a blanketing effect similar to that of greenhouse gases. However, clouds are also bright reflectors of solar radiation and thus also act to cool the surface. Although on average there is strong cancellation between the two opposing effects of short-wave and long-wave cloud heating, the net global effect of clouds in our current climate, as determined by space-based measurements, is a small cooling of the surface. A key issue is how clouds will change as climate changes. This issue is complicated by the fact that clouds are also strongly influenced by particulate pollution, which tends to make more smaller cloud droplets, and thus makes clouds brighter and more reflective of solar radiation. These effects may also influence precipitation. If cloud tops become higher, the radiation to space from clouds is at a colder temperature and this produces a warming. However, more extensive low clouds would be likely to produce cooling because of the greater influence on solar radiation.

3. *REGIONAL PATTERNS*

The annual mean absorbed solar radiation (ASR) and outgoing long-wave radiation (OLR) are shown in Fig. 3. Most of the atmosphere is relatively transparent to solar radiation, with the most notable exception being clouds. At the surface, snow and ice have a high albedo and consequently absorb little incoming radiation. Therefore, the main departures in the ASR from what would be expected simply from the sun–earth geometry are the signatures of persistent clouds. Bright clouds occur over Indonesia and Malaysia, across the Pacific near 10°N, and over the Amazon in the southern summer, contributing to the relatively low values in these locations, whereas dark oceanic cloud-free regions along and south of the equator in the Pacific and Atlantic and in the subtropical anticyclones absorb most solar radiation.

The OLR, as noted previously, is greatly influenced by water vapor and clouds but is generally more uniform with latitude than ASR in Fig. 3. Nevertheless, the signature of high-top and therefore cold clouds is strongly evident in the OLR. Similarly, the dry, cloud-free regions are where the most surface radiation escapes to space. There is a remarkable cancellation between much of the effects of clouds on

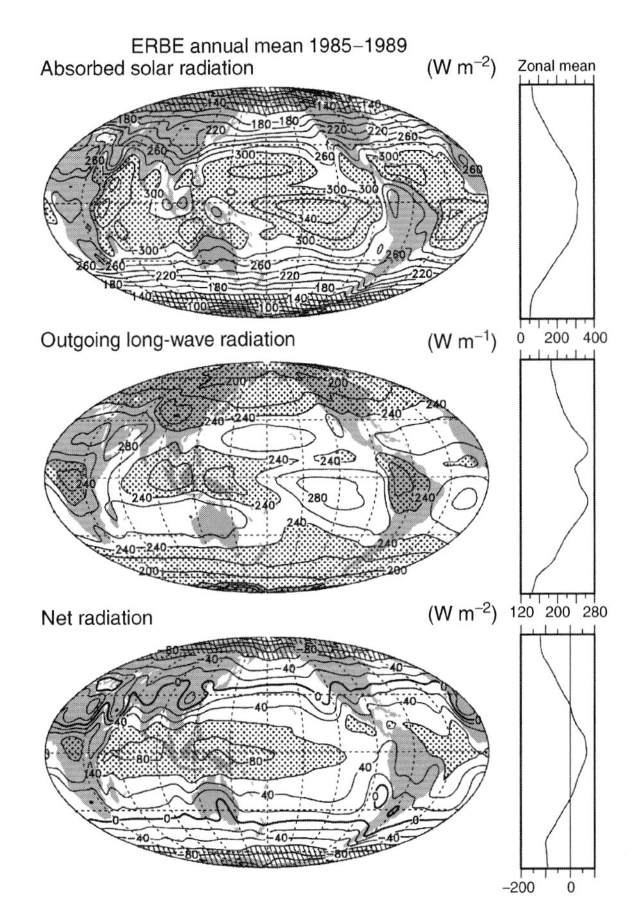

FIGURE 3 Maps of the annual mean absorbed solar radiation, outgoing long-wave radiation, and net radiation at the top of the atmosphere during the period between April 1985 and February 1989, when the Earth Radiation Budget Experiment was under way. From Trenberth and Stepaniak (2003).

the net radiation (Fig. 3). In particular, the high convective clouds are bright and reflect solar radiation but are also cold and hence reduce OLR. The main remaining signature of clouds in the net radiation from Earth is seen from the low stratocumulus cloud decks that persist above cold ocean waters, most notably off the west coasts of California and Peru. Such clouds are also bright, but because they have low tops they radiate at temperatures close to those at the surface, resulting in a cooling of the planet. Note that the Sahara desert has a high OLR, consistent with dry, cloud-free, and warm conditions, but it is also bright and reflects solar radiation, and it stands out as a region of net radiation deficit.

For the earth, on an annual mean basis there is an excess of solar over outgoing long-wave radiation in the tropics and a deficit at mid- to high latitudes (Fig. 1) that set up an equator-to-pole temperature gradient. These result, along with the earth's rotation, in a broad band of westerlies and a jet stream in each hemisphere in the troposphere. Embedded within the midlatitude westerlies are large-scale weather systems that, along with the ocean, act to transport heat polewards to achieve an overall energy balance.

4. THE ATMOSPHERE

In the atmosphere, phenomena and events are loosely classified into the realms of "weather" and "climate." Climate is usually defined as average weather and thus is thought of as the prevailing weather, which includes not just average conditions but also the range of variations. Climate involves variations in which the atmosphere is influenced by and interacts with other parts of the climate system and the external forcings. The large fluctuations in the atmosphere from hour to hour or day to day constitute the weather but occur as part of much larger scale organized weather systems that arise mainly from atmospheric instabilities driven by heating patterns from the sun.

Much of the incoming solar radiation penetrates through the relatively transparent atmosphere to reach the surface, and so the atmosphere is mostly heated from below. The decreasing density of air with altitude also means that air expands as it rises, and consequently it cools. In the troposphere, this means that temperatures generally decrease with height, and warming from below causes convection that transports heat upwards. Convection gives rise to the clouds and thunderstorms, driven by solar heating at the earth's surface that produces buoyant thermals that rise, expand and cool, and produce rain and cloud.

Another example is the cyclones (low-pressure areas or systems) and anticyclones (high-pressure systems) and their associated cold and warm fronts, which arise from the equator-to-pole temperature differences and distribution of heating (Fig. 1). The atmosphere attempts to reduce these temperature gradients by producing these weather systems, which have, in the Northern Hemisphere, southerly winds to carry warm air polewards and cold northerly winds to reduce temperatures in lower latitudes. In the Southern Hemisphere, it is the southerlies that are cold and the northerly winds are warm. These weather systems migrate, develop, evolve, mature, and decay over periods of days to weeks and constitute a form of atmospheric turbulence.

Energy in the atmosphere comes in several forms. The incoming radiant energy is transformed when it is absorbed into sensible energy. On the surface of land, the heat may be manifested as increases in temperature or as increases in evaporation, and the partitioning depends on the available moisture and nature of vegetative ground cover. Increased evaporation adds moisture to the atmosphere, and this is often referred to as latent energy because it is realized as latent heating when the moisture is condensed in subsequent precipitation. Increases in temperature increase the internal energy of the atmosphere, which also causes it to expand. Consequently, it changes the altitude of the air and increases the potential energy. Therefore, there is a close relationship between internal and potential energy in the atmosphere, which are combined into the concept of enthalpy or sensible heat. Once air starts to move, usually because temperature gradients give rise to pressure gradients that in turn cause wind, some energy is converted into kinetic energy. The sum of the internal, potential, kinetic, and latent energies is a constant in the absence of a transfer of heat. However, the possibilities for conversions among all these forms of atmospheric energy are a key part of what provides the richness of atmospheric phenomena.

In terms of poleward transport of energy, the main transports at middle and higher latitudes occur in the form of sensible heat and potential energy, which are often combined as dry static energy, whereas at low to middle latitudes, latent energy also plays a major role. In the tropics, much of the movement of energy occurs through large-scale overturning of the atmosphere. The classic examples are the monsoon circulations and the Hadley circulation. At low levels, the moisture in the atmosphere is transported toward areas where it is forced to rise and hence cool, resulting in strong latent heating that drives the upward branch of the overturning cells and constitutes the monsoon rains. In the subtropics, the downward branch of the circulation suppresses clouds and is very dry as the "freeze-dried air" from aloft slowly descends, and the air warms as it descends and expands. Hence, in these regions the surface can radiate to space without much interference from clouds and water vapor greenhouse effects. This is what happens over the Sahara. However, another key part of the cooling in the subtropics that drives the monsoons is the link to midlatitude weather systems. The latter transport both sensible and latent heat polewards in the cyclones and anticyclones and hence cool the subtropics while warming the higher latitudes.

Thus, the atmosphere is primarily responsible for the energy transports (Fig. 1) that compensate for the

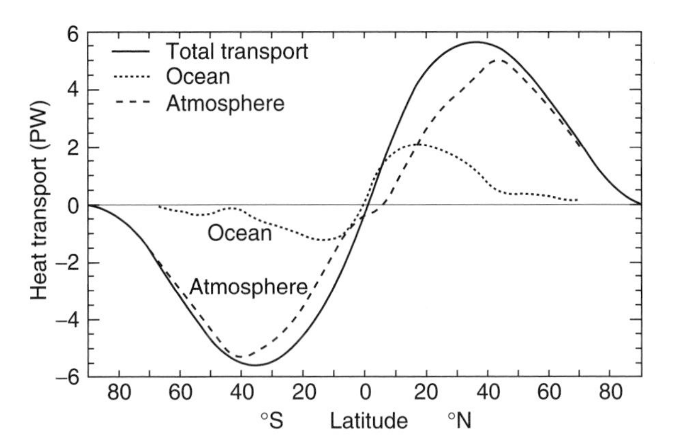

FIGURE 4 Poleward heat transports by the atmosphere and ocean and the total transport. From Trenberth and Caron (2001).

net radiation imbalances. Annual mean poleward transports of atmospheric energy of 5.0 PW peak at 43°N and with similar values near 40°S. Poleward ocean heat transports are dominant only between 0 and 17°N. At 35° latitude, where the peak total poleward transport in each hemisphere occurs, the atmospheric transport accounts for 78% of the total in the Northern Hemisphere and 92% in the Southern Hemisphere (Fig. 4).

5. THE HYDROLOGICAL CYCLE

Driven by exchanges of energy, the hydrological cycle involves the transfer of water from the oceans to the atmosphere, to the land, and back to the oceans both on top of and beneath the land surface. In the tropics in summer, land warms relative to the ocean and sets the stage for monsoon development. Water is evaporated from the ocean surface, cooling the ocean. As water vapor, it is transported perhaps thousands of kilometers before it is involved in clouds and weather systems and precipitated out as rain, snow, hail, or some other frozen pellet back to the earth's surface. During this process, it heats the atmosphere. Over land, soil moisture and surface waters can act through evaporative cooling to moderate temperatures. Some precipitation infiltrates or percolates into soils and some runs off into streams and rivers. Ponds and lakes or other surface water may evaporate moisture into the atmosphere and can also freeze so that water can become locked up for awhile. The surface water weathers rocks, erodes the landscape, and replenishes subterranean aquifers. Over land, plants transpire moisture into the atmosphere.

An estimate of the global hydrological cycle in units of $10^3\,km^3$ per year is that evaporation over the oceans (436) exceeds precipitation (399), leaving a net of 37 units of moisture transported onto land as water vapor. On average, this flow must be balanced by a return flow over and beneath the ground through river and stream flows and subsurface ground water flow. Consequently, precipitation over land exceeds evapotranspiration by this same amount (37). The average precipitation rate over the oceans exceeds that over land by 72% (allowing also for the differences in areas). It has been estimated that, on average, more than 80% of the moisture precipitated out comes from locations more than 1000 km distant, highlighting the important role of the winds in moving moisture.

6. THE OCEANS

The oceans cover 70.8% of the surface of the earth, although there is a much greater fraction in the Southern Hemisphere (80.9% of the area) than the Northern Hemisphere (60.7%), and through their fluid motions and high heat capacity they have a central role in shaping the earth's climate and its variability. The average depth of the ocean is 3795 m. The oceans are stratified opposite to the atmosphere, with warmest waters near the surface. Consequently, in the ocean, convection arises from cooling at the surface and transport of heat upwards occurs through colder and denser waters sinking and being replaced by lighter and more buoyant waters. Another vital factor in convection is the salinity of the water because this also affects density. Consequently, the densest waters are those that are cold and salty, and these are found at high latitudes where there is a deficit in radiation and atmospheric temperatures are low. The formation of sea ice also leads to a rejection of brine and increases the salinity of the surrounding waters. The cold, deep, abyssal ocean turns over only very slowly on timescales of hundreds to thousands of years.

The most important characteristic of the oceans is that they are wet. Water vapor, evaporated from the ocean surface, provides latent heat energy to the atmosphere when precipitated out. Wind blowing on the sea surface drives the large-scale ocean circulation in its upper layers. The ocean currents carry heat and salt along with fresh water around the globe (Fig. 4). The oceans therefore store heat, absorbed at the surface, for varying durations and release it in different places, thereby ameliorating temperature changes over nearby land and contributing substantially to variability of climate on many timescales.

Additionally, the ocean thermohaline circulation, which is the circulation driven by changes in seawater density arising from temperature (thermal) or salt (haline) effects, allows water from the surface to be carried into the deep ocean, where it is isolated from atmospheric influence and hence it may sequester heat for periods of 1000 years or more.

The main energy transports in the ocean are those of heat associated with overturning circulations as cold waters flow equatorwards at some depth while the return flow is warmer near the surface. There is a small poleward heat transport by gyre circulations whereby western boundary currents, such as the Gulf Stream in the north Atlantic or the Kuroshio in the north Pacific, move warm waters polewards while part of the return flow is of colder currents in the eastern oceans. However, a major part of the Gulf Stream return flow is at depth, and the most pronounced thermohaline circulation is in the Atlantic Ocean. In contrast, the Pacific Ocean is fresher and features shallower circulations. A key reason for the differences lies in salinity. Because there is a net atmospheric water vapor transport in the tropical easterly trade winds from the Atlantic to the Pacific across the central American isthmus, the north Atlantic is much saltier than the north Pacific.

7. THE LAND

The heat penetration into land is limited and slow, as it occurs mainly through conduction, except where water plays a role. Temperature profiles taken from bore holes into land or ice caps provide a coarse estimate of temperatures in years long past. Consequently, surface air temperature changes over land occur much faster and are much larger than those over the oceans for the same heating, and because we live on land, this directly affects human activities. The land surface encompasses an enormous variety of topographical features and soils, differing slopes (which influence runoff and radiation received), and water capacity. The highly heterogeneous vegetative cover is a mixture of natural and managed ecosystems that vary on very small spatial scales. Changes in soil moisture affect the disposition of heat at the surface and whether it results in increases in air temperature or increased evaporation of moisture. The latter is complicated by the presence of plants, which can act to pump moisture out of the root zone into the leaves, where it can be released into the atmosphere as the plants participate in photosynthesis; this process is called transpiration. The behavior of land ecosystems can be greatly influenced by changes in atmospheric composition and climate. The availability of surface water and the use of the sun's energy in photosynthesis and transpiration in plants influence the uptake of carbon dioxide from the atmosphere as plants transform the carbon and water into usable food. Changes in vegetation alter how much sunlight is reflected and how rough the surface is in creating drag on the winds, and the land surface and its ecosystems play an important role in the carbon cycle and fluxes of water vapor and other trace gases.

8. ICE

Major ice sheets, such as those over Antarctica and Greenland, have a large heat capacity but, like land, the penetration of heat occurs primarily through conduction so that the mass involved in changes from year to year is small. Temperature profiles can be taken directly from bore holes into ice and it is estimated that terrestrial heat flow is 51 mW/m^2. On century timescales, however, the ice sheet heat capacity becomes important. Unlike land, the ice can melt, which has major consequences through changes in sea level on longer timescales.

Sea ice is an active component of the climate system that is important because it has a high albedo. A warming that reduces sea ice, reduces the albedo and hence enhances the absorption of solar radiation, amplifying the original warming. This is known as the ice albedo feedback effect. Sea ice varies greatly in areal extent with the seasons, but only at higher latitudes. In the Arctic, where sea ice is confined by the surrounding continents, mean sea ice thickness is 3 or 4 m deep and multiyear ice can be present. Around Antarctica, the sea ice is unimpeded and spreads out extensively but, as a result, the mean thickness is typically 1 or 2 m.

9. THE ROLE OF HEAT STORAGE

The different components of the climate system contribute on different timescales to climate variations and change. The atmosphere and oceans are fluid systems and can move heat through convection and advection in which the heat is carried by the currents, whether small-scale short-lived eddies or large-scale atmospheric jet streams or ocean currents. Changes in phase of water, from ice to liquid to water vapor, affect the storage of heat. However, even ignoring these complexities, many facets of the climate are determined simply by the heat capacity of the different components of the climate system. The total heat capacity considers the mass involved as well as its capacity for holding heat, as measured by the specific heat of each substance.

The atmosphere does not have much capability to store heat. The heat capacity of the global atmosphere corresponds to that of only 3.2 m of the ocean. However, the depth of ocean actively involved in climate is much greater. The specific heat of dry land is approximately a factor of $4\frac{1}{2}$ less than that of seawater (for moist land the factor is probably closer to 2). Moreover, heat penetration into land is limited and only approximately the top 2 m typically plays an active role (as a depth for most of the variations on annual timescales). Accordingly, land plays a much smaller role in the storage of heat and in providing a memory for the climate system. Similarly, the ice sheets and glaciers do not play a strong role, whereas sea ice is important where it forms.

The seasonal variations in heating penetrate into the ocean through a combination of radiation, convective overturning (in which cooled surface waters sink while warmer, more buoyant waters below rise), and mechanical stirring by winds through what is called the "mixed layer" and on average involve approximately 90 m of ocean. The thermal inertia of the 90-m layer would add a delay of approximately 6 years to the temperature response to an instantaneous change. This value corresponds to an exponential time constant in which there is a 63% response toward a new equilibrium value following an abrupt change. Thus, the actual change is gradual. The total ocean, however, with a mean depth of approximately 3800 m, if rapidly mixed would add a delay of 230 years to the response. Mixing is not a rapid process for most of the ocean, so in reality the response depends on the rate of ventilation of water between the mixed upper layers of the ocean and the deeper, more isolated layers through the thermocline (region of temperature gradient). Such mixing is not well-known and varies greatly geographically. An overall estimate of the delay in surface temperature response caused by the oceans is 10–100 years. The slowest response should be in high latitudes, where deep mixing and convection occur, and the fastest response is expected in the tropics. Consequently, the oceans are a great moderating effect on climate variations, especially changes such as those involved with the annual cycle of the seasons.

Generally, the observed variability of temperatures over land is a factor of two to six greater than that over the oceans. At high latitudes over land in winter, there is often a strong surface temperature inversion. In this situation, the temperature increases with altitude because of the cold land surface, and it makes for a very stable layer of air that can trap pollutants. The strength of an inversion is very sensitive to the amount of stirring in the atmosphere. Such wintertime inversions are greatly affected by human activities; for instance, an urban heat island effect exceeding 10°C has been observed during strong surface inversion conditions in Fairbanks, Alaska. Strong surface temperature inversions over midlatitude continents also occur in winter. In contrast, over the oceans, surface fluxes of heat into the atmosphere keep the air temperature within a narrow range. Thus, it is not surprising that over land, month-to-month persistence in surface temperature anomalies is greatest near bodies of water. Consequently, for a given heating perturbation, the response over land should be much greater than that over the oceans; the atmospheric winds are the reason why the observed factor is only in the two to six range.

Another example is the contrast between the Northern Hemisphere (NH) (60.7% water) and Southern Hemisphere (SH) (80.9% water) mean annual cycle of surface temperature. The amplitude of the 12-month cycle between 40 and 60° latitude ranges from <3°C in the SH to ~ 12°C in the NH. Similarly, in midlatitudes, the average lag in temperature response relative to the sun for the annual cycle is 33 days in the NH versus 44 days in the SH, again reflecting the difference in thermal inertia.

10. ATMOSPHERE–OCEAN INTERACTION: EL NIÑO

The climate system becomes more involved as the components interact. A striking example is a phenomenon that would not occur without interactions between the atmosphere and ocean, El Niño, which consists of a warming of the surface waters of the tropical Pacific Ocean. It takes place from the International Dateline to the west coast of South America and results in changes in the local and regional ecology. Historically, El Niños have occurred approximately every 3–7 years and alternated with the opposite phases of below average temperatures in the tropical Pacific, called La Niña. In the atmosphere, a pattern of change called the Southern

Oscillation is closely linked with these ocean changes so that scientists refer to the total phenomenon as ENSO. El Niño is the warm phase of ENSO and La Niña is the cold phase.

El Niño develops as a coupled ocean–atmosphere phenomenon, and the amount of warm water in the tropics is redistributed and depleted during El Niño and restored during La Niña. There is often a mini global warming following an El Niño as a consequence of heat from the ocean affecting the atmospheric circulation and changing temperatures throughout the world. Consequently, interannual variations occur in the energy balance of the combined atmosphere–ocean system and are manifested as important changes in weather regimes and climate throughout the world.

11. ANTHROPOGENIC CLIMATE CHANGE

11.1 Human Influences

Climate can vary for many reasons, and in particular, human activities can lead to changes in several ways. However, to place human influences in perspective, it is worthwhile to compare the output from a power plant with that from the sun. The largest power plants that exist are on the order of 1000 MW, and these service human needs for electricity in appliances and heating that use power in units of kilowatts. Figure 2 shows that the energy received at the top of the atmosphere is $342\,\mathrm{W\,m^{-2}}$ on an annual mean basis averaged over the globe. This is equivalent to 175 PW, although only approximately 120 PW is absorbed. One petawatt is 1 billion MW or 1 million huge power plants. Consequently, it is easily recognized that human generation of heat is only a tiny fraction of the energy available from the sun. This also highlights the fact that the main way in which human activities can compete with nature is if they somehow interfere with the natural flow of energy from the sun, through the climate system, and back out to space.

There have been major changes in land use during the past two centuries. Conversion of forest to cropland, in particular, has led to a higher albedo in places such as the eastern and central United States and changes in evapotranspiration, both of which have probably cooled the region in summer by perhaps 1°C and in autumn by more than 2°C, although global effects are less clear. In cities, the building of "concrete jungles" allows heat to be soaked up and stored during the day and released at night, moderating nighttime temperatures and contributing to an urban heat island. Space heating also contributes to this effect. Urbanization changes also affect runoff of water, leading to drier conditions unless compensated by water usage and irrigation, which can cool the area. However, these influences, although real changes in climate, are quite local. Widespread irrigation on farms can have more regional effects.

Combustion of fossil fuels not only generates heat but also generates particulate pollution (e.g., soot and smoke) as well as gaseous pollution that can become particulates (e.g., sulfur dioxide and nitrogen dioxide, which get oxidized to form tiny sulfate and nitrate particles). Other gases are also formed in burning, such as carbon dioxide; thus, the composition of the atmosphere is changing. Several other gases, notably methane, nitrous oxide, the chlorofluorocarbons (CFCs), and tropospheric ozone, are also observed to have increased due to human activities (especially from biomass burning, landfills, rice paddies, agriculture, animal husbandry, fossil fuel use, leaky fuel lines, and industry), and these are all greenhouse gases. However, the observed decreases in lower stratospheric ozone since the 1970s, caused principally by human-introduced CFCs and halons, contribute to a small cooling.

11.2 The Enhanced Greenhouse Effect

The amount of carbon dioxide in the atmosphere has increased by more than 30% in the past two centuries since the beginning of the industrial revolution, an increase that is known to be in part due to combustion of fossil fuels and the removal of forests. Most of this increase has occurred since World War II. Because carbon dioxide is recycled through the atmosphere many times before it is finally removed, it has a long lifetime exceeding 100 years. Thus, emissions lead to a buildup in concentrations in the atmosphere. In the absence of controls, it is projected that the rate of increase in carbon dioxide may accelerate and concentrations could double from pre-industrial values within approximately the next 60 years.

If the amount of carbon dioxide in the atmosphere were suddenly doubled, with other things remaining the same, the outgoing long-wave radiation would be reduced by approximately $4\,\mathrm{W\,m^{-2}}$ and instead trapped in the atmosphere, resulting in an enhanced greenhouse effect. To restore the radiative balance, the atmosphere must warm up and, in the absence of other changes, the warming at the surface and throughout the troposphere would be approximately 1.2°C. In reality, many other factors will change, and various feedbacks come into play, so that the best estimate of the average global warming for doubled carbon dioxide is 2.5°C. In other words, the net effect of the feedbacks is positive and approximately doubles the response otherwise expected.

Increased heating therefore increases global mean temperatures and also enhances evaporation of surface moisture. It follows that naturally occurring droughts are likely to be exacerbated by enhanced drying. After the land is dry, all the solar radiation goes into increasing temperature, causing heat waves. Temperature increases signify that the water-holding capacity of the atmosphere increases and, with enhanced evaporation, the atmospheric moisture increases, as is currently observed in many areas. The presence of increased moisture in the atmosphere implies stronger moisture flow converging into all precipitating weather systems, whether they are thunderstorms, extratropical rain, or snow storms. This leads to the expectation of enhanced rainfall or snowfall events, which are also observed to be happening. Globally, there must be an increase in precipitation to balance the enhanced evaporation and hence there is an enhanced hydrological cycle.

The main positive feedback comes from water vapor. As the amount of water vapor in the atmosphere increases as the earth warms, and because water vapor is an important greenhouse gas, it amplifies the warming. However, increases in clouds may act either to amplify the warming through the greenhouse effect of clouds or reduce it by the increase in albedo; which effect dominates depends on the height and type of clouds and varies greatly with geographic location and time of year. Decreases in sea ice and snow cover, which have high albedo, decrease the radiation reflected back to space and thus produce warming, which may further decrease the sea ice and snow cover, known as ice albedo feedback. However, increased open water may lead to more evaporation and atmospheric water vapor, thereby increasing fog and low clouds, offsetting the change in surface albedo.

Other, more complicated feedbacks may involve the atmosphere and ocean. For example, cold waters off the western coasts of continents (such as California or Peru) encourage development of extensive low stratocumulus cloud decks that block the sun and this helps keep the ocean cold. A warming of the waters, such as during El Niño, eliminates the cloud deck and leads to further sea surface warming through solar radiation.

11.3 Effects of Aerosols

Aerosols occur in the atmosphere from natural causes; for instance, they are blown off the surface of deserts or dry regions. As a result of the eruption of Mt. Pinatubo in the Philippines in June 1991, considerable amounts of aerosol were added to the stratosphere that, for approximately 2 years, scattered solar radiation leading to a loss of radiation and a cooling at the surface.

As noted previously, human activities also affect the amount of aerosol in the atmosphere. The main direct effect of aerosols, such as sulfate particles, is the scattering of some solar radiation back to space, which tends to cool the earth's surface. Aerosols can also influence the radiation budget by directly absorbing solar radiation, leading to local heating of the atmosphere, and, to a lesser extent, by absorbing and emitting thermal radiation. A further influence of aerosols is that many of them act as nuclei on which cloud droplets condense. A changed concentration therefore affects the number and size of droplets in a cloud and hence alters the reflection and the absorption of solar radiation by the cloud.

Because man-made aerosols typically remain in the atmosphere for only a few days before they are washed out by precipitation, they tend to be concentrated near their sources, such as industrial regions, adding complexity to climate change because they can help mask, at least temporarily, any global warming arising from increased greenhouse gases.

12. OBSERVED AND PROJECTED TEMPERATURES

Changes in climate have occurred in the distant past as the distribution of continents and their landscapes have changed, as the so-called Milankovitch changes in the orbit of the earth and the earth's tilt relative to the ecliptic plane have varied the insolation received on earth, and as the composition of the atmosphere has changed, all through natural processes. Recent evidence obtained from ice cores drilled through the Greenland ice sheet indicates that changes in climate may often have been quite rapid and major and not associated with any known external forces.

Observations of surface temperature show a global mean warming of approximately 0.7°C during the past 100 years. The warming became noticeable from the 1920s to the 1940s, leveled off from the 1950s to the 1970s, and increased again in the late 1970s. The calendar year 1998 is the warmest on record. The 1990s are the warmest decade on record. Information from paleodata further indicates that these years are the warmest in at least the past 1000 years, which is as far back as a hemispheric estimate of temperatures can be made. The melting of glaciers throughout most of the world and rising sea levels confirm the reality of the global temperature increases.

Climate modeling suggests that solar variability has contributed to some of the warming of the 20th century, perhaps 0.2°C, up to approximately 1950. Changes in aerosols in the atmosphere, both from volcanic eruptions and from visible pollutants and their effects on clouds, have also contributed to reduced warming, perhaps by a couple of tenths of a degree. A temporary cooling in 1991 and 1992 followed the eruption of Mount Pinatubo in June 1991. Heavy industrialization and associated pollution following World War II may have contributed to the plateau in global temperatures from approximately 1950 to 1970. Interactions between the atmosphere and the oceans, including El Niño, have contributed to natural fluctuations of perhaps two-tenths of a degree. It is only after the late 1970s that global warming from increases in greenhouse gases has emerged as a clear signal in global temperatures.

Projections have been made of future global warming effects based on model results. Because the actions of humans are not predictable in any deterministic sense, future projections necessarily contain a "what if" emissions scenario. In addition, for a given scenario, the rate of temperature increase depends on the model and features such as how clouds are depicted so that there is a range of possible outcomes. Various projections in which the concentrations of carbon dioxide double 1990 values by the year 2100 indicate global mean temperature increases ranging from 2 to 4°C higher than 1990 values. Uncertainties in the projections of carbon dioxide and aerosols in the atmosphere add to this range. However, a major concern is that the rates of climate change as projected exceed anything seen in nature in the past 10,000 years.

Acknowledgments

The National Center for Atmospheric Research is sponsored by the National Science Foundation.

SEE ALSO THE FOLLOWING ARTICLES

Climate Change and Energy, Overview • Climate Change: Impact on the Demand for Energy • Climate Protection and Energy Policy • Ecosystems and Energy: History and Overview • Environmental Gradients and Energy • Greenhouse Gas Emissions from Energy Systems, Comparison and Overview • Lithosphere, Energy Flows in • Ocean, Energy Flows in

Further Reading

Bond, G., *et al.* (1997). A pervasive millennial-scale cycle in North Atlantic Holocene and glacial climates. *Science* **278,** 1257–1266.

Dahl-Jensen, D., Mosegaard, K., Gundestrup, N., Clow, G. D., Johnsen, S. J., Hansen, A. W., and Balling, N. (1998). Past temperatures directly from the Greenland Ice Sheet. *Science* **282,** 268–271.

Hansen, J., *et al.* (1996). A Pinatubo climate modeling investigation. *In* "The Mount Pinatubo Eruption: Effects on the Atmosphere and Climate" (G. Fiocco, D. Fua, and G. Visconti, Eds.), NATO ASI Series I, No. 42, pp. 233–272. Springer-Verlag, Heidelberg.

Intergovernmental Panel on Climate Change (2001). *In* "Climate Change 2001. The Scientific Basis" (J. T. Houghton *et al.*, Eds.). Cambridge Univ. Press, Cambridge, UK.

Kiehl, J. T., and Trenberth, K. E. (1997). Earth's annual global mean energy budget. *Bull. Am. Meteorol. Soc.* **78,** 197–208.

Levitus, S., Antonov, J. I., Wang, J., Delworth, T. L., Dixon, K. W., and Broccoli, A. J. (2001). Anthropogenic warming of the earth's climate system. *Science* **292,** 267–270.

Pollack, H. N., Huang, S., and Shen, P. -Y. (1998). Climate change record in subsurface temperatures: A global perspective. *Science* **282,** 279–281.

Trenberth, K. E. (1983). What are the seasons? *Bull. Am. Meteor. Soc.* **64,** 1276–1282.

Trenberth, K. E. (1998). Atmospheric moisture residence times and cycling: Implications for rainfall rates with climate change. *Climate Change* **39,** 667–694.

Trenberth, K. E. (2001). Stronger evidence for human influences on climate: The 2001 IPCC Assessment. *Environment* **43**(4), 8–19.

Trenberth, K. E., and Caron, J. M. (2001). Estimates of meridional atmosphere and ocean heat transports. *J. Climate* **14,** 3433–3443.

Trenberth, K. E., and Hoar, T. J. (1997). El Niño and climate change. *Geophys. Res. Lett.* **24,** 3057–3060.

Trenberth, K. E., and Stepaniak, D. P. (2003). Seamless poleward atmospheric energy transports and implications for the Hadley circulation. *J. Climate* **16,** 3705–3721.

Trenberth, K. E., Houghton, J. T. and Meira Filho, L. G. (1996). The climate system: An overview. In "Climate Change 1995. The Science of Climate Change" (J. T. Houghton, L. G. Meira Filho, B. Callander, N. Harris, A. Kattenberg, and K. Maskell, Eds.), Contribution of WG 1 to the Second Assessment Report of the Intergovernmental Panel on Climate Change, pp. 51–64. Cambridge Univ. Press, Cambridge, UK.

Easter Island: Resource Depletion and Collapse

JAMES A. BRANDER
University of British Columbia
Vancouver, British Columbia, Canada

1. The Mystery of Easter Island
2. First Discovery and Early History
3. Resource Depletion and Collapse
4. Malthusian Population Dynamics
5. A Predator–Prey System with Unfortunate Parameter Values
6. The Role of Property Rights and Open Access
7. Aftermath
8. Lessons of Easter Island

Glossary

core sample A specimen obtained in a long cylinder, taken from a physical site for subsequent study; used in many fields, including oil exploration and archaeology. In archaeology, elements in the core sample can be carbon dated, allowing a reconstruction of the implicit historical record.

Malthusian population dynamics A demographic theory originated by Thomas Malthus, suggesting that the human population has a natural tendency to grow sufficiently rapidly to offset and perhaps even overshoot productivity gains in food production. Malthus also made the related assertion that population growth will therefore be sensitive to per capita consumption—tending to increase when consumption rises and to decrease when consumption falls. Hence, the term refers to a situation in which population growth is positively related to per capita consumption. It is noteworthy that the term does not describe societies that have gone through the so-called demographic transition in which fertility tends to fall at sufficiently high income levels, it does seem to describe societies at the level of development of Easter Island prior to European contact.

market failure A situation in which markets fail to achieve economic efficiency. Open-access resources and, more broadly, difficulty in establishing property rights constitute an important source of market failure.

open-access resource A resource to which any potential user can gain access. Various communities have sometimes had land to which everyone had access, e.g., a "common," consisting of grazing land. It was commonly observed that such common land was often overgrazed. The tendency to overuse open-access resources is sometimes referred to as the "open-access" problem, which is an extreme version of "incomplete property rights," arising when the owner or controller of an economic resource can exclude other users only at significant cost.

predator–prey system A system in which a "prey" forms an important part of the consumption for a "predator." It has often been observed that predators and prey (such as lynx and rabbits) sometimes go through dramatic population cycles. In the early 20th century, Lotka and Volterra independently studied the formal properties of simple linear two-equation dynamic systems in an effort to shed light on predator–prey systems. They discovered that population cycles are a natural outcome of such systems for particular parameter values. Extinction and monotonic convergence to a steady state are also possible dynamic trajectories in predator–prey models.

Easter island (or rapa nui) is remarkable for being a small geographic entity that has generated enormous academic and popular interest. This interest arose primarily from the enormous stone statues, or moai, positioned at various locations around the island. These moai are sometimes referred to as "the mystery of Easter Island." The mystery of the moai remained unsolved until the last decade of the 20th century, when it became clear that the statues were intimately connected to a story of resource depletion and collapse.

1. THE MYSTERY OF EASTER ISLAND

1.1 Early European Contact

First European contact with Easter Island occurred on Easter Day of 1722, when three Dutch ships under the command of Jacob Rogaveen stopped for a 1-day visit at the very isolated South Pacific island, just over 3000 km west of Chile. The visitors observed a small, treeless island populated by what they thought were about 3000 islanders and, to their surprise, by a large number of giant statues. The outside world had no previous knowledge of Easter Island and the islanders apparently had no awareness of the outside world. The visit must have been a shock to the Easter Islanders, but no systematic account of their reaction survives and no further contact with Europeans occurred until 1770, when a Spanish vessel made a brief stop.

The first systematic study of Easter Island dates from the 1774 visit of James Cook. At least one member of Cook's crew could understand the Polynesian dialect spoken by the Easter Islanders. This allowed Cook to infer that the Easter Island population was in fact Polynesian. In addition, Cook learned that the islanders believed that the statues had been built and moved to their then-current locations by some supernatural force. Certainly the islanders of 1774 had no concept of how to carve or move the statues. Cook estimated that the island's population was about 2000, which would have been insufficient to move the statues using any technology available on the island.

1.2 Explanations of the Moai

The mystery of how the statues came into existence remained a topic of increasingly fanciful speculation from Cook's day forward. Among the more striking theories advanced to explain the mystery include Thor Heyerdahl's view that the statues had been built by an advanced South American civilization that had discovered and colonized Easter Island before somehow being displaced by a less advanced Polynesian society. In an effort to support his theory, Heyerdahl undertook a famous voyage to Easter Island, in a reed boat named the Kon Tiki, starting from off the coast of Chile. An alternative and even more exotic theory was that Easter Island was the remnant of the hypothesized archipelago civilization of "Mu" that, similar in concept to Atlantis, had sunk beneath the ocean due to geological events. Perhaps the most remarkable theory was that of Eric von Daniken, who wrote a best-selling book (*Chariots of the Gods*) and produced a popular 1970s television series, which proposed that the statues had been built by extraterrestrials.

These imaginative explanations have been displaced by scientific evidence, including carbon dating of core samples taken from Easter Island and a genetic study of skeletal remains on Easter Island. This evidence has allowed the development of a rather different understanding of Easter Island's past. This understanding, based on resource depletion and collapse, explains both the statues and the decline of the Easter Island civilization.

2. FIRST DISCOVERY AND EARLY HISTORY

2.1 The Virgin Forest

Although there is some uncertainty about dates, it now seems that Easter Island was first discovered by a small group of Polynesians sometime between 400 and 700 AD. The striking surprise that emerged from analysis of core samples is that Easter Island was at this time covered by a dense forest of *Jubaea chilensis* (the Chilean wine palm). This particular palm is a large, slow-growing tree that grows in temperate climates.

2.2 The Easter Island Economy

Following first discovery of Easter Island, the new inhabitants developed an economy based on the wine palm. Harvested trees were the raw material for canoes that could be used for fishing, leading to an early diet that depended very heavily on fish, gradually supplemented from other sources, including yams grown on deforested sections of the island, and domestic fowl. The population grew steadily and the civilization grew more sophisticated, allowing a substantial diversion of labor into ceremonial and other nonsubsistence activity. However, as time progressed, the forest stock gradually diminished. At first, liberation of forested land for agricultural uses, for dwelling space, and for other purposes would have been seen as a benefit. The rate of forest decline was slow. In a typical lifetime (somewhere between 30 and 40 years for those who survived early childhood), it would have been hard to notice a dramatic change in the forest cover because the stock would not have declined by more than about 5 or 6% over a 30- to 40-year period, until relatively near the end of viable forest.

3. RESOURCE DEPLETION AND COLLAPSE

3.1 An End to the Golden Age

By the turn of the millennium in 1000 AD, the population was still rising and a statue-carving culture had developed. The statues were carved in Easter Island's lone quarry and then transported to desired locations using logs as rollers. The forest stock would have been down to perhaps two-thirds of its original level and the islanders probably would have regarded as apocryphal any claims that the island was once fully forested. However, this much loss of forest cover would have begun to reduce rainfall (because low clouds could pass overhead more easily) and would also have reduced the capacity of the soil to retain water. These trends continued for the next several hundred years. The peak population (of probably between 10,000 and 14,000), the greatest intensity of statue carving, and the virtual elimination of the forest stock all occurred in the neighborhood of 1400 AD.

3.2 Collapse of a Civilization

The subsequent 100 years on Easter Island must have been very difficult. Statue carving declined and soon ceased entirely. A new weapon, almost certainly used for hand-to-hand combat, enters the archaeological record at this time. Chickens were moved into fortified enclosures, much of the population moved into caves, and there is evidence of cannibalism from this period. The population began to decline steadily. The population was apparently still decreasing in the 18th century, because the 1774 population estimate by Cook was significantly less than Rogaveen's estimate of 50 years earlier. In addition, Cook noted that many of the statues had been pushed over on their faces, but no such phenomenon was observed by Rogaveen. It seems likely that the middle of the 18th century was a turbulent time on Easter Island. As has been frequently observed, the Easter Island civilization temporarily surpassed its limits and crashed devastatingly.

4. MALTHUSIAN POPULATION DYNAMICS

4.1 A Cautionary Tale?

The basic facts regarding Easter Island are no longer in serious dispute. It is, however, important to understand why Easter Island suffered an internally generated collapse based on resource overuse. The real question is whether Easter Island is an unusual and isolated case or whether it is a cautionary tale for the modern world. It is therefore important to uncover any general principles that might be inferred from the Easter Island experience.

4.2 Malthusian Population Dynamics and Overshooting

One important idea that might shed light on Easter Island derives from Malthusian population dynamics. Polynesian societies (along with virtually all other societies at comparable stages of development) were Malthusian in the sense that fertility was positively related to consumption (or real income) and mortality was negatively related to consumption. Correspondingly, in times of increasing hardship, fertility would decrease and mortality would increase. Depending on the precise responsiveness of fertility and mortality to real income (or food supply), and on other aspects of the environment, it is possible for Malthusian populations to "overshoot." Such populations would tend to increase too rapidly to be sustained, leading to an ultimate population "crash" enforced by some of the more unpleasant instruments of mortality, such as disease, starvation, and violent conflict. This possible feature of Malthusian models led subsequent readers of economist Thomas Malthus to refer to economics as the "dismal science."

4.3 Mitigating Factors and Reinforcing Factors

Various factors can mitigate Malthusian population dynamics (as Malthus recognized), including technological progress and the possibility of a "demographic transition" under which fertility declines at sufficiently high incomes. (The combination of technological progress and a demographic transition has, of course, allowed most of the modern world to avoid the Malthusian trap over the two centuries that have passed since the time of Malthus.) On the other hand, as Malthus also realized, there are factors that tend to exacerbate Malthusian dynamics, including reliance on open-access resources and lack of property rights more generally.

4.4 Easter Island and Malthusian Overshooting

At a superficial level, Easter Island might look like a simple example of Malthusian population overshooting. The mitigating factors were not present, in that little technological progress occurred and the level of income was never high enough for a demographic transition to be relevant. In addition, the Easter Island economy did suffer from heavy reliance on a renewable resource that very likely was characterized by open-access problems or at least by incomplete property rights. However, Malthusian overshooting cannot be the whole story, for most Polynesian societies did not experience the boom-and-bust pattern exhibited by Easter Island, even though these societies were similar to the Easter Island culture in most important respects. Thus, for example, islands such as Tahiti and New Zealand's North Island appeared to be in something approaching a long-run steady state at the time of first European contact. Therefore, there must be more to the story than simple Malthusian overshooting.

5. A PREDATOR–PREY SYSTEM WITH UNFORTUNATE PARAMETER VALUES

5.1 Easter Island as a Predator–Prey System

As reported in 1998 by Brander and Taylor, the understanding of Easter Island can be greatly advanced by using formal mathematical analysis. The formal description provided here relies on that analysis. One key point is that the Easter Island economy can be viewed as a "predator–prey" system. The forest resource can be considered as the prey and the Easter Islanders as the predators. One important aspect of even relatively simple predator–prey systems is that small to moderate changes in one or more key parameters can give rise to major qualitative changes in the dynamic pattern of population growth and decline. In particular, depending on parameter values, it is possible for a given model structure to give rise to smooth, or "monotonic," movement toward a steady state or, alternatively, to a boom-and-bust cycle. One possible behavior is long-run stability of the type apparently observed in Tahiti. Another possible behavior is the rise and fall that characterizes Easter Island. More broadly, it is quite possible that the same general model might describe a variety of Polynesian societies, with plausible variations in parameter values giving rise to the observed differences in demographic and economic experience.

5.2 Modeling the Renewable Resource Stock

The economy and ecology of Easter Island were relatively simple, but obtaining a tractable mathematical representation still requires significant abstraction and approximation. At a minimum, it is necessary to model the behavior of two "stocks" or "state variables": the forest stock and the population. In fact, it is an oversimplification to focus just on the forest stock. Even in the absence of the forest stock, grass, shrubs, and certain vegetables could be grown and this biomass was sufficient to support a diminished but significant population. Rather than introduce a third "stock" representing other biomass, it is simpler to aggregate the renewable resource base growing on the land into a single resource stock, denoted as S. For simplicity, sometimes just the term "forest" is used, but it should be kept in mind that other components of the flora biomass are also relevant.

5.3 Logistic Growth

The starting point for describing the evolution of a renewable resource stock is the logistic growth function. Using t to denote time, a simple logistic growth function has the form $G(t) = rS(1-S/K)$. The variable r is the intrinsic growth rate and K is the environmental carrying capacity, or maximum possible size of the resource stock. $G(t)$ is the growth rate defined in biomass units and G/S is the proportional growth rate (i.e., a number such as 0.1 or 10%). If the forest stock S reaches a level equal to carrying capacity K, then $G = 0$ and no further growth occurs. Similarly, if there is no forest stock, then $S = 0$ and no growth occurs. If the forest stock is small relative to carrying capacity (but positive), then S/K is negligible and $G = rS$. In this case, proportional growth G/S approaches r. Thus, we can think of r as the proportional growth rate in the absence of congestion effects. As the forest stock increases, there is congestion (competition for space, rainfall, sunlight, etc.), with the result that the proportional growth rate tends to decline. However, the base to which that growth rate is applied increases as S increases. Therefore, the absolute growth rate in biomass units increases with S up to a critical point beyond which the congestion effect dominates the increasing base effect and the absolute growth rate declines. A logistic growth function is illustrated in Fig. 1. It is clear from Fig. 1 that the maximum sustainable yield occurs at the maximum

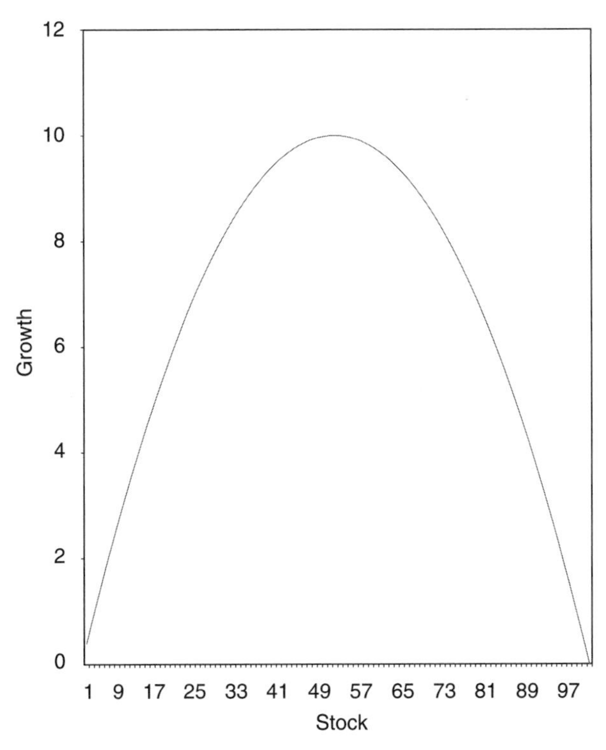

FIGURE 1 Logistic growth.

growth rate. In other words, it would be possible to maintain a stock of $K/2$ and harvest at this maximum growth rate in steady state.

5.4 Other Growth Functions

The logistic growth function is no doubt far from being a perfect approximation to the growth of the *Jubaea chilensis* forest on Easter Island. Various extensions to the logistic growth function are possible. For example, it is quite possible that at very low values of the stock, the proportional growth rate actually tends to decline. This is referred to as "depensation" or the "Allee effect." The logistic growth function can be augmented to allow for depensation, but that will not be pursued here. The basic logistic growth function does allow for the critical elements of biological renewable resource growth that are needed to capture at least the qualitative experience of Easter Island.

5.5 Open Access, Property Rights, and Harvesting Behavior

The natural growth of the forest stock provides part of the story in characterizing the evolution of the forest stock. The other part of the story depends on the harvesting behavior of the Easter Islanders. The key economic point is that under conditions of incomplete property rights, it is natural to expect that, for any given stock, the amount harvested at any given time period, $H(t)$, is increasing with population size. In other words, for any given set of external conditions and any given forest stock, we expect more harvesting to occur if there is a larger population seeking to use the forest stock. Similarly, for a given population, harvesting is easier if the stock is larger, and therefore more harvesting would be likely to occur. In the case of pure open access (no enforceable property rights) and standard demand and production functions, the harvesting function can be shown to take on the form $H = aSP$, where a is a parameter reflecting demand conditions and the harvesting technology and P is the population. This functional form is used here for concreteness, although other related functional forms would have similar properties. The differential equation that governs the evolution of the forest stock can therefore be written as

$$dS/dt = G(t) - H(t) = rS(1 - S/K) - aSP, \quad (1)$$

where state variables S and P are functions to time t.

5.6 Population Growth

To describe population growth, we adopt the linear form used by Lotka and Volterra in their original predator–prey formulations. The first component is a base net fertility rate, n, that would apply in the absence of a resource stock (and would presumably be negative in our environment). In addition, there is a second term that, from our point of view, captures the Malthusian idea that net fertility rises when per capita consumption rises. This implies that proportional net fertility should rise when, other things equal, the resource stock is larger. The effect follows because a higher stock (for a given population) would normally give rise to higher per capita consumption. This reasoning leads to the following population growth function:

$$dP/dt = nP + bPS, \quad (2)$$

where b is a parameter that reflects economic conditions. It can be seen from Eq. (2) that proportional population growth rate $(dP/dt)/P$ equals the base rate plus some increment related to the per capita resource stock. The functional form could be changed to include a "congestion effect," similar to the congestion effect that arises in the logistic growth function that is used for the resource stock, but

Eq. (2) should allow a reasonable approximation to actual events.

5.7 A Dynamic System

Equations (1) and (2) form an interdependent dynamic system similar to the Lotka–Volterra predator–prey model. Equation (1) shows the evolution of prey (the forest stock) and has the key property that the prey tends to diminish more quickly when there are more predators (humans). Equation (2) shows the evolution of the predator and has the property that the predator increases more rapidly when prey are more numerous. Figure 2 illustrates the relationship between the principal "stocks" and "flows" in the model using a formal flow diagram.

It is possible to examine the properties of this model analytically. Solutions for steady states can be determined and the dynamic path leading from any initial conditions can be characterized. Depending on parameter values, three steady states are possible, one with no humans and the forest at environmental carrying capacity, one with no humans and no resource stock, and one "interior" steady state with positive stocks of humans and forest biomass. The dynamic evolution of the model can be either monotonic or cyclical, once again depending on parameter values.

5.8 Simulation of Easter Island

Perhaps the most revealing method of analysis of this dynamic system is through simulation. It is possible to estimate plausible values for the parameters in Eqs. (1) and (2) and simulate the evolution of Easter Island. The necessary parameters are a, b, n, r, and environmental carrying capacity K. It is also necessary to specify the size $P(0)$ of the initial founding population.

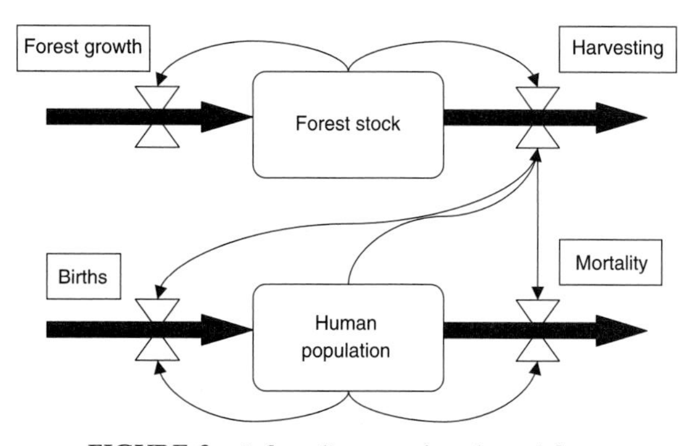

FIGURE 2 A flow diagram of stocks and flows.

The environmental carrying capacity and the initial forest stock are taken to be the same. Of course, the size of this stock can be normalized to any value. For scaling purposes, it is convenient if the resource stock is normalized to be of approximately the same magnitude as the maximum population, so we assume the initial resource stock (and environmental carrying capacity) is 12,000 biomass units. The initial population probably arrived in one or a few large ocean-going canoes that had traveled a considerable distance in search of new land. Take the founding population to be 40, although it makes very little difference if the founding population were taken to be twice or three times as large or, for that matter, if it were smaller. The net fertility rate, n, is the proportional (or percentage) rate of growth in the absence of the resource stock. In the simulation, discrete periods of 10 years will be used. Using a value of -0.1 for n indicates that, without the biomass in place, the population would tend to decline at a rate of 10% per decade. Parameters a and b reflect particular underlying economic conditions related to harvesting technology and to the preferences of the Easter Islanders. Because Eqs. (1) and (2) are being used as the starting points for the formal analysis here, parameters a and b could be viewed as essentially arbitrary parameters that we choose appropriately to "fit" the model to the experience of Easter Island. However, Eqs. (1) and (2) can be derived from more fundamental economic structures that suggest plausible magnitudes for a and b. A plausible magnitude for a is 4×10^{-5} and a plausible magnitude for b is 1.6×10^{-4}.

The key parameter of interest is the intrinsic growth rate, r. As it happens, *Jubaea chilensis* is a well-known species. It prefers temperate climate and grows slowly, normally requiring 40 to 60 years from first planting to the time it bears its first fruit. An intrinsic growth rate on the order of 4% per decade would be plausible. The Easter Island flora biomass consisted of more than just *J. chilensis*, but the forest was by far the most important part of the resource stock, so 4% (or 0.04) is used in the base case simulation for Easter Island.

To obtain simulations for Easter Island, Eqs. (1) and (2) are converted to discrete format. Thus, the two simulated equations are as follows:

$$S_{t+1} = S_t + rS_t(1 - St/K) - aS_tP_t, \qquad (1A)$$

$$P_{t+1} = nP_t + bP_tS_t. \qquad (2A)$$

The units of time are decades and the model is run for about 1200 years. The results are shown in Fig. 3.

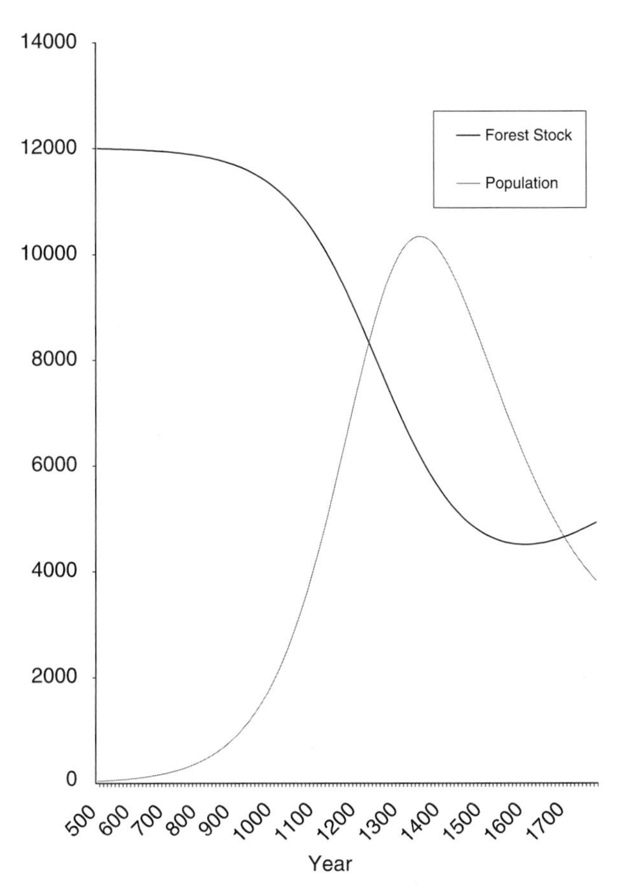

FIGURE 3 Easter Island simulation.

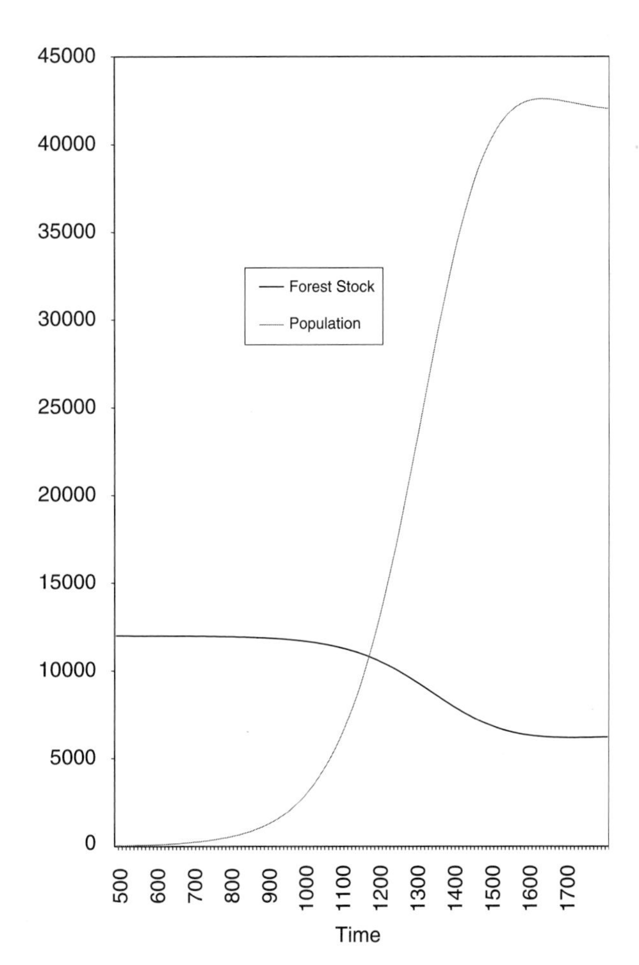

FIGURE 4 A fast-growing resource.

As can be seen in Fig. 3, this relatively simply model tracks the known facts about Easter Island quite well. If it is assumed that first colonization occurred in 500 AD, then peak population occurs in the 14th century at just over 10,000. From there, the population decreases and is down to a little over 4000 by the time of first European contact in 1722—a little high relative to actual estimates, but quite close. The resource stock is down to less than half its initial level by 1400 AD. On the actual Easter Island, this was about the time the wine palm forest completely disappeared, so for this to be accurate it would need to be the case that the flora biomass in the absence of the palm forest was about half of the biomass with the forest. This is certainly high, because the original palm forest had far more than twice the biomass of whatever shrubs, grass, and other vegetation was left after the forest was gone. Still, the general qualitative pattern shown in Fig. 3 is a reasonable representation of the Easter Island experience.

A particularly interesting variation of this simulation arises from making one change in a model parameter. Most of Polynesia did have palm forests. However, the palms growing on other islands grew much faster than the Chilean wine palm. The two most common large palms are the *Cocos* (coconut palm) and the *Pritchardia* (Fijian fan palm). These palms reach the fruit-growing stage after 7 to 10 years (much less than the 40- to 60-year period for the wine palm) and would have an intrinsic growth rate on the order of 35% per decade (rather than 4%). The resulting simulation is shown in Fig. 4, which illustrates a dynamic pattern strikingly different from that of Fig. 3. No boom-and-bust pattern is discernible. In fact, there is a very slight cycle that would not disappear entirely unless the intrinsic growth rate exceeded 71% per decade, but, even at 35%, the very gentle cycle would be much too damped to be evident to archaeologists and the approach to steady state is virtually monotonic. The forest stock declines smoothly to a new steady state that would correspond to having some standing forest in steady state. Low intrinsic growth rates, on the other hand, produce sharp cyclical fluctuations.

Therefore, in the predator–prey model described here, an island with a slow-growing resource base will exhibit overshooting and collapse. An otherwise identical island with a more rapidly growing resource base will exhibit a near-monotonic adjustment of population and resource stocks toward steady-state values. This fact alone can explain the sharp difference in experience between Easter Island and other Polynesian islands.

There are 12 so-called mystery islands in Polynesia that had been settled by Polynesians but were unoccupied at the time of first European contact. All these islands but one had relatively small carrying capacities. Applying our model, we see that if K (carrying capacity) is too small, then there is no interior steady state, implying that a colonizing population might arrive and expand but would eventually be driven to extinction.

6. THE ROLE OF PROPERTY RIGHTS AND OPEN ACCESS

A presumption underlying the analysis here is that property rights were incomplete, perhaps not to the point of full open access, but at least to the point where it was hard to prevent increased harvesting as the population increased. Specifically, in Eq. (1) harvesting is assumed to be increasing with population size for any given stock. This need not be the case if property rights are complete, because an efficient owner or manager of the resource would restrict access beyond some optimal point.

It is unlikely that Easter Island was characterized by complete open access, but it is very likely that property rights were incomplete in the sense that different competing harvesters probably had access to some common part of the forest. Incomplete property rights give rise to market failure, with the result that overharvesting occurs. In other words, the individual incentives faced by individual harvesters (or by small kinship groups) would lead to overharvesting from the collective point of view. As shown in Fig. 4, even with incomplete property rights, it is still possible to have monotonic adjustment to a steady state if parameter values are suitable. However, it would be even better if the society could enforce strict property rights. If a dictator or "social planner" could impose strict property rights, that person could, for example, set harvesting at the level of maximum sustainable yield and keep harvesting at that level independent of population size. However, by all accounts, that level of control was very difficult to impose and maintain in Polynesian and in most other societies.

In addition to the property rights problem, there was also a significant intertemporal market failure problem. Specifically, no one who planted or nurtured a young wine palm could expect to make personal use of that tree in his or lifetime, because life expectancy was short relative to the time taken for trees to reach maturity. Even the planter's children would be unlikely to benefit from the such activity. It would be the generation of one's grandchildren who would benefit from current planting activity. Thus, the incentives to undertake investments in planting and growing were weak. In modern societies, someone who plants a stand of trees can sell the stand at any time to someone else willing to hold the tree lot for some period of time—secure in the knowledge that it can be sold to still another person in the future. This mechanism of overlapping ownership connected through markets allows, in effect, future generations to "pay" for current investments. This is necessary for economic efficiency. However, in the Easter Island context, there would be no such markets. Thus, there would be a "market failure" leading to under investment in planting and growing trees. In primitive societies, social norms often take the place of market transactions in enforcing particular economic investments, but social norms are highly unreliable. In the Easter Island case, it is quite possible that the statue-carving norm dominated any conservation norms and worsened rather than mitigated the overharvesting problem.

7. AFTERMATH

It is useful to consider the history of Easter Island after the 18th century. The story is a very sad one—much worse than the endogenous boom-and-bust cycle that had already occurred on Easter Island. From the time of James Cook in the late 18th century through the mid-19th century, conditions on Easter Island gradually improved and the population increased gradually. Population was reliably estimated at something exceeding 3000 in 1862. In 1862 and 1863, slave traders from the Spanish community in Peru invaded the island and took about one-third of the islanders as slaves. Most of these slaves died of smallpox and other infectious diseases within a few years. A few returned to Easter Island, inadvertently causing a smallpox epidemic that killed most of the remaining Islanders.

By 1877, the population reached its low point of 111. Some of the deaths would have been due to causes other than smallpox, but, even allowing for these other causes, these numbers imply that the death rate from smallpox exceeded 90% of the base population—probably the highest death rate ever observed from an infectious disease invading a population of significant size. Although this has nothing to do with the pattern of resource overuse and collapse, it illustrates an important force in human history: the devastating effect that a new infectious disease can have on a long-isolated population.

From 1877, the population gradually increased through natural increase, immigration from other Polynesian islands (especially Tahiti), and immigration from the South American mainland. In 1888, the island was annexed by Chile, which maintains possession of the island at present. Population on the island has risen to something approaching 3000 and is devoted primarily to agriculture and to tourism. There is some mild tension between the predominantly Polynesian traditional or indigenous peoples and the South American population in Easter Island, but, on the whole, Easter Island is a peaceful and pleasant island that has become a major tourist destination of particular interest to "cultural tourists" who wish to see the now carefully maintained and protected moai.

8. LESSONS OF EASTER ISLAND

Some observers like to see Easter Island as a metaphor for the modern world. However, it is important to be aware of the differences. The first point, as has been emphasized already, is that societies of the Easter Island type do not necessarily have a dramatic decline. In the case of Easter Island, the civilization was "unlucky" to be on an island where the main resource was very slow growing. If the resource had been faster growing, then the boom-and-bust cycle probably would have been avoided, as it was in most of Polynesia. Therefore, even if Earth as a whole could be viewed as Easter Island writ large, there would no presumption about any inevitability of the imminent decline and fall of modern civilization.

On the other hand, Easter Island should not be viewed as an isolated case. As modern archaeology has made increasing use of sophisticated scientific methods, it has become increasingly clear that resource degradation has played an important role in the rise and fall of many civilizations. The truly unique feature of Easter Island was its isolation, which prevented migration as a response to resource degradation. In other parts of the world, overshooting population and resource degradation have led to substantial out-migration that mitigated, but did not eliminate, the population losses and the cultural losses associated with the boom-and-bust cycle. The lesson here is that resource management is very difficult, especially in conditions approaching open access, and overshooting is a genuine concern. This applies to the modern world just as it did to Easter Island. Thus, for example, major fisheries, major wildlife resources, and major forest resources around the world have been significantly compromised already. It is quite possible that various populations around the world, especially in areas with high population growth, such as Africa and parts of the Middle East, are on an overshooting trajectory.

Another lesson to be drawn from Easter Island concerns the danger of making simple linear projections based on a short time series. If an Easter Islander had, in the year 1300, extrapolated trends in population and real income based on the previous few centuries, that person would have failed to anticipate the "turning point" in population and real income that was coming. It is a characteristic of resource systems and predator–prey systems more broadly that cyclical patterns are common.

Even the two centuries of remarkable economic performance that have occurred in the time since Malthus should not encourage complacency about avoiding future Malthusian adjustment, especially if major renewable resource stocks (fish, forests, and soil) continue to decline. The world's current population growth rate is about 1.3% per year, significantly lower than its peak of 2.2% (reached in 1963), but still high by historical standards. Current population growth rates imply a population doubling in just over 50 years. Current resource stocks, even if degradation could be slowed or halted, would have a hard time supporting such a population at anything like current levels of real income, even adjusting for likely technological improvements over the next 50 years. A subsequent doubling in the latter half of the 21st century would seem completely infeasible. Therefore, it seems that population growth will have to fall dramatically from current levels over the next 100 years. Whether this is achieved through a benign demographic transition or through the more unpleasant Malthusian mechanisms of disease, famine, and violent conflict is still very much an open question at this stage.

SEE ALSO THE FOLLOWING ARTICLES

Biomass Resource Assessment • Depletion and Valuation of Energy Resources • Ecological Footprints and Energy • Ecological Risk Assessment Applied to Energy Development • Ecosystem Health: Energy Indicators • Population Growth and Energy • Sociopolitical Collapse, Energy and

Further Reading

Bahn, P., and Flenley, J. (1992). "Easter Island, Earth Island." Thames and Hudson, London.

Brander, J. A., and Taylor, M. S. (1998). The simple economics of Easter Island: A Ricardo–Malthus model of renewable resource use. *Am. Econ. Rev.* 88(1), 119–138.

Clark, C. W. (1990). "Mathematical Bioeconomics, The Optimal Management of Renewable Resources." 2nd ed. Wiley, New York.

Dark, K. R. (1995). "Theoretical Archaeology." Cornell University Press, Ithaca, N.Y.

Lotka, A. J. (1925). "Elements of Physical Biology." Williams & Wilkins, Baltimore.

Malthus, T. R. (1798). "An Essay on the Theory of Population." Oxford University Press, Oxford.

Ostrum, E. (1990). "Governing the Commons, The Evolution of Institutions for Collective Action." Cambridge University Press, Cambridge.

von Daniken, E. (1970). "Chariots of the Gods? Unsolved Mysteries of the Past." Putnam, New York.